T0207309

ALLE ZEIT WACH
1842

Originally published as
ÉLÉMENTS DE MATHÉMATIQUE,
ALGÈBRE 1–3

Mathematics Subject Classification (1991): 00A05

Distribution rights worldwide:
Springer-Verlag Berlin Heidelberg New York London Paris Tokyo

ISBN 3-540-64243-9 Springer-Verlag Berlin Heidelberg New York
ISBN 3-540-19373-1 2nd printing Springer-Verlag Berlin Heidelberg New York

Softcover edition of the 2nd printing 1989

Library of Congress Cataloging-in-Publication Data
Bourbaki, Nicolas. [Algèbre. English] Algebra / Nicolas Bourbaki. p. cm.-(Elements of mathematics / Nicolas Bourbaki) Translation of: Algèbre. Includes index. Contents:1. Chapters 1-3.
ISBN 0-387-19373-1 (v. 1: U.S.)
1. Algebra. I. Title. II. Series: Bourbaki, Nicolas. Éléments de mathématique. English.
QA155.B68213 1988 512-dc 19 88-31211

Printed in Germany
SPIN 11002123
41/3111-5 4 3 2 1
Printed on acid-free paper

NICOLAS BOURBAKI

ELEMENTS OF MATHEMATICS

Algebra I

Chapters 1–3

Springer-Verlag
Berlin Heidelberg New York
London Paris Tokyo

TO THE READER

1. This series of volumes, a list of which is given on pages ix and x, takes up mathematics at the beginning, and gives complete proofs. In principle, it requires no particular knowledge of mathematics on the readers' part, but only a certain familiarity with mathematical reasoning and a certain capacity for abstract thought. Nevertheless, it is directed especially to those who have a good knowledge of at least the content of the first year or two of a university mathematics course.

2. The method of exposition we have chosen is axiomatic and abstract, and normally proceeds from the general to the particular. This choice has been dictated by the main purpose of the treatise, which is to provide a solid foundation for the whole body of modern mathematics. For this it is indispensable to become familiar with a rather large number of very general ideas and principles. Moreover, the demands of proof impose a rigorously fixed order on the subject matter. It follows that the utility of certain considerations will not be immediately apparent to the reader unless he has already a fairly extended knowledge of mathematics; otherwise he must have the patience to suspend judgment until the occasion arises.

3. In order to mitigate this disadvantage we have frequently inserted examples in the text which refer to facts the reader may already know but which have not yet been discussed in the series. Such examples are always placed between two asterisks: * . . . *. Most readers will undoubtedly find that these examples will help them to understand the text, and will prefer not to leave them out, even at a first reading. Their omission would of course have no disadvantage, from a purely logical point of view.

4. This series is divided into volumes (here called "Books"). The first six Books are numbered and, in general, every statement in the text assumes as known only those results which have already been discussed in the preceding

volumes. This rule holds good within each Book, but for convenience of exposition these Books are no longer arranged in a consecutive order. At the beginning of each of these Books (or of these chapters), the reader will find a precise indication of its logical relationship to the other Books and he will thus be able to satisfy himself of the absence of any vicious circle.

5. The logical framework of each chapter consists of the *definitions*, the *axioms*, and the *theorems* of the chapter. These are the parts that have mainly to be borne in mind for subsequent use. Less important results and those which can easily be deduced from the theorems are labelled as "propositions", "lemmas", "corollaries", "remarks", etc. Those which may be omitted at a first reading are printed in small type. A commentary on a particularly important theorem appears occasionally under the name of "scholium".

To avoid tedious repetitions it is sometimes convenient to introduce notations or abbreviations which are in force only within a certain chapter or a certain section of a chapter (for example, in a chapter which is concerned only with commutative rings, the word "ring" would always signify "commutative ring"). Such conventions are always explicitly mentioned, generally at the beginning of the chapter in which they occur.

6. Some passages in the text are designed to forewarn the reader against serious errors. These passages are signposted in the margin with the sign

Z ("dangerous bend").

7. The Exercises are designed both to enable the reader to satisfy himself that he has digested the text and to bring to his notice results which have no place in the text but which are nonetheless of interest. The most difficult exercises bear the sign ¶.

8. In general, we have adhered to the commonly accepted terminology, except where there appeared to be good reasons for deviating from it.

9. We have made a particular effort always to use rigorously correct language. without sacrificing simplicity. As far as possible we have drawn attention in the text to *abuses of language*, without which any mathematical text runs the risk of pedantry, not to say unreadability.

10. Since in principle the text consists of the dogmatic exposition of a theory, it contains in general no references to the literature. Bibliographical references are gathered together in *Historical Notes*, usually at the end of each chapter. These notes also contain indications, where appropriate, of the unsolved problems of the theory.

The bibliography which follows each historical note contains in general only those books and original memoirs which have been of the greatest importance in the evolution of the theory under discussion. It makes no sort of pretence to completeness; in particular, references which serve only to determine questions of priority are almost always omitted.

As to the exercises, we have not thought it worthwhile in general to indicate their origins, since they have been taken from many different sources (original papers, textbooks, collections of exercises).

11. References to a part of this series are given as follows:

a) If reference is made to theorems, axioms, or definitions presented *in the same section*, they are quoted by their number.
b) If they occur *in another section of the same chapter*, this section is also quoted in the reference.
c) If they occur *in another chapter in the same Book*, the chapter and section are quoted.
d) If they occur *in another Book*, this Book is first quoted by its title.

The *Summaries of Results* are quoted by the letter R; thus *Set Theory*, R signifies "*Summary of Results of the Theory of Sets*".

CONTENTS
OF
THE ELEMENTS OF MATHEMATICS SERIES

Lie groups and lie algebras

Commutative algebra

Spectral theory

Differential and Analytic Manifolds

CONTENTS

INTRODUCTION

Algebra is essentially concerned with *calculating*, that is, performing, on elements of a set, "algebraic operations", the most well-known example of which is provided by the "four rules" of elementary arithmetic.

This is not the place to retrace the slow process of progressive abstraction by which the notion of algebraic operation, at first restricted to natural numbers and measurable magnitudes, little by little enlarged its domain, whilst the notion of "number" received a parallel generalization, until, passing beyond the latter, it came to be applied to elements with no "numerical" character, for example permutations of a set (see the Historical Note to Chapter I). It is no doubt the possibility of these successive extensions, in which the *form* of the calculations remained the same, whereas the *nature* of the mathematical entities subjected to these calculations varied considerably, which was responsible for the gradual isolation of the guiding principle of modern mathematics, namely that mathematical entities in themselves are of little importance; what matters are their *relations* (see Book I). It is certain in any case that Algebra attained this level of abstraction well before the other branches of Mathematics and it has long been accustomed to being considered as the study of algebraic operations, independent of the mathematical entities to which they can be applied.

Deprived of any specific character, the common notion underlying the usual algebraic operations is very simple: performing an algebraic operation on two elements a, b of the same set E, means associating with the ordered pair (a, b) a well-defined third element c of the set E. In other words, there is nothing more in this notion than that of *function*: to be given an algebraic operation is to be given a function defined on $\mathrm{E} \times \mathrm{E}$ and taking its values in E; the only peculiarity lies in the fact that the set of definition of the function is the product of two sets identical with the set where the function takes its values; to such a function we give the name *law of composition*.

Alongside these "internal laws", mathematicians have been led (chiefly under the influence of Geometry) to consider another type of "law of composition"; these are the "laws of action", where, besides the set E (which remains as it were in the foreground), an auxiliary set Ω occurs, whose elements are called *operators*: the law this time associating with an ordered pair (α, a) consisting of an operator $\alpha \in \Omega$ and an element $a \in E$ a second element b of E. For example a homothety of given centre on the Euclidean space E associates with a real number k (the "homothety ratio", which is here the operator) and a point A of E another point A′ of E: this is a law of action on E.

In conformity with the general definitions (*Set Theory*, IV, § 1, no. 4), being given on a set E one or several laws of composition or laws of action defines a *structure* on E; for the structures defined in this way we reserve precisely the name *algebraic structures* and it is the study of these which constitutes Algebra.

There are several *species* (*Set Theory*, IV, § 1, no. 4) of algebraic structures, characterized on the one hand by the laws of composition or laws of action which define them and on the other by the *axioms* to which these laws are subjected. Of course, these axioms have not been chosen arbitrarily, but are just the properties of most of the laws which occur in applications, such as associativity, commutativity, etc. Chapter I is essentially devoted to the exposition of these axioms and the general consequences which follow from them; also there is a more detailed study of the two most important species of algebraic structure, that of a *group* (in which only *one* law of composition occurs) and that of a *ring* (with *two* laws of composition), of which a *field* structure is a special case.

In Chapter I are also found the definitions of *groups with operators* and *rings with operators*, where, besides the laws of composition, there occur one or several laws of action. The most important groups with operators are *modules*, of which *vector spaces* are a particular example, which play an important role both in Classical Geometry and in Modern Analysis. The study of module structures derives its origin from that of *linear equations*, whence its name *Linear Algebra*; the results on this subject are to be found in Chapter II.

Similarly, the rings with operators which occur most often are called *algebras* (or *hypercomplex systems*). In Chapters III and IV a detailed study is made of two particular types of algebras: *exterior algebras*, which with the theory of determinants which is derived from them is a valuable tool of Linear Algebra; and *polynomial algebras*, which are fundamental to the theory of algebraic equations.

In Chapter V there is an exposition of the general theory of *commutative fields* and their classification. The origin of this theory is the study of algebraic equations in one unknown; the questions which gave birth to this today have little more than a historical interest, but the theory of commutative fields remains fundamental to Algebra, being basic to the theory of algebraic numbers on the one hand and Algebraic Geometry on the other.

As the set of natural numbers has two laws of composition, addition and multiplication, Classical Arithmetic (or Number Theory), which is the study of natural numbers, is subordinate to Algebra. However, related to the algebraic structure defined by these two laws is the structure defined by the *order relation* "a divides b"; and the distinguishing aspect of Classical Arithmetic is precisely the study of the relations between these two associated structures. This is not the only example where an order structure is thus associated with an algebraic structure by a "divisibility" relation: this relation plays just as important a role in polynomial rings. A general study will therefore be made of this in Chapter VI; this study will be applied in Chapter VII to determining the structure of modules over certain particularly simple rings and in particular to the theory of "elementary divisors".

Chapters VIII and IX are devoted to more particular theories, but which have many applications in Analysis: on the one hand, the theory of *semisimple modules and rings*, closely related to that of *linear representations of groups*; on the other, the theory of *quadratic forms* and *Hermitian forms*, with the study of the groups associated with them. Finally, the *elementary geometries* (affine, projective, euclidean, etc.) are studied in Chapters II, VI and IX from a purely algebraic point of view: here there is little more than a new language for expressing results of Algebra already obtained elsewhere, but it is a language which is particularly well adapted to the later developments of Algebraic Geometry and Differential Geometry, to which this chapter serves as an introduction.

CHAPTER I

Algebraic Structures

§ 1. LAWS OF COMPOSITION; ASSOCIATIVITY; COMMUTATIVITY

1. LAWS OF COMPOSITION

DEFINITION 1. *Let* E *be a set. A mapping f of* $E \times E$ *into* E *is called a law of composition on* E. *The value* $f(x, y)$ *of* f *for an ordered pair* $(x, y) \in E \times E$ *is called the composition of* x *and* y *under this law. A set with a law of composition is called a magma.*

The composition of x and y is usually denoted by writing x and y in a definite order and separating them by a characteristic symbol of the law in question (a symbol which it may be agreed to omit). Among the symbols most often used are $+$ and ., the usual convection being to omit the latter if desired; with these symbols the composition of x and y is written respectively as $x + y$ and $x.y$ or xy. A law denoted by the symbol $+$ is usually called *addition* (the composition $x + y$ being called the *sum* of x and y) and we say that it is *written additively*; a law denoted by the symbol . is usually called *multiplication* (the composition $x.y = xy$ being called the *product* of x and y) and we say that it is *written multiplicatively*. In the general arguments of paragraphs 1 to 3 of this chapter we shall generally use the symbols $\top$ and $\perp$ to denote arbitrary laws of composition.

By an abuse of language, a mapping of a *subset* of $E \times E$ into E is sometimes called a law of composition *not everywhere defined* on E.

Examples. (1) The mappings $(X, Y) \mapsto X \cup Y$ and $(X, Y) \mapsto X \cap Y$ are laws of composition on the set of subsets of a set E.

(2) On the set **N** of natural numbers, addition, multiplication and exponentiation are laws of composition (the compositions of $x \in \mathbf{N}$ and $y \in \mathbf{N}$ under these laws being denoted respectively by $x + y$, xy or $x.y$ and x^y) (*Set Theory*, III, § 3, no. 4).

(3) Let E be a set; the mapping $(X, Y) \mapsto X \circ Y$ is a law of composition on the set of subsets of $E \times E$ (*Set Theory*, II, § 3, no. 3, Definition 6); the mapping $(f, g) \mapsto f \circ g$ is a law of composition on the set of mappings of E into E (*Set Theory*, II, § 5, no. 2).

(4) Let E be a lattice (*Set Theory*, III, § 1, no. 11); if $\sup(x, y)$ denotes the least upper bound of the set $\{x, y\}$, the mapping $(x, y) \mapsto \sup(x, y)$ is a law of

composition on E. Similarly for the greatest lower bound $\inf(x, y)$. Example 1 above is a particular case of this with $\mathfrak{P}(E)$ ordered by inclusion.

(5) Let $(E_\iota)_{\iota \in I}$ be a family of magmas. Let $\top_\iota$ denote the law of composition on E_ι. The mapping

$$((x_\iota), (y_\iota)) \mapsto ((x_\iota \top_\iota y_\iota))$$

is a law of composition on the product $E = \prod_{\iota \in I} E_\iota$, called the *product* of the laws $\top_\iota$. The set E with this law is called the *product magma* of the magmas E_ι. In particular, if all the magmas E_ι are equal to the same magma M, we obtain the *magma of mappings of* I *from* M.

Let $(x, y) \mapsto x \top y$ be a law of composition on a set E. Given any two subsets X, Y of E, $X \top Y$ (provided this notation does not lead to confusion†) will denote the set of elements $x \top y$ in E, where $x \in X$, $y \in Y$ (in other words, the image of $X \times Y$ under the mapping $(x, y) \mapsto x \top y$).

If $a \in E$ we usually write $a \top Y$ instead of $\{a\} \top Y$ and $X \top a$ instead of $X \top \{a\}$. The mapping $(X, Y) \mapsto X \top Y$ is a law of composition on the set of subsets of E.

DEFINITION 2. *Let* E *be a magma and* $\top$ *denote its law of composition. The law of composition* $(x, y) \mapsto y \top x$ *on* E *is called the opposite of the above. The set* E *with this law is called the opposite magma of* E.

Let E and E′ be two magmas; we shall denote their laws by the same symbol $\top$. Conforming with the general definitions (*Set Theory*, IV, § 1, no. 5), a bijective mapping f of E onto E′ such that

$$f(x \top y) = f(x) \top f(y) \tag{1}$$

for every ordered pair $(x, y) \in E \times E$ is called an *isomorphism of* E *onto* E′. E and E′ are said to be *isomorphic* if there exists an isomorphism of E onto E′.

More generally:

DEFINITION 3. *A mapping f of* E *into* E′ *such that relation* (1) *holds for every ordered pair* $(x, y) \in E \times E$ *is called a homomorphism, or morphism, of* E *into* E′*; if* E = E′, *f is called an endomorphism of* E.

The identity mapping of a magma E is a homomorphism, the composition of two homomorphism is a homomorphism.

---

† The following is an example where this principle would lead to confusion and should therefore not be used. Suppose that the law of composition in question is the law $(A, B) \mapsto A \cup B$ between subsets of a set E; a law of composition

$$(\mathfrak{A}, \mathfrak{B}) \mapsto F(\mathfrak{A}, \mathfrak{B})$$

is derived between subsets of $\mathfrak{P}(E)$, $F(\mathfrak{A}, \mathfrak{B})$ being the set of $A \cup B$ with $A \in \mathfrak{A}$, $B \in \mathfrak{B}$; but $F(\mathfrak{A}, \mathfrak{B})$ should not be denoted by $\mathfrak{A} \cup \mathfrak{B}$, as this notation already has a different meaning (the union of $\mathfrak{A}$ and $\mathfrak{B}$ considered as subsets of $\mathfrak{P}(E)$).

For a mapping f of E into E′ to be an isomorphism, it is necessary and sufficient that it be a bijective homomorphism and $\overset{-1}{f}$ is then an isomorphism of E′ onto E.

2. COMPOSITION OF AN ORDERED SEQUENCE OF ELEMENTS

Recall that a *family* of elements of a set E is a mapping $\iota \mapsto x_\iota$ of a set I (called an indexing set) into E; a family $(x_\iota)_{\iota \in I}$ is called *finite* if the indexing set is *finite*.

A finite family $(x_\iota)_{\iota \in I}$ of elements of E whose indexing set I is *totally ordered* is called an *ordered sequence* of elements of E.

In particular, every finite sequence $(x_i)_{i \in H}$, where H is a finite subset of the set **N** of natural numbers, can be considered as an ordered sequence if H is given the order relation induced by the relation $m \leqslant n$ between natural numbers.

Two ordered sequences $(x_i)_{i \in I}$ and $(y_\kappa)_{\kappa \in K}$ are called *similar* if there exists an ordered set isomorphism ϕ of I onto K such that $y_{\phi(i)} = x_i$ for all $i \in I$.

Every ordered sequence $(x_\alpha)_{\alpha \in A}$ is similar to a suitable finite sequence. For there exists an increasing bijection of A onto an interval $[0, n]$ of **N**.

DEFINITION 4. *Let $(x_\alpha)_{\alpha \in A}$ be an ordered sequence of elements in a magma E whose indexing set A is non-empty. The composition (under the law $\top$) of the ordered sequence $(x_\alpha)_{\alpha \in A}$, denoted by $\mathop{\top}\limits_{\alpha \in A} x_\alpha$, is the element of E defined by induction on the number of elements in A as follows:*

(1) *if* $A = \{\beta\}$ *then* $\mathop{\top}\limits_{\alpha \in A} x_\alpha = x_\beta$;

(2) *if A has $p > 1$ elements, β is the least element of A and $A' = A - \{\beta\}$, then* $\mathop{\top}\limits_{\alpha \in A} x_\alpha = x_\beta \top \left(\mathop{\top}\limits_{\alpha \in A'} x_\alpha\right)$.

It follows immediately (by induction on the number of elements in the indexing sets) that the compositions of two *similar* ordered sequences are *equal*; in particular, the composition of any ordered sequence is equal to the composition of a finite sequence. If $A = \{\lambda, \mu, \nu\}$ has three elements ($\lambda < \mu < \nu$) the composition $\mathop{\top}\limits_{\alpha \in A} x_\alpha$ is $x_\lambda \top (x_\mu \top x_\nu)$.

> *Remark.* Note that there is a certain arbitrariness about the definition of the composition of an ordered sequence; the induction we introduced proceeds "from right to left". If we proceeded "from left to right", the composition of the above ordered sequence $(x_\lambda, x_\mu, x_\nu)$ would be $(x_\lambda \top x_\mu) \top x_\nu$.

As a matter of notation, the composition of an ordered sequence $(x_\alpha)_{\alpha \in A}$ is written $\mathop{\perp}\limits_{\alpha \in A}$ for a law denoted by $\perp$; for a law written additively it is usually denoted by $\sum\limits_{\alpha \in A} x_\alpha$ and called the *sum* of the ordered sequence $(x_\alpha)_{\alpha \in A}$ (the x_α being called the *terms* of the sum); for a law written multiplicatively it is usually denoted by $\prod_{\alpha \in A} x_\alpha$ and called the *product* of the ordered sequence (x_α) (the x_α being called the *factors* of the product).†

When there is no possible confusion over the indexing set (nor over its ordering) it is often dispensed with in the notation for the composition of an ordered sequence and we then write, for example for a law written additively, $\sum_\alpha x_\alpha$ instead of $\sum_{\alpha \in A} x_\alpha$; similarly for the other notations.

For a law denoted by $\top$ the composition of a *sequence* (x_i) with indexing set a non-empty interval $[p, q]$ of $\mathbf{N}$ is denoted by $\mathop{\top}\limits_{p \leqslant i \leqslant q} x_i$ or $\mathop{\top}\limits_{i=p}^{q} x_i$; similarly for laws denoted by other symbols.

Let E and F be two magmas whose laws of composition are denoted by $\top$ and f a homomorphism of E into F. For every ordered sequence $(x_\alpha)_{\alpha \in A}$ of elements of E

$$f\Big(\mathop{\top}\limits_{\alpha \in A}\Big) = \mathop{\top}\limits_{\alpha \in A} f(x_\alpha). \tag{2}$$

3. ASSOCIATIVE LAWS

DEFINITION 5. *A law of composition* $(x, y) \mapsto x \top y$ *on a set* E *is called associative if, for all elements* x, y, z *in* E,

$$(x \top y) \top z = x \top (y \top z).$$

A magma whose law is associative is called an associative magma.

The opposite law of an associative law is associative.

Examples. (1) Addition and multiplication of natural numbers are associative laws of composition on $\mathbf{N}$ (*Set Theory*, III, § 3, no. 3, Corollary to Proposition 5)

(2) The laws cited in Examples (1), (3) and (4) of no. 1 are associative.

THEOREM 1 (Associativity theorem). *Let* E *be an associative magma whose law is denoted by* $\top$. *Let* A *be a totally ordered non-empty finite set, which is the union of an ordered sequence of non-empty subsets* $(B_\iota)_{\iota \in I}$ *such that the relations* $\alpha \in B_\iota$, $\beta \in B_\kappa$, $\iota < \kappa$ *imply* $\alpha < \beta$; *let* $(x_\alpha)_{\alpha \in A}$ *be an ordered sequence of elements in* E *with* A *as indexing set. Then*

$$\mathop{\top}\limits_{\alpha \in A} x_\alpha = \mathop{\top}\limits_{\iota \in I}\Big(\mathop{\top}\limits_{\alpha \in B_\iota} x_\alpha\Big) \tag{3}$$

† The use of this terminology and the notation $\prod_{\alpha \in A} x_\alpha$ must be avoided if there is any risk of confusion with the product of the sets x_α defined in the theory of sets (*Set Theory*, II, § 5, no. 3). However, if the x_α are cardinals and addition (resp. multiplication) is the cardinal sum (resp. the cardinal product), the cardinal denoted by $\sum_{\alpha \in A} x_\alpha$ $\left(\text{resp. } \prod_{\alpha \in A} x_\alpha\right)$ in the above notation is the cardinal sum (resp. cardinal product) of the family $(x_\alpha)_{\alpha \in A}$ (*Set Theory*, III, § 3, no. 3).

We prove the theorem by induction on the cardinal n of A. Let p be the cardinal of I and h its least element; let $J = I - \{h\}$. If $n = 1$, as the B_i are non-empty, of necessity $p = 1$ and the theorem is obvious. Otherwise, assuming the theorem holds for an indexing set with at most $n - 1$ elements, we distinguish two cases:

(a) B_h has a single element β. Let $C = \bigcup_{i \in J} B_i$. The left-hand side of (3) is equal, by definition, to $x_\beta \top \left(\mathop{\top}_{\alpha \in C} x_\alpha\right)$; the right-hand side is equal, by definition, to

$$x_\beta \top \left(\mathop{\top}_{i \in J} \left(\mathop{\top}_{\alpha \in B_i} x_\alpha\right)\right);$$

the result follows from the fact that the theorem is assumed true for C and $(B_i)_{i \in J}$.

(b) Otherwise, let β be the least element of A (and hence of B_h); let $A' = A - \{\beta\}$ and let $B'_i = A' \cap B_i$ for $i \in I$; then $B'_i = B_i$ for $i \in J$. The set A′ has $n - 1$ elements and the conditions of the theorem hold for A′ and its subsets B'_i; hence by hypothesis:

$$\mathop{\top}_{\alpha \in A'} x_\alpha = \left(\mathop{\top}_{\alpha \in B'_h} x_\alpha\right) \top \left(\mathop{\top}_{i \in J}\left(\mathop{\top}_{\alpha \in B_i} x_\alpha\right)\right).$$

Forming the composition of x_β with each side, we have on the left-hand side side, by definition, $\mathop{\top}_{\alpha \in A} x_\alpha$ and on the right-hand side, using the associativity,

$$\left(x_\beta \top \left(\mathop{\top}_{\alpha \in B'_h} x_\alpha\right)\right) \top \left(\mathop{\top}_{i \in J}\left(\mathop{\top}_{\alpha \in B_i} x_\alpha\right)\right)$$

which is equal, by Definition 3, to the right-hand side of formula (3).

For an associative law denoted by $\top$ the composition $\mathop{\top}_{p \leqslant i \leqslant q} x_i$ of a sequence $(x_i)_{i \in [p, q]}$ is also denoted (since no confusion can arise) by

$$x_p \top \cdots \top x_q.$$

A particular case of Theorem 1 is the formula

$$x_0 \top x_1 \top \cdots \top x_n = (x_0 \top x_1 \top \cdots \top x_{n-1}) \top x_n.$$

Consider an ordered sequence of n terms all of whose terms are equal to the same element $x \in E$. The composition of this sequence is denoted by $\overset{n}{\top}\, x$ for a law denoted by $\top$, $\overset{n}{\bot}\, x$ for a law denoted by $\bot$. For a law written multiplicatively the composition is denoted by x^n and called the n-th *power* of x. For a law written additively the composition is usually denoted by nx. The associativity theorem applied to an ordered sequence all of whose terms are equal gives the equation

$$\overset{n_1+n_2+\cdots+n_p}{\top} x = \left(\overset{n_1}{\top} x\right) \top \left(\overset{n_2}{\top} x\right) \top \cdots \top \left(\overset{n_p}{\top} x\right).$$

In particular, if $p = 2$,

$$\overset{m+n}{\top} x = \left(\overset{m}{\top} x\right) \top \left(\overset{n}{\top} x\right) \tag{4}$$

and if $n_1 = n_2 = \cdots = n_p = m$,

$$\overset{pm}{\top} x = \overset{p}{\top} \left(\overset{m}{\top} x\right). \tag{5}$$

If X is a subset of E, we sometimes denote, in conformity with the above notation, by $\overset{p}{\top}$ X the set $X_1 \top X_2 \top \cdots \top X_p$, where

$$X_1 = X_2 = \cdots = X_p = X;$$

it is thus the set of all compositions $x_1 \top x_2 \top \cdots \top x_p$ with $x_1 \in X$, $x_2 \in X, \ldots$, $x_p \in X$.

It is important not to confuse this set with the set of $\overset{p}{\top} x$, where x runs through X.

4. STABLE SUBSETS. INDUCED LAWS

DEFINITION 6. *A subset* A *of a set* E *is called stable under the law of composition* $\top$ *on* E *if the composition of two elements of* A *belongs to* A. *The mapping* $(x, y) \mapsto x \top y$ *of* A $\times$ A *into* A *is then called the law induced on* A *by the law* $\top$. *The set* A *with the law induced by* $\top$ *is called a submagma of* E.

In other words, for A to be stable under a law $\top$ it is necessary and sufficient that $A \top A \subset A$. A stable subset of E and the corresponding submagma are often identified.

The intersection of a family of stable subsets of E is stable; in particular there exists a smallest stable subset A of E containing a given subset X; it is said to be *generated* by X and X is called a *generating system* of A or a *generating set* of A. The corresponding submagma is also said to be *generated* by X.

PROPOSITION 1. *Let* E *and* F *be two magmas and* f *a homomorphism of* E *into* F.

(i) *The image under* f *of a stable subset of* E *is a stable subset of* F.

(ii) *The inverse image under* f *of a stable subset* F *is a stable subset of* E.

(iii) *Let* X *be a subset of* E. *The image under* f *of the stable subset of* E *generated by* X *is the stable subset of* F *generated by* $f(X)$.

(iv) *If* g *is a second homomorphism of* E *from* F *the set of elements* x *of* E *such that* $f(x) = g(x)$ *is a stable subset of* E.

Assertions (i), (ii) and (iv) are obvious; we prove (iii). Let $\overline{X}$ be the stable subset of E generated by X and $\overline{f(X)}$ the stable subset of F generated by $f(X)$. By (i) $\overline{f(X)} \subset f(\overline{X})$ and by (ii) $\overline{X} \subset \overset{-1}{f}(\overline{f(X)})$, whence $f(\overline{X}) \subset \overline{f(X)}$.

Proposition 2. *Let* E *be an associative magma and* X *a subset of* E. *Let* X' *be the set of* $x_1 \top x_2 \top \cdots \top x_n$, *where* $n \geqslant 1$ *and where* $x_i \in X$ *for* $1 \leqslant i \leqslant n$. *The stable subset generated by* X *is equal to* X'.

It follows immediately by induction on n that the composition of an ordered sequence of n terms belonging to X belongs to the stable subset generated by X; it is therefore sufficient to verify that X' is stable. Now if u and v are elements of X' they are of the form $u = x_0 \top x_1 \top \cdots \top x_{n-1}$, $v = x_n \top x_{n+1} \top \cdots \top x_{n+p}$ with $x_i \in X$ for $0 \leqslant i \leqslant n + p$; then (Theorem 1) $u \top v = x_0 \top x_1 \top \cdots \top x_{n+p}$ belongs to X'.

Examples. (1) In the set **N** of natural numbers the stable subset under addition generated by $\{1\}$ is the set of integers $\geqslant 1$; under multiplication the set $\{1\}$ is stable.

(2) Given a law $\top$ on a set E, for a subset $\{h\}$ consisting of a single element to be stable under the law $\top$ it is necessary and sufficient that $h \top h = h$; h is then said to be *idempotent*. For example, every element of a lattice is idempotent for each of the laws sup and inf.

(3) For an associative law $\top$ on a set E the stable subset generated by a set $\{a\}$ consisting of a single element is the set of elements $\overset{n}{\top}\, a$, where n runs through the set of integers > 0.

5. PERMUTABLE ELEMENTS. COMMUTATIVE LAWS

Definition 7. *Let* E *be a magma whose law is denoted by* $\top$. *Two elements* x *and* y *of* E *are said to* commute (*or to be* permutable) *if* $y \top x = x \top y$.

Definition 8. *A law of composition on a set* E *is called* commutative *if any two elements of* E *commute under this law. A magma whose law of composition is commutative is called a* commutative magma.

A commutative law is equal to its opposite.

Examples. (1) Addition and multiplication of natural numbers are commutative laws on **N** (*Set Theory*, III, § 3, no. 3, Corollary to Proposition 5).

(2) On a lattice the laws sup and inf are commutative; so, in particular, are the laws $\cup$ and $\cap$ between subsets of a set E.

(3) Let E be a set of cardinal > 1. The law $(f, g) \mapsto f \circ g$ between mappings of E into E is not commutative as is seen by taking f and g to be distinct constant mappings, but the identity mapping is permutable with every mapping.

(4) Let $(x, y) \mapsto x \top y$ be a commutative law on E; the law

$$(X, Y) \mapsto X \top Y$$

between subsets of E is commutative.

Definition 9. *Let* E *be a magma and* X *a subset of* E. *The set of elements of* E *which commute with each of the elements of* X *is called the* centralizer *of* X.

Let X and Y be two subsets of E and X′ and Y′ their respective centralizers. If $X \subset Y$, then $Y' \subset X'$.

Let $(X_i)_{i \in I}$ be a family of subsets of E and for all $i \in I$ let X'_i be the centralizer of X_i. The centralizer of $\bigcup_{i \in I} X_i$ is $\bigcap_{i \in I} X'_i$.

Let X be a subset of E and X′ the centralizer of X. The centralizer X″ of X′ is called the *bicentralizer* of X. Then $X \subset X''$. The centralizer X‴ of X″ is equal to X′. For X′ is contained in its bicentralizer X‴ and the relation $X \subset X''$ implies $X''' \subset X'$.

Proposition 3. *Let* E *be an associative magma whose law is denoted by* $\top$. *If an element* x *of* E *commutes with each of the elements* y *and* z *of* E, *it commutes with* $y \top z$.

For

$$x \top (y \top z) = (x \top y) \top z = (y \top x) \top z$$
$$= y \top (x \top z) = y \top (z \top x) = (y \top z) \top x.$$

Corollary. *Let* E *be an associative magma. The centralizer of any subset of* E *is a stable subset of* E.

Definition 10. *The centralizer of a magma* E *is called the* centre *of* E. *An element of the centre of* E *is called a* central *element of* E.

If E is an associative magma its centre is a stable subset by the Corollary to Proposition 3 and the law induced on its centre is commutative.

Proposition 4. *Let* E *be an associative magma,* X *and* Y *two subsets of* E. *If every element of* X *commutes with every element of* Y *every element of the stable subset generated by* X *commutes with every element of the stable subset generated by* Y.

Let X′ and X″ be the centralizer and bicentralizer of X. They are stable subsets of E. Now $X \subset X''$ and $Y \subset X'$ and hence X″ (resp. X′) contains the stable subset of E generated by X (resp. Y). As every element of X″ commutes with every element of X′, the proposition follows.

Corollary 1. *If* x *and* y *are permutable under the associative law* $\top$ *so are* $\overset{m}{\top} x$ *and* $\overset{n}{\top} y$ *for all integers* $m > 0$ *and* $n > 0$; *in particular* $\overset{m}{\top} x$ *and* $\overset{n}{\top} x$ *are permutable for all* x *and all integers* $m > 0$ *and* $n > 0$.

COROLLARY 2. *If all pairs of elements of a subset X are permutable under an associative law* $\top$, *the law induced by* $\top$ *on the stable subset generated by* X *is associative and commutative.*

THEOREM 2 (Commutativity theorem). *Let* $\top$ *be an associative law of composition on* E; *let* $(x_\alpha)_{\alpha \in A}$ *be a non-empty finite family of elements of* E *which are pairwise permutable; let* B *and* C *be two totally ordered sets with* A *as underlying set. Then* $\top_{\alpha \in B} x_\alpha = \top_{\alpha \in C} x_\alpha$.

Since the theorem is true if A has a single element, we argue by induction on the number p of elements in A. Let p be an integer > 1 and suppose the theorem is true when Card A $< p$. We prove it for Card A $= p$. It may be assumed that A is the interval $[0, p-1]$ in $\mathbf{N}$; the composition of the ordered sequence $(x_\alpha)_{\alpha \in A}$ defined by the natural order relation on A is $\top_{i=0}^{p-1} x_i$.

Let A be given another total ordering and let h be the least element of A under this ordering and A′ the set of other elements of A (totally ordered by the induced ordering). Suppose first $0 < h < p - 1$ and let $P = \{0, 1, \ldots, h-1\}$ and $Q = \{h+1, \ldots, p-1\}$; the theorem being assumed true for A′, applying the associativity theorem, we obtain (since $A' = P \cup Q$)

$$\top_{\alpha \in A'} x_\alpha = \left(\top_{i=0}^{h-1} x_i\right) \top \left(\top_{i=h+1}^{p-1} x_i\right)$$

whence, composing x_h with both sides and repeatedly applying the commutativity and associativity of $\top$:

$$\begin{aligned}\top_{\alpha \in A} x_\alpha = x_h \top \left(\top_{\alpha \in A'} x_\alpha\right) &= x_h \top \left(\top_{i=0}^{h-1} x_i\right) \top \left(\top_{i=h+1}^{p-1} x_i\right)\\ &= \left(\top_{i=0}^{h-1} x_i\right) \top x_h \top \left(\top_{i=h+1}^{p-1} x_i\right) = \top_{i=0}^{p-1} x_i,\end{aligned}$$

which proves the theorem for this case. If $h = 0$ or $h = p - 1$, the same result follows, but more simply, the terms arising from P or the terms arising from Q not appearing in the formulae.

Under a commutative associative law on a set E the *composition* of a *finite family* $(x_\alpha)_{\alpha \in A}$ of elements of E is by definition the common value of the composition of all the *ordered sequences* obtained by totally ordering A in all possible ways. This composition will still be denoted by $\top_{\alpha \in A} x_\alpha$ under a law denoted by $\top$; similarly for other notations.

THEOREM 3. *Let* $\top$ *be an associative law on* E *and* $(x_\alpha)_{\alpha \in A}$ *a non-empty finite family of elements of* E *which are pairwise permutable. If* A *is a union of non-empty subsets* $(B_i)_{i \in I}$ *which are pairwise disjoint, then*

$$\top_{\alpha \in A} x_\alpha = \top_{i \in I}\left(\top_{\alpha \in B_i} x_\alpha\right). \tag{6}$$

This follows from Theorem 2 if A and I are totally ordered so that the B_i satisfy the conditions of Theorem 1.

We single out two important special cases of this theorem:

1. If $(x_{\alpha\beta})_{(\alpha,\beta)\in A\times B}$ is a finite family of permutable elements of an associative magma whose indexing set is the product of two non-empty finite sets A, B (a "double family"), then

$$(7) \qquad \mathop{\mathrm{T}}_{(\alpha,\beta)\in A\times B} x_{\alpha\beta} = \mathop{\mathrm{T}}_{\alpha\in A}\Big(\mathop{\mathrm{T}}_{\beta\in B} x_{\alpha\beta}\Big) = \mathop{\mathrm{T}}_{\beta\in B}\Big(\mathop{\mathrm{T}}_{\alpha\in A} x_{\alpha\beta}\Big)$$

as follows from Theorem 3 by considering $A \times B$ as the union of the sets $\{\alpha\} \times B$ on the one hand and of the sets $A \times \{\beta\}$ on the other.

For example, if B has n elements and for each $\alpha \in A$ all the $x_{\alpha\beta}$ have the same value x_α, then

$$(8) \qquad \mathop{\mathrm{T}}_{\alpha\in A}\Big(\overset{n}{\mathrm{T}}\, x_\alpha\Big) = \overset{n}{\mathrm{T}}\Big(\mathop{\mathrm{T}}_{\alpha\in A} x_\alpha\Big).$$

If B has two elements, we obtain the following results: let $(x_\alpha)_{\alpha\in A}$, $(y_\alpha)_{\alpha\in A}$ be two non-empty families of elements of E. If the x_α and the y_β are pairwise permutable, then

$$(9) \qquad \mathop{\mathrm{T}}_{\alpha\in A} x_\alpha \,\mathrm{T}\, y_\alpha = \Big(\mathop{\mathrm{T}}_{\alpha\in A} x_\alpha\Big)\,\mathrm{T}\,\Big(\mathop{\mathrm{T}}_{\alpha\in A} y_\alpha\Big).$$

Because of formula (7) the composition of a double sequence (x_{ij}) whose indexing set is the product of two intervals (p, q) and (r, s) in **N** is often denoted for a commutative associative law written additively by

$$\sum_{i=p}^{q}\sum_{j=r}^{s} x_{ij} \quad\text{or}\quad \sum_{j=r}^{s}\sum_{i=p}^{q} x_{ij}$$

and similarly for laws denoted by other symbols.

2. Let n be an integer >0 and let A be the set of ordered pairs of integers (i, j) such that $0 \leqslant i \leqslant n$, $0 \leqslant j \leqslant n$ and $i < j$; the composition of a family $(x_{ij})_{(i,j)\in A}$ (under a commutative associative law) is also denoted by $\mathop{\mathrm{T}}_{0\leqslant i<j\leqslant n} x_{ij}$ (or simply $\mathop{\mathrm{T}}_{i<j} x_{ij}$, if no confusion arises); Theorem 3 here gives the formulae

$$(10) \qquad \mathop{\mathrm{T}}_{0\leqslant i<j\leqslant n} x_{ij} = \mathop{\mathrm{T}}_{i=0}^{n-1}\Big(\mathop{\mathrm{T}}_{j=i+1}^{n} x_{ij}\Big) = \mathop{\mathrm{T}}_{j=1}^{n}\Big(\mathop{\mathrm{T}}_{i=0}^{j-1} x_{ij}\Big).$$

There are analogous formulae to (7) for a family whose indexing set is the product of more than two sets and analogous formulae to (10) for a family whose indexing set is the set S_p of *strictly increasing sequences* $(i_k)_{1\leqslant k\leqslant p}$ of p integers such that $0 \leqslant i_k \leqslant n$ $(p \leqslant n+1)$: in the latter case the composition of the family $(x_{i_1 i_2 \ldots i_p})_{(i_1,\ldots,i_p)\in S_p}$ is denoted by

$$\mathop{\mathrm{T}}_{0\leqslant i_1<i_2<\cdots<i_p<n} x_{i_1 i_2\ldots i_p}, \quad\text{or simply}\quad \mathop{\mathrm{T}}_{i_1<i_2<\cdots<i_p} x_{i_1 i_2\ldots i_p}.$$

Proposition 5. *Let* E *and* F *be magmas whose laws are denoted by* $\top$ *and let* f *and* g *be homomorphisms of* E *into* F. *Let* $f \top g$ *be the mapping* $x \mapsto f(x) \top g(x)$ *of* E *into* F. *If* F *is associative and commutative,* $f \top g$ *is a homomorphism.*

For all elements x and y of E:

$$\begin{aligned}(f \top g)(x \top y) &= f(x \top y) \top g(x \top y) = f(x) \top f(y) \top g(x) \top g(y)\\ &= f(x) \top g(x) \top f(y) \top g(y) = ((f \top g)(x)) \top ((f \top g)(y)).\end{aligned}$$

6. QUOTIENT LAWS

Definition 11. *Let* E *be a set. A law of composition* $\top$ *and an equivalence relation* R *on* E *are said to be compatible if the relations* $x \equiv x' \pmod{R}$ *and* $y \equiv y' \pmod{R}$ *(for* $x, x', y, y' \in E$*) imply* $x \top y \equiv x' \top y' \pmod{R}$; *the law of composition on the quotient set* E/R *which maps the equivalence classes of* x *and* y *to the equivalence class of* $x \top y$ *is called the quotient law of the law* $\top$ *with respect to* R. *The set* E/R *with the quotient law is called the quotient magma of* E *with respect to* R.

To say that an equivalence relation R on E is compatible with the internal law of composition f: $E \times E \to E$ on E means that the mapping f is compatible (in the sense of *Set Theory*, II, § 6, no. 5) with the product equivalence relation $R \times R$ on $E \times E$ and the equivalence relation R on E. (*Set Theory*, II, § 6, no. 8). This also means that the graph of R is a submagma of $E \times E$.

If the law $\top$ is associative (resp. commutative) so is the quotient law (more briefly we say that *associativity, or commutativity, is preserved when passing to the quotient*).

The canonical mapping from the magma E to the magma E/R is a homomorphism.

For a mapping g of E/R into a magma F to be a homomorphism it is necessary and sufficient that the composition of g with the canonical mapping of E onto E/R be a homomorphism.

The two following propositions are immediate from the definitions:

Proposition 6. *Let* E *and* F *be two magmas and* f *a homomorphism of* E *into* F. *Let* $R\{x, y\}$ *denote the relation* $f(x) = f(y)$ *between elements* x, y *of* E. *Then* R *is an equivalence relation on* E *compatible with the law on* E *and the mapping of* E/R *onto* $f(E)$ *derived from* f *by passing to the quotient is an isomorphism of the quotient magma* E/R *onto the submagma* $f(E)$ *of* F.

Proposition 7. *Let* E *be a magma and* R *an equivalence relation on* E *compatible with the law on* E. *For an equivalence relation* S *on* E/R *to be compatible with the quotient law it is necessary and sufficient that* S *be of the form* T/R *where* T *is an equivalence relation on* E *implied by* R *and compatible with the law on* E. *The canonical mapping of* E/T *onto* (E/R)/(T/R) (*Set Theory*, II, § 6, no. 7) *is then a magma isomorphism.*

Proposition 8. *Let* E *be a magma,* A *a stable subset of* E *and* R *an equivalence relation on* E *compatible with the law on* E. *The saturation* B *of* A *with respect to* R (*Set*

Theory, II, § 6, no. 5) *is a stable subset. The equivalence relations* R_A *and* R_B *induced by* R *on* A *and* B *respectively are compatible with the induced laws and the mapping derived from the canonical injection of* A *into* B *by passing to the quotients is a magma isomorphism of* A/R_A *onto* B/R_B.

Let $\top$ denote the law on E. If x and y are two elements of B there exist two elements x' and y' of A such that $x \equiv x'$ (mod R) and $y \equiv y'$ (mod R); then $x \top y \equiv x' \top y'$ (mod R) and $x' \top y' \in A$, whence $x \top y \in B$. Thus B is a stable subset of E and the other assertions are obvious.

Let M be a magma and $((u_\alpha, v_\alpha))_{\alpha \in I}$ a family of elements of M $\times$ M. Consider all the equivalence relations S on M which are compatible with the law on M and such that $u_\alpha \equiv v_\alpha$ (mod S) for all $\alpha \in I$. The intersection of the graphs of these relations is the graph of an equivalence relation R which is compatible with the law on M and such that $u_\alpha \equiv v_\alpha$ (mod R). Hence R is the *finest* (*Set Theory*, III, § 1, nos. 3 and 7) equivalence relation with these two properties. It is called the equivalence relation compatible with the law on M *generated* by the (u_α, v_α).

PROPOSITION 9. *Preserving the above notation, let* f *be a homomorphism of* M *into a magma such that* $f(u_\alpha) = f(v_\alpha)$ *for all* $\alpha \in I$. *Then* f *is compatible with* R.

Let T be the equivalence relation associated with f. Then $u_\alpha \equiv v_\alpha$ (mod T) for all $\alpha \in I$ and T is compatible with the law on M, hence T is coarser than R; this proves the proposition.

§ 2. IDENTITY ELEMENT; CANCELLABLE ELEMENTS; INVERTIBLE ELEMENTS

1. IDENTITY ELEMENT

DEFINITION 1. *Under a law of composition* $\top$ *on a set* E *an element* e *of* E *is called an* identity element *if, for all* $x \in E$, $e \top x = x \top e = x$.

There exists at most one identity element under a given law $\top$, for if e and e' are identity elements then $e = e \top e' = e'$. An identity element is permutable with every element: it is a central element.

DEFINITION 2. *A magma with an identity element is called a unital magma. If* E, E′ *are unital magmas, a homomorphism of the magma* E *into the magma* E′ *which maps the identity element of* E *to the identity element of* E′ *is called a unital homomorphism* (or *morphism*) *of* E *into* E′. *An associative unital magma is called a monoid.*

If E, E′ are monoids, a *unital* morphism of E into E′ is called a *monoid homomorphism* or a *monoid morphism* of E into E′.

Examples. (1) In the set **N** of natural numbers 0 is an identity element under addition and 1 is an identity element under multiplication. Each of these two laws gives **N** a commutative monoid structure (*Set Theory*, III, § 3, no. 3).

(2) In the set of subsets of a set E, $\varnothing$ is an identity element under the law $\cup$ and E under the law $\cap$. More generally, in a lattice the least element, if it exists, is identity element under the law sup; conversely, if there exists an identity element under this law it is the least element of the set. Similarly for the greatest element and the law inf.

(3) The set **N** has no identity element under the law $(x, y) \mapsto x^y$. Under the law $(X, Y) \mapsto X \circ Y$ between subsets of $E \times E$ the diagonal Δ is the identity element. Under the law $(f, g) \mapsto f \circ g$ between mappings of E into E the identity mapping of E onto E is the identity element.

(4) Let E be a magma and R an equivalence relation on E compatible with the law on E (§ 1, no. 6). If e is an identity element of E the canonical image of e in E/R is an identity element of the magma E/R.

The identity element of a unital magma is a unital homomorphism; the composition of two unital homomorphisms is also one. For a mapping to be a unital magma isomorphism it is necessary and sufficient that it be a bijective unital homomorphism and the inverse mapping is then a unital homomorphism. Let E and E′ be unital magmas and e' the identity element of E′; the constant mapping of E into E′ mapping E to e' is a unital homomorphism, called a *trivial homomorphism.*

The product of a family of unital magmas (resp. monoids) is a unital magma (resp. monoid).

Every quotient magma of a unital magma (resp. monoid) is a unital magma (resp. monoid).

Let E be a unital magma and e its identity element. A submagma A of E such that $e \in A$ is called a *unital submagma* of E. Clearly e is the identity element of the magma A. Every intersection of unital submagmas of E is a unital submagma of E. If X is a subset of E then there exists a smallest unital submagma of E containing X; it is called the *unital submagma of* E *generated by* X; it is equal to $\{e\}$ if X is empty. If E is a monoid, a unital submagma of E is called a *submonoid* of E.

If F is a magma without identity element a submagma of F may possess an identity element. For example, if F is associative and h is an idempotent element of F (I, § 1, no. 4), the set of $h \top x \top h$, where x runs through F, is a submagma of F with h as identity element.

If E is a magma with identity element e, a submagma A of E such that $e \notin A$ may still possess an identity element.

Definition 3. *Let* E *be a unital magma. The identity element of* E *is called the composition of the empty family of elements of* E.

If $(x_\alpha)_{\alpha\in\varnothing}$ is the empty family of elements of E its composition e is also denoted by $\mathop{\top}\limits_{\alpha\in\varnothing} x_\alpha$. For example, we write

$$\mathop{\top}\limits_{q\leqslant i\leqslant p} x_i = e$$

when $p < q$ ($p, q \in \mathbf{N}$). Similarly we write $\overset{0}{\top} x = e$ for arbitrary x. *With these definitions Theorems* 1 *and* 3 *of* § 1 *remain true if the hypothesis that the sets* A *and* B_i *are non-empty is suppressed.* Similarly the formulae $\overset{m+n}{\top} x = \left(\overset{m}{\top} x\right) \top \left(\overset{n}{\top} x\right)$ and $\overset{mn}{\top} x = \overset{m}{\top}\left(\overset{n}{\top} x\right)$ are then true for $m \geqslant 0, n \geqslant 0$.

Let E be a unital magma whose law is denoted by $\top$ and e its identity element. The *support* of a family $(x_i)_{i\in I}$ of elements of E is the set of indices $i \in I$ such that $x_i \neq e$. Let $(x_i)_{i\in I}$ be a family of elements of E *with finite support*. We shall define the composition $\mathop{\top}\limits_{i\in I} x_i$ in the two following cases:

(a) the set I is totally ordered;
(b) E is associative and the x_i are pairwise permutable.

In these two cases let S be the support of the family (x_i). If J is a finite subset of I containing S, then $\mathop{\top}\limits_{i\in J} x_i = \mathop{\top}\limits_{i\in S} x_i$, as is seen by induction on the number of elements in J, applying Theorem 1 of § 1 in case (a) and Theorem 3 of § 1 in case (b). Let $\mathop{\top}\limits_{i\in I} x_i$ denote the common value of the compositions $\mathop{\top}\limits_{i\in J} x_i$ for all finite subsets of I containing S. When I is the interval $[p, \rightarrow[$ of $\mathbf{N}$, we also write $\mathop{\top}\limits_{i=p}^{\infty} x_i$.

With these definitions and notation, Theorems 1 and 3 of § 1 and the remarks following Theorem 3 extend to families with finite support.

The identity element of a law written *additively* is usually denoted by 0 and called *zero* or the *null element* (or sometimes the *origin*). Under a law written *multiplicatively* it is usually denoted by 1 and called the *unit element* (or *unit*).

2. CANCELLABLE ELEMENTS

DEFINITION 4. *Given a law of composition* $\top$ *on a set* E, *the mapping* $x \mapsto a \top x$ (resp. $x \mapsto x \top a$) *of* E *into itself is called left translation* (resp. *right translation*) *by an element* $a \in$ E.

On passing to the opposite law, left translations become right translations and conversely.

Let γ_a, δ_a (or $\gamma(a)$, $\delta(a)$) denote the left and right translations by $a \in$ E; then

$$\gamma_a(x) = a \top x, \qquad \delta_a(x) = x \top a.$$

PROPOSITION 1. *If the law* $\top$ *is associative, then for all* $x \in E$ *and* $y \in E$

$$\gamma_{x \top y} = \gamma_x \circ \gamma_y, \qquad \delta_{x \top y} = \delta_y \circ \delta_x.$$

For all $z \in E$:

$$\gamma_{x \top y}(z) = (x \top y) \top z = x \top (y \top z) = \gamma_x(\gamma_y(z))$$
$$\delta_{x \top y}(z) = z \top (x \top y) = (z \top x) \top y = \delta_y(\delta_x(z))$$

In other words, the mapping $x \mapsto \gamma_x$ is a homomorphism from the magma E to the set E^E of mappings of E into itself with the law $(f, g) \mapsto f \circ g$; the mapping $x \mapsto \delta_x$ is a homomorphism of E into the set E^E with the opposite law. If E is a monoid, these homomorphisms are unital.

DEFINITION 5. *An element a of a magma* E *is called* left (resp. right) cancellable (*or* regular) *if left* (resp. *right*) *translation by a is injective. A left and right cancellable element is called a cancellable* (*or regular*) *element.*

In other words, for a to be cancellable under the law $\top$, it is necessary and sufficient that each of the relations $a \top x = a \top y$, $x \top a = y \top a$ imply $x = y$ (it is said that "a can be cancelled" from each of these equalities). If there exists an identity element e under the law $\top$, it is cancellable under this law: the translations γ_e and δ_e are then the identity mapping of E onto itself.

Examples. (1) Every natural number is cancellable under addition; every natural number $\neq 0$ is cancellable under multiplication.

(2) In a lattice there can be no cancellable element under the law sup other than the identity element (least element) if it exists; similarly for inf. In particular, in the set of subsets of a set E, $\varnothing$ is the only cancellable element under the law $\cup$ and E the only cancellable element under the law $\cap$.

PROPOSITION 2. *The set of cancellable* (resp. *left cancellable*, resp. *right cancellable*) *elements of an associative magma is a submagma.*

If γ_x and γ_y are injective so is $\gamma_{x \top y} = \gamma_x \circ \gamma_y$ (Proposition 1). Similarly for $\delta_{x \top y}$.

3. INVERTIBLE ELEMENTS

DEFINITION 6. *Let* E *be a unital magma,* $\top$ *its law of composition, e its identity element and x and x′ two elements of* E. *x′ is called a* left inverse (resp. right inverse, resp. inverse) *of x if* $x' \top x = e$ (resp. $x \top x' = e$, resp. $x' \top x = x \top x' = e$).

An element x of E *is called* left invertible (resp. right invertible, resp. invertible) *if it has a left inverse* (resp. *right inverse*, resp. *inverse*).

A monoid all of whose elements are invertible is called a group.

Symmetric and *symmetrizable* are sometimes used instead of *inverse* and *invertible*. When the law on E is written additively, we generally say *negative* instead of *inverse*.

Examples. (1) An identity element is its own inverse.

(2) In the set of mappings of E into E an element f is left invertible (resp. right invertible) if f is a surjection (resp. injection). The left inverses (resp. right inverses) are then the retractions (resp. sections) associated with f (*Set Theory*, II, § 3, no. 8, Definition 11). For f to be invertible it is necessary and sufficient that f be a bijection. Its unique inverse is then the inverse bijection of f.

Let E and F be two unital magmas and f a unital homomorphism of E into F. If x' is the inverse of x in E, $f(x')$ is the inverse of $f(x)$ in F. Hence, if x is an invertible element of E, $f(x)$ is an invertible element of F.

In particular, if R is an equivalence relation compatible with the law on a unital magma E, the canonical image of an invertible element of E is invertible in E/R.

PROPOSITION 3. *Let* E *be a monoid and* x *an element of* E.

(i) *For* x *to be left* (resp. *right*) *invertible it is necessary and sufficient that the right* (resp. *left*) *translation by* x *be surjective.*

(ii) *For* x *to be invertible it is necessary and sufficient that it be left and right invertible. In that case* x *has a unique inverse, which is also its unique left* (resp. *right*) *inverse.*

If x' is a left inverse of x then (Proposition 1)

$$\boldsymbol{\delta}_x \circ \boldsymbol{\delta}_{x'} = \boldsymbol{\delta}_{x' \top x} = \boldsymbol{\delta}_e = \mathrm{Id}_E$$

and $\boldsymbol{\delta}_x$ is surjective. Conversely, if $\boldsymbol{\delta}_x$ is surjective, there exists an element x' of E such that $\boldsymbol{\delta}_x(x') = e$ and x' is the left inverse of x. The other assertions of (i) follow similarly.

If x' (resp. x'') is a left (resp. right) inverse of x, then

$$x' = x' \top e = x' \top (x \top x'') = (x' \top x) \top x'' = e \top x'' = x'',$$

whence (ii).

Remark. Let E be a monoid and x an element of E. If x is left invertible it is left cancellable; for, if x' is a left inverse of x, then

$$\boldsymbol{\gamma}_{x'} \circ \boldsymbol{\gamma}_x = \boldsymbol{\gamma}_{x' \top x} = \boldsymbol{\gamma}_e = \mathrm{Id}_E$$

and $\boldsymbol{\gamma}_x$ is injective. In particular, if x is left invertible, the left and right translations by x are *bijective*. Conversely, suppose that $\boldsymbol{\gamma}_x$ is bijective; there exists $x' \in E$ such that $xx' = \boldsymbol{\gamma}_x(x') = e$; then $\boldsymbol{\gamma}_x(x'x) = (xx')\,x = x = \boldsymbol{\gamma}_x(e)$ and hence $x'x = e$, so that x is invertible. We see similarly that if $\boldsymbol{\delta}_x$ is bijective, x is invertible.

PROPOSITION 4. *Let* E *be a monoid and* x *and* y *invertible elements of* E *with inverses* x' *and* y' *respectively. Then* $y' \top x'$ *is the inverse of* $x \top y$.

This follows from the relation

$$(y' \top x') \top (x \top y) = y' \top (x' \top x) \top y = y' \top y = e$$

and the analogous calculations for $(x \top y) \top (y' \top x')$.

COROLLARY 1. *Let* E *be a monoid; if each of the elements* x_α *of an ordered sequence* $(x_\alpha)_{\alpha \in A}$ *of elements of* E *has an inverse* x'_α, *the composition* $\top_{\alpha \in A} x_\alpha$ *has inverse* $\top_{\alpha \in A'} x'_\alpha$, *where* A' *is the totally ordered set derived from* A *by replacing the order on* A *by the opposite order.*

This corollary follows from Proposition 4 by induction on the number of elements in A.

In particular, if x and x' are inverses, $\overset{n}{\top}\, x$ and $\overset{n}{\top}\, x'$ are inverses for every integer $n \geqslant 0$.

COROLLARY 2. *In a monoid the set of invertible elements is stable.*

PROPOSITION 5. *If in a monoid* x *and* x' *are inverses and* x *commutes with* y, *so does* x'.

From $x \top y = y \top x$, we deduce $x' \top (x \top y) \top x' = x' \top (y \top x) \top x'$ and hence $(x' \top x) \top (y \top x') = (x' \top y) \top (x \top x')$, that is $y \top x' = x' \top y$.

COROLLARY 1. *Let* E *be a monoid,* X *a subset of* E *and* X' *the centralizer of* X. *The inverse of every invertible element of* X' *belongs to* X'.

COROLLARY 2. *In a monoid the inverse of a central invertible element is a central element.*

4. MONOID OF FRACTIONS OF A COMMUTATIVE MONOID

In this no., e will denote the identity element of a monoid and x^* the inverse of an invertible element x of E.

Let E be a *commutative* monoid, S a subset of E and S' the submonoid of E generated by S.

Lemma 1. *In* E × S' *the relation* R$\{x, y\}$ *defined by:*

"there exist $a, b \in$ E and $p, q, s \in$ S' such that $x = (a, p)$, $y = (b, q)$ and $aqs = bps$"

is an equivalence relation compatible with the law on the product monoid E × S'.

It is immediate that R is reflexive and symmetric. Let $x = (a, p)$, $y = (b, q)$ and $z = (c, r)$ be elements of E × S' such that R$\{x, y\}$ and R$\{y, z\}$ hold. Then there exist two elements s and t of S' such that

$$aqs = bps, \qquad brt = cqt,$$

whence it follows that

$$ar(stq) = bpsrt = cp(stq)$$

and hence $R\{x, z\}$ holds, for stq belongs to S'. The relation R is therefore transitive.

Further, let $x = (a, p), y = (b, q), x' = (a', p')$, and $y' = (b', q')$ be elements of $E \times S'$ such that $R\{x, y\}$ and $R\{x', y'\}$ hold. There exist s and s' in S' such that

$$aqs = bps, \quad a'q's' = b'p's',$$

whence it follows that $(aa')(qq')(ss') = (bb')(pp')(ss')$ and hence $R\{xx', yy'\}$ for $ss' \in S'$. The equivalence relation R is therefore compatible with the law of composition on $E \times S'$.

The quotient magma $(E \times S')/R$ is a commutative monoid.

DEFINITION 7. *Let* E *be a* commutative *monoid,* S *a subset of* E *and* S' *the submonoid of* E *generated by* S. *The quotient monoid* $(E \times S')/R$, *where the equivalence relation* R *is as described in Lemma* 1, *is denoted by* E_S *and is called the monoid of fractions*† *of* E *associated with* S (*or with denominators in* S).

For $a \in E$ and $p \in S'$ the class of (a, p) modulo R is in general denoted by a/p and called the *fraction* with *numerator* a and *denominator* p. Then by definition $(a/p).(a'/p') = aa'/pp'$. The fractions a/p and a'/p' are equal if and only if there exists s in S' with $spa' = sp'a$; if so, there exist σ and σ' in S' with $a\sigma = a'\sigma'$ and $p\sigma = p'\sigma'$. In particular, $a/p = sa/sp$ for $a \in E$ and s, p in S'. The identity element of E_S is the fraction e/e.

Let $a/e = \varepsilon(a)$ for all $a \in E$. The above shows that ε is a homomorphism of E into E_S, called *canonical*. For all $p \in S'$, $(p/e).(e/p) = e/e$ and hence e/p is the inverse of $\varepsilon(p) = p/e$; every element of $\varepsilon(S')$ is therefore invertible. Then $a/p = (a/e).(e/p)$, whence

$$a/p = \varepsilon(a)\varepsilon(p)^* \tag{1}$$

for $a \in E$ and $p \in S'$, the monoid E_S is therefore generated by $\varepsilon(E) \cup \varepsilon(S)^*$.

PROPOSITION 6. *The notation is that of Definition* 7 *and* ε *denotes the canonical homomorphism of* E *into* E_S.

(i) *Let* a *and* b *be in* E; *in order that* $\varepsilon(a) = \varepsilon(b)$, *it is necessary and sufficient that there exist* $s \in S'$ *with* $sa = sb$.

(ii) *For* ε *to be injective it is necessary and sufficient that every element of* S *be cancellable.*

(iii) *For* ε *to be bijective it is necessary and sufficient that every element of* S *be invertible.*

Assertion (i) is clear and shows that ε is injective if and only if every element of S' is cancellable; but since the set of cancellable elements of E is a submonoid of E (no. 2, Proposition 2), it amounts to the same to say that every element of S is cancellable.

† It is also called *monoid of differences* if the law on E is written additively.

If ε is bijective every element of S is invertible, for $\varepsilon(S)$ is composed of invertible elements of E_S. Conversely, suppose every element of S is invertible; then every element of S′ is invertible (no. 3, Corollary 2 to Proposition 4) and hence cancellable. Then ε is injective by (ii) and $a/p = \varepsilon(a.p^*)$ by (1), hence ε is surjective.

THEOREM 1. *Let* E *be a commutative monoid,* S *a subset of* E, E_S *the monoid of fractions associated with* S *and* $\varepsilon: E \to E_S$ *the canonical homomorphism. Further let f be a homomorphism of* E *into a monoid* F *(not necessarily commutative) such that every element of* $f(S)$ *is invertible in* F. *There exists one and only one homomorphism* $\bar{f}$ *of* E_S *into* F *such that* $f = \bar{f} \circ \varepsilon$.

If $\bar{f}$ is a homomorphism of E_S into F such that $f = \bar{f} \circ \varepsilon$, then

$$\bar{f}(a/p) = \bar{f}(\varepsilon(a)\varepsilon(p)^*) = \bar{f}(\varepsilon(a))\,\bar{f}(\varepsilon(p))^* = f(a)\,f(p)^*$$

for $a \in E$ and $p \in S'$, whence the uniqueness of $\bar{f}$.

Let g be the mapping of $E \times S'$ into F defined by $g(a, p) = f(a).f(p)^*$. We show that g is a homomorphism of $E \times S'$ into F. First of all,

$$g(e, e) = f(e)f(e)^* = e.$$

Let (a, p) and (a', p') be two elements of $E \times S'$; as a' and p commute in E, $f(a')$ and $f(p)$ commute in F, whence $f(a')f(p)^* = f(p)^*f(a')$ by no. 3, Proposition 5. Moreover $f(pp')^* = f(p'p)^* = (f(p')f(p))^* = f(p)^*f(p')^*$ by no. 3, Proposition 4, whence

$$\begin{aligned} g(aa', pp') &= f(aa')f(pp')^* = f(a)f(a')f(p)^*f(p')^* = f(a)f(p)^*f(a')f(p')^* \\ &= g(a, p)\,g(a', p'). \end{aligned}$$

We show that g is compatible with the equivalence relation R on $E \times S'$: if (a, p) and (a', p') are congruent mod. R, there exists $s \in S'$ with $spa' = sap'$, whence $f(s)f(p)f(a') = f(s)f(a)f(p')$. As $f(s)$ is invertible, it follows that $f(p)f(a') = f(a)f(p')$ and then by left multiplication by $f(p)^*$ and right multiplication by $f(p')^*$

$$g(a', p') = f(a')f(p')^* = f(p)^*f(a) = f(a)f(p)^* = g(a, p).$$

Hence there exists a homomorphism $\bar{f}$ of E_S into F such that $\bar{f}(a/p) = g(a, p)$, whence $\bar{f}(\varepsilon(a)) = \bar{f}(a/e) = f(a)f(e)^* = f(a)$. Hence $\bar{f} \circ \varepsilon = f$.

COROLLARY. *Let* E *and* F *be two commutative monoids,* S *and* T *subsets of* E *and* F *respectively, f a homomorphism of* E *into* F *such that* $f(S) \subset T$ *and* $\varepsilon: E \to E_S$, $\eta: F \to F_T$ *the canonical homomorphisms. There exists one and only one homomorphism* $g: E_S \to F_T$ *such that* $g \circ \varepsilon = \eta \circ f$.

The homomorphism $\eta \circ f$ of E into F_T maps every element of S to an invertible element of F_T.

Remarks. (1) Theorem 1 can also be expressed by saying that (E_S, ε) is the solution of the universal mapping problem for E, relative to monoids, monoid homomorphisms and homomorphisms of E into monoids which map the elements of S to invertible elements (*Set Theory*, IV, § 3, no. 1). It follows (*loc. cit.*) that every other solution of this problem is isomorphic in a unique way to (E_S, ε).

(2) For the existence of a solution to the above universal mapping problem it is unnecessary to assume that the monoid E is commutative, as follows from *Set Theory*, IV, § 3, no. 2 (cf. Exercise 17).

We mention two important special cases of monoids of fractions.

(a) Let $\bar{E} = E_E$. As the monoid $\bar{E}$ is generated by the set $\varepsilon(E) \cup \varepsilon(E)^*$, which is composed of invertible elements, every element of $\bar{E}$ is invertible (no. 3, Corollary 2 to Proposition 4). In other words, $\bar{E}$ is a commutative group. Moreover, by Theorem 1 every homomorphism f of E into a group G can be uniquely factorized in the form $f = \bar{f} \circ \varepsilon$, where $\bar{f}\colon \bar{E} \to G$ is a homomorphism. $\bar{E}$ is called the ***group of fractions of*** E (or ***group of differences of*** E in the case of additive notation).

(b) Let $\Phi = E_\Sigma$, where Σ consists of the cancellable elements of E. By Proposition 6, (ii), the canonical homomorphism of E into Φ is injective; it will be profitable to identify E with its image in Φ. Hence E is a submonoid of Φ, every cancellable element of E has an inverse in Φ and every element of Φ is of the form $a/p = a.p^*$ with $a \in E$ and $p \in \Sigma$; then $a/p = a'/p'$ if and only if $ap' = pa'$. It is easily seen that the invertible elements of Φ are the fractions a/p, where a and p are cancellable, and p/a is the inverse of a/p.

Now let S be a set of cancellable elements of E and S′ be the submonoid of E generated by S. If a/p and a'/p' are two elements of E_S, then $a/p = a'/p'$ if and only if $ap' = pa'$ (for $sap' = spa'$ implies $ap' = pa'$ for all $s \in S'$). E_S may therefore be identified with the submonoid of Φ generated by $E \cup S^*$.

If every element of E is cancellable, then $\Phi = \bar{E}$ and E is a submonoid of the commutative group Φ. Conversely, if E is isomorphic to a submonoid of a group, every element of E is cancellable.

5. APPLICATIONS: I. RATIONAL INTEGERS

Consider the commutative monoid **N** of natural numbers with law of composition addition; all the elements of **N** are cancellable under this law (*Set Theory*, III, § 5, no. 2, Corollary 3). The group of differences of **N** is denoted by **Z**; its elements are called the ***rational integers***; its law is called ***addition of rational integers*** and also denoted by +. The canonical homomorphism from **N** to **Z** is injective and we shall identify each element of **N** with its image in **Z**. The elements of **Z** are by definition the equivalence classes determined in **N** × **N** by the relation between (m_1, n_1) and (m_2, n_2) which is written $m_1 + n_2 = m_2 + n_1$; an element m of **N** is identified with the class consisting of the elements $(m + n, n)$, where

$n \in \mathbf{N}$; it admits as negative in $\mathbf{Z}$ the class of elements $(n, m + n)$. Every element (p, q) of $\mathbf{N} \times \mathbf{N}$ may be written in the form $(m + n, n)$ if $p \geqslant q$ or in the form $(n, m + n)$ if $p \leqslant q$; it follows that $\mathbf{Z}$ is *the union of* $\mathbf{N}$ *and the set of negatives of the elements of* $\mathbf{N}$. The identity element 0 is the only element of $\mathbf{N}$ whose negative belongs to $\mathbf{N}$.

For every natural number m, $-m$ denotes the negative rational integer of m and $-\mathbf{N}$ denotes the set of elements $-m$ for $m \in \mathbf{N}$. Then

$$\mathbf{Z} = \mathbf{N} \cup (-\mathbf{N}) \quad \text{and} \quad \mathbf{N} \cap (-\mathbf{N}) = \{0\}.$$

for $m \in \mathbf{N}$, $m = -m$ if and only if $m = 0$.

Let m and n be two natural numbers;

(a) if $m \geqslant n$, then $m + (-n) = p$, where p is the element of $\mathbf{N}$ such that $m = n + p$;

(b) if $m \leqslant n$, then $m + (-n) = -p$, where p is the element of $\mathbf{N}$ such that $m + p = n$;

(c) $(-m) + (-n) = -(m + n)$.

Properties (b) and (c) follow from no. 3, Proposition 4; as $\mathbf{Z} = \mathbf{N} \cup (-\mathbf{N})$, addition in $\mathbf{N}$ and properties (a), (b) and (c) describe completely addition in $\mathbf{Z}$.

More generally $-x$ is used to denote the negative of an arbitrary rational integer x; the composition $x + (-y)$ is abbreviated to $x - y$ (cf. no. 8).

The order relation $\leqslant$ between natural numbers is characterized by the following property: $m \leqslant n$ if and only if there exists an integer $p \in \mathbf{N}$ such that $m + p = n$ (*Set Theory*, III, § 3, no. 6, Proposition 13 and § 5, no. 2, Proposition 2). The relation $y - x \in \mathbf{N}$ between rational integers x and y is a *total order* relation on $\mathbf{Z}$ which extends the order relation $\leqslant$ on $\mathbf{N}$. For, for all $x \in \mathbf{Z}$, $x - x = 0 \in \mathbf{N}$; if $y - x \in \mathbf{N}$ and $z - y \in \mathbf{N}$, then

$$z - x = (z - y) + (y - x) \in \mathbf{N},$$

for $\mathbf{N}$ is stable under addition; if $y - x \in \mathbf{N}$ and $x - y \in \mathbf{N}$, then $y - x = 0$, for 0 is the only element of $\mathbf{N}$ whose negative belongs to $\mathbf{N}$; for arbitrary rational integers x and y, $y - x \in \mathbf{N}$ or $x - y \in \mathbf{N}$, for $\mathbf{Z} = \mathbf{N} \cup (-\mathbf{N})$; finally, if x and y are natural numbers, then $y - x \in \mathbf{N}$ if and only if there exists $p \in \mathbf{N}$ such that $x + p = y$. This order relation is also denoted by $\leqslant$.

Henceforth when $\mathbf{Z}$ is considered as an ordered set, it will always be, unless otherwise mentioned, with the ordering that has just been defined, the natural numbers are identified with the integers $\geqslant 0$; they are also called *positive* integers; the integers $\leqslant 0$, negatives of the positive integers, are called *negative* integers; the integers > 0 (resp. < 0) are called *strictly positive* (resp. *strictly negative*); the set of integers > 0 is sometimes denoted by $\mathbf{N}^*$.

Let x, y and z be three rational integers; then $x \leqslant y$ if and only if

$x + z \leqslant y + z$. For $x - y = (x + z) - (y + z)$. This property is expressed by saying that the order relation on **Z** is *invariant under translation.*

6. APPLICATIONS: II. MULTIPLICATION OF RATIONAL INTEGERS

Lemma 2. *Let* E *be a monoid and* x *an element of* E.

(i) *There exists a unique homomorphism* f *of* **N** *into* E *with* $f(1) = x$ *and* $f(n) = \overset{n}{\top} x$ *for all* $n \in \mathbf{N}$.

(ii) *If* x *is invertible, there exists a unique homomorphism* g *of* **Z** *into* E *such that* $g(1) = x$ *and* g *coincides with* f *on* **N**.

Writing $f(n) = \overset{n}{\top} x$ for all $n \in \mathbf{N}$, the formulae

$$\overset{0}{\top} x = e \quad \text{and} \quad \left(\overset{m}{\top} x\right) \top \left(\overset{n}{\top} x\right) = \overset{m+n}{\top} x$$

(no. 1) express the fact that f is a homomorphism of **N** into E and obviously $f(1) = x$. If f' is a homomorphism of **N** into E such that $f'(1) = x$, then $f = f'$, by § 1, no. 4, Proposition 1, (iv).

Suppose now that x is invertible. By no. 3, Corollary 2 to Proposition 4, $f(n) = \overset{n}{\top} x$ is invertible for every integer $n \geqslant 0$. By construction, **Z** is the group of differences of **N** and hence (no. 4, Theorem 1) f extends uniquely to a homomorphism g of **Z** into E. If g' is a homomorphism of **Z** into E with $g'(1) = x$, the restriction f' of g' to **N** is a homomorphism of **N** into E with $f'(1) = x$. Hence $f' = f$, whence $g' = g$.

We shall apply Lemma 2 to the case where the monoid E is **Z**; for every integer $m \in \mathbf{Z}$ there therefore exists an endomorphism f_m of **Z** characterized by $f_m(1) = m$. If m is in **N**, the mapping $n \mapsto mn$ of **N** into **N** is an endomorphism of the magma **N** (*Set Theory*, III, § 3, no. 3, Corollary to Proposition 5); hence $f_m(n) = mn$ for all m, n in **N**.

Multiplication on **N** can therefore be extended to multiplication on **Z** by the formula $mn = f_m(n)$ for m, n in **Z**. We shall establish the formulae:

$$xy = yx \tag{2}$$

$$(xy)z = x(yz) \tag{3}$$

$$x(y + z) = xy + xz \tag{4}$$

$$(x + y)z = xz + yz \tag{5}$$

$$0.x = x.0 = 0 \tag{6}$$

$$1.x = x.1 = x \tag{7}$$

$$(-1).x = x.(-1) = -x \tag{8}$$

for x, y, z in **Z**. (*In other words, **Z** is a commutative ring.*) The formulae $x(y + z) = xy + xz$ and $x.0 = 0$ express the fact that f_x is an endomorphism of the additive monoid $\mathbf{Z}$ and $f_x(1) = x$ may be written $x.1 = x$. The endomorphism $f_x \circ f_y$ of $\mathbf{Z}$ maps 1 to xy and hence is equal to f_{xy}, whence (3). Now $f_x(-y) = -f_x(y)$, that is $x(-y) = -xy$; similarly, the endomorphism $y \mapsto -xy$ of $\mathbf{Z}$ maps 1 to $-x$, whence $(-x).y = -xy$ and therefore

$$(-x)(-y) = -(x(-y)) = -(-xy) = xy.$$

For m, n in $\mathbf{N}$, $mn = nm$ (*Set Theory*, III, § 3, no. 3, Corollary to Proposition 5), whence $(-m).n = n(-m)$ and $(-m)(-n) = (-n)(-m)$; as $\mathbf{Z} = \mathbf{N} \cup (-\mathbf{N})$, $xy = yx$ for x, y in $\mathbf{Z}$; and this formula means that (5) follows from (4) and completes the proof of formula (6) to (8).

7. APPLICATIONS: III. GENERALIZED POWERS

Let E be a monoid with identity element e and law of composition denoted by $\top$. If x is invertible in E, let g_x be the homomorphism of $\mathbf{Z}$ into E mapping 1 to x. Let $g_x(n) = \overset{n}{\top}\, x$ for all $n \in \mathbf{Z}$; by Lemma 2 this notation is compatible for $n \in \mathbf{N}$ with the notation introduced earlier. Then

$$\overset{m+n}{\top}\, x = \left(\overset{m}{\top}\, x\right) \top \left(\overset{n}{\top}\, x\right) \tag{9}$$

$$\overset{0}{\top}\, x = e \tag{10}$$

$$\overset{1}{\top}\, x = x \tag{11}$$

for x invertible in E and m, n in $\mathbf{Z}$. Further, if $y = \overset{m}{\top}\, x$, the mapping $n \mapsto g_x(mn)$ of $\mathbf{Z}$ into E is a homomorphism mapping 1 to y, whence $g_x(mn) = g_y(n)$, that is

$$\overset{mn}{\top}\, x = \overset{n}{\top}\left(\overset{m}{\top}\, x\right). \tag{12}$$

As -1 is the negative of 1 in $\mathbf{Z}$, $\overset{-1}{\top}\, x$ is the inverse of $x = \overset{1}{\top}\, x$ in E. If we write $n = -m$ in (9), it is seen that $\overset{-m}{\top}\, x$ is the inverse of $\overset{m}{\top}\, x$.

8. NOTATION

(a) As a general rule the law of a commutative monoid is written additively. It is then a convention that $-x$ denotes the negative of x. The notation $x + (-y)$ is abbreviated to $x - y$ and similarly

$$x + y - z, \quad x - y - z, \quad x - y + z - t, \quad \text{etc.} \ldots$$

represent respectively

$$x + y + (-z), \quad x + (-y) + (-z), \quad x + (-y) + z + (-t), \quad \text{etc.} \ldots$$

For $n \in \mathbf{Z}$ the notation $\overset{n}{\top} x$ is replaced by nx. Formulae (9) to (12) then become

(13) $$(m + n).x = m.x + n.x$$

(14) $$0.x = 0$$

(15) $$1.x = x$$

(16) $$m.(n.x) = (mn).x$$

where m and n belong to $\mathbf{N}$ or even to $\mathbf{Z}$ if x admits a negative. Also in the latter case the relation $(-1).x = -x$ holds. We also note the formula

(17) $$n.(x + y) = n.x + n.y.$$

(b) Let E be a monoid written multiplicatively. For $n \in \mathbf{Z}$ the notation $\overset{n}{\top} x$ is replaced by x^n. We have the relations

$$x^{m+n} = x^m.x^n$$
$$x^0 = 1$$
$$x^1 = x$$
$$(x^m)^n = x^{mn}$$

and also $(xy)^n = x^n y^n$ if x and y commute.

When x has an inverse, this is precisely x^{-1}. The notation $\frac{1}{x}$ is also used instead of x^{-1}. Finally, when the monoid E is commutative, $\frac{x}{y}$ or x/y is also used for xy^{-1}.

§ 3. ACTIONS

1. ACTIONS

DEFINITION 1. *Let* Ω *and* E *be two sets. A mapping of* Ω *into the set* E^E *of mappings of* E *into itself is called an* action of Ω on E.

Let $\alpha \mapsto f_\alpha$ be an action of Ω on E. The mapping $(\alpha, x) \mapsto f_\alpha(x)$ (resp. $(x, \alpha) \mapsto f_\alpha(x)$) is called the *law of left* (resp. *right*) *action of* Ω *on* E† *associated with the given action of* Ω *on* E. Given a mapping g of $\Omega \times E$ (resp. $E \times \Omega$) into E, there exists one and only one action $\alpha \mapsto f_\alpha$ of Ω on E such that the associated law of left (resp. right) action is g (*Set Theory*, II, § 5, no. 2, Proposition 3).

In this chapter we shall say, for the sake of abbreviation, "law of action"

† Or sometimes the *external law of composition* on E with Ω as operating set.

instead of "law of left action". The element $f_\alpha(x)$ of E (for $\alpha \in \Omega$ and $x \in$ E) is sometimes called the *transform* of x under α or the *composition* of α and x. It is often denoted by left (resp. right) multiplicative notation $\alpha . x$ (resp. $x . \alpha$), the dot may be omitted; the composition of α and x is then called the *product* of α and x (resp. x and α). The exponential notation x^α is also used. In the arguments of the following paragraphs we shall generally use the notation $\alpha \perp x$. The elements of Ω are often called *operators*.

Examples. (1) Let E be an associative magma written multiplicatively. The mapping which associates with a strictly positive integer n the mapping $x \mapsto x^n$ of E into itself is an action of $\mathbf{N}^*$ on E. If E is a group, the mapping which associates with a rational integer a the mapping $x \mapsto x^a$ of E into E is an action of $\mathbf{Z}$ on E.

(2) Let E be a magma with law denoted by $\top$. The mapping which associates with $x \in$ E the mapping $A \mapsto x \top A$ of the set of subsets of E into itself is an action of E on $\mathfrak{P}(E)$.

(3) Let E be a set. The identity mapping of E^E is an action of E^E on E, called the *canonical action*. The corresponding law of action is the mapping $(f, x) \mapsto f(x)$ of $E^E \times E$ into E.

(4) Let $(\Omega_\iota)_{\iota \in I}$ be a family of sets. For all $i \in I$, let $f_i \colon \Omega_\iota \to E^E$ be an action of Ω_ι on E. Let Ω be the sum of the Ω_ι (*Set Theory*, II, § 4, no. 8). The mapping f of Ω onto E^E, extending the f_ι, is an action of Ω on E. This allows us to reduce the study of a family of actions to that of a single action.

(5) Given an action of Ω on E with law denoted by $\perp$, a subset Ξ of Ω and a subset X of E, $\Xi \perp X$ denotes the set of $\alpha \perp x$ with $\alpha \in \Xi$ and $x \in X$; when Ξ consists of a single element α, we generally write $\alpha \perp X$ instead of $\{\alpha\} \perp X$. The mapping which associates with $\alpha \in \Omega$ the mapping $X \mapsto \alpha \perp X$ is an action of Ω on $\mathfrak{P}(E)$, which is said to be *derived* from the given action by extension to the set of subsets.

(6) Let $\alpha \mapsto f_\alpha$ be an action of Ω on E. Let g be a mapping of Ω' into Ω. Then the mapping $\beta \mapsto f_{g(\beta)}$ is an action of Ω' on E.

(7) Let $f \colon E \times E \to E$ be a law of composition on a set E. The mapping $\gamma \colon x \mapsto \gamma_x$ (resp. $\delta \colon x \mapsto \delta_x$) (§ 2, no. 2) which associates with the element $x \in$ E left (resp. right) translation by x is an action of E on itself; it is called the *left* (resp. *right*) *action* of E on itself *derived* from the given law. When f is commutative, these two actions coincide.

The law of left (resp. right) action associated with γ is f (resp. the opposite law to f). The law of right (resp. left) action associated with δ is f (resp. the opposite law to f).

Let Ω, E, F be sets, $\alpha \mapsto f_\alpha$ an action of Ω on E and $\alpha \mapsto g_\alpha$ an action of Ω on F. An Ω-*morphism of* E *into* F, or *mapping of* E *into* F *compatible with the action of* Ω, is a mapping h of E into F such that

$$g_\alpha(h(x)) = h(f_\alpha(x))$$

for all $\alpha \in \Omega$ and $x \in E$. The composition of two Ω-morphisms is an Ω-morphism.

Let Ω, Ξ, E, F be sets, $\alpha \mapsto f_\alpha$ an action of Ω on E, $\beta \mapsto g_\beta$ an action of Ξ on F and ϕ a mapping of Ω into Ω'. A *ϕ-morphism* of E into F is a mapping h of E into F such that

$$g_{\phi(\alpha)}(h(x)) = h(f_\alpha(x))$$

for all $\alpha \in \Omega$ and $x \in E$.

2. SUBSETS STABLE UNDER AN ACTION. INDUCED ACTION

DEFINITION 2. *A subset* A *of a set* E *is called* stable *under an action* $\alpha \mapsto f_\alpha$ *of* Ω *on* E *if* $f_\alpha(A) \subset A$ *for all* $\alpha \in \Omega$. *An element* x *of* E *is called invariant under an element* α *of* Ω *if* $f_\alpha(x) = x$.

The intersection of a family of stable subsets of E under a given action is stable. There therefore exists a smallest stable subset of E containing a given subset X of E; it is said to be *generated* by X; it consists of the elements $(f_{\alpha_1} \circ f_{\alpha_2} \circ \cdots \circ f_{\alpha_n})(x)$, where $x \in X$, $n \geqslant 0$, $\alpha_i \in \Omega$ for all i.

Remark. Let E be a magma with law denoted by $\top$. It should be noted that a subset A of E which is stable under the left action on E on itself is not necessarily stable under the right action of E on itself; a subset A of E stable under the left (resp. right) action of E on itself is stable under the law on E but the converse is not in general true. More precisely, A is stable under the law on E if and only if $A \top A \subset A$ whereas A is stable under the left (resp. right) action on E on itself if and only if $E \top A \subset A$ (resp. $A \top E \subset A$).

Example. Take the magma E to be the set **N** with multiplication. The set $\{1\}$ is stable under the internal law of **N**, but the stable subset under the action of **N** on itself generated by $\{1\}$ is the whole of **N**.

DEFINITION 3. *Let* $\alpha \mapsto f_\alpha$ *be an action of* Ω *on* E *and* A *a stable subset of* E. *The mapping which associates with an element* $\alpha \in \Omega$ *the restriction of* f_α *to* A (*considered as a mapping of* A *into itself*) *is an action of* Ω *on* A *said to be* induced *by the given action*.

3. QUOTIENT ACTION

DEFINITION 4. *Let* $\alpha \mapsto f_\alpha$ *be an action of a set* Ω *on a set* E. *An equivalence relation* R *on* E *is said to be compatible with the given action if, for all elements* x *and* y *of* E *such that* $x \equiv y \pmod{R}$ *and all* $\alpha \in \Omega$, $f_\alpha(x) \equiv f_\alpha(y) \pmod{R}$. *The mapping which associates with an element* $\alpha \in \Omega$ *the mapping of* E/R *into itself derived from* f_α *by passing to the quotients is an action of* Ω *on* E/R *called the quotient of the action of* Ω *on* E.

Let E be a magma and R an equivalence relation on E. R is said to be *left* (resp. *right*) *compatible* with the law on E if it is compatible with the left (resp.

right) action of E on itself derived from the law on E. For R to be compatible with the law on E it is necessary and sufficient that it be left and right compatible with the law on E.

We leave to the reader the statement and proof of the analogues of Propositions 6, 7 and 8 of § 1, no. 6.

4. DISTRIBUTIVITY

DEFINITION 5. *Let* $E_1, \ldots, E_n$ *and* F *be sets and u a mapping of* $E_1 \times \cdots \times E_n$ *into* F. *Let* $i \in [1, n]$. *Suppose* E_i *and* F *are given the structures of magmas. u is said to be distributive relative to the index variable i if the partial mapping*

$$x_i \mapsto u(a_1, \ldots, a_{i-1}, x_i, a_{i+1}, \ldots, a_n)$$

is a homomorphism of E_i *into* F *for all fixed* a_j *in* E_j *and* $j \neq i$.

If $\top$ denotes the internal laws on E_i and F, the distributivity of u is given by the equations

$$(1)\quad u(a_1, \ldots, a_{i-1}, x_i \top x'_i, a_{i+1}, a_n) = u(a_1, \ldots, a_{i-1}, x_i, a_{i+1}, \ldots, a_n) \top u(a_1, \ldots, a_{i-1}, x'_i, a_{i+1}, \ldots, a_n)$$

for $i = 1, 2, \ldots, n$, $a_1 \in E_1, \ldots, a_{i-1} \in E_{i-1}$, $x_i \in E_i$, $x'_i \in E_i$, $a_{i+1} \in E_{i+1}, \ldots$, $a_n \in E_n$.

Example. Let E be a monoid (resp. group) written multiplicatively. The mapping $(n, x) \mapsto x^n$ of $\mathbf{N} \times E$ (resp. $\mathbf{Z} \times E$) into E is distributive with respect to the first variable by the equation $x^{m+n} = x^m x^n$ (with addition as law on $\mathbf{N}$). If E is commutative, this mapping is distributive with respect to the second variable by the equation $(xy)^n = x^n y^n$.

PROPOSITION 1. *Let* $E_1, E_2, \ldots, E_n$ *and* F *be commutative monoids written additively and let u be a mapping of* $E_1 \times \cdots \times E_n$ *into* F, *which is distributive with respect to all the variables. For* $i = 1, 2, \ldots, n$, *let* L_i *be a non-empty finite set and* $(x_{i,\lambda})_{\lambda \in L_i}$ *a family of elements of* E_i. *Let* $y_i = \sum_{\lambda \in L_i} x_{i,\lambda}$ *for* $i = 1, 2, \ldots, n$. *Then*

$$(2)\qquad u(y_1, \ldots, y_n) = \sum_{\alpha} u(x_{1,\alpha_1}, \ldots, x_{n,\alpha_n})$$

the sum being taken over all sequences $\alpha = (\alpha_1, \ldots, \alpha_n)$ *belonging to* $L_1 \times \cdots \times L_n$.

We argue by induction on n, the case $n = 1$ following from formula (2) of § 1, no. 2. From the same reference

$$(3)\qquad u(y_1, \ldots, y_{n-1}, y_n) = \sum_{\alpha_n \in L_n} u(y_1, \ldots, y_{n-1}, x_{n,\alpha_n})$$

for $y_n = \sum_{\alpha_n \in L_n} x_{n,\alpha_n}$ and the mapping $z \mapsto u(y_1, \ldots, y_{n-1}, z)$ of E_n into F is a magma homomorphism. By the induction hypothesis applied to the distributive mappings $(z_1, \ldots, z_{n-1}) \mapsto u(z_1, \ldots, z_{n-1}, x_{n,\alpha_n})$ from $E_1 \times \cdots \times E_{n-1}$ to F,

$$(4) \qquad u(y_1, \ldots, y_{n-1}, x_{n,\alpha_n}) = \sum_{\alpha_1, \ldots, \alpha_{n-1}} u(x_{1,\alpha_1}, \ldots, x_{n-1,\alpha_{n-1}}, x_{n,\alpha_n}),$$

the sum being taken over the sequences $(\alpha_1, \ldots, \alpha_{n-1})$ belonging to $M = L_1 \times \cdots \times L_{n-1}$. Now $L_1 \times \cdots \times L_n = M \times L_n$; writing

$$t_{\alpha_1, \ldots, \alpha_n} = u(x_{1,\alpha_1}, \ldots, x_{n,\alpha_n}),$$

we have

$$(5) \qquad \sum_{\alpha_1, \ldots, \alpha_n} t_{\alpha_1, \ldots, \alpha_n} = \sum_{\alpha_n} \Big(\sum_{\alpha_1, \ldots, \alpha_n} t_{\alpha_1, \ldots, \alpha_{n-1}, \alpha_n} \Big)$$

by formula (7) of § 1, no. 5. (2) follows immediately from (3), (4) and (5).

Remark. If $u(a_1, \ldots, a_{i-1}, 0, a_{i+1}, \ldots, a_n) = 0$ for $i = 1, 2, \ldots, n$ and $a_j \in E_j$ $(j \neq i)$, then formula (2) remains true for families $(x_{i,\lambda})_{\lambda \in L_i}$ *of finite support.*

A special case of Definition 5 is that where u is the law of action associated with the action of a set Ω on a magma E. If u is distributive with respect to the second variable, it is also said that the action of Ω on the magma E is distributive. In other words:

DEFINITION 6. *An action $\alpha \mapsto f_\alpha$ of a set Ω on a magma* E *is said to be* distributive *if, for all $\alpha \in \Omega$, the mapping f_α is an endomorphism of the magma* E.

If $\top$ denotes the law of the magma E and $\perp$ the law of action associated with the action of Ω on E, the distributivity of the latter is then expressed by the formula

$$(6) \qquad \alpha \perp (x \top y) = (\alpha \perp x) \top (\alpha \perp y) \quad (\text{for } \alpha \in \Omega \text{ and } x, y \in E).$$

By an abuse of language, it is also said that the law $\perp$ is distributive (or right distributive) with respect to the law $\top$.

Formula (2) of § 1, no. 2 shows that then, for every ordered sequence $(x_\lambda)_{\lambda \in L}$ of elements of E and every $\alpha \in \Omega$,

$$(7) \qquad \alpha \perp \Big(\mathop{\top}_{\lambda \in L} x_\lambda \Big) = \mathop{\top}_{\lambda \in L} (\alpha \perp x_\lambda).$$

If an action $\alpha \mapsto f_\alpha$ is distributive and an equivalence relation R on E is compatible with the law of composition of E and the action $\alpha \mapsto f_\alpha$, the quotient action on E/R is distributive.

When the law on E is written multiplicatively, we often use the exponential notation x^α for a law of action which is distributive with respect to this multiplication, so that distributivity is expressed by the identity $(xy)^\alpha = x^\alpha y^\alpha$. If the law on E is written additively, we often use left (resp. right) multiplicative notation $\alpha.x$ (resp. $x.\alpha$) for a law of action which is distributive with respect to this addition, the distributivity being expressed by the identity

$$\alpha(x + y) = \alpha x + \alpha y \quad (\text{resp. } (x + y)\alpha = x\alpha + y\alpha).$$

We may also consider the case where Ω has an internal law, denoted by $\overline{\top}$, and the law of action is distributive with respect to the first variable, which means that

$$(\alpha \,\overline{\top}\, \beta) \perp x = (\alpha \perp x) \top (\beta \perp x) \tag{8}$$

for all $\alpha, \beta \in \Omega$ and $x \in \mathrm{E}$. Then, by formula (2) of § 1, no. 2

$$\Big(\mathop{\overline{\top}}_{\lambda \in \mathrm{L}} \alpha_\lambda\Big) \perp x = \mathop{\top}_{\lambda \in \mathrm{L}} (\alpha_\lambda \perp x) \tag{9}$$

for every ordered sequence $(\alpha_\lambda)_{\lambda \in \mathrm{L}}$ of elements of Ω and all $x \in \mathrm{E}$.

5. DISTRIBUTIVITY OF ONE INTERNAL LAW WITH RESPECT TO ANOTHER

DEFINITION 7. *Let $\top$ and $\perp$ be two internal laws on a set* E. *The law $\perp$ is said to be distributive with respect to the law $\top$ if*

$$x \perp (y \top z) = (x \perp y) \top (x \perp z) \tag{10}$$

$$(x \top y) \perp z = (x \perp z) \top (y \perp z) \tag{11}$$

for all x, y, z in E.

Note that (10) and (11) are equivalent if the law $\perp$ is commutative. In general, one of the laws is written additively and the other multiplicatively; if multiplication is distributive with respect to addition, then:

$$x.(y + z) = x.y + x.z \tag{12}$$

$$(x + y).z = x.z + y.z \tag{13}$$

Examples. (1) In the set $\mathfrak{P}(\mathrm{E})$ of subsets of a set E, each of the internal laws $\cap$ and $\cup$ is distributive with respect to itself and the other. This follows from formulae of the form

$$\mathrm{A} \cap (\mathrm{B} \cup \mathrm{C}) = (\mathrm{A} \cap \mathrm{B}) \cup (\mathrm{A} \cap \mathrm{C})$$
$$\mathrm{A} \cup (\mathrm{B} \cap \mathrm{C}) = (\mathrm{A} \cup \mathrm{B}) \cap (\mathrm{A} \cup \mathrm{C}).$$

(2) In **Z** (and more generally, in any totally ordered set) each of laws sup and inf is distributive with respect to the other and with respect to itself.

(3) In **Z** (*and more generally in any ring$_*$) multiplication is distributive with respect to addition.

(4) In **N** addition and multiplication are distributive with respect to the laws sup and inf.

§ 4. GROUPS AND GROUPS WITH OPERATORS

1. GROUPS

Recall the following definition (§ 2, no. 3, Definition 6).

DEFINITION 1. *A set with an associative law of composition, possessing an identity element and under which every element is invertible, is called a* group.

In other words, a group is a *monoid* (§ 2, no. 1, Definition 1) in which every element is invertible. A law of composition on a set which determines a group structure on it is called a *group law*. If G and H are two groups, a magma homomorphism of G into H is also called a *group homomorphism*. Such a homomorphism f maps identity element to identity element; for, let e (resp. e') be the identity element of G (resp. H); writing the group laws of G and H multiplicatively, $e.e = e$, whence $f(e).f(e) = f(e)$ and, multiplying by $f(e)^{-1}$, $f(e) = e'$. Hence f is unital. It then follows from no. 3 of § 2 that $f(x^{-1}) = f(x)^{-1}$ for all $x \in G$.

Example. In any monoid E the set of invertible elements with the structure induced by that on E is a group. In particular, the set of bijective mappings of a set F onto itself (or set of *permutations* of F) is a group under the law $(f, g) \mapsto f \circ g$, called the *symmetric group of the set* F and denoted by $\mathfrak{S}_F$.

In this paragraph, unless otherwise indicated, the law of composition of a group will always be written *multiplicatively* and e will denote the identity element of such a group law.

A group G is called *finite* if the underlying set of G is finite; otherwise it is called *infinite*; the cardinal of a group is called the *order* of the group.

If a law of composition on G determines a group structure on G, so does the opposite law. The mapping of a group G onto itself which associates with each $x \in G$ the inverse of x is an *isomorphism* of G onto the opposite group (§ 2, no. 3, Proposition 4).

Following our general conventions (*Set Theory*, II, § 3, no. 1), we shall denote by A^{-1} the image of a subset A of G under the mapping $x \mapsto x^{-1}$. But it is important to note that, in spite of the analogy of notation, A^{-1} is definitely not the inverse element of A under the law of composition $(X, Y) \mapsto XY$ between subsets of G (recall that XY is the set of xy with $x \in X$, $y \in Y$): the identity element under this law is $\{e\}$ and the only invertible elements of $\mathfrak{P}(G)$ under this law are the sets A consisting of a single element (such an A, moreover, certainly has inverse A^{-1}). The identity

$(AB)^{-1} = B^{-1}A^{-1}$ holds for $A \subset G$, $B \subset G$. A is called a symmetric subset of G if $A = A^{-1}$. For all $A \subset G$, $A \cup A^{-1}$, $A \cap A^{-1}$ and AA^{-1} are symmetric.

2. GROUPS WITH OPERATORS

DEFINITION 2. *Let Ω be a set. A group* G *together with an action of Ω on* G *which is distributive with respect to the group law, is called a* group with operators in Ω.

In what follows x^α will denote the composition of $\alpha \in \Omega$ and $x \in G$. Distributivity is then expressed by the identity $(xy)^\alpha = x^\alpha y^\alpha$.

In a group with operators G, each operator defines an *endomorphism* of the underlying *group* structure; these endomorphisms will sometimes be called the *homotheties* of the group with operators G.

A group with operators G is called *commutative* (or *Abelian*) if its group law is commutative.

In what follows a group G will be identified with the group with operators in $\varnothing$ obtained by giving G the unique action of $\varnothing$ on G. This allows us to consider groups as special cases of groups with operators and to apply to them the definitions and results relating to the latter which we shall state.

Example. In a commutative group G, written multiplicatively, $(xy)^n = x^n y^n$ for all $n \in \mathbf{Z}$ (§ 2, no. 8, equation (1)); the action $n \mapsto (x \mapsto x^n)$ of $\mathbf{Z}$ on G therefore defines, together with the group law, the structure of a group with operators on G.

DEFINITION 3. *Let* G *and* G′ *be groups with operators in Ω. A* homomorphism of groups with operators of G *into* G′ *is a homomorphism of the group* G *into the group* G′ *such that*

$$f(x^\alpha) = (f(x))^\alpha$$

for all $\alpha \in \Omega$ and all $x \in G$.

An *endomorphism* of the group with operators G is an endomorphism of the group G which is *permutable with all the homotheties of* G.

Z As two homotheties of a group with operators G are not necessarily permutable, *a homothety of* G *is not in general an endomorphism of the group with operators* G.

The identity mapping of a group with operators is a homomorphism of groups with operators; the composition of two homomorphisms of groups with operators is also one. For a mapping to be an isomorphism of groups with operators, it is necessary and sufficient that it be a bijective homomorphism of groups with operators and the inverse mapping is then an isomorphism of groups with operators.

More generally, let G (resp. G′) be a group with operators in Ω (resp. Ω'). Let ϕ be a mapping of Ω into Ω'. A *ϕ-homomorphism* of G into G′ is a homomorphism of the group G into the group G′ such that

$$f(x^\alpha) = (f(x))^{\phi(\alpha)}$$

for all $\alpha \in \Omega$ and all $x \in G$.

In the rest of this paragraph we shall be given a set Ω. Unless otherwise mentioned the groups with operators considered will admit Ω as set of operators.

3. SUBGROUPS

DEFINITION 4. *Let* G *be a group with operators. A stable subgroup of* G *is a subset* H *of* G *with the following properties:*

(i) $e \in H$;
(ii) $x, y \in H$ *implies* $xy \in H$;
(iii) $x \in H$ *implies* $x^{-1} \in H$;
(iv) $x \in H$ *and* $\alpha \in \Omega$ *imply* $x^\alpha \in H$.

If H is a stable subgroup of G, the structure induced on H by the structure of a group with operators on G is the structure of a group with operators and the canonical injection of H into G is a homomorphism of groups with operators.

Let G be a group. A stable subgroup of G with the action of $\varnothing$ (no. 2), which is a subset of G satisfying conditions (i), (ii), (iii) of Definition 4, is called a *subgroup* of G. When we speak of a subgroup of a group of operators we shall always mean a subgroup of the underlying group of G. A subgroup of a group with operators G is not necessarily a stable subgroup of G.

Example (1). Let Σ be a species of structure (*Set Theory*, IV, § 1, no. 4) and S a structure of species Σ on a set E (*loc. cit.*). The set of *automorphisms* of S is a subgroup of $\mathfrak{S}_E$.

PROPOSITION 1. *Let* G *be a group with operators and* H *a subset of* G *which is stable under the homotheties of* G. *The following conditions are equivalent:*

(a) H *is a stable subgroup of* G.
(b) H *is non-empty and the relations* $x \in H$, $y \in H$ *imply* $xy \in H$ *and* $x^{-1} \in H$.
(c) H *is non-empty and the relations* $x \in H$, $y \in H$ *imply* $xy^{-1} \in H$.
(d) H *is stable under the law on* G *and the law of composition induced on* H *by the law of composition on* G *is a group law.*

Clearly (a) implies (b). We show that (b) implies (a). It suffices to show that H contains the identity element of G. As the subset H is non-empty, let $x \in H$. Then $x^{-1} \in H$ and $e = xx^{-1} \in H$. Clearly (b) implies (c). We show that (c) implies (b). First of all since H is non-empty it contains an element x. Hence

$xx^{-1} = e$ is an element of H. For every element x of H, $x^{-1} = ex^{-1}$ belongs to H; hence the relations $x \in H$, $y \in H$ imply $x(y^{-1})^{-1} = xy \in H$. Clearly (a) implies (d). We show that (d) implies (a): the canonical injection of H into G is a group homomorphism; hence $e \in H$ and the relation $x \in H$ implies $x^{-1} \in H$ (no. 1).

Remarks. (1) Similarly it can be shown that condition (b) is equivalent to the condition

(c′) $H \neq \varnothing$ and the relations $x \in H$ and $y \in H$ imply $y^{-1}x \in H$.

(2) For every subgroup H of G there are the following relations

$$H.H = H \quad \text{and} \quad H^{-1} = H. \tag{1}$$

For $H.H \subset H$ and $H^{-1} \subset H$ by (b). As $e \in H$, $H.H \supset e.H = H$ and taking inverses transforms the inclusion $H^{-1} \subset H$ into $H \subset H^{-1}$, whence formulae (1).

If H is a stable subgroup of G and K is a stable subgroup of H, clearly K is a stable subgroup of G.

The set $\{e\}$ is the smallest stable subgroup of G. The intersection of a family of stable subgroups of G is a stable subgroup. There is therefore a smallest stable subgroup H of G containing a given subset X of G; it is called the *stable subgroup generated by* X and X is called a *generating system* (or *generating set*) of H.

PROPOSITION 2. *Let* X *be a non-empty subset of a group with operators* G *and* $\hat{X}$ *the stable subset under the action of* Ω *on* G *generated by* X. *The stable subgroup generated by* X *is the stable subset under the law on* G *generated by the set* $Y = \hat{X} \cup \hat{X}^{-1}$.

The latter subset Z is the set of compositions of finite sequences all of whose terms are elements of $\hat{X}$ or inverse of elements of $\hat{X}$: the inverse of such a composition is a composition of the same form (§ 2, no. 3, Corollary 1 to Proposition 5) and Z is stable under the action of Ω, as is seen by applying § 3, no. 4, Proposition 1 to the homotheties of G, hence (Proposition 1) Z is a *stable* subgroup of G. Conversely, every stable subgroup containing X obviously contains Y and hence Z.

COROLLARY 1. *Let* G *be a group with operators and* X *a subset of* G *which is stable under the action of* Ω. *The subgroup generated by* X *and the stable subgroup generated by* X *coincide.*

COROLLARY 2. *Let* G *be a group and* X *a subset of* G *consisting of pairwise permutable elements. The subgroup of* G *generated by* X *is commutative.*

The set $Y = X \cup X^{-1}$ consists of pairwise permutable elements (§ 2, no. 3, Proposition 5) and the law induced on the stable subset generated by Y is commutative (§ 1, no. 5, Corollary 2).

If G is a group with operators, the *stable* subgroup generated by a subset of G consisting of pairwise permutable elements is not necessarily commutative.

Corollary 3. *Let $f\colon G \to G'$ be a homomorphism of groups with operators and X a subset of G. The image under f of the stable subgroup of G generated by X is the stable subgroup of G' generated by $f(X)$.*

Let $X' = f(X)$. Then $\hat{X}' = f(\hat{X})$ and $X'^{-1} = f(X^{-1})$. Hence

$$f(\hat{X} \cup \hat{X}^{-1}) = \hat{X}' \cup \hat{X}'^{-1}.$$

The corollary then follows from § 1, no. 4, Proposition 1.

Example (2). Let G be a group and x an element of G. The subgroup generated by $\{x\}$ (called more simply the subgroup generated by x) is the set of x^n, $n \in \mathbf{Z}$. The stable subset (under the law on G) generated by $\{x\}$ is the set of x^n where $n \in \mathbf{N}^*$. These two sets are in general distinct.

Thus, in the additive group $\mathbf{Z}$, the subgroup generated by an element x is the set $x.\mathbf{Z}$ of xn, $n \in \mathbf{Z}$, and the stable subset generated by x is the set of xn, $n \in \mathbf{N}^*$. These two sets are always distinct if $x \neq 0$.

The union of a *right directed* family of stable subgroups of G is obviously a stable subgroup. It follows that, if P is a subset of G and H a stable subgroup of G not meeting P, the set of stable subgroups of G containing H and not meeting P, ordered by inclusion, is inductive (*Set Theory*, III, § 2, no. 4). Applying Zorn's Lemma (*Set Theory*, III, § 2, no. 4), we obtain the following result:

Proposition 3. *Let G be a group with operators, P a subset of G and H a stable subgroup G not meeting P. The set of stable subgroups of G containing H and not meeting P has a maximal element.*

4. QUOTIENT GROUPS

Theorem 1. *Let R be an equivalence relation on a group with operators G; if R is left* (resp. *right*) *compatible* (§ 3, no. 3) *with the group law on G and compatible with the action of Ω, the equivalence class of e is a stable subgroup H of G and the relation R is equivalent to* $x^{-1}y \in H$ (resp. $yx^{-1} \in H$). *Conversely, if H is a stable subgroup of G, the relation* $x^{-1}y \in H$ (resp. $yx^{-1} \in H$) *is an equivalence relation which is left* (resp. *right*) *compatible with the group law on G and compatible with the action of Ω and under which H is the equivalence class of e.*

We restrict our attention to the case where the relation R is left compatible with the law on G (the case of a right compatible relation follows by replacing the law on G by the opposite law). The relation $y \equiv x \pmod{R}$ is equivalent to $x^{-1}y \equiv e \pmod{R}$, for $y \equiv x$ implies $x^{-1}y \equiv x^{-1}x = e$ and conversely $x^{-1}y \equiv e$ implies $y = x(x^{-1}y) \equiv x$. If H denotes the equivalence class of e, the relation R is then equivalent to $x^{-1}y \in H$. We show that H is a stable subgroup of G. For every operator α, the relation $x \equiv e$ implies $x^\alpha \equiv e^\alpha = e$, hence $H^\alpha \subset H$ and H is stable under the action of Ω. It suffices to establish (Proposition 1) that $x \in H$ and $y \in H$ imply $x^{-1}y \in H$, that is $x \equiv e$ and $y \equiv e$ imply $x \equiv y$, which is a consequence of the transitivity of R.

Conversely, let H be a stable subgroup of G; the relation $x^{-1}y \in H$ is reflexive since $x^{-1}x = e \in H$; it is symmetric since $x^{-1}y \in H$ implies $y^{-1}x = (x^{-1}y)^{-1} \in H$; it is transitive, for $x^{-1}y \in H$ and $y^{-1}z \in H$ imply $x^{-1}z = (x^{-1}y)(y^{-1}z) \in H$; it is left compatible with the law of composition on G, for $x^{-1}y = (zx)^{-1}(zy)$ for all $z \in G$; finally, for every operator α, the relation $y \in xH$ implies $y^\alpha \in x^\alpha H^\alpha \subset x^\alpha H$ and hence the equivalence relation $x^{-1}y \in H$ is compatible with the action of Ω on G.

Let G be a group and H a subgroup of G; the relation $x^{-1}y \in H$ (resp. $yx^{-1} \in H$) is also written in the equivalent form $y \in xH$ (resp. $y \in Hx$). Every subgroup H of G thus defines two equivalent relations on G, namely $y \in xH$ and $y \in Hx$: the equivalent classes under these relations are respectively the sets xH, which are called *left cosets of* H (or *modulo* H), and the sets Hx, which are called *right cosets of* H (or *modulo* H). By *saturating* a subset $A \subset G$ with respect to these relations (*Set Theory*, II, § 6, no. 4), we obtain respectively the sets AH and HA. The mapping $x \mapsto x^{-1}$ transforms left cosets modulo H into right cosets modulo H and conversely.

The cardinal of the set of left cosets (mod. H) is called the *index* of the subgroup H with respect to G and is denoted by $(G:H)$; it is also equal to the cardinal of the set of right cosets.

If a subgroup K of G contains H, it is a union of left (or right) cosets of H. Since a left coset of K is obtained from K by left translation, the set of left cosets of H contained in a left coset of K has cardinal independent of the latter. Hence (*Set Theory*, III, § 5, no. 8, Proposition 9):

PROPOSITION 4. *Let* H *and* K *be two subgroups of a group* G *such that* $H \subset K$. *Then*

$$(G:H) = (G:K)(K:H). \tag{2}$$

COROLLARY. *If* G *is a finite group of order* g *and* H *is a subgroup of* G *of order* h, *then*

$$h.(G:H) = g \tag{3}$$

(in particular, the order and index of H are *divisors* of the order of G).

Theorem 1 allows us to determine the equivalence relations compatible with the laws on a group with operators G: if R is such a relation, it is both left and right compatible with the group law on G and with the action of Ω. Hence, if H is the class of e (mod. R), H is a stable subgroup such that the relations $y \in xH$ and $y \in Hx$ are equivalent (since both are equivalent to R); hence $xH = Hx$ for all $x \in G$. Conversely, if this is so, one or other of the equivalent relations $y \in x\mathrm{H}$, $y \in \mathrm{H}x$ is compatible with the group law, since it is both left and right compatible with this law (§ 3, no. 4) and is compatible with the action of Ω. Since the equation $x\mathrm{H} = \mathrm{H}x$ is equivalent to $x\mathrm{H}x^{-1} = \mathrm{H}$, we make the following definition:

DEFINITION 5. *Let* G *be a group with operators. A stable subgroup* H *of* G *is called a* normal (*or* invariant) stable subgroup *of* G *if* $x\mathrm{H}x^{-1} = \mathrm{H}$ *for all* $x \in \mathrm{G}$.

If $\Omega = \varnothing$, a normal stable subgroup of G is called a *normal* (or *invariant*) *subgroup* of G. In a commutative group every subgroup is normal.

To verify that a stable subgroup H is normal, it suffices to show that $x\mathrm{H}x^{-1} \subset \mathrm{H}$ for all $x \in \mathrm{G}$; for if so then $x^{-1}\mathrm{H}x \subset \mathrm{H}$ for all $x \in \mathrm{G}$, that is $\mathrm{H} \subset x\mathrm{H}x^{-1}$, and hence $\mathrm{H} = x\mathrm{H}x^{-1}$.

Let H be a normal stable subgroup of G and R the equivalence relation $y \in x\mathrm{H}$ defined by H; on the quotient set G/R, the internal law, the quotient by R of the law of the group G, is associative; the class of e is the identity element of this quotient law; the classes of two inverse elements in G are inverses under the quotient law and the action of Ω, the quotient by R of the action of Ω on G, is distributive with respect to the internal law on G/R (§ 3, no. 5). Hence, summarizing the results obtained:

THEOREM 2. *Let* G *be a group with operators. For an equivalence relation* R *on* G *to be compatible with the group law and the action of* Ω, *it is necessary and sufficient that it be of the form* $x^{-1}y \in \mathrm{H}$, *where* H *is a normal stable subgroup of* G (*the relation* $x^{-1}y \in \mathrm{H}$ *being moreover equivalent to* $yx^{-1} \in \mathrm{H}$ *for such a subgroup*). *The law of composition on* G/R *the quotient of that on* G *and the action of* Ω *on* G/R *the quotient of that of* Ω *on* G *by such a relation* R *give* G/R *the structure of a group with operators, called the quotient structure, and the canonical mapping of the passage to the quotient is a homomorphism of groups with operators.*

DEFINITION 6. *The quotient of a group with operators* G *by the equivalence relation defined by a normal subgroup* H *of* G, *with the quotient structure, is called the* quotient group with operators *of* G *by* H *and is denoted by* G/H. *The canonical mapping* $\mathrm{G} \rightarrow \mathrm{G/H}$ *is called a* canonical homomorphism

Let G be a group and H a normal subgroup of G. The quotient G/H, with its group structure, is called the *quotient group* of G by H. For a mapping from G/H to a group with operators to be a homomorphism of groups with operators, it is necessary and sufficient that its composition with the canonical mapping of G onto G/H be one: this justifies the name "quotient group" (*Set Theory*, IV, § 2, no. 6).

The equivalence relation defined by a normal stable subgroup of G is denoted by $x \equiv y$ (mod. H) or $x \equiv y(\mathrm{H})$.

Proposition 5. *Let* $f: G \to G'$ *be a homomorphism of groups with operators and* H *and* H' *normal stable subgroups of* G *and* G' *respectively such that* $f(H) \subset H'$. *The mapping* f *is compatible with the equivalence relations defined by* H *and* H'. *Let* $\pi: G \to G/H$ *and* $\pi': G' \to G'/H'$ *be the canonical homomorphisms. The mapping* $\bar{f}: G/H \to G'/H'$ *derived from* f *by passing to the quotients is a homomorphism.*

If $x \equiv y$ (mod. H), then $x^{-1}y \in H$, whence

$$f(x)^{-1}f(y) = f(x^{-1})f(y) = f(x^{-1}y) \in f(H) \subset H'$$

and hence $f(x) \equiv f(y)$ (mod. H'). The second assertion follows from the universal property of quotient laws (§ 1, no. 6).

Remarks. (1) If A is any subset of a group G and H is a normal subgroup of G, then AH = HA; this set is obtained by saturating A with respect to the relation $x \equiv y$ (mod. H).

(2) If H is a normal subgroup of G of finite index, the quotient group G/H is a finite group of order (G:H).

Z Note that if H is a normal subgroup of a group G and K is a normal subgroup of H, K is not necessarily a normal subgroup of G (I, § 5, Exercise 10).

Let G be a group with operators. The intersection of every family of normal stable subgroups of G is a normal stable subgroup. Hence, for every subset X of G, there exists a smallest normal stable subgroup containing X, called the normal stable subgroup *generated by* X.

In a group with operators G, the stable subgroups G and $\{e\}$ are normal.

Definition 7. *A group with operators* G *is called* simple *if* $G \neq \{e\}$ *and there exists no normal stable subgroup of* G *other than* G *and* $\{e\}$.

5. DECOMPOSITION OF A HOMOMORPHISM

Proposition 6. *Let* G *be a group with operators and* G' *a magma with an action by* Ω, *written exponentially. Let* $f: G \to G'$ *be a homomorphism of the magna* G *into the magma* G' *such that, for all* $\alpha \in \Omega$ *and all* $x \in G$, $f(x^\alpha) = f(x)^\alpha$. *Then* $f(G)$ *is a stable subset of* G' *under the law on* G' *and the action of* Ω; *the set* $f(G)$ *with the induced laws is a group with operators and the mapping* $x \mapsto f(x)$ *of* G *into* $f(G)$ *is a homomorphism of groups with operators.*

By virtue of § 1, no. 4, Proposition 1, $f(G)$ is a stable subset of G' under the internal law on G'. For every element $x \in G$ and for every operator α, $f(x)^\alpha = f(x^\alpha) \in f(G)$ and therefore $f(G)$ is stable under the action of Ω on G'. Writing the internal law of G' multiplicatively,

$$(f(x)f(y))f(z) = f(xy)f(z) = f((xy)z) = f(x(yz)) = f(x)f(yz) = f(x)(f(y)f(z))$$

for all elements x, y, z in G; therefore the induced law on $f(G)$ is associative.

Let e be the identity element of G. Its image $f(e)$ is an identity element of $f(G)$ (§ 2, no. 1). Every element of $f(G)$ is invertible in $f(G)$ (§ 2, no. 3). Therefore the law induced on $f(G)$ by the internal law on G' is a group law. For all elements x and y in G and every operator α,

$$(f(x)f(y))^\alpha = (f(xy))^\alpha = f((xy)^\alpha) = f(x^\alpha y^\alpha) = f(x^\alpha)f(y^\alpha) = (f(x))^\alpha(f(y))^\alpha$$

which shows that the action of Ω is distributive with respect to the group law on $f(G)$. Therefore $f(G)$ with the induced laws is a group with operators and clearly the mapping $x \mapsto f(x)$ is a homomorphism of groups with operators.

DEFINITION 8. *Let $f: G \to G'$ be a homomorphism of groups with operators. The inverse image of the identity element of G' is called the* kernel of f.

The kernel of f is often denoted by $\mathrm{Ker}(f)$ and the image $f(G)$ of f is sometimes denoted by $\mathrm{Im}(f)$.

THEOREM 3. *Let $f: G \to G'$ be a homomorphism of groups with operators.*

(a) $\mathrm{Ker}(f)$ *is a normal stable subgroup of* G;

(b) $\mathrm{Im}(f)$ *is a stable subgroup of* G';

(c) *the mapping f is compatible with the equivalence relation defined on* G *by* $\mathrm{Ker}(f)$;

(d) *the mapping $\bar{f}: G/\mathrm{Ker}(f) \to \mathrm{Im}(f)$ derived from f by passing to the quotient is an isomorphism of groups with operators;*

(e) $f = \iota \circ \bar{f} \circ \pi$, *where ι is the canonical injection of* $\mathrm{Im}(f)$ *into* G' *and π is the canonical homomorphism of* G *onto* $G/\mathrm{Ker}(f)$.

Assertion (b) follows from Proposition 6. The equivalence relation $f(x) = f(y)$ on G is compatible with the group with operators structure on G. By Theorem 2 (no. 4), it is therefore of the form $y \in xH$, where H is a normal stable subgroup of G and H is the class of the identity element, whence $H = \mathrm{Ker}(f)$. Assertions (a), (c) and (d) then follow. Assertion (e) is obvious (*Set Theory*, II, § 6, no. 5).

6. SUBGROUPS OF A QUOTIENT GROUP

PROPOSITION 7. *Let* G *and* H *be two groups with operators, f a homomorphism of* G *into* H *and* N *the kernel of f.*

(a) *Let* H' *be a stable subgroup of* H. *The inverse image* $G' = \overset{-1}{f}(H')$ *is a stable subgroup of* G *and* G' *is normal in* G *if* H' *is normal in* H. *Further,* N *is a normal subgroup of* G'. *If f is surjective, then* $H' = f(G')$ *and f defines an isomorphism of* G'/N *onto* H' *on passing to the quotient.*

(b) *Let* G' *be a stable subgroup of* G. *The image* $H' = f(G')$ *is a stable subgroup of* H *and* $\overset{-1}{f}(H') = G'N = NG'$. *In particular,* $\overset{-1}{f}(H') = G'$ *if and only if* $N \subset G'$. *If f is surjective and* G' *is normal in* G, *then* H' *is normal in* H.

(a) Let x and y be in G' and $\alpha \in \Omega$; then $f(x) \in H'$ and $f(y) \in H'$, whence

$f(xy^{-1}) = f(x)f(y)^{-1} \in H'$, that is $xy^{-1} \in G'$; hence G' is a subgroup of G. Now $f(x^\alpha) = f(x)^\alpha \in H'$, whence $x^\alpha \in G'$ and therefore G' is stable. Suppose H' is normal in H and let $x \in G'$, $y \in G$; then $f(x) \in H'$ and

$$f(yxy^{-1}) = f(y)f(x)f(y)^{-1} \in H'$$

whence $yxy^{-1} \in G'$; hence G' is normal in G. For all $n \in N$, $f(n) = e \in H'$, whence $N \subset G'$; as N is normal in G, it is normal in G'. Finally, if f is surjective, $f(\overset{-1}{f}(A)) = A$ for every subset A of H, whence $H' = f(G')$; the restriction of f to G' is a homomorphism f' of G' onto H' of kernel N, hence f' defines on passing to the quotient an isomorphism of G'/N onto H'.

(b) Let a and b be in H' and α in Ω; there exist x, y in G' with $a = f(x)$ and $b = f(y)$, whence $ab^{-1} = f(xy^{-1}) \in H'$, hence H' is a subgroup of H which is stable, for $a^\alpha = f(x^\alpha) \in H'$. Let $x \in G$; then $x \in \overset{-1}{f}(H')$ if and only if $f(x) \in H' = f(G')$, that is if and only if there exists y in G' with $f(x) = f(y)$; the relation $f(x) = f(y)$ is equivalent to the existence of $n \in N$ with $x = yn$; finally, $x \in \overset{-1}{f}(H')$ is equivalent to $x \in G'N = NG'$. Clearly the relation $G' = G'N$ is equivalent to $G' \supset N$. Suppose finally that f is surjective and G' is normal in G; let $a \in H'$ and $b \in H$; there exist $x \in G'$ and $y \in G$ with $a = f(x)$ and $b = f(y)$, whence $bab^{-1} = f(yxy^{-1}) \in f(G') = H'$. Hence H' is normal in H.

COROLLARY 1. *Suppose that f is surjective. Let $\mathfrak{G}$* (resp. $\mathfrak{G}'$) *be the set of stable* (resp. *normal stable*) *subgroups of* G *containing* N *and* $\mathfrak{H}$ (resp. $\mathfrak{H}'$) *the set of stable* (resp. *normal stable*) *subgroups of* H, *these sets being ordered by inclusion. The mapping* $G' \mapsto f(G')$ *is an ordered set isomorphism* $\Phi: \mathfrak{G} \to \mathfrak{H}$; *the inverse isomorphism* $\Psi: \mathfrak{H} \to \mathfrak{G}$ *is the mapping* $H' \mapsto \overset{-1}{f}(H')$. *Further* Φ *and* Ψ *induce isomorphisms* $\Phi': \mathfrak{G}' \to \mathfrak{H}'$ *and* $\Psi': \mathfrak{H}' \to \mathfrak{G}'$.

COROLLARY 2. *Let $f: G \to H$ be a homomorphism of groups with operators,* N *the kernel of f,* G' *a stable subgroup of* G *and* L *a normal stable subgroup of* G'. *Then* LN, $L.(G' \cap N)$ *and* $f(L)$ *are normal stable subgroups of* $G'N$, G' *and* $f(G')$ *respectively and the three quotient groups with operators* $G'N/LN$, $G'/L.(G' \cap N)$ *and* $f(G')/f(L)$ *are isomorphic.*

Let $H' = f(G')$ and f' denote the homomorphism of G' onto H' which coincides with f on G'; the kernel of f' is $G \cap N$ and $f'(L) = f(L)$; by Proposition 7, $f'(L)$ is a normal stable subgroup of H' and

$$\overset{-1}{f'}(f'(L)) = L.(G' \cap N)$$

is a normal stable subgroup of G'. Let λ be the canonical homomorphism of H' onto $H'/f'(L) = f(G')/f(L)$: as $\lambda \circ f'$ is surjective with kernel

$$\overset{-1}{f'}(f'(L)) = L.(G' \cap N),$$

it defines an isomorphism of $G'/L.(G' \cap N)$ onto $f(G')/f(L)$. By Proposition 7, (b), $\overset{-1}{f}(H') = G'N$; if f'' is the homomorphism of $G'N$ onto H' which coincides with f on $G'N$, the homomorphism $\lambda \circ f''$ of $G'N$ onto $f(G')/f(L)$ is surjective with kernel $\overset{-1}{f}(f(L)) = LN$; this proves that LN is a normal stable subgroup of $G'N$ and that $\lambda \circ f''$ defines an isomorphism of $G'N/LN$ onto $f(G')/f(L)$.

Corollary 3. *Let $f: G \to H$ be a homomorphism of groups with operators, N its kernel, X a subset of G such that $f(X)$ generates H and Y a subset of N which generates N. Then $X \cup Y$ generates N.*

Let G' be the stable subgroup of G generated by $X \cup Y$. As $Y \subset G'$, $N \subset G'$. As $f(X) \subset f(G')$, $f(G') = H$, whence $G' = \overset{-1}{f}(H) = G$.

Remark. In the notation of Proposition 7, the fact that the inverse image of a subgroup of H is a subgroup of G follows from the following more general fact.

If A and B are subsets of H and f is surjective, then

$$\overset{-1}{f}(A.B) = \overset{-1}{f}(A).\overset{-1}{f}(B), \quad \overset{-1}{f}(A^{-1}) = \overset{-1}{f}(A)^{-1}.$$

Obviously $\overset{-1}{f}(A).\overset{-1}{f}(B) \subset \overset{-1}{f}(A.B)$; on the other hand, if $z \in \overset{-1}{f}(A.B)$, there exists $a \in A$ and $b \in B$ such that $f(z) = ab$; as f is surjective, there exists $x \in G$ such that $f(x) = a$; writing $y = x^{-1}z$, $f(y) = a^{-1}f(z) = b$ and $z = xy$, whence $z \in \overset{-1}{f}(A).\overset{-1}{f}(B)$. The relation $x \in \overset{-1}{f}(A^{-1})$ is equivalent to $f(x) \in A^{-1}$, hence to $f(x^{-1}) \in A$, that is to $x^{-1} \in \overset{-1}{f}(A)$ and finally to $x \in \overset{-1}{f}(A)^{-1}$.

Proposition 8. *Let G be a group with operators and A and B two stable subgroups of G. Suppose that the relations $a \in A$ and $b \in B$ imply $aba^{-1} \in B$ *(in other words, A normalizes B)*. Then $AB = BA$ is a stable subgroup of G, $A \cap B$ is a normal stable subgroup of A and B is a normal stable subgroup of AB. The canonical injection of A into AB defines on passing to the quotient an isomorphism of $A/(A \cap B)$ onto AB/B.*

The formulae

$$(ab)(a'b') = aa'(a'^{-1}ba'.b')$$
$$(ab)^{-1} = a^{-1}(ab^{-1}a^{-1})$$
$$(ab)^{\alpha} = a^{\alpha}b^{\alpha}$$

for $a, a' \in A$, $b, b' \in B$ and every operator α on G, show that AB is a stable subgroup of G. Let $a \in A$ and $x \in A \cap B$; then $axa^{-1} \in B$ by the hypotheses made on A and B and clearly axa^{-1} belongs to A, hence $A \cap B$ is normal in A. Let $a \in A$ and b, b' be in B; the formula $(ab)b'(ab)^{-1} = a(bb'b^{-1})a^{-1}$ shows that B is normal in AB. Let ϕ be the restriction to A of the canonical homomorphism of AB onto AB/B; then $\phi(a) = a\mathrm{B}$ and hence the kernel of ϕ is equal to $\mathrm{A} \cap \mathrm{B}$. Clearly ϕ is surjective and hence defines an isomorphism of $\mathrm{A}/(\mathrm{A} \cap \mathrm{B})$ onto AB/B.

THEOREM 4. *Let* G *be a group with operators and* N *a normal stable subgroup of* G.

(a) *The mapping* $\mathrm{G}' \mapsto \mathrm{G}'/\mathrm{N}$ *is a bijection of the set of stable subgroups of* G *containing* N *onto the set of stable subgroups of* G/N.

(b) *Let* G′ *be a stable subgroup of* G *containing* N. *For* G′/N *to be normal in* G/N, *it is necessary and sufficient that* G′ *be normal in* G *and the groups* G/G′ *and* (G/N)/(G′/N) *are then isomorphic.*

(c) *Let* G′ *be a stable subgroup of* G. *Then* G′N *is a stable subgroup of* G *and* N *is normal in* G′N. *Further* $\mathrm{G}' \cap \mathrm{N}$ *is normal in* G′ *and the groups* $\mathrm{G}'/(\mathrm{G}' \cap \mathrm{N})$ *and* G′N/N *are isomorphic.*

Let f denote the canonical homomorphism of G onto G/N. For all $x \in \mathrm{G}$, $f(x) \in x\mathrm{N}$; therefore, $f(\mathrm{G}') = \mathrm{G}'/\mathrm{N}$ for every subgroup G′ of G containing N. As f is surjective, assertion (a) follows from Corollary 1 to Proposition 7; similarly for the equivalence "G′ normal" $\Leftrightarrow$ "G′/N normal". Suppose that G′ is a normal stable subgroup of G containing N. By no. 4, Proposition 5 applied to Id_{G}, there exists a homomorphism u of G/N into G/G′ defined by $u(x\mathrm{N}) = x\mathrm{G}'$ for all $x \in \mathrm{G}$. It is immediate that u is surjective with kernel G′/N whence the desired isomorphism of (G/N)/(G′/N) onto G/G′. Finally, (c) follows immediately from Proposition 8.

7. THE JORDAN-HÖLDER THEOREM

DEFINITION 9. *A composition series of a group with operators* G *is a finite sequence* $(\mathrm{G}_i)_{0 \leqslant i \leqslant n}$ *of stable subgroups of* G, *with* $\mathrm{G}_0 = \mathrm{G}$ *and* $\mathrm{G}_n = \{e\}$ *and such that* G_{i+1} *is a normal subgroup of* G_i *for* $0 \leqslant i \leqslant n - 1$. *The quotients* $\mathrm{G}_i/\mathrm{G}_{i+1}$ *are called the quotients of the series. A composition series* Σ' *is said to be finer than a composition series* Σ *if* Σ *is a series taken from* Σ'.

If $(\mathrm{G}_i)_{0 \leqslant i \leqslant n}$ *and* $(\mathrm{H}_j)_{0 \leqslant j \leqslant m}$ *are respectively composition series of two groups with operators* G *and* H, *they are said to be equivalent if* $m = n$ *and there exists a permutation* ϕ *of the interval* $[0, n-1]$ *of* **N**, *such that the groups with operators* $\mathrm{G}_i/\mathrm{G}_{i+1}$ *and* $\mathrm{H}_{\phi(i)}/\mathrm{H}_{\phi(i)+1}$ *are isomorphic for all* i.

Note that in general a series taken from a composition series (G_i) is not a composition series, since for $j > i + 1$, G_j is not in general a normal subgroup of G_i.

THEOREM 5 (Schreier). *Given two composition series* Σ_1, Σ_2 *of a group with operators* G, *there exist two equivalent composition series* Σ'_1, Σ'_2, *finer respectively than* Σ_1 *and* Σ_2.

Let $\Sigma_1 = (\mathrm{H}_i)_{0 \leqslant i \leqslant n}$ and $\Sigma_2 = (\mathrm{K}_j)_{0 \leqslant j \leqslant p}$ be the two given composition series with respectively $n + 1$ and $p + 1$ terms; we shall see that the composition series Σ'_1 can be formed by inserting $p - 1$ subgroups $\mathrm{H}'_{i,j}$

$(1 \leqslant j \leqslant p-1)$ between H_i and H_{i+1} for $0 \leqslant i \leqslant n-1$ and the series Σ'_2 by inserting $n-1$ subgroups $K'_{j,i}$ $(1 \leqslant i \leqslant n-1)$ between K_j and K_{j+1} for $0 \leqslant j \leqslant p-1$; thus two series of $pn+1$ stable subgroups of G will be obtained; by choosing suitably the inserted stable subgroups, we shall show that these series are equivalent composition series.

To this end note that $H_i \cap K_j$ is a stable subgroup of H_i and of K_j and hence (Theorem 4) $H_{i+1}.(H_i \cap K_j)$ is a stable subgroup of H_i containing H_{i+1} and $K_{j+1}.(H_i \cap K_j)$ is a stable subgroup of K_j containing K_{j+1}. If we write $H'_{i,j} = H_{i+1}.(H_i \cap K_j)$ and $K'_{j,i} = K_{j+1}.(H_i \cap K_j)$, $H'_{i,j+1}$ is a stable subgroup of $H'_{i,j}$ $(0 \leqslant j \leqslant p-1)$ and $K'_{j,i+1}$ is a stable subgroup of $K'_{j,i}$ $(0 \leqslant i \leqslant n-1)$. Moreover $H'_{i,0} = H_i$, $H'_{i,p} = H_{i+1}$, $K'_{j,0} = K_j$ and $K'_{j,p} = K_{j+1}$. To show the theorem, it suffices to show that $H'_{i,j+1}$ (resp. $K'_{j,i+1}$) is a normal stable subgroup of $H'_{i,j}$ (resp. $K'_{j,i}$) and that the quotient groups $H'_{i,j}/H'_{i,j+1}$ and $K'_{j,i}/K'_{j,i+1}$ are isomorphic $(0 \leqslant i \leqslant n-1,$ $0 \leqslant j \leqslant p-1)$. This follows from the following lemma by taking $H = H_i$, $H' = H_{i+1}$, $K = K_j$, $K' = K_{j+1}$.

Lemma 1 (Zassenhaus). *Let* H *and* K *be two stable subgroups of a group with operators* G *and* H′ *and* K′ *normal stable subgroups of* H *and* K *respectively; then* $H'.(H \cap K')$ *is a normal stable subgroup of* $H'.(H \cap K)$, $K'.(K \cap H')$ *is a normal stable subgroup of* $K'.(K \cap H)$ *and the quotient groups with operators* $(H'.(H \cap K))/(H'.(H \cap K'))$ *and* $(K'.(K \cap H))/(K'.(K \cap H'))$ *are isomorphic.*

By Theorem 4, $H' \cap K = H' \cap (H \cap K)$ is a normal stable subgroup of $H \cap K$; similarly $K' \cap H$ is a normal stable subgroup of $K \cap H$; hence (no. 6, Corollary 2) $(H' \cap K)(K' \cap H)$ is a normal stable subgroup of $H \cap K$. By Theorem 4 applied to the group H,

$$H'.(H' \cap K).(K' \cap H) = H'.(H \cap K')$$

is a normal stable subgroup of $H'.(H \cap K)$ and the quotient group

$$(H'.(H \cap K))/(H'.(H \cap K))$$

is isomorphic to

$$(H \cap K)/((H' \cap K).(K' \cap H)).$$

In the last quotient, H and H′ on the one hand and K and K′ on the other appear symmetrically; permuting them the stated result is obtained.

Definition 10. *A Jordan-Hölder series of a group with operators* G *is a strictly decreasing decomposition series* Σ *such that there exists no strictly decreasing decomposition series distinct from* Σ *and finer than* Σ.

Proposition 9. *For a decomposition series of* G *to be a Jordan-Hölder series it is necessary and sufficient that all the quotients of the series be simple.*

A decomposition series is strictly decreasing if and only if none of its successive quotients is reduced to the identity element. If a strictly decreasing composition series Σ is not a Jordan-Hölder series, there exists a strictly decreasing composition series Σ' which is finer than Σ and distinct from Σ. There are therefore two consecutive terms G_i, G_{i+1} of Σ which are not consecutive in Σ'; let H be the first term which follows G_i in Σ'; H is a normal stable subgroup of G_i, containing G_{i+1} and distinct from the latter; hence H/G_{i+1} is a normal stable subgroup of G_i/G_{i+1}, distinct from the latter and from the identity element; therefore G_i/G_{i+1} is not simple. Conversely, if Σ is a strictly decreasing composition series one of whose quotients G_i/G_{i+1} is not simple, this quotient contains a normal stable subgroup other than itself and $\{e\}$, whose inverse image in G_i is a normal stable subgroup H of G_i, distinct from G_i and G_{i+1} (Theorem 4); it suffices to insert H between G_i and G_{i+1} to obtain a strictly decreasing composition series distinct from Σ and finer than Σ.

THEOREM 6 (Jordan-Hölder). *Two Jordan-Hölder series of a group with operators are equivalent.*

Let Σ_1, Σ_2 be two Jordan-Hölder series of a group with operators G; by applying Theorem 5 two *equivalent* composition series Σ'_1, Σ'_2 are obtained which are respectively finer than Σ_1 and Σ_2; since the latter are Jordan-Hölder series, Σ'_1 is identical with Σ_1 or is derived from it by repeating certain terms; the series of quotients of Σ'_1 is derived from that for Σ_1 by inserting a number of terms isomorphic to the group $\{e\}$; since Σ_1 is strictly decreasing, the series of quotients of Σ_1 is derived from that of Σ'_1 by suppressing in the latter *all* the terms isomorphic to $\{e\}$. Similarly for Σ_2 and Σ'_2. As the series of quotients of Σ'_1 and Σ'_2 differ (up to isomorphism) only in the order of the terms, the same is true of Σ_1 and Σ_2; the theorem is proved.

COROLLARY. *Let* G *be a group with operators in which there exists a Jordan-Hölder series. If* Σ *is any strictly decreasing composition series of* G, *there exists a Jordan-Hölder series finer than* Σ.

Let Σ_0 be a Jordan-Hölder series of G; by Theorem 5, there exist two equivalent composition series, Σ' and Σ'_0 respectively finer than Σ and Σ_0; the argument of Theorem 6 shows that, by suppressing from Σ' the repetitions, a sequence Σ'' is obtained which is equivalent to Σ_0 and hence a Jordan-Hölder series, since all its quotients are simple (Proposition 9). As Σ is strictly decreasing, Σ'' is finer than Σ, whence the corollary.

Remark. A group with operators does not always possess a Jordan-Hölder series; an example is given by the additive group $\mathbf{Z}$ of rational integers: the sequence $(2^n.\mathbf{Z})_{n \geqslant 0}$ is a strictly decreasing infinite sequence of (normal) subgroups of $\mathbf{Z}$; for all p, the first p terms of this sequence form with the group $\{0\}$ a strictly decreasing composition series; if there existed a Jordan-Hölder series for $\mathbf{Z}$, it would have at least $p + 1$ terms, by the Corollary to Theorem 6; absurd, since p is arbitrary.

On the other hand, there exists a Jordan-Hölder series in every *finite* group with operators G: if $G \neq \{e\}$, among the normal stable subgroups of G distinct from G, let H_1 be a maximal subgroup; similarly define H_{n+1} by induction as a maximal element in the set of normal subgroups of H_n distinct from H_n, when $H_n \neq \{e\}$; the sequence of orders of the H_n is strictly decreasing, hence there exists n such that $H_n = \{e\}$ and the sequence consisting of G and the H_i $(1 \leqslant i \leqslant n)$ is by its formation a Jordan-Hölder series.

DEFINITION 11. *Let* G *be a group with operators; the length of* G *is the upper bound of the integers n such that there exists a strictly decreasing composition series of* G $(G_i)_{0 \leqslant i \leqslant n}$.

If G admits a Jordan-Hölder series, the length of G is the number of successive quotients of this series, as follows from the Corollary to Theorem 6. If G does not admit a Jordan-Hölder series, its length is infinite; by Proposition 9, for every strictly decreasing series of G there exists a strictly finer strictly decreasing series. The group consisting of the identity element is the only group with operators of length zero. A group with operators is simple if and only if it is of length 1.

Let G be a group with operators, H a normal stable subgroup of G, K the quotient G/H and $\pi\colon G \to K$ the canonical homomorphism. Let

$$\Sigma' = (H_i)_{0 \leqslant i \leqslant n}$$

be a decomposition series of H and $\Sigma'' = (K_j)_{0 \leqslant j \leqslant p}$ be a composition series of K. Writing $G_i = \overset{-1}{\pi}(K_i)$ for $0 \leqslant i \leqslant p$ and $G_i = H_{i-p}$ for $p \leqslant i \leqslant n + p$, a composition series $\Sigma = (G_i)_{0 \leqslant i \leqslant n+p}$ of G is obtained. The sequence of quotients of Σ is obtained by juxtaposing the sequence of quotients of Σ'' and the sequence of quotients of Σ'. If Σ' and Σ'' are Jordan-Hölder series, Σ is a Jordan-Hölder series of G, by Proposition 9. If H or K admits composition series of arbitrary length, so does G. We have proved:

PROPOSITION 10. *Let* G *be a group with operators and* H *a normal stable subgroup of* G. *The length of* G *is the sum of the lengths of* H *and* G/H.

COROLLARY. *Let* G *be a group with operators and* $(G_i)_{0 \leqslant i \leqslant n}$ *a composition series of* G. *The length of* G *is the sum of the lengths of the* G_i/G_{i+1}, $0 \leqslant i \leqslant n - 1$.

If G and G′ are isomorphic groups with operators and G admits a Jordan-Hölder series, so does G′ and the Jordan-Hölder series of G and G′ are equivalent. However, non-isomorphic groups can have equivalent Jordan-Hölder series; such is true for $\mathbf{Z}/4\mathbf{Z}$ and $(\mathbf{Z}/2\mathbf{Z}) \times (\mathbf{Z}/2\mathbf{Z})$, cf. no. 10.

8. PRODUCTS AND FIBRE PRODUCTS

Let $(G_i)_{i\in I}$ be a family of groups with operators. Let G be the product monoid of the G_i. Consider the action of Ω on G defined by

$$((x_i)_{i\in I})^\alpha = (x_i^\alpha)_{i\in I} \quad (\alpha \in \Omega, x_i \in G_i).$$

With this structure G is a group with operators. For all $i \in I$, the projection mapping $\mathrm{pr}_i: G \to G_i$ is a homomorphism of groups with operators.

DEFINITION 12. *The group with operators* $G = \prod_{i\in I} G_i$ *defined above is the* product group with operators *of the* G_i. *The mappings* $\mathrm{pr}_i: G \to G_i$ *are called the* projection homomorphisms.

A particular case of the product of groups with operators is the group G^E consisting of the mappings of a set E into a group with operators G, the laws being defined by:

$$(fg)(x) = f(x)g(x) \quad (f, g \in G^E, x \in E)$$
$$f^\alpha(x) = f(x)^\alpha \quad (f \in G^E, \alpha \in \Omega, x \in E).$$

Let $(\phi_i: H \to G_i)_{i\in I}$ be a family of homomorphisms of groups with operators. The mapping $h \mapsto (\phi_i(h))_{i\in I}$ of H into $\prod_{i\in I} G_i$ is a homomorphism of groups with operators. It is the only homomorphism $\Phi: H \to \prod_{i\in I} G_i$ satisfying $\mathrm{pr}_i \circ \Phi = \phi_i$ for all i. This justifies the name "product group with operators" (*Set Theory*, IV, §2, no. 4).

Let $(\phi_i: H_i \to G_i)_{i\in I}$ be a family of homomorphisms of groups with operators. The mapping $\prod_{i\in I} \phi_i: (h_i)_{i\in I} \mapsto (\phi_i(h_i))_{i\in I}$ of $\prod_{i\in I} H_i$ into $\prod_{i\in I} G_i$ is a homomorphism of groups with operators.

PROPOSITION 11. *Let* $(\phi_i: H_i \to G_i)_{i\in I}$ *be a family of homomorphisms of groups with operators and let* $\Phi = \prod_{i\in I} \phi_i$. *Then:*

(a) $\mathrm{Ker}(\Phi) = \prod_{i\in I} \mathrm{Ker}(\phi_i)$; *in particular, if all the* ϕ_i *are injective,* Φ *is injective.*

(b) $\mathrm{Im}(\Phi) = \prod_{i\in I} \mathrm{Im}(\phi_i)$; *in particular, if all the* ϕ_i *are surjective,* Φ *is surjective.*

This is immediate.

In particular, let $(G_i)_{i\in I}$ be a family of groups with operators and, for all i, let H_i be a stable (resp. normal stable) subgroup of G_i. The product $\prod_{i \in I} H_i$ is a stable (resp. normal stable) subgroup of $\prod_{i \in I} G_i$ and the canonical mapping of $\prod_{i \in I} G_i$ onto $\prod_{i \in I} (G_i/H_i)$ defines when passing to the quotient an isomorphism of

$$\Big(\prod_{i \in I} G_i\Big)\Big/\Big(\prod_{i \in I} H_i\Big)$$

onto $\prod_{i \in I} (G_i/H_i)$. For example, let J be a subset of I. The subgroup G_J of $\prod_{i \in I} G_i$ consisting of the $(x_i)_{i \in I}$ such that $x_i = e_i$ for $i \notin J$ is a normal stable subgroup. The mapping ι_J which associates with $x = (x_j)_{j \in J}$ the element $y = (y_i)_{i \in I}$ such that $y_i = e_i$ for $i \notin J$ and $y_i = x_i$ for $i \in J$, is an isomorphism of $\prod_{j \in J} G_j$ onto G_J. The mapping pr_{I-J} defines when passing to the quotient an isomorphism θ_J of the quotient group G/G_J onto $\prod_{i \in I-J} G_i$. The composition $\mathrm{pr}_J \circ \iota_J$ is the identity mapping of $\prod_{j \in J} G_j$. G/G_J is often identified with $\prod_{i \in I-J} G_i$ because of θ_J and $\prod_{i \in J} G_i$ with G_J because of ι_J.

If J_1 and J_2 are disjoint subsets of I, it follows from the definitions that every element of G_{J_1} commutes with every element of G_{J_2}.

DEFINITION 13. *Let* G *be a group with operators and* $(H_i)_{i \in I}$ *a family of normal stable subgroups of* G. *Let* $p_i: G \to G/H_i$ *be the canonical homomorphism.* G *is called the internal product* (or *product*) *of the family of quotient groups* (G/H_i) *if the homomorphism* $g \mapsto (p_i(g))_{i \in I}$ *is an isomorphism of* G *onto* $\prod_{i \in I} G/H_i$.

Let G and H be groups with operators and let ϕ and ψ be two homomorphisms of G into H. The set of elements x in G such that $\phi(x) = \psi(x)$ is a stable subgroup of G, called the *coincidence group* of ϕ and ψ. In particular, let $\phi_1: G_1 \to H$ and $\phi_2: G_2 \to H$ be homomorphisms of groups with operators; the coincidence group of the homomorphisms $\phi_1 \circ \mathrm{pr}_1$ and $\phi_2 \circ \mathrm{pr}_2$ of $G_1 \times G_2$ into H is called the *fibre product* of G_1 and G_2 over H relative to ϕ_1 and ϕ_2. It is denoted by $G_1 \times_H G_2$ when there is no ambiguity about ϕ_1 and ϕ_2 and the restrictions p_1 and p_2 of pr_1 and pr_2 to $G_1 \times_H G_2$ are also called projection homomorphisms. Then $\phi_1 \circ p_1 = \phi_2 \circ p_2$. The elements of $G_1 \times_H G_2$ are the ordered pairs $(g_1, g_2) \in G_1 \times G_2$ such that $\phi_1(g_1) = \phi_2(g_2)$. If f_i is a homomorphism of a group with operators K into G_i $(i = 1, 2)$ and $\phi_1 \circ f_1 = \phi_2 \circ f_2$, there exist one and only one homomorphism f of K into $G_1 \times_H G_2$ such that $f_i = p_i \circ f$ for $i = 1, 2$.

9. RESTRICTED SUMS

Let $(G_i)_{i \in I}$ be a family of groups with operators and, for $i \in I$, let H_i be a stable subgroup of G_i. The subset of $\prod_{i \in I} G_i$ consisting of the $(x_i)_{i \in I}$ such that the set of $i \in I$ with $x_i \notin H_i$ is *finite* is a stable subgroup of $\prod_{i \in I} G_i$ equal to $\prod_{i \in I} G_i$ if I is finite. It is called the *restricted sum of the* G_i *with respect to the* H_i. When, for all i except for a finite number, H_i is a normal stable subgroup of G_i, the restricted sum is a normal stable subgroup of the product. When, for all i, the subgroup H_i is reduced to the identity element of G_i, the direct sum of the G_i with respect to the H_i is called simply the *restricted sum of the* G_i and is sometimes denoted by $\coprod_{i \in I} G_i$. For all $i_0 \in I$, the mapping $\iota_{i_0}: G_{i_0} \to \coprod_{i \in I} G_i$ defined by $\iota_{i_0}(x) = (x_i)_{i \in I}$, where $x_{i_0} = x$ and $x_i = e_i$ if $i \neq i_0$, is an injective homomorphism of groups with operators called the *canonical injection*. G_i is identified with the stable subgroup $\text{Im}(\iota_i)$. The subgroups G_i are normal. For $i \neq j$, the elements of G_i and G_j commute and $G_i \cap G_j = \{e\}$. The group $\coprod_{i \in I} G_i$ is generated by the set $\bigcup_{i \in I} G_i$.

Proposition 12. *Let* $(\phi_i: G_i \to K)_{i \in I}$ *be a family of homomorphisms of groups with operators such that, for all* $i \in I$ *and* $j \in I$ *with* $i \neq j$, $x \in G_i$, $y \in G_j$, *the elements* $\phi_i(x)$ *and* $\phi_j(y)$ *of* K *commute; there exists one and only one homomorphism of groups with operators* Φ *of* $\coprod_{i \in I} G_i$ *into* K *such that* $\phi_i = \Phi \circ \iota_i$ *for all* $i \in I$. *For every element* $x = (x_i)_{i \in I}$ *of* $\coprod_{i \in I} G_i$, $\Phi(x) = \prod_{i \in I} \phi_i(x_i)$.

If Φ and Φ are solutions to the problem, they coincide on $\bigcup_{i \in I} G_i$ and hence on $\coprod_{i \in I} G_i$, whence the uniqueness of Φ. We now show the existence of Φ: for every element $x = (x_i)_{i \in I}$ of $\coprod_{i \in I} G_i$, let $\Phi(x) = \prod_{i \in I} \phi_i(x_i)$ (§ 1, no. 5). Clearly $\Phi \circ \iota_i = \phi_i$ for all i and Φ commutes with homotheties; the formula $\Phi(xy) = \Phi(x)\Phi(y)$ follows from § 1, no. 5, formula (9).

Definition 14. *Let* G *be a group with operators and* $(H_i)_{i \in I}$ *a family of stable subgroups of* G. G *is called the* internal restricted sum (*or* restricted sum) *of the family of subgroups* (H_i) *if every element of* H_i *is permutable with every element of* H_j *for* $j \neq i$ *and the unique homomorphism of* $\coprod_{i \in I} H_i$ *into* G *whose restriction to each of the* H_i *is the canonical injection is an isomorphism.*

When I is finite, we also say, by an abuse of language, *internal direct product* (or *direct product*, or *product*) instead of internal restricted sum. Every stable subgroup H of G for which there exists a stable subgroup H′ of G such that G is the direct product of H and H′ is called a *direct factor* of G.

PROPOSITION 13. *Let* G *be a group with operators and* $(H_\iota)_{\iota \in I}$ *a family of stable subgroups of* G *such that every element of* H_ι *is permutable with every element of* H_j *for* $j \neq i$. *For* G *to be the restricted sum of the family of subgroups* $(H_\iota)_{\iota \in I}$, *it is necessary and sufficient that every element* x *of* G *be expressible uniquely in the form* $\prod_{\iota \in I} y_\iota$, *where* $(y_\iota)_{\iota \in I}$ *is a family with finite support of elements of* G *with* $y_\iota \in H_\iota$ *for all* i.

Obvious.

PROPOSITION 14. *Let* G *be a group with operators and* $(H_\iota)_{\iota \in I}$ *a finite family of stable subgroups of* G. *For* G *to be the restricted sum of the family of subgroups* (H_ι), *it is necessary and sufficient that each* H_ι *be normal and that* G *be the product of the quotient groups* (G/H^ι), *where* H^ι *is the subgroup generated by the* H_j *for* $j \neq i$.

The condition is obviously necessary. Conversely, suppose G is the product of the $K_\iota = G/H^\iota$ and let G be identified with the product of the K_ι. Then H_ι is identified with a subgroup of K_ι, so that, for $i \neq j$, every element of H_ι is permutable with every element of H_j; on the other hand, H^ι is identified with the product of the K_j for $j \neq i$, hence $H_\iota = K_\iota$ for all i and G is the direct product of the H_ι.

PROPOSITION 15. *Let* G *be a group with operators and* $(H_\iota)_{1 \leqslant \iota \leqslant n}$ *a sequence of normal stable subgroups of* G *such that*

$$(H_1 H_2 \ldots H_\iota) \cap H_{\iota+1} = \{e\} \quad \text{for } 1 \leqslant i \leqslant n-1,$$

the set $H_1 H_2 \ldots H_n$ *is a normal stable subgroup of* G, *the restricted sum of the* H_ι.

By induction on n, this is immediately reduced to proving the proposition for $n = 2$. We show first that, if $x \in H_1$ and $y \in H_2$, x and y are *permutable*; for $xyx^{-1}y^{-1} = (xyx^{-1})y^{-1} = x(yx^{-1}y^{-1})$ and hence (since H_1 and H_2 are normal) $xyx^{-1}y^{-1} \in H_1 \cap H_2$, that is $xyx^{-1}y^{-1} = e$, by the hypothesis. Moreover H_1H_2 is a subset of G which is stable under the homotheties of G. It follows (by no. 3, Proposition 1) that H_1H_2 is a stable subgroup of G and it is immediately verified that this subgroup is normal. Suppose finally that $xy = x'y'$, with $x \in H_1$, $x' \in H_1$, $y \in H_2$, $y' \in H_2$; then $x'^{-1}x = y'y^{-1}$, hence $x'^{-1} \in H_1 \cap H_2 = \{e\}$, $x' = x$ and similarly $y' = y$; H_1H_2 is thus the direct product of H_1 and H_2.

When the group considered are commutative, the term *direct sum* is used instead of direct product.

10. MONOGENOUS GROUPS

Let $a \in \mathbf{Z}$; since $a\mathbf{Z}$ is a subgroup of $\mathbf{Z}$, the relation between elements x, y of $\mathbf{Z}$ which states "there exists $z \in \mathbf{Z}$ such that $x - y = az$" is an equivalence relation, which we agree, once and for all, to write as $x \equiv y \pmod{a}$ or $x \equiv y(a)$ and which is called *congruence modulo a*. Replacing a by $-a$ an equivalent relation is obtained, hence it may be supposed that $a \geqslant 0$; for $a = 0$, $x \equiv y(0)$ means $x = y$, hence a relation distinct from equality is obtained only if $a \neq 0$: we shall therefore suppose in what follows that $a > 0$ unless otherwise indicated.

For $a > 0$, the quotient of $\mathbf{Z}$ by the congruence $x \equiv y(a)$, that is the group $\mathbf{Z}/a\mathbf{Z}$, is called the *group of rational integers modulo a*.

PROPOSITION 16. *Let a be an integer* >0. *The integers r such that* $0 \leqslant r < a$ *form a system of representatives of the equivalence relation* $x \equiv y \pmod{a}$ *on* $\mathbf{Z}$.

If x is an integer $\geqslant 0$, there exist (*Set Theory*, III, § 5, no. 6) integers q and r such that $x = aq + r$ and $0 \leqslant r < a$ and $x \equiv r \pmod{a}$. If x is an integer $\leqslant 0$, the integer $-x$ is $\geqslant 0$ and by the above there exists an integer r such that $0 \leqslant r < a$ and $-x \equiv r \pmod{a}$. Writing $r' = 0$ if $r = 0$ and $r' = a - r$ if $r > 0$, then

$$x \equiv -r \equiv r' \pmod{a}$$

and $0 \leqslant r' < a$. We now show that if $0 \leqslant r < r' < a$, then $r \not\equiv r' \pmod{a}$. Now $r' - r < na$ for $n \geqslant 1$ and $r' - r > na$ for $n \leqslant 0$, whence $r' - r \notin a\mathbf{Z}$.

COROLLARY. *Let a be an integer* >0. *The group* $\mathbf{Z}/a\mathbf{Z}$ *of rational integers modulo a is a group of order a.*

PROPOSITION 17. *Let* H *be a subgroup of* $\mathbf{Z}$. *There exists one and only one integer* $a \geqslant 0$ *such that* $\mathrm{H} = a\mathbf{Z}$.

If $\mathrm{H} = \{0\}$, then $\mathrm{H} = 0\mathbf{Z}$. Suppose that $\mathrm{H} \neq \{0\}$. The subgroup H has an element $x \neq 0$. Then $x > 0$ or $-x > 0$, and therefore H has elements >0. Let a be the smallest element >0 in H. The subgroup $a\mathbf{Z}$ generated by a is contained in H; we show that $\mathrm{H} \subset a\mathbf{Z}$. Let $y \in \mathrm{H}$. By Proposition 16, there exists an integer r such that $y \equiv r \pmod{a}$ and $0 \leqslant r < a$. *A fortiori* $y \equiv r$ (mod. H), whence $r \in \mathrm{H}$. But this is only possible if $r = 0$ and therefore $y \in a\mathbf{Z}$. The integer a is unique: if $\mathrm{H} = \{0\}$, then necessarily $a = 0$, and if $\mathrm{H} \neq \{0\}$, the integer a is the order of $\mathbf{Z}/\mathrm{H}$.

DEFINITION 15. *A group is called* monogenous *if it admits a system of generators consisting of a single element. A finite monogenous group is called* cyclic.

Every monogenous group is commutative (no. 3, Corollary 2 to Proposition 2). Every quotient group of a monogenous group is monogenous (no. 3, Corollary 3 to Proposition 2).

The additive group $\mathbf{Z}$ is monogenous: it is generated by $\{1\}$. For every positive integer a, the group $\mathbf{Z}/a\mathbf{Z}$ is monogenous, for it is a quotient of $\mathbf{Z}$.

PROPOSITION 18. *A finite monogenous group of order a is isomorphic to* $\mathbf{Z}/a\mathbf{Z}$. *An infinite monogenous group is isomorphic to* $\mathbf{Z}$.

Let G be a monogenous group (written multiplicatively) and x a generator of G. The identity $x^m x^n = x^{m+n}$ (§ 1, no. 3, formula (1)) shows that the mapping $n \mapsto x^n$ is a homomorphism of $\mathbf{Z}$ into G. Its image is a subgroup of G containing x and hence it is G. By no. 5, Theorem 3, the group G is isomorphic to the quotient of $\mathbf{Z}$ by a subgroup, which is necessarily of the form $a\mathbf{Z}$ (Proposition 17). If $a > 0$, the group G is finite of order a and if $a = 0$, the group G is isomorphic to $\mathbf{Z}$.

PROPOSITION 19. *Let a be an integer* >0. *Let* H *be a subgroup of* $\mathbf{Z}/a\mathbf{Z}$, *b the order of* H *and c its index in* $\mathbf{Z}/a\mathbf{Z}$. *Then* $a = bc$, $\mathrm{H} = c\mathbf{Z}/a\mathbf{Z}$ *and* H *is isomorphic to* $\mathbf{Z}/b\mathbf{Z}$.

Conversely, let b and c be two integers >0 *such that* $a = bc$. *Then* $a\mathbf{Z} \subset c\mathbf{Z}$ *and* $c\mathbf{Z}/a\mathbf{Z}$ *is a subgroup of* $\mathbf{Z}/a\mathbf{Z}$, *of order b and index c.*

$a = bc$ by no. 4, Corollary to Proposition 4. By no. 7, Theorem 4, H is of the form $\mathrm{H}'/a\mathbf{Z}$, where H' is a subgroup of $\mathbf{Z}$ and Z/H' is isomorphic to $(\mathbf{Z}/a\mathbf{Z})/\mathrm{H}$ and hence of order c. By Proposition 17 and the Corollary to Proposition 16, $\mathrm{H}' = c\mathbf{Z}$ and hence H is monogenous. Finally, H is isomorphic to $\mathbf{Z}/b\mathbf{Z}$ by Proposition 18. Conversely, if $a = bc$, then $a\mathbf{Z} \subset c\mathbf{Z}$ for $a \in c\mathbf{Z}$: the quotient group $(\mathbf{Z}/a\mathbf{Z})/(c\mathbf{Z}/a\mathbf{Z})$ is isomorphic to $\mathbf{Z}/c\mathbf{Z}$ (no. 7, Theorem 4) and hence of order c (no. 4, Corollary to Proposition 4) and index b (no. 4, Corollary to Proposition 4).

COROLLARY. *Every subgroup of a monogenous group is monogenous.*

Let a and b be two integers $\neq 0$. The relation $b \in a\mathbf{Z}$ is also written: *b is a multiple of a*, and also *a divides b* or *a is a divisor of b*.

DEFINITION 16. *An integer* $p > 0$ *is called prime if* $p \neq 1$ *and it admits no divisor* >1 *other than p.*

PROPOSITION 20. *An integer* $p > 0$ *is prime if and only if the group* $\mathbf{Z}/p\mathbf{Z}$ *is a simple group.*

This follows from Proposition 19.

COROLLARY. *Every commutative simple group is cyclic of prime order.*

Let G be such a group. Then $\mathrm{G} \neq \{e\}$; let $a \neq e$ be an element of G. The subgroup generated by a is normal since G is commutative, it is not reduced to $\{e\}$ and hence is equal to G. Therefore G is monogenous and hence isomorphic to a group of the form $\mathbf{Z}/p\mathbf{Z}$ with $p > 0$, since $\mathbf{Z}$ is not simple, and p is necessarily prime.

Remark. A finite group G of prime order is necessarily cyclic. G admits no subgroup other than G and $\{e\}$ and hence it is generated by every element $\neq e$.

Lemma 2. *Let a be an integer >0. By associating with every composition series $(H_i)_{0 \leqslant i \leqslant n}$ of the group $\mathbf{Z}/a\mathbf{Z}$ the sequence $(s_i)_{1 \leqslant i \leqslant n}$, where s_i is the order of H_{i-1}/H_i, a one-to-one correspondence is obtained between the composition series of $\mathbf{Z}/a\mathbf{Z}$ and the finite sequences (s_i) of integers >0 such that $a = s_1 \dots s_n$. The composition series $(H_i)_{0 \leqslant i \leqslant n}$ is a Jordan-Hölder series if and only if the s_i are prime.*

If $(H_i)_{0 \leqslant i \leqslant n}$ is a composition series of $\mathbf{Z}/a\mathbf{Z}$, it follows, by induction on n, from no. 4, Proposition 4, that $a = \prod_{i=1}^{n} (H_{i-1}:H_i)$.

Conversely, let $(s_i)_{1 \leqslant i \leqslant n}$ be a sequence of integers >0 such that $a = s_1 \dots s_n$. If $(H_i)_{0 \leqslant i \leqslant n}$ is a composition series of $\mathbf{Z}/a\mathbf{Z}$ such that $(H_{i-1}:H_i) = s_i$ for $1 \leqslant i \leqslant n$, then necessarily $((\mathbf{Z}/a\mathbf{Z}):H_i) = \prod_{1 \leqslant j \leqslant i} s_j$ as is seen by induction on i, whence $H_i = \left(\prod_{j=1}^{i} s_j\right)\mathbf{Z}/a\mathbf{Z}$ (Proposition 19), which shows the injectivity of the mapping in question. We now show its surjectivity. Writing $H_i = \left(\prod_{j=1}^{i} s_j\right)\mathbf{Z}/a\mathbf{Z}$ for $0 \leqslant i \leqslant n$, a composition series of $\mathbf{Z}/a\mathbf{Z}$ is obtained such that $(H_{i-1}:H_i) = s_i$ (Proposition 19). The second assertion follows from Proposition 20 and no. 7, Proposition 9.

Let $\mathfrak{P}$ denote the set of prime numbers.

THEOREM 7 (decomposition into prime factors). *Let a be a strictly positive integer. There exists one and only one family $(v_p(a))_{p \in \mathfrak{P}}$ of integers >0 such that the set of $p \in \mathfrak{P}$ with $v_p(a) \neq 0$ is finite and*

$$a = \prod_{p \in \mathfrak{P}} p^{v_p(a)}. \tag{5}$$

As the group $\mathbf{Z}/a\mathbf{Z}$ is finite, it admits a Jordan-Hölder series. Lemma 2 then implies that a is a product of prime integers, whence the existence of the family $(v_p(a))$; further, for every family $(v_p(a))_{p \in \mathfrak{P}}$ satisfying the conditions of Theorem 7, the integer $v_p(a)$ is, for all $p \in \mathfrak{P}$, equal to the number of factors of a Jordan-Hölder series of $\mathbf{Z}/a\mathbf{Z}$ isomorphic to $\mathbf{Z}/p\mathbf{Z}$ (Lemma 2). The uniqueness of the family $(v_p(a))$ therefore follows from the Jordan-Hölder theorem (no. 7, Theorem 6).

COROLLARY. *Let a and b be two integers >0. Then $v_p(ab) = v_p(a) + v_p(b)$. For a to divide b, it is necessary and sufficient that $v_p(a) \leqslant v_p(b)$ for every prime number p.*

In any group G, if the (monogenous) subgroup generated by an element $x \in G$ is of finite order d, x is called an element of *order* d; the number d is therefore the least integer >0 such that $x^d = e$; if the subgroup generated by x is infinite, x is said to be of *infinite order*. These definitions, together with Proposition 4 (no. 4), imply in particular that in a finite group G the order of every element of G is a *divisor* of the order of G.

PROPOSITION 21. *In a finite group* G *of order* n, $x^n = e$ *for all* $x \in$ G.

If p is the order of x, then $n = pq$, with q an integer, and hence $x^n = (x^p)^q = e$.

§ 5. GROUPS OPERATING ON A SET

1. MONOID OPERATING ON A SET

DEFINITION 1. *Let* M *be a monoid, with law written multiplicatively and identity element denoted by* e, *and* E *a set. An action* $\alpha \mapsto f_\alpha$ *of* M *on* E *is called* a left (*resp.* right) operation *of* M *on* E *iff* $f_e = \mathrm{Id}_E$ *and* $f_{\alpha\beta} = f_\alpha \circ f_\beta$ (*resp.* $f_{\alpha\beta} = f_\beta \circ f_\alpha$) *for all* $\alpha, \beta \in$ M.

In other words, a left (resp. right) operation of a monoid M on a set E is a *monoid homomorphism* of M into the monoid E^E (resp. the opposite monoid of E^E) with composition of mappings. If the law of action corresponding to the action of M is denoted by left (resp. right) multiplication, the fact that this action is a left (resp. right) operation may be expressed by the formulae

$$(1) \qquad e.x = x;\ \alpha.(\beta.x) = (\alpha\beta).x \quad \text{for } \alpha, \beta \in M \text{ and } x \in E.$$
$$(\text{resp. } x.e = x;\ (x.\alpha).\beta = x.(\alpha\beta) \quad \text{for } \alpha, \beta \in M \text{ and } x \in E).$$

Under these conditions, it is also said that M *operates* on E *on the left* (resp. *right*) and that the corresponding laws of action are *laws of left* (resp. *right*) *operation* of the monoid M on E.

Let M be a monoid; a set E with a left (resp. right) operation of M on E is called a left (resp. right) M-*set*. The monoid M is said to operate on the left (resp. right) *faithfully* if the mapping $\alpha \mapsto f_\alpha$ of M into E^E is injective.

Examples. (1) Let E be a set; the canonical action of E^E on E (§ 3, no. 1, *Example* 3) is a left operation.

(2) Let M be a monoid. The left (resp. right) action of M on itself derived from the law on M (§ 3, no. 3, *Example* 7) is a left (resp. right) operation of M on itself. When considering this operation, we say that M operates on itself by *left* (resp. *right*) *translation*.

Let E be a left (resp. right) M-set and M^0 the opposite monoid to M. Under the same action, the monoid M^0 operates on E on the right (resp. left). The M^0-set obtained is called *opposite* to the M-set E. The definitions and results relating to left M-sets carry over to right M^0-sets when passing to the opposite structures.

In the rest of this paragraph, we shall consider, unless otherwise stated, only left M-sets which we shall call simply M-sets. Their law of action will be denoted by left multiplication.

Let E be a set. Let G be a group operating on E. For all α in G, the element of E^E defined by α is a permutation of E (§ 2, no. 3, *Example* 2). Being given an operation of G on E therefore amounts to being given a homomorphism of G into $\mathfrak{S}_E$.

In conformity with § 3, no. 3, we make the following definition:

DEFINITION 2. *Let* M *be a monoid and* E *and* E′ M-*sets. A mapping* f *of* E *into* E′ *such that, for all* $x \in E$ *and all* $\alpha \in M$, $f(\alpha . x) = \alpha . f(x)$ *is called an* M-set homomorphism (*or* M-morphism, *or* mapping compatible with the operations of M).

The identity mapping of an M-set is an M-morphism. The composition of two M-morphisms is an M-morphism. For a mapping of one M-set into another to be an isomorphism, it is necessary and sufficient that it be a bijective M-morphism and the inverse mapping is then an M-morphism.

Let $(E_i)_{i \in I}$ be a family of M-sets and let E be the product set of E_i. The monoid M operates on E by $\alpha . (x_i)_{i \in I} = (\alpha . x_i)_{i \in I}$ and E, with this action, is an M-set; let E′ be an M-set; a mapping f of E′ into E is an M-morphism if and only if $\mathrm{pr}_i \circ f$ is an M-morphism of E′ into E_i for all $i \in I$.

Let E be an M-set and F a stable subset of E under the action of M; with the induced law, F is an M-set and the canonical injection $F \rightarrow E$ is an M-morphism.

Let E be an M-set and R an equivalence relation on E compatible with the action of M; the quotient E/R with the quotient action is an M-set and the canonical mapping $E \rightarrow E/R$ is an M-morphism.

Let $\phi : M \rightarrow M'$ be a monoid homomorphism, E an M-set and E′ an M′-set. A mapping f of E into E′ such that, for all $x \in E$ and $\alpha \in M$,

$$f(\alpha . x) = \phi(\alpha) . f(x)$$

is called a ϕ-*morphism* of E into E′ (cf. § 3, no. 1).

Extension of a law of operation. Given (for example) three sets F_1, F_2, F_3, permutations f_1, f_2, f_3 of F_1, F_2, F_3 respectively and an echelon F on the base sets F_1, F_2, F_3 (*Set Theory*, IV, § 1, no. 1), we can define, proceeding step by step on the construction of the echelon F, a permutation of F called the *canonical extension* of f_1, f_2, f_3 to F (*Set Theory*, IV, § 1, no. 2); we shall denote it by $\phi_F(f_1, f_2, f_3)$.

Then let G be a group and h_i a homomorphism of G into the symmetric group of F_i ($i = 1, 2, 3$), in other words an operation of G on F_i. The mapping $x \mapsto x_F = \phi_F(h_1(x), h_2(x), h_3(x))$ is a homomorphism of G into $\mathfrak{S}_F$, in other words an operation of G on F, called the *extension* of h_1, h_2, h_3 to F. Let P be a subset of F such that, for all $x \in G$, $x_F(P) = P$; let x_P be the restriction of x_F to

P; then the mapping $x \mapsto x_P$ is an operation of G on P, also called the *extension* of h_1, h_2, h_3 to P.

For example, let K and L be two echelons on F_1, F_2, F_3; take F to be the set of subsets of K × L and P to be the set of mappings of K into L, identified with their graphs. If $w \in P$ and $x \in G$, $x_P(w)$ is the mapping $k \mapsto x_L(w(x_K^{-1}(k)))$ of K into L.

2. STABILIZER, FIXER

DEFINITION 3. *Let* M *be a monoid operating on a set* E *and* A *and* B *subsets of* E. *The set of* $\alpha \in M$ *such that* $\alpha A \subset B$ (*resp.* $\alpha A = B$) *is called the* transporter (*resp.* strict transporter) *of* A *to* B. *The transporter* (resp. *strict transporter*) *of* A *to* A *is called the* stabilizer (*resp.* strict stabilizer) *of* A. *The set of* $\alpha \in M$ *such that* $\alpha a = a$ *for all* $a \in A$ *is called the* fixer *of* A.

An element α of M is said to stabilize (resp. stabilize strictly, resp. fix) a subset A of E if α belongs to the stabilizer (resp. strict stabilizer, resp. fixer) of A. A subset P of M is said to stabilize (resp. stabilize strictly, resp. fix) a subset A of E if all the elements of P stabilize (resp. strictly stabilize, resp. fix) A. The fixer of A is contained in the strict stabilizer of A which itself is contained in the stabilizer of A.

PROPOSITION 1. *Let* M *be a monoid operating on a set* E *and* A *a subset of* E.

(a) *The stabilizer, strict stabilizer and fixer of* A *are submonoids of* M.

(b) *Let* α *be an invertible element of* M; *if* α *belongs to the strict stabilizer* (resp. *fixer*) *of* A, *so does* α^{-1}.

Let e be the identity element of M; then $ea = a$ for every element $a \in A$ and therefore e belongs to the fixer of A. Let α and β be elements of E which stabilize A. Then $(\alpha\beta)A = \alpha(\beta A) \subset \alpha A \subset A$ and therefore the stabilizer of A is a submonoid of M. Similarly for the strict stabilizer and fixer of A, whence (a). If $\alpha A = A$, then $A = \alpha^{-1}(\alpha A) = \alpha^{-1}A$. If for all $a \in A$, $\alpha a = a$, then $a = \alpha^{-1}(\alpha a) = \alpha^{-1}a$, whence (b).

COROLLARY. *Let* G *be a group operating on a set* E *and* A *be a subset of* E. *The strict stabilizer* S *and the fixer* F *of* A *are subgroups of* G *and* F *is a normal subgroup of* S.

The first assertion follows from the proposition and F is the kernel of the homomorphism of S into $\mathfrak{S}_A$ associated with the operation of S on A.

A group G operates faithfully on a set E if and only if the fixer of E consists of the identity element of G. The fixer of E is the kernel of the given homomorphism of G into $\mathfrak{S}_E$; this homomorphism is injective if and only if its kernel consists of the identity element (§ 4, no. 5, Theorem 3).

Let M be a monoid, E an M-set and a an element of E. The fixer, strict stabilizer and stabilizer of $\{a\}$ are equal; this monoid is called equally the fixer or stabilizer of a. The fixer of a subset A of E is the intersection of the fixers of the elements of A. a is called an *invariant* element of E if the fixer of a is the monoid M. M is said to operate *trivially* on E if every element of E is invariant.

Proposition 2. *Let* G *be a group operating on a set* E *and, for all* $x \in E$, *let* S_x *be the stabilizer of* x. *For all* $\alpha \in G$, $S_{\alpha x} = \alpha S_x \alpha^{-1}$.

If $s \in S_x$, then $\alpha s \alpha^{-1}(\alpha x) = \alpha s x = \alpha x$, whence $\alpha S_x \alpha^{-1} \subset S_{\alpha x}$. As $x = \alpha^{-1}(\alpha x)$, $\alpha^{-1} S_{\alpha x} \alpha \subset S_x$, whence $S_{\alpha x} \subset \alpha S_x \alpha^{-1}$.

It is seen similarly that, if A and B are two subsets of E and T is the transporter (resp. strict transporter) of A to B, then the transporter (resp. strict transporter) of αA to αB is equal to $\alpha T \alpha^{-1}$.

3. INNER AUTOMORPHISMS

Let G be a group. The set Aut(G) of automorphisms of the group G is a subgroup of $\mathfrak{S}_G$ (§ 4, no. 1, *Example* 2).

Proposition 3. *Let* G *be a group. For every element* x *of* G, *the mapping* $\operatorname{Int}(x) : y \mapsto xyx^{-1}$ *of* G *into itself is an automorphism of* G. *The mapping* $\operatorname{Int} : x \mapsto \operatorname{Int}(x)$ *of* G *into* Aut(G) *is a group homomorphism, whose kernel is the centre of* G *and whose image is a normal subgroup of* Aut(G).

If x, y and z are elements of G, then $(xyx^{-1})(xzx^{-1}) = xyzx^{-1}$ and hence $\operatorname{Int}(x)$ is an endomorphism of G. For x and y elements of G,

$$\operatorname{Int}(x) \circ \operatorname{Int}(y) = \operatorname{Int}(xy):$$

for all $z \in G$, $x(yzy^{-1})x^{-1} = (xy)z(xy)^{-1}$. On the other hand, $\operatorname{Int}(e)$ is the identity mapping of G. The mapping Int is therefore a monoid homomorphism from G to the monoid End(G) of endomorphisms of the group G. As the elements of G are invertible, the mapping Int takes its values in the set Aut(G) of invertible elements of End(G) (§ 2, no. 3). Now $xyx^{-1} = y$ if and only if x and y commute and hence $\operatorname{Int}(x)$ is the identity mapping of G if and only if x is a central element. Finally, let α be an automorphism of G and let $x \in G$; then

$$\operatorname{Int}(\alpha(x)) = \alpha \circ \operatorname{Int}(x) \circ \alpha^{-1}. \tag{2}$$

For $y \in G$,

$$\alpha(x).y.\alpha(x)^{-1} = \alpha(x).\alpha(\alpha^{-1}(y)).\alpha(x)^{-1} = \alpha(x.\alpha^{-1}(y).x^{-1}).$$

Hence $\alpha.\operatorname{Int}(G).\alpha^{-1} \subset \operatorname{Int}(G)$.

Definition 4. *Let* G *be a group and* $x \in G$. *The automorphism* $y \mapsto xyx^{-1}$ *is called the* inner automorphism *of* G *defined by* x *and is denoted by* Int x.

For $x, y \in G$, we also write $x^y = y^{-1}xy = (\operatorname{Int} y^{-1})(x)$.

A subgroup of G is normal if and only if it is stable under all inner automorphisms of G (§ 4, no. 4, Definition 5). A subgroup of G is called *charac-*

teristic if it stable under all automorphisms of G. The centre of a group G is a characteristic subgroup (formula (2)).

The centre of a group G is not necessarily stable under all endomorphisms of G (Exercise 22). In particular, the centre of a group with operators is not necessarily a stable subgroup.

PROPOSITION 4. *Let* G *be a group,* H *a characteristic* (resp. *normal*) *subgroup of* G *and* K *a characteristic subgroup of* H. *Then* K *is a characteristic* (resp. *normal*) *subgroup of* G.

The restriction to H of an automorphism (resp. inner automorphism) of G is an automorphism of H and therefore leaves K invariant.

Let G be a group, $A \subset G$ and $b \in G$. b is said to *normalize* A if $bAb^{-1} = A$; b is said to *centralize* A if, for all $a \in A$, $bab^{-1} = a$. Let A and B be subsets of G; B is said to *normalize* (resp. *centralize*) A if every element of B normalizes (resp. centralizes) A.

The set of $g \in G$ which normalize (resp. centralize) A is called the *normalizer* (resp. *centralizer*) of A (cf. § 1, no. 5, Definition 9); it is often denoted by $N_G(A)$ or simply $N(A)$ (resp. $C_G(A)$ or $C(A)$). It is a subgroup of G. When A is a subgroup of G, $N_G(A)$ may be characterized as the largest subgroup of G which contains A and in which A is normal.

Remarks. (1) The normalizer (resp. centralizer) of A is the strict stabilizer (resp. fixer) of A when G operates on itself by inner automorphisms. In particular the centralizer is a normal subgroup of the normalizer.

(2) The set of elements $b \in G$ such that $bAb^{-1} \subset A$ is a submonoid of G. Even when A is a subgroup of G, this set is not necessarily a subgroup of G (Exercise 27).

4. ORBITS

DEFINITION 5. *Let* G *be a group,* E *a* G*-set and* $x \in G$. *An element* $y \in E$ *is* conjugate *to* x *under the operation of* G *if there exists an element* $\alpha \in G$ *such that* $y = \alpha x$. *The set of conjugate elements of* x *is called the* orbit of x *in* E.

The relation "*y is conjugate to x*" is an equivalence relation. For $x = ex$; if $y = \alpha x$, then $x = \alpha^{-1}y$; if $y = \alpha x$ and $z = \beta y$, then $z = \beta\alpha x$. The orbits are the equivalence classes under this relation.

A subset X of E is stable if and only if it is saturated with respect to the relation of conjugation.

The mapping $\alpha \mapsto \alpha x$ of G into E is sometimes called the *orbital mapping* defined by x. It is a G-morphism of G (with the operation of G on itself by left translation) into E. The image $G.x$ of G under this mapping is the orbit of x.

G is said to operate *freely* on E if, for all $x \in E$, the orbital mapping defined by x is injective or also if the mapping $(g, x) \mapsto (gx, x)$ of $G \times E$ into $E \times E$ is injective.

Examples. (1) Let G be a group and consider the operation of G on itself by inner automorphisms. Two elements of G which are conjugate under this operation are called conjugate under inner automorphisms or simply *conjugate*. The orbits are called *conjugacy classes*. Similarly, two subsets H and H′ of G are called *conjugate* if there exists an element $\alpha \in G$ such that $H' = \alpha . H . \alpha^{-1}$, that is if they are conjugate under the extension to $\mathfrak{P}(G)$ of the operation of G on itself by inner automorphisms.

(2) *In the space $\mathbf{R}^n$, the orbit of a point x under the action of the orthogonal group $\mathbf{O}(n, \mathbf{R})$ is the Euclidean sphere of radius $\|x\|$.*

The stabilizers of two conjugate elements of E are conjugate subgroups of G (no. 2, Proposition 2).

The quotient set of E under the relation of conjugation is the set of orbits of E; it is sometimes denoted by E/G or G\E. (Sometimes the notation E/G is reserved for the case where E is a right G-set and the notation G\E for the case where E is a left G-set.)

Let G be a group operating on a set E on the right. Let H be a normal subgroup of G. The group G operates on E/H on the right, the corresponding law of right action being $(xH, g) \mapsto xHg = xgH$; under this operation, H operates trivially, whence a right operation of G/H on E/H. Let ϕ be the canonical mapping of E/H onto E/G; the inverse images under ϕ of the points of E/G are the orbits of G (or of G/H) in E/H. Hence ϕ defines when passing to the quotient a bijection, called *canonical*, of $(E/H)/G = (E/H)/(G/H)$ onto E/G.

Let G (resp. H) be a group operating on a set E on the left (resp. right). Suppose that the actions of G and H on E *commute*, that is

$$(g.x).h = g.(x.h) \quad \text{for } g \in G,\ x \in E \text{ and } h \in H.$$

The action of H on E is also a left operation of the opposite group H^0 to H. It then follows from § 4, no. 9, Proposition 12 that the mapping which associates with the element $(g, h) \in G \times H^0$ the mapping $x \mapsto g.x.h$ of E into itself is a left operation of $G \times H^0$ on E. The orbit of an element $x \in E$ under this operation is the set GxH. The set of these orbits is denoted by G\E/H. On the other hand, the operation of G (resp. H) is compatible with the relation of conjugation with respect to the operation of H (resp. G) and the set of orbits G\(E/H) (resp. (G\E)/H) is identified with G\E/H: in the diagram

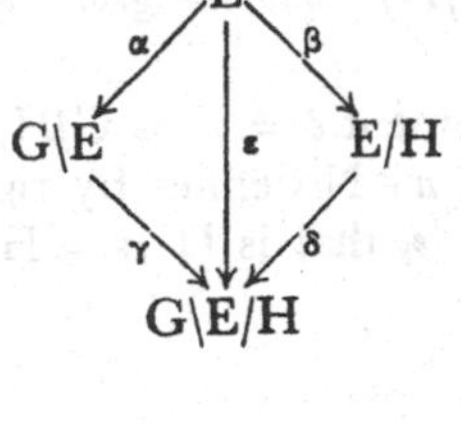

$$\begin{array}{ccccc} & & E & & \\ & \overset{\alpha}{\swarrow} & \downarrow \varepsilon & \overset{\beta}{\searrow} & \\ G\backslash E & & & & E/H \\ & \underset{\gamma}{\searrow} & \downarrow & \underset{\delta}{\swarrow} & \\ & & G\backslash E/H & & \end{array}$$

(where α, β, γ, δ, ε denote the canonical mappings of taking quotients), $\gamma \circ \alpha = \delta \circ \beta = \varepsilon$.

Let G be a group and H a subgroup of G. Consider the *right* operation of H on G by right translation (no. 1, *Example* 2). The set of orbits G/H is the set of *left cosets* modulo H; note that G operates on G/H *on the left* by the law $(g, x\mathrm{H}) \mapsto gx\mathrm{H}$ (cf. no. 5). Similarly, the set of right cosets modulo H is the set $\mathrm{H}\backslash\mathrm{G}$ of orbits of the left operation of H on G by left translation. If K is a subgroup of G containing H and Γ is a left (resp. right) coset modulo H, then $\Gamma\mathrm{K}$ (resp. $\mathrm{K}\Gamma$) is a left (resp. right) coset modulo K. The mapping $\Gamma \mapsto \Gamma\mathrm{K}$ (resp. $\Gamma \mapsto \mathrm{K}\Gamma$) is called the *canonical mapping* of G/H into G/K (resp. of $\mathrm{H}\backslash\mathrm{G}$ into $\mathrm{K}\backslash\mathrm{G}$). It is surjective.

Let G be a group and H and K two subgroups of G. Let H operate on G on the left by left translation and K on the right by right translation; these two operations commute, which allows us to consider the set $\mathrm{H}\backslash\mathrm{G}/\mathrm{K}$. The elements of $\mathrm{H}\backslash\mathrm{G}/\mathrm{K}$ are called the *double cosets* of G modulo H and K. When $\mathrm{K} = \mathrm{H}$, we simply say double cosets modulo H. For the canonical mapping of G/H onto $\mathrm{H}\backslash\mathrm{G}/\mathrm{H}$ to be a bijection, it is necessary and sufficient that H be a normal subgroup of G.

5. HOMOGENEOUS SETS

DEFINITION 6. *Let* G *be a group. An operation of* G *on a set* E *is called* transitive *if there exists an element* $x \in \mathrm{E}$ *whose orbit is* E. *A* G*-set* E *is called* homogeneous *if the operation of* G *on* E *is transitive.*

It is also said that G *operates transitively* on E; or that E is a *homogeneous set under* G. It amounts to the same to say that E *is non-empty* and that, for all elements x and y of E, there exists an element $\alpha \in \mathrm{G}$ such that $\alpha . x = y$.

Example. If E is a G-set, each orbit of E, with the induced operation, is a homogeneous set under G.

Let G be a group and H a subgroup of G. Consider the set G/H of left cosets modulo H. The group G operates on G/H *on the left* by $(g, x\mathrm{H}) \mapsto gx\mathrm{H}$. Let N be the normalizer of H. The group N operates on G/H on the right by $(x\mathrm{H}, n) \mapsto x\mathrm{H}n = xn\mathrm{H}$. This operation induces on H the trivial operation and hence, on passing to the quotient, N/H operates on G/H *on the right*. Let $\phi\colon (\mathrm{N}/\mathrm{H})^0 \to \mathfrak{S}_{\mathrm{G}/\mathrm{H}}$ be the homomorphism corresponding to this operation.

PROPOSITION 5. *With the above notation,* G/H *is a homogeneous* G*-set. The mapping* ϕ *induces an isomorphism of* $(\mathrm{N}/\mathrm{H})^0$ *onto the group of automorphisms of the* G*-set* G/H.

The orbit in G/H of the element $\dot{e} = \mathrm{H}$ is G/H, whence the first assertion. We now prove the second. If $n \in \mathrm{N}$ defines by right translation the identity mapping on G/H, then $\dot{e}.n = \dot{e}$, that is $\mathrm{H}.n = \mathrm{H}$, whence $n \in \mathrm{H}$. Therefore

N/H operates faithfully on G/H on the right and ϕ is injective. The left operation of G and right operation of N/H on G/H commute and hence the operators of N/H define G-morphisms of G/H into itself, which are necessarily G-automorphisms since they are bijective. Therefore ϕ takes its values in the group Φ of G-automorphisms of G/H. We show that the image of ϕ is Φ. Let $f \in \Phi$. By transporting the structure, the stabilizer of $f(\dot{e})$ in G is equal to the stabilizer of $\dot{e}$ and hence to H. Let $n \in G$ be such that $f(\dot{e}) = n\dot{e}$. The stabilizer of $n\dot{e}$ in G is nHn^{-1} (no. 2, Proposition 2), whence $nHn^{-1} = H$ and $n \in N$. For every element xH of G/H, $f(xH) = f(x.\dot{e}) = x.f(\dot{e}) = xnH = xHn$ and f coincides with the mapping defined by n.

Remarks. (1) Let G be a group, H a subgroup of G and $\phi\colon G \to \mathfrak{S}_{G/H}$ the homomorphism corresponding to the operation of G on G/H. The kernel of ϕ is the intersection of the conjugates of H (no. 2, Proposition 2). It is also the largest normal subgroup contained in H (no. 3). In particular, G operates faithfully on G/H if and only if the intersection of the conjugates of H reduces to $\{e\}$.

(2) Let G be a group and H and K subgroups such that H is a normal subgroup of K. Then K/H operates on the G-set G/H on the right and the canonical mapping of G/H onto G/K defines on passing to the quotient a G-set isomorphism $(G/H)/(K/H) \to G/K$ (cf. no. 4).

PROPOSITION 6. *Let* G *be a group,* E *a homogeneous* G-*set,* $a \in E$, H *the stabilizer of* a *and* K *a subgroup of* G *contained in* H. *There exists one and only one* G-*morphism* f *of* G/K *into* E *such that* $f(e.K) = a$. *If* K = H, f *is an isomorphism.*

If f is a solution, then $f(x.K) = x.a$ for all x in G, whence the uniqueness; we show the existence. The orbital mapping defined by a is compatible with the equivalence relation $y \in xK$ on G. For, if $y = xk$, $k \in K$, then

$$y.a = xk.a = x.a.$$

A mapping f is thus derived of G/K into H which satisfies $f(x.K) = x.a$ for all x in G. This mapping is a G-morphism and $f(K) = a$. This mapping is surjective for its image is a non-empty stable subset of E. Suppose now that K = H and let us show that f is injective. If $f(x.H) = f(y.H)$, then $x.a = y.a$, whence $x^{-1}y.a = a$ and $x^{-1}y \in H$, whence $x.H = y.H$.

THEOREM 1. *Let* G *be a group.*

(a) *Every homogeneous* G-*set is isomorphic to a homogeneous* G-*set of the form* G/H *where* H *is a subgroup of* G.

(b) *Let* H *and* H′ *be two subgroups of* G. *The* G-*sets* G/H *and* G/H′ *are isomorphic if and only if* H *and* H′ *are conjugate.*

As a homogeneous G-set is non-empty, assertion (a) follows from Proposition 6. We show (b). Let $f\colon G/H \to G/H'$ be a G-set isomorphism. The subgroup H is the stabilizer of H and hence, by transport of structure, the stabilizer of an element of G/H'. The subgroups H and H' are therefore conjugate (no. 2, Proposition 2). If $H' = \alpha H \alpha^{-1}$, H' is the stabilizer of the element $\alpha . H$ of G/H (no. 2, Proposition 2) and hence G/H' is isomorphic to G/H (Proposition 6).

Examples. (1) Let E be a non-empty set. The group $\mathfrak{S}_E$ operates transitively on E. If x and y are two elements of E, the mapping $\tau\colon E \to E$ such that $\tau(x) = y$, $\tau(y) = x$ and $\tau(z) = z$ for $z \neq x, y$, is a permutation of E. Let $a \in E$. The stabilizer of a is identified with $\mathfrak{S}_F$, where $F = E - \{a\}$. The homogeneous G-set E is thus isomorphic to $\mathfrak{S}_E/\mathfrak{S}_F$.

(2) Let E be a set of n elements and $(p_i)_{i \in I}$ a finite family of integers >0 such that $\sum_i p_i = n$. Let X be the set of partitions $(F_i)_{i \in I}$ of E such that $\mathrm{Card}(F_i) = p_i$ for all i. The group $\mathfrak{S}_E$ operates transitively on X. The stabilizer H of an element $(F_i)_{i \in I}$ of X is canonically isomorphic to $\prod_{i \in I} \mathfrak{S}_{F_i}$ and is hence of order $\prod_{i \in I} p_i!$. Applying Theorem 1 and § 4, no. 4, Corollary to Proposition 4, a new proof is obtained of the fact that

$$\mathrm{Card}(X) = \frac{n!}{\prod_{i \in I} p_i!}.$$

In particular, take $I = \{1, 2, \ldots, r\}$, $E = \{1, 2, \ldots, n\}$,

$$F_i = \{p_1 + \cdots + p_{i-1} + 1, \ldots, p_1 + \cdots + p_i\}$$

for $1 \leqslant i \leqslant r$. Let S be the set of $\tau \in \mathfrak{S}_E$ such that $\tau|_{F_i}$ is increasing for $1 \leqslant i \leqslant r$. If $(G_1, \ldots, G_r) \in X$ there exists one and only one $\tau \in S$ which maps $(F_1, \ldots, F_r)$ to $(G_1, \ldots, G_r)$. In other words, each left coset of $\mathfrak{S}_E$ modulo H meets S in one and only one point.

(3) *Let n be an integer $\geqslant 1$. The orthogonal group $\mathbf{O}(n, \mathbf{R})$ operates transitively on the unit sphere $\mathbf{S}_{n-1}$ in $\mathbf{R}^n$. The stabilizer of the point $(0, \ldots, 0, 1)$ is identified with the orthogonal group $\mathbf{O}(n-1, \mathbf{R})$. The homogeneous $\mathbf{O}(n, \mathbf{R})$-set $\mathbf{S}_{n-1}$ is thus isomorphic to $\mathbf{O}(n, \mathbf{R})/\mathbf{O}(n-1, \mathbf{R})$.*

6. HOMOGENEOUS PRINCIPAL SETS

DEFINITION 7. *Let* G *be a group. An operation of* G *on a set* E *is called* simply transitive *if there exists an element* x *of* E *such that the orbital mapping defined by* x *is a bijection. A set* E, *together with a simply transitive left action of* G *on* E, *is called a* left homogeneous principal G-set (*or* left homogeneous principal set under G).

It amounts to the same to say that G operates freely and transitively on E, or also that there exists an element $x \in E$ such that the orbital mapping defined by x is an isomorphism of the G-set G (where G operates by left translation) onto E; or also that the two following conditions are satisfied:

(i) E is non-empty.

(ii) for all elements x and y of E, there exists one and only one element $\alpha \in G$ such that $\alpha x = y$.

Condition (ii) is also equivalent to the following condition:

(iii) the mapping $(\alpha, x) \mapsto (\alpha x, x)$ is a bijection of $G \times E$ onto $E \times E$.

We leave to the reader the task of defining right homogeneous principal G-sets.

Examples. (1) Let G operate on itself by left (resp. right) translation. Thus a left (resp. right) homogeneous principal G-set structure is defined on G, which is sometimes denoted by G_s (resp. G_d).

(2) Let E be a homogeneous set under a *commutative* group G. If G operates faithfully on E, the latter is a homogeneous principal G-set.

(3) Let E and F be two isomorphic sets with structures of the same species and let Isom(E, F) be the set of isomorphisms of E onto F (with the given structures). The group Aut(E) of automorphisms of E (with the given structure) operates on Isom(E, F) on the right by the law $(\sigma, f) \mapsto f \circ \sigma$ and Isom(E, F) is a right homogeneous principal Aut(E)-set. Similarly, the group Aut(F) operates on Isom(E, F) on the left by the law $(\sigma, f) \mapsto \sigma \circ f$ and Isom(E, F) is a left homogeneous principal Aut(F)-set.

(4) *A homogeneous principal set under the additive group of a vector space is called an *affine space* (cf. II, § 9, no. 1).*

The group of automorphisms of the homogeneous principal G-set G_s (Example 1) is the group of right translations of G which is identified with G^0 (no. 5, Proposition 5). Let E be a homogeneous principal G-set and a an element of E. The orbital mapping ω_a defined by a is an isomorphism of the G-set G_s onto E. By transporting the structure an isomorphism ψ_a is derived of G^0 onto Aut(E). It should be noted that ψ_a *in general depends on* a; more precisely, for $\alpha \in G$,

$$\psi_{\alpha a} = \psi_a \circ \mathrm{Int}_{G^0}(\alpha) = \psi_a \circ \mathrm{Int}(\alpha^{-1}). \tag{3}$$

For, writing δ_a for the translation $x \mapsto x\alpha$ on G,

$$\omega_{\alpha a} = \phi\omega_a \circ \delta_\alpha$$

and

$$\psi_a(x) = \omega_a \circ \delta_x \circ \omega_a^{-1}, \quad x \in G,$$

whence

$$\psi_{\alpha a}(x) = \omega_a \circ \delta_\alpha \circ \delta_x \circ \delta_\alpha^{-1} \circ \omega_\alpha^{-1} = \omega_a \circ \delta_{\alpha^{-1} x \alpha} \circ \omega_\alpha^{-1} = \psi_a(\alpha^{-1} x \alpha).$$

7. PERMUTATION GROUPS OF A FINITE SET

If E is a finite set with n elements, the symmetric group $\mathfrak{S}_E$ (§ 4, no. 1) is a finite group of order $n!$. When E is the interval $[1, n]$ of the set $\mathbf{N}$ of natural numbers, the corresponding symmetric group is denoted by $\mathfrak{S}_n$; the symmetric group of any set with n elements is isomorphic to $\mathfrak{S}_n$.

DEFINITION 8. *Let* E *be a finite set,* $\zeta \in \mathfrak{S}_E$ *a permutation of* E *and* $\bar{\zeta}$ *the subgroup of* $\mathfrak{S}_E$ *generated by* ζ. ζ *is called a* cycle *if, under the operation of* $\bar{\zeta}$ *on* E, *there exists one and only one orbit which is not reduced to a single element. This orbit is called the* support of ζ.

Let ζ be a cycle. The support $\mathrm{supp}(\zeta)$ of ζ is the set of $x \in E$ such that $\zeta(x) \neq x$.

The *order* of a cycle ζ is equal to the cardinal of its support. The subgroup $\bar{\zeta}$ generated by ζ operates transitively and faithfully on $\mathrm{supp}(\zeta)$. As ζ is commutative, $\mathrm{supp}(\zeta)$ is a principal set under $\bar{\zeta}$ (no. 6, Example 2) and hence $\mathrm{Card}(\mathrm{supp}(\zeta)) = \mathrm{Card}(\bar{\zeta})$.

Lemma 1. *Let* $(\zeta_\iota)_{\iota \in I}$ *be a family of cycles whose supports are pairwise disjoint. Then the* ζ_ι *are pairwise permutable. Let* $\sigma = \prod_{\iota \in I} \zeta_\iota$ *and* $\bar{\sigma}$ *be the subgroup generated by* σ. *Then* $\sigma(x) = \zeta_\iota(x)$ *for* $x \in S_\iota$, $\iota \in I$, *and* $\sigma(x) = x$ *for* $x \notin \bigcup_{\iota \in I} S_\iota$. *The mapping* $\iota \mapsto S_\iota$ *is a bijection of* I *onto the set of* $\bar{\sigma}$*-orbits not consisting of a single element.*

Let ζ and ζ' be two cycles whose supports are disjoint. If

$$x \notin \mathrm{supp}(\zeta) \cup \mathrm{supp}(\zeta'),$$

then $\zeta\zeta'(x) = \zeta'\zeta(x) = x$. If x belongs to the support of ζ, then $\zeta'(x) = x$ and $\zeta(x)$ belongs to the support of ζ, whence $\zeta\zeta'(x) = \zeta'\zeta(x) = \zeta(x)$. Similarly when x belongs to the support of ζ', then $\zeta'\zeta(x) = \zeta\zeta'(x) = \zeta'(x)$. Hence $\zeta\zeta' = \zeta'\zeta$. Therefore the ζ_ι are pairwise permutable and, for $\iota \in I$ and $x \in S_\iota$, $\sigma(x) = \zeta_\iota(x) \in S_\iota$. The mappings σ and ζ_ι coincide on S_ι, hence S_ι is stable under σ and the subgroup of $\mathfrak{S}_{S_\iota}$ generated by the restriction of σ to S_ι operates transitively on S_ι; therefore S_ι is a $\bar{\sigma}$-orbit. As the S_ι are non-empty and are pairwise disjoint, the mapping $\iota \mapsto S_\iota$ is injective. As $\bigcup_\iota S_\iota$ is the set of x such that $\sigma(x) \neq x$, every $\bar{\sigma}$-orbit not consisting of a single element is one of the S_ι.

PROPOSITION 7. *Let* E *be a finite set and* σ *a permutation of* E. *There exists one and only one finite set* C *of cycles satisfying the two following conditions:*

(a) *the supports of the elements of* C *are pairwise disjoints;*

(b) $\sigma = \prod_{\zeta \in C} \zeta$ (the elements of C being pairwise permutable by Lemma 1).

Let $\bar{\sigma}$ be the subgroup generated by σ and let S be the set of $\bar{\sigma}$-orbits not consisting of a single element. For $s \in S$, write $\zeta_s(x) = \sigma(x)$ if $x \in s$ and $\zeta_s(x) = x$ if $x \notin s$. For all $s \in S$, ζ_s is a cycle whose support is s and $\sigma = \prod_{s \in S} \zeta_s$, as is seen by applying the two sides to any element of E. The uniqueness of C follows from Lemma 1.

DEFINITION 9. *A cycle of order* 2 *is called a* transposition.

Let x and y be two *distinct* elements of E. Let $\tau_{x,y}$ denote the unique transposition with support $\{x, y\}$.

For every permutation σ of E the permutation $\sigma.\tau_{x,y}.\sigma^{-1}$ is a transposition whose support is $\{\sigma(x), \sigma(y)\}$. Hence:

$$\sigma.\tau_{x,y}.\sigma^{-1} = \tau_{\sigma(x),\sigma(y)}. \tag{4}$$

Transpositions thus form a conjugacy class in the group $\mathfrak{S}_E$.

PROPOSITION 8. *Let* E *be a finite set. The group* $\mathfrak{S}_E$ *is generated by the transpositions.*

For every permutation σ let F_σ be the set of $x \in E$ such that $\sigma(x) = x$. We show by descending induction on p that every permutation σ such that $\mathrm{Card}(F_\sigma) = p$ is a product of transpositions. If $p \geqslant \mathrm{Card}(E)$, the permutation σ is the identity mapping of E; it is the product of the empty family of transpositions. If $p < \mathrm{Card}(E)$, suppose that the property is proved for every permutation σ' such that $\mathrm{Card}(F_{\sigma'}) > p$. Now $E - F_\sigma \neq \varnothing$; let $x \in E - F_\sigma$ and $y \in \sigma(x)$. Then $y \neq x$ and $y \in E - F_\sigma$. Let $\sigma' = \tau_{x,y}.\sigma$. The set $F_{\sigma'}$ contains F_σ and x and hence $\mathrm{Card}(F_{\sigma'}) > \mathrm{Card}(F_\sigma) = p$. By the induction hypothesis σ' is a product of transpositions and hence $\sigma = \tau_{x,y}.\sigma'$ is a product of transpositions.

PROPOSITION 9. *Let* n *be an integer* $\geqslant 0$. *The group* $\mathfrak{S}_n$ *is generated by the transpositions* $(\tau_{i,i+1})_{1 \leqslant i \leqslant n-1}$.

By virtue of Proposition 8, it suffices to show that every transposition $\tau_{p,q}$, $1 \leqslant p < q \leqslant n$, belongs to the subgroup H generated by the $\tau_{i,i+1}$, $1 \leqslant i \leqslant n-1$. We show this by induction on $q - p$. For $q - p = 1$, it is obvious. If $q - p > 1$, then (formula (4)) $\tau_{p,q} = \tau_{q-1,q}\tau_{p,q-1}\tau_{q-1,q}$. By the induction hypothesis $\tau_{p,q-1} \in H$ and therefore $\tau_{p,q} \in H$.

If $\sigma \in \mathfrak{S}_n$, every ordered pair (i, j) of elements of $[1, n]$ such that $i < j$ and $\sigma(i) > \sigma(j)$ is called an *inversion* of σ. Let $\nu(\sigma)$ denote the number of inversions of σ.

Let P be the additive group of mappings from $\mathbf{Z}^n$ to $\mathbf{Z}$. For $f \in P$ and $\sigma \in \mathfrak{S}_n$, let σf be the element of P defined by

$$\sigma f(z_1, \ldots, z_n) = f(z_{\sigma(1)}, \ldots, z_{\sigma(n)}). \tag{5}$$

The action of $\mathfrak{S}_n$ on P thus defined is an operation; for all $\sigma, \tau \in \mathfrak{S}_n$ and $f \in P$, $ef = f$ and

$$\begin{aligned}(\tau(\sigma f))(z_1, \ldots, z_n) &= \sigma f(z_{\tau(1)}, \ldots, z_{\tau(n)}) = f(z_{\tau\sigma(1)}, \ldots, z_{\tau\sigma(n)}) \\ &= ((\tau\sigma)f)(z_1, \ldots, z_n).\end{aligned}$$

Formula (5) shows that $\sigma(-f) = -\sigma f$ for $\sigma \in \mathfrak{S}_n$ and $f \in P$.

Let p be the element of P defined by

$$p(z_1, \ldots, z_n) = \prod_{i<j} (z_j - z_i). \tag{6}$$

Lemma 2. $p \neq 0$ *and* $\sigma p = (-1)^{\nu(\sigma)}$ *for* $\sigma \in \mathfrak{S}_n$.

$p(1, 2, \ldots, n) = \prod_{i<j} (j - i) \neq 0$ and hence $p \neq 0$. On the other hand, if $\sigma \in \mathfrak{S}_n$, then

$$\sigma p(z_1, \ldots, z_n) = p(z_{\sigma(1)}, \ldots, z_{\sigma(n)}) = \prod_{i<j} (z_{\sigma(j)} - z_{\sigma(i)}).$$

Let C be the set of ordered pairs (i, j) such that $1 \leqslant i \leqslant n$, $1 \leqslant j \leqslant n$, $i < j$. A permutation θ is defined on C by setting $\theta(i, j) = (\sigma(i), \sigma(j))$ if (i, j) is not an inversion, $\theta(i, j) = (\sigma(j), \sigma(i))$ if (i, j) is an inversion. This implies $\sigma p = (-1)^{\nu(\sigma)}p$.

THEOREM 2. *Let* E *be a finite set. There exists one and only one homomorphism* ε *from* $\mathfrak{S}_n$ *to the multiplicative group* $\{-1, +1\}$ *such that* $\varepsilon(\tau) = -1$ *for every transposition* τ.

The uniqueness follows from Proposition 8. We show the existence. By transporting the structure, it may be assumed that $E = [1, n]$. Using the above notation, let $\varepsilon(\sigma) = (-1)^{\nu(\sigma)}$. Then (Lemma 2)

$$\sigma(\sigma' p) = \sigma(\varepsilon(\sigma')p) = \varepsilon(\sigma')(\sigma p) = \varepsilon(\sigma')\varepsilon(\sigma)p.$$

On the other hand,

$$\sigma(\sigma' p) = (\sigma\sigma')p = \varepsilon(\sigma\sigma')p.$$

As $p \neq -p$, it follows that $\varepsilon(\sigma\sigma') = \varepsilon(\sigma)\varepsilon(\sigma')$ and thus ε is a homomorphism. We now show that, for every transposition τ, $\varepsilon(\tau) = -1$. $\nu(\tau_{n-1,n}) = 1$, whence $\varepsilon(\tau_{n-1,n}) = -1$. As every transposition τ is conjugate to $\tau_{n-1,n}$ and the group $\{-1, +1\}$ is commutative, $\varepsilon(\tau) = \varepsilon(\tau_{n-1,n}) = -1$.

DEFINITION 10. *In the notation of Theorem* 2, *the number* $\varepsilon(\sigma)$ (*also denoted* ε_σ) *is called the* signature *of the permutation* σ. *The kernel of the homomorphism* ε *is called the* alternating group of E.

σ is called *even* (resp. *odd*) if $\varepsilon(\sigma) = 1$ (resp. $\varepsilon(\sigma) = -1$). The alternating group of E is denoted by $\mathfrak{A}_E$. It is a normal subgroup of $\mathfrak{S}_E$. When $E = [1, n]$, it is simply denoted by $\mathfrak{A}_n$. When the cardinal n of E is $\geqslant 2$, it is a subgroup of index 2 and hence of order $n!/2$. It can be shown that, for $n = 3$ or $n \geqslant 5$, the group $\mathfrak{A}_n$ is a simple group (cf. Exercise 16).

Example. If σ is a cycle of order d, then

$$\varepsilon(\sigma) = (-1)^{d-1}.$$

The number of inversions of the permutation

$$(1, 2, 3, \ldots, d) \mapsto (d, 1, 2, \ldots, d-1)$$

is equal to $d - 1$.

§6. EXTENSIONS, SOLVABLE GROUPS, NILPOTENT GROUPS

Throughout this paragraph, the group laws are, unless expressly mentioned otherwise, written multiplicatively.

1. EXTENSIONS

DEFINITION 1. *Let* F *and* G *be two groups. An* extension *of* G *by* F *is a triple* $\mathscr{E} = (\mathrm{E}, i, p)$, *where* E *is a group,* i *an injective homomorphism of* F *into* E *and* p *a surjective homomorphism of* E *onto* G *such that* $\mathrm{Im}(i) = \mathrm{Ker}(p)$. *A homomorphism* $s\colon \mathrm{G} \to \mathrm{E}$ (*resp.* $r\colon \mathrm{E} \to \mathrm{F}$) *such that* $p \circ s = \mathrm{Id}_{\mathrm{G}}$ (*resp.* $r \circ i = \mathrm{Id}_{\mathrm{F}}$) *is called a* section (*resp.* retraction) *of the extension* $\mathscr{E}$.

An extension $\mathscr{E} = (\mathrm{E}, i, p)$ of G by F is often denoted by the diagram $\mathscr{E}\colon \mathrm{F} \xrightarrow{i} \mathrm{E} \xrightarrow{p} \mathrm{G}$, in which i and p are sometimes omitted if no confusion can arise. It is sometimes said simply that the group E is an extension of G by F.

For a group E to be an extension of G by F, it is necessary and sufficient that it contain a normal subgroup F′ isomorphic to F such that the quotient group E/F′ is isomorphic to G.

An extension $\mathscr{E}\colon \mathrm{F} \xrightarrow{i} \mathrm{E} \xrightarrow{p} \mathrm{G}$ is called *central* if the image $i(\mathrm{F})$ is contained in the centre of E; this is only possible if F is commutative.

Let $\mathscr{E}\colon \mathrm{F} \xrightarrow{i} \mathrm{E} \xrightarrow{p} \mathrm{G}$ and $\mathscr{E}'\colon \mathrm{F} \xrightarrow{i'} \mathrm{E}' \xrightarrow{p'} \mathrm{G}$ be two extensions of G by F. A *morphism* of $\mathscr{E}$ into $\mathscr{E}'$ is a homomorphism $u\colon \mathrm{E} \to \mathrm{E}'$ such that $p' \circ u = p$ and $u \circ i = i'$, or, in other words, such that the following diagram is commutative:

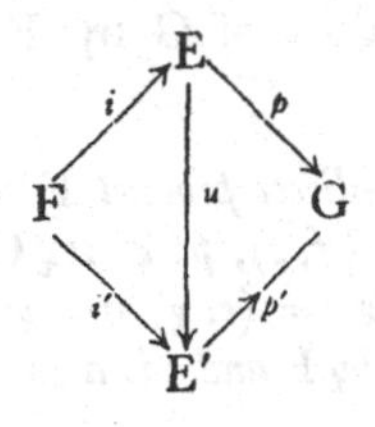

PROPOSITION 1. *Let* $\mathscr{E}: F \xrightarrow{i} E \xrightarrow{p} G$ *and* $\mathscr{E}': F \xrightarrow{i'} E' \xrightarrow{p'} G$ *be extensions of* G *by* F. *If* $u: E \to E'$ *is a morphism of* $\mathscr{E}$ *into* $\mathscr{E}'$, *u is an isomorphism of* E *onto* E' *and* u^{-1} *is a morphism of* $\mathscr{E}'$ *into* $\mathscr{E}$.

Let $x \in E$ be such that $u(x) = e$. Then $p(x) = p'(u(x)) = e$, whence $x \in i(F)$. Let $y \in F$ be such that $x = i(y)$; then $i'(y) = u(i(y)) = e$. As i' is injective, $y = e$ and $x = e$. Therefore u is injective. By virtue of § 4, no. 6, Corollary 1 to Proposition 7, u is surjective since $u(i(F)) = i'(F)$. The last assertion is immediate.

In other words, the extensions $\mathscr{E}$ and $\mathscr{E}'$ are *isomorphic* if and only if there exists a morphism of $\mathscr{E}$ into $\mathscr{E}'$.

Let F and G be two groups and write $E_0 = F \times G$; let $i: F \to E_0$ be the canonical injection and $p: E_0 \to G$ be the canonical projection. Every extension of G by F isomorphic to the extension $\mathscr{E}_0: F \xrightarrow{i} E_0 \xrightarrow{p} G$ is called a *trivial extension.*

PROPOSITION 2. *Let* $\mathscr{E}: F \xrightarrow{i} E \xrightarrow{p} G$ *be an extension of* G *by* F. *The following conditions are equivalent:*

(i) $\mathscr{E}$ *is a trivial extension;*

(ii) $\mathscr{E}$ *has a retraction* r*;*

(iii) $\mathscr{E}$ *has a section* s *such that* $s(G)$ *is contained in the centralizer of* $i(F)$.

Clearly (i) implies (ii) and (iii). If (ii) holds, the mapping $(r, p): E \to F \times G$ is a morphism of $\mathscr{E}$ into $\mathscr{E}_0$, whence (i). If (iii) holds, the homomorphism of $F \times G$ into E corresponding to (i, s) (§ 4, no. 9, Proposition 12) is a morphism of $\mathscr{E}_0$ into $\mathscr{E}$, whence (i).

Z It may be that an extension $\mathscr{E}: F \to E \to G$ is not trivial and yet the group E is isomorphic to $F \times G$ (Exercise 6).

DEFINITION 2. *Let* F *and* G *be two groups and* τ *a homomorphism of* G *into the automorphism group of* F. *Write* $\tau(g)(f) = {}^{g}f$ *for* $g \in G$ *and* $f \in F$. *The set* $F \times G$ *with the law of composition*

$$((f, g), (f', g')) \mapsto (f, g) \cdot_{\tau} (f', g') = (f \cdot {}^{g}f', gg') \tag{1}$$

is called the external semi-direct product of G by F relative to τ.

The external semi-direct product of G by F relative to τ is denoted by $F \times_{\tau} G$.

PROPOSITION 3. *The external semi-direct product* $F \times_{\tau} G$ *is a group. The mappings* $i: F \to F \times_{\tau} G$ *defined by* $i(f) = (f, e)$, $p: F \times_{\tau} G \to G$ *defined by* $p(f, g) = g$, *and* $s: G \to F \times_{\tau} G$ *defined by* $s(g) = (e, g)$ *are group homomorphisms. The triple* $(F \times_{\tau} G, i, p)$ *is an extension of* G *by* F *and* s *is a section of the extension.*

We have:

$$((f, g) \cdot_\tau (f', g')) \cdot_\tau (f'', g'') = (f.{}^g f', gg') \cdot_\tau (f'', g'') = (f.{}^g f'.{}^{gg'}f'', gg'g'');$$

$$(f, g) \cdot_\tau ((f', g') \cdot_\tau (f'', g'')) = (f, g) \cdot_\tau (f'{}^{g'}f'', g'g'') = (f.{}^g(f'.{}^{g'}f''), gg'g'').$$

Now ${}^g(f'.{}^{g'}f'') = {}^g f'.{}^{gg'}f''$, which shows that the law of composition defined by (1) is associative. The element (e, e) is the identity under this law. The element (f, g) admits as inverse $({}^{g^{-1}}f^{-1}, g^{-1})$. Hence the law of composition on $F \times_\tau G$ is a group law. The other assertions are immediate.

Using the notation of Proposition 3, $\mathscr{E}_\tau$ will denote the extension

$$F \xrightarrow{i} F \times_\tau G \xrightarrow{p} G.$$

Let $\mathscr{E}' : F \xrightarrow{i'} E' \xrightarrow{p'} G$ be an extension of G by F and $s' : G \to E'$ a section of $\mathscr{E}'$. We define an operation τ of G on the group F by:

(2) $$i'(\tau(g, f)) = s'(g)i'(f)s'(g)^{-1} = \text{Int}(s'(g))(i'(f)).$$

PROPOSITION 4. *With the above notation, there exists one and only one isomorphism u of $\mathscr{E}_\tau$ onto $\mathscr{E}'$ such that $u \circ s = s'$.*

$(f, g) = (f, e) \cdot_\tau (e, g) = i(f) \cdot_\tau s(g)$. Therefore, if u is a solution to the problem, of necessity $u(f, g) = i'(f).s'(g)$, whence the uniqueness of u. We prove the existence. We write $u(f, g) = i'(f).s'(g)$. Then

$$\begin{aligned} u(f, g).u(f', g') &= i'(f)s'(g)i'(f')s'(g') \\ &= i'(f)(s'(g)i'(f')s'(g)^{-1})s'(g)s'(g') \\ &= i'(f)i'(\tau(g, f')).s'(g)s'(g') \\ &= i'(f.\tau(g, f')).s'(gg') \\ &= u((f, g) \cdot_\tau (f', g')). \end{aligned}$$

Therefore, u is a homomorphism of $F \times_\tau G$ into E'. Obviously $u \circ i = i'$, $p' \circ u = p$ and $u \circ s = s'$.

Remark. The definition of the operation τ by formula (2) depends on the extension $\mathscr{E}'$ and the section s'. When F is commutative, the operation τ does not depend on s'. For $\text{Int}(s'(g)) \mid i'(F)$ depends then only on the coset of $s'(g)$ mod. $i'(F)$.

More generally, let $\mathscr{E}: F \to E \to G$ be an extension of G by a commutative group F (it is not assumed that $\mathscr{E}$ admits a section). The group E operates on F by inner automorphisms, this image is trivial on the image of F and hence defines an operation of G on F. If $\mathscr{E}$ admits a section, this operation is that defined by formula (2).

COROLLARY. *Let* G *be a group and* H *and* K *two subgroups of* G *such that* H *is normal,* $H \cap K = \{e\}$ *and* $H.K = G$. *Let* τ *be the operation of* K *on* H *by inner automorphisms of* G. *The mapping* $(h, k) \mapsto hk$ *is an isomorphism of* $H \times_\tau K$ *onto* G.

Under the hypotheses of this corollary, G is said to be the *semi-direct product* of K by H.

Examples. (1) Let G be a group and E a homogeneous principal G-set; let Γ denote the automorphism group of G. Let A be the set of permutations f of E with the following property:

There exists $\gamma \in \Gamma$ such that f is a γ-morphism of E into E (that is, $f(gb) = \gamma(g)f(b)$ for $b \in E$ and $g \in G$).

The above formula $f(gb) = \gamma(g)f(b)$ shows that if $f \in A$ there exists a *unique* $\gamma \in \Gamma$ such that f is a γ-morphism, we shall denote it by $p(f)$.

Let f, f' be in A, $\gamma = p(f)$, $\gamma' = p(f')$. Then, for all $b \in E$ and all $g \in G$,

$$(f' \circ f)(gb) = f'(\gamma(g)f(b)) = \gamma'(\gamma(g))f'(f(b))$$

which proves that $f' \circ f \in A$ and that $p(f' \circ f) = p(f')p(f)$. On the other hand, $f(\gamma^{-1}(g)f^{-1}(b)) = gb$, whence $f^{-1}(gb) = \gamma^{-1}(g)f^{-1}(b)$ and $f^{-1} \in A$. Thus A is a subgroup of $\mathfrak{S}_E$ and p is a homomorphism of A into Γ. The kernel of p is the set $\mathrm{Aut}_G(E)$ of automorphisms of the G-set E.

We fix $a \in E$. We have defined in § 5, no. 6 an isomorphism ψ_a of G^0 onto $\mathrm{Aut}_G(E)$ such that $\psi_a(x)(ga) = gxa$ for all g, x in G. On the other hand, for $\gamma \in \Gamma$, let $s_a(\gamma)$ be the permutation of E defined by $s_a(\gamma)(ga) = \gamma(g)a$ for all $g \in G$; it is immediately verified that s_a is a homomorphism of Γ into A such that $p \circ s_a = \mathrm{Id}_\Gamma$. Thus $G^0 \xrightarrow{\psi_a} A \xrightarrow{p} \Gamma$ is an *extension* of Γ by G^0 and s_a is a *section* of this extension. This extension and this section define an operation of Γ on G^0, $s_a(\Gamma)$ acting on $\psi_a(G^0)$ by inner automorphisms; we write this operation exponentially. We show that *this operation is the natural operation* (§ 3, no. 1, *Example* 3): for x, g in G and $\gamma \in \Gamma$,

$$\begin{aligned}(\psi_a({}^\gamma x))(ga) &= (s_a(\gamma) \circ \psi_a(x) \circ s_a(\gamma)^{-1})(ga) \\ &= (s_a(\gamma) \circ \psi_a(x))(\gamma^{-1}(g)a) = s_a(\gamma)(\gamma^{-1}(g)xa) \\ &= g\gamma(x)a = \psi_a(\gamma(x))ga\end{aligned}$$

whence ${}^\gamma x = \gamma(x)$.

Proposition 4 then shows that A is isomorphic to the semidirect product of $\Gamma = \mathrm{Aut}(G)$ by G^0 under the natural operation of Aut(G) on G^0. Note that the isomorphism which we have constructed depends in general on the choice of the element $a \in E$.

(2) *Let A be a commutative ring. The upper triangular group $T(n, A)$ is the semi-direct product of the diagonal subgroup $D(n, A)$ by the upper strict triangular group $T_1(n, A)$.*

2. COMMUTATORS

DEFINITION 3. *Let* G *be a group and* x *and* y *two elements of* G. *The element* $x^{-1}y^{-1}xy$ *of* G *is called the* commutator *of* x *and* y.

(x, y) is used to denote the commutator of x and y. Then obviously

$$(y, x) = (x, y)^{-1}.$$

For x and y to commute it is necessary and sufficient that $(x, y) = e$. More generally,

$$xy = yx(x, y).$$

On the other hand we write

$$x^y = y^{-1}xy = x(x, y) = (y, x^{-1})x. \tag{3}$$

As the mapping $x \mapsto x^y$ is the inner automorphism $\mathrm{Int}(y^{-1})$, $(x, y)^z = (x^z, y^z)$ for all $x, y, z \in \mathrm{G}$.

For $x, y, z \in \mathrm{G}$, we prove the following relations:

(4) $\quad (x, yz) = (x, z).(x, y)^z = (x, z).(z, (y, x)).(x, y)$

(4 bis) $\quad (xy, z) = (x, z)^y.(y, z) = (x, z).((x, z), y).(y, z)$

(5) $\quad (x^y, (y, z)).(y^z, (z, y)).(z^x, (x, y)) = e$

(6) $\quad (x, yz).(z, xy).(y, zx) = e$

(6 bis) $\quad (xy, z).(yz, x).(zx, y) = e.$

Now

$$\begin{aligned}(x, yz) &= x^{-1}z^{-1}y^{-1}xyz = (x, z)z^{-1}x^{-1}y^{-1}xyz = (x, z)(x, y)^z \\ &= (x, z)(z, (x, y)^{-1})(x, y)\end{aligned}$$

by (3), which proves (4). Formula (4 bis) follows similarly. On the other hand,

$$\begin{aligned}(x^y, (y, z)) &= (x^y)^{-1}(z, y)(x^y)(y, z) \\ &= y^{-1}x^{-1}yz^{-1}y^{-1}zyy^{-1}xyy^{-1}z^{-1}yz \\ &= (yzy^{-1}xy)^{-1}(zxz^{-1}yz).\end{aligned}$$

Then writing $u = yzy^{-1}xy$, $v = zxz^{-1}xy$ and $w = xyx^{-1}zx$, we obtain

$$(x^y, (y, z)) = u^{-1}v.$$

By cyclically permuting x, y, z, we deduce $(y^z, (z, x)) = v^{-1}w$ and

$$(z^x, (x, y)) = w^{-1}u,$$

which immediately imply (5). Finally, (6) follows by multiplying together the two sides in the three formulae obtained by cyclically permuting x, y, z in the formula $(x, yz) = x^{-1}z^{-1}y^{-1}xyz = (yzx)^{-1}(xyz)$, and similarly for (6 bis).

If A and B are two subgroups of G, (A, B) denotes the *subgroup generated* by the commutators (a, b) with $a \in \mathrm{A}$ and $b \in \mathrm{B}$.† Then $(\mathrm{A}, \mathrm{B}) = \{e\}$ if and only if A centralizes B. $(\mathrm{A}, \mathrm{B}) \subset \mathrm{A}$ if and only if B normalizes A. If A and B are normal (resp. characteristic), so is (A, B).

† We here reject the notational convention made in § 1, no 1 of extending a law of composition to subsets.

Proposition 5. *Let* A, B, C *be three subgroups of* G.

(i) *The subgroup* A *normalizes the subgroup* (A, B).

(ii) *If the subgroup* (B, C) *normalizes* A, *the subgroup* (A, (B, C)) *is generated by the elements* $(a, (b, c))$ with $a \in A$, $b \in B$ *and* $c \in C$.

(iii) *If* A, B *and* C *are normal, then*

$$(A, (B, C)) \subset (C, (B, A)).(B, (C, A)).$$

By (4 bis), for $a, a' \in A$ and $b \in B$,

$$(a, b)^{a'} = (aa', b).(a', b)^{-1},$$

whence (i). Suppose now that (B, C) normalizes A. For $a \in A$, $b \in B$, $c \in C$ and $x \in G$, (4) implies

$$(a, (b, c).x) = (a, x).(x, ((b, c), a))(a, (b, c))$$

and $((b, c), a) \in A$ since (B, C) normalizes A, whence by induction on p the fact that $\left(a, \prod_{i=1}^{p} (b_i, c_i)\right)$, for $b_i \in B$, $c_i \in C$, belongs to the subgroup generated by the elements of the form $(a, (b, c))$. If finally A, B and C are normal, so are the subgroups (A, (B, C)), (C, (B, A)) and (B, (C, A)). It therefore suffices by (ii) to show that

$$(a, (b, c)) \in (C, (B, A)).(B, (C, A))$$

for all $a \in A$, $b \in B$ and $c \in C$. Now by (5), writing $a^{b^{-1}} = u$

$$(a, (b, c)) = ((u^b), (b, c)) = (c^u, (u, b))^{-1}.(b^c, (c, u))^{-1},$$

whence (iii).

Definition 4. *Let* G *be a group. The subgroup generated by the commutators of elements of* G *is called the* derived group *of* G.

The derived group of G is thus the subgroup (G, G). It is also denoted by D(G). By an abuse of language, it is sometimes called the *commutator group* of G although it is in general distinct from the set of commutators of elements of G (Exercise 16). $D(G) = \{e\}$ if and only if G is commutative.

Proposition 6. *Let* $f: G \to G'$ *be a group homomorphism. Then* $f(D(G)) \subset D(G')$. *If* f *is surjective, the homomorphism of* D(G) *into* D(G') *the restriction of* f *is surjective.*

The image under f of a commutator of elements of G is a commutator of elements of G'. If f is surjective, the image under f of the set of commutators of G is the set of commutators of G'. The proposition thus follows from § 4, no. 3, Corollary 3 to Proposition 2.

Corollary 1. *The derived group of a group* G *is a characteristic subgroup of* G. *In particular it is a normal subgroup of* G.

COROLLARY 2. *Let G be a group. The quotient group G/D(G) is commutative. Let $\pi: G \rightarrow G/D(G)$ be the canonical homomorphism. Every homomorphism f of G into a commutative group G′ can be expressed uniquely in the form $f = \bar{f} \circ \pi$, where $\bar{f}: G/D(G) \rightarrow G'$ is a homomorphism.*

Now $\pi(D(G)) = \{e\}$. As π is surjective, it follows that $D(G/D(G)) = \{e\}$, whence the first assertion. The second follows from § 4, no. 4, Proposition 5.

COROLLARY 3. *Let H be a subgroup of G. The following conditions are equivalent:*

(i) $H \supset D(G)$;

(ii) *H is a normal subgroup and G/H is commutative.*

(ii) ⇒ (i) by Corollary 2 and (i) ⇒ (ii) by § 4, no. 7, Theorem 4, since every subgroup of a commutative group is normal.

COROLLARY 4. *Let G be a group and X a subset of G which generates G. The group D(G) is the* normal *subgroup of G generated by the commutators of elements of X.*

Let H be the normal subgroup of G generated by the commutators of elements of X and $\phi: G \rightarrow G/H$ the canonical homomorphism. The set $\phi(X)$ generates G/H. The elements of $\phi(X)$ are pairwise permutable and hence H is commutative (§ 4, no. 3, Corollary 2 to Proposition 2). Hence (Corollary 3) H contains D(G). On the other hand, obviously $H \subset D(G)$.

Remarks. (1) Corollary 2 can also be expressed by saying that G/D(G), together with π, is a solution of the universal mapping problem for G, relative to commutative groups and homomorphisms from G to commutative groups.

(2) Under the hypotheses of Corollary 4, the subgroup generated by the commutators of elements of X is contained in D(G) but is not in general equal to D(G) (cf. Exercise 15e).

Examples. (1) If G is a non-commutative simple group, then D(G) = G. Therefore every homomorphism of G into a commutative group is trivial.

(2) The derived group of the symmetric group $\mathfrak{S}_n$ is the alternating group $\mathfrak{A}_n$. For $\mathfrak{A}_n$ is generated by the products of two transpositions; if $\tau = \tau_{x,y}$ and $\tau' = \tau_{x',y'}$ are two transpositions, let σ be a permutation such that $\sigma(x') = x$ and $\sigma(y') = y$. Then $\tau' = \sigma^{-1}\tau\sigma$ and $\tau\tau' = \tau^{-1}\tau' = \tau^{-1}\sigma^{-1}\tau\sigma$ is a commutator. Hence $\mathfrak{A}_n \subset D(\mathfrak{S}_n)$. As $\mathfrak{S}_n/\mathfrak{A}_n$ is commutative, $\mathfrak{A}_n \supset D(\mathfrak{S}_n)$ (Corollary 3).

3. LOWER CENTRAL SERIES, NILPOTENT GROUPS

Let G be a group, H a subgroup of G and K a normal subgroup of G. The image of H in G/K is contained in the centre of G/K if and only if $(G, H) \subset K$.

DEFINITION 5. *Let G be a group. The* lower central series *of G is the sequence $(C^n(G))_{n \geqslant 1}$ of subgroups of G defined inductively by:*

$$C^1(G) = G, \qquad C^{n+1}(G) = (G, C^n(G)).$$

Let $f\colon G \to G'$ be a group homomorphism. It is seen, by induction on n, that $f(C^n(G)) \subset C^n(G')$ and that, if f is surjective, $f(C^n(G)) = C^n(G')$. In particular, for all $n \geqslant 1$, $C^2(G)$ is a characteristic (and hence normal) subgroup of G. For all $n \geqslant 1$, $C^n(G)/C^{n+1}(G)$ is contained in the centre of $G/C^{n+1}(G)$.

Let $(G_1, G_2, \ldots)$ be a decreasing sequence of normal subgroups of G such that (1) $G_1 = G$; (2) for all i, G_i/G_{i+1} is contained in the centre of G/G_{i+1}. Then $C^i(G) \subset G_i$, as is seen by induction on i.

Now

$$(C^m(G), C^n(G)) \subset C^{m+n}(G). \tag{7}$$

For, if this relation is denoted by $(F_{m,n})$, it follows from $(F_{m,n})$, by no. 2, Proposition 5, that

$$\begin{aligned}(C^m(G), C^{n+1}(G)) &\subset (G, (C^m(G), C^n(G))).(C^n(G), (G, C^m(G)))\\ &\subset C^{m+n+1}(G).(C^{m+1}(G), C^n(G)).\end{aligned}$$

Hence $((F_{m,n})$ and $(F_{m+1,n})) \Rightarrow (F_{m,n+1})$. As $(F_{m,1})$ and $(F_{1,n})$ are obvious $(F_{m,n})$ follows by induction.

DEFINITION 6. *A group* G *is called nilpotent if there exists an integer n such that* $C^{n+1}(G) = \{e\}$. *The least integer n such that* $C^{n+1}(G) = \{e\}$ *is called the* nilpotency class *of a nilpotent group* G.

If $n \in \mathbf{N}$, a group of nilpotency class n is called a nilpotent group of class n. It is sometimes said that the nilpotency class of a group G is finite if G is nilpotent.

Examples. (1) A group is nilpotent of class 0 (resp. $\leqslant 1$) if and only if it consists of the identity element (resp. is commutative).

(2) *For every commutative ring A and every integer $n \geqslant 1$, the upper strict triangular group $T_1(n, A)$ is nilpotent of class $\leqslant n - 1$ (and exactly of class $n - 1$ if $A \neq \{0\}$).*

(3) Let G be a nilpotent group of class n. Every subgroup (resp. every quotient group) of G is nilpotent of class $\leqslant n$. For, if H is a subgroup of G, then $C^n(H) \subset C^n(G)$. If G′ is a quotient group of G and $\pi\colon G \to G'$ is the canonical homomorphism, then $C^n(G') = \pi(C^n(G))$.

(4) A finite product of nilpotent groups is nilpotent.

PROPOSITION 7. *Let* G *be a group and n an integer. The following conditions are equivalent:*

(a) G *is nilpotent of class* $\leqslant n$.

(b) *There exists a series of subgroups of* G:

$$G = G^1 \supset G^2 \supset \cdots \supset G^{n+1} = \{e\}$$

such that $(G, G^k) \subset G^{k+1}$ *for all* $k \in [1, n]$.

(c) *There exists a subgroup* A *of* G *contained in the centre of* G *such that* G/A *is nilpotent of class* $\leqslant n - 1$.

(a) ⇒ (b): it suffices to take $G^k = C^k(G)$.
(b) ⇒ (a): by induction on k, $C^k(G) \subset G^k$.
(a) ⇒ (c): it suffices to take $A = C^n(G)$.
(c) ⇒ (a): let $\pi: G \to G/A$ be the canonical homomorphism; then $\pi(C^n(G)) = C^n(G/A) = \{e\}$ and hence $C^n(G) \subset A$, whence $C^{n+1}(G) = \{e\}$.

More briefly: a group is nilpotent of class $\leqslant n$ if it can be obtained from the group $\{e\}$ by n successive *central* extensions.

Corollary. *A central extension of a nilpotent group (by a necessarily commutative group) is nilpotent.*

Proposition 8. *Let* G *be a nilpotent group of class* $\leqslant n$ *and let* H *be a subgroup of* G. *There exists a sequence of subgroups*

$$G = H^1 \supset H^2 \supset \ldots \supset H^{n+1} = H,$$

such that H^{k+1} *is normal in* H^k *and* H^k/H^{k+1} *is commutative for all* $k \leqslant n$.

Choose a sequence (G^k) of subgroups of G satisfying the conditions of Proposition 7 (b) for all k; G^k is normal in G. Write:

$$H^k = H.G^k.$$

It is necessary to verify that H^{k+1} is normalized by $H^k = H.G^k$; as it is normalized by H, it suffices to verify that it is by G^k. Now, if $s \in G^k$ and $h \in H$,

$$shs^{-1} = shs^{-1}h^{-1}.h \in (G, G^k).H$$

and $(G, G^k).H$ is contained in $G^{k+1}.H = H^{k+1}$; hence $s.H^{k+1}.s^{-1} = H^{+k1}$, which shows that H^{k+1} is normal in H^k.

Finally, the canonical homomorphism $G^k/G^{k+1} \to H^k/H^{k+1}$ is obviously surjective; as the first group is commutative, so is the second.

Corollary 1. *Let* G *be a nilpotent group and* H *a subgroup of* G. *If* H *is distinct from* G, *the normalizer* $N_G(H)$ *of* H *in* G *is distinct from* H.

Let k be the largest index such that $H^k \neq H$. The group H^k normalizes H and is distinct from H.

Corollary 2. *Let* G *be a nilpotent group and* H *a subgroup of* G. *If* H *is distinct from* G, *there exists a normal subgroup* N *of* G, *containing* H, *distinct from* G *and such that* G/N *is commutative.*

Let k be the least index such that $H^k \neq G$. The group H^k satisfies the required conditions.

COROLLARY 3. *Let* G *be a nilpotent group and* H *a subgroup of* G. *If* G = H.(G, G), *then* G = H.

Every subgroup N of G which contains H and such that G/N is commutative contains H.(G, G). Corollary 3 thus follows from Corollary 2.

Corollary 3 can also be formulated thus: let X be a subset of G. For X to generate G, it is necessary and sufficient that the image of X in G/D(G) generate G/D(G).

COROLLARY 4. *Let* f: G' → G *be a group homomorphism. Suppose that*

(a) G *is nilpotent.*

(b) *The homomorphism* f_1: G'/(G', G') → G/(G, G), *derived from* f *by passing to the quotients, is surjective.*

Then f *is surjective.*

This follows from Corollary 3 applied to the subgroup H = f(G').

PROPOSITION 9. *Let* G *be a nilpotent group of class* $\leqslant n$ *and let* N *be a normal subgroup of* G. *There exists a series of subgroups*

$$N = N^1 \supset N^2 \supset \dots \supset N^{n+1} = \{e\}$$

such that $(G, N^k) \subset N^{k+1}$ *for* $k = 1, \dots, n$.

If (G^k) satisfies condition (b) of Proposition 7, then take

$$N^k = G^k \cap N.$$

COROLLARY 1. *Let* G *be a nilpotent group,* Z *the centre of* G *and* N *a normal subgroup of* G. *If* $N \neq \{e\}$, *then* $N \cap Z \neq \{e\}$.

Let k be the largest index such that $N^k \neq \{e\}$. The group N^k is contained in N. On the other hand, $(G, N^k) \subset N^{k+1} = \{e\}$; hence N^k is contained in the centre Z of G.

COROLLARY 2. *Let* f *be a homomorphism from a nilpotent group* G *to a group* G'. *If the restriction of* f *to the centre of* G *is injective,* f *is injective.*

This is Corollary 1 applied to Ker(f).

4. DERIVED SERIES, SOLVABLE GROUPS

DEFINITION 7. *Let* G *be a group. The* derived series *of* G *is the series* $(D^n(G))_{n \in \mathbf{N}}$ *defined inductively by:*

$$D^0(G) = G; \qquad D^{n+1}(G) = D(D^n(G)) \quad \textit{for } n \in \mathbf{N}.$$

Then $D^0(G) = C^1(G) = G$, $D^1(G) = C^2(G) = D(G) = (G, G)$. For all

$n \in \mathbf{N}$, $D^n(G) \subset C^{2^n}(G)$, as is seen by induction on n using formula (7) of no. 3.

Let $f: G \to G'$ be a group homomorphism. It is seen, by induction on n, that $f(D^n(G)) \subset D^n(G')$ and that, if f is surjective, $f(D^n(G)) = D^n(G')$. In particular, for all $n \in \mathbf{N}$, $D^n(G)$ is a characteristic (and therefore normal) subgroup of G. For all $n \in \mathbf{N}$, the group $D^n(G)/D^{n+1}(G)$ is a commutative normal (but not in general central) subgroup of $G/D^{n+1}(G)$.

Let $(G_0, G_1, \ldots)$ be a decreasing sequence of subgroups of G such that: (1) $G_0 = G$; (2) for all i, G_{i+1} is normal in G_i and G_i/G_{i+1} is commutative. Then $D^i(G) \subset G_i$ for all i, as is seen by induction on i.

DEFINITION 8. *A group* G *is called solvable if there exists an integer n such that* $D^n(G) = \{e\}$. *If* G *is a solvable group, the least integer n such that* $D^n(G) = \{e\}$ *is called the* solvability class *of* G.

A solvable group of solvability class n is called a solvable group of class n. A group is sometimes said to be of finite solvability class if it is solvable.

Examples. (1) A group is solvable of class 0 (resp. $\leqslant 1$) if and only if it is reduced to $\{e\}$ (resp. is commutative).

(2) Every nilpotent group of class $\leqslant 2^n - 1$ is solvable of class $\leqslant n$; this follows from the relation $D^n(G) \subset C^{2^n}(G)$ proved above.

(3) Let G be a solvable group of class $\leqslant n$. Every subgroup (resp. quotient group) of G is solvable of class $\leqslant n$ (proof analogous to that of no. 3, Example 3).

(4) If G is a solvable group of class p and F is a solvable group of class q, every extension E of G by F is a solvable group of class $\leqslant p + q$. For, let $\pi: E \to G$ be the projection; then $\pi(D^p(E)) \subset D^p(G) = \{e\}$ and therefore $D^p(E) \subset F$; it follows that $D^{p+q}(E) = D^q(D^p(E)) \subset D^q(F) = \{e\}$.

(5) The symmetric group $\mathfrak{S}_n$ is solvable if and only if $n < 5$ (cf. § 5, Exercises 10 and 16).

(6) *If A is a commutative ring, the upper triangular group T(n, A) is solvable but not in general nilpotent.*

PROPOSITION 10. *Let* G *be a group and n an integer. The following conditions are equivalent:*

(i) G *is solvable of class* $\leqslant n$.

(ii) *There exists a series of normal subgroups of* G

$$G = G^0 \supset G^1 \supset \cdots \supset G^n = \{e\}$$

such that the groups G^k/G^{k+1} *are commutative.*

(iii) *There exists a series of subgroups of* G

$$G = G^0 \supset G^1 \supset \cdots \supset G^n = e$$

such that, for all k, G^{k+1} *is a normal subgroup of* G^k *and* G^k/G^{k+1} *is commutative.*

(iv) *There exists a normal commutative subgroup* A *of* G *such that* G/A *is solvable of class* $\leqslant n - 1$.

For (i) $\Rightarrow$ (ii) it suffices to take G^k equal to $D^k(G)$. (ii) $\Rightarrow$ (iii) trivially. (iii) $\Rightarrow$ (i) for $D^k(G)$ is necessarily contained in G^k. The equivalence of (ii) and (iv) is immediate by induction on n.

More briefly: a group is solvable of class $\leqslant n$ if it can be obtained by successive extensions of n commutative groups.

COROLLARY. *Let* G *be a finite group and*

$$G = G^0 \supset G^1 \supset \cdots \supset G^n = \{e\}$$

a Jordan-Hölder series of G *For* G *to be solvable, it is necessary and sufficient that the quotients* G^k/G^{k+1} *be cyclic of prime order.*

If the quotients of a composition series of G are cyclic and hence commutative, G is solvable by Proposition 10. Conversely, if G is solvable, the group G^k/G^{k+1} is, for all k, solvable and simple (§ 4, no. 7, Proposition 9). Now, every solvable simple group H is cyclic of prime order. For D(H) is a normal subgroup of H; D(H) = H is impossible for in that case $D^k(H) = H$ for all k; then $D(H) = \{e\}$ and H is commutative. The corollary then follows from § 4, no. 10, Corollary to Proposition 20.

5. p-GROUPS

In this number and the following, the letter p *denotes a prime number* (§ 4, no 10, Proposition 16).

DEFINITION 9. *A finite group whose order is a power of* p *is called a* p-*group.*

Let G be a p-group of order p^r. Every divisor of p^r is a power of p (§ 4, no. 10, Corollary to Theorem 7). Therefore every subgroup and every quotient group of G is a p-group (§ 4, no. 4, Corollary to Proposition 4); the cardinal of every homogeneous space of G is a power of p (§ 5, no. 5, Theorem 1).

An extension of a p-group by a p-group is a p-group.

**Examples.* (1) A commutative p-group is isomorphic to a product of cyclic groups $\mathbf{Z}/p^n\mathbf{Z}$ (cf. Exercise 19 and also VII, § 4, no. 7, Proposition 7).

(2) Let k be a finite field of characteristic p. The strict triangular group $T_1(n, k)$ is a p-group.

(3) The quaternionic group $\{\pm 1, \pm i, \pm j, \pm k\}$ is a 2-group (cf. Exercise 4).*

PROPOSITION 11. *Let* E *be a finite set and* G *a* p-*group operating on* E. *Let* E^G *denote the set of* $x \in E$ *such that* $gx = x$ *for all* $g \in G$ (the fixed points). *Then*

$$\text{Card}(E^G) \equiv \text{Card}(E) \quad (\text{mod. } p).$$

$E - E^G$ is a disjoint union of orbits not reduced to a point. The cardinal of such an orbit is a power of p distinct from $p^0 = 1$ and hence a multiple of p.

COROLLARY. *Let* G *be a* p*-group. If* G *is not reduced to* e*, its centre is not reduced to* e*.*

Let G operate on itself by inner automorphisms. The set of fixed points is the centre Z of G. By Proposition 11,

$$\mathrm{Card}(Z) \equiv \mathrm{Card}(G) \equiv 0 \pmod{p},$$

whence $\mathrm{Card}(Z) \neq 1$ and $Z \neq \{e\}$.

THEOREM 1. *Let* G *be a* p*-group and* p^r *its order. There exists a sequence of subgroups of* G

$$G = G^1 \supset G^2 \supset \dots \supset G^{r+1} = \{e\}$$

such that $(G, G^k) \subset G^{k+1}$, $1 \leqslant k \leqslant r$, and G^k/G^{k+1}, $1 \leqslant k \leqslant r$, *is cyclic of order* p.

The theorem is true for $G = \{e\}$. We prove it by induction on Card(G). Let Z be the centre of G, $x \neq e$ an element of Z (Corollary to Proposition 11) and p^s, $s \neq 0$, the order of x. Then $x^{p^{s-1}}$ is an element of order p and therefore Z contains a subgroup G^r which is cyclic of order p. By the induction hypothesis, the group $G' = G/G^r$ has a series of subgroups $(G'^k)_{1 \leqslant k \leqslant r}$ with the required properties. Let $\pi\colon G \to G'$ be the canonical homomorphism. The sequence of subgroups of G defined by $G^k = \pi^{-1}(G'^k)$, $1 \leqslant k \leqslant r$, $G^{r+1} = \{e\}$ is a solution for G^k/G^{k+1} is isomorphic to G'^k/G'^{k+1} for $1 \leqslant k \leqslant r$ (§ 4, no. 7, Theorem 4).

COROLLARY. *Every* p*-group is nilpotent.*

This follows from no. 3, Proposition 7.

PROPOSITION 12. *Let* G *be a* p*-group and* H *a subgroup of* G *distinct from* G*. Then:*

(a) *The normalizer* $N_G(H)$ *of* H *in* G *is distinct from* G.
(b) *There exists a normal subgroup* N *of* G *of index* p *in* G*, which contains* H.

Assertion (a) follows from no. 3, Corollary 1 to Proposition 8. We prove (b). By no. 3, Corollary 2 to Proposition 8, there exists a normal subgroup N′ of G containing H, distinct from G and such that G/N′ is commutative. Let N be a maximal subgroup distinct from G containing N′. Then N is normal (no. 2, Corollary 3 to Proposition 6) and G/N is a simple commutative p-group and hence cyclic of order p (§ 4, no. 10, Corollary to Proposition 20).

COROLLARY. *Let* G *be a* p*-group. Every subgroup of* G *of index* p *is normal.*

6. SYLOW SUBGROUPS

DEFINITION 10. *Let G be a finite group. A Sylow p-subgroup of G is any subgroup P of G satisfying the two following conditions:*

(a) *P is a p-group.*
(b) *(G:P) is not a multiple of p.*

If the order of G is written in the form $p^r m$, where m is not a multiple of p, conditions (a) and (b) are equivalent to $\mathrm{Card}(P) = p^r$.

Examples. (1) In the group $\mathfrak{S}_p$ let ζ be a cycle of order p. The subgroup generated by ζ is a Sylow p-subgroup of $\mathfrak{S}_p$ for p does not divide $(p - 1)!$.

(2) *Let k be a finite field of characteristic p and let n be a positive integer. The strict triangular group $T_1(n, k)$ is a Sylow p-subgroup of the group $\mathbf{GL}(n, k)$.*

THEOREM 2. *Every finite group contains a Sylow p-subgroup.*

The proof depends on the following lemma.

Lemma. Let $n = p^r m$, where m is an integer which is not a multiple of p. Then

$$\binom{n}{p^r} \not\equiv 0 \pmod{p}.$$

Let S be a group of order p^r (for example $\mathbf{Z}/p^r\mathbf{Z}$) and T a set with m elements. Write $X = S \times T$ and let E be the set of subsets of X with p^r elements. Then $\mathrm{Card}(X) = n$, whence $\mathrm{Card}(E) = \binom{n}{p^r}$ (*Set Theory*, III, § 5, no. 8, Corollary 1 to Proposition 11). Let S operate on X by $s.(x, y) = (sx, y)$ $(s, x \in S, y \in T)$ and consider the canonical extension of this operation to E. In the notation of no. 5, Proposition 11, the set E^S is the set of orbits of X, that is the set of subsets $Y \subset X$ of the form $S \times \{t\}$, $t \in T$, whence $\mathrm{Card}(E^S) = m$. By no. 5, Proposition 11,

$$\binom{n}{p^r} = \mathrm{Card}(E) \equiv \mathrm{Card}(E^S) = m \not\equiv 0 \pmod{p},$$

which proves the lemma.

We now prove the theorem. Let G be a finite group and n its order; we write $n = p^r m$, where m is not a multiple of p. Let E be the set of subsets of G with p^r elements. Then

$$\mathrm{Card}(E) = \binom{n}{p^r};$$

whence, by virtue of the lemma, $\mathrm{Card}(E) \not\equiv 0 \pmod{p}$. Consider the extension to E of the operation of G on itself by left translation. There exists $X \in E$ whose orbit has non-zero cardinal mod. p. If H_X denotes the stabilizer of X, then

$(G:H_X) \not\equiv 0$ (mod. p), which means that p^r divides $\mathrm{Card}(H_X)$. But H_X consists of the $s \in G$ such that $sX = X$; if $x \in X$, then $H_X \subset X.x^{-1}$, whence $\mathrm{Card}(H_X) \leqslant \mathrm{Card}(X) = p^r$. Hence $\mathrm{Card}(H_X) = p^r$.

COROLLARY. *If the order of* G *is divisible by* p, *the group* G *contains an element of order* p.

By virtue of Theorem 2, this is reduced to the case where G is a p-group $\neq \{e\}$; if $x \in G$ is different from e, the cyclic group generated by x is then of order p^n with $n \geqslant 1$ and it therefore contains a subgroup of order p.

Remark. For every prime number q dividing Card(G), let P_q be a Sylow q-subgroup of G. Then the subgroup H of G generated by the P_q is of order a multiple of $\mathrm{Card}(P_q)$ for each q and of order a divisor of Card(G), hence it is equal to G.

THEOREM 3. *Let* G *be a finite group.*

(a) *The Sylow* p-*subgroups of* G *are conjugate to one another. Their number is congruent to* 1 mod. p.

(b) *Every subgroup of* G *which is a* p-*group is contained in a Sylow* p-*subgroup.*

Let P be a Sylow p-subgroup of G (Theorem 2) and let H be a p-subgroup of G. Let $E = G/P$ and consider the operation of H on G/P. As $\mathrm{Card}(E) \not\equiv 0$ mod. p, Proposition 11 of no. 5, shows that there exists $x \in G/P$ such that $hx = x$ for all $h \in H$. If g is a representative of x in G, this means that $H \subset gPg^{-1}$, whence assertion (b).

If H is a Sylow p-subgroup, then $\mathrm{Card}(H) = \mathrm{Card}(P) = \mathrm{Card}(gPg^{-1})$, whence $H = gPg^{-1}$, which proves the first assertion of (a).

We now prove the second assertion of (a). Let $\mathscr{S}$ be the set of Sylow p-subgroups of G and let P operate on $\mathscr{S}$ by inner automorphisms. The element $P \in \mathscr{S}$ is a fixed point under this operation, we show that it is the only one. Let $Q \in \mathscr{S}$ be a fixed point; Q is a Sylow subgroup of G normalized by P and hence P is contained in the normalizer N of Q. The groups P and Q are Sylow p-subgroups of N; hence there exists $n \in N$ such that $P = nQn^{-1} = Q$. By no. 5, Proposition 11, $\mathrm{Card}(\mathscr{S}) \equiv \mathrm{Card}(\mathscr{S}^P) = 1$ (mod. p).

COROLLARY 1. *Let* P *be a Sylow* p-*subgroup of* G, *let* N *be its normalizer in* G *and let* M *be a subgroup of* G *containing* N. *The normalizer of* M *in* G *is equal to* M.

Let $s \in G$ be such that $sMs^{-1} = M$. The subgroup sPs^{-1} of M is a Sylow p-subgroup of M. There thus exists $t \in M$ such that $sPs^{-1} = tPt^{-1}$; then $t^{-1}s \in N$, whence $s \in tN \subset M$.

COROLLARY 2. *Let* $f: G_1 \to G_2$ *be a homomorphism of finite groups. For every Sylow* p-*subgroup* P_1 *of* G_1 *there exists a Sylow* p-*subgroup* P_2 *of* G_2 *such that* $f(P_1) \subset P_2$.

This follows from Theorem 3 (b) applied to the subgroup $f(P_1)$ of G_2.

COROLLARY 3. (a) *Let* H *be a subgroup of* G. *For every Sylow p-subgroup* P *of* H *there exists a Sylow p-subgroup* Q *of* G *such that* $P = Q \cap H$.

(b) *Conversely, if* Q *is a Sylow p-subgroup of* G *and* H *is normal in* G, *the group* $Q \cap H$ *is a Sylow p-subgroup of* H.

(a) The p-group P is contained in a Sylow p-subgroup Q of G and $Q \cap H$ is a p-subgroup of H containing P and is hence equal to P.

(b) Let P′ be a Sylow p-subgroup of H. There exists an element $g \in G$ such that $gP'g^{-1} \subset Q$. As H is normal, $P = gP'g^{-1}$ is contained in H and hence in $Q \cap H$. As $Q \cap H$ is a p-subgroup of H and P is a Sylow p-subgroup of H, $P = Q \cap H$.

COROLLARY 4. *Let* N *be a normal subgroup of* G. *The image in* G/N *of a Sylow p-subgroup of* G *is a Sylow p-subgroup of* G/N *and every Sylow p-subgroup of* G/N *is obtained in this way.*

Let $G' = G/N$ and P′ be the image in G′ of a Sylow p-subgroup P of G. The group G operates transitively on G′/P′ and hence G′/P′ is equipotent to G/S, where S is a subgroup of G containing P. Therefore $(G' : P')$ divides $(G : P)$, is thus not a multiple of p and the p-group P′ is a Sylow p-subgroup of G′. Let Q′ be another Sylow p-subgroup of G′; then $Q' = g'P'g'^{-1}$ for some $g' \in G'$; if $g \in G$ is a representative of g', the group Q′ is the image of $Q = gPg^{-1}$.

7. FINITE NILPOTENT GROUPS

THEOREM 4. *Let* G *be a finite group. The following conditions are equivalent:*

(a) G *is nilpotent.*

(b) G *is a product of p-groups.*

(c) *For every prime number p there exists a normal Sylow p-subgroup of* G.

(b) ⇒ (a) (no. 5, Corollary to Theorem 1).

Suppose (a) holds and let P be a Sylow p-subgroup of G. If N is the normalizer of P in G, Corollary 1 to Theorem 3 shows that N is its own normalizer. By § 6, no. 3, Corollary to Proposition 8, this shows that $N = G$. Hence (a) ⇒ (c).

Suppose (c) holds and let I be the set of prime numbers dividing Card(G). For all $p \in I$, let P_p be a normal Sylow p-subgroup of G. For all $p \neq q$, $P_p \cap P_q$ is reduced to e for it is both a p-group and a q-group, hence P_p and P_q centralize one another (§ 4, no. 9, Proposition 15). Let ϕ be the canonical homomorphism (§ 4, no. 9, Proposition 12) of $\prod_{p \in I} P_p$ into G. The homomorphism ϕ is surjective by the *Remark* of no. 6. As $\text{Card}\left(\prod_{p \in I} P_p\right) = \text{Card}(G)$, it follows that ϕ is bijective.

Remarks. (1) Let G be a finite group and p a prime number. By no. 6, Theorem 3 (a) and no. 6, Theorem 2, the following conditions are equivalent:

(i) there exists a normal Sylow p-subgroup of G;
(ii) every Sylow p-subgroup of G is normal;
(iii) there exists only one Sylow p-subgroup of G.

(2) Let G be a nilpotent finite group. Let I be the set of prime divisors of Card(G). By Theorem 4 and *Remark* 1, $G = \prod_{p \in I} G_p$, where G_p is the unique Sylow p-group of G.

(3) Applied to commutative groups, Theorem 4 gives the decomposition, of commutative finite groups as a product of primary components, which will be studied from another point of view in Chapter VII.

Example. The group $\mathfrak{S}_3$ is of order 6. It contains a normal Sylow 3-subgroup of order 3: the group $\mathfrak{A}_3$. It contains three Sylow 2-subgroups of order 2: the groups $\{e, \tau\}$, where τ is a transposition. The group $\mathfrak{S}_3$ is thus not nilpotent.

§ 7. FREE MONOIDS, FREE GROUPS

In this paragraph X *will denote a set. Unless otherwise mentioned, the identity element of a monoid will be denoted by e.*

1. FREE MAGMAS

A sequence of sets $M_n(X)$ is defined by induction on the integer $n \geqslant 1$ as follows: writing $M_1(X) = X$, for $n \geqslant 2$, $M_n(X)$ is the set the sum of the sets $M_p(X) \times M_{n-p}(X)$ for $1 \leqslant p \leqslant n - 1$. The set the sum of the family $(M_n(X))_{n \geqslant 1}$ is denoted by M(X); each of the sets $M_n(X)$ is identified with its canonical image in M(X). For every element w of M(X) there exists a unique integer n such that $w \in M_n(X)$; it is called the *length* of w and denoted by $l(w)$. The set X consists of the elements in M(X) of length 1.

Let w and w' be in M(X); write $p = l(w)$ and $q = l(w')$. The image of (w, w') under the canonical injection of $M_p(X) \times M_q(X)$ into the sum set $M_{p+q}(X)$ is called the *composition* of w and w' and is denoted by ww' or $w.w'$. Then $l(w.w') = l(w) + l(w')$ and every element of M(X) of length $\geqslant 2$ can be written uniquely in the form $w'w''$ with w', w'' in M(X).

The set M(X) with the law of composition $(w, w') \mapsto w.w'$ is called the *free magma constructed on* X (§ 1, no. 1, Definition 1).

Proposition 1. *Let* M *be a magma. Every mapping f of* X *into* M *may be extended in a unique way to a morphism of* M(X) *into* M.

By induction on $n \geqslant$, mappings f_n: $M_n(X) \to M$ are defined as follows: let

$f_1 = f$; for $n \geqslant 2$, the mapping f_n is defined by $f_n(w.w') = f_p(w).f_{n-p}(w')$ for $p = 1, 2, \ldots, n-1$ and (w, w') in $M_p(X) \times M_{n-p}(X)$. Let g be the mapping of M(X) into M which induces f_n on $M_n(X)$ for every integer $n \geqslant 1$. Clearly g is the unique morphism of M(X) into M which extends f.

Let u be a mapping of X into a set Y. By Proposition 1, there exists one and only one homomorphism of M(X) into M(Y) which coincides with u on X. It will be denoted by $M(u)$. If v is a mapping of Y into a set Z, the homomorphism $M(v) \circ M(u)$ of M(X) into M(Z) coincides with $v \circ u$ on X, whence

$$M(v) \circ M(u) = M(v \circ u).$$

PROPOSITION 2. *Let* $u: X \to Y$ *be a mapping. If* u *is injective* (resp. *surjective, bijective*), *so is* $M(u)$.

Suppose u is injective. When X is empty, M(X) is empty, hence $M(u)$ is injective. If X is non-empty, there exists a mapping v of Y into X such that $v \circ u$ is the identity mapping of X (*Set Theory*, II, § 3, no. 8, Proposition 8); the mapping $M(v) \circ M(u) = M(v \circ u)$ is the identity mapping of M(X) and hence $M(u)$ is injective.

When u is surjective, there exists a mapping w of Y into X such that $u \circ w$ is the identity mapping of Y (*Set Theory*, II, § 3, no. 8, Proposition 8). Then $M(u) \circ M(w) = M(u \circ w)$ is the identity mapping of M(Y) and hence $M(u)$ is surjective.

Finally, if u is bijective, it is injective and surjective and hence $M(u)$ has the same properties.

Let S be a subset of X. By Proposition 2 the injection of S into X can be extended to an isomorphism of M(S) onto a submagma M'(S) of M(X). The magmas M(S) and M'(S) are identified by means of this isomorphism. Then M(S) is the submagma of M(X) generated by S.

Let X be a set and $(u_\alpha, v_\alpha)_{\alpha \in I}$ be a family of ordered pairs of elements of M(X). Let R be the equivalence relation on M(X) compatible with the law of M(X) and generated by the (u_α, v_α) (§ 1, no. 6). The magma M(X)/R is called the *magma defined by* X *and the relators* $(u_\alpha, v_\alpha)_{\alpha \in I}$. Let h be the canonical morphism of M(X) onto M(X)/R. Then M(X)/R is generated by $h(X)$.

Let N be a magma and $(n_x)_{x \in X}$ a family of elements of N. Let k be the morphism from M(X) to N such that $k(x) = n_x$ for all $x \in X$ (Proposition 1). If $k(u_\alpha) = k(v_\alpha)$ for all $\alpha \in I$, there exists one and only one morphism $f: M(X)/R \to N$ such that $f(h(x)) = n_x$ for all $x \in X$ (§ 1, no. 6, Proposition 9).

2. FREE MONOIDS

Any finite sequence $w = (x_i)_{1 \leqslant i \leqslant n}$ of elements of X indexed by an interval $(1, n)$ of **N** (possibly empty) is called a *word* constructed on X. The integer n is called the *length* of the word w and denoted by $l(w)$. There is a unique word of length 0, namely the empty sequence e. X will be identified with the set of words of length 1.

Let $w = (x_i)_{1 \leqslant i \leqslant m}$ and $w' = (x'_j)_{1 \leqslant j \leqslant n}$ be two words. The composition of w and w' is the word $u = (y_k)_{1 \leqslant k \leqslant m+n}$ defined by

$$(1) \qquad y_k = \begin{cases} x_k & \text{for } 1 \leqslant k \leqslant m \\ x'_{k-m} & \text{for } m+1 \leqslant k \leqslant m+n. \end{cases}$$

In other words, the sequence w'' is obtained by first writing the elements of the sequence w and then those of w'. The composition of w and w' is generally denoted by ww' or $w.w'$; it is sometimes said that it is obtained by *juxtaposition* of w and w'. Then by construction $l(w.w') = l(w) + l(w')$.

The relation $we = ew = w$ is immediately established for every word w. Let $w = (x_i)_{1 \leqslant i \leqslant m}$, $w' = (x'_j)_{1 \leqslant j \leqslant n}$ and $w'' = (x''_k)_{1 \leqslant k \leqslant p}$ be three words; clearly the words $w(w'w'')$ and $(ww')w''$ are both equal to the word $(y_l)_{1 \leqslant l \leqslant m+n+p}$ defined by

$$(2) \qquad y_l = \begin{cases} x_l & \text{if } 1 \leqslant l \leqslant m \\ x_{l-m} & \text{if } m+1 \leqslant l \leqslant m+n \\ x''_{l-m-n} & \text{if } m+n+1 \leqslant l \leqslant m+n+p. \end{cases}$$

The above shows that the set of words constructed on X with the law of composition $(w, w') \mapsto w.w'$ is a monoid with identity element e. It is denoted by Mo(X) and called the *free monoid constructed on* X. It follows immediately from the definition of product of words that every word $w = (x_i)_{1 \leqslant i \leqslant n}$ is equal to the product $\prod_{i=1}^{n} x_i$. A word may therefore be written in the form $x_1 \ldots xn$.

Proposition 3. *Let* M *be a monoid. Every mapping* f *of* X *into* M *extends uniquely to a homomorphism of* Mo(X) *into* M.

Let g be a homomorphism of Mo(X) into M extending f. If $w = (x_i)_{1 \leqslant i \leqslant n}$ is a word, then $w = \prod_{i=1}^{n} x_i$ in the monoid Mo(X), whence

$$g(w) = \prod_{i=1}^{n} g(x_i) = \prod_{i=1}^{n} f(x_i)$$

in the monoid M (§ 1, no. 2, formula (2)). This proves the uniqueness of g.

Let $h(w) = \prod_{i=1}^{n} f(x_i)$ for every word $w = (x_i)_{1 \leqslant i \leqslant n}$. The associativity theorem (§ 1, no. 3, Theorem 1) and the definition of product in Mo(X) imply $h(ww') = h(w)h(w')$. By convention the empty product $h(e)$ is the identity element of M and $h(x) = f(x)$ for $x \in$ X. Hence h is a homomorphism of Mo(x) into M extending f.

Let u: X $\rightarrow$ Y be a mapping. By Proposition 3, there exists one and only one homomorphism of Mo(X) into Mo(Y) which coincides with u on X; it is denoted by Mo(u). It maps a word $(x_i)_{1 \leq i \leq n}$ to the word $(u(x_i))_{1 \leq i \leq n}$. As in the case of magmas (no. 1), the equation $\text{Mo}(v \circ u) = \text{Mo}(v) \circ \text{Mo}(u)$ is established for every mapping $v: Y \to Z$ and it can be shown that Mo(u) is injective (resp. surjective, bijective) if u is. For every subset S of X, Mo(S) is identified with the submonoid of Mo(X) generated by S.

Let X be a set and $(u_\alpha, v_\alpha)_{\alpha \in I}$ be a family of ordered pairs of elements of Mo(X). Let R be the equivalence relation on Mo(X) compatible with the law on Mo(X) and generated by the (u_α, v_α) (§ 1, no. 6). The monoid Mo(X)/R is called the *monoid defined by* X *and the relators* $(u_\alpha, v_\alpha)_{\alpha \in I}$. Let h be the canonical morphism of Mo(X) onto Mo(X)/R. Then Mo(X) is generated by h(X).

Let N be a monoid and $(n_x)_{x \in X}$ a family of elements of N. Let k be the morphism of Mo(X) into N such that $k(x) = n_x$ for all $x \in X$ (Proposition 3). If $k(u_\alpha) = k(v_\alpha)$ for all $\alpha \in I$, there exists one and only one magma morphism $f: \text{Mo}(X)/R \to N$ such that $f(h(x)) = n_x$ for all $x \in X$ (§ 1, no. 6, Proposition 9); as k is unital, f is a monoid morphism.

3. AMALGAMATED SUM OF MONOIDS

Let $(M_i)_{i \in I}$ *denote a family of monoids and* e_i *the identity element of* M_i. *We are given a monoid* A *and a family of homomorphisms* $h_i: A \to M_i$ (for $i \in I$).

The set S the sum of the family $(M_i)_{i \in I}$ has elements the ordered pairs (i, x) with $i \in I$ and $x \in M_i$. For every triple $\alpha = (i, x, x')$ with $i \in I$, x, x' in M_i, write $u_\alpha = (i, xx')$ and $v_\alpha = (i, x).(i, x')$; for every triple $\lambda = (i, j, a)$ in $I \times I \times A$, write $p_\lambda = (i, h_i(a))$ and $q_\lambda = (j, h_j(a))$; for all $i \in I$, write $\varepsilon_i = (i, e_i)$. The monoid M defined by S and the relators (u_α, v_α), (p_λ, q_λ) and (ε_i, e) is called the *sum of the family* $(M_i)_{i \in I}$ *amalgamated by* A. Let ϕ denote the canonical homomorphism of Mo(S) onto M and write $\phi_i(x) = \phi(i, x)$ for $(i, x) \in S$. It is said that ϕ_i is the *canonical mapping* of M_i into M. For all $a \in A$, the element $\phi(i, h_i(a))$ is independent of i and denoted by $h(a)$.†

The universal property of monoids defined by generators and relators (no. 2) implies the following result:

PROPOSITION 4. (a) *For all* $i \in I$, *the mapping* ϕ_i *is a homomorphism of* M_i *into* M *and* $\phi_i \circ h_i = h$ *for all* $i \in I$. *Further,* M *is generated by* $\bigcup_{i \in I} \phi_i(M_i)$.

(b) *Let* M' *be a monoid and* $f'_i: M_i \to M'$ (*for* $i \in I$) *homomorphisms such that* $f_i \circ h_i$ *is independent of* $i \in I$. *There exists one and only one homomorphism* $f: M \to M'$ *such that* $f_i = f \circ \phi_i$ *for all* $i \in I$.

In what follows we shall make the following hypothesis:

(A) *For all* $i \in I$, *there exists a subset* P_i *of* M_i *containing* e_i *such that the mapping* $(a, p) \mapsto h_i(a).p$ *of* $A \times P_i$ *into* M_i *is bijective.*

† When I is empty, $M = \{e\}$ and $h(a) = e$ for all $a \in A$.

It implies that the homomorphisms h_i are injective. Let $x \in M$; every finite sequence $\sigma = (a: i_1, \ldots, i_n; p_1, \ldots, p_n)$ with $a \in A$, $i_\alpha \in I$ and $p_\alpha \in P_{i_\alpha}$ for $1 \leqslant \alpha \leqslant n$, satisfying

$$(3) \qquad x = h(a).\prod_{\alpha=1}^{n} \phi_{i_\alpha}(p_\alpha)$$

is called a *decomposition of* x. The integer $n \geqslant 0$ is called the *length* of the decomposition σ and is denoted by $l(\sigma)$; the sequence (e) is a decomposition of length 0 of the identity element of M. The decomposition σ is called *reduced* if $i_\alpha \neq i_{\alpha+1}$ for $1 \leqslant \alpha < n$ and $p_\alpha \neq e_{i_\alpha}$ for $1 \leqslant \alpha \leqslant n$.

PROPOSITION 5. *Under hypothesis* (A) *every element x of* M *admits a unique reduced decomposition σ. Every decomposition $\sigma' \neq \sigma$ of x satisfies $l(\sigma') > l(\sigma)$.*

(A) *Uniqueness of a reduced decomposition:*

Let Σ denote the set of sequences $\sigma = (a; i_1, \ldots, i_n; p_1, \ldots, p_n)$ with $n \geqslant 0$, $a \in A$, $i_\alpha \in I$ and $p_\alpha \in P_{i_\alpha} - \{e_{i_\alpha}\}$ for $1 \leqslant \alpha \leqslant n$, such that $i_\alpha \neq i_{\alpha+1}$ for $1 \leqslant \alpha < n$. Let Φ denote the mapping of Σ into M defined by

$$(4) \qquad \Phi(a; i_1, \ldots, i_n; p_1, \ldots, p_n) = h(a).\prod_{\alpha=1}^{n} \phi_{i_\alpha}(p_\alpha).$$

A reduced decomposition of $x \in M$ is an element σ of Σ such that $\Phi(\sigma) = x$.

For all $i \in I$, let Σ_i be the subset of Σ consisting of the sequences $(e; i_1, \ldots, i_n; p_1, \ldots, p_n)$ with $i \neq i_1$ when $n > 0$. Let

$$\sigma = (e; i_1, \ldots, i_n; p_1, \ldots, p_n)$$

be in Σ_i and ξ in M_i; let $\xi = h_i(a).p$ with $a \in A$ and $p \in P_i$, and

$$(5) \qquad \Psi_i(\xi, \sigma) = \begin{cases} (a; i_1, \ldots, i_n; p_1, \ldots, p_n) & \text{if } p = e_i \\ (a; i, i_1, \ldots, i_n; p, p_1, \ldots, p_n) & \text{if } p \neq e_i. \end{cases}$$

It is immediate that Ψ_i is a bijection of $M_i \times \Sigma_i$ onto Σ.

Let $i \in I$ and $x \in M_i$; as Ψ_i is bijective, a mapping $f_{i,x}$ of Σ into itself is defined by

$$(6) \qquad f_{i,x}(\Psi_i(\xi, \sigma)) = \Psi_i(x\xi, \sigma) \quad (\xi \in M_i, \sigma \in \Sigma_i).$$

Further, for $a \in A$, f_a denotes the mapping of Σ into itself defined by

$$(7) \qquad f_a(a'; i_1, \ldots, i_n; p_1, \ldots, p_n) = (aa'; i_1, \ldots, i_n; p_1, \ldots, p_n).$$

Clearly f_{i,e_i} is the identity mapping of Σ and $f_{i,xx'} = f_{i,x} \circ f_{i,x'}$ for x, x' in M_i and $f_{i,h_i(a)} = f_a$ for $a \in A$ and $i \in I$.

Then Proposition 4 may be applied to the case where M′ is the monoid of mappings of Σ into itself with law of composition $(f, f') \mapsto f \circ f'$ and where

f_i is the homomorphism $x \mapsto f_{i,x}$ of M_i into M'; then there exists a homomorphism f of M into M′ such that $f_{i,x} = f(\phi_i(x))$ for $i \in I$ and $x \in M_i$. Let

$$\sigma = (a; i_1, \dots, i_n; p_1, \dots, p_n)$$

be in Σ. Formulae (5) to (7) imply by induction on n the relation

$$\begin{aligned}\sigma &= (f_a \circ f_{i_1, p_1} \circ \cdots \circ f_{i_n, p_n})(e)\\ &= f(h(a)\phi_{i_1}(p_1) \dots \phi_{i_n}(p_n))(e),\end{aligned}$$

that is $\sigma = f(\Phi(\sigma))(e)$. This proves that Φ is injective.

(B) *Existence of a decomposition:*

Let D be the set of elements of M admitting a decomposition. Then $e \in D$ and M is generated by $\bigcup_{i \in I} \phi_i(M_i)$ and hence by $h(A) \cup \bigcup_{i \in I} \phi_i(P_i)$. Then $D.\phi_i(P_i) \subset D$ for all $i \in I$; to prove that $D = M$, it thus suffices to prove the relation $D.h(A) \subset D$. This follows from the following more precise lemma:

Lemma 1. *Let $i_1, \dots, i_n$ be in* I *and p_α in P_{i_α} for $1 \leqslant \alpha \leqslant n$. For all $a \in A$ there exists $a' \in A$ and a sequence $(p'_\alpha)_{1 \leqslant \alpha \leqslant n}$ with $p'_\alpha \in P_{i_\alpha}$ such that*

$$\phi_{i_1}(p_1) \dots \phi_{i_n}(p_n) h(a) = h(a') \phi_{i_1}(p'_1) \dots \phi_{i_n}(p'_n).$$

$h(a) = \phi_{i_n}(h_{i_n}(a))$ and there exists $a_n \in A$ and $p'_n \in P_{i_n}$ with

$$p_n . h_{i_n}(a) = h_{i_n}(a_n) . p'_n.$$

It follows that $\phi_{i_n}(p_n) h(a) = h(a_n) \phi_{i_n}(p'_n)$, whence

$$\phi_{i_1}(p_1) \dots \phi_{i_{n-1}}(p_{n-1}) \phi_{i_n}(p_n) h(a) = \phi_{i_1}(p_1) \dots \phi_{i_{n-1}}(p_{n-1}) h(a_n) \phi_{i_n}(p'_n);$$

the lemma follows from this by induction on n.

(C) *End of the proof:*

Let $x \in M$ and let n be the minimum of the lengths of decompositions of x. We shall prove that every decomposition σ of x of length n is reduced. This will establish the existence of a reduced decomposition of x; the uniqueness of the reduced decomposition then implies $l(\sigma') > l(\sigma)$ for every decomposition $\sigma' \neq \sigma$ of x.

The case $n = 0$ being trivial, suppose $n > 0$. Let

$$\sigma = (a; i_1, \dots, i_n; p_1, \dots, p_n)$$

be a decomposition of x of length n. If there existed an integer α with $1 \leqslant \alpha \leqslant n$ and $p_\alpha = e_{i_\alpha}$, the sequence

$$(a; i_1, \dots, i_{\alpha-1}, i_{\alpha+1}, \dots, i_n; p_1, \dots, p_{\alpha-1}, p_{\alpha+1}, \dots, p_n)$$

would be a decomposition of x of length $n-1$, which is excluded. Suppose that there exists an integer α with $1 \leqslant \alpha < n$ and $i_\alpha = i_{\alpha+1}$ and let

$$p_\alpha p_{\alpha+1} = h_{i_\alpha}(a').p'_\alpha$$

with $a' \in \mathrm{A}$ and $p'_\alpha \in \mathrm{P}_{i_\alpha}$; by Lemma 1 there exists elements $a'' \in \mathrm{A}$,

$$p'_1 \in \mathrm{P}_{i_1}, \ldots, p'_{\alpha-1} \in \mathrm{P}_{i_{\alpha-1}}$$

such that

$$\phi_{i_1}(p_1)\ldots\phi_{i_{\alpha-1}}(p_{\alpha-1})h(a') = h(a'')\phi_{i_1}(p'_1)\ldots\phi_{i_{\alpha-1}}(p'_{\alpha-1})$$

and the sequence

$$(aa''; i_1, \ldots, i_{\alpha-1}, i_\alpha, i_{\alpha+2}, \ldots, i_n; p'_1, \ldots, p'_{\alpha-1}, p'_\alpha, p_{\alpha+2}, \ldots, p_n)$$

is a decomposition of x of length $n-1$, which is a contradiction.

We have thus proved that σ is reduced.

Corollary. *Under hypothesis* (A) *the homomorphisms* ϕ_i *and* h *are injective. For* $i \neq j$ *in* I, $\phi_i(\mathrm{M}_i) \cap \phi_j(\mathrm{M}_j) = h(\mathrm{A})$.

First h is injective: if $h(a) = h(a')$, then (a) and (a') are two reduced decompositions of the same element of M, whence $a = a'$. Let $i \in \mathrm{I}$; then $h(\mathrm{A}) = \phi_i(h_i(\mathrm{A})) \subset \phi_i(\mathrm{M}_i)$; the uniqueness of reduced decompositions implies

$$h(\mathrm{A}) \cap \phi_i(\mathrm{M}_i - h_i(\mathrm{A})) = \varnothing,$$

whence $\phi_i(\mathrm{M}_i - h_i(\mathrm{A})) = \phi_i(\mathrm{M}_i) - h(\mathrm{A})$.

The injectivity of the homomorphisms ϕ_i and the relation

$$\phi_i(\mathrm{M}_i) \cap \phi_j(\mathrm{M}_j) \subset h(\mathrm{A})$$

for $i \neq j$ are then consequences of the following fact: for i, j in I, x in $\mathrm{M}_i - h_i(\mathrm{A})$ and y in $\mathrm{M}_j - h_j(\mathrm{A})$, the relation $\phi_i(x) = \phi_j(y)$ implies $i = j$ and $x = y$. Let $x = h_i(a).p$ and $y = h_j(b).q$ with a, b in A, p in $P_i - \{e_i\}$ and y in $\mathrm{P}_j - \{e_j\}$. Then $\phi_i(x) = h(a)\phi_i(p)$ and $\phi_j(y) = h(b)\phi_j(q)$ and hence $(a; i; p)$ and $(b; j; q)$ are two reduced decompositions of the same element of M. It follows that $i = j$, $a = b$ and $p = q$, whence $x = h_i(a)p = h_j(b)q = y$.

When hypothesis (A) is fulfilled, we shall identify each monoid M_i with a submonoid of M by means of ϕ_i; similarly, we shall identify A with a submonoid of M by h. Then M is generated by $\bigcup_{i \in \mathrm{I}} \mathrm{M}_i$ and $\mathrm{M}_i \cap \mathrm{M}_j = \mathrm{A}$ for $i \neq j$. Every element of M can be written uniquely in the form $a.\prod_{\alpha=1}^{n} p_\alpha$ with $a \in \mathrm{A}$, $p_1 \in \mathrm{P}_{i_1} - \{e\}, \ldots, p_n \in \mathrm{P}_{i_n} - \{e\}$ and $i_\alpha \neq i_{\alpha+1}$ for $1 \leqslant \alpha < n$. Finally, if M′ is a monoid and $(f_i: \mathrm{M}_i \to \mathrm{M}')$ (for $i \in I$) a family of homomorphisms whose restrictions to A are the same homomorphism of A into M′, there exists one and only one homomorphism $f: \mathrm{M} \to \mathrm{M}'$ inducing f_i on M_i for all $i \in \mathrm{I}$.

Hypothesis (A) is satisfied in two important cases:

(a) $A = \{e\}$. In this case, there is a family $(M_\iota)_{\iota \in I}$ of monoids and M is called the *monoidal sum* of this family. Each M_ι is identified with a submonoid of M and M is generated by $\bigcup_{\iota \in I} M_\iota$; further, $M_i \cap M_j = \{e\}$ for $i \neq j$. Every element of M may be written uniquely in the form $x_1 \ldots x_n$ with

$$x_1 \in M_{\iota_1} - \{e\}, \ldots, x_n \in M_{\iota_n} - \{e\}$$

and $\iota_\alpha \neq \iota_{\alpha+1}$ for $1 \leqslant \alpha \leqslant n$. Finally, for every family of homomorphisms $(f_\iota \colon M_\iota \to M')$, there exists a unique homomorphism $f \colon M \to M'$ whose restriction to M_ι is f_ι for all $\iota \in I$.

(b) There is a family of groups $(G_\iota)_{\iota \in I}$ containing as subgroup the same group A and h_ι is the injection of A into G_ι. The sum of the family $(G_\iota)_{\iota \in I}$ amalgamated by A is then a *group* G: the monoid G is generated by $\bigcup_{\iota \in I} \phi_\iota(G_\iota)$ and every element of $\bigcup_{\iota \in I} \phi_\iota(G_\iota)$ admits an inverse in G (cf. § 2, no. 3, Corollary 1 to Proposition 4); it is denoted by $\ast_A G_\iota$ or $G_1 *_A G_2$ when $I = \{1, 2\}$. When A consists of the identity element, it is also said that G is the *free product of the family* $(G_\iota)_{\iota \in I}$ of groups and it is denoted by $\ast G_\iota$ (or $G_1 * G_2$ if $I = \{1, 2\}$).†

4. APPLICATION TO FREE MONOIDS

Lemma 2. *Let* M *be the monoidal sum of the family* $(M_x)_{x \in X}$ *defined by* $M_x = \mathbf{N}$ *for all* $x \in X$ *and let* ϕ_x *denote the canonical homomorphism of* M_x *into* M. *The mapping* $x \mapsto \phi_x(1)$ *of* X *into* M *extends to an isomorphism h of* Mo(X) *onto* M.

Let h be the homomorphism of Mo(X) into M characterized by $h(x) = \phi_x(1)$. For every integer $n \geqslant 0$, $\phi_x(n) = \phi_x(1)^n = h(x)^n$ and as M is generated by $\bigcup_{x \in X} \phi_x(\mathbf{N})$, it is also generated by $h(X)$. Hence h is *surjective.* Moreover, for all x in X, the mapping $n \mapsto x^n$ is a homomorphism of $\mathbf{N} = M_x$ into Mo(X); there thus exists (no. 3, Proposition 4) a homomorphism h' of M into Mo(X) such that $h'(\phi_x(n)) = x^n$ for $x \in X$ and $n \in \mathbf{N}$; in particular, $h'(h(x)) = x$ for $x \in X$ and hence $h' \circ h$ is the identity homomorphism of Mo(X). Therefore h is *injective.* It has thus been proved that h is bijective.

PROPOSITION 6. *Let w be an element of* Mo(X).

(a) *There exist an integer* $n \geqslant 0$, *elements* x_α *of* X *and integers* $m(\alpha) > 0$ (*for*

† Note that $G_1 * G_2$ is not the "product" of G_1 and G_2 in the sense of *Set Theory*, IV, § 2, no. 4 (nor in the sense of the "theory of categories"; in the context of this theory, $G_1 * G_2$ is the "sum" of G_1 and G_2).

$1 \leqslant \alpha \leqslant n$) *such that* $x_\alpha \neq x_{\alpha+1}$ *for* $1 \leqslant \alpha < n$ *and* $w = \prod_{\alpha=1}^{n} x_\alpha^{m(\alpha)}$. *The sequence* $(x_\alpha, m(\alpha))_{1 \leqslant \alpha \leqslant n}$ *is determined uniquely by these conditions.*

(b) *Let* p *be a positive integer,* x'_β *in* X *and* $m'(\beta)$ *in* **N** *for* $1 \leqslant \beta \leqslant p$ *such that* $w = \prod_{\beta=1}^{p} x'^{m'(\beta)}$. *Then* $p \geqslant n$. *If* $p = n$, *then* $x'_\beta = x_\beta$ *and* $m'(\beta) = m(\beta)$ *for* $1 \leqslant \beta \leqslant p$.

In the notation of Lemma 2, $\overset{-1}{h}(\phi_x(n)) = x^n$ for $x \in \mathrm{X}$ and $n \in \mathbf{N}$. Proposition 6 then follows from no. 3, Proposition 5.

5. FREE GROUPS

Let $\mathrm{G}_x = \mathbf{Z}$ for all $x \in \mathrm{X}$. The free product of the family $(\mathrm{G}_x)_{x \in \mathrm{X}}$ is called the *free group constructed on* X and is denoted by F(X). Let ϕ_x denote the canonical homomorphism of $\mathrm{G}_x = \mathbf{Z}$ into F(X). By no. 3, Corollary to Proposition 5, the mapping $x \mapsto \phi_x(1)$ of X into F(X) is injective; we shall identify X with its image in F(X) under this mapping. Then X generates F(X) and $e \notin \mathrm{X}$.

Applying no. 3, Proposition 5, we obtain the following result:

PROPOSITION 7. *Let* g *be an element of the free group* F(X). *There exist an integer* $n \geqslant 0$ *and a sequence* $(x_\alpha, m(\alpha))_{1 \leqslant \alpha \leqslant n}$ *determined uniquely by the relations* $x_\alpha \in \mathrm{X}$, $x_\alpha \neq x_{\alpha+1}$ *for* $1 \leqslant \alpha < n$, $m(\alpha) \in \mathbf{Z}$, $m(\alpha) \neq 0$ *for* $1 \leqslant \alpha \leqslant n$, *and* $g = \prod_{\alpha=1}^{n} x_\alpha^{m(\alpha)}$.

The free group F(X) enjoys the following universal property:

PROPOSITION 8. *Let* G *be a group and* f *a mapping of* X *into* G. *There exists one and only one homomorphism* $\bar{f}$ *of* F(X) *into* G *which extends* f.

The uniqueness of $\bar{f}$ follows from the fact that the group F(X) is generated by X. For all x in X, let f_x be the homomorphism $n \mapsto f(x)^n$ of **Z** into G. By no. 3, Proposition 4, there exists a homomorphism $\bar{f}$ of F(X) into G such that $\bar{f}(x^n) = f_x(n)$ for $x \in \mathrm{X}$ and $n \in \mathbf{Z}$; in particular, $\bar{f}(x) = f_x(1) = f(x)$ for all $x \in \mathrm{X}$ and hence $\bar{f}$ extends f.

Let $u\colon \mathrm{X} \to \mathrm{Y}$ be a mapping. By Proposition 8 there exists one and only one homomorphism of F(X) into F(Y) which coincides with u on X; it is denoted by F(u). As in the case of magmas (no. 1) the formula

$$\mathrm{F}(v \circ u) = \mathrm{F}(v) \circ \mathrm{F}(u)$$

is established for every mapping $v\colon \mathrm{Y} \to \mathrm{Z}$ and it is shown that F(u) is injective (resp. surjective, bijective) if u is. For every subset S of X, F(S) will be identified with the subgroup of F(X) generated by S.

Let I be a set. In certain cases it is of interest not to identify i in I with its canonical image $\phi_i(1)$ in the free group $F(I)$; the latter will be denoted by T_i (or T'_i, X_i, ... as the case may be) and called the *indeterminate* of index i. The free group $F(I)$ is then denoted by $F((T_i)_{i \in I})$ or $F(T_1, \ldots, T_n)$ if $I = \{1, 2, \ldots, n\}$.

Let G be a group and $\mathbf{t} = (t_i)_{i \in I}$ a family of elements of G. By Proposition 8 there exists a homomorphism $f_{\mathbf{t}}$ of $F((T_i)_{i \in I})$ into G characterized by $f_{\mathbf{t}}(T_i) = t_i$ for all $i \in I$. The image of an element w of $F((T_i)_{i \in I})$ under $f_{\mathbf{t}}$ will be denoted by $w(\mathbf{t})$ or $w(t_1, \ldots, t_n)$ if $I = \{1, 2, \ldots, n\}$; $w(\mathbf{t})$ is said to result from the *substitution* $T_i \mapsto t_i$ in w. In particular, if we take $G = F((T_i)_{i \in I})$ and $(t_i) = (T_i) = \mathbf{T}$, $f_{\mathbf{T}}$ is the identity homomorphism of G, whence $w(\mathbf{T}) = w$; for $I = \{1, 2, \ldots, n\}$, then $w(T_1, \ldots, T_n) = w$.

Let G and G′ be two groups, u a homomorphism of G into G′ and $\mathbf{t} = (t_1, \ldots, t_n)$ a finite sequence of elements in G. Let $\mathbf{t}' = (u(t_1), \ldots, u(t_n))$; the homomorphism $u \circ f_{\mathbf{t}}$ of $F(T_1, \ldots, T_n)$ into G′ maps T_i to $u(t_i)$ for $1 \leqslant i \leqslant n$ and hence is equal to $f_{\mathbf{t}'}$; for w in $F(T_1, \ldots, T_n)$, then

$$u(w(t_1, \ldots, t_n)) = w(u(t_1), \ldots, u(t_n)). \tag{8}$$

Let w be given in $F(T_1, \ldots, T_n)$ and elements $v_1, \ldots, v_n$ in the free group $F(T'_1, \ldots, T'_m)$. The substitution $T_i \mapsto v_i$ defines an element $w' = w(v_1, \ldots, v_n)$ of $F(T'_1, \ldots, T'_m)$. Let G be a group, $t_1, \ldots, t_m$ elements of G and u the homomorphism of $F(T'_1, \ldots, T'_m)$ into G characterized by $u(T'_j) = t_j$ for $1 \leqslant j \leqslant m$. Then $u(v_i) = v_i(t_1, \ldots, t_m)$ and $u(w') = w(t_1, \ldots, t_m)$; formula (8) thus implies

$$w'(t_1, \ldots, t_m) = w(v_1(t_1, \ldots, t_m), \ldots, v_n(t_1, \ldots, t_m)). \tag{9}$$

This justifies the "functional notation" $w(t_1, \ldots, t_n)$. The reader is left to extend formulae (8) and (9) to the case of arbitrary indexing sets.

6. PRESENTATIONS OF A GROUP

Let G be a group and $\mathbf{t} = (t_i)_{i \in I}$ a family of elements of G. Let $f_{\mathbf{t}}$ be the unique homomorphism of the free group $F(I)$ into G which maps i to t_i. The image of $f_{\mathbf{t}}$ is the subgroup generated by the elements t_i of G. The elements of the kernel of $f_{\mathbf{t}}$ are called the *relators* of the family $\mathbf{t}$. $\mathbf{t}$ is called *generating* (resp. *free*, *basic*) if $f_{\mathbf{t}}$ is surjective (resp. injective, bijective).

Let G be a group. A *presentation* of G is an ordered pair $(\mathbf{t}, \mathbf{r})$ consisting of a generating family $\mathbf{t} = (t_i)_{i \in I}$ and a family $\mathbf{r} = (r_j)_{j \in J}$ of relators such that the kernel $N_{\mathbf{t}}$ of $f_{\mathbf{t}}$ is generated by the elements $g r_j g^{-1}$ for $g \in F(I)$ and $j \in J$. It amounts to the same to say that $N_{\mathbf{t}}$ is the normal subgroup of $F(I)$ generated by the r_j for $j \in J$ (in other words, the smallest normal subgroup of $F(I)$ containing the elements r_j $(j \in J)$, cf. § 4, no. 4). By an abuse of language the generators t_i and the relations $r_j(\mathbf{t}) = e$ are said to constitute a *presentation* of the group G.

Let I be a set and $\mathbf{r} = (r_j)_{j \in J}$ a family of elements of the free group F(I). Let N($\mathbf{r}$) be the normal subgroup of F(I) generated by the r_j for $j \in J$. Let $F(I, \mathbf{r}) = F(I)/N(\mathbf{r})$ and τ_i denote the class of i modulo N($\mathbf{r}$). The ordered pair $(\boldsymbol{\tau}, \mathbf{r})$ with $\boldsymbol{\tau} = (\tau_i)_{i \in I}$ is a presentation of the group F(I, $\mathbf{r}$); if G is a group and $(\mathbf{t}, \mathbf{r})$ is a presentation of G with $\mathbf{t} = (t_i)_{i \in I}$, there exists a unique isomorphism u of F(I, $\mathbf{r}$) onto G such that $u(\tau_i) = t_i$ for all $i \in I$. The group F(I, $\mathbf{r}$) is said to be defined by the *generators* τ_i and the *relators* r_j, or by an abuse of language that it is *defined by the generators* τ_i *and the relations* $r_j(\boldsymbol{\tau}) = e$. When $I = [1, n]$ and $J = [1, m]$, it is said that F(I, $\mathbf{r}$) is defined by the presentation

$$\langle \tau_1, \ldots, \tau_n; r_1, \ldots, r_m \rangle.$$

If $r_j = u_j^{-1} v_j$ with u_j and v_j in F(I), this presentation is equally denoted by the symbol

$$\langle \tau_1, \ldots, \tau_n; u_1 = v_1, \ldots, u_m = v_m \rangle.$$

Examples. (1) The group defined by the presentation $\langle \tau; \tau^q = e \rangle$ is cyclic of order q.

(2) The group defined by the presentation $\langle x, y; xy = yx \rangle$ is isomorphic to $\mathbf{Z} \times \mathbf{Z}$.

PROPOSITION 9. *Let* G *be a group,* $\mathbf{t} = (t_i)_{i \in I}$ *a generating family of* G *and* $\mathbf{r} = (r_j)_{j \in J}$ *a family of relators of* $\mathbf{t}$. *The following conditions are equivalent:*

(a) *The ordered pair* $(\mathbf{t}, \mathbf{r})$ *is a presentation of* G.

(b) *Let* G' *be a group and* $\mathbf{t}' = (t'_i)_{i \in I}$ *a family of elements of* G'. *If* $r_j(\mathbf{t}') = e$ *for all* $j \in J$, *there exists a homomorphism* u *of* G *into* G' *such that* $u(t_i) = t'_i$ *for all* $i \in I$.

(c) *Let* $\overline{G}$ *be a group and* $\bar{\mathbf{t}} = (\bar{t}_i)_{i \in I}$ *a generating family of* $\overline{G}$ *such that* $r_j(\bar{\mathbf{t}}) = e$ *for all* $j \in J$. *Every homomorphism of* $\overline{G}$ *into* G *which maps* $\bar{t}_i$ *to* t_i *for all* $i \in I$ *is an isomorphism.*

Let f denote the homomorphism of F(I) into G which maps i to t_i for all $i \in I$ and N the kernel of f.

(a) $\Rightarrow$ (b): Suppose that $(\mathbf{t}, \mathbf{r})$ is a presentation of G and let $\mathbf{t}' = (t'_i)_{i \in I}$ be a family of elements of a group G' with $r_j(\mathbf{t}') = e$ for all $j \in J$. Let f' be the homomorphism of F(I) into G' characterized by $f'(i) = t'_i$ for all $i \in I$. By hypothesis $f'(r_j) = e$ for all $j \in J$ and, as N is generated by the elements $g r_j g^{-1}$ for $j \in J$ and $g \in F(I)$, $f'(N) = \{e\}$. As the homomorphism $f: F(I) \to G$ is surjective with kernel N, there exists a homomorphism $u: G \to G'$ such that $f' = u \circ f$. Then $u(t_i) = u(f(i)) = f'(i) = t'_i$.

(b) $\Rightarrow$ (c): Suppose condition (b) holds. Let $\bar{\mathbf{t}} = (\bar{t}_i)_{i \in I}$ be a generating family of a group $\overline{G}$ such that $r_j(\bar{\mathbf{t}}) = e$ for all $j \in J$ and let v be a homomorphism of $\overline{G}$ into G such that $v(\bar{t}_i) = t_i$ for all $i \in I$. As the family $(t_i)_{i \in I}$ generates G, the homomorphism v is *surjective*. By property (b) there exists a homomorphism $u: G \to \overline{G}$ such that $u(t_i) = \bar{t}_i$ for all $i \in I$. Then $u(v(\bar{t}_i)) = \bar{t}_i$ for all $i \in I$ and hence $u \circ v$ is the identity on $\overline{G}$, which proves that v is *injective*. Hence v is an isomorphism and condition (c) holds.

(c) $\Rightarrow$ (a): Suppose condition (c) holds. Let t'_i be the canonical image of i in $F(I, \mathbf{r})$ and $\mathbf{t}' = (t'_i)_{i \in I}$; then $r_j(\mathbf{t}') = e$ for all $j \in J$. As $r_j(\mathbf{t}) = e$ for all $j \in J$, there exists one and only one homomorphism v of $F(I, \mathbf{r})$ into G such that $v(\bar{t}_i) = t_i$ for all $i \in I$. By (c), v is an isomorphism of $F(I, \mathbf{r})$ onto G which transforms the presentation $(\mathbf{t}', \mathbf{r})$ of $F(I, \mathbf{r})$ into a presentation $(\mathbf{t}, \mathbf{r})$ of G.

7. FREE COMMUTATIVE GROUPS AND MONOIDS

The set $\mathbf{Z}^X$ of all mappings of X into $\mathbf{Z}$ is a commutative group under the law defined by $(\alpha + \beta)(x) = \alpha(x) + \beta(x) (\alpha, \beta \in \mathbf{Z}^X, x \in X)$; the elements of $\mathbf{Z}^X$ are sometimes called *multiindices*. The identity element, denoted by 0, is the constant mapping with value 0. For $\alpha \in \mathbf{Z}^X$, the set S_α of $x \in X$ such that $\alpha(x) \neq 0$ is called the *support* of α; then $S_0 = \varnothing$ and $S_{\alpha+\beta} \subset S_\alpha \cup S_\beta$ for α, β in $\mathbf{Z}^X$. Therefore the set $\mathbf{Z}^{(X)}$ of mappings $\alpha: X \rightarrow \mathbf{Z}$ of finite support is a subgroup of $\mathbf{Z}^X$ called the *free commutative group constructed on* X.

For all $x \in X$, let δ_x denote the element of $\mathbf{Z}^{(X)}$ defined by

$$\delta_x(y) = \begin{cases} 1 & \text{if } y = x \\ 0 & \text{if } y \neq x. \end{cases} \tag{10}$$

Also, for $\alpha \in \mathbf{Z}^{(X)}$, the integer $|\alpha|$, the *length* of α is defined by the formula

$$|\alpha| = \sum_{x \in X} \alpha(x). \tag{11}$$

The following relations are immediately established:

$$\alpha = \sum_{x \in X} \alpha(x) . \delta_x \tag{12}$$

$$|\delta_x| = 1, \qquad |0| = 0 \tag{13}$$

$$|\alpha + \beta| = |\alpha| + |\beta| \tag{14}$$

for α, β in $\mathbf{Z}^{(X)}$ and x in X.

The order relation $\alpha \leqslant \beta$ is defined in $\mathbf{Z}^{(X)}$ by $\alpha(x) \leqslant \beta(x)$ for all $x \in X$. The relations $\alpha \leqslant \beta$ and $\alpha' \leqslant \beta'$ imply $\alpha + \alpha' \leqslant \beta + \beta'$, $|\alpha| \leqslant |\beta|$ and $-\alpha \geqslant -\beta$; further, the relation $\alpha \leqslant \beta$ is equivalent to $\beta - \alpha \geqslant 0$. The set of elements $\alpha \geqslant 0$ in $\mathbf{Z}^{(X)}$ is denoted by $\mathbf{N}^{(X)}$; it is the set of mappings of X into $\mathbf{N}$ of finite support and it is a submonoid of $\mathbf{Z}^{(X)}$ called the *free commutative monoid constructed on* X. The elements of length 1 are the minimal elements in $\mathbf{N}^{(X)} - \{0\}$ and constitute the set of δ_x $(x \in X)$.

The monoid $\mathbf{N}^{(X)}$ and the group $\mathbf{Z}^{(X)}$ enjoy the following universal property.

PROPOSITION 10. *Let* M *be a commutative monoid* (resp. *group*) *and f a mapping of* X *into* M. *There exists one and only one homomorphism of* $\mathbf{N}^{(X)}$ (resp. $\mathbf{Z}^{(X)}$) *into M such that* $g(\delta_x) = f(x)$ *for all* $x \in \mathrm{X}$. *If* M *is written additively, then* $g(\alpha) = \sum_{x \in \mathrm{X}} \alpha(x).f(x)$ *for all* α *in* $\mathbf{N}^{(\mathrm{X})}$ (resp. $\mathbf{Z}^{(\mathrm{X})}$).

Let g be a homomorphism of $\mathbf{N}^{(\mathrm{X})}$ (resp. $\mathbf{Z}^{(\mathrm{X})}$) into M such that $g(\delta_x) = f(x)$ for all $x \in \mathrm{X}$. For all α in $\mathbf{N}^{(\mathrm{X})}$ (resp. $\mathbf{Z}^{(\mathrm{X})}$), it follows from (12) that

$$g(\alpha) = \sum_{x \in \mathrm{X}} \alpha(x).g(\delta_x) = \sum_{x \in \mathrm{X}} \alpha(x).f(x),$$

whence the uniqueness of g.

For all α in $\mathbf{N}^{(\mathrm{X})}$ (resp. $\mathbf{Z}^{(\mathrm{X})}$) we write $g(\alpha) = \sum_{x \in \mathrm{X}} \alpha(x).f(x)$. Then obviously $g(0) = 0$; for α, β in $\mathbf{N}^{(\mathrm{X})}$ (resp. $\mathbf{Z}^{(\mathrm{X})}$),

$$\begin{aligned} g(\alpha + \beta) &= \sum_{x \in \mathrm{X}} (\alpha(x) + \beta(x)).f(x) \\ &= \sum_{x \in \mathrm{X}} [\alpha(x).f(x) + \beta(x).f(x)] \\ &= \sum_{x \in \mathrm{X}} \alpha(x).f(x) + \sum_{x \in \mathrm{X}} \beta(x).f(x) \\ &= g(\alpha) + g(\beta) \end{aligned}$$

and hence g is a homomorphism of $\mathbf{N}^{(\mathrm{X})}$ (resp. $\mathbf{Z}^{(\mathrm{X})}$) into M. Also, for y in X,

$$g(\delta_y) = \sum_{x \in \mathrm{X}} \delta_y(x).f(x);$$

now $\delta_y(x).f(x) = 0$ for $x \neq y$ and $\delta_y(y).f(y) = f(y)$, whence $g(\delta_y) = f(y)$.

Let $u: \mathrm{X} \to \mathrm{Y}$ be a mapping. By Proposition 10 there exists one and only one homomorphism of $\mathbf{Z}^{(\mathrm{X})}$ into $\mathbf{Z}^{(\mathrm{Y})}$ which maps δ_x to $\delta_{u(x)}$ for all $x \in \mathrm{X}$. It is denoted by $\mathbf{Z}^{(u)}$; it is immediately seen that it maps $\alpha \in \mathbf{Z}^{(\mathrm{X})}$ to the element $\beta \in \mathbf{Z}^{(\mathrm{Y})}$ defined by

$$\beta(y) = \sum_{x \in u^{-1}(y)} \alpha(x). \tag{15}$$

As in the case of magmas (no. 1), the formula $\mathbf{Z}^{(v \circ u)} = \mathbf{Z}^{(v)} \circ \mathbf{Z}^{(u)}$ is established for every mapping $v: \mathrm{Y} \to \mathrm{Z}$; it is also shown that $\mathbf{Z}^{(u)}$ is injective (resp. surjective, bijective) if u is.

Let S be a subset of X; if i is the injection of S into X, the mapping $f = \mathbf{Z}^{(i)}$ is an isomorphism of $\mathbf{Z}^{(\mathrm{S})}$ onto the subgroup H of $\mathbf{Z}^{(\mathrm{X})}$ generated by the elements δ_s for $s \in \mathrm{S}$. By (15),

$$(f(\alpha))(x) = \begin{cases} \alpha(x) & \text{if } x \in \mathrm{S} \\ 0 & \text{if } x \in \mathrm{X} - \mathrm{S} \end{cases}$$

and therefore H is the set of elements of $\mathbf{Z}^{(X)}$ of support contained in S. Henceforth $\mathbf{Z}^{(S)}$ will be identified with H by means of f.

Formula (15) shows that the restriction of $\mathbf{Z}^{(u)}$ to $\mathbf{N}^{(X)}$ induces a homomorphism $\mathbf{N}^{(u)}$ of $\mathbf{N}^{(X)}$ into $\mathbf{N}^{(Y)}$. Then $\mathbf{N}^{(v \circ u)} = \mathbf{N}^{(v)} \circ \mathbf{N}^{(u)}$ for every mapping $v: Y \to Z$; further, $\mathbf{N}^{(u)}$ is injective (resp. surjective, bijective) if u is. If S is a subset of X,

$$\mathbf{N}^{(S)} = \mathbf{Z}^{(S)} \cap \mathbf{N}^{(X)}.$$

Remark. Let M be the multiplicative monoid of strictly positive integers and let $\mathfrak{P}$ be the set of prime numbers (§ 4, no. 10, Definition 15). By Proposition 10 there exists a homomorphism u of $\mathbf{N}^{(\mathfrak{P})}$ into M characterized by $u(\delta_p) = p$ for every prime number p. Then $u(\alpha) = \prod_{p \in \mathfrak{P}} p^{\alpha(p)}$ for α in $\mathbf{N}^{(\mathfrak{P})}$ and Theorem 7 of § 4, no. 10 shows that u is an isomorphism of $\mathbf{N}^{(\mathfrak{P})}$ onto H.

8. EXPONENTIAL NOTATION

Let M be a monoid, written multiplicatively, and $\mathbf{u} = (u_x)_{x \in X}$ a family of elements of M, commuting in pairs. Let α be in $\mathbf{N}^{(X)}$; the elements $u_x^{\alpha(x)}$ and $u_y^{\alpha(y)}$ of M commute for x, y in X and there exists a finite subset S of X such that $u_x^{\alpha(x)} = 1$ for x in X − S. We may therefore write:

$$\mathbf{u}^\alpha = \prod_{x \in X} u_x^{\alpha(x)}. \tag{16}$$

Let M′ be the submonoid of M generated by the family $(u_x)_{x \in X}$; it is commutative (§ 1, no. 5, Corollary 2 to Proposition 4). There thus exists (no. 7, Proposition 10) a unique homomorphism f of $\mathbf{N}^{(X)}$ into M′ such that $f(\delta_x) = u_x$ for all $x \in X$ and $f(\alpha) = \mathbf{u}^\alpha$ for all α in $\mathbf{N}^{(X)}$. We deduce the following formulae

$$\mathbf{u}^{\alpha+\beta} = \mathbf{u}^\alpha . \mathbf{u}^\beta \tag{17}$$

$$\mathbf{u}^0 = 1 \tag{18}$$

$$\mathbf{u}^{\delta_x} = u_x \tag{19}$$

for α, β in $\mathbf{N}^{(X)}$ and x in X.

Let $\mathbf{v} = (v_x)_{x \in X}$ be another family of elements of M; suppose that $v_x v_y = v_y v_x$ and $u_x v_y = v_y u_x$ for x, y in X. Then there exists (§ 1, no. 5, Corollary 2 to Proposition 4) a commutative submonoid L of M such that $u_x \in L$ and $v_x \in L$ for all $x \in X$. The mapping $\alpha \mapsto u^\alpha . v^\alpha$ of $\mathbf{N}^{(X)}$ into L is then a homomorphism (§ 1, no. 5, Proposition 5) mapping δ_x to $u_x . v_x$. Thus we have the formula

$$\mathbf{u}^\alpha . \mathbf{v}^\alpha = (\mathbf{u} . \mathbf{v})^\alpha, \tag{20}$$

where $\mathbf{u} . \mathbf{v}$ is the family $(u_x . v_x)_{x \in X}$.

When M is commutative, $\mathbf{u}^\alpha$ can be defined for every family $\mathbf{u}$ of elements of M and formulae (15) to (20) hold without restriction.

9. RELATIONS BETWEEN THE VARIOUS FREE OBJECTS

As the free monoid Mo(X) is a magma, Proposition 1, of no. 1 shows the existence of a homomorphism $\lambda: M(X) \to Mo(X)$ whose restriction to X is the identity. Similarly, as the free group F(X) is a monoid the identity mapping of X extends to a homomorphism $\mu: Mo(X) \to F(X)$ (no. 2, Proposition 3). By no. 4, Proposition 6 and no. 5, Proposition 7, μ is injective. Similarly Proposition 10 of no. 7 and Proposition 8 of no. 5 show the existence of homomorphisms $\nu: Mo(X) \to \mathbf{N}^{(X)}$ and $\pi: F(X) \to \mathbf{Z}^{(X)}$ characterized by $\nu(x) = \delta_x$ and $\pi(x) = \delta_x$ for all $x \in X$. If ι is the injection of $\mathbf{N}^{(X)}$ into $\mathbf{Z}^{(X)}$, the two homomorphisms $\iota \circ \nu$ and $\pi \circ \mu$ of Mo(X) into $\mathbf{Z}^{(X)}$ coincide on X, whence $\iota \circ \nu = \pi \circ \mu$. The situation may be summarized by the following commutative diagram:

$$\begin{array}{ccccc} M(X) & \xrightarrow{\lambda} & Mo(X) & \xrightarrow{\nu} & \mathbf{N}^{(X)} \\ & & \downarrow{\scriptstyle\mu} & & \uparrow{\scriptstyle\iota} \\ & & F(X) & \xrightarrow{\pi} & \mathbf{Z}^{(X)}. \end{array}$$

The homomorphisms λ, μ, ν and π will be called *canonical.*

Let w be in M(X); it is immediately shown by induction on $l(w)$ that the length of the word $\lambda(w)$ is equal to that of w. Moreover

$$\nu(x_1 \ldots x_n) = \sum_{i=1}^{n} \delta_{x_i} \tag{21}$$

for $x_1, \ldots, x_n$ in X, whence $|\nu(x_1 \ldots x_n)| = n$ by (13) and (14). In other words,

$$|\nu(u)| = l(u) \quad (u \in Mo(X)). \tag{22}$$

PROPOSITION 11. *The canonical homomorphism* ν *of* Mo(X) *into* $\mathbf{N}^{(X)}$ *is surjective. Let* $w = x_1 \ldots x_n$ *and* $w' = x'_1 \ldots x'_m$ *be two elements of* Mo(X); *in order that* $\nu(w) = \nu(w')$, *it is necessary and sufficient that* $m = n$ *and that there exist a permutation* $\sigma \in \mathfrak{S}_n$ *with* $x'_i = x_{\sigma(i)}$ *for* $1 \leqslant i \leqslant n$.

The image of ν is a submonoid I of $\mathbf{N}^{(X)}$ containing the elements δ_x (for $x \in X$). Formula (12) (no. 7) shows that $\mathbf{N}^{(X)}$ is generated by the family $(\delta_x)_{x \in X}$, where $I = \mathbf{N}^{(X)}$. Therefore ν is surjective.

If $m = n$ and $x'_i = x_{\sigma(i)}$ for $1 \leqslant i \leqslant n$, then

$$\nu(w') = \sum_{i=1}^{n} \delta_{x'_i} = \sum_{i=1}^{n} \delta_{x_{\sigma(i)}} = \sum_{i=1}^{n} \delta_{x_i} = \nu(w)$$

by formula (21) and the commutativity theorem (§ 1, no. 5, Theorem 2).

Conversely, suppose that $\nu(w)$ and $\nu(w')$ are equal to the same element α of $\mathbf{N}^{(X)}$; by formula (22), $n = |\alpha| = m$. For all $x \in X$, let I_x (resp. I'_x) be the set of integers i such that $1 \leqslant i \leqslant n$ and $x_i = x$ (resp. $x'_i = x$). Hence $(I_x)_{x \in X}$ and $(I'_x)_{x \in X}$ are partitions of the interval $(1, n)$ of $\mathbf{N}$; further, the formula $\alpha = \sum_{i=1}^{n} \delta_{x_i}$ shows that $\alpha(x)$ is the cardinal of I_x; similarly the formula $\alpha = \sum_{i=1}^{n} \delta_{x'_i}$ shows that $\alpha(x)$ is the cardinal of I'_x. There thus exists a permutation σ of $(1, n)$ such that $\sigma(I'_x) = I_x$ for all $x \in X$, that is $x'_i = x_{\sigma(i)}$ for $i = 1, \ldots, n$.

Remark. Let S be a subset of X. Recall that we have identified M(S) with a submagma of M(X), Mo(S) with a submonoid of Mo(X) and $\mathbf{N}^{(S)}$ with a submonoid of $\mathbf{N}^{(X)}$. Then

(23) $$M(S) = \lambda^{-1}(Mo(S)).$$

Clearly $\lambda(M(S)) \subset Mo(S)$. Let $w \in \lambda^{-1}(Mo(S))$; we show by induction on $l(w)$ that $w \in M(S)$. It is obvious if $l(w) = 1$. If $l(w) > 1$, we may write $w = w_1 w_2$ with $w_1, w_2 \in M(X)$, $l(w_1) < l(w)$, $l(w_2) < l(w)$. Then $\lambda(w_1)\lambda(w_2) \in Mo(S)$, hence $\lambda(w_1) \in Mo(S)$ and $\lambda(w_2) \in Mo(S)$, whence $w_1 \in M(S)$ and $w_2 \in M(S)$ by the induction hypothesis and finally $w \in M(S)$.

Also

(24) $$Mo(S) = \nu^{-1}(\mathbf{N}^{(S)}).$$

This follows immediately from formula (21).

Further, $\mathbf{N}^{(S)}$ is the set of elements of $\mathbf{N}^{(X)}$ whose support is contained in S; if $(S_i)_{i \in I}$ is a family of subsets of X of intersection S, then $\mathbf{N}^{(S)} = \bigcap_{i \in I} \mathbf{N}^{(S_i)}$ and formulae (23) and (24) imply

(25) $$M(S) = \bigcap_{i \in I} M(S_i), \qquad Mo(S) = \bigcap_{i \in I} Mo(S_i).$$

§ 8. RINGS

1. RINGS

DEFINITION 1. *A ring is a set* A *with two laws of composition called respectively addition and multiplication, satisfying the following axioms:*

(AN I) *Under addition* A *is a commutative group.*

(AN II) *Multiplication is associative and possesses an identity element.*

(AN III) *Multiplication is distributive with respect to addition.*

The ring A *is said to be commutative if its multiplication is commutative.*

In what follows, $(x, y) \mapsto x + y$ denotes addition and $(x, y) \mapsto xy$ multiplication; 0 denotes the identity element for addition and 1 that for multiplication. Finally, $-x$ denotes the negative of x under addition. The axioms of a ring are therefore expressed by the following identities:

(1) $x + (y + z) = (x + y) + z$ (associativity of addition)
(2) $x + y = y + x$ (commutativity of addition)
(3) $0 + x = x + 0 = x$ (zero)
(4) $x + (-x) = (-x) + x = 0$ (negative)
(5) $x(yz) = (xy)z$ (associativity of multiplication)
(6) $x.1 = 1.x = x$ (unit element)
(7) $(x + y).z = xz + yz$
(8) $x.(y + z) = xy + xz$ (distributivity)

Finally the ring A is commutative if $xy = yx$ for x, y in A.

With addition alone A is a commutative group called the *additive group* of A. For all $x \in A$, we define the left homothety γ_x and right homothety δ_x by $\gamma_x(y) = xy$, $\delta_x(y) = yx$. By formulae (7) and (8), γ_x and δ_x are endomorphisms of the additive group of A and thus map zero to zero and negative to negative. Therefore

(9) $$x.0 = 0.x = 0$$

(10) $$x.(-y) = (-x).y = -xy;$$

it follows that $(-x)(-y) = -((-x).y) = -(-xy)$, whence

(11) $$(-x)(-y) = xy.$$

Formulae (10) and (11) constitute the *sign rule*. It follows that

$$-x = (-1)x = x(-1)$$

and $(-1)(-1) = 1$.

From (11) it follows by induction on n that

(12) $$(-x)^n = \begin{cases} x^n & \text{if } n \text{ is even} \\ -x^n & \text{if } n \text{ is odd.} \end{cases}$$

When we speak of *cancellable* elements, *invertible* elements, *permutable* elements, *central* elements, *centralizer* or *centre* of a ring A, all these notions will refer to the multiplication on A. If $x, y \in A$ and y is invertible, the element xy^{-1} of A is also denoted by x/y when A is commutative. The set of invertible elements of A is stable under multiplication. Under the law induced by multiplication it is a group called the *multiplicative group* of A, sometimes denoted by A*.

Let x, y be in A. x is said to be a *left* (resp. *right*) *multiple* of y if there exists $y' \in A$ such that $x = y'y$ (resp. $x = yy'$); it is also said that y is a *right* (resp.

left) *divisor* of x. When A is commutative, there is no need to distinguish between "left" and "right".

In conformity with the above terminology, every element $y \in A$ would be considered as a right and left divisor of 0; but, by an abuse of language, in general the term "*right* (resp. *left*) *divisor of* 0" is reserved for elements y such that there exists $x \neq 0$ in A satisfying the relation $xy = 0$ (resp. $yx = 0$). In other words, the right (resp. left) divisors of zero are the right (resp. left) non-cancellable elements.

Let $x \in A$. x is called *nilpotent* if there exists an integer $n > 0$ with $x^n = 0$. The element $1 - x$ is then invertible, with inverse equal to

$$1 + x + x^2 + \cdots + x^{n-1}.$$

As A is a commutative group under addition, the element nx for $n \in \mathbf{Z}$ and $x \in A$ has been defined (§ 2, no. 8). As γ_x and δ_x are endomorphisms of the additive group A, $\gamma_x(ny) = n\gamma_x(y)$ and $\delta_y(nx) = n\delta_y(x)$, whence

$$x.(ny) = (nx).y = n.(xy).$$

In particular, $nx = (n.1)x$.

A set A with addition and multiplication satisfying the axioms of a ring with the exception of that assuring the existence of the identity element under multiplication, is called a *pseudo-ring*.

2. CONSEQUENCES OF DISTRIBUTIVITY

Distributivity of multiplication with respect to addition allows us to apply Proposition 1 of § 3, no. 4, which gives

$$\prod_{i=1}^{n}\Big(\sum_{\lambda \in L_i} x_{i,\lambda}\Big) = \sum_{\alpha_1,\ldots,\alpha_n} \prod_{i=1}^{n} x_{i,\alpha_i} \tag{13}$$

where the sum extends over all sequences $(\alpha_1, \ldots, \alpha_n)$ belonging to $L_1 \times \cdots \times L_n$ and for $i = 1, \ldots, n$ the family $(x_{i,\lambda})_{\lambda \in L_i}$ of elements of the ring A is of finite support.

Proposition 1. *Let* A *be a* commutative *ring and* $(x_\lambda)_{\lambda \in L}$ *a finite family of elements of* A. *For every family of positive integers* $\beta = (\beta_\lambda)_{\lambda \in L}$, *let* $|\beta| = \sum_{\lambda \in L} \beta_\lambda$. *Then*

$$\Big(\sum_{\lambda \in L} x_\lambda\Big)^n = \sum_{|\beta| = n} \frac{n!}{\prod_{\lambda \in L} \beta_\lambda !} \prod_{\lambda \in L} x_\lambda^{\beta_\lambda}. \tag{14}$$

We apply formula (13) with $L_i = L$ and $x_{i,\lambda} = x_\lambda$ for $1 \leqslant i \leqslant n$. Then

$$\Big(\sum_{\lambda \in L} x_\lambda\Big)^n = \sum_{\alpha_1,\ldots,\alpha_n} x_{\alpha_1} \ldots x_{\alpha_n}, \tag{15}$$

the sum extending over all sequences $\alpha = (\alpha_1, \ldots, \alpha_n) \in L^n$.

Let α be in L^n; for all $\lambda \in L$ let U_λ^α denote the set of integers i such that $1 \leqslant i \leqslant n$ and $\alpha_i = \lambda$ and let $\Phi(\alpha) = (U_\lambda^\alpha)_{\lambda \in L}$. It is immediate that Φ is a bijection of L^n onto the set of partitions of $\{1, 2, \dots, n\}$ indexed by L. For all $\beta \in \mathbf{N}^L$ such that $|\beta| = n$, let L_β^n denote the set of $\alpha \in L^n$ such that $\operatorname{Card} U_\beta^\alpha = \beta_\lambda$ for all $\lambda \in L$. It follows that the family $(L_\beta^n)_{|\beta|=n}$ is a partition of L^n and that

$$\operatorname{Card} L_\beta^n = \frac{n!}{\prod_{\lambda \in L} \beta_\lambda!}$$

(§ 5, no. 5).

Finally, for $\alpha \in L_\beta^n$,

$$x_{\alpha_1} \dots x_{\alpha_n} = \prod_{\lambda \in L} \prod_{i \in U_\lambda^\alpha} x_{\alpha_i} = \prod_{\lambda \in L} \prod_{i \in U_\lambda^\alpha} x_\lambda = \prod_{\lambda \in L} x_\lambda,$$

whence

$$\begin{aligned}\sum_{(\alpha_1, \dots, \alpha_n)} x_{\alpha_1} \dots x_{\alpha_n} &= \sum_{|\beta|=n} \sum_{\alpha \in L_\beta^n} x_{\alpha_1} \dots x_{\alpha_n} \\ &= \sum_{|\beta|=n} \sum_{\alpha \in L_\beta^n} \prod_{\lambda \in L} x_\lambda^{\beta_\lambda} \\ &= \sum_{|\beta|=n} \frac{n!}{\prod_{\lambda \in L} \beta_\lambda!} x_\lambda^{\beta_\lambda}\end{aligned}$$

and formula (14) thus follows from (15).

Corollary 1 (binomial formula). *Let x and y be two elements of a commutative ring* A. *Then:*

$$(x + y)^n = \sum_{p=0}^{n} \binom{n}{p} x^p y^{n-p}.$$

Formula (14) applied to $L = \{1, 2\}$, $x_1 = x$ and $x_2 = y$ gives

$$(x + y)^n = \sum_{p+q=n} \frac{n!}{p!\,q!} x^p y^q,$$

the sum extending over ordered pairs of positive integers p, q with $p + q = n$. The binomial formula follows immediately from this (*Set Theory*, III, § 5, no. 8).

Corollary 2. *Let* A *be a commutative ring,* X *a set,* $\mathbf{u} = (u_x)_{x \in X}$ *and* $\mathbf{v} = (v_x)_{x \in X}$ *two families of elements of* A. *Let* $\mathbf{u} + \mathbf{v}$ *denote the family* $(u_x + v_x)_{x \in X}$. *For all* $\lambda \in \mathbf{N}^{(X)}$ *we write* $\lambda! = \prod_{x \in X} \lambda(x)!$. *Then for all* $\alpha \in \mathbf{N}^{(X)}$, *in the notation of* § 7, no. 8,

$$(\mathbf{u} + \mathbf{v})^\alpha = \sum_{\beta+\gamma=\alpha} \frac{\alpha!}{\beta!\,\gamma!} \mathbf{u}^\beta \mathbf{v}^\gamma.$$

For $x \in X$,

$$(u_x + v_x)^{\alpha(x)} = \sum_{m+n=\alpha(x)} \frac{\alpha(x)!}{m!\,n!} u_x^m v_x^n$$

by Corollary 1. Taking the product of these equations for $x \in X$ and using (13), we obtain the corollary.

PROPOSITION 2. *Let* A *be a ring,* $x_1, \ldots, x_n$ *elements of* A *and* $I = \{1, 2, \ldots, n\}$. *For* $H \subset I$, *we write* $x_H = \sum_{i \in H} x_i$. *Then*

$$(16) \qquad (-1)^n \sum_{\sigma \in \mathfrak{S}_n} x_{\sigma(1)} \ldots x_{\sigma(n)} = \sum_{H \subset I} (-1)^{\operatorname{Card} H} (x_H)^n.$$

In particular, if A *is commutative,*

$$(-1)^n n! x_1 x_2 \ldots x_n = \sum_{H \subset I} (-1)^{\operatorname{Card} H} (x_H)^n.$$

Let C be the set of mappings of I into $\{0, 1\}$. If each $H \subset I$ is mapped to its characteristic function, a bijection is obtained of $\mathfrak{P}(I)$ onto C. The right hand side of (16) is thus equal to:

$$\begin{aligned}
\sum_{a \in C} (-1)^{a(1)+\cdots+a(n)} \Big(\sum_{i \in I} a(i) x_i\Big)^n
&= \sum_{a \in C} (-1)^{a(1)+\cdots+a(n)} \sum_{(i_1, \ldots, i_n) \in I^n} a(i_1) \ldots a(i_n) x_{i_1} \ldots x_{i_n} \\
&= \sum_{(i_1, \ldots, i_n) \in I^n} c_{i_1 \ldots i_n} x_{i_1} \ldots x_{i_n}
\end{aligned}$$

where

$$c_{i_1 \ldots i_n} = \sum_{a \in C} (-1)^{a(1)+\cdots+a(n)} a(i_1) \ldots a(i_n).$$

(1) Suppose that $(i_1, \ldots, i_n)$ is not a permutation of I. There exists a $j \in I$ distinct from $i_1, \ldots, i_n$. Let C′ be the set of $a \in C$ such that $a(j) = 0$. For all $a \in C'$, let a^* be the sum of a and the characteristic function of $\{j\}$. Then $a^*(1) + \cdots + a^*(n) = a(1) + \cdots + a(n) + 1$ and hence

$$\begin{aligned}
c_{i_1 \ldots i_n} &= \sum_{a \in C'} (-1)^{a(1)+\cdots+a(n)} a(i_1) \ldots a(i_n) + (-1)^{a^*(1)+\cdots+a^*(n)} a^*(i_1) \ldots a^*(i_n) \\
&= \sum_{a \in C'} ((-1)^{a(1)+\cdots+(n)} + (-1)^{a(1)+\cdots+a(n)+1}) a(i_1) \ldots a(i_n) = 0.
\end{aligned}$$

(2) Suppose that there exists $\sigma \in \mathfrak{S}_n$ such that $i_1 = \sigma(1), \ldots, i_n = \sigma(n)$. Then $a(i_1) \ldots a(i_n) = 0$ unless a only takes the value 1. Thus $c_{i_1 \ldots i_n} = (-1)^n$.

3. EXAMPLES OF RINGS

I. *Zero ring.* Let A be a ring. For $0 = 1$ in A, it is necessary and sufficient that A consist of a single element. The condition is obviously sufficient. On the other hand, if $0 = 1$, then, for all $x \in A$, $x = x.1 = x.0 = 0$. Such a ring is called a zero ring.

II. *Ring of rational integers.* With the addition defined in § 2, no. 5, and the multiplication defined in § 2, no. 6, $\mathbf{Z}$ is a commutative ring. The notation 0, $1, -x$ is in accordance with the notation introduced earlier.

*III. *Ring of real-valued functions.* Let I be an interval in the set $\mathbf{R}$ of real numbers and let A be the set of continuous functions defined on I with real values. The sum $f + g$ and product $f.g$ of two functions f and g are defined by

$$(f + g)(t) = f(t) + g(t), \qquad (fg)(t) = f(t)g(t) \quad (t \in \mathrm{I}).$$

A commutative ring is obtained whose unit element is the constant 1.$_*$

*IV. *Convolution pseudo-ring.* Let E be the set of real-valued continuous functions on $\mathbf{R}$, which are zero outside a bounded interval. The sum of two functions is defined as in III, but the product is now defined by

$$(fg)(t) = \int_{-\infty}^{\infty} f(s)g(t - s)\, ds$$

("convolution product"). Thus a commutative pseudo-ring is obtained which is not a ring (cf. *Integration*, VIII, §4).$_*$

V. *Opposite ring of a ring* A. Let A be a ring. The set A with the same addition as A and the multiplication $(x, y) \mapsto yx$ is often denoted by A^0. It is a ring (called the opposite ring of A) with the same zero and same unit as A and which coincides with A if and only if A is commutative.

VI. *Endomorphism ring of a commutative group.* Let G be a commutative group written additively. Let E denote the set of endomorphisms of G. Given f and g in E, the mappings $f + g$ and fg of G into G are defined by

$$(f + g)(x) = f(x) + g(x), \qquad (fg)(x) = f(g(x)) \quad (x \in \mathrm{G}).$$

By § 1, no. 5, Proposition 5, $f + g$ is an endomorphism of G and so obviously also is $fg = f \circ g$. By § 4, no. 8, E is a (commutative) group under addition. Multiplication is obviously associative and has identity element Id_G. Also for f, g and h in E, we write $\phi = f.(g + h)$; for all $x \in G$,

$$\phi(x) = f((g + h)(x)) = f(g(x) + h(x)) = f(g(x)) + f(h(x))$$

for f is an endomorphism of G; hence $\phi = fg + fh$ and clearly

$$(g + h)f = gf + hf.$$

Therefore E is a (not in general commutative) ring called the *endomorphism ring of* G.

VII. *Pseudo-ring of zero square.* A pseudo-ring A is said to be of zero square if $xy = 0$ for all $x, y \in A$. Let G be a commutative group. If the set G is given the addition of the group G and multiplication $(x, y) \mapsto 0$, a pseudo-ring of zero square is obtained. It is only a ring if $G = \{0\}$, in which case it is the zero ring.

4. RING HOMOMORPHISMS

DEFINITION 2. *Let* A *and* B *be two rings. A morphism, or homomorphism, of* A *into* B *is any mapping* f *of* A *into* B *satisfying the relations:*

$$f(x + y) = f(x) + f(y), \qquad f(xy) = f(x).f(y), \qquad f(1) = 1, \tag{17}$$

for all x, y *in* A.

The composition of two ring homomorphisms is a ring homomorphism. Let A and B be two rings and f a mapping of A into B; for f to be an isomorphism, it is necessary and sufficient that it be a bijective homomorphism; in that case, $\overset{-1}{f}$ is a homomorphism of B into A. A homomorphism of a ring A into itself is called an *endomorphism* of A.

Let $f\colon A \to B$ be a ring homomorphism. The mapping f is a homomorphism of the additive group of A into the additive group of B; in particular, $f(0) = 0$ and $f(-x) = -f(x)$ for all $x \in A$. The image under f of an invertible element of A is an invertible element of B and f induces a homomorphism of the multiplicative group of A into the multiplicative group of B.

Examples. (1) Let A be a ring. It is immediately seen that the mapping $n \mapsto n.1$ of $\mathbf{Z}$ into A is the unique homomorphism of $\mathbf{Z}$ into A. In particular, the identity mapping of $\mathbf{Z}$ is the unique endomorphism of the ring $\mathbf{Z}$.

In particular, take A to be the endomorphism ring of the additive group $\mathbf{Z}$ (no. 3, Example VI). The mapping $n \mapsto n.1$ of $\mathbf{Z}$ into A is an isomorphism of $\mathbf{Z}$ onto A by the very construction of multiplication in $\mathbf{Z}$ (§ 2, no. 6).

(2) Let a be an invertible element of a ring A. The mapping $x \mapsto axa^{-1}$ is an endomorphism of A for

$$a(x + y)a^{-1} = axa^{-1} + aya^{-1},$$
$$a(xy)a^{-1} = (axa^{-1})(aya^{-1}).$$

It is bijective, for the relation $x' = axa^{-1}$ is equivalent to $x = a^{-1}x'a$. It is therefore an automorphism of the ring A, called the *inner automorphism* associated with a.

5. SUBRINGS

DEFINITION 3. *Let* A *be a ring. A subring of* A *is any subset* B *of* A *which is a subgroup of* A *under addition, which is stable under multiplication and which contains the unit of* A.

The above conditions may be written as follows

$$0 \in B, \quad B + B \subset B, \quad -B \subset B, \quad B.B \subset B, \quad 1 \in B.$$

If B is a subring of A, it is given the addition and multiplication induced by those on A, which make it into a ring. The canonical injection of B into A is a ring homomorphism.

Examples. (1) Every subgroup of the additive group **Z** which contains 1 is equal to **Z**. Thus **Z** is the only subring of **Z**.

(2) Let A be a ring and $(A_\iota)_{\iota \in I}$ a family of subrings of A; it is immediate that $\bigcap_{\iota \in I} A_\iota$ is a subring of A. In particular, the intersection of the subrings of A containing a subset X of A is a subring called the *subring of* A *generated by* X.

(3) Let X be a subset of a ring A. The centralizer of X in A is a subring of A. In particular, the centre of A is a subring of A.

(4) Let G be a commutative group with operators; let Ω denote the set of operators and $\alpha \mapsto f_\alpha$ the action of Ω on G. Let E be the endomorphism ring of the group without operators G and F the set of endomorphisms of the group with operators G. By definition, F consists of the endomorphisms ϕ of G such that $\phi.f_\alpha = f_\alpha.\phi$ for all $\alpha \in \Omega$. Therefore F is a subring of the ring E. F is called *the endomorphism ring of the group with operators* G (cf. II, § 1, no. 2). Let F_1 be the subring of E generated by the f_α. Then F is the centralizer of F_1 in E.

6. IDEALS

DEFINITION 4. *Let* A *be a ring. A subset* $\mathfrak{a}$ *of* A *is called a left* (resp. *right*) *ideal if it is a subgroup of the additive group of* A *and the relations* $a \in A$, $x \in \mathfrak{a}$ *imply* $ax \in \mathfrak{a}$ (resp. $xa \in \mathfrak{a}$). $\mathfrak{a}$ *is called a two-sided ideal of* A *if it is both a left ideal and a right ideal of* A.

The definition of a left ideal may be expressed by the relations

$$0 \in \mathfrak{a}, \quad \mathfrak{a} + \mathfrak{a} \subset \mathfrak{a}, \quad A.\mathfrak{a} \subset \mathfrak{a}$$

the relation $-\mathfrak{a} \subset \mathfrak{a}$ following from the formula $(-1).x = -x$ and $A.\mathfrak{a} \subset \mathfrak{a}$. For all $x \in A$, let γ_x be the mapping $a \mapsto xa$ of A into A; the action $x \mapsto \gamma_x$ gives the additive group A^+ of A the structure of a group with operators with A as set of operators. The left ideals of A are just the subgroups of A^+ which are stable under this action.

The left ideals in the ring A are just the right ideals of the opposite ring A^0. In a commutative ring the three species of ideals are the same; they are simply called *ideals*.

Examples. (1) Let A be a ring. The set A is a two-sided ideal of A; so is the set consisting of 0, which is called the *zero* ideal and sometimes denoted by 0 or (0) instead of {0}.

(2) For every element a of A, the set $A.a$ of left multiples of a is a left ideal; similarly the set $a.A$ is a right ideal. When a is in the centre of A, $A.a = a.A$; this ideal is called *the principal ideal* generated by a and is denoted by (a). $(a) = A$ if and only if a is invertible.

(3) Let M be a subset of A. The set of elements $x \in A$ such that $xy = 0$ for all $y \in M$ is a left ideal of A called the *left annihilator* of M. The right annihilator of M is defined similarly.

(4) Every intersection of left (resp. right, two-sided) of A is a left (resp. right, two-sided) ideal. Given a subset X of A, there thus exists a smallest left (resp. right, two-sided) ideal containing X; it is called the left (resp. right, two-sided) ideal *generated by* X.

Let $\mathfrak{a}$ be a left ideal of A. The conditions $1 \notin \mathfrak{a}$, $\mathfrak{a} \neq A$ are obviously equivalent.

DEFINITION 5. *Let* A *be a ring. By an abuse of language, a left ideal* $\mathfrak{a}$ *is said to be maximal if it is a maximal element of the set of left ideals* distinct from A.

In other words, $\mathfrak{a}$ is maximal if $\mathfrak{a} \neq A$ and the only left ideals of A containing $\mathfrak{a}$ are $\mathfrak{a}$ and A.

THEOREM 1 (Krull). *Let* A *be a ring and* $\mathfrak{a}$ *a left ideal of* A *distinct from* A. *There exists a maximal ideal* $\mathfrak{m}$ *of* A *containing* $\mathfrak{a}$.

Consider A as operating on the additive group A^+ of A by left multiplication. Then the left ideals of A are the stable subgroups of A^+. The theorem thus follows from § 4, no. 3, Proposition 3 applied to the subset $P = \{1\}$ of A^+.

PROPOSITION 3. *Let* A *be a ring,* $(x_\lambda)_{\lambda \in L}$ *a family of elements of* A *and* $\mathfrak{a}$ (resp. $\mathfrak{b}$) *the set of sums* $\sum_{\lambda \in L} a_\lambda x_\lambda$ *where* $(a_\lambda)_{\lambda \in L}$ *is a family with finite support of elements of* A (resp. $\sum_{\lambda \in L} a_\lambda x_\lambda b_\lambda$ *where* $(a_\lambda)_{\lambda \in L}$, $(b_\lambda)_{\lambda \in L}$ *are families with finite support of elements of* A). *Then* $\mathfrak{a}$ (resp. $\mathfrak{b}$) *is the left* (resp. *two-sided*) *ideal of* A *generated by the elements* x_λ.

The formulae

$$0 = \sum_{\lambda \in L} 0.x_\lambda \tag{18}$$

$$\sum_{\lambda \in L} a_\lambda x_\lambda + \sum_{\lambda \in L} a'_\lambda x_\lambda = \sum_{\lambda \in L} (a_\lambda + a'_\lambda) x_\lambda \tag{19}$$

$$a.\sum_{\lambda \in L} a_\lambda x_\lambda = \sum_{\lambda \in L} (a a_\lambda) x_\lambda \tag{20}$$

prove that $\mathfrak{a}$ is a left ideal. Let $\mathfrak{a}'$ be a left ideal such that $x_\lambda \in \mathfrak{a}'$ for all $\lambda \in L$ and let $(a_\lambda)_{\lambda \in L}$ be a family with finite support in A. Then $a_\lambda x_\lambda \in \mathfrak{a}'$ for all $\lambda \in L$, whence $\sum_{\lambda \in L} a_\lambda x_\lambda \in \mathfrak{a}'$; hence $\mathfrak{a} \subset \mathfrak{a}'$. Hence $\mathfrak{a}$ is the left ideal of A generated by the x_λ. The argument for $\mathfrak{b}$ is analogous.

PROPOSITION 4. *Let A be a ring and $(\mathfrak{a}_\lambda)_{\lambda \in L}$ a family of left ideals of A. The left ideal generated by $\bigcup_{\lambda \in L} \mathfrak{a}_\lambda$ consists of the sums $\sum_{\lambda \in L} y_\lambda$ where $(y_\lambda)_{\lambda \in L}$ is a family with finite support such that $y_\lambda \in \mathfrak{a}_\lambda$ for all $\lambda \in L$.*

Let $\mathfrak{a}$ be the set of sums $\sum_{\lambda \in L} y_\lambda$ with $y_\lambda \in \mathfrak{a}_\lambda$ for all $\lambda \in L$. The formulae $\sum_{\lambda \in L} x_\lambda + \sum_{\lambda \in L} y_\lambda = \sum_{\lambda \in L} (x_\lambda + y_\lambda)$ and $a.\sum_{\lambda \in L} x_\lambda = \sum_{\lambda \in L} ax_\lambda$ shows that $\mathfrak{a}$ is a left ideal of A. Let $\lambda \in L$ and $x \in \mathfrak{a}_\lambda$; write $y_\lambda = x$ and $y_\mu = 0$ for $\mu \neq \lambda$; then $x = \sum_{\lambda \in L} y_\lambda$, whence $x \in \mathfrak{a}$ and finally $\mathfrak{a}_\lambda \subset \mathfrak{a}$. If a left ideal $\mathfrak{a}'$ contains $\mathfrak{a}_\lambda$ for all $\lambda \in L$, it obviously contains $\mathfrak{a}$ and hence $\mathfrak{a}$ is generated by $\bigcup_{\lambda \in L} \mathfrak{a}_\lambda$.

The ideal $\mathfrak{a}$ generated by $\bigcup_{\lambda \in L} \mathfrak{a}_\lambda$ is called *the sum of the left ideals* $\mathfrak{a}_\lambda$ and is denoted by $\sum_{\lambda \in L} \mathfrak{a}_\lambda$ (cf. II, § 1, no. 7). In particular, the sum $\mathfrak{a}_1 + \mathfrak{a}_2$ of the two left ideals consists of the sums $a_1 + a_2$ where $a_1 \in \mathfrak{a}_1$ and $a_2 \in \mathfrak{a}_2$.

7. QUOTIENT RINGS

Let A be a ring. If $\mathfrak{a}$ is a two-sided ideal of A, two elements x and y of A are said to be *congruent modulo* $\mathfrak{a}$, written $x \equiv y \pmod{\mathfrak{a}}$ or $x \equiv y(\mathfrak{a})$, if $x - y \in \mathfrak{a}$. This is an equivalence relation on A. The relations $x \equiv y(\mathfrak{a})$ and $x' \equiv y'(\mathfrak{a})$ imply $x + x' \equiv y + y'(\mathfrak{a})$, $xx' \equiv xy'(\mathfrak{a})$ for $\mathfrak{a}$ is a left ideal and $xy' \equiv yy'(\mathfrak{a})$ for $\mathfrak{a}$ is a right ideal, whence $xx' \equiv yy'(\mathfrak{a})$. Conversely, if R is an equivalence relation on A compatible with addition and multiplication, the set $\mathfrak{a}$ of $x \equiv 0$ mod. R is a two-sided ideal and $x \equiv y$ mod. R is equivalent to $x \equiv y$ mod. $\mathfrak{a}$.

Let A be a ring and $\mathfrak{a}$ a two-sided ideal of A. $A/\mathfrak{a}$ denotes the quotient set of A by the equivalence relation $x \equiv y(\mathfrak{a})$, with addition and multiplication the quotients of those on A (§ 1, no. 6, Definition 11). We show that $A/\mathfrak{a}$ is a ring:

(a) Under addition, $A/\mathfrak{a}$ is the quotient commutative group of the additive group of A by the subgroup $\mathfrak{a}$.

(b) Under multiplication, $A/\mathfrak{a}$ is a monoid (§ 2, no. 1).

(c) Let ξ, η, ζ be in $A/\mathfrak{a}$ and let $\pi: A \to A/\mathfrak{a}$ be the canonical mapping; we choose elements x, y, z in A such that $\pi(x) = \xi$, $\pi(y) = \eta$ and $\pi(z) = \zeta$. Then

$$\begin{aligned}\xi(\eta + \zeta) &= \pi(x)\pi(y + z) = \pi(x(y + z)) = \pi(xy + xz) \\ &= \pi(x)\pi(y) + \pi(x)\pi(z) = \xi\eta + \xi\zeta\end{aligned}$$

and the relation $(\xi + \eta)\zeta = \xi\zeta + \eta\zeta$ is established similarly.

DEFINITION 6. *Let* A *be a ring and* $\mathfrak{a}$ *a two-sided ideal of* A. *The quotient ring of* A *by* $\mathfrak{a}$, *denoted by* $A/\mathfrak{a}$, *is the quotient set of* A *by the equivalence relation* $x \equiv y(\mathfrak{a})$, *with addition and multiplication the quotients of those on* A.

The ring $A/\{0\}$ is isomorphic to A and A/A is a zero ring.

THEOREM 2. *Let* A *be a ring and* $\mathfrak{a}$ *a two-sided ideal of* A.

(a) *The canonical mapping* π *of* A *onto* $A/\mathfrak{a}$ *is a ring homomorphism.*

(b) *Let* B *be a ring and* f *a homomorphism of* A *into* B. *If* $f(\mathfrak{a}) = \{0\}$, *there exists one and only one homomorphism* $\bar{f}$ *of* $A/\mathfrak{a}$ *into* B *such that* $f = \bar{f} \circ \pi$.

By construction, $\pi(x + y) = \pi(x) + \pi(y)$ and $\pi(xy) = \pi(x)\pi(y)$ for x, y in A; also $\pi(1)$ is the unit ε of $A/\mathfrak{a}$, whence (a).

Let A^+ be the additive group of A and B^+ that of B; as f is a homomorphism of A^+ into B^+, zero on the subgroup $\mathfrak{a}$ of A^+, there exists (§ 4, no. 4, Proposition 5) one and only one homomorphism $\bar{f}$ of $A^+/\mathfrak{a}$ into B^+ such that $f = \bar{f} \circ \pi$. Let ξ, η be in $A/\mathfrak{a}$; choose x, y in A with $\pi(x) = \xi$ and $\pi(y) = \eta$; then $\xi\eta = \pi(xy)$, whence

$$\bar{f}(\xi\eta) = \bar{f}(\pi(xy)) = f(xy) = f(x).f(y) = \bar{f}(\xi).\bar{f}(\eta)$$

and $\bar{f}(\varepsilon) = f(\pi(1)) = f(1)$, hence $\bar{f}$ is a ring homomorphism.

THEOREM 3. *Let* A *and* B *be rings and* f *a homomorphism of* A *into* B.

(a) *The kernel* $\mathfrak{a}$ *of* f *is a two-sided ideal of* B.

(b) *The image* $B' = f(B)$ *of* f *is a subring of* B.

(c) *Let* $\pi: A \to A/\mathfrak{a}$ *and* $i: B' \to B$ *be the canonical morphisms. There exists one and only one morphism* $\bar{f}$ *of* $A/\mathfrak{a}$ *into* B' *such that* $f = i \circ \bar{f} \circ \pi$ *and* $\bar{f}$ *is an isomorphism.*

As f is a morphism of the additive group of A into that of B, $\mathfrak{a}$ is a subgroup of A. If $x \in \mathfrak{a}$ and $a \in A$, then $f(ax) = f(a)f(x) = 0$, hence $ax \in \mathfrak{a}$ and similarly $xa \in \mathfrak{a}$; hence $\mathfrak{a}$ is a two-sided ideal of A. Assertion (b) is obvious. As f is zero on $\mathfrak{a}$, there exists a morphism $\bar{f}$ of $A/\mathfrak{a}$ into B' such that $f = i \circ \bar{f} \circ \pi$ (Theorem 2). The uniqueness of $\bar{f}$ and the fact that $\bar{f}$ is an isomorphism follow from *Set Theory*, II, § 6, no. 4.

8. SUBRINGS AND IDEALS IN A QUOTIENT RING

PROPOSITION 5. *Let* A *and* A' *be two rings,* f *a homomorphism of* A *into* A' *and* $\mathfrak{a}$ *the kernel of* f.

(a) *Let* B' *be a subring of* A'. *Then* $B = \overset{-1}{f}(B')$ *is a subring of* A *containing* $\mathfrak{a}$. *If* f *is surjective, then* $f(B) = B'$ *and* $f \mid B$ *defines when passing to the quotient an isomorphism of* $B/\mathfrak{a}$ *onto* B'.

(b) *Let* $\mathfrak{b}'$ *be a left* (resp. *right, two-sided*) *ideal of* A'. *Then* $\mathfrak{b} = \overset{-1}{f}(\mathfrak{b}')$ *is a left* (resp. *right, two-sided*) *ideal of* A *containing* $\mathfrak{a}$.

(c) *If* $\mathfrak{b}'$ *is a two-sided ideal of* A', *the composite mapping of the canonical morphism* $A' \to A'/\mathfrak{b}'$ *and* $f: A \to A'$ *defines, when passing to the quotient, an injective morphism* $\bar{f}$ *of* $A/\mathfrak{b}$ *into* $A'/\mathfrak{b}'$. *If* f *is surjective,* $\bar{f}$ *is an isomorphism of* $A/\mathfrak{b}$ *onto* $A'/\mathfrak{b}'$.

(d) *Suppose* f *is surjective. Let* Φ *be the set of subrings* (resp. *left ideals, right ideals, two-sided ideals*) *of* A *containing* $\mathfrak{a}$. *Let* Φ' *be the set of subrings* (resp. *left ideals, right ideals, two-sided ideals*) *of* A'. *The mappings* $B \to f(B)$ *and* $B' \mapsto \overset{-1}{f}(B')$ *are inverse bijections of* Φ *onto* Φ' *and* Φ' *onto* Φ.

(a) and (b) are obvious, except the last assertion of (a) which follows from no. 7, Theorem 3.

The composite morphism $g: A \to A' \to A'/\mathfrak{b}'$ considered in (c) has kernel $\mathfrak{b}$ and hence $\bar{f}$ is an injective morphism of $A/\mathfrak{b}$ into $A'/\mathfrak{b}'$ (§ 8, no. 7, Theorem 3). If f is surjective, g is surjective and hence $\bar{f}$ is surjective.

Suppose f is surjective. By the above, the mapping $\theta: B' \mapsto \overset{-1}{f}(B')$ is a mapping of Φ' into Φ. Clearly the mapping $\eta: B \mapsto f(B)$ is a mapping of Φ into Φ'. Then $\theta \circ \eta = \mathrm{Id}_{\Phi}$, $\eta \circ \theta = \mathrm{Id}_{\Phi'}$, whence (d).

Remark. In the above notation, θ and η are ordered set isomorphisms (Φ and Φ' being ordered by inclusion).

COROLLARY. *Let* A *be a ring and* $\mathfrak{a}$ *a two-sided ideal of* A.

(a) *Every left* (resp. *right, two-sided*) *ideal of* $A/\mathfrak{a}$ *can be written uniquely in the form* $\mathfrak{b}/\mathfrak{a}$, *where* $\mathfrak{b}$ *is a left* (resp. *right, two-sided*) *ideal of* A *containing* $\mathfrak{a}$.

(b) *If* $\mathfrak{b}$ *is two-sided, the composite homomorphism* $A \to A/\mathfrak{a} \to (A/\mathfrak{a})/(\mathfrak{b}/\mathfrak{a})$ *defines when passing to the quotient an isomorphism of* $A/\mathfrak{b}$ *onto* $(A/\mathfrak{a})/(\mathfrak{b}/\mathfrak{a})$.

It suffices to apply Proposition 5 to the canonical morphism of A onto $A/\mathfrak{a}$.

9. MULTIPLICATION OF IDEALS

Let A be a ring and $\mathfrak{a}$ and $\mathfrak{b}$ two-sided ideals of A. The set of elements of the form $x_1y_1 + \cdots + x_ny_n$ with $n \geqslant 0$, $x_i \in \mathfrak{a}$ and $y_i \in \mathfrak{b}$ for $1 \leqslant i \leqslant n$, is obviously a two-sided ideal of A, which is denoted by $\mathfrak{ab}$ and called the *product* of the two-sided ideals $\mathfrak{a}$ and $\mathfrak{b}$. Under this multiplication, the set of two-sided ideals of A is a monoid with unit element the two-sided ideal A. If $\mathfrak{a}$, $\mathfrak{b}$, $\mathfrak{c}$ are two-sided ideals of A, then $\mathfrak{a}(\mathfrak{b} + \mathfrak{c}) = \mathfrak{ab} + \mathfrak{ac}$, $(\mathfrak{b} + \mathfrak{c})\mathfrak{a} = \mathfrak{ba} + \mathfrak{ca}$. If A is commutative, multiplication of ideals is commutative.

$\mathfrak{ab} \subset \mathfrak{a}A \subset \mathfrak{a}$ and $\mathfrak{ab} \subset A\mathfrak{b} \subset \mathfrak{b}$, hence

$$\mathfrak{ab} \subset \mathfrak{a} \cap \mathfrak{b}. \tag{21}$$

PROPOSITION 6. *Let* $\mathfrak{a}, \mathfrak{b}_1, \ldots, \mathfrak{b}_n$ *be two-sided ideals of* A. *If* $A = \mathfrak{a} + \mathfrak{b}_i$ *for all* i, *then* $A = \mathfrak{a} + \mathfrak{b}_1\mathfrak{b}_2\ldots\mathfrak{b}_n = \mathfrak{a} + (\mathfrak{b}_1 \cap \mathfrak{b}_2 \cap \cdots \cap \mathfrak{b}_n)$.

By (21) it suffices to prove that $A = \mathfrak{a} + \mathfrak{b}_1\mathfrak{b}_2 \dots \mathfrak{b}_n$. By induction, it suffices to consider the case where $n = 2$. By hypothesis, there exist $a, a' \in \mathfrak{a}$, $b_1 \in \mathfrak{b}_1$, $b_2 \in \mathfrak{b}_2$ such that $1 = a + b_1 = a' + b_2$. Then

$$1 = a' + (a + b_1)b_2 = (a' + ab_2) + b_1b_2 \in \mathfrak{a} + \mathfrak{b}_1\mathfrak{b}_2,$$

whence $A = \mathfrak{a} + \mathfrak{b}_1\mathfrak{b}_2$.

PROPOSITION 7. *Let $\mathfrak{b}_1, \dots, \mathfrak{b}_n$ be two-sided ideals of A such that $\mathfrak{b}_i + \mathfrak{b}_j = A$ for $i \neq j$. Then $\mathfrak{b}_1 \cap \mathfrak{b}_2 \cap \dots \cap \mathfrak{b}_n = \sum_{\sigma \in \mathfrak{S}_n} \mathfrak{b}_{\sigma(1)}\mathfrak{b}_{\sigma(2)} \dots \mathfrak{b}_{\sigma(n)}$. In particular, if A is commutative, $\mathfrak{b}_1 \cap \mathfrak{b}_2 \cap \dots \cap \mathfrak{b}_n = \mathfrak{b}_1\mathfrak{b}_2 \dots \mathfrak{b}_n$* (cf. Exercise 2).

Suppose first $n = 2$. There exist $a_1 \in \mathfrak{b}_1$, $a_2 \in \mathfrak{b}_2$ such that $a_1 + a_2 = 1$. If $x \in \mathfrak{b}_1 \cap \mathfrak{b}_2$, then $x = x(a_1 + a_2) = xa_1 + xa_2 \in \mathfrak{b}_2\mathfrak{b}_1 + \mathfrak{b}_1\mathfrak{b}_2$. Hence $\mathfrak{b}_1 \cap \mathfrak{b}_2 = \mathfrak{b}_1\mathfrak{b}_2 + \mathfrak{b}_2\mathfrak{b}_1$.

Suppose now the equation of the proposition is established for all integers $< n$. By Proposition 6, $\mathfrak{b}_n + (\mathfrak{b}_1\mathfrak{b}_2 \dots \mathfrak{b}_{n-1}) = A$ and hence

$$\begin{aligned}\mathfrak{b}_1 \cap \mathfrak{b}_2 \cap \dots \cap \mathfrak{b}_n &= (\mathfrak{b}_1 \cap \mathfrak{b}_2 \cap \dots \cap \mathfrak{b}_{n-1})\mathfrak{b}_n + \mathfrak{b}_n(\mathfrak{b}_1 \cap \mathfrak{b}_2 \cap \dots \cap \mathfrak{b}_{n-1}) \\ &= \Big(\sum_{\sigma \in \mathfrak{S}_{n-1}} \mathfrak{b}_{\sigma(1)}\mathfrak{b}_{\sigma(2)} \dots \mathfrak{b}_{\sigma(n-1)}\Big)\mathfrak{b}_n \\ &\quad + \mathfrak{b}_n\Big(\sum_{\sigma \in \mathfrak{S}_{n-1}} \mathfrak{b}_{\sigma(1)}\mathfrak{b}_{\sigma(2)} \dots \mathfrak{b}_{\sigma(n-1)}\Big) \\ &\subset \sum_{\tau \in \mathfrak{S}_n} \mathfrak{b}_{\tau(1)}\mathfrak{b}_{\tau(2)} \dots \mathfrak{b}_{\tau(n)} \subset \mathfrak{b}_1 \cap \mathfrak{b}_2 \cap \dots \cap \mathfrak{b}_n.\end{aligned}$$

10. PRODUCT OF RINGS

Let $(A_\iota)_{\iota \in I}$ be a family of rings. Let A be the product set $\prod_{\iota \in I} A_\iota$. On A addition and multiplication are defined by the formulae

$$(x_\iota) + (y_\iota) = (x_\iota + y_\iota), \qquad (x_\iota)(y_\iota) = (x_\iota y_\iota). \tag{22}$$

It is immediately verified that A is a ring called the *product* of the rings A_ι, with zero the element $0 = (0_\iota)_{\iota \in I}$, where 0_ι is the zero of A_ι, and unit $1 = (1_\iota)_{\iota \in I}$, where 1_ι is the unit of A_ι. If the A_ι are commutative, so is A. If C_ι is the centre of A_ι, the centre of A is $\prod_{\iota \in I} C_\iota$.

For all $\iota \in I$, the projection pr_ι of A onto A_ι is a ring homomorphism. If B is a ring and $f_\iota \colon B \to A_\iota$ is a family of homomorphisms, there exists a unique homomorphism $f \colon B \to A$ such that $f_\iota = \mathrm{pr}_\iota \circ f$ for all $\iota \in I$; it is given by $f(b) = (f_\iota(b))_{\iota \in I}$.

For all $\iota \in I$, let $\mathfrak{a}_\iota$ be a left ideal of A_ι. Then $\mathfrak{a} = \prod_{\iota \in I} \mathfrak{a}_\iota$ is a left ideal of A. There is an analogous result for right ideals, two-sided ideals and subrings.

Suppose that $\mathfrak{a}_\iota$ is a two-sided ideal for all $\iota \in I$ and let f_ι denote the canonical mapping of A_ι onto $A_\iota/\mathfrak{a}_\iota$. Then the mapping $f: (x_\iota)_{\iota\in I} \mapsto (f_\iota(x_\iota))_{\iota\in I}$ of $\prod_{\iota\in I} A_\iota$ onto $\prod_{\iota\in I}(A_\iota/\mathfrak{a}_\iota)$ is a ring homomorphism of kernel $\prod_{\iota\in I}\mathfrak{a}_\iota$ and hence defines when passing to the quotient an isomorphism of $\left(\prod_{\iota\in I} A_\iota\right)/\left(\prod_{\iota\in I}\mathfrak{a}_\iota\right)$ onto $\prod_{\iota\in I}(A_\iota/\mathfrak{a}_\iota)$.

Let $(I_\lambda)_{\lambda\in L}$ be a partition of I. The canonical bijection of $\prod_{\iota\in I} A_\iota$ onto $\prod_{\lambda\in L}\left(\prod_{\iota\in I_\lambda} A_\iota\right)$ is a ring isomorphism, under which these two rings are identified.

Let $J \subset I$. Let e_J denote the element $(x_\iota)_{\iota\in I}$ of A defined by $x_\iota = 1_\iota$ for $\iota \in J$, $x_\iota = 0_\iota$ for $\iota \in I - J$. Then e_J is a central idempotent (§ 1, no. 4) of A. The following formulae follow immediately:

$$e_I = 1;$$
$$e_\varnothing = 0;$$
$$e_{J\cap K} = e_J e_K \quad \text{for } J \subset I, K \subset I;$$
$$e_{J\cup K} = e_J + e_K \quad \text{for } J \subset I, K \subset I, J \cap K = \varnothing;$$
$$\sum_\lambda e_{J_\lambda} = 1 \quad \text{if } (J_\lambda) \text{ is a finite partition of I.}$$

Let $A_J = \prod_{\iota\in J} A_\iota$. Let η_J be the canonical projection of A onto A_J. For $x = (x_\iota)_{\iota\in J} \in A_J$, let $\varepsilon_J(x)$ be the element $(y_\iota)_{\iota\in I}$ of A defined by $y_\iota = x_\iota$ for $\iota \in J$, $y_\iota = 0_\iota$ for $\iota \in I - J$. Then η_J is a ring homomorphism of A onto A_J, ε_J is an injective homomorphism of the additive group of A_J onto A and in the diagram

$$A_J \xrightarrow{\varepsilon_J} A \xrightarrow{\eta_{I-J}} A_{I-J}$$

the kernel $\mathfrak{a}_J$ of η_{I-J} is equal to the image of ε_J. Then $\varepsilon_J(xx') = \varepsilon_J(x)\varepsilon_J(x')$ for all $x, x' \in A_J$; but ε_J is not in general a ring homomorphism for $\varepsilon_J(1) = e_J$. Clearly $\mathfrak{a}_J = e_J A = A e_J$.

Let $e_{\{\iota\}} = e_\iota$ and $\mathfrak{a}_\iota = \mathfrak{a}_{\{\iota\}} = e_\iota A = A e_\iota$ for all $\iota \in I$. Then $e_\iota^2 = e_\iota$,

$$e_\iota e_\kappa = e_\kappa e_\iota = 0$$

for $\iota \neq \kappa$. If I is finite, then $\sum_{\iota\in I} e_\iota = 1$, the additive group A is the direct sum of the two-sided ideals $\mathfrak{a}_\iota$ and if $x \in A$ its component in $\mathfrak{a}_\iota$ is xe_ι. The following proposition is immediately deduced.

PROPOSITION 8. *Suppose* I *is finite. If* $\mathfrak{b}$ *is a left or right ideal of* A, $\mathfrak{b}$ *is the direct sum of the* $\mathfrak{b} \cap \mathfrak{a}_\iota$.

11. DIRECT DECOMPOSITION OF A RING

Let A be a ring and $(\mathfrak{b}_i)_{i \in I}$ a family of two-sided ideals of A. We shall call the homomorphism

$$x \mapsto (\phi_i(x))_{i \in I},$$

where ϕ_i is the canonical homomorphism of A onto $A/\mathfrak{b}_i$, the canonical homomorphism of A into $\prod_{i \in I} (A/\mathfrak{b}_i)$.

Proposition 9. *Let* A *be a ring and* $(\mathfrak{b}_1, \ldots, \mathfrak{b}_n)$ *two-sided ideals of* A *such that* $\mathfrak{b}_i + \mathfrak{b}_j = A$ *for* $i \neq j$. *The canonical homomorphism of* A *into* $\prod_{i=1}^{n} (A/\mathfrak{b}_i)$ *is surjective of kernel* $\bigcap_{i=1}^{n} \mathfrak{b}_i = \sum_{\sigma \in \mathfrak{S}_n} \mathfrak{b}_{\sigma(1)}\mathfrak{b}_{\sigma(2)} \ldots \mathfrak{b}_{\sigma(n)}$.

Clearly the kernel is $\bigcap_{i=1}^{n} \mathfrak{b}_i$. To prove the surjectivity, it is necessary to show that, for every family $(x_i)_{1 \leqslant i \leqslant n}$ of elements of A, there exists $x \in A$ such that $x \equiv x_i \ (\mathfrak{b}_i)$ for all $1 \leqslant i \leqslant n$. We prove this assertion by induction on the cardinal n of I, the case $n \leqslant 1$ being trivial. By the induction hypothesis, there exists $y \in A$ such that $y \equiv x_i \ (\mathfrak{b}_i)$ for $1 \leqslant i \leqslant n - 1$. We look for an x of the form $y + z$ with $z \in A$. Of necessity $z \equiv 0 \ (\mathfrak{b}_i)$ for $i < n$, that is $z \in \mathfrak{b} = \bigcap_{i=1}^{n-1} \mathfrak{b}_i$, and on the other hand $z \equiv x_n - y \ (\mathfrak{b}_n)$. Now $\mathfrak{b}_n + \mathfrak{b} = A$ by no. 9, Proposition 6, whence the existence of z. Finally, the second expression of the kernel follows from no. 9, Proposition 7.

Definition 7. *Let* A *be a ring. A finite family* $(\mathfrak{b}_i)_{i \in I}$ *of two-sided ideals of* A *such that the canonical homomorphism of* A *into* $\prod_{i \in I} (A/\mathfrak{b}_i)$ *is an isomorphism is called a direct decomposition of* A.

Proposition 10. *Let* A *be a ring,* A′ *its centre and* $(\mathfrak{b}_i)_{i \in I}$ *a finite family of two-sided ideals of* A. *The following conditions are equivalent:*

(a) *the family* $(\mathfrak{b}_i)_{i \in I}$ *is a direct decomposition of* A;

(b) *there exists a family* $(e_i)_{i \in I}$ *of idempotents of* A′ *such that* $e_i e_j = 0$ *for* $i \neq j$, $1 = \sum_{i \in I} e_i$ *and* $\mathfrak{b}_i = A(1 - e_i)$ *for* $i \in I$;

(c) $\mathfrak{b}_i + \mathfrak{b}_j = A$ *for* $i \neq j$ *and* $\bigcap_{i \in I} \mathfrak{b}_i = \{0\}$;

(d) $\mathfrak{b}_i + \mathfrak{b}_j = A$ *for* $i \neq j$ *and* $\prod_{i \in I} \mathfrak{b}_i = \{0\}$ *for every total order on* I;

(e) *there exists a direct decomposition* $(\mathfrak{b}'_i)_{i \in I}$ *of* A′ *such that* $\mathfrak{b}_i = A\mathfrak{b}'_i$ *for* $i \in I$.

(a) ⇒ (b). If condition (a) holds, A may be identified with the ring $\prod_{i \in I} (A/\mathfrak{b}_i)$

and $\mathfrak{b}_i$ with the kernel of pr_i. The existence of the e_i with the properties in (b) then follows from no. 10.

(b) ⇒ (d). Suppose that the e_i exist with the properties in (b). For $i \neq j$, $1 - e_i \in \mathfrak{b}_i$, $e_i = e_i(1 - e_j) \in \mathfrak{b}_j$, hence $1 \in \mathfrak{b}_i + \mathfrak{b}_j$ and $A = \mathfrak{b}_i + \mathfrak{b}_j$. On the other hand, if I is given a total ordering and $(x_i)_{i \in I}$ is a family of elements of A, then, since the e_i are central,

$$\prod_{i \in I} x_i(1 - e_i) = \Big(\prod_{i \in I} x_i\Big)\Big(\prod_{i \in I} (1 - e_i)\Big) = \Big(\prod_{i \in I} x_i\Big)\Big(1 - \prod_{i \in I} e_i\Big) = 0$$

hence $\prod_{i \in I} \mathfrak{b}_i = \{0\}$.

(d) ⇒ (c). This follows from no. 9, Proposition 7.

(c) ⇒ (a). This follows from Proposition 9.

Thus conditions (a), (b), (c) and (d) are equivalent. Suppose they hold. By (b) ⇒ (a), the family of $\mathfrak{b}'_i = A'(1 - e_i)$ is a direct decomposition of A'. Then $\mathfrak{b}_i = A(1 - e_i) = A\mathfrak{b}'_i$ for all $i \in I$. Hence condition (e) holds.

Finally, suppose condition (e) holds. By (a) ⇒ (b), there exists a family $(e_i)_{i \in I}$ of idempotents of A' such that $e_i e_j = 0$ for $i \neq j$, $1 = \sum_{i \in I} e_i$ and $\mathfrak{b}'_i = A'(1 - e_i)$ for $i \in I$. Then $\mathfrak{b}_i = A\mathfrak{b}'_i = A(1 - e_i)$ for $i \in I$, hence condition (b) holds.

Remark. Let A be a ring. Let $(\mathfrak{a}_i)_{i \in I}$ be a finite family of subgroups of the additive group A^+ of A such that A^+ is the direct sum of the $\mathfrak{a}_i$. Suppose $\mathfrak{a}_i\mathfrak{a}_i \subset \mathfrak{a}_i$ for $i \in I$ and $\mathfrak{a}_i\mathfrak{a}_j = \{0\}$ for $i \neq j$. Then $\mathfrak{a}_i$ is for all $i \in I$ a two-sided ideal of A. With addition and multiplication induced by those on A, $\mathfrak{a}_i$ is a ring with unit element the component of $1 \in A$ in $\mathfrak{a}_i$. If $\mathfrak{b}_i = \sum_{j \neq i} \mathfrak{a}_j$, clearly the $\mathfrak{b}_i$ satisfy condition (c) of Proposition 10 and hence $(\mathfrak{b}_i)_{i \in I}$ is a direct decomposition of A, which is said to be *defined by* $(\mathfrak{a}_i)_{i \in I}$.

Example: Ideals and quotient rings of $\mathbf{Z}$

An ideal of $\mathbf{Z}$ is an additive subgroup of $\mathbf{Z}$ and hence of the form $n.\mathbf{Z}$ with $n \geqslant 0$; conversely, for every integer $n \geqslant 0$, the set $n.\mathbf{Z}$ is an ideal, the principal ideal (n). Thus every ideal of $\mathbf{Z}$ is principal and is represented uniquely in the form $n\mathbf{Z}$ with $n \geqslant 0$. The ideal (1) is equal to $\mathbf{Z}$, the ideal (0) consists of 0 and the ideals distinct from $\mathbf{Z}$ and $\{0\}$ are therefore of the form $n\mathbf{Z}$ with $n > 1$. If $m \geqslant 1$ and $n \geqslant 1$, $m\mathbf{Z} \supset n\mathbf{Z}$ if and only if $n \in m.\mathbf{Z}$, that is m divides n. Therefore, for the ideal $n\mathbf{Z}$ to be maximal, it is necessary and sufficient that there exist no integer $m > 1$ distinct from n and dividing n; in other words, *the maximal ideals of* $\mathbf{Z}$ *are the ideals of the form* $p\mathbf{Z}$ *where* p *is a prime number* (§ 4, no. 10, Definition 16).

Let m and n be two integers $\geqslant 1$. The ideal $m\mathbf{Z} + n\mathbf{Z}$ is principal, whence there is an integer $d \geqslant 1$ characterized by $d\mathbf{Z} = m\mathbf{Z} + n\mathbf{Z}$; for every integer

$r \geqslant 1$, the relation "r divides d" is equivalent to $r\mathbf{Z} \supset d\mathbf{Z}$ and hence to "$r\mathbf{Z} \supset m\mathbf{Z}$ and $r\mathbf{Z} \supset n\mathbf{Z}$", that is to "$r$ divides m and n". It is thus seen that the common divisors of m and n are the divisors of d and that d is *the greatest* of the divisors $\geqslant 1$ common to m and n; d is called *the greatest common divisor* (abbreviated to g.c.d.) of m and n. As $d\mathbf{Z} = m\mathbf{Z} + n\mathbf{Z}$, there exist two integers x and y such that $d = mx + ny$. m and n are said to be *relatively prime* if their g.c.d. is equal to 1. It amounts to the same to assume that there exist integers x and y with $mx + ny = 1$.

The intersection of the ideals $m\mathbf{Z}$ and $n\mathbf{Z}$ is non-zero for it contains mn and hence is of the form $r\mathbf{Z}$ with $r \geqslant 1$. Arguing as above, it is seen that the multiples of r are the common multiples of m and n and that r is *the least* of the integers $\geqslant 1$ which are common multiples of m and n; it is called *the least common multiple* (l.c.m.) of m and n.

The product of the ideals $m\mathbf{Z}$ and $n\mathbf{Z}$ is the set of $\sum_{i=1}^{r} mx_i ny_i = mn\left(\sum_{i=1}^{r} x_i y_i\right)$ for $x_1, \ldots, y_r \in \mathbf{Z}$ and hence is equal to $mn\mathbf{Z}$.

For every integer $n \geqslant 1$, the quotient ring $\mathbf{Z}/n\mathbf{Z}$ is called the *ring of integers modulo* n; it has n elements, which are the classes modulo n of the integers $0, 1, 2, \ldots, n - 1$. For $n = 1$, we obtain the zero ring.

PROPOSITION 11. *Let $n_1, \ldots, n_d$ be integers $\geqslant 1$ which are relatively prime in pairs and $n = n_1 \ldots n_r$. The canonical homomorphism of* $\mathbf{Z}$ *into the product ring* $\prod_{i=1}^{r} \mathbf{Z}/n_i\mathbf{Z}$ *is surjective of kernel* $n\mathbf{Z}$ *and defines a ring isomorphism of* $\mathbf{Z}/n\mathbf{Z}$ *onto* $\prod_{i=1}^{r} \mathbf{Z}/n_i\mathbf{Z}$.

Let $\mathfrak{a}_i = n_i\mathbf{Z}$ for $i = 1, \ldots, r$. By hypothesis, $\mathfrak{a}_i + \mathfrak{a}_j = \mathbf{Z}$ for $i \neq j$. The proposition then follows from Proposition 9.

The above results, as also those concerning decomposition into prime factors, will be generalized in Chapter VII, § 1, which is devoted to the study of principal ideal domains and in *Commutative Algebra*, Chapter VII, § 3, which is devoted to the study of factorial domains.

12. RINGS OF FRACTIONS

THEOREM 4. *Let* A *be a commutative ring and* S *a subset of* A. *Let* A_S *be the monoid of fractions of* A (*provided only with multiplication*) *with denominators in* S (§ 2, no. 4). *Let* $\varepsilon : A \to A_S$ *be the canonical morphism. There exists on* A_S *one and only one addition satisfying the following conditions:*

(a) A_S, *with this addition and its multiplication, is a commutative ring;*

(b) ε *is a ring homomorphism.*

Suppose an addition has been found for A_S satisfying conditions (a) and (b).

Let $x, y \in A_S$. Let S' be the stable multiplicative submonoid of A generated by S. There exist $a, b \in A$ and $p, q \in S'$ such that $x = a/p$, $y = b/q$. Then

$$x = \varepsilon(aq)\varepsilon(pq)^{-1}, \qquad y = \varepsilon(bp)\varepsilon(pq)^{-1},$$

whence

$$\begin{aligned} x + y &= (\varepsilon(aq) + \varepsilon(bp))\varepsilon(pq)^{-1} \\ &= \varepsilon(aq + bp)\varepsilon(pq)^{-1} \\ &= (aq + bp)/pq. \end{aligned} \tag{23}$$

This proves the uniqueness of the addition.

We now *define* an addition on A_S by setting $x + y = (aq + bp)/pq$. It is necessary to show that this definition does not depend on the choice of a, b, p, q. Now, if $a', b' \in A$, $p', q' \in S'$ are such that $x = a'/p', y = b'/q'$, there exist s and t in S' such that $ap's = a'ps$, $bq't = b'qt$, whence

$$(aq + bp)(p'q')(st) = (a'q' + b'p')(pq)(st)$$

and hence

$$(aq + bp)/pq = (a'q' + b'p')/p'q'.$$

It is easily verified that addition in A_S is associative and commutative, that $0/1$ is identity element for addition, that $(-a)/p$ is the negative of a/p and that $x(y + z) = xy + xz$ for all $x, y, z \in A_S$. If $a, b \in A$, then

$$\varepsilon(a + b) = (a + b)/1 = a/1 + b/1 = \varepsilon(a) + \varepsilon(b)$$

and hence ε is a ring homomorphism.

DEFINITION 8. *The ring defined in Theorem 4 is called the ring of fractions associated with* S, *or with denominators in* S, *and is denoted by* $A[S^{-1}]$.

The zero of $A[S^{-1}]$ is $0/1$, the unit of $A[S^{-1}]$ is $1/1$.

We shall return to the properties of $A[S^{-1}]$ in *Commutative Algebra*, Chapter II, § 2.

If S is the set of cancellable elements of A, the ring $A[S^{-1}]$ is called the total ring of fractions of A. A is then identified with a subring of $A[S^{-1}]$ by means of the mapping ε, which is then injective (I, § 2, no. 4, Proposition 6).

THEOREM 5. *Let* A *be a commutative ring,* S *a subset of* A, B *a ring and* f *a homomorphism of* A *into* B *such that every element of* $f(S)$ *is invertible. There exists one and only one* $\bar{f}$ *of* $A[S^{-1}]$ *into* B *such that* $f = \bar{f} \circ \varepsilon$.

We know (§ 2, no. 4, Theorem 1) that there exists one and only one morphism $\bar{f}$ of the multiplicative monoid $A[S^{-1}]$ into the multiplicative monoid B such that $f = \bar{f} \circ \varepsilon$. Let $a, b \in A$, $p, q \in S'$ (stable multiplicative submonoid of A generated by S). As the elements of $f(A)$ commute in pairs,

$$\begin{aligned}\bar{f}(a/p + b/q) &= \bar{f}((aq + bp)/pq) = f(aq + bp)f(pq)^{-1} \\ &= (f(a)f(q) + f(b)f(p))f(p)^{-1}f(q)^{-1} \\ &= f(a)f(p)^{-1} + f(b)f(q)^{-1} \\ &= \bar{f}(a/p) + \bar{f}(b/q).\end{aligned}$$

Hence $\bar{f}$ is a ring homomorphism.

§ 9. FIELDS

1. FIELDS

Definition 1. *A ring* K *is called a field if it does not consist only of* 0 *and every non-zero element of* K *is invertible.*

The set of non-zero elements of the field K, with multiplication, is a group, which is precisely the multiplicative group K* of the ring K (§ 8, no. 1). The opposite ring of a field is a field. A field is called *commutative* if its multiplication is commutative; such a field is identified with its opposite. A non-commutative field is sometimes called a *skew field.*

Examples. *(1) We shall define in no. 4 the field of *rational numbers*; In *General Topology* the field of real numbers will be defined (*General Topology*, IV, § 1, no. 3), also the field of complex numbers (*General Topology*, VIII, § 1, no. 1) and the field of quaternions (*General Topology*, VIII, § 1, no. 4). These fields are commutative with the exception of the field of quaternions.*

(2) The ring $\mathbf{Z}/2\mathbf{Z}$ is obviously a field.

Let K be a field. Every subring L of K which is a field is called a *subfield* of K and K is then called an *extension field* of L; it amounts to the same to say that L is a subring of K and that $x^{-1} \in L$ for every non-zero element x of L. If $(L_\iota)_{\iota \in I}$ is a family of subfields of K, then $\bigcap_{\iota \in I} L_\iota$ is a subfield of K; for every subset X of K there thus exists a smallest subfield of K containing X; it is said to be *generated by* X.

Proposition 1. *Let* K *be a field. For every subset* X *of* K, *the centralizer* (§ 8, no. 5, *Example* 3) X′ *of* X *is a subfield of* K.

We know (*loc. cit.*) that X′ is a subring of K. On the other hand, if $x \neq 0$ is permutable with $z \in X$, so is x^{-1} (I, § 2, no. 3, Proposition 5) and hence X′ contains the inverse of every non-zero element of X′.

Corollary. *The centre of a field* K *is a* (commutative) *subfield of* K.

THEOREM 1. *Let A be a ring. The following conditions are equivalent:*

(a) A *is a field;*
(b) A *is not reduced to* 0 *and the only left ideals of* A *are* 0 *and* A.

Suppose that A is a field. Then A is not reduced to 0. Let $\mathfrak{a}$ be a left ideal of A distinct from 0. There exists a non-zero a belonging to $\mathfrak{a}$. For all $x \in A$, $x = (xa^{-1})a \in \mathfrak{a}$; hence $\mathfrak{a} = A$.

Suppose that A satisfies condition (b). Let $x \neq 0$ be in A. We need to prove that x is invertible. The left ideal Ax contains x and hence is not zero, whence $Ax = A$. There thus exists $x' \in A$ such that $x'x = 1$. Then $x' \neq 0$ since $1 \neq 0$. Applying the above result to x', it is seen that there exists $x'' \in A$ such that $x''x' = 1$. Then $x'' = x''.1 = x''x'x = 1.x = x$, hence $xx' = 1$. Hence x' is the inverse of x.

Remark. In Theorem 1, left ideals may be replaced by right ideals. In Chapter VIII, § 5, no. 2, we shall study non-zero rings A which have no *two-sided* ideal distinct from 0 and A; such rings (called *quasi-simple*) are not necessarily fields *(for example, the ring $\mathbf{M}_2(\mathbf{Q})$ of square matrices of order 2 with rational coefficients is quasi-simple but is not a field)*.

COROLLARY 1. *Let* A *be a ring and* $\mathfrak{a}$ *a two-sided ideal of* A. *For the ring* $A/\mathfrak{a}$ *to be a field, it is necessary and sufficient that* $\mathfrak{a}$ *be a maximal left ideal of* A.

The left ideals of $A/\mathfrak{a}$ are of the form $\mathfrak{b}/\mathfrak{a}$ where $\mathfrak{b}$ is a left ideal of A containing $\mathfrak{a}$ (§ 8, no. 8, Corollary to Proposition 5). To say that $A/\mathfrak{a} \neq 0$ means that $\mathfrak{a} \neq A$. Under this hypothesis, $A/\mathfrak{a}$ is a field if and only if the only left ideals of A containing $\mathfrak{a}$ are $\mathfrak{a}$ and A (Theorem 1), whence the corollary.

COROLLARY 2. *Let* A *be a commutative ring which is not* 0. *There exists a homomorphism of* A *onto a commutative field.*

By Krull's Theorem (§ 8, no. 6, Theorem 1), there exists in A a maximal ideal $\mathfrak{a}$. Then $A/\mathfrak{a}$ is a field (Corollary 1).

COROLLARY 3. *Let a be an integer* $\geqslant 0$. *For the ring* $\mathbf{Z}/a\mathbf{Z}$ *to be a field, it is necessary and sufficient that a be prime.*

This follows from Corollary 1 and § 8, no. 11.

For p prime the field $\mathbf{Z}/p\mathbf{Z}$ is denoted by $\mathbf{F}_p$.

THEOREM 2. *Let* K *be a field and* A *a non-zero ring. If f is a homomorphism of* K *into* A, *then the subring* $f(K)$ *of* A *is a field and f defines an isomorphism of* K *onto* $f(K)$.

Let $\mathfrak{a}$ be the kernel of f. Then $1 \notin \mathfrak{a}$ for $f(1) = 1 \neq 0$ in A and, as $\mathfrak{a}$ is a left ideal of K, $\mathfrak{a} = \{0\}$ by Theorem 1. Therefore f is injective and hence an isomorphism of K onto the subring $f(K)$ of A; the latter ring is therefore a field.

2. INTEGRAL DOMAINS

DEFINITION 2. *A ring* A *is called an integral domain (or a domain of integrity) if it is commutative, non-zero, and the product of two non-zero elements of* A *is non-zero.*

The ring **Z** of rational integers is an integral domain; it is commutative and non-zero; the product of two integers >0 is non-zero; every non-zero integer is of the form a or $-a$ with $a > 0$ and $(-a)b = -ab$, $(-a)(-b) = ab$ for $a > 0$, $b > 0$, whence our assertion.

Every commutative field is an integral domain. A subring of an integral domain is an integral domain. In particular, a subring of a commutative field is an integral domain. We shall show that conversely every integral domain A is isomorphic to a subring of a commutative field. Recall (§ 8, no. 12) that A has been identified with a subring of its total ring of fractions. Our assertion then follows from the following proposition:

PROPOSITION 2. *If* A *is an integral domain, the total ring of fractions* K *of* A *is a commutative ring.*

The ring K is commutative. It is non-zero since $A \neq \{0\}$. As A is an integral domain, every non-zero element of A is cancellable and K consists of the fractions a/b with $b \neq 0$. Now $a/b \neq 0$ implies $a \neq 0$ and the fraction b/a is then inverse of a/b.

The total ring of fractions of an integral domain is called its *field of fractions*. Such a ring is identified with its image in its field of fractions.

PROPOSITION 3. *Let* B *be a non-zero ring and* A *a commutative subring of* B *such that every non-zero element of* A *is invertible in* B.

(a) A *is an integral domain.*

(b) *Let* A′ *be the field of fractions* A. *The canonical injection of* A *into* B *can be extended uniquely to an isomorphism f of* A′ *onto a subfield of* B.

(c) *The elements of $f(A')$ are the xy^{-1} where $x \in A$, $y \in A$, $y \neq 0$.*

Assertion (a) is obvious. The canonical injection of A into B extends uniquely to a homomorphism f of A′ into B (§ 8, no. 12, Theorem 5). Assertion (b) then follows from no. 1, Theorem 2. The elements of A′ are the fractions x/y with $x \in A$, $y \in A$, $y \neq 0$ and $f(x/y) = xy^{-1}$, whence (c).

3. PRIME IDEALS

PROPOSITION 4. *Let* A *be a commutative ring and* $\mathfrak{p}$ *an ideal of* A. *The following conditions are equivalent:*

(a) *the ring* $A/\mathfrak{p}$ *is an integral domain;*

(b) $A \neq \mathfrak{p}$ *and, if* $x \in A - \mathfrak{p}$ *and* $y \in A - \mathfrak{p}$, *then* $xy \in A - \mathfrak{p}$.

(c) $\mathfrak{p}$ *is the kernel of a homomorphism of* A *into a field.*

The implications (c) ⇒ (b) ⇒ (a) are obvious. If $A/\mathfrak{p}$ is an integral domain, let f be the canonical injection of $A/\mathfrak{p}$ into its field of fractions and g the canonical homomorphism of A onto $A/\mathfrak{p}$; then $\mathfrak{p}$ is the kernel of $f \circ g$, whence the implication (a) ⇒ (c).

DEFINITION 3. *In a commutative ring* A, *an ideal* $\mathfrak{p}$ *satisfying the conditions of Proposition 4 is called a prime ideal.*

Examples. (1) Let A be a commutative ring. If $\mathfrak{m}$ is a maximal ideal of A, $\mathfrak{m}$ is prime; the ring $A/\mathfrak{m}$ is a field (no. 1, Corollary 1 to Theorem 1).

(2) If A is an integral domain, the ideal $\{0\}$ of A is prime (but not maximal in general, as the example of the ring $\mathbf{Z}$ proves).

4. THE FIELD OF RATIONAL NUMBERS

DEFINITION 4. *The field of fractions of the ring* $\mathbf{Z}$ *of rational integers is called the field of rational numbers and is denoted by* $\mathbf{Q}$. *The elements of* $\mathbf{Q}$ *are called rational numbers.*

Every rational number is thus of the form a/b where a and b are rational integers with $b \neq 0$ (and we may even take $b > 0$ as the relation

$$a/b = (-a)/(-b)$$

proves). $\mathbf{Q}_+$ is used to denote the set of rational numbers of the form a/b with $a \in \mathbf{N}$ and $b \in \mathbf{N}^*$.

We have the relations:

$$\mathbf{Q}_+ + \mathbf{Q}_+ = \mathbf{Q}_+ \tag{1}$$

$$\mathbf{Q}_+ . \mathbf{Q}_+ = \mathbf{Q}_+ \tag{2}$$

$$\mathbf{Q}_+ \cap (-\mathbf{Q}_+) = \{0\} \tag{3}$$

$$\mathbf{Q}_+ \cup (-\mathbf{Q}_+) = \mathbf{Q} \tag{4}$$

$$\mathbf{Q}_+ \cap \mathbf{Z} = \mathbf{N}. \tag{5}$$

The first two follow from the formulae $a/b + a'/b' = (ab' + a'b)/bb'$, $(a/b)(a'/b') = aa'/bb'$, $0 \in \mathbf{Q}_+$, $1 \in \mathbf{Q}_+$ and the fact that $\mathbf{N}$ is stable under addition and multiplication and $\mathbf{N}^*$ under multiplication. Obviously $0 \in \mathbf{Q}_+$, whence $0 \in (-\mathbf{Q}_+)$; let x be in $\mathbf{Q}_+ \cap (-\mathbf{Q}_+)$; then there exist positive integers a, b, a', b' with $b \neq 0$, $b' \neq 0$ and $x = a/b = -a'/b'$; then $ab' + ba' = 0$, whence $ab' = 0$ (for $ab' \geqslant 0$ and $ba' \geqslant 0$) and therefore $a = 0$ since $b' \neq 0$; in other words, $x = 0$. Finally, obviously $\mathbf{N} \subset \mathbf{Z}$ and $\mathbf{N} \subset \mathbf{Q}_+$. Conversely, if x belongs to $\mathbf{Z} \cap \mathbf{Q}_+$, it is a rational integer; there exist two rational integers a and b with $a \geqslant 0$, $b > 0$ and $x = a/b$, whence $a = bx$; if $x \notin \mathbf{N}$, then $-x > 0$, whence $-a = b(-x) > 0$ and consequently $a < 0$ contrary to the hypothesis.

Given two rational numbers x and y, we write $x \leqslant y$ if $y - x \in \mathbf{Q}_+$. It is easily deduced from formulae (1), (3) and (4) that $x \leqslant y$ is a total ordering on

$\mathbf{Q}$, from (5) that this relation induces the usual order relation on $\mathbf{Z}$. Finally, from (1) it follows that the relations $x \leqslant y$ and $x' \leqslant y'$ imply $x + x' \leqslant y + y'$ and from (2) that the relation $x \leqslant y$ implies $xz \leqslant yz$ for all $z \geqslant 0$ and $xz \geqslant yz$ for $z \leqslant 0$ (cf. VI, § 2, no. 1).

Let x be a rational number. x is said to be *positive* if $x \geqslant 0$, *strictly positive* if $x > 0$, *negative* if $x \leqslant 0$ and *strictly negative* if $x < 0$.† The set of positive rational numbers is $\mathbf{Q}_+$ and that of negative numbers is $-\mathbf{Q}_+$. If $\mathbf{Q}^*$ denotes the set of non-zero rational numbers, the set $\mathbf{Q}^*_+$ of strictly positive numbers is equal to $\mathbf{Q}^* \cap \mathbf{Q}_+$ and $-\mathbf{Q}^*_+$ is the set of strictly negative numbers.

The sets $\mathbf{Q}^*_+$ and $\{1, -1\}$ are subgroups of the multiplicative group $\mathbf{Q}^*$. Every rational number $x \neq 0$ can be expressed in one and only one way in the form $1 . y$, $(-1) . y$, where $y > 0$; hence the multiplicative group $\mathbf{Q}^*$ is the product of the subgroups $\mathbf{Q}^*_+$ and $\{1, -1\}$, the component of x in $\mathbf{Q}^*_+$ is called the *absolute value* of x and is denoted by $|x|$; the component of x in $\{-1, 1\}$ (equal to 1 if $x > 0$, to -1 if $x < 0$) is called the *sign* of x and denoted by sgn x.

Usually these two functions are extended to the whole of $\mathbf{Q}$ by setting $|0| = 0$ and sgn $0 = 0$.

§ 10. INVERSE AND DIRECT LIMITS

Throughout this paragraph, I will denote a non-empty preordered set, $\alpha \leqslant \beta$ the preordering on I. The notion of inverse (resp. direct) system of sets relative to the indexing set I is defined in *Set Theory*, III, § 7, no. 1 (resp. *Set Theory*, III, § 7, no. 5, under the hypothesis that I is right directed).

1. INVERSE SYSTEMS OF MAGMAS

DEFINITION 1. *An inverse system of magmas relative to the indexing set* I *is an inverse system of sets* $(E_\alpha, f_{\alpha\beta})$ *relative to* I, *each* E_α *having a magma structure and each* $f_{\alpha\beta}$ *being a magma homomorphism.*

Let $(E_\alpha, f_{\alpha\beta})$ be an inverse system of magmas whose laws are written multiplicatively. The inverse limit set $E = \varprojlim E_\alpha$ is a subset of the product magma $\prod_{\alpha \in I} E_\alpha$ consisting of the families $(x_\alpha)_{\alpha \in I}$ such that $x_\alpha = f_{\alpha\beta}(x_\beta)$ for $\alpha \leqslant \beta$. If (x_α) and (y_α) belong to E, then for $\alpha \leqslant \beta$, $x_\alpha = f_{\alpha\beta}(x_\beta)$ and $y_\alpha = f_{\alpha\beta}(y_\beta)$, hence $x_\alpha y_\alpha = f_{\alpha\beta}(x_\beta) f_{\alpha\beta}(y_\beta) = f_{\alpha\beta}(x_\beta y_\beta)$; hence E is a submagma of $\prod_{\alpha \in I} E_\alpha$. E will be given the law induced by that on $\prod_{\alpha \in I} E_\alpha$; the magma obtained is called the

† We avoid the current terminology where positive means > 0.

inverse limit magma of the magmas E_α. It enjoys the following universal property:

(a) For all $\alpha \in I$, the canonical mapping f_α of E into E_α is a magma homomorphism of E into E_α. $f_\alpha = f_{\alpha\beta} \circ f_\beta$ for $\alpha \leqslant \beta$.

(b) Suppose a magma F is given and homomorphisms $u_\alpha: F \to E_\alpha$ such that $u_\alpha = f_{\alpha\beta} \circ u_\beta$ for $\alpha \leqslant \beta$. There exists one and only one homomorphism $u: F \to E$ such that $u_\alpha = f_\alpha \circ u$ for all $\alpha \in I$ (namely $x \mapsto u(x) = (u_\alpha(x))_{\alpha \in I}$).

If the magmas E_α are associative (resp. commutative), so is E. Suppose that each magma E_α admits an identity element e_α and that the homomorphisms $f_{\alpha\beta}$ are unital. Then $e = (e_\alpha)_{\alpha \in I}$ belongs to E for $e_\alpha = f_{\alpha\beta}(e_\beta)$ for $\alpha \leqslant \beta$ and it is an identity element of the magma E; with the above notation, the homomorphisms f_α are unital and if the u_α are unital then u is unital. Further, an element $x = (x_\alpha)_{\alpha \in I}$ of E is invertible if and only if each of the x_α is invertible in the corresponding magma E_α and $x^{-1} = (x_\alpha^{-1})_{\alpha \in I}$; this follows from the formula $f_{\alpha\beta}(x_\beta^{-1}) = f_{\alpha\beta}(x_\beta)^{-1} = x_\alpha^{-1}$ for $\alpha \leqslant \beta$.

From these remarks it can be deduced that if the magmas E_α are monoids (resp. groups) and the $f_{\alpha\beta}$ are monoid homomorphisms, then the magma E is a monoid (resp. a group). In this case we speak of an inverse system of monoids (resp. groups). The universal property goes over immediately to this case.

It is left to the reader to define an inverse system of rings $(E_\alpha, f_{\alpha\beta})$ and to verify that $E = \varprojlim E_\alpha$ is a subring of the product ring $\prod_{\alpha \in I} E_\alpha$, called the *inverse limit ring of the rings* E_α; it can be verified that the universal property extends to this case.

Let $\mathfrak{E} = (E_\alpha, f_{\alpha\beta})$ and $\mathfrak{E}' = (E'_\alpha, f'_{\alpha\beta})$ be two inverse systems of magmas (resp. monoids, groups, rings) relative to the same indexing set. A homomorphism of $\mathfrak{E}$ into $\mathfrak{E}'$ is an inverse system $(u_\alpha)_{\alpha \in I}$ of mappings $u_\alpha: E_\alpha \to E'_\alpha$ such that each u_α is a homomorphism. Under these conditions, the mapping $u = \varprojlim u_\alpha$ of $\varprojlim E_\alpha$ into $\varprojlim E'_\alpha$ is a homomorphism (cf. *Set Theory*, III, § 7, no. 2).

2. INVERSE LIMITS OF ACTIONS

Suppose there are given two inverse systems of sets $(\Omega_\alpha, \phi_{\alpha\beta})$ and $(E_\alpha, f_{\alpha\beta})$ relative to the same indexing set I. Suppose there is given for all $\alpha \in I$ an action of Ω_α on E_α such that

$$f_{\alpha\beta}(\omega_\beta x_\beta) = \phi_{\alpha\beta}(\omega_\beta) . f_{\alpha\beta}(x_\beta) \tag{1}$$

for $\alpha \leqslant \beta$, $x_\beta \in E_\beta$, $\omega_\beta \in \Omega_\beta$. Then the family of actions considered is said to be an *inverse system of actions*. Let $\Omega = \varprojlim \Omega_\alpha$ and $E = \varprojlim E_\alpha$; if $x = (x_\alpha)_{\alpha \in I}$ belongs to E and $\omega = (\omega_\alpha)_{\alpha \in I}$ belongs to Ω, then $\omega . x = (\omega_\alpha . x_\alpha)_{\alpha \in I}$ belongs to E by (1). Thus an action of Ω on E is defined called the *inverse limit of the actions of the* Ω_α *on the* E_α.

The above applies especially in the case where the Ω_α are monoids and each action of Ω_α on E_α is an operation. Then the inverse limit of these operations is an operation of the monoid on E.

It is left to the reader to define the inverse limit of an inverse system of groups with operators and to verify that this limit is a group with operators.

3. DIRECT SYSTEMS OF MAGMAS

In this no. and the following we shall assume that I is *right directed.*

DEFINITION 2. *A direct system of magmas relative to the indexing set* I *is a direct system of sets* $(E_\alpha, f_{\beta\alpha})$ *relative to* I, *each* E_α *having a magma structure and each* $f_{\beta\alpha}$ *being a magma homomorphism.*

Let $(E_\alpha, f_{\beta\alpha})$ be a direct system of magmas. E will denote the direct limit set $\varinjlim E_\alpha$ and f_α the canonical mapping of E_α into E. Recall that

$$f_\beta \circ f_{\beta\alpha} = f_\alpha \quad \text{for } \alpha \leqslant \beta, \tag{2}$$

$$E = \bigcup_{\alpha \in I} f_\alpha(E_\alpha). \tag{3}$$

By (2), also

$$f_\alpha(E_\alpha) \subset f_\beta(E_\beta) \quad \text{for } \alpha \leqslant \beta. \tag{4}$$

If $x_\alpha, y_\alpha \in E_\alpha$ are such that $f_\alpha(x_\alpha) = f_\alpha(y_\alpha)$, there exists a $\beta \geqslant \alpha$ such that $f_{\beta\alpha}(x_\alpha) = f_{\beta\alpha}(y_\alpha)$.

PROPOSITION 1. *There exists on* E *one and only one magma structure for which the mappings* $f_\alpha \colon E_\alpha \to E$ *are homomorphisms. If the magmas* E_α *are associative* (resp. *commutative*), *so is* E. *If the magmas* E_α *and the homomorphism* $f_{\beta\alpha}$ *are unital, so are the magma* E *and the homomorphisms* f_α.

The magmas E_α will be written multiplicatively.

Let x, y be in E. There exist α in I and x_α, y_α in E_α such that $x = f_\alpha(x_\alpha)$ and $y = f_\alpha(y_\alpha)$. If there exists a magma structure on E for which f_α is a homomorphism, then $x.y = f_\alpha(x_\alpha y_\alpha)$, whence the *uniqueness* of this magma structure.

To prove the existence, we must prove that for α, β in I, x_α, y_α in E_α and x'_β, y'_β in E_β, the relations

$$f_\alpha(x_\alpha) = f_\beta(x'_\beta), \qquad f_\alpha(y_\alpha) = f_\beta(y'_\beta) \tag{5}$$

imply $f_\alpha(x_\alpha y_\alpha) = f_\beta(x'_\beta y'_\beta)$. For $\gamma \geqslant \alpha$ and $\gamma \geqslant \beta$, we set $x_\gamma = f_{\gamma\alpha}(x_\alpha)$,

$y_\gamma = f_{\gamma\alpha}(y_\alpha)$, $x'_\gamma = f_{\gamma\beta}(x'_\beta)$, $y'_\gamma = f_{\gamma\beta}(y'_\beta)$. By the definition of direct limit, there exists γ in I with $\gamma \geqslant \alpha$, $\gamma \geqslant \beta$, $x_\gamma = x'_\gamma$, $y_\gamma = y'_\gamma$. Then

$$\begin{aligned} f_\alpha(x_\alpha y_\alpha) &= f_\gamma(f_{\gamma\alpha}(x_\alpha y_\alpha)) = f_\gamma(x_\gamma y_\gamma) = f_\gamma(x'_\gamma y'_\gamma) = f_\gamma(f_{\gamma\beta}(x'_\beta y'_\beta)) \\ &= f_\beta(x'_\beta y'_\beta). \end{aligned}$$

Suppose the magmas E_α are associative. Let x, y, z be in E. There exist $\alpha \in I$ and elements $x_\alpha, y_\alpha, z_\alpha$ in E_α such that

$$x = f_\alpha(x_\alpha), \quad y = f_\alpha(y_\alpha), \quad z = f_\alpha(z_\alpha).$$

Then $xy = f_\alpha(x_\alpha y_\alpha)$, whence $(xy)z = f_\alpha((x_\alpha y_\alpha)z_\alpha)$; similarly

$$x(yz) = f_\alpha(x_\alpha(y_\alpha z_\alpha)),$$

whence $(xy)z = x(yz)$ for $(x_\alpha y_\alpha)z_\alpha = x_\alpha(y_\alpha z_\alpha)$. The case of commutative magmas is treated analogously.

Suppose finally that each magma E_α has an identity element e_α and that $f_{\beta\alpha}(e_\alpha) = e_\beta$ for $\alpha \leqslant \beta$. For α, β in I, there exists γ in I with $\gamma \geqslant \alpha$ and $\gamma \geqslant \beta$, whence

$$f_\alpha(e_\alpha) = f_\gamma(f_{\gamma\alpha}(e_\alpha)) = f_\gamma(e_\gamma) = f_\gamma(f_{\gamma\beta}(e_\beta)) = f_\beta(e_\beta)$$

and there thus exists an element e in E such that $f_\alpha(e_\alpha) = e$ for all $\alpha \in I$. Let x be in E; choose $\alpha \in I$ and $x_\alpha \in E_\alpha$ such that $x = f_\alpha(x_\alpha)$. Then

$$ex = f_\alpha(e_\alpha).f_\alpha(x_\alpha) = f_\alpha(e_\alpha.x_\alpha) = f_\alpha(x_\alpha) = x$$

and similarly $x.e = x$, hence e is the identity element of E.

The magma E is called the *direct limit of the magmas* E_α.

PROPOSITION 2. *Let* $(E_\alpha, f_{\beta\alpha})$ *be a direct system of magmas and let* E *be its direct limits* $f_\alpha: E_\alpha \to E$ *the canonical homomorphisms. Suppose a magma* F *and homomorphisms* $u_\alpha: E_\alpha \to F$ *are given such that* $u_\alpha = u_\beta \circ f_{\beta\alpha}$ *for* $\alpha \leqslant \beta$. *There exists one and only one homomorphism* $u: E \to F$ *such that* $u_\alpha = u \circ f_\alpha$ *for all* $\alpha \in I$. *If the magmas* E_α *and* F *and the homomorphisms* $f_{\beta\alpha}$ *and* u_α *are unital, the homomorphism* u *is unital.*

We know (*Set Theory*, III, § 7, no. 6, Proposition 6) that there exists one and only one mapping $u: E \to F$ such that $u_\alpha = u \circ f_\alpha$ for all $\alpha \in I$. We verify that u is a homomorphism: let x, y be in E, α in I and x_α, y_α in E_α such that $x = f_\alpha(x_\alpha)$ and $y = f_\alpha(y_\alpha)$. Then $xy = f_\alpha(x_\alpha y_\alpha)$, whence

$$\begin{aligned} u(xy) &= u(f_\alpha(x_\alpha y_\alpha)) = u_\alpha(x_\alpha y_\alpha) = u_\alpha(x_\alpha)u_\alpha(y_\alpha) \\ &= u(f_\alpha(u_\alpha))u(f_\alpha(y_\alpha)) = u(x)u(y). \end{aligned}$$

We consider now the unital case and let e_α denote the identity element of E_α, e that of E and e' that of F. Choose $\alpha \in I$, then $e = f_\alpha(e_\alpha)$, whence

$$u(e) = u(f_\alpha(e_\alpha)) = u_\alpha(e_\alpha) = e'$$

for u_α is unital. Hence u is unital.

By analogy with the notion of direct system of magmas, that of a direct system of monoids or groups can be formulated. Proposition 1 shows that the magma E which is the limit of a direct system of monoids $(E_\alpha, f_{\beta\alpha})_{\alpha, \beta \in I}$ is a monoid. We show that E is a group if the E_α are groups: let $x \in E$, $\alpha \in I$ and $x_\alpha \in E_\alpha$ be such that $x = f_\alpha(x_\alpha)$; the element $y = f_\alpha(x_\alpha^{-1})$ of E is the inverse of x (§ 2, no. 3). The universal property of Proposition 2 goes over immediately in the case of a direct system of monoids or groups.

The reader is left to define a direct system of rings. Let $(A_\alpha, f_{\beta\alpha})$ be such a direct system; let $A = \varinjlim A_\alpha$ and $f_\alpha: A_\alpha \to A$ the canonical homomorphisms. There exists (Proposition 2) on A one and only one addition and multiplication characterized by $x + y = f_\alpha(x_\alpha + y_\alpha)$, $xy = f_\alpha(x_\alpha y_\alpha)$ for $\alpha \in I$, x_α, y_α in A_α and $x = f_\alpha(x_\alpha)$, $y = f_\alpha(y_\alpha)$. Under addition A is a commutative group and multiplication is associative and unital. Finally, for x, y, z in A, choose α in I and x_α, y_α and z_α in A_α with

$$x = f_\alpha(x_\alpha), \quad y = f_\alpha(y_\alpha) \text{ and } z = f_\alpha(z_\alpha).$$

Then

$$\begin{aligned}(x + y).z &= f_\alpha(x_\alpha + y_\alpha)f_\alpha(z_\alpha) = f_\alpha((x_\alpha + y_\alpha)z_\alpha) \\ &= f_\alpha(x_\alpha z_\alpha + y_\alpha z_\alpha) = f_\alpha(x_\alpha z_\alpha) + f_\alpha(y_\alpha z_\alpha) = xz + yz\end{aligned}$$

and the relation $x(y + z) = xy + xz$ is proved analogously. In other words, A has the structure of a ring, characterized by the fact that f_α is a ring homomorphism for all $\alpha \in I$.

The ring A is called the *direct limit of the rings* A_α. Proposition 2 extends immediately to the case of rings.

PROPOSITION 3. (a) *If the* A_α *are non-zero,* A *is non-zero.*

(b) *If the* A_α *are integral domains,* A *is an integral domain.*

(c) *If the* A_α *are fields,* A *is a field.*

Let 0_α, 1_α be the zero and unit of A_α and 0, 1 the zero and unit of A. There exists $\alpha \in I$ such that $f_\alpha(0_\alpha) = 0$, $f_\alpha(1_\alpha) = 1$. If $0 = 1$, there exists $\beta \geqslant \alpha$ such that $f_{\beta\alpha}(0_\alpha) = f_{\beta\alpha}(1_\alpha)$, that is $0_\beta = 1_\beta$. This proves (a).

Suppose that A_α are integral domains. Then A is commutative and non-zero by (a). Let x, y be elements of A such that $xy = 0$. There exists $\alpha \in I$ and $x_\alpha, y_\alpha \in A_\alpha$ such that $x = f_\alpha(x_\alpha)$, $y = f_\alpha(y_\alpha)$. Then $f_\alpha(x_\alpha y_\alpha) = xy = 0 = f_\alpha(0_\alpha)$. Hence there exists $\beta \geqslant \alpha$ such that $f_{\beta\alpha}(x_\alpha y_\alpha) = f_{\beta\alpha}(0_\alpha)$. As A_β is an integral domain, it follows that $f_{\beta\alpha}(x_\alpha) = 0_\beta$ or $f_{\beta\alpha}(y_\alpha) = 0_\beta$, hence $x = 0$ or $y = 0$. This proves (b).

Suppose that the A_α are fields. Then $A \neq \{0\}$ by (a). Let x be a non-zero element of A. There exist $\alpha \in I$ and $x_\alpha \in A_\alpha$ such that $x = f_\alpha(x_\alpha)$. Then $x_\alpha \neq 0$ and $f_\alpha(x_\alpha^{-1})$ is the inverse of x in A. This proves (c).

Let $\mathfrak{E} = (E_\alpha, f_{\beta\alpha})$ and $\mathfrak{E}' = (E'_\alpha, f'_{\beta\alpha})$ be two direct systems of magmas (resp. monoids, groups, rings). A homomorphism of $\mathfrak{E}$ into $\mathfrak{E}'$ is a direct system $(u_\alpha)_{\alpha\in I}$ of mappings $u_\alpha\colon E_\alpha \to E'_\alpha$ such that each u_α is a homomorphism. Under these conditions, the mapping $u = \varinjlim u_\alpha$ from $E = \varinjlim E_\alpha$ to $E' = \varinjlim E'_\alpha$ is a homomorphism (cf. *Set Theory*, III, § 7, no. 6).

4. DIRECT LIMIT OF ACTIONS

Suppose that there are given two direct systems of sets $(\Omega_\alpha, \phi_{\beta\alpha})$ and $(E_\alpha, f_{\beta\alpha})$ relative to the same indexing set I and for each $\alpha \in I$ an action of Ω_α on E_α. Suppose that

$$f_{\beta\alpha}(\omega_\alpha . x_\alpha) = \phi_{\beta\alpha}(\omega_\alpha) . f_{\beta\alpha}(x_\alpha)$$

for $\alpha \leqslant \beta$, $\omega_\alpha \in \Omega_\alpha$ and $x_\alpha \in E_\alpha$. Then the family of actions under consideration is said to be a *direct system of actions*. It is easily verified as in Proposition 2 that there exists an action h of $\Omega = \varinjlim \Omega_\alpha$ on $E = \varinjlim E_\alpha$ which is described as follows: let $\omega \in \Omega$ and $x \in E$; choose $\alpha \in I$ and $\omega_\alpha \in \Omega_\alpha$, $x_\alpha \in E_\alpha$ such that $\omega = \phi_\alpha(\omega_\alpha)$ and $x = f_\alpha(x_\alpha)$ ($\phi_\alpha\colon \Omega_\alpha \to \Omega$ and $f_\alpha\colon E_\alpha \to E$ denote the canonical mappings); then $\omega . x = f_\alpha(\omega_\alpha . x_\alpha)$. The action of Ω on E is called the *direct limit of the actions* of the Ω_α on the E_α.

If the Ω_α are monoids and each action of Ω_α on E_α is an operation, the direct limit action is an operation.

It is left to the reader to define the direct limit of a direct system of groups with operators and to verify that this limit is a group with operators.

EXERCISES

§ 1

1. Let E be a set, $A \subset E \times E$ and $(x, y) \mapsto x \top y$, $(x, y) \in A$, an internal law of composition not everywhere defined on E. Given two subsets X and Y of E, let $X \top Y$ denote the set of elements of E of the form $x \top y$ with $x \in X$, $y \in Y$ and $(x, y) \in A$. A law of composition is thus defined which is *everywhere defined* on $\mathfrak{P}(E)$.

If $(X_\alpha)_{\alpha \in A}$ and $(Y_\beta)_{\beta \in B}$ are any two families of subsets of E, then

$$\Big(\bigcup_{\alpha \in A} X_\alpha\Big) \top \Big(\bigcup_{\beta \in B} Y_\beta\Big) = \bigcup_{(\alpha, \beta) \in A \times B} (X_\alpha \top Y_\beta).$$

2. Let $\top$ be a law of composition not everywhere defined on a set E. Let E′ be the subset of $\mathfrak{P}(E)$ consisting of the sets $\{x\}$ where x runs through E and the empty subset $\varnothing$ of E; show that E′ is a stable subset of $\mathfrak{P}(E)$ under the law $(X, Y) \mapsto X \top Y$ (Exercise 1); deduce that, if $\overline{E}$ denotes the set obtained by *adjoining* (*Set Theory*, II, § 4, no. 8) to E an element ω, the law $\top$ can be extended to $\overline{E} \times \overline{E}$, so that the law $\top$ is identical with the law induced on E by this extended law.

3. Let $\top$ be a law of composition not everywhere defined on E.

(*a*) For the law $(X, Y) \mapsto X \top Y$ (Exercise 1) between subsets of E to be associative, it is necessary and sufficient that, for all $x \in E$, $y \in E$, $z \in E$, if *one* of the two sides of the formula $(x \top y) \top z = x \top (y \top z)$ is defined, so is the other and is equal to it (use Exercise 1 to show that the condition is sufficient).

(*b*) If this condition holds, show that Theorem 1 of no. 3 can be generalized as follows: if one of the two sides of formula (3) is defined, the other is defined and is equal to it.

4. (*a*) Given a set E, let Φ be the set of mappings into E of any subset of E; if f, g, h are three elements of Φ, show that if the composition $(f \circ g) \circ h$ is defined so is $f \circ (g \circ h)$, but that the converse is not true; if these two compositions are defined they are equal.

(*b*) Let $\mathfrak{F}$ be a family of non-empty subsets of E such that no two have an element in common and let Ψ be the subset of Φ consisting of the bijective mappings of a set of $\mathfrak{F}$ *onto* a set of $\mathfrak{F}$. Show that, under the law induced by the law $f \circ g$ on Ψ, the condition of Exercise 3(*a*) holds.

5. Show that the only triples (m, n, p) of natural numbers $\neq 0$ such that $(m^n)^p = m^{n^p}$ are: $(1, n, p)$, n and p being arbitrary; $(m, n, 1)$, m and n arbitrary and $(m, 2, 2)$ where m is arbitrary.

6. Let $\top$ be a law of composition on a set E; let A be the subset of E consisting of the elements x such that $x \top (y \top z) = (x \top y) \top z$ for all $y \in E$, $z \in E$; show that A is a stable subset of E and that the law induced on A by $\top$ is associative.

7. If $\top$ is an associative law on E and a and b two elements of E, the sets $\{a\} \top E$, $E \top \{b\}$, $\{a\} \top E \top \{b\}$, $E \top \{a\} \top E$ are stable subsets of E under the law $\top$.

8. Let $\top$ be an associative law on E and a an element of E; for all $x \in E$, $y \in E$, we write $x \perp y = x \top a \top y$: show that the law $\perp$ is associative.

9. On a set E the mappings $(x, y) \mapsto x$ and $(x, y) \mapsto y$ are opposite associative laws of composition.

10. Let X and Y be any two subsets of a set E and let $X \top Y = X \cup Y$ if $X \cap Y = \varnothing$, $X \top Y = E$ if $X \cap Y \neq \varnothing$; show that the law of composition thus defined on $\mathfrak{P}(E)$ is associative and commutative.

11. Let $\top$ be an associative law on E and A and B two subsets of E which are stable under this law; show that, if $B \top A \subset A \top B$, $A \top B$ is a stable subset of E.

12. The only distinct natural numbers $\neq 0$ which are permutable under the law $(x, y) \mapsto x^y$ are 2 and 4.

13. Show that under the law $(X, Y) \mapsto X \circ Y$ between subsets of $E \times E$ the centre is the set consisting of $\varnothing$ and the diagonal.

14. Show that under the law of composition $f \circ g$ between mappings of E into E the centre consists of the identity mapping.

15. A law $\top$ on a set E is called *idempotent* if all the elements of E are idempotent (no. 4) under this law, that is if $x \top x = x$ for all $x \in E$. Show that, if a law $\top$ on E is associative, commutative and idempotent, the relation $x \top y = y$ is an order relation on E; if we write $x \leqslant y$, any two elements x, y of E admit a least upper bound (under this order relation) equal to $x \top y$. Obtain the converse.

16. Let $\top$ be an associative and commutative law of composition on a set E. Let n be an integer >0. The mapping $x \mapsto \overset{n}{\top}\, x$ of E into E is a morphism.

§ 2

1. Let $\top$ be a law of composition on a set E. Denoting by F the sum set (*Set Theory*, II, § 4, no. 8) of E and a set $\{e\}$ consisting of a single element and identifying E and $\{e\}$ with the corresponding subsets of F, show that it is possible in one and only one way to define on F a law of composition $\bar{\top}$ which induces on E the law $\top$ and under which e is identity element; if $\top$ is associative, the law $\bar{\top}$ is associative. (F is said to be derived from E "by adjoining an identity element".)

2. Let $\top$ be a law everywhere defined on E.
(*a*) For $\top$ to be associative, it is necessary and sufficient that every left translation γ_x be permutable with every right translation δ_y in the set of mappings of E into E (with the law $f \circ g$).
(*b*) Suppose that E has an identity element. For $\top$ to be associative and commutative, it is necessary and sufficient that the mapping $(x, y) \mapsto x \top y$ be a morphism of the magma $E \times E$ into the magma E.

3. In the set F of mappings of E into E, for the relation $f \circ g = f \circ h$ to imply $g = h$, it is necessary and sufficient that f be injective; for the relation $g \circ f = h \circ f$ to imply $g = h$, it is necessary and sufficient that f be surjective; for f to be cancellable (under the law $\circ$), it is necessary and sufficient that f be bijective.

4. For there to exist on a set E a law of composition such that every permutation of E is an isomorphism of E onto itself under this law, it is necessary and sufficient that E have 0, 1 or 3 elements.

5. For $2 \leqslant n \leqslant 5$, determine on a set E with n elements all the laws everywhere defined admitting an identity element and under which all the elements of E are cancellable; for $n = 5$ show that there exist non-associative laws satisfying these conditions.

(N.B. Exercises 6 to 17 inclusive refer to associative laws written *multiplicatively*; e denotes the identity when it exists, γ_a and δ_a the translations by $a \in E$; for every subset X of E we write $\gamma_a(X) = aX$, $\delta_a(X) = Xa$.)

6. For an associative law on a *finite* set, every cancellable element is invertible (use no. 3, Proposition 3).

7. Given an associative law on a set E and an element $x \in E$, let A be the set of x^n for $n \in \mathbf{N}^*$; if there is an identity element, let B be the set of x^n for $n \in \mathbf{N}$; if further x is invertible, let C be the set of x^a for $a \in \mathbf{Z}$. Show that, if A (resp.

B, C) is infinite, it is isomorphic (with the law induced by the given law on E) to $\mathbf{N}^*$ (resp. $\mathbf{N}$, $\mathbf{Z}$) with addition.

8. In the notation of Exercise 7, suppose A is finite; show that A contains one and only one idempotent (§ 1, no. 4) (observe that if x^p and x^q are idempotents, then $x^p = x^{pq}$, $x^q = x^{pq}$, hence $x^p = x^q$); if $h = x^p$, the set of x^n for $n \geqslant p$ is a stable subset D of E such that under the law induced on D, D is a group.

9. Under a multiplicative law on a set E let a be a left cancellable element of E.

(*a*) If there exists an element u such that $au = a$, show that $ux = x$ for all $x \in \mathrm{E}$; in particular, if $xu = x$ for all $x \in \mathrm{E}$, u is identity element.

(*b*) If there exist $u \in \mathrm{E}$ such that $au = a$ and $b \in \mathrm{E}$ such that $ab = u$, show that $ba = u$ (form aba); in particular, if there exists an identity element e and an element b such that $ab = e$, b is the inverse of a.

10. If a and b are two elements of E such that ba is left cancellable, show that a is left cancellable. Deduce that under an associative and commutative law on E the set S of non-cancellable elements of E is such that $\mathrm{ES} \subset \mathrm{S}$ (and in particular is stable).

¶ 11. E is called a *left semi-group* if every element x of E is left cancellable.

(*a*) If u is an idempotent (§ 1, no. 4), then $ux = x$ for all $x \in \mathrm{E}$ (use Exercise 9(*a*)); u is identity element for the law induced on Eu.

(*b*) If u and v are two distinct idempotents of E, then $\mathrm{E}u \cap \mathrm{E}v = \varnothing$ and the sets Eu and Ev (with the laws induced by that on E) are isomorphic.

(*c*) Let R be the complement of the union of the sets Eu, where u runs through the set of idempotents of E. Show that $\mathrm{ER} \subset \mathrm{R}$; it follows that R is a stable subset of E. If R is non-empty, then $a\mathrm{R} \neq \mathrm{R}$ for all $a \in \mathrm{R}$ (use Exercise 9(*a*) to prove that, in the contrary case, R would contain an idempotent); in particular, R is then infinite. R is called the *residue* of the left semi-group E.

(*d*) If R is non-empty and there exists at least one idempotent u in E (that is $\mathrm{E} \neq \mathrm{R}$), for all $x \in \mathrm{RE}u$, $x\mathrm{E}u \neq \mathrm{E}u$; in particular, no element of REu is invertible in Eu and REu is an infinite set (use Exercise 8).

(*e*) If E has an identity element e, e is the only idempotent in E and R is empty (note that $\mathrm{E} = \mathrm{E}e$).

(*f*) If there exists a right cancellable element a in E (in particular if the given law on E is commutative), either E has an identity element or $\mathrm{E} = \mathrm{R}$ (note that if there exists an idempotent u, then $xa = xua$ for all $x \in \mathrm{E}$). For example, if E is the set of integers $\geqslant 1$, with addition, E is a commutative semi-group such that $\mathrm{E} = \mathrm{R}$.

(*g*) If there exists $a \in \mathrm{E}$ such that δ_a is surjective, either E has an identity element or $\mathrm{E} = \mathrm{R}$ (examine separately the case where $a \in \mathrm{R}$ and the case where $a \in \mathrm{E}u$ for an idempotent u).

¶ 12. Under a multiplicative law on E, let $a \in \mathrm{E}$ be such that γ_a is surjective.

(*a*) Show that, if there exists u such that $ua = a$, then $ux = x$ for all $x \in \mathrm{E}$.

(*b*) For an element $b \in \mathrm{E}$ to be such that ba is left cancellable, it is necessary and sufficient that γ_a be surjective and that b be left cancellable.

¶ 13. Suppose that, for *all* $x \in \mathrm{E}$, γ_x is surjective. Show that if, for an element $a \in \mathrm{E}$, γ_a is a bijective mapping, γ_x is a bijective mapping for all $x \in \mathrm{E}$ (use Exercise 12(*b*)); this holds in particular if there exist two elements a, b of E such that $ab = b$ (use Exercise 12(*a*)); E is then a left semi-group whose residue R is empty; moreover, for every idempotent u, every element of Eu is invertible in Eu.

¶ 14. For an associative law on a *finite* set E there exist *minimal* subsets of the form $a\mathrm{E}$ (that is minimal elements of the set of subsets of E of this form, ordered by inclusion).

(*a*) If $\mathrm{M} = a\mathrm{E}$ is minimal, then $x\mathrm{M} = x\mathrm{E} = \mathrm{M}$ for all $x \in \mathrm{M}$; with the law induced by that on E, M is a left semi-group (Exercise 11) in which every left translation is bijective (cf. Exercise 13).

(*b*) If $\mathrm{M} = a\mathrm{E}$ and $\mathrm{M}' = a'\mathrm{E}$ are minimal and distinct, then $\mathrm{M} \cap \mathrm{M}' = \varnothing$; for all $b \in \mathrm{M}$, the mapping $x' \mapsto bx'$ of M′ into M is a bijective mapping of M′ onto M. Deduce that there exist an idempotent $u' \in \mathrm{M}'$ such that $bu' = b$ and an idempotent $u \in \mathrm{M}$ such that $uu' = u$ (take u to be the idempotent such that $bu = b$). Show that $u'u = u'$ (consider $uu'u$) and that every $y' \in \mathrm{M}'$ such that $y'u = y'$ belongs to $\mathrm{M}'u'$.

(*c*) Show that the mapping $x' \mapsto ux'$ of $\mathrm{M}'u'$ into M is an isomorphism of $\mathrm{M}'u'$ onto $\mathrm{M}u$; deduce that M and M′ are isomorphic left semi-groups.

(*d*) Let M_i $(1 \leqslant i \leqslant r)$ be the distinct minimal subsets of the form $a\mathrm{E}$: deduce from (*b*) that the idempotents of each left semi-group M_i can be arranged in a sequence (u_{ij}) $(1 \leqslant j \leqslant s)$ so that $u_{ij}u_{kj} = u_{ij}$ for all i, j, k. If K is the union of the M_i, show that $\mathrm{E}u_{ij} \subset \mathrm{K}$ (note that, for all $x \in \mathrm{E}$, $xu_{ij}\mathrm{E}$ is a minimal subset of the form $a\mathrm{E}$); deduce that $\mathrm{E}u_{ij}$ is the union of the $u_{kj}\mathrm{E}u_{kj}$ for $1 \leqslant k \leqslant r$ (show that $(\mathrm{E}u_{ij}) \cap \mathrm{M}_k = u_{kj}\mathrm{E}u_{kj}$) and that $\mathrm{E}u_{ij}\mathrm{E} = \mathrm{K}$. Prove finally that every minimal subset of the form Eb is identical with one of the s sets $\mathrm{E}u_{ij}$ (note, using (*a*), that $\mathrm{E}bu_{ij}b = \mathrm{E}b$ and deduce that $\mathrm{E}b \subset \mathrm{K}$).

15. Let $(x_i)_{1 \leqslant i \leqslant n}$ be a finite sequence of left cancellable elements.

(*a*) Show that the relation $x_1x_2\ldots x_n = e$ implies all the relations $x_{i+1}\ldots x_nx_1x_2\ldots x_i = e$ which are obtained from one another by "cyclic permutation" for $1 \leqslant i \leqslant n$.

(*b*) Deduce that if the composition of the sequence (x_i) is invertible each of the x_i is invertible.

¶ 16. Let $\top$ be an associative law on a set E and E* the set of cancellable elements of E; suppose that E* is non-empty and that every cancellable element is a central element. Let $\mathfrak{F}$ denote the set of subsets X of E with the following property: there exists $y \in E^*$ such that $\delta_y(E) \subset X$.

(*a*) Show that the intersection of two sets of $\mathfrak{F}$ belong to $\mathfrak{F}$.

(*b*) Let Φ be the set of functions f defined on a set of $\mathfrak{F}$ taking their values in E and such that, for $X \in \mathfrak{F}$, $\overset{-1}{f}(X)$ belongs to $\mathfrak{F}$. Let R denote the relation between elements f and g of Φ: "there exists a set $X \in \mathfrak{F}$ such that the restrictions of f and g to X are identical". Show that R is an equivalence relation; let $\Psi = \Phi/R$ be the quotient set of Φ under this relation.

(*c*) Let f and g be two elements of Φ, $A \in \mathfrak{F}$ and $B \in \mathfrak{F}$ the sets where f and g are respectively defined; there exists $X \subset B$ and belonging to $\mathfrak{F}$ such that $g(X) \subset A$; if g_X is the restriction of g to X, show that the mapping $f \circ g_X$ belongs to Φ and that its class (mod. R) depends only on the classes of f and g (and not on X); this class is called the composition of that of f and that of g; show that the law of composition thus defined on Ψ is associative and has an identity element.

(*d*) For all $a \in E$, show that the left translation γ_a belongs to Φ; let ϕ_a be its class (mod. R). Show that the mapping $x \mapsto \phi_x$ is an isomorphism of E onto a submonoid of Ψ and that, if $x \in E^*$, ϕ_x is invertible in Ψ (consider the inverse mapping of γ_x, show that it belongs to Φ and that its class (mod. R) is the inverse of ϕ_x). Deduce a generalization of Theorem 1 of no. 4.

17. Let E be a monoid and S a stable subset of E with the following properties:

(α) For all $s \in S$ and all $a \in E$, there exist $t \in S$ and $b \in E$ such that $sb = at$.

(β) For all $a, b \in E$ and all $s \in S$ such that $sa = sb$, there exist $t \in S$ such that $at = bt$.

(*a*) In $E \times S$ let $(a, s) \sim (b, t)$ denote the relation: "there exist s' and t' in S such that $tt' = ss'$ and $as' = bt'$."

Show that $\sim$ is an equivalence relation. We write $\overline{E} = E \times S/\sim$, denote by as^{-1} the equivalence class of (a, s) (mod $\sim$) and by $\varepsilon: E \to \overline{E}$ the mapping $a \mapsto ae^{-1}$.

(*b*) Show that there exists on $\overline{E}$ one and only one monoid structure such that ε is a unital homomorphism and such that, for all $s \in S$, $\varepsilon(s)$ is invertible.

(*c*) Show that $(\overline{E}, \varepsilon)$ has the universal property described in Theorem 1 (no. 4).

(*d*) Show that ε is injective if and only if the elements of S are cancellable.

§ 3

1. Let $\alpha \mapsto f_\alpha$ be an action of Ω on E. Let F be the image of Ω in E^E and G the stable subset of E^E under the law $(f, g) \mapsto f \circ g$ generated by F and the identity mapping of E.

(*a*) Show that every subset of E which is stable under the action of Ω is also stable under the restriction to G of the canonical action of E^E on E (no. 1, *Example* 3).

(*b*) Let X be a subset of E; show that the stable subset of E generated by X is the set of $f(x)$ where f runs through G and x runs through X.

2. Let $\perp$ be an internal law on E which is doubly distributive with respect to the associative internal law $\top$; show that, if $x \perp x'$ and $y \perp y'$ are cancellable under the law $\top$, $x \perp y'$ is permutable with $y \perp x'$ under the law $\top$ (calculate the composition $(x \top y) \perp (x' \top y')$ in two different ways). In particular, if the law $\perp$ has an identity element, two cancellable elements under the law $\top$ are permutable under this law; if all the elements of E are cancellable under the law $\top$, this law is commutative.

3. Let $\top$ and $\perp$ be two internal laws on a set E such that $\perp$ is right distributive with respect to $\top$.

(*a*) If the law $\top$ has an identity element e, $x \perp e$ is idempotent under the law $\top$ for all $x \in E$; if further there exists y such that $x \perp y$ is cancellable under the law $\top$, then $x \perp e = e$.

(*b*) If the law $\perp$ has an identity element u and there exists $z \in E$ which is cancellable for both laws $\perp$ and $\top$, u is cancellable under the law $\top$.

¶ 4. Let $\top$ and $\perp$ be two internal laws on E, each with an identity element; if the left action of E on itself derived from each of the laws is distributive with respect to the other, every element of E is idempotent under both these laws (if e is the identity element for $\top$, u the identity element for $\perp$, prove first that $e \perp e = e$, noting that $e = e \perp (u \top e)$).

5. On a set E suppose three internal laws of composition are given: an addition (not necessarily associative nor commutative), a multiplication (not necessarily associative) and a law denoted by $\top$. Suppose that the multiplication has an identity element e, that the law $\top$ is right distributive with respect to multiplication and the law $\top$ is left distributive with respect to addition. Show that, if there exist x, y, z such that $x \top z$, $y \top z$ and $(x + y) \top z$ are cancellable under multiplication, then $e + e = e$ (use Exercise 3(*a*)).

¶ 6. An internal law of composition $\top$ on E is said to determine on E a *quasi-group* structure if, for all $x \in E$, the left and right translations γ_x and δ_x are bijective mappings of E onto itself. A quasi-group is said to be *distributive* if the law $\top$ is doubly distributive with respect to itself.

(*a*) Determine all the distributive quasi-group structures on a set of n elements for $2 \leqslant n \leqslant 6$.

(*b*) *Show that the set $\mathbf{Q}$ of rational numbers, with the law of composition $(x, y) \mapsto \frac{1}{2}(x + y)$, is a distributive quasi-group.*

(*c*) In a distributive quasi-group E, every element is idempotent. Deduce that, if E has more than one element, the law $\top$ can neither have an identity element nor be associative.

(*d*) The right and left translations of a distributive quasi-group E are automorphisms of E.

(*e*) If E is a *finite* distributive quasi-group, the structure induced on every stable subset of E is a distributive quasi-group structure.

(*f*) Let E be a distributive quasi-group. If R is an equivalence relation which is left (resp. right) compatible (no. 3) with the law $\top$, the classes mod. R are stable subsets of E. If E is finite, all these classes are obtained one from another by left (resp. right) translation; under the same conditions, if R is compatible with the law $\top$, the quotient structure on E/R is a distributive quasi-group structure.

(*g*) Let E be a distributive quasi-group. The set A_a of elements of E which are permutable with a given element a is stable; if E is finite, for all $x \in A_a$, $A_x = A_a$ (note, using (*e*), that there exists $y \in A_a$ such that $x = a \top y$) and when x runs through E the sets A_x are identical with the equivalence classes with respect to a relation compatible with the law $\top$.

(*h*) If E is a finite distributive quasi-group and the law $\top$ is commutative, the number of elements of E is odd (consider the ordered pairs (x, y) of elements of E such that $x \top y = y \top x = a$, where a is given).

7. Suppose given on a set E an (associative and commutative) addition under which all the elements of E are invertible and an (associative) multiplication which is doubly distributive with respect to addition *(in other words, a ring structure)*; write $[x, y] = xy - yx$; the law $(x, y) \mapsto [x, y]$ is doubly distributive with respect to addition. For x and y to be permutable under multiplication, it is necessary and sufficient that $[x, y] = 0$; and we have the identities

$$[x,y] = -[y,x]; \qquad [x,[y,z]] + [y,[z,x]] + [z,[x,y]] = 0$$

(the second is known as the "Jacobi identity"). The second identity may also be written

$$[x,[y,z]] - [[x,y],z] = [[z,x],y]$$

which expresses the "deviation from associativity" of the law $[x, y]$.

8. With the same hypotheses as in Exercise 7, let $x \top y = xy + yx$; then the law $\top$ is commutative and doubly distributive with respect to addition but not in general associative.

(*a*) For all $x \in E$, show that $\overset{m+n}{\top}\, x = \left(\overset{m}{\top}\, x\right) \top \left(\overset{n}{\top}\, x\right)$.

(*b*) If we write $[x, y, z] = (x \top y) \top z - x \top (y \top z)$ (deviation from associativity of the law $\top$), prove the identities

$$[x,y,z] + [z,y,x] = 0$$

$$[x, y, z] + [y, z, x] + [z, x, y] = 0$$
$$[x \top y, u, z] = [u, [(x \top y), z)]]$$

(the notation $[x, y]$ meaning the same as in Exercise 7)

$$[x \top y, u, z] + [y \top z, u, x] + [z \top x, u, y] = 0.$$

¶ 9. Suppose given on E an (associative and commutative) addition under which all the elements of E are invertible and a multiplication which is *non-associative*, but commutative and doubly distributive with respect to addition. Suppose further that $n \in \mathbf{Z}$, $n \neq 0$ and $nx = 0$ imply $x = 0$ in E. Show that if, writing $[x, y, z] = (xy)z - x(yz)$, the identity

$$[xy, u, z] + [yz, u, z] + [zx, u, y] = 0$$

holds, then $x^{m+n} = x^m x^n$ for all x (show, by induction on p, that the identity $[x^q, y, x^{p-q}] = 0$ holds for $1 \leqslant q < p$).

10. Let E be a commutative monoid whose law is written additively and whose identity element is denoted by 0. Let $\top$ be an internal law on E which is distributive with respect to addition and such that $0 \top x = x \top 0 = 0$ for all $x \in$ E. Let S be a subset of E which is stable under addition and under each of the external laws derived from $\top$; let $\overline{\mathrm{E}}$ denote the monoid of differences of E with respect to S under addition and ε the canonical homomorphism of E into $\overline{\mathrm{E}}$. Then there exists on $\overline{\mathrm{E}}$ one and only one law $\overline{\top}$ which is distributive with respect to addition and such that ε is a homomorphism for the laws $\top$ and $\overline{\top}$. If the law $\top$ is associative (resp. commutative), so is the law $\overline{\top}$.

§ 4

1. Determine all the group structures on a set of n elements for $2 \leqslant n \leqslant 6$ (cf. § 2, Exercise 5). Determine the subgroups and quotient groups of these groups and also their Jordon-Hölder series.

¶ 2. (*a*) An associative law $(x, y) \mapsto xy$ on a set E is a group law if there exists $e \in$ E such that, for all $x \in$ E, $ex = x$ and if, for all $x \in$ E, there exists x' such that $x'x = e$ (show that $xx' = e$ by considering the composition $x'xx'$; deduce that e is the identity element).

(*b*) Show that the same is true if, for *all* $x \in$ E, the left translation γ_x is a mapping of E onto E and if there exists *some* $a \in$ E such that the right translation δ_a is a mapping of E onto E.

3. In a group G every *finite* non-empty stable subset H is a subgroup of G (cf. § 2, Exercise 6).

4. Let A and B be two subgroups of a group G.

(*a*) Show that the least subgroup containing A and B (that is the subgroup generated by $A \cup B$) is identical with the set of compositions of the sequences $(x_i)_{1 \leqslant i \leqslant 2n+1}$ of an (arbitrary) odd number of elements such that $x_i \in A$ for i odd and $x_i \in B$ for i even.

(*b*) For AB to be a subgroup of G (in which case it is the subgroup generated by $A \cup B$), it is necessary and sufficient that A and B be permutable, that is that $AB = BA$.

(*c*) If A and B are permutable and C is a subgroup containing A, A is permutable with $B \cap C$ and $A(B \cap C) = C \cap (AB)$.

5. If a subgroup of a group G has index 2 it is normal in G.

6. Let (G_α) be a family of normal subgroups of a group G such that $\bigcap_\alpha G_\alpha = \{e\}$; show that G is isomorphic to a subgroup of the product group $\prod_\alpha (G/G_\alpha)$.

7. If G is the direct product of two subgroups A and B and H is a subgroup of G such that $A \subset H$, H is the direct product of A and $H \cap B$.

8. Let A and A′ be two groups and G a subgroup of $A \times A'$. We write:

$$N = G \cap (A \times \{e\}), \qquad H = \mathrm{pr}_1(G)$$

$$N' = G \cap (\{e\} \times A'), \qquad H' = \mathrm{pr}_2(G).$$

(*a*) Show that N is normal in H and N′ is normal in H′; define isomorphisms

$$H/N \to G/(N \times N') \to H'/N'.$$

(*b*) Suppose that $H = A$, $H' = A'$, that the groups A and A′ are of finite length and that no quotient of a Jordan-Hölder series of A is isomorphic to a quotient of a Jordan-Hölder series of A′. Show that $G = A \times A'$.

9. Let G be a group which does not consist only of the identity element. Suppose that there exists a finite subset S of G such that G is generated by S (resp. by the elements gsg^{-1}, $g \in G$, $s \in S$). Let $\mathfrak{N}$ be the set of subgroups of G (resp. of normal subgroups of G) distinct from G. Show that $\mathfrak{N}$, ordered by inclusion, is inductive. Deduce the existence of a normal subgroup H of G such that G/H is simple.

10. Let H be a normal subgroup of a group G contained in the centre of G. Show that if G/H is a monogenous group G is commutative.

11. If all the elements of a group G other than the identity element are of order 2, G is commutative; if G is finite, its order n is a power of 2 (argue by induction on n).

12. Let G be a group such that, for a fixed integer $n > 1$, $(xy)^n = x^n y^n$ for all $x \in G$, $y \in G$. If $G^{(n)}$ denotes the set of x^n, where x runs through G, and $G_{(n)}$ the set of $x \in G$ such that $x^n = e$, show that $G^{(n)}$ and $G_{(n)}$ are normal subgroups of G; if G is finite, the order of $G^{(n)}$ is equal to the index of $G_{(n)}$. Show that, for all x, y in G, also $x^{1-n}y^{1-n} = (xy)^{1-n}$ and deduce that $x^{n-1}y^n = y^n x^{n-1}$; conclude from this that the set of elements of G of the form $x^{n(n-1)}$ generates a commutative subgroup of G.

13. Let G be a group. If S is a simple group, S is said to occur in G if there exist two subgroups H, H' of G with H a normal subgroup of H' such that H'/H is isomorphic to S. Let in(G) denote the set of isomorphism classes of simple groups occurring in G.

(*a*) Show that $\mathrm{in}(G) = \varnothing \Leftrightarrow G = \{e\}$.

(*b*) If H is a subgroup of G, show that $\mathrm{in}(H) \subset \mathrm{in}(G)$; if moreover H is normal, then

$$\mathrm{in}(G) = \mathrm{in}(H) \cup \mathrm{in}(G/H).$$

(*c*) Let G_1 and G_2 be two groups. Show the equivalence of the two following properties:

(i) $\mathrm{in}(G_1) \cap \mathrm{in}(G_2) = \varnothing$;

(ii) Every subgroup of $G_1 \times G_2$ is of the form $H_1 \times H_2$ with $H_1 \subset G_1$ and $H_2 \subset G_2$.

(Use Exercise 8.)

*When G_1 and G_2 are finite groups, show that (i) and (ii) are equivalent to:

(iii) $\mathrm{Card}(G_1)$ and $\mathrm{Card}(G_2)$ are relatively prime.*

14. Let G be a group and H a subgroup of G. Any subset T of G which meets every left coset mod. H in one and only one point is called a representative system of the left cosets mod. H (cf. *Set Theory*, II, § 6, no. 2); this condition is equivalent to the following:

The mapping $(x, y) \mapsto xy$ is a bijection of $T \times H$ onto G.

(*a*) Let T be such a representative system; for all $g \in G$ and all $t \in T$, let $x(g, t) \in T$ and $y(g, t) \in H$ be such that $gt = x(g, t)y(g, t)$. Let S be a subset of G generating G. Show that the elements $y(g, t)$, $g \in S$, $t \in T$ generate H. (If H' denotes the subgroup of H generated by these elements, show that T.H' is stable under the γ_g, $g \in S$, hence also under the γ_g, $g \in G$, and deduce that $T.H' = G$, whence $H' = H$.)

(*b*) Suppose that (G:H) is finite. Show that G can be generated by a finite subset if and only if H can.

¶ 15. Let $\mathfrak{F}$ be a set of stable subgroups of a group with operators G; $\mathfrak{F}$ is said to satisfy the maximal condition (resp. minimal condition) if every subset of $\mathfrak{F}$, ordered by inclusion, has a maximal (resp. minimal) element.

Suppose that the set of *all* stable subgroups of a group G satisfies the minimal condition.

(*a*) Prove that there exists no stable subgroup of G isomorphic to G and distinct from G (argue by *reductio ad absurdum* by showing that the hypothesis would imply the existence of a strictly decreasing infinite sequence of stable subgroups of G).

(*b*) The minimal elements of the set of normal stable subgroups of G not consisting only of e are called *minimal* normal stable subgroups. Let $\mathfrak{M}$ be a set of minimal normal stable subgroups of G and S the smallest stable subgroup of G containing all the subgroups belonging to $\mathfrak{M}$; show that S is the direct product of a *finite* number of minimal normal stable subgroups of G (let (M_n) be a sequence of minimal normal stable subgroups of G belonging to $\mathfrak{M}$ and such that M_{n+1} is not contained in the stable subgroup generated by the union of $M_1, M_2, \ldots, M_n$; let S_k be the stable subgroup generated by the union of the M_n of index $n \geqslant k$; show that $S_{k+1} = S_k$ after a certain rank and therefore that (M_n) is a finite sequence; finally use no. 9, Proposition 15).

(*c*) If G is a group without operators, show that every minimal normal subgroup M of G is the direct product of a finite number of simple groups isomorphic to one another (let N be a minimal normal subgroup *of* M; show that M is the smallest subgroup of G containing all the subgroups aNa^{-1} where a runs through G, and then apply (*b*) to the group M).

16. If the set of stable subgroups of a group with operators G satisfies the maximal and minimal conditions (Exercise 15), G has a Jordan-Hölder series (consider, for a subgroup H of G, a maximal element of the set of normal stable subgroups of H distinct from H).

¶ 17. Let G be a group with operators; a composition series (G_i) of G is called *normal* if all the G_i are normal stable subgroups *of* G; a normal series Σ is called *principal* if it is strictly decreasing and there exists no normal series distinct from Σ, finer than Σ and strictly decreasing.

(*a*) If (G_i) and (H_j) are two normal series of G, show that there exist two equivalent normal series finer respectively than (G_i) and (H_j) (apply Schreier's Theorem by considering a suitable domain of operators for G). Give a second proof of this proposition by "inserting" the subgroups $G'_{ij} = G_i \cap (G_{i+1}H_j)$ and $H'_{ji} = H_j \cap (H_{j+1}G_i)$ in the sequences (G_i) and (H_j) respectively.

(*b*) If G has a principal series, any two principal series of G are equivalent; for every strictly decreasing normal series Σ there exists a principal series finer than Σ. Deduce that, for G to possess a principal series, it is necessary and sufficient that the set of normal stable subgroups of G satisfy the maximal and minimal conditions.

(*c*) If G is a group without operators and possesses a principal series and if the set of subgroups of G satisfies the minimal condition, every quotient group

G_i/G_{i+1} is the direct product of a finite number of simple subgroups isomorphic to one another (use Exercise 15).

¶ 18. (*a*) Let G be a group with operators generated by the union of a family $(H_\iota)_{\iota \in I}$ of *simple* normal stable subgroups of G. Show that there exists a subset J of I such that G is the restricted sum of the family $(H_\iota)_{\iota \in J}$ (apply Zorn's Lemma to the set of subsets K of I such that the subgroup generated by the union of the H_ι for $\iota \in K$ is the restricted sum of this family).

(*b*) Let A be a normal stable subgroup of G. Show that there exists a subset J_A of I such that G is the restricted sum of A and the H_ι for $\iota \in J_A$. Deduce that A is isomorphic to the restricted sum of a subfamily of the H_ι.

¶ 19. (*a*) Let G be a group such that every normal subgroup of G distinct from G is contained in a direct factor of G distinct from G. Show that G is the restricted sum of a family of simple subgroups. (Let K be the subgroup of G generated by the union of all the simple subgroups of G. Suppose $K \neq G$; if $x \in G - K$, consider a normal subgroup M of G which is maximal amongst those containing K and not containing x. If $G = M' \times S$, where M' contains M and $S \neq \{e\}$, show that $x \notin M'$ and therefore $M' = M$; on the other hand, if N is a normal subgroup of S, show that $M \times N = G$ and conclude that S is simple, whence a contradiction. Conclude with the aid of Exercise 18.) Obtain the converse (cf. Exercise 18).

(*b*) In order that there exist in a group G a family (N_α) of normal simple subgroups such that the G/N_α are simple and $\bigcap_\alpha N_\alpha = \{e\}$, it is necessary and sufficient that, for every normal subgroup $N \neq \{e\}$ of G, there exist a normal subgroup $N' \neq G$ such that $NN' = G$. (To see that the condition is necessary, consider an $x \in N$ distinct from e and an N_α not containing x. To see that the condition is sufficient, for all $x \neq e$ in G, consider the normal subgroup N generated by x and a normal subgroup $N' \neq G$ such that $NN' = G$. Show that if a normal subgroup M of G is maximal amongst the normal subgroups of G containing N' and not containing x, it is maximal in the set of all normal subgroups $\neq G$ and deduce the conclusion.)

20. (*a*) Let G be a group, the direct product of two subgroups A and B. Let C_A be the centre of A and let N be a normal subgroup of G such that $N \cap A = \{1\}$. Show that $N \subset C_A \times B$.

(*b*) Let $(S_\iota)_{\iota \in I}$ be a finite family of non-commutative simple groups. Show that every normal subgroup of $\prod_{\iota \in I} S_\iota$ is equal to one of the partial products $\prod_{\iota \in J} S_\iota$ where J is a subset of I.

21. Let G be a group of finite length. Show the equivalence of the following properties:

(*a*) G is a product of simple groups isomorphic to one another.

(*b*) No subgroup of G distinct from $\{1\}$ and G is stable under all the automorphisms of G.

¶ 22. Let E be a quasi-group (§ 3, Exercise 6) written multiplicatively, possessing an idempotent e and such that $(xy)(zt) = (xz)(yt)$ for all x, y, z, t. Let $u(x)$ denote the element of E such that $u(x)e = x$ and $v(x)$ the element of E such that $ev(x) = x$; show that the law of composition $(x, y) \mapsto u(x)v(y)$ is a commutative group law on E under which e is identity element, that the mappings $x \mapsto xe$ and $y \mapsto ey$ are permutable endomorphisms of this group structure and that $xy = u(xe)v(ey)$ (start by establishing the identities $e(xy) = (ex)(ey)$, $(xy)e = (xe)(ye)$, $e(xe) = (ex)e$; then note that the relations $x = y$, $ex = ey$ and $xe = ye$ are equivalent). Obtain the converse.

¶ 23. Consider on a set E an internal law, *not everywhere defined*, written multiplicatively and satisfying the following conditions:

(1) if one of the compositions $(xy)z$, $x(yz)$ is defined, so is the other and they are equal;

(2) if x, x', y are such that xy and $x'y$ (resp. yx and yx') are defined and equal, then $x = x'$;

(3) for all $x \in E$, there exist three elements e_x, e'_x and x^{-1} such that $e_x x = x$, $xe'_x = x$, $x^{-1}x = e'_x$; e_x is called the left unit of x, e'_x the right unit of x, x^{-1} the inverse of x (by an abuse of language).

(*a*) Show that the compositions xx^{-1}, $x^{-1}e_x$, $e'_x x^{-1}$, $e_x e_x$, $e'_x e'_x$ are defined and that $xx^{-1} = e_x$, $x^{-1}e_x = e'_x x^{-1} = x^{-1}$, $e_x e_x = e_x$, $e'_x e'_x = e'_x$.

(*b*) Every idempotent e of E (§ 1, no. 4) is a left unit for all the x such that ex is defined and right unit for all the y such that ye is defined.

(*c*) For the composition xy to be defined, it is necessary and sufficient that the right unit of x be the same as the left unit of y (to see that the condition is sufficient, use the relation $e_y = yy^{-1}$); if $xy = z$, then $x^{-1}z = y$, $zy^{-1} = x$, $y^{-1}x^{-1} = z^{-1}$, $z^{-1}x = y^{-1}$, $yz^{-1} = x^{-1}$ (the compositions appearing on the left-hand sides of these relations being defined).

(*d*) For any two idempotents e, e' of E, let $G_{e,e'}$ denote the set of $x \in E$ such that e is left unit and e' is right unit of x. Show that, if $G_{e,e}$ is non-empty, it is a group under the law induced by that on E.

E is called a *groupoid* if it also satisfies the following condition:

(4) for all idempotents e, e', $G_{e,e'}$ is non-empty.

(*e*) In a groupoid E, if $a \in G_{e,e'}$, show that $x \mapsto xa$ is a bijective mapping of $G_{e,e}$ onto $G_{e,e'}$, $y \mapsto ay$ a bijective mapping of $G_{e',e'}$ onto $G_{e,e'}$ and $x \mapsto a^{-1}xa$ an isomorphism of the group $G_{e,e}$ onto the group $G_{e',e'}$.

(*f*) Show that the law defined in § 1, Exercise 4(*b*) determines a groupoid structure on the set Ψ, if all the sets of the family $\mathfrak{F}$ are equipotent.

24. (*a*) Given any set E, consider in the set E × E the law of composition, not everywhere defined, written multiplicatively, under which the composition

of (x, y) and (y', z) is only defined if $y' = y$ and in this case has the value (x, z). Show that $E \times E$, with this law of composition, is a groupoid (Exercise 23).

(*b*) Let $(x, y) \equiv (x', y')$ (R) be an equivalence relation compatible with the above composition (that is such that $(x, y) \equiv (x', y')$ and $(y, z) \equiv (y', z')$ imply $(x, z) \equiv (x', z')$); suppose further that the relation R satisfies the following condition: for all x, y, z there exists one and only one $t \in E$ such that $(x, y) \equiv (z, t)$ and there exists a $u \in E$ such that $(x, y) \equiv (u, z)$.

Show that under these conditions the quotient structure by R of the groupoid structure on $E \times E$ is a group structure (prove first that the quotient law is everywhere defined, since, if $\dot{x}, \dot{y}, \dot{z}$ are three classes such that $\dot{x}\dot{y} = \dot{x}\dot{z}$, then $\dot{y} = \dot{z}$; establish finally that, for all $x \in E$, $y \in E$, $(x, x) \equiv (y, y)$).

(*c*) Let G be the group obtained by giving the quotient of $E \times E$ by R the above structure. Let a be any element of E; if, for all $x \in E$, $f_a(x)$ denotes the class mod. R of (a, x), show that f_a is a bijective mapping of E onto G and that the relation $(x, y) \equiv (x', y')$ is equivalent to the relation

$$f_a(x)(f_a(x'))^{-1} = f_a(y)(f_a(y'))^{-1}.$$

For G to be commutative, it is necessary and sufficient that the relation $(x, y) \equiv (x', y')$ imply $(x, x') \equiv (y, y')$.

¶ 25. Let E be a set and f a mapping of E^m into E; we write

$$f(x_1, x_2, \ldots, x_m) = x_1 x_2 \ldots x_m;$$

suppose that f satisfies the following conditions:

(1) $$(x_1 x_2 \ldots x_m) x_{m+1} \ldots x_{2m-1} = x_1 (x_2 x_3 \ldots x_{m+1}) x_{m+2} \ldots x_{2m-1}$$

identically;

(2) For all $a_1, a_2, \ldots, a_{m-1}$, the mappings

$$x \mapsto x a_1 a_2 \ldots a_{m-1}$$

$$x \mapsto a_1 a_2 \ldots a_{m-1} x$$

are bijections of E onto E.

(*a*) Show that identically

$$(x_1 x_2 \ldots x_m) x_{m+1} \ldots x_{2m-1} = x_1 x_2 \ldots x_i (x_{i+1} \ldots x_{i+m}) x_{i+m+1} \ldots x_{2m-1}$$

for every index i such that $1 \leqslant i \leqslant m - 1$ (argue by induction on i, by considering the element

$$((x_1 x_2 \ldots x_m) x_{m+1} \ldots x_{2m-1}) a_1 a_2 \ldots a_{m-1}).$$

(*b*) For all $a_1, a_2, \ldots, a_{m-2}$, there exists $u \in E$ such that for all x

$$x = x a_1 a_2 \ldots a_{m-2} u = u a_1 a_2 \ldots a_{m-2} x.$$

(*c*) In the set E^k of sequences $(u_1, u_2, \ldots, u_k)$ of k elements of E ($1 \leqslant k \leqslant m - 1$) consider the equivalence relation R_k which states: for all $x_1, x_2, \ldots, x_{m-k}$, $u_1u_2\ldots u_kx_1x_2\ldots x_{m-k} = v_1v_2\ldots v_kx_1x_2\ldots x_{m-k}$; let E_k denote the quotient set E^k/R_k and G the sum set (*Set Theory*, II, § 4, no. 8) of $E_1 = E, E_2, \ldots, E_{m-1}$. Let $\alpha \in E_i$, $\beta \in E_j$; if $(u_1, u_2, \ldots, u_i)$ is a sequence of class α, $(v_1, \ldots, v_j)$ a sequence of class β, consider the sequence

$$(u_1, u_2, \ldots, u_i, v_1, \ldots, v_j)$$

in E^{i+j} if $i + j < m$, the sequence

$$(u_1, u_2, \ldots, u_{i+j-m}, (u_{i+j-m+1}\ldots u_iv_1v_2\ldots v_j))$$

in $E^{i+j-m+1}$ if $i + j \geqslant m$; show that the class of this sequence in E_{i+j} (resp. $E_{i+j-m+1}$) depends only on the classes α and β; it is denoted by $\alpha . \beta$. Show that a group law is thus defined on G, that $H = E_{m-1}$ is a normal subgroup of G and that the quotient group G/H is cyclic of order $m - 1$; prove finally that E is identical with a class (mod. H) which generates G/H and that $x_1x_2\ldots x_m$ is just the composition, in the group G, of the sequence $(x_1, x_2, \ldots, x_m)$.

¶ 26. Let G, G′ be two groups, $f: G \rightarrow G'$ a mapping such that, for two arbitrary elements x, y of G, $f(xy) = f(x)f(y)$ *or* $f(xy) = f(y)f(x)$. It is proposed to prove that f is a homomorphism of G into G′ *or* a homomorphism of G into the opposite group G'^0 (in other words, either $f(xy) = f(x)f(y)$ for *every* ordered pair (x, y) or $f(xy) = f(y)f(x)$ for *every* ordered pair (x, y)).

(*a*) Show that the set $N = \overset{-1}{f}(e')$ (e' the identity element of G′) is a normal subgroup of G and that f factors into $G \xrightarrow{p} G/N \xrightarrow{g} G'$, where g is injective. Attention may therefore be confined to the case where f is *injective*, which will be assumed throughout what follows.

(*b*) Show that if $xy = yx$ then $f(x)f(y) = f(y)f(x)$ (consider $f(x^2y)$ and $f(x^2y^2)$, expressing them in several ways).

(*c*) Show that if $f(xy) = f(x)f(y)$ then $f(yx) = f(y)f(x)$. (Show with the aid of (*b*) that attention may be confined to the case where $xy \neq yx$.)

(*d*) Show that $f(xyx) = f(x)f(y)f(x)$. (Use (*a*) or (*b*) according to whether $xy = yx$ or $xy \neq yx$; in the second case, consider $f(x^2y)$ and $f(yx^2)$.)

(*e*) Let A be the set of $x \in G$ such that $f(xy) = f(x)f(y)$ for *all* $y \in G$ and B the set of $x \in G$ such that $f(xy) = f(y)f(x)$ for *all* $y \in G$. Show that $A \neq G$, $B \neq G$ and $A \cup B = G$ is an impossible situation. (By virtue of (*b*) there would then exist x, y in A such that $f(xy) = f(x)f(y) \neq f(y)f(x)$ and u, v in B such that $f(uv) = f(v)f(u) \neq f(u)f(v)$. Deduce from this a contradiction by considering successively the two possibilities $xu \in A$, $xu \in B$; in the first case consider $f(xuv)$ and in the second $f(yxu)$.)

The rest of the proof consists of proving that $A \cup B \neq G$ is impossible, arguing by *reductio ad absurdum*. If $A \cup B \neq G$, there exist a, b, c in G such that

$$f(ab) = f(a)f(b) \neq f(b)f(a) \quad \text{and} \quad f(ac) = f(c)f(a) \neq f(a)f(c).$$

(f) Show that $f(c)f(a)f(b) = f(b)f(a)f(c)$. (Consider two cases according to whether $bc \neq cb$ or $bc = cb$. In the first case, consider $f(bac)$; in the second, use (b) and (c) and consider $f(abc)$, $f(bca)$ and $f(bac)$.)

(g) Consider $f(abac)$ and obtain a contradiction.

§ 5

1. In a finite group G, show that the number of conjugates (no. 4) of an element $a \in G$ is equal to the index of the normalizer of a and is therefore a divisor of the order of G.

¶ 2. *If G is a finite group of order n, the number of automorphisms of G is $\leqslant n^{\log n/\log 2}$ (show that there exists a generating set $\{a_1, a_2, \ldots, a_m\}$ of G such that a_i does not belong to the subgroup generated by $a_1, a_2, \ldots, a_{i-1}$ for $2 \leqslant i \leqslant m$; deduce that $2^m \leqslant n$ and that the number of automorphisms of G is $\leqslant n^m$).*

3. Let Γ be the group of automorphisms of a group G and Δ the group of inner automorphisms of G; show that Δ is a normal subgroup of Γ. For an automorphism σ of G to be permutable with all the inner automorphisms of G, it is necessary and sufficient that, for all $x \in G$, $x^{-1}\sigma(x)$ belong to the centre of G; deduce that, if the centre of G is reduced to the identity element, so is the centralizer of Δ in Γ.

¶ 4. (a) Let G be a non-commutative simple group, Γ the automorphism group of G and Δ the group of inner automorphisms of G (isomorphic to G). If s is an automorphism of the group Γ, show that $s(\Delta) = \Delta$ (using Exercise 3 above and § 4, no. 9, Proposition 15, note that the intersection $\Delta \cap s(\Delta)$ cannot consist only of the identity element of Γ).

(b) Show that the only automorphism of Γ leaving invariant each of the elements of Δ is the identity element (use the fact that this automorphism leaves invariant α and $\sigma\alpha\sigma^{-1}$ for all $\alpha \in \Delta$ and $\sigma \in \Gamma$ and use Exercise 3).

(c) Let s be an automorphism of Γ, ϕ the isomorphism $x \mapsto \mathrm{Int}(x)$ of G onto Δ, ψ the inverse isomorphism and σ the automorphism $\psi \circ s \circ \phi$ of G; show that the automorphism $\xi \mapsto \sigma^{-1}s(\xi)\sigma$ of Γ is the identity automorphism (use (b), noting that, for all $x \in G$, $s(\mathrm{Int}(x)) = \mathrm{Int}(\sigma(x))$). Deduce that every automorphism of Γ is an inner automorphism.

5. Let G be a group.

(a) If H is a subgroup of G of finite index n, show that the intersection N of the conjugates of H is of index a divisor of $n!$ (note that G/N is isomorphic to a subgroup of $\mathfrak{S}_n$).

(b) G is called *residually finite*, if the intersection of its subgroups of finite index is $\{e\}$. Show that this is equivalent to saying that G is isomorphic to a subgroup of a product of finite groups.

(*c*) Suppose that G can be generated by a finite set. Show that, for every integer n, the set P_n of subgroups of G of index n is finite.

(*d*) Under the hypothesis of (*c*), consider an endomorphism f of G which is surjective. Show that, for every integer n, the mapping $H \mapsto \overset{-1}{f}(H)$ is a bijection of P_n onto itself (observe that it is an injection). Deduce that the kernel of f is contained in every subgroup of G of finite index. In particular, if G is residually finite, f is bijective.

6. Let H be a subgroup of finite index in a group G. Suppose that G is the union of the conjugates of H. Show that H = G. (Reduce it to the case where G is finite by means of Exercise 5. In the latter case, note that:

$$\mathrm{Card}\Big(\bigcup_{x \in G} xHx^{-1}\Big) \leqslant 1 + (\mathrm{Card}(H) - 1)\mathrm{Card}(G/H).)$$

7. Let p be a prime number.

(*a*) Show that $n^p \equiv n \pmod{p}$ for all $n \in \mathbf{Z}$. (Reduce it to the case where n is positive and argue by induction on n using the binomial formula.)

(*b*) Let x, y be two elements of a group G. Suppose that

$$yxy^{-1} = x^n, \quad \text{with } n \in \mathbf{Z}, \quad \text{and that} \quad x^p = 1.$$

Show that $y^pxy^{-p} = x^n$; deduce that y^{p-1} commutes with x.

(*c*) Let G be a group all of whose elements $\neq 1$ are of order p and are conjugate to one another. Show that Card(G) is equal to 1 or 2. (Use (*b*) to prove that G is commutative.)

8. Let A and B be subgroups of a finite group G, N_A and N_B the normalizers of A and B, ν_A and ν_B the indices of N_A and N_B in G, r_A the number of conjugates of A which contain B and r_B the number of conjugates of B which contain A; show that $\nu_A r_B = \nu_B r_A$.

¶ 9. Let G be a group of permutations of a finite set E; for all $s \in G$, let $\chi(s)$ denote the cardinal of the set of fixed points of s.

(*a*) Show that $\sum_{s \in G} \chi(s) = Nt$, where N = Card(G) and t is the number o orbits of G in E (evaluate in two ways the number of ordered pairs $(s, x) \in G \times E$ such that $s(x) = x$).

(*b*) Suppose that, for all $a \in E$, the stabilizer H_a of a is not reduced to e. Show that, if $\chi(s)$ is independent of s for $s \neq e$ and is equal to an integer k, then

$$k \leqslant t \leqslant 2k.$$

(Note that $\mathrm{Card}(E) \leqslant Nk$.)

In the particular case where $k = 2$, find the possible orders of the groups H_a; show that, if $t = 3$, the order of H_a cannot be $\geqslant 3$ for elements of two of the three orbits unless N has one of the values 12, 24 or 60.

(*c*) Suppose that G operates transitively on E and let H denote the stabilizer of an element of E; show that $\sum_{s \in G} \chi(s)^2$ is equal to $N . t_H$, where t_H is the number of orbits of H in E. (For every (s, u), write $\chi(s, u) = \chi(s)$. Evaluate in two ways the sum $\sum_{(s, u) \in R} \chi(s, u)$, where R is the set of ordered pairs (s, u) such that $usu^{-1} \in H$.)

10. (*a*) Show that the elements $\tau_{12}\tau_{34}, \tau_{13}\tau_{24}, \tau_{14}\tau_{23}$ of the alternating group $\mathfrak{A}_4$ form with the identity element a commutative subgroup H of $\mathfrak{A}_4$. Show that H is normal in $\mathfrak{A}_4$ and in $\mathfrak{S}_4$ and that $\mathfrak{A}_4/H$ is cyclic of order 3.

(*b*) Show that the centralizer of H in $\mathfrak{S}_4$ is equal to H. Deduce that the mapping $s \mapsto (h \mapsto shs^{-1})$ of $\mathfrak{S}_4$ into Aut(H) defines when passing to the quotient an isomorphism of $\mathfrak{S}_4/H$ onto Aut(H) and that the latter group is isomorphic to $\mathfrak{S}_3$.

(*c*) Let K be a subgroup of H of order 2. Show that K is not normal in $\mathfrak{A}_4$.

11. Let E be a finite set, $\zeta \in \mathfrak{S}_E$ a cycle of support E and τ a transposition. Show that $\mathfrak{S}_E$ is generated by ζ and τ.

¶ 12. (*a*) Show that every permutation $\sigma \in \mathfrak{A}_n$ is a product of cycles of order 3 (which are not in general component cycles of σ). (Prove it for a product of two transpositions and use § 5, no. 7, Proposition 9.)

(*b*) If $a_1, \ldots, a_p$ are p distinct elements of $[1, n]$, let $(a_1 a_2 \ldots a_p)$ denote the cycle of order p whose support is $\{a_1, a_2, \ldots, a_p\}$ and which maps a_i to a_{i+1} for $1 \leqslant i \leqslant p-1$ and a_p to a_1. Show that $\mathfrak{A}_n$ is generated by the $n-2$ cycles $(1\,2\,3), (1\,2\,4), \ldots, (1\,2\,n)$. (Use (*a*).)

(*c*) Deduce that, if n is odd, $\mathfrak{A}_n$ is generated by $(1\,2\,3)$ and $(1\,2\,\ldots\,n)$ and, if n is even, by $(1\,2\,3)$ and $(2\,3\,\ldots\,n)$.

(*d*) Show that, if a normal subgroup of $\mathfrak{A}_n$ contains a cycle of order 3, it is identical with $\mathfrak{A}_n$ (prove that such a subgroup contains all the cycles $(1\,2\,k)$, $3 \leqslant k \leqslant n$).

¶ 13. Let G be a transitive group of permutations of a set X and H the stabilizer of an element $x \in X$. Show the equivalence of the following properties:

(*a*) Every subgroup H′ of G containing H is equal to H or G.

(*b*) Every subset Y of X such that, for all $g \in G$, gY is either contained in Y or disjoint from Y, is equal to X or consists of a single element.

(If H′ satisfies (*a*) and $Y = H'x$, then $gY = Y$ for all $g \in H'$ and $gY \cap Y = \varnothing$ for all $g \notin H'$. Conversely, if Y enjoys the property in (*b*), the set H′ of $g \in G$ such that $gY = Y$ is a subgroup of G containing H.)

A transitive permutation group G satisfying (*a*) and (*b*) and not consisting only of the identity element is called *primitive*.

14. A group of permutations Γ of a set E is called *r-ply transitive* if, for any two sequences $(a_1, a_2, \ldots, a_r)$, $(b_1, b_2, \ldots, b_r)$ of r distinct elements of E, there exists a permutation $\sigma \in \Gamma$ such that $\sigma(a_i) = b_i$ for $1 \leqslant i \leqslant r$, this property not holding for at least one ordered pair of sequences of $r + 1$ distinct elements of E.

(*a*) Show that an r-ply transitive group is primitive if $r > 1$.

(*b*) The order of a permutation group Γ operating on n objects which is r-ply transitive is of the form $n(n-1)\ldots(n-r+1)d$, where d is a divisor of $(n-r)!$ (consider a subgroup of the permutations of Γ leaving invariant r elements and calculate its index).

¶ 15. Let Γ be an r-ply transitive group of permutations of a set E with n elements; for a permutation $\sigma \in \Gamma$ distinct from the identity permutation, let $n - s$ be the number of elements of E invariant under σ. If $s > r$, show that there exists a permutation $\tau \in \Gamma$ such that $\sigma^{-1}\tau\sigma\tau^{-1}$ is distinct from the identity permutation and the number of elements of E it leaves invariant is $\geqslant n - 2(s - r + 1)$ (use the decomposition of σ into its component cycles). If $s = r$, show similarly that there exists $\tau \in \Gamma$ such that $\sigma^{-1}\tau\sigma\tau^{-1}$ is a cycle of order 3. Deduce that, if $r \geqslant 3$ and Γ does not contain the alternating group $\mathfrak{A}_n$, then $s \geqslant 2r - 2$ for every permutation in Γ (use Exercise 10). Conclude finally that, if Γ is not identical with $\mathfrak{A}_n$ or $\mathfrak{S}_n$, then $r \leqslant n/3 + 1$.

¶ 16. (*a*) Show that the alternating group $\mathfrak{A}_n$ is $(n-2)$-ply transitive.

(*b*) Show that $\mathfrak{A}_n$ is a non-commutative simple group for $n \geqslant 5$. (Use (*a*), the method of Exercise 15 and Exercise 12(*d*) to prove that $\mathfrak{A}_n$ is simple if $n > 6$; examine analogously the cases $n = 5$ and $n = 6$.)

(*c*) Show that, for $n \geqslant 5$, the only normal subgroups of $\mathfrak{S}_n$ are $\{e\}$, $\mathfrak{A}_n$ and $\mathfrak{S}_n$.

17. Let G be a transitive group of permutations of a set X; suppose G is primitive (Exercise 13). Show that every normal subgroup of G distinct from $\{e\}$ is transitive.

18. Let Γ be a group of permutations of a set E. Let A and B be two subsets of E, complements one of the other and stable under Γ. Let Γ_A and Γ_B denote the groups consisting of the restrictions of the permutations in Γ to A and B respectively and Δ_A and Δ_B the subgroups of Γ which leave invariant respectively every element of A and every element of B. Show that Δ_A and Δ_B are normal subgroups of Γ and that Γ_A is isomorphic to Γ/Δ_A and Γ_B to Γ/Δ_B; if Δ_{AB} (resp. Δ_{BA}) is the group consisting of the restrictions of the permutations in Δ_A (resp. Δ_B) to B (resp. A), show that the quotient groups Γ_A/Δ_{BA}, Γ_B/Δ_{AB} and $\Gamma/(\Delta_A\Delta_B)$ are isomorphic (use Exercise 8).

19. Let G be a group, H a subgroup of G and G/H the set of left cosets (mod. H) in G. Let $r\colon G/H \to G$ be a section associated with the canonical projection $G \to G/H$. Then for all $X \in G/H$, $X = r(X).H$. An internal law of composition is defined on G/H by setting $X \top Y = r(X)r(Y).H$.

(*a*) Show that $X \top H = X$ for all X and that under the law $\top$ every left translation is bijective. If G' is the subgroup of G generated by the set of elements $r(X)$ and $H' = H \cap G'$, the internal law defined analogously on G'/H' by the mapping r determines on this set a structure isomorphic to that determined on G/H by the law $\top$.

(*b*) For the law $\top$ to be associative, it is necessary and sufficient that H' be a normal subgroup of G' in which case the structure determined by $\top$ is isomorphic to the structure of the quotient group G'/H' (to see that the condition is sufficient, show first, with the aid of § 4, Exercise 2(*a*), that if $\top$ is associative it determines on G/H a group structure; denoting by K the largest normal subgroup of G' contained in H', show then, using the associativity condition on $\top$, that $(r(X \top Y))^{-1}r(X)r(Y) \in K$ for all X, Y; deduce that the mapping $X \mapsto r(X).K$ is an isomorphism of the group G/H (with the law $\top$) onto the quotient group G'/K; conclude that $H' = K$, noting that H' is a union of cosets mod. K).

(*c*) Conversely, suppose given on a set E an internal law of composition $\top$ such that every left translation is bijective and there exists $e \in E$ such that $x \top e = x$ for all $x \in E$. Let Γ be the group of permutations of E generated by the left translations γ_x and Δ the subgroup of permutations in Γ leaving e invariant; show that to every left coset X modulo Δ there corresponds one and only one element $x \in E$ such that $\gamma_x \in X$; if we write $r(X) = \gamma_x$, the mapping $x \mapsto \gamma_x$ is an isomorphism of the set E with the law $\top$ onto the set Γ/Δ with the law $(X, Y) \mapsto r(X)r(Y).\Delta$.

20. Let G be a simple group and H a subgroup of G of finite index $n > 1$.
(*a*) Show that G is finite and that its order divides $n!$ (use Exercise 5).
(*b*) Show that G is commutative if $n \leqslant 4$ (use Exercise 10).

21. Let $p(n)$ be the number of conjugacy classes of the symmetric group $\mathfrak{S}_n$. Show that $p(n)$ is equal to the number of families $(x_1, \ldots, x_n)$ of integers $\geqslant 0$ such that $\sum_{i=1}^{n} i.x_i = n$.

Deduce the identity $\sum_{n=0}^{\infty} p(n)T^n = \prod_{m=1}^{\infty} \frac{1}{1 - T^m}$.

22. Let $A = \mathbf{Z}/2\mathbf{Z}$, $B = \mathfrak{S}_3$ and $G = A \times B$. If s is an element of B of order 2, let ϕ be the endomorphism of G whose kernel is equal to B and whose image is $\{1, s\}$. Show that the centre of G is not stable under ϕ.

¶ 23. (*a*) Let s be an element of order 2 in the group $\mathfrak{S}_n$ which is a product of k transpositions $t_1, \ldots, t_k$ with mutually disjoint supports. Show that the centralizer C_s of s in $\mathfrak{S}_n$ is isomorphic to $\mathfrak{S}_{n-2k} \times A$, where A admits a normal subgroup of order 2^k such that A/B is isomorphic to $\mathfrak{S}_k$ (observe that every element of C_s permutes the fixed points of s and also the supports of $t_1, \ldots, t_k$).

(*b*) Show that, if $n = 5$ or $n \geqslant 7$, every element of order 2 in $\mathfrak{S}_n$ whose centralizer is isomorphic to that of a transposition is a transposition (compare the Jordan-Hölder quotients of these centralizers using Exercise 16). Deduce that every automorphism of $\mathfrak{S}_n$ permutes the transpositions.

(*c*) If a, b are distinct elements of $[1, n]$, let F_a (resp. F_{ab}) denote the stabilizer of a (resp. the fixer of $\{a, b\}$) in $\mathfrak{S}_n$ and $(a\ b)$ the transposition $\tau_{a,b}$. Let a, b, c, d be distinct elements of $[1, n]$. Show that, if $n \geqslant 5$, the group generated by F_{ab} and F_{cd} is $\mathfrak{S}_n$ (show that this group contains all the transpositions; for this use the fact that, if $x \in [1, n] - \{a, b, c. d\}$, then $(a\ c) = (c\ x)(a\ x)(c\ x)$). Show that the group generated by F_{ab} and F_{ac} is F_a (use the same equation as in the above case).

(*d*) Show that an automorphism α of $\mathfrak{S}_n$ such that $\alpha(F_a) = F_a$ for all $a \in [1, n]$ is the identity (show that α maps every transposition to itself). Show that, if there exists $a \in [1, n]$ such that $\alpha(F_a) = F_a$, α is an inner automorphism (show that there exists an inner automorphism β such that $\beta \circ \alpha$ maps each of the groups F_b, $b \in [1, n]$, to itself).

(*e*) Let a, b be two distinct elements of $[1, n]$ and let C_{ab} be the centralizer of (a, b) in $\mathfrak{S}_n$. Show that, if $n \geqslant 5$, the derived group of C_{ab} is of index 4 in C_{ab} (use the fact that the derived group of $\mathfrak{S}_{n-2}$ is $\mathfrak{A}_{n-2}$). Show that F_{ab} is the unique subgroup of C_{ab} of index 2 which does not contain $(a\ b)$ and is not contained in $\mathfrak{A}_n$.

(*f*) Show that, if $n \geqslant 5$, an automorphism α of $\mathfrak{S}_n$ which leaves fixed a transposition $(a\ b)$ is inner. (Show by means of (*e*) that $\alpha(F_{ab}) = F_{ab}$; show equally that, if $c \notin \{a, b\}$, $\alpha(F_{ac})$ is either of the form F_{ax}, $x \notin \{a, b\}$, or of the form F_{bx}, $x \notin \{a, b\}$; in the first case conclude that $\alpha(F_a) = F_a$ and apply (*d*); reduce the second case to the first by multiplying α by the inner automorphism of $\mathfrak{S}_n$ defined by $(a\ b)$.)

(*g*) Deduce from (*b*) and (*f*) that every automorphism of $\mathfrak{S}_n$ is inner if $n = 5$ or $n \geqslant 7$.

24. (*a*) Show that there exists a subgroup H of $\mathfrak{S}_6$ which is isomorphic to $\mathfrak{S}_5$ and leaves fixed no element of $[1, 6]$ (let $\mathfrak{S}_5$ operate by inner automorphisms on the set of its subgroups of order 5, which has 6 elements).

(*b*) Show that there exists an automorphism σ of $\mathfrak{S}_6$ which maps $(1\ 2)$ to $(1\ 2)(3\ 4)(5\ 6)$. (Let $\mathfrak{S}_6$ operate on $\mathfrak{S}_6/H$, where H is chosen as above.)

(*c*) Show that the group of inner automorphisms of $\mathfrak{S}_6$ is of index 2 in the group of all automorphism of $\mathfrak{S}_6$. (Let α be an automorphism of $\mathfrak{S}_6$; show that α maps a transposition either to a transposition or to a conjugate of $(1\ 2)(3\ 4)(5\ 6)$. Using Exercise 23, show that in the first (resp. second) case α (resp. $\alpha \circ \sigma$) is an inner automorphism.)

25. Let n be an integer $\geqslant 5$. Show that the subgroups of $\mathfrak{S}_n$ of order $(n-1)!$ are isomorphic to $\mathfrak{S}_{n-1}$ and form a single conjugacy class (resp. two conjugacy classes) if $n \neq 6$ (resp. if $n = 6$). (Let H be such a subgroup and $x_1, \ldots, x_n$ the

elements of $\mathfrak{S}_n/H$. Show that the action of $\mathfrak{S}_n$ on the x_i defines an automorphism of $\mathfrak{S}_n$ and apply Exercises 23 and 24.)

¶ 26. (*a*) Let G be a group operating on two finite sets E_1 and E_2. If $s \in G$, let s_1 (resp. s_2) denote the permutation of E_1 (resp. E_2) defined by s and E_1^s (resp. E_2^s) the set of elements invariant under s_1 (resp. s_2).

Show the equivalence of the following properties:

(i) For all $s \in G$, $\mathrm{Card}(E_1^s) = \mathrm{Card}(E_2^s)$.

(ii) For all $s \in G$, the orders of the component cycles of s_1 (cf. no. 7, Proposition 7) are the same (to within a permutation) as those of s_2.

(iii) For all $s \in G$, there exists a bijection $f_s\colon E_1 \to E_2$ such that $s_2 \circ f_s = f_s \circ s_1$.

If these properties hold, the G-sets E_1 and E_2 are said to be weakly equivalent.

(*b*) Give an example of two weakly equivalent G-sets which are not isomorphic (take G to be a non-cyclic group of order 4).

(*c*) Let H_1 and H_2 be two subgroups of a finite group G. Show the equivalence of the following properties:

(1) For every conjugacy class C of G,

$$\mathrm{Card}(C \cap H_1) = \mathrm{Card}(C \cap H_2).$$

(2) The G-sets G/H_1 and G/H_2 are weakly equivalent.

(*d*) Show that, if H_1 and H_2 satisfy properties (1) and (2) above and H_1 is normal in G, then $H_2 = H_1$.

(*e*) Let $G = \mathfrak{A}_6$ and let

$$H_1 = \{e, (1\,2)(3\,4), (1\,3)(2\,4), (1\,4)(2\,3)\}$$
$$H_2 = \{e, (1\,2)(3\,4), (1\,2)(5\,6), (3\,4)(5\,6)\}.$$

Show that H_1 and H_2 satisfy properties (1) and (2) and that H_1 and H_2 are not conjugate in $\mathfrak{A}_6$ (nor in $\mathfrak{S}_6$).

27. Let Y be a subset of a set X and let A be the fixer of Y in the permutation group $\mathfrak{S}_X$ of X. Let M be the submonoid of $\mathfrak{S}_X$ consisting of the elements s such that $sAs^{-1} \subset A$. Show that M is a subgroup of $\mathfrak{S}_X$ if and only if one of the sets Y and X − Y is finite.

28. Let G be a group, A and B two subgroups of G and ϕ an isomorphism of A onto B. Show that there exists a group G_1 containing G such that ϕ is the restriction of an *inner* automorphism of G_1 (use the permutation group of the set G). Show that, if G is finite, G_1 can be chosen to be finite.

29. Let X be an infinite set. Let G be the subgroup of $\mathfrak{S}_X$ consisting of the permutations σ which enjoy the following property: there exists a finite subset Y_σ of X such that $\sigma x = x$ if $x \in X - Y_\sigma$ and the restriction of σ to Y_σ is even.

(*a*) Show that G is a non-commutative simple group. Show that every subset of G which generates G is equipotent to X.

(*b*) Show that every finite group is isomorphic to a subgroup of G (use the fact that $\mathfrak{S}_n$ is isomorphic to a subgroup of $\mathfrak{A}_{n+2}$).

§ 6

1. Let G be a group and H a normal subgroup of G such that G/H is cyclic of finite order n. Let x be an element of G whose image $\bar{x}$ in G/H generates G/H. Let ϕ be the automorphism $h \mapsto xhx^{-1}$ of H and let $y = x^n$; then $y \in \mathrm{H}$. Show that $\phi(y) = y$ and that ϕ^n is the inner automorphism of H defined by y.

Let τ be the operation of $\mathbf{Z}$ on H defined by $(m, h) \mapsto \phi^m(h)$ and let $\mathrm{E} = \mathrm{H} \times_\tau \mathbf{Z}$ be the corresponding semi-direct product. Show that the element (y^{-1}, n) of E generates a central subgroup C_y of E and that the quotient E/C_y is isomorphic to G.

2. Show that every central extension of $\mathbf{Z}$ is trivial.

3. Define extensions (cf. § 5, Exercise 10)

$$\mathbf{Z}/2\mathbf{Z} \times \mathbf{Z}/2\mathbf{Z} \to \mathfrak{A}_4 \to \mathfrak{A}_3$$

$$\mathbf{Z}/2\mathbf{Z} \times \mathbf{Z}/2\mathbf{Z} \to \mathfrak{S}_4 \to \mathfrak{S}_3.$$

Show that these extensions are non-trivial, that the first admits a section and that the second does not.

¶ 4. (*a*) Let G be a finite group of order mn such that there exists a normal subgroup H of G which is cyclic of order m, the quotient group G/H being cyclic of order n. Show that G is generated by two elements a, b such that $a^m = e$, $b^n = a^r$, $bab^{-1} = a^s$, where r and s are two integers such that $r(s-1)$ and $s^n - 1$ are multiples of m (take a to be an element generating H, b an element of a coset generating G/H; express the elements $b^h a^k b^{-h}$ as powers of a and apply this in particular to the cases $h = n, k = 1$ and $h = 1, k = r$).

(*b*) *Conversely, let $\mathrm{G}(m, n, r, s)$ be the group defined by the presentation

$$\langle a, b;\ a^m = e,\ b^n = a^r,\ bab^{-1} = a^s \rangle,$$

where m and n are two integers $\geqslant 0$ and r and s arbitrary integers (cf. § 7, no. 6); show that, if m, $r(s-1)$ and $s^n - 1$ are not all zero, $\mathrm{G}(m, n, r, s)$ is a finite group of order qn, where q is the greatest common divisor of m, $|r(s-1)|$ and $|s^n - 1|$; in this group, the subgroup H generated by a is a normal subgroup of order q and G/H is a cyclic group of order n (prove that every element of $\mathrm{G}(m, n, r, s)$ can be written in the form $a^x b^y$, where x and y are two integers such that $0 \leqslant x \leqslant q - 1$, $0 \leqslant y \leqslant n - 1$, and $\mathrm{G}(m, n, r, s)$ is isomorphic to the group consisting of the ordered pairs (x, y) of integers subjected to the above conditions, with law of composition

$$(x, y).(x', y') = \begin{cases} (x + x's^y, y + y') & \text{if } y + y' \leqslant n - 1 \\ (x + x's^y + r, y + y' - n) & \text{if } y + y' \geqslant n \end{cases}$$

the first coordinate of the right-hand side being a sum modulo q). Examine the cases where $m = r(s - 1) = s^n - 1 = 0$.

The group $G(n, 2, 0, -1)$ is called the *dihedral group* of order $2n$ and is denoted by $\mathbf{D}_n$; the group $G(4, 2, 2, -1)$ is a group of order 8 called the *quaternionic group* and denoted by $\mathfrak{Q}$. Show that in $\mathfrak{Q}$ every subgroup is normal and that the intersection of the subgroups distinct from $\{e\}$ is a subgroup distinct from $\{e\}$. Prove that $\mathbf{D}_4$ is not isomorphic to $\mathfrak{Q}$.*

5. Let F be a group whose centre is $\{e\}$ and A its automorphism group. F is identified with a subgroup of A by means of the homomorphism $\text{Int}: F \rightarrow A$; let $\Gamma = A/F$.

(*a*) Show that the extension $F \rightarrow A \rightarrow \Gamma$ is trivial only if $\Gamma = \{e\}$ (note that the centralizer of F in A is $\{e\}$).

(*b*) Suppose that $\Gamma = \{e\}$. Show that every extension

$$F \rightarrow E \rightarrow G$$

of a group G by the group F is trivial.

6. Let I be a set; write $F = \mathbf{Z}$, $E = \mathbf{Z} \times (\mathbf{Z}/2\mathbf{Z})^I$ and $G = \mathbf{Z}/2\mathbf{Z} \times (\mathbf{Z}/2\mathbf{Z})^I$.

(*a*) Define a non-trivial extension $F \xrightarrow{i} E \xrightarrow{p} G$.

(*b*) Show that, if I is infinite, E is isomorphic to $F \times G$.

7. Let G and A be two groups, A being commutative. Let τ be a homomorphism of G into the automorphism group Aut(A) of A; if $g \in G$, $a \in A$, we write ${}^g a = \tau(g)(a)$. A *crossed homomorphism* of G into A is any mapping $\phi: G \rightarrow A$ such that

$$\phi(gg') = \phi(g) + {}^g\phi(g'),$$

the group A being written additively. The crossed homomorphisms of G into A form a group under addition denoted by Z(G, A).

(*a*) If $a \in A$, the mapping $g \mapsto {}^g a - a$ is denoted by θ_a. Show that $a \mapsto \theta_a$ is a homomorphism θ of A into Z(G, A). The kernel of θ is the subgroup of A consisting of the elements invariant under G. The image of θ is denoted by B(G, A).

(*b*) Let $X = A \times_\tau G$ be the semi-direct product of G by A. If ϕ is a mapping of G into A, show that $g \mapsto (\phi(g), g)$ is a *section* of X if and only if ϕ belongs to Z(G, A). For the sections corresponding to $\phi_1, \phi_2 \in Z(G, A)$ to be conjugate by an element of A, it is necessary and sufficient that $\phi_1 \equiv \phi_2$ mod. B(G, A).

(*c*) Let $F \xrightarrow{i} E \xrightarrow{p} G$ be an extension of G by a group F whose centre is equal to A. Suppose that, if $y \in E$ and $x = p(y)$, then

$$i({}^{x}f) = y.i(f).y^{-1}$$

for all $f \in A$. If $\phi \in Z(G, A)$, let u_{ϕ} be the mapping of E into E given by the formula $u_{\phi}(y) = i(\phi(p(y))).y$. Show that $\phi \mapsto u_{\phi}$ is an isomorphism of Z(G, A) onto the automorphism group Aut(E) of the extension E. This isomorphism maps B(G, A) to the group of automorphisms $\mathrm{Int}_{E}(x)$, where x runs through A.

¶ 8. With the notation of the above exercise, C(G, A) is used to denote the group of mappings of G into A. G operates on C(G, A) by the formula

$$({}^{g}\phi)(g') = {}^{g}(\phi(g^{-1}g')).$$

If σ denotes the corresponding homomorphism of G into Aut(C(G, A)), the semi-direct product $C(G, A) \times_{\sigma} G$ is denoted by E_0.

(*a*) Let $\varepsilon: A \to C(G, A)$ be the mapping which associates with each $a \in A$ the constant mapping equal to a. Show that ε is an injective homomorphism, compatible with the action of G.

(*b*) Let $A \xrightarrow{i} E \xrightarrow{p} G$ be an extension of G by A such that $i({}^{g}a) = xi(a)x^{-1}$ if $a \in A$, $x \in E$ and $g = p(x)$. Let ρ be a mapping $G \to E$ such that $p \circ \rho = \mathrm{Id}_G$. For all $x \in E$, let ϕ_x be the mapping of G into A such that

$$x\rho(p(x^{-1})g) = i(\phi_x(g))\rho(g) \quad \text{for all } g \in G.$$

Show that

$$\phi_{xy} = \phi_x + {}^{p(x)}\phi_y \quad \text{if } x, y \in E.$$

Deduce that the mapping $\Phi: E \to E_0$ defined by

$$\Phi(x) = (\phi_x, p(x))$$

is a homomorphism which makes the following diagram commutative:

$$\begin{array}{ccccc} A & \xrightarrow{i} & E & \xrightarrow{p} & G \\ \downarrow{\scriptstyle\varepsilon} & & \downarrow{\scriptstyle\Phi} & & \downarrow{\scriptstyle\mathrm{Id}_G} \\ C(G, A) & \longrightarrow & E_0 & \longrightarrow & G \end{array}.$$

(*c*) A G-*mean* on A is a homomorphism

$$m: C(G, A) \to A$$

satisfying the two following conditions:

$$(c_1) \quad m \circ \varepsilon = \mathrm{Id}_A$$

$$(c_2) \quad m({}^{g}\phi) = {}^{g}m(\phi) \quad \text{if } g \in G, \phi \in C(G, A).$$

Let m be a G-mean on A. Show that, if ϕ is a crossed homomorphism and $a = m(\phi)$, then $\phi = \theta_{-a}$ (cf. Exercise 7). In particular,

$$Z(G, A) = B(G, A).$$

Show that the mapping $(\phi, g) \mapsto (m(\phi), g)$ is a homomorphism of E_0 into $A \times_\tau G$. Deduce that every extension E of G by A satisfying the condition in (*b*) is isomorphic to $A \times_\tau G$ (use the composition $E \xrightarrow{\Phi} E_0 \to A \times_\tau G$) and hence admits a section. Show that two such sections are transformed one into the other by an inner automorphism of E defined by an element of A (use Exercise 7 and the fact that $Z(G, A) = B(G, A)$).

(*d*) Suppose that G is finite of order n and that the mapping $\alpha \mapsto n\alpha$ is an automorphism of A. For all $\phi \in C(G, A)$, let $m(\phi)$ denote the unique element of A such that

$$n.m(\phi) = \sum_{g \in G} \phi(g).$$

Show that m is a G-mean on A.

This applies in particular when A is finite of order prime to n.

¶ 9. Let $F \to E \to G$ be an extension of finite groups. Suppose no prime number divides both the order of F and the order of G.

(*a*) Suppose F is solvable. Show that there exists a section $s: G \to E$ and that two such sections are conjugate by an element of F. (Argue by induction on the solvability class of F; where F is commutative, use the preceding Exercise.)

(*b*) Show the existence of a section† $s: G \to E$ without assuming that F is solvable. (Argue by induction on the order of G. If p divides the order of F, choose a Sylow p-subgroup P of F and consider its normalizer N in E. The image of N in G is equal to G, cf. Exercise 25. If $N \neq E$, conclude by means of the induction hypothesis; if $N = E$, use (*a*) and the induction hypothesis applied to $F/P \to E/P \to G$.)

¶ 10. Let G be a solvable finite group of order mn, where m and n have no common prime factor. Show that there exists a subgroup H of G of order m and that every subgroup of G whose order divides m is contained in a conjugate of H ("*Hall's Theorem*").

(Argue by induction on the order of G. If $mn \neq 1$, choose a commutative subgroup A of G which is normal and not equal to $\{e\}$ and whose order a is a power of a prime number. Apply this induction hypothesis to G/A and to $(m/a, n)$ or $(m, n/a)$ according to whether a divides m or n. In the second case, use Exercise 9 to pass from G/A to G.)

---

† Here again it can be shown that two such sections are conjugate, cf. FEIT and THOMPSON, *Proc. Nat. Acad. Sci., U.S.A.*, **48**, 1962, 968–970.

11. Show that the simple group $\mathfrak{A}_5$ of order 60 contains no subgroup of order 15.

¶ 12. Let G be a nilpotent group and H the set of elements of G of finite order.

(*a*) Show that H is a subgroup of G.
(*b*) Show that every finite subset of H generates a finite subgroup.
(*c*) Show that two elements of H of relatively prime orders commute.

13. Let G be a nilpotent group. Show that G is finite (resp. countable) if and only if G/D(G) is.

14. Let G be a finite group. Suppose that, for all $x, y \in G$, the subgroup of G generated by $\{x, y\}$ is nilpotent. Show that G is nilpotent (apply the hypothesis to x and y of orders $p_1^{n_1}$ and $p_2^{n_2}$, where p_1 and p_2 are distinct prime numbers).

15. Let G be a group and X a subset of G generating G.

(*a*) Write $X_1 = X$; for $n \geqslant 2$, define X_n, by induction on n, as the set of commutators (x, y), $x \in X$, $y \in X_{n-1}$. Show that $X_n = \{e\}$ if and only if G is nilpotent of class $\leqslant n$. (Argue by induction on n. If $X_n = \{e\}$, show that the subgroup H of G generated by X_{n-1} is contained in the centre of G and apply the induction hypothesis to G/H.)

(*b*) Show that $C^n(G)$ is a normal subgroup of G generated by X_n. Deduce that the image of X_n in $C^n(G)/C^{n+1}(G)$ generates the group $C^n(G)/C^{n+1}(G)$.

(*c*) Take $G = \mathfrak{S}_m$ $(m \geqslant 4)$ and $X = \{s, t\}$, where s is a transposition and t a cycle of order m. Show that X_2 does not generate the group $C^2(G) = \mathfrak{A}_m$.

16. *Let V be a vector space over a commutative field k and

$$W = \bigwedge^2 V$$

(cf. III, § 7, no. 1). On $E = V \times W$ a law of composition is defined by the formula

$$(v_1, w_1) . (v_2, w_2) = (v_1 + v_2, w_1 + w_2 + v_1 \wedge v_2).$$

(*a*) Show that this law gives E a group structure, that of a central extension of V by W.

(*b*) Suppose the characteristic of k is different from 2. Show that the derived group D(E) is the set of $(0, w)$, $w \in W$; it is a group isomorphic to W. Show that, if $\dim(V) \geqslant 4$, there exist elements of D(E) which are not commutators.*

17. Let G be a finite commutative group. Show the equivalence of the following conditions:

(*a*) G is cyclic.
(*b*) Every Sylow subgroup of G is cyclic.

(*c*) For every prime number p, the number of elements $x \in G$ such that $x^p = e$ is $\leqslant p$.

18. Let n be an integer $\geqslant 1$ and let G be a commutative group written additively, such that $nx = 0$ for all $x \in G$. Let H be a subgroup of G and f a homomorphism of H into $\mathbf{Z}/n\mathbf{Z}$.

(*a*) Let $x \in G$ and let q be the order of the image of x in G/H; then $qx \in H$. Show that there exists $\alpha \in \mathbf{Z}/n\mathbf{Z}$ such that $q\alpha = f(x)$. Deduce the existence of an extension of f to the subgroup of G generated by H and x.

(*b*) Show that f extends to a homomorphism of G into $\mathbf{Z}/n\mathbf{Z}$ (use Zorn's Lemma).

19. Let G be a finite commutative group.

(*a*) Show that there exists an element $x \in G$ whose order n is a multiple of the orders of the elements of G. (Use the decomposition of G as a product of p-groups.) Show that the subgroup H generated by such an element is a direct factor of G (apply the above exercise to construct a homomorphism $G \to H$ whose restriction to H is the identity).

(*b*) Show that G is a direct product of cyclic groups. (Argue by induction on Card(G) using (*a*).) (Cf. VII, § 4, no. 6.)

20. Let G be a finite commutative group written additively. Let $x = \sum_{g \in G} g$ and let G_2 be the subgroup of G consisting of the elements g such that $2g = 0$.

(*a*) Show that $x = \sum_{g \in G_2} g$.

(*b*) Show that $x = 0$ if $\mathrm{Card}(G_2) \neq 2$. If $\mathrm{Card}(G_2) = 2$, show that x is the unique non-zero element of G_2.

(*c*) Show that $\mathrm{Card}(G_2) = 2$ if and only if the Sylow 2-subgroup of G is cyclic and $\neq 0$.

(*d*) *Let p be a prime number. Show that

$$(p-1)! \equiv -1 \pmod{p}.$$

(Apply (*b*) to the multiplicative group of the field $\mathbf{Z}/p\mathbf{Z}$.)

21. Let p and q be two prime numbers such that $p > q$.

(*a*) Show that every finite group G of order pq is an extension of $\mathbf{Z}/q\mathbf{Z}$ by $\mathbf{Z}/p\mathbf{Z}$ (note that every Sylow p-subgroup of G is normal). If further $p \not\equiv 1 \pmod{q}$, show that G is cyclic.

(*b*) Deduce that the centre of a group cannot be of index 69.

(*c*) Show that, if $p \equiv 1 \pmod{q}$, there exist a group of order pq whose centre is $\{e\}$.

(*d*) Show that every non-commutative group of order 6 is isomorphic to $\mathfrak{S}_3$.

22. Let G be a finite group of order $n > 1$ and let p be the least prime

number dividing n. Let P be a Sylow p-subgroup of G and N its normalizer. Show that, if P is cyclic, P is contained in the centre of N (show that the order of N/P is prime to the order of the automorphism group of P).

¶ 23. Let σ be an automorphism of a group G.

(*a*) Show the equivalence of the following conditions:

(i) $\sigma(x) = x$ implies $x = e$.

(ii) The mapping $x \mapsto x^{-1}\sigma(x)$ is injective.

When these conditions are satisfied, σ is said (by an abuse of language) to be *without fixed point*.

(*b*) Suppose G is finite and σ without fixed point. The mapping $x \mapsto x^{-1}\sigma(x)$ is then bijective. If H is a normal subgroup of G stable under σ, show that the automorphism of G/H defined by σ is without fixed point.

(*c*) Under the hypotheses of (*b*), let p be a prime number. Show that there exists a Sylow p-subgroup P of G which is stable under σ. (If P_0 is a Sylow p-subgroup, there exists $y \in G$ such that $\sigma(P_0) = yP_0y^{-1}$; write y^{-1} in the form $x^{-1}\sigma(x)$ and take $P = xP_0x^{-1}$.) Show that such a subgroup is unique and contains every p-subgroup of G which is stable under σ.

(*d*) Suppose further that σ is of order 2. Show that $\sigma(g) = g^{-1}$ for all $g \in G$ (write g in the form $x^{-1}\sigma(x)$ with $x \in G$). Deduce that G is commutative and of odd order.

24. Let G be a finite group, H a Sylow subgroup of G and N the normalizer of H. Let X_1, X_2 be two subsets of the centre of H and $s \in G$ such that $sX_1s^{-1} = X_2$. Show that there exists $n \in N$ such that $nxn^{-1} = sxs^{-1}$ for all $x \in X_1$ (apply the theorem on the conjugacy of Sylow subgroups to the centralizer of X_1). Deduce that two central elements of H are conjugate in G if and only if they are so in N.

25. Let $\phi: G \to G'$ be a surjective homomorphism of finite groups; let P be a Sylow p-subgroup of G.

(*a*) Let P_1 be a Sylow p-subgroup of G such that $\phi(P) = \phi(P_1)$. Show that there exists an element x of the kernel of ϕ such that $P_1 = xPx^{-1}$.

(*b*) Let N (resp. N') be the normalizer in G (resp. G') of P (resp. $\phi(P)$). Show that $\phi(N) = N'$. In particular, if the order of G' is not divisible by p, then $\phi(P) = \{e\}$ and $\phi(N) = G'$.

¶ 26. Let G be a finite group. G is called *supersolvable* if there exists a composition series $(G_i)_{0 \leqslant i \leqslant n}$ of G consisting of normal subgroups of G such that the quotients G_i/G_{i+1} are cyclic.

(*a*) Show that every subgroup, every quotient group and every finite product of supersolvable groups is supersolvable.

(*b*) Show that nilpotent $\Rightarrow$ supersolvable $\Rightarrow$ solvable and that the converse implications are false.

(*c*) Suppose G is supersolvable. Show that, if $G \neq \{1\}$, there exists a normal subgroup C of G which is cyclic of prime order. Show that the derived group D(G) of G is nilpotent. Show that every subgroup of G distinct from G which is maximal is of prime index.

(*d*) A finite group G is *bicyclic* if there exist two elements a, b of G such that, if C_a and C_b denote the cyclic subgroups generated by a and b respectively, then $G = C_a C_b = C_b C_a$. Show that every bicyclic group is supersolvable. (Reduce it to proving that there exists in G a normal cyclic subgroup $N \neq \{1\}$. Attention may be confined to the case where $C_a \cap C_b = \{1\}$; let $m \geqslant n > 1$ be the orders of a and b respectively; by considering the elements $b^{-1}a^k$, where $0 \leqslant k \leqslant m - 1$, show that there exists an integer s such that $a^s \neq 1$ and $b^{-1}a^s b \in C_a$; the set of a^s with this property is the desired cyclic subgroup N.)

27. Let S be a finite group. S is called a *minimal simple* group if S is simple and non-commutative and if every subgroup of S distinct from S is solvable.

(*a*) Show that the alternating group $\mathfrak{A}_n$ is minimal simple if and only if $n = 5$.

(*b*) Let G be a finite group. Show that, if G is not solvable, there exist two subgroups H and K of G with H normal in K such that K/H is a minimal simple group.

¶ 28. Let G be a finite group and p a prime number. An element $s \in G$ is called p-unipotent (resp. p-regular) if its order is a power of p (resp. not divisible by p).

(*a*) Let $x \in G$. Show that there exists a unique ordered pair (u, t) of elements of G satisfying the following conditions: u is p-unipotent, t is p-regular, $x = ut = tu$. (Consider first the case where G is the cyclic group generated by x.)

(*b*) Let P be a Sylow p-group of G, C its centralizer and E the set of p-regular elements of G. Show that

$$\mathrm{Card}(E) \equiv \mathrm{Card}(E \cap C) \pmod{p}.$$

Deduce that $\mathrm{Card}(E) \not\equiv 0 \pmod{p}$. (Argue by induction on Card(G) and reduce it to the case where $C = G$; then use (*a*) to show that $\mathrm{Card}(E) = (G:P)$.)

¶ 29. Let G be a finite group of even order and let H be a Sylow 2-subgroup of G. For all $s \in G$, let $\varepsilon(s)$ be the signature of the permutation $x \mapsto sx$ of G.

(*a*) Let $s \in H$. Show that $\varepsilon(s) = -1$ if and only if s generates H.

(*b*) Show that ε is surjective if and only if H is cyclic.

(*c*) Suppose that H is cyclic. Show that there exists one and only one normal subgroup D of G such that G is the semi-direct product of H and D. (Argue by induction on the order of H.) Show that the normalizer N of H in G is the direct product of H and $N \cap D$.

(*d*) Show that the order of a non-commutative simple group is either odd† or divisible by 4.

¶ 30. Let p be a prime number, r and t integers $\geqslant 0$, S a cyclic group of order p^{r+t} and T the subgroup of S of order p^t.

(*a*) If S operates on a finite set E, show that

$$\mathrm{Card}(\mathrm{E}) \equiv \mathrm{Card}(\mathrm{E}^{\mathrm{T}}) \pmod{p^{r+1}},$$

where E^{T} denotes the set of elements of E invariant under T.

(*b*) Let $m, n \geqslant 0$. Show that

$$\binom{p^{r+t}m}{p^t m} \equiv \binom{p^r m}{n} \pmod{p^{r+1}}.$$

(Let M be a set with m elements and E the set of subsets of $\mathrm{S} \times \mathrm{M}$ with $p^t n$ elements. Define an operation of S on $\mathrm{S} \times \mathrm{M}$ such that there exists a bijection of E^{T} onto the set of subsets of $(\mathrm{S}/\mathrm{T}) \times \mathrm{M}$ with n elements; apply (*a*).)

31. Let G be a finite group and S a Sylow p-subgroup of G. Let r, t be integers $\geqslant 0$ such that $\mathrm{Card}(\mathrm{S}) = p^{r+t}$. Let E be the set of subgroups of G of order p^t.

(*a*) Suppose that S is cyclic. Show that $\mathrm{Card}(\mathrm{E}) \equiv 1 \pmod{p^{r+1}}$. (Let S operate on E by conjugation and use Exercise 30.) Deduce that, if the subgroup of S of order p^t is not normal in G, there are at least $1 + p^{r+1}$ Sylow p-subgroups in G.

(*b*) Show that $\mathrm{Card}(\mathrm{E}) \equiv 1 \pmod{p}$ even if S is not cyclic. (Let G operate by translation on the set F of subsets of G with p^t elements. Show that the elements of E give distinct orbits with $(\mathrm{G}:\mathrm{S})p^r$ elements and that all the other orbits have a number of elements divisible by p^{r+1}. Apply Exercise 30.)

¶ 32. Let p be a prime number and G a p-group. Let $\mathrm{G}^* = \mathrm{G}^p\mathrm{D}(\mathrm{G})$ be the subgroup of G generated by D(G) and the x^p, $x \in \mathrm{G}$.

(*a*) Show that G^* is the intersection of the kernels of the homomorphisms of G into $\mathbf{Z}/p\mathbf{Z}$.

(*b*) Show that a subset S of G generates G if and only if the image of S in G/G^* generates G/G^*. Deduce that, if $(\mathrm{G}:\mathrm{G}^*) = p^n$, the integer n is the minimum number of elements in a subset of G which generates G; in particular, G is cyclic if and only if $n \leqslant 1$.

(*c*) Let u be an automorphism of G of order prime to p and let $\bar{u}$ be the corresponding automorphism of G/G^*. Show that $\bar{u} = 1$ implies $u = 1$.

¶ 33. Let G be a p-group and A its automorphism group.

(*a*) Let $(\mathrm{G}_i)_{0 \leqslant i \leqslant n}$ be a composition series of G such that $(\mathrm{G}_i:\mathrm{G}_{i+1}) = p$ for

† In fact this case is impossible, as has been shown by Feit and Thompson, *Pac. J. of Math.*, **13** (1963), 775–1029.

$0 \leqslant i \leqslant n-1$. Let P be the subgroup of A consisting of the automorphisms u such that, for all i and all $x \in G_i$, $u(x)x^{-1} \in G_{i+1}$. Show that, if an element $u \in P$ is of order prime to p, then $u = 1$ (argue by induction on n). Deduce that P is a p-group.

(*b*) Conversely, let P be a p-subgroup of A. Show that there exists a composition series $(G_i)_{0 \leqslant i \leqslant n}$ of G which is stable under P and such that $(G_i : G_{i+1}) = p$ for $0 \leqslant i \leqslant n-1$; show that the G_i may be chosen to be normal in G.

34. Let p be a prime number. Show that every group of order p^2 is commutative.

35. Let G be a group such that the derived group $D = D(G)$ is contained in the centre C of G.

(*a*) Show that the mapping $(x, y) \mapsto (x, y)$ induces a mapping

$$\phi : (G/C) \times (G/C) \to D$$

such that, writing these groups additively:

$$\phi(\alpha + \beta, \gamma) = \phi(\alpha, \gamma) + \phi(\beta, \gamma) \qquad \phi(\alpha, \beta) = -\phi(\beta, \alpha)$$
$$\phi(\alpha, \beta + \gamma) = \phi(\alpha, \beta) + \phi(\alpha, \gamma) \qquad \phi(\alpha, \alpha) = 0$$

for all $\alpha, \beta, \gamma \in G/C$.

(*b*) Show that, for every integer n,

$$(yx)^n = y^n x^n (x, y)^{n(n-1)/2}$$

for $x, y \in G$. Deduce that if n is odd and $d^n = 1$ for all $d \in D$, the mapping $x \mapsto x^n$ induces a homomorphism $\theta : G/D \to C$.

¶ 36. Let p be a prime number and G a non-commutative group of order p^3.

(*a*) Show that, with the notation of the preceding exercise, $C = D$ and that G/D is isomorphic to the product of two groups of order p.

(*b*) Suppose that p is odd and that the homomorphism θ of the preceding exercise is non-zero. Show that G is the semi-direct product of a group of order p by a cyclic group of order p^2. Show that there exist elements x, y generating G such that

$$x^{p^2} = 1, \qquad y^p = 1, \quad (x, y) = x^p,$$

that G is characterized up to isomorphism by this property (and by the fact that it is of order p^3) and that such a group G exists.

(*c*) For $p = 2$, consider the same question as in (*b*) with the hypothesis $\theta \neq 0$ replaced by the hypothesis that G contains a non-central element of order 2. Show that such a group is isomorphic to the dihedral group $\mathbf{D}_4$ (cf. Exercise 4) and also to a Sylow 2-subgroup of $\mathfrak{S}_4$.

(*d*) Suppose that p is odd and that the homomorphism θ of the preceding exercise is zero. Show that there are elements x, y, z generating G such that

$$(x, y) = z; \qquad x^p = y^p = z^p = (x, z) = (y, z) = 1,$$

that G is characterized by this property and that such a group exists.

*Show that G is isomorphic to the multiplicative group of matrices of the form

$$\begin{pmatrix} 1 & a & b \\ 0 & 1 & c \\ 0 & 0 & 1 \end{pmatrix}$$

with coefficients a, b, c in the field with p elements.*

(*e*) Suppose that $p = 2$ and that no non-central element of G is of order 2 (hence that they are all of order 4). Show that G is generated by elements x, y such that

$$x^2 = y^2 = (x, y); \qquad (x, y)^2 = 1,$$

that G is characterized by this property and is isomorphic to the quaternionic group (cf. Exercise 4).

(*f*) *Is the group of matrices

$$\begin{pmatrix} 1 & a & b \\ 0 & 1 & c \\ 0 & 0 & 1 \end{pmatrix},$$

where a, b, c run through the field with 2 elements, of type (*c*) or of type (*e*)?*

37. (*a*) Let G be a finite group, H a normal subgroup of G, p a prime number not dividing the order of H and P a Sylow p-subgroup of G. Show that HP is the semi-direct product of P by H.

(*b*) We say that a finite group G has property (ST) if there exists a numbering $p_1, \ldots, p_s$ of distinct prime numbers dividing the order

$$n = p_1^{\alpha_1} \ldots p_s^{\alpha_s}$$

of G and Sylow p_i-subgroups $P_1, \ldots, P_s$ of G such that, for $1 \leqslant i \leqslant s$, the set $G_i = P_1 P_2 \ldots P_i$ is a normal subgroup of G. Show that, if this is so, G_i does not depend on the choice of the P_i; it is the unique subgroup of G of order $p_1^{\alpha_1} \ldots p_i^{\alpha_i}$; further, G_i is the semi-direct product of P_i by G_{i-1}.

(*c*) Show that a finite group G has property (ST) if and only if there exists a Sylow subgroup P of G which is normal in G and such that G/P has property (ST).

(*d*) Show that every subgroup, every quotient group and every central extension of a group with property (ST) has property (ST).

¶ 38. Let G be a finite group of order $n = p^a m$, where p is a prime number not dividing m, and let X be the set of Sylow p-subgroups of G. Let $P \in X$, let N be the normalizer of P and $r = \text{Card}(X)$.

(*a*) Show that $r = (G:N)$. Deduce that r belongs to the set R of positive divisors of m which are congruent to 1 (mod p). In particular, if R is $\{1\}$, P is normal in G.

(*b*) If $a = 1$, show that the number of elements of G of order p is $r(p - 1)$. If further $N = P$, show that the elements of G of order different from p are equal in number to m; if further m is a power of a prime number q, deduce that a Sylow q-subgroup Q of G contains all the elements of G of order different from p and hence is normal in G.

(*c*) Let H be a group of order 210. Show that H contains a normal subgroup G of order 105 (cf. Exercise 29). Let $p = 3$ (resp. 5, 7). Show that, if G has no normal Sylow p-subgroup, the group G contains at least 14 (resp. 84, 90) elements of order p. Conclude that H and G have property (ST) (cf. the preceding exercise).

(*d*) Show that every group of order $n \leqslant 100$ has property (ST), except possibly if $n = 24, 36, 48, 60, 72, 96$ (same method as in (*c*)).

(*e*) Show that the group $\mathfrak{S}_4$, of order 24, does not have property (ST). Construct analogous examples for $n = 48, 60, 72, 96$. (For $n = 36$, see the following exercise.)

¶ 39. Let G be a finite group, $n = \text{Card}(G)$, p a prime number, X the set of Sylow p-subgroups of G and $r = \text{Card}(X)$. Then G operates transitively on X by conjugation; let $\phi: G \to \mathfrak{S}_X \approx \mathfrak{S}_r$ be the corresponding homomorphism ($A \approx B$ denotes the relation "A is isomorphic to B") and K its kernel.

(*a*) Show that K is the intersection of the normalizers of the $P \in X$ and that the Sylow p-subgroup of K is the intersection of the $P \in X$. Deduce that if $r > 1$ the order k of K divides n/rp.

(*b*) Show that if G has no subgroup of index 2, then $\phi(G) \subset \mathfrak{A}_X$ and that if $n = kr!$ (resp. $\frac{1}{2}kr!$), then $\phi(G) = \mathfrak{S}_X$ (resp. $\mathfrak{A}_X$).

(*c*) Using (*a*), (*b*) and the preceding exercise, show the following facts:

If $n = 12$ and a Sylow 3-subgroup of G is not normal, then $G \approx \mathfrak{A}_4$.

If $n = 24$ and G does not have property (ST), then $G \approx \mathfrak{S}_4$.

If $n = 36$, then G has property (ST). (Show that if a Sylow 3-subgroup of G is not normal, then G contains a subgroup K of order 3 such that $G/K \approx \mathfrak{A}_4$ and that K is central.)

If $n = 48$ and G does not have property (ST), then G contains a normal subgroup K of order 2 such that $G/K \approx \mathfrak{S}_4$.

If $n = 60$ and G does not have property (ST), then $G \approx \mathfrak{A}_5$. (Show that such a group G has no normal subgroup of order 5, nor of order 2, nor of index 2. Deduce that G is isomorphic to a subgroup of $\mathfrak{S}_6$ and, arguing as in Exercise 25 of § 5, that $G \approx \mathfrak{A}_5$.)

If $n = 72$ and a Sylow 3-subgroup of G is not normal, then G contains a normal subgroup K such that G/K is isomorphic either to $\mathfrak{S}_4$ or to $\mathfrak{A}_4$.

If $n = 96$ and a Sylow 2-subgroup of G is not normal, then G contains a normal subgroup K such that $G/K \approx \mathfrak{S}_3$.

(*d*) Show that a group G of order $\leqslant 100$ is solvable unless $G \approx \mathfrak{A}_5$.

40. Let G_1, G_2 be two groups and G a subgroup of $G_1 \times G_2$ such that $\mathrm{pr}_1 G = G_1$, $\mathrm{pr}_2 G = G_2$; identifying G_1 and G_2 with $G_1 \times \{e_2\}$ and $\{e_1\} \times G_2$ respectively, $H_1 = G \cap G_1$ and $H_2 = G \cap G_2$ are normal subgroups of G such that G/H_1 is isomorphic to G_2 and G/H_2 isomorphic to G_1.

(*a*) Show that for G to be normal in $G_1 \times G_2$, it is necessary and sufficient that G contain the commutator group $D(G_1 \times G_2) = D(G_1) \times D(G_2)$ (prove that G contains $D(G_1)$ by writing $xzx^{-1}z^{-1} \in G$ for $x \in G_1$ and $z \in G$). Deduce that G_1/H_1 is commutative.

(*b*) If G is normal in $G_1 \times G_2$, show that a homomorphism is defined of $G_1 \times G_2$ onto G_1/H_1 by associating with each ordered pair (s_1, s_2) the coset of $s_1 t_1^{-1}$ modulo H_1, where $t_1 \in G_1$ is such that $(t_1, s_2) \in G$. Deduce that $(G_1 \times G_2)/G$ is isomorphic to $G/(H_1 \times H_2)$.

41. Let x, y be two elements of a group G.

(*a*) For there to exist a, b in G such that $bay = xab$, it is necessary and sufficient that xy^{-1} be a commutator.

(*b*) For there to exist $2n + 1$ elements $a_1, a_2, \ldots, a_{2n+1}$ in G such that

$$x = a_1 a_2 \ldots a_{2n+1} \quad \text{and} \quad y = a_{2n+1} a_{2n} \ldots a_1,$$

it is necessary and sufficient that xy^{-1} be a product of n commutators (argue by induction on n).

§ 7

1. Enumerate the elements of length $\leqslant 4$ in M(X) when X consists of a single element.

2. Let X be a set and M one of the magmas M(X), Mo(X) or $\mathbf{N}^{(X)}$. Show that every automorphism of M leaves X stable. Deduce an isomorphism of $\mathfrak{S}_X$ onto the group Aut(M).

Show that the endomorphisms of M correspond bijectively to the mappings of X into M.

3. Let X be a set and M_a the free magma constructed on a set $\{a\}$ with one element. Let ρ denote the homomorphism of M(X) into M_a which maps every element of X onto a; on the other hand let $\lambda\colon M(X) \to Mo(X)$ be the homomorphism defined in no. 9. Show that the mapping $w \mapsto (\lambda(w), \rho(w))$ is an isomorphism of M(X) onto the submagma of $Mo(X) \times M_a$ consisting of the ordered pairs (u, v) where u and v have the same length.

¶ 4. Let X be a set and c an element not belonging to X. We write $Y = X \cup \{c\}$. Let M be the magma with underlying set Mo(Y) and law of composition given by $(u, v) \mapsto cuv$. Let L be the submagma of M generated by X and let ε be the mapping from Y to **Z** equal to -1 on X and to 1 at c.

(*a*) Show that L is the subset of M consisting of the words $w = y_1 \ldots y_p$, $y_i \in Y$, satisfying the two following conditions:

(i) $\varepsilon(y_1) + \cdots + \varepsilon(y_p) = -1$

(ii) $\varepsilon(y_1) + \cdots + \varepsilon(y_i) \geqslant 0$ for $1 \leqslant i \leqslant p - 1$.

Show that every element of L − X may be written uniquely in the form cuv with $u, v \in L$.

(The elements w satisfying (i) and (ii) form a submagma L′ of M containing X. If $w = y_1 \ldots y_p$ belongs to L′, let k be the least integer such that $\varepsilon(y_1) + \cdots + \varepsilon(y_k) = 0$; show that $y_1 = c$ and that the elements $u = y_2 \ldots y_k$ and $v = y_{k+1} \ldots y_p$ belong to L′; then $w = cuv$, whence by induction on p the relation $w \in L$. Then show that the relations $u', v' \in L$ and $w = cu'v'$ imply $u' = u$, $v' = v$.)

(*b*) Show that the injection of X into L extends to an isomorphism of M(X) onto L.

5. Let X be a set with one element. If n is an integer $\geqslant 1$, let u_n denote the number of elements in M(X) of length n.

(*a*) For every set Y, show that

$$\operatorname{Card}(M_n(Y)) = u_n . \operatorname{Card}(Y)^n.$$

(Use Exercise 3.)

(*b*) Establish the relation

$$u_n = \sum_{p=1}^{n-1} u_p u_{n-p} \quad \text{for } n \geqslant 2.$$

(*c*) *Let $f(T)$ be the formal power series $\sum_{n=1}^{\infty} u_n T^n$. Show that

$$f(T) = T + f(T)^2.$$

Deduce the formulae

$$f(T) = \tfrac{1}{2}(1 - (1 - 4T)^{1/2}), \qquad u_n = \frac{2^{n-1}}{n!} 1.3.5 \ldots (2n - 3) \quad \text{for } n \geqslant 2.{}_*$$

Obtain the last result using Exercise 11 of *Set Theory*, III, § 5.

¶ 6. Let X be a set.

(*a*) Let N be a submagma of M(X) and Y = N − (N.N). The injection $Y \to N$ extends to a homomorphism u: $M(Y) \to N$. Show that u is an isomorphism. (In other words, every submagma of a free magma is free.)

(*b*) If $x \in \mathrm{M(X)}$, let M_x denote the submagma of M(X) generated by x; by (*a*), M_x is identified with the free magma constructed on x. Show that, if $x, y \in \mathrm{M(X)}$, then either $\mathrm{M}_x \subset \mathrm{M}_y$ or $\mathrm{M}_y \subset \mathrm{M}_x$ or $\mathrm{M}_x \cap \mathrm{M}_y = \varnothing$. (If $\mathrm{M}_x \cap \mathrm{M}_y \neq \varnothing$, let z be an element of $\mathrm{M}_x \cap \mathrm{M}_y$ of minimum length; show that either $z = x$ or $z = y$.)

¶ 7. Let M_x be the free magma constructed on a set $\{x\}$ with one element.

(*a*) For all $y \in \mathrm{M}_x$, show that there exists a unique endomorphism f_y of M_x such that $f_y(x) = y$; it is an isomorphism of M_x onto the submagma M_y generated by y (cf. Exercise 6).

(*b*) If $y, z \in \mathrm{M}_x$, we write $y \circ z = f_y(z)$. Show that the law of composition $(y, z) \mapsto y \circ z$ makes M_x into a monoid with identity element x which is isomorphic to the monoid $\mathrm{End}(\mathrm{M}_x)$ of endomorphisms of M_x.

(*c*) Let N be the set of monogenous submagmas of M_x distinct from M_x. An element y of M_x is called *primitive* if M_y is a maximal element of N. Show that xx, $x(x(xx))$ are primitive whereas x, $(xx)(xx)$ are not.

(*d*) Let P be the set of primitive elements of M_x. Show that, if $y, z \in \mathrm{P}, y \neq z$, then $\mathrm{M}_y \cap \mathrm{M}_z = \varnothing$ (use Exercise 6).

(*e*) Let Mo(P) be the free monoid constructed on P. If M_x is given the monoid structure defined in (*b*), the injection $\mathrm{P} \to \mathrm{M}_x$ can be extended to a homomorphism $p\colon \mathrm{Mo(P)} \to \mathrm{M}_x$. Show that p is an isomorphism.

(If $z \in \mathrm{M}_x$, show by induction on $l(z)$ that $z \in \mathrm{Im}(p)$. On the other hand, if $y_1, \ldots, y_n, z_1, \ldots, z_m \in \mathrm{P}$ are such that $p(y_1 \ldots y_n) = p(z_1 \ldots z_m)$, then $\mathrm{M}_{y_1} \cap \mathrm{M}_{z_1} \neq \varnothing$, whence $y_1 = z_1$ and $f_{y_1}(p(y_2 \ldots y_n)) = f_{y_1}(p(z_2 \ldots z_m))$; whence $p(y_2 \ldots y_n) = p(z_2 \ldots z_m)$ and it follows that $y_1 \ldots y_n = z_1 \ldots z_m$ arguing by induction on $\sup(n, m)$.)

8. Let X be a set; for every integer $q \geqslant 0$ let $\mathrm{Mo(X)}_q$ be the set of elements of Mo(X) of length q and let $\mathrm{Mo}^{(q)}(\mathrm{X})$ be the union of $\mathrm{Mo(X)}_{nq}$, $n \in \mathbf{N}$. The injection $\mathrm{Mo(X)}_q \to \mathrm{Mo}^{(q)}(\mathrm{X})$ extends to a homomorphism

$$\mathrm{Mo}(\mathrm{Mo(X)}_q) \to \mathrm{Mo}^{(q)}(\mathrm{X});$$

show that this homomorphism is bijective if $q \geqslant 1$.

9. Let $x, y \in \mathrm{X}$ with $x \neq y$. Show that the submonoid of Mo(X) generated by $\{x, xy, yx\}$ is not isomorphic to a free monoid.

10. Show that the group defined by the presentation

$$\langle x, y;\ xy^2 = y^3x,\ yx^2 = x^3y \rangle$$

reduces to the identity element.

(The first relation implies $x^2y^8x^{-2} = y^{18}$ and $x^3y^8x^{-3} = y^{27}$; use the second relation to deduce that $y^{18} = y^{27}$, whence $y^9 = e$; the fact that y^2 is conjugate to y^3 then implies $y = e$, whence $x = e$.)

11. The group defined by the presentation $\langle x, y; x^2, y^3, (xy)^2\rangle$ is isomorphic to $\mathfrak{S}_3$.

¶ 12. The group defined by the presentation $\langle x, y; x^3, y^2, (xy)^3\rangle$ is isomorphic to $\mathfrak{A}_4$. (Show that the conjugates of y form, with the identity element, a normal subgroup of order $\leqslant 4$ and index $\leqslant 3$.)

13. Let G be a group and X a subset of G disjoint from X^{-1}. Let $S = X \cup X^{-1}$. Show the equivalence of the two following properties:
(i) X is a free family in G.
(ii) No product $s_1 \ldots s_n$, with $n \geqslant 1$, $s_i \in S$ and $s_i s_{i+1} \neq e$ for $1 \leqslant i \leqslant n-1$, is equal to e.

14. A group G is called *free* if it possesses a basic family and hence is isomorphic to a free group F(X) constructed on a set X.
(*a*) Show that, if $(t_i)_{i \in I}$ and $(t'_j)_{j \in J}$ are two such families, then Card(I) = Card(J). (When Card(I) is infinite, show that Card(J) is also and that both are equal to Card(G). When Card(I) is an integer d, show that the number of subgroups of index 2 in G is $2^d - 1$.)
The cardinal of a basic family of G is called the *rank* of G.
(*b*) A free group is of rank 0 (resp. 1) if and only if it reduces to $\{e\}$ (resp. is isomorphic to $\mathbf{Z}$).
(*c*) Let G be a free group of finite rank d and $(x_1, \ldots, x_d)$ a generating family of d elements of G. Show that this family is basic. (Use Exercise 34 and § 5, Exercise 5.)
(*d*) Show that a free group of rank $\geqslant 2$ contains a free subgroup of given rank d for every cardinal $d \leqslant \aleph_0$. (Use Exercise 22.)

¶ 15. Let G be a group with presentation $(\mathbf{t}, \mathbf{r})$, where $\mathbf{t} = (t_i)_{i \in I}$ and $\mathbf{r} = (r_j)_{j \in J}$; let $F((T_i)_{i \in I})$ denote the free group F(I) and $\mathbf{T} = (T_i)$. On the other hand, let $\mathbf{x} = (x_\alpha)_{\alpha \in A}$ be a generating family of G; write

$$F(A) = F((X_\alpha)_{\alpha \in X}),$$

$\mathbf{X} = (X_\alpha)$ and $R_{\mathbf{x}}$ the set of relators of $\mathbf{x}$; it is a normal subgroup of F(A). For all $i \in I$, let $\phi_i(\mathbf{X})$ be an element of F(A) such that $t_i = \phi_i(\mathbf{x})$ in G; for all $\alpha \in A$, let $\psi_\alpha(\mathbf{T})$ be an element of F(I) such that $x_\alpha = \psi_\alpha(\mathbf{t})$ in G. Show that $R_{\mathbf{x}}$ is the normal subgroup of F(A) generated by the elements $X_\alpha^{-1}\psi_\alpha(\phi_i(\mathbf{X})_{i \in I})$, $\alpha \in A$ and $r_j(\phi_i(\mathbf{X})_{i \in I})\, j \in J$. (If $R'_{\mathbf{x}}$ denotes the normal subgroup of F(A) generated by the elements in question, then $R'_{\mathbf{x}} \subset R_{\mathbf{x}}$; on the other hand, show that the homomorphism $F(I) \to F(A)$ defined by the ϕ_i gives, when passing to the quotient, a homomorphism $G \to F(A)/R'_{\mathbf{x}}$ the inverse of the canonical homomorphism $F(A)/R'_{\mathbf{x}} \to F(A)/R_{\mathbf{x}} = G$.)

16. A group is called *finitely generated* (resp. *finitely presented*) if it admits a

presentation $\langle (l_i)_{i\in I}, (r_j)_{j\in J}\rangle$ where I is a finite set (resp. where I and J are finite sets).

(*a*) Let G be a finitely presented group and $\mathbf{x} = (x_\alpha)_{\alpha\in A}$ a finite generating family of G. Show that there exists a finite family $\mathbf{s} = (s_k)_{k\in K}$ of relators of the family $\mathbf{x}$ such that $(\mathbf{x}, \mathbf{s})$ is a presentation of G. (Use the preceding exercise.)

(*b*) Show that a finite group is finitely presented.

(*c*) Give an example of a finitely presented group with a subgroup which is not finitely generated.

(*d*) If G is a finitely presented group and H a normal subgroup of G, show the equivalence of the following properties:

(i) G/H is finitely presented.

(ii) There exists a finite subset X of G such that H is the normal subgroup of G generated by X.

(iii) If $(H_i)_{i\in I}$ is a right directed family of normal subgroup of G whose union is equal to H, there exists $i \in I$ such that $H_i = H$.

(Use (*a*) to prove that (i) implies (ii).)

(*e*) If G is a finitely presented group, show that $G/C^r(G)$ is for all $r \geqslant 1$. (Use § 6, Exercise 15.)

(*f*) Let $(G_\alpha)_{\alpha\in A}$ be a family of groups and let $G = \prod_{\alpha\in A} G_\alpha$ be the product of the G_α. Show that G is finitely generated (resp. finitely presented) if and only if each of the G_α is finitely generated (resp. finitely presented) and $G_\alpha = \{e\}$ for all but a finite number of α.

¶ 17. (*a*) Show that every subgroup of a finitely generated nilpotent group is finitely generated. (In the commutative case, argue by induction on the minimum number of generators of the group; then proceed by induction on the nilpotency class of the group.)

Give an example of a finitely generated solvable group containing a subgroup which is not finitely generated.

(*b*) Show that every finitely generated nilpotent group is finitely presented. (Write the group in the form F(X)/R, with X finite, and choose r such that $R \supset C^r(F(X))$; use (*a*) to show that $R/C^r(F(X))$ is finitely generated; observe that $F(X)/C^r(F(X))$ is finitely presented, cf. Exercise 16.)

18. Let S be a symmetric subset of a group G. Two elements g_1, g_2 of G are called S-*neighbours* if $g_2^{-1}g_1$ belongs to S. A sequence $(g_1, \ldots, g_n)$ of elements of G is called an S-*chain* if g_i and g_{i+1} are S-neighbours for $1 \leqslant i \leqslant n-1$; the elements g_1 and g_n are called respectively the beginning and the end of the chain.

(*a*) Let Y be a subset of G and $R_Y\{a, b\}$ the relation "there exists an S-chain in Y beginning at a and ending at b". Show that R_Y is an equivalence relation on Y (cf. *Set Theory*, II, § 6, Exercise 10); the equivalence classes under this relation are called the *connected* S-*components* of Y. Y is called S-*connected* if it has at most one connected component.

(*b*) Show that G is S-connected if and only if S generates G.

(*c*) Suppose that no element $s \in S$ satisfies the relation $s^2 = e$; choose a subset X of S such that S is the disjoint union of X and X^{-1}. Show the equivalence of the following conditions:

(i) The family X is free.

(ii) There exists no S-chain $(g_1, \ldots, g_n)$ in G consisting of distinct elements $n \geqslant 3$ in number such that g_1 and g_n are S-neighbours.

(Use Exercise 13.)

¶ 19. Let X be a set, $G = F(X)$ the free group constructed on X and $S = X \cup X^{-1}$.

(*a*) If $g \in G$, every sequence $(s_1, \ldots, s_n)$ of elements of S such that $g = s_1 \ldots s_n$ and $s_i s_{i+1} \neq e$ for $1 \leqslant i < n$ is called a *reduced decomposition* of g. Show that every element of G admits one and only one reduced decomposition; the integer n is the *length* of g (cf. Exercise 26).

(*b*) Let Y be a subset of G containing e. Show the equivalence of the following conditions:

(b_1) Y is S-connected (cf. Exercise 18).

(b_2) For all $g \in Y$, if $(s_1, \ldots, s_n)$ is the reduced decomposition of g, then $s_1 \ldots s_i \in Y$ for $1 \leqslant i \leqslant n$.

(*c*) Let Y_1 and Y_2 be two non-empty S-connected subsets of G such that $Y_1 \cap Y_2 = \varnothing$. Show that there exists at most one ordered pair

$$(y_1, y_2) \in Y_1 \times Y_2$$

such that y_1 and y_2 are S-neighbours. There exists one if and only if $Y_1 \cup Y_2$ is S-connected; Y_1 and Y_2 are then said to be *neighbours*.

(*d*) Let $Y_1, \ldots, Y_n$ be a sequence of disjoint non-empty S-connected subsets of G. Suppose that Y_i and Y_{i+1} are neighbours for $1 \leqslant i < n$. Show that, if $n \geqslant 3$, Y_1 and Y_n are not neighbours.

¶ 20. We preserve the notation of the preceding exercise. Let H be a subgroup of G.

(*a*) Let R_H be the set of S-connected subsets T of G such that $hT \cap T = \varnothing$ if $h \in H$, $h \neq e$. Show that R_H is inductive for the relation of inclusion.

(*b*) Let Y be a maximal element of R_H. Show that $G = \bigcup_{h \in H} hY$, in other words Y is a system of representatives of the right cosets modulo H. (If $G' = \bigcup hY$ is distinct from G, show that there exist $x \in G'$ and $y \in Y$ which are S-neighbours and deduce that $Y \cup \{x\}$ belongs to R_H.)

(*c*) Let S_H be the set of elements $h \in H$ such that Y and hY are neighbours. It is a symmetric subset of H. Show that an element $h \in H$ belongs to S_H if and only if $h \neq e$ and there exist $y, y' \in Y$, $s \in S$ with $ys = hy'$.

Show that S_H generates H. (If H' is the subgroup of H generated by S_H, prove that $\bigcup_{h \in H'} hY$ is a connected S-component of G.) Obtain this result by means of Exercise 14 of § 4.

(*d*) Let X_H be a subset of S_H such that S_H is the disjoint union of X_H and X_H^{-1} (show that such a subset exists).

Show that X_H is a basic family for H and in particular that H is a free group ("*Nielsen-Schreier Theorem*"). (Apply the criterion of Exercise 18(*c*) to H together with S_H, noting that two elements h, h' of H are S_H-neighbours if and only if hY and $h'Y$ are neighbouring subsets of G. Use Exercise 19(*d*) to prove that S_H satisfies condition (ii) in Exercise 18(*c*).)

(*e*) Let $d = (G:H) = \text{Card}(Y)$. Suppose d is finite. Show that:

$$\text{Card}(X_H) = \text{Card}(X) = \text{Card}(G) \quad \text{if Card}(X) \text{ is infinite}$$
$$\text{Card}(X_H) = 1 + d(\text{Card}(X) - 1) \quad \text{if Card}(X) \text{ is finite.}$$

(Let T be a non-empty S-connected finite subset of G and T′ (resp. T″) the set of ordered pairs of S-neighbouring elements whose first element (resp. both of whose elements) belongs (resp. belong) to T. $\text{Card}(T') = 2\text{Card}(X)\text{Card}(T)$ and it should be shown by induction on Card(T) that

$$\text{Card}(T'') = 2\text{Card}(T) - 2.$$

Apply the result to T = Y, noting that S_H is equipotent to T′ — T″.)

21. Let H be a subgroup of a group G of finite index. Show that H is finitely presented if and only if G is. (It may be assumed that G is finitely generated, cf. § 4, Exercise 14. Write G in the form G = F/R with F free and finitely generated; the inverse image F′ of H in F is free and finitely generated, cf. Exercise 20, and H = F′/R. If R is generated (as a normal subgroup of F) by $(r_j)_{j \in J}$, show that it is generated (as a normal subgroup of F′) by the yr_jy^{-1}, $y \in Y$, $j \in J$, where Y denotes a representative system in F of the right cosets modulo F′.)

22. (*a*) Let $(y_n)_{n \in \mathbf{Z}}$ be a basic family of a group F. Let ϕ be the automorphism of F which maps y_n to y_{n+1}; this automorphism defines as operation τ of $\mathbf{Z}$ on F; let $E = F \times_\tau \mathbf{Z}$ be the corresponding semi-direct product. Show that the elements $x = (e, 1)$ and $y = (y_0, 0)$ form a basic family of E. (Let $F_{x,y}$ be the free group constructed on $\{x, y\}$ and let f be the canonical homomorphism of $F_{x,y}$ into E. Show that there exists a homomorphism $g: E \to F_{x,y}$ such that $g((0, 1)) = x$ and $g((y_n, 0)) = x^n y x^{-n}$ and f and g are inverses of one another.)

(*b*) Deduce that the normal subgroup of $F_{x,y}$ generated by y is a free group with basic family $(x^n y x^{-n})_{n \in \mathbf{Z}}$. Obtain this result by applying Exercise 20 to the representative system consisting of the x^n, $n \in \mathbf{Z}$.

(*c*) Extend the above to free groups of arbitrary rank.

23. Let F be a free group with basic family (x, y), $x \neq y$. Show that the

derived group of F has as basic family the family of commutators (x^i, y^j), $i \in \mathbf{Z}$, $j \in \mathbf{Z}$, $i \neq 0, j \neq 0$.

(Use Exercise 20 or Exercise 32.)

24. Let G be a group, $S = G - \{e\}$ and $F = F(S)$. For all $s \in S$, let x_s denote the corresponding element of F. If $s, t \in S$, let $r_{s,t}$ denote the element of F defined by:

$$r_{s,t} = x_s x_t x_{st}^{-1} \quad \text{if } st \neq e \text{ in G}$$
$$r_{s,t} = x_s x_t \quad \text{if } st = e \text{ in G}.$$

Let ϕ be the homomorphism of F onto G which maps x_s to s and let R be the kernel of ϕ.

Show that the family $(r_{s,t})$ for $s, t \in S \times S$, is a basic family of the group R. (Apply Exercise 20 to the system of representatives of G in F consisting of e and the x_s.)

25. Let G be a group. Show the equivalence of the following properties:

(*a*) G is a free group.

(*b*) For every group H and extension E of G by H there exists a section $G \to E$.

(To prove that (*b*) implies (*a*), take E to be a free group and use Exercise 20.)

26. Let X be a set and g an element of the free group F(X). Let $g = \prod_{\alpha=1}^{n} x_\alpha^{m(\alpha)}$ be the canonical decomposition of g as a product of powers of elements of X (cf. no. 5, Proposition 7). The integer $l(g) = \sum_\alpha |m(\alpha)|$ is called the *length* of g.

(*a*) g is called *cyclically reduced* if either $x_1 \neq x_n$ or $x_1 = x_n$ and $m(1)m(n) > 0$. Show that every conjugacy class of F(X) contains one and only one cyclically reduced element; it is the element of minimum length of the class in question.

(*b*) An element $x \in F(X)$ is called *primitive* if there does not exist an integer $n \geqslant 2$ and an element $g \in F(X)$ such that $x = g^n$. Show that every element $z \neq e$ of F(X) can be written uniquely in the form x^n with $n \geqslant 1$ and x primitive. (Reduce it to the case where z is cyclically reduced and observe that, if $z = x^n$, the element x is also cyclically reduced.)

Show that the centralizer of z is the same as that of x; it is the cyclic subgroup generated by x.

27. Let $G = \ast_A G_i$ be the sum of a family $(G_i)_{i \in I}$ of groups amalgamated by a common subgroup A. For all $i \in I$, choose a subset P_i of G_i satisfying condition (A) of no. 3.

(*a*) Let $x \in G$ and let $(a; i_1, \ldots, i_n; p_1, \ldots, p_n)$ be a reduced decomposition of x; then

$$x = a \prod_{\alpha=1}^{n} p_\alpha, \quad \text{with} \quad p_\alpha \in P_{i_\alpha} - \{e_{i_\alpha}\}, \; i_\alpha \neq i_{\alpha+1}.$$

Show that, if $i_1 \neq i_n$, the order of x is infinite. Show that, if $i_1 = i_n$, x is conjugate to an element with a decomposition of length $\leqslant n - 1$.

(*b*) Deduce that every element of G of finite order is conjugate to an element of one of the G_ι. In particular, if the G_ι have no element of finite order except e, the same is true of G.

28. We preserve the notation of the preceding exercise.

(*a*) Let N be a subgroup of A. Suppose that, for all $i \in I$, the group N is normal in G_ι. Show that N is normal in $G = \mathop{\ast}_A G_\iota$ and that the canonical homomorphism of $\mathop{\ast}_{A/N} G_\iota/N$ into G/N is an isomorphism.

(*b*) For all $i \in I$, let H_ι be a subgroup of G_ι and let B be a subgroup of A. Suppose that $H_\iota \cap A = B$ for all i. Show that the canonical homomorphism of $\mathop{\ast}_B H_\iota$ into G is injective; its image is the subgroup of G generated by the H_ι.

¶ 29. Let $G = G_1 *_A G_2$ be the sum of two groups G_1 and G_2 amalgamated by a subgroup A.

(*a*) Show that G is finitely generated if G_1 and G_2 are finitely generated.

(*b*) Suppose G_1 and G_2 are finitely presented. Show the equivalence of the two following properties:

(1) G is finitely presented.

(2) A is finitely generated.

(Show first that (1) implies (2). On the other hand, if A is not finitely generated, there exists a right directed family $(A_\iota)_{\iota \in I}$ of subgroups of A with $\bigcup_{\iota \in I} A_\iota = A$ and $A_\iota \neq A$ for all i. Let H_ι (resp. H) be the kernel of the canonical homomorphism of $G_1 * G_2$ onto $G_1 *_{A_\iota} G_2$ (resp. onto G). Show that H is the union of the H_ι and that $H_\iota \neq H$ for all i; then apply Exercise 16(*d*).)

(*c*) Deduce an example of a finitely generated group which is not finitely presented. (Take G_1 and G_2 to be free groups of rank 2 and A a free group of finite rank, cf. Exercise 22.)

30. Let A and B be two groups and $G = A * B$ their free product.

(*a*) Let Z be an element of A\G/A distinct from A. Show that Z contains one and only one element z of the form

$$b_1 a_1 b_2 a_2 \ldots b_{n-1} a_{n-1} b_n,$$

with $n \geqslant 1$, $a_i \in A - \{e\}$, $b_i \in B - \{e\}$. Show that every element of Z can be written uniquely in the form aza' with $a, a' \in A$.

(*b*) Let $x \in G - A$. Let f_x be the homomorphism of $A * A$ into G whose restriction to the first (resp. second) factor of $A * A$ is the identity (resp. the mapping $a \mapsto xax^{-1}$). Show that f_x is injective. (Write x in the form aza' as above and reduce it thus to the case $x = z$; use the uniqueness of reduced decompositions in G.)

In particular, $A \cap xAx^{-1} = \{e\}$ and the normalizer of A in G is A.

(*c*) Let T be an element of A\G/B. Show that T contains one and only one element t of the form

$$b_1a_1b_2a_2\dots b_{n-1}a_{n-1},$$

with $n \geqslant 1$, $a_i \in A - \{e\}$, $b_i \in B - \{e\}$. Show that every element of T can be written uniquely in the form atb with $a \in A$, $b \in B$.

(*d*) Let $x \in G$ and let h_x be the unique endomorphism of G such that $h_x(a) = a$ if $a \in A$ and $h_x(b) = xbx^{-1}$ if $b \in B$. Show that h_x is injective. (Same method as in (*b*), using the decomposition $x = atb$ of (*c*).) In particular, $A \cap xBx^{-1} = \{e\}$.

Give an example where h_x is not surjective.

31. Let G be the free product of a family $(G_\iota)_{\iota \in I}$ of groups. Let S be the set of subgroups of G which are conjugate to one of the G_ι and Θ the union of the elements of S.

(*a*) Let H, H′ ∈ S, with H ≠ H′, and let $x \in H - \{e\}$, $x' \in H' - \{e\}$. Show that $xx' \in G - \Theta$. (Reduce it to the case where H is one of the G_i; write H′ in the form yG_jy^{-1} and proceed as in the preceding exercise.)

(*b*) Let C be a subgroup of G contained in Θ. Show that there exists H ∈ S such that C ⊂ H.

(*c*) Let C be a subgroup of G all of whose elements are of finite order. Show that C is contained in Θ (cf. Exercise 27); deduce that C is contained in a conjugate of one of the G_ι.

¶ 32. Let A and B be two groups and $G = A * B$ their free product. Let X be the set of commutators (a, b) with $a \in A - \{e\}$ and $b \in B - \{e\}$ and $S = X \cup X^{-1}$.

(*a*) Show that $X \cap X^{-1} = \varnothing$ and that, if x is an element of X, there exists a unique ordered pair a, b in $(A - \{e\}) \times (B - \{e\})$ such that $(a, b) = x$.

(*b*) Let $n \geqslant 1$ and let $s_1, \dots, s_n$ be a sequence of elements of S such that $s_is_{i+1} \neq e$ for $1 \leqslant i < n$; let $a_n \in A - \{e\}$, $b_n \in B - \{e\}$, $\varepsilon(n) = \pm 1$ be such that $s_n = (a_n, b_n)^{\varepsilon(n)}$. Let $g = s_1\dots s_n$. Show, by induction on n, that the reduced decomposition of g is of length $\geqslant n + 3$ and terminates either with a_nb_n if $\varepsilon(n) = 1$ or with b_na_n if $\varepsilon(n) = -1$.

(*c*) Let R be the kernel of the canonical projection $A * B \to A \times B$ corresponding (no. 3) to the canonical injections $A \to A \times B$ and $B \to A \times B$. Show that R is a free group with basic family X. (Use (*b*) to prove that X is free; if R_X is the subgroup of G generated by X, show that R_X is normal; deduce that it coincides with R.)

33. Let $E = G *_A G'$ be the sum of two groups amalgamated by a common subgroup A. Let P (resp. P′) be a subset of G (resp. G′) satisfying condition (A) of no. 3.

Let n be an integer $\geqslant 1$. Let L_n denote the set of $x \in E$ whose reduced decomposition is of length $\leqslant 2n - 1$. Let M_n denote the set of $x \in E$ of the form $ap_1p'_1p_2p'_2 \ldots p_np'_n$, with $a \in A$, $p_i \in P - \{e\}$, $p'_i \in P' - \{e\}$; similarly, let M'_n denote the set of elements of the form $ap'_1p_1p'_2p_2 \ldots p'_np_n$, where a, p_i, p'_i satisfy the same conditions.

(*a*) Show that L_n, M_n and M'_n are disjoint and that their union is the set of elements of E whose reduced decomposition is of length $\leqslant 2n$.

(*b*) Construct a bijection ε of M_n onto M'_n such that $\varepsilon(ax) = a\varepsilon(x)$ if $a \in A$, $x \in M_n$.

(*c*) Let $X_n = L_n \cup M_n$ and $X'_n = L_n \cup M'_n$; show that $G.X_n = X_n$ and $G'.X'_n = X'_n$.

(*d*) ε extends to a bijection of X_n onto X'_n by setting $\varepsilon(x) = x$ if $x \in L_n$. Show that E can operate on X_n so that $g(x) = gx$ if $g \in G$, $x \in X_n$ and $g'(x) = \varepsilon^{-1}(g' \in (x))$ if $g' \in G'$, $x \in X_n$. Show that, if $g \in X_n$, then $g(e) = g$.

(*e*) Let $\tau_n : E \rightarrow \mathfrak{S}_{X_n}$ be the homomorphism defined by the action of E on X_n described above and let R_n be the kernel of τ_n. Show that $R_n \cap X_n = \{e\}$; deduce that the intersection of the R_n reduces to e.

(*f*) When G and G′ are finite, show that E is residually finite (cf. § 5, Exercise 5). (Note that the R_n are then subgroups of finite index.)

¶ 34. (*a*) Let (H_ι) be a right directed family of normal subgroups of a group G. Suppose that $\bigcap_\iota H_\iota = \{e\}$. Show that, for every group G′, the intersection of the kernels of the canonical homomorphisms $G' * G \rightarrow G' * (G/H_\iota)$ is equal to $\{e\}$.

(*b*) Let G be the free product of a family (G_α) of groups. Show that, if all the G_α are residually finite (cf. § 5, Exercise 5), so is G. (Reduce it first to the case where the family is finite, then to the case where it has two elements; by (*a*) it may be assumed that the G_α are finite, then apply the preceding exercise.)

(*c*) Deduce that every free group is residually finite.

35. *Let G (resp. G′) be the additive group of fractions of the form a/b with $a \in \mathbf{Z}$, $b \in \mathbf{Z}$ and b not divisible by 2 (resp. by 3). Let $E = G *_{\mathbf{Z}} G'$ be the sum of G and G′ amalgamated by their common subgroup $\mathbf{Z}$. Show that G and G′ are residually finite, but that E is not. (Prove that the only subgroup of E of finite index is E.)*

36. Let $(G_1, \ldots, G_n)$ and $(A_1, \ldots, A_{n-1})$ be two sequences of groups; for each i, let $\phi_i : A_i \rightarrow G_i$ and $\psi_i : A_i \rightarrow G_{i+1}$ be two injective homomorphisms; by these homomorphisms A_i can be identified both with a subgroup of G_i and with a subgroup of G_{i+1}.

(*a*) Let $H_1 = G_1, H_2 = H_1 *_{A_1} G_2, \ldots, H_n = H_{n-1} *_{A_{n-1}} G_n$. The group H_n is called the sum of the groups (G_i) amalgamated by the subgroups (A_i); it is denoted by $G_1 *_{A_1} G_2 *_{A_2} \cdots *_{A_{n-1}} G_n$. If T is an arbitrary group, define a bijection between the homomorphisms of H_n into T and the families

$(t_1, \ldots, t_n)$ of homomorphisms $t_i : G_i \to T$ such that $t_i \circ \phi_i = t_{i+1} \circ \psi_i$ for $1 \leqslant i \leqslant n-1$.

(*b*) Each G_i is identified with the corresponding subgroup of H_n. Show that $G_i \cap G_{i+1} = A_i$ and that $G_i \cap G_{i+2} = A_i \cap A_{i+1}$ (the latter intersection being taken in G_{i+1}).

¶ 37. Let G be a group and S a finite symmetric subset of G generating G. Let $\mathfrak{T}$ denote the set of finite subsets of G, ordered by inclusion; it is a directed set.

(*a*) Let $T \in \mathfrak{T}$ and let B_T be the set of connected S-components of $G - T$ (cf. Exercise 18); let $B_T^\infty = B_T - (\mathfrak{T} \cap B_T)$. Show that B_T is finite (note that, if $T \neq \varnothing$ and $Z \in B_T$, there exists $t \in T$ and $z \in Z$ which are S-neighbours).

(*b*) Let $T, T' \in \mathfrak{T}$ with $T \subset T'$. If $Z' \in B_{T'}$, show that there exists a unique element $Z \in B_T$ such that $Z' \subset Z$; the mapping $f_{TT'} : B_{T'} \to B_T$ such that $f_{TT'}(Z') = Z$ maps $B_{T'}^\infty$ onto B_T^∞. The inverse limit of the B_T, with T running through $\mathfrak{T}$, is denoted by B. Show that $B = \varprojlim B_T^\infty$ and that the image of B in B_T is B_T^∞. The set B is called the set of *ends* of G; it is non-empty if and only if G is infinite.

(*c*) Let T be a non-empty S-connected finite subset of G and let $Z \in B_T^\infty$. Show that there exist $g \in G$ such that $gT \cap T = \varnothing$ and $gT \cap Z \neq \varnothing$ and that then $gT \subset Z$. Show that there exists $Z_1 \in B_T$ such that gZ_1 contains T and all the $Z' \in B_T$ such that $Z' \neq Z$ (observe that $T \cup \bigcup_{Z' \neq Z} Z'$ is S-connected and does not meet gT); deduce that, if $Z' \in B_T$ is different from Z_1, then $gZ' \subset Z$. Show that, writing $T' = T \cup gT$, the inverse image of Z in $B_{T'}^\infty$ consists of at least $n-1$ elements, where $n = \text{Card}(B_T^\infty)$.

(*d*) Let S' be a symmetric finite subset of G generating G and B' the set of ends of G relative to S'. Define a bijection of B onto B' (reduce it to the case where $S \subset S'$). The cardinal of B is thus independent of the choice of S; it is called the *number of ends* of G.

(*e*) Show, using (*c*), that the number of ends of G is equal to 0, 1, 2 or $2^{\aleph_0}$.

(*f*) Show that the number of ends of $\mathbf{Z}$ is 2 and that that of $\mathbf{Z}^n$ ($n \geqslant 2$) is 1.

¶ 38. Let X be a finite set, F(X) the free group constructed on X and $S = X \cup X^{-1}$.

(*a*) Let S_n be the set of products $s_1 \ldots s_m$, $s_i \in S$, $m \leqslant n$. Show that every connected S-component (cf. Exercise 18) of $F(X) - S_n$ contains one and only one element of the form $s_1 \ldots s_{n+1}$ with $s_i \in S$ and $s_i s_{i+1} \neq e$ for $1 \leqslant i \leqslant n$.

(*b*) Deduce from (*a*) a bijection of the set B of ends of F(X) (cf. Exercise 37) onto the set of infinite sequences $(s_1, \ldots, s_n, \ldots)$ of elements of S such that $s_i s_{i+1} \neq e$ for all *i*. In particular, the number of ends of F(X) is equal to $2^{\aleph_0}$ if $\text{Card}(X) \geqslant 2$.

39. *Let G be a finitely generated group and $\mathfrak{F}$ the set of symmetric finite subsets of G containing *e*, which generate G.

(*a*) If $X \in \mathfrak{F}$ and $n \in \mathbf{N}$, let $d_n(X)$ denote the number of elements in G of the form $x_1 \ldots x_n$ with $x_i \in X$. If $X, Y \in \mathfrak{F}$, show that there exists an integer $a \geqslant 1$ such that

$$d_n(X) \leqslant d_{an}(Y) \quad \text{for all } n \in \mathbf{N}.$$

Deduce that $\limsup_{n \to \infty} \dfrac{\log(d_n(X))}{\log(n)}$ does not depend on the choice of X in $\mathfrak{F}$; this limit (finite or infinite) is denoted by $e(G)$.

(*b*) If $X \in \mathfrak{F}$ and G is infinite, show that $d_n(X) < d_{n+1}(X)$ for all n. Deduce that $e(G)$ is $\geqslant 1$.

(*c*) Let H be a finitely generated group. Show that $e(H) \leqslant e(G)$ if H is isomorphic to a subgroup (resp. a quotient group) of G. If E is an extension of G by H, show that $e(E) \geqslant e(H) + e(G)$, with equality when $E = G \times H$.

(*d*) Show that $e(\mathbf{Z}^n) = n$.

(*e*) Let $X \in \mathfrak{F}$. Consider the following property:

(i) There exists a real number $a > 1$ such that $d_n(X) \geqslant a^n$ for all sufficiently large n.

Show that, if property (i) holds for one element of $\mathfrak{F}$, it holds for every element of $\mathfrak{F}$. G is then said to be *of exponential type;* this implies $e(G) = +\infty$.

(*f*) Let G_1 and G_2 be two groups containing the same subgroup A and let $H = G_1 *_A G_2$ the corresponding amalgamated sum. Suppose G_1 and G_2 are finitely generated (hence also H is) and $G_i \neq A$ for $i = 1, 2$. Show that H is of exponential type if at least one of the indices $(G_1 : A)$, $(G_2 : A)$ is $\geqslant 3$.

(*g*) Show that a free group of rank $\geqslant 2$ is of exponential type.$_*$

40. *Let n be an integer $\geqslant 2$ and T_n the group of upper triangular matrices of order n over $\mathbf{Z}$ with all the diagonal terms equal to 1. Let $e(T_n)$ be the invariant of the group T_n defined in Exercise 39. Show that

$$e(T_n) \leqslant \sum_{i=1}^{n} i(n-i).{}_*$$

§ 8

1. Determine all the ring structures on a set of n elements for $2 \leqslant n \leqslant 7$ and also the ideals of these rings.

2. Let A be a ring.

(*a*) The mapping which associates with $a \in A$ the left homothety $x \mapsto ax$ is an isomorphism of A onto the ring of endomorphisms of the additive group of A which commute with right homotheties.

(*b*) Let M be the set of sequences (a, b, c, d) of four elements of A which may be represented in the form of a table or matrix $\begin{pmatrix} a & b \\ c & d \end{pmatrix}$. With every element

$\begin{pmatrix} a & b \\ c & d \end{pmatrix}$ of M we associate the mapping $(x, y) \mapsto (ax + by, cx + dy)$ of $A \times A$ into itself. Show that a bijection is thus obtained of M onto the ring B of endomorphisms of the commutative group $A \times A$ which commute with the operations $(x, y) \mapsto (xr, yr)$ for all $r \in A$. Henceforth M is identified with the ring B. Calculate the product of two matrices.

(*c*) Show that the subset T of M consisting of the matrices of the form $\begin{pmatrix} a & b \\ 0 & c \end{pmatrix}$ is the subring of M which leaves stable the subgroup $A \times 0$ of $A \times A$. The mappings $\begin{pmatrix} a & b \\ 0 & c \end{pmatrix} \mapsto a$ and $\begin{pmatrix} a & b \\ 0 & c \end{pmatrix} \mapsto c$ are homomorphisms of T into A; let $\mathfrak{a}$ and $\mathfrak{b}$ be their kernels. Show that $\mathfrak{a} + \mathfrak{b} = T$, that $\mathfrak{a}\mathfrak{b} = \{0\}$ and that $\mathfrak{b}\mathfrak{a} = \mathfrak{b} \cap \mathfrak{a}$ is $\neq \{0\}$ if $A \neq \{0\}$ (cf. no. 9, Proposition 6).

(*d*) If $A = \mathbf{Z}/2\mathbf{Z}$, T is not commutative, but the multiplicative group of invertible elements of T is commutative.

3. Let A be a ring and $a \in A$. If there exists one *and only one* $a' \in A$ such that $aa' = 1$, a is invertible and $a' = a^{-1}$ (show first that a is left cancellable, then consider the product $aa'a$).

4. Let A be a pseudo-ring in which every additive subgroup of A is a left ideal of A.

(*a*) If A is a ring, A is canonically isomorphic to $\mathbf{Z}/n\mathbf{Z}$ for some suitable integer n.

(*b*) If A has no unit element and every non-zero element is cancellable, A is isomorphic to a pseudo-subring of $\mathbf{Z}$.

5. Let M be a monoid, $\mathbf{Z}^{(M)}$ the free commutative group with basis M and $u: M \to \mathbf{Z}^{(M)}$ the canonical mapping.

(*a*) Show that there exists a unique multiplication on $\mathbf{Z}^{(M)}$ such that $\mathbf{Z}^{(M)}$ is a ring and such that u is a monoid morphism of the monoid M into the multiplicative monoid of the ring $\mathbf{Z}^{(M)}$.

(*b*) Let A be a ring and $v: M \to A$ a monoid morphism of the monoid M into the multiplicative monoid of A. There exists a unique ring morphism $f: \mathbf{Z}^{(M)} \to A$ such that

$$f \circ u = v.$$

6. Let A be a commutative ring and M a submonoid of the additive group of A. Let n be an integer >0 such that $m^n = 0$ for $m \in M$.

(*a*) If $n!$ is not a divisor of zero in A, the product of a family of n elements of M is zero.

(*b*) Show that the conclusion in (*a*) is not necessarily true if $n!$ is a divisor of zero in A.

7. For all $n \in \mathbf{N}^*$,

$$n! = \sum_{\nu=0}^{n-1} (-1)^{\nu}\binom{n}{\nu}(n-\nu)^n$$

(use no. 2, Proposition 2).

8. In a ring, the right ideal generated by a left ideal is a two-sided ideal.

9. In a ring, the right annihilator of a right ideal is a two-sided ideal.

10. In a ring A, the two-sided ideal generated by the elements $xy - yx$, where x and y run through A, is the smallest of the two-sided ideals $\mathfrak{a}$ such that $A/\mathfrak{a}$ is commutative.

¶ 11. In a ring A, a two-sided ideal $\mathfrak{a}$ is said to be *irreducible* if there exists no ordered pair of two-sided ideals $\mathfrak{b}, \mathfrak{c}$ distinct from $\mathfrak{a}$ and such that $\mathfrak{a} = \mathfrak{b} \cap \mathfrak{c}$.

(*a*) Show that the intersection of all the irreducible two-sided ideals of A reduces to 0 (note that the set of two-sided ideals containing no element $a \neq 0$ is inductive and apply Zorn's Lemma).

(*b*) Deduce that every two-sided ideal $\mathfrak{a}$ of A is the intersection of all the irreducible ideals which contain it.

12. Let A be a ring and $(\mathfrak{a}_\iota)_{\iota \in I}$ a finite family of left ideals of A such that the additive group of A is the direct sum of the additive groups $\mathfrak{a}_\iota$. If $1 = \sum_{\iota \in I} e_\iota (e_\iota \in \mathfrak{a}_\iota)$, then $e_\iota^2 = e_\iota$, $e_\iota e_j = 0$ if $\iota \neq j$ and $\mathfrak{a}_\iota = Ae_\iota$ (write $x = x \, . \, 1$ for all $x \in A$). Conversely, if $(e_\iota)_{\iota \in I}$ is a finite family of idempotents of A such that $e_\iota e_j = 0$ for $\iota \neq j$ and $1 = \sum_\iota e_\iota$, then A is the direct sum of the left ideals Ae_ι. For the ideals Ae_ι to be two-sided, it is necessary and sufficient that the e_ι be in the centre of A.

¶ 13. Let A be a ring and e an idempotent of A.

(*a*) Show that the additive group of A is the direct sum of the left ideal $\mathfrak{a} = Ae$ and the left annihilator $\mathfrak{b}$ of e (note that, for all $x \in A$, $x - xe \in \mathfrak{b}$).

(*b*) Every right ideal $\mathfrak{d}$ of A is the direct sum of $\mathfrak{d} \cap \mathfrak{a}$ and $\mathfrak{d} \cap \mathfrak{b}$.

(*c*) If $Ae = eA$, $\mathfrak{a}$ and $\mathfrak{b}$ are two-sided ideals of A and define a direct decomposition of A.

14. Let A be a commutative ring in which there is only a finite number n of divisors of zero. Show that A has at most $(n + 1)^2$ elements. (Let $a_1, \ldots, a_n$ be the divisors of zero. Show that the annihilator $\mathfrak{Z}_1$ of a_1 has at most $n + 1$ elements and that the quotient ring $A/\mathfrak{Z}_1$ has at most $n + 1$ elements.)

15. A *ringoid* is a set E with two laws of composition: (*a*) an associative

multiplication xy; (*b*) a law written additively, *not everywhere defined* (§ 1, no. 1) and satisfying the following conditions:

(1) it is *commutative* (in other words, if $x + y$ is defined, so is $y + x$ and $x + y = y + x$; x and y are then said to be *addible*);

(2) if x and y are addible, for $x + y$ and z to be addible it suffices that x and y on the one hand, and y and z on the other, be addible; then x and $y + z$ are addible and $(x + y) + z = x + (y + z)$;

(3) there exists an identity element 0;

(4) if x and z on the one hand, y and z on the other, are addible and $x + z = y + z$, then $x = y$;

(5) if x, y are addible, so are xz and yz (resp. zx and zy) and

$$(x + y)z = xz + yz$$

(resp. $z(x + y) = zx + zy$) for all $z \in E$.

Every ring is a ringoid.

(*a*) Examine how the definitions and results of § 8 and the above exercises extend to ringoids (a left ideal of a ringoid E is a subset $\mathfrak{a}$ of E which is stable under addition and such that $E.\mathfrak{a} \subset \mathfrak{a}$).

(*b*) Let G be a group with operators and f and g two endomorphisms of G. For the mapping $x \mapsto f(x)g(x)$ to be an endomorphism of G, it is necessary and sufficient that every element of the subgroup $f(G)$ be permutable with every element of $g(G)$; if this endomorphism is then denoted by $f + g$ and fg is the composite endomorphism $x \mapsto f(g(x))$, show that the set E of endomorphisms of G with these laws of composition is a *ringoid* with unit element; for E to be a ring, it is necessary and sufficient that G be commutative.

For an element $f \in E$ to be addible to all the elements of E, it is necessary and sufficient that $f(G)$ be contained in the centre of G; the set N of these endomorphisms is a *pseduo-ring* called the *kernel* of the ringoid E.

An endomorphism f of G is called *normal* if it is permutable with all the inner automorphisms of G; for every normal stable subgroup H of G, $f(G)$ is then a normal stable subgroup of G. Show that the set D of normal endomorphisms of G forms a subringoid of E and that the kernel N is a two-sided ideal in D.†

16. Give examples of ideals $\mathfrak{a}, \mathfrak{b}, \mathfrak{c}$ in the ring **Z** such that $\mathfrak{a}\mathfrak{b} \neq \mathfrak{a} \cap \mathfrak{b}$ and $(\mathfrak{a} + \mathfrak{b})(\mathfrak{a} + \mathfrak{c}) \neq \mathfrak{a} + \mathfrak{b}\mathfrak{c}$.

§ 9

1. Which are the field structures amongst the ring structures determined in § 8, Exercise 1?

† See H. Fitting, Die Theorie der Automorphismenringe Abelscher Gruppen und ihre Analogen bei nicht kommutativen Gruppen, *Math. Ann.*, **107** (1933), 514.

2. A *finite* ring in which every non-zero element is cancellable is a field (cf. § 2, Exercise 6).

3. Let G be a commutative group with operators which is simple (§ 4, no. 4, Definition 7). Show that the endomorphism ring A of G is a field.

4. Let K be a field. Show that there exists a smallest subfield F of K and that F is canonically isomorphic either to the field $\mathbf{Q}$ or to the field $\mathbf{Z}/p\mathbf{Z}$ for some prime number p. In the first case (resp. second case) K is said to be of *characteristic* 0 (resp. p).

5. Let K be a commutative field of characteristic $\neq 2$; let G be a subgroup of the additive group of K such that, if H denotes the set consisting of 0 and the inverses of the elements $\neq 0$ of G, H is also a subgroup of the additive group of K. Show that there exists an element $a \in K$ and a subfield K' of K such that $G = aK'$ (establish first that, if x and y are elements of G such that $y \neq 0$, then $x^2/y \in G$; deduce that, if x, y, z are elements of G such that $z \neq 0$, then $xy/z \in G$).

6. Let K be a commutative field of characteristic $\neq 2$; let f be a mapping of K into K such that $f(x + y) = f(x) + f(y)$ for all x and y and for all $x \neq 0$ $f(x)f(1/x) = 1$. Show that f is an isomorphism of K onto a subfield of K (prove that $f(x^2) = (f(x))^2$).

7. Let A be a commutative ring.

(*a*) Every prime ideal is irreducible (§ 8, Exercise 11).

(*b*) The set of prime ideals is inductive for the relations $\subset$ and $\supset$.

(*c*) Let $\mathfrak{a}$ be an ideal of A distinct from A; let $\mathfrak{b}$ be the set of $x \in A$ such that there exists an integer $n \geqslant 0$ with $x^n \in \mathfrak{a}$ (n depends on x). Show that $\mathfrak{b}$ is an ideal and is equal to the intersection of the prime ideals of A containing $\mathfrak{a}$.

¶ 8. A ring A is called a *Boolean ring* if each of its elements is idempotent (in other words, if $x^2 = x$ for all $x \in A$).

(*a*) In the set $\mathfrak{P}(E)$ of subsets of a set E, show that a Boolean ring structure is defined by setting $AB = A \cap B$ and $A + B = (A \cap \complement B) \cup (B \cap \complement A)$. This ring is isomorphic to the ring K^E of mappings of E into the ring $K = \mathbf{Z}/(2)$ of integers modulo 2 (consider for each subset $X \subset E$ its "characteristic function" ϕ_X such that $\phi_X(x) = 1$ if $x \in X$, $\phi_X(x) = 0$ if $x \notin X$).

(*b*) Every Boolean ring A is commutative and such that $x + x = 0$ for $x \in A$ (write $x + x$ as an idempotent and then $x + y$ as an idempotent).

(*c*) If a Boolean ring A contains no divisor of 0, it is reduced to 0 or is isomorphic to $\mathbf{Z}/(2)$ (if x and y are any two elements of A, show that $xy(x + y) = 0$). Deduce that in a Boolean ring every prime ideal is maximal.

(*d*) In a Boolean ring A, every ideal $\mathfrak{a} \neq A$ is the intersection of the prime ideals containing $\mathfrak{a}$ (apply Exercise 7(*c*)). Deduce that every irreducible ideal is maximal (in other words, that the notions of irreducible ideal, prime ideal and maximal ideal coincide in a Boolean ring).

(*e*) Show that every Boolean ring is isomorphic to a subring of a product ring K^E, where $K = \mathbf{Z}/(2)$ (use (*d*)).

(*f*) If $\mathfrak{p}_i$ $(1 \leqslant i \leqslant n)$ are n distinct maximal ideals in a Boolean ring A and $\mathfrak{a} = \bigcap_{1 \leqslant i \leqslant n} \mathfrak{p}_i$, show that the quotient ring $A/\mathfrak{a}$ is isomorphic to the product ring K^n. Deduce that every finite Boolean ring is of the form K^n.

(*g*) In a Boolean ring A, the relation $xy = x$ is an order relation; if it is denoted by $x \leqslant y$, show that with this relation A is a distributive lattice (*Set Theory*, III, § 1, Exercise 16), has a least element α and that, for every ordered pair (x, y) of elements such that $x \leqslant y$, there exists an element $d(x, y)$ such that $\inf(x, d(x, y)) = \alpha$, $\sup(x, d(x, y)) = y$. Conversely, if an ordered set A has these three properties, show that the laws of composition $xy = \inf(x, y)$ and $x + y = d(\sup(x, y), \inf(x, y))$ define a Boolean ring structure on A.

9. Let A be a ring such that $x^3 = x$ for all $x \in A$. It is proposed to prove that A is commutative.

(*a*) Show that $6A = \{0\}$ and that 2A and 3A are two-sided ideals such that $2A + 3A = A$ and $2A \cap 3A = \{0\}$. Deduce that it can be assumed that either $2A = \{0\}$ or $3A = \{0\}$.

(*b*) If $2A = \{0\}$, calculate $(1 + x)^3$, deduce that $x^2 = x$ for all $x \in A$ and conclude by means of Exercise 8.

(*c*) If $3A = \{0\}$, calculate $(x + y)^3$ and $(x - y)^3$, show that $x^2y + xyx + yx^2 = 0$ and left multiply by x; deduce that $xy - yx = 0$.

(*d*) Let A be a ring such that $3A = \{0\}$ and $x^3 = x$ for all $x \in A$; define a set I and an injective homomorphism of A into $(\mathbf{Z}/3\mathbf{Z})^I$ (same method as in Exercise 8).

10. Let A be a commutative ring and $\mathfrak{R}$ the intersection of its maximal ideals.

(*a*) Show that the canonical homomorphism $A^* \to (A/\mathfrak{R})^*$ is surjective and that its kernel is $1 + \mathfrak{R}$.

(*b*) Suppose that the set M of maximal ideals of A is finite and that A^* is finite. Deduce that A is finite (apply Proposition 8 of § 8, no. 10 to $A/\mathfrak{R}$).

(*c*) Deduce that the set of prime numbers is infinite (note that $\mathbf{Z}^* = \{1, -1\}$).

(*d*) Suppose that A^* is finite and that $A/\mathfrak{m}$ is finite for every maximal ideal $\mathfrak{m}$. Can it be deduced that A is finite, that A is countable?

11. Let P be the set of prime numbers and A the product ring of the fields $\mathbf{Z}/p\mathbf{Z}$, $p \in P$. Let $\mathfrak{a}$ be the subset of A consisting of the elements $(a_p)_{p \in P}$ such that $a_p \neq 0$ only for a finite number of indices p.

(*a*) $\mathfrak{a}$ is an ideal of A.

(*b*) Let $B = A/\mathfrak{a}$. For every integer $n > 0$ and every $b \neq 0$ in B, there exists one and only one element b' of B such that $nb' = b$ (note that if $p \in P$ does not divide n, multiplication by n in $\mathbf{Z}/p\mathbf{Z}$ is bijective). Deduce that B contains a subfield isomorphic to $\mathbf{Q}$.

12. Let A be a commutative ring and $a, b \in A$. Show that the canonical image of ab in $A/(a - a^2b)$ is an idempotent. Give an example where this idempotent is distinct from 0 and 1.

13. Determine the endomorphisms of the ring $\mathbf{Z}$ and the ring $\mathbf{Z} \times \mathbf{Z}$. More generally, if I and J are finite sets, determine the homomorphisms of the ring $\mathbf{Z}^I$ into the ring $\mathbf{Z}^J$.

14. Let A and B be two rings and f and g two homomorphisms of A into B. Is the set of $x \in A$ such that $f(x) = g(x)$ a subring of A, an ideal of A?

¶ 15. Let A be a non-commutative ring with no divisors of 0. A is said to admit a *field of left fractions* if it is isomorphic to a subring B of a field K such that every element of K is of the form $x^{-1}y$, where $x \in B$, $y \in B$.

(*a*) Let A′ be the set of elements $\neq 0$ in A. For A to admit a field of left fractions, it is necessary that the following condition be fulfilled:

(G) for all $x \in A$, $x' \in A'$, there exist $u \in A'$ and $v \in A$ such that $ux = vx'$.

(*b*) Suppose conversely that condition (G) is fulfilled. Show that in the set $A \times A'$ the relation R between (x, x') and (y, y') which states "for every ordered pair (u, v) of elements $\neq 0$ such that $ux' = vy'$, $ux = vy$" is an equivalence relation.

Let (x, x'), (y, y') be two elements of $A \times A'$, ξ and η their respective classes (mod. R). For every ordered pair $(u, u') \in A \times A'$ such that $u'x = uy'$, show that the class (mod. R) of $(uy, u'y')$ depends only on the classes ξ and η; if it is denoted by $\xi\eta$, a law of composition is defined on the set $K = (A \times A')/R$. If K′ is the set of elements of K distinct from the class 0 of elements $(0, x')$ of $A \times A'$, K′, with the law induced by the above law, is a group.

For every element $x \in A$, the elements $(x'x, x')$, where x' runs through A′, define an isomorphism of A (with multiplication alone) onto a subring of K. Identifying A with its image under this isomorphism, the class (mod. R) of an ordered pair $(x, x') \in A \times A'$ is identified with the element $x'^{-1}x$.

This being so, if $\xi = x'^{-1}x$ is an element of K and 1 denotes the unit element of K′, then $\xi + 1$ denotes the element $x'^{-1}(x + x')$, which does not depend on the representation of ξ in the form $x'^{-1}x$. We then write $\xi + 0 = \xi$ and, for $\eta \neq 0$, $\xi + \eta = \eta(\eta^{-1}\xi + 1)$. Show that the addition and multiplication thus defined on K determine on that set a field structure which extends the ring structure on A; in other words, condition (G) is *sufficient* for A to admit a field of left fractions.

16. Let A be a ring in which every non-zero element is cancellable. For A to admit a field of left fractions (Exercise 15), it is necessary and sufficient that in A the intersection of two left ideals distinct from 0 never reduce to 0.

17. Let $(K_i)_{i \in I}$ be a family of fields. For every subset J of I, let e_J denote the element of the product $K = \prod_{i \in I} K_i$ whose i-th component is equal to 0 if $i \in J$ and to 1 if $i \in I - J$.

(*a*) Let $x = (x_i)$ be an element of K and $J(x)$ the set of elements $i \in I$ such that $x_i = 0$. Show that there exists an invertible element u in K such that $x = ue_{J(x)} = e_{J(x)}u$.

(*b*) Let $\mathfrak{a}$ be a left (resp. right) ideal of K distinct from K. Let $\mathfrak{F}_\mathfrak{a}$ be the set of subsets J of I such that $e_J \in \mathfrak{a}$. Show that $\mathfrak{F}_\mathfrak{a}$ has the following properties:

(b_1) If $J \in \mathfrak{F}_\mathfrak{a}$ and $J' \supset J$, then $J' \in \mathfrak{F}_\mathfrak{a}$.

(b_2) If $J_1, J_2 \in \mathfrak{F}_\mathfrak{a}$, then $J_1 \cap J_2 \in \mathfrak{F}_\mathfrak{a}$.

(b_3) $\varnothing \notin \mathfrak{F}_\mathfrak{a}$.

Show that $x \in \mathfrak{a}$ is equivalent to $J(x) \in \mathfrak{F}_\mathfrak{a}$ (use (*a*)). Deduce that $\mathfrak{a}$ is a two-sided ideal.

(*c*) A set of subsets of I is called a *filter* if it satisfies properties (b_1), (b_2), (b_3) above (*cf. *General Topology*, I, § 6*). Show that the mapping $\mathfrak{a} \mapsto \mathfrak{F}_\mathfrak{a}$ is a strictly increasing bijection of the set of ideals of K distinct from K onto the set of filters of I. *Show that $\mathfrak{a}$ is maximal if and only if $\mathfrak{F}_\mathfrak{a}$ is an ultrafilter.*

(*d*) *Let $\mathfrak{F}$ be a non-trivial ultrafilter on the set P of prime numbers, let $\mathfrak{a}$ be the corresponding ideal of the ring $K = \prod_{p \in P} \mathbf{Z}/p\mathbf{Z}$ and let $k = K/\mathfrak{a}$. Show that k is a field of characteristic 0.*

¶ 18. (*a*) Let K be a field and a, b two *non-permutable* elements of K. Show that

(1) $a = (b - (a-1)^{-1}b(a-1))(a^{-1}ba - (a-1)^{-1}b(a-1))^{-1}$

(2) $a = (1 - (a-1)^{-1}b^{-1}(a-1)b)(a^{-1}b^{-1}ab - (a-1)^{-1}b^{-1}(a-1)b)^{-1}$.

(*b*) Let K be a non-commutative field and x an element not belonging to the centre of K. Show that K is generated by the set of conjugates axa^{-1} of x. (Let K_1 be the subfield of K generated by the set of conjugates of x. Deduce from (*a*) that, if $K_1 \neq K$ and $a \in K_1$ and $b \notin K_1$, then necessarily $ab = ba$. Deduce a contradiction by considering in K_1 two non-permutable elements a, a' and an element $b \notin K_1$ and noting that $ba \notin K_1$.)

(*c*) Deduce from (*b*) that if Z is the centre of K and H is a subfield of K such that $Z \subset H \subset K$ and $aHa^{-1} = H$ for all $a \neq 0$ in K, then $H = Z$ or $H = K$ (*Cartan-Brauer-Hua Theorem*).

(*d*) Deduce from (*a*) that in a non-commutative field K, the set of commutators $aba^{-1}b^{-1}$ of elements $\neq 0$ in K generates the field K.

19. Let A be a non-zero pseudo-ring such that, for all $a \neq 0$ in A and all $b \in A$, the equation $ax + ya = b$ has solutions consisting of ordered pairs (x, y) of elements of A; in other words, $aA + Aa = A$ for $a \neq 0$ in A.

(*a*) Show that A is the only non-zero two-sided ideal in A.

(*b*) Show that the relation $a \neq 0$ implies $a^2 \neq 0$ (note that if $a^2 = 0$, it would follow that $aAa = \{0\}$).

(*c*) Show that if $ab = 0$ and $a \neq 0$ (resp. $b \neq 0$) then $b = 0$ (resp. $a = 0$). (Note that $(ba)^2 = 0$ and deduce that $bAb = \{0\}$.)

20. Let A be a non-zero pseudo-ring such that there exists $a \in A$ for which, for all $b \in A$, one of the equations $ax = b$, $xa = b$ admits a solution $x \in A$. Show that if A has no divisor of 0 other than 0, A admits a unit element.

21. Let A be a non-zero pseudo-ring in which, for all $a \neq 0$ and all $b \in A$, one of the equations $ax = b$, $xa = b$ admits a solution $x \in A$. Show that A is a field (use Exercises 19 and 20).

§ 10

1. Let X be the direct limit of a direct system $(X_\iota, f_{j\iota})$ of sets relative to a right directed set I. Let F_ι (resp. F) be the free group constructed on X_ι (resp. on X); each $f_{j\iota}$ extends to a homomorphism $\phi_{j\iota} : F_\iota \to F_j$.

(*a*) Show that $(F_\iota, \phi_{j\iota})$ is a direct system of groups.

(*b*) Let ε_ι be the composite mapping $X_\iota \to F_\iota \to \varinjlim F_\iota$. Show that $\varepsilon_j \circ \phi_{j\iota} = \varepsilon_\iota$ if $j \geqslant \iota$. Let ε be the mapping of X into $\varinjlim F_\iota$ defined by the ε_ι. Show that ε extends to an isomorphism of F onto $\varinjlim F_\iota$ (so that $F(\varinjlim X_\iota)$ can be identified with $\varinjlim F(X_\iota)$).

(*c*) State and prove analogous results for free magmas, free monoids, free commutative groups, free commutative monoids.

2. Show that a direct limit of simple groups is either a simple group or a group with one element.

HISTORICAL NOTE

(Numbers in brackets refer to the bibliography
at the end of this Note.)

There are few notions in Mathematics more primitive than that of a law of composition: it seems inseparable from the first rudiments of calculations on natural numbers and measurable quantities. The most ancient documents that have come down to us on the Mathematics of the Egyptians and Babylonians show that they already had a complete system of rules for calculating with natural numbers >0, rational numbers >0, lengths and areas; although the preserved texts deal only with problems in which the given quantities have explicit numerical values,† they leave us in no doubt concerning the generality they attributed to the rules employed and show a quite remarkable technical ability in the manipulation of equations of the first and second degree ([1], p. 179 et seq.). On the other hand there is not the slightest trace of a concern to justify the rules used nor even of precise definitions of the operations which occur: the latter as the former arise in a purely empirical manner.

Just such a concern, however, is already very clearly manifest in the works of the Greeks of the classical age; true, an axiomatic treatment is not found of the theory of natural numbers (such an axiomatization was only to appear at the end of the nineteenth century; see the Historical Note to *Set Theory*, IV); but in Euclid's *Elements* there are numerous passages giving formal proofs of rules of calculation which are just as intuitively "obvious" as those of calculation with integers (for example, the commutativity of the product of two rational numbers). The most remarkable proofs of this nature are those

† It must not be forgotten that Viete (16th century) was the first to use letters to denote all the elements (known and unknown) occurring in an algebraic problem. Until then the only problems solved in treatises on Algebra had numerical coefficients; when an author stated a general rule for treating analogous equations, he did so (as best he could) in ordinary language; in the absence of an explicit statement of this type, the way in which the calculations were carried out in the numerical cases dealt with rendered the possession of such a rule more or less credible.

relating to the *theory of magnitudes*, the most original creation of Greek Mathematics (equivalent, as is known, to our theory of real numbers >0; see the Historical Note to *General Topology*, IV); here Euclid considers amongst other things the product of two ratios of magnitudes and shows that it is independent of the form of presentation of these ratios (the first example of a "quotient" of a law of composition by an equivalence relation in the sense of § 1, no. 6) and that they commute ([2], Book V, Prop. 22–23).†

It must not however be concealed that this progress towards rigour is accompanied in Euclid by a stagnation and even in some ways by a retrogression as far as the technique of algebraic calculations is concerned. The overwhelming preponderance of Geometry (in the light of which the theory of magnitudes was obviously conceived) paralyses any independent development of algebraic notation: the elements entering into the calculations must always be "represented" geometrically; moreover the two laws of composition which occur are not defined on the same set (addition of ratios is not defined in a general way and the product of two lengths is not a length but an area); the result is a lack of flexibility making it almost impracticable to manipulate algebraic relations of degree greater than the second.

Only with the decline of classical Greek Mathematics does Diophantus return to the tradition of the "logisticians" or professional calculators, who had continued to apply such rules as they had inherited from the Egyptians and Babylonians: no longer encumbering himself with geometric representations of the "numbers" he considers, he is naturally led to the development of rules of abstract algebraic calculation; for example, he gives rules which (in modern language) are equivalent to the formula $x^{m+n} = x^m x^n$ for small values (positive or negative) of m and n ([3], vol. I, pp. 8–13); a little later (pp. 12–13), the "rule of signs" is stated, the beginnings of calculating with negative numbers‡; finally, Diophantus uses for the first time a letter to represent an unknown in an equation. In contrast, he seems little concerned to attach a general significance to the methods he applies for the solution of his problems; as for the axiomatic conception of laws of composition as begun by Euclid, this appears foreign to Diophantus' thought as to that of his immediate successors; it only reappears in Algebra at the beginning of the 19th century.

There were first needed, during the intervening centuries, on the one hand the development of a system of algebraic notation adequate for the expression

† Euclid gives at this point, it is true, no formal definition of the product of two ratios and the one which is found a little later on in the *Elements* (Book VI, Definition 5) is considered to be interpolated: he has of course a perfectly clear conception of this operation and its properties.

‡ It seems that Diophantus was not acquainted with negative numbers; this rule can therefore only be interpreted as relating to the calculus of polynomials and allowing the "expansion" of such products as $(a - b)(c - d)$.

of abstract laws and on the other a broadening of the notion of "number" sufficient to give it, through the observation of a wide enough variety of special cases, a general conception. For this purpose, the axiomatic theory of ratios of magnitudes created by the Greeks was insufficient, for it did no more than make precise the intuitive concept of a real number >0 and the operations on these numbers, which were already known to the Babylonians in a more confused form; now "numbers" would be considered of which the Greeks had had no concept and which to begin with had no sensible "representation": on the one hand, zero and negative numbers which appeared in the High Middle Ages in Hindu Mathematics; on the other, imaginary numbers, the creation of Italian algebraists of the 16th century.

With the exception of zero, introduced first as a separating sign before being considered as a number (see the Historical Note to *Set Theory*, III), the common character of these extensions is that (at least to begin with) they were purely "formal". By this it must be understood that the new "numbers" appeared first of all as the result of operations applied to situations where, keeping to their strict definition, they had no meaning (for example, the difference $a - b$ of two natural numbers when $a < b$): whence the names "false", "fictive", "absurd", "impossible", "imaginary", etc., numbers attributed to them. For the Greeks of the Classical Period, enamoured above all of clear thought, such extensions were inconceivable; they could only arise with calculators more disposed than were the Greeks to display a somewhat mystic faith in the power of their methods ("the generality of Analysis" as the 18th century would say) and to allow themselves to be carried along by the mechanics of their calculations without investigating whether each step was well founded; a confidence moreover usually justified *a posteriori*, by the exact results to which the extension led, with these new mathematical entities, of rules of calculation uniquely valid, in all rigour, for the numbers which were already known. This explains how little by little these generalizations of the concept of number, which at first occurred only as intermediaries in a sequence of operations whose starting and finishing points were genuine "numbers", came to be considered for their own sakes (independent of any application to concrete calculations); once this step had been taken, more or less tangible interpretations were sought, new objects which thus acquired the right to exist in Mathematics.†

On this subject, the Hindus were already aware of the interpretation to be given to negative numbers in certain cases (a debt in a commercial problem, for example). In the following centuries, as the methods and results of Greek

† This search was merely a transitory stage in the evolution of the concepts in question; from the middle of the 19th century there has been a return, this time fully deliberate, to a formal conception of the various extensions of the notion of number, a conception which has finally become integrated in the "formalistic" and axiomatic point of view which dominates the whole of modern mathematics.

and Hindu mathematics spread to the West (via the Arabs), the manipulation of these numbers became more familiar and they took on other "representations" of a geometric or kinematic character. With the progressive improvement in algebraic notation, this was the only notable progress in Algebra during the close of the Middle Ages.

At the beginning of the 16th century, Algebra took on a new impulse thanks to the discovery by mathematicians of the Italian school of the solution "by radicals" of equations of the 3rd and then of the 4th degree (of which we shall speak in more detail in the Historical Note to *Algebra*, V); it is at this point that, in spite of their aversion, they felt compelled to introduce into their calculations imaginary numbers; moreover, little by little confidence is born in the calculus of these "impossible" numbers, as in that of negative numbers, even though no "representation" of them was conceived for more than two centuries.

On the other hand, algebraic notation was decisively perfected by Viete and Descartes; from the latter onwards, algebraic notation is more or less that which we use today.

From the middle of the 17th to the end of the 18th century, it seems that the vast horizons opened by the creation of Infinitesimal Calculus resulted in something of a neglect of Algebra in general and especially of mathematical reflection on laws of composition and on the nature of real and complex numbers.† Thus the composition of forces and the composition of velocities, well known in Mechanics from the end of the 17th century, had no repercussions in Algebra, even though they already contained the germ of vector calculus. In fact it was necessary to await the movement of ideas which, about 1800, led to the geometric representation of complex numbers (see the Historical Note to *General Topology*, VIII) to see addition of vectors used in Pure Mathematics.‡

It was round about the same time that for the first time in Algebra the notion of law of composition was extended in two different directions to elements which have only distant analogies with "numbers" (in the broadest sense of the word up till then). The first of these extensions is due to C. F. Gauss occasioned by his arithmetical researches on quadratic forms $ax^2 + bxy + cy^2$ with integer coefficients. Lagrange had defined on the set of forms with the

† The attempts of Leibniz should be excepted, on the one hand to give an algebraic form to the arguments of formal logic, on the other to found a "geometric calculus" operating directly on geometric objects without using coordinates ([4], vol. V, p. 141). But these attempts remained as sketches and found no echo in his contemporaries; they were only to be taken up again during the 19th century (see below).

‡ This operation was moreover introduced without reference to Mechanics; the connection between the two theories was only explicitly recognized by the founders of Vector Calculus in the second quarter of the 19th century.

same discriminant an equivalence relation† and had on the other hand proved an identity which provided this set with a commutative law of composition (not everywhere defined); starting from these results, Gauss showed that this law is compatible (in the sense of § 1, no. 6) with a refined form of the above equivalence relation ([5], vol. I, p. 272): "*This shows*", he then says, "*what must be understood by the composite class of two or several classes*". He then proceeds to study the "quotient" law, establishes essentially that it is (in modern language) a commutative group law and does so with arguments whose generality for the most part goes far beyond the special case considered by Gauss (for example, the argument by which he proves the uniqueness of the inverse element is identical with the one we have given for an arbitrary law of composition in § 2, no 3 (*ibid.*, p. 273)). Nor does he stop there: returning a little later to this question, he recognizes the analogy between composition of classes and multiplication of integers modulo a prime number‡ (*ibid.*, p. 371), but proves also that the group of classes of quadratic forms of given discriminant is not always a cyclic group; the indications he gives on this subject show clearly that he had recognized, at any rate in this particular case, the general structure of finite Abelian groups, which we shall study in Chapter VII ([5], vol. I, p. 374 and vol. II, p. 266).

The other series of research of which we wish to speak also ended in the concept of a group and its definitive introduction into Mathematics: this is the "theory of substitutions", the development of the ideas of Lagrange, Vandermonde and Gauss on the solution of algebraic equations. We do not give here the detailed history of this subject (see the Historical Note to *Algebra*, V); we must refer to the definition by Ruffini and then Cauchy ([6], (2), vol. I, p. 64) of the "product" of two permutations of a finite set,§

† Two forms are equivalent when one of them transforms into the other by a "change of variables" $x' = \alpha x + \beta y$, $y' = \gamma x + \delta y$, where $\alpha, \beta, \gamma, \delta$ are integers such that $\alpha\delta - \beta\gamma = 1$.

‡ It is quite remarkable that Gauss writes composition of classes of quadratic forms *additively*, in spite of the analogy which he himself points out and in spite of the fact that Lagrange's identity, which defines the composition of two forms, much more naturally suggests multiplicative notation (to which moreover all Gauss's successors returned). This indifference in notational matters must further bear witness to the generality at which Gauss had certainly arrived in his concept of laws of composition. Moreover he did not restrict his attention to commutative laws, as is seen from a fragment not published in his lifetime but dating from the years 1819–1820, where he gives, more than twenty years before Hamilton, the formulae for multiplication of quaternions ([5], vol. VIII, p. 357).

§ The notion of composite function was of course known much earlier, at any rate for functions of real or complex variables; but the algebraic aspect of this law of composition and the relation with the product of two permutations only came to light in the works of Abel ([7], vol. I, p. 478) and Galois.

and the first notions concerning finite permutation groups: transitivity, primitivity, identity element, permutable elements, etc. But these first results remain, as a whole, quite superficial and Evariste Galois must be considered as the true initiator of the theory: having, in his memorable work [8], reduced the study of algebraic equations to that of permutation groups associated with them, he took the study of the latter considerably deeper, both with regard to the general properties of groups (Galois was the first to define the notion of normal subgroup and to recognize its importance) and with regard to the determination of groups with particular properties (where the results he obtained are still today considered to be amongst the most subtle in the theory). The first idea of the "linear representation of groups" also goes back to Galois† and this fact shows clearly that he possessed the notion of *isomorphism* of two group structures, independently of their "realizations".

However, if the ingenious methods of Gauss and Galois incontestably led them to a very broad conception of the notion of law of composition, they did not develop their ideas particularly on this point and their works had no immediate effect on the evolution of Abstract Algebra.‡ The clearest progress towards abstraction was made in a third direction: after reflecting on the nature of imaginary numbers (whose geometric representation had brought about, at the beginning of the 19th century, a considerable number of works), algebraists of the English school were the first, between 1830 and 1850, to isolate the abstract notion of law of composition and immediately they broadened the field of Algebra by applying this notion to a host of new mathematical entities: the algebra of Logic with Boole (see the Historical Note to *Set Theory*, IV), vectors and quaternions with Hamilton, general hypercomplex systems, matrices and non-associative laws with Cayley ([10], vol. I, pp. 127 and 301 and vol. II, pp. 185 and 475). A parallel evolution was pursued independently on the Continent, notably in Vector Calculus (Möbius, Bellavitis), Linear Algebra and hypercomplex systems (Grassmann), of which we shall speak in more detail in the Historical Note to *Algebra*, III.§

From all this ferment of original and fertile ideas which revitalized Algebra in the first half of the 19th century the latter emerged thoroughly renewed. Previously methods and results had gravitated round the central problem of the solution of algebraic equations (or Diophantine equations in the Theory

---

† Here it is that Galois, by a daring extension of the "formalism" which had led to complex numbers, considers "imaginary roots" of a congruence modulo a prime number and thus discovers *finite fields*, which we shall study in *Algebra*, V.

‡ Moreover those of Galois remained unknown until 1846 and those of Gauss only exercised a direct influence on the Theory of Numbers.

§ The principal theories developed during this period are to be found in a remarkable contemporary exposition of H. Hankel [11], where the abstract notion of law of composition is conceived and presented with perfect clarity.

of Numbers): "*Algebra*", said Serret in the Introduction to his Course of Higher Algebra [12], "*is, properly speaking, the Analysis of equations*". After 1850, if the treatises on Algebra still gave preeminence to the theory of equations for some time to come, new research works were no longer dominated by the concern for immediate applications to the solution of numerical equations but were oriented more and more towards what today is considered to be the essential problem of Algebra, the study of algebraic structures for their own sake.

These works fall quite neatly into three main currents which extend respectively the three movements of ideas which we have analysed above and pursue parallel courses without noticeably influencing one another until the last years of the 19th century.†

The first of these arose out of the work of the German school of the 19th century (Dirichlet, Kummer, Kronecker, Dedekind, Hilbert) in building up the theory of algebraic numbers, issuing out of the work of Gauss who was responsible for the first study of this type, that of the numbers $a + bi$ (a and b rational). We shall not follow the evolution of this theory here (see the Historical Note to *Commutative Algebra*, VII): we need only mention, for our purposes, the abstract algebraic notions which arose here. Right from the first successors of Gauss, the idea of a *field* (of algebraic numbers) is fundamental to all the works on this subject (as also to the works of Abel and Galois on algebraic equations); its field of application was enlarged when Dedekind and Weber [13] modelled the theory of algebraic functions of one variable on that of algebraic numbers. Dedekind [14] was also responsible for the notion of an *ideal* which provides a new example of a law of composition between *sets* of elements; he and Kronecker were responsible for the more and more important role played by commutative groups and modules in the theory of algebraic fields; we shall return to this in the Historical Notes to *Algebra*, III, V and VII.

We shall also refer in the Historical Notes to *Algebra*, III and VIII to the development of Linear Algebra and hypercomplex systems which was pursued without introducing any new algebraic notion during the end of the 19th and the beginning of the 20th century, in England (Sylvester, W. Clifford) and America (B. and C. S. Pierce, Dickson, Wedderburn) following the path traced by Hamilton and Cayley, in Germany (Weierstrass, Dedekind, Frobenius, Molien) and in France (Laguerre, E. Cartan) independently of the Anglo-Saxons and using quite different methods.

---

† We choose here to leave aside all consideration of the development, during this period, of algebraic geometry and the theory of invariants which is closely related to it; these two theories developed following their own methods oriented towards Analysis rather than Algebra and it is only recently that they have found their place in the vast edifice of Modern Algebra.

As far as group theory is concerned, this developed first of all mainly as the theory of finite permutation groups, following the publication of the works of Galois and their diffusion through the works of Serret [12] and above all the great "Traité des Substitutions" of C. Jordan [15]. The latter there developed, greatly improving them, the works of his predecessors on the particular properties of permutation groups (transitivity, primitivity, etc.), obtaining results most of which have not been superseded since; he made an equally profound study of a very important particular class of groups, linear groups and their subgroups; moreover, he it was who introduced the fundamental notion of a homomorphism of one group into another and also (a little later) that of quotient group and who proved a part of the theorem known as the "Jordan-Hölder Theorem".† Finally Jordan was the first to study *infinite* groups [16], which S. Lie on the one hand and F. Klein and H. Poincaré on the other were to develop considerably in two different directions several years later.

In the meantime, what is essential in a group, that is its law of composition and not the nature of the objects which constitute the group, slowly come to be realized (see for example [10], vol. II, pp. 123 and 131 and [14], vol. III, p. 439). However, even the research works on finite abstract groups had for a long time been conceived as the study of permutation groups and only around 1880 was the independent theory of finite groups beginning to develop consciously. We cannot pursue further the history of this theory which is only touched on very superficially in this Treatise; we shall just mention two of the tools which are still today among the most used in the study of finite groups and which both go back to the 19th century: the Sylow theorems‡ on p-groups which date from 1872 [17] and the theory of characters created in the last years of the century by Frobenius [19]. We refer the reader who wishes to go deeply into the theory of finite groups and the many difficult problems it raises, to the monographs of Burnside [20], Speiser [23], Zassenhaus [24] and Gorenstein [27].

Around 1880 too the systematic study of group *presentations* was begun; previously only presentations of particular groups had been encountered, for example the alternating group $\mathfrak{A}_5$ in a work of Hamilton [9 *bis*] or the monodromy groups of linear differential equations and Riemann surfaces (Schwarz, Klein, Schläfli). W. Dyck was the first [18] to define (without giving it a name) the free group generated by a finite number of generators,

† Jordan had only established the invariance (up to order) of the *orders* of the quotient groups of a "Jordan-Hölder series" for a finite group; it was O. Hölder who showed that the quotient groups themselves are (up to order) independent of the series considered.

‡ The existence of a subgroup of order p^n in a group whose order is divisible by p^n is mentioned without proof in the papers of Galois [8].

but he was less interested in it for its own sake than as a "universal" object allowing a precise definition of a group given "by generators and relations". Developing an idea first set forth by Cayley and visibly influenced on the other hand by the works of Riemann's school mentioned above, Dyck described an interpretation of a group of given presentation, where each generator is represented by a product of two inversions with respect to tangent or intersecting circles (see also Burnside [20], Chapter XVIII). A little later, after the development of the theory of automorphic functions by Poincaré and his successors, then the introduction by Poincaré of the tools of Algebraic Topology, the first studies of the fundamental group would be on an equal footing with those of group presentations (Dehn, Tietze), the two theories lending one another mutual support. Moreover it was a topologist, J. Nielsen, who in 1924 introduced the term *free group* in the first deep study of its properties [21], almost immediately afterwards E. Artin (always with reference to questions in topology) introduced the notion of the free product of groups and 0. Schreier defined more generally the free product with amalgamated subgroups [22] (cf. I, § 7, no. 3, Exercise 20). Here again we cannot go into the history of the later developments in this direction, referring the reader to the work [26] for more details.

There is no further room to speak of the extraordinary success, since the end of the 19th century, of the idea of group (and that of *invariant* which is intimately related with it) in Analysis, Geometry, Mechanics and Theoretical Physics. An analogous invasion by this notion and the algebraic notions related to it (groups with operators, rings, ideals, modules) into parts of Algebra which until then seemed quite unrelated is the distinguishing mark of the last period of evolution which we retrace here and which ended with the synthesis of the three branches which we have followed above. This unification is above all the work of the German school of the years 1900–1930: begun by Dedekind and Hilbert in the last years of the 19th century, the work of axiomatization of Algebra was vigorously pursued by E. Steinitz, then, starting in 1920, under the impulse of E. Artin, E. Noether and the algebraists of their school (Hasse, Krull, O. Schreier, van der Waerden). The treatise by van der Waerden [25], published in 1930, brought together these works for the first time in a single exposition, opening the way and serving as a guide to many research works in Algebra during more than twenty years.

Bibliography

1. O. Neugebauer, *Vorlesungen über Geschichte der antiken Mathematik*, Vol. I: Vorgriechische Mathematik, Berlin (Springer), 1934.

2. *Euclidis Elementa*, 5 vols., ed. J. L. Heiberg, Lipsiae (Teubner), 1883–88.

2 *bis*. T. L. Heath, *The thirteen books of Euclid's Elements* . . ., 3 vols., Cambridge, 1908.

3. *Diophanti Alexandrini Opera Omnia . . .*, 2 vols., ed. P. Tannery, Lipsiae (Teubner), 1893–95.
3 *bis*. *Diophante d'Alexandrie*, trad. P. Ver Eecke, Bruges (Desclée-de Brouwer), 1926.
4. G. W. Leibniz, *Mathematische Schriften*, ed. C. I. Gerhardt, vol. V, Halle (Schmidt), 1858.
5. C. F. Gauss, *Werke*, vols. I (Göttingen, 1870), II (*ibid.*, 1863) and VIII (*ibid.*, 1900).
6. A. L. Cauchy, *Oeuvres complètes* (2), vol. I, Paris (Gauthier-Villars), 1905.
7. N. H. Abel, *Oeuvres*, 2 vol., ed. Sylow and Lie, Christiania, 1881.
8. E. Galois, *Ecrits et mémoires mathématiques*, ed. R. Bourgne and J. Y. Azra, Paris (Gauthier-Villars), 1962.
9. W. R. Hamilton, *Lectures on Quaternions*, Dublin, 1853.
9 *bis*. W. R. Hamilton, Memorandum respecting a new system of roots of unity, *Phil. Mag.* (4), **12** (1856), p. 446.
10. A. Cayley, *Collected mathematical papers*, vols. I and II, Cambridge (University Press), 1889.
11. H. Hankel, *Vorlesungen über die complexen Zahlen und ihre Functionen*, 1st part: Theorie der complexen Zahlensysteme, Leipzig (Voss), 1867.
12. J. A. Serret, *Cours d'Algèbre supérieure*, 3rd ed., Paris (Gauthier-Villars), 1866.
13. R. Dedekind and H. Weber, Theorie der algebraischen Funktionen einer Veränderlichen, *Crelle's J.*, **92** (1882), pp. 181–290.
14. R. Dedekind, *Gesammelte mathematische Werke*, 3 vols., Braunschweig (Vieweg), 1932.
15. C. Jordan, *Traité des substitutions et des équations algébriques*, Paris (Gauthier-Villars), 1870 (reprinted, Paris (A. Blanchard), 1957).
16. C. Jordan, Mémoire sur les groups des mouvements, *Ann. di Mat.* (2), **2** (1868), pp. 167–215 and 322–345 (= *Oeuvres*, vol. IV, pp. 231–302, Paris (Gauthier-Villars), 1964).
17. L. Sylow, Théorèmes sur les groups de substitutions, *Math. Ann.*, **5** (1872), pp. 584–594.
18. W. Dyck, Gruppentheoretische Studien, *Math. Ann.*, **20** (1882), pp. 1–44.
19. G. Frobenius, Über Gruppencharaktere, *Berliner Sitzungsber.*, 1896, pp. 985–1021 (= *Gesammelte Abhandlungen*, ed. J. P. Serre, Berlin–Heidelberg–New York (Springer), vol. III (1968), pp. 1–37).
20. W. Burnside, *Theory of groups of finite order*, 2nd ed., Cambridge, 1911.
21. J. Nielsen, Die Isomorphismengruppen der freien Gruppen, *Math. Ann.*, **91** (1924), pp. 169–209.
22. O. Schreier, Die Untergruppen der freien Gruppe, *Abh. Hamb.*, **5** (1927), pp. 161–185.

23. A. Speiser, *Theorie der Gruppen von endlicher Ordnung*, 3rd ed., Berlin (Springer), 1937.
24. H. Zassenhaus, *Lehrbuch der Gruppentheorie*, vol. I, Leipzig–Berlin (Teubner), 1937.
25. B. L. van der Waerden, *Moderne Algebra*, 2nd ed., vol. I, Berlin (Springer), 1937; vol. II (*ibid.*), 1940.
26. W. Magnus, A. Karrass and D. Solitar, *Combinatorial group theory*, New York (Interscience), 1966.
27. D. Gorenstein, *Finite groups*, New York (Harper and Row), 1968.

CHAPTER II

Linear Algebra

This chapter is essentially devoted to the study of a particular type of *commutative groups with operators* (I, § 4, no. 2) called *modules*. Certain properties stated in §§ 1 and 2 for modules extend to all commutative groups with operators; these will be indicated as they occur. Moreover it will be seen in Chapter III, § 2, no. 6 that the study of a commutative group with operators is always equivalent to that of a module suitably associated with the group with operators in question.

§ 1. MODULES

1. MODULES; VECTOR SPACES; LINEAR COMBINATIONS

DEFINITION 1. *Given a ring* A, *a left module over* A (or *left* A-*module*) *is a set* E *with an algebraic structure defined by giving:*

(1) *a commutative group law on* E (*written additively in what follows*);

(2) *a law of action* $(\alpha, x) \mapsto \alpha \top x$, *whose domain of operators is the ring* A *and which satisfies the following axioms:*

(M_I) $\alpha \top (x + y) = (\alpha \top x) + (\alpha \top y)$ *for all* $\alpha \in A$, $x \in E$, $y \in E$;

(M_{II}) $(\alpha + \beta) \top x = (\alpha \top x) + (\beta \top x)$ *for all* $\alpha \in A$, $\beta \in A$, $x \in E$;

(M_{III}) $\alpha \top (\beta \top x) = (\alpha\beta) \top x$ *for all* $\alpha \in A$, $\beta \in A$, $x \in E$;

(M_{IV}) $1 \top x = x$ *for all* $x \in E$.

Axiom (M_I) means that the external law of a left A-module E is *distributive* with respect to addition on E; a module is thus a commutative group with operators.

If in Definition 1, axiom (M_{III}) is replaced by

(M'_{III}) $\alpha \top (\beta \top x) = (\beta\alpha) \top x$ *for all* $\alpha \in A$, $\beta \in A$, $x \in E$,

E with the algebraic structure thus defined is a *right module over* A or a *right* A-*module.*

When speaking of A-modules (left or right), the elements of the ring A are often called *scalars.*

Usually the external law of composition of a left A-module (resp. right A-module) is written *multiplicatively*, with the operator written on the left (resp. right); condition (M_{III}) is then written $\alpha(\beta x) = (\alpha\beta)x$, condition ($M'_{III}$) is written $(x\beta)\alpha = x(\beta\alpha)$.

If A^0 denotes the *opposite* ring to A (I, § 8, no. 3), every *right* module E over the ring A is a *left* module over the ring A^0. It follows that an exposition can be given of the properties of modules whilst systematically confining attention either to left modules or to right modules; in §§ 1 and 2 we shall in general give this exposition for *left* modules and when we speak of a *module* (without specifying which type) we shall mean a left module whose external law will be written multiplicatively. When the ring A is *commutative* the notions of right module and left module with respect to A are identical.

For all $\alpha \in A$, the mapping $x \mapsto \alpha x$ of an A-module E into itself is called the *homothety of ratio* α of E (I, § 4, no. 2); by (M_I) a homothety is an endomorphism of the commutative group structure (*without operators*) on E, but *not* in general of the module structure on E (I, § 4, no. 2; cf. II, § 1, no. 2 and no. 13). Hence $\alpha 0 = 0$ and $\alpha(-x) = -(\alpha x)$; if α is an *invertible* element of A, the homothety $x \mapsto \alpha x$ is an *automorphism* of the commutative group structure (without operators) on E, for the relation $y = \alpha x$ implies by virtue of (M_{IV}), $x = \alpha^{-1}(\alpha x) = \alpha^{-1}y$.

Similarly, by virtue of (M_{II}), for all $x \in E$, the mapping $\alpha \mapsto \alpha x$ is a homomorphism of the additive group A into the commutative group (without operators) E; hence $0x = 0$ and $(-\alpha)x = -(\alpha x)$; moreover, by (M_{IV}), for every integer $n \in \mathbf{Z}$, $n.x = (n.1)x$.

When the ring A consists only of the element 0, *every* A-module E consists only of the element 0, for then $1 = 0$ in A, whence, for all $x \in E$, $x = 1.x = 0.x = 0$.

Examples. (1) Let ϕ be a homomorphism of a ring A into a ring B; the mapping $(a, x) \mapsto \phi(a)x$ (resp. $(a, x) \mapsto x\phi(a)$) of A × B into B defines on B a left (resp. right) A-module structure. In particular if ϕ is taken to be the identity mapping on A, a canonical left (resp. right) A-module structure is obtained on A; to avoid confusion, the set A with this structure is denoted by A_s (resp. A_d).

(2) On a commutative group G (written additively) the group with operators structure defined by the external law $(n, x) \mapsto n.x$ (I, § 3, no. 1) is a module structure over the ring **Z** of rational integers.

(3) Let E be a commutative group written additively, $\mathscr{E}$ the *endomorphism*

ring of E (I, § 8, no. 3: recall that the product fg of two endomorphisms is by definition the composite endomorphism $f \circ g$). The external law $(f, x) \mapsto f(x)$ between operators $f \in \mathcal{E}$ and elements $x \in E$ defines on E a canonical left $\mathcal{E}$-module structure.

Consider now a ring A and suppose there is given on E a left (resp. right) A-module structure; for all $\alpha \in A$, the homothety $h_\alpha : x \mapsto \alpha x$ (resp. $x \mapsto x\alpha$) belongs to $\mathcal{E}$; the mapping $\phi : \alpha \mapsto h_\alpha$ is a *homomorphism* of the ring A (resp. the opposite ring A^0) into the ring $\mathcal{E}$ and by definition $\alpha x = (\phi(\alpha))(x)$ (resp. $x\alpha = (\phi(\alpha))(x)$). Conversely, giving a ring homomorphism $\phi : A \to \mathcal{E}$ (resp. $\phi : A^0 \to \mathcal{E}$) defines a left (resp. right) A-module structure on E by the above formulae. In other words, being given a left (resp. right) A-module structure on an additive group E with additive law the given group law is equivalent to being given a ring homomorphism $A \to \mathcal{E}$ (resp. $A^0 \to \mathcal{E}$).

DEFINITION 2. *A left* (resp. *right*) *vector space over a field* K *is a left* (resp. *right*) K-*module.*

The elements of a vector space are sometimes called *vectors.*

Examples. (4) A field is both a left and a right vector space with respect to any of its subfields.

(5) The real number space of three dimensions $\mathbf{R}^3$ is a vector space with respect to the field of real numbers $\mathbf{R}$, the product tx of a real number t and a point x with coordinates x_1, x_2, x_3 being the point with coordinates tx_1, tx_2, tx_3. Similarly, the set of real-valued functions defined on an arbitrary set F is a vector space with respect to $\mathbf{R}$, the product tf of a real number t and a function f being the real-valued function $x \mapsto tf(x)$.

For two families $(x_\iota)_{\iota \in I}$, $(y_\iota)_{\iota \in I}$ of elements of an A-module E of finite support (I, § 2, no. 1), the following equations hold:

$$\sum_{\iota \in I} (x_\iota + y_\iota) = \sum_{\iota \in I} x_\iota + \sum_{\iota \in I} y_\iota \tag{1}$$

$$\alpha . \sum_{\iota \in I} x_\iota = \sum_{\iota \in I} (\alpha x_\iota) \quad \text{for all } \alpha \in A; \tag{2}$$

the equations are immediately reduced to the analogous equations for finite sums by considering a finite subset H of I containing the supports of (x_ι) and (y_ι).

DEFINITION 3. *An element x of an* A-*module* E *is said to be a linear combination with coefficients in* A *of a family $(a_\iota)_{\iota \in I}$ of elements of* E *if there exists a family $(\lambda_\iota)_{\iota \in I}$ of elements of* A, *of finite support, such that $x = \sum_{\iota \in I} \lambda_\iota a_\iota$.*

In general there are several distinct families (λ_ι) satisfying this condition (cf. no. 11).

Note that 0 is the only linear combination of the *empty family* of elements of E (by the convention of I, § 2, no. 1).

2. LINEAR MAPPINGS

DEFINITION 4. *Let* E *and* F *be two (left) modules with respect to the same ring* A. *A linear mapping* (or A-*linear mapping*, or *homomorphism*, or A-*homomorphism*) *of* E *into* F *is any mapping* $u: E \rightarrow F$ *such that:*

$$(3) \qquad u(x + y) = u(x) + u(y) \quad \textit{for } x \in E, y \in E;$$

$$(4G) \qquad u(\lambda . x) = \lambda . u(x) \quad \textit{for } \lambda \in A, x \in E.$$

If E and F are two *right* A-modules, a linear mapping $u: E \rightarrow F$ is a mapping satisfying (3) and

$$(4D) \qquad u(x . \lambda) = u(x) . \lambda \quad \text{for } \lambda \in A, x \in E.$$

Remark. When E and F are two commutative groups considered as modules over the ring **Z** (no. 1), every homomorphism u of the group E (without operators) into the group F (without operators) is also a linear mapping of E into F, since for n an integer > 0, the relation $u(n.x) = n.u(x)$ follows from $u(x + y) = u(x) + u(y)$ by induction on n and for $n = -m < 0$,

$$u(n.x) = u(-(m.x)) = -u(m.x) = -(m.u(x)) = n.u(x).$$

Examples. (1) Let E be an A-module and a an element of E; the mapping $\lambda \mapsto \lambda a$ of the A-module A_s into E is a linear mapping θ_a such that $\theta_a(1) = a$.

*(2) Let I be an open interval of the real line **R**, E the vector space of differentiable real-valued functions on I and F the vector space of all real-valued functions defined on I. The mapping $x \mapsto x'$ which associates with every differentiable function x its derivative is a linear mapping from E to F.$_*$

Note that a homothety $x \mapsto \alpha x$ on an A-module E *is not necessarily a linear mapping*: in other words, the relation $\alpha(\lambda x) = \lambda(\alpha x)$ does not necessarily hold for all $\lambda \in A$ and $x \in E$. This relation is however true when α belongs to the *centre* of A; $x \mapsto \alpha x$ is then called a *central homothety* (cf. no. 13).

If $u: E \rightarrow F$ is a linear mapping, then, for every family $(x_\iota)_{\iota \in I}$ of elements of E and every family $(\lambda_\iota)_{\iota \in I}$ of elements of A such that the support of the family $(\lambda_\iota x_\iota)_{\iota \in I}$ is finite,

$$(5) \qquad u\left(\sum_{\iota \in I} \lambda_\iota x_\iota\right) = \sum_{\iota \in I} \lambda_\iota u(x_\iota)$$

as follows immediately from (3) and (4G) by induction on the cardinal of the support of the family $(\lambda_\iota x_\iota)$.

PROPOSITION 1. *Let* E, F, G *be three* A-*modules*, u *a linear mapping of* E *into* F *and* v *a linear mapping of* F *into* G. *Then the composite mapping* $v \circ u$ *is linear.*

PROPOSITION 2. *Let* E, F *be two* A-*modules.*

(1) *If* $u: E \to F$ *and* $v: F \to E$ *are two linear mappings such that* $v \circ u$ *is the identity mapping on* E *and* $u \circ v$ *is the identity mapping on* F, u *is an isomorphism of* E *onto* F *and* v *is the inverse isomorphism.*

(2) *Every bijective linear mapping* $u: E \to F$ *is an isomorphism of* E *onto* F.

These propositions follow immediately from Definition 4.

Propositions 1 and 2 show that linear mappings can be taken as the *morphisms* for the species of A-module structures (*Set Theory*, IV, § 2, no. 1); we shall henceforth always assume that this choice of morphisms has been made.

Given two left (resp. right) A-modules E and F, let Hom(E, F) or $\mathrm{Hom}_A(E, F)$ denote the set of linear mappings of E into F.

The set Hom(E, F) is a *commutative ring*, a subgroup of the product commutative group F^E of all mappings of E into F (I, § 4, no. 8); recall that for two elements u, v of F^E and for all $x \in E$,

$$(u + v)(x) = u(x) + v(x), \qquad (-u)(x) = -u(x)$$

whence it follows immediately that, if u and v are linear, so are $u + v$ and $-u$. If G is a third left (resp. right) A-module, f, f_1, f_2 elements of Hom(E, F) and g, g_1, g_2 elements of Hom(F, G), the following relations are immediately verified:

$$g \circ (f_1 + f_2) = g \circ f_1 + g \circ f_2 \tag{6}$$

$$(g_1 + g_2) \circ f = g_1 \circ f + g_2 \circ f \tag{7}$$

$$g \circ (-f) = (-g) \circ f = -(g \circ f). \tag{8}$$

In particular, the law of composition $(f, g) \mapsto f \circ g$ on Hom(E, E) defines with the above additive group structure a *ring* structure on Hom(E, E) whose unit element, denoted by 1_E or Id_E, is the identity mapping on E; the linear mappings of E into itself are also called *endomorphisms* of the A-module E and the ring Hom(E, E) is also denoted by End(E) or $\mathrm{End}_A(E)$. The *automorphisms* of the A-module E are just the *invertible* elements of End(E) (Proposition 2); they form a multiplicative *group*, denoted by Aut(E) or **GL**(E) which is also called the *linear group* relative to E.

It follows from (6) and (7) that, for two A-modules E, F, Hom(E, F) has the canonical structure of a *left module* over the ring Hom(F, F) and of a *right module* over Hom(E, E).

Let E, F, E′, F′ be four (left) A-modules, $u: E' \to E$ and $v: F \to F'$ A-linear mappings. If every element $f \in \mathrm{Hom}(E, F)$ is associated with the element $v \circ f \circ u \in \mathrm{Hom}(E', F')$, a mapping

$$\mathrm{Hom}(E, F) \to \mathrm{Hom}(E', F')$$

is defined, which is **Z**-*linear* and which is denoted by $\operatorname{Hom}(u, v)$ or $\operatorname{Hom}_A(u, v)$. If u, u_1, u_2 belong to $\operatorname{Hom}(E', E)$ and v, v_1, v_2 to $\operatorname{Hom}(F, F')$, then

$$(9) \qquad \begin{cases} \operatorname{Hom}(u_1 + u_2, v) = \operatorname{Hom}(u_1, v) + \operatorname{Hom}(u_2, v) \\ \operatorname{Hom}(u, v_1 + v_2) = \operatorname{Hom}(u, v_1) + \operatorname{Hom}(u, v_2) \end{cases}$$

Let E'', F'' be two A-modules, $u' : E'' \to E'$ and $v' : F' \to F''$ linear mappings. Then

$$(10) \qquad \operatorname{Hom}(u \circ u', v' \circ v) = \operatorname{Hom}(u', v') \circ \operatorname{Hom}(u, v).$$

If u is an isomorphism of E' onto E and v an isomorphism of F onto F', $\operatorname{Hom}(u, v)$ is an isomorphism of $\operatorname{Hom}(E, F)$ onto $\operatorname{Hom}(E', F')$ whose inverse isomorphism is $\operatorname{Hom}(u^{-1}, v^{-1})$ by (10).

If h (resp. k) is an endomorphism of E (resp. F), $\operatorname{Hom}(h, 1_F)$ (resp. $\operatorname{Hom}(1_E, k)$) is just the homothety with ratio h (resp. k) for the right (resp. left) module structure on the ring $\operatorname{End}(E)$ (resp. $\operatorname{End}(F)$) defined above.

3. SUBMODULES; QUOTIENT MODULES

Let E be an A-module and M a subset of E; for the A-module structure on E to induce an A-module structure on M, it is necessary and sufficient that M be a *stable* subgroup of E (I, § 4, no. 3), for if this is so, the structure of a group with operators induced on M obviously satisfies axioms (M_{II}), (M_{III}) and (M_{IV}); then M with this structure (or, by an abuse of language, the set M itself) is called a *submodule* of E; the canonical injection $M \to E$ is a linear mapping. When E is a *vector space*, its submodules are called *vector subspaces* (or simply *subspaces* if no confusion can arise).

Examples. (1) In any module E the set consisting of 0 is a submodule (the *zero* submodule, often denoted by 0, by an abuse of notation).

(2) Let A be a ring. The submodules of A_s (resp. A_d) are just the *left ideals* (resp. *right ideals*) of the ring A.

(3) Let E be an A-module, x an element of E and $\mathfrak{a}$ a left ideal of A. The set of elements αx, where α runs through $\mathfrak{a}$, is a submodule of E, denoted by $\mathfrak{a}x$.

(4) In a commutative group G considered as a **Z**-module (no. 1), every subgroup of G is also a submodule.

*(5) Let I be an open interval of the real line **R**; the set C of real-valued functions defined and *continuous* on I is a vector subspace of the vector space $\mathbf{R}^I$ of all real-valued functions defined on I. Similarly, the set D of *differentiable* functions on I is a vector subspace of C.*

Let E be an A-module. Every equivalence relation *compatible* (I, § 1, no. 6) with the module structure on E is of the form $x - y \in M$, where M is a stable subgroup of E (I, § 4, no. 4), that is a *submodule* of E. It is immediately verified that the structure of a group with operators on the quotient group E/M (I, § 4, no. 4) is an A-module structure, under which the canonical mapping $E \to E/M$ is linear; with this structure, E/M is called the *quotient module* of E by the submodule M. A quotient module of a vector space E is called a *quotient vector space* (or simply *quotient space*) of E.

Example (6). Every left ideal $\mathfrak{a}$ in a ring A defines a quotient module $A_s/\mathfrak{a}$ of the left A-module A_s; by an abuse of notation, this quotient module is often denoted by $A/\mathfrak{a}$.

Let E, F be two A-modules. It follows from the general properties of groups with operators (I, § 4, no. 5) (or directly from the definitions) that if $u: E \to F$ is a linear mapping, the image under u of any submodule of E is a submodule of F and the inverse image under u of any submodule of F is a submodule of E. In particular, the *kernel* $N = \overset{-1}{u}(0)$ is a submodule of E and the image $u(E)$ of E under u is a submodule of F (I, § 4, no. 6, Proposition 7); by an abuse of language $u(E)$ is called the *image of u*. The quotient module E/N is also called the *coimage* of u and the quotient module $F/u(E)$ the *cokernel* of u. In the *canonical decomposition* of u (I, § 4, no. 5)

$$u: E \xrightarrow{p} E/N \xrightarrow{v} u(E) \xrightarrow{i} F \tag{11}$$

v is an *isomorphism* of the coimage of u onto the image of u (no. 2, Proposition 2). For u to be *injective*, it is necessary and sufficient that its kernel be zero; for u to be *surjective*, it is necessary and sufficient that its cokernel be zero.

The kernel, image, coimage and cokernel of u are denoted respectively by Ker u, Im u, Coim u, Coker u.

Remark. Let M be a submodule of an A-module E and $\phi: E \to E/M$ the canonical homomorphism. For an A-linear mapping $u: E \to F$ to be of the form $v \circ \phi$, where v is a linear mapping of E/M into F, it is necessary and sufficient that $M \subset \mathrm{Ker}(u)$; for, if this condition holds, the relation $x - y \in M$ implies $u(x) = u(y)$, hence u is compatible with this equivalence relation and clearly the mapping $v: E/M \to F$ derived from u by taking quotients is linear.

4. EXACT SEQUENCES

DEFINITION 5. *Let* F, G, H *be three* A*-modules; let f be a homomorphism of* F *into* G *and g a homomorphism of* G *into* H. *The ordered pair* (f, g) *is called an exact sequence if*

$$\overset{-1}{g}(0) = f(F)$$

in other words, if the kernel of g is equal to the image of f.

The diagram

$$F \xrightarrow{f} G \xrightarrow{g} H \tag{12}$$

is also called an *exact sequence.*

We consider similarly a diagram consisting of four A-modules and three homomorphisms:

$$E \xrightarrow{f} F \xrightarrow{g} G \xrightarrow{h} H. \tag{13}$$

This diagram is called *exact at* F if the diagram $E \xrightarrow{f} F \xrightarrow{g} G$ is exact; it is called *exact at* G if $F \xrightarrow{g} G \xrightarrow{h} H$ is exact. If diagram (13) is *exact at* F *and at* G, it is simply called *exact*, or an *exact sequence.* Exact sequences with an arbitrary number of terms are defined similarly.

Remark. (1) If the ordered pair (f, g) is an exact sequence, then $g \circ f = 0$; but of course this property does not characterize exact sequences, for it only means that the image of f is *contained* in the kernel of g.

In the statements below, E, F, G denote A-modules, 0 the A-module reduced to its identity element; the arrows represent A-module homomorphisms. As there is only one homomorphism from the module 0 to a module E (resp. of E to 0), there is no point in giving these homomorphisms a name in the exact sequences where they appear.

PROPOSITION 3. (a) *For*

$$0 \longrightarrow E \xrightarrow{f} F$$

to be an exact sequence, it is necessary and sufficient that f be injective.

(b) *For*

$$E \xrightarrow{f} F \longrightarrow 0$$

to be an exact sequence, it is necessary and sufficient that f be surjective.

(c) *For*

$$0 \longrightarrow E \xrightarrow{f} F \longrightarrow 0$$

to be an exact sequence, it is necessary and sufficient that f be bijective (in other words (no. 2, Proposition 2) that f be an *isomorphism* of E onto F).

(d) *If* F *is a submodule of* E *and* $i: F \rightarrow E$ *is the canonical injection,* $p: E \rightarrow E/F$ *the canonical homomorphism, the diagram*

$$0 \longrightarrow F \xrightarrow{i} E \xrightarrow{p} E/F \longrightarrow 0 \tag{14}$$

is an exact sequence.

(e) *If* $f: E \to F$ *is a homomorphism, the diagram*

$$0 \longrightarrow \overset{-1}{f}(0) \xrightarrow{i} E \xrightarrow{f} F \xrightarrow{p} F/f(E) \to 0$$

(*where* i *is the canonical injection and* p *the canonical surjection*) *is an exact sequence.*

The proposition follows immediately from the definitions and Proposition 2 of no. 2.

Remarks. (2) To say that there is an exact sequence

$$0 \longrightarrow E \xrightarrow{f} F \xrightarrow{g} G \longrightarrow 0$$

means that f is injective, g surjective and that the canonical bijection associated with g is an *isomorphism* of $F/f(E)$ onto G. It is also said that the triple (F, f, g) is an *extension of the module* G *by the module* E (I, § 6, no. 7).

(3) If there is an exact sequence with 4 terms

$$E \xrightarrow{f} F \xrightarrow{g} G \xrightarrow{h} H$$

the cokernel of f is $F/f(E) = F/\overset{-1}{g}(0)$ and the kernel of h is $g(F)$; the canonical bijection associated with g is therefore an *isomorphism*

$$\operatorname{Coker} f \to \operatorname{Ker} h.$$

(4) Consider an ordered pair of A-module homomorphisms

$$E \xrightarrow{f} F \xrightarrow{g} G. \tag{15}$$

For diagram (15) to be an exact sequence, it is necessary and sufficient that there exist two A-modules S, T and homomorphisms $a: E \to S$, $b: S \to F$, $c: F \to T$, $d: T \to G$ such that the three sequences

$$\begin{cases} E \xrightarrow{a} S \longrightarrow 0 \\ 0 \longrightarrow S \xrightarrow{b} F \xrightarrow{c} T \longrightarrow 0 \\ 0 \longrightarrow T \xrightarrow{d} G \end{cases} \tag{16}$$

are *exact* and $f = b \circ a$ and $g = d \circ c$.

If (15) is an exact sequence, then take $S = f(E) = \overset{-1}{g}(0)$ and $T = g(F)$, b and d being the canonical injections and a (resp. c) the homomorphism with the same graph as f (resp. g). Conversely, if S, T, a, b, c, d satisfy the above conditions, then $f(E) = b(a(E)) = b(S)$ and $\overset{-1}{g}(0) = \overset{-1}{c}(\overset{-1}{d}(0)) = \overset{-1}{c}(0)$, hence the exactness of (16) shows that $f(E) = \overset{-1}{g}(0)$.

The use of explicit letters to denote homomorphisms in an exact sequence is often dispensed with when it is not necessary for the arguments.

Remark. (5) The definition of an exact sequence extends immediately to *groups* which are not necessarily commutative; in this case of course multiplicative notation is used with 0 replaced by 1 in the formulae (if no confusion arises). Parts (*a*), (*b*), (*c*) of Proposition 3 are still valid and also (*d*) when F is a *normal* subgroup of E. *Remark* 2 and Proposition 3(*e*) hold on condition that $f(E)$ is a normal subgroup of F; *Remarks* 3 and 4 are valid without modification.

5. PRODUCTS OF MODULES

Let $(E_\iota)_{\iota \in I}$ be a family of modules over the same ring A. It is immediately verified that the *product* of the module structures on the E_ι (I, § 4, no. 8) is an A-module structure on the product set $E = \prod_{\iota \in I} E_\iota$. With this structure the set E is called the *product module* of the modules E_ι; if $x = (x_\iota)$, $y = (y_\iota)$ are two elements of E, then

$$(17) \quad \begin{cases} x + y = (x_\iota + y_\iota) \\ \lambda . x = (\lambda x_\iota) \qquad \text{for all } \lambda \in A. \end{cases}$$

Formulae (17) express the fact that the projections $\mathrm{pr}_\iota : E \to E_\iota$ are *linear* mappings; these mappings are obviously surjective.

Recall that if the indexing set I is *empty*, the product set $\prod_{\iota \in I} E_\iota$ then consists of a single element; the product module structure on this set is then that under which this unique element is 0.

Proposition 4. *Let* $E = \prod_{\iota \in I} E_\iota$ *be the product of a family of* A-*modules* $(E_\iota)_{\iota \in I}$. *For every* A-*module* F *and every family of linear mappings* $f_\iota : F \to E_\iota$ *there exists one and only one mapping* f *of* F *into* E *such that* $\mathrm{pr}_\iota \circ f = f_\iota$ *for all* $\iota \in I$ *and this mapping is linear.*

This follows directly from the definitions.

Product of modules is "associative": if $(J_\lambda)_{\lambda \in L}$ is a partition of I, the canonical mapping

$$\prod_{\iota \in I} E_\iota \to \prod_{\lambda \in L} \Big(\prod_{\iota \in J_\lambda} E_\iota \Big)$$

is an isomorphism.

Proposition 5. (i) *Let* $(E_\iota)_{\iota \in I}$, $(F_\iota)_{\iota \in I}$ *be two families of* A-*modules with the same indexing set* I; *for every family of linear mappings* $f_\iota : E_\iota \to F_\iota$ $(\iota \in I)$, *the mapping* $f : (x_\iota) \to (f_\iota(x_\iota))$ *of* $\prod_\iota E_\iota$ *into* $\prod_\iota F_\iota$ $\Big($*sometimes denoted by* $\prod_\iota f_\iota\Big)$ *is linear.*

(ii) *Let* $(G_\iota)_{\iota \in I}$ *be a third family of* A-*modules with* I *as indexing set and, for all*

$\iota \in I$, *let* $g_\iota : F_\iota \to G_\iota$ *be a linear mapping; let* $g = \prod_\iota g_\iota$. *For each of the sequences* $E_\iota \xrightarrow{f_\iota} F_\iota \xrightarrow{g_\iota} G_\iota$ *to be exact, it is necessary and sufficient that the sequence*

$$\prod_\iota E_\iota \xrightarrow{f} \prod_\iota F_\iota \xrightarrow{g} \prod_\iota G_\iota$$

be exact.

Assertion (i) follows immediately from the definitions. On the other hand, to say that $y = (y_\iota)$ belongs to $\operatorname{Ker}(g)$ means that $g_\iota(y_\iota) = 0$ for all $\iota \in I$ and hence that $y_\iota \in \operatorname{Ker}(g_\iota)$ for all $\iota \in I$; similarly, to say that y belongs to $\operatorname{Im}(f)$ means that there exists $x = (x_\iota) \in \prod_\iota E_\iota$ such that $y = f(x)$, which is equivalent to saying that $y_\iota = f_\iota(x_\iota)$ for all $\iota \in I$ or also that $y_\iota \in \operatorname{Im}(f_\iota)$ for all $\iota \in I$; whence (ii).

Corollary. *Under the conditions of Proposition* 5, (i),

$$\operatorname{Ker}(f) = \prod_{\iota \in I} \operatorname{Ker}(f_\iota), \qquad \operatorname{Im}(f) = \prod_{\iota \in I} \operatorname{Im}(f_\iota) \tag{18}$$

and there are canonical isomorphisms

$$\operatorname{Coim}(f) \to \prod_{\iota \in I} \operatorname{Coim}(f_\iota), \qquad \operatorname{Coker}(f) = \prod_{\iota \in I} \operatorname{Coker}(f_\iota)$$

obtained by respectively associating with the class of an element $x = (x_\iota)$ *of* $\prod_\iota E_\iota$, mod $\operatorname{Ker}(f)$, (resp. *with the class of an element* $y = (y_\iota)$ *of* $\prod_\iota F_\iota$, mod. $\operatorname{Im}(f)$) *the family of classes of the* x_ι mod. $\operatorname{Ker}(f_\iota)$ (resp. *the family of classes of the* y_ι mod. $\operatorname{Im}(f)$).

In particular, for f to be injective (resp. *surjective, bijective, zero*) *it is necessary and sufficient that, for all $\iota \in I$, f_ι be injective* (resp. *surjective, bijective, zero*).

If, for all $\iota \in I$, we consider a submodule F_ι of E_ι, the module $\prod_{\iota \in I} F_\iota$ is a submodule of $\prod_{\iota \in I} E_\iota$ and by virtue of the Corollary to Proposition 5, there is a canonical isomorphism

$$\prod_{\iota \in I} (E_\iota / F_\iota) \to \Big(\prod_{\iota \in I} E_\iota\Big) \Big/ \Big(\prod_{\iota \in I} F_\iota\Big). \tag{19}$$

An important example of a product of modules is that where all the factor modules are equal to the same module F; their product F^I is then just the set of *mappings of* I *into* F. The *diagonal* mapping $F \to F^I$ mapping $x \in F$ to the constant function equal to x on I is linear. If $(E_\iota)_{\iota \in I}$ is a family of A-modules and, for all $\iota \in I$, $f_\iota : F \to E_\iota$ is a linear mapping, then the linear mapping $x \mapsto (f_\iota(x))$ of F into $\prod_{\iota \in I} E_\iota$ is the composition of the mapping $(x_\iota) \mapsto (f_\iota(x_\iota))$ of F^I into $\prod_{\iota \in I} E_\iota$ and the diagonal mapping $F \to F^I$.

6. DIRECT SUM OF MODULES

Let $(E_\iota)_{\iota \in I}$ be a family of A-modules and $F = \prod_{\iota \in I} E_\iota$ their product. The set E of $x \in F$ such that $\text{pr}_\iota\, x = 0$ except for a *finite* number of indices is obviously a *submodule* of F, called the *external direct sum* (or simply *direct sum*) of the family of modules (E_ι) and denoted by $\bigoplus_{\iota \in I} E_\iota$ (I, § 4, no. 9). When I is *finite*, then $\bigoplus_{\iota \in I} E_\iota = \prod_{\iota \in I} E_\iota$; if $I = [p, q]$ (interval of $\mathbf{Z}$), we also write

$$\bigoplus_{\iota \in I} E_\iota = E_p \oplus E_{p+1} \oplus \cdots \oplus E_q.$$

For all $\kappa \in I$, let j_κ be the mapping $E_\kappa \to F$ which associates with each $x_\kappa \in E_\kappa$ the element of F such that $\text{pr}_\iota(j_\kappa(x_\kappa)) = 0$ for $\iota \neq \kappa$ and $\text{pr}_\kappa(j_\kappa(x_\kappa)) = x_\kappa$; it is immediate that j_κ is an injective linear mapping of E_κ into the *direct sum* E of the E_ι, which we shall call the *canonical injection*; the submodule $j_\kappa(E_\kappa)$ of E, isomorphic to E_κ, is called the *component* submodule of E of index κ. It is often identified with E_κ by means of j_κ.

For all $x \in E = \bigoplus_{\iota \in I} E_\iota$, we have therefore

$$x = \sum_{\iota \in I} j_\iota(\text{pr}_\iota\, x). \tag{20}$$

Proposition 6. *Let* $(E_\iota)_{\iota \in I}$ *be a family of* A-*modules,* M *an* A-*module and, for all* $\iota \in I$, *let* $f_\iota : E_\iota \to M$ *be a linear mapping. Then there exists one and only one linear mapping* $g : \bigoplus_{\iota \in I} E_\iota \to M$ *such that, for all* $\iota \in I$:

$$g \circ j_\iota = f_\iota. \tag{21}$$

By virtue of (20), if g exists, then necessarily, for all $x \in \bigoplus_{\iota \in I} E_\iota$,

$$g(x) = \sum_\iota g(j_\iota(\text{pr}_\iota(x))) = \sum_\iota f_\iota(\text{pr}_\iota(x)),$$

whence the uniqueness of g. Conversely, setting $g(x) = \sum_\iota f_\iota(\text{pr}_\iota(x))$ for all $x \in \bigoplus_{\iota \in I} E_\iota$, it is immediately verified that a linear mapping has been defined satisfying the conditions of the statement.

When no confusion can arise, we write $g = \sum_{\iota \in I} f_\iota$ (which is contrary to the conventions of I, § 2, no. 1, when the family (f_ι) is not of finite support).

In particular, if J is any subset of I, the canonical injections j_ι for $\iota \in J$ define a canonical linear mapping $j_J: \bigoplus_{\iota \in J} E_\iota \to \bigoplus_{\iota \in I} E_\iota$, which associates with each $(x_\iota)_{\iota \in I}$ the element $(x'_\iota)_{\iota \in I}$ such that $x'_\iota = x_\iota$ for $\iota \in J$, $x'_\iota = 0$ for $\iota \notin J$; this mapping is obviously injective. Moreover, if $(J_\lambda)_{\lambda \in L}$ is a partition of I, the mapping $i: \bigoplus_{\lambda \in L} \left(\bigoplus_{\iota \in J_\lambda} E_\iota\right) \to \bigoplus_{\iota \in I} E_\iota$ corresponding to the family (j_{J_λ}) by Proposition 6 is an *isomorphism* called canonical ("associativity" of direct sums).

COROLLARY 1. *Let* $(E_\iota)_{\iota \in I}$, $(F_\lambda)_{\lambda \in L}$ *be two families of A-modules. The mapping*

$$\operatorname{Hom}_A\left(\bigoplus_{\iota \in I} E_\iota, \prod_{\lambda \in L} F_\lambda\right) \to \prod_{(\iota, \lambda) \in I \times L} \operatorname{Hom}_A(E_\iota, F_\lambda) \tag{22}$$

which associates with each $g \in \operatorname{Hom}_A\left(\bigoplus_{\iota \in I} E_\iota, \prod_{\lambda \in L} F_\lambda\right)$ *the family* $(\mathrm{pr}_\lambda \circ g \circ j_\iota)$, *is a* **Z**-*module isomorphism* (called *canonical*).

This follows from Proposition 6 and no. 5, Proposition 4.

COROLLARY 2. Let $(E_\iota)_{\iota \in I}$ *be a family of A-modules, F an A-module and, for each* $\iota \in I$, *let* $f_\iota: E_\iota \to F$ *be a linear mapping. For* $f = \sum_{\iota \in I} f_\iota$ *to be an isomorphism of* $E = \bigoplus_{\iota \in I} E_\iota$ *onto F, it is necessary and sufficient that there exist for each* $\iota \in I$ *a linear mapping* $g_\iota: F \to E_\iota$ *with the following properties:*

(1) $g_\iota \circ f_\iota = 1_{E_\iota}$ *for all* $\iota \in I$.

(2) $g_\iota \circ f_\kappa = 0$ *for* $\iota \neq \kappa$.

(3) *For all* $y \in F$, *the family* $(g_\iota(y))$ *has finite support and*

$$y = \sum_{\iota \in I} f_\iota(g_\iota(y)). \tag{23}$$

Note that if I is finite, the last condition may also be written as

$$\sum_{\iota \in I} f_\iota \circ g_\iota = 1_F. \tag{24}$$

Obviously the conditions are necessary for they are satisfied by the $g_\iota = \mathrm{pr}_\iota \circ \overset{-1}{f}$. Conversely if they hold, for all $y \in F$, $g(y) = \sum_\iota j_\iota(g_\iota(y))$ is defined and it is immediate that g is a linear mapping of F into E. For all $y \in F$, $f(g(y)) = \sum_{\iota \in I} f_\iota(g_\iota(y)) = y$ by hypothesis. On the other hand, for all $x \in E$, $g_\kappa(f(x)) = g_\kappa\left(\sum_\iota f_\iota(\mathrm{pr}_\iota(x))\right) = g_\kappa(f_\kappa(\mathrm{pr}_\kappa(x))) = \mathrm{pr}_\kappa(x)$ by hypothesis;

therefore $g(f(x)) = \sum_\iota j_\iota(g_\iota(f(x))) = \sum_\iota j_\iota(\mathrm{pr}_\iota(x)) = x$, which proves the corollary.

PROPOSITION 7. (i) *Let* $(\mathrm{E}_\iota)_{\iota \in \mathrm{I}}$, $(\mathrm{F}_\iota)_{\iota \in \mathrm{I}}$ *be two families of* A-*modules with the same indexing set* I; *for every family of linear mappings* $f_\iota : \mathrm{E}_\iota \to \mathrm{F}_\iota$ $(\iota \in \mathrm{I})$, *the restriction to* $\bigoplus_{\iota \in \mathrm{I}} \mathrm{E}_\iota$ *of the linear mapping* $(x_\iota) \mapsto (f_\iota(x_\iota))$ *is a linear mapping* $f : \bigoplus_{\iota \in \mathrm{I}} \mathrm{E}_\iota \to \bigoplus_{\iota \in \mathrm{I}} \mathrm{F}_\iota$ *denoted by* $\bigoplus_{\iota \in \mathrm{I}} f_\iota$ *or* $\bigoplus_\iota f_\iota$ *(where* $f = f_p \oplus f_{p+1} \oplus \cdots \oplus f_q$ *if* $\mathrm{I} = [p, q]$ *is an interval in* **Z**).

(ii) *Let* $(\mathrm{G}_\iota)_{\iota \in \mathrm{I}}$ *be a third family of* A-*modules with* I *as indexing set and, for all* $\iota \in \mathrm{I}$, *let* $g_\iota : \mathrm{F}_\iota \to \mathrm{G}_\iota$ *be a linear mapping; we write* $g = \bigoplus_\iota g_\iota$. *For each of the sequences* $\mathrm{E}_\iota \xrightarrow{f_\iota} \mathrm{F}_\iota \xrightarrow{g_\iota} \mathrm{G}_\iota$ *to be exact, it is necessary and sufficient that the sequence*

$$\bigoplus_\iota \mathrm{E}_\iota \xrightarrow{f} \bigoplus_\iota \mathrm{F}_\iota \xrightarrow{g} \bigoplus_\iota \mathrm{G}_\iota$$

be exact.

Obviously, for all $(x_\iota) \in \bigoplus_\iota \mathrm{E}_\iota$, the family $(f_\iota(x_\iota))$ has finite support, whence (i). On the other hand, to say that an element $y = (y_\iota)$ of $\bigoplus_\iota \mathrm{F}_\iota$ belongs to $\mathrm{Ker}(g)$ means that $y_\iota \in \mathrm{Ker}(g_\iota)$ for all $\iota \in \mathrm{I}$ (no. 5, Proposition 5); similarly, if $y_\iota \in \mathrm{Im}(f_\iota)$ for all $\iota \in \mathrm{I}$, there exists for each $\iota \in \mathrm{I}$ an $x_\iota \in \mathrm{E}_\iota$ such that $y_\iota = f_\iota(x_\iota)$ and when $y_\iota = 0$, it may be supposed that $x_\iota = 0$; hence $y \in \mathrm{Im}(f)$ and the converse is obvious.

COROLLARY 1. *Under the conditions of Proposition* 7, (i),

$$\mathrm{Ker}(f) = \bigoplus_{\iota \in \mathrm{I}} \mathrm{Ker}(f_\iota), \qquad \mathrm{Im}(f) = \bigoplus_{\iota \in \mathrm{I}} \mathrm{Im}(f_\iota) \tag{25}$$

and there are canonical isomorphisms

$$\mathrm{Coim}(f) \to \bigoplus_{\iota \in \mathrm{I}} \mathrm{Coim}(f_\iota), \qquad \mathrm{Coker}(f) \to \bigoplus_{\iota \in \mathrm{I}} \mathrm{Coker}(f_\iota)$$

defined as in no. 5, *Corollary to Proposition* 5. *In particular, for f to be injective* (resp. *surjective, bijective, zero*), *it is necessary and sufficient that each of the* f_ι *be injective* (resp *surjective, bijective, zero*).

If, for all $\iota \in \mathrm{I}$, we consider a submodule F_ι of E_ι, the module $\bigoplus_{\iota \in \mathrm{I}} \mathrm{F}_\iota$ is a submodule of $\bigoplus_{\iota \in \mathrm{I}} \mathrm{E}_\iota$ and, by virtue of Corollary 1 to Proposition 7, there is a canonical isomorphism

$$\bigoplus_{\iota \in \mathrm{I}} (\mathrm{E}_\iota/\mathrm{F}_\iota) \to \Big(\bigoplus_{\iota \in \mathrm{I}} \mathrm{E}_\iota\Big)\Big/\Big(\bigoplus_{\iota \in \mathrm{I}} \mathrm{F}_\iota\Big). \tag{26}$$

COROLLARY 2. *Let* $(E_\iota)_{\iota\in I}$, $(E'_\iota)_{\iota\in I}$, $(F_\lambda)_{\lambda\in L}$, $(F'_\lambda)_{\lambda\in L}$ *be four families of* A-*modules and, for each* $\iota \in I$ (*resp. each* $\lambda \in L$), $u_\iota : E'_\iota \to E_\iota$ (*resp.* $v_\lambda : F_\lambda \to F'_\lambda$) *a linear mapping. Then the diagram*

$$\begin{array}{ccc}
\mathrm{Hom}\Big(\bigoplus_{\iota\in I} E'_\iota, \prod_{\lambda\in L} F'_\lambda\Big) & \xrightarrow{\phi'} & \prod_{(\iota,\lambda)\in I\times L} \mathrm{Hom}(E'_\iota, F'_\lambda) \\
\uparrow {\scriptstyle \mathrm{Hom}(\bigoplus_\iota u_\iota, \prod_\lambda v_\lambda)} & & \uparrow {\scriptstyle \prod \mathrm{Hom}(u_\iota, v_\lambda)} \\
\mathrm{Hom}\Big(\bigoplus_{\iota\in I} E_\iota, \prod_{\lambda\in L} F_\lambda\Big) & \xrightarrow[\phi]{} & \prod_{(\iota,\lambda)\in I\times L} \mathrm{Hom}(E_\iota, F_\lambda)
\end{array}$$

(*where* ϕ *and* ϕ' *are the canonical isomorphisms defined in Corollary* 1 *to Proposition* 6) *is commutative.*

The verification follows immediately from the definitions.

When all the E_ι are equal to the same A-module E, the direct sum $\bigoplus_{\iota\in I} E_\iota$ is also denoted by $E^{(I)}$: its elements are the mappings of I into E with finite support. If, for all ι, f_ι is taken to be the identity mapping $E \to E$, by Proposition 6, a linear mapping $E^{(I)} \to E$ is obtained, called *codiagonal*, which associates with every family $(x_\iota)_{\iota\in I}$ of elements of E, of finite support, its *sum* $\sum_{\iota\in I} x_\iota$.

Remark. Recall that the definition of direct sum extends immediately to a family $(E_\iota)_{\iota\in I}$ of *groups* which are not necessarily commutative, multiplicative notation of course replacing the additive notation; we then say "*restricted product*" or "*restricted sum*" instead of "*direct sum*" (I, § 4, no. 9). Note that E is a *normal* subgroup of the product $F = \prod_{\iota\in I} E_\iota$ and that each of the $j_\kappa(E_\kappa)$ is a *normal* subgroup of F; moreover, for two distinct indices λ, μ, every element of $j_\lambda(E_\lambda)$ is *permutable* with every element of $j_\mu(E_\mu)$. Proposition 6 extends to the general case with the hypothesis that, for two distinct indices λ, μ, every element of $f_\lambda(E_\lambda)$ is *permutable* in M with every element of $f_\mu(E_\mu)$ (I, § 4, no. 9, Proposition 12). The property of "associativity" of the restricted sum follows immediately from this. Proposition 7 and its Corollaries 1 and 2 hold without modification.

7. INTERSECTION AND SUM OF SUBMODULES

For every family $(M_\iota)_{\iota\in I}$ of submodules of an A-module E, the intersection $\bigcap_{\iota\in I} E_\iota$ is a submodule of E. If, for each $\iota \in I$, ϕ_ι denotes the canonical homomorphism $E \to E/M_\iota$, $\bigcap_{\iota\in I} M_\iota$ is the *kernel* of the homomorphism $\phi : x \mapsto (\phi_\iota(x))$ of E into $\prod_{\iota\in I} (E/M_\iota)$, in other words, there is an *exact sequence*

$$0 \longrightarrow \bigcap_{\iota\in I} M_\iota \longrightarrow E \xrightarrow{\phi} \prod_{\iota\in I} (E/M_\iota). \tag{27}$$

The linear mapping ϕ and the mapping

$$E/\Big(\bigcap_{\iota\in I} M_\iota\Big) \to \prod_{\iota\in I} (E/M_\iota)$$

which is obtained by passing to the quotient, are called *canonical*.

In particular:

PROPOSITION 8. *If a family* $(M_\iota)_{\iota\in I}$ *of submodules of* E *has intersection reduced to* 0 E *is canonically isomorphic to a submodule of* $\prod_{\iota\in I} (E/M_\iota)$.

Given a subset X of an A-module E, the intersection F of the submodules of E containing X is called the submodule *generated* by X and X is called a *generating set* (or *generating system*) of F (I, § 4, no. 3); for a family $(a_\iota)_{\iota\in I}$ of elements of E, the submodule generated by the set of a_ι is called the submodule generated by the family (a_ι).

An A-module is called *finitely generated* if it has a *finite* generating set.

PROPOSITION 9. *The submodule generated by a family* $(a_\iota)_{\iota\in I}$ *of elements of an A-module* E *is the set of linear combinations of the family* (a_ι).

Every submodule of E which contains all the a_ι also contains the linear combinations of the a_ι. Conversely, formulae (1) and (2) of no. 1 prove that the set of linear combinations of the a_ι is a submodule of E which obviously contains all the a_ι and is therefore the smallest submodule containing them.

COROLLARY 1. *Let* $u: E \to F$ *be a linear mapping,* S *a subset of* E *and* M *the submodule of* E *generated by* S. *Then* $u(M)$ *is the submodule of* F *generated by* $u(S)$.

In particular, the image under u of any finitely generated submodule of E is a finitely generated submodule of F.

Remark. If $u(x) = 0$ for all $x \in S$, then also $u(x) = 0$ for all $x \in M$. We shall sometimes refer to this result as the "*principle of extension of linear identities*" or "*principle of extension by linearity*".

In particular, to verify that a linear mapping $u: E \to F$ is of the form $v \circ \phi$, where $v: E/M \to F$ is linear and $\phi: E \to E/M$ is the canonical projection, it suffices to verify that $u(S) = 0$.

COROLLARY 2. *The submodule generated by the union of a family* $(M_\iota)_{\iota\in I}$ *of submodules of a module* E *is identical with the set of sums* $\sum_{\iota\in I} x_\iota$, *where* $(x_\iota)_{\iota\in I}$ *runs through the set of families of elements of* E *of finite support such that* $x_\iota \in M_\iota$ *for all* $\iota \in I$.

Clearly every linear combination of elements of $\bigcup_{\iota\in I} M_\iota$ is of the above form: the converse is obvious.

The submodule of E generated by the union of a family $(M_\iota)_{\iota \in I}$ of submodules of E is called the *sum* of the family (M_ι) and is denoted by $\sum_{\iota \in I} M_\iota$. If for all $\iota \in I$, h_ι is the canonical injection $M_\iota \to E$ and $h:(x_\iota) \mapsto \sum_\iota h_\iota(x_\iota)$ the linear mapping of $\bigoplus_{\iota \in I} M_\iota$ into E corresponding to the family (h_ι) (no. 6, Proposition 6), $\sum_{\iota \in I} M_\iota$ is the *image* of h; in other words, there is an *exact sequence*

$$\bigoplus_{\iota \in I} M_\iota \xrightarrow{h} E \longrightarrow E\Big/\Big(\sum_{\iota \in I} M_\iota\Big) \longrightarrow 0. \tag{28}$$

COROLLARY 3. *If $(M_\lambda)_{\lambda \in L}$ is a right directed family of submodules of an* A*-module* E, *the sum $\sum_{\lambda \in L} M_\lambda$ is identical with the union $\bigcup_{\lambda \in L} M_\lambda$.*

$\bigcup_{\lambda \in L} M_\lambda \subset \sum_{\lambda \in L} M_\lambda$ is always true without hypothesis; on the other hand, for every finite subfamily $(M_\lambda)_{\lambda \in J}$ of $(M_\lambda)_{\lambda \in L}$, there exists by hypothesis a $\mu \in L$ such that $M_\lambda \subset M_\mu$ for all $\lambda \in J$, hence $\sum_{\lambda \in L} M_\lambda \subset M_\mu$ and it thus follows from Corollary 2 that $\sum_{\lambda \in L} M_\lambda \subset \bigcup_{\lambda \in L} M_\lambda$.

COROLLARY 4. *Let $0 \to E \xrightarrow{f} F \xrightarrow{g} G \to 0$ be an exact sequence of* A*-modules,* S *a generating system of* E, T *a generating system of* G. *If* T′ *is a subset of* F *such that* $g(T') = T$, $T' \cup f(S)$ *is a generating system of* F.

The submodule F′ of F generated by $T' \cup f(S)$ contains $f(E)$ and, as $g(F')$ contains T, $g(F') = G$; hence $F' = F$.

COROLLARY 5. *In an exact sequence $0 \to E \to F \to G \to 0$ of* A*-modules, if* E *and* G *are finitely generated, so is* F.

PROPOSITION 10. *Let* M, N *be two submodules of an* A*-module* E. *Then there are two exact sequences*

$$0 \longrightarrow M \cap N \xrightarrow{u} M \oplus N \xrightarrow{i-j} M + N \longrightarrow 0 \tag{29}$$

$$0 \longrightarrow E/(M \cap N) \xrightarrow{v} (E/M) \oplus (E/N) \xrightarrow{p-q} E/(M + N) \longrightarrow 0 \tag{30}$$

where $i: M \to M + N$, $j: N \to M + N$ are the canonical injections,

$$p: E/M \to E/(M + N) \quad \textit{and} \quad q: E/N \to E/(M + N)$$

the canonical surjections and where the homomorphisms u and v are defined as follow:

if $f: \mathrm{M} \cap \mathrm{N} \to \mathrm{M} \to \mathrm{M} \oplus \mathrm{N}$ *and* $g: \mathrm{M} \cap \mathrm{N} \to \mathrm{N} \to \mathrm{M} \oplus \mathrm{N}$ *are the canonical injections,* $u = f + g$, *and if* $r: \mathrm{E}/(\mathrm{M} \cap \mathrm{N}) \to \mathrm{E}/\mathrm{M} \to (\mathrm{E}/\mathrm{M}) \oplus (\mathrm{E}/\mathrm{N})$ *and*

$$s: \mathrm{E}/(\mathrm{M} \cap \mathrm{N}) \to \mathrm{E}/\mathrm{N} \to (\mathrm{E}/\mathrm{M}) \oplus (\mathrm{E}/\mathrm{N})$$

are the canonical mappings, $v = r + s$.

We prove the exactness of (29): obviously $i - j$ is surjective and u is injective. On the other hand, to say that $(i - j)(x, y) = 0$, where $x \in \mathrm{M}$ and $y \in \mathrm{N}$, means that $i(x) - j(y) = 0$, hence $i(x) = j(y) = z \in \mathrm{M} \cap \mathrm{N}$, whence by definition $x = f(z)$, $y = g(z)$, which proves that $\mathrm{Ker}(i - j) = \mathrm{Im}\, u$.

We prove the exactness of (30): clearly $p - q$ is surjective. On the other hand, to say that $v(t) = 0$ for $t \in \mathrm{E}/(\mathrm{M} \cap \mathrm{N})$ means that $r(t) = s(t) = 0$, hence t is the class mod. $(\mathrm{M} \cap \mathrm{N})$ of an element $z \in \mathrm{E}$ whose classes mod. M and mod. N are zero, which implies $z \in \mathrm{M} \cap \mathrm{N}$ and $t = 0$. Finally, to say that $(p - q)(x, y) = 0$, where $x \in \mathrm{E}/\mathrm{M}$, $y \in \mathrm{E}/\mathrm{N}$ means that $p(x) = q(y)$, or also that there exist two elements z', z'' of E whose classes mod. M and mod. N are respectively x and y and which are such that $z' - z'' \in \mathrm{M} + \mathrm{N}$. Hence there are $t' \in \mathrm{M}$, $t'' \in \mathrm{N}$ such that $z' - z'' = t' - t''$, whence

$$z' - t' = z'' - t'' = z.$$

Let w be the class mod. $(\mathrm{M} \cap \mathrm{N})$ of z; $r(w)$ is the class mod. M of z and hence also that of z', that is x; similarly $s(w) = y$, which completes the proof that $\mathrm{Ker}(p - q) = \mathrm{Im}\, v$.

8. DIRECT SUMS OF SUBMODULES

DEFINITION 6. *An* A-*module* E *is said to be the direct sum of a family* $(\mathrm{M}_\iota)_{\iota \in \mathrm{I}}$ *of submodules of* E *if the canonical mapping* $\bigoplus_{\iota \in \mathrm{I}} \mathrm{M}_\iota \to \mathrm{E}$ (no. 6) *is an isomorphism.*

It amounts to the same to say that every $x \in \mathrm{E}$ can be written *in a unique way* in the form $x = \sum_{\iota \in \mathrm{I}} x_\iota$, where $x_\iota \in \mathrm{E}_\iota$ for all $\iota \in \mathrm{I}$; the element x_ι thus corresponding to x is called the *component* of x in E_ι; the mapping $x \mapsto x_\iota$ is *linear*.

Remark. (1) Let $(\mathrm{M}_\iota)_{\iota \in \mathrm{I}}$, $(\mathrm{N}_\iota)_{\iota \in \mathrm{I}}$ be two families of submodules of a module E, with the same indexing set; suppose that E is *both* the direct sum of the family (M_ι) and of the family (N_ι) and *that* $\mathrm{N}_\iota \subset \mathrm{M}_\iota$ for all $\iota \in \mathrm{I}$. Then $\mathrm{N}_\iota = \mathrm{M}_\iota$ *for all* $\iota \in \mathrm{I}$, as follows immediately from no. 6, Corollary 1 to Proposition 7 applied to the canonical injections $f_\iota: \mathrm{N}_\iota \to \mathrm{M}_\iota$.

PROPOSITION 11. *Let* $(\mathrm{M}_\iota)_{\iota \in \mathrm{I}}$ *be a family of submodules of an* A-*module* E. *The following properties are equivalent:*

(a) *The submodule* $\sum_{\iota \in \mathrm{I}} \mathrm{M}_\iota$ *is the direct sum of the family* $(\mathrm{M}_\iota)_{\iota \in \mathrm{I}}$.

(b) *The relation* $\sum_{\iota \in I} x_\iota = 0$, *where* $x_\iota \in M_\iota$ *for all* $\iota \in I$, *implies* $x_\iota = 0$ *for all* $\iota \in I$.

(c) *For all* $\kappa \in I$, *the intersection of* M_κ *and* $\sum_{\iota \neq \kappa} M_\iota$ *reduces to* 0.

It is immediate that (a) and (b) are equivalent, since the relation $\sum_\iota x_\iota = \sum_\iota y_\iota$ is equivalent to $\sum_\iota (x_\iota - y_\iota) = 0$. On the other hand, by virtue of Definition 6, (a) implies (c) by the uniqueness of the expression of an element of $\bigoplus_{\iota \in I} M_\iota$ as a direct sum of elements $x_\iota \in M_\iota$. Finally, the relation $\sum_\iota x_\iota = 0$, where $x_\iota \in M_\iota$ for all ι, can be written, for all $\kappa \in I$, $x_\kappa = \sum_{\iota \neq \kappa} (-x_\iota)$; condition (c) then implies $x_\kappa = 0$ for all $\kappa \in I$, hence (c) implies (b).

DEFINITION 7. *An endomorphism e of an* A*-module* E *is called a projector if* $e \circ e = e$ (in other words, if e is an *idempotent* in the ring End(E)). *In* End(E) *a family* $(e_\lambda)_{\lambda \in L}$ *of projectors is called orthogonal if* $e_\lambda \circ e_\mu = 0$ *for* $\lambda \neq \mu$.

PROPOSITION 12. *Let* E *be an* A*-module.*

(i) *If* E *is the direct sum of a family* $(M_\lambda)_{\lambda \in L}$ *of submodules and, for all* $x \in E$, $e_\lambda(x)$ *is the component of* x *in* M_λ, (e_λ) *is an orthogonal family of projectors such that* $x = \sum_{\lambda \in L} e_\lambda(x)$ *for all* $x \in E$.

(ii) *Conversely, if* $(e_\lambda)_{\lambda \in L}$ *is an orthogonal family of projectors in* End(E) *such that* $x = \sum_{\lambda \in L} e_\lambda(x)$ *for all* $x \in E$, E *is the direct sum of the family of submodules* $M_\lambda = e_\lambda(E)$.

Property (i) follows from the definitions and (ii) is a special case of no. 6, Corollary 2 to Proposition 6, applied to the canonical injections $M_\lambda \to E$ and the mappings $e_\lambda : E \to M_\lambda$.

Note that when L is finite the condition $x = \sum_{\lambda \in L} e_\lambda(x)$ for all $x \in E$ may also be written in End(E).

$$1_E = \sum_{\lambda \in L} e_\lambda. \tag{31}$$

COROLLARY. *For every projector e of* E, E *is the direct sum of the image* $M = e(E)$ *and the kernel* $N = \overset{-1}{e}(0)$ *of e; for all* $x = x_1 + x_2 \in E$ *with* $x_1 \in M$ *and* $x_2 \in N$, $x_1 = e(x)$; $1 - e$ *is a projector of* E *of image* N *and kernel* M.

$(1 - e)^2 = 1 - 2e + e^2 = 1 - e$ in End(E) and hence $1 - e$ is a projector; as also $e(1 - e) = (1 - e)e = e - e^2 = 0$, E is the direct sum of the images M and N of e and $1 - e$ by Proposition 12. Finally, for all $x \in E$, the relation $x \in M$ is equivalent to $x = e(x)$; for $x = e(x)$ implies by definition $x \in M$ and, conversely, if $x = e(x')$ with $x' \in E$, then $e(x) = e^2(x') = e(x') = x$;

this shows therefore that M is the kernel of $1 - e$ and, exchanging the roles of e and $1 - e$, it is similarly seen that N is the kernel of e.

Remark. (2) Let E, F be two A-modules such that E is the direct sum of a *finite* family $(M_i)_{1 \leqslant i \leqslant m}$ of submodules and F the direct sum of a *finite* family $(N_j)_{1 \leqslant j \leqslant n}$ of submodules. Then it is known (no. 6, Corollary 1 to Proposition 6) that $\mathrm{Hom}_A(E, F)$ is canonically identified with the product $\prod_{i,j} \mathrm{Hom}_A(M_i, N_j)$; to be precise, to a family (u_{ji}), where $u_{ji} \in \mathrm{Hom}_A(M_i, N_j)$ there corresponds the linear mapping $u: E \to F$ defined as follows. It suffices to define the restriction of u to each of the M_i and for each $x_i \in M_i$, $u(x_i) = \sum_{j=1}^{n} u_{ji}(x_i)$.

Now let G be a third A-module, the direct sum of a *finite* family $(P_k)_{1 \leqslant k \leqslant p}$ of submodules; let v be a linear mapping of F into G and let $(v_{kj}) \in \prod_{j,k} \mathrm{Hom}_A(N_j, P_k)$ be the family corresponding to it canonically. For all $x_i \in M_i$,

$$v(u(x_i)) = \sum_{j=1}^{n} v(u_{ji}(x_i)) = \sum_{k=1}^{p} \sum_{j=1}^{n} v_{kj}(u_{ji}(x_i)).$$

Thus it is seen that if we write

$$w_{ki} = \sum_{j=1}^{n} v_{kj} \circ u_{ji} \in \mathrm{Hom}_A(M_i, P_k) \tag{32}$$

the family (w_{ki}) corresponds canonically to the *composite* linear mapping $w = v \circ u$ from E to G (cf. § 10, no. 5).

9. SUPPLEMENTARY SUBMODULES

DEFINITION 8. *In an* A-*module* E, *two submodules* M_1, M_2 *are said to be supplementary if* E *is the direct sum of* M_1 *and* M_2.

Proposition 11 of no. 8 shows that, for M_1 and M_2 to be supplementary, it is necessary and sufficient that $M_1 + M_2 = E$ and $M_1 \cap M_2 = \{0\}$ (cf. I, § 4, no. 9, Proposition 15).

PROPOSITION 13. *Let* M_1, M_2 *be two supplementary submodules in an* A-*module* E. *The restriction to* M_1 *of the canonical mapping* $E \to E/M_2$ *is an isomorphism of* M_1 *onto* E/M_2.

This linear mapping is surjective since $M_1 + M_2 = E$ and it is injective since its kernel is the intersection of M_1 and the kernel M_2 of $E \to E/M_2$ and hence reduces to $\{0\}$.

COROLLARY. *If* M_2 *and* M'_2 *are two supplements of the same submodule* M_1 *of* E, *the set of ordered pairs* $(x, x') \in M_2 \times M'_2$ *such that* $x - x' \in M_1$ *is the graph of an isomorphism of* M_2 *onto* M'_2.

It is immediate that it is the graph of the composite isomorphism $M_2 \to E/M_1 \to M'_2$.

Definition 9. *A submodule* M *of an* A*-module* E *is called a direct factor of* E *if it has a supplementary submodule in* E.

When this is so, E/M is isomorphic to a supplement of M (Proposition 13).

A submodule does not necessarily admit a supplement (Exercise 11). When a submodule is a direct factor, it has in general several distinct supplements; these supplements are however canonically isomorphic to one another (Corollary to Proposition 13).

Proposition 14. *For a submodule* M *of a module* E *to be a direct factor, it is necessary and sufficient that there exist a projector of* E *whose image is* M *or a projector of* E *whose kernel is* M.

This follows immediately from no. 8, Proposition 12 and Corollary.

Proposition 15. *Given an exact sequence of* A*-modules*

$$(33) \qquad 0 \longrightarrow E \xrightarrow{f} F \xrightarrow{g} G \longrightarrow 0$$

the following propositions are equivalent:

(a) *The submodule* $f(E)$ *of* F *is a direct factor.*

(b) *There exists a linear retraction* $r : F \to E$ *associated with* f (*Set Theory*, II, § 3, no. 8, Definition 11).

(c) *There exists a linear section* $s : G \to F$ *associated with* g (*Set Theory*, II, § 3, no. 8, Definition 11).

When this is so, $f + s : E \oplus G \to F$ *is an isomorphism.*

If there exists a projector e in End(F) such that $e(F) = f(E)$, the homomorphism $\overset{-1}{f} \circ e : F \to E$ is a linear retraction associated with f; conversely, if there exists such a retraction r, it is immediate that $f \circ r$ is a projector in F whose image is $f(E)$, hence (a) and (b) are equivalent (Proposition 14). If $f(E)$ admits a supplement E′ in F and $j : E' \to F$ is the canonical injection, $g \circ j$ is an isomorphism of E′ onto G and the inverse isomorphism, considered as a mapping of G into F, is a linear section associated with g. Conversely, if such a section s exists, $s \circ g$ is a projector in F whose kernel is $f(E)$, hence (a) and (c) are equivalent (Proposition 14). Moreover s is a bijection of G onto $s(G)$ and as $s(G)$ is supplementary to $f(E)$, $f + s$ is an isomorphism.

Note that being given r (resp. s) is equivalent to being given a supplement of $f(E)$ in F, namely the kernel of r (resp. image of s).

When the exact sequence (33) satisfies the conditions of Proposition 15, it is said to *split* or (F, f, g) is said to be a *trivial* extension of G by E (I, § 6, no. 1).

COROLLARY 1. *Let* $u: \mathrm{E} \to \mathrm{F}$ *be a linear mapping. For there to exist a linear mapping* $v: \mathrm{F} \to \mathrm{E}$ *such that* $u \circ v = 1_{\mathrm{F}}$ (the case where u is said to be *right invertible* and v is said to be the *right inverse of* u), *it is necessary and sufficient that* u *be surjective and that its kernel be a direct factor in* E. *The submodule* $\mathrm{Im}(v)$ *of* E *is then a supplement of* $\mathrm{Ker}(u)$.

It is obviously necessary that u be surjective; as v is then a section associated with u, the conclusion follows from Proposition 15.

COROLLARY 2. *Let* $u: \mathrm{E} \to \mathrm{F}$ *be a linear mapping. For there to exist a linear mapping* $v: \mathrm{F} \to \mathrm{E}$ *such that* $v \circ u = 1_{\mathrm{E}}$ (the case where u is said to be *left invertible* and v is said to be the *left inverse of* u), *it is necessary and sufficient that* u *be injective and that its image be a direct factor in* F. *The submodule* $\mathrm{Ker}(v)$ *of* F *is then a supplement of* $\mathrm{Im}(u)$.

It is obviously necessary that u be injective; as v is then a retraction associated with u, the conclusion also follows from Proposition 15.

Remarks. (1) Let M, N be two supplementary submodules in an A-module E, p, q the projectors of E onto M and N respectively, corresponding to the decomposition of E as a direct sum of M and N. It is known (no. 6, Corollary 1 to Proposition 6) that, for every A-module F, the mapping $(u, v) \mapsto u \circ p + v \circ q$ is an isomorphism of

$$\mathrm{Hom}_{\mathrm{A}}(\mathrm{M}, \mathrm{F}) \oplus \mathrm{Hom}_{\mathrm{A}}(\mathrm{N}, \mathrm{F})$$

onto $\mathrm{Hom}_{\mathrm{A}}(\mathrm{E}, \mathrm{F})$. The image of $\mathrm{Hom}_{\mathrm{A}}(\mathrm{M}, \mathrm{F})$ under this isomorphism is the set of linear mappings $w: \mathrm{E} \to \mathrm{F}$ *such that* $w(x) = 0$ *for all* $x \in \mathrm{N}$.

(2) If M, N are two submodules of E such that $\mathrm{M} \cap \mathrm{N}$ is a direct factor of M and N, then $\mathrm{M} \cap \mathrm{N}$ is also a direct factor of $\mathrm{M} + \mathrm{N}$: if P (resp. Q) is a supplement of $\mathrm{M} \cap \mathrm{N}$ in M (resp. N), $\mathrm{M} + \mathrm{N}$ is the direct sum of $\mathrm{M} \cap \mathrm{N}$, P and Q, as is immediately verified.

10. MODULES OF FINITE LENGTH

Recall (I, § 4, no. 4, Definition 7) that an A-module M is called *simple* if it is not reduced to 0 and it contains no submodule distinct from M and $\{0\}$. An A-module M is said to be of *finite length* if it has a Jordan-Hölder series $(\mathrm{M}_i)_{0 \leqslant i \leqslant n}$ and the number n of quotients of this series (which does not depend on the Jordan-Hölder series of M considered) is then called the *length* of M (I, § 4, no. 7, Definition 11); we shall denote it by $\mathrm{long}(\mathrm{M})$ or $\mathrm{long}_{\mathrm{A}}(\mathrm{M})$. An A-module which is reduced to 0 is of length 0; if M is a non-zero A-module of finite length then $\mathrm{long}(\mathrm{M}) > 0$.

PROPOSITION 16. *Let* M *be an* A-*module and* N *a submodule of* M; *for* M *to be of finite length, it is necessary and sufficient that* N *and* M/N *be so, and then*

$$\mathrm{long}(\mathrm{N}) + \mathrm{long}(\mathrm{M}/\mathrm{N}) = \mathrm{long}(\mathrm{M}). \tag{34}$$

The proof has been given in I, § 4, no. 7, Proposition 10.

COROLLARY 1. *Let* M *be an* A*-module of finite length; for a submodule* N *of* M *to be equal to* M, *it is necessary and sufficient that* long(N) = long(M).

COROLLARY 2. *Let* $u: \mathrm{M} \to \mathrm{N}$ *be an* A*-module homomorphism. If* M *or* N *is of finite length, so is* Im(u). *If* M *is of finite length, so is* Ker(u) *and*

$$\text{(35)} \qquad \operatorname{long}(\operatorname{Im}(u)) + \operatorname{long}(\operatorname{Ker}(u)) = \operatorname{long}(\mathrm{M}).$$

If N *is of finite length, so is* Coker(u) *and*

$$\text{(36)} \qquad \operatorname{long}(\operatorname{Im}(u)) + \operatorname{long}(\operatorname{Coker}(u)) = \operatorname{long}(\mathrm{N}).$$

COROLLARY 3. *Let* $(\mathrm{M}_i)_{0 \leqslant i \leqslant n}$ *be a finite family of* A*-modules of finite length. If there exists an exact sequence of linear mappings*

$$\text{(37)} \quad 0 \longrightarrow \mathrm{M}_0 \xrightarrow{u_0} \mathrm{M}_1 \xrightarrow{u_1} \mathrm{M}_2 \longrightarrow \cdots \longrightarrow \mathrm{M}_{n-1} \xrightarrow{u_{n-1}} \mathrm{M}_n \longrightarrow 0$$

then

$$\text{(38)} \qquad \sum_{k=0}^{n} (-1)^k \operatorname{long}(\mathrm{M}_k) = 0.$$

The corollary is obvious for $n = 1$ and is just Proposition 16 for $n = 2$; we argue by induction on n. If $\mathrm{M}'_{n-1} = \operatorname{Im}(u_{n-2})$, then, by the induction hypothesis,

$$\sum_{k=0}^{n-2} (-1)^k \operatorname{long}(\mathrm{M}_k) + (-1)^{n-1} \operatorname{long}(\mathrm{M}'_{n-1}) = 0.$$

On the other hand, the exact sequence $0 \to \mathrm{M}'_{n-1} \to \mathrm{M}_{n-1} \to \mathrm{M}_n \to 0$ gives

$$\operatorname{long}(\mathrm{M}'_{n-1}) + \operatorname{long}(\mathrm{M}_n) = \operatorname{long}(\mathrm{M}_{n-1}),$$

whence relation (38).

COROLLARY 4. *Let* M *and* N *be two submodules of finite length of an* A*-module* E; *then* M + N *is of finite length and*

$$\text{(39)} \qquad \operatorname{long}(\mathrm{M} + \mathrm{N}) + \operatorname{long}(\mathrm{M} \cap \mathrm{N}) = \operatorname{long}(\mathrm{M}) + \operatorname{long}(\mathrm{N}).$$

It suffices to apply Corollary 3 to the exact sequence (29) (no. 7).

$$0 \to \mathrm{M} \cap \mathrm{N} \to \mathrm{M} \oplus \mathrm{N} \to \mathrm{M} + \mathrm{N} \to 0$$

using the fact that $\operatorname{long}(\mathrm{M} \oplus \mathrm{N}) = \operatorname{long}(\mathrm{M}) + \operatorname{long}(\mathrm{N})$ by (34).

COROLLARY 5. *Let* M *be an* A*-module the sum of a family* (N_ι) *of submodules of finite length. Then* M *is of finite length and*

$$\text{(40)} \qquad \operatorname{long}(\mathrm{M}) \leqslant \sum_\iota \operatorname{long}(\mathrm{N}_\iota).$$

Moreover, for the two sides of (40) *to be equal, it is necessary and sufficient that* M *be the direct sum of the* N_ι.

It has been seen (no. 7, formula (28)) that there is a canonical surjective linear mapping $h: \bigoplus_\iota N_\iota \to M$; hence the corollary follows from (35).

COROLLARY 6. *Let* M *and* N *be two submodules of an* A*-module* E *such that* E/M *and* E/N *are modules of finite length; then* $E/(M \cap N)$ *is of finite length and*

$$\text{(41)} \quad \text{long}(E/(M \cap N)) + \text{long}(E/(M + N)) = \text{long}(E/M) + \text{long}(E/N).$$

It suffices to apply Corollary 3 to the exact sequence (30)

$$0 \to E/(M \cap N) \to (E/M) \oplus (E/N) \to E/(M + N) \to 0$$

using the fact that

$$\text{long}((E/M) \oplus (E/N)) = \text{long}(E/M) + \text{long}(E/N).$$

COROLLARY 7. *Let* (M_ι) *be a finite family of submodules of an* A*-module* E *such that the* E/M_ι *are modules of finite length. Then* $E/\left(\bigcap_\iota M_\iota\right)$ *is of finite length and*

$$\text{(42)} \qquad \text{long}\left(E/\left(\bigcap_\iota M_\iota\right)\right) \leqslant \sum_\iota \text{long}(E/M_\iota).$$

It has been seen (formula (27)) that there is a canonical injective linear mapping $E/\left(\bigcap_\iota M_\iota\right) \to \bigoplus_\iota (E/M_\iota)$.

Remark. With the exception of no. 7, Proposition 9, *all* the results of nos. 2 to 10 are valid for arbitrary *commutative groups with operators*, submodules (resp. quotient modules) being replaced in the statements by stable subgroups (resp. quotient groups by stable subgroups); we also make the convention of calling homomorphisms of groups with operators "linear mappings". The corollaries to no. 7, Proposition 9 are also valid for commutative groups with operators: this is obvious for Corollaries 4 and 5, as also for Corollary 2, since $\alpha\left(\sum_{\iota \in I} x_\iota\right) = \sum_{\iota \in I} \alpha x_\iota$ for every operator α, and Corollary 3 follows from it. As for Corollary 1, it suffices to note that if N is a stable subgroup of F containing $u(S)$, $\overset{-1}{u}(N)$ is a stable subgroup of E containing S, hence $\overset{-1}{u}(N)$ contains M and therefore $u(M) \subset N$.

11. FREE FAMILIES. BASES

Let A be a ring, T a set and consider the A-module $A_s^{(T)}$. By definition, it is the external direct sum of a family $(M_t)_{t \in T}$ of A-modules all equal to A_s and for all $t \in T$ there is a canonical injection $j_t: A_s \to A_s^{(T)}$ (no. 6). We write

$j_t(1) = e_t$ so that $e_t = (\delta_{tt'})_{t' \in T}$, where $\delta_{tt'}$ is equal to 0 if $t' \neq t$, to 1 if $t' = t$ ("*Kronecker symbol*"; $(t, t') \mapsto \delta_{tt'}$ is just the characteristic function of the diagonal of $T \times T$); every $x = (\xi_t)_{t \in T} \in A_s^{(T)}$ may then be written uniquely: $x = \sum_{t \in T} \xi_t e_t$. The mapping $\phi : t \mapsto e_t$ of T into $A_s^{(T)}$ is called *canonical*; it is *injective* if A is non-zero. We shall see that the ordered pair $(A_s^{(T)}, \phi)$ is a solution of a *universal mapping problem* (*Set Theory*, IV, § 3, no. 1).

Proposition 17. *For every* A-*module* E *and every mapping* $f : T \to E$, *there exists one and only one* A-*linear mapping* $g : A_s^{(T)} \to E$ *such that* $f = g \circ \phi$.

The condition $f = g \circ \phi$ means that $g(e_t) = f(t)$ for all $t \in T$, which is equivalent to $g(\xi e_t) = \xi f(t)$ for all $\xi \in A$ and all $t \in T$ and also means that $g \circ j_t$ is the linear mapping $\xi \mapsto \xi f(t)$ of A_s into E for all $t \in T$; the proposition is therefore a special case of no. 6, Proposition 6.

The linear mapping g is said to be *determined* by the family $(f(t))_{t \in T}$ of elements of E; by definition

$$g\Big(\sum_{t \in T} \xi_t e_t\Big) = \sum_{t \in T} \xi_t f(t). \tag{43}$$

The kernel R of g is the set of $(\xi_t) \in A_s^{(T)}$ such that $\sum_t \xi_t f(t) = 0$; it is sometimes said that the module R is *the module of linear relations between the elements of the family* $(f(t))_{t \in T}$. The exact sequence

$$0 \longrightarrow R \longrightarrow A_s^{(T)} \xrightarrow{g} E \tag{44}$$

is said to be *determined* by the family $(f(t))_{t \in T}$.

Corollary 1. *Let* T, T' *be two sets*, $g : T \to T'$ *a mapping. Then there exists one and only one* A-*linear mapping* $f : A^{(T)} \to A^{(T')}$ *which renders commutative the diagram*

$$\begin{array}{ccc} T & \xrightarrow{g} & T' \\ \phi \downarrow & & \downarrow \phi' \\ A^{(T)} & \xrightarrow{f} & A^{(T')} \end{array}$$

where ϕ *and* ϕ' *are the canonical mappings.*

It suffices to apply Proposition 17 to the composite mapping $T \xrightarrow{g} T' \xrightarrow{\phi'} A^{(T')}$.

Corollary 2. *For a family* $(a_t)_{t \in T}$ *of elements of an* A-*module* E *to be a generating system of* E, *it is necessary and sufficient that the linear mapping* $A_s^{(T)} \to E$ *determined by this family be surjective.*

This is just another way of expressing Proposition 9 of no. 7.

DEFINITION 10. *A family* $(a_t)_{t \in T}$ *of elements of an* A*-module* E *is called a* free family (resp. *a* basis *of* E) *if the linear mapping* $A_s^{(T)} \to E$ *determined by this family is injective* (resp. *bijective*). *A module is called free if it has a basis.*

In particular, a commutative group G is called *free* if G (written additively) is a *free* **Z***-module* (cf. I, § 7, no. 7).

Definition 10, together with Corollary 2 to Proposition 17, show that a basis of an A-module E is a *free generating family* of E. Every free family of elements of E is thus a basis of the submodule it generates.

By definition, the A-module $A_s^{(T)}$ is free and the family $(e_t)_{t \in T}$ is a basis (called *canonical*) of this A-module. When $A \neq \{0\}$, T is often identified with the set of e_t by the canonical bijection $t \mapsto e_t$; this amounts to writing $\sum_{t \in T} \xi_t . t$ nstead of $\sum_{t \in T} \xi_t a_t$ for the elements of $A_s^{(T)}$. When this convention is adopted, the elements of $A_s^{(T)}$ are called *formal linear combinations* (with coefficients in A) *of the elements of* T.

Definition 10 and Proposition 17 give immediately the following result:

COROLLARY 3. *Let* E *be a free* A*-module,* $(a_t)_{t \in T}$ *a basis of* E, F *an* A*-module and* $(b_t)_{t \in T}$ *a family of elements of* F. *There exists one and only one linear mapping* $f: E \to F$ *such that*

$$f(a_t) = b_t \quad \textit{for all } t \in T. \tag{45}$$

For f to be injective (resp. *surjective*), *it is necessary and sufficient that* (b_t) *be a free family in* F (resp. *a generating system of* F).

When a family $(a_t)_{t \in T}$ is not free, it is called *related.* Definition 10 may also be expressed as follows: to say that the family $(a_t)_{t \in T}$ is *free* means that the relation $\sum_{t \in T} \lambda_t a_t = 0$ (where the family (λ_i) is of finite support) implies $\lambda_t = 0$ for all $t \in T$; to say that $(a_t)_{t \in T}$ is a *basis* of E means that every $x \in E$ can be written in one and only one way in the form $x = \sum_{t \in T} \xi_t a_t$; for all $t \in T$, ξ_t is then called the *component* (or *coordinate*) *of x of index t with respect to the basis* (a_t); the mapping $x \to \xi_t$ from E to A_s is *linear.*

Suppose $A \neq \{0\}$; then, in an A-module E, two elements of a free family $(a_t)_{t \in T}$ whose indices are distinct are themselves *distinct*: for if $a_{t'} = a_{t''}$ for $t' \neq t''$, then $\sum_{t \in T} \lambda_t a_t = 0$ with $\lambda_{t'} = 1$, $\lambda_{t''} = -1$ and $\lambda_t = 0$ for the elements of T distinct from t' and t''. A subset S of E will be called a *free subset* (resp. a *basis* of E) if the family defined by the identity mapping of S onto itself is free (resp. a basis of E); every family defined by a bijective mapping of an indexing set onto S is then free (resp. a basis). The elements of a free subset of E are also called *linearly independent.*

If a subset of E is not free, it is called *related* or a *related system* and its elements are said to be *linearly dependent.*

Every subset of a free subset is free; in particular, the empty subset is free and is a basis of the submodule $\{0\}$ of E.

PROPOSITION 18. *For a family* $(a_t)_{t \in T}$ *of a module* E *to be free, it is necessary and sufficient that every finite subfamily of* $(a_t)_{t \in T}$ *be free.*

This follows immediately from the definition.

Proposition 18 shows that the set of free subsets of E, ordered by inclusion, is *inductive* (*Set Theory*, III, § 2, no. 4); as it is non-empty (since $\varnothing$ belongs to it), it has a *maximal* element $(a_\iota)_{\iota \in I}$ by Zorn's Lemma. It follows (if $A \neq \{0\}$) that for all $x \in E$ there exist an element $\mu \neq 0$ of A and a family (ξ_ι) of element A such that $\mu x = \sum_\iota \xi_\iota a_\iota$ (cf. § 7, no. 1).

PROPOSITION 19. *Let* E *be an* A*-module, the direct sum of a family* $(M_\lambda)_{\lambda \in L}$ *of submodules. If, for each* $\lambda \in L$, S_λ *is a free subset* (resp. *generating set, basis*) of M_λ, *then* $S = \bigcup_{\lambda \in L} S_\lambda$ *is a free subset* (resp. *generating set, basis*) *of* E.

The proposition follows from the definitions and the relation $A_s^{(S)} = \bigoplus_{\lambda \in L} A_s^{(S_\lambda)}$ (associativity of direct sums, cf. no. 6).

Remark. (1) By Definition 10, if $A \neq \{0\}$ and $(a_\iota)_{\iota \in I}$ is a free family, no element a_κ can be equal to a linear combination of the a_ι of index $\iota \neq \kappa$. But conversely, a family (a_ι) satisfying this condition is not necessarily a free family. For example, let A be an integral domain and a, b two distinct non-zero elements; in A, considered as an A-module, a and b form a related system, since $(-b)a + ab = 0$. But in general there does not exist an element $x \in A$ such that $a = xb$ of $b = xa$ (cf. however § 7, no. 1, *Remark*).

An element x of a module E is called *free* if $\{x\}$ is a free subset, that is if the relation $\alpha x = 0$ implies $\alpha = 0$. Every element of a free subset is free and in particular 0 cannot belong to any free subset when $A \neq \{0\}$.

Remarks. (2) A free module can have elements $\neq 0$ which are not free: for example, the A-module A_s is free but the right divisors of zero in A are not free elements of A_s.

(3) In the additive group $\mathbf{Z}/(n)$ (n an integer $\geqslant 2$) considered as a $\mathbf{Z}$-module, no element is free and *a fortiori* $\mathbf{Z}/(n)$ is not a free module.

(4) It can happen that every element $\neq 0$ of an A-module is free without the module being free. For example, the field $\mathbf{Q}$ of rational numbers is a $\mathbf{Z}$-module with this property, for two elements $\neq 0$ of $\mathbf{Q}$ always form a

related system and a basis of $\mathbf{Q}$ could therefore only contain a single element a; but the elements of $\mathbf{Q}$ are not all of the form na with $n \in \mathbf{Z}$ (cf. VII, § 3).

PROPOSITION 20. *Every* A-*module* E *is isomorphic to a quotient module of a free* A-*module.*

If T is a generating set of E, there exists a surjective linear mapping $A_s^{(T)} \to E$ (Corollary 2 to Proposition 17), and if R is the kernel of this mapping, E is isomorphic to $A_s^{(T)}/R$.

In particular we may take T = E; then there is a surjective linear mapping $A_s^{(E)} \to E$, called *canonical.*

In particular, to say that an A-module E is *finitely generated* (no. 7) means that it is isomorphic to a quotient of a free A-module with a *finite basis* or also that there exists an exact sequence of the form

$$A_s^n \to E \to 0 \quad (n \text{ an integer} > 0).$$

Note that if $A \neq \{0\}$ every basis of a *finitely generated* free module E is necessarily *finite*, for if S is a finite generating system and B a basis of E, each element of S is a linear combination of a finite number of elements of B and if B′ is the set of all the elements of B which figure thus in the expression for the elements of S, B′ is finite and every $x \in E$ is a linear combination of elements of B′, hence B′ = B.

PROPOSITION 21. *Every exact sequence of* A-*modules*

$$0 \longrightarrow G \xrightarrow{g} E \xrightarrow{f} F \longrightarrow 0$$

in which F *is a* free A-*module, splits* (no. 9). *To be precise, if* $(b_\lambda)_{\lambda \in L}$ *is a basis of* F *and, for each* $\lambda \in L$, a_λ *is an element of* E *such that* $f(a_\lambda) = b_\lambda$, *the family* $(a_\lambda)_{\lambda \in L}$ *is free and generates a supplementary submodule to* $g(G)$.

There exists one and only one linear mapping $h: F \to E$ such that $h(b_\lambda) = a_\lambda$ for all $\lambda \in L$ (Corollary 3 to Proposition 17). As h is a linear section associated with f, the proposition follows from I, § 4, no. 9, Proposition 15.

Remark. (5) Let $(a_i)_{1 \leq i \leq n}$ be a *basis* of an A-module E and let $(b_i)_{1 \leq i \leq n}$ be a family of elements of E given by the relations

$$b_i = \lambda_{1i}a_1 + \cdots + \lambda_{ii}a_i \quad (1 \leq i \leq n) \tag{46}$$

where λ_{ii} is *invertible* in A; then $(b_i)_{1 \leq i \leq n}$ is a *basis* of E. It suffices to argue by induction on n, the proposition being obvious for $n = 1$. If E′ is the submodule of E generated by the family $(a_i)_{1 \leq i \leq n-1}$, it follows from the induction hypothesis that $(b_i)_{1 \leq i \leq n-1}$ is a basis of E′; on the other hand, it follows from (46) that if $\mu b_n \in E'$ with $\mu \in A$, then also $\mu\lambda_{nn}a_n \in E'$, whence $\mu = 0$ since λ_{nn} is invertible. The family $(b_\iota)_{1 \leq \iota \leq n}$ is thus free and, as

$$a_n = -\lambda_{nn}^{-1}\lambda_{1n}a_1 - \cdots - \lambda_{nn}^{-1}\lambda_{n-1,n}a_{n-1} + \lambda_{nn}^{-1}b_n$$

it is seen that $(b_\iota)_{1 \leq \iota \leq n}$ is a generating system of E, which completes the proof. This result is easily generalized to a family $(a_\iota)_{\iota \in I}$ whose indexing set I is well ordered.

12. ANNIHILATORS. FAITHFUL MODULES. MONOGENOUS MODULES

DEFINITION 11. *The annihilator of a subset* S *of an* A*-module* E *is the set of elements* $\alpha \in A$ *such that* $\alpha x = 0$ *for all* $x \in S$.

The annihilator of S is usually denoted by Ann(S); for a subset S consisting of a single element x, we write Ann(x) instead of Ann($\{x\}$) and call Ann(x) *the annihilator of* x.

The relation $\alpha x = 0$ may also be expressed by saying that x *is annihilated by* α.

It is immediate that the annihilator of an arbitrary subset S of E is a *left ideal* of A; for it to be equal to A, it is necessary and sufficient (by virtue of (M_{IV})) that $S = \{0\}$. If two subsets S, T of E are such that $S \subset T$, the annihilator of T is contained in the annihilator of S. If $(S_\iota)_{\iota \in I}$ is an arbitrary family of subsets of E, the annihilator of the union $\bigcup_\iota S_\iota$ is the intersection of the annihilators of the S_ι. In particular, the annihilator of a subset S of E is the intersection of the annihilators of the elements of S. To say that an element of E is *free* is equivalent to saying that its annihilator is $\{0\}$. For all $x \in E$ and all $\alpha \in A$, the annihilator of αx is the set of $\beta \in A$ such that $\beta\alpha \in \text{Ann}(x)$.

The annihilator of a *submodule* M of E is a *two-sided ideal* of A; for, if $\alpha x = 0$ for all $x \in M$, then also $\alpha(\beta x) = 0$ for all $x \in M$ and all $\beta \in A$, hence $\alpha\beta$ belongs to the annihilator of M for all $\beta \in A$. In particular, the annihilator of E is a two-sided ideal of A.

For all $\alpha \in A$, let h_α be the homothety $x \mapsto \alpha x$; it is known that the mapping $\alpha \mapsto h_\alpha$ of A into the endomorphism ring $\mathscr{E} = \text{Hom}_{\mathbf{Z}}(E, E)$ of the commutative group (without operators) E, is a *ring homomorphism* (§ 2, no. 5). The inverse image of 0 under this homomorphism is the *annihilator* $\mathfrak{a}$ of E; the image of A under the homomorphism $\alpha \mapsto h_\alpha$ is therefore isomorphic to the quotient ring $A/\mathfrak{a}$. The module E is called *faithful* if its annihilator $\mathfrak{a}$ reduces to 0.

Let E be any A-module, $\mathfrak{a}$ a two-sided ideal of A contained in Ann(E) and let $\dot{\alpha}$ be an element of the quotient ring $A/\mathfrak{a}$; for all $x \in E$, the element αx is the same for all the $\alpha \in A$ belonging to the class $\dot{\alpha}$ mod. $\mathfrak{a}$; if this element is denoted by $\dot{\alpha}x$, it is immediately seen that the mapping

$(\dot{\alpha}, x) \mapsto \dot{\alpha}x$ defines (with addition on E) an $(A/\mathfrak{a})$-module structure on E. When $\mathfrak{a} = \text{Ann}(E)$, the $(A/\mathfrak{a})$-module E thus defined is *faithful*; we shall say that it is the faithful module *associated* with the A-module E. Observe that every submodule of an A-module E is also a submodule of the associated faithful module and conversely.

DEFINITION 12. *A module is called monogenous if it is generated by a single element.*

Proposition 9 of no. 7 shows that, if E is a monogenous A-module and a is an element generating E, E is identical with the set $A.a$ of ξa, where ξ runs through A.

Examples. (1) Every monogenous group, being commutative (I, § 4, no. 10, Proposition 18), is a monogenous $\mathbf{Z}$-module.

(2) If A is a commutative ring, the monogenous submodules of the A-module A are just the *principal ideals* (I, § 8, no. 6) of the ring A.

(3) Every *simple* A-module E is monogenous, since the submodule of E generated by an element $\neq 0$ of E is necessarily equal to E.

PROPOSITION 22. *Let* A *be a ring. Every quotient module of* A_s *is monogenous. Conversely, let* E *be a monogenous* A*-module,* c *a generator of* E *and* $\mathfrak{a}$ *its annihilator; the linear mapping* $\xi \mapsto \xi c$ *defines, when passing to the quotient, an isomorphism of* $A_s/\mathfrak{a}$ *onto* E.

As A_s is itself monogenous, being generated by 1, the first assertion follows from no. 7, Corollary 1 to Proposition 9. The second is obvious, since $\xi \mapsto \xi c$ is by hypothesis surjective and has kernel $\mathfrak{a}$.

Note that, if A is not commutative, the annihilators of two distinct generators c, c' of a monogenous A-module E are in general *distinct* and are also distinct from the annihilator of the module E. On the other hand, if A is *commutative*, the annihilator of a generator c of E is contained in the annihilator of every element of E and hence is the annihilator of the whole of E.

COROLLARY. *Every submodule of a monogenous* A*-module* E *is isomorphic to a quotient module* $\mathfrak{b}/\mathfrak{a}$ *where* $\mathfrak{a}$ *and* $\mathfrak{b}$ *are two left-ideals of* A *such that* $\mathfrak{a} \subset \mathfrak{b}$. *Every quotient module of a monogenous* A*-module is monogenous.*

The second assertion is immediate and the first follows from Proposition 22 and I, § 4, no. 6, Theorem 4.

Note on the other hand that a submodule of a monogenous module is not necessarily monogenous. For example, if A is a commutative ring in which there exist non-principal ideals (VII, § 1, no. 1), these ideals are non-monogenous submodules of the monogenous A-module A.

It follows from the definitions that a submodule of an A-module E generated by a family (a_ι) of elements of E is the *sum* of the monogenous submodules

Aa_ι of E; for (a_ι) to be a *basis* of E, it is necessary and sufficient that each of the a_ι be a *free* element of E and that the sum of the Aa_ι be *direct*.

PROPOSITION 23. *Let* E *be an* A-*module, the direct sum of an* infinite *family* $(M_\iota)_{\iota \in I}$ *of non-zero submodules. For every generating system* S *of* E, $\mathrm{Card}(S) \geqslant \mathrm{Card}(I)$.

For all $x \in S$, let C_x be the finite set of indices $\iota \in I$ such that the component of x in M_ι is $\neq 0$ and let $C = \bigcup_{x \in S} C_x$. Every $x \in S$ belongs by definition to the submodule of E the direct sum of the M_ι for $\iota \in C$ and the hypothesis that S generates E implies therefore that $C = I$; as I is by hypothesis infinite, so is S (*Set Theory*, III, § 5, no. 1, Corollary 1 to Proposition 1); therefore $\mathrm{Card}(I) = \mathrm{Card}(C) \leqslant \mathrm{Card}(S)$ (*Set Theory*, III, § 6, no. 3, Corollary 3 to Theorem 2).

COROLLARY 1. *Under the hypotheses of Proposition* 23, *suppose that each* M_ι *is monogenous and that* E *is the direct sum of a second family* $(N_\lambda)_{\lambda \in L}$ *of non-zero monogenous submodules. Then* $\mathrm{Card}(L) = \mathrm{Card}(I)$.

If b_λ is a generator of N_λ, the set of b_λ is a generating system of E, hence $\mathrm{Card}(L) \geqslant \mathrm{Card}(I)$. In particular L is infinite and, exchanging the roles of (M_ι) and (N_λ), similarly $\mathrm{Card}(I) \geqslant \mathrm{Card}(L)$, whence the corollary.

COROLLARY 2. *If a module* E *admits an infinite basis* B, *every generating system of* E *has cardinal* $\geqslant \mathrm{Card}(B)$ *and every basis of* E *is equipotent to* B.

13. CHANGE OF RING OF SCALARS

Let A, B be two rings and ρ a homomorphism of the ring B into the ring A. For every A-module E, the external law $(\beta, x) \mapsto \rho(\beta)x$ defines (with addition) a B-*module* structure said to be *associated* with ρ and the A-module structure on E; this B-module is denoted by $\rho_*(E)$ or $E_{[B]}$ (and even simply E if no confusion can arise). In particular, if B is a *subring* of A and $\rho: B \to A$ is the canonical injection, $E_{[B]}$ is called the B-module obtained by *restricting* the ring of scalars A *to* B; by an abuse of language, this expression is also used when the homomorphism ρ is arbitrary.

If F is a submodule of the A-module E, $\rho_*(F)$ is a submodule of $\rho_*(E)$ and $\rho_*(E/F)$ is equal to $\rho_*(E)/\rho_*(F)$.

Let E, F be two A-modules; every A-linear mapping $u: E \to F$ is also a B-linear mapping $E_{[B]} \to F_{[B]}$ denoted by $\rho_*(u)$; in other words, there is a *canonical injection* of $\mathbf{Z}$-modules

$$\mathrm{Hom}_A(E, F) \to \mathrm{Hom}_B(E_{[B]}, F_{[B]}). \tag{47}$$

This mapping is not necessarily bijective; in other words a B-linear mapping $E_{[B]} \to F_{[B]}$ is not necessarily A-linear. For example, a sub-B-module of $E_{[B]}$ is not necessarily a sub-A-module of E: if A is a field and B a subfield of A, the vector subspace B_s of the vector B-space $(A_s)_{[B]}$ is not a vector sub-A-space if $B \neq A$.

It is immediate that, for every family $(E_\iota)_{\iota \in I}$ of A-modules, the B-module $\rho_*\left(\prod_{\iota \in I} E_\iota\right)$ $\left(\text{resp. } \rho_*\left(\bigoplus_{\iota \in I} E_\iota\right)\right)$ is equal to $\prod_{\iota \in I} \rho_*(E_\iota)$ $\left(\text{resp. } \bigoplus_{\iota \in I} \rho_*(E_\iota)\right)$. Every generating system of $\rho_*(E)$ is a generating system of E but the converse is not necessarily true.

Proposition 24. *Let* A, B *be two rings and* $\rho : B \to A$ *a ring homomorphism.*

(i) *If* ρ *is surjective, the canonical mapping* (47) *is bijective. For every* A*-module* E, *every sub-*B*-module of* $\rho_*(E)$ *is a sub-*A*-module of* E; *every generating system of* E *is a generating system of* $\rho_*(E)$.

(ii) *If* ρ *is injective, every free family in the* A*-module* E *is a free family in the* B*-module* $\rho_*(E)$.

The proposition follows immediately from the definitions.

Note that even if ρ is injective, a free family in $\rho_*(E)$ is not necessarily free in E.

*For example, 1 and $\sqrt{2}$ do not form a free system in $\mathbf{R}$ considered as a vector $\mathbf{R}$-space, although they form a free system in $\mathbf{R}$ considered as a vector $\mathbf{Q}$-space (cf. *Remark* 1).*

Proposition 25. *Let* A, B *be two rings,* $\rho : B \to A$ *a ring homomorphism and* E *an* A*-module. Let* $(\alpha_\lambda)_{\lambda \in L}$ *be a generating system* (resp. *free family of elements, basis*) *of* A *considered as a left* B*-module. Let* $(a_\mu)_{\mu \in M}$ *be a generating system* (resp. *free family of elements, basis*) *of the* A*-module* E. *Then* $(\alpha_\lambda a_\mu)_{(\lambda, \mu) \in L \times M}$ *is a generating system* (resp. (*when* ρ *is injective*) *free family of elements, basis*) *of the* B*-module* $\rho_*(E)$.

If $x = \sum_{\mu \in M} \gamma_\mu a_\mu$, where $\gamma_\mu \in A$ and (α_λ) is a generating system of A, we may write $\gamma_\mu = \sum_{\lambda \in L} \rho(\beta_{\lambda\mu})\alpha_\lambda$, with $\beta_{\lambda\mu} \in B$, for all $\mu \in M$, whence $x = \sum_{\lambda, \mu} \rho(\beta_{\lambda\mu})\alpha_\lambda a_\mu$. On the other hand, if (α_λ) and (a_μ) are free families, a relation $\sum_{\lambda, \mu} \rho(\beta_{\lambda\mu})\alpha_\lambda a_\mu = 0$, with $\beta_{\lambda\mu} \in B$, may be written $\sum_{\mu \in M} \left(\sum_{\lambda \in L} \rho(\beta_{\lambda\mu})\alpha_\lambda\right) a_\mu = 0$; it thus implies $\sum_{\lambda \in L} \rho(\beta_{\lambda\mu})\alpha_\lambda = 0$ for all $\mu \in M$ and therefore $\beta_{\lambda\mu} = 0$ for all λ, μ if ρ is injective.

Corollary. *If* A *is a finitely generated left* B*-module and* E *a finitely generated left* A*-module,* $\rho_*(E)$ *is a finitely generated left* A*-module.*

Let C be a third ring, $\rho' : C \to B$ a ring homomorphism and $\rho'' = \rho \circ \rho'$ the composite homomorphism. It follows immediately from the definitions that

$\rho''_*(E) = \rho'_*(\rho_*(E))$ for every A-module E. In particular, if ρ is an *isomorphism* of B onto A, then $E = \rho'_*(\rho_*(E))$, where ρ' denotes the inverse isomorphism of ρ.

Remarks. (1) Let K be a field and A a subring of K with the following property: for every finite family $(\xi_i)_{1 \leqslant i \leqslant n}$ of elements of K, there exists a $\gamma \in A$ which is non-zero and such that $\gamma\xi_i \in A$ for $1 \leqslant i \leqslant n$ (a hypothesis which is always satisfied when A is *commutative* and K is the field of fractions of A). Let E be a vector space over K and $E_{[A]}$ the A-module obtained by restricting the ring of scalars to A. Then, if a family $(x_\lambda)_{\lambda \in L}$ is *free in* $E_{[A]}$ it is also *free in* E. Attention may be confined to the case where $L = [1, n]$; if there were a relation $\sum_{i=1}^{n} \xi_i x_i = 0$ with $\xi_i \in K$, the ξ_i not all zero, it would follow that for all $\beta \in A$, $\sum_{i=1}^{n} (\beta\xi_i)x_i = 0$. By hypothesis we can suppose $\beta \neq 0$ in A such that $\beta\xi_i = \alpha_i$ belongs to A for all i; but the relation $\sum_{i=1}^{n} \alpha_i x_i = 0$ is contrary to the hypothesis, the α_i being not all zero.

(2) If the ring homomorphism $\rho : B \to A$ is surjective and $\mathfrak{b}$ is its kernel (so that A is canonically identified with $B/\mathfrak{b}$), then, for every A-module E, $\mathfrak{b}$ is contained in the annihilator of $\rho_*(E)$ and E is the A-module derived from $\rho_*(E)$ by the process defined in no. 12.

Let A, B be two rings and $\rho : B \to A$ a homomorphism. Let E be an A-module and F a B-module; a B-*linear* mapping $u : F \to \rho_*(E)$ (also called a B-*linear mapping of* F *into* E if no confusion arises) is also called a *semi-linear mapping* (relative to ρ) of the B-module F into the A-module E; it is also said that the ordered pair (u, ρ) is a *dimorphism* of F into E; this therefore means that, for $x \in F$, $y \in F$ and $\beta \in B$,

$$\text{(48)} \qquad \begin{cases} u(x + y) = u(x) + u(y) \\ u(\beta x) = \rho(\beta)u(x). \end{cases}$$

The set $\mathrm{Hom}_B(F, \rho_*(E))$ of B-linear mappings of F into E is also written as $\mathrm{Hom}_B(F, E)$ if no confusion can arise.

When ρ is an *isomorphism* of B onto A, the relation $u(\beta x) = \rho(\beta)u(x)$ for all $\beta \in B$ may also be written as $u(\rho'(\alpha)x) = \alpha x$ for all $\alpha \in A$, where ρ' denotes the inverse isomorphism of ρ; to say that u is semi-linear for ρ is then equivalent to saying that u is an A-*linear* mapping *of* $\rho'_*(F)$ *into* E.

Example. It has been seen (no. 1) that a homothety $h_\alpha : x \mapsto \alpha x$ on an A-module E is not necessarily a linear mapping. But if α is *invertible*, h_α is a *semi-linear* mapping (which is moreover bijective) relative to the inner automorphism $\xi \mapsto \alpha\xi\alpha^{-1}$ of A, for $\alpha(\lambda x) = (\alpha\lambda\alpha^{-1})(\alpha x)$.

Let C be a third ring, $\rho' : C \to B$ a homomorphism and G a C-module. If $v : G \to F$ is a semi-linear mapping relative to ρ', the composition $w = u \circ v$ is a semi-linear mapping of G into E relative to the homomorphism $\rho'' = \rho \circ \rho'$. If ρ is an *isomorphism* and $u : F \to E$ is a *bijective* semi-linear mapping relative to ρ, the inverse mapping $u' : E \to F$ is a semi-linear mapping *relative to the inverse isomorphism* $\rho' : A \to B$ of ρ.

It is thus seen that, for the species of structure defined by giving on an ordered pair (A, E) of sets a ring structure on A and a left A-module structure on E, the *dimorphisms* (u, ϕ) can be taken as *morphisms* (*Set Theory*, IV, § 2, no. 1); we shall always assume in what follows that this choice of morphisms has been made.

Remark (3). Let A_1, A_2 be two rings, $A = A_1 \times A_2$ their product and let $e_1 = (1, 0)$, $e_2 = (0, 1)$ in A, so that A_1 and A_2 are canonically identified with the two-sided ideals Ae_1 and Ae_2 of A. For every A-module E, e_1E and e_2E are sub-A-modules E_1, E_2 of E, annihilated respectively by e_2 and e_1, so that, canonically identifying A/Ae_2 with A_1 and A/Ae_1 with A_2, E_1 (resp. E_2) is given an A_1-module (resp. A_2-module) structure. Moreover, E is the *direct sum* of E_1 and E_2, for every $x \in E$ can be written as $x = e_1x + e_2x$ and the relation $e_1x = e_2y$ implies $e_1x = e_1^2x = e_1e_2y = 0$. Conversely, for every ordered pair consisting of an A_1-module F_1 and an A_2-module F_2, let E_1 be the A-module $(p_1)_*(F_1)$, E_2 the A-module $(p_2)_*(F_2)$, p_1 and p_2 being the projections of A onto A_1 and A_2 respectively; then in the A-module $E = E_1 \oplus E_2$, $E_1 = e_1E$, $E_2 = e_2E$. The study of A-modules is thus reduced to that of A_1-modules and that of A_2-modules. In particular, every submodule M of E is of the form $M_1 \oplus M_2$, where $M_1 = e_1M$ and $M_2 = e_2M$.

14. MULTIMODULES

Let A, B be two rings and consider on a set E two left module structures with the *same* additive law and whose ring of operators are respectively A and B; let $\mathscr{E}$ be the endomorphism ring of the additive group E and for all $\alpha \in A$ (resp. $\beta \in B$) let h_α (resp. h'_β) denote the element $x \mapsto \alpha x$ (resp. $x \mapsto \beta x$) of $\mathscr{E}$. Clearly the three following properties are equivalent: (a) $h_\alpha \circ h'_\beta = h'_\beta \circ h_\alpha$ for all α and β; (b) the image of A under the homomorphism $a \mapsto h_\alpha$ is *contained in* $\mathrm{Hom}_B(E, E)$; (c) the image of B under the homomorphism $\beta \mapsto h'_\beta$ is *contained in* $\mathrm{Hom}_A(E, E)$. When the A-module (resp. B-module) structure in question is a right module structure, the ring A (resp. B) must be replaced in (b) (resp. (c)) by A^0 (resp. B^0). The above properties can be expressed by saying that the two (left or right) module structures defined on E are *compatible*.

DEFINITION 13. *Let* $(A_\lambda)_{\lambda \in L}$, $(B_\mu)_{\mu \in M}$ *be two families of rings; an* $((A_\lambda), (B_\mu))$-*multimodule* (or *multimodule over the families of rings* $(A_\lambda)_{\lambda \in L}$, $(B_\mu)_{\mu \in M}$) *is a set* E *with, for each* $\lambda \in L$, *a left* A_λ-*module structure and, for each* $\mu \in M$, *a right* B_μ-*module structure, all these module structures being compatible with one another.*

When the family (B_μ) (resp. (A_λ)) is empty, E is called a *left* (resp. *right*) *multimodule*. When Card(L) + Card(M) = 2, we say "*bimodule*" instead of "multimodule"; it is then often convenient to consider (as can always be done by replacing a ring of operators by its opposite, cf. no. 1) a bimodule as having a *left module* structure with respect to a ring A and a *right module* structure with respect to a ring B, the permutability of the laws then being expressed by the relation

$$(49) \qquad \alpha(x\beta) = (\alpha x)\beta \quad \text{for } x \in E, \alpha \in A, \beta \in B.$$

It is then also said that E is an (A, B)-*bimodule*.

Two *multimodule* structures on a set E are said to be *compatible* if all the module structures on E which define one or the other of these multimodule structures are compatible with one another.

Examples. (1) On a ring A the module structures of A_s and A_d are compatible and A can therefore be considered canonically as an (A, A)-bimodule.

(2) A left A-module E has a canonical left module structure over the ring $\mathrm{End}_A(E)$ and the A-module and $\mathrm{End}_A(E)$-module structures on E are *compatible*.

Clearly when E is a multimodule over two families $(A_\lambda)_{\lambda \in L}$, $(B_\mu)_{\mu \in M}$ of rings, E is also a multimodule over any two subfamilies $(A_\lambda)_{\lambda \in L'}$, $(B_\mu)_{\mu \in M'}$, where the A_λ-module and B_μ-module structures for $\lambda \in L'$ and $\mu \in M'$ are those initially given.

Since multimodules are particular examples of commutative groups with operators, the results of nos. 2 to 10 (cf. no. 10, *Remark*) can be applied to them; in particular, if E, F are two $((A_\lambda), (B_\mu))$-multimodules, a *homomorphism* $u: E \to F$ is a mapping which is an A_λ-homomorphism for all $\lambda \in L$ and a B_μ-homomorphism for all $\mu \in M$. The stable subgroups of an $((A_\lambda), (B_\mu))$-multimodule are $((A_\lambda), (B_\mu))$-multimodules (called *submultimodules*), as also are the quotients by such subgroups (called *quotient multimodules*); similarly for products and direct sums.

Let E be an $((A_\lambda), (B_\mu))$-multimodule and for each $\lambda \in L$ (resp. $\mu \in M$) let $\phi_\lambda: A'_\lambda \to A_\lambda$ (resp. $\psi_\mu: B'_\mu \to B_\mu$) be a ring homomorphism; clearly the A'_λ-module structures associated with the ϕ_λ and the A_λ-module structures given on E and the B'_μ-module structures associated with the ψ_μ and the B_λ-module structures given on E are compatible with one another and hence define on E an $((A'_\lambda), (B'_\mu))$-multimodule structure, said to be *associated* with the given $((A_\lambda), (B_\mu))$-multimodule structure and the ϕ_λ and ψ_μ.

If E, F are two $((A_\lambda), (B_\mu))$-multimodules, the additive group of homomorphisms of E into F is denoted by $\mathrm{Hom}_{(A_\lambda),(B_\mu)}(E, F)$ (or simply Hom(E, F)). Formulae (6) to (8) of no. 2 are obviously valid for $((A_\lambda), (B_\mu))$-multimodule homomorphisms and, in particular, Hom(E, E) = End(E) has a *ring* structure;

moreover Hom(E, F) has a canonical *left* End(F)-module structure and *right* End(E)-module structure, these two structures being compatible; in other words, Hom(E, F) has a canonical (End(F), End(E))-*bimodule* structure.

Suppose now that E has a multimodule structure whose rings of left (resp. right) operators are on the one hand the A_λ for $\lambda \in L$ (resp. the B_μ for $\mu \in M$) and on the other the rings of another family $(A'_{\lambda'})_{\lambda' \in L'}$ (resp. $(B'_{\mu'})_{\mu' \in M'}$). Suppose similarly that F has a multimodule structure whose rings of left (resp. right) operators are on the one hand the A_λ for $\lambda \in L$ (resp. the B_μ for $\mu \in M$) and on the other the rings of another family $(A''_{\lambda''})_{\lambda'' \in L''}$ (resp. $(B''_{\mu''})_{\mu'' \in M''}$); to abbreviate we shall say that E is an $((A_\lambda), (A'_{\lambda'}); (B_\mu), (B'_{\mu'}))$-multimodule and F an $((A_\lambda), (A''_{\lambda''}); (B_\mu), (B''_{\mu''}))$-multimodule. Consider E and F as $((A_\lambda), (B_\mu))$-multimodules, thus *restricting* the operators to the subfamilies (A_λ) and (B_μ). By what was said at the beginnings of this no., the multimodule structures given on E and F define canonically ring homomorphisms

$$A'_{\lambda'} \to \mathrm{End}_{(A_\lambda),(B_\mu)}(E), \qquad B'^{0}_{\mu'} \to \mathrm{End}_{(A_\lambda),(B_\mu)}(E),$$
$$A''_{\lambda''} \to \mathrm{End}_{(A_\lambda),(B_\mu)}(F), \qquad B''_{\mu''} \to \mathrm{End}_{(A_\lambda),(B_\mu)}(F);$$

moreover, two elements of $\mathrm{End}_{(A_\lambda),(B_\mu)}(E)$ (resp. $\mathrm{End}_{(A_\lambda),(B_\mu)}(F)$) respective images of elements of two distinct rings among the $A'_{\lambda'}$ or the $B'^{0}_{\mu'}$ (resp. the $A''_{\lambda''}$ or the $B''^{0}_{\mu''}$) are permutable; it follows that the above homomorphisms define on $\mathrm{Hom}_{(A_\lambda),(B_\mu)}(E, F)$ a *multimodule* structure whose rings of *left* operators are the $A''_{\mu''}$ $(\lambda'' \in L'')$ and the $B'_{\mu'}$ $(\mu' \in M')$ and whose rings of *right* operators are the $A'_{\lambda'}$ $(\lambda' \in L')$ and the $B''_{\mu''}$ $(\mu'' \in M'')$.

If now E′ is an $((A_\lambda), (A'_{\lambda'}); (B_\mu), (B'_{\mu'}))$-multimodule and F′ an $((A_\lambda), (A''_{\lambda''}); (B_\mu), (B''_{\mu''}))$-multimodule, $\mathrm{Hom}_{(A_\lambda),(B_\mu)}(E', F')$ is an

$$((A''_{\lambda''}), (B'_{\mu'}) : (A'_{\lambda'}), (B''_{\mu''}))\text{-multimodule};$$

if $u: E' \to E$, $v: F \to F'$ are multimodule homomorphisms,

$$\mathrm{Hom}(u, v): \mathrm{Hom}_{(A_\lambda),(B_\mu)}(E, F) \to \mathrm{Hom}_{(A_\lambda),(B_\mu)}(E', F')$$

is defined as in no. 2 and is a *multimodule* homomorphism.

Remarks. (1) Let F be an A-module and C the *centre* of the ring A; as central homotheties commute with all homotheties, F has a *bimodule* structure whose rings of left operators are A and C. If E is another A-module, $\mathrm{Hom}_A(E, F)$ has therefore a canonical C-*module* structure (where, for $f \in \mathrm{Hom}_A(E, F)$ and $\gamma \in C$, γf is the homomorphism $x \mapsto \gamma f(x)$); if E′, F′ are two A-modules, $u: E' \to E$, $v: F \to F'$ two A-homomorphisms, the mapping Hom(u, v) is C-*linear*.

(2) Let E be a left A-module; as A has a canonical (A, A)-bimodule structure, so has the direct sum $A^{(T)}$ for any indexing set T; by the above,

$\mathrm{Hom}_A(A_s^{(T)}, E)$ has a canonical *left* A-*module* structure arising from the *right* A-module structure on $A_s^{(T)}$: for $f \in \mathrm{Hom}_A(A_s^{(T)}, E)$ and $\alpha \in A$, αf is the linear mapping $x \mapsto f(x\alpha)$. Corollary 2 to Proposition 17 of no. 11 defines a canonical mapping $j_{E,T}$ from the product module E^T to $\mathrm{Hom}_A(A_s^{(T)}, E)$, the image under $j_{E,T}$ of a family $(x_t)_{t \in T}$ being the linear mapping $f: A_s^{(T)} \to E$ such that $f(e_t) = x_t$ for all $t \in T$ (where (e_t) is the canonical basis of $A_s^{(T)}$); it is known (*loc. cit.*) that $j_{E,T}$ is *bijective* and it follows from the definition given above of the A-module structure on $\mathrm{Hom}_A(A_s^{(T)}, E)$ that $j_{E,T}$ is A-*linear*. Finally, if $u: E \to F$ is an A-module homomorphism, the diagram

$$\begin{array}{ccc} E^T & \xrightarrow{j_{E,T}} & \mathrm{Hom}_A(A_s^{(T)}, E) \\ {\scriptstyle u^T}\downarrow & & \downarrow{\scriptstyle \mathrm{Hom}(1,u)} \\ F^T & \xrightarrow[j_{F,T}]{} & \mathrm{Hom}_A(A_s^{(T)}, F) \end{array} \tag{50}$$

is *commutative*.

Note that when T consists of a single element, $j_E: E \to \mathrm{Hom}_A(A_s, E)$ is just the mapping $x \mapsto \theta_x$ defined in no. 2, *Example* 1.

§ 2. MODULES OF LINEAR MAPPINGS. DUALITY

1. PROPERTIES OF $\mathrm{Hom}_A(E, F)$ RELATIVE TO EXACT SEQUENCES

THEOREM 1. *Let* A *be a ring,* E′, E, E″ *three* A-*modules and* $u: E' \to E$, $v: E \to E''$ *two homomorphisms. For the sequence*

$$E' \xrightarrow{u} E \xrightarrow{v} E'' \longrightarrow 0 \tag{1}$$

to be exact, it is necessary and sufficient that, for every A-*module* F, *the sequence*

$$0 \longrightarrow \mathrm{Hom}(E'', F) \xrightarrow{\bar{v}} \mathrm{Hom}(E, F) \xrightarrow{\bar{u}} \mathrm{Hom}(E', F) \tag{2}$$

(*where* $\bar{u} = \mathrm{Hom}(u, 1_F)$, $\bar{v} = \mathrm{Hom}(v, 1_F)$) *be exact.*

Suppose that sequence (1) is exact. If $w \in \mathrm{Hom}(E'', F)$ and $\bar{v}(w) = w \circ v = 0$, then $w = 0$ since v is surjective. Sequence (2) is therefore exact at Hom(E″, F). We show that it is exact at Hom(E, F). $\bar{u} \circ \bar{v} = \mathrm{Hom}(v \circ u, 1_F)$ (§ 1, no. 2, formula (10)) and $v \circ u = 0$ since sequence (1) is exact at E. Therefore $\bar{u} \circ \bar{v} = 0$, that is $\mathrm{Im}(\bar{v}) \subset \mathrm{Ker}(\bar{u})$. On the other hand, if $w \in \mathrm{Ker}(\bar{u})$, then $w \circ u = 0$ and hence $\mathrm{Ker}(w) \supset \mathrm{Im}(u)$. But as sequence (1) is exact at E, $\mathrm{Im}(u) = \mathrm{Ker}(v)$ and hence $\mathrm{Ker}(w) \supset \mathrm{Ker}(v)$; as v is surjective, it follows from § 1, no. 3, *Remark* that there exists a $w' \in \mathrm{Hom}(E'', F)$ such that

$w = w' \circ v = \bar{v}(w')$. Therefore $\mathrm{Ker}(\bar{u}) \subset \mathrm{Im}(\bar{v})$, which completes the proof that sequence (2) is exact.

Conversely, suppose that (2) is exact for every A-module F. As $\bar{u} \circ \bar{v} = \mathrm{Hom}(v \circ u, 1_F) = 0$, $w \circ v \circ u = 0$ for every homomorphism $w: E'' \to F$. Taking $F = E''$ and $w = 1_{E''}$, it is seen first that $v \circ u = 0$ and hence $u(E') \subset \mathrm{Ker}(v)$. Now take $F = \mathrm{Coker}(u)$ and let $\phi: E \to F = E/u(E')$ be the canonical mapping. Then $\bar{u}(\phi) = \phi \circ u = 0$ by definition and hence there exists a $\psi \in \mathrm{Hom}(E'', F)$ such that $\phi = \bar{v}(\psi) = \psi \circ v$; this obviously implies $u(E') = \mathrm{Ker}(\phi) \supset \mathrm{Ker}(v)$, which proves that sequence (1) is exact at E. Finally, let θ be the canonical homomorphism of E'' onto $F = E''/v(E)$; then $\bar{v}(\theta) = \theta \circ v = 0$, hence $\theta = 0$; therefore, $F = \{0\}$ and v is surjective. Sequence (1) is therefore exact at E''.

COROLLARY. *For an* A-*linear mapping* $u: E \to F$ *to be surjective* (resp. *bijective*, resp. *zero*), *it is necessary and sufficient that, for every* A-*module* G, *the mapping* $\mathrm{Hom}(u, 1_G): \mathrm{Hom}(F, G) \to \mathrm{Hom}(E, G)$ *be injective* (resp. *bijective*, resp. *zero*).

It suffices to apply Theorem 1 to the case where $E'' = \{0\}$ (resp. $E' = \{0\}$ resp. $E'' = E$ and $v = 1_E$).

Note that starting from an exact sequence

$$0 \longrightarrow E' \xrightarrow{u} E \xrightarrow{v} E'' \longrightarrow 0$$

the corresponding sequence

$$0 \longrightarrow \mathrm{Hom}(E'', F) \xrightarrow{\bar{v}} \mathrm{Hom}(E, F) \xrightarrow{\bar{u}} \mathrm{Hom}(E', F) \longrightarrow 0$$

is not necessarily exact, in other words, the homomorphism $\bar{u}$ is not necessarily surjective. If E' is identified with a submodule of E, this means that a linear mapping of E' into F cannot always be extended to a linear mapping of E into F (Exercises 11 and 12). However:

PROPOSITION 1. *If the exact sequence of linear mappings*

$$(3) \qquad 0 \longrightarrow E' \xrightarrow{u} E \xrightarrow{v} E'' \longrightarrow 0$$

splits (in other words, if $u(E')$ is a *direct factor* of E) *the sequence*

$$(4) \qquad 0 \longrightarrow \mathrm{Hom}(E'', F) \xrightarrow{\bar{v}} \mathrm{Hom}(E, F) \xrightarrow{\bar{u}} \mathrm{Hom}(E', F) \longrightarrow 0$$

is exact and splits. Conversely, if, for every A-module F, *sequence* (4) *is exact, sequence* (3) *splits.*

If the exact sequence (3) splits, there exists a linear retraction $u': E \to E'$ associated with u (§ 1, no. 9, Proposition 15); if

$$\bar{u}' = \mathrm{Hom}(u', 1_F): \mathrm{Hom}(E', F) \to \mathrm{Hom}(E, F),$$

the fact that $u' \circ u$ is the identity implies that $\bar{u} \circ \bar{u}'$ is the identity (§ 1, no. 2, formula (10)) and hence the first assertion follows from § 1, no. 9, Proposition 15. Conversely, suppose sequence (4) is exact for $F = E'$. Then there exists an element $f \in \mathrm{Hom}(E, E')$ such that $f \circ u = 1_{E'}$, and the conclusion follows from § 1, no. 9, Proposition 15.

Note that the first assertion of Proposition 1 can also be considered as a special case of § 1, no. 6, Corollary 1 to Proposition 6, canonically identifying $\mathrm{Hom}(E', F) \oplus \mathrm{Hom}(E'', F)$ with $\mathrm{Hom}(E' \oplus E'', F)$ by means of the **Z**-linear mapping $\mathrm{Hom}(p', 1_F) + \mathrm{Hom}(p'', 1_F)$, where $p' : E' \oplus E'' \to E'$ and $p'' : E' \oplus E'' \to E''$ are the canonical projections.

THEOREM 2. *Let* A *be a ring*, F', F, F'' *three* A-*modules and* $u : F' \to F$, $v : F \to F''$ *two homomorphisms. For the sequence*

$$(5) \qquad 0 \longrightarrow F' \xrightarrow{v} F \xrightarrow{u} F''$$

to be exact, it is necessary and sufficient that, for every A-*module* E, *the sequence*

$$(6) \qquad 0 \longrightarrow \mathrm{Hom}(E, F') \xrightarrow{\bar{u}} \mathrm{Hom}(E, F) \xrightarrow{\bar{v}} \mathrm{Hom}(E, F'')$$

(where $\bar{u} = \mathrm{Hom}(1_E, u)$, $\bar{v} = \mathrm{Hom}(1_E, v)$) *be exact.*

Suppose that the sequence (5) is exact. Note first that

$$\bar{v} \circ \bar{u} = \mathrm{Hom}(1_E, v \circ u) = 0$$

(II, § 1, no. 2, formula (10)) since $v \circ u = 0$. The image of $\mathrm{Hom}(E, F')$ under $\bar{u}$ is therefore contained in the kernel N of $\bar{v}$; let f be the homomorphism of the **Z**-module $\mathrm{Hom}(E, F')$ into N whose graph is equal to that of $\bar{u}$; it is necessary to prove that f is *bijective* and hence to define a mapping $g : N \to \mathrm{Hom}(E, F')$ such that $f \circ g$ and $g \circ f$ are the identity mappings. For this, let w be an element of N, that is a linear mapping $w : E \to F$ such that $v \circ w = 0$. The latter relation is equivalent to $w(E) \subset \mathrm{Ker}(v) = u(F')$ by hypothesis, hence, since u is injective, there exists one and only one linear mapping $w' : E \to F'$ such that $w = u \circ w'$ and we take $g(w) = w'$; it is immediately verified that g satisfies the desired conditions.

Conversely, suppose that sequence (6) is exact for every A-module E. As $\mathrm{Hom}(1_E, v \circ u) = \bar{v} \circ \bar{u} = 0$, then $v \circ u \circ w = 0$ for every homomorphism $w : E \to F'$. Taking $E = F'$ and $w = 1_{F'}$, it is seen first that $v \circ u = 0$ and hence $u(F') \subset \mathrm{Ker}(v)$. Now we take $E = \mathrm{Ker}(v)$ and let $\phi : E \to F$ be the canonical injection. Then $\bar{v}(\phi) = v \circ \phi = 0$ by definition and hence there exists $\psi \in \mathrm{Hom}(E, F')$ such that $\phi = \bar{u}(\psi) = u \circ \psi$, which obviously implies $\mathrm{Ker}(v) \subset u(F')$ and completes the proof of the exactness of (5) at F. Finally, if θ is the identity mapping of $\mathrm{Ker}\, u$, then $\bar{u}(\theta) = 0$, hence $\theta = 0$ and $\mathrm{Ker}\, u = \{0\}$, which proves the exactness of (5) at F'.

Remark. (1) Theorem 2 allows us, for every submodule F′ of F, to identify Hom(E, F′) with a sub-**Z**-module of Hom(E, F). When this identification is made, then, for every family (M_λ) of submodules of F

$$\mathrm{Hom}\Big(E, \bigcap_\lambda M_\lambda\Big) = \bigcap_\lambda \mathrm{Hom}(E, M_\lambda)$$

for if $u \in \mathrm{Hom}(E, F)$ belongs to each of the $\mathrm{Hom}(E, M_\lambda)$, then, for all $x \in E$, $u(x) \in M_\lambda$ for all λ and hence u maps E into $\bigcap_\lambda M_\lambda$.

Corollary. *For an* A*-linear mapping* $u: E \to F$ *to be injective, it is necessary and sufficient that, for every* A*-module* G, *the mapping* $\mathrm{Hom}(1_G, u): \mathrm{Hom}(G, E) \to \mathrm{Hom}(G, F)$ *be injective.*

It suffices to apply Theorem 2 to the case where $F' = \{0\}$.

Starting from an exact sequence

$$0 \longrightarrow F' \xrightarrow{u} F \xrightarrow{v} F'' \longrightarrow 0$$

the corresponding sequence

$$0 \longrightarrow \mathrm{Hom}(E, F') \xrightarrow{\bar{u}} \mathrm{Hom}(E, F) \xrightarrow{\bar{v}} \mathrm{Hom}(E, F'') \longrightarrow 0$$

is not necessarily exact, in other words $\bar{v}$ is not necessarily surjective. If F′ is identified with a submodule of F and F″ with the quotient module F/F′, this means that a linear mapping of E into F″ is not necessarily of the form $v \circ w$, where w is a linear mapping of E into F. However:

Proposition 2. *If the exact sequence*

$$0 \longrightarrow F' \xrightarrow{u} F \xrightarrow{v} F'' \longrightarrow 0 \tag{7}$$

splits (in other words, if $u(F')$ is a *direct factor* of F), *the sequence*

$$0 \longrightarrow \mathrm{Hom}(E, F') \xrightarrow{\bar{u}} \mathrm{Hom}(E, F) \xrightarrow{\bar{v}} \mathrm{Hom}(E, F'') \longrightarrow 0 \tag{8}$$

is exact and splits. Conversely, if sequence (8) *is exact for every* A*-module* E, *the exact sequence* (7) *splits.*

The first assertion follows from the fact that

$$\mathrm{Hom}(E, F') \oplus \mathrm{Hom}(E, F'')$$

is canonically identified with $\mathrm{Hom}(E, F' \oplus F'')$ by means of the **Z**-linear mapping $\mathrm{Hom}(1_E, j') + \mathrm{Hom}(1_E, j'')$, $j': F' \to F' \oplus F''$ and $j'': F'' \to F' \oplus F''$ being the canonical injections (§ 1, no. 6, Corollary 1 to Proposition 6). Conversely, if sequence (8) is exact for $E = F''$, there is an element $g \in \mathrm{Hom}(F'', F)$ such that $v \circ g = 1_{F''}$ and the conclusion follows from § 1, no. 9, Proposition 15.

Remark (2). The results of this no. are valid without modification for *all* commutative groups with operators.

2. PROJECTIVE MODULES

DEFINITION 1. *An* A-*module* P *is called projective if, for every exact sequence* $F' \to F \to F''$ *of* A-*linear mapping, the sequence*

$$\text{Hom}(P, F') \to \text{Hom}(P, F) \to \text{Hom}(P, F'')$$

is exact.

PROPOSITION 3. *For an* A-*module* P, *the direct sum of a family of submodules* (M_ι), *to be projective, it is necessary and sufficient that each of the* M_ι *be projective.*

For every A-module homomorphism $u : E \to F$,

$$\text{Hom}(1_P, u) : \text{Hom}(P, E) \to \text{Hom}(P, F)$$

is identified with $\prod_\iota \text{Hom}(1_{M_\iota}, u)$ (§ 1, no. 6, Corollary 1 to Proposition 6); the conclusion thus follows from Definition 1 and § 1, no. 5, Proposition 5 (ii).

COROLLARY. *Every free* A-*module is projective.*

It suffices by Proposition 3 to show that A_s is projective, which follows immediately from the commutativity of diagram (50) of § 1, no. 14.

PROPOSITION 4. *Let* P *be an* A-*module. The following properties are equivalent:*

(a) P *is projective.*

(b) *For every exact sequence* $0 \to F' \to F \to F'' \to 0$ *of* A-*linear mappings, the sequence*

$$0 \to \text{Hom}(P, F') \to \text{Hom}(P, F) \to \text{Hom}(P, F'') \to 0$$

is exact.

(c) *For every surjective* A-*module homomorphism* $u : E \to E''$ *and every homomorphism* $f : P \to E''$, *there exists a homomorphism* $g : P \to E$ *such that* $f = u \circ g$ (it is said that f can be "lifted" to a homomorphism of P into E).

(d) *Every exact sequence* $0 \to E' \to E \xrightarrow{v} P \to 0$ *of* A-*linear mappings splits* (and therefore P is isomorphic to a *direct factor* of E).

(e) P *is isomorphic to a direct factor of a free* A-*module.*

It is trivial that (a) implies (b). To see that (b) implies (c), it suffices to apply (b) to the exact sequence $0 \to E' \to E \xrightarrow{u} E'' \to 0$, where $E' = \text{Ker}(u)$, since (c) expresses the fact that

$$\text{Hom}(1_P, u) : \text{Hom}(P, E) \to \text{Hom}(P, E'')$$

is surjective. To see that (c) implies (d), it suffices to apply (c) to the surjective homomorphism $v: E \to P$ and the homomorphism $1_P: P \to P$; the existence of a homomorphism $g: P \to E$ such that $1_P = v \circ g$ implies that the sequence

$$0 \longrightarrow E' \longrightarrow E \xrightarrow{v} P \longrightarrow 0$$

splits (§ 1, no. 9, Proposition 15). As for every A-module M there exist a free A-module L and an exact sequence $0 \to R \to L \to M \to 0$ (§ 1, no. 11, Proposition 20), clearly (d) implies (e). Finally (e) implies (a) by virtue of Proposition 3 and its Corollary.

COROLLARY 1. *For an* A-*module to be projective and finitely generated, it is necessary and sufficient that it be a direct factor of a free* A-*module with a finite basis.*

The condition is obviously sufficient; conversely, a finitely generated projective module E is isomorphic to a quotient of a free module F with a finite basis (§ 1, no. 11) and E is isomorphic to a direct factor of F by virtue of Proposition 4 (d).

COROLLARY 2. *Let* C *be a commutative ring and* E, F *two finitely generated projective* C-*modules; then* $\mathrm{Hom}_C(E, F)$ *is a finitely generated projective* C-*module.*

It may be assumed that there are two finitely generated free C-modules M, N such that $M = E \oplus E'$, $N = F \oplus F'$; it follows from § 1, no. 6, Corollary 1 to Proposition 6 that $\mathrm{Hom}_C(M, N)$ is finitely generated and free and on the other hand that $\mathrm{Hom}_C(M, N)$ is isomorphic to

$$\mathrm{Hom}_C(E, F) \oplus \mathrm{Hom}_C(E', F) \oplus \mathrm{Hom}_C(E, F') \oplus \mathrm{Hom}(E', F'),$$

whence the corollary.

3. LINEAR FORMS; DUAL OF A MODULE

Let E be a *left* A-module. As A is an (A, A)-bimodule, $\mathrm{Hom}_A(E, A_s)$ has a canonical *right* A-module structure (§ 1, no. 14).

DEFINITION 2. *For every right* A-*module* E, *the right* A-*module* $\mathrm{Hom}_A(E, A_s)$ *is called the dual module of* E (or simply the *dual*† of E) *and its elements are called the linear forms on* E.

If E is a *right* A-module, the set $\mathrm{Hom}_A(E, A_d)$ with its canonical *left* A-module structure is likewise called the *dual* of E and its elements are called *linear forms* on E.

---

† In *Topological Vector Spaces*, IV, we shall define, for vector spaces with a *topology*, a notion of "dual space" which will depend on this topology and will be distinct from the one defined here. The reader must beware of incautiously applying to the "topological" dual space the properties of the "algebraic" dual which are established in this paragraph.

In this chapter, E* will be used to denote the dual of a (left or right) A-module E.

Example. *On the vector space (with respect to the field $\mathbf{R}$) of continuous real-valued functions on an interval $[a, b]$ of $\mathbf{R}$, the mapping $x \mapsto \int_a^b x(t)\,dt$ is a linear form.*

Let E be a left A-module and E* its dual; for every ordered pair of elements $x \in E$, $x^* \in E^*$, the element $x^*(x)$ of A is denoted by $\langle x, x^* \rangle$. Then the relations

$$\langle x + y, x^* \rangle = \langle x, x^* \rangle + \langle y, x^* \rangle \tag{9}$$

$$\langle x, x^* + y^* \rangle = \langle x, x^* \rangle + \langle x, y^* \rangle \tag{10}$$

$$\langle \alpha x, x^* \rangle = \alpha \langle x, x^* \rangle \tag{11}$$

$$\langle x, x^* \alpha \rangle = \langle x, x^* \rangle \alpha \tag{12}$$

hold for x, y in E, x^*, y^* in E* and $\alpha \in A$. The mapping $(x, x^*) \mapsto \langle x, x^* \rangle$ of $E \times E^*$ into A is called the *canonical bilinear form* on $E \times E^*$ (the notion of bilinear form will be defined generally in IX, § 1). Every linear form x^* on E can be considered as the *partial* mapping $x \mapsto \langle x, x^* \rangle$ corresponding to the canonical bilinear form.

When E is a *right* A-module, the value $x^*(x)$ of a linear form $x^* \in E^*$ at an element $x \in E$ is denoted by $\langle x^*, x \rangle$ and the formulae corresponding to (11) and (12) are written as

$$\langle x^*, x\alpha \rangle = \langle x^*, x \rangle \alpha$$

$$\langle \alpha x^*, x \rangle = \alpha \langle x^*, x \rangle.$$

When A is commutative, either notation is permissible.

PROPOSITION 5. *For every ring* A, *the mapping which associates with every* $\xi \in A$ *the linear form* $\eta \mapsto \eta\xi$ *on* A_s *is an isomorphism of* A_d *onto the dual of* A_s.

It is the particular case of the canonical isomorphism $E \to \mathrm{Hom}_A(A_s, E)$ of § 1, no. 14, *Remark* 2, corresponding to $E = A_s$; the commutativity of diagram (50) of § 1, no. 14, shows that we have here an isomorphism of right A-modules.

If A_d is identified with the dual of A_s by means of the isomorphism of Proposition 5, the canonical bilinear form on $A_s \times A_d$ is then expressed by

$$\langle \xi, \xi^* \rangle = \xi\xi^* \quad \text{for } \xi, \xi^* \text{ in A.} \tag{13}$$

Similarly, the dual of A_d is canonically identified with A_s, the canonical bilinear form on $A_d \times A_s$ being expressed by

$$\langle \xi^*, \xi \rangle = \xi^*\xi \quad \text{for } \xi, \xi^* \text{ in A.} \tag{14}$$

4. ORTHOGONALITY

DEFINITION 3. *Let* E *be an* A*-module and* E^* *its dual; an element* $x \in E$ *and an element* $x^* \in E^*$ *are called orthogonal if* $\langle x, x^* \rangle = 0$.

A subset M of E and a subset M′ of E^* are called *orthogonal sets* if, for all $x \in M$, $x^* \in M'$, x and x^* are orthogonal. In particular, $x^* \in E^*$ (resp. $x \in E$) is called orthogonal to M (resp. M′) if it is orthogonal to every element of M (resp. M′). If x^* and y^* are orthogonal to M, so is $x^* + y^*$ and $x^*\alpha$ for all $\alpha \in A$ by virtue of (10) and (12) (no. 3), which justifies the following definition:

DEFINITION 4. *Given a subset* M *of* E (resp. *a subset* M′ *of* E^*), *the set of* $x^* \in E^*$ (resp. *the set of* $x \in E$) *which are orthogonal to* M (resp. M′) *is called the submodule totally orthogonal to* M (resp. M′) (or simply the submodule *orthogonal to* M (resp. M′) if no confusion can arise).

By definition of a linear form, the submodule of E^* orthogonal to E is reduced to 0; the submodule of E^* orthogonal to $\{0\}$ is identical with E^*.

PROPOSITION 6. *Let* M, N *be two subsets of* E *such that* $M \subset N$; *if* M′ *and* N′ *are the submodules of* E^* *orthogonal to* M *and* N *respectively, then* $N' \subset M'$.

PROPOSITION 7. *Let* (M_ι) *be a family of subsets of* E; *the submodule orthogonal to the union of the* M_ι *is the intersection of the submodules* M'_ι *which are respectively orthogonal to the* M_ι; *this submodule is also the submodule orthogonal to the submodule of* E *generated by the union of the* M_ι.

These results are immediate consequences of the definitions.

There is an analogous proposition (which we shall leave to the reader to state) for submodules of E orthogonal to subsets of E^*.

If M is a submodule of E, M′ the submodule of E^* orthogonal to M and M″ the submodule of E orthogonal to M′, then $M \subset M''$ but it may be that $M \neq M''$ (Exercise 9). Note however that if M‴ is the orthogonal of M″ in E^*, then $M''' = M'$; for $M' \subset M'''$ and on the other hand the relation $M \subset M''$ implies $M''' \subset M'$.

5. TRANSPOSE OF A LINEAR MAPPING

Let E, F be two left A-modules; for every linear mapping $u: E \to F$, the mapping $\operatorname{Hom}(u, 1_{A_s})$ is a linear mapping of the right A-module F^* into the right A-module E^* (§ 1, no. 2), called the *transpose* of u.

In other words:

DEFINITION 5. *For every linear mapping* u *of an* A*-module* E *into an* A*-module* F, *the linear mapping* $y^* \mapsto y^* \circ u$ *of the dual* F^* *of* F *into the dual* E^* *of* E *is called the transpose of* u *and is denoted by* ${}^t u$.

The transpose ${}^tu$ is therefore defined by the relation

(15) $\langle u(x), y^*\rangle = \langle x, {}^tu(y^*)\rangle$ for all $x \in \mathrm{E}$ and all $y^* \in \mathrm{F}^*$.

Definition 5 applies without alteration to *right* A-modules and is then equivalent to the relation

$$\langle y^*, u(x)\rangle = \langle {}^tu(y^*), x\rangle \quad \text{for all } x \in \mathrm{E} \text{ and all } y^* \in \mathrm{F}^*.$$

Formulae (9) and (10) of § 1, no. 2 here give

(16) $${}^t(u_1 + u_2) = {}^tu_1 + {}^tu_2$$

for two elements u_1, u_2 of $\mathrm{Hom}_A(\mathrm{E}, \mathrm{F})$ and

(17) $${}^t(v \circ u) = {}^tu \circ {}^tv$$

for $u \in \mathrm{Hom}_A(\mathrm{E}, \mathrm{F})$ and $v \in \mathrm{Hom}_A(\mathrm{F}, \mathrm{G})$, G being a third A-module; finally, clearly

(18) $${}^t1_{\mathrm{E}} = 1_{\mathrm{E}^*}.$$

Remark. From (17) and (18) it follows that if u is *left* (resp. *right*) *invertible*, ${}^tu$ is *right* (resp. *left*) *invertible.*

PROPOSITION 8. *Let* $u : \mathrm{E} \to \mathrm{F}$ *be an* A*-linear mapping,* M *a submodule of* E *and* M′ *the orthogonal of* M *in* E*; *the orthogonal of* $u(\mathrm{M})$ *in* F* *is the inverse image* ${}^tu^{-1}(\mathrm{M}')$.

This follows immediately from (15).

COROLLARY. *The orthogonal of the image* $u(\mathrm{E})$ *in* F* *is the kernel* ${}^tu^{-1}(0)$ *of* ${}^tu$.

The orthogonal of E in E* is 0.

If $u : \mathrm{E} \to \mathrm{F}$ is an isomorphism, ${}^tu : \mathrm{F}^* \to \mathrm{E}^*$ is an isomorphism and if $v : \mathrm{F} \to \mathrm{E}$ is the inverse isomorphism of u, ${}^tv$ is the inverse isomorphism of ${}^tu$ (formulae (17) and (18)).

DEFINITION 6. *Given an isomorphism* u *of an* A*-module* E *onto an* A*-module* F, *the transpose of the inverse isomorphism of* u (equal to the inverse isomorphism of the transpose of u) *is called the contragredient isomorphism of* u *and denoted by* $\check{u}$.

The isomorphism $\check{u}$ is thus characterized by the relation

(19) $\langle u(x), \check{u}(y^*)\rangle = \langle x, x^*\rangle$ for $x \in \mathrm{E}$, $x^* \in \mathrm{E}^*$.

If $v : \mathrm{F} \to \mathrm{G}$ is an isomorphism, the contragredient isomorphism of $v \circ u$ is $\check{v} \circ \check{u}$.

In particular, the mapping $u \mapsto \check{u}$ is an *isomorphism* of the linear group $\mathbf{GL}(\mathrm{E})$ onto a subgroup of the linear group $\mathbf{GL}(\mathrm{E}^*)$.

Let $\sigma: A \to B$ be an *isomorphism* of a ring A onto a ring B, E a left A-module, F a left B-module and $u: E \to F$ a *semi-linear* mapping (§ 1, no. 13) *relative to* σ. Let σ^{-1} be the inverse isomorphism of σ; for all $y^* \in F^*$, the mapping $x \mapsto \langle u(x), y^* \rangle^{\sigma^{-1}}$ of E into A is a *linear form*; if it is also denoted by ${}^t u(y^*)$, a mapping ${}^t u: F^* \to E^*$ is defined which is also called the *transpose* of the semi-linear mapping u; it is thus characterized by the identity

$$\langle u(x), y^* \rangle = \langle x, {}^t u(y^*) \rangle^{\sigma} \tag{20}$$

for $x \in E$, $y^* \in F^*$. It is immediately verified that ${}^t u$ is a *semi-linear* mapping *relative to* σ^{-1}. If v denotes the mapping u considered as an A-*linear* mapping of E into $\sigma_*(F)$ (§ 1, no. 13), we may write $u = \phi \circ v$, where ϕ is the identity mapping $\sigma_*(F) \to F$, considered as a semi-linear mapping relative to σ. It is immediate that ${}^t u = {}^t v \circ {}^t \phi$ and $({}^t \phi, \sigma^{-1})$ is a di-isomorphism of F^* onto $(\sigma_*(F))^*$ relative to the isomorphism σ^{-1}; this relation allows us immediately to extend the properties of transposes of linear mappings to transposes of semi-linear mappings.

6. DUAL OF A QUOTIENT MODULE. DUAL OF A DIRECT SUM. DUAL BASES

We apply Theorem 1 of no. 1 to the case where $F = A_s$:

PROPOSITION 9. *Let* E′, E, E″ *be* A-*modules and*

$$E' \xrightarrow{u} E \xrightarrow{v} E'' \longrightarrow 0 \tag{21}$$

an exact sequence of linear mappings. Then the sequence of transpose mappings

$$0 \longrightarrow E''^* \xrightarrow{{}^t v} E^* \xrightarrow{{}^t u} E'^*$$

is exact.

COROLLARY. *Let* M *be a submodule of an* A-*module* E *and* $\phi: E \to E/M$ *the canonical homomorphism. Then* ${}^t \phi$ *is an isomorphism of the dual of* E/M *onto the submodule* M′ *of* E* *orthogonal to* M.

If $j: M \to E$ is the canonical injection, the kernel of ${}^t j$ is by definition the orthogonal of M in E*.

Moreover, in the notation of the corollary, an *injective* homomorphism $E^*/M' \to M^*$ is obtained from ${}^t j$ when passing to the quotient.

PROPOSITION 10. *Let* $(E_\iota)_{\iota \in I}$ *be a family of* A-*modules and for all* $\iota \in I$ *let* $j_\iota: E_\iota \to E = \bigoplus_{\iota \in I} E_\iota$ *be the canonical injection. Then the product mapping* $x^* \mapsto ({}^t j_\iota(x^*))$ *is an isomorphism of the dual* E* *of* E *onto the product* $\prod_{\iota \in I} E_\iota^*$.

This is a particular case of § 1, no. 6, Corollary 1 to Proposition 6, applied to the case where $\prod_\lambda F_\lambda = A_s$.

If, by means of the canonical injections j_ι, the E_ι are identified with submodules of their direct sum E and if, by means of the product mapping $x^* \mapsto ({}^t j_\iota(x^*))$, E^* is identified with $\prod_{\iota \in I} E_\iota^*$, it can then be said that $\prod_{\iota \in I} E_\iota^*$ is the *dual of* $\bigoplus_{\iota \in I} E_\iota$, the canonical bilinear form being given by

$$\langle (x_\iota), (x_\iota^*) \rangle = \sum_{\iota \in I} \langle x_\iota, x_\iota^* \rangle. \tag{22}$$

COROLLARY. *Let* M, N *be two supplementary submodules in an* A*-module* E *and* $p: E \to M$, $q: E \to N$ *the corresponding projectors; then* ${}^t p + {}^t q: M^* \oplus N^* \to E^*$ *is an isomorphism and* ${}^t p$ (resp. ${}^t q$) *is an isomorphism of* M^* (resp. N^*) *onto the submodule of* E^* *orthogonal to* N (resp. M). *Moreover, if* $i: M \to E$ *and* $j: N \to E$ *are the canonical injections,* ${}^t p \circ {}^t i$ *and* ${}^t q \circ {}^t j$ *are the projectors* $E^* \to {}^t p(M^*)$, $E^* \to {}^t q(N^*)$ *corresponding to the decomposition of* E^* *as the direct sum of* ${}^t p(M^*)$ *and* ${}^t q(N^*)$.

$p \circ i = 1_M$, $q \circ j = 1_N$, $p \circ j = q \circ i = 0$, $i \circ p + j \circ q = 1_E$, whence, by transposition (no. 5, formulae (16), (17) and (18)), ${}^t i \circ {}^t p = 1_{M^*}$, ${}^t j \circ {}^t q = 1_{N^*}$, ${}^t j \circ {}^t p = {}^t i \circ {}^t q = 0$. ${}^t p \circ {}^t i + {}^t q \circ {}^t j = 1_{E^*}$ and the proposition follows from § 1, no. 6, Corollary 2 to Proposition 6.

Under the hypotheses of the Corollary, M^* (resp. N^*) is often identified with the orthogonal ${}^t p(M^*)$ (resp. ${}^t q(N^*)$) of N (resp. M) in E^*, thus identifying every linear form u on M (resp. N) with the linear form on E extending u and which is zero on N (resp. M).

When an A-module E admits a *basis* $(e_t)_{t \in T}$, it has been seen that giving this basis defines canonically an isomorphism $u: A_s^{(T)} \to E$. By virtue of Proposition 10 and no. 3, Proposition 5, the dual of $A_s^{(T)}$ is canonically identified with the product A_d^T; consider the contragredient isomorphism $\check{u}: A_d^T \to E^*$. If, for all $t \in T$, f_t is the element of A_d^T all of whose projections are zero with the exception of that of index t, which is equal to 1, and if we write $e_t^* = \check{u}(f_t)$, the elements e_t^* of E^* are, by (19) and (22), characterized by the relations

$$\langle e_t, e_{t'}^* \rangle = \begin{cases} 0 & \text{for } t' \neq t \\ 1 & \text{for } t' = t. \end{cases} \tag{23}$$

It amounts to the same to say that, for all $x = \sum_{t \in T} \xi_t e_t \in E$, $e_t^*(x) = \xi_t$; also e_t^* is called the *coordinate form* of index t on E. It follows from (23) that (e_t^*) is a *free system* in E^*.

In particular, if T is *finite*, the e_t^* form a *basis* of E^*, the f_t then forming the canonical basis of A_d^T. Hence:

PROPOSITION 11. *The dual of a free module with a basis of n elements is a free module with a basis of n elements.*

Note that the dual of a free module with an infinite basis is not necessarily a free module (VII, § 3, Exercise 10).

DEFINITION 7. *If* E *is a free module with a finite basis* (e_t), *the basis* (e_t^*) *of the dual* E^* *of* E *defined by relations* (23) *is called the dual basis of* (e_t).

Relations (23) can also be written in the form

$$\langle e_t, e_{t'}^* \rangle = \delta_{tt'} \tag{24}$$

where $\delta_{tt'}$ is the *Kronecker symbol* on $T \times T$.

Note that if T is finite and (e_t^*) is the dual basis of (e_t), then, for $x = \sum_{t \in T} \xi_t e_t \in E$, $x^* = \sum_{t \in T} \xi_t^* e_t^* \in E^*$,

$$\langle x, x^* \rangle = \sum_{t \in T} \xi_t \xi_t^*. \tag{25}$$

The dual basis of a finite basis of a *right* A-module is of course defined similarly.

COROLLARY. *The dual of a finitely generated projective module is a finitely generated projective module.*

A finitely generated projective left A-module can be identified with a direct factor M of a free A-module A_s^n with a finite basis (no. 2, Corollary 1 to Proposition 4). Then (Proposition 11 and Corollary to Proposition 10) M^* is isomorphic to a direct factor of A_d^n, whence the corollary.

PROPOSITION 12. *Let* E *be an* A*-module and* $(a_t)_{t \in T}$ *a generating system of* E. *The following conditions are equivalent:*

(a) E *is a projective* A*-module.*

(b) *There exists a family* $(a_t^*)_{t \in T}$ *of linear forms on* E *such that, for all* $x \in E$, *the family* $(\langle x, a_t^* \rangle)_{t \in T}$ *has finite support and*

$$x = \sum_{t \in T} \langle x, a_t^* \rangle a_t. \tag{26}$$

There exists a surjective homomorphism $u: L \to E$, where $L = A_s^{(T)}$, such that if $(e_t)_{t \in T}$ is the canonical basis of L then $u(e_t) = a_t$ (§ 1, no. 11, Proposition 17); for E to be projective, it is necessary and sufficient that there exist a linear mapping $v: E \to L$ such that $u \circ v = 1_E$ (no. 2, Proposition 4 and § 1, no. 9, Proposition 15). If such a mapping exists and we write ${}^t v(e_t^*) = a_t^*$, then $\langle x, a_t^* \rangle = \langle x, {}^t v(e_t^*) \rangle = \langle v(x), e_t^* \rangle$, hence the family $(\langle x, a_t^* \rangle)$ has finite support and $x = u\left(\sum_{t \in T} \langle (x), e_t^* \rangle e_t\right) = \sum_{t \in T} \langle x, a_t^* \rangle a_t$ for all $x \in E$. Conversely, if condition (b) of the statement is fulfilled, the sum $\sum_{t \in T} \langle x, a_t^* \rangle e_t$ is defined for all $x \in E$ and $x \to \sum_{t \in T} \langle x, a_t^* \rangle e_t$ is a linear mapping $v: E \to L$ such that $u \circ v = 1_E$.

7. BIDUAL

Let E be a left A-module. The dual E^{**} of the dual E^* of E is called the *bidual* of E; it is also a *left* A-module (no. 3). For all $x \in E$, it follows from no. 3, formulae (10) and (12), that the mapping $x^* \mapsto \langle x, x^* \rangle$ is a *linear form* on the right A-module E^*, in other words an element of the bidual E^{**}, which we shall denote by $\tilde{x}$; moreover, it follows immediately from (9) and (11) (no. 3) that the mapping $c_E: x \mapsto \tilde{x}$ of E into E^{**} is *linear*; this mapping will be called *canonical*; in general, it is netither injective nor surjective, even when E is finitely generated (cf. Exercise 9(e) and § 7, no. 5, Theorem 6).

An A-module E is called *reflexive* if the canonical homomorphism $c_E: E \to E^{**}$ is *bijective*.

Let F be a second left A-module; for every linear mapping $u: E \to F$, the mapping ${}^t({}^t u): E^{**} \to F^{**}$, which will also be denoted by ${}^{tt}u$, is linear and the diagram

$$\begin{array}{ccc} E & \xrightarrow{u} & F \\ {\scriptstyle c_E}\downarrow & & \downarrow{\scriptstyle c_F} \\ E^{**} & \xrightarrow[{}^{tt}u]{} & F^{**} \end{array} \tag{27}$$

is commutative, as follows immediately from the definitions and formula (15) giving the transpose of a linear mapping.

PROPOSITION 13. *If* E *is a free module* (resp. *a free module with a finite basis*), *the canonical mapping* $c_E: E \to E^{**}$ *is injective* (resp. *bijective*).

Let $(e_t)_{t \in T}$ be a basis of E and let (e_t^*) be the family of corresponding coordinate forms; by definition, if $x \in E$ is such that $\tilde{x} = 0$, then $\langle x, e_t^* \rangle = 0$ for all $t \in T$, in other words all the coordinates of x are zero, hence $x = 0$. Suppose further that T is finite; since $\langle \tilde{e}_t, e_{t'}^* \rangle = \delta_{tt'}$, $(\tilde{e}_t)$ is the dual basis of (e_t^*) in E^{**} and, as c_E transforms a basis of E into a basis of E^{**}, c_E is bijective (§ 1, no. 11, Corollary 3 to Proposition 17). We have moreover proved:

COROLLARY 1. *Let* E *be a free* A*-module with a finite basis; for every basis* (e_t) *of* E, $(c_E(e_t))$ *is the dual basis of the basis* (e_t^*) *of* E^* *dual to* (e_t).

In this case it is said that (e_t) and (e_t^*) are two *dual* bases *of one another*.

COROLLARY 2. *If* E *is a free* A-*module with a finite basis, every finite basis of* E^* *is the dual basis of a basis of* E.

It suffices to consider in E^{**} the dual basis of the given basis and canonically to identify E with E^{**}.

COROLLARY 3. *Let* E, F *be two* A-*modules each with a finite basis,* E (resp. F) *being canonically identified with its bidual* E^{**} (resp. F^{**}). *Then, for every linear mapping* $u: E \to F$, ${}^{tt}u = u$.

This follows immediately from the commutativity of diagram (27).

COROLLARY 4. *If* P *is a projective module* (resp. *a finitely generated projective module*) *the canonical mapping* $c_P: P \to P^{**}$ *is injective* (resp. *bijective*).

We shall use the following lemma:

Lemma 1. *Let* M, N *be two supplementary submodules in an* A-*module* E *and* $i: M \to E$, $j: N \to E$ *the canonical injections. Then the diagram*

$$(28) \qquad \begin{array}{ccc} M \oplus N & \xrightarrow{c_M \oplus c_N} & M^{**} \oplus N^{**} \\ \Big\downarrow{\scriptstyle i+j} & & \Big\downarrow{\scriptstyle {}^{tt}i + {}^{tt}j} \\ E & \xrightarrow[c_E]{} & E^{**} \end{array}$$

is commutative.

By definition, for $x \in M$, $y \in N$, $z^* \in E^*$,

$$\begin{aligned} \langle c_E(i(x) + j(y)), z^* \rangle &= \langle i(x) + j(y), z^* \rangle \\ &= \langle i(x), z^* \rangle + \langle j(y), z^* \rangle \\ &= \langle x, {}^t i(z^*) \rangle + \langle y, {}^t j(z^*) \rangle \\ &= \langle c_M(x), {}^t i(z^*) \rangle + \langle c_N(y), {}^t j(z^*) \rangle \\ &= \langle {}^{tt}i(c_M(x)) + {}^{tt}j(c_N(y)), z^* \rangle. \end{aligned}$$

This being so, if E is a free module (resp. a free module with a finite basis), c_E is injective (resp. bijective); on the other hand, it follows from no. 6, Proposition 10, that ${}^{tt}i \oplus {}^{tt}j$ is bijective; the commutativity of diagram (28) then implies that $c_M \oplus c_N$ is injective (resp. bijective) and so therefore are c_M and c_N (§ 1, no. 6, Corollary 1 to Proposition 7), whence the corollary, taking account of no. 2, Proposition 4.

8. LINEAR EQUATIONS

Let E, F be two A-modules. Every equation of the form $u(x) = y_0$, where $u: E \to F$ is a given linear mapping, y_0 a given element of F and the unknown x is subjected to the condition that it take its values in E, is called a *linear equation*; y_0 is called the *right hand side* of the equation; if $y_0 = 0$, the equation is called *homogeneous linear.*

Every element $x_0 \in E$ such that $u(x_0) = y_0$ is called a *solution of the linear equation* $u(x) = y_0$.†

It is often said, loosely speaking, that a problem is *linear* if it is equivalent to determining the solutions of a linear equation.

Given a linear equation $u(x) = y_0$, the equation $u(x) = 0$ is called the *homogeneous linear equation associated with* $u(x) = y_0$.

PROPOSITION 14. *If x_0 is a solution of the linear equation $u(x) = y_0$, the set of solutions of this equation is equal to the set of elements $x_0 + z$, where z runs through the set of solutions of the associated homogeneous equation $u(x) = 0$.*

The relation $u(x) = y_0$ may be written as $u(x) = u(x_0)$, which is equivalent to $u(x - x_0) = 0$.

In other words, if the equation $u(x) = y_0$ has at least one solution x_0, the set of its solutions is the set $x_0 + \overset{-1}{u}(0)$, obtained by translation from the kernel $\overset{-1}{u}(0)$ of u. Observe that $\overset{-1}{u}(0)$, being a submodule, is never empty, since it contains 0 (called the *zero solution*, or *trivial solution*, of the homogeneous equation $u(x) = 0$).

By virtue of Proposition 14, for a linear equation $u(x) = y_0$ to have *exactly one solution*, it is necessary and sufficient that it have at least one solution and that $\overset{-1}{u}(0) = \{0\}$ (in other words, that the associated homogeneous equation have no non-zero solution, or also that u be *injective*); in this case, for *all* $y \in F$, the equation $u(x) = y$ has *at most* one solution.

PROPOSITION 15. *Let u be a linear mapping of a module E into a module F. If the equation $u(x) = y_0$ has at least one solution, y_0 is orthogonal to the kernel of ${}^t u$.*

To say that $u(x) = y_0$ admits a solution means that $y_0 \in u(E)$ and the proposition follows from no. 5, Corollary to Proposition 8.

---

† This is in fact an abuse of language; from the logical point of view, we are not here defining the word "solution", but simply the sentence "x_0 is a solution of the equation $u(x) = y_0$" as equivalent to the relation "$x_0 \in E$ and $u(x_0) = y_0$". Observe that in a mathematical theory $\mathcal{T}$ where the relation "A is a ring, E and F are A-modules, u is a homomorphism of E into F, y_0 an element of F" is a theorem, every *term* T of $\mathcal{T}$ such that the relation "$T \in E$ and $u(T) = y_0$" is true in $\mathcal{T}$ is a *solution* of the equation $u(x) = y_0$ in the sense of *Set Theory*, I, § 5, no. 2; this justifies the above abuse of language.

Observe that the necessary criterion for the existence of a solution of $u(x) = y_0$, given by Proposition 15, is sufficient when A is a *field* (§ 7, no. 6, Proposition 12), but *not in general* (Exercise 10).

Remarks. (1) Let E be an A-module, $(F_\iota)_{\iota \in I}$ a family of A-modules and for all $\iota \in I$ let $u_\iota : E \to F_\iota$ be a linear mapping. Every system of linear equations

$$u_\iota(x) = y_\iota \quad (\iota \in I) \tag{29}$$

where the $y_\iota \in F_\iota$ are given, is equivalent to *a single* linear equation $u(x) = y$, where u is the mapping $x \mapsto (u_\iota(x))$ of E into $F = \prod_{\iota \in I} F_\iota$ and $y = (y_\iota)$. The system (29) is called *homogeneous* if $y_\iota = 0$ for all $\iota \in I$.

(2) Suppose that E admits a *basis* $(a_\lambda)_{\lambda \in L}$; if we set $u(a_\lambda) = b_\lambda$ for all $\lambda \in L$, to say that $x = \sum_{\lambda \in L} \xi_\lambda a_\lambda$ satisfies the equation $u(x) = y_0$ is equivalent to saying that the family (of finite support) $(\xi_\lambda)_{\lambda \in L}$ of elements of A satisfies the relation

$$\sum_{\lambda \in L} \xi_\lambda b_\lambda = y_0. \tag{30}$$

Conversely, looking for families $(\xi_\lambda)_{\lambda \in L}$ of elements of A of finite support satisfying (30), is equivalent to solving the linear equation $u(x) = y_0$, where u is the unique linear mapping of E into F such that $u(a_\lambda) = b_\lambda$ for all $\lambda \in L$ (§ 1, no. 11, Corollary 3 to Proposition 17).

(3) A linear equation $u(x) = y_0$ is called *scalar* when $F = A_s$ and therefore u is a *linear form* on E and y_0 a *scalar*. If E admits a basis $(a_\lambda)_{\lambda \in L}$, it follows from *Remark* (2) that such an equation may also be written as

$$\sum_{\lambda \in L} \xi_\lambda \alpha_\lambda = y_0 \in A \tag{31}$$

where the family of scalars (α_λ) is arbitrary and where it is understood that the family (ξ_λ) must have finite support. In general, by the *solution* (in A) of a system of scalar linear equations

$$\sum_{\lambda \in L} \xi_\lambda \alpha_{\lambda\iota} = \eta_\iota \quad (\iota \in I) \tag{32}$$

where $\alpha_{\lambda\iota} \in A$ and $\eta_\iota \in A$, is understood a family $(\xi_\lambda)_{\lambda \in L}$ of elements of A of *finite* support and satisfying (32); the $\alpha_{\lambda\iota}$ are called the *coefficients* of the system of equations and the η_ι the *right hand sides*. The solution of such a system is equivalent to that of the equation $u(x) = y$, where $y = (\eta_\iota)$ and $u : A_s^{(L)} \to A_s^I$ is the linear mapping

$$(\xi_\lambda) \mapsto \Big(\sum_{\lambda \in L} \xi_\lambda \alpha_{\lambda\iota}\Big).$$

(4) A linear system (32) is also called a *system of left scalar linear equations* when it is necessary to avoid confusion. A system of equations

$$\sum_{\lambda \in L} \alpha_{\lambda\iota} \xi_\lambda = \eta_\iota \quad (\iota \in I) \tag{33}$$

is likewise called a system of *right scalar linear equations*; such a system can immediately be transformed into a system (32) by considering the ξ_λ, η_ι and $\alpha_{\lambda\iota}$ as belonging to the *opposite* ring A^0 to A.

§ 3. TENSOR PRODUCTS

1. TENSOR PRODUCT OF TWO MODULES

Let G_1, G_2 be two **Z**-modules; a mapping u of the set $G = G_1 \times G_2$ into a **Z**-module is called *biadditive* (or **Z**-*bilinear*) if $u(x_1, x_2)$ is "additive with respect to x_1 and with respect to x_2"; to be precise, this means that, for x_1, y_1 in G_1, x_2, y_2 in G_2,

$$u(x_1 + y_1, x_2) = u(x_1, x_2) + u(y_1, x_2)$$
$$u(x_1, x_2 + y_2) = u(x_1, x_2) + u(x_1, y_2).$$

Note that this implies in particular that $u(0, x_2) = u(x_1, 0) = 0$ for all $x_1 \in G_1$, $x_2 \in G_2$.

Let A be a ring, E a *right* A-module and F a *left* A-module. We are going to consider the *universal mapping problem* (*Set Theory*, IV, § 3, no. 1) where Σ is the species of **Z**-*module* structure (the morphisms then being **Z**-linear mappings, in other words, additive group homomorphisms) and the α-mappings are the mappings f of E × F into a **Z**-module G which are **Z**-*bilinear* and further satisfy, for all $x \in E$, $y \in F$ and $\lambda \in A$

$$f(x\lambda, y) = f(x, \lambda y). \tag{1}$$

We show that this problem admits a solution. For this we consider the **Z**-module $C = \mathbf{Z}^{(E \times F)}$ of formal linear combinations of the elements of E × F with coefficients in **Z** (§ 1, no. 11), a basis of which can be considered to consist of the ordered pairs (x, y), where $x \in E$ and $y \in F$. Let D be the sub-**Z**-module of C *generated* by the elements of one of the following types:

$$\begin{cases} (x_1 + x_2, y) - (x_1, y) - (x_2, y) \\ (x, y_1 + y_2) - (x, y_1) - (x, y_2) \\ (x\lambda, y) - (x, \lambda y) \end{cases} \tag{2}$$

where x, x_1, x_2 are in E, y, y_1, y_2 are in F and λ is in A.

DEFINITION 1. *The tensor product of the right* A-*module* E *and the left* A-*module* F, *denoted by* $E \underset{A}{\otimes} F$ *or* $E \otimes_A F$ (or simply $E \otimes F$ if no confusion is to be feared) *is the quotient* **Z**-*module* C/D (the quotient of the **Z**-module C of formal linear combinations of elements of $E \times F$ with coefficients in $\mathbf{Z}$, by the submodule D generated by the elements of one of the types (2)). *For $x \in E$ and $y \in F$, the element of $E \otimes_A F$ which is the canonical image of the element (x, y) of $C = \mathbf{Z}^{(E \times F)}$ is denoted by $x \otimes y$ and called the tensor product of x and y.*

The mapping $(x, y) \mapsto x \otimes y$ of $E \times F$ into $E \otimes_A F$ is called *canonical.* It is a $\mathbf{Z}$-*bilinear* mapping which satisfies conditions (1).

We show that the tensor product $E \otimes_A F$ and the above canonical mapping form a solution of the universal mapping problem posed earlier. To be precise:

PROPOSITION 1. (a) *Let g be a $\mathbf{Z}$-linear mapping of $E \otimes_A F$ into a $\mathbf{Z}$-module G. The mapping $(x, y) \mapsto f(x, y) = g(x \otimes y)$ of $E \times F$ into G is $\mathbf{Z}$-bilinear and satisfies conditions* (1).

(b) *Conversely, let f be a $\mathbf{Z}$-bilinear mapping of $E \times F$ into a $\mathbf{Z}$-module G satisfying conditions* (1). *Then there exists one and only one $\mathbf{Z}$-linear mapping g of $E \otimes_A F$ into G such that $f(x, y) = g(x \otimes y)$ for $x \in E$, $y \in F$.*

If ϕ denotes the canonical mapping of $E \times F$ into $E \otimes_A F$, then $f = g \circ \phi$; whence (a). To show (b), we note that, in the notation of Definition 1, f extends to a $\mathbf{Z}$-linear mapping $\bar{f}$ of C into G (§ 1, no. 11, Proposition 17). By virtue of relations (1), $\bar{f}$ is zero for all the elements of C of one of the types (2) and hence on D. There therefore exists a $\mathbf{Z}$-linear mapping g of $C/D = E \otimes_A F$ into G such that $\bar{f} = g \circ \psi$, where $\psi: C \to C/D$ is the canonical homomorphism (§ 1, no. 8, *Remark*). The uniqueness of g is immediate since $E \otimes_A F$ is generated, as a $\mathbf{Z}$-module, by the elements of the form $x \otimes y$.

Proposition 1 defines a *canonical isomorphism* of the $\mathbf{Z}$-module of $\mathbf{Z}$-bilinear mappings f of $E \times F$ into G, satisfying conditions (1), onto the $\mathbf{Z}$-module $\mathrm{Hom}_{\mathbf{Z}}(E \otimes_A F, G)$.

When $A = \mathbf{Z}$, conditions (1) are automatically satisfied for *every* $\mathbf{Z}$-bilinear mapping f and the submodule D of C is already generated by the elements of the first two types in (2).

If now we return to the general case and E' and F' denote the underlying $\mathbf{Z}$-modules of E and F respectively, the above remark and Definition 1 show immediately that the $\mathbf{Z}$-module $E \otimes_A F$ can be canonically identified with the *quotient* of the $\mathbf{Z}$-module $E' \otimes_{\mathbf{Z}} F'$ by the sub-$\mathbf{Z}$-module generated by the elements of the form $(x\lambda) \otimes y - x \otimes (\lambda y)$, where x runs through E, y runs through F and λ runs through A.

COROLLARY 1. *Let H be a $\mathbf{Z}$-module and $h: E \times F \to H$ a $\mathbf{Z}$-bilinear mapping satisfying conditions* (1) *and such that H is generated by $h(E \times F)$. Suppose that for every $\mathbf{Z}$-module G and every $\mathbf{Z}$-bilinear mapping f of $E \times F$ into G satisfying* (1) *there exists a* **Z**-*linear mapping* $g: \mathrm{H} \to \mathrm{G}$ *such that* $f = g \circ h$. *Then, if* ϕ *denotes the canonical mapping of* $\mathrm{E} \times \mathrm{F}$ *into* $\mathrm{E} \otimes_{\mathrm{A}} \mathrm{F}$, *there exists one and only one isomorphism* θ *of* $\mathrm{E} \otimes_{\mathrm{A}} \mathrm{F}$ *onto* H *such that* $h = \theta \circ \phi$.

The hypothesis that $h(\mathrm{E} \times \mathrm{F})$ generates H implies the uniqueness of g; the corollary is then just the general uniqueness property of a solution of a universal mapping problem (*Set Theory*, IV, § 3, no. 1).

Corollary 2. *Let* E^0 (resp. F^0) *denote the module* E (resp. F) *considered as a left* (resp. *right*) *module over the opposite ring* A^0; *then there exists one and only one* **Z**-*module isomorphism* $\sigma: \mathrm{E} \otimes_{\mathrm{A}} \mathrm{F} \to \mathrm{F}^0 \otimes_{\mathrm{A}^0} \mathrm{E}^0$ *such that* $\sigma(x \otimes y) = y \otimes x$ *for* $x \in \mathrm{E}$ *and* $y \in \mathrm{F}$ ("commutativity" of tensor products).

By definition of the A^0-module structures on E^0 and F^0, the mapping $(x, y) \mapsto y \otimes x$ of $\mathrm{E} \times \mathrm{F}$ into $\mathrm{F}^0 \otimes_{\mathrm{A}^0} \mathrm{E}^0$ is **Z**-bilinear and satisfies conditions (1), whence the existence and uniqueness of the **Z**-linear mapping σ. Similarly a **Z**-linear mapping $\tau: \mathrm{F}^0 \otimes_{\mathrm{A}^0} \mathrm{E}^0 \to \mathrm{E} \otimes_{\mathrm{A}} \mathrm{F}$ is defined such that $\tau(y \otimes x) = x \otimes y$ and clearly σ and τ are inverse isomorphisms.

Remark. The tensor product of non-zero modules may be zero: for example, taking the two **Z**-modules $\mathrm{E} = \mathbf{Z}/2\mathbf{Z}$ and $\mathrm{F} = \mathbf{Z}/3\mathbf{Z}$, $2x = 0$ and $3y = 0$ for all $x \in \mathrm{E}$ and $y \in \mathrm{F}$; therefore, in $\mathrm{E} \otimes_{\mathbf{Z}} \mathrm{F}$,

$$x \otimes y = 3(x \otimes y) - 2(x \otimes y) = x \otimes (3y) - (2x) \otimes y = 0$$

for all x and y (cf. no. 6, Corollary 4 to Proposition 6).

2. TENSOR PRODUCT OF TWO LINEAR MAPPINGS

Let A be a ring, E, E′ two right A-modules, F, F′ two left A-modules and $u: \mathrm{E} \to \mathrm{E}'$ and $v: \mathrm{F} \to \mathrm{F}'$ two A-linear mappings. It is easily verified that the mapping

$$(x, y) \mapsto u(x) \otimes v(y)$$

of $\mathrm{E} \times \mathrm{F}$ into $\mathrm{E}' \oplus_{\mathrm{A}} \mathrm{F}'$ is **Z**-bilinear and satisfies conditions (1) of no. 1. By Proposition 1 of no. 1 there thus exists one and only one **Z**-linear mapping $w: \mathrm{E} \otimes_{\mathrm{A}} \mathrm{F} \to \mathrm{E}' \otimes_{\mathrm{A}} \mathrm{F}'$ such that

$$w(x \otimes y) = u(x) \otimes v(y) \tag{3}$$

for $x \in \mathrm{E}$, $y \in \mathrm{F}$. This mapping is denoted by $u \otimes v$ (when no confusion can arise) and is called the *tensor product* of the linear mappings u and v.

It follows immediately from (3) that $(u, v) \mapsto u \otimes v$ is a **Z**-*bilinear* mapping called *canonical*

$$\mathrm{Hom}_{\mathrm{A}}(\mathrm{E}, \mathrm{E}') \times \mathrm{Hom}_{\mathrm{A}}(\mathrm{F}, \mathrm{F}') \to \mathrm{Hom}_{\mathbf{Z}}(\mathrm{E} \otimes_{\mathrm{A}} \mathrm{F}, \mathrm{E}' \otimes_{\mathrm{A}} \mathrm{F}').$$

There corresponds to it by Proposition 1 of no. 1 a $\mathbf{Z}$-linear mapping called *canonical*

$$(4) \qquad \mathrm{Hom_A}(E, E') \otimes_{\mathbf{Z}} \mathrm{Hom_A}(F, F') \rightarrow \mathrm{Hom_{\mathbf{Z}}}(E \otimes_A F, E' \otimes_A F')$$

which associates with every element $u \otimes v$ *of the tensor product* the *linear mapping* $u \otimes v: E \otimes_A F \rightarrow E' \otimes_A F'$. Note that the canonical mapping (4) *is not necessarily injective nor surjective.* The notation $u \otimes v$ can therefore lead to confusion and it will be necessary for the context to indicate whether it denotes an element of the tensor product or a linear mapping.

Further, let E″ be a right A-module, F″ a left A-module and $u': E' \rightarrow E''$, $v': F' \rightarrow F''$ A-linear mappings; it follows from (3) that

$$(5) \qquad (u' \circ u) \otimes (v' \circ v) = (u' \otimes v') \circ (u \otimes v).$$

3. CHANGE OF RING

PROPOSITION 2. *Let* A, B *be two rings,* $\rho: B \rightarrow A$ *a ring homomorphism and* E (resp. F) *a right* (resp. *left*) *A-module. Then there exists one and only one* $\mathbf{Z}$*-linear mapping*

$$(6) \qquad \phi: \rho_*(E) \otimes_B \rho_*(F) \rightarrow E \otimes_A F$$

such that, for all $x \in E$ *and* $y \in F$, *the image under* ϕ *of the element* $x \otimes y$ *of* $\rho_*(E) \otimes_B \rho_*(F)$ *is the element* $x \otimes y$ *of* $E \otimes_A F$; *this* $\mathbf{Z}$*-linear mapping is surjective.*

We consider the mapping $(x, y) \mapsto x \otimes y$ of $\rho_*(E) \times \rho_*(F)$ into $E \otimes_A F$; it is $\mathbf{Z}$-bilinear and, for all $\beta \in B$, by definition $(x\rho(\beta)) \otimes y = x \otimes (\rho(\beta)y)$, hence conditions (1) of no. 1 hold, whence the existence and uniqueness of ϕ (no. 1, Proposition 1). The latter assertion follows from the fact that the elements $x \otimes y$ generate the $\mathbf{Z}$-module $E \otimes_A F$.

The mapping (6) is called *canonical.*

COROLLARY. *Let* $\mathfrak{I}$ *be a two-sided ideal of* A *such that* $\mathfrak{I}$ *is contained in the annihilator of* E *and in the annihilator of* F, *so that* E (resp. F) *has a canonical left* (resp. *right*) $(A/\mathfrak{I})$*-module structure* (§ 1, no. 12). *Then the canonical homomorphism* (6)

$$\phi: E \otimes_A F \rightarrow E \otimes_{A/\mathfrak{I}} F$$

corresponding to the canonical homomorphism $\rho: A \rightarrow A/\mathfrak{I}$ *is the identity.*

For all $\bar{\alpha} \in A/\mathfrak{I}$, all $x \in E$ and all $y \in F$, $x\bar{\alpha} = x\alpha$ (resp. $\bar{\alpha}y = \alpha y$) for all α such that $\rho(\alpha) = \bar{\alpha}$. If $C = \mathbf{Z}^{(E \times F)}$, the submodule of C generated by the elements $(x\alpha, y) - (x, \alpha y)$ is then equal to the submodule generated by the elements $(x\bar{\alpha}, y) - (x, \bar{\alpha}y)$.

With the hypotheses and notation of Proposition 2, let E′ be a right B-module, F′ a left B-module and consider two *semi-linear* mappings $u: E' \rightarrow E$,

$v: F' \to F$ relative to the homomorphism $\rho: B \to A$; u (resp. v) can be considered as a B-linear mapping $E' \to \rho_*(E)$ (resp. $F' \to \rho_*(F)$), whence a **Z**-linear mapping $w: E' \otimes_B F' \to \rho_*(E) \otimes_B \rho_*(F)$ such that $w(x' \otimes y') = u(x') \otimes v(y')$ for $x' \in E'$, $y' \in F'$; composing the canonical mapping (6) with this mapping, a **Z**-linear mapping $w': E' \otimes_B F' \to E \otimes_A F$ is hence obtained such that $w'(x' \otimes y') = u(x') \otimes v(y')$ for $x' \in E'$, $y' \in F'$; this is the mapping which will normally be denoted by $u \otimes v$ if no confusion can arise. Clearly $(u, v) \mapsto u \otimes v$ is a **Z**-bilinear mapping

$$\mathrm{Hom}_B(E', \rho_*(E)) \times \mathrm{Hom}_B(F', \rho_*(F)) \to \mathrm{Hom}_{\mathbf{Z}}(E' \otimes_B F', E \otimes_A F).$$

Moreover, if C is a third ring, $\sigma: C \to B$ a homomorphism, E'' a right C-module, F'' a left C-module, $u': E'' \to E'$ and $v': F'' \to F'$ semi-linear mappings relative to σ, then

$$(u \circ u') \otimes (v \circ v') = (u \otimes v) \circ (u' \otimes v').$$

4. OPERATORS ON A TENSOR PRODUCT; TENSOR PRODUCTS AS MULTIMODULES

With the hypotheses and notation of no. 1, for every endomorphism u (resp. v) of the A-module E (resp. F), $u \otimes 1_F$ (resp. $1_E \otimes v$) is an endomorphism of the **Z**-module $E \otimes_A F$; it follows immediately from (5) (no. 2) that the mapping $u \mapsto u \otimes 1_F$ (resp. $v \mapsto 1_E \otimes v$) is a *ring homomorphism*

$$\mathrm{End}_A(E) \to \mathrm{End}_{\mathbf{Z}}(E \otimes_A F)$$

(resp. $\mathrm{End}_A(F) \to \mathrm{End}_{\mathbf{Z}}(E \otimes_A F)$); moreover,

$$(7) \qquad (u \otimes 1_F) \circ (1_E \otimes v) = (1_E \otimes v) \circ (u \otimes 1_F) = u \otimes v$$

and therefore (§ 1, no. 14) $E \otimes_A F$ has a canonical *left bimodule* structure with respect to the rings $\mathrm{End}_A(E)$ and $\mathrm{End}_A(F)$.

This being so, suppose given on E a $((B'_i); A, (C'_j))$-*multimodule* structure and on F a $(A, (B''_h); (C''_k))$-*multimodule* structure (§ 1, no. 14); it amounts to the same to say that ring homomorphisms $B'_i \to \mathrm{End}_A(E)$, $C'^0_j \to \mathrm{End}_A(E)$ are given with pairwise permutable images, and ring homomorphisms $B''_h \to \mathrm{End}_A(F)$, $C''^0_k \to \mathrm{End}_A(F)$ with pairwise permutable images. If these homomorphisms are composed respectively with the canonical homomorphisms $\mathrm{End}_A(E) \to \mathrm{End}_{\mathbf{Z}}(E \otimes_A F)$ and $\mathrm{End}_A(F) \to \mathrm{End}_{\mathbf{Z}}(E \otimes_A F)$ defined above, it is seen (taking account of (7)) that ring homomorphisms

$$B'_i \to \mathrm{End}_{\mathbf{Z}}(E \otimes_A F), \qquad C'^0_j \to \mathrm{End}_{\mathbf{Z}}(E \otimes_A F)$$
$$B''_h \to \mathrm{End}_{\mathbf{Z}}(E \otimes_A F), \qquad C''^0_k \to \mathrm{End}_{\mathbf{Z}}(E \otimes_A F)$$

are defined with *pairwise permutable* images; in other words, there has been defined on $E \otimes_A F$ a $((B'_i), (B''_h); (C'_j), (C''_k))$-*multimodule* structure; it is this multimodule which is also called the *tensor product* (relative to A) *of the*

$((B'_i); A, (C'_j))$-*multimodule* E *and the* $(A, (B''_h); (C''_k))$-*multimodule* F. This multimodule is the solution of a universal mapping problem analogous to that considered in no. 1; to be precise:

PROPOSITION 3. *Let* G *be a* $((B'_i), (B''_h); (C'_j), (C''_k))$-*multimodule.*

(a) *Let* g *be a linear mapping of the multimodule* $E \otimes_A F$ *into* G. *The mapping* $f:(x, y) \mapsto g(x \otimes y)$ *of* $E \times F$ *into* G *is* **Z**-*bilinear and satisfies relations* (1) *of* no. 1 *and the conditions*

$$(8) \quad \begin{cases} f(\mu'_i x, y) = \mu'_i f(x, y), & f(x\nu'_j, y) = f(x, y)\nu'_j \\ f(x, \mu''_h y) = \mu''_h f(x, y), & f(x, y\nu''_k) = f(x, y)\nu''_k \end{cases}$$

for all $x \in E$, $y \in F$, $\mu'_i \in B'_i$, $\nu'_j \in C'_j$, $\mu''_h \in B''_h$, $\nu''_k \in C''_k$, i, j, h, k *arbitrary.*

(b) *Conversely, let* f *be a* **Z**-*bilinear mapping of* $E \times F$ *into* G *satisfying conditions* (1) (no. 1) *and* (8). *Then there exists one and only one linear mapping* g *of the multimodule* $E \otimes_A F$ *into the multimodule* G *such that* $f(x, y) = g(x \otimes y)$ *for* $x \in E$, $y \in F$.

Assertion (a) follows immediately from the definition of the multimodule structure on $E \otimes_A F$, since for example $(x \otimes y)\nu'_j = (x\nu'_j) \otimes y$. To prove (b), we note first that Proposition 1 of no. 1 gives the existence and uniqueness of a **Z**-linear mapping g such that $g(x \otimes y) = f(x, y)$ for $x \in E$, $y \in F$; all that is needed is to verify that g is linear for the *multimodule* structures. As the elements $x \otimes y$ generate the **Z**-module $E \otimes_A F$, it suffices to verify the relations $g(\mu'(x \otimes y)) = \mu'_i g(x \otimes y)$ and their analogues; but this follows immediately from the formula $g(x \otimes y) = f(x, y)$ and relations (8).

SCHOLIUM. An element of $E \otimes_A F$ may in general be written in several ways in the form $\sum_i (x_i \otimes y_i)$, where $x_i \in E$ and $y_i \in F$; but to define a linear mapping g of the multimodule $E \otimes_A F$ into a multimodule G, there is no need to verify that, if $\sum_i (x_i \otimes y_i) = \sum_j (x'_j \otimes y'_j)$, then $\sum_i g(x_i \otimes y_i) = \sum_j g(x'_j \otimes y'_j)$; it suffices to be given $g(x \otimes y)$ for $x \in E$ and $y \in F$ and to verify that $(x, y) \mapsto g(x \otimes y)$ is **Z**-bilinear and satisfies conditions (1) (no. 1) and (8).

Let E′ be a $((B'_i), A, (C'_j))$-multimodule, F′ an $(A, (B''_h); (C''_k))$-multimodule and $u: E \to E'$, $v: F \to F'$ linear mappings of *multimodules*; it follows immediately from the definitions (no. 2) that $u \otimes v$ is a linear mapping of the multimodule $E \otimes_A F$ into the multimodule $E' \otimes_A F'$.

With E always denoting a right A-module, let ${}_sA_d$ denote the ring A considered as an (A, A)-bimodule (§ 1, no. 14, *Example* 1); by the above, the tensor product $E \otimes_A ({}_sA_d)$ has a canonical *right* A-*module* structure such that $(x \otimes \lambda)\mu = x \otimes (\lambda\mu)$ for $x \in E$, $\lambda \in A$, $\mu \in A$. The mapping $(x, \lambda) \mapsto x\lambda$ of $E \times ({}_sA_d)$ into E is **Z**-bilinear and satisfies conditions (1) (no. 1) and (8) (where, in the latter, the B'_i, C'_j and B''_h are absent and the family (C''_k) reduces to A); hence (Proposition 3), there exists an A-*linear* mapping g (called *canonical*) of $E \otimes_A ({}_sA_d)$ into E such that $g(x \otimes \lambda) = x\lambda$ for $x \in E$, $\lambda \in A$.

PROPOSITION 4. *If* E *is a right* A-*module, the mapping* $h : x \mapsto x \otimes 1$ *of* E *into* $E \otimes_A ({}_sA_d)$ *is a right* A-*module isomorphism, whose inverse isomorphism* g *is such that* $g(x \otimes \lambda) = x\lambda$ *for* $x \in E$, $\lambda \in A$.

If g is the canonical mapping, $g \circ h$ is the identity mapping 1_E and $h \circ g$ coincides with the identity mapping of $E \otimes_A ({}_sA_d)$ onto itself for elements of the form $x \otimes y$, which generate the latter **Z**-module; hence the conclusion.

We shall normally write $E \otimes_A A$ instead of $E \otimes_A ({}_sA_d)$ and shall often identify $E \otimes_A A$ with E by means of the above canonical isomorphisms. Observe that, if E also has a (left or right) B-module structure which is compatible with its right A-module structure, g and h are also isomorphisms for the B-module structures on E and $E \otimes_A A$ (and hence multimodule isomorphisms).

Now let F be a left A-module; $({}_sA_d) \otimes_A F$ (also denoted by $A \otimes_A F$) then has a canonical left A-module structure and as in Proposition 4 a canonical isomorphism is defined of $A \otimes_A F$ onto F mapping $\lambda \otimes x$ to λx, and its inverse isomorphism is $x \mapsto 1 \otimes x$.

In particular there exists a canonical isomorphism of the (A, A)-bimodule $({}_sA_d) \otimes_A ({}_sA_d)$ onto ${}_sA_d$ which maps $\lambda \otimes \mu$ to $\lambda\mu$.

5. TENSOR PRODUCT OF TWO MODULES OVER A COMMUTATIVE RING

Let C be a *commutative* ring; for every C-module E, the module structure on E is *compatible with itself* (§ 1, no. 14). If E and F are two C-modules, the considerations of no. 4 then allow us to define *two* C-module structures on the tensor product $E \otimes_C F$, respectively such that $\gamma(x \otimes y) = (\gamma x) \otimes y$ and such that $\gamma(x \otimes y) = x \otimes (\gamma y)$; but as, by Definition 1 of no. 1, in this case $(\gamma x) \otimes y = x \otimes (\gamma y)$, these two structures are *the same*. Henceforth when we speak of $E \otimes_C F$ as a C-*module*, we shall mean with the structure thus defined, unless otherwise stated. The canonical isomorphism.

$$\sigma : F \otimes_C E \to E \otimes_C F$$

(no. 1, Corollary 2 to Proposition 1) is then a C-module isomorphism.

It follows from this definition that, if $(a_\lambda)_{\lambda \in L}$ (resp. $(b_\mu)_{\mu \in M}$) is a *generating system* of the C-module E (resp. F), $(a_\lambda \otimes b_\mu)$ is a *generating system* of the C-module $E \otimes_C F$; in particular, if E and F are *finitely generated* C-modules, so is $E \otimes_C F$.

For every C-module G, the **Z**-bilinear mappings f of $E \times F$ into G for which

$$(9) \qquad f(\gamma x, y) = f(x, \gamma y) = \gamma f(x, y) \quad \text{for } x \in E, y \in F, \gamma \in C$$

are then called C-*bilinear* and form a C-*module* denoted by $\mathscr{L}_2(E, F; G)$; Proposition 3 (no. 4) defines a *canonical* C-*module isomorphism* (cf. § 1, no. 14, *Remark* 1).

$$\mathscr{L}_2(E, F; G) \to \mathrm{Hom}_C(E \otimes_C F, G). \tag{10}$$

Let E', F' be two C-modules and $u: E \to E'$, $v: F \to F'$ two C-linear mappings; then (no. 4) $u \otimes v$ is a C-*linear* mapping of $E \otimes_C F$ into $E' \otimes_C F'$. Further, it is immediate that $(u, v) \mapsto u \otimes v$ is a C-*bilinear* mapping of $\mathrm{Hom}_C(E, E') \times \mathrm{Hom}_C(F, F')$ into $\mathrm{Hom}_C(E \otimes_C F, E' \otimes_C F')$; hence there canonically corresponds to it a C-*linear* mapping, called *canonical*:

$$\mathrm{Hom}_C(E, E') \otimes_C \mathrm{Hom}_C(F, F') \to \mathrm{Hom}_C(E \otimes_C F, E' \otimes_C F') \tag{11}$$

which associates with every *element* $u \otimes v$ *of the tensor product*

$$\mathrm{Hom}_C(E, E') \otimes_C \mathrm{Hom}_C(F, F')$$

the linear mapping $u \otimes v$. Note that the canonical mapping (11) *is not necessarily injective nor surjective* (§ 4, Exercise 2).

Remarks. (1) Let A, B be two *commutative* rings, $\rho: B \to A$ a ring homomorphism and E and F two A-modules; then the canonical mapping (6) of no. 3 is a B-*linear* mapping

$$\rho_*(E) \otimes \rho_*(F) \to \rho_*(E \otimes_A F). \tag{12}$$

(2) What has been said in this no. can be generalized to the following case: let E be a right A-module, F a left A-module, C a commutative ring and $\rho: C \to A$ a homomorphism of C into A such that $\rho(C)$ is contained in the *centre* of A (cf. III, § 1, no. 3). We may then consider the C-modules $\rho_*(E)$ and $\rho_*(F)$ and the hypothesis on ρ implies that these C-module structures are compatible respectively with the A-module structures on E and F (§ 1, no. 14). The tensor product $E \otimes_A F$ is thus (by virtue of no. 4) given two C-module structures such that $\gamma(x \otimes y) = (x\rho(\gamma)) \otimes y$ and

$$\gamma(x \otimes y) = x \otimes (\rho(\gamma)y)$$

respectively for $\gamma \in C$, $x \in E$, $y \in F$ and Definition 1 (no. 1) shows also that these two structures are *identical*. If E' (resp. F') is a right (resp. left) A-module and $u: E \to E'$, $v: F \to F'$ are two A-linear mappings, then $u \otimes v: E \otimes_A F \to E' \otimes_A F'$ is C-*linear* for the C-module structures just defined; the mapping $(u, v) \mapsto u \otimes v$:

$$\mathrm{Hom}_A(E, E') \times \mathrm{Hom}_A(F, F') \to \mathrm{Hom}_C(E \otimes_A F, E' \otimes_A F')$$

is C-*bilinear* (for the C-module structures on $\mathrm{Hom}_A(E, E')$ and $\mathrm{Hom}_A(F, F')$

defined in § 1, no. 14, *Remark* 1), whence we also derive a C-*linear* mapping, called *canonical*

$$\text{(13)} \qquad \mathrm{Hom}_A(E, E') \otimes_C \mathrm{Hom}_A(F, F') \to \mathrm{Hom}_C(E \otimes_A F, E' \otimes_A F').$$

(3) Let A be an integral domain and K its field of fractions. If E and F are two vector K-spaces, the canonical mapping

$$(E_{[A]}) \otimes_A (F_{[A]}) \to E \otimes_K F$$

(no. 3 and § 1, no. 13) is *bijective*. It suffices (no. 4) to prove that if f is an A-*bilinear* mapping of $E \times F$ into a vector K-space G, f is also K-*bilinear*. Now, for all $\alpha \neq 0$ in A.

$$\alpha f(\alpha^{-1}x, y) = f(x, y) = \alpha f(x, \alpha^{-1}y)$$

whence

$$f(\alpha^{-1}x, y) = f(x, \alpha^{-1}y) = \alpha^{-1}f(x, y)$$

since G is a vector K-space.

6. PROPERTIES OF $E \otimes_A F$ RELATIVE TO EXACT SEQUENCES

PROPOSITION 5. *Let* E, E′, E″ *be right* A-*modules*, F *a left* A-*module and*

$$\text{(14)} \qquad E' \xrightarrow{u} E \xrightarrow{v} E'' \longrightarrow 0$$

an exact sequence of linear mappings. Writing $\bar{u} = u \otimes 1_F$, $\bar{v} = v \otimes 1_F$, *the sequence*

$$\text{(15)} \qquad E' \otimes_A F \xrightarrow{\bar{u}} E \otimes_A F \xrightarrow{\bar{v}} E'' \otimes_A F \longrightarrow 0$$

of **Z**-*homomorphisms is exact.*

By virtue of no. 2, formula (5), $\bar{v} \circ \bar{u} = (v \circ u) \otimes 1_F = 0$; the image $H = \bar{u}(E' \otimes F)$ is contained in the kernel $L = \mathrm{Ker}(\bar{v})$; by passing to the quotient, we therefore derive from $\bar{v}$ a **Z**-linear mapping f of the cokernel $M = (E \otimes F)/H$ of $\bar{u}$ into $E'' \otimes F$; it must be proved that f is *bijective* and it will hence suffice to define a **Z**-linear mapping $g: E'' \otimes F \to M$ such that $g \circ f$ and $f \circ g$ are the identity mappings.

Let $x'' \in E''$, $y \in F$; by hypothesis there exists $x \in E$ such that $v(x) = x''$. We show that, if x_1, x_2 are two elements of E such that $v(x_1) = v(x_2) = x''$ and $\phi: E \otimes F \to M$ is the canonical mapping, then $\phi(x_1 \otimes y) = \phi(x_2 \otimes y)$. It suffices to prove that if $v(x) = 0$ then $\phi(x \otimes y) = 0$, which follows from the fact that $x = u(x')$ with $x' \in E'$, whence $x \otimes y = u(x') \otimes y = \bar{u}(x' \otimes y) \in H$. If (x'', y) is mapped to the unique value of $\phi(x \otimes y)$ for all $x \in E$ such that $v(x) = x''$, a mapping is defined of $E'' \times F$ into M; this mapping is **Z**-bilinear and satisfies conditions (1) (no. 1), since $v(x\lambda) = x''\lambda$ and $(x\lambda) \otimes y = x \otimes (\lambda y)$ for $x \in E$; hence there is a **Z**-linear mapping g of $E'' \otimes F$ into M such that $g(x'' \otimes y) = \phi(x \otimes y)$ for $y \in F$, $x \in E$ and $x'' = v(x)$. This definition further proves that $f \circ g$ coincides with the identity mapping for the elements of

$E'' \otimes F$ of the form $x'' \otimes y$ and hence $f \circ g$ is the identity mapping of $E'' \otimes F$; on the other hand, for $x \in E$ and $y \in F$, $f(\phi(x \otimes y)) = v(x) \otimes y$ by definition, hence $g(f(\phi(x \otimes y))) = \phi(x \otimes y)$ and, as the elements of the form $\phi(x \otimes y)$ generate M, $g \circ f$ is the identity mapping of M.

COROLLARY. *Let* F, F', F'' *be left* A-*modules*, E *a right* A-*module and*

$$F' \xrightarrow{s} F \xrightarrow{t} F'' \longrightarrow 0 \tag{16}$$

an exact sequence of linear mappings. Writing $\bar{s} = 1_E \otimes s$, $\bar{t} = 1_E \otimes t$, *the sequence of* **Z**-*homomorphisms*

$$E \otimes_A F' \xrightarrow{\bar{s}} E \otimes_A F \xrightarrow{\bar{t}} E \otimes_A F'' \longrightarrow 0 \tag{17}$$

is exact.

When E (resp. F) is considered as a left (resp. right) A^0-module, $F \otimes_{A^0} E$ is identified with $E \otimes_A F$ and there are analogous identifications for $F' \otimes_{A^0} E$ and $F'' \otimes_{A^0} E$ (no. 1, Corollary 2 to Proposition 1); the corollary then follows immediately from Proposition 5.

Remark. Note that in general, if E' is a submodule of a right A-module E and $j: E' \to E$ the canonical injection, the canonical mapping

$$j \otimes 1_F : E' \otimes F \to E \otimes F$$

is not necessarily injective. In other words, for an exact sequence

$$0 \longrightarrow E' \xrightarrow{u} E \xrightarrow{v} E'' \longrightarrow 0 \tag{18}$$

it cannot in general be concluded that the sequence

$$0 \longrightarrow E' \otimes F \xrightarrow{\bar{u}} E \otimes F \xrightarrow{\bar{v}} E'' \otimes F \longrightarrow 0 \tag{19}$$

is exact.

Take for example $A = \mathbf{Z}$, $E = \mathbf{Z}$, $E' = 2\mathbf{Z}$, $F = \mathbf{Z}/2\mathbf{Z}$. As E' is isomorphic to E, $E' \otimes F$ is isomorphic to $E \otimes F$ which is itself isomorphic to F (no. 4, Proposition 4). But for all $x' = 2x \in E'$ (where $x \in E$) and all $y \in F$, $j(x') \otimes y = (2x) \otimes y = x \otimes (2y) = 0$, since $2y = 0$, and the canonical image of $E' \otimes F$ in $E \otimes F$ reduces to 0.

In other words, care must be taken to distinguish, for a submodule E' of E and an element $x \in E'$, between the element $x \otimes y$ "calculated in $E' \otimes F$" and the element $x \otimes y$ "calculated in $E \otimes F$" (in other words, the element $j(x) \otimes y$).

We shall study later, under the name of *flat* modules, the modules F such that the sequence (19) is exact for every exact sequence (18) (*Commutative Algebra*, I, § 2).

PROPOSITION 6. *Given two exact sequences* (14) *and* (16), *the homomorphism* $v \otimes t : E \otimes_A F \to E'' \otimes_A F''$ *is surjective and its kernel is equal to*

$$\operatorname{Im}(u \otimes 1_F) + \operatorname{Im}(1_E \otimes s)$$

Now $v \otimes t = (v \otimes 1_{F''}) \circ (1_E \otimes t)$ (no. 2, formula (5)) and $v \otimes t$ is therefore surjective, being the composition of two surjective homomorphisms by virtue of Proposition 5 and its Corollary. On the other hand, for $z \in E \otimes F$ to be in the kernel of $v \otimes t$, it is necessary and sufficient that $(1_E \otimes t)(z)$ belong to the kernel of $v \otimes 1_{F''}$, that is, by virtue of (15) to the image of

$$u \otimes 1_{F''}: E' \otimes F'' \to E \otimes F''.$$

But as the homomorphism $t: F \to F''$ is surjective, so is

$$1_{E'} \otimes t: E' \otimes F \to E' \otimes F''$$

by the Corollary to Proposition 5, hence the condition on z reduces to the existence of an $a \in E' \otimes F$ such that

$$(1_E \otimes t)(z) = (u \otimes t)(a).$$

Let $b = z - (u \otimes 1_F)(a)$; then $(1_E \otimes t)(b) = 0$, and by virtue of (17), b belongs to the image of $1_E \otimes s$, which proves the proposition.

In other words:

COROLLARY 1. *Let* E' *be a submodule of a right* A-*module* E, F' *a submodule of a left* A-*module* F *and* $\mathrm{Im}(E' \otimes_A F)$ *and* $\mathrm{Im}(E \otimes_A F')$ *the sub-*$\mathbf{Z}$*-modules of* $E \otimes_A F$, *the respective images of the canonical mappings* $E' \otimes_A F \to E \otimes_A F$,

$$E \otimes_A F' \to E \otimes_A F.$$

Then there is a canonical $\mathbf{Z}$*-module isomorphism*

$$\pi: (E/E') \otimes_A (F/F') \to (E \otimes_A F)/(\mathrm{Im}(E' \otimes_A F) + \mathrm{Im}(E \otimes_A F')) \tag{20}$$

such that, for $\xi \in E/E'$, $\eta \in F$, $\pi(\xi \otimes \eta)$ *is the class of all elements* $x \otimes y \in E \otimes_A F$ *such that* $x \in \xi$ *and* $y \in \eta$.

Note that when E is a $((B'_i); A, (C'_j))$-multimodule, F a $(A, (B''_h); (C''_k))$-multimodule and E' and F' *submultimodules* of E and F respectively, the isomorphism (20) is an isomorphism for the $((B'_i), (B''_h); (C'_j), (C''_k))$-multimodule structures of the two sides (no. 3).

COROLLARY 2. *Let* $\mathfrak{a}$ *be a right ideal of* A, F *a left* A-*module and* $\mathfrak{a}F$ *the sub-*$\mathbf{Z}$*-module of* F *generated by the elements of the form* λx, *where* $\lambda \in \mathfrak{a}$ *and* $x \in F$. *Then there is a canonical* $\mathbf{Z}$*-module isomorphism*

$$\pi: (A/\mathfrak{a}) \otimes_A F \to F/\mathfrak{a}F \tag{21}$$

such that, for all $\bar{\lambda} \in A/\mathfrak{a}$ *and all* $x \in F$, $\pi(\bar{\lambda} \otimes x)$ *is the class* mod. $\mathfrak{a}F$ *of* λx, *where* $\lambda \in \bar{\lambda}$.

In particular, for $A = \mathbf{Z}$, it is seen that for every integer n and every $\mathbf{Z}$-module F, $(\mathbf{Z}/n\mathbf{Z}) \otimes_{\mathbf{Z}} F$ is canonically identified with the quotient $\mathbf{Z}$-module F/nF.

COROLLARY 3. *Let* A *be a commutative ring,* $\mathfrak{a}$ *an ideal of* A *and* E *and* F *two* A*-modules such that* $\mathfrak{a}$ *is contained in the annihilator of* F. *Then the* $(A/\mathfrak{a})$*-modules* $E \otimes_A F$ *and* $(E/\mathfrak{a}E) \otimes_{A/\mathfrak{a}} F$ *are canonically isomorphic.*

F and $E \otimes_A F$ are annihilated by $\mathfrak{a}$ and hence have canonical $(A/\mathfrak{a})$-module structures (§ 1, no. 12) and if we write $E' = \mathfrak{a}E$, then $\mathrm{Im}(E' \otimes_A F) = 0$; then there is a canonical isomorphism (20) of $E \otimes_A F$ onto $(E/\mathfrak{a}E) \otimes_A F$ and the latter is itself identical with $(E/\mathfrak{a}E) \otimes_{A/\mathfrak{a}} F$ (no. 3, Corollary to Proposition 2).

COROLLARY 4. *Let* $\mathfrak{a}$, $\mathfrak{b}$ *be two ideals in a commutative ring* C; *the* C*-module* $(C/\mathfrak{a}) \otimes_C (C/\mathfrak{b})$ *is then canonically isomorphic to* $C(\mathfrak{a} + \mathfrak{b})$.

7. TENSOR PRODUCTS OF PRODUCTS AND DIRECT SUMS

Let $(E_\lambda)_{\lambda \in L}$ be a family of right A-modules, $(F_\mu)_{\mu \in M}$ a family of left A-modules and consider the product modules $C = \prod_{\lambda \in L} E_\lambda$, $D = \prod_{\mu \in M} F_\mu$. The mapping $((x_\lambda), (y_\mu)) \mapsto (x_\lambda \otimes y_\mu)$ of $C \times D$ into the product **Z**-module

$$\prod_{(\lambda, \mu) \in L \times M} (E_\lambda \otimes_A F_\mu)$$

is **Z**-bilinear and obviously satisfies conditions (1) (no. 1). Thus there exists (no. 1, Proposition 1) a **Z**-linear mapping, called *canonical*

$$f : \Big(\prod_{\lambda \in L} E_\lambda\Big) \otimes_A \Big(\prod_{\mu \in M} F_\mu\Big) \to \prod_{(\lambda, \mu) \in L \times M} (E_\lambda \otimes_A F_\mu) \tag{22}$$

such that $f((x_\lambda) \otimes (y_\mu)) = (x_\lambda \otimes y_\mu)$.

When $C = R^L$, $D = S^M$, R (resp. S) being a right (resp. left) A-module, the canonical mapping (22) associates with every tensor product $u \otimes v$, where u is a mapping of L into R and v a mapping of M into S, the mapping $(\lambda, \mu) \mapsto u(\lambda) \otimes v(\mu)$ of $L \times M$ into $R \otimes_A S$; even in this case the canonical mapping (22) is in general *neither injective nor surjective* (Exercise 3; cf. Corollary 3 to Proposition 7).

When the E_λ are $((B'_i); A, (C'_j))$-multimodules and the F_μ $(A, (B''_h); (C''_k))$-multimodules, the homomorphism (22) is also a homomorphism for the $((B'_i), (B''_h); (C'_j), (C''_k))$-multimodule structures of the two sides.

Consider now the submodule $E = \bigoplus_{\lambda \in L} E_\lambda$ (resp. $\bigoplus_{\mu \in M} F_\mu$) of C (resp. D); the canonical injections $E \to C$, $F \to D$ define canonically a **Z**-linear mapping $E \otimes_A F \to C \otimes_A D$ which, composed with the mapping (22), gives a **Z**-linear mapping g of $E \otimes_A F$ into $\prod_{\lambda, \mu} (E_\lambda \otimes_A F_\mu)$ such that

$$g((x_\lambda) \otimes (y_\mu)) = (x_\lambda \otimes y_\mu);$$

moreover, as the families (x_λ) and (y_μ) have finite support, so does $(x_\lambda \otimes y_\mu)$ and hence finally g is a canonical homomorphism

$$(23) \qquad g:\left(\bigoplus_{\lambda \in L} E_\lambda\right) \otimes_A \left(\bigoplus_{\mu \in M} F_\mu\right) \to \bigoplus_{(\lambda, \mu) \in L \times M} (E_\lambda \otimes_A F_\mu),$$

which is a multimodule homomorphism under the same conditions as (22).

PROPOSITION 7. *The canonical mapping* (23) *is bijective.*

To prove this it suffices to define a **Z**-linear mapping h of the direct sum $G = \bigoplus_{(\lambda, \mu) \in L \times M} (E_\lambda \otimes_A F_\mu)$ into $E \otimes_A F$ such that $g \circ h$ and $h \circ g$ are the identity mappings. But, to define a **Z**-linear mapping of G into $E \otimes_A F$, it suffices (§ 1, no. 6, Proposition 6) to define a **Z**-linear mapping

$$h_{\lambda\mu}: E_\lambda \otimes_A F_\mu \to E \otimes_A F$$

for every ordered pair (λ, μ), and we take $h_{\lambda\mu} = i_\lambda \otimes j_\mu$, where $i_\lambda: E_\lambda \to E$ and $j_\mu: F_\mu \to F$ are the canonical injections. Then clearly $h \circ g$ coincides with the identity mapping for the elements of the form $\left(\sum_\lambda x_\lambda\right) \otimes \left(\sum_\mu y_\mu\right)$ which generate the **Z**-module $E \otimes_A F$; similarly $g \circ h$ coincides with the identity mapping for the elements of the form $\sum_{\lambda, \mu} (x_\lambda \otimes y_\mu)$ which generate the **Z**-module G, since for each ordered pair (λ, μ) the products

$$x_\lambda \otimes y_\mu \quad (x_\lambda \in E_\lambda, y_\mu \in F_\mu)$$

generate the **Z**-module $E_\lambda \otimes_A F_\mu$. Whence the proposition.

Let $u_\lambda: E_\lambda \to E'_\lambda$, $v_\mu: F_\mu \to F'_\mu$ be A-homomorphisms; clearly the diagram

$$\begin{array}{ccc}
\left(\bigoplus_\lambda E_\lambda\right) \otimes_A \left(\bigoplus_\mu F_\mu\right) & \longrightarrow & \bigoplus_{\lambda, \mu} (E_\lambda \otimes_A (F_\mu) \\
\Big\downarrow {\scriptstyle (\oplus u_\lambda) \otimes (\oplus v_\mu)} & & \Big\downarrow {\scriptstyle \oplus (u_\lambda \otimes v_\mu)} \\
\left(\bigoplus_\lambda E'_\lambda\right) \otimes_A \left(\bigoplus_\mu F'_\mu\right) & \longrightarrow & \bigoplus_{\lambda, \mu} (E'_\lambda \otimes_A F'_\mu)
\end{array}$$

is commutative.

COROLLARY 1. *If the left* A*-module* F *admits a basis* $(b_\mu)_{\mu \in M}$, *every element of* $E \otimes_A F$ *can be written uniquely in the form* $\sum_\mu (x_\mu \otimes b_\mu)$, *where* $x_\mu \in E$ *and the family* (x_μ) *has finite support. The* **Z***-module* $E \otimes_A F$ *is isomorphic to* $E^{(M)}$ *considered as a* **Z***-module.*

The basis (b_μ) defines an isomorphism of F onto $\bigoplus_{\mu \in M} Ab_\mu$, whence there is an isomorphism $E \otimes_A F \to \bigoplus_{\mu \in M} (E \otimes_A Ab_\mu)$ by virtue of Proposition 7; as

$\xi \mapsto \xi b_\mu$ is an isomorphism of A_s onto Ab_μ, $x \mapsto x \otimes b_\mu$ is an isomorphism of E onto $E \otimes_A (Ab_\mu)$ by virtue of Proposition 4 of no. 4, whence the corollary.

If E is a $((B'_i); A, (C'_j))$-multimodule, the canonical isomorphism $E \otimes_A F \to E^{(M)}$ is a $((B'_i); (C'_j))$-multimodule isomorphism.

In particular, if also E admits a basis $(a_\lambda)_{\lambda \in L}$, every $z \in E \otimes_A F$ may be written in one and only one way in the form $\sum_{\lambda, \mu} (a_\lambda \xi_{\lambda\mu}) \otimes b_\mu$, where the $\xi_{\lambda\mu}$ belong to A (and form a family of finite support); the mapping $z \mapsto (\xi_{\lambda\mu})_{(\lambda, \mu) \in L \times M}$ is an isomorphism of $E \otimes_A F$ onto $A^{(L \times M)}$ for the **Z**-module structures (and even the module structures over the *centre* of A). More particularly:

COROLLARY 2. *If* E *and* F *are two free modules over a commutative ring* C *and* (a_λ) (resp. (b_μ)) *is a basis of the* C*-module* E (resp. F), *then* $(a_\lambda \otimes b_\mu)$ *is a basis of the* C*-module* $E \otimes_C F$.

By an abuse of language, the basis $(a_\lambda \otimes b_\mu)$ is sometimes called the *tensor product* of the bases (a_λ) and (b_μ).

Remark (1). Let E be a free right A-module, F a free left A-module, $(a_\lambda)_{\lambda \in L}$ a basis of E and $(b_\mu)_{\mu \in M}$ a basis of F. Every element $z \in E \otimes_A F$ may be written uniquely as $\sum_\lambda a_\lambda \otimes y_\lambda$, where $y_\lambda \in F$, and also uniquely as $\sum_\mu x_\mu \otimes b_\mu$, where $x_\mu \in E$. If we write $y_\lambda = \sum_\mu \eta_{\lambda\mu} b_\mu$, $x_\mu = \sum_\lambda a_\lambda \xi_{\lambda\mu}$, where the $\xi_{\lambda\mu}$ and $\eta_{\lambda\mu}$ belong to A, then $\xi_{\lambda\mu} = \eta_{\lambda\mu}$ for all (λ, μ), for

$$\sum_\lambda \Big(a_\lambda \otimes \Big(\sum_\mu \eta_{\lambda\mu} b_\mu\Big)\Big) = \sum_{\lambda, \mu} ((a_\lambda \eta_{\lambda\mu}) \otimes b_\mu) = \sum_\mu \Big(\Big(\sum_\lambda a_\lambda \eta_{\lambda\mu}\Big) \otimes b_\mu\Big).$$

COROLLARY 3. *Let* $(E_\lambda)_{\lambda \in L}$ *be a family of right* A*-modules and* F *a free* (resp. *finitely generated free*) *left* A*-module. Then the canonical mapping* (22)

$$\Big(\prod_{\lambda \in L} E_\lambda\Big) \otimes_A F \to \prod_{\lambda \in L} (E_\lambda \otimes_A F)$$

is injective (resp. *bijective*).

If (b_μ) is a basis of F, every element of $\Big(\prod_{\lambda \in L} E_\lambda\Big) \otimes_A F$ can be written uniquely as $z = \sum_\mu ((x_\lambda^{(\mu)}) \otimes b_\mu)$ (Corollary 1); to say that its canonical image is zero means that, for all $\lambda \in L$, $\sum (x_\lambda^{(\mu)} \otimes b_\mu) = 0$, hence $x_\lambda^{(\mu)} = 0$ for all $\lambda \in L$ and all μ (Corollary 1) and therefore $z = 0$.

Showing that the canonical mapping is bijective when F admits a finite basis is immediately reduced, by virtue of Proposition 7, to the case where

$F = A_s$; but then the two sides are canonically identified with $\prod_{\lambda \in L} E_\lambda$ (no. 4, Proposition 4) and after these identifications the canonical mapping (22) becomes the identity.

COROLLARY 4. *Let A be a ring with no divisor of zero, E a free right A-module and F a free left A-module. Then the relation $x \otimes y = 0$ in $E \otimes_A F$ implies $x = 0$ or $y = 0$.*

Let (a_λ) be a basis of E, (b_μ) a basis of F and let $x = \sum_\lambda a_\lambda \xi_\lambda$, $y = \sum_\mu \eta_\mu b_\mu$; then $x \otimes y = \sum_{\lambda, \mu} ((a_\lambda \xi_\lambda \eta_\mu) \otimes b_\mu$ and the relation $x \otimes y = 0$ implies $\xi_\lambda \eta_\mu = 0$ for every ordered pairs of indices (λ, μ) (Corollary 1). Hence, if $x \neq 0$, that is $\xi_\lambda \neq 0$ for at least one λ, it follows that $\eta_\mu = 0$ for all μ, whence $y = 0$.

COROLLARY 5. *Let E be a right A-module, F a left A-module, M a submodule of E and N a submodule of F. If M is a direct factor of E and N a direct factor of F, the canonical homomorphism $M \otimes_A N \to E \otimes_A F$ is injective and the image of $M \otimes_A N$ under this homomorphism is a direct factor of the* **Z**-*module $E \otimes_A F$.*

This follows immediately from Proposition 7.

Note that if E is a $((B'_i); A, (C'_j))$-multimodule and F an $(A, (B''_h); (C''_k))$-multimodule and M and N direct factors in these multimodules, $M \otimes N$ is a direct factor of the $((B'_i), (B''_h); (C'_j), (C''_k))$-multimodule $E \otimes F$.

COROLLARY 6. *Let P be a projective left A-module and E, F two right A-modules. For every injective homomorphism $u: E \to F$, the homomorphism*

$$u \otimes 1_P : E \otimes_A P \to F \otimes_A P$$

is injective.

There exists a left A-module Q such that $L = P \oplus Q$ is free (§ 2, no. 2, Proposition 4) and $u \otimes 1_L$ is identified (Proposition 7) with

$$(u \otimes 1_P) \oplus (u \otimes 1_Q);$$

it therefore suffices to prove the corollary when P is *free* (§ 1, no. 6, Corollary 1 to Proposition 7). The same argument reduces the problem to the case where $P = A_s$, which follows immediately from no. 4, Proposition 4.

COROLLARY 7. *Let C be a commutative ring. If E and F are two projective C-modules, $E \otimes_C F$ is a projective C-module.*

This follows immediately from Corollary 5 and the fact that the tensor product of two free C-modules is a free C-module (Corollary 2).

Remark 2. Under the hypotheses of Proposition 7, let E'_λ be a submodule of E_λ, F'_μ a submodule of F_μ and let $E' = \bigoplus_{\lambda \in L} E'_\lambda$, $F' = \bigoplus_{\mu \in M} F'_\mu$. Let $\mathrm{Im}(E' \otimes_A F')$

(resp. $\mathrm{Im}(E'_\lambda \otimes_A F'_\mu)$) denote the image of $E' \otimes_A F'$ (resp. $E'_\lambda \otimes_A F'_\mu$) in $E \otimes_A F$ (resp. $E_\lambda \otimes_A F_\mu$) under the canonical mapping; then the isomorphism (23) identifies the sub-**Z**-modules

$$\mathrm{Im}(E' \otimes_A F') \quad \text{and} \quad \bigoplus_{(\lambda,\mu)\in I\times M} \mathrm{Im}(E'_\lambda \otimes_A F'_\mu);$$

this follows immediately from the commutativity of the diagram

$$\begin{array}{ccc} \left(\bigoplus_\lambda E'_\lambda\right) \otimes_A \left(\bigoplus_\mu F'_\mu\right) & \longrightarrow & \left(\bigoplus_\lambda E_\lambda\right) \otimes_A \left(\bigoplus_\mu F_\mu\right) \\ \downarrow & & \downarrow \\ \bigoplus_{\lambda,\mu} (E'_\lambda \otimes_A F'_\mu) & \longrightarrow & \bigoplus_{\lambda,\mu} (E_\lambda \otimes_A F_\mu) \end{array}$$

where the vertical arrows are the canonical isomorphisms.

8. ASSOCIATIVITY OF THE TENSOR PRODUCT

PROPOSITION 8. *Let* A, B *be two rings,* E *a right* A-*module,* F *an* (A, B)-*bimodule and* G *a left* B-*module. Then* $E \otimes_A F$ *is a right* B-*module,* $F \otimes_B G$ *a left* A-*module and there exists one and only one* **Z**-*linear mapping*

$$\phi : (E \otimes_A F) \otimes_B G \to E \otimes_A (F \otimes_B G)$$

such that $\phi((x \otimes y) \otimes z) = x \otimes (y \otimes z)$ *for* $x \in E$, $y \in F$, $z \in G$; *moreover this* **Z**-*linear mapping is bijective* ("associativity" of the tensor product).

The right B-module structure on $E \otimes_A F$ and left A-module structure on $F \otimes_B G$ have been defined in no. 4. The uniqueness of ϕ is obvious since the elements $(x \otimes y) \otimes z$ generate the **Z**-module $(E \otimes_A F) \otimes_B G$. To show the existence of ϕ, we note that, for all $z \in G$, $h_z : y \mapsto y \otimes z$ is an A-linear mapping of the left A-module F into the left A-module $F \otimes_B G$. We write $g_z = 1_E \otimes h_z$, which is therefore a **Z**-linear mapping of $E \otimes_A F$ into $E \otimes_A (F \otimes_B G)$ and consider the mapping $(t, z) \mapsto g_z(t)$ from $(E \otimes_A F \times G$ into $E \otimes_A (F \otimes_B G)$; as $h_{z+z'} = h_z + h_{z'}$ for $z \in G$, $z' \in G$, it is immediate that the above mapping is **Z**-bilinear. Further, we show that for all $\mu \in B$, $g_{\mu z}(t) = g_z(t\mu)$; it is obviously sufficient to do this for $t = x \otimes y$ where $x \in E$ and $y \in F$; now

$$g_{\mu z}(x \otimes y) = x \otimes (y \otimes \mu z)$$

and

$$g_z((x \otimes y)\mu) = g_z(x \otimes y\mu) = x \otimes (y\mu \otimes z).$$

Proposition 1 (no. 1) then proves the existence of a **Z**-linear mapping

$$\phi : (E \otimes_A F) \otimes_B G \to E \otimes_A (F \otimes_B G)$$

such that $\phi(t \otimes z) = g_z(t)$, hence $\phi((x \otimes y) \otimes z) = x \otimes (y \otimes z)$. Similarly a **Z**-linear mapping

$$\psi : E \otimes_A (F \otimes_B G) \to (E \otimes_A F) \otimes_B G$$

is defined such that $\psi(x \otimes (y \otimes z)) = (x \otimes y) \otimes z$ and clearly $\psi \circ \phi$ and $\phi \circ \psi$ are the identity mappings of $(E \otimes_A F) \otimes_B G$ and $E \otimes_A (F \otimes_B G)$ respectively, since they reduce to the identity mappings on generating systems of these **Z**-modules.

It is immediate that, if E is a $((C'_i); A, (D'_j))$-multimodule, F an $(A, (C''_h); B, (D''_k))$-multimodule and G a $(B, (C'''_l); (D'''_m))$-multimodule, the canonical isomorphism defined in Proposition 8 is a $((C'_i), (C''_h), (C'''_l); (D'_j), (D''_k), (D'''_m))$-*multimodule* isomorphism. In particular, if C is a *commutative* ring and E, F, G three C-*modules*, there is a canonical C-*module* isomorphism

$$(E \otimes_C F) \otimes_C G \to E \otimes_C (F \otimes_C G).$$

We shall see below that, under certain conditions, the definition of tensor product can be generalized to a family of multimodules, which will in particular give us under the hypotheses of Proposition 8 a **Z**-module $E \otimes_A F \otimes_B G$, which is canonically isomorphic to each of the **Z**-modules $(E \otimes_A F) \otimes_B G$ and $E \otimes_A (F \otimes_B G)$ and with which the latter will be identified.

9. TENSOR PRODUCT OF FAMILIES OF MULTIMODULES

Let $(G_\lambda)_{\lambda \in L}$ be a family of **Z**-*modules*; a mapping u of the set $G = \prod_{\lambda \in L} G_\lambda$ into a **Z**-module is said to be *multiadditive* (or **Z**-*multilinear*) if $(x_\lambda) \mapsto u((x_\lambda))$ is additive with respect to each of the variables x_λ; to be precise, this means that, for all $\mu \in L$ and every element $(a_\lambda) \in \prod_{\lambda \neq \mu} G_\lambda$, canonically identifying G with $G_\mu \times \prod_{\lambda \neq \mu} G_\lambda$,

$$u(x_\mu + y_\mu, (a_\lambda)) = u(x_\mu, (a_\lambda)) + u(y_\mu, (a_\lambda)) \quad \text{for } x_\mu, y_\mu \text{ in } G_\mu. \tag{24}$$

This implies in particular that $u((x_\lambda)) = 0$ if one of the x_λ is zero.

We consider also the *universal mapping problem* where Σ is the species of **Z**-module structure and the α-mappings are the multiadditive mappings of G into a **Z**-module. A solution is still obtained by considering the **Z**-module $C = \mathbf{Z}^{(G)}$ of formal linear combinations of elements of G with coefficients in **Z** and the sub-**Z**-module D of C generated by the elements of the form

$$(x_\mu + y_\mu, (z_\lambda)_{\lambda \neq \mu}) - (x_\mu, (z_\lambda)_{\lambda \neq \mu}) - (y_\mu, (z_\lambda)_{\lambda \neq \mu})$$

where $\mu \in L$, $x_\mu \in G_\mu$, $y_\mu \in G_\mu$ and the $z_\lambda \in G_\lambda$ $(\lambda \neq \mu)$ are arbitrary. The quotient **Z**-module C/D is called the *tensor product* (*over* **Z**) *of the family* $(G_\lambda)_{\lambda \in L}$ *of* **Z**-*modules* and is denoted by $\bigotimes_{\lambda \in L} G_\lambda$; for every element $(x_\lambda)_{\lambda \in L}$ of G which is an element of the canonical basis of C, $\bigotimes_{\lambda \in L} x_\lambda$ denotes the canonical image of this element in C/D. It follows from the above definitions that the mapping $\phi:(x_\lambda)\mapsto\bigotimes_{\lambda\in L}x_\lambda$ of G into $\bigotimes_{\lambda\in L}G_\lambda$ is **Z**-multilinear and that, for every **Z**-multilinear mapping f of G into a **Z**-module H, there exists one and only one **Z**-linear mapping $g:\bigotimes_{\lambda\in L}G_\lambda\to H$ such that $f=g\circ\phi$; the ordered pair $\left(\bigotimes_{\lambda\in L}G_\lambda,\phi\right)$ thus resolves the universal mapping problem in question.

Let $(G'_\lambda)_{\lambda\in L}$ be another family of **Z**-modules and, for all $\lambda\in L$, let $v_\lambda:G_\lambda\to G'_\lambda$ be a **Z**-linear mapping (in other words a homomorphism of commutative groups). Then the mapping

$$(x_\lambda)\mapsto\bigotimes_{\lambda\in L}v_\lambda(x_\lambda)$$

of G into $\bigotimes_{\lambda\in L}G'_\lambda$ is **Z**-*multilinear* and hence defines canonically a **Z**-linear mapping of $\bigotimes_{\lambda\in L}G_\lambda$ into $\bigotimes_{\lambda\in L}G'_\lambda$ denoted by $\bigotimes_{\lambda\in L}v_\lambda$ and such that

$$\left(\bigotimes_{\lambda\in L}v_\lambda\right)\left(\bigotimes_{\lambda\in L}x_\lambda\right)=\bigotimes_{\lambda\in L}v_\lambda(x_\lambda). \tag{25}$$

In particular, we consider, for some $\mu\in L$, an endomorphism θ of G_μ; we denote by $\tilde\theta$ the endomorphism of $\bigotimes_{\lambda\in L}G_\lambda$ equal to $\bigotimes_{\lambda\in L}v_\lambda$ where $v_\mu=\theta$ and $v_\lambda=1_{G_\lambda}$ for $\lambda\neq\mu$.

Then we suppose given a set Ω, a mapping

$$c:\omega\mapsto(\rho(\omega),\sigma(\omega))$$

of Ω into $L\times L$ and, for all $\omega\in\Omega$, an endomorphism p_ω of $G_{\rho(\omega)}$ and an endomorphism q_ω of $G_{\sigma(\omega)}$; there correspond to them two endomorphisms $\tilde p_\omega$ and $\tilde q_\omega$ of $P=\bigotimes_{\lambda\in L}G_\lambda$. Let R be the sub-**Z**-module of P *generated by the union of the images of the endomorphisms* $\tilde p_\omega-\tilde q_\omega$ when ω runs through Ω. The quotient **Z**-module P/R is called the *tensor product of the family* $(G_\lambda)_{\lambda\in L}$ *relative to* c, p, q and is denoted by $\bigotimes_{(c,p,q)}G_\lambda$; composing the canonical homomorphism $P\to P/R$ with the mapping $\phi:G\to\bigotimes_{\lambda\in L}G_\lambda$ defined above, we obtain a **Z**-multilinear mapping $\phi_{(c,p,q)}:G\to\bigotimes_{(c,p,q)}G_\lambda$ and write $\phi_{(c,p,q)}((x_\lambda))=\bigotimes_{(c,p,q)}x_\lambda$ or simply $\bigotimes_{(c)}x_\lambda$. The ordered pair consisting of $\bigotimes_{(c,p,q)}x_\lambda$ and $\phi_{(c,p,q)}$ resolves the following *universal mapping problem*: let $\bar p_\omega$ (resp. $\bar q_\omega$) denote the mapping

$$(x_{\rho(\omega)},(x_\lambda)_{\lambda\neq\rho(\omega)})\mapsto(p_\omega(x_{\rho(\omega)}),(x_\lambda)_{\lambda\neq\rho(\omega)})$$

$$(\text{resp. }(x_{\sigma(\omega)},(x_\lambda)_{\lambda\neq\sigma(\omega)})\mapsto(q_\omega(x_{\sigma(\omega)}),(x_\lambda)_{\lambda\neq\sigma(\omega)}))$$

of G into itself. Then Σ is taken to be the species of **Z**-module structure and the

α-mappings the **Z**-multilinear mappings u of G into a **Z**-module which further satisfy the conditions

$$u \circ \bar{p}_\omega = u \circ \bar{q}_\omega \tag{26}$$

for all $\omega \in \Omega$. The proof is obvious from the above definitions.

This construction recovers in particular that of $E \otimes_A F$ described in no. 1: in this case we take $L = \{1, 2\}$, $G_1 = E$, $G_2 = F$, $\Omega = A$; further, for all $\omega \in A$, we must have $\rho(\omega) = 1$, $\sigma(\omega) = 2$, $_\omega$ is the endomorphism $x \mapsto x\omega$ of the **Z**-module E and q_ω the endomorphism $y \mapsto \omega y$ of the **Z**-module F.

Let $(G'_\lambda)_{\lambda \in L}$ be a second family of **Z**-modules; keeping the mapping c the same, suppose given for all $\omega \in \Omega$, an endomorphism p'_ω of $G'_{\rho(\omega)}$ and an endomorphism q'_ω of $G'_{\sigma(\omega)}$. For all $\lambda \in L$, then let $v_\lambda : G_\lambda \to G'_\lambda$ be a **Z**-linear mapping such that, for all $\omega \in \Omega$,

$$v_{\rho(\omega)} \circ p_\omega = p'_\omega \circ v_{\rho(\omega)} \quad \text{and} \quad v_{\sigma(\omega)} \circ q_\omega = q'_\omega \circ v_{\sigma(\omega)} \tag{27}$$

(in other words, for all $\lambda \in L$, v_λ is a morphism for the laws of action on G_λ (resp. G'_λ) defined on the p_ξ and q_η (resp. p'_ξ and q'_η) with ξ and η such that $\rho(\xi) = \lambda$ and $\sigma(\eta) = \lambda$). Then the mapping

$$u : (x_\lambda) \mapsto \bigotimes_{(c, p', q')} v_\lambda(x_\lambda)$$

of G into $\bigotimes_{(c, p', q')} G'_\lambda$ is **Z**-multilinear and satisfies conditions (26) and hence defines a **Z**-*linear* mapping of $\bigotimes_{(c, p, q)} G_\lambda$ into $\bigotimes_{(c, p', q')} G'_\lambda$, which we shall denote simply by $\bigotimes_{(c)} v_\lambda$ if no confusion can arise.

We shall now give an "*associativity*" property for the general tensor products thus defined. Let $(L_i)_{1 \leqslant i \leqslant n}$ be a *finite* partition of L; for every index i, let Ω_i denote the subset of Ω consisting of the elements such that $\rho(\omega) \in L_i$ *and* $\sigma(\omega) \in L_i$; clearly the Ω_i are *pairwise disjoint*; we set $\Omega' = \Omega - \left(\bigcup_i \Omega_i\right)$. For each index i, $c^{(i)}$ will denote the mapping $\omega \mapsto (\rho(\omega), \sigma(\omega))$ of Ω_i into $L_i \times L_i$; for $\omega \in \Omega_i$, we write $p^{(i)}_\omega$ and $q^{(i)}_\omega$ instead of p_ω and q_ω. Then for each i there is a "partial" tensor product

$$F_i = \bigotimes_{(c^{(i)}, p^{(i)}, q^{(i)})} G_\lambda.$$

We shall further make the following "permutability" hypothesis:

(P) *If* $\omega \in \Omega'$, p_ω (resp. q_ω) *permutes with each of the endomorphisms* p_ξ *and* q_η *of* $G_{\rho(\omega)}$ (resp. $G_{\sigma(\omega)}$) *such that* $\xi \notin \Omega'$, $\eta \notin \Omega'$ *and* $\rho(\omega) = \rho(\xi) = \sigma(\eta)$ (resp. $\sigma(\omega) = \rho(\xi) = \sigma(\eta)$).

For each $\omega \in \Omega'$, let i be the index such that $\rho(\omega) \in L_i$; then consider the family $(v_\lambda)_{\lambda \in L_i}$ where $v_{\rho(\omega)} = p_\omega$ and $v_\lambda = 1_{G_\lambda}$ for $\lambda \neq \rho(\omega)$; hypothesis (P) implies that the family (v_λ) satisfies conditions (27) (where p' and p must be replaced by $p^{(i)}$, q' and q by $q^{(i)}$, ω by an element ξ running through Ω_i); thus an endomorphism $\bigotimes_{(c^{(i)})} v_\lambda = r_\omega$ is derived of the $\mathbf{Z}$-module F_i. Similarly, an endomorphism s_ω is defined of the $\mathbf{Z}$-module F_j starting with q_ω, j being the index such that $\sigma(\omega) \in L_j$; finally let $d(\omega) = (i, j)$. Then we can define the *tensor product* $\bigotimes_{(d, r, s)} F_i$ and the corresponding canonical mapping

$$\phi_{(d, r, s)}: \prod_{i=1}^{n} F_i \to \bigotimes_{(d, r, s)} F_i.$$

On the other hand, for each i, the canonical mapping

$$\psi_i = \phi_{(c^{(i)}, p^{(i)}, q^{(i)})}: \prod_{\lambda \in L_i} G_\lambda \to F_i;$$

using the associativity of the product of sets, a $\mathbf{Z}$-multilinear mapping $\psi = \phi_{(d, r, s)} \circ (\psi_i)$ of G into $\bigotimes_{(d, r, s)} F_i$ is derived. We show that the *ordered pair* $\left(\bigotimes_{(d, r, s)} F_i, \psi\right)$ *is a solution of the same universal problem as* $\left(\bigotimes_{(c, p, q)} G_\lambda, \phi_{(c, p, q)}\right)$, whence will follow the existence of a *unique* $\mathbf{Z}$*-module isomorphism*

$$\theta: \bigotimes_{(c, p, q)} G_\lambda \to \bigotimes_{(d, r, s)} F_i$$

such that $\psi = \theta \circ \phi_{(c, p, q)}$ (*Set Theory*, IV, § 3, no. 1).

By induction on n, the proof is reduced to the case $n = 2$; for simplicity we write $F_1 \underset{(d)}{\otimes} F_2$ and $y_1 \underset{(d)}{\otimes} y_2$ instead of $\bigotimes_{(d, r, s)} F_i$ and $\bigotimes_{(d, r, s)} y_i$. Consider the mapping from G to $F_1 \underset{(d)}{\otimes} F_2$

$$h: (x_\lambda) \to \left(\bigotimes_{(c^{(1)})} x_\lambda\right) \underset{(d)}{\otimes} \left(\bigotimes_{(c^{(2)})} x_\lambda\right).$$

It is obviously $\mathbf{Z}$-multilinear; we show that it satisfies conditions (26) for all $\omega \in \Omega$. This is obvious if $\omega \in \Omega_1$ or $\omega \in \Omega_2$; otherwise, supposing, in order to fix the ideas, that $\rho(\omega) \in L_1$ and $\sigma(\omega) \in L_2$, the values of $h \circ \bar{p}_\omega$ and $h \circ \bar{q}_\omega$ at (x_λ) are respectively

$$\left(r_\omega\left(\bigotimes_{(c^{(1)})} x_\lambda\right)\right) \underset{(d)}{\otimes} \left(\bigotimes_{(c^{(2)})} x_\lambda\right) \quad \text{and} \quad \left(\bigotimes_{(c^{(1)})} x_\lambda\right) \underset{(d)}{\otimes} \left(s_\omega\left(\bigotimes_{(c^{(2)})} x_\lambda\right)\right)$$

which are also equal by definition of $F_1 \underset{(d)}{\otimes} F_2$.

This being so, let u be a $\mathbf{Z}$-multilinear mapping of G into a $\mathbf{Z}$-module H, satisfying conditions (26); we shall define a $\mathbf{Z}$-linear mapping $v: F_1 \underset{(d)}{\otimes} F_2 \to H$

such that $u = v \circ h$ and that will prove our assertion (repeating the argument of no. 1, Corollary 1 to Proposition 1). For all $z_2 = (x_\lambda)_{\lambda \in L_2}$ we consider the "partial" linear mapping of $\prod_{\lambda \in L_1} G_\lambda$ into H

$$u(.\,, z_2) : (x_\lambda)_{\lambda \in L_1} \mapsto u((x_\lambda)_{\lambda \in L_1}, z_2) = u((x_\lambda)_{\lambda \in L}). \tag{28}$$

Clearly it is **Z**-multilinear and satisfies conditions (26) for $\omega \in \Omega_1$; by definition there thus exists a **Z**-linear mapping $y_1 \mapsto w_1(y_1, x_2)$ of F_1 into H such that

$$w_1\Big(\bigotimes_{(c^{(1)})} x_\lambda, z_2\Big) = u((x_\lambda)_{\lambda \in L_1}, z_2) \tag{29}$$

We next consider the mapping

$$u_2 : (x_\lambda)_{\lambda \in L_2} \mapsto w_1(.\,, (x_\lambda)_{\lambda \in L_2})$$

of $\prod_{\lambda \in L_2} G_\lambda$ into $\mathrm{Hom}_{\mathbf{Z}}(F_1, H)$; it is obviously **Z**-multilinear and satisfies conditions (26) for $\omega \in \Omega_2$, by virtue of the hypothesis on u and relations (28) and (29) and taking account of the fact that the elements of the form $\bigotimes_{(c^{(1)})} x_\lambda$ generate the **Z**-module F_1. Hence there is a **Z**-linear mapping

$$w_2 : F_2 \to \mathrm{Hom}_{\mathbf{Z}}(F_1, H)$$

such that

$$w_2\Big(\bigotimes_{(c^{(2)})} x_\lambda\Big) = u_2((x_\lambda)_{\lambda \in L_2})$$

or also

$$\Big(w_2\Big(\bigotimes_{(c^{(2)})} x_\lambda\Big)\Big)\Big(\bigotimes_{(c^{(1)})} x_\lambda\Big) = u((x_\lambda)_{\lambda \in L}). \tag{30}$$

We now consider, for $y_1 \in F_1$, $y_2 \in F_2$, the element of H

$$w(y_1, y_2) = (w_2(y_2))(y_1). \tag{31}$$

Clearly w is a **Z**-*bilinear* mapping of $F_1 \times F_2$ into H. We show further that, for all $\omega \in \Omega'$, (assuming, to fix the ideas, that $\rho(\omega) \in L_1$ and $\sigma(\omega) \in L_2$)

$$w(r_\omega(y_1), y_2) = w(y_1, s_\omega(y_2)). \tag{32}$$

It suffices to verify this relation when y_1 (resp. y_2) is of the form $\bigotimes_{(c^{(1)})} x_\lambda$ (resp. $\bigotimes_{(c^{(2)})} x_\lambda$), since these elements generate the **Z**-module F_1 (resp. F_2). But by definition, $r_\omega\Big(\bigotimes_{(c^{(1)})} x_\lambda\Big) = \bigotimes_{(c^{(1)})} x'_\lambda$, where $x'_{\rho(\omega)} = p_\omega(x_{\rho(\omega)})$ and $x'_\lambda = x_\lambda$ for

$\lambda \neq \rho(\omega)$ in L_1; similarly $s_\omega\left(\bigotimes_{(c^{(2)})} x_\lambda\right) = \bigotimes_{(c^{(2)})} x''_\lambda$, where $x''_{\sigma(\omega)} = q_\omega(x'_{\sigma(\omega)})$ and $x''_\lambda = x_\lambda$ for $\lambda \neq \sigma(\omega)$ in L_2; using (30) and (31), relation (32) then follows from (26). Hence there exists a **Z**-*linear* mapping v of $F_1 \underset{(d)}{\otimes} F_2$ into H such that $v(y_1 \underset{(d)}{\otimes} y_2) = w(y_1, y_2)$ and it then follows from (30) and (31) that $v \circ h = u$.

The most important special case of the general tensor product defined above is the following: we start with a family $(A_i)_{1 \leqslant i \leqslant n-1}$ of rings and a family $(E_i)_{1 \leqslant i \leqslant n}$, where E_1 is a right A_1-module, E_n is a left A_{n-1} module and for $2 \leqslant i \leqslant n-1$, E_i is an (A_{i-1}, A_i)-*bimodule*. Then the above definition is applied as follows: L is the set $[1, n]$, $G_i = E_i$, Ω is the set the *sum* of the A_i $(1 \leqslant i \leqslant n-1)$. For $\omega \in A_i$ $(1 \leqslant i \leqslant n-1)$, take $\rho(\omega) = i$, $\sigma(\omega) = i + 1$, p_ω is the endomorphism $x \mapsto x\omega$ of the **Z**-module E_i and q_ω the endomorphism $y \mapsto \omega y$ of the **Z**-module E_{i+1}; the corresponding tensor product is denoted by

$$E_1 \otimes_{A_1} E_2 \otimes_{A_2} E_3 \otimes \cdots \otimes_{A_{n-2}} E_{n-1} \otimes_{A_{n-1}} E_n \tag{33}$$

(a notation where the A_i may occasionally be suppressed) and the elements $\bigotimes_{(c, p, q)} x_i$ of this tensor product, for a family (x_i) such that $x_i \in E_i$ for $1 \leqslant i \leqslant n$, may be written $x_1 \otimes x_2 \otimes \cdots \otimes x_n$ if no confusion can arise; an analogous notation is used for a **Z**-linear mapping $\underset{(c)}{\otimes} v_i$. Hypothesis (P) holds for *every* partition of $[1, n]$, since the E_i are *bimodules* for $2 \leqslant i \leqslant n-1$. When $n = 3$, we have thus defined the **Z**-module $E \otimes_A F \otimes_B G$ alluded to in no. 8, and recovered Proposition 8 (no. 8).

When each of the E_i is a *multimodule* (with, for $2 \leqslant i \leqslant n-1$, A_{i-1} one of the rings operating on the left and A_i one of the rings operating on the right and analogous conditions for $i = 1$ and $i = n$), as in no. 4, a *multimodule* structure is defined on $E_1 \otimes_{A1} E_2 \otimes \cdots \otimes_{An-1} E_n$ with respect to *all* the rings except the A_i which operate on the E_i $(1 \leqslant i \leqslant n)$.

In particular, let C be a *commutative* ring, $(E_i)_{1 \leqslant i \leqslant n}$ a family of C-*modules*. By giving E_1 and E_n *two* C-module structures identical with the given structure and E_i for $2 \leqslant i \leqslant n-1$ *three* C-module structures identical with the given structure, we define on the tensor product

$$E_1 \otimes_C E_2 \otimes_C E_3 \otimes \cdots \otimes_C E_{n-1} \otimes_C E_n \tag{34}$$

n C-modules structures which are compatible with one another and which are in fact *identical*, since for all $\gamma \in C$ and $(x_i) \in \prod_{i=1}^{n} E_i$, by definition

$$(\gamma x_1) \otimes x_2 \otimes \cdots \otimes x_n$$
$$= x_1 \otimes (\gamma x_2) \otimes \cdots \otimes x_n = \cdots = x_1 \otimes x_2 \otimes \cdots \otimes (\gamma x_n).$$

When we speak of the tensor product (34) as a C-*module*, we shall always mean with this structure, unless otherwise mentioned, and the tensor product (34) is also denoted by $\bigotimes_{1 \leqslant i \leqslant n} E_i$ if no confusion can arise. For every C-module G, the **Z**-multilinear mappings of $\prod_{i=1}^{n} E_i$ into G which, for every index i, satisfy the relation

$$(35) \qquad f(x_1, \ldots, x_{i-1}, \gamma x_i, x_{i+1}, \ldots, x_n) = \gamma f(x_1, \ldots, x_n)$$

for $\gamma \in C$ and $(x_i) \in \prod_i E_i$ are then called C-*multilinear* and form a C-*module* denoted by $\mathscr{L}_n(E_1, \ldots, E_n; G)$; the universal property of the tensor product (34) then allows us to define a *canonical* C-*module isomorphism*

$$(36) \qquad \mathscr{L}_n(E_1, \ldots, E_n; G) \to \operatorname{Hom}_C(E_1 \otimes_C E_2 \otimes \cdots \otimes_C E_n, G)$$

which associates with every C-multilinear mapping f the C-linear mapping g such that

$$f(x_1, \ldots, x_n) = g(x_1 \otimes x_2 \otimes \cdots \otimes x_n).$$

A C-multilinear mapping of $E_1 \times \cdots \times E_n$ into C is also called an *n-linear form.*

Let $(F_i)_{1 \leqslant i \leqslant n}$ be another family of C-modules; for every system of n C-*linear* mappings $u_i : E_i \to F_i$, $u_1 \otimes u_2 \otimes \cdots \otimes u_n$ (also denoted by $\bigotimes_{1 \leqslant i \leqslant n} u_i$) is a C-*linear* mapping of

$$E_1 \otimes_C E_2 \otimes \cdots \otimes_C E_n \quad \text{into } F_1 \otimes_C F_2 \otimes \cdots \otimes_C F_n.$$

Further, $(u_1, \ldots, u_n) \mapsto u_1 \otimes u_2 \otimes \cdots \otimes u_n$ is a C-*multilinear* mapping of $\prod_i \operatorname{Hom}_C(E_i, F_i)$ into

$$\operatorname{Hom}_C(E_1 \otimes_C E \otimes \cdots \otimes_C E_n, F_1 \otimes_C F_2 \otimes \cdots \otimes_C F_n).$$

Hence there corresponds canonically to the latter mapping a C-*linear* mapping called *canonical*

$$(37) \quad \operatorname{Hom}_C(E_1, F_1) \otimes_C \operatorname{Hom}_C(E_2, F_2) \otimes \cdots \otimes_C \operatorname{Hom}_C(E_n, F_n) \\ \to \operatorname{Hom}_C(E_1 \otimes_C E_2 \otimes \cdots \otimes_C E_n, F_1 \otimes_C F_2 \otimes \cdots \otimes_C F_n)$$

generalizing that defined in no. 5 for $n = 2$.

The general associativity property seen earlier can be specialized here as follows. Given a partition $(J_k)_{1 \leqslant k \leqslant m}$ of the interval $[1, n]$ of **N**, let, for each k, F_k be the tensor product $E_{i_1} \otimes_C E_{i_2} \otimes \cdots \otimes_C E_{i_r}$, where $(i_1, \ldots, i_r)$ is the strictly increasing sequence of elements of J_k. Then there is a canonical C-module isomorphism (called "*associativity isomorphism*")

$$F_1 \otimes_C F_2 \otimes \cdots \otimes_C F_m \to E_1 \otimes_C E_2 \otimes \cdots \otimes_C E_n$$

which, in the above notation, maps the tensor product

$$y_1 \otimes y_2 \otimes \cdots \otimes y_m, \quad \text{where } y_k = x_{i_1} \otimes x_{i_2} \otimes \cdots \otimes x_{i_r},$$

to the tensor product $x_1 \otimes x_2 \otimes \cdots \otimes x_n$ (where $x_i \in E_i$ for all i).

In particular, if π is a *permutation* of $[1, n]$, writing $J_k = \{\pi(k)\}$ for $1 \leqslant k \leqslant n$, we obtain a canonical ("*commutativity*") isomorphism

$$E_{\pi(1)} \otimes_C E_{\pi(2)} \otimes \cdots \otimes_C E_{\pi(n)} \to E_1 \otimes_C E_2 \otimes \cdots \otimes_C E_n$$

which maps $x_{\pi(1)} \otimes x_{\pi(2)} \otimes \cdots \otimes x_{\pi(n)}$ to $x_1 \otimes x_2 \otimes \cdots \otimes x_n$. We shall often identify the various tensor products which correspond to one another under these canonical isomorphisms.

For $1 \leqslant i \leqslant n$, suppose that E_i admits a basis $(b^{(i)}_{\lambda_i})_{\lambda_i \in L_i}$; by induction on n, it follows from no. 7, Corollary 2 to Proposition 7 that the family $(b^{(1)}_{\lambda_1} \otimes b^{(2)}_{\lambda_2} \otimes \cdots \otimes b^{(n)}_{\lambda_n})$, where $(\lambda_1, \ldots, \lambda_n)$ runs through $\prod_{1 \leqslant i \leqslant n} L_i$, is a basis of $\bigoplus_{1 \leqslant i \leqslant n} E_i$, sometimes called the *tensor product* of the bases $(b^{(1)}_{\lambda_1})$ in question.

Remarks. (1) The above remarks concerning the case of modules over a commutative ring generalize as in no. 5, *Remark* 2 when there is a tensor product $E_1 \otimes_{A_1} E_2 \otimes \cdots \otimes_{A_{n-1}} E_n$ where the rings A_i are not necessarily commutative and where, for each i, there is a homomorphism $\rho_i : C \to A$ of the same *commutative* ring C such that: (1) $\rho_i(C)$ is contained in the *centre* of A_i; (2) for $2 \leqslant i \leqslant n - 1$, the C-module structures on E_i obtained using the homomorphisms ρ_{i-1} and ρ_i *coincide*. Then a C-*module* structure is obtained on $E_1 \otimes_{A_1} E_2 \otimes \cdots \otimes_{A_{n-1}} E_n$ and canonical mappings analogous to (13) (no. 5), which will be left to the reader to describe.

(2) Let A, B be two rings, E a right A-module, E′ a left A-module, F a right B-module and F′ a left B-module. The **Z**-*bilinear* mappings of $(E \otimes_A E') \times (F \otimes_B F')$ into a **Z**-module G are then in *one-to-one correspondence* with the **Z**-*multilinear* mappings f of $E \times E' \times F \times F'$ into G satisfying the conditions

$$\left\{\begin{aligned} f(x\lambda, x', y, y') &= f(x, \lambda x', y, y') \\ f(x, x', y\mu, y') &= f(x, x', y, \mu y') \end{aligned}\right. \tag{38}$$

for $\lambda \in A$, $\mu \in B$, $x \in E$, $x' \in E'$, $y \in F$, $y' \in F'$. The general constructions given in this no. reduce the proof of this to defining a canonical **Z**-module isomorphism between $(E \otimes_A E') \otimes_{\mathbf{Z}} (F \otimes_B F')$ and $E \otimes_A E' \otimes_{\mathbf{Z}} F \otimes_B F'$, which follows from the associativity property of tensor products of the form (33).

§4. RELATIONS BETWEEN TENSOR PRODUCTS AND HOMOMORPHISM MODULES

1. THE ISOMORPHISMS $\mathrm{Hom}_B(E \otimes_A F, G) \to \mathrm{Hom}_A(F, \mathrm{Hom}_B(E, G))$ AND $\mathrm{Hom}_C(E \otimes_A F, G) \to \mathrm{Hom}_A(E, \mathrm{Hom}_C(F, G))$

Let E be a right A-module, F a left A-module, G a **Z**-module and H the **Z**-module of mappings $f: E \times F \to G$ which are **Z**-bilinear and satisfy

$$f(x\lambda, y) = f(x, \lambda y) \quad \text{for } x \in E, y \in F, \lambda \in A. \tag{1}$$

It has been seen (§3, no. 1, Proposition 1) that there exists a canonical **Z**-module homomorphism

$$H \to \mathrm{Hom}_{\mathbf{Z}}(E \otimes_A F, G). \tag{2}$$

On the other hand, a left A-module structure has been defined on $\mathrm{Hom}_{\mathbf{Z}}(E, G)$ and a right A-module structure on $\mathrm{Hom}_{\mathbf{Z}}(F, G)$ (§3, no. 3); we may therefore consider the **Z**-modules $\mathrm{Hom}_A(E, \mathrm{Hom}_{\mathbf{Z}}(F, G))$ and $\mathrm{Hom}_A(F, \mathrm{Hom}_{\mathbf{Z}}(E, G))$. A mapping f of $E \times F$ into G is canonically identified with a mapping of E into the set G^F of mappings of F into G (*Set Theory*, II, §5, no. 2); by expressing the fact that the latter mapping belongs to $\mathrm{Hom}_A(E, \mathrm{Hom}_{\mathbf{Z}}(F, G))$, we obtain precisely the fact that f is biadditive and conditions (1); whence there is a canonical isomorphism

$$H \to \mathrm{Hom}_A(E, \mathrm{Hom}_{\mathbf{Z}}(F, G)) \tag{3}$$

and similarly there is defined a canonical isomorphism

$$H \to \mathrm{Hom}_A(F, \mathrm{Hom}_{\mathbf{Z}}(E, G)). \tag{4}$$

Suppose now that E and G also have left (resp. right) B-module structures and that the A-module and B-module structures on E are compatible. Then $E \otimes_A F$ has canonically a left (resp. right) B-module structure (§3, no. 4) and on the other hand $\mathrm{Hom}_B(E, G)$ has canonically a left A-module structure (§1, no. 14). We may therefore consider the **Z**-modules $\mathrm{Hom}_B(E \otimes_A F, G)$ and $\mathrm{Hom}_A(F, \mathrm{Hom}_B(E, G))$, which are submodules of $\mathrm{Hom}_{\mathbf{Z}}(E \otimes_A F, G)$ and $\mathrm{Hom}_A(F, \mathrm{Hom}_{\mathbf{Z}}(E, G))$ respectively (§2, no. 1, Theorem 2). We examine under what condition a mapping $f \in H$ has as image under the isomorphisms (2) and (4) an element of $\mathrm{Hom}_B(E \otimes_A F, G)$ and an element of $\mathrm{Hom}_A(E, \mathrm{Hom}_B(F, G))$ respectively; in each of the two cases we find the *same* condition

$$f(\beta x, y) = \beta f(x, y)$$
$$(\text{resp. } f(x\beta, y) = f(x, y)\beta)$$

for $x \in E, y \in F, \beta \in B$.

Similarly, suppose that F and G are left (resp. right) C-modules and that the A-module and C-module structures on F are compatible. Then, for a mapping $f \in \mathrm{H}$ to have as image under (2) or (3) an element of $\mathrm{Hom}_{\mathrm{C}}(\mathrm{E} \otimes_{\mathrm{A}} \mathrm{F}, \mathrm{G})$ or $\mathrm{Hom}_{\mathrm{A}}(\mathrm{E}, \mathrm{Hom}_{\mathrm{C}}(\mathrm{F}, \mathrm{G}))$ respectively, it is necessary and sufficient that it satisfy the same condition

$$f(x, \gamma y) = \gamma f(x, y)$$

$$(\text{resp. } f(x, y\gamma) = f(x, y)\gamma)$$

for $x \in \mathrm{E}$, $y \in \mathrm{F}$, $\gamma \in \mathrm{C}$.

We have therefore established the following result (in the notation introduced above):

Proposition 1. (a) *Let* E *be a* (B, A)*-bimodule,* F *a left* A*-module and* G *a left* B*-module. For every mapping* $g \in \mathrm{Hom}_{\mathrm{B}}(\mathrm{E} \otimes_{\mathrm{A}} \mathrm{F}, \mathrm{G})$, *let* g' *be the mapping of* F *into* $\mathrm{Hom}_{\mathrm{B}}(\mathrm{E}, \mathrm{G})$ *defined by* $(g'(y))(x) = g(x \otimes y)$ *for* $x \in \mathrm{E}$, $y \in \mathrm{F}$. *The mapping* $g \mapsto g'$ *is an isomorphism*

$$\beta : \mathrm{Hom}_{\mathrm{B}}(\mathrm{E} \otimes_{\mathrm{A}} \mathrm{F}, \mathrm{G}) \rightarrow \mathrm{Hom}_{\mathrm{A}}(\mathrm{F}, \mathrm{Hom}_{\mathrm{B}}(\mathrm{E}, \mathrm{G})). \tag{5}$$

(b) *Let* E *be a right* A*-module,* F *an* (A, C)*-bimodule and* G *a right* C*-module. For every mapping* $h \in \mathrm{Hom}_{\mathrm{C}}(\mathrm{E} \otimes_{\mathrm{A}} \mathrm{F}, \mathrm{G})$, *let* h' *be the mapping of* E *into* $\mathrm{Hom}_{\mathrm{C}}(\mathrm{F}, \mathrm{G})$ *defined by* $(h'(x))(y) = h(x \otimes y)$ *for* $x \in \mathrm{E}$, $y \in \mathrm{F}$. *The mapping* $h \mapsto h'$ *is an isomorphism*

$$\gamma : \mathrm{Hom}_{\mathrm{C}}(\mathrm{E} \otimes_{\mathrm{A}} \mathrm{F}, \mathrm{G}) \rightarrow \mathrm{Hom}_{\mathrm{A}}(\mathrm{E}, \mathrm{Hom}_{\mathrm{C}}(\mathrm{F}, \mathrm{G})). \tag{6}$$

In particular B and C may be taken to be a subring Γ of the *centre* of the ring A; then for every Γ-module G, the three Γ-modules

$$\mathrm{Hom}_{\Gamma}(\mathrm{E} \otimes_{\mathrm{A}} \mathrm{F}, \mathrm{G}), \qquad \mathrm{Hom}_{\mathrm{A}}(\mathrm{E}, \mathrm{Hom}_{\Gamma}(\mathrm{F}, \mathrm{G})), \qquad \mathrm{Hom}_{\mathrm{A}}(\mathrm{F}, \mathrm{Hom}_{\Gamma}(\mathrm{E}, \mathrm{G}))$$

are canonically isomorphic to the Γ-module of Γ-*bilinear* mappings of $\mathrm{E} \times \mathrm{F}$ into G which satisfy (1). More particularly:

Corollary. *If* C *is a commutative ring and* E, F, G *three* C*-modules, then the* C*-modules*

$$\mathrm{Hom}_{\mathrm{C}}(\mathrm{E} \otimes_{\mathrm{C}} \mathrm{F}, \mathrm{G}), \qquad \mathrm{Hom}_{\mathrm{C}}(\mathrm{E}, \mathrm{Hom}_{\mathrm{C}}(\mathrm{F}, \mathrm{G})),$$
$$\mathrm{Hom}_{\mathrm{C}}(\mathrm{F}, \mathrm{Hom}_{\mathrm{C}}(\mathrm{E}, \mathrm{G})), \qquad \mathscr{L}_2(\mathrm{E}, \mathrm{F}; \mathrm{G})$$

are canonically isomorphic.

2. THE HOMOMORPHISM $\mathrm{E}^* \otimes_{\mathrm{A}} \mathrm{F} \rightarrow \mathrm{Hom}_{\mathrm{A}}(\mathrm{E}, \mathrm{F})$

Let A, B be two rings, E a left A-module, F a left B-module and G an (A, B)-*bimodule*. The **Z**-module $\mathrm{Hom}_{\mathrm{A}}(\mathrm{E}, \mathrm{G})$ has canonically a *right* B*-module* structure (§ 1, no. 14) such that $(u\beta)(x) = u(x)\beta$ for $\beta \in B$, $u \in \mathrm{Hom}_A(E, G)$, $x \in E$. On the other hand, $G \otimes_B F$ has canonically a *left* A-*module* structure (§ 3, no. 4). We shall define a *canonical* **Z**-*homomorphism*

$$\nu : \mathrm{Hom}_A(E, G) \otimes_B F \to \mathrm{Hom}_A(E, G \otimes_B F). \tag{7}$$

To this end, we consider, for all $y \in F$ and all $u \in \mathrm{Hom}_A(E, G)$, the mapping $\nu'(u, y) : x \mapsto u(x) \otimes y$ of E onto $G \otimes_B F$. It is immediately verified that $\nu'(u, y)$ is A-linear and that ν' is a **Z**-bilinear mapping of $\mathrm{Hom}_A(E, G) \times F$ into $\mathrm{Hom}_A(E, G \otimes_B F)$; moreover, for all $\beta \in B$, $\nu'(u\beta, y)$ and $\nu'(u, \beta y)$ are equal, for $(u(x)\beta) \otimes y = u(x) \otimes (\beta y)$. We conclude (§ 3, no. 1, Proposition 1) the existence of the desired homomorphism ν such that $\nu(u \otimes y)$ is the A-linear mapping $x \mapsto u(x) \otimes y$.

It is immediately verified that, if E is an $(A, (C'_i); (D'_j))$-multimodule, F a $(B, (C''_h); (D''_k))$-multimodule and G an $(A, (C'''_l); B, (D'''_m))$-multimodule, the mapping (7) is a $((D'_j), (C''_h), (C'''_l); (C'_i), (D''_k), (D'''_m))$-multimodule homomorphism.

PROPOSITION 2. (i) *When* F *is a projective* (resp. *finitely generated projective*) B-*module, the canonical homomorphism* (7) *is injective* (resp. *bijective*).

(ii) *When* E *is a finitely generated projective* A-*module, the canonical homomorphism* (7) *is bijective.*

(i) Fixing E and G, for *every* left B-module F, we write

$$T(F) = \mathrm{Hom}_A(E, G) \otimes_B F, \quad T'(F) = \mathrm{Hom}_A(E, G \otimes_B F);$$

for every left B-module homomorphism $u : F \to F'$, we write $T(u) = 1 \otimes u$ (1 here denoting the identity mapping of $\mathrm{Hom}_A(E, G)$),

$$T'(u) = \mathrm{Hom}(1_E, 1_G \otimes u);$$

on the other hand we write ν_F instead of ν. Then we have the following lemmas:

Lemma 1. *For every homomorphism* $u : F \to F'$, *the diagram*

$$\begin{array}{ccc} T(F) & \xrightarrow{\nu_F} & T'(F) \\ {\scriptstyle T(u)}\downarrow & & \downarrow{\scriptstyle T'(u)} \\ T(F') & \xrightarrow[\nu_{F'}]{} & T'(F') \end{array} \tag{8}$$

is commutative.

The verification is immediate.

Lemma 2. *Let* M, N *be two supplementary submodules in* F *and* $i: \mathrm{M} \to \mathrm{F}$, $j: \mathrm{N} \to \mathrm{F}$ *the canonical injections. The diagram*

$$\begin{array}{ccc} \mathrm{T(M)} \oplus \mathrm{T(N)} & \xrightarrow{\nu_{\mathrm{M}} \oplus \nu_{\mathrm{N}}} & \mathrm{T'(M)} \oplus \mathrm{T'(N)} \\ \downarrow{\scriptstyle \mathrm{T}(i)+\mathrm{T}(j)} & & \downarrow{\scriptstyle \mathrm{T}'(i)+\mathrm{T}'(j)} \\ \mathrm{T(F)} & \xrightarrow[\nu_{\mathrm{F}}]{} & \mathrm{T'(F)} \end{array} \tag{9}$$

is commutative and the vertical arrows are bijective.

The commutativity follows from Lemma 1, the other assertions from § 1, no. 6, Corollary 2 to Proposition 6 and § 3, no. 7, Proposition 7.

Lemma 3. *Under the hypotheses of Lemma* 2, *for* ν_{F} *to be injective* (resp. *surjective*), *it is necessary and sufficient that* ν_{M} *and* ν_{N} *be so.*

This follows from Lemma 2 and § 1, no. 6, Corollary 1 to Proposition 6.

Then Lemma 3, together with § 2, no. 2, Proposition 4, shows that it suffices to consider the case where F is a *free* module. But, if (b_μ) is a basis of F, every element of $\mathrm{Hom}_{\mathrm{A}}(\mathrm{E}, \mathrm{G}) \otimes_{\mathrm{B}} \mathrm{F}$ may then be written uniquely as $\sum_\mu u_\mu \otimes b_\mu$, where $u_\mu \in \mathrm{Hom}_{\mathrm{A}}(\mathrm{E}, \mathrm{G})$ (§ 3, no. 7, Corollary 1 to Proposition 7); the image of this element under ν is the A-linear mapping $x \mapsto \sum_\mu u_\mu(x) \otimes b_\mu$; it cannot be zero for all $x \in \mathrm{E}$ unless $u_\mu(x) = 0$ for all $x \in \mathrm{E}$ and all μ, which is equivalent to saying that $u_\mu = 0$ for all μ; hence ν is injective. When also F admits a *finite basis*, Lemma 3 shows (by induction on the number of elements in the basis of F) that to prove that ν is surjective, it suffices to do so when $\mathrm{F} = \mathrm{B}_s$; but in this case the two sides of (7) are canonically identified with $\mathrm{Hom}_{\mathrm{A}}(\mathrm{E}, \mathrm{G})$ (§ 3, no. 4, Proposition 4) and ν becomes the identity.

(ii) To show the proposition when E is projective and finitely generated, this time we fix F and G and write, for *every* left A-module E,

$$\mathrm{T(E)} = \mathrm{Hom}_{\mathrm{A}}(\mathrm{E}, \mathrm{G}) \otimes_{\mathrm{B}} \mathrm{F}, \qquad \mathrm{T'(E)} = \mathrm{Hom}_{\mathrm{A}}(\mathrm{E}, \mathrm{G} \otimes_{\mathrm{B}} \mathrm{F})$$

and, for every left A-module homomorphism $v: \mathrm{E} \to \mathrm{E}'$,

$$\mathrm{T}(v) = \mathrm{Hom}(v, 1_{\mathrm{G}}) \otimes 1_{\mathrm{F}}, \qquad \mathrm{T}'(v) = \mathrm{Hom}(v, 1_{\mathrm{G}} \otimes 1_{\mathrm{F}});$$

on the other hand, we write ν_{E} instead of ν. Then we have the two lemmas:

Lemma 4. *For every homomorphism* $v: \mathrm{E} \to \mathrm{E}'$, *the diagram*

$$\begin{array}{ccc} \mathrm{T(E')} & \xrightarrow{\nu_{\mathrm{E}'}} & \mathrm{T'(E')} \\ \downarrow{\scriptstyle \mathrm{T}(v)} & & \downarrow{\scriptstyle \mathrm{T}'(v)} \\ \mathrm{T(E)} & \xrightarrow[\nu_{\mathrm{E}}]{} & \mathrm{T'(E)} \end{array} \tag{10}$$

is commutative.

Lemma 5. *Let* M *and* N *be two supplementary submodules in* E *and* $p: E \to M$, $q: E \to N$ *the canonical projections. The diagram*

$$\begin{array}{ccc} T(M) \oplus T(N) & \xrightarrow{\nu_M \oplus \nu_N} & T'(M) \oplus T'(N) \\ \downarrow {\scriptstyle T(p)+T(q)} & & \downarrow {\scriptstyle T'(p)+T'(q)} \\ T(E) & \xrightarrow[\nu_E]{} & T'(E) \end{array}$$

is commutative and the vertical arrows are bijective.

They are proved as Lemmas 1 and 2, taking account of § 1, no. 6, Corollary 2 to Proposition 6, § 2, no. 1, Proposition 1 and § 3, no. 7, Proposition 7.

The remainder of the proof then proceeds as in (i) and is reduced to the case where $E = A_s$; the two sides of (7) are then canonically identified with $G \otimes_B F$ and ν becomes the identity.

In particular take B = A and G the (A, A)-bimodule ${}_sA_d$ (§ 3, no. 4), so that the right A-module $\mathrm{Hom}_A(E, {}_sA_d)$ is just the *dual* E^* of E and $({}_sA_d) \otimes_A F$ is canonically identified with F (§ 3, no. 4, Proposition 4). Homomorphism (7) then becomes a canonical **Z**-homomorphism

$$(11) \qquad \theta: E^* \otimes_A F \to \mathrm{Hom}_A(E, F)$$

and $\theta(x^* \otimes y)$ is the linear mapping of E into F

$$x \mapsto \langle x, x^* \rangle y.$$

Remark (1). The characterization of *projective* A-modules given in § 2, no. 6, Proposition 12, can also be expressed as follows: for a left A-module E to be projective, it is necessary and sufficient that the canonical homomorphism

$$\theta_E: E^* \otimes_A E \to \mathrm{Hom}_A(E, E) = \mathrm{End}_A(E)$$

be such that 1_E *belongs to the image of* θ_E.

Corollary. (i) *When* F *is a projective* (resp. *finitely generated projective*) *module, the canonical homomorphism* (11) *is injective* (resp. *bijective*).

(ii) *When* E *is a finitely generated projective module, the canonical homomorphism* (11) *is bijective.*

Even when E and F are both finitely generated, θ is not necessarily surjective, as is shown by the example $A = \mathbf{Z}$, $E = F = \mathbf{Z}/2\mathbf{Z}$; the right hand side of (11) is non-zero but $E^* = 0$. On the other hand, examples can be given where E is *free*, but (11) is neither injective nor surjective (Exercise 3(*b*)).

When E admits a finite basis (e_i), the inverse isomorphism θ^{-1} of θ can be found explicitly as follows. Let(e_i^*) be the dual basis of (e_i) (§ 2, no. 6); for all $u \in \mathrm{Hom}(E, F)$ and all $x = \sum_i \xi_i e_i$ with $\xi_i \in A$,

$$u(x) = \sum_i \xi_i u(e_i) = \sum_i \langle x, e_i^* \rangle u(e_i)$$

and therefore $u = \sum_i \theta(e_i^* \otimes u(e_i))$, in other words

$$\theta^{-1}(u) = \sum_i e_i^* \otimes u(e_i). \tag{12}$$

In particular, if further $F = E$, it is seen that the image under θ_E^{-1} of the identity mapping 1_E is the element $\sum_i e_i^* \otimes e_i$, which is therefore *independent* of the basis (e_i) considered in E.

Note on the other hand that when E is a finitely generated projective module the *ring* structure on $\mathrm{End}_A(E)$ can be transported by θ_E^{-1} to $E^* \otimes_A E$; it is immediately verified that, for x, y in E, x^*, y^* in E^*, in the ring $\mathrm{End}_A(E)$,

$$\theta_E(x^* \otimes x) \circ \theta_E(y^* \otimes y) = \theta_E((y^* \langle y, x^* \rangle) \otimes x). \tag{13}$$

Remark (2). Let E be a *right* A-module; replacing E by E^* in (11), we obtain a canonical **Z**-homomorphism

$$E^{**} \otimes_A F \to \mathrm{Hom}_A(E^*, F). \tag{14}$$

On the other hand, there is a canonical A-homomorphism $c_E : E \to E^{**}$, whence there is a **Z**-homomorphism $c_E \otimes 1_F : E \otimes_A F \to E^{**} \otimes_A F$; composing the latter with the homomorphism (14), we hence obtain a canonical **Z**-homomorphism

$$\theta' : E \otimes_A F \to \mathrm{Hom}_A(E^*, F) \tag{15}$$

such that $\theta'(x \otimes y)$ is the linear mapping

$$x^* \mapsto \langle x, x^* \rangle y.$$

If E *and* F are *projective* modules, the mapping (15) is *injective*. For c_E is then injective (§ 2, no. 7, Corollary 4 to Proposition 13) and as F is projective, the **Z**-homomorphism $c_E \otimes 1_F : E \otimes_A F \to E^{**} \otimes_A F$ is also injective (§ 3, no. 7, Corollary 6 to Proposition 7); finally, it has been seen (Proposition 2) that the homomorphism (14) is injective, whence the conclusion.

If E is *projective* and *finitely generated*, the mapping (15) is *bijective* for the two mappings of which it is composed are then bijective (§ 2, no. 7, Corollary 4 to Proposition 13 and Proposition 2 above).

3. TRACE OF AN ENDOMORPHISM

Let C be a *commutative* ring and E a C-module. The mapping $(x^*, x) \mapsto \langle x, x^* \rangle$ of $E^* \times E$ into C is then *C-bilinear*, since, for all $\gamma \in C$, $\langle \gamma x, x^* \rangle = \gamma \langle x, x^* \rangle$ and $\langle x, x^* \gamma \rangle = \langle x, x^* \rangle \gamma$; we derive a canonical *C-linear* mapping

$$\tau : E^* \otimes_C E \to C \tag{16}$$

such that $\tau(x^* \otimes x) = \langle x, x^* \rangle$ (§ 3, no. 5). Suppose now also that E is a *finitely generated projective* C-module; the canonical isomorphism (11) of no. 2 is then a C-*module* isomorphism and we can therefore define by transporting the structure a *canonical linear form* $\mathrm{Tr} = \tau \circ \theta_E^{-1}$ on the C-module $\mathrm{End}_C(E)$. For all $u \in \mathrm{End}_C(E)$ the scalar $\mathrm{Tr}(u)$ is called the *trace* of the endomorphism u; every $u \in \mathrm{End}_C(E)$ can be written (in general in an infinity of ways) in the form $x \mapsto \sum_i \langle x, x_i^* \rangle y_i$ where $x_i^* \in E^*$ and $y_i \in E$, by virtue of no. 2, Corollary to Proposition 2; then

$$\mathrm{Tr}(u) = \sum_i \langle y_i, x_i^* \rangle \quad \text{(cf. § 10, no. 11).} \tag{17}$$

By definition,

$$\mathrm{Tr}(u + v) = \mathrm{Tr}(u) + \mathrm{Tr}(v) \tag{18}$$

$$\mathrm{Tr}(\gamma u) = \gamma \mathrm{Tr}(u) \tag{19}$$

for u, v in $\mathrm{End}_C(E)$ and $\gamma \in C$. Moreover:

PROPOSITION 3. *Let* C *be a commutative ring,* E, F *two finitely generated projective* C-*modules and* $u : E \to F$ *and* $v : F \to E$ *two linear mappings; then*

$$\mathrm{Tr}(v \circ u) = \mathrm{Tr}(u \circ v). \tag{20}$$

The two mappings $(u, v) \mapsto \mathrm{Tr}(u \circ v)$, $(u, v) \mapsto \mathrm{Tr}(v \circ u)$ of

$$\mathrm{Hom}_C(E, F) \times \mathrm{Hom}_C(F, E)$$

into C are *C-bilinear*; it therefore suffices to verify (20) when u is of the form $x \mapsto \langle x, a^* \rangle b$ and v of the form $y \mapsto \langle y, b^* \rangle a$, with $a \in E$, $a^* \in E^*$, $b \in F$, $b^* \in F^*$. But then $v \circ u$ is the mapping $x \mapsto \langle x, a^* \rangle \langle b, b^* \rangle a$ and $u \circ v$ the mapping $y \mapsto \langle y, b^* \rangle \langle a, a^* \rangle b$. Formula (17) shows that the values of the two sides of (20) are equal to $\langle a, a^* \rangle \langle b, b^* \rangle$.

COROLLARY. *If* $u_1, \dots, u_p$ *are endomorphisms of* E, *then*

$$\mathrm{Tr}(u_1 \circ u_2 \circ \cdots \circ u_p) = \mathrm{Tr}(u_i \circ u_{i+1} \circ \cdots \circ u_p \circ u_1 \circ \cdots \circ u_{i-1})$$

for $1 \leqslant i \leqslant p$ ("invariance of the trace under cyclic permutation").

It suffices to apply (20) to the product

$$(u_1 \circ u_2 \circ \cdots \circ u_{i-1}) \circ (u_i \circ u_{i+1} \circ \cdots \circ u_p).$$

Note that on the other hand it is not necessarily true that

$$\mathrm{Tr}(u \circ v \circ w) = \mathrm{Tr}(u \circ w \circ v)$$

for three endomorphisms u, v, w of E.

4. THE HOMOMORPHISM $\mathrm{Hom}_C(E_1, F_1) \otimes_C \mathrm{Hom}_C(E_2, F_2) \to \mathrm{Hom}_C(E_1 \otimes_C E_2, F_1 \otimes_C F_2)$

Let C be a *commutative* ring and E_1, E_2, F_1, F_2 four C-modules; in § 3, no. 5, formula (13) we defined a canonical C-module homomorphism

$$(21) \qquad \lambda : \mathrm{Hom}(E_1, F_1) \otimes \mathrm{Hom}(E_2, F_2) \to \mathrm{Hom}(E_1 \otimes E_2, F_1 \otimes F_2).$$

PROPOSITION 4. *When one of the ordered pairs* (E_1, E_2), (E_1, F_1), (E_2, F_2) *consists of finitely generated projective* C*-modules, the canonical homomorphism* (21) *is bijective.*

It is obviously sufficient to perform the proof for the ordered pairs (E_1, F_1) and (E_1, E_2).

We consider first the case of the ordered pair (E_1, F_1); we fix E_2, F_1, F_2 and write for *every* C-module $T(E) = \mathrm{Hom}(E, F_1) \otimes_C \mathrm{Hom}(E_2, F_2)$ and $T'(E) = \mathrm{Hom}(E \otimes E_2, F_1 \otimes F_2)$ and, for every C-homomorphism $v : E \to E'$,

$$T(v) = \mathrm{Hom}(v, 1_{F_1}) \otimes 1_{\mathrm{Hom}(E_2, F_2)}$$

and

$$T'(v) = \mathrm{Hom}(v \otimes 1_{E_2}, 1_{F_1 \otimes F_2}).$$

Then *Lemmas* 4 *and* 5 (no. 2) (where ν is replaced by λ) *are valid* and are proved by completely analogous methods.

We next fix E_2 and F_2 and this time write, for every C-module F, $T(F) = \mathrm{Hom}(C, F) \otimes_C \mathrm{Hom}(E_2, F_2)$ and $T'(F) = \mathrm{Hom}(C \otimes E_2, F \otimes F_2)$ and, for every C-homomorphism $u : F \to F'$,

$$T(u) = \mathrm{Hom}(1_C, u) \otimes 1_{\mathrm{Hom}(E_2, F_2)}$$

and

$$T'(u) = \mathrm{Hom}(1_C \otimes 1_{E_2}, u \otimes 1_{F_2}).$$

This time it is immediately verified that *Lemmas* 1 *and* 2 (no. 2) (where λ always replaces ν) *are valid.*

This being so, we show the proposition first when $E_1 = C$ and F_1 is projective and finitely generated. The argument of no. 2 (which rests on Lemmas 1 and 2), together with the above remarks, reduces this to proving the proposition when also $F_1 = C$; then $\mathrm{Hom}(E_1, F_1)$, $E_1 \otimes E_2$ and $F_1 \otimes F_2$ are identified with C, E_2 and F_2 respectively (§ 3, no. 4, Proposition 4); the two sides of (21) are then both canonically identified with $\mathrm{Hom}(E_2, F_2)$ and, after these identifications, it is verified that λ becomes the identity.

We now suppose that F_1 is projective and finitely generated; the argument of no. 2 (depending this time on Lemmas 4 and 5) reduces the proof for E_1 any finitely generated projective module to the case where $E_1 = C$, that is to the first case dealt with.

For the ordered pair (E_1, E_2), the procedure is similar, this time applying Lemmas 4 and 5 twice; we leave the details to the reader.

Note that when $E_1 = C^{(I)}$, $E_2 = C^{(J)}$ are free (finitely generated or not), then $\mathrm{Hom}(E_1, F_1) = F_1^I$, $\mathrm{Hom}(E_2, F_2 = F_2^J$ and

$$\mathrm{Hom}(E_1 \otimes E_2, F_1 \otimes F_2) = (F_1 \otimes F_2)^{I \times J}$$

to within canonical isomorphisms and (21) is then identical with a special case of the canonical homomorphism (22) of § 3, no. 7.

When $E_2 = C$, the canonical homomorphism (21) gives, after identifying $\mathrm{Hom}(E_2, F_2)$ with F_2 and $E_1 \otimes E_2$ with E_1, a canonical homomorphism

$$\mathrm{Hom}(E, F) \otimes G \to \mathrm{Hom}(E, F \otimes G) \tag{22}$$

for any three C-modules E, F, G which is just the homomorphism (7) of no. 2 for $A = B = C$.

Note that when $F = C$ the canonical homomorphism (22) again gives (11) (no. 2) for the case of a commutative ring.

Suppose now that $F_1 = F_2 = C$; as $F_1 \otimes F_2$ is identified with C, there is this time a canonical homomorphism

$$\mu : E^* \otimes F^* \to (E \otimes F)^* \tag{23}$$

for two C-modules E, F; for $x^* \in E^*$, $y^* \in F^*$, the image of $x^* \otimes y^*$ under the canonical homomorphism (23) is the linear form u on $E \otimes F$ such that

$$u(x \otimes y) = \langle x, x^* \rangle \langle y, y^* \rangle. \tag{24}$$

Moreover, if E_1, E_2, F_1, F_2 are four C-modules, $f : E_1 \to E_2$, $g : F_1 \to F_2$ two linear mappings, it follows immediately from (24) that the diagram

$$\begin{array}{ccc} E_2^* \otimes F_2^* & \xrightarrow{\mu} & (E_2 \otimes F_2)^* \\ {\scriptstyle {}^t f \otimes {}^t g}\downarrow & & \downarrow{\scriptstyle {}^t(f \otimes g)} \\ E_1^* \otimes F_1^* & \xrightarrow[\mu]{} & (E_1 \otimes F_1)^* \end{array} \tag{25}$$

is *commutative*.

COROLLARY 1. *If one of the modules* E, F *is projective and finitely generated, the canonical homomorphism* (23) *is bijective.*

COROLLARY 2. *Let* E_1, E_2 *be two finitely generated projective* C-*modules*, u_1 *an endomorphism of* E_1 *and* u_2 *an endomorphism of* E_2; *then*

$$\mathrm{Tr}(u_1 \otimes u_2) = \mathrm{Tr}(u_1)\mathrm{Tr}(u_2). \tag{26}$$

By linearity, it suffices to consider the case where u_1 is of the form $x_1 \mapsto \langle x_1, x_1^* \rangle y_1$ and u_2 of the form $x_2 \mapsto \langle x_2, x_2^* \rangle y_2$; then the image of $x_1 \otimes x_2$ under $u_1 \otimes u_2$ is by definition

$$\langle x_1, x_1^* \rangle \langle x_2, x_2^* \rangle (y_1 \otimes y_2) = \langle x_1 \otimes x_2, x_1^* \otimes x_2^* \rangle (y_1 \otimes y_2)$$

$x_1^* \otimes x_2^*$ being canonically identified under μ with an element of $(E_1 \otimes E_2)^*$. As $\langle y_1 \otimes y_2, x_1^* \otimes x_2^* \rangle = \langle y_1, x_1^* \rangle \langle y_2, x_2^* \rangle$, formula (26) follows in this case from (17).

Remark. If E, F, G are any three C-modules, it is immediately verified that the diagram

$$\begin{array}{ccc} E^* \otimes F^* \otimes G^* & \xrightarrow{\mu \otimes 1} & (E \otimes F)^* \otimes G^* \\ {\scriptstyle 1 \otimes \mu}\downarrow & & \downarrow{\scriptstyle \mu} \\ E^* \otimes (F \otimes G)^* & \xrightarrow[\mu]{} & (E \otimes F \otimes G)^* \end{array} \tag{27}$$

is *commutative*, by virtue of formula (24).

We note also that, without any hypothesis on the C-modules E, F, there are canonical *isomorphisms*

$$(E \otimes F)^* \to \mathrm{Hom}(E, F^*) \tag{28}$$

$$(E \otimes F)^* \to \mathrm{Hom}(F, E^*) \tag{29}$$

which are just the isomorphism (6) and (5) of no. 1 for $G = C$, $A = B = C$.

Thus a canonical one-to-one correspondence has been defined between the *bilinear forms* on $E \times F$, the *homomorphisms of* E *into* F^* and the *homomorphisms of* F *into* E^*: if u (resp. v) is a homomorphism of E into F^* (resp. of F into E^*), the corresponding bilinear form is given by

$$(x, y) \mapsto \langle y, u(x) \rangle \quad (\text{resp. } (x, y) \mapsto \langle x, v(y) \rangle).$$

§ 5. EXTENSION OF THE RING OF SCALARS

1. EXTENSION OF THE RING OF SCALARS OF A MODULE

Let A, B be two rings and $\rho: A \to B$ a ring homomorphism; we consider the *right* A-*module* $\rho^*(B_d)$ defined by this homomorphism (§ 1, no. 13); this A-module also has a *left* B-*module* structure, namely that of B_s and, as $b'(b\rho(a)) = (b'b)\rho(a)$ for $a \in A$, b, b' in B, these two module structures on B are *compatible* (§ 1, no. 14). This allows us, for every *left* A-module E, to define a *left* B-*module* structure on the tensor product $\rho_*(B_d) \otimes_A E$ such that $\beta'(\beta \otimes x) = (\beta'\beta) \otimes x$ for β, β' in B and $x \in E$ (§ 3, no. 3). This left B-module is said to be *derived from* E *by extending the ring of scalars to* B *by means of* ρ and it is denoted by $\rho^*(E)$ or $E_{(B)}$ if no confusion arises.

PROPOSITION 1. *For every left* A-*module* E, *the mapping* $\phi: x \mapsto 1 \otimes x$ *of* E *into the* A-*module* $\rho_*(\rho^*(E))$ *is* A-*linear and the set* $\phi(E)$ *generates the* B-*module* $\rho^*(E)$. *Further, for every left* B-*module* F *and every* A-*linear mapping* f *of* E *into the* A-*module* $\rho_*(F)$, *there exists one and only one* B-*linear mapping* $\bar{f}$ *of* $\rho^*(E)$ *into* F *such that* $\bar{f}(1 \otimes x) = f(x)$ *for all* $x \in E$.

B can be considered as a (B, A)-bimodule by means of ρ; then there is a canonical **Z**-module isomorphism

$$\mathrm{Hom}_B(B \otimes_A E, F) \to \mathrm{Hom}_A(E, \mathrm{Hom}_B(B_s, F)) \tag{1}$$

as has been seen in § 4, no. 1, Proposition 1. But the left A-module $\mathrm{Hom}_B(B_s, F)$ is canonically identified with $\rho^*(F)$: for, by definition (§ 1, no. 14), there corresponds to an element $y \in F$ the homomorphism $\theta(y): B_s \to F$ such that $(\theta(y))(1) = y$; for all $\lambda \in A$, there thus corresponds to $\rho(\lambda)y \in F$ the homomorphism $\mu \mapsto \mu\rho(\lambda)y$ of B_s into F, which is just $\lambda\theta(y)$ for the left A-module structure on $\mathrm{Hom}_B(B_s, F)$ (§ 1, no. 14). Using this identification, we obtain therefore a *canonical* **Z**-module *isomrophism*, the inverse of (1)

$$\delta: \mathrm{Hom}_A(E, \rho_*(F)) \to \mathrm{Hom}_B(\rho^*(E), F) \tag{2}$$

and it follows immediately from the definitions that if $\delta(f) = \bar{f}$ then $\bar{f}(1 \otimes x) = f(x)$ for all $x \in E$. In particular, the mapping $\phi_E: x \mapsto 1 \otimes x$ is just

$$\phi_E = \delta^{-1}(1_{\rho^*(E)}). \tag{3}$$

Proposition 1 is therefore proved. The mapping $\phi_E: E \to \rho_*(\rho^*(E))$ is called *canonical.*

Remarks. (1) Proposition 1 shows that the ordered pair consisting of $E_{(B)}$ and ϕ_E is a solution of the *universal mapping problem* (*Set Theory*, IV, § 3, no. 1), where Σ is the species of left B-module structures (the morphisms being B-linear mappings) and the α-mappings are the A-linear mappings of E into a B-module.

(2) If E is an $(A, (C'_i); (D'_j))$-multimodule and F a $(B, (C''_h); (D''_k))$-multimodule, then the isomorphism (2) is linear with respect to the $((D'_j, (C''_h); (C'_i), (D''_k))$-multimodule structures of the two sides (§ 1, no. 14 and § 3, no. 4).

(3) Let E be a left A-module, $\mathfrak{a}$ a two-sided ideal of A and $\rho: A \to A/\mathfrak{a}$ the canonical homomorphism. In the notation of § 3, no. 6, Corollary 2 to Proposition 6, the A-module $E/\mathfrak{a}E$ is annihilated by $\mathfrak{a}$ and therefore has canonically a left $(A/\mathfrak{a})$-module structure (§ 1, no. 12); it is immediate that the canonical mapping $\pi: \rho^*(E) \to E/\mathfrak{a}E$ defined in § 3, no. 6, Corollary 2 to Proposition 6 is an isomorphism for the $(A/\mathfrak{a})$-module structures.

COROLLARY. *Let* E, E′ *be two left* A*-modules; for every* A*-linear mapping* $u: E \to E'$, $v = 1_B \otimes u$ *is the unique* B*-linear mapping which renders commutative the diagram*

$$\begin{array}{ccc} E & \xrightarrow{\phi_E} & E_{(B)} \\ {\scriptstyle u}\downarrow & & \downarrow{\scriptstyle v} \\ E' & \xrightarrow[\phi_{E'}]{} & E'_{(B)} \end{array}$$

where ϕ_E *and* $\phi_{E'}$ *are the canonical mappings.*

It suffices to apply Proposition 1 to the A-homomorphism $\phi_{E'} \circ u: E \to E'_{(B)}$.

The mapping v defined in the above corollary is denoted by $\rho^*(u)$ or $u_{(B)}$. If E″ is a third left A-module and $v: E' \to E''$ an A-linear mapping, it is immediate that

$$(v \circ u)_{(B)} = v_{(B)} \circ u_{(B)}.$$

Extending the ring of operators of a module is a *transitive* operation; to be precise:

PROPOSITION 2. *Let* $\rho: A \to B$, $\sigma: B \to C$ *be ring homomorphisms. For every left* A*-module* E, *there exists one and only one* C*-homomorphism*

$$\sigma^*(\rho^*(E)) \to (\sigma \circ \rho)^*(E) \tag{4}$$

mapping $1 \otimes (1 \otimes x)$ *to* $1 \otimes x$ *for all* $x \in E$ *and this homomorphism is bijective.*

The underlying **Z**-modules of $\sigma^*(\rho^*(E))$ and $(\sigma \circ \rho)^*(E)$ are respectively $C \otimes_B (B \otimes_A E)$ and $C \otimes_A E$. There exists a canonical **Z**-isomorphism $C \otimes_B (B \otimes_A E) \to (C \otimes_B B) \otimes_A E$ (§ 3, no. 8, Proposition 8), which is also a C-isomorphism for the left C-module structures on both sides. Moreover, the

C-module $C \otimes_B B$ is canonically identified with the C-module C_s under the isomorphism which maps $\gamma \otimes \beta$ to $\gamma\sigma(\beta)$ (§ 3, no. 4, Proposition 4) and this isomorphism is also an isomorphism for the right A-module structure on $C \otimes_B B$ defined by ρ and the right A-module structure on C defined by $\sigma \circ \rho$. Thus a canonical isomorphism

$$(C \otimes_B B) \otimes_A E \to C \otimes_A E$$

is obtained and, composing it with the isomorphism

$$C \otimes_B (B \otimes_A E) \to (C \otimes_B B) \otimes_A E$$

defined earlier, the desired canonical isomorphism is obtained.

If ϕ, ϕ' and ϕ'' denote the canonical mappings $E \to \rho^*(E)$, $\rho^*(E \to) \sigma^*(\rho^*(E))$ and $E \to (\sigma \circ \rho)^*(E)$, $\phi' \circ \phi$ is identified with ϕ'' under the canonical isomorphism of Proposition 2.

PROPOSITION 3. *Let* A, B *be two commutative rings,* $\rho: A \to B$ *a ring homomorphism and* E, E' *two* A*-modules. There exists one and only one* B*-homomorphism*

$$E_{(B)} \otimes_B E'_{(B)} \to (E \otimes_A E')_{(B)} \tag{5}$$

mapping $(1 \otimes x) \otimes (1 \otimes x')$ *to* $1 \otimes (x \otimes x')$ *for* $x \in E$, $x' \in E'$, *and this homomorphism is bijective.*

The left hand side of (5) may be written $(B \otimes_A E) \otimes_B (B \otimes_A E')$ and is identified with $(E \otimes_A B) \otimes_B (B \otimes_A E')$ since A and B are commutative; the latter product is identified successively with $E \otimes_A (B \otimes_B B) \otimes_A E'$, $E \otimes_A (B \otimes_A E')$, $E \otimes_A (E' \otimes_A B)$ and finally $(E \otimes_A E') \otimes_A B$, using the associativity of the tensor product (§ 3, no. 8, Proposition 8), Proposition 4, § 3, no. 4, and the commutativity of A and B. The desired isomorphism is the composition of the successive canonical isomorphisms.

Clearly if S is a generating system of E, the image of S under the canonical mapping $E \to E_{(B)}$ is a generating system of $E_{(B)}$; in particular, if E is a finitely generated A-module, $E_{(B)}$ is a finitely generated B-module.

PROPOSITION 4. *Let* E *be an* A*-module admitting a basis* $(a_\lambda)_{\lambda \in L}$; *if* $\phi: x \mapsto 1 \otimes x$ *is the canonical mapping of* E *into* $\rho^*(E)$, *then* $(\phi(a_\lambda))_{\lambda \in L}$ *is a basis of* $\rho^*(E)$. *If* ρ *is injective, so is* ϕ.

The first assertion follows immediately from § 3, no. 7, Corollary 1 to Proposition 7. Also, for every family $(\xi_\lambda)_{\lambda \in L}$ of elements of A of finite support, $\phi\left(\sum_{\lambda \in L} \xi_\lambda a_\lambda\right) = \sum_{\lambda \in L} \rho(\xi_\lambda)\phi(a_\lambda)$ and the relation $\phi\left(\sum_{\lambda \in L} \xi_\lambda a_\lambda\right) = 0$ is therefore equivalent to $\rho(\xi_\lambda) = 0$ for all $\lambda \in L$, whence the second assertion.

COROLLARY. *For every projective* A*-module* E, *the* B*-module* $\rho^*(E)$ *is projective. If further* ρ *is injective, the canonical mapping of* E *into* $\rho^*(E)$ *is injective.*

By hypothesis there exists a free A-module M containing E and in which E admits a supplement F. It follows immediately from § 3, no. 7, Proposition 7 that $M_{(B)}$ is identified with the direct sum of $E_{(B)}$ and $F_{(B)}$ and if ϕ and ψ are the canonical mappings $E \to E_{(B)}$ and $F \to F_{(B)}$, the canonical mapping $M \to M_{(B)}$ is just $x + y \mapsto \phi(x) + \psi(y)$. The corollary follows immediately from Proposition 4 applied to the A-module M.

When E is a *right* A-module, we write similarly $\rho^*(E) = E \otimes_A \rho_*(B_s)$, B being considered this time as an (A, B)-bimodule and the right B-module structure on $\rho^*(E)$ being such that $(x \otimes \beta)\beta' = x \otimes (\beta\beta')$ for $\beta \in B'$ $\beta' \in B$ and $x \in E$. We leave to the reader the task of stating for right modules the results corresponding to those of this no. and the following.

Remark (4). Consider the *left* A-module $\rho_*(B_s)$ defined by ρ and for every left A-module E, consider the **Z**-module

$$\tilde{\rho}(E) = \mathrm{Hom}_A(\rho_*(B_s), E). \tag{6}$$

As $\rho_*(B_s)$ has a *right* B-*module* structure, a *left* B-*module* structure is derived on $\tilde{\rho}(E)$ (§ 1, no. 14) such that, if $u \in \tilde{\rho}(E)$ and $b' \in B$, $b'u$ is the homomorphism $b \mapsto u(bb')$ of $\rho_*(B_s)$ into E. We further define an A-*linear* mapping, called *canonical*

$$\eta : \rho_*(\tilde{\rho}(E)) \to E \tag{7}$$

associating with every homomorphism $u \in \tilde{\rho}(E)$ the element $u(1)$ in E. As B can be considered as an (A, B)-bimodule by means of ρ, for every left B-module F, there is a canonical **Z**-module isomorphism

$$\mathrm{Hom}_A(\rho_*(B_s) \otimes_B F, E) \to \mathrm{Hom}_B(F, \mathrm{Hom}_A(\rho_*(B_s), E))$$

(§ 1, no. 1, Proposition 1). As the left A-module $\rho^*(B_s) \otimes_B F$ is canonically identified with $\rho_*(F)$ by virtue of § 3, no. 4, Proposition 4, we obtain a canonical **Z**-module isomorphism, the inverse of the above

$$\mathrm{Hom}_B(F, \tilde{\rho}(E)) \to \mathrm{Hom}_A(\rho_*(F), E) \tag{8}$$

which associates with every B-linear mapping g of F into $\tilde{\rho}(E)$ the composite mapping $\eta \circ g$, considered as an A-linear mapping of $\rho_*(F)$ into E. In particular, under the hypotheses of Proposition 2, if F is replaced by $\sigma_*(C_s)$, we obtain a canonical C–isomorphism

$$\tilde{\sigma}(\tilde{\rho}(E)) \to (\sigma \circ \rho)^{\sim}(E). \tag{9}$$

2. RELATIONS BETWEEN RESTRICTION AND EXTENSION OF THE RING OF SCALARS

Let $\rho : A \to B$ be a ring homomorphism. For every left A-module E, a canonical A-linear mapping

$$\phi_E : E \to \rho_*(\rho^*(E)) \tag{10}$$

was defined in no. 1 such that $\phi_E(x) = 1 \otimes x$. We consider now a left B-module F and apply Proposition 1 (no. 1) to the A homomorphism $1_{\rho_*(F)}: \rho_*(F) \to \rho_*(F)$: we obtain a B-linear mapping

$$\psi_F: \rho^*(\rho_*(F)) \to F \tag{11}$$

equal to $\delta(1_{\rho_*(F)})$ and such therefore that, for all $y \in F$ and all $\beta \in B$, $\psi_F(\beta \otimes y) = \beta y$.

Proposition 5. *Let E be a left A-module and F a left B-module; the composite mappings*

$$\rho^*(E) \xrightarrow[\rho^*(\phi_E)]{} \rho^*(\rho_*(\rho^*(E))) \xrightarrow[\psi_{\rho^*(E)}]{} \rho^*(E) \tag{12}$$

$$\rho_*(E) \xrightarrow[\phi_{\rho_*(F)}]{} \rho_*(\rho^*(\rho_*(F))) \xrightarrow[\rho_*(\psi_F)]{} \rho_*(F) \tag{13}$$

are respectively equal to the identity mappings of $\rho^*(E)$ *and* $\rho_*(F)$.

We give the proof, for example, for (12); for all $x \in E$, the mapping $\rho^*(\phi_E)$ maps $1 \otimes x$ to the element $1 \otimes (1 \otimes x)$ and the mapping $\psi_{\rho^*(E)}$ maps $1 \otimes (1 \otimes x)$ to the element $1 \otimes x$; the conclusion follows from the fact that the elements of the form $1 \otimes x$ generate the B-module $\rho^*(E)$; the proof is even simpler for (13).

Corollary. *The mappings* $\rho^*(\phi_E)$ *and* $\phi_{\rho^*(F)}$ *are injective and respectively identify* $\rho^*(E)$ *with a direct factor of* $\rho^*(\rho_*(\rho^*(E)))$ *and* $\rho_*(F)$ *with a direct factor of* $\rho_*(\rho^*(\rho_*(F)))$.

This is a consequence of Proposition 5 and § 1, no. 9, Corollary 2 to Proposition 15.

Proposition 6. *Let E be a left A-module and F a right B-module. There exists one and only one* **Z**-*homomorphism*

$$\rho_*(F) \otimes_A E \to F \otimes_B \rho^*(E) \tag{14}$$

mapping $y \otimes x$ *to* $y \otimes (1 \otimes x)$ *for all* $x \in E$ *and all* $y \in F$ *and this homomorphism is bijective.*

By definition the right hand side of (14) is $F \otimes_B (B \otimes_A E)$, where B is considered as a (B, A)-bimodule, and there is a canonical **Z**-isomorphism $(F \otimes_B B) \otimes_A E \to F \otimes_B (B \otimes_A E)$ defined in § 3, no. 8, Proposition 8; on the other hand, the canonical isomorphism $F \to F \otimes_B B$ of § 3, no. 4, Proposition 4 is an isomorphism for the right A-module structures on the two sides, defined by ρ. Whence the desired isomorphism.

When A and B are *commutative*, the isomorphism (14) is an A-*module* isomorphism

$$\rho_*(F) \otimes_A E \to \rho_*(F \otimes_B \rho^*(E)).$$

3. EXTENSION OF THE RING OF OPERATORS OF A HOMOMORPHISM MODULE

Let A be a *commutative* ring, B a ring, $\rho: A \to B$ a ring homomorphism and E, F two A-modules; as B is an (A, A)-bimodule (by means of ρ) and F can be considered as an (A, A)-bimodule, there are on the **Z**-module $B \otimes_A F$ *two* A-module structures, under which respectively $a(b \otimes y) = (\rho(a)b) \otimes y$ and $a(b \otimes y) = b \otimes (ay)$ for $a \in A$, $b \in B$, $y \in F$. We shall denote the two A-modules thus defined by G′ and G″; G′ is moreover just the A-module $\rho_*(\rho^*(F))$.

This being so, in the definition of the canonical homomorphism of § 4, no. 2, formula (7), we replace B by A, the B-module F by the ring B considered as an A-module by means of ρ and G by F considered as an (A, A)-bimodule; as A is commutative, we may write the *canonical* **Z**-*homomorphism* obtained as

$$(15) \qquad B \otimes_A \operatorname{Hom}_A(E, F) \to \operatorname{Hom}_A(E, G'').$$

On the other hand (no. 1, formula (2)), there is a *canonical* **Z**-*isomorphism*

$$(16) \qquad \operatorname{Hom}_A(E, G') = \operatorname{Hom}_A(E, \rho_*(\rho^*(F))) \to \operatorname{Hom}_B(\rho^*(E), \rho^*(F)).$$

Suppose now that $\rho(A)$ is contained in the *centre* of B, in which case ρ is also called a *central* homomorphism *(or ρ is said to define an A-*algebra* structure on B, cf. III, § 1, no. 3)*. Then the A-module structures of G′ and G″ are *identical* and composing the homomorphisms (16) and (15) we thus obtain a canonical **Z**-homomorphism

$$(17) \qquad \omega: B \otimes_A \operatorname{Hom}_A(E, F) \to \operatorname{Hom}_B(E_{(B)}, F_{(B)})$$

which is characterized by the fact that, for all $u \in \operatorname{Hom}_A(E, F)$ and all $b \in B$

$$(18) \qquad \omega(b \otimes u) = r_b \otimes u,$$

where r_b denotes right multiplication by b in B.

Moreover, the hypothesis that ω is a *central* homomorphism implies that $(bb')\rho(a) = b\rho(a))b'$ for b, b' in B and $a \in A$; in other words the *right* B-module structure of B_d is *compatible* with its A-module structure; it thus defines on $B \otimes_A \operatorname{Hom}_A(E, F)$ a *right* B-*module* structure (§ 3, no. 4) and also on $F_{(B)} = B \otimes_A F$, and finally, as the left and right B-module structures on $F_{(B)}$ are *compatible*, a *right* B-*module* structure is also obtained on $\operatorname{Hom}_B(E_{(B)}, F_{(B)})$ (§ 1, no. 14). Then it is immediately verified that (17) is a *right* B-*module homomorphism* for these structures.

PROPOSITION 7. *Let* A *be a commutative ring,* B *a ring,* $\rho: A \to B$ *a central homomorphism and* E, F *two* A-*modules.*

(i) *If* B *is a projective* (resp. *finitely generated projective*) A*-module, the homomorphism* (17) *is injective* (resp. *bijective*).

(ii) *If* E *is a finitely generated projective* A*-module, the homomorphism* (17) *is bijective.*

As (16) is bijective, the proposition follows from § 4, no. 2. Proposition 2, applied to the canonical homomorphism (15).

4. DUAL OF A MODULE OBTAINED BY EXTENSION OF SCALARS

Let A, B be two rings, $\rho: A \to B$ a ring homomorphism, E a left A-module and E^* its dual. We shall define a canonical B-*linear* mapping

$$\upsilon_E : (E^*)_{(B)} \to (E_{(B)})^*. \tag{19}$$

The left hand side of (19) may be written as $\mathrm{Hom}_A(E, A) \otimes_A \rho_*(B_s)$, where, in $\mathrm{Hom}_A(E, A)$, A is considered as an (A, A)-bimodule. Then there is a canonical **Z**-homomorphism (§ 4, no. 2, formula (7))

$$\nu : \mathrm{Hom}_A(E, A) \otimes_A \rho_*(B_s) \to \mathrm{Hom}_A(E, A \otimes_A \rho_*(B_s)) = \mathrm{Hom}_A(E, \rho_*(B_s))$$

with the identification given by the canonical isomorphism of § 3, no. 4, Proposition 4. On the other hand, the right hand side of (19) may be written as $\mathrm{Hom}_B(\rho_*(B_d) \otimes_A E, B_s)$; as B is a (B, A)-bimodule, there is a canonical **Z**-isomorphism (§ 4, no. 1, Proposition 1)

$$\beta : \mathrm{Hom}_B(\rho_*(B_d) \otimes_A E, B_s) \to \mathrm{Hom}_A(E, \mathrm{Hom}_B(B_s, B_s))$$

and $\mathrm{Hom}_B(B_s, B_s)$ is canonically identified, as an A-module, with $\rho_*(B_s)$ (see the proof of no. 1, Proposition 1). Taking account of these identifications, we obtain the homomorphism υ_E; it is easily verified that this homomorphism is characterized by the equation

$$\langle \xi \otimes x, \upsilon_E(x^* \otimes \eta) \rangle = \xi \rho(\langle x, x^* \rangle)\eta, \tag{20}$$

for $x \in E$, $x^* \in E^*$, ξ, η in B, which shows immediately that υ_E is B-*linear*. Moreover, for every A-linear mapping $u: E \to F$ the diagram

$$\begin{array}{ccc} (F^*)_{(B)} & \xrightarrow{\upsilon_F} & (F_{(B)})^* \\ \downarrow {\scriptstyle ({}^t u)_{(B)}} & & \downarrow {\scriptstyle {}^t(u_{(B)})} \\ (E^*)_{(B)} & \xrightarrow[\upsilon_E]{} & (E_{(B)})^* \end{array}$$

is commutative.

PROPOSITION 8. *If one of the* A*-modules* E, $\rho_*(B_s)$ *is projective and finitely generated, the homomorphism* υ_E *is bijective.*

This follows from the above and § 4, no. 2, Proposition 2.

Suppose in particular that E is a *finitely generated free* A-module and let $(e_i)_{1 \leqslant i \leqslant n}$ be a basis of E and (e_i^*) the dual basis; then the canonical isomorphism (19) maps the basis $(e_i^* \otimes 1)$ of $(E^*)_{(B)}$ to the dual basis of the basis $(1 \otimes e_i)$ of $E_{(B)}$.

5. A CRITERION FOR FINITENESS

Proposition 9. *Let* B *be a ring,* A *a subring of* B *and* P *a projective left* A*-module. Then, if* $P_{(B)}$ *is a finitely generated* B*-module,* P *is itself a finitely generated* A*-module.*

We know (§ 2, no. 6, Proposition 12) that there exists a family $(a_\lambda)_{\lambda \in L}$ of elements of P and a family $(a_\lambda^*)_{\lambda \in L}$ of elements of the dual P^* such that, for all $x \in P$, the family $(\langle x, a_\lambda^* \rangle)$ is of finite support and $x = \sum_\lambda \langle x, a_\lambda^* \rangle a_\lambda$. Since $P_{(B)}$ is finitely generated, there exists a finite family $(y_\iota)_{\iota \in I}$ of elements of P such that $P_{(B)}$ is generated by the elements $1 \otimes y_\iota$. For each index ι, the family $(\langle y_\iota, a_\lambda^* \rangle)$ has finite support. Hence there exists a finite subset H of L such that $\langle y_\iota, a_\lambda^* \rangle = 0$ for $\iota \in I$ and $\lambda \notin H$. Since

$$\langle 1 \otimes y_\iota, 1d_B \otimes a_\lambda^* \rangle = \langle y_\iota, a_\lambda^* \rangle,$$

it follows that $1d_B \otimes a_\lambda^* = 0$ for $\lambda \notin H$. Hence, for all $x \in P$,

$$\langle x, a_\lambda^* \rangle = \langle 1 \otimes x, 1_B \otimes a_\lambda^* \rangle = 0$$

for $\lambda \notin H$. This shows that the A-module P is generated by the a_λ such that $\lambda \in H$.

§ 6. INVERSE AND DIRECT LIMITS OF MODULES

Throughout this paragraph, I *will denote a non-empty preordered set and* $\alpha \leqslant \beta$ *the preorder relation on* I*. Unless otherwise mentioned, the inverse and direct systems have indexing set* I.

1. INVERSE LIMITS OF MODULES

Let $(A_\alpha, \phi_{\alpha\beta})$ be an inverse system of rings (I, § 10, no. 1), $(E_\alpha, f_{\alpha\beta})$ an inverse system of commutative groups (written additively) (I, § 10, no. 1) and suppose that each E_α has a *left* A_α*-module* structure; moreover suppose that for $\alpha \leqslant \beta$ $(f_{\alpha\beta}, \phi_{\alpha\beta})$ is a *dimorphism* of E_β into E_α (§ 1, no. 13), in other words that

$$f_{\alpha\beta}(\lambda_\beta x_\beta) = \phi_{\alpha\beta}(\lambda_\beta) f_{\alpha\beta}(x_\beta), \tag{1}$$

for $x_\beta \in E_\beta$, $\lambda_\beta \in A_\beta$; then it follows from I, § 10, no. 2 that $E = \varprojlim E_\alpha$ has a *left module* structure over $A = \varprojlim A_\alpha$. For all $\alpha \in I$, let $f : E \to E_\alpha$, $\phi_\alpha : A \to A_\alpha$ be the canonical mappings; then (f_α, ϕ_α) is a *dimorphism* of E into E_α. We shall say that $(E_\alpha, f_{\alpha\beta})$ is an *inverse system of left* A_α-*modules* and that the A-module E is its *inverse limit.*

Let $(E'_\alpha, f'_{\alpha\beta})$ be another inverse system of left A_α-modules and, for all α, let $u_\alpha : E'_\alpha \to E_\alpha$ be an A_α-*linear* mapping, these mappings forming an *inverse system*; then $u = \varprojlim u_\alpha$ is an A-*linear* mapping of $\varprojlim E'_\alpha$ into $\varprojlim E_\alpha$.

Moreover:

PROPOSITION 1. *Let* $(E_\alpha, f_{\alpha\beta})$, $(E'_\alpha, f'_{\alpha\beta})$, $(E''_\alpha, f''_{\alpha\beta})$ *be three inverse systems of* A_α-*modules and* (u_α), (v_α) *two inverse systems of* A_α-*linear mappings such that the sequences*

$$0 \longrightarrow E'_\alpha \xrightarrow{u_\alpha} E_\alpha \xrightarrow{v_\alpha} E''_\alpha$$

are exact for all α. *Then, writing* $u = \varprojlim u_\alpha$, $v = \varprojlim v_\alpha$, *the sequence*

$$0 \longrightarrow \varprojlim E'_\alpha \xrightarrow{u} \varprojlim E_\alpha \xrightarrow{v} \varprojlim E''_\alpha$$

is exact.

As $u_\alpha^{-1}(0) = \{0\}$ for all α, it follows from *Set Theory*, III, § 7, no. 2, Proposition 2 that $u^{-1}(0) = \{0\}$, hence u is injective; further, the $u_\alpha(E'_\alpha)$ form an inverse system of subsets of the E_α and thus $u(\varprojlim E'_\alpha) = \varprojlim u_\alpha(E'_\alpha)$. As $u_\alpha(E'_\alpha) = v_\alpha^{-1}(0)$ by hypothesis, $v^{-1}(0) = \varprojlim u_\alpha(E'_\alpha) = u(\varprojlim E'_\alpha)$ (*Set Theory*, III, § 7, no. 2, Proposition 2), which completes the proof.

Remarks. (1) Proposition 1 and its proof are valid for arbitrary *groups*, except for change of notation.

(2) Note that if there are exact sequences

$$0 \longrightarrow E'_\alpha \xrightarrow{u_\alpha} E_\alpha \xrightarrow{v_\alpha} E''_\alpha \longrightarrow 0$$

it does not necessarily follow that the sequence

$$0 \longrightarrow \varprojlim E'_\alpha \xrightarrow{u} \varprojlim E_\alpha \xrightarrow{v} \varprojlim E''_\alpha \longrightarrow 0$$

is exact; in other words, the inverse limit of an inverse system of surjective linear mappings is not necessarily surjective (cf. Exercise 1).

Suppose now that the A_α are equal to the *same ring* A and the $\phi_{\alpha\beta}$ to 1_A; then for every inverse system $(E_\alpha, f_{\alpha\beta})$ of A-modules, $E = \varprojlim E_\alpha$ is an A-module. Let F be an A-module and, for all α, let $u_\alpha : F \to E_\alpha$ be an A-linear mapping such that (u_α) is an inverse system of mappings; then $u = \varprojlim u_\alpha$ is an A-linear mapping of F into $\varprojlim E_\alpha$. *Conversely*, for every A-linear mapping $v : F \to \varprojlim E_\alpha$, the family of $v_\alpha = f_\alpha \circ v$ is an inverse system of A-linear mappings such that $v = \varprojlim v_\alpha$. We note on the other hand that for $\alpha \leqslant \beta$ the mapping

$$\mathrm{Hom}(1_F, f_{\alpha\beta}) = \bar{f}_{\alpha\beta} : \mathrm{Hom}_A(F, E_\beta) \to \mathrm{Hom}_A(F, E_\alpha)$$

is a **Z**-module homomorphism such that $(\mathrm{Hom}_A(F, E_\alpha), \bar{f}_{\alpha\beta})$ is an *inverse system of* **Z**-*modules*; as $\bar{f}_{\alpha\beta}(v_\beta) = f_{\alpha\beta} \circ v_\beta$, the above remarks can therefore be expressed as follows:

PROPOSITION 2. *For every inverse system* $(E_\alpha, f_{\alpha\beta})$ *of* A-*modules and every* A-*module* F, *the canonical mapping* $u \mapsto (f_\alpha \circ u)$ *is a* **Z**-*module isomorphism*

$$l_F : \mathrm{Hom}_A(F, \varprojlim E_\alpha) \to \varprojlim \mathrm{Hom}_A(F, E_\alpha). \tag{2}$$

COROLLARY. *For every* A-*module homomorphism* $v : F \to F'$, *the*

$$\bar{v}_\alpha = \mathrm{Hom}(v, 1_{E_\alpha}) : \mathrm{Hom}(F', E_\alpha) \to \mathrm{Hom}(F, E_\alpha)$$

form an inverse system of **Z**-*linear mappings and the diagram*

$$\begin{array}{ccc} \mathrm{Hom}(F', \varprojlim E_\alpha) & \xrightarrow{l_{F'}} & \varprojlim \mathrm{Hom}(F', E_\alpha) \\ {\scriptstyle \mathrm{Hom}(v, 1_E)} \downarrow & & \downarrow {\scriptstyle \varprojlim \bar{v}_\alpha} \\ \mathrm{Hom}(F, \varprojlim E_\alpha) & \xrightarrow[l_F]{} & \varprojlim \mathrm{Hom}(F, E_\alpha) \end{array} \tag{3}$$

is commutative.

For all $u \in \mathrm{Hom}(F', \varprojlim E_\alpha)$, $l_F(u \circ v) = (f_\alpha \circ u \circ v)$ by definition and the commutativity of diagram (3) follows immediately from the definitions.

2. DIRECT LIMITS OF MODULES

Henceforth I *is assumed to be right directed.*

Let $(A_\alpha, \phi_{\beta\alpha})$ be a direct system of rings (I, § 10, no. 3), $(E_\alpha, f_{\beta\alpha})$ a direct system of commutative groups (written additively) (I, § 10, no. 3) and suppose that each E_α has a *left* A_α-*module* structure; further, suppose that, for $\alpha \leqslant \beta$, $(f_{\beta\alpha}, \phi_{\beta\alpha})$ is a *dimorphism* of E_α into E_β (§ 1, no. 13), in other words that

$$f_{\beta\alpha}(\lambda_\alpha x_\alpha) = \phi_{\beta\alpha}(\lambda_\alpha) f_{\beta\alpha}(x_\alpha) \tag{4}$$

for $x_\alpha \in E_\alpha$, $\lambda_\alpha \in A_\alpha$; then $E = \varinjlim E_\alpha$ has a *left module* structure over $A = \varinjlim A_\alpha$ (I, § 10, no. 4). For all $\alpha \in I$, let $f_\alpha : E_\alpha \to E$, $\phi_\alpha : A_\alpha \to A$ be the canonical mappings; then (f_α, ϕ_α) is a *dimorphism* of E_α into E. We shall say that $(E_\alpha, f_{\beta\alpha})$ is a *direct system of left* A_α-*modules* and that the A-module E is its *direct limit.*

Let $(E'_\alpha, f'_{\beta\alpha})$ be another direct system of left A_α-modules and, for all α, let $u_\alpha : E'_\alpha \to E_\alpha$ be an A_α-*linear* mapping, these mappings forming a *direct system*; then $u = \varinjlim u_\alpha$ is an A-*linear* mapping of $\varinjlim E'_\alpha$ into $\varinjlim E_\alpha$.

Moreover:

PROPOSITION 3. *Let* $(E_\alpha, f_{\beta\alpha})$, $(E'_\alpha, f'_{\beta\alpha})$, $(E''_\alpha, f''_{\beta\alpha})$ *be three direct systems of* A_α-*modules and* (u_α), (v_α) *two direct systems of* A_α-*linear mappings such that the sequences*

$$E'_\alpha \xrightarrow{u_\alpha} E_\alpha \xrightarrow{v_\alpha} E''_\alpha$$

are exact for all α. *Then, writing* $u = \varinjlim u_\alpha$, $v = \varinjlim v_\alpha$, *the sequence*

$$\varinjlim E'_\alpha \xrightarrow{u} \varinjlim E_\alpha \xrightarrow{v} \varinjlim E''_\alpha$$

is exact.

$u(\varinjlim E'_\alpha) = \varinjlim u_\alpha(E'_\alpha)$ and $v^{-1}(0) = \varinjlim v_\alpha^{-1}(0)$ (*Set Theory*, III, § 7, no. 6, Corollary to Proposition 7).

Loosely speaking, Proposition 3 can also be expressed by saying that *passing to the direct limit preserves exactness.*

PROPOSITION 4. *Let* $(E_\alpha, f_{\beta\alpha})$ *be a direct system of* A_α-*modules,* $E = \varinjlim E_\alpha$ *its direct limit and* $\phi_\alpha : A_\alpha \to A$ *and* $f_\alpha : E_\alpha \to E$ *the canonical mappings for all* $\alpha \in I$. *If, for all* $\alpha \in I$, S_α *is a generating system of* E_α, *then* $S = \bigcup_{\alpha \in I} f_\alpha(S_\alpha)$ *is a generating system of* E.

Every $x \in E$ is of the form $f_\alpha(x_\alpha)$ for some $\alpha \in I$ and some $x_\alpha \in E_\alpha$ and by hypothesis $x = \sum_i \lambda_\alpha^{(i)} y_\alpha^{(i)}$, where $\lambda_\alpha^{(i)} \in A_\alpha$ and $y_\alpha^{(i)} \in S_\alpha$; writing $\lambda^{(i)} = \phi_\alpha(\lambda_\alpha^{(i)})$, $y^{(i)} = f_\alpha(y_\alpha^{(i)})$, we obtain $x = \sum_i \lambda^{(i)} y^{(i)}$.

PROPOSITION 5. *With the hypotheses and notation of Proposition 4, suppose that for all* $\alpha \in I$, E_α *is the direct sum of a family* $(M_\alpha^\lambda)_{\alpha \in L}$ *of submodules (the indexing set* L *being independent of* α*) and that* $f_{\beta\alpha}(M_\alpha^\lambda) \subset M_\alpha^\lambda$ *for* $\alpha \leqslant \beta$ *and for all* $\lambda \in L$. *Then* E *is the direct sum of the submodules* $M^\lambda = \varinjlim_\alpha M_\alpha^\lambda$ $(\lambda \in L)$.

It follows from Proposition 4 that E is the sum of the M^λ. Let $(y^\lambda)_{\lambda \in L}$ be a family such that $y^\lambda \in M^\lambda$ for all $\lambda \in L$ and whose support is finite and suppose that $\sum_\lambda y^\lambda = 0$. By virtue of *Set Theory*, III, § 7, no. 5, Lemma 1, there exist an $\alpha \in I$ and a family $(x_\alpha^\lambda)_{\alpha \in L}$ of finite support consisting of elements of E_α such that $x_\alpha^\lambda \in M_\alpha^\lambda$ and $y^\lambda = f_\alpha(x_\alpha^\lambda)$ for all $\lambda \in L$. The relation $f_\alpha\left(\sum_{\lambda \in L} x_\alpha^\lambda\right) = 0$ implies the existence of a $\beta \geqslant \alpha$ such that $f_{\beta\alpha}\left(\sum_{\lambda \in L} x_\alpha^\lambda\right) = 0$ (*Set Theory*, III,

§ 7, no. 5, Lemma 1), which may be written as $\sum_{\lambda \in L} x_\beta^\lambda = 0$, where $x_\beta^\lambda = f_{\beta\alpha}(x_\alpha^\lambda) \in M_\beta^\lambda$ by hypothesis; hence $x_\beta^\lambda = 0$ for all $\lambda \in L$ and therefore $y^\lambda = f_\beta(x_\beta^\lambda) = 0$ for all $\lambda \in L$, which proves that the sum of the M^λ is direct.

COROLLARY. *Let* (P_α) *be a direct system of subsets of* E_α *and let* $P = \varinjlim P_\alpha$. *If, for all* $\alpha \in I$, P_α *is a free subset* (resp. *basis*) *of* E_α, *then* P *is a free subset* (resp. *basis*) *of* E.

The second assertion follows immediately from the first and Proposition 4. It is therefore sufficient to prove that if the P_α are free every subset $\{y^{(i)}\}_{1 \leqslant i \leqslant n}$ consisting of distinct elements of P, is free. There exists an $\alpha \in I$ and elements $x_\alpha^{(i)} \in P_\alpha$ such that $y^{(i)} = f_\alpha(x_\alpha^{(i)})$ for $1 \leqslant i \leqslant n$ (*Set Theory*, III, § 7, no. 5, Lemma 1); if $\sum_i \lambda^{(i)} y^{(i)} = 0$, it may be assumed that $\lambda^{(i)} = \phi_\alpha(\lambda_\alpha^{(i)})$ for $1 \leqslant i \leqslant n$ and hence $f_\alpha\left(\sum_i \lambda_\beta^{(i)} x_\beta^{(i)}\right) = 0$; this implies $\sum_i \lambda_\beta^{(i)} x_\beta^{(i)} = 0$ for some $\beta \geqslant \alpha$, where $\lambda_\beta^{(i)} = \phi_{\beta\alpha}(\lambda_\alpha^{(i)})$, $x_\beta^{(i)} = f_{\beta\alpha}(x_\alpha^{(i)})$ and the $x_\beta^{(i)}$ belong to P_β and are distinct since $y^{(i)} = f_\beta(x_\beta^{(i)})$; then $\lambda_\beta^{(i)} = 0$ for $1 \leqslant i \leqslant n$, whence

$$\lambda^{(i)} = \phi_\beta(\lambda_\beta^{(i)}) = 0$$

for $1 \leqslant i \leqslant n$.

Suppose now that all the rings A_α are equal to the *same ring* A and the $\phi_{\beta\alpha}$ to 1_A; then, for every direct system $(E_\alpha, f_{\beta\alpha})$ of A-modules, $E = \varinjlim E_\alpha$ is an A-module. Let F be an A-module and for all α let $u_\alpha : E_\alpha \to F$ be an A-linear mapping such that (u_α) is a direct system of mappings; then $u = \varinjlim u_\alpha$ is an A-linear mapping of E into F. *Conversely*, for every A-linear mapping $v : \varinjlim E_\alpha \to F$, the family of $v_\alpha = v \circ f_\alpha$ is a direct system of A-linear mappings such that $v = \varinjlim v_\alpha$. On the other hand we note that for $\alpha \leqslant \beta$ the mapping

$$\mathrm{Hom}(f_{\beta\alpha}, 1_F) = \bar{f}_{\alpha\beta} : \mathrm{Hom}_A(E_\beta, F) \to \mathrm{Hom}_A(E_\alpha, F)$$

is a **Z**-module homomorphism such that $(\mathrm{Hom}_A(E_\alpha, F), \bar{f}_{\alpha\beta})$ is an *inverse system of* **Z**-*modules*; as $\bar{f}_{\alpha\beta}(v_\beta) = v_\beta \circ f_{\beta\alpha}$, the above remarks can be expressed as follows:

PROPOSITION 6. *For every direct system* $(E_\alpha, f_{\beta\alpha})$ *of* A-*modules and every* A-*module* F, *the canonical mapping* $u \mapsto (u \circ f_\alpha)$ *is a* **Z**-*module isomorphism*

$$d_F : \mathrm{Hom}_A(\varinjlim E_\alpha, F) \to \varprojlim \mathrm{Hom}_A(E_\alpha, F). \tag{5}$$

COROLLARY 1. *For every* A-*module homomorphism* $v : F \to F'$, *the*

$$\bar{v}_\alpha = \mathrm{Hom}(1_{E_\alpha}, v) : \mathrm{Hom}(E_\alpha, F) \to \mathrm{Hom}(E_\alpha, F')$$

form an inverse system of **Z**-*linear mappings and the diagram*

$$
\text{(6)}\qquad
\begin{array}{ccc}
\mathrm{Hom}(\varinjlim \mathrm{E}_\alpha, \mathrm{F}) & \xrightarrow{d_\mathrm{F}} & \varprojlim \mathrm{Hom}(\mathrm{E}_\alpha, \mathrm{F}) \\
\Big\downarrow{\scriptstyle \mathrm{Hom}(1_\mathrm{E}, v)} & & \Big\downarrow{\scriptstyle \varprojlim \bar{v}_\alpha} \\
\mathrm{Hom}(\varinjlim \mathrm{E}_\alpha, \mathrm{F}') & \xrightarrow[d_{\mathrm{F}'}]{} & \varprojlim \mathrm{Hom}(\mathrm{E}_\alpha, \mathrm{F}')
\end{array}
$$

is commutative.

For all $u \in \mathrm{Hom}(\varinjlim \mathrm{E}_\alpha, \mathrm{F})$, $d_{\mathrm{F}'}(v \circ u) = (v \circ u \circ f_\alpha)$ by definition and the commutativity of diagram (6) then follows immediately from the definitions.

Corollary 2. *If* $(\mathrm{E}_\alpha, f_{\beta\alpha})$ *is a direct system of left* A-*modules and* $\mathrm{E} = \varinjlim \mathrm{E}_\alpha$, $(\mathrm{E}_\alpha^*, {}^t f_{\beta\alpha})$ *is an inverse system of right* A-*modules and* $\varprojlim \mathrm{E}_\alpha^*$ *is canonically isomorphic to* E^*.

Remark. Let E be an A-module and $(\mathrm{M}_\alpha)_{\alpha \in \mathrm{I}}$ an increasing family of submodules of E such that E is the *union* of the M_α; if $j_{\beta\alpha}: \mathrm{M}_\alpha \to \mathrm{M}_\beta$ (for $\alpha \leqslant \beta$) and $j_\alpha: \mathrm{M}_\alpha \to \mathrm{E}$ are the canonical injections, it is immediate that $j = \varinjlim j_\alpha$ is an isomorphism of $\varinjlim \mathrm{M}_\alpha$ onto E (*Set Theory*, III, § 7, no. 6, *Remark* 1). In particular, every A-module is the direct limit of the right directed family of its *finitely generated* submodules.

3. TENSOR PRODUCT OF DIRECT LIMITS

Let $(\mathrm{A}_\alpha, \rho_{\beta\alpha})$ be a direct system of rings and $(\mathrm{E}_\alpha, f_{\beta\alpha})$ (resp. $(\mathrm{F}_\alpha, g_{\beta\alpha})$) be a direct system of right (resp. left) A_α-modules. For $\alpha \leqslant \beta$, there is a **Z**-module homomorphism

$$f_{\beta\alpha} \otimes g_{\beta\alpha}: \mathrm{E}_\alpha \otimes_{\mathrm{A}_\alpha} \mathrm{F}_\alpha \to (\mathrm{E}_\beta)_{[\mathrm{A}_\alpha]} \otimes_{\mathrm{A}_\alpha} (\mathrm{F}_\beta)_{[\mathrm{A}_\alpha]}$$

and on the other hand there is a canonical **Z**-module homomorphism

$$(\mathrm{E}_\beta)_{[\mathrm{A}_\alpha]} \otimes_{\mathrm{A}_\alpha} (\mathrm{F}_\beta)_{[\mathrm{A}_\alpha]} \to \mathrm{E}_\beta \otimes_{\mathrm{A}_\beta} \mathrm{F}_\beta$$

corresponding to the ring homomorphism $\rho_{\beta\alpha}$ (§ 3, no. 3, Proposition 2); whence by composition we obtain a **Z**-module homomorphism

$$h_{\beta\alpha}: \mathrm{E}_\alpha \otimes_{\mathrm{A}_\alpha} \mathrm{F}_\alpha \to \mathrm{E}_\beta \otimes_{\mathrm{A}_\beta} \mathrm{F}_\beta$$

which maps the tensor product $x_\alpha \otimes y_\alpha$ to $f_{\beta\alpha}(x_\alpha) \otimes g_{\beta\alpha}(y_\alpha)$. Clearly

$$(\mathrm{E}_\alpha \otimes_{\mathrm{A}_\alpha} \mathrm{F}_\alpha, h_{\beta\alpha})$$

is a *direct system* of **Z**-modules. Let $\mathrm{A} = \varinjlim \mathrm{A}_\alpha$, $\mathrm{E} = \varinjlim \mathrm{E}_\alpha$, $\mathrm{F} = \varinjlim \mathrm{F}_\alpha$ and let $\rho_\alpha : A_\alpha \to A$, $f_\alpha : E_\alpha \to E$, $g_\alpha : F_\alpha \to F$ be the canonical mappings. As above, a **Z**-linear mapping $\pi_\alpha : E_\alpha \otimes_{A_\alpha} F_\alpha \to E \otimes_A F$ is defined, which maps the tensor product $x_\alpha \otimes y_\alpha$ to $f_\alpha(x_\alpha) \otimes g_\alpha(y_\alpha)$, and it is immediate that these mappings form a direct system. Thus we obtain a **Z**-linear mapping

$$\pi = \varinjlim \pi_\alpha : \varinjlim (E_\alpha \otimes_{A_\alpha} F_\alpha) \to E \otimes_A F. \tag{7}$$

PROPOSITION 7. *The* **Z**-*linear mapping* (7) *is bijective.*

We set $P = \varinjlim(E_\alpha \otimes_{A_\alpha} F_\alpha)$ and, for all $\alpha \in I$, let $h_\alpha : E_\alpha \otimes_{A_\alpha} F_\alpha \to P$ be the canonical mapping. On the other hand, for all $\alpha \in I$, let

$$t_\alpha : E_\alpha \times F_\alpha \to E_\alpha \otimes_{A_\alpha} F_\alpha$$

be the canonical **Z**-bilinear mapping; for $\alpha \leqslant \beta$,

$$t_\beta(f_{\beta\alpha}(x_\alpha), g_{\beta\alpha}(y_\alpha)) = f_{\beta\alpha}(x_\alpha) \otimes g_{\beta\alpha}(y_\alpha) = h_{\beta\alpha}(t_\alpha(x_\alpha, y_\alpha))$$

and hence (t_α) is a direct system of mappings. Canonically identifying $\varinjlim(E_\alpha \times F_\alpha)$ with $E \times F$ (*Set Theory*, III, § 7, no. 7, Proposition 10), we derive a mapping $t = \varinjlim t_\alpha : E \times F \to P$ such that

$$t(f_\alpha(x_\alpha), g_\alpha(y_\alpha)) = h_\alpha(t_\alpha(x_\alpha, y_\alpha)) = h_\alpha(x_\alpha \otimes y_\alpha).$$

Taking account of *Set Theory*, III, § 7, no. 5, Lemma 1, it is immediately seen that t is **Z**-bilinear; moreover, for $x \in E$, $y \in F$, $\lambda \in A$, there exists $\alpha \in I$ such that $x = f_\alpha(x_\alpha)$, $y = g_\alpha(y_\alpha)$, $\lambda = \rho_\alpha(\lambda_\alpha)$ with $\lambda_\alpha \in A_\alpha$, $x_\alpha \in E_\alpha$, $y_\alpha \in F_\alpha$ (*Set Theory*, III, § 3, no. 7, Lemma 1); whence

$$t(x\lambda, y) = h_\alpha((x_\alpha\lambda_\alpha) \otimes y_\alpha) = h_\alpha(x_\alpha \otimes (\lambda_\alpha y_\alpha)) = t(x, \lambda y).$$

Hence there exists one and only one **Z**-*linear* mapping $\pi' : E \otimes_A F \to P$ such that $\pi'(x \otimes y) = t(x, y)$ (§ 3, no. 1, Proposition 1). Moreover, by definition

$$\pi'(\pi(h_\alpha(x_\alpha \otimes y_\alpha))) = \pi'(f_\alpha(x_\alpha) \otimes g_\alpha(y_\alpha)) = h_\alpha(x_\alpha \otimes y_\alpha)$$

$$\pi(\pi'(f_\alpha(x_\alpha) \otimes g_\alpha(y_\alpha))) = \pi(h_\alpha(x_\alpha \otimes y_\alpha)) = f_\alpha(x_\alpha) \otimes g_\alpha(y_\alpha)$$

and as the elements of the form $f_\alpha(x_\alpha) \otimes g_\alpha(y_\alpha)$ (resp. $h_\alpha(x_\alpha \otimes y_\alpha)$) generate the **Z**-module $E \otimes_A F$ (resp. P), $\pi' \circ \pi$ and $\pi \circ \pi'$ are the identity mappings.

Loosely speaking, Proposition 7 may be expressed by saying that *tensor products commute with direct limits* and usually the two sides of (7) are identified by means of the isomorphism π.

COROLLARY 1. *Let* $(E'_\alpha, f'_{\beta\alpha})$ (resp. $(F'_\alpha, g'_{\alpha\beta})$) *be another direct system of right* (resp. *left*) A_α-*modules; for all* $\alpha \in I$, *let* $u_\alpha : E_\alpha \to E'_\alpha$ (resp. $v_\alpha : F_\alpha \to F'_\alpha$) *be an*

A_α-linear mapping such that (u_α) (resp. (v_α)) is a direct system. Then $(u_\alpha \oplus v_\alpha)$ is a direct system of $\mathbf{Z}$-linear mappings and the diagram

$$(8)\qquad \begin{array}{ccc} \varinjlim(E_\alpha \otimes_{A_\alpha} F_\alpha) & \longrightarrow & (\varinjlim E_\alpha) \otimes_A (\varinjlim F_\alpha) \\ \Big\downarrow {\scriptstyle \varinjlim(u_\alpha \otimes v_\alpha)} & & \Big\downarrow {\scriptstyle (\varinjlim u_\alpha) \otimes (\varinjlim v_\alpha)} \\ \varinjlim(E'_\alpha \otimes_{A_\alpha} F'_\alpha) & \longrightarrow & (\varinjlim E'_\alpha) \otimes_A (\varinjlim F'_\alpha) \end{array}$$

is commutative.

The verification is immediate.

Let $(A'_\alpha, \rho'_{\beta\alpha})$ be another direct system of rings and suppose that each E_α is an (A'_α, A_α)-bimodule, the $f_{\beta\alpha}$ being (A'_α, A_α)-linear for $\alpha \leqslant \beta$. Then if we write $A' = \varinjlim A'_\alpha$, the isomorphism (7) is *linear* with respect to the left A'-module structures on the two sides by virtue of Corollary 1. This can be immediately generalized to arbitrary multimodules.

In particular, if the A_α are *commutative*, $A = \varinjlim A_\alpha$ is commutative and isomorphism (7) is an *A-module isomorphism.*

COROLLARY 2. *Let $(E_\alpha, f_{\beta\alpha})$ be a direct system of right A_α-modules and let $E'_\alpha = E_\alpha \otimes_{A_\alpha} A$ be the A-module obtained by extending the ring of scalars to $A = \varinjlim A_\alpha$ by means of the canonical homomorphism $\rho_\alpha : A_\alpha \to A$. Then $(E'_\alpha, f_{\beta\alpha} \otimes 1_A)$ is a direct system of right A-modules, whose direct limit is canonically isomorphic to $\varinjlim E_\alpha$.*

It suffices to apply Proposition 7 with F_α the ring A considered as an (A_α, A)-bimodule by means of ρ_α.

COROLLARY 3. *Let A be a ring, $(E_\alpha, f_{\beta\alpha})$ a direct system of right A-modules and F a left A-module. Then the $\mathbf{Z}$-modules $\varinjlim (E_\alpha \otimes_A F)$ and $(\varinjlim E_\alpha) \otimes_A F$ are canonically isomorphic.*

It suffices to take $A_\alpha = A$ and $F_\alpha = F$ for all $\alpha \in I$ in Proposition 7.

In particular, if $\rho : A \to B$ is a ring homomorphism, $\varinjlim \rho^*(E_\alpha)$ and $\rho^*(\varinjlim E_\alpha)$ are canonically isomorphic.

COROLLARY 4. *Let M be a right A-module, N a left A-module, $(x_i)_{1 \leqslant i \leqslant n}$ a family of elements of M, $(y_i)_{1 \leqslant i \leqslant n}$ a family of elements of N, such that $\sum_i (x_i \otimes y_i) = 0$ in $M \otimes_A N$. Then there exists a finitely generated submodule M_1 (resp. N_1) of M (resp. N) containing the x_i (resp. the y_i) and such that $\sum_i (x_i \otimes y_i) = 0$ in $M_1 \otimes_A N_1$.*

M (resp. N) is canonically identified with the direct limit of the right directed family of its finitely generated submodules containing the x_i (resp. the y_i) and it suffices to apply *Set Theory*, III, § 7, no. 5, Lemma 1.

§ 7. VECTOR SPACES

1. BASES OF A VECTOR SPACE

THEOREM 1. *Every vector space over a field* K *is a free* K*-module.*

It must be proved that every vector space admits a *basis*; this will follow from the following more precise theorem:

THEOREM 2. *Given a generating system* S *of a vector space* E *over a field* K *and a free subset* L *of* E *contained in* S, *there exists a basis* B *of* E *such that* $L \subset B \subset S$.

Theorem 1 follows from this statement by taking $L = \varnothing$.

To prove Theorem 2, we note that the set $\mathfrak{L}$ of free subsets of E contained in S, ordered by inclusion, is *an inductive set* (*Set Theory*, III, § 2, no. 4), by virtue of § 1, no. 11; so is the set $\mathfrak{M}$ of free subsets containing L and contained in S. By Zorn's Lemma, $\mathfrak{M}$ admits a maximal element B and it suffices to prove that the vector subspace of E generated by B is equal to E. This follows immediately from the definition of B and the following lemma:

Lemma 1. *Let* $(a_\iota)_{\iota \in I}$ *be a free family of elements of* E; *if* $b \in E$ *does not belong to the subspace* F *generated by* (a_ι), *the subset of* E *consisting of the* a_ι *and* b *is free.*

Suppose that there were a relation $\mu b + \sum_\iota \lambda_\iota a_\iota = 0$ with $\mu \in K$ and $\lambda_\iota \in K$ for all $\iota \in I$, the family (λ_ι) having finite support; if $\mu \neq 0$, it would follow that $b = -\sum_\iota (\mu^{-1}\lambda_\iota)a_\iota$ and hence $b \in F$ contrary to the hypothesis; hence $\mu = 0$ and the relation becomes $\sum_\iota \lambda_\iota a_\iota = 0$, which implies $\lambda_\iota = 0$ for all $\iota \in I$ by hypothesis; whence the lemma.

COROLLARY. *For a subset* B *of a vector space* E, *the following properties are equivalent:*

(a) B *is a basis of* E.
(b) B *is a maximal free subset of* E.
(c) B *is a minimal generating system of* E.

This follows immediately from Theorem 2.

Example. Given a ring A and a *subfield* K of A, A is a (right or left) vector space over K and therefore admits a basis; in particular, every *extension field* of a field K has a basis as a left (resp. right) vector space over K. *Thus the field **R** of real numbers admits an (infinite) basis as a vector space over the field **Q** of rational numbers; such a basis of **R** is called a *Hamel basis*.*

Remark. For a family $(a_\iota)_{\iota \in I}$ of elements of a vector space E over a field K to be *free*, it is necessary and sufficient that, for all $\kappa \in I$, a_κ belong to no subspace of E generated by the a_ι of index $\iota \neq \kappa$. We know that this condition is necessary in any module (§ 1, no. 11, *Remark* 1). It is sufficient by virtue of Lemma 1, as is seen immediately arguing by *reductio ad absurdum* and considering a minimal related subfamily of (a_ι).

2. DIMENSION OF VECTOR SPACES

THEOREM 3. *Two bases of the same vector space* E *over a field* K *are equipotent.*

We note first that if E admits an *infinite* basis B, it follows from § 1, no. 12, Corollary 2 to Proposition 23 that every other basis of E is equipotent to B. We may therefore confine our attention to the case where E has a finite basis of n elements. We note that every *monogenous* vector space over K, not reduced to 0, is a *simple* K-module (I, § 4, no. 4, Definition 7), for it is generated by each of its elements $\neq 0$, by virtue of the relation $\mu a = (\mu\lambda)(\lambda^{-1}a)$ for $\mu \in K$, $\lambda \in K$ and $\lambda \neq 0$. Hence if $(a_i)_{1 \leqslant i \leqslant n}$ is a basis of E, then $E = \bigoplus_{i=1}^{n} Ka_i$ up to isomorphism and the subspaces $E_k = \bigoplus_{i=1}^{k} Ka_i$ for $0 \leqslant k \leqslant n$ form a *Jordan-Hölder series* of E, E_k/E_{k-1} being isomorphic to Ka_k. Theorem 3 then follows in this case from the Jordan-Hölder Theorem (I, § 4, no. 7, Theorem 6).

A proof can be given independent of the Jordan-Hölder Theorem, by showing by induction on n that, if E admits a basis of n elements, every other basis B′ has *at most n* elements. The proposition is obvious for $n = 0$. If $n \geqslant 1$, B′ is non-empty; then let $a \in B'$. By Theorem 2 (no. 1) there exists a subset C of B such that $\{a\} \cup C$ is a basis of E and $a \notin C$, since $\{a\} \cup B$ is obviously a generating system of E. As B is a basis of E, $C = B$ is impossible (no. 1, Corollary to Theorem 2) and hence C has at most $n - 1$ elements. Let V be the subspace generated by C and V′ the subspace generated by $B' - \{a\}$; V and V′ are both supplementary to the subspace Ka of E and hence are isomorphic (§ 1, no. 10, Proposition 13). As V admits a basis with at most $n - 1$ elements, $B' - \{a\}$ has at most $n - 1$ elements by the induction hypothesis and hence B′ has at most n elements.

DEFINITION 1. *The dimension of a vector space* E *over a field* K, *denoted by* $\dim_K E$ *or* $[E:K]$ (*or simply* dim E) *is the cardinal of any of the bases of* E. *If* M *is a subset of* E, *the rank of* M (*over* K), *denoted by* rg M *or* rg_K M, *is the dimension of the vector subspace of* E *generated by* M.

To say that E is finite-dimensional is equivalent to saying that E is a K-module of *finite length* and $\dim_K E = \mathrm{long}_K E$.

COROLLARY. *For every subset* M *of* E, *the rank of* M *is at most equal to* dim E.

If V is the vector subspace of E generated by M, M contains a basis B′ of V (no. 1, Theorem 2) and as B′ is a free subset of E, it is contained in a basis B of E (no. 1, Theorem 2); then Card(B′) $\leqslant$ Card(B), whence the corollary.

Theorems 2 and 3 immediately imply the following proposition:

PROPOSITION 1. (i) *For a left vector space over* K *to be of finite dimension n, it is necessary and sufficient that it be isomorphic to* K_s^n.

(ii) *For two vector spaces* K_s^m *and* K_s^n *to be isomorphic* (*m and n integers* $\geqslant 0$), *it is necessary and sufficient that* $m = n$.

(iii) *In a vector space* E *of finite dimension n, every generating system has at least n elements; a generating system of* E *with n elements is a basis of* E.

(iv) *In a vector space* E *of finite dimension n, every free subset has at most n elements; a free subset with n elements is a basis of* E.

PROPOSITION 2. *Let* $(E_\iota)_{\iota\in I}$ *be a family of vector spaces over* K. *Then*

$$\dim_K\Big(\bigoplus_{\iota\in I} E_\iota\Big) = \sum_{\iota\in I} \dim_K E_\iota. \tag{1}$$

If the E_ι are canonically identified with subspaces of $E = \bigoplus_{\iota\in I} E_\iota$ and B_ι is a basis of E_ι $(\iota \in I)$, then $B = \bigcup_{\iota\in I} B_\iota$ is a basis of E (§ 1, no. 11, Proposition 19); whence relation (1) since the B_ι are pairwise disjoint.

Remark. (1) Examples can be given of modules admitting two finite bases not having the same number of elements (§ 1, Exercise 16(*c*)). However:

PROPOSITION 3. *Let* A *be a ring such that there exists a homomorphism* ρ *of* A *into a field* D; *then for every free* A*-module* E, *any two bases of* E *are equipotent.*

Consider the vector space $\rho^*(E) = D \otimes_A E$ over D obtained by extending the ring of scalars to D (§ 5, no. 1) and let $\phi: x \mapsto 1 \otimes x$ be the canonical mapping of E into $\rho^*(E)$; if (a_λ) is a basis of E, $(\phi(a_\lambda))$ is a basis of $\rho^*(E)$ (§ 5, no. 1, Proposition 4); the proposition then follows from Theorem 3.

COROLLARY. *If* A *is a commutative ring* $\neq 0$ *and* E *a free* A*-module, any two bases of* E *are equipotent.*

There exists in A at least one maximal ideal $\mathfrak{m}$ (I, § 8, no. 6, Theorem 1) and, as $A/\mathfrak{m}$ is a field, the conditions of Proposition 3 are fulfilled.

Remarks. (2) When a free A-module E is such that any two bases of E are equipotent, the cardinal of an arbitrary basis of E over A is also called the *dimension* or *rank* of E and denoted by $\dim_A E$ or dim E.

(3) Let A be a ring such that any two bases of a free A-module are equipotent and let K be a subfield of A, so that A can be considered as a *left vector space* over K by restricting the scalars. Every free A-module E can similarly be considered as a left vector space over K and it then follows from § 1, no. 13, Proposition 25 that

$$\dim_K E = \dim_K E . \dim_K A_s. \tag{2}$$

(4) In Chapter VIII we shall see examples of rings satisfying the conclusion of Proposition 3 but not the hypothesis.

3. DIMENSION AND CODIMENSION OF A SUBSPACE OF A VECTOR SPACE

PROPOSITION 4. *Every subspace* F *of a vector space* E *is a direct factor of* E *and*

$$\dim F + \dim(E/F) = \dim E. \tag{3}$$

As the quotient vector space E/F is a free module, we know (§ 1, no. 11, Proposition 21) that F is a direct factor of E; relation (3) is then a special case of formula (1) (no. 2).

COROLLARY 1. *If* E, F, G *are vector spaces over a field* K, *every exact sequence of linear mappings* $0 \to E \to F \to G \to 0$ *splits.*

This is another way of expressing Proposition 4 (§ 1, no. 9).

COROLLARY 2. *Let* $(E_i)_{0 \leqslant i \leqslant n}$ *be a finite family of vector spaces over a field* K. *If there exists an exact sequence of linear mappings*

$$0 \longrightarrow E_0 \xrightarrow{u_0} E_1 \xrightarrow{u_1} E_2 \longrightarrow \cdots \longrightarrow E_{n-1} \xrightarrow{u_{n-1}} E_n \xrightarrow{u_n} 0 \tag{4}$$

the relation

$$\sum_{2k+1 \leqslant n} \dim E_{2k+1} = \sum_{2k \leqslant n} \dim E_{2k} \tag{5}$$

holds, or, if all the spaces are finite dimensional,

$$\sum_{i=1}^{n} (-i)^i \dim E_i = 0. \tag{6}$$

Let $I_k = \operatorname{Im} u_k = \operatorname{Ker} u_{k+1}$ for $0 \leqslant k \leqslant n-1$; I_{k+1} is therefore isomorphic to E_{k+1}/I_k, hence (formula (3)) $\dim I_k + \dim I_{k+1} = \dim E_{k+1}$ for $0 \leqslant k \leqslant n-2$ and moreover $\dim I_0 = \dim E_0$ and $I_{n-1} = E_n$, hence $\dim I_{n-1} = \dim E_n$. Replacing $\dim E_i$ by its expression as a function of the $\dim I_k$ in the two sides of (5), we obtain on each side $\sum_{k=0}^{n-1} \dim I_k$, whence the corollary.

Corollary 3. *If* M *and* N *are two subspaces of a vector space* E, *then*

$$\dim(M + N) + \dim(M \cap N) = \dim M + \dim N. \tag{7}$$

It suffices to apply Corollary 2 to the exact sequence

$$0 \to M \cap N \to M \oplus N \to M + N \to 0$$

(§ 1, no. 8, Proposition 10) taking account of the fact that

$$\dim(M \oplus N) = \dim M + \dim N$$

(no. 2, Proposition 2).

Corollary 4. *For every subspace* F *of a vector space* E, $\dim F \leqslant \dim E$; *if* E *is* finite *dimensional, the relation* $\dim F = \dim E$ *is equivalent to* $F = E$.

The first assertion is obvious from (3); further, if $\dim E$ is finite, the relation $\dim F = \dim E$ implies $\dim(E/F) = 0$ by (3) and a vector space of dimension 0 reduces to 0.

Corollary 5. *If a vector space* E *is the sum of a family* (F_ι) *of vector subspaces, then*

$$\dim E \leqslant \sum_\iota \dim F_\iota. \tag{8}$$

If further $\dim E$ *is finite, the two sides of* (8) *are equal if and only if* E *is the direct sum of the family* (F_ι).

The inequality (8) follows from (3) and the fact that E is isomorphic to a quotient of $\bigoplus_\iota F_\iota$ (§ 1, no. 7, formula (28)). The second assertion is a particular case of § 1, no. 10, Corollary 5 to Proposition 16, for the equality of the two sides of (8) implies that $\dim F_\iota = 0$ except for a finite number of indices.

Definition 2. *Given a vector space* E, *the codimension (with respect to* E) *of a subspace* F *of* E, *denoted by* $\operatorname{codim}_E F$, *or simply* $\operatorname{codim} F$, *is the dimension of* E/F (equal to that of any supplement of F in E).

Relation (3) may then be written

$$\dim F + \operatorname{codim} F = \dim E. \tag{9}$$

Proposition 5. *Let* F, F′ *be two subspaces of a vector space* E, *such that* $F \subset F'$. *Then* $\operatorname{codim}_E F' \leqslant \operatorname{codim}_E F \leqslant \dim E$. *If* $\operatorname{codim}_E F$ *is finite, the relation* $\operatorname{codim}_E F' = \operatorname{codim}_E F$ *implies* $F = F'$.

The inequality $\operatorname{codim}_E F \leqslant \dim E$ is obvious from (9) and if $\dim E$ is finite the relation $\operatorname{codim}_E F = \dim E$ implies $\dim F = 0$ and hence $F = \{0\}$. The remainder of the proposition follows from this, for

$$\operatorname{codim}_E F' = \operatorname{codim}_{E/F}(F'/F),$$

since E/F' is canonically isomorphic to $(E/F)/(F'/F)$ (I, § 4, no. 7, Theorem 4).

PROPOSITION 6. *If* M *and* N *are two subspaces of a vector space* E, *then*

$$\operatorname{codim}(M + N) + \operatorname{codim}(M \cap N) = \operatorname{codim} M + \operatorname{codim} N. \tag{10}$$

It suffices to apply Corollary 2 of Proposition 4 to the exact sequence

$$0 \to E/(M \cap N) \to (E/M) \oplus (E/N) \to E/(M + N) \to 0$$

(§ 1, no. 7, Proposition 10) and use no. 2, Proposition 2.

Note that if E is finite dimensional, (10) is a consequence of (7) and (9).

PROPOSITION 7. *If* (F_ι) *is a* finite *family of subspaces of a vector space* E, *then* $\operatorname{codim}\left(\bigcap_\iota F_\iota\right) \leqslant \sum_\iota \operatorname{codim} F_\iota$.

If $F = \bigcap_\iota F_\iota$, E/F is isomorphic to a subspace of the direct sum of the E/F_ι (§ 1, no. 7, formula (27)).

Vector subspaces of dimension 1 (resp. dimension 2) of a vector space E are often called *lines passing through* 0 (resp. *planes passing through* 0) (or simply *lines* (resp. *planes*) if no confusion arises (cf. § 9, no. 3)), by analogy with the language of Classical Geometry; a subspace of E is called a *hyperplane passing through* 0 (or simply a *hyperplane*) if it is of codimension 1. Hyperplanes can also be defined as the *maximal* elements of the set $\mathfrak{S}$ of vector subspaces of E *distinct* from E, ordered by inclusion. There is a one-to-one correspondence between the subspaces of E containing a subspace H and the subspaces of E/H (I, § 4, no. 7, Theorem 4); if E is of dimension $\geqslant 1$, $\mathfrak{S}$ is non-empty and to say that H is maximal in $\mathfrak{S}$ means that E/H contains no subspace distinct from $\{0\}$ and E/H, which implies that E/H is generated by any of its elements $\neq 0$, in other words it is of dimension 1.

In a vector space of finite dimension $n \geqslant 1$, the hyperplanes are the subspaces *of dimension* $n - 1$ by formula (3).

PROPOSITION 8. *In a vector space* E *over a field* K, *every vector subspace* F *is the intersection of the hyperplanes which contain it.*

It suffices to show that for all $x \notin F$ there exists a hyperplane H containing F and not containing x. By hypothesis $F \cap Kx = \{0\}$ and hence the sum M of F and Kx is direct. Let N be supplementary to M in E; E is then the direct sum of $H = F + N$ and Kx and H is therefore a hyperplane with the desired property.

Z

Remark. Most of the properties proved in this no. for subspaces of a vector space do not hold for submodules of a free module whose *dimension* (§ 2, no. 7, *Remark* 2) is defined. *For example, an ideal of a commutative ring does not necessarily admit a basis, for there are integral domains A in which certain ideals are not principal (VII, § 1, no. 1) and any two elements of such a ring are linearly dependent (§ 1, no. 11, *Remark* 1).* A submodule of a free A-module E may be free, distinct from E and have the same dimension as E, as is shown by principal ideals in an integral domain A; the same example proves moreover that a free submodule of a free A-module does not necessarily admit a supplement.

4. RANK OF A LINEAR MAPPING

DEFINITION 3. *Let* E, F *be two vector spaces over a field* K. *For every linear mapping* u *of* E *into* F, *the dimension of the subspace* $u(\mathrm{E})$ *of* F *is called the rank of* u *and denoted by* $\mathrm{rg}(u)$.

If $\mathrm{N} = \mathrm{Ker}(u)$, E/N is isomorphic to $u(\mathrm{E})$, whence the relation

$$\mathrm{rg}(u) = \mathrm{codim}_{\mathrm{E}}(\mathrm{Ker}(u)) \tag{11}$$

and therefore

$$\mathrm{rg}(u) + \dim(\mathrm{Ker}(u) = \dim \mathrm{E}. \tag{12}$$

Moreover, by formula (3)

$$\mathrm{rg}(u) + \dim(\mathrm{Coker}(u)) = \dim \mathrm{F}. \tag{13}$$

PROPOSITION 9. *Let* E, F *be two vector spaces over a field* K *and* $u:\mathrm{E} \rightarrow \mathrm{F}$ *a linear mapping.*

(i) $\mathrm{rg}(u) \leqslant \inf(\dim \mathrm{E}, \dim \mathrm{F})$.

(ii) *Suppose that* E *is finite-dimensional; in order that* $\mathrm{rg}(u) = \dim \mathrm{E}$, *it is necessary and sufficient that* u *be injective.*

(iii) *Suppose that* F *is finite-dimensional; in order that* $\mathrm{rg}(u) = \dim \mathrm{F}$, *it is necessary and sufficient that* u *be surjective.*

This follows immediately from relations (12) and (13).

COROLLARY. *Let* E *be a vector space of finite dimension* n *and* u *an endomorphism of* E. *The following properties are equivalent:*

(a) u *is bijective;*
(b) u *is injective;*
(c) u *is surjective;*
(d) u *is right invertible;*
(e) u *is left invertible;*
(f) u *is of rank* n.

If E is an infinite-dimensional vector space, there are injective (resp. surjective) endomorphisms of E which are not bijective (Exercise 9).

Let K, K′ be two fields, $\sigma: K \to K'$ an *isomorphism* of K onto K′, E a vector K-space, E′ a vector K′-space and $u: K \to K'$ a *semi-linear* mapping relative to σ (§ 1, no. 13); the dimension of the subspace $u(E)$ of E′ is also called the *rank* of u. It is also the rank of u considered as a linear mapping of E into $\sigma_*(E')$, for every basis of $u(E)$ is also a basis of $\sigma_*(u(E))$.

5. DUAL OF A VECTOR SPACE

THEOREM 4. *The dimension of the dual* E^* *of a vector space* E *is at least equal to the dimension of* E. *For* E^* *to be finite-dimensional, it is necessary and sufficient that* E *be so, and then* $\dim E^* = \dim E$.

If K is the field of scalars of E, E is isomorphic to a space $K_s^{(I)}$ and therefore E^* is isomorphic to K_d^I (§ 2, no. 6, Proposition 10). As $K_s^{(I)}$ is a subspace of K_d^I, $\dim E = \mathrm{Card}(I) \leqslant \dim E^*$ (no. 3, Corollary 4 to Proposition 4); further, if I is finite, $K_d^I = K_d^{(I)}$ (cf. Exercise 3(d)).

COROLLARY. *For a vector space* E, *the relations* $E = \{0\}$ *and* $E^* = \{0\}$ *are equivalent.*

THEOREM 5. *Given two exact sequences of vector spaces (over the same field* K*) and linear mappings*

$$0 \to E' \to E \to E'' \to 0$$
$$0 \to F' \to F \to F'' \to 0$$

and two vector spaces G, H *over* K, *the corresponding sequences*

$$0 \to \mathrm{Hom}(E'', G) \to \mathrm{Hom}(E, G) \to \mathrm{Hom}(E', G) \to 0$$
$$0 \to \mathrm{Hom}(H, F') \to \mathrm{Hom}(H, F) \to \mathrm{Hom}(H, F'') \to 0$$

are exact and split.

This follows from the fact that every vector subspace is a direct factor (no. 3, Proposition 4) and from § 2, no. 1, Propositions 1 and 2.

COROLLARY. *For every exact sequence*

$$0 \longrightarrow E' \xrightarrow{u} E \xrightarrow{v} E'' \longrightarrow 0$$

of vector spaces over the same field K *and linear mappings the sequence*

$$0 \longrightarrow E''^* \xrightarrow{{}^t v} E^* \xrightarrow{{}^t u} E'^* \longrightarrow 0$$

is exact and splits.

It follows in particular that for every vector subspace M of E the canonical homomorphism $E^*/M' \to M^*$, where M' is the subspace of E^* orthogonal to M (§ 2, no. 4), is *bijective.*

THEOREM 6. *For every vector space* E *over a field* K, *the canonical mapping* $c_E : E \to E^{**}$ (§ 2, no. 7) *is injective; for it to be bijective, it is necessary and sufficient that* E *be finite-dimensional.*

The first assertion and the fact that if E is finite-dimensional c_E is bijective are special cases of § 2, no. 7, Proposition 14. Suppose that E is infinite-dimensional, so that we may assume that $E = K_s^{(L)}$, where L is an infinite set and therefore $E^* = K_d^L$. Let $(e_\lambda)_{\lambda \in L}$ be the canonical basis of E and let $(e_\lambda^*)_{\lambda \in L}$ be the corresponding family of coordinate forms in E^* (§ 2, no. 6); the vector subspace of E^* generated by the e_λ^* is just the direct sum $F' = K_d^{(L)}$ and the hypothesis that L is infinite implies that $F' \neq E^*$. Then there exists a hyperplane H' of E^* containing F' (no. 3, Proposition 8) and, as E^*/H' is non-zero, its dual is also non-zero (Corollary to Theorem 4), which is identified with the orthogonal H'' of H' in E^{**} (§ 2, no. 6, Corollary to Proposition 9). But $H'' \cap c_E(E)$ is contained in the image under c_E of the orthogonal of F' in E, which is by definition 0; hence $c_E(E) = E^{**}$ is impossible.

E will usually be *identified* with the subspace of E^{**} the image of c_E.

Let E, F be two vector spaces over a field K and $u : E \to F$ a linear mapping. We shall define *canonical isomorphisms*:

(1) *Of the dual of* $\mathrm{Im}(u) = u(E)$ *onto* $\mathrm{Im}({}^t u) = {}^t u(F^*)$.

(2) *Of the dual of* $\mathrm{Ker}(u) = \overset{-1}{u}(0)$ *onto* $\mathrm{Coker}({}^t u) = E^*/{}^t u(F^*)$.

(3) *Of the dual of* $\mathrm{Coker}(u) = F/u(E)$ *onto* $\mathrm{Ker}({}^t u) = {}^t u^{-1}(0)$.

We write $I = \mathrm{Im}(u)$, $N = \mathrm{Ker}(u)$, $C = \mathrm{Coker}(u)$; from the exact sequences

$$(14) \quad 0 \longrightarrow N \longrightarrow E \xrightarrow{p} I \longrightarrow 0, \qquad 0 \longrightarrow I \xrightarrow{j} F \longrightarrow C \longrightarrow 0$$

we derive, by transposition (Corollary to Theorem 5), the exact sequences

$$(15) \qquad \begin{array}{l} 0 \longrightarrow I^* \xrightarrow{{}^t p} E^* \longrightarrow N^* \longrightarrow 0, \\ 0 \longrightarrow C^* \longrightarrow F^* \xrightarrow{{}^t j} I^* \longrightarrow 0. \end{array}$$

Moreover, as $u = j \circ p$, ${}^t u = {}^t p \circ {}^t j$; the exact sequences (15) thus define canonical isomorphisms of C^* onto $\mathrm{Ker}({}^t u)$, of I^* onto $\mathrm{Im}({}^t u)$ and of N^* onto $\mathrm{Coker}({}^t u)$, since ${}^t p$ is injective and ${}^t j$ surjective. To be precise, let $y \in \mathrm{Im}(u)$, $z \in \mathrm{Ker}(u)$, $t \in \mathrm{Coker}(u)$, $y' \subset \mathrm{Im}({}^t u)$, $z' \in \mathrm{Coker}({}^t u)$, $t' \in \mathrm{Ker}({}^t u)$; when y', z', t' are canonically identified with linear forms on $\mathrm{Im}(u)$, $\mathrm{Ker}(u)$ and $\mathrm{Coker}(u)$ respectively, then

$$(16) \qquad \langle y, y' \rangle = \langle x, y' \rangle \quad \text{for all } x \in E \text{ such that } u(x) = y;$$

(17) $\langle z, z' \rangle = \langle z, x^* \rangle$ for all $x^* \in E^*$ whose class mod. ${}^t u(F)$ is equal to z';

(18) $\langle t, t' \rangle = \langle s, t' \rangle$ for all $s \in F$ whose class mod. $u(E)$ is equal to t.

In particular we derive from these results:

PROPOSITION 10. *Let* E, F *be two vector spaces over the same field* K *and* $u: E \to F$ *a linear mapping.*

(i) *For* u *to be injective* (resp. *surjective*), *it is necessary and sufficient that* ${}^t u$ *be surjective* (resp. *injective*).

(ii) $\operatorname{rg}(u) \leqslant \operatorname{rg}({}^t u)$ *and* $\operatorname{rg}(u) = \operatorname{rg}({}^t u)$ *if* $\operatorname{rg}(u)$ *is finite.*

The second assertion follows from the above and Theorem 4.

THEOREM 7. *Let* E *be a vector space over a field* K, F *a subspace of* E *and* F′ *the orthogonal of* F *in* E*.

(i) $\dim F' \geqslant \operatorname{codim}_E F$; *for* $\dim F'$ *to be finite, it is necessary and sufficient that* $\operatorname{codim}_E F$ *be finite, and then* $\dim F' = \operatorname{codim}_E F$.

(ii) *The orthogonal of* F′ *in* E *is equal to* F.

(iii) *Every finite-dimensional subspace* G′ *of* E* *is the orthogonal of some subspace of* E, *necessarily equal to the orthogonal of* G′ *in* E *and of finite codimension.*

(i) We know that F′ is isomorphic to the dual $(E/F)^*$ (§ 2, no. 6, Corollary to Proposition 9) and hence the assertion follows from Theorem 4, since $\dim(E/F) = \operatorname{codim}_E F$ by definition.

(ii) Let F_1 be the orthogonal of F′ in E; clearly $F \subset F_1$ and the orthogonal F'_1 of F_1 is equal to F′ (§ 2, no. 4); the canonical linear mapping $(E/F_1)^* \to (E/F)^*$, the transpose of $E/F \to E/F_1$ is therefore bijective (§ 2, no. 6, Corollary to Proposition 10); it then follows from Proposition 10 that the canonical mapping $E/F \to E/F_1$ is bijective, which implies $F_1 = F$.

(iii) Let G′ be a subspace of E* of finite dimension p and let F be its orthogonal in E; then $\operatorname{codim}_E F \leqslant \dim G'$. For, if $(a_i^*)_{1 \leqslant i \leqslant p}$ is a basis of G′, F is the kernel of the linear mapping $x \mapsto (\langle x, a_i^* \rangle)$ from E to K_s^p whose rank is at most p (no. 4, Proposition 9), whence the conclusion (no. 4). Then let F′ be the orthogonal of F in E*; it follows from (i) that $\dim F' \leqslant \dim G'$; but on the other hand obviously $G' \subset F'$, whence $F' = G'$ (§ 2, no. 3, Corollary 4 to Proposition 4).

Remark. An *infinite*-dimensional subspace G′ of E* is not necessarily the orthogonal of a subspace of E, in other words, if F is the orthogonal of G′ in E, the orthogonal F′ of F in E* may be distinct from G′ (Exercise 20(*b*)).†

---

† By giving E and E* suitable *topologies* and only considering in E and E* subspaces *closed* with respect to these topologies, it is possible to re-establish a perfect symmetry between the properties of E and E* when E is infinite-dimensional (cf. *Topological vector spaces,* II, § 6).

COROLLARY 1. *Let* $(x_i^*)_{1 \leqslant i \leqslant p}$ *be a finite sequence of linear forms on* E *and let* F *be the subspace of* E *consisting of the* x *such that*

$$\langle x, x_i^* \rangle = 0 \quad \textit{for } 1 \leqslant i \leqslant p.$$

Then $\operatorname{codim}_E F$ *is equal to the rank of the set of the* x_i^* *and every linear form on* E *which is zero on* F *is a linear combination of the* x_i^*. *Then* $\operatorname{codim}_E F \leqslant p$ *and in order that* $\operatorname{codim}_E F = p$, *it is necessary and sufficient that the* x_i^* *be linearly independent.*

The set G′ of linear combinations of the x_i^* is a subspace of E^* and F is the orthogonal of G′ in E, hence $\operatorname{codim}_E F = \dim G'$ by Theorem 7; further $\dim G' \leqslant p$ and the relation $\dim G' = p$ means that (x_i^*) is a free system (no. 2, Proposition 1); hence the corollary.

COROLLARY 2. (i) *Let* $(x_i^*)_{1 \leqslant i \leqslant p}$ *be a finite sequence of linear forms on* E. *For* (x_i^*) *to be a free system, it is necessary and sufficient that there exist a sequence* $(x_i)_{1 \leqslant i \leqslant p}$ *of elements of* E *such that* $\langle x_i, x_j^* \rangle = \delta_{ij}$ (Kronecker index).

(ii) *Let* $(x_i)_{1 \leqslant i \leqslant p}$ *be a finite sequence of elements of* E. *For* (x_i) *to be a free system, it is necessary and sufficient that there exist a sequence* $(x_i^*)_{1 \leqslant i \leqslant p}$ *of linear forms on* E *such that* $\langle x_i, x_j^* \rangle = \delta_{ij}$.

Clearly (ii) follows from (i) by considering E as identified with a subspace of E^{**} by means of c_E (Theorem 6). Let G′ be the subspace of E^* generated by the x_i^* and F its orthogonal in E; E/F and G′ can each be canonically identified with the dual of the other; if the family (x_i^*) is free, there is in E/F a basis $(\dot{x}_i)$ the dual of (x_i^*) and every representative system (x_i) of the classes $\dot{x}_i$ has the required properties. Conversely, the existence of the system (x_i) such that $\langle x_i, x_j^* \rangle = \delta_{ij}$ implies that for all i the subspace of E^* orthogonal to $K.x_i$ contains the x_j^* of index $j \neq i$ but does not contain x_i^*, hence the system $(x_i^*)_{1 \leqslant i \leqslant p}$ is free.

COROLLARY 3. *Let* S *be a set and* V *a vector subspace of the right vector* K*-space* K_d^S *of mappings of* S *into* K. *In order that* $\dim V \geqslant p$ *(where* p *is an integer), it is necessary and sufficient that there exist* p *elements* s_i *of* S *and* p *elements* f_i *of* V $(1 \leqslant i \leqslant p)$ *such that* $f_i(s_j) = \delta_{ij}$.

The space K_d^S is canonically identified with the dual of $E = K_s^{(S)}$ and $f(s) = \langle e_s, f \rangle$ for $s \in S$ and $f \in K_s^S$, $(e_s)_{s \in S}$ being the canonical basis of E. Corollary 2 then shows that the condition is sufficient. Conversely, suppose that $\dim V \geqslant p$, so that there exists a subspace G′ of V of dimension p; let F be the orthogonal of G′ in E, so that $\dim(E/F) = p$. It follows from no. 1, Theorem 2 that there exist p elements $s_i \in S$ such that the e_{s_i} $(1 \leqslant i \leqslant p)$ form a basis of a supplement F of E (applying Theorem 2 (no. 1) to a free subset consisting of a basis of E and the generating system the union of this free subset and the canonical basis of E); we then take the f_ι to be the elements of a basis of G' dual to the basis of E/F consisting of the classes of the e_{s_ι} mod. F.

COROLLARY 4. *Let* E *be a vector space and* M, N *two subspaces of* E *of finite codimension; if* M', N' *are the orthogonals of* M *and* N *in* E*, *the orthogonal of* M ∩ N *in* E* *is* M' + N'.

As M (resp. N) is the orthogonal of M' (resp. N') in E (Theorem 7), M ∩ N is the orthogonal of M' + N' in E and hence M' + N' is the orthogonal of M ∩ N in E* (Theorem 7 (iii)).

COROLLARY 5. *Let* E *be a vector space of finite dimension* n. *For every subspace* F *of* E *of dimension* p, *the orthogonal* F' *of* F *in* E* *is of dimension* $n - p$. *For every subspace* G' *of* E* *of dimension* q, *the orthogonal* G *of* G' *in* E *is of dimension* $n - q$ *and* G' *is the orthogonal of* G *in* E*.

Theorem 7 gives another characterization of *hyperplanes* in E:

PROPOSITION 11. *For every hyperplane* H *in a vector space* E, *there exists a linear form* x_0^* *on* E *such that* $H = x_0^{*-1}(0)$. *Given such a form* x_0^*, *for a linear form* x^* *on* E *to satisfy* $H = x^{*-1}(0)$, *it is necessary and sufficient that* $x^* = x_0^*\alpha$, *where* α *is a scalar* $\neq 0$. *Conversely, for every linear form* $x^* \neq 0$ *on* E, *the subspace* $x^{*-1}(0)$ *is a hyperplane of* E.

This statement merely expresses Theorem 7 for subspaces of E of codimension 1 and subspaces of E* of dimension 1.

If H is a hyperplane and x_0^* a linear form such that $H = x^{*-1}(0)$, the relation

$$\langle x, x_0^* \rangle = 0$$

which characterizes the elements $x \in H$, is called *an equation of* H.

More generally, if (x_ι^*) is a family of linear forms on E and F denotes the vector subspace the intersection of the hyperplanes $x_\iota^{*-1}(0)$, the relation "for all ι, $\langle x, x_\iota^* \rangle = 0$" characterizes the elements of F; the relations

$$\langle x, x_\iota^* \rangle = 0 \quad \text{for all } \iota$$

form *a system of equations* of the subspace F. Theorem 7 (ii) expresses the fact that *every vector subspace of* E *can be defined by a system of equations*.

Theorem 7 (i) and (ii) proves moreover that a subspace F of *finite* codimension p can be defined by a system of p equations

$$\langle x, x_i^* \rangle = 0, \quad 1 \leqslant i \leqslant p, \tag{19}$$

where the forms x_i^* are *linearly independent*. Conversely, Corollary 1 to Theorem 7 shows that a subspace F defined by a system of p equations (19) is of codimension $\leqslant p$ and that it is of codimension p if and only if the x_i^* are linearly independent; it amounts to the same to say that F *cannot be defined by a system consisting of at most* $p - 1$ *of the equations* (19).

6. LINEAR EQUATIONS IN VECTOR SPACES

PROPOSITION 12. *Let* E, F *be two vector spaces over a field* K *and* $u: E \to F$ *a linear mapping. For the linear equation*

$$u(x) = y_0 \tag{20}$$

to have at least one solution $x \in E$, *it is necessary and sufficient that* y_0 *be orthogonal to the kernel of the transpose mapping* ${}^t u$.

The orthogonal of $u(E)$ in F^* is ${}^t u^{-1}(0)$ (§ 2, no. 5, Corollary to Proposition 8) and the orthogonal of ${}^t u^{-1}(0)$ in F is therefore $u(E)$ (no. 5, Theorem 7 (ii)).

We shall obtain a more convenient criterion for *systems of scalar linear equations*

$$\langle x, x_\iota^* \rangle = \eta_\iota \quad (\iota \in I) \tag{21}$$

where the unknown x takes its values in a vector space E over a field K, the x_ι^* are linear forms on E and the right hand sides η_ι elements of K.

If we consider a basis $(a_\lambda)_{\lambda \in L}$ of E, the system (21) is equivalent to the system of equations

$$\sum_{\lambda \in L} \xi_\lambda \langle a_\lambda, x_\iota^* \rangle = \eta_\iota \quad (\iota \in I) \tag{22}$$

with $x = \sum_{\lambda \in L} \xi_\lambda a_\lambda$, the solutions of (22) being of necessity families (ξ_λ) of elements of K *of finite support.*

DEFINITION 4. *The dimension of the subspace of* E^* *generated by the family* (x_ι^*) *is called the rank of the system* (21).

PROPOSITION 13. *For the system* (21) *to be of* finite *rank* r, *it is necessary and sufficient that the linear mapping* $u: x \mapsto (\langle x, x_\iota^* \rangle)$ *of* E *into* K_s^I *be of rank* r.

If F' is the subspace of E^* generated by the x_ι^*, the kernel of u is the orthogonal F of F' in E; if F' is of dimension r, F is of codimension r and conversely (no. 5, Theorem 7) and $\mathrm{rg}(u) = \mathrm{codim}_E F$ (no. 4, formula (11)).

THEOREM 8. *Let*

$$\langle x, x_\iota^* \rangle = \eta_\iota \quad (\iota \in I) \tag{21}$$

be a system of scalar linear equations on a vector space E *over a field* K. *For this system* *to have at least one solution, it is necessary that, for every family* (ρ_ι) *of scalars of finite support such that* $\sum_\iota x_\iota^* \rho_\iota = 0$, $\sum_\iota \eta_\iota \rho_\iota = 0$. *If the rank of the system* (21) *is finite, this condition is also sufficient.*

The condition is obviously necessary. It says that, if F' is the subspace of E* generated by the family (x_ι^*), there exists a linear mapping $f: F' \to K_d$ such that $f(x_\iota^*) = \eta_\iota$ for all $\iota \in I$. If F' is of finite dimension r, F' is the orthogonal of a subspace F of E of codimension r (no. 5, Theorem 7) and F' is identified with the dual of E/F (§ 2, no. 6, Corollary to Proposition 9); f is therefore an element of the bidual $(E/F)^{**}$. As E/F is finite-dimensional there exists one and only one element $y \in E/F$ such that $f(x^*) = \langle y, x^* \rangle$ for all $x^* \in F'$ (no. 5, Theorem 6). The solutions of (21) are then the $x \in E$ whose canonical image in E/F is y.

Remark. When the rank of the system (21) is *infinite*, the condition of Theorem 8 is no longer sufficient. For example, suppose that the x_ι^* are the *coordinate forms* on the space $E = K_s^{(I)}$, I being infinite (§ 2, no. 6); as the x_ι^* are linearly independent, the condition of Theorem 8 holds for every family (η_ι) but the system (21) only has solutions if the family (η_ι) has finite support.

A system (21) is always of finite rank if it has only a *finite number of equations* and its rank is then *at most equal* to the number of equations (no. 2, Proposition 1). Similarly, if E is of finite dimension n (which for a system (22) corresponds to the case where there is only a *finite number n of unknowns*), its dual E* is of dimension n and hence the rank of system (21) is at most equal to n (no. 3, Corollary 4 to Proposition 4). From this we deduce:

Corollary 1. *A system of scalar linear equations in a vector space, consisting of a finite number of equations whose left hand sides are linearly independent forms, always admits solutions.*

Corollary 2. *For a homogeneous linear system* (22) *of equations in n unknowns with coefficients in a field* K *to admit non-trivial solutions consisting of elements of* K, *it is necessary and sufficient that its rank be* $< n$.

This will always be the case if the number of equations is finite and $< n$.

Corollary 3. *For a linear system* (22) *with coefficients and right hand sides in a field* K, *consisting of n equations in n unknowns, to have one and only one solution consisting of elements of* K, *it is necessary and sufficient that the associated homogeneous system have no non-trivial solution* (or, what amounts to the same, that the left hand sides of this system be *linearly independent* forms).

7. TENSOR PRODUCTS OF VECTOR SPACES

The results of §§ 3, 4 and 5 relating to free or projective modules apply in particular to vector spaces and give the following properties:

PROPOSITION 14. *Given an exact sequence*

$$0 \rightarrow E' \rightarrow E \rightarrow E'' \rightarrow 0 \tag{23}$$

of right vector spaces over a field K *and linear mappings and a left vector space* F *over* K, *the corresponding sequence of* **Z**-*linear mappings*

$$0 \rightarrow E' \otimes_K F \rightarrow E \otimes_K F \rightarrow E'' \otimes_K F \rightarrow 0$$

is exact and splits.

As the sequence (23) splits, this is a particular case of § 3, no. 7, Corollary 5 to Proposition 7 and § 3, no. 6, Proposition 5.

Because of Proposition 14, when E′ is a vector subspace of E, $j: E' \rightarrow E$ the canonical injection, $E' \otimes_K F$ is usually *identified* with a *sub-*$\mathbf{Z}$*-module* of $E \otimes_K F$ by means of the injection $j \otimes 1_F$. With this convention:

COROLLARY. *Let* K *be a field,* E *a right vector space over* K, F *a left vector space over* K, $(M_\alpha)_{\alpha \in A}$ *a family of vector subspaces of* E *and* $(N_\beta)_{\beta \in B}$ *a family of vector subspaces of* F. *Then*

$$\Big(\bigcap_{\alpha \in A} M_\alpha\Big) \otimes_K \Big(\bigcap_{\beta \in B} N_\beta\Big) = \bigcap_{(\alpha, \beta) \in A \times B} (M_\alpha \otimes_K N_\beta). \tag{24}$$

It is obviously sufficient to prove the particular case

$$\Big(\bigcap_{\alpha \in A} M_\alpha\Big) \otimes_K F = \bigcap_{\alpha \in A} (M_\alpha \otimes_K F). \tag{25}$$

Clearly the left hand side of (25) is contained in the right hand side. To prove the converse, we consider a basis $(f_\lambda)_{\lambda \in L}$ of F. Every element of $E \otimes_K F$ can then be expressed uniquely in the form $\sum_{\lambda \in L} x_\lambda \otimes f_\lambda$, where $x_\lambda \in E$ (§ 3, no. 7, Corollary 1 to Proposition 7); if E′ is a vector subspace of E, the relation $\sum_{\lambda \in L} x_\lambda \otimes f_\lambda \in E' \otimes_K F$ is equivalent, by Proposition 14, to $x_\lambda \in E'$ for all $\lambda \in L$. To say that $\sum_{\lambda \in L} x_\lambda \otimes f_\lambda$ belongs to each of the $M_\alpha \otimes_K F$ thus means that for all $\lambda \in L$ and all $\alpha \in A$, $x_\lambda \in M_\alpha$, that is $x_\lambda \in \bigcap_{\alpha \in A} M_\alpha$ for all $\lambda \in L$, which proves that the right hand side of (25) is contained in the left hand side.

PROPOSITION 15. *If* $(E_\lambda)_{\lambda \in L}$ *is a family of right vector spaces over a field* K *and* $(F_\mu)_{\mu \in M}$ *a family of left vector spaces over* K, *the canonical mapping*

$$\Big(\prod_{\lambda \in L} E_\lambda\Big) \otimes_K \Big(\prod_{\mu \in M} F_\mu\Big) \rightarrow \prod_{(\lambda, \mu) \in L \times M} (E_\lambda \otimes_K F_\mu) \tag{26}$$

(§ 3, no. 7, formula (22)) *is injective.*

We write $F = \prod_{\mu \in M} F_\mu$; the mapping (26) is the composition of the canonical mappings

$$\left(\prod_{\lambda \in L} E_\lambda\right) \otimes_K F \to \prod_{\lambda \in L} (E_\lambda \otimes_K F)$$

and

$$\prod_{\lambda \in L} (E_\lambda \otimes_K F) \to \prod_{\lambda \in L} \left(\prod_{\mu \in M} (E_\lambda \otimes_K F_\mu)\right);$$

as F and the E_λ are vector spaces over K, this reduces to § 3, no. 7, Corollary 3 to Proposition 7.

When the conditions of Proposition 15 are fulfilled, the tensor product $\left(\prod_{\lambda \in L} E_\lambda\right) \otimes_K \left(\prod_{\mu \in M} F_\mu\right)$ is often identified with its canonical image in

$$\prod_{\lambda, \mu} (E_\lambda \otimes_K F_\mu).$$

With this convention:

COROLLARY. *Let* F *be a left vector space over* K; *for every set* X, *the left vector space* $K_d^X \otimes_K F$ *is identified with the subspace of the space* F^X *of all mappings of* X *into* F, *consisting of the mappings* u *such that* $u(X)$ *is of finite rank in* F.

If (f_λ) is a basis of F, the element $\sum_{\lambda \in L} v_\lambda \otimes f_\lambda$ of $K_d^X \otimes_K F$ is identified under (26) with the mapping $x \mapsto \sum_\lambda v_\lambda(x) f_\lambda$. As $v_\lambda = 0$ except for the indices λ belonging to a finite subset H of L, the image of X under the above mapping is contained in the finite-dimensional subspace of F generated by the f_λ of indices $\lambda \in H$. Conversely, let $u : X \to F$ be a mapping such that $u(X)$ is contained in a finite-dimensional subspace G of F and let $(b_i)_{1 \leqslant i \leqslant n}$ be a basis of G. For all $x \in X$, we may write $u(x) = \sum_{i=1}^{n} v_i(x) b_i$, where the $v_i(x)$ are well determined elements of K; thus n mappings $v_i : X \to K$ are defined and clearly u is then identified with the element $\sum_{i=1} v_i \otimes b_i$.

Similarly, for a right vector space E over K and a set Y, $E \otimes_K K_s^Y$ is identified with a subspace of the space E^Y, consisting of the mappings $v : Y \to E$ such that $v(Y)$ is of finite rank. More particularly, for every field K, $K_d^X \otimes_K K_s^Y$ is identified with a subspace of the space $K^{X \times Y}$ of mappings of $X \times Y$ into K (K being considered as a (K, K)-bimodule); an element $\sum_i u_i \otimes v_i$, where u_i is a mapping of X into K and v_i a mapping of Y into K, is identified with the mapping $(x, y) \mapsto \sum_i u_i(x) v_i(y)$ of $X \times Y$ into K.

Proposition 16. (i) *Let* K, L *be two fields*, E *a left vector space over* K, F *a left vector space over* L *and* G *a* (K, L)-*bimodule. Then the canonical* **Z**-*homomorphism*

$$\nu : \mathrm{Hom}_{K}(E, G) \otimes_{L} F \to \mathrm{Hom}_{K}(E, G \otimes_{L} F) \tag{27}$$

(§ 4, no. 2, formula (7)) *is injective; it is bijective when one of the vector spaces* E, F *is finite-dimensional.*

(ii) *Let* E_1, E_2, F_1, F_2 *be four vector spaces over a commutative field* K; *then the canonical* K-*homomorphism*

$$\lambda : \mathrm{Hom}(E_1, F_1) \otimes \mathrm{Hom}(E_2, F_2) \to \mathrm{Hom}(E_1 \otimes E_2, F_1 \otimes F_2) \tag{28}$$

(§ 4, no. 4, formula (21)) *is injective; it is bijective if one of the ordered pairs* (E_1, E_2), (E_1, F_1), (E_2, F_2) *consists of finite-dimensional spaces.*

Assertion (i) is a particular case of § 4, no. 2, Proposition 2. Similarly the second assertion of (ii) is a particular case of § 4, no. 4, Proposition 4. Finally, to see that the homomorphism (28) is always injective, observe that $\mathrm{Hom}(E_i, F_i)$ is a vector subspace of $F_i^{E_i}$ $(i = 1, 2)$ and that

$$\mathrm{Hom}(E_1 \otimes E_2, F_1 \otimes F_2)$$

is canonically identified with a vector subspace of the space $(F_1 \otimes F_2)^{E_1 \times E_2}$ (II, § 3, no. 1, Proposition 1); when these identifications are made and also the left hand side of (28) is identified with a subspace of $F_1^{E_1} \otimes F_2^{E_2}$ (Proposition 14), the canonical mapping (28) becomes the restriction to this subspace of the canonical mapping (26) and it has been seen (Proposition 15) that the latter is injective.

Corollary 1. *Let* E *and* F *be two vector spaces over a field* K; *the canonical mapping*

$$E^* \otimes_K F \to \mathrm{Hom}_K(E, F)$$

(§ 4, no. 2, formula (11)) *is injective; it is bijective when* E *or* F *is finite-dimensional.*

This is a special case of Proposition 16 (i).

Corollary 2. *Let* E *be a right vector space and* F *a left vector space over the same field* K; *the canonical mapping*

$$E \otimes_K F \to \mathrm{Hom}_K(E^*, F) \tag{29}$$

(§ 4, no. 2, formula (15)) *is injective; it is bijective when* E *is finite-dimensional.*

This is a special case of § 4, no. 2, *Remark* 2.

Remarks. (1) Let K be a *commutative* field, E, F two vector spaces over K,

(a_λ) a basis of E and (b_μ) a basis of F; then $(a_\lambda \otimes b_\mu)$ is a basis of the vector K-space $E \otimes_K F$ (II, § 3, no. 7, Corollary 2 to Proposition 2) and therefore

$$\dim_K(E \otimes_K F) = \dim_K E . \dim_K F. \tag{30}$$

(2) Let K be a *commutative* field, E_1, E_2, F_1, F_2 four vector spaces over K and $u: E_1 \to F_1$, $v: E_2 \to F_2$ two linear mappings; then

$$\operatorname{rg}(u \otimes v) = \operatorname{rg}(u) . \operatorname{rg}(v). \tag{31}$$

It is immediate that $(u \otimes v)(E_1 \otimes E_2)$ is the canonical image of $u(E_1) \otimes v(E_2)$ in $F_1 \otimes F_2$ and hence (Proposition 14) is isomorphic to $u(E_1) \otimes v(E_2)$; the conclusion then follows from (30).

(3) Under the same hypotheses as in *Remark* 1,

$$\dim_K(\operatorname{Hom}_K(E, F)) \geqslant \dim_K E . \dim_K F. \tag{32}$$

If E is isomorphic to $K^{(I)}$, Hom(E, F) is isomorphic to $(\operatorname{Hom}(K, F))^I$ (§ 1, no. 6, Corollary 1 to Proposition 6) and hence to F^I (§ 1, no. 14); as $F^{(I)}$ is a subspace of F^I and $\dim(F^{(I)}) = \operatorname{Card}(I) . \dim F = \dim E . \dim F$ (no. 2, Proposition 2), the inequality (32) follows from no. 3, Corollary 4 to Proposition 4. The same argument shows that the two sides of (32) are equal when dim E is *finite* (cf. § 10, nos. 3 and 4).

8. RANK OF AN ELEMENT OF A TENSOR PRODUCT

Let E be a right vector space and F a left vector space over the same field K; to every element $u \in E \otimes_K F$ there corresponds canonically under (29) a homomorphism $u_1 \in \operatorname{Hom}_K(E^*, F)$; if $u = \sum_i x_i \otimes y_i$ with $x_i \in E$, $y_i \in F$, the element u_1 is the linear mapping

$$x^* \mapsto \sum_i \langle x^*, x_i \rangle y_i. \tag{33}$$

On the other hand, $E \otimes_K F$ is canonically identified with $F \otimes_{K^0} E$, where E is considered as a left vector space and F as a right vector space over the opposite field K^0; thus there corresponds canonically to u a homomorphism $u_2 \in \operatorname{Hom}_K(F^*, E)$ given by

$$y^* \mapsto \sum_i x_i \langle y_i, y^* \rangle; \tag{34}$$

u_1 (resp. u_2), considered as a mapping of E^* into F^{**} (resp. of F^* into E^{**}) is just the *transpose* of u_2 (resp. u_1). The *ranks* of u_1 and u_2 are thus *equal* to the same *finite* number r, the common dimension of the subspaces $u_1(E^*)$ of F and $u_2(F^*)$ of E, each of which is canonically isomorphic to the dual of the other (no. 5); we shall say that r (denoted $\operatorname{rg}(u)$) is the *rank* of the element u of $E \otimes_K F$ and that $u_1(E^*)$ and $u_2(F^*)$ are the subspaces (of F and E respectively) *associated* with u.

PROPOSITION 17. *Let u be an element of $E \otimes_K F$ and $M \subset E$ and $N \subset F$ its associated subspaces. For every expression $u = \sum_{i=1}^{s} x_i \otimes y_i$ of u, where $x_i \in E$ and $y_i \in F$ for $1 \leqslant i \leqslant s$, the subspace* M (resp. N) *is contained in the subspace of* E (resp. F) *generated by the x_i* (resp. *the y_i*). *Moreover, the following properties are equivalent:*

(a) *The integer s is equal to the rank of u.*
(b) *The family $(x_i)_{1 \leqslant i \leqslant s}$ is a basis of* M.
(c) *The family $(y_i)_{1 \leqslant i \leqslant s}$ is a basis of* N.
(d) *The families $(x_i)_{1 \leqslant i \leqslant s}$ and $(y_i)_{1 \leqslant i \leqslant s}$ are both free.*

By (33) (resp. (34)) each element of $N = u_1(E^*)$ (resp. $M = u_2(F^*)$) is a linear combination of the y_i (resp. the x_i); whence the first assertion. If $s = r$, the subspace generated by the x_i (resp. y_i) with dimension $\leqslant \dim M$ (resp. $\leqslant \dim N$) and containing M (resp. N) is identical with it and hence (a) implies (b) and (c) and *a fortiori* (d). Conversely, each of conditions (b) and (c) implies (a) by definition of $\text{rg}(u)$. Finally if (d) holds, there exists a family $(x_i^*)_{1 \leqslant i \leqslant s}$ of elements of E^* such that $\langle x_i, x_j^* \rangle = \delta_{ij}$ (no. 5, Corollary 1 to Theorem 7) and hence it follows from (33) that (y_i) is a basis of N, which completes the proof.

COROLLARY 1. *The rank of u is the smallest integer s such that there exists an expression $u = \sum_{i=1}^{s} x_i \otimes y_i$, where $x_i \in E$ and $y_i \in F$ for $1 \leqslant i \leqslant s$.*

This follows immediately from Proposition 17 and no. 2, Proposition 1.

COROLLARY 2. *Let* K *be a commutative field,* E, F *two vector spaces over* K *and* L *a commutative extension field of* K. *Let u be an element of $E \otimes_K F$,* M *and* N *its associated subspaces and u' the canonical image of u in $(E \otimes_K F)_{(L)}$ (canonically identified with $E_{(L)} \otimes_L F_{(L)}$,* cf. § 5, no. 1, Proposition 3); *then $\text{rg}(u') = \text{rg}(u)$ and the subspaces associated with u' are canonically identified with $M_{(L)}$ and $N_{(L)}$.*

If $u = \sum_{i=1}^{r} x_i \otimes y_i$, where the families (x_i) and (y_i) are free, then $u' = \sum_{i=1}^{r} (1 \otimes x_i) \otimes (1 \otimes y_i)$ and the families $(1 \otimes x_i)$ and $(1 \otimes y_i)$ are free in $E_{(L)}$ and $F_{(L)}$ respectively (§ 5, no. 1, Proposition 4).

9. EXTENSION OF SCALARS FOR A VECTOR SPACE

Recall (I, § 9, no. 1, Theorem 2) that a homomorphism of a field K into a non-zero ring A is necessarily *injective*.

PROPOSITION 18. *Let* ρ *be a homomorphism of a field* K *into a ring* A. *For every exact sequence of vector* K-*spaces and* K-*linear mappings*

$$E' \xrightarrow{u} E \xrightarrow{v} E''$$

the sequence

$$E'_{(A)} \xrightarrow{u_{(A)}} E_{(A)} \xrightarrow{v_{(A)}} E''_{(A)}$$

is exact.

This is a special case of no. 7, Proposition 14, taking account of § 1, no. 4, *Remark* 4.

COROLLARY. *For every* K-*linear mapping* $f: E' \rightarrow E$, $\operatorname{Im}(f_{(A)}) = (\operatorname{Im}(f))_{(A)}$, $\operatorname{Ker}(f_{(A)}) = (\operatorname{Ker}(f))_{(A)}$, $\operatorname{Coker}(f_{(A)}) = (\operatorname{Coker}(f))_{(A)}$, *to within canonical isomorphisms.*

PROPOSITION 19. *Let* ρ *be an injective homomorphism of a field* K *into a ring* A. *For every left vector space* E *over* K, *the canonical mapping* $\phi: E \rightarrow \rho^*(E) = A \otimes_K E$ *is injective. Moreover, for every vector subspace* E' *of* E, $\rho^*(E') = A \otimes_K E'$ *is canonically identified with a direct factor sub-A-module of* $A \otimes_K E$ *and, with this identification,*

$$(A \otimes_K E') \cap \phi(E) = \phi(E'). \tag{35}$$

The first assertion is a special case of § 5, no. 1, Proposition 4; the second is a special case of no. 7, Proposition 14; finally, to show (35), it suffices to take in A (considered as a right K-module) a basis $(a_\lambda)_{\lambda \in L}$ such that $a_{\lambda_0} = 1$ for some index λ_0 (no. 1, Theorem 2); the elements of $A \otimes_K E$ can be written uniquely as $\sum_\lambda a_\lambda \otimes x_\lambda$ with $x_\lambda \in E$ and, for such an element to belong to $A \otimes_K E'$, it is necessary and sufficient that $x_\lambda \in E'$ for all λ. On the other hand, the elements of $\phi(E)$ are those for which $x_\lambda = 0$ for $\lambda \neq \lambda_0$; for an element $\sum_\lambda a_\lambda \otimes x_\lambda$ to belong to $(A \otimes_K E') \cap \phi(E)$, it is necessary and sufficient that $x_\lambda = 0$ for $\lambda \neq \lambda_0$ and $x_{\lambda_0} \in E'$, whence the conclusion.

COROLLARY. *Let* ρ *be an injective homomorphism of a field* K *into a ring* A. *For a* K-*linear mapping* $f: E \rightarrow F$ (*where* E *and* F *are two vector spaces over* K) *to be injective* (resp. *surjective, zero*), *it is necessary and sufficient that* $f_{(A)}: E_{(A)} \rightarrow F_{(A)}$ *be injective* (resp. *surjective, zero*).

This follows immediately from Proposition 19 and the Corollary to Proposition 18.

PROPOSITION 20. *Let* ρ *be a homomorphism of a field* K *into a ring* A. *For every left vector space* E *over* K, *the canonical right* A*-module homomorphism*

$$\upsilon : (\mathrm{E}^*)_{(\mathrm{A})} \to (\mathrm{E}_{(\mathrm{A})})^*$$

(§ 5, no. 4) *is injective; it is bijective when* E *is finite-dimensional.*

The second assertion follows from § 5, no. 4, Proposition 8. To prove the first, we note that every element of $(\mathrm{E}^*)_{(\mathrm{A})}$ may be written as $\sum_i x_i^* \otimes \alpha_i$, where $\alpha_i \in \mathrm{A}$ and $(x_i^*)_{1 \leqslant i \leqslant n}$ is a *free* family in E^*; there corresponds to it in $(\mathrm{E}_{(\mathrm{A})})^*$ the linear form y^* such that $y^*(1 \otimes x) = \sum_i \rho(\langle x, x_i^* \rangle)\alpha_i$ for all $x \in \mathrm{E}$. But, there exists in E a family $(x_i)_{1 \leqslant i \leqslant n}$ such that $\langle x_i, x_j^* \rangle = \delta_{ij}$ (no. 5, Corollary 2 to Theorem 7), whence $y^*(1 \otimes x_i) = \alpha_i$; the relation $y^* = 0$ therefore implies $\alpha_i = 0$ for all i, which proves our assertion.

PROPOSITION 21. *Let* K *be a field and* L *an extension field of* K.

(i) *For every vector space* E *over* K, $\dim_{\mathrm{L}}(\mathrm{E}_{(\mathrm{L})}) = \dim_{\mathrm{K}} \mathrm{E}$.

(ii) *For every* K*-linear mapping* $u : \mathrm{E} \to \mathrm{F}$, *where* E *and* F *are vector spaces over* K, $\mathrm{rg}(u_{(\mathrm{L})}) = \mathrm{rg}(u)$.

If $(e_\iota)_{\iota \in \mathrm{I}}$ is a basis of E over K, $(1 \otimes e_\iota)_{\iota \in \mathrm{I}}$ is a basis of $\mathrm{E}_{(\mathrm{L})}$ over L (§ 5, no. 1, Proposition 4), whence the first assertion; the second follows from the first and the fact that $u_{(\mathrm{L})}(\mathrm{E}_{(\mathrm{L})})$ is canonically identified with $(u(\mathrm{E}))_{(\mathrm{L})}$ by the Corollary to Proposition 18.

PROPOSITION 22. *Let* K *be a commutative field,* $\rho : \mathrm{K} \to \mathrm{A}$ *an injective central homomorphism and* E, F *two vector spaces over* K. *Then the canonical homomorphism*

$$\omega : \mathrm{A} \otimes_{\mathrm{K}} \mathrm{Hom}(\mathrm{E}, \mathrm{F}) \to \mathrm{Hom}_{\mathrm{A}}(\mathrm{E}_{(\mathrm{A})}, \mathrm{F}_{(\mathrm{A})}) \tag{36}$$

(§ 5, no. 3, formula (17)) *is injective; it is bijective if* A *or* E *is a finite-dimensional vector space over* K.

This is a particular case of § 5, no. 3, Proposition 7.

10. MODULES OVER INTEGRAL DOMAINS

PROPOSITION 23. *In a module* E *over an integral domain* A, *the set* T *of non-free elements is a submodule of* E.

If x and y are not free, there exist two non-zero elements α, β in A such that $\alpha x = 0$ and $\beta y = 0$. Then $\alpha\beta \neq 0$ since A is an integral domain and $\alpha\beta(\lambda x + \mu y) = 0$ for all λ and μ in A since A is commutative, hence $\lambda x + \lambda y$ is not free.

Remark. Let E be a module over any commutative ring A. If x is a non-free element of E, every element of the submodule Ax is non-free. On the other hand, if A contains divisors of 0, the sum of two non-free elements of E may be free; for example, in $\mathbf{Z}/6\mathbf{Z}$ considered as a module over itself, 3 and 4 are not free, but $3 + 4 = 1$ is free.

Proposition 23 leads to the following definition:

DEFINITION 5. *In a module* E *over an integral domain* A, *the torsion submodule of* E *is the submodule of* E *consisting of the non-free elements* (also called the *torsion elements* of E).

When E is equal to its torsion submodule (that is when every element of E is annihilated by an element $\neq 0$ of A) E is called a *torsion module*. When the torsion submodule of E is reduced to 0 (that is every non-zero element of E is *free*) E is called (by an abuse of language) a *torsion-free module*.

Every submodule of a free A-module (and in particular every *projective* A-module) is torsion-free. The $\mathbf{Z}$-module $\mathbf{Q}$ is torsion-free.

PROPOSITION 24. *Let* A *be an integral domain. For every* A-*module* E, *let* T(E) *denote the torsion submodule of* E. *Let* $f: E \to E'$ *be an* A-*linear mapping,* E *and* E' *being* A-*modules.*

(i) $f(T(E)) \subset T(E')$.

(ii) *If* f *is injective,* $f(T(E)) = T(E') \cap f(E)$.

(iii) *If* f *is surjective and* $\operatorname{Ker}(f) \subset T(E)$, *then* $f(T(E)) = T(E')$.

Assertions (i) and (ii) are obvious. On the other hand, if f is surjective and $x' \in T(E')$, then $x' = f(x)$, where $x \in E$, and by hypothesis there exists $\alpha \neq 0$ in A such that $f(\alpha x) = \alpha x' = 0$; whence $\alpha x \in \operatorname{Ker}(f)$ and by the hypothesis there exists $\beta \neq 0$ in A such that $\beta(\alpha x) = 0$; as $\beta\alpha \neq 0$, $x \in T(E)$.

COROLLARY 1. *For every* A-*module* E, E/T(E) *is torsion-free.*

If $f: E \to E'$ is an A-linear mapping, let f_T denote the mapping $T(E) \to T(E')$ with the same graph as the restriction of f to T(E). With this notation:

COROLLARY 2. *For every exact sequence of* A-*modules and* A-*linear mappings*

$$0 \longrightarrow E' \xrightarrow{f} E \xrightarrow{g} E''$$

the sequence

$$0 \longrightarrow T(E') \xrightarrow{f_T} T(E) \xrightarrow{g_T} T(E'')$$

is exact.

$$\operatorname{Ker}(g_T) = \operatorname{Ker}(g) \cap T(E) = f(E') \cap T(E) = f(T(E')) = \operatorname{Im}(f_T).$$

PROPOSITION 25. *Let* A *be an integral domain and* (E_ι) *a family of* A-*modules; then*

$$T\Big(\bigoplus_\iota E_\iota\Big) = \bigoplus_\iota T(E_\iota). \tag{37}$$

Let (x_ι) be an element of $\bigoplus_\iota E_\iota$ such that $x_\iota \in T(E_\iota)$ for all ι; then each of the x_ι is annihilated by an element $\alpha_\iota \neq 0$ of A and it may be assumed that $\alpha_\iota = 1$ when $x_\iota = 0$; as the family (x_ι) has finite support, the element $\alpha = \prod_\iota \alpha_\iota$ of A is defined and $\neq 0$; it obviously annihilates $\bigoplus_\iota x_\iota$ and hence $\bigoplus_\iota T(E_\iota) \subset T\left(\bigoplus_\iota E_\iota\right)$; the converse is immediate.

If E and F are two A-modules, clearly $T(E \otimes_A F)$ contains the canonical images of $T(E) \otimes_A F$ and $E \otimes_A T(F)$; but examples can be given of *torsion-free* A-modules E, F such that $T(E \otimes_A F) \neq 0$ (Exercise 31).

Note that an *infinite* product of torsion modules is not necessarily a torsion module; for example, in the **Z**-module $\prod_{n=1}^{\infty} (\mathbf{Z}/p^n\mathbf{Z})$ (p an integer > 1), the element all of whose coordinates are 1 is free.

PROPOSITION 26. *Let* A *be an integral domain,* K *its field of fractions,* E *an* A-*module and* $E_{(K)} = K \otimes_A E$ *the vector space over* K *obtained by extending the ring of operators; let* ϕ *denote the canonical* A-*linear mapping* $x \mapsto 1 \otimes x$ *of* E *into* $E_{(K)}$.

(i) *Every element of* $E_{(K)}$ *is of the form* $\lambda^{-1}\phi(x)$ *for* $\lambda \in A$, $\lambda \neq 0$ *and* $x \in E$.

(ii) *The kernel of* ϕ *is the torsion submodule* T(E) *of* E.

(i) Every element of $E_{(K)}$ is of the form $z = \sum_{i=1}^{n} \xi_i \phi(x_i)$ with $\xi_i \in K$ and $x_i \in E$; for all i, there exists $\alpha_i \in A$ such that $\alpha_i \neq 0$ and $\alpha_i \xi_i \in A$; if $\alpha = \prod_{i=1}^{n} \alpha_i$, then $\alpha \neq 0$ and $\alpha\xi_i = \beta_i \in A$ for all i, whence, in $E_{(K)}$,

$$z = \alpha^{-1}(\alpha z) = \alpha^{-1} \sum_{i=1}^{n} \beta_i \phi(x_i) = \alpha^{-1}\phi\left(\sum_{i=1}^{n} \beta_i x_i\right)$$

since ϕ is A-linear.

(ii) If $x \neq 0$ is not free in E, there exists $\alpha \neq 0$ in A such that $\alpha x = 0$, whence $\alpha\phi(x) = \phi(\alpha x) = 0$ in $E_{(K)}$, which implies $\phi(x) = 0$. Conversely, suppose that, for some $x \in E$, $1 \otimes x = 0$ in $E_{(K)}$; we show that x is a torsion element of E. We consider the set $\mathfrak{M}$ of *monogenous* sub-A-modules of K; this is a right directed set under the relation of inclusion, for any two elements α, β of K can be written as $\alpha = \zeta^{-1}\xi$, $\beta = \zeta^{-1}\eta$, where ξ, η, ζ belong to A and $\zeta \neq 0$, hence $A.\alpha \subset A.\zeta^{-1}$ and $A.\beta \subset A.\zeta^{-1}$. Moreover K is the union of the modules $M \in \mathfrak{M}$ and can therefore be considered as the *direct limit* of the direct system defined by the modules $M \in \mathfrak{M}$ and the canonical injections (§ 6, no. 2, *Remark*). Hence also, to within a canonical isomorphism, $E_{(K)} = \varinjlim (M \otimes_A E)$ (§ 6, no. 3, Proposition 7) and the relation $1 \otimes x = 0$ in $E_{(K)}$ implies that there exists an

$M \in \mathfrak{M}$ such that $1 \in M$ and $1 \otimes x = 0$ *in the tensor product* $M \otimes_A E$ (*Set Theory*, III, § 7, no. 5, Lemma 1). It may further be supposed (replacing if need be M by a monogenous submodule $M' \supset M$ of K) that $M = A.\gamma^{-1}$, where $\gamma \in A$ and $\gamma \neq 0$. Now the mapping $\xi \mapsto \gamma\xi$ is an isomorphism of M onto the A-module A; on the other hand, the canonical isomorphism $A \otimes_A E \to E$ (§ 3, no. 4, Proposition 4) maps $\xi \otimes x$ to the element ξx of E; thus there exists an *isomorphism* $M \otimes_A E \to E$ which maps the tensor product $\xi \otimes x$ to the element $(\gamma\xi)x$ of E. The hypothesis $1 \otimes x = 0$ in $M \otimes_A E$ thus implies $\gamma x = 0$.

Remark. Let $\alpha^{-1}\phi(x)$, $\beta^{-1}\phi(y)$ be two elements of $E_{(K)}$, with $\alpha \in A$, $\beta \in A$, $x \in E$, $y \in E$, $\alpha\beta \neq 0$. In order that $\alpha^{-1}\phi(x) = \beta^{-1}\phi(y)$, it is necessary and sufficient that $\beta x - \alpha y$ be a *torsion element* of E, for this relation is equivalent to $\beta\phi(x) = \alpha\phi(y)$, which may also be written $\phi(\beta x - \alpha y) = 0$.

COROLLARY 1. *If* E *is a torsion-free* A*-module, the canonical mapping* $\phi: E \to E_{(K)}$ *is injective.*

Recall (§ 5, no. 1) that for every A-linear mapping of E into a vector space F over K, there exists one and only one K-linear mapping $\bar{f}: E_{(K)} \to F$ such that $f = \bar{f} \circ \phi$; we shall say that $\bar{f}$ is *associated* with f.

COROLLARY 2. *Let f be an* A*-linear mapping of* E *into a vector space* F *over* K; *if* $\operatorname{Ker}(f) \subset T(E)$, *the* K*-linear mapping* $\bar{f}$ *associated with* f *is injective.*

We write an element of $\operatorname{Ker}(\bar{f})$ in the form $\lambda^{-1}\phi(x)$, where $\lambda \in A$, $\lambda \neq 0$, $x \in E$; the relation $\bar{f}(\lambda^{-1}\phi(x)) = 0$ is equivalent to $\lambda^{-1}\bar{f}(\phi(x)) = 0$ in F and hence to $f(x) = \bar{f}(\phi(x)) = 0$. By hypothesis, this implies $x \in T(E)$ and hence $\phi(x) = 0$, which proves the corollary.

COROLLARY 3. *Let* E *be an* A*-module and g an* A*-linear mapping of* E *into a vector space* F *over* K *such that* $g(E)$ *generates* F *and* $\operatorname{Ker}(g) \subset T(E)$. *Then the* K*-linear mapping* $\bar{g}$ *associated with* g *is an isomorphism of* $E_{(K)}$ *onto* F.

$\bar{g}$ is injective by Corollary 2 and the hypothesis that $g(E)$ generates F implies that $\bar{g}$ is surjective.

For every A-module E, the vector space $E_{(K)}$ is said to be *associated* with E. For every subset S of E the *rank* of S over K (or by an abuse of language, the *rank* of S) is the rank of the canonical image $\phi(S)$ of S in $E_{(K)}$, in other words (no. 2, Definition 1) the dimension over K of the vector subspace of $E_{(K)}$ generated by $\phi(S)$.

When E is a *torsion-free* A-module, it is usually identified with its canonical image $\phi(E)$ in $E_{(K)}$. With this convention, every generating system of E contains a *basis* of $E_{(K)}$ (no. 1, Theorem 2). In particular:

COROLLARY 4. *Every finitely generated* A*-module is of finite rank.*

Note that the converse of this corollary is not necessarily true; for example $\mathbf{Q}$ is a $\mathbf{Z}$-module of rank 1 but is not finitely generated over $\mathbf{Z}$.

Recall (§ 5, no. 1) that for every linear mapping $f: E \to E'$ (where E and E' are A-modules), $f_{(K)}$ denotes the K-linear mapping $1_K \otimes f: E_{(K)} \to E'_{(K)}$.

PROPOSITION 27. *For every exact sequence*

$$E' \xrightarrow{f} E \xrightarrow{g} E''$$

of A*-linear mappings, the corresponding sequence of* K*-linear mappings*

$$E'_{(K)} \xrightarrow{f_{(K)}} E_{(K)} \xrightarrow{g_{(K)}} E''_{(K)}$$

is exact.

Suppose that $g_{(K)}(\lambda^{-1} \otimes x) = 0$, with $\lambda \in A$, $\lambda \neq 0$, $x \in E$; this is equivalent to $\lambda^{-1} \otimes g(x) = 0$ in $E''_{(K)}$ and hence also to

$$1 \otimes g(x) = \lambda(\lambda^{-1} \otimes g(x)) = 0;$$

by Proposition 26, there exists $\alpha \neq 0$ in A such that $\alpha g(x) = 0$ in E'', or also $g(\alpha x) = 0$. By hypothesis, there is therefore an $x' \in E'$ such that $\alpha x = f(x')$ and therefore $\lambda^{-1} \otimes x = f_{(K)}(\alpha^{-1}\lambda^{-1} \otimes x')$, which proves the proposition.

COROLLARY 1. *If* E' *is a submodule of* E, $E'_{(K)}$ *is canonically identified with a vector subspace of* $E_{(K)}$ *and* $(E/E')_{(K)}$ *with* $E_{(K)}/E'_{(K)}$.

It suffices to apply Proposition 27 to the exact sequence

$$0 \to E' \to E \to E/E' \to 0.$$

COROLLARY 2. *For every* A*-linear mapping* $f: E \to F$, $\mathrm{Ker}(f_{(K)}) = (\mathrm{Ker}(f))_{(K)}$, $\mathrm{Im}(f_{(K)}) = (\mathrm{Im}(f))_{(K)}$, $\mathrm{Coker}\,(f_{(K)}) = (\mathrm{Coker}(f))_{(K)}$ *to within canonical isomorphisms. In particular, for* $f_{(K)}$ *to be injective* (resp. *surjective*, resp. *zero*), *it is necessary and sufficient that* $\mathrm{Ker}(f) \subset T(E)$ (resp. *that* $\mathrm{Coker}(f)$ *be a torsion module*, resp. *that* $\mathrm{Im}(f) \subset T(F)$).

This follows from Corollary 1 and Proposition 26.

COROLLARY 3. *Let* E *be an* A*-module and* $(x_\lambda)_{\lambda \in L}$ *a family of elements of* E. *For* (x_λ) *to be a free family, it is necessary and sufficient that in the vector* K*-space* $E_{(K)}$ *the family* $(1 \otimes x_\lambda)$ *be free.*

The family (x_λ) defines an A-linear mapping $f: A^{(L)} \to E$ such that $f(e_\lambda) = x_\lambda$ for all $\lambda \in L$ ((e_λ) being the canonical basis of $A^{(L)}$) and to say that (x_λ) is free means that f is injective. It suffices to apply Corollary 2 to f, observing that $A^{(L)}$ is torsion-free (Proposition 25).

§8. RESTRICTION OF THE FIELD OF SCALARS IN VECTOR SPACES

Throughout this paragraph, K *denotes a field and* K′ *a subfield of* K. *On a set* V, *a right* (resp. *left*) *vector space structure over* K *defines, by restriction of scalars, a right* (resp. *left*) *vector space structure over* K′.

1. DEFINITION OF K′-STRUCTURES

PROPOSITION 1. *Let* V *be a right vector space over* K *and* V′ *a subset of* V *which is a vector subspace over* K′. *The following conditions are equivalent:*

(a) *The* K*-linear mapping* λ *of* $V'_{(K)} = V' \otimes_{K'} K$ *into* V *such that* $\lambda(x' \otimes \xi) = x'\xi$ *for all* $x' \in V'$, $\xi \in K$, *is bijective.*

(b) *Every* K′*-linear mapping* f' *of* V′ *into a vector* K*-space* W *can be extended uniquely to a* K*-linear mapping* f *of* V *into* W.

(c) *Every basis of* V′ *over* K′ *is a basis of* V *over* K.

(d) *There exists a basis of* V′ *over* K′ *which is also a basis of* V *over* K.

(e) *The vector* K*-space* V *is generated by* V′ *and every subset of* V′ *free over* K′ *is free over* K.

We know (§5, no. 1) that $V'_{(K)}$ has a right vector K-space structure under which $(x' \otimes \xi)\eta = x' \otimes (\xi\eta)$ (ξ, η in K, $x' \in V'$) and that for every K′-linear mapping f' of V′ into a vector K-space W, there exists one and only one K-linear mapping $\bar{f}'$ of $V'_{(K)}$ into W such that $\bar{f}'(x' \otimes 1) = f'(x')$ for $x' \in V'$. If j is the canonical injection of V′ into V, λ is just the corresponding K-linear mapping $\bar{j}$. If λ is bijective, then for every K′-linear mapping $f': V' \to W$, $\bar{f}' \circ \lambda^{-1}$ is the unique K-linear mapping of V into W extending f'; in other words, (a) implies (b). Conversely, if (b) holds, there exists in particular a K-linear mapping μ of V into $V'_{(K)}$ such that $\mu(x') = x' \otimes 1$ for all $x' \in V'$; it is immediate that $\mu \circ \lambda = 1_{V'_{(K)}}$; on the other hand, $\lambda(\mu(x')) = x'$ for all $x' \in V'$ and, as by hypothesis $j: V' \to V$ can be extended uniquely to an endomorphism of V, of necessity $\lambda \circ \mu = 1_V$, which completes the proof that (a) and (b) are equivalent.

For every basis B′ of V′ over K′, the set B of elements $v' \otimes 1$ of $V'_{(K)}$, where v' runs through B′, is a basis of $V'_{(K)}$ over K (§5, no. 1, Proposition 4) and $\lambda(B) = B'$. For λ to be bijective, it is necessary that the image under λ of every basis of $V'_{(K)}$ be a basis of V over K and it is sufficient that it be so for a single basis of $V'_{(K)}$ (§1, no. 11, Corollary 2 to Proposition 17). This proves the equivalence of (a), (c) and (d).

As every subset of V′ which is free over K′ is contained in a basis of V′ over K′ (§7, no. 1, Theorem 2), (c) implies (e). Finally, suppose that (e) holds; if B′ is a basis of V′ over K′, it is a free subset of V over K; on the other hand B′

generates V′ over K′ and hence generates V over K by hypothesis; therefore B′ is a basis of V over K, which proves that (e) implies (c).

DEFINITION 1. *Let* V *be a right vector space over a field* K *and* K′ *a subfield of* K. *Every vector sub-*K′*-space* V′ *of* V *satisfying the equivalent conditions of Proposition* 1 *is called a* K′*-structure on* V.

Example. Let B be a basis of V over K. For *every* subfield K′ of K, the vector sub-K′-space of V generated by B admits B as a basis over K′ and hence is a K′-structure on V. *For example, if K is commutative and V is taken to be the polynomial K-algebra $K[X_1, \ldots, X_n]$, then, for every subfield K′ of K, $K'[X_1, \ldots, X_n]$ is a K′-structure on V.*

2. RATIONALITY FOR A SUBSPACE

DEFINITION 2. *Let* V *be a right vector space over* K, *with a* K′*-structure* V′. *A vector of* V *is said to be rational over* K′ *if it belongs to* V′. *A vector sub-*K*-space* W *of* V *is said to be rational over* K′ *if it is generated* (*over* K) *by vectors rational over* K′.

Let $(v'_\iota)_{\iota \in I}$ be a basis of V′ over K′, which is therefore also a basis of V over K (no. 1, Proposition 1). For a vector $x = \sum_\iota v'_\iota \xi_\iota$ of V to be rational over K′, it is necessary and sufficient that $\xi_\iota \in K'$ for all $\iota \in I$.

If W is a vector sub-K-space of V which is *rational* over K′, it follows from Definition 2 that $W' = W \cap V'$ is a vector sub-K′-space of W, which *generates* W over K; on the other hand every subset of W′ which is free over K′ is also free over K since it is contained in V′ (no. 1, Proposition 1). It follows (no. 1, Proposition 1) that W′ is a K′-*structure* on W said to be *induced* by the K′-structure V′ on V.

For every vector sub-K′-space W′ of V′, we shall denote by W′.K the vector sub-K-space of V consisting of the linear combinations of elements of W′ with coefficients in K.

PROPOSITION 2. *Let* V *be a right vector space over* K *and* V′ *a* K′*-structure on* V. *The mapping* $W' \mapsto W'.K$ *is a bijection of the set of vector sub-*K′*-spaces of* V′ *onto the set of vector sub-*K*-spaces of* V *which are rational over* K′ *and the inverse bijection is* $W \mapsto W \cap V'$.

Clearly the bijection $\lambda^{-1}: V \to V' \otimes_{K'} K$, the inverse of the bijection λ defined in no. 1, Proposition 1, maps every vector sub-K′-space W′ of V′ onto its image under the canonical injection $x' \mapsto x' \otimes 1$ and W′.K onto $W' \otimes_{K'} K$; the assertions of Proposition 2 are thus consequences of Definition 2 and § 7, no. 9, Proposition 19.

COROLLARY 1. *Every sum and every intersection of vector sub-*K*-spaces of* V *which are rational over* K′ *is a rational subspace over* K′.

The assertion concerning sums is obvious. On the other hand, if $(W'_\iota)_{\iota \in I}$ is a family of vector sub-K'-spaces of V', then $\left(\bigcap_{\iota \in I} W'_\iota\right) \otimes_{K'} K = \bigcap_{\iota \in I} (W_\iota \otimes_{K'} K)$ (§ 7, no. 7, Corollary to Proposition 14), which proves the corollary.

A basis B of V over K is said to be *rational over* K' if it consists of vectors which are rational over K'.

COROLLARY 2. *Every basis of* V *over* K *which is rational over* K' *is a basis of* V' *over* K'.

If W' is the vector sub-K'-space of V' generated by B, then $W'.K = V = V'.K$, whence $V' = W'$ by virtue of Proposition 2.

3. RATIONALITY FOR A LINEAR MAPPING

DEFINITION 3. *Let* V_1, V_2 *be two right vector spaces over* K *with* K'*-structures* V'_1, V'_2 *respectively. A* K*-linear mapping* $f: V_1 \to V_2$ *is said to be rational over* K' *if* $f(V'_1) \subset V'_2$.

If V_3 is a third right vector space over K, with a K'-structure V'_3 and a K-linear mapping $g: V_2 \to V_3$ is rational over K', clearly $g \circ f: V_1 \to V_3$ is rational over K'.

PROPOSITION 3. *Let* V_1, V_2 *be two right vector spaces over* K *and* V'_1, V'_2 K'*-structures on* V_1, V_2 *respectively.* V_1 (resp. V_2) *is canonically identified with* $V'_1 \otimes_{K'} K$ (resp. $V'_2 \otimes_{K'} K$) (no. 1, Proposition 1).

(i) *The mapping* $f' \mapsto f' \otimes 1_K = f'_{(K)}$ *is a bijection of* $\mathrm{Hom}_{K'}(V'_1, V'_2)$ *onto the set of* K*-linear mappings of* V_1 *into* V_2 *which are rational over* K'*; the inverse bijection associates with every* K*-linear mapping* $f: V_1 \to V_2$ *rational over* K' *the* K'*-linear mapping* $f': V'_1 \to V'_2$ *with the same graph as the restriction of* f *to* V'_1.

(ii) *For every* K*-linear mapping* $f: V_1 \to V_2$ *rational over* K',

$$f(V'_1) = f(V_1) \cap V'_2 \quad \text{and} \quad \overset{-1}{f}(V'_2) = V'_1 + \mathrm{Ker}(f).$$

(i) Clearly with the above identifications, if $f': V'_1 \to V'_2$ is a K'-linear mapping, $f'_{(K)} = f' \otimes 1_K$ is rational over K' and f' is the mapping with the same graph as the restriction of $f'_{(K)}$ to V'_1. Conversely, if $f: V_1 \to V_2$ is a K-linear mapping rational over K' and $f': V'_1 \to V'_2$ has the same graph as the restriction of f to V'_1, f and $f'_{(K)}$ coincide on V'_1, which is a generating system of V_1 over K, hence $f = f'_{(K)}$.

(ii) If $f = f' \otimes 1_K$, then $f(V_1) = f(V'_1 \otimes_{K'} K) = f'(V'_1) \otimes_{K'} K$ and, as $f'(V'_1) \subset V'_2, f(V'_1) = f'(V'_1) = f(V_1) \cap V'_2$ (§ 7, no. 9, Proposition 19); the formula $\overset{-1}{f}(V'_2) = V'_1 + \mathrm{Ker}(f)$ follows immediately.

COROLLARY 1. *In the notation of Proposition* 3, $\mathrm{Im}(f) = (\mathrm{Im}(f'))_{(K)}$,

$$\mathrm{Ker}(f) = (\mathrm{Ker}(f'))_{(K)}, \qquad \mathrm{Coker}(f) = (\mathrm{Coker}(f'))_{(K)}.$$

In particular, for f to be injective (resp. *surjective, zero*), *it is necessary and sufficient that f' be so. If f is bijective, its inverse mapping is rational over* K'.

This is a particular case of § 7, no. 9, Corollary to Proposition 18.

COROLLARY 2. *Let $f: V_1 \to V_2$ be a K-linear mapping rational over* K'. *For every vector sub-K-space W_1 of V_1* (resp. W_2 *of* V_2) *rational over* K', $f(W_1)$ (resp. $\overset{-1}{f}(W_2)$) *is a vector sub-K-space of* V_2 (resp. V_1) *rational over* K'.

In the notation of Proposition 3, for every vector sub-K'-space W'_1 of V'_1, $f_{(K)}(W'_1 \otimes_{K'} K) = f'(W'_1) \otimes_{K'} K$; whence the assertion relating to W_1 (no. 2, Proposition 2). On the other hand, let W'_2 be a vector sub-K'-space of V'_2 and let g' be the canonical K'-linear mapping $V'_2 \to V'_2/W'_2$; then $\overset{-1}{f'}(W'_2) = \mathrm{Ker}(g' \circ f')$; hence, by Corollary 1,

$$\overset{-1}{f'_{(K)}}(W'_2 \otimes_{K'} K) = \overset{-1}{f'}(W'_2) \otimes_{K'} K,$$

whence the assertion relating to W_2.

Let V_1, V_2 be two right vector K-spaces with K'-structures V'_1, V'_2 respectively. It is immediate that $V'_1 \times V'_2$ is a *K'-structure* on $V_1 \times V_2$, called the *product* of the K'-structures V'_1 and V'_2.

PROPOSITION 4. *For a K-linear mapping $f: V_1 \to V_2$ to be rational over K', it is necessary and sufficient that its graph Γ be rational over K' for the product K'-structure on $V_1 \times V_2$.*

Let g be the mapping $x_1 \mapsto (x_1, f(x_1))$ from V_1 to $V_1 \times V_2$; this is a K-linear mapping such that $\Gamma = g(V_1)$; if f is rational over K', so is Γ by virtue of Corollary 2 to Proposition 3. Conversely, suppose that Γ is rational over K' and let it have the K'-structure induced by that on $V_1 \times V_2$; it follows immediately from the definitions that the restrictions p_1, p_2 to Γ of the projections pr_1, pr_2, are K-linear mappings, rational over K' of Γ into V_1 and V_2 respectively. As p_1 is bijective, its inverse mapping q_1 is rational over K' (Corollary 1 to Proposition 3) and hence so is $f = p_2 \circ q_1$.

4. RATIONAL LINEAR FORMS

Let V be a right vector space over K with a K'-structure V'. As K'_d is a K'-structure on the right vector K-space K_d, we may define *linear forms* $x^* \in V^*$, *rational over* K', as the linear mappings of V into K_d, rational over K' for the

K′-structures on V and K_d. By virtue of no. 3, Proposition 3, the set R′ of these linear forms is the image of the dual V'^* of V′ under the composite mapping

$$V'^* \xrightarrow{\phi} K \otimes_{K'} V'^* \xrightarrow{\upsilon} V^* \tag{1}$$

where $\phi(x'^*) = 1 \otimes x'^*$ and $\upsilon(\xi \otimes x'^*)$ is the linear form y^* on V such that $y^*(x') = \xi\langle x'^*, x'\rangle$ for all $x' \in V'$ (§ 5, no. 4). We know that this mapping is injective (§ 7, no. 9, Propositions 20 and 19) and clearly R′ is a left vector sub-K′-space of V^*; moreover every subset of R′ free over K′ is free over K. But in general R′ *does not necessarily generate* V^* over K and does not therefore define a K′-structure on V^* (Exercise 2). However, if V is of *finite* dimension *n* over K, V'^* is of dimension *n* over K′ and R′ then defines canonically a K′-structure on V^*.

PROPOSITION 5. *Let* V *be a right vector space over* K, V′ *a* K′*-structure on* V *and* W *a vector sub-*K*-space of* V. *For* W *to be rational over* K′, *it is necessary and sufficient that there exist a set* $H \subset V^*$ *of rational linear forms over* K′ *such that* W *is the orthogonal of* H *in* V (§ 2, no. 4).

Let H be a subset of V^* whose elements are rational linear forms over K′. For all $x^* \in H$, the kernel of x^* is a vector sub-K-space of V, rational over K′ (no. 3, Corollary 2 to Proposition 3); the intersection of these kernels is therefore also a vector sub-K-space of V, rational over K′ (no. 2, Corollary 1 to Proposition 2).

Conversely, let W be a vector sub-K-space of V rational over K′, so that W is identified with $W' \otimes_{K'} K$, where $W' = W \cap V'$ (no. 2, Proposition 2). For a linear form $x'^* \in V'^*$ to be zero on W′, it is necessary and sufficient that the linear form $x^* \in V^*$ which corresponds to it under (1) be zero on W, for by no. 3, Corollary I to Proposition 3,

$$\operatorname{Ker}(x^*) = (\operatorname{Ker}(x'^*)) \otimes_{K'} K \quad \text{and} \quad \operatorname{Ker}(x^{*\prime}) = (\operatorname{Ker}(x^*)) \cap V'.$$

Let H′ be the orthogonal of W′ in V'^*; we know (§ 7, no. 5, Theorem 7) that W′ is the orthogonal of H′ in V′; if H is the image of H′ in V^* under the mapping (1), it follows from the above that W is the orthogonal of H in V, taking account of no. 7, Corollary to Proposition 14.

5. APPLICATION TO LINEAR SYSTEMS

PROPOSITION 6. (i) *Given a system of homogeneous linear equations*

$$\sum_{\iota \in I} \alpha_{\mu\iota}\xi_\iota = 0 \quad (\mu \in M) \tag{2}$$

whose coefficients $\alpha_{\mu\iota}$ *belong to* K′, *every solution* (ξ_ι) *of this system consisting of elements of* K *is a linear combination with coefficients in* K *of solutions* (ξ'_ι) *of* (2) *consisting of elements of* K′.

(ii) *Given a system of linear equations*

$$\sum_{\iota \in I} \alpha_{\mu\iota} \xi_\iota = \beta_\mu \quad (\mu \in M) \tag{3}$$

whose coefficients $\alpha_{\mu\iota}$ *and right hand sides* β_μ *belong to* K′, *if there exists a solution to the system consisting of elements of* K, *there also exists a solution consisting of elements of* K′.

(i) For every set S, let the right vector K-space $K_d^{(S)}$ be given the K′-structure $K_d'^{(S)}$. Let f be the K-linear mapping of $K_d^{(I)}$ into $K_d^{(M)}$ mapping every vector $(\xi_\iota)_{\iota \in I}$ to the vector $(\zeta_\mu)_{\mu \in M}$ defined by $\zeta_\mu = \sum_{\iota \in I} \alpha_{\mu\iota} \xi_\iota$ for all $\mu \in M$. Clearly f is rational over K′; its kernel V, which is the set of solutions in K of the system (2), is a subspace of $K_d^{(I)}$ rational over K′ (no. 3, Corollary 2 to Proposition 3) and hence generated by the solutions of (2) in K′.

(ii) We consider K as a left vector K′-space; there exists a K′-linear projector p of K onto its vector subspace K_s' (§ 7, no. 3, Proposition 4); if (ξ_ι) is a solution of (3) in K, then $\sum_{\iota \in I} \alpha_{\mu\iota} p(\xi_\iota) = p\left(\sum_{\iota \in I} \alpha_{\mu\iota} \xi_\iota\right) = p(\beta_\mu) = \beta_\mu$, which proves that $(p(\xi_\iota))$ is a solution of (3) in K′.

A ring K is called (left) *faithfully flat* over a subring K′ if Proposition 6 holds for K and K′; we shall study this notion in detail later (*Commutative Algebra*, I, § 3).

6. SMALLEST FIELD OF RATIONALITY

Let V be a right vector K-space with a K′-structure V′. For every field L such that $K' \subset L \subset K$, we write $V_L = V'.L$; clearly every basis of V′ over K′ is a basis of V over K and a basis of V_L over L. Hence V_L is an L-structure on V and V′ a K′-structure on V_L.

PROPOSITION 7. (i) *Let* V *be a right vector* K-*space with a* K′-*structure* V′. *For every vector* $x \in V$ (resp. *every vector sub-K-space* W *of* V), *the set of subfields* L *of* K *containing* K′ *and such that* x (resp. W) *is rational over* L *has a least element* K′(x) (resp. K′(W)).

(ii) *Let* V_1, V_2 *be two right vector* K-*spaces with* K′-*structures* V_1', V_2' *respectively. For every* K-*linear mapping* f *of* V_1 *into* V_2, *the set of subfields* L *of* K *containing* K′ *and such that* f *is rational over* L *has a least element* K′(f).

We first prove assertion (i) for a vector $x \in V$. Let B be a basis of V rational over K′; B is a basis of V′ over K′ and a basis of V_L over L for every field L such that $K' \subset L \subset K$; for $x = \sum_{b \in B} b\xi_b$ to be rational over L, it is necessary and sufficient that the ξ_b belong to L (no. 2) and hence the smallest field L with this property is the field *generated* by K', and the ξ_b for $b \in B$.

We next prove (ii). Let B_1, B_2 be bases of V_1, V_2 respectively, which are rational over K' and write, for all $b_1 \in B_1$, $f(b_1) = \sum_{b_2 \in B_2} b_2 \alpha_{b_2 b_1}$ (*the family $(\alpha_{b_2 b_1})$ is just the *matrix* of f with respect to the bases B_1 and B_2; cf. § 10, no. 4$_*$). As B_1 (resp. B_2) is a basis of $(V_1)_L$ (resp. $(V_2)_L$) over L for every field such that $K' \subset L \subset K$, for f to be rational over L, it is necessary and sufficient that the $\alpha_{b_2 b_1}$ belong to L; the smallest field with this property is therefore the field *generated* by K' and the $\alpha_{b_2 b_1}$ for $b_1 \in B_1$, $b_2 \in B_2$.

Finally, to establish assertion (i) for a subspace W of V, we shall first prove the following lemma:

Lemma 1. *Let* V *be a right vector* K*-space with a* K'*-structure* V' *and* W *a vector sub-*K*-space of* V. *There exist two vector sub-*K*-spaces* W_1, W_2 *of* V, *rational over* K', *such that* V *is the direct sum of* W_1 *and* W_2 *and, if* V *is identified with* $W_1 \times W_2$, W *is the graph of a* K*-linear mapping* g *of* W_1 *into* W_2.

Let B be a basis of V rational over K'. Applying Theorem 2 of § 7, no. 1, to a basis of W over K, considered as a free subset of V, and the generating system the union of this free subset and B, it is seen that there exists a subset C of B such that V is the direct sum of W and the subspace W_2 of V generated by C. Also let W_1 be the subspace of V generated by $B - C$. As $B \subset V'$, clearly W_1 and W_2 are rational over K'. Moreover, for all $x \in W_1$, there exists one and only one vector $g(x)$ of W_2 such that $x + g(x) \in W$, since V is the direct sum of W and W_2; then W is the graph of g and g is K-linear since W is a vector sub-K-space of V.

Having proved this lemma, it is known that W is rational over a subfield L of K containing K' if and only if g is rational over L (no. 3, Proposition 4). The smallest field $K'(g)$ such that g is rational over $K'(g)$ is therefore also the smallest field over which W is rational.

7. CRITERIA FOR RATIONALITY

For every subfield L of K, let $\mathrm{End}_L(K)$ denote the endomorphism ring of K considered as a *left vector space* over L; if L contains K', $\mathrm{End}_L(K)$ is a subring of $\mathrm{End}_{K'}(K)$. For every subset $\mathscr{M}$ of $\mathrm{End}_{K'}(K)$, there exists a *largest subfield* L of K containing K' and such that $\mathscr{M}$ is contained in $\mathrm{End}_L(K)$, namely the set of $\xi \in K$ such that $\phi(\xi\eta) = \xi.\phi(\eta)$ for all $\eta \in K$ and all $\phi \in \mathscr{M}$ (it is immediately verified that this set is a ring and, on the other hand, replacing η by $\xi^{-1}\eta$ in the preceding relation, we obtain $\phi(\xi^{-1}\eta) = \xi^{-1}.\phi(\eta)$ when $\xi \neq 0$). We shall call this field the *centralizer* of $\mathscr{M}$ in K and denote it by $\chi(\mathscr{M})$.

Now let V be a right vector K-space with a K'-structure V'. For all $\phi \in \mathrm{End}_{K'}(K)$, there exists one and only one endomorphism ϕ_V of the **Z**-

module V such that $\phi_V(x'.\xi) = x'.\phi(\xi)$ for $x' \in V'$ and $\xi \in K$: for, in no. 1, a **Z**-isomorphism λ of $V' \otimes_{K'} K$ onto V was defined which maps $x' \otimes \xi$ to $x'.\xi$ and ϕ_V is necessarily equal to $\lambda \circ (1_{V'} \otimes \phi) \circ \lambda^{-1}$.

THEOREM 1. *Let* $\mathscr{M}$ *be a subset of* $\mathrm{End}_{K'}(K)$ *and* $L = \chi(\mathscr{M})$ *the subfield of* K *the centralizer of* $\mathscr{M}$.

(i) *Let* V *be a right vector* K*-space with a* K'*-structure. For a vector* $x \in V$ *to be rational over* L, *it is necessary and sufficient that* $\phi_V(x.\eta) = x.\phi(\eta)$ *for all* $\phi \in \mathscr{M}$ *and all* $\eta \in K$. *For a vector sub-*K*-space* W *of* V *to be rational over* L, *it is necessary and sufficient that* $\phi_V(W) \subset W$ *for all* $\phi \in \mathscr{M}$.

(ii) *Let* V_1, V_2 *be two right vector* K*-spaces each with a* K'*-structure. For a* K*-linear mapping* f *of* V_1 *into* V_2 *to be rational over* L, *it is necessary and sufficient that* $f(\phi_{V_1}(x_1)) = \phi_{V_2}(f(x_1))$ *for all* $x_1 \in V_1$ *and all* $\phi \in \mathscr{M}$.

We first prove assertion (i) for x. Let B be a basis of V rational over K' and write $x = \sum_{b \in B} b.\xi_b$; then, for $\phi \in \mathscr{M}$ and $\eta \in K$,

$$\phi_V(x.\eta) - x.\phi(\eta) = \sum_{b \in B} b.(\phi(\xi_b\eta) - \xi_b.\phi(\eta))$$

and therefore, the relations

"for all $\phi \in \mathscr{M}$ and all $\eta \in K$, $\phi_V(x.\eta) = x.\phi(\eta)$"

and

"for all $\phi \in \mathscr{M}$, all $b \in B$ and all $\eta \in K$, $\phi(\xi_b\eta) = \xi_b.\phi(\eta)$"

are equivalent. The second of these relations means that for all $b \in B$, $\xi_b \in \chi(\mathscr{M})$, which proves the first assertion of (i).

We next prove (ii). For f to be rational over L, it is necessary and sufficient that, for all $x'_1 \in V_1$ rational over K', $f(x'_1)$ is a vector in V_2 rational over L; this will imply that $f(x_1)$ is rational over L for every vector x_1 of V_1 rational over L, such a vector being a linear combination with coefficients in L of vectors rational over K'. The above condition is equivalent, by the first part of the argument, to the relation

$$f(x'_1).\phi(\eta) = \phi_{V_2}(f(x'_1).\eta) \quad \text{for } \phi \in \mathscr{M} \text{ and } \eta \in K \tag{4}$$

which may also be written

$$f(\phi_{V_1}(x'_1.\eta)) = \phi_{V_2}(f(x'_1.\eta)) \quad \text{for } \phi \in \mathscr{M} \text{ and } \eta \in K. \tag{5}$$

As every element of V_1 is a linear combination with coefficients in K of elements of V_1 rational over K', condition (5) is equivalent to

$$f(\phi_{V_1}(x_1)) = \phi_{V_2}(f(x_1))$$

for all $x_1 \in V_1$ and all $\phi \in \mathscr{M}$.

Finally, to prove the second assertion in (i), we use no. 6, Lemma 1: W is the graph of a K-linear mapping $g: W_1 \to W_2$ and W is rational over if and only if the mapping g is rational over L (no. 3, Proposition 4). By (ii), for g to be rational over L, it is necessary and sufficient that $g(\phi_{W_1}(x_1)) = \phi_{W_2}(g(x_1))$ for all $x_1 \in W_1$ and all $\phi \in \mathscr{M}$; as $\phi_V = \phi_{W_1} \times \phi_{W_2}$, the above condition means that the graph W of g is stable under ϕ_V for all $\phi \in \mathscr{M}$.

§ 9. AFFINE SPACES AND PROJECTIVE SPACES

1. DEFINITION OF AFFINE SPACES

DEFINITION 1. *Given a left* (resp. *right*) *vector space* T *over a field* K, *an affine space attached to* T *is any homogeneous space* E *of the additive group* T (I, § 5, no. 5) *such that* 0 *is the only operator in* T *leaving invariant all the elements of* E (that is, T *operates faithfully and transitively on* E). *Under these conditions,* T *is called the translation space of* E *and its elements are called the translations of* E (or *free vectors of* E).

In what follows we shall confine our attention to the case where T is a left vector space over K. The dimension (over K) of the translation vector space T of an affine space E is called the *dimension* of E (over K) and is denoted by dim E or $\dim_K$ E. An affine space of dimension one (resp. two) is called an *affine line* (resp. an *affine plane*). The elements of an affine space are also called *points*.

Under the conditions of Definition 1, for $\mathbf{t} \in T$ and $a \in E$ we shall denote by $\mathbf{t} + a$ or $a + \mathbf{t}$ the transform of the point a under $\mathbf{t}$. Then the relations

$$(1) \qquad \mathbf{s} + (\mathbf{t} + a) = (\mathbf{s} + \mathbf{t}) + a, \qquad 0 + a = a$$

hold for $\mathbf{s} \in T$, $\mathbf{t} \in T$, $a \in E$. The mapping $x \mapsto x + \mathbf{t}$ is a bijection of E onto itself, which we identify with $\mathbf{t}$. Definition 1 moreover implies that, for all $a \in E$, the mapping $\mathbf{t} \mapsto \mathbf{t} + a$ is a *bijection* of T onto E. In other words, given two points a, b of E, there exists one and only one translation $\mathbf{t}$ such that $b = \mathbf{t} + a$; we shall denote this translation by $b - a$; then the formulae

$$(2) \qquad a - a = 0, \qquad a - b = -(b - a), \qquad b = (b - a) + a$$
$$(c - b) + (b - a) = c - a$$

hold for $a \in E$, $b \in E$, $c \in E$. If four points a, b, a', b' of E are such that $b - a = b' - a'$, the formula

$$b' = (b' - b) + (b - a) + a = (b' - a') + (a' - a) + a$$

and the commutativity of addition in T show that $b' - b = a' - a$.

Given a point $a \in E$, the mapping $x \mapsto x - a$ is a bijection of E onto T; when E is identified with T under this mapping, we say that E is considered as the vector space obtained *by taking a as origin* in E. Conversely, every vector space T has canonically the structure of an affine space attached to T, namely the homogeneous space structure corresponding to the subgroup $\{0\}$ of T (I, § 5, no. 6).

Remark. The definitions of this no. and some of the results which follow extend immediately to the case where, instead of a vector space T, we consider an arbitrary *commutative group with operators* T.

2. BARYCENTRIC CALCULUS

PROPOSITION 1. *Let* $(x_\iota)_{\iota \in I}$ *be a family of points in an affine space* E *and* $(\lambda_\iota)_{\iota \in L}$ *a family of elements of* K *of finite support such that* $\sum_{\iota \in I} \lambda_\iota = 1$ $\left(\text{resp. } \sum_{\iota \in I} \lambda_\iota = 0\right)$ *If a is any point of* E, *the point* $x \in E$ *defined by*

$$x - a = \sum_{\iota \in I} \lambda_\iota (x_\iota - a)$$

(resp. *the free vector* $\sum_{\iota \in I} \lambda_\iota (x_\iota - a)$) *is independent of the point considered.*

If a' is another point of E, then

$$\sum_\iota \lambda_\iota (x_\iota - a') = \sum_\iota \lambda_\iota ((x_\iota - a) + (a - a')) = \sum_\iota \lambda_\iota (x - a) + \left(\sum_\iota \lambda_\iota\right)(a - a').$$

If $\sum_\iota \lambda_\iota = 1$, then $\sum_\iota \lambda_\iota (x_\iota - a) = (x - a) + (a - a') = x - a'$; if $\sum_\iota \lambda_\iota = 0$, then $\sum_\iota \lambda_\iota (x - a') = \sum_\iota \lambda_\iota (x - a)$; whence the proposition.

Under the conditions of Proposition 1, the point x defined by

$$x - a = \sum_{\iota \in I} \lambda_\iota (x_\iota - a)$$

$\left(\text{resp. the free vector } \sum_{\iota \in I} \lambda_\iota (x_\iota - a)\right)$ is denoted by $\sum_{\iota \in I} \lambda_\iota x_\iota$.

Thus in particular the notation $b - a$ introduced in no. 1 is recovered. When $\sum_\iota \lambda_\iota = 1$, the point $x = \sum_\iota \lambda_\iota x_\iota$ is called the *barycentre of the points* x_ι *given the masses* λ_ι.

Given m points $a_1, \dots, a_m$ of E, whose number m is not a multiple of the characteristic of K (V, § 1), the point $g = \sum_{i=1}^{m} \frac{1}{m} a_i$ is called (by an abuse of language) the *barycentre of the points* a_i $(1 \leqslant i \leqslant m)$ (for $m = 2$, we say "midpoint" instead of "barycentre"); it is characterized by the relation $\sum_{i=1}^{m} (a_i - g) = 0$.

3. AFFINE LINEAR VARIETIES

DEFINITION 2. *Given an affine space* E, *a subset* V *of* E *is called an affine linear variety* (or simply a *linear variety* or an *affine subset* of E) *if, for every family* $(x_\iota)_{\iota \in I}$ *of points of* V *and every family* $(\lambda_\iota)_{\iota \in I}$ *of elements of* K *of finite support such that* $\sum_{\iota \in I} \lambda_\iota = 1$, *the barycentre* $\sum_{\iota \in I} \lambda_\iota x_\iota$ *belongs to* V.

It amounts to the same to say that the condition of Definition 2 holds for every *finite* family of points of V.

The empty set is a linear variety; every intersection of linear varieties is a linear variety.

Let V be a non-empty subset of E and a a point of V; the relation

$$x - a = \sum_{i=1}^{n} \lambda_i (x_i - a)$$

means that x is a barycentre $\sum_{i=1}^{n} \lambda_i x_i + \left(1 - \sum_{i=1}^{n} \lambda_i\right) a$ of the family consisting of the x_i and a. Therefore:

PROPOSITION 2. *For a non-empty subset* V *of an affine space* E *to be a linear variety, it is necessary and sufficient that* V *be a vector subspace for the vector space structure on* E *obtained by taking a point of* V *as origin.*

In particular, the non-empty affine linear varieties of a vector space T (considered as an affine space) are just the *translates* of the vector subspaces of T; the vector subspaces of T are therefore the linear varieties containing 0.

Let V be a non-empty linear variety of the affine space E; the set of free vectors $x - y$, where x and y run through V, is a vector subspace D of the translation space T of E called the *direction* of V: for, if $a \in V$, then

$$x - y = (x - a) - (y - a)$$

and our assertion follows from Proposition 2. It is immediate that D operates faithfully and transitively on V, which therefore has canonically the structure of an *affine space attached to* D. By the *dimension* of the affine variety V, we mean the dimension of V with this affine space structure, that is the dimension of the vector space D. The linear varieties of dimension 0 are the points of E; those of dimension 1 (resp. 2) are called *lines* (resp. *planes*) of E.

Every vector $\neq 0$ belonging to the direction of a line is called a *direction vector* of this line; its components with respect to a basis of T form what is called a system of *direction parameters* of the line in question.

The *codimension* of a linear variety V in E is the codimension of its direction D in T; a linear variety of codimension 1 in E is called an (affine) *hyperplane* of E.

Two linear varieties with the same direction are called *parallel*; it amounts to the same to say that one is derived from the other by translation. If V is a linear variety in T (considered as an affine space), its direction is the linear variety parallel to V and containing 0.

PROPOSITION 3. *Given a family* $(a_\iota)_{\iota \in I}$ *of points of an affine space* E, *the set* V *of barycentres* $\sum_{\iota \in I} \lambda_\iota a_\iota \left((\lambda_\iota)\right.$ *of finite support,* $\left.\sum_{\iota \in I} \lambda_\iota = 1\right)$ *is a linear variety of* E.

If the family (a_ι) is empty, then $V = \varnothing$ because of the condition $\sum_\iota \lambda_\iota = 1$. It may therefore be assumed that the family (a_ι) is non-empty and in this case the proposition is obvious, taking one of the a_ι as origin in E.

The variety V is obviously the smallest linear variety containing the a_ι; it is said to be *generated* by the family (a_ι) and this family is called a *generating system* of V.

In the notation of Proposition 3, assuming the family (a_ι) is non-empty, for the expression for every point $x \in V$ in the form $x = \sum_\iota \lambda_\iota a_\iota$ to be *unique,* it is necessary and sufficient that, denoting an arbitrary index of I by κ, the family of vectors $a_\iota - a_\kappa$, where ι runs through the set of indices $\neq \kappa$, be free in T. Then the family $(a_v)_{\iota \in I}$ of points of E is said to be *affinely free* (or that its elements form an *affinely free system*, or are *affinely independent*) and that λ_ι is the *barycentric coordinate* of x of index ι with respect to the affinely free family (a_ι).

A family $(a_\iota)_{\iota \in I}$ of points of E which is not affinely free is said to be *affinely related.*

PROPOSITION 4. *For a non-empty family* $(a_\iota)_{\iota \in I}$ *of points in an affine space* E *to be affinely related, it is necessary and sufficient that there exist a family* $(\lambda_\iota)_{\iota \in I}$ *of elements not all zero in* K, *of finite support, such that* $\sum_{\iota \in I} \lambda_\iota = 0$ *and* $\sum_{\iota \in I} \lambda_\iota a_\iota = 0$.

Given an index $\kappa \in I$, to say that the family of vectors $(a_\iota - a_\kappa)$, where ι runs through the set of indices $\neq \kappa$, is related in T, means that there exists a family of scalars $(\lambda_\iota)_{\iota \neq \kappa}$ not all zero such that $\sum_{\iota \neq \kappa} \lambda_\iota(a_\iota - a_\kappa) = 0$, which may also be written $\sum_{\iota \in I} \lambda_\iota a_\iota = 0$, with $\lambda_\kappa = -\sum_{\iota \neq \kappa} \lambda_\iota$, in other words $\sum_{\iota \in I} \lambda_\iota = 0$.

PROPOSITION 5. *For a non-empty family $(a_\iota)_{\iota \in I}$ of points of an affine space* E *to be affinely free, it is necessary and sufficient that, for every index* $\kappa \in I$, a_κ *do not belong to the linear variety generated by the* a_ι *of index* $\neq \kappa$.

The proposition is obvious if I has only a single element. Otherwise, taking as origin in E one of the a_ι of index $\neq \kappa$, the proposition follows from § 7, no. 1, *Remark*.

4. AFFINE LINEAR MAPPINGS

DEFINITION 3. *Given two affine spaces* E, E′ *attached to two vector spaces* T, T′ *over the same field* K, *a mapping u of* E *into* E′ *is called an affine linear mapping* (or *an affine mapping*) *if, for every family* $(x_\iota)_{\iota \in I}$ *of points of* E *and every family* $(\lambda_\iota)_{\iota \in I}$ *such that* $\sum_{\iota \in I} \lambda_\iota = 1$,

$$u\Big(\sum_{\iota \in I} \lambda_\iota x_\iota\Big) = \sum_{\iota \in I} \lambda_\iota u(x_\iota). \tag{3}$$

PROPOSITION 6. *Let u be an affine mapping of* E *into* E′. *There exists one and only one linear mapping v of* T *into* T′ *such that*

$$u(x + \mathbf{t}) = u(x) + v(\mathbf{t})$$

for all $x \in E$, $\mathbf{t} \in T$.

Let a be any point of E. The mapping

$$\mathbf{t} \mapsto u(a + \mathbf{t}) - u(a)$$

is a linear mapping v_a of T into T′, for we may write

$$a + \lambda\mathbf{t} = \lambda(a + \mathbf{t}) + (1 - \lambda)a$$
$$a + \mathbf{s} + \mathbf{t} = (a + \mathbf{s}) + (a + \mathbf{t}) - a$$

and it follows from (3) that $v_a(\lambda\mathbf{t}) = \lambda v_a(\mathbf{t})$ and $v_a(\mathbf{s} + \mathbf{t}) = v_a(\mathbf{s}) + v_a(\mathbf{t})$. Moreover, if b is another point of E, then $v_a = v_b$; for the relation

$$(a + \mathbf{t}) - a + b = b + \mathbf{t}$$

implies

$$u(a + \mathbf{t}) - u(a) + u(b) = u(b + \mathbf{t})$$

that is $u(a + \mathbf{t}) - u(a) = u(b + \mathbf{t}) - u(b)$. Whence the existence of v; the uniqueness is immediate.

v is called the linear mapping of T into T′ *associated with u*. Conversely, for every linear mapping v of T into T′ and every ordered pair of points $a \in E$, $a' \in E'$, it is immediately verified that

$$x \mapsto a' + v(x - a)$$

is an affine mapping of E into E′ whose associated linear mapping is v. To say that u is an affine mapping of E into E′ therefore also means that, if an arbitrary point a in E and the point $u(a)$ in E′ are taken as origins, u is a *linear* mapping for the two vector spaces thus obtained.

Let E″ be a third affine space, T″ its translation space, u' an affine mapping of E′ into E″ and v' the linear mapping of T′ into T″ associated with u'. Clearly $u' \circ u$ is an affine mapping of E into E″; moreover, for $a \in \mathrm{E}$ and $\mathbf{t} \in \mathrm{T}$,

$$u'(u(a + \mathbf{t})) = u'(u(a) + v(\mathbf{t})) = u'(u(a)) + v'(v(\mathbf{t}))$$

and hence $v' \circ v$ is the linear mapping of T into T″ associated with $u' \circ u$. For an affine mapping u to be bijective, it is necessary and sufficient that the associated linear mapping v be so, and u^{-1} is then an affine mapping whose associated linear mapping is v^{-1}.

In particular, the affine bijections of E onto itself form a group G, called the *affine group* of E. The mapping which associates with $u \in \mathrm{G}$ the linear mapping v associated with u is, by the above, a *homomorphism* of G *onto the linear group* $\mathbf{GL}(\mathrm{T})$. If u is a translation, v is the identity and conversely. Hence, the kernel of the above homomorphism is the translation group T of E which is therefore a *normal subgroup* of G.

If $u \in \mathrm{G}$, the automorphism $\mathbf{t} \mapsto u\mathbf{t}u^{-1}$ of T (where $\mathbf{t}$ is identified with the translation $x \mapsto x + \mathbf{t}$) is the linear mapping v associated with u. For $x \in \mathrm{E}$ and $\mathbf{t} \in \mathrm{T}$, by definition

$$x + u\mathbf{t}u^{-1} = u(u^{-1}(x) + \mathbf{t}) = u(u^{-1}(x)) + v(\mathbf{t}) = x + v(\mathbf{t})$$

and hence $u\mathbf{t}u^{-1} = v(\mathbf{t})$.

Let $a \in \mathrm{E}$ and G_a be the subgroup of G consisting of the $u \in \mathrm{G}$ such that $u(a) = a$. If E is identified with T by taking a as origin, G_a is identified with $\mathbf{GL}(\mathrm{T})$. Every $u \in \mathrm{G}$ can be expressed uniquely in the form $u = \mathbf{t}_1 u_1$ (resp. in the form $u = u_2\mathbf{t}_2$), where u_1, u_2 are in G_a and $\mathbf{t}_1$, $\mathbf{t}_2$ in T: for, writing $\mathbf{t}_1 = u(a) - a$, $u^{-1}\mathbf{t}_1 \in \mathrm{G}_a$, whence the existence of u_1 and $\mathbf{t}_1$; the existence of u_2 and $\mathbf{t}_2$ is obtained analogously. The uniqueness follows from the fact that $\mathrm{G}_a \cap \mathrm{T}$ reduces to the identity element of G. Moreover

$$\mathbf{t}_1 u_1 = u_1(u_1^{-1}\mathbf{t}_1 u_1)$$

whence $u_2 = u_1$, $\mathbf{t}_2 = u_1^{-1}\mathbf{t}_1 u_1$. Finally, the linear mappings associated with u and u_1 are the same and hence, if as above G_a is identified with $\mathbf{GL}(\mathrm{T})$, u_1 is the linear mapping from T to itself associated with u. It is thus seen that G is the *semi-direct product* of G_a by T (I, § 6, no. 1).

Let E, E′ be two affine spaces over K. The direct (resp. inverse) image of a linear variety of E (resp. E′) under an affine mapping u of E into E′ is a linear variety of E′ (resp. E); the *rank* of u is by definition the dimension of $u(\mathrm{E})$; it is equal to the rank of the linear mapping associated with u. If V, V′ are linear varieties of the same finite dimension m in E, E′ respectively, there exists an affine mapping u of E into E' such that $u(V) = V'$: taking as origins in E and E' points of V and V' respectively, then taking in E (resp. E') a basis whose first m vectors form a basis of V (resp. V'), the proposition follows immediately from § 1, no. 11, Corollary 3 to Proposition 17.

As the field K has canonically a left vector space structure (of dimension 1) over K, it can be considered as an affine space of dimension 1. An affine mapping of an affine space D (over K) into the affine space K is also called an *affine linear function* (or an *affine function*). If a point a is taken as origin in E, every affine function on E can then be written uniquely as $x \mapsto \alpha + v(x)$, where $\alpha \in K$ and v is a linear form on the vector space E thus obtained; the affine functions on E therefore form a *right vector space over* K of dimension $1 + \dim E$. If u is a non-constant affine function on E and $\lambda \in K$, the set of $x \in E$ satisfying the equation $u(x) = \lambda$ is a hyperplane; conversely, for every hyperplane H in E, there exists an affine function u_0 on E such that $H = \overset{-1}{u_0}(0)$ and every affine function u such that $H = \overset{-1}{u}(0)$ is of the form $u_0\mu$, where $\mu \in K$ (§ 7, no. 5, Proposition 11). If u is an affine function on E, the hyperplanes with equations $u(x) = \alpha$ and $u(x) = \beta$ are parallel.

5. DEFINITION OF PROJECTIVE SPACES

DEFINITION 4. *Given a left* (resp. *right*) *vector space* V *over a field* K, *the left* (resp. *right*) *projective space derived from* V, *denoted by* $\mathbf{P}(V)$, *is the quotient of the complement* $V - \{0\}$ *of* $\{0\}$ *in* V *by the equivalence relation* $\Delta(V)$ "*there exists* $\lambda \neq 0$ *in* K *such that* $y = \lambda x$ (resp. $y = x\lambda$) *between* x *and* y *in* $V - \{0\}$.

When $V = K_s^{n+1}$, we also write $\mathbf{P}_n(K)$ instead of $\mathbf{P}(K_s^{n+1})$ and $\Delta_n(K)$ instead of $\Delta(V)$.

Definition 4 can also be expressed by saying that $\mathbf{P}(V)$ is the set of lines (passing through 0) in V with the origin removed; $\mathbf{P}(V)$ is therefore canonically identified with the set of lines (passing through 0) in V. The elements of a projective space are called the *points* of that space.

When V is of dimension n, the integer $n - 1$ is called the *dimension* of the projective space $\mathbf{P}(V)$ if n is finite, and the cardinal n otherwise; this cardinal is denoted by $\dim_K \mathbf{P}(V)$ or $\dim \mathbf{P}(V)$. Thus a projective space of dimension -1 is empty and a projective space of dimension 0 is a single point. A projective space of dimension 1 (resp. 2) is called a *projective line* (resp. *projective plane*).

Henceforth we shall only consider left projective spaces.

6. HOMOGENEOUS COORDINATES

Let V be a vector space of finite dimension $n + 1$ over K, $\mathbf{P}(V)$ the projective space of dimension n derived from V and $(e_i)_{0 \leq i \leq n}$ a basis of V. Let π denote the canonical mapping of $V - \{0\}$ onto the quotient set $\mathbf{P}(V)$. For every point $x = \sum_{i=0}^{n} \xi_i e_i$ of $V - \{0\}$, $(\xi_0, \xi_1, \ldots, \xi_n)$ is called a *system of homogeneous coordinates* of the point $\pi(x)$ with respect to the basis (e_i) of V. Every system (ξ_i) of $n + 1$ elements *not all zero* of K is therefore a system of homogeneous coordinates of a point of $\mathbf{P}(V)$ with respect to (e_i); for two such systems (ξ_i), (ξ'_i) to be systems of homogeneous coordinates of the same point of $\mathbf{P}(V)$ with respect to the same basis (e_i), it is necessary and sufficient that there exists an element $\lambda \neq 0$ of K such that $\xi'_i = \lambda\xi_i$ for $0 \leqslant i \leqslant n$.

This definition is immediately generalized to the case where V is infinite-dimensional.

Given another basis $(\bar{e}_i)$ of V such that $e_i = \sum_{j=0}^{n} \alpha_{ij}\bar{e}_j$ $(0 \leqslant i \leqslant n)$ and a system (ξ_i) of homogeneous coordinates of $\pi(x)$ with respect to the basis (e_i), for a system $(\bar{\xi}_i)$ of $n + 1$ elements of K to be a system of homogeneous coordinates of $\pi(x)$ with respect to the basis $(\bar{e}_i)$, it is necessary and sufficient that there exist $\lambda \neq 0$ in K such that

$$\lambda\bar{\xi}_i = \sum_{j=0}^{n} \xi_j\alpha_{ji} \quad \text{for } 0 \leqslant i \leqslant n.$$

In particular, if $e_i = \gamma_i\bar{e}_i$ with $\gamma_i \neq 0$ $(0 \leqslant i \leqslant n)$, then $\bar{\xi}_i = \mu\xi_i\gamma_i$ where $\mu \neq 0$.

7. PROJECTIVE LINEAR VARIETIES

Let W be a vector subspace of a vector space V; the canonical image of $W - \{0\}$ in the projective space $\mathbf{P}(V)$ derived from V is called a *projective linear variety* (or simply a *linear variety* when no confusion is to be feared); as the equivalence relation $\Delta(W)$ on $W - \{0\}$ is induced by the relation $\Delta(V)$, the projective linear variety the image of $W - \{0\}$ in $\mathbf{P}(V)$ can be identified with the projective space $\mathbf{P}(W)$ derived from W and hence we may speak of the dimension of such a variety. In a projective space $\mathbf{P}(V)$, the canonical image of a hyperplane (with the origin removed) of V is a linear variety called a *projective hyperplane* (or simply a *hyperplane*); if $\mathbf{P}(V)$ is of finite dimension n the hyperplanes in $\mathbf{P}(V)$ are the linear varieties of dimension $n - 1$.

Every proposition concerning vector subspaces of a vector space goes over to a proposition concerning projective linear varieties. For example, if a projective space $\mathbf{P}(V)$ is of finite dimension n and $(e_i)_{0 \leqslant i \leqslant n}$ is a basis of V, every linear variety $L \subset \mathbf{P}(V)$ of dimension r can be defined by a system of $n - r$ homogeneous linear equations

$$\sum_{i=0}^{n} \xi_i\alpha_{ij} = 0 \qquad (1 \leqslant j \leqslant n - r) \tag{4}$$

between the homogeneous coordinates ξ_i $(0 \leqslant i \leqslant n)$ of a point of $\mathbf{P}(V)$ with respect to the basis (e_i), the left hand sides of (4) being linearly independent forms on V. In particular, a projective hyperplane is defined by a single homogeneous linear equation with coefficients not all zero. Conversely, the points of $\mathbf{P}(V)$ satisfying an arbitrary system of homogeneous linear equations with respect to the ξ_i form a linear variety L; if the system considered consists of $k \leqslant n + 1$ equations, L is of dimension $\geqslant n - k$.

Every intersection of linear varieties of $\mathbf{P}(V)$ is a linear variety; for every subset A of $\mathbf{P}(V)$, there exists a smallest linear variety L containing A; it is called the linear variety *generated by* A and A is called a *generating system* of L. If W is the vector subspace of V generated by $\overset{-1}{\pi}(A)$, then $L = \mathbf{P}(W)$.

If L and M are any two linear varieties in $\mathbf{P}(V)$ and N the variety generated by $L \cup M$, then (§ 7, no. 3, Corollary 3 to Proposition 4)

$$\dim L + \dim M = \dim(L \cap M) + \dim N. \tag{5}$$

In particular, if $\mathbf{P}(V)$ is finite-dimensional and $\dim L + \dim M \geqslant \dim \mathbf{P}(V)$, it follows from (5) that $L \cap M$ is non-empty.

Let (x_ι), (y_ι) be two families of points in the vector space V with the same indexing set, such that $y_\iota = \lambda_\iota x_\iota$, where $\lambda_\iota \neq 0$ for all ι. If the family (x_ι) is free, so is (y_ι) and conversely; then it is said that the family of points $\pi(x_\iota)$ of $\mathbf{P}(V)$ is *projectively free* (or simply *free*). It amounts to the same to say that for every index κ, the point $\pi(x_\kappa)$ does not belong to the linear variety generated by the $\pi(x_\iota)$ for $\iota \neq \kappa$. A family of points of $\mathbf{P}(V)$ which is not projectively free is called *projectively related* (or simply *related*).

For a family (x_ι) of points of $V - \{0\}$ to be such that the family $(\pi(x_\iota))$ is projectively free and generates $\mathbf{P}(V)$, it is necessary and sufficient that (x_ι) be a basis of V. If $\mathbf{P}(V)$ is of dimension n the number of elements in such a family is therefore $n + 1$. Note that giving such a family $(\pi(x_\iota))$ in $\mathbf{P}(V)$ does not determine (even to within a left factor) the homogeneous coordinates of a given point of $\mathbf{P}(V)$ with respect to a basis (y_ι) of V such that $\pi(y_\iota) = \pi(x_\iota)$ for all ι (cf. no. 6).

8. PROJECTIVE COMPLETION OF AN AFFINE SPACE

Let V be a (left) vector space over a field K and consider the vector space $K_s \times V$ over K; the projective space $\mathbf{P}(K_s \times V)$ is called the projective space *canonically associated* with the vector space V. If V is of dimension n, $\mathbf{P}(K_s \times V)$ is of the same dimension n. Consider in $K_s \times V$ the affine hyperplane $V_1 = \{1\} \times V$, whose direction (no. 3) is the subspace $V_0 = \{0\} \times V$; if a line (passing through 0) of $K_s \times V$ is not contained in V_0, it contains a point (α, x) with $\alpha \neq 0$ and $x \in V$, hence it contains also the point $\alpha^{-1}(\alpha, x) = (1, \alpha^{-1}x)$ of V_1; the converse is immediate and it is seen that there is a one-to-one correspondence between the points of V_1 and the lines (passing through 0) of $K_s \times V$ not contained in V_0, each of the latter meeting V_1 in one and only one point. It follows that the mapping $x \mapsto \phi(x) = \pi(1, x)$ is an injection (called *canonical*) of V into the projective space $\mathbf{P}(K_s \times V)$; V is often identified with its image under this injection. The complement of $\phi(V)$ in $\mathbf{P}(K_s \times V)$ is the projective hyperplane $\mathbf{P}(V_0)$ called the *hyperplane at infinity* of $\mathbf{P}(K_s \times V)$ (or of V, by an abuse of language); its points are also called the "points at infinity" of $\mathbf{P}(K_s \times V)$ (or of V). If (a_ι) is a basis of V and in $K_s \times V$ the basis is taken consisting of the elements $e_\iota = (0, a_\iota)$ and the element $e_\omega = (1, 0)$, the points at infinity in $\mathbf{P}(K_s \times V)$ are those whose homogeneous coordinate of index ω is 0.

Let M be an affine linear variety in V (no. 3) and D its direction; the canonical image $\phi(M)$ of M in $\mathbf{P}(K_s \times V)$ is contained in the canonical image $\overline{M} = \pi(M_2)$ of the vector subspace M_2 of $K_s \times V$ generated by the affine variety $M_1 = \{1\} \times M$ of $K_s \times V$. More precisely, if (a_ι) is an affinely free system of M generating M, the elements $(1, a_\iota)$ form a basis of M_2 and therefore $\overline{M}$ is just the *projective linear variety generated by* $\phi(M)$; if M is finite-dimensional, $\overline{M}$ has the same dimension as M. The complement of $\phi(M)$ in $\overline{M}$ is the intersection of $\overline{M}$ and the hyperplane at infinity and is equal to the canonical image $\pi(M_0)$, where $M_0 = \{0\} \times D$.

Conversely, let N be a projective linear variety not contained in the hyperplane at infinity and let $R = \overset{-1}{\pi}(N)$; $R \cap V_1$ is an affine linear variety of $K \times V$ of the form $\{1\} \times M$, where M is an affine linear variety of V and it is immediately seen that N is the affine linear variety $\overline{M}$ generated by $\phi(M)$.

There is therefore a one-to-one correspondence between the affine linear varieties of V and the projective linear varieties of $\mathbf{P}(K_s \times V)$ not contained in the hyperplane at infinity; for two affine linear varieties of V to be *parallel*, it is necessary and sufficient that the projective linear varieties which they generate have the same intersection with the hyperplane at infinity (which is sometimes expressed by saying that the two affine linear varieties in question have the same points at infinity).

9. EXTENSION OF RATIONAL FUNCTIONS

If the results of no. 8 are applied to the vector space $V = K_s$ of dimension 1, it is seen that there exists a canonical injection ϕ of K_s into the projective line $\mathbf{P}_1(K) = \mathbf{P}(K_s \times K_s)$; for all $\xi \in K$, $\phi(\xi)$ is the point with homogeneous coordinates $(1, \xi)$ with respect to the canonical basis (§ 1, no. 11) of $K_s \times K_s$. The complement of $\phi(K)$ in $\mathbf{P}_1(K)$ consists of the single point with homogeneous coordinates $(0, 1)$ with respect to the above basis; it is called the "point at infinity". $\mathbf{P}_1(K)$ is also called the *projective field* associated with K and denoted by $\tilde{K}$, the point at infinity in $\tilde{K}$ being denoted by ∞.

*Consider in particular the case where K is a *commutative* field and let

$f \in K(X)$ be a rational function in one indeterminate over K (IV, § 4); if $f \neq 0$, there is a unique expression $f = \alpha p/q$, where $\alpha \in K^*$ and p and q are two relatively prime monic polynomials (VII, § 1); let m and n be their respective degrees and let $r = \sup(m, n)$. We write

$$p_1(T, X) = T^r p(X/T), \qquad q_1(T, X) = T^r q(X/T);$$

p_1 and q_1 are two homogeneous polynomials of degree r over K such that $p(X) = p_1(1, X)$, $q(X) = q_1(1, X)$. Hence, for every element $\xi \in K$ which is not a zero of $q(X)$, $f(\xi) = \alpha p(\xi)/q(\xi)$ is defined and we may write

$$f(\xi) = \alpha p_1(1, \xi)/q_1(1, \xi) = \alpha p_1(\lambda, \lambda\xi)/q_1(\lambda, \lambda\xi)$$

for all $\lambda \neq 0$ in K. Consider then the mapping

$$(\eta, \xi) \mapsto (q_1(\eta, \xi), p_1(\eta, \xi))$$

of K^2 into itself; it is compatible with the equivalence relation $\Delta(K^2)$ and therefore defines, when passing to the quotients, a mapping $\tilde{f}$ of $\tilde{K}$ into itself which coincides with $\xi \mapsto f(\xi)$ at the points where this rational function is defined; it is said, by an abuse of language, that $\tilde{f}$ is the *canonical extension* of f to $\tilde{K}$.

For example, if $f = 1/X$, then $\tilde{f}(0) = \infty$ and $\tilde{f}(\infty) = 0$; if

$$f = (aX + b)/(cX + d)$$

with $ad - bc \neq 0$, then $\tilde{f}(-d/c) = \infty$, $\tilde{f}(\infty) = a/c$ if $c \neq 0$, $\tilde{f}(\infty) = \infty$ if $c = 0$. If $f = a_0X^n + \cdots + a_n$ is a polynomial of degree $n > 0$, then $\tilde{f}(\infty) = \infty$. ¶

10. PROJECTIVE LINEAR MAPPINGS

Let V, V′ be two left vector spaces over a field K, f a linear mapping of V into V′ and $N = \overset{-1}{f}(0)$ its kernel. It is immediate that the image under f of a line (passing through 0) in V not contained in N is a line (passing through 0) in V′; hence, on passing to the quotients, f defines a mapping g of $\mathbf{P}(V) - \mathbf{P}(N)$ into $\mathbf{P}(V')$. Such a mapping is called a *projective linear mapping* (or, simply, a *projective mapping*); although it is defined on $\mathbf{P}(V) - \mathbf{P}(N)$ and not on $\mathbf{P}(V)$ (when $N \neq \{0\}$), we shall say by an abuse of language that g is a projective mapping of $\mathbf{P}(V)$ into $\mathbf{P}(V')$. The projective linear variety $\mathbf{P}(N)$, where g is not defined, is called the *centre* of g.

Note that, when g is defined on the whole of $\mathbf{P}(V)$ (that is when $N = \{0\}$), g is an *injection* of $\mathbf{P}(V)$ into $\mathbf{P}(V')$.

When bases $(a_\lambda)_{\lambda \in L}$, $(b_\mu)_{\mu \in M}$ are given in V and V′ respectively, a projective mapping of $\mathbf{P}(V)$ into $\mathbf{P}(V')$ maps a point of $\mathbf{P}(V)$ with homogeneous coordinates ξ_λ ($\lambda \in L$) to a point of $\mathbf{P}(V')$ with a system of homogeneous coordinates η_μ ($\mu \in M$) of the form

$$\eta_\mu = \sum_{\lambda \in L} \xi_\lambda \alpha_{\lambda\mu} \qquad (\alpha_{\lambda\mu} \in K). \tag{6}$$

The centre of g is the linear variety defined by the equations

$$\sum_{\lambda \in L} \xi_\lambda \alpha_{\lambda\mu} = 0 \qquad (\mu \in M).$$

If C is the centre of g and M is a linear variety of $\mathbf{P}(V)$, the image under g of $M - (M \cap C)$ is a linear variety of $\mathbf{P}(V')$ denoted (by an abuse of language) by $g(M)$. Then

$$\dim g(M) + \dim(M \cap C) + 1 = \dim M \tag{7}$$

(§ 7, no. 4, formula (12)). If M′ is a linear variety of $\mathbf{P}(V')$, $\overset{-1}{g}(M') \cup C$ is a linear variety of $\mathbf{P}(V)$ and

$$\dim(\overset{-1}{g}(M') \cup C) = \dim C + \dim(M' \cap g(\mathbf{P}(V))) + 1. \tag{8}$$

It is said, by an abuse of language, that $\overset{-1}{g}(M') \cup C$ is the *inverse image* of M′ under g.

As the values taken by a linear mapping on a basis (e_ι) of V can be chosen arbitrarily in V′, it is seen that there exists a projective linear mapping of $\mathbf{P}(V)$ into $\mathbf{P}(V')$ taking *arbitrary* values at the points $\pi(e_\iota)$. But (even when g is everywhere defined) giving $g(\pi(e_\iota))$ does not determine g uniquely (Exercise 10).

The composition of two projective mappings which are bijections is a projective mapping; so is the inverse mapping of such a bijection. The bijective projective mappings of a projective space $\mathbf{P}(V)$ onto itself thus form a group, called the *projective group* of $\mathbf{P}(V)$ and denoted by $\mathbf{PGL}(V)$; we write $\mathbf{PGL}_n(K)$ or $\mathbf{PGL}(n, K)$ instead of $\mathbf{PGL}(K_s^n)$.

Remark. In a projective space $\mathbf{P}(V)$ over a field K, let $H = \mathbf{P}(W)$ be a hyperplane. There exists a bijective linear mapping f of V onto $K_s \times W$ such that $f(W) = W$; let g be the projective mapping obtained from f by passing to the quotients. It has been seen (no. 8) that the complement of $\mathbf{P}(W)$ in $\mathbf{P}(K_s \times W)$ can be identified with an affine space whose translation space is W. When $\mathbf{P}(V)$ is identified with $\mathbf{P}(K_s \times W)$ by means of g, it is said that H *has been taken as hyperplane at infinity* in $\mathbf{P}(V)$; the complement of H in P(V) is then identified with an affine space whose translation space is W.

11. PROJECTIVE SPACE STRUCTURE

Given a set E and a field K, a (*left*) *projective space structure* on E with respect to the field K is defined by giving a non-empty set Φ of *bijections* of subsets of the projective space $\mathbf{P}(K_s^{(E)})$ *onto* E satisfying the following axioms:

(EP_I) *The set of definition of every mapping $f \in \Phi$ is a linear variety of $\mathbf{P}(K_s^{(E)})$.*

(EP_{II}) *For every ordered pair of elements f, g of Φ defined respectively on the linear varieties $\mathbf{P}(V)$ and $\mathbf{P}(W)$, the bijection $h = \overset{-1}{g} \circ f$ of $\mathbf{P}(V)$ onto $\mathbf{P}(W)$ is a projective mapping.*

(EP_{III}) *Conversely, if $f \in \Phi$ is defined on the linear variety $\mathbf{P}(V)$ and h is a bijective projective mapping of $\mathbf{P}(V)$ onto a linear variety $\mathbf{P}(W) \subset \mathbf{P}(K_s^{(E)})$, then $f \circ h^{-1} \in \Phi$.*

Let E be a set, $(V_\lambda)_{\lambda \in L}$ a family of vector spaces over K and suppose given for each $\lambda \in L$ a bijection f_λ of $\mathbf{P}(V_\lambda)$ onto E such that, for every ordered pair of indices λ, μ, $\overset{-1}{f_\lambda} \circ f_\mu$ is a *projective mapping* of $\mathbf{P}(V_\mu)$ onto $\mathbf{P}(V_\lambda)$. Then we can define on E a projective space structure with respect to K as follows: let $(e_\iota)_{\iota \in I}$ be a basis of a space V_λ and write $a_\iota = f_\lambda(\pi(e_\iota))$; let b_ι be the element of index a_ι in the canonical basis of $K_s^{(E)}$ (§ 1, no. 11). The relation $\iota \neq \kappa$ implies $b_\iota \neq b_\kappa$ because of the hypothesis that f_λ is bijective; hence the b_ι form a basis of a vector subspace W_0 of $K_s^{(E)}$ and there therefore exists a bijective projective mapping h of $\mathbf{P}(W_0)$ onto $\mathbf{P}(V_\lambda)$ such that $h(\pi(b_\iota)) = \pi(e_\iota)$ for all $\iota \in I$. If Φ is taken to be the set of all bijective projective mappings $f_\lambda \circ h \circ g^{-1}$, where g runs through the set of all bijective projective mappings $\mathbf{P}(W) \subset \mathbf{P}(K_s^{(E)})$, it is immediately verified that Φ satisfies axioms (EP_I), (EP_{II}) and (EP_{III}). It is moreover immediate that Φ depends neither on the choice of index $\lambda \in L$, nor on the choice of basis (e_ι) in V_λ, nor on the choice of h.

In particular (taking L to consist of a single element), every projective space $\mathbf{P}(V)$ derived from a vector space V (no. 5, Definition 4) thus has a well determined "projective space structure" in the sense of the definition given in this no. Hence any set with a projective space structure can be called a *projective space*.

With the same notation, a *linear variety* in a projective space E is a subset M of E such that, for at least *one* bijection $f \in \Phi$ defined on $\mathbf{P}(V) \subset \mathbf{P}(K_s^{(E)})$, $\overset{-1}{f}(M)$ is a linear variety in $\mathbf{P}(V)$ in the sense of no. 7 (this property then holds for *all* $f \in \Phi$). It follows from the above that every linear variety in a projective space has canonically a projective space structure.

A projective space E is said to be *of dimension n* if, for all $f \in \Phi$, $\overset{-1}{f}(E)$ is a linear variety of dimension n (it suffices that this hold for *one* mapping $f \in \Phi$).

§ 10. MATRICES

1. DEFINITION OF MATRICES

DEFINITION 1. *Let* I, K, H *be three sets; a matrix of type* (I, K) *with elements in* H (or *a matrix of type* (I, K) *over* H) *is any family* $M = (m_{\iota\kappa})_{(\iota, \kappa) \in I \times K}$ *of elements of* H *whose indexing set is the product* I × K. *For all* $\iota \in I$, *the family* $(m_{\iota\kappa})_{\kappa \in K}$ *is called the row of* M *of index* ι; *for all* $\kappa \in K$, *the family* $(m_{\iota\kappa})_{\iota \in I}$ *is called the column of* M *of index* κ.

If I (resp. K) is finite, M is said to be a matrix with a finite number of rows (resp. columns). The set of matrices of type (I, K) over H is identified with the product $H^{I \times K}$.

The names "row" and "column" arise from the fact that, in the case where I and K are intervals $(1, p)$, $(1, q)$ of **N**, the elements of the matrix are envisaged as set out in a rectangular array with p rows (arranged horizontally) and q columns (arranged vertically):

$$\begin{pmatrix} m_{11} & m_{12} & \dots & m_{1q} \\ m_{21} & m_{22} & \dots & m_{2q} \\ . & . & . & . \\ m_{p1} & m_{p2} & \dots & m_{pq} \end{pmatrix}$$

When p and q are explicit integers sufficiently small for it to be practicable, it is a convention that the above array is a symbol effectively denoting the matrix in question; this notation enables us to dispense with the use of indices, it being understood that the indices of an element are determined by its place in the array; for example, when we speak of the matrix

$$\begin{pmatrix} a & b & c \\ d & e & f \end{pmatrix}$$

we mean the matrix $(m_{ij})_{1 \leqslant i \leqslant 2, 1 \leqslant j \leqslant 3}$ such that

$$m_{11} = a,\ m_{12} = b,\ m_{13} = c,\ m_{21} = d,\ m_{22} = e,\ m_{23} = f.$$

Instead of matrix of type $((1, p), (1, q))$, we also say matrix *of type* (p, q), or matrix *with* p *rows and* q *columns*, if no confusion arises; the set of matrices of type (p, q) over H is sometimes denoted by $\mathbf{M}_{p,q}(H)$.

Every matrix over H for which one of the indexing sets I, K is empty is identical with the empty family of elements of H; it is also called the *empty matrix*. When $I = \{i_0\}$ (resp. $K = \{k_0\}$) is a set consisting of a single element, M is called a *row matrix* (resp. *column matrix*) and the row (resp. column) index can then be suppressed in the notation; when I and K are both sets with one element, a matrix of type (I, K) is often identified with the unique element in this matrix.

A subfamily $M' = (m_{\iota\kappa})_{(\iota,\kappa)\in J\times L}$ of a matrix $M = (m_{\iota\kappa})_{(\iota,\kappa)\in I\times K}$, whose indexing set is the product of a subset J of I and a subset L of K, is called a *submatrix* of the matrix M; it is said to be obtained by *suppressing* in M the rows of index $\iota \notin J$ and the columns of index $\kappa \notin L$; conversely, M is said to be obtained by *bordering* M' with the rows of index $\iota \notin J$ and the columns of index $\kappa \notin L$.

DEFINITION 2. *The transpose of a matrix* $M = (m_{\iota\kappa})_{(\iota,\kappa)\in I\times K}$, *denoted by* ${}^tM$, *is the matrix* $(m'_{\kappa\iota})_{(\kappa,\iota)\in K\times I}$ *over* H *given by* $m'_{\kappa\iota} = m_{\iota\kappa}$ *for all* $(\iota, \kappa) \in K \times L$.

It follows from this definition that the transpose of a matrix of type (I, K) is a matrix of type (K, I) and that

$$ {}^t({}^tM) = M. \tag{1} $$

2. MATRICES OVER A COMMUTATIVE GROUP

Let G be a commutative group (written additively). The set of matrices over G, with the given indexing sets I, K, has a *commutative group* structure since it is the set of mappings from I × K to G; this group is written additively, so that if $M = (m_{\iota\kappa})$ and $M' = (m'_{\iota\kappa})$ are two of its elements, then

$$ M + M' = (m_{\iota\kappa} + m'_{\iota\kappa}); $$

the identity element of this group is therefore the matrix all of whose elements are *zero* (called the *zero matrix*). Clearly

$$ {}^t(M + M') = {}^tM + {}^tM'. \tag{2} $$

The sum of two matrices is thus only defined if the indexing sets of the rows and the columns are the *same* for the two matrices.

Let H′, H″ be two sets, G a commutative group (written additively) and $f : (h', h'') \mapsto h'h''$ a mapping from H′ × H″ to G. Given two matrices

$$ M' = (m'_{ik})_{(i,k)\in I\times K}, \qquad M'' = (m''_{kl})_{(k,l)\in K\times L} $$

over H′ and H″ respectively such that the indexing set K of the columns of M′ is *finite* and equal to the indexing set of the rows of M'', the *product of* M' *and* M'' *via* f, denoted by $M'M''$ or $f(M', M'')$, is the matrix

$$ \Big(\sum_{k\in K} m'_{ik}m''_{kl}\Big)_{(i,l)\in I\times L} \tag{3} $$

over G.

The above definition supposes that the indexing set of the columns of M' is equal to the indexing set of the rows of M''; in particular the product $M''M'$ *has no meaning if* I ≠ L. In formula (3) the elements of the *same* row of M' figure multiplied on the right by the elements of the *same* column of M''; the multiplication is said to be made "rows by columns".

Let f^0 be the mapping $(h'', h') \mapsto h'h''$ of $H'' \times H'$ to G; it follows immediately from the definitions that

$$(4) \qquad {}^t(M'M'') = {}^tM''.{}^tM'$$

where the product on the left (resp. right) hand side is calculated via f (resp. via f^0).

When H' and H'' are themselves commutative groups (written additively) and f is **Z**-*bilinear* (§ 3, no. 1), the distributivity formulae

$$(5) \qquad \begin{cases}(M' + N')M'' = M'M'' + N'M'' \\ M'(M'' + N'') = M'M'' + M'N''\end{cases}$$

are immediately verified, the indexing sets being such that the sums and products appearing are defined.

Now let $H_1, H_2, H_3, H_{12}, H_{23}$ and H be commutative groups (written additively), $f_{12}: H_1 \times H_2 \rightarrow H_{12}$, $f_{23}: H_2 \times H_3 \rightarrow H_{23}$ mappings and

$$f_3: H_{12} \times H_3 \rightarrow H, \qquad f_1: H_1 \times H_{23} \rightarrow H$$

Z-bilinear mappings; suppose further that, for all $x_i \in H_i$ $(i = 1, 2, 3)$

$$f_3(f_{12}(x_1, x_2), x_3) = f_1(x_1, f_{23}(x_2, x_3))$$

(which may also be written as above $(x_1x_2)x_3 = x_1(x_2x_3)$); then, if $M' = (m'_{rs})$, $M'' = (m''_{st})$, $M''' = (m'''_{tu})$ are matrices over H_1, H_2, H_3 respectively,

$$(6) \qquad (M'M'')M''' = M'(M''M''')$$

when the products on the two sides (calculated respectively via f_{12}, f_3, f_{23} and f_1) are defined; for

$$\sum_t \Big(\sum_s m'_{rs}m''_{st}\Big)m'''_{tu} = \sum_t \sum_s (m'_{rs}m''_{st})m'''_{tu} = \sum_s \sum_t m'_{rs}(m''_{st}m'''_{tu})$$
$$= \sum_s m'_{rs}\Big(\sum_t m''_{st}m'''_{tu}\Big)$$

by virtue of the hypotheses made.

The two sides of (6) are also denoted by $M'M''M'''$. Analogous conventions are made for products of more than three factors.

Remark. The above formulae extend to a more general situation. To be precise:

(a) Suppose $H = \bigcup_{(\iota, \kappa) \in I \times K} G_{\iota\kappa}$ where each $G_{\iota\kappa}$ is a commutative group written additively; then the sum $M + M'$ may be defined when, for each ordered pair (ι, κ), $m_{\iota\kappa} \in G_{\iota\kappa}$ and $m'_{\iota\kappa} \in G_{\iota\kappa}$.

(b) Let I, K, L be three sets with K assumed finite and let $H' = \bigcup_{(i,k)\in I\times K} H'_{ik}$, $H'' = \bigcup_{(k,l)\in K\times L} H''_{kl}$, $H = \bigcup_{(i,l)\in I\times L} H_{il}$ be three sets; suppose that each H_{il} is a commutative group written additively and for each triple (i, k, l) let

$$f_{ikl}: H'_{ik} \times H''_{kl} \to H_{il}$$

be a mapping. Then if $M' = (m'_{ik})_{(i,k)\in I\times K}$, $M'' = (m''_{kl})_{(k,l)\in K\times L}$ are matrices such that $m'_{ik} \in H'_{ik}$ and $m''_{kl} \in H''_{kl}$ for all i, k, l we can define the product $M'M''$ via the f_{ikl}. We leave to the reader the task of writing down and proving the formulae analogous to (4), (5) and (6).

3. MATRICES OVER A RING

The most important matrices in Mathematics are matrices over a *ring* A. The set $A^{I\times K}$ of matrices over A corresponding to indexing sets I, K then has canonically an (A, A)-*bimodule* structure (§ 1, no. 14).

For every ordered pair $(i, k) \in I \times K$, let E_{ik} be the matrix (a_{jl}) such that $a_{ik} = 1$ and $a_{jl} = 0$ for $(j, l) \neq (i, k)$; the E_{ik} are called the *matrix units* in the set of matrices $A^{I\times K}$; if I and K are finite, they form the *canonical basis* of this set for its left or right A-module structure (§ 1, no. 11). Clearly

$${}^tE_{ik} = E_{ki}.$$

Unless otherwise mentioned, the *product* $M'M''$ of two matrices over A (assumed to be defined) will always be understood to be relative to the multiplication $(x, y) \mapsto xy$ *in* A (or, as is also said, will be "calculated in A"). Then we have (no. 2) the associativity and distributivity formulae

$$(XY)Z = X(YZ) \tag{7}$$

$$\begin{cases} X(Y + Z) = XY + XZ \\ (X + Y)Z = XZ + YZ \end{cases} \tag{8}$$

for three matrices X, Y, Z over A, whenever the sums and products appearing in these formulae are defined.

In particular, if E_{ik} (resp. E'_{kl}, E''_{il}) are the matrix units in $A^{I\times K}$ (resp. $A^{K\times L}$, $A^{I\times L}$) respectively, with $I = (1, p)$, $K = (1, q)$, $L = (1, r)$, we obtain the formulae

$$\begin{cases} E_{ik}E'_{jl} = 0 \quad \text{if } k \neq j \\ E_{ik}E'_{kl} = E''_{il}. \end{cases} \tag{9}$$

Let A^0 be the *opposite* ring of A and let $a * b$ $(= ba)$ denote the product of a and b in A^0; then, for two matrices X, Y over A whose product is defined,

$${}^t(XY) = {}^tY * {}^tX \tag{10}$$

where on the right hand side ${}^tY$ and ${}^tX$ are considered as matrices with elements in A^0; when A is *commutative*, then

$$(11) \qquad {}^t(XY) = {}^tY.{}^tX$$

PROPOSITION 1. *Let* A, B *be two rings and* $M = (m_{\iota\kappa})_{(\iota, \kappa) \in I \times K}$ *and*

$$M' = (m'_{\iota\kappa})_{(\iota, \kappa) \in I \times K}$$

two matrices with finite indexing sets over an (A, B)*-bimodule* G. *Suppose that for every matrix unit* $L = (a_\iota)_{\iota \in I}$ *with one row and elements in* A *and every matrix unit* $C = (b_\kappa)_{\kappa \in K}$ *with one column and elements in* B, $L.M.C = L.M'.C$ *(the products being calculated via the external laws of the* (A, B)*-module* G); *then* $M = M'$.

If L is taken to be the matrix unit (a_s) with $a_i = 1$, $a_s = 0$ for $s \neq i$, and C the matrix unit (b_t) with $b_k = 1$, $b_t = 0$ for $t \neq k$, the products $L.M.C$ and $L.M'.C$ are matrices with a single element respectively equal to m_{ik} and m'_{ik}.

Let A, B be two rings and $\sigma : A \to B$ a homomorphism.

For every matrix $M = (m_{\iota\kappa})$ over A, we shall denote by $\sigma(M)$ the matrix $(\sigma(m_{\iota\kappa}))$ over B; clearly $\sigma(aM) = \sigma(a)\sigma(M)$, $\sigma(Ma) = \sigma(M)\sigma(a)$ for $a \in A$, also $\sigma({}^tM) = {}^t(\sigma(M))$ and

$$(12) \qquad \begin{cases} \sigma(M + M') = \sigma(M) + \sigma(M') \\ \sigma(MM') = \sigma(M)\sigma(M') \end{cases}$$

when the operations considered are defined, the products on the left and right hand sides of (12) being calculated in A and B respectively. When σ is denoted by $x \mapsto x^\sigma$, we write M^σ instead of $\sigma(M)$.

Consider in particular an *anti-endomorphism* σ of A, that is a homomorphism of A to the opposite ring A^0, or a mapping of A into itself such that

$$\sigma(a + a') = \sigma(a) + \sigma(a'), \qquad \sigma(aa') = \sigma(a')\sigma(a)$$

for all a, a' in A; then, for two matrices M, M' over A whose product MM' is defined,

$$(13) \qquad \sigma(MM') = {}^t(\sigma({}^tM').\sigma({}^tM))$$

where the products on the two sides are calculated *in* A; this follows immediately from (10) and (12).

4. MATRICES AND LINEAR MAPPINGS

Let A be a ring and E a (*right or left*) A-module admitting a basis $(e_\iota)_{\iota \in I}$. For every element $x \in E$, the *matrix of* x *with respect to the basis* (e_ι), denoted by $M(x)$ or $\mathbf{x}$ (or sometimes simply x when no confusion can arise), is the *column matrix* consisting of the components x_ι ($\iota \in I$) of x with respect to (e_ι) (§ 1, no. 11); in calculations it will sometimes be convenient, in order to remember that the index i is a row index, to adjoin to it a column index taking only one value and write the matrix $M(x)$ as (x_{i0}).

We now consider two (left or right) A-modules E and F with bases $(e_i)_{i \in I}$ and $(f_k)_{k \in K}$ respectively; let (f_k^*) be the family of *coordinate forms* corresponding to (f_k). For a linear mapping u of E into F, we shall define *the matrix of u with respect to the bases (e_i), (f_k)* in each of the following cases:

(D) E and F are right A-modules, u is A-linear.
(G) E and F are left A-modules, u is A-linear.

In what follows, we shall attach the letter (D) (resp. (G)) to formulae applying to right (resp. left) modules.

DEFINITION 3. *In each of the above two cases, the matrix of u with respect to the bases (e_i), (f_k) is the matrix $M(u) = (u_{ki})_{(k, i) \in K \times I}$ such that*

$$u_{ki} = f_k^*(u(e_i)) \tag{14}$$

which is written respectively as

$$u_{ki} = \langle f_k^*, u(e_i)\rangle \tag{14 D}$$

$$u_{ki} = \langle u(e_i), f_k^*\rangle. \tag{14 G}$$

The *column* of $M(u)$ of index i is therefore equal to $M(u(e_i))$.

Clearly if u, v are two linear mappings of E into F and $M(u)$, $M(v)$ their matrices with respect to the same bases, then

$$M(u + v) = M(u) + M(v) \tag{15}$$

and

$$M(\gamma u) = \gamma M(u) \tag{16}$$

for every element γ of the *centre* Γ of A. In other words, once the bases (e_i), (f_k) are fixed, the mapping $u \mapsto M(u)$ is a *Γ-module isomorphism* of $\mathrm{Hom}_A(E, F)$ onto a subset of the set $A^{K \times I}$, equal to $A^{K \times I}$ if K is *finite*.

PROPOSITION 2. *Suppose* I *and* K *are finite. For every element $x \in E$, the matrix $M(u(x))$ with respect to the basis (f_k) is given by the formula*

$$M(u(x)) = M(u).M(x) \tag{17 D}$$

$$^tM(u(x)) = {}^tM(x).{}^tM(u). \tag{17 G}$$

We verify for example (17 G). Let $x = \sum_i x_{i0}e_i$, $u(x) = \sum_k y_{k0}f_k$ with $x_{i0} \in A$, $y_{k0} \in A$; then $u(x) = u\Big(\sum_i x_{i0}e_i\Big) = \sum_i x_{i0}u(e_i) = \sum_{i,k} x_{i0}u_{ki}f_k$; whence

$y_{k0} = \sum_i x_{i0} u_{ki}$. In order to bring the two indices i along side one another, we consider the transpose matrices ${}^tM(x) = (x'_{0i})$, where $x'_{0i} = x_{i0}$ and

$$ {}^tM(u) = (u'_{ik}), $$

where $u'_{ik} = u_{ki}$; then $y_{k0} = \sum_i x'_{0i} u'_{ik}$ and the right hand side is the element of index k of the matrix with one row ${}^tM(x).{}^tM(u)$, whence (17 G).

When A is commutative, (17 G) reduces to (17 D) by formula (4) of no. 2.

COROLLARY. *Let* E, F, G *be three right* (resp. *left*) *modules over a ring* A, $(e_i)_{i \in I}$, $(f_k)_{k \in K}$, $(g_l)_{l \in L}$ *respective finite bases of* E, F, G, $u: E \to F$, $v: F \to G$ *two linear mappings*, $M(u)$ *the matrix of* u *relative to the bases* (e_i), (f_k), $M(v)$ *the matrix of* v *relative to the bases* (f_k), (g_l) *and* $M(v \circ u)$ *the matrix of* $v \circ u$ *relative to the bases* (e_i), (g_l)*; then*

$$ M(v \circ u) = M(v)M(u) \tag{18 D} $$

$$ {}^tM(v \circ u) = {}^tM(u){}^tM(v). \tag{18 G} $$

We prove for example (18 G). For all $x \in E$, by (17 G):
$$ {}^tM(x).{}^tM(v \circ u) = {}^tM(v(u(x))) = {}^tM(u(x)).{}^tM(v) = {}^tM(x).{}^tM(u).{}^tM(v) $$
by associativity; the corollary then follows from no. 3, Proposition 1 since the matrix ${}^tM(x)$ with one row is arbitrary.

Remark (1). Formula (17 D) can be considered as a special case of (18 D). For there corresponds canonically to every $x \in E$ the linear mapping $\theta_x: A_d \to E$ mapping every $\alpha \in A$ to $x\alpha$ (§ 2, no. 1). It is immediate that the matrix $M(\theta_x)$ with respect to the basis 1 of A_d and the basis (e_i) of E is just the matrix $M(x)$; similarly $M(\theta_{u(x)}) = M(u(x))$ and formula (17 D) can therefore be considered as a translation of the relation

$$ \theta_{u(x)} = u \circ \theta_x. $$

PROPOSITION 3. *Let* E, F *be two right* (resp. *left*) A*-modules and* $(e_i)_{i \in I}$, $(f_k)_{k \in K}$ *finite bases of* E *and* F *respectively. For every linear mapping* u *of* E *into* F, *let* $M(u)$ *be the matrix of* u *with respect to the bases* (e_i) *and* (f_k). *Then the matrix of* ${}^tu: F^* \to E^*$ *with respect to the dual bases* (f_k^*) *and* (e_i^*) *is equal to* ${}^tM(u)$.

E is canonically identified with its bidual E** and (e_i) with the dual basis of (e_i^*); then (supposing for example that E and F are right modules)

$$ \langle {}^tu(f_k^*), e_i \rangle = \langle f_k^*, u(e_i) \rangle, $$

whence the proposition.

Remarks. (2) Let E and F be two left A-modules with bases $(e_\iota)_{\iota \in I}$ and $(f_\kappa)_{\kappa \in K}$ respectively. For every A-linear mapping $u: E \to F$, by (14 G), $u(e_\iota) = \sum_\kappa u_{\kappa\iota} f_\kappa$; these relations can also be interpreted by saying that the *column matrix* $(u(e_\iota)_{\iota \in I})$ with elements in F is equal to the product ${}^t\mathrm{M}(u) . (f_\kappa)$, where $(f_\kappa)_{\kappa \in K}$ is considered as a *column matrix* with elements in F and the product is calculated for the mapping $A \times F \to F$ defining the law of action on the A-module F (no. 2).

(3) Let A, B be two *commutative* rings and $\sigma: A \to B$ a ring homomorphism. In the notation of Proposition 3, $(e_\iota \otimes 1)$ and $(f_\kappa \otimes 1)$ are respective bases of $E_{(B)} = E \otimes_A B$ and $F_{(B)} = F \otimes_A B$ (§ 5, no. 1, Proposition 4); moreover, if (e_ι^*) and (f_κ^*) are respectively the dual bases of (e_ι) and (f_κ), then $(e_\iota^* \otimes 1)$ and $(f_\kappa^* \otimes 1)$ are respectively the dual bases of $(e_\iota \otimes 1)$ and $(f_\kappa \otimes 1)$ (§ 5, no. 4). For every A-linear mapping $u: E \to F$, let $M(u)$ and $M(u \otimes 1)$ be the matrix of u with respect to (e_ι) and (f_κ) and the matrix of the B-linear mapping $u \otimes 1$ with respect to $(e_\iota \otimes 1)$ and $(f_\kappa \otimes 1)$. It follows from § 5, no. 4, formula (20) that

$$M(u \otimes 1) = \sigma(M(u)).$$

Consider a system of a *finite* number of right scalar linear equations in a *finite* number of unknowns

$$\sum_{\iota \in I} a_{\kappa\iota} x_\iota = b_\kappa \qquad (\kappa \in K) \tag{19}$$

with $a_{\kappa\iota}$, x_ι, b_κ in A.

Let $(e_\iota)_{\iota \in I}$, $(f_\kappa)_{\kappa \in K}$ be the canonical bases of $E = A_d^I$ and $F = A_d^K$; the system (19) is equivalent to the equation $u(x) = b$, where $x = \sum_\iota e_\iota x_\iota$, $b = \sum_\kappa f_\kappa b_\kappa$ and $u: E \to F$ is the linear mapping such that the matrix $M(u)$ with respect to the bases (e_ι) and (f_κ) are equal to $A = (a_{\kappa\iota})_{(\kappa,\iota) \in K \times L}$. This matrix is called *the matrix of the system of linear equations* (19). Recall (§ 2, no. 8, *Remarks* 2 and 3), that, writing $c_\iota = \sum_\kappa f_\kappa a_{\kappa\iota}$, the system (19) is equivalent to the unique vector equation

$$\sum_\iota c_\iota x_\iota = b, \tag{20}$$

and as c_ι is the *column* of index i in the matrix A, we see that to say that the system (19) admits a solution amounts to saying that the matrix $b = (b_{\kappa 0})$ with one column is a *linear combination* of the columns of the matrix A.

We leave to the reader the task of formulating the analogous definitions and remarks for systems of left linear equations.

5. BLOCK PRODUCTS

The definitions of no. 4 can be generalized as follows. Let E be a (right or left) A-module, the *direct sum* of a family $(E_\iota)_{\iota \in I}$ of submodules. For all $x \in E$, let $x = \sum_{\iota \in I} x_\iota$ with $x_\iota \in E_\iota$ for all $i \in I$; we shall say that the *column matrix* $M(x) = (x_\iota)_{\iota \in I}$ with elements in E is *the matrix of x with respect to the decomposition* $(E_\iota)_{\iota \in I}$ *of* E *as a direct sum.*

Let F be another A-module (E and F being both right A-modules or both left A-modules) and suppose that F is the *direct sum* of a family $(F_k)_{k \in K}$ of submodules. For all $u \in \mathrm{Hom}(E, F)$ and all $x_\iota \in E_\iota$, let $u(x_\iota) = \sum_k u_{k\iota}(x_\iota)$ with $u_{k\iota}(x_\iota) \in F_k$ for all $k \in K$; then $u_{k\iota} \in \mathrm{Hom}(E_\iota, F_k)$; we shall say that the matrix $M(u) = (u_{k\iota})_{(k, \iota) \in K \times I}$ of type (K, I) with elements in the set H the *sum* of the $\mathrm{Hom}(E_\iota, F_k)$ is *the matrix of u with respect to the decompositions* (E_ι) *and* (F_k) *of* E *and* F *as direct sums.*

With these definitions, it is obvious that if u, v are two A-linear mappings of E into F then, for matrices with respect to the same decompositions as direct sums

$$(21) \qquad M(u + v) = M(u) + M(v), \qquad M(\gamma u) = \gamma M(u)$$

for every element γ of the centre of A (no. 2, *Remark*).

Moreover, the definition of the $u_{k\iota}$ shows that, if K is finite, we can write

$$(22) \qquad M(u(x)) = M(u) . M(x)$$

where $M(u(x))$ is the matrix of $u(x)$ with respect to the decomposition (F_k), the product on the right hand side of (22) being calculated for the mappings $(t, z) \mapsto t(z)$ of $\mathrm{Hom}(E_\iota, F_k) \times E_\iota$ into F_k (no. 2, *Remark*).

Let G be a third A-module, the direct sum of a family $(G_l)_{l \in L}$ of submodules, so that there corresponds to every A-linear mapping $v: F \to G$ a matrix $M(v) = (v_{lk})$ with respect to the decompositions (F_k) and (G_l). If I, K and L are *finite*, then

$$(23) \qquad M(v \circ u) = M(v) . M(u)$$

where the left hand side is the matrix $(w_{l\iota})$ of $w = v \circ u$ with respect to the decomposition (E_ι) and (G_l) and the product on the left hand side is calculated for the mappings $(t, s) \mapsto t \circ s$ of $\mathrm{Hom}(F_k, G_l) \times \mathrm{Hom}(E_\iota, F_k)$ into $\mathrm{Hom}(E_\iota, G_l)$ (no. 2, *Remark*). This is just formula (32) of § 1, no. 8, expressed in terms of matrices.

Finally, if I and K are assumed to be finite, E^* (resp. F^*) is canonically identified with the direct sum of the modules E_ι^* (resp. F_k^*) (§ 2, no. 6, Proposition 10). Then it is immediately verified that the matrix of ${}^t u$ with respect to the decompositions (F_k^*) and (E_ι^*) is just $({}^t u_{k\iota})_{(k, \iota) \in K \times I}$.

Suppose now that I and K are finite and further that each of the E_i (resp. F_k) admits a *finite* basis. It amounts to the same to say that E (resp. F) admits a basis $(e_r)_{r \in R}$ (resp. $(f_s)_{s \in S}$) and that R (resp. S) admits a partition $(R_i)_{i \in I}$ (resp. $(S_k)_{k \in K}$) such that for all $i \in I$ (resp. $k \in K$), $(e_r)_{r \in R_i}$ is a basis of E_i (resp. $(f_s)_{s \in S_k}$ is a basis of F_k). Then, if $X = M(u)$ is the matrix of u with respect to the bases $(e_r)_{r \in R}$ and $(f_s)_{s \in S}$, the matrix $X_{ki} = M(u_{ki})$ with respect to the bases $(e_r)_{r \in R_i}$ and $(f_s)_{s \in S_k}$ is just the *submatrix* of X obtained by suppressing the rows of index $s \notin S_k$ and the columns of index $r \notin R_i$. Thus we define a *one-to-one correspondence*

$$X \mapsto (X_{ki})_{(k,i) \in K \times I} \tag{24}$$

between the set of matrices of type (S, R) with elements in A and the set of *matrices of matrices* $(X_{ki})_{(k,i) \in K \times I}$ of type $K \times I$, where each X_{ki} is a matrix over A of type (S_k, R_i). Suppose that further G admits a finite basis $(g_t)_{t \in T}$ and that $T = (T_l)_{l \in L}$ is a partition of T such that, for each $l \in L$, $(g_t)_{t \in T_l}$ is a basis of G_l; let $Y = M(v)$ be the matrix of v with respect to the bases $(f_s)_{s \in S}$ and $(g_t)_{t \in T}$, $Y_{lk} = M(v_{lk})$ that of v_{lk} with respect to the bases $(f_s)_{s \in S_k}$ and $(g_t)_{t \in T_l}$, $Z = M(w)$ the matrix of $w = v \circ u$ with respect to the bases $(e_r)_{r \in R}$ and $(g_t)_{t \in T}$ and $Z_{li} = M(w_{li})$ that of w_{li} with respect to the bases $(e_r)_{r \in R_i}$ and $(g_t)_{t \in T_l}$; then it follows from (23) that the submatrices Z_{li} of $Z = YX$ are given by

$$Z_{li} = \sum_k Y_{lk} X_{ki} \tag{25}$$

in other words, the one-to-one correspondence (24) *transforms products into products* when all the products in question are defined (products of matrices of matrices being defined in the sense of no. 2, *Remark*); when the submatrices Z_{li} of the product YX are calculated thus, this product is said to be carried out "*in blocks*".

This name arises from the fact that, when $I = [1, p]$ and $K = [1, q]$, the table representing the matrix X is envisaged as divided into "blocks" forming an "array of matrices"

$$\begin{pmatrix} X_{11} & X_{12} & \dots & X_{1p} \\ X_{21} & X_{22} & \dots & X_{2p} \\ \cdot & \cdot & \cdot & \cdot \\ X_{q1} & X_{q2} & \dots & X_{qp} \end{pmatrix}$$

which is considered as a symbol denoting X when p and q are specific integers sufficiently small for this to be practicable.

6. MATRIX OF A SEMI-LINEAR MAPPING

Let A, B be two rings, $\sigma: A \to B$ a homomorphism of A into B, E a right (resp. left) A-module with basis $(e_i)_{i \in I}$ and F a right (resp. left) B-module with basis $(f_\kappa)_{\kappa \in K}$. Let $u: E \to F$ be a *semi-linear* mapping relative to σ and $u(e_\iota) = \sum_{\kappa \in K} f_\kappa u_{\kappa\iota}$ $\left(\text{resp. } u(e_\iota) = \sum_{\kappa \in K} u_{\kappa\iota} f_\kappa\right)$, where the $u_{\kappa\iota}$ are therefore elements of B; by definition, the matrix $M(u) = (u_{\kappa\iota})$ of type $K \times I$ is also called the *matrix of u with respect to the bases* (e_ι) and (f_κ). By the same calculation as in Proposition 2 of no. 4, it is immediately verified that for all $x \in E$, if I and K are *finite*,

$$M(u(x)) = M(u).\sigma(M(x)) \tag{26 D}$$

(resp.

$$^tM(u(x)) = \sigma(^tM(x)).^tM(u)). \tag{26 G}$$

Let C be a third ring, $\tau: B \to C$ a homomorphism, G a right (resp. left) C-module with basis $(g_\lambda)_{\lambda \in L}$ and v a semi-linear mapping of F into G relative to τ; if $M(v)$ is the matrix of v with respect to (f_κ) and (g_λ) and $M(v \circ u)$ the matrix of $v \circ u$ relative to (e_ι) and (g_λ), then, if I, K and L are finite,

$$M(v \circ u) = M(v).\tau(M(u)) \tag{27 D}$$

(resp.

$$^tM(v \circ u) = \tau(^tM(u)).^tM(v)). \tag{27 G}$$

To show for example (27 D), note that for all $x \in E$, by (26 D),

$$\begin{aligned} M(v \circ u).\tau(\sigma(M(x))) &= M(v(u(x))) \\ &= M(v).\tau(M(u(x))) = M(v).\tau(M(u)).\tau(\sigma(M(x))), \end{aligned}$$

whence (27 D) by Proposition 1 of no. 3.

Suppose finally that $\sigma: A \to B$ is an *isomorphism*; then recall that $^tu: F^* \to E^*$ is a semi-linear mapping relative to σ^{-1} (§ 2, no. 5); when I and K are finite, the matrix ^tu with respect to the dual bases (f_κ^*) and (e_ι^*) is given by

$$M(^tu) = \sigma^{-1}(^tM(u)) \tag{28}$$

for, by definition, supposing for example that E and F are right modules, $\langle {}^tu(f_\kappa^*), e_\iota \rangle^\sigma = \langle f_\kappa^*, u(e_\iota) \rangle$ when σ is denoted by $x \mapsto x^\sigma$.

Remark. Let A be a ring and σ an *anti-endomorphism* of A (no. 3); consider the two following situations:

(GD) E is a left A-module, F a right A-module and u a $\mathbf{Z}$-linear mapping of E into F such that $u(ax) = u(x)\sigma(a)$ for $a \in A$, $x \in E$; in other words, u is a *semi-linear* mapping relative to σ of the right A^0-module E into the right A-module F.

(DG) E is a right A-module, F a left A-module and u a $\mathbf{Z}$-linear mapping of E into F such that $u(xa) = \sigma(a)u(x)$ for $a \in A$, $x \in E$; in other words, u is a *semi-linear* mapping relative to σ of the left A^0-module E into the left A-module F.

In the two cases, the matrix $M(u)$ of u relative to bases of E and F has its elements in A; if these bases are finite, then, for all $x \in E$, we have the respective formulae

(17 GD) $$M(u(x)) = M(u).\sigma(M(x))$$

(17 DG) $${}^tM(u(x)) = \sigma({}^tM(x)).{}^tM(u),$$

the products on the two sides being calculated *in* A. This follows immediately from (26 D) and (26 G) respectively.

7. SQUARE MATRICES

DEFINITION 4. *A matrix whose rows and columns have the same indexing set is called a square matrix.*

A square matrix with n rows and n columns is called a *matrix of order n.*

Remark. It should be noted that a matrix for which the indexing sets of the rows and columns have the *same cardinal* but *are not identical*, must not be considered as a square matrix; in particular, the product of two such matrices over a ring *is not defined.*

Clearly addition and multiplication of square matrices over A with a finite set as indexing set of the rows and columns, define on the set of these matrices a *ring* structure because of formulae (7), (8) and (9) (no. 3); the matrix (δ_{ij}), where δ_{ij} is the Kronecker index (for $i \in I$, $j \in I$), is the unit element of this ring and is denoted by I_n or 1_n when I has n elements. When $I = [1, n]$, the ring of matrices thus defined is denoted simply by $\mathbf{M}_n(A)$; the group of invertible elements of $\mathbf{M}_n(A)$ is denoted by $\mathbf{GL}_n(A)$ or $\mathbf{GL}(n, A)$.

For a square matrix $U = (a_{ij})$ of order n over A to be right (resp. left) invertible, it is necessary and sufficient that, for every system $(b_i)_{1 \leqslant i \leqslant n}$ of elements of A, the system of n equations in n unknowns

$$\sum_{j=1}^{n} a_{ij}x_j = b_i \quad (1 \leqslant i \leqslant n)$$

$$\left(\text{resp. } \sum_{j=1}^{n} x_j a_{ji} = b_i\right)$$

have *one solution* (x_i) in A.

Let I be a finite indexing set, A a ring and E a right (resp. left) A-module with basis $(e_i)_{i \in I}$. For every *endomorphism* u of E, the matrix $M(u)$ of u with respect to the *two bases identical with* (e_i) is a square matrix; more briefly, it is called the matrix of *u with respect to the basis* (e_i).

Suppose that $I = [1, n]$. The mapping $u \mapsto M(u)$ (resp. $u \mapsto {}^tM(u)$) is an *isomorphism* of the ring $\text{End}_A(E)$ onto $\mathbf{M}_n(A)$ (resp. onto the opposite ring of $\mathbf{M}_n(A)$, as follows from formulae (18 D) (resp. 18 G)) (no. 4). The invertible elements of the ring $\mathbf{M}_n(A)$ called *invertible matrices*, correspond under the mapping $u \mapsto M(u)$ (resp. $u \mapsto {}^tM(u)$) to the *automorphisms* of E; the group $\mathbf{GL}(n, A)$ is therefore canonically identified with the group $\mathbf{GL}(A_d^n)$.

If u is an automorphism of E, its *contragredient* $\check{u}$ is an automorphism of the left (resp. right) A-module E^*, such that $\check{u} = ({}^tu)^{-1} = {}^t(u^{-1})$ (§ 2, no. 5, Definition 6); if $M(\check{u})$ is the matrix of $\check{u}$ with respect to the dual basis (e_i^*), then, by virtue of Proposition 3 (no. 4),

$$M(\check{u}) = ({}^tM(u))^{-1} = {}^tM(u^{-1}). \tag{29}$$

For every invertible matrix X, it therefore follows that ${}^t(X^{-1}) = ({}^tX)^{-1}$; this matrix is also denoted by ${}^tX^{-1}$ and called the *contragredient* of the matrix X.

Let σ be an *automorphism* of the ring A; for every *semi-linear* mapping $u: E \to E$ relative to σ, the matrix $M(u)$ of this mapping with respect to a basis (e_i) of E is also a square matrix. It follows immediately from (27 D) (no. 6) that, if u is bijective, then

$$M(u^{-1}) = (\sigma^{-1}(M(u)))^{-1}.$$

Let E be an A-module which is the *direct sum* of a finite family $(E_i)_{i \in I}$ of submodules; for every endomorphism u of E, the matrix $M(u) = (u_{ki})$ of u with respect to the two decompositions of E identical with (E_i) (no. 5) is a *square matrix of linear mappings*. In order that $u(E_i) \subset E_i$ for all $i \in I$, it is necessary and sufficient that $u_{ki} = 0$ for $k \neq i$. When $I = [1, n]$, the relations

$$u(E_i) \subset E_i + E_{i+1} + \cdots + E_n \quad (1 \leqslant i \leqslant n)$$

are equivalent to the relations $u_{ki} = 0$ for $k < i$.

Examples of square matrices. I. *Diagonal matrices.* In a square matrix

$$M = (m_{\iota\kappa})_{(\iota, \kappa) \in I \times I},$$

the elements both of whose indices are equal are called *diagonal elements* and the family $(m_{\iota\iota})_{\iota \in I}$ is called the *diagonal* of M; a square matrix $M = (m_{\iota\kappa})$ over a ring, whose elements other than the diagonal elements are zero, is called a *diagonal matrix*. For every family $(a_\iota)_{\iota \in I}$ of elements of a ring A, the diagonal matrix $(m_{\iota\kappa})$ such that $m_{\iota\iota} = a_\iota$ for all $\iota \in I$ is denoted by $\operatorname{diag}(a_\iota)_{\iota \in I}$ (or $\operatorname{diag}(a_1, a_2, \ldots, a_n)$ when $I = [1, n]$). In the set $\mathbf{M}_n(A)$ of square matrices of order n over A, the unit matrix I_n is a diagonal matrix and also every multiple $aI_n = I_n a$ of this matrix by a scalar a (the diagonal matrix (called *scalar*) all of whose diagonal elements are equal to a).

For every family $(d_i)_{1 \leqslant i \leqslant n}$ of elements of A and every matrix $X = (x_{ij})$ of type (n, q) (resp. (p, n)) over A, writing $D = \operatorname{diag}(d_i)$,

$$\begin{cases} DX = (d_i x_{ij}) \\ XD = (x_{ij} d_j). \end{cases} \tag{30}$$

In particular, for two diagonal matrices of order n,

$$\text{(31)} \qquad \begin{aligned} \operatorname{diag}(a_i) + \operatorname{diag}(b_i) &= \operatorname{diag}(a_i + b_i) \\ \operatorname{diag}(a_i) . \operatorname{diag}(b_i) &= \operatorname{diag}(a_i b_i). \end{aligned}$$

The diagonal matrices therefore form a *subring* of $\mathbf{M}_n(A)$ isomorphic to the product ring A^n; the scalar matrices form a subring isomorphic to A.

II. *Permutation matrices; monomial matrices.* Let π be any *permutation* of a finite set I and let $(e_i)_{i \in I}$ be the canonical basis of the A-module $E = A_d^I$; there exists one and only one endomorphism u_π of E such that, for all $i \in I$, $u_\pi(e_i) = e_{\pi(i)}$ (§ 1, no. 11, Corollary 3 to Proposition 17). For all $i \in I$, the column of index i in the matrix $M(u_\pi)$ with respect to the basis (e_i) has all its elements zero except the one in the row of index $\pi(i)$, which is equal to 1. By an abuse of language, $M(u_\pi)$ is called *the matrix of the permutation* π. It is immediate that for any two permutations σ, τ of I, $u_{\sigma\tau} = u_\sigma \circ u_\tau$ and that for the identity permutation ε, u_ε is the identity; the mapping $\pi \mapsto M(u_\pi)$ is therefore an *isomorphism* of the symmetric group $\mathfrak{S}_I$ onto the group of permutation matrices.

Each row and each column of a permutation matrix contains only a single element $\neq 0$. A finite square matrix R over a non-zero ring A, with this property, is called a *monomial matrix*; let r_i be the unique element $\neq 0$ in the column of R of index i and let $\pi(i)$ be the index of the row where this element is; clearly π is a permutation of the indexing set I and $R = M(u_\pi)D$, where $D = \operatorname{diag}(r_i)$.

III. *Triangular matrices.* In the ring $\mathbf{M}_n(A)$ of square matrices of order n over a ring A, any matrix (a_{ij}) such that $a_{ij} = 0$ for $i > j$ (resp. $i < j$) is called an *upper* (resp. *lower*) *triangular matrix*; it is also said that such a matrix *has only zeros below* (resp. *above*) *its diagonal*. It is immediately established that the upper (resp. lower) triangular matrices form a subring S (resp. T) of $\mathbf{M}_n(A)$, $S \cap T$ being obviously the ring of diagonal matrices.

The set S′ (resp. T′) of matrices in S (resp. T) whose diagonal elements are *invertible* is a multiplicative *group* of matrices called the *upper* (resp. *lower*) *total triangular group*, this follows immediately from § 1, no. 11, *Remark* 5. The set S_1 (resp. T_1) of matrices in S (resp. T) whose diagonal elements are all equal to 1 is a *subgroup* of the above group, called the *upper* (resp. *lower*) *strict triangular group*, and every matrix $M \in S'$ (resp. $M \in T'$) whose diagonal is (d_i), may be written as $M = DM_1 = M_1'D$, where $D = \operatorname{diag}(d_i)$ and M_1 and M_1' matrices belonging to S_1 (resp. T_1).

IV. *Diagonal and triangular matrices of matrices.* Let $(I_k)_{1 \leqslant k \leqslant p}$ be a partition of the finite set I; every square matrix over a ring A with indexing set I can be written in the form of a *square matrix of matrices* corresponding to *the same partition* (I_k) of the indexing set of the rows and the indexing set of the columns (no. 5)

$$(32)\qquad \begin{pmatrix} X_{11} & X_{12} & \dots & X_{1p} \\ X_{21} & X_{22} & \dots & X_{2p} \\ \cdot & \cdot & \cdot & \cdot \\ X_{p1} & X_{p2} & \dots & X_{pp} \end{pmatrix}$$

where each X_{kk} is a square matrix with I_k as indexing set of the rows and columns.

With this notation, (32) will be called a *diagonal* (resp. *upper triangular*, resp. *lower triangular*) *matrix of matrices* if all the matrices X_{ij} such that $i \neq j$ (resp. $i > j$, resp. $i < j$) are *zero*. The interpretation of endomorphisms u whose matrix is a diagonal, resp. triangular, matrix of matrices has been seen earlier, by considering the corresponding matrix $M(u)$ of linear mappings. The lower triangular (resp. upper triangular, diagonal) matrices of matrices for a given partition (I_k) of I form *subrings* of the ring of matrices $A^{I \times I}$. In particular, the ring of diagonal matrices of matrices relative to the partition (I_k) is isomorphic to the product $\prod_{k=1}^{p} \mathrm{End}_A(E_k)$.

8. CHANGE OF BASES

PROPOSITION 4. *Let* E *be a right* A*-module with finite basis* $(e_i)_{1 \leqslant i \leqslant n}$ *of n elements. For a family of n elements* $e'_i = \sum_{j=1}^{n} e_j a_{ji}$ $(1 \leqslant i \leqslant n)$ *to be a basis of* E, *it is necessary and sufficient that the square matrix* $P = (a_{ji})$ *of order n be invertible.*

P is just the matrix, with respect to the basis (e_i) of the endomorphism u of E defined by $u(e_i) = e'_i$ $(1 \leqslant i \leqslant n)$. Now, for u to be an automorphism of E, it is necessary and sufficient that $(u(e_i))$ be a basis of E (§ 1, no. 11, Corollary 3 to Proposition 17); whence the proposition.

The invertible matrix P is called the *matrix of passage from the basis* (e_i) *to the basis* (e'_i). It can also be interpreted as the matrix of the identity mapping 1_E with respect to the bases (e'_i) and (e_i) *(in that order)*; then clearly the matrix of passage *from the basis* (e'_i) *to the basis* (e_i) is the *inverse* P^{-1} of P.

PROPOSITION 5. *Let* (e_i), (e'_i) *be two bases of n elements of* E *and P the matrix of passage from* (e_i) *to* (e'_i). *If* (e^*_i) *and* (e'^*_i) *are the respective dual bases of* (e_i) *and* (e'_i), *the matrix of passage from* (e^*_i) *to* (e'^*_i) *is the contragredient* ${}^tP^{-1}$ *of P.*

The transpose of the identity mapping 1_E is the identity mapping 1_{E^*}; by Proposition 3, no. 4, the matrix of 1_{E^*} with respect to the bases (e'^*_i) and (e^*_i) (in that order) is the transpose of the matrix of 1_E with respect to the bases (e_i) and (e'_i) (in that order), that is the transpose of P^{-1}.

PROPOSITION 6. *Let* E *and* F *be two right* A*-modules,* (e_i) *and* (e'_i) *two bases of* E *with n elements, (f_j) and (f'_j) two bases of* F *with m elements, P the matrix of passage from (e_i) to (e'_i) and Q the matrix of passage from (f_j) to (f'_j). For every linear mapping u of* E *into* F, *let $M(u)$ be the matrix of u with respect to the bases (e_i) and (f_j) and $M'(u)$ the matrix of u with respect to the bases (e'_i) and (f'_i); then*

$$M'(u) = Q^{-1}M(u)P. \tag{33 D}$$

We may write $u = 1_F \circ u \circ 1_E$. Formula (33) follows immediately from no. 4, Corollary to Proposition 2 when the matrix of 1_E is taken with respect to (e'_i) and (e_i), that of u with respect to (e_i) and (f_i) and that of 1_F with respect to (f_i) and (f'_j).

Corollary 1. *If u is an endomorphism of* E *and $M(u)$ and $M'(u)$ its matrices with respect to the bases (e_i) and (e'_i) respectively, then*

$$M'(u) = P^{-1}M(u)P. \tag{34 D}$$

Corollary 2. *If* $M(x)$ and $M'(x)$ *are the matrices with one column of the same element $x \in$* E *with respect to the bases (e_i) and (e'_i) respectively, then*

$$M(x) = P.M'(x). \tag{35 D}$$

This is a special case of Proposition 6, applied to the mapping $\theta_x : a \mapsto xa$ of A_d to E (no. 4, *Remark* 1).

Formula (35) is equivalent to

$$x_i = \sum_{j=1}^{n} a_{ij}x'_j \quad (1 \leqslant i \leqslant n) \tag{36 D}$$

for the elements x_i and x'_i of the matrices $M(x)$ and $M'(x)$ respectively. Formulae (36 D) are called *formulae of change of coordinates*. Observe that they express the components of x relative to the "old" basis (e_i) as functions of the components of x relative to the "new" basis (e'_i) and the elements of P, that is the components of the "new" basis relative to the "old" basis.

Remarks. (1) We now start with a *left* A-module E with two bases (e_i), (e'_i), each with n elements; if we write $e'_i = \sum_{j=1}^{n} a_{ji}e_i$, $P = (a_{ji})$ is also called the *matrix of passing* from (e_i) to (e'_i); it is also the matrix of the automorphism of E such that $u(e_i) = e'_i$, with respect to the basis (e_i) and also the matrix of 1_E with respect to the bases (e'_i) and (e_i) *in that order*. The above results then hold with only the following modifications: formulae (33 D) to (36 D) are respectively replaced by

$$^tM'(u) = {}^tP.{}^tM(u).{}^tQ^{-1} \tag{33 G}$$

$$^tM'(u) = {}^tP.{}^tM(u).{}^tP^{-1} \tag{34 G}$$

$$^tM(u) = {}^tM'(u).{}^tP. \tag{35 G}$$

(36 G) $$x_i = \sum_{j=1}^{n} x'_j a_{ij} \quad (1 \leqslant i \leqslant n).$$

(2) Under the hypotheses of Proposition 4, consider an element $x^* \in E^*$; as the matrix of passage from (e_i^*) to $(e_i'^*)$ is ${}^tP^{-1}$ (Proposition 5), for the matrices $M(x^*)$ and $M'(x^*)$ of x^* with respect to these two bases respectively,

$$ {}^tM(x^*) = {}^tM'(x^*).P^{-1}$$

or also

(37 D) $$ {}^tM'(x^*) = {}^tM(x^*).P$$

which is equivalent to the system of equations

(38 D) $$x_i'^* = \sum_{j=1}^{n} x_j^* a_{ji} \quad (1 \leqslant i \leqslant n)$$

for the elements (x_i^*) and $(x_i'^*)$ of the matrices $M(x^*)$ and $M'(x^*)$. The corresponding formulae for a left A-module E are

(37 G) $$M'(x^*) = {}^tP.M(x^*)$$

(38 G) $$x_i'^* = \sum_{j=1}^{n} a_{ji} x_j^* \quad (1 \leqslant i \leqslant n).$$

(3) Let A, B be two rings, $\sigma: A \to B$ a *homomorphism* of A into B, E a right (resp. left) A-module, (e_i), (e'_i) two bases with n elements of E, F a right (resp. left) B-module, (f_j), (f'_j) two bases with m elements of F and P (resp. Q) the matrix of passage from (e_i) to (e'_i) (resp. from (f_j) to (f'_j)).

For every *semi-linear* mapping $u: E \to F$, relative to σ, let $M(u)$ be the matrix of u with respect to (e_i) and (f_j) and $M'(u)$ its matrix with respect to (e'_i) and (f'_j). Then

(39 D) $$M'(u) = Q^{-1}M(u)\sigma(P)$$

(resp.

(39 G) $$ {}^tM'(u) = \sigma({}^tP).{}^tM(u).{}^tQ^{-1}).$$

The proof is the same as that for (33 D) and (33 G), this time using formulae (27 D) and (27 G) (no. 6).

9. EQUIVALENT MATRICES; SIMILAR MATRICES

DEFINITION 5. *Two matrices X, X' with m rows and n columns over a ring are called equivalent if there exists an invertible square matrix P of order m and an invertible square matrix Q of order n such that*

(40) $$X' = PXQ.$$

Clearly the relation "X and X' are equivalent" is an *equivalence relation* (*Set Theory*, II, § 6, no. 1) on the set A^{mn} of matrices of type (m, n) over A, which justifies the terminology.

With this definition, Proposition 6 of no. 8 can be stated by saying that when the bases are changed in two right A-modules E, F (with finite bases), the matrix of a linear mapping $u: E \to F$ with respect to the new bases is *equivalent* to the matrix of u with respect to the old bases.

Conversely, if relation (40) holds and $u: A_d^n \to A_d^m$ is a linear mapping whose matrix is X with respect to the respective canonical bases (e_i) and (f_j) of A_d^n and A_d^m, then X' is the matrix of u with respect to the bases (e'_i) and (f'_j) such that Q is the matrix of passage from (e_i) to (e'_i) and P^{-1} the matrix of passage from (f_j) to (f'_j).

Examples of equivalent matrices. (1) Two matrices $X = (x_{ij})$ and $X' = (x'_{ij})$ with m rows and n columns "*differ only in the order of their rows*" if there exists a permutation σ of the interval $(1, m)$ of **N**, such that for every ordered pair of indices (i, j), $x'_{ij} = x_{\sigma(i), j}$ (we also say that X' is obtained by performing the permutation σ^{-1} on the rows of X). The matrices X and X' are then *equivalent*, for $X' = PX$, where P is the matrix of the permutation σ^{-1} (cf. no. 7, Example II).

Similarly X and X' are said to *differ only in the order of their columns* if there exists a permutation τ of $(1, n)$ such that $x'_{ij} = x_{i, \tau(j)}$ for every ordered pair of indices (i, j), X and X' are also *equivalent*, for $X' = XQ$ where Q is the matrix of the permutation τ.

Note that in the above notation P is the matrix of passage from a basis $(f_j)_{1 \leqslant j \leqslant m}$ to the basis $(f_{\sigma^{-1}(j)})_{1 \leqslant j \leqslant m}$ and Q the matrix of passage from a basis $(e_i)_{1 \leqslant i \leqslant n}$ to the basis $(e_{\tau(i)})_{1 \leqslant i \leqslant n}$.

(2) Let j, k be *distinct* elements of $(1, n)$ and let $a \in A$.

Suppose that for $1 \leqslant i \leqslant m$, $x'_{ij} = x_{ij} + x_{ik}a$ and $x'_{il} = x_{il}$ for $j \neq l$ and $1 \leqslant i \leqslant m$; X' is said to be derived from X by *adding to the column of X of index j the column of index k multiplied on the right by a.* In this case X and X' are also equivalent: for if $Q = I_n + aE_{kj}$ (an invertible triangular matrix, as seen in no. 7), then $X' = XQ$.

Similarly, let h, i be two *distinct* elements of $(1, m)$ and a an element of A; if X' is derived from X by adding to the *row* of X of index i the row of index h multiplied on the *left* by a, X and X' are equivalent, for $X' = PX$, where $P = I_m + aE_{ih}$.

(3) Finally, if, for a given index j, $x'_{ij} = x_{ij}c$ for $1 \leqslant i \leqslant m$, where c is *invertible* and $x'_{il} = x_{il}$ for $1 \leqslant i \leqslant m$ and $l \neq j$, X and X' are equivalent; for $X' = XQ$, where Q is the matrix $\operatorname{diag}(a_k)$ with $a_j = c$, $a_k = 1$ for $k \neq j$. Then X' is said to be derived from X by *multiplying the column of X of index j on the right by a.*

Similarly, if X' is derived from X' by multiplying the row of X of index i on the *left* by an *invertible* element $c \in A$, X' and X are equivalent, for $X' = PX$ where P is the matrix $\operatorname{diag}(b_h)$ with $b_i = c$, $b_h = 1$ for $h \neq i$.

DEFINITION 6. *Two square matrices X, X' of order n over a ring A are called similar if there exists an invertible square matrix P of order n such that*

$$X' = PXP^{-1} \tag{41}$$

Clearly the relation "X and X' are similar" is an *equivalence relation* on $\mathbf{M}_n(A)$ meaning that X and X' are transformed into one another by an *inner automorphism* of this ring.

With this definition, Corollary 1 to Proposition 6 of no. 8 can be stated by saying that when the basis of an A-module E (with a finite basis) is changed, the matrix of an endomorphism u with respect to the new basis is *similar* to the matrix of u with respect to the old basis.

Remarks. (1) Two square matrices which differ only in the order of their rows (or the order of their columns) are equivalent, but in general *not similar.* A matrix similar to a square matrix $X = (x_{ij})$ can be obtained by performing *the same permutation* σ^{-1} on the rows and columns, that is by considering the matrix $X' = (x'_{ij})$, where $x'_{ij} = x_{\sigma(i),\sigma(j)}$ for every ordered pair of indices; for if X is the matrix of an endomorphism u of A_d^n with respect to a basis $(e_i)_{1 \leqslant i \leqslant n}$, X' is the matrix of u with respect to the basis $(e_{\sigma(i)})_{1 \leqslant i \leqslant n}$.

(2) Let X and X' be two square matrices of order n which can be written in the form of diagonal matrices of square matrices (no. 7, Example IV):

$$X = \begin{pmatrix} X_1 & 0 & \dots & 0 \\ 0 & X_2 & \dots & 0 \\ \cdot & \cdot & \cdot & \cdot \\ 0 & 0 & \dots & X_p \end{pmatrix} \qquad X' = \begin{pmatrix} X'_1 & 0 & \dots & 0 \\ 0 & X'_2 & \dots & 0 \\ \cdot & \cdot & \cdot & \cdot \\ 0 & 0 & \dots & X'_p \end{pmatrix}$$

corresponding to the *same* partition of the indexing set $[1, n]$ for X and X'. If, for $1 \leqslant i \leqslant p$, X_i and X'_i are equivalent (resp. similar), then X and X' are equivalent (resp. similar): for, if $X'_i = P_i X_i Q_i$ for $1 \leqslant i \leqslant p$, then $X' = PXQ$ where

$$P = \begin{pmatrix} P_1 & 0 & \dots & 0 \\ 0 & P_2 & \dots & 0 \\ \cdot & \cdot & \cdot & \cdot \\ 0 & 0 & \dots & P_p \end{pmatrix} \qquad Q = \begin{pmatrix} Q_1 & 0 & \dots & 0 \\ 0 & Q_2 & \dots & 0 \\ \cdot & \cdot & \cdot & \cdot \\ 0 & 0 & \dots & Q_p \end{pmatrix}$$

as follows from calculating the "block" product (no. 6). Moreover, if $Q_i = P_i^{-1}$ for all i, then $Q = P^{-1}$.

10. TENSOR PRODUCT OF MATRICES OVER A COMMUTATIVE RING

Let C be a commutative ring, E, F, U, V four C-modules and $\phi: E \to U$, $\psi: F \to V$ two C-linear mappings. Suppose that E, F, U, V have respectively finite bases $(e_\lambda)_{\lambda \in L}$, $(f_\mu)_{\mu \in M}$, $(u_\rho)_{\rho \in R}$, $(v_\sigma)_{\sigma \in S}$; let $A = (a_{\rho\lambda})$ be the matrix of ϕ with respect to (e_λ) and (u_ρ), $B = (b_{\sigma\mu})$ that of ψ with respect to (f_μ) and (v_σ). For every ordered pair $(\lambda, \mu) \in L \times M = N$, let $g_{\lambda\mu} = e_\lambda \otimes f_\mu$; for every ordered pair $(\rho, \sigma) \in R \times S = T$ let $w_{\rho\sigma} = u_\rho \otimes v_\sigma$; the $g_{\lambda\mu}$ then form a basis of $E \otimes F$ and the $w_{\rho\sigma}$ a basis of $U \otimes V$ (§ 3, no. 6, Corollary 2 to Proposition 7). The *tensor product* of A by B, denoted by $A \otimes B$, is the matrix $X = (x_{\tau\nu})_{(\tau, \nu) \in T \times N}$ whose elements are given by

$$x_{(\rho, \sigma), (\lambda, \mu)} = a_{\rho\lambda} b_{\sigma\mu}. \tag{42}$$

Then $A \otimes B$ is *the matrix of* $\phi \otimes \psi$ *with respect to the bases* $(g_{\lambda\mu})$ *and* $(w_{\rho\sigma})$. By definition (§ 3, no. 2, formula (3))

$$\begin{aligned}(\phi \otimes \psi)(g_{\lambda\mu}) &= (\phi \otimes \psi)(e_\lambda \otimes f_\mu) = \phi(e_\lambda) \otimes \psi(f_\mu) \\ &= \sum_{\rho, \sigma} a_{\rho\lambda} b_{\sigma\mu}(u_\rho \otimes v_\sigma) = \sum_{\rho, \sigma} a_{\rho\lambda} b_{\sigma\mu} w_{\rho\sigma}.\end{aligned}$$

Definition (42) of the elements of $A \otimes B$ shows that this matrix corresponds bijectively with the matrix of matrices $(a_{\rho\lambda}B)_{(\rho, \lambda) \in R \times L}$ and also the matrix $(Ab_{\sigma\mu})_{(\sigma, \mu) \in S \times M}$ (no. 5).

The fact that $(\phi, \psi) \mapsto \phi \otimes \psi$ is a C-bilinear mapping and formula (9) of § 3, no. 5, can be expressed by the identities

$$\begin{cases} A \otimes (B_1 + B_2) = A \otimes B_1 + A \otimes B_2 \\ (A_1 + A_2) \otimes B = A_1 \otimes B + A_2 \otimes B \end{cases} \tag{43}$$

$$(cA) \otimes B = A \otimes (cB) = c(A \otimes B) \quad \text{for } c \in C \tag{44}$$

$$(A_1 \otimes B_1)(A_2 \otimes B_2) = (A_1 A_2) \otimes (B_1 B_2) \tag{45}$$

when the operations appearing are defined. The transpose of a tensor product of matrices is given by

$${}^t(A \otimes B) = ({}^tA) \otimes ({}^tB). \tag{46}$$

If A and B are invertible square matrices over C, $A \otimes B$ is invertible and

$$(A \otimes B)^{-1} = (A^{-1}) \otimes (B^{-1}). \tag{47}$$

Let $(e'_\lambda)_{\lambda \in L}$ be another basis of E and $(f'_\mu)_{\mu \in M}$ another basis of F; if P is the matrix of passage from the basis (e_λ) to the basis (e'_λ) and Q the matrix of passage from the basis (f_μ) to the basis (f'_μ), the matrix of passage from the basis $(e_\lambda \otimes f_\mu)$ to the basis $(e'_\lambda \otimes f'_\mu)$ is $P \otimes Q$. If A' is *equivalent* (resp. *similar*) to A and B' *equivalent* (resp. *similar*) to B, then $A' \otimes B'$ is *equivalent* (resp. *similar*) to $A \otimes B$.

The definition of tensor product of matrices can be generalized in an obvious way to an arbitrary finite number of matrices over C; in particular we have the associativity formula

$$(48) \qquad \Big(\bigotimes_{\iota \in I_1} X_\iota\Big) \otimes \Big(\bigotimes_{\iota \in I_2} X_\iota\Big) = \bigotimes_{\iota \in I} X_\iota$$

for every *partition* (I_1, I_2) of the finite indexing set I.

11. TRACE OF A MATRIX

Let C be a *commutative* ring; for every square matrix $X = (x_{\iota\kappa})$ over C corresponding to the finite indexing set I, the *trace* of X is the element

$$(49) \qquad \operatorname{Tr}(X) = \sum_{\iota \in I} x_{\iota\iota}.$$

Let E be a C-module admitting a finite basis $(e_\iota)_{\iota \in I}$; for every endomorphism u of E,

$$(50) \qquad \operatorname{Tr}(u) = \operatorname{Tr}(M(u))$$

$M(u)$ being the matrix u with respect to the basis (e_ι); this follows immediately from § 4, no. 3, formula (17), when this formula is applied to the endomorphism $x \mapsto \langle x, e_i^* \rangle e_j$ (where e_i^*) is the dual basis of (e_ι)); from this we pass to the general case by linearity. Formula (49) shows that

$$(51) \qquad \operatorname{Tr}(u) = \sum_\iota \langle u(e_\iota), e_\iota^* \rangle$$

for every basis (e_ι) of E (cf. § 4, no. 3, formula (17)).

If X is a matrix of type (m, n) over C and Y a matrix of type (n, m) over C, then

$$(52) \qquad \operatorname{Tr}(XY) = \operatorname{Tr}(YX)$$

as follows from the above and Proposition 3 of § 4, no. 3; (52) can also be obtained directly, for if $X = (x_{ij})$, $Y = (y_{ji})$ $(1 \leqslant i \leqslant m, 1 \leqslant j \leqslant n)$, then

$$(53) \qquad \operatorname{Tr}(XY) = \sum_{i,j} x_{ij} y_{ji}$$

by (49). The latter formula proves moreover:

PROPOSITION 7. *Let C be a commutative ring and for every matrix $P \in \mathbf{M}_n(C)$ let f_P be the linear form $X \mapsto \operatorname{Tr}(PX)$ on $\mathbf{M}_n(C)$; the mapping $P \mapsto f_P$ is a C-linear bijection of $\mathbf{M}_n(C)$ onto its dual.*

PROPOSITION 8. *If g is a linear form on the C-module $\mathbf{M}_n(C)$ such that $g(XY) = g(YX)$ for all matrices X, Y in $\mathbf{M}_n(C)$, there exists one and only one scalar $c \in C$ such that $g(X) = c.\operatorname{Tr}(X)$ for every matrix $X \in \mathbf{M}_n(C)$.*

Since the proposition is trivial for $n = 1$, attention may be confined to the case where $n \geqslant 2$. Taking $X = E_{ij}$, $Y = E_{jk}$ with $i \neq k$, we obtain $g(E_{ik}) = 0$; then taking $X = E_{ij}$, $Y = E_{ji}$ with $i \neq j$, we find $g(E_{ii}) = g(E_{jj})$; the proposition follows immediately since the E_{ij} form a basis of $\mathbf{M}_n(\mathrm{C})$.

12. MATRICES OVER A FIELD

The finite matrices with m rows and n columns over a field K are in one-to-one correspondence with the linear mappings of the right vector space $\mathrm{E} = \mathrm{K}_d^n$ into the right vector space K_d^m when the matrices of these mappings are taken with respect to the canonical bases of E and F. By definition, the *rank* of such a matrix X is the rank of the linear mapping $u:\mathrm{E} \to \mathrm{F}$ corresponding to it; as this number is by definition the dimension of the subspace $u(\mathrm{E})$ of F, it amounts to the same (identifying the columns of X with the images under u of the canonical basis of E) to give the following definition:

DEFINITION 7. *Given a matrix X with m rows and n columns over a field* K, *the dimension of the subspace of* K_d^m *generated by the n columns of X is called the rank of X with respect to* K *and denoted by* $\mathrm{rg}(X)$.

It can also be said that the rank of X is the *maximum number of linearly independent columns of X* (as elements of K_d^m). Obviously $\mathrm{rg}(X) \leqslant \inf(m, n)$; for every submatrix Y of X, $\mathrm{rg}(Y) \leqslant \mathrm{rg}(X)$.

If E and F are two finite-dimensional vector spaces over K and u a linear mapping of E into F, the rank of the matrix $M(u)$ with respect to any two bases is equal to the rank of u.

PROPOSITION 9. *If the elements of a matrix X with m rows and n columns belong to a subfield* K_0 *of a field* K, *the rank of X with respect to* K_0 *is equal to the rank of X with respect to* K.

Let F_0 be the right vector K_0-space generated by the canonical basis of the right vector K-space $\mathrm{E} = \mathrm{K}_d^m$; by hypothesis the columns of X belong to E_0. Let V_0 (resp. V) be the vector sub-K_0-space of F_0 (resp. the vector sub-K-space of E) generated by these columns. Then $\mathrm{V} = \mathrm{V}_0 \otimes_{\mathrm{K}_0} \mathrm{K}$ (§ 8, no. 2, Proposition 2) and hence $\dim_{\mathrm{K}} \mathrm{V} = \dim_{\mathrm{K}_0} \mathrm{V}_0$.

PROPOSITION 10. *The rank of a matrix X over a field* K *is equal to the rank of its transpose* ${}^tX$ *over the opposite field* K^0.

In the notation introduced before Definition 7, the rank of u is equal to that of ${}^tu$ (§ 7, no. 5, Proposition 10) and the proposition therefore follows from no. 4, Proposition 3.

It is thus seen that the rank of X can also be defined as the *maximum number of linearly independent rows of X* (considering them as elements of the left vector K-space K_s^n).

The square matrices of order n over a field K correspond bijectively with the endomorphisms of $\mathrm{E} = \mathrm{K}_d^n$ and form a ring isomorphic to the ring

$\mathrm{End}_K(E)$ (no. 7); corresponding to the automorphisms of E are the invertible square matrices.

PROPOSITION 11. *Let X be a square matrix of order n over a field* K. *The following properties are equivalent:*

(a) *X is invertible in $\mathbf{M}_n(K)$.*
(b) *X is right invertible in $\mathbf{M}_n(K)$.*
(c) *X is left invertible in $\mathbf{M}_n(K)$.*
(d) *X is of rank n.*

This is just a translation of § 7, no. 4, Corollary to Proposition 9.

PROPOSITION 12. *For a system of m linear equations in n unknowns*

$$\sum_{j=1}^{n} a_{ij}x_j = b_i \quad (1 \leqslant i \leqslant m) \tag{54}$$

over a field K *to have at least one solution, it is necessary and sufficient that the matrix $A = (a_{ij})$ of the system and the matrix B, obtained by bordering A with an $(n+1)$-th column equal to (b_i), be matrices of the same rank.*

It has been seen (no. 4) that the existence of a solution of (54) is equivalent to the fact that the column (b_i) is a linear combination of the columns of A and the proposition therefore follows from § 7, no. 3, Corollary 4 to Proposition 4.

Note that the condition of Proposition 12 is always fulfilled when $m = n$ and A is invertible, that is of rank n (Proposition 11). If x and b then denote the matrices with one column (x_i) and (b_i) respectively, system (54) is equivalent to $A.x = b$ and its unique solution is $x = A^{-1}.b$.

13. EQUIVALENCE OF MATRICES OVER A FIELD

PROPOSITION 13. *Let* E, F *be two finite-dimensional vector spaces over a field* K. *If $u: E \to F$ is a linear mapping of rank r, there exist bases of* E *and* F *such that, with respect to these bases,*

$$M(u) = \begin{pmatrix} I_r & 0 \\ 0 & 0 \end{pmatrix}. \tag{55}$$

Every matrix of type (m, n) over K *and of rank r is equivalent to a matrix of the form* (55).

The second assertion is trivially equivalent to the first. To show the latter, let $\dim E = n$, $\dim F = m$. The kernel $N = \overset{-1}{u}(0)$ is of dimension $n - r$ (§ 7, no. 4, formula (11)); let V be a supplementary subspace of N in E and $(e_i)_{1 \leqslant i \leqslant n}$ a basis of E such that $(e_i)_{1 \leqslant i \leqslant r}$ is a basis of V and $(e_i)_{r+1 \leqslant i \leqslant n}$ a basis of N. Then the $u(e_j)$ $(1 \leqslant j \leqslant r)$ form a basis of $u(E)$; hence there exists a basis $(f_j)_{1 \leqslant j \leqslant m}$ of F such that $f_j = u(e_j)$ for $1 \leqslant j \leqslant r$ (§ 7, no. 1, Theorem 2) and clearly with respect to the bases (e_i) and (f_j) the matrix $M(u)$ is given by (55).

COROLLARY. *For two matrices over a field, of type (m, n), to be equivalent, it is necessary and sufficient that they have the same rank.*

We shall now recover Proposition 13 by another more explicit method. For every ring A, every $\lambda \in A$, every integer $m > 1$ and every ordered pair of *distinct* integers i, j in $[1, m]$, we write

$$B_{ij}(\lambda) = I_m + \lambda E_{ij} \tag{56}$$

an invertible matrix of order m by no. 8.

Lemma 1. *Let $X = (\xi_{ij})$ be a matrix of type (m, n) over a ring* A. *Suppose that $m \geqslant 2$ and that there exists an element ξ_{i1} in the first column of X which is invertible in* A. *Then there exist two invertible square matrices $P \in \mathbf{M}_m(A)$, $Q \in \mathbf{M}_n(A)$ and a matrix Y of type $(m-1, n-1)$ over* A *such that P* (resp. Q) *is a product of matrices of the form $B_{ij}(\lambda)$ of order m* (resp. n) *and*

$$PXQ = \begin{pmatrix} 1 & 0 & \dots & 0 \\ 0 & & & \\ \vdots & & Y & \\ 0 & & & \end{pmatrix}. \tag{57}$$

The matrix $B_{ij}(\lambda)X$ is obtained by adding to the row of X of index i the row of index j multiplied on the left by λ (no. 9, *Example* 2); if ξ_{i1} is invertible, then there exists $\lambda \in A$ such that, for the matrix $X' = B_{1i}(\lambda)X = (\xi'_{kl})$, $\xi'_{11} = 1$; multiplying X' on the left by suitably chosen matrices $B_{k1}(\mu_k)$ of order m (for $1 \leqslant k \leqslant m$), a matrix $X'' = (\xi''_{kl})$ is obtained such that $\xi''_{11} = 1$, $\xi''_{k1} = 0$ for $k \neq 1$. Then the matrix obtained is multiplied successively *on the right* by suitable matrices $B_{1j}(\nu_j)$ of order n $(2 \leqslant j \leqslant n)$ and a matrix is obtained of the form (57).

PROPOSITION 14. *Let X be a matrix of type (m, n) over a field* K. *If X is of rank r, there exist two invertible square matrices $P \in \mathbf{M}_m(K)$, $Q \in M_n(K)$ such that P* (resp. Q) *is a product of matrices of order m* (resp. n) *of the form $B_{ij}(\lambda)$ and*

$$PXQ = \begin{pmatrix} 1 & 0 & \dots & 0 & 0 & \dots & 0 \\ 0 & 1 & \dots & 0 & 0 & \dots & 0 \\ \cdot & \cdot & \cdot & \cdot & \cdot & \cdot & \cdot \\ 0 & 0 & \dots & \delta_r & 0 & \dots & 0 \\ 0 & 0 & \dots & 0 & 0 & \dots & 0 \\ \cdot & \cdot & \cdot & \cdot & \cdot & \cdot & \cdot \\ 0 & 0 & \dots & 0 & 0 & \dots & 0 \end{pmatrix} \tag{58}$$

(a matrix (η_{ij}) *all of whose terms are zero except the* η_{ii} *for* $1 \leqslant i \leqslant r$, *with* $\eta_{ii} = 1$ *for* $1 \leqslant i \leqslant r - 1$, $\eta_{rr} = \delta_r \neq 0$). *If* $r \neq m$ or $r \neq n$, *it may also be assumed that* $\delta_r = 1$.

The proposition is obvious if $X = 0$; suppose therefore $X \neq 0$. If $m = n = 1$ the proposition is obvious (with $P = I_m$, $Q = I_n$, $\delta_1 \neq 0$ arbitrary). If $n = 1$, $m \geqslant 2$, we can apply Lemma 1 (since $X \neq 0$), which gives the desired form (58) with $r = 1$, $\delta_r = 1$. We argue by induction on $n > 1$; there exists an element $\xi_{ij} \neq 0$ in X; if $j = 1$, Lemma 1 can be applied and reduces the problem to the case where X has the form (57). The induction hypothesis then applies to Y and there are therefore invertible matrices

$$P' \in \mathbf{M}_{m-1}(\mathrm{K}), \qquad Q' \in \mathbf{M}_{n-1}(\mathrm{K})$$

which are products of matrices of the form $B_{ij}(\lambda)$ of order $m - 1$ (resp. $n - 1$), such that $P'YQ'$ is of the form (58). But, if $B_{ij}(\lambda)$ belongs for example to $\mathbf{M}_{m-1}(\mathrm{K})$, then

$$\begin{pmatrix} 1 & 0 \\ 0 & B_{ij}(\lambda) \end{pmatrix} = B_{i+1,j+1}(\lambda);$$

formula (58) then follows from the formula for block products writing

$$P = \begin{pmatrix} 1 & 0 \\ 0 & P' \end{pmatrix} \quad \text{and} \quad Q = \begin{pmatrix} 1 & 0 \\ 0 & Q' \end{pmatrix}.$$

If finally $j \neq 1$, it would be sufficient to consider the matrix $XB_{j1}(1)$ to reduce it to the above case.

Proposition 14 recovers Proposition 13 immediately.

Corollary 1. *If X is an invertible square matrix of order n over a field* K, *there exist three invertible matrices P, Q, D of order n such that $X = PDQ$, P and Q being products of matrices of the form $B_{ij}(\lambda)$ and D a diagonal matrix of the form*

$$D = \operatorname{diag}(1, 1, \ldots, \delta),$$

where $\delta \neq 0$ (cf. Exercise 13).

Corollary 2. *For every field* K, *the group of invertible matrices* $\mathbf{GL}(n, \mathrm{K})$ *is generated by the permutation matrices* (no. 7, *Example* 2), *the diagonal matrices* $\operatorname{diag}(a, 1, \ldots, 1)$ ($a \neq 0$ in K) *and the matrices* $B_{12}(\lambda)$ ($\lambda \in \mathrm{K}$).

It has been seen (no. 9) that the right (resp. left) product of a matrix by the matrix of a suitable transposition exchanges any two columns (resp. rows). Then the matrix $\operatorname{diag}(1, \ldots, 1, a)$ is equal to the product of $\operatorname{diag}(a, 1, \ldots, 1)$ and permutation matrices and every matrix $B_{ij}(\lambda)$ is equal to the product of $B_{12}(\lambda)$ and permutation matrices, whence the corollary.

Remarks. (1) In Chapter III, we shall see that, if $m = n = r$ and K is *commutative*, then, for all choices of P and Q satisfying the conditions of Proposition 14, the element δ_r is always the same and equal to the *determinant* of X (III, § 8, no. 6).

(2) The argument of Proposition 14, slightly modified, shows that there is a permutation matrix R such that (with the same conditions on P)

$$PXR = \begin{pmatrix} I_r & N \\ 0 & 0 \end{pmatrix}$$

if $m = n = r$ does not hold, and

$$PXR = \mathrm{diag}(1, \ldots, 1, \delta)$$

otherwise. Observe also that the method of proof gives an explicit determination of the matrices P, Q, R when X is given explicitly.

§ 11. GRADED MODULES AND RINGS

From no. 2 of this paragraph onwards, Δ *will denote a commutative monoid* (I, § 2, no. 1), *written additively, with an identity element denoted by* 0.

1. GRADED COMMUTATIVE GROUPS

We are going to translate into another language the definitions concerning direct sums (§ 1, no. 8).

DEFINITION 1. *Given a commutative group* G *written additively and a set* Δ, *a graduation of type* Δ *on* G *is a family* $(G_\lambda)_{\lambda \in L}$ *of subgroups of* G, *of which* G *is the direct sum. The set* G, *with the structure defined by its group law and its graduation, is called a graded (commutative) group of type* Δ.

Δ is called the *set of degrees* of G. An element $x \in G$ is called *homogeneous* if it belongs to one of the G_λ, *homogeneous of degree* λ if $x \in G_\lambda$. The element 0 is therefore homogeneous of all degrees; but if $x \neq 0$ is homogeneous, it belongs to only one of the G_λ; the index λ such that $x \in G_\lambda$ is then called *the degree of* x (or sometimes the *weight* of x) and is sometimes denoted by $\deg(x)$. Every $y \in G$ may be written uniquely as a sum $\sum_\lambda y_\lambda$ of homogeneous elements with $y_\lambda \in G_\lambda$; y_λ is called the *homogeneous component of degree* λ (or simply the *component of degree* λ) of y. When the word "weight" is used instead of "degree", the adjective "homogeneous" is replaced by "*isobaric*".

Examples. (1) Given any commutative monoid Δ (with identity element 0) and a commutative group G, a graduation $(G_\lambda)_{\lambda \in \Delta}$ is defined on G by taking $G_0 = G$ and $G_\lambda = \{0\}$ for $\lambda \neq 0$; this graduation is called *trivial*.

(2) Let Δ, Δ' be two sets and ρ a mapping of Δ into Δ'. Let $(G_\lambda)_{\lambda \in \Delta}$ be a graduation of type Δ on a commutative group G; for $\mu \in \Delta'$, let G'_μ be the sum of the G_λ such that $\rho(\lambda) = \mu$; clearly $(G'_\mu)_{\mu \in \Delta'}$ is a graduation of type Δ' on G, said to be *derived* from (G_λ) by means of the mapping ρ.

When Δ is a commutative group written additively and ρ the mapping $\lambda \mapsto -\lambda$ of Δ onto itself, (G'_μ) is called the *opposite* graduation of (G_λ).

(3) If $\Delta = \Delta_1 \times \Delta_2$ is a product of two sets, a graduation of type Δ is called a *bigraduation* of types Δ_1, Δ_2. For all $\lambda \in \Delta_1$, let $G'_\lambda = \bigoplus_{\mu \in \Delta_2} G_{\lambda\mu}$ and, for all $\mu \in \Delta_2$, let $G''_\mu = \bigoplus_{\lambda \in \Delta_1} G_{\lambda\mu}$; clearly $(G'_\lambda)_{\lambda \in \Delta_1}$ is a graduation of type Δ_1 and $(G''_\mu)_{\mu \in \Delta_2}$ a graduation of type Δ_2 on G; these graduations are called the *partial graduations* derived from the bigraduation $(G_{\lambda\mu})$. Note that $G_{\lambda\mu} = G'_\lambda \cap G''_\mu$; conversely, if $(G'_\lambda)_{\lambda \in \Delta_1}$ and $(G''_\mu)_{\mu \in \Delta_2}$ are two graduations on G such that G is the direct sum of the $G_{\lambda\mu} = G'_\lambda \cap G''_\mu$, these subgroups form a bigraduation of types Δ_1, Δ_2 on G, of which (G'_λ) and (G''_μ) are the partial graduations. We leave to the reader the task of generalizing this to the case where Δ is a finite product of sets.

(4) Let Δ_0 be a commutative monoid written additively, with identity element denoted by 0; let I be any set and $\Delta_0^{(I)} = \Delta$ denote the submonoid of the product Δ_0^I consisting of the families $(\lambda_\iota)_{\iota \in I}$ of finite support. Let $\rho: \Delta \to \Delta_0$ be the surjective (codiagonal) homomorphism of Δ into Δ_0 defined by $\rho((\lambda_\iota)) = \sum_{\iota \in I} \lambda_\iota$. From every graduation of type Δ a graduation of type Δ_0 is derived by means of ρ (*Example* 2); it is called the *total graduation* associated with the given "multigraduation" of type Δ.

The definitions and examples of this no. extend immediately to the case where G is a group which is *not necessarily commutative*; it is simply necessary to replace everywhere the notion of direct sum by that of "restricted sum" (§ 1, no. 6, *Remark*). Note that in this case the G_λ are normal subgroups of G and that for $\lambda \neq \mu$ every element of G_λ is permutable with every element of G_μ.

2. GRADED RINGS AND MODULES

DEFINITION 2. *Given a ring* A *and a graduation* (A_λ) *of type* Δ *on the additive group* A, *this graduation is said to be compatible with the ring structure on* A *if*

$$(1) \qquad A_\lambda A_\mu \subset A_{\lambda+\mu} \quad \textit{for all } \lambda, \mu \textit{ in } \Delta.$$

The ring A *with this graduation is then called a graded ring of type* Δ.

PROPOSITION 1. *If every element of* Δ *is cancellable and* (A_λ) *is a graduation of type* Δ *compatible with the structure of a ring* A, A_0 *is a subring of* A (*and in particular* $1 \in A_0$).

As $A_0A_0 \subset A_0$ by definition, it suffices to prove that $1 \in A_0$. Let $1 = \sum_{\lambda \in \Delta} e_\lambda$ be the decomposition of 1 into its homogeneous components. If $x \in A_\mu$, then $x = x.1 = \sum_{\lambda \in \Delta} xe_\lambda$; comparing the components of degree μ, (since $\mu + \lambda = \mu$ implies $\lambda = 0$) $x = xe_0$. Since this relation is true for every homogeneous element of A, it is true for all $x \in A$; in particular $1 = 1.e_0 = e_0 \in A_0$.

DEFINITION 3. *Let* A *be a graded ring of type* Δ, (A_λ) *its graduation and* M *a left* (resp. *right*) A-*module; a graduation* (M_λ) *of type* λ *on the additive group* M *is compatible with the* A-*module structure on* M *if*

$$(2) \qquad A_\lambda M_\mu \subset M_{\lambda+\mu} \quad (\text{resp. } M_\mu A_\lambda \subset M_{\lambda+\mu})$$

for all λ, μ *in* Δ. *The module* M *with this graduation is then called a left* (resp. *right*) *graded module of type* Δ *over the graded ring* A.

When the elements of Δ are cancellable, it follows from (2) and Proposition 1 that the M_λ are A_0-*modules.*

Clearly if A is a graded ring of type Δ, the left A-module A_s (resp. the right A-module A_d) is graded of type Δ.

Examples. (1) On any ring A the trivial graduation of type Δ is compatible with the ring structure. If A is graded by the trivial graduation, for a graduation (M_λ) of type Δ on an A-module M to be compatible with the A-module structure, it is necessary and sufficient that the M_λ be *submodules* of M.

(2) Let A be a graded ring of type Δ, M a graded A-module of type Δ and ρ a homomorphism of Δ into a commutative monoid Δ' whose identity element is denoted by 0. Then A is a graded ring of type Δ' and M a graded module of type Δ' for the graduations of type Δ' derived from ρ and the graduations of type Δ on A and M by the procedure of no. 1, *Example* 1: this follows immediately from the relation $\rho(\lambda + \mu) = \rho(\lambda) + \rho(\mu)$.

In particular, if $\Delta = \Delta_1 \times \Delta_2$ is a product of two commutative monoids, the projections pr_1 and pr_2 are homomorphisms and the corresponding graduations are just the *partial graduations* derived from the graduations of type Δ (no. 1, *Example* 3); these partial graduations are thus compatible with the ring structure of A and the module structure of M.

Similarly, if $\Delta = \Delta_0^{(I)}$ (where Δ_0 is a commutative monoid with identity element denoted by 0), the *total graduation* (no. 1, Example 4) of type Δ_0 derived from the graduation of type Δ on A (resp. M) by means of the codiagonal homomorphism is compatible with the ring structure on A (resp. with the module structure on M).

(3) Let A be a graded ring of type Δ, M a graded A-module of type Δ and λ_0 an element of Δ; for all $\lambda \in \Delta$, let $M'_\lambda = M_{\lambda+\lambda_0}$ and let M′ be the **Z**-module $\bigoplus_{\lambda \in \Delta} M'_\lambda$. As $A_\lambda M'_\mu \subset M_{\lambda+\mu+\lambda_0} = M'_{\lambda+\mu}$, M′ is an A-module and the M'_λ form

on M′ a graduation of type Δ compatible with the A-module structure of M′; the graded A-module M′ of type Δ thus defined is said to be obtained by *shifting by* λ_0 the graduation of M and it is denoted by $M(\lambda_0)$. When Δ is a *group*, the underlying A-module of the graded A-module M′ is identified with M.

*(4) Let B be a commutative ring. The polynomial ring B[X] in one indeterminate is graded of type **N** by the subgroups BX^n ($n \geqslant 0$) (cf. III, § 2, no. 9 and IV).*

*(5) Let B be a commutative ring, E a B-module, Q a quadratic form on E and C(Q) the Clifford algebra of Q (cf. IX, § 9). The sub-B-modules $C^+(Q)$ and $C^-(Q)$ form on C(Q) a graduation of type **Z**/2**Z** compatible with the ring structure on C(Q).*

Remarks. (1) The graduations most often used are of type **Z** or of type $\mathbf{Z}^n$; when we speak of *graded* (resp. bigraded, trigraded, etc.) modules and rings without mentioning the type, it is understood that we mean graduations of type **Z** (resp. $\mathbf{Z}^2$, $\mathbf{Z}^3$, etc.); a graded ring (resp. module) of type **N** is also called a *graded* ring (resp. module) *with positive degrees.*

(2) The graded **Z**-modules of type Δ, when **Z** has the trivial graduation, are just the graded commutative groups (whose set of degrees is a commutative monoid) of Definition 1 (no. 1).

DEFINITION 4. *Let* A, A′ *be two graded rings of the same type* Δ *and* (A_λ), (A'_λ) *their respective graduations. A ring homomorphism* $h\colon A \to A'$ *is called graded if* $h(A_\lambda) \subset A'_\lambda$ *for all* $\lambda \in \Delta$.

Let M, M′ *be two graded modules of type* Δ *over a graded ring of type* Δ. *Let* $u\colon M \to M'$ *be an* A*-homomorphism and* δ *an element of* Δ; u *is called graded of degree* δ *if* $u(M_\lambda) \subset M_{\lambda+\delta}$ *for all* $\lambda \in \Delta$.

Let A be a graded ring of type Δ, A′ a graded ring of type Δ' and $\rho\colon \Delta \to \Delta'$ a homomorphism. A ring homomorphism $h\colon A \to A'$ is called *graded* if h is a graded homomorphism of graded rings of type Δ' when A is given the graduation of type Δ' derived from its graduation of type Δ by means of ρ (no. 1, *Example* 2); this therefore means that $h(A_\lambda) \subset A'_{\rho(\lambda)}$ for all $\lambda \in \Delta$.

An A-homomorphism $u\colon M \to M'$ is called *graded* if there exists $\delta \in \Delta$ such that u is graded of degree δ. If $u \neq 0$ and every element of Δ is *cancellable*, the degree δ of u is then determined *uniquely*.

If $h\colon A \to A'$, $h'\colon A' \to A''$ are two graded homomorphisms of graded rings of type Δ, so is $h' \circ h\colon A \to A''$; for a mapping $h\colon A \to A'$ to be a graded ring *isomorphism*, it is necessary and sufficient that h be bijective and that h and the inverse mapping h' be graded homomorphisms; it also suffices for this that h be a bijective graded homomorphism. Thus it is seen that graded homomorphisms can be taken as the *morphisms* of the species of graded ring structure of type Δ (*Set Theory*, IV, § 2, no. 1).

Similarly, if $u\colon \mathrm{M} \to \mathrm{M}'$ and $u'\colon \mathrm{M}' \to \mathrm{M}''$ are two graded homomorphisms of graded A-modules of type Δ, of respective degrees δ and δ', $u' \circ u\colon \mathrm{M} \to \mathrm{M}''$ is a graded homomorphism of degree $\delta + \delta'$. If δ admits an inverse $-\delta$ in Δ and $u\colon \mathrm{M} \to \mathrm{M}'$ is a bijective graded homomorphism of degree δ, the inverse mapping $u'\colon \mathrm{M}' \to \mathrm{M}$ is a bijective graded homomorphism of degree $-\delta$. It follows as above that the *graded homomorphisms of degree* 0 can be taken as the *morphisms* of the species of graded A-module of type Δ. But a bijective graded homomorphism $u\colon \mathrm{M} \to \mathrm{N}$ of degree $\neq 0$ is not a graded A-module isomorphism if M and N are non-zero and the elements of Δ are cancellable.

Examples. (6) If M is a graded A-module and $\mathrm{M}(\lambda_0)$ is a graded A-module obtained by shifting (no. 2, *Example* 3), the **Z**-linear mapping of $\mathrm{M}(\lambda_0)$ into M which coincides with the canonical injection on each $\mathrm{M}_{\lambda+\lambda_0}$ is a graded homomorphism of degree λ_0 (which is bijective when Δ is a *group*).

(7) If a is a homogeneous element of degree δ belonging to the centre of A, the homothety $x \mapsto ax$ of any graded A-module M is a graded homomorphism of degree δ.

Remark (3) A graded A-module M is called a *graded free* A-*module* if there exists a basis $(m_\iota)_{\iota\in\mathrm{I}}$ of M consisting of *homogeneous* elements. Suppose it is and Δ is a commutative *group*; let λ_ι be the degree of m_ι and consider for each ι the shifted A-module $\mathrm{A}(-\lambda_\iota)$ (no. 2, *Example* 3); if e_ι denotes the element 1 of A considered as an element *of degree* λ_ι in $\mathrm{A}(-\lambda_\iota)$, the A-linear mapping $u\colon \bigoplus_{\iota\in\mathrm{I}} \mathrm{A}(-\lambda_\iota) \to \mathrm{M}$ such that $u(e_\iota) = m_\iota$ for all ι, is a *graded* A-*module isomorphism*.

Assuming always that Δ is a commutative group, now let N be a graded A-module, $(n_\iota)_{\iota\in\mathrm{I}}$ a system of *homogeneous* generators of N and suppose that n_ι is of degree μ_ι. Then the A-linear mapping $v\colon \bigoplus_{\iota\in\mathrm{I}} \mathrm{A}(-\mu_\iota) \to \mathrm{N}$ such that $u(e_\iota) = n_\iota$ for all ι is a *surjective graded* A-*module homomorphism of degree* 0. If N is a *finitely generated* graded A-module, there is always a finite system of homogeneous generators of N and hence there is a surjective homomorphism of the above type with I *finite*.

3. GRADED SUBMODULES

PROPOSITION 2. *Let* A *be a graded ring of type* Δ, M *a graded* A-*module of type* Δ, (M_λ) *its graduation and* N *a sub*-A-*module of* M. *The following properties are equivalent:*

(a) N *is the sum of the family* $(\mathrm{N} \cap \mathrm{M}_\lambda)_{\lambda\in\Delta}$.
(b) *The homogeneous components of every element of* N *belong to* N.
(c) N *is generated by homogeneous elements.*

Every element of N can be written uniquely as a sum of elements of the M_λ and hence it is immediate that (a) and (b) are equivalent and that (a) implies (c). We show that (c) implies (b). Then let $(x_\iota)_{\iota \in I}$ be a family of homogeneous generators $\neq 0$ of N and let $\delta(\iota)$ be the degree of x_ι. Every element of N can be written as $\sum_\iota a_\iota x_\iota$ with $a_\iota \in A$; if $a_{\iota,\lambda}$ is the component of a_ι of degree λ, the conclusion follows from the relation

$$\sum_{\iota \in I}\Big(\sum_{\mu \in \Delta} a_{\iota,\mu}x_\iota\Big) = \sum_{\lambda \in \Delta}\Big(\sum_{\mu + \delta(\iota) = \lambda} a_{\iota,\mu}x_\iota\Big).$$

Remark (1) In the above notation, the relation $\sum_{\iota \in I} a_\iota x_\iota = 0$ is therefore equivalent to the system of relations $\sum_{\mu + \delta(\iota) = \lambda} a_{\iota,\mu}x_\iota = 0$. When Δ is a group, these relations can be written $\sum_{\iota \in I} a_{\iota,\lambda - \delta(\iota)}x_\iota = 0$.

When a submodule N of M has the equivalent properties stated in Proposition 2, clearly the $N \cap M_\lambda$ form a graduation compatible with the A-module structure of N, called the *graduation induced* by that on M; N with this graduation is called a *graded submodule* of M.

COROLLARY 1. *If* N *is a graded submodule of* M *and* (x_ι) *is a generating system of* N, *the homogeneous components of the* x_ι *form a generating system of* N.

COROLLARY 2. *If* N *is a finitely generated submodule of* M, N *admits a finite generating system consisting of homogeneous elements.*

It suffices to apply Corollary 1 noting that an element of M has only a finite number of homogeneous components $\neq 0$.

A graded submodule of A_s (resp. A_d) is called a *graded left* (resp. *right*) *ideal* of the graded ring A. For every subring B of A$(B \cap A_\lambda)(B \cap A_\mu) \subset B \cap A_{\lambda+\mu}$; if B is a *graded sub-$\mathbf{Z}$-module* of A, the graduation induced on B by that on A is therefore compatible with the ring structure on B; B is then called a *graded subring* of A.

Clearly if N (resp. B) is a graded sub-A-module of M (resp. a graded subring of A), the canonical injection $N \to M$ (resp. $B \to A$) is a graded module homomorphism of degree 0 (resp. a graded ring homomorphism).

If N is a graded submodule of a graded A-module M and $(M_\lambda)_{\lambda \in \Delta}$ the graduation of M, the submodules $(M_\lambda + N)/N$ of M/N form a *graduation* compatible with the structure of this quotient module. For, if $N_\lambda = M_\lambda \cap N$, $(M_\lambda + N)/N$ is identified with M_λ/N_λ and it follows from Proposition 2 and § 1, no. 6, formula (26) that M/N is their direct sum. Moreover,

$$A_\lambda(M_\mu + N) \subset A_\lambda M_\mu + N \subset M_{\lambda+\mu} + N$$

and hence $A_\lambda((M_\mu + N)/N) \subset (M_{\lambda+\mu} + N)/N$, which establishes our assertion. The graduation $((M_\lambda + N)/N)_{\lambda \in \Delta}$ is called the *quotient graduation* of that on M by N and the quotient module M/N with this graduation is called the *graded quotient module* of M by the graded submodule N; the canonical homomorphism $M \to M/N$ is a graded homomorphism of degree 0 for this graduation.

If $\mathfrak{b}$ is a graded two-sided ideal of A, the quotient graduation on $A/\mathfrak{b}$ is compatible with the ring structure on $A/\mathfrak{b}$; the ring $A/\mathfrak{b}$ with this graduation is called the *quotient graded ring* of A by $\mathfrak{b}$; the canonical homomorphism $A \to A/\mathfrak{b}$ is a homomorphism of graded rings for this graduation.

PROPOSITION 3. *Let* A *be a graded ring of type* Δ, M, N *two graded* A*-modules of type* Δ *and* $u: M \to N$ *a graded* A*-homomorphism of degree* δ. *Then:*

(i) $\mathrm{Im}(u)$ *is a graded submodule of* N.

(ii) *If* δ *is a regular element of* Δ, $\mathrm{Ker}(u)$ *is a graded submodule of* M.

(iii) *If* $\delta = 0$, *the bijection* $M/\mathrm{Ker}(u) \to \mathrm{Im}(u)$ *canonically associated with* u *is an isomorphism of graded modules.*

Assertion (i) follows immediately from the definitions and Proposition 2(c). If x is an element of M such that $u(x) = 0$ and $x = \sum_\lambda x_\lambda$ is its decomposition into homogeneous components (where x_λ is of degree λ), then

$$\sum_\lambda u(x_\lambda) = u(x) = 0$$

and $u(x_\lambda)$ is of degree $\lambda + \delta$; if δ is regular the relation $\lambda + \delta = \mu + \delta$ implies $\lambda = \mu$, hence the $u(x_\lambda)$ are the homogeneous components of $u(x)$ and necessarily $u(x_\lambda) = 0$ for all $\lambda \in \Delta$, which proves (ii). The bijection $v: M/\mathrm{Ker}(u) \to \mathrm{Im}(u)$ canonically associated with u is then a graded homomorphism of degree δ, as follows from the definition of the quotient graduation; whence (iii) when $\delta = 0$.

COROLLARY. *Let* A, B *be two graded rings of type* Δ *and* $u: A \to B$ *a graded homomorphism of graded rings. Then* $\mathrm{Im}(u)$ *is a graded submodule of* B, $\mathrm{Ker}(u)$ *a graded two-sided ideal of* A *and the bijection* $A/\mathrm{Ker}(u) \to \mathrm{Im}(u)$ *canonically associated with* u *is an isomorphism of graded rings.*

It suffices to apply Proposition 3 to u considered as a homomorphism of degree 0 of graded **Z**-modules.

PROPOSITION 4. *Let* A *be a graded ring of type* Δ *and* M *a graded* A*-module of type* Δ.

(i) *Every sum and every intersection of graded submodules of* M *is a graded submodule.*

(ii) *If* x *is a homogeneous element of* M *of degree* μ *which is cancellable in* Δ, *the annihilator of* x *is a graded left ideal of* A.

(iii) *If all the elements of* Δ *are cancellable, the annihilator of a graded submodule of* M *is a graded two-sided ideal of* A.

If (N_ι) is a family of graded submodules of M, property (c) of Proposition 2 shows that the sum of the N_ι is generated by homogeneous elements and property (b) of Proposition 2 proves that the homogeneous components of every element of $\bigcap_\iota N_\iota$ belongs to $\bigcap_\iota N_\iota$; whence (i).

To prove (ii), it suffices to note that $\text{Ann}(x)$ is the kernel of the homomorphism $a \mapsto ax$ of the A-module A_s into M and that this homomorphism is graded of degree μ; the conclusion follows from Proposition 3(ii). Finally (iii) is a consequence of (i) and (ii) for the annihilator of a graded submodule N of M is the intersection of the annihilators of the homogeneous elements of N, by virtue of Proposition 2.

Remark 2. Let M be a graded A-module and E a submodule of M; it follows from Proposition 4(i) that there exists a *largest* graded submodule N′ of M contained in E and a *smallest* graded submodule N″ of M containing E; N′ is the set of $x \in E$ all of whose homogeneous components belong to E and N″ is the submodule of M generated by the homogeneous components of a generating system of E.

PROPOSITION 5. *Let* A *be a graded ring of type* Δ. *If every element of* Δ *is cancellable, then, for every homogeneous element* $a \in A$, *the centralizer of* a *in* A (I, § 1, no. 5) *is a graded subring of* A.

Suppose that a is of degree δ; let $b = \sum_\lambda b_\lambda$ be an element permutable with a, b_λ being the homogeneous component of b of degree λ for all $\lambda \in \Delta$. Then by hypothesis $\sum_\lambda (ab_\lambda - b_\lambda a) = 0$ and $ab_\lambda - b_\lambda a$ is homogeneous of degree $\lambda + \delta$; as δ is cancellable, it follows that $ab_\lambda = b_\lambda a$ for all λ, which proves our assertion.

COROLLARY. *If every element of* Δ *is cancellable, the centralizer of the graded subring* B *of* A (and in particular the *centre* of A) *is a graded subring of* A.

It is the intersection of the centralizers of the homogeneous elements of B.

Remark (3) A *direct system* $(A_\alpha, \phi_{\beta\alpha})$ *of graded rings of type* Δ (resp. a *direct system* $(M_\alpha, f_{\beta\alpha})$ *of graded* A_α*-modules of type* Δ) is a direct system of rings (resp. A_α-modules) such that each A_α (resp. M_α) is graded of type Δ and each $\phi_{\beta\alpha}$ (resp. $f_{\beta\alpha}$) is a *homomorphism of graded rings* (resp. an A_α*-homomorphism of degree* 0 *of graded modules*). If $(A_\alpha^\lambda)_{\lambda \in \Delta}$ (resp. $(M_\alpha^\lambda)_{\lambda \in \Delta}$) be the graduation of A_α (resp. M_α) and we write

$$A = \varinjlim A_\alpha, \quad A^\lambda = \varinjlim_\alpha A_\alpha^\lambda \qquad (\text{resp. } M = \varinjlim M_\alpha,\ M^\lambda = \varinjlim_\alpha M_\alpha^\lambda),$$

it follows from § 6, no. 2, Proposition 5 that (A^λ) (resp. (M^λ)) is a graduation of A (resp. M) and it follows from I, § 10, nos. 3 and 4 that this graduation is compatible with the ring structure on A (resp. the A-module structure on M). The *graded* ring A (resp. *graded* A-module M) is called the *direct limit* of the direct system of graded rings $(A_\alpha, \phi_{\beta\alpha})$ (resp. graded modules $(M_\alpha, f_{\beta\alpha})$). If $\phi_\alpha: A_\alpha \to A$ (resp. $f_\alpha: M_\alpha \to M$) is the canonical mapping, ϕ_α (resp. f_α) is a homomorphism of graded rings (resp. a homomorphism of degree 0 of graded A_α-modules).

4. CASE OF AN ORDERED GROUP OF DEGREES

An order structure (denoted by $\leqslant$) on a commutative group Δ written additively is said to be *compatible* with the group structure if, for all $\rho \in \Delta$, the relation $\lambda \leqslant \mu$ implies $\lambda + \rho \leqslant \mu + \rho$. The group Δ with this order structure is then called an *ordered group*. We shall study these groups in detail in VI, § 1; here we restrict ourselves to the remark that in such a group the relation $\lambda > 0$ implies $\lambda + \mu > \mu$ for all μ, for it implies $\lambda + \mu \geqslant \mu$ by definition and the relation $\xi + \mu = \mu$ is equivalent to $\xi = 0$.

Let Δ be an ordered commutative group, A a graded ring of type Δ and (A_λ) its graduation and suppose that the relation $A_\lambda \neq \{0\}$ implies $\lambda \geqslant 0$; then it follows from the definitions that $\mathfrak{F}_0 = \sum_{\lambda > 0} A_\lambda$ is a *graded two-sided ideal* of A, by virtue of the remark made above.

Proposition 6. *Let Δ be an ordered commutative group,* A *a graded ring of type* Δ, (A_λ) *its graduation,* M *a graded* A*-module of type* Δ *and* (M_λ) *its graduation. Suppose that the relation* $A_\lambda \neq \{0\}$ *implies* $\lambda \geqslant 0$ *and that there exists* λ_0 *such that* $M_{\lambda_0} \neq \{0\}$ *and* $M_\lambda = \{0\}$ *for* $\lambda < \lambda_0$. *Then, if* $\mathfrak{F}_0 = \sum_{\lambda > 0} A_\lambda$, $\mathfrak{F}_0 M \neq M$.

Let x be a non-zero element of M_{λ_0}; suppose that $x \in \mathfrak{F}_0 M$. Then $x = \sum_i a_i x_i$, where the a_i are homogeneous elements $\neq 0$ of $\mathfrak{F}_0$ and the x_i homogeneous elements $\neq 0$ of M with $\deg(x) = \deg(a_i) + \deg(x_i)$ for all i (no. 2). But, as $\deg(a_i) > 0$, $\lambda_0 = \deg(a_i) + \deg(x_i) > \deg(x_i)$, which contradicts the hypothesis.

Corollary 1. *With the hypotheses on* Δ *and* A *of Proposition* 6, *if M is a finitely generated graded* A*-module such that* $\mathfrak{F}_0 M = M$, *then* $M = \{0\}$.

Suppose $M \neq \{0\}$. Let λ_0 be a minimal element of the set of degrees of a finite generating system of M consisting of homogeneous elements $\neq 0$; then the hypotheses of Proposition 6 would be fulfilled, which implies a contradiction.

Corollary 2. *With the hypotheses on* Δ *and* A *of Proposition* 6, *let* M *be a finitely generated graded* A*-module and* N *a graded submodule of* M *such that* $N + \mathfrak{J}_0 M = M$*; then* $N = M$.

M/N is a finitely generated graded A-module and the hypothesis implies that $\mathfrak{J}_0 . (M/N) = M/N$; hence $M/N = 0$.

COROLLARY 3. *With the hypotheses on* Δ *and* A *of Proposition* 6, *let* $u: M \to N$ *be a graded homomorphism of graded right* A*-modules, where* N *is assumed to be finitely generated. If the homomorphism*

$$u \otimes 1 : M \otimes_A (A/\mathfrak{J}_0) \to N \otimes_A (A/\mathfrak{J}_0)$$

is surjective, then u is surjective.

$u(M)$ is a graded submodule of N and the $(A/\mathfrak{J}_0)$-module

$$(N/u(M)) \otimes_A (A/\mathfrak{J}_0)$$

is isomorphic to $(N \otimes_A (A/\mathfrak{J}_0))/\mathrm{Im}(u \otimes 1)$ (§ 3, no. 6, Proposition 6). The hypothesis therefore implies $(N/u(M)) \otimes_A (A/\mathfrak{J}_0) = 0$ and hence $N = u(M)$ by Corollary 1.

Remark. It follows from the proof of Corollary 1 that Corollaries 1 and 2 (resp. Corollary 3) are still valid when, instead of assuming that M (resp. N) is finitely generated, the following hypothesis is made: there exists a subset Δ^+ of Δ satisfying the following conditions:

(1) for $\lambda \notin \Delta^+$, $M_\lambda = \{0\}$ (resp. $N_\lambda = \{0\}$);
(2) every non-empty subset of Δ^+ has a least element.

This will be the case if $\Delta = \mathbf{Z}$ and M (resp. N) is a graded module with *positive* degrees.

PROPOSITION 7. *Suppose that* $\Delta = \mathbf{Z}$. *With the hypotheses on* A *and* M *of Proposition* 6, *consider the graded* A_0*-module* $N = M/\mathfrak{J}_0 M$ *and suppose the following conditions hold:*

(i) *each of the* N_λ *considered as an* A_0*-module admits a basis* $(y_{\iota\lambda})_{\iota \in I_\lambda}$*;*
(ii) *the canonical homomorphism* $\mathfrak{J}_0 \otimes_A M \to M$ *is injective.*

Then M *is a graded free* A*-module* (no. 2, *Remark* 3) *and, to be precise, if* $x_{\iota\lambda}$ *is an element of* M_λ *whose image in* N_λ *is* $y_{\iota\lambda}$, *the family* $(x_{\iota\lambda})_{(\iota, \lambda) \in I}$ (*where* I *is the set the sum of the* I_λ) *is a basis of* M.

We know (no. 2, *Remark* 3) that there is a graded free A-module L (of graduation (L_λ)) and a surjective homomorphism $p: L \to M$ of degree 0 such that $p(e_{\iota\lambda}) = x_{\iota\lambda}$ for all $(\iota, \lambda) \in I$ ($(e_{\iota\lambda})_{(\iota, \lambda) \in I}$ being a basis of L consisting of homogeneous elements $e_{\iota\lambda} \in L_\lambda$). It follows from the above *Remark* that p is *surjective.* Consider the graded A-module $R = \mathrm{Ker}(p)$ and note that $R_\lambda = \{0\}$ for $\lambda < \lambda_0$ by definition; we need to prove that $R = \{0\}$ and by Proposition 6 it will suffice to show that $\mathfrak{J}_0 R = R$. Consider the commutative diagram (§ 3, no. 6, Proposition 5)

$$\begin{array}{ccccccc} & & \mathfrak{J}_0 \otimes R & \xrightarrow{1 \otimes j} & \mathfrak{J}_0 \otimes L & \xrightarrow{1 \otimes p} & \mathfrak{J}_0 \otimes M & \longrightarrow 0 \\ & & \downarrow a & & \downarrow b & & \downarrow c & \\ 0 & \longrightarrow & R & \xrightarrow{f} & L & \xrightarrow{p} & M & \longrightarrow 0 \end{array}$$

where j is the canonical injection, a, b, c deriving from the canonical injection $\mathfrak{J}_0 \to A$ (§ 3, no. 4, Proposition 4); it must be shown that a is *surjective*. Note that, as L is free, b is *injective* (§ 3, no. 7, Corollary 6 to Proposition 7) and c is injective by hypothesis. Then let t be an element of R and $\bar{t}$ its class in $R/\mathfrak{J}_0 R$; then there is an exact sequence (§ 3, no. 6, Proposition 5 and Corollary 2 to Proposition 6)

$$R/\mathfrak{J}_0 R \xrightarrow{\bar{j}} L/\mathfrak{J}_0 L \xrightarrow{\bar{p}} M/\mathfrak{J}_0 M \longrightarrow 0$$

where $\bar{j}$ and $\bar{p}$ derive from j and p when passing to the quotients and $\bar{p}$ is by hypothesis a *bijection*; then $\bar{j}(\bar{t}) = 0$, in other words $j(t) \in \mathfrak{J}_0 L$. Then there is an element $z \in \mathfrak{J}_0 \otimes L$ such that $j(t) = b(z)$; as $p(b(z)) = 0$, $c((1 \otimes p)(z)) = 0$ and, as c is injective, $(1 \otimes p)(z) = 0$. In other words, z is the image of an element $t' \in \mathfrak{J}_0 \otimes R$ under $1 \otimes j$ and then $j(a(t')) = b(z) = j(t)$; as j is injective, this implies $t = a(t')$.

We shall show later (*Commutative Algebra*, II, § 3, no. 2, Proposition 5) how this proposition can be extended to non-graded modules.

Lemma 1. *For a commutative group* Δ *to be such that there exists on* Δ *a total ordering compatible with the group structure of* Δ, *it is necessary and sufficient that* Δ *be torsion-free.*

If there exists such an order structure on Δ and if $\lambda > 0$, then $\lambda + \mu > 0$ for all $\mu \geqslant 0$ and in particular, by induction on the integer $n > 0$, $n.\lambda > 0$, which proves that Δ is torsion-free (since every element $\neq 0$ of Δ is either >0 or <0). Conversely, if Δ is torsion-free, Δ is a sub-$\mathbf{Z}$-module of a vector $\mathbf{Q}$-space (§ 7, no. 10, Corollary 1 to Proposition 26) which may be assumed of the form $\mathbf{Q}^{(I)}$; if I is given a well-ordering (*Set Theory*, III, § 2, no. 3, Theorem 1) and $\mathbf{Q}$ its usual ordering, the set $\mathbf{Q}^{(I)}$ with the *lexicographical ordering* is totally ordered (*Set Theory*, III, § 2, no. 6); it is immediate that this ordering is compatible with the additive group structure of $\mathbf{Q}^{(I)}$.

PROPOSITION 8. *Let* Δ *be a torsion-free commutative group and* A *a graded ring of type* Δ. *If the product in* A *of two homogeneous elements* $\neq 0$ *is* $\neq 0$, *the ring* A *has no divisor of* 0.

Let Δ be given a total ordering compatible with its group structure (Lemma

1) and let $x = \sum_{\lambda \in \Delta} x_\lambda$, $y = \sum_{\lambda \in \Delta} y_\lambda$ be two non-zero elements of A (x_λ and y_λ being homogeneous of degree λ for all $\lambda \in \Delta$); let α (resp. β) be the greatest of the elements $\lambda \in \Delta$ such that $x_\lambda \neq 0$ (resp. $y_\lambda \neq 0$); it is immediate that if $\lambda \neq \alpha$ or $\mu \neq \beta$, either $x_\lambda y_\mu = 0$ or $\deg(x_\lambda y_\mu) < \alpha + \beta$; the homogeneous component of xy of degree $\alpha + \beta$ is therefore $x_\alpha y_\beta$, which is non-zero by hypothesis; whence $xy \neq 0$.

5. GRADED TENSOR PRODUCT OF GRADED MODULES

Let Δ be a commutative monoid with its identity element denoted by 0, A a graded ring of type Δ and M (resp. N) a graded right (resp. left) A-module of type Δ. Let (A_λ) (resp. (M_λ), (N_λ)) be the graduation of A (resp. M, N); the tensor product $M \otimes_{\mathbf{Z}} N$ of the **Z**-modules M and N is the direct sum of the $M_\lambda \otimes_{\mathbf{Z}} N_\mu$ (§ 3, no. 7, Proposition 7) and hence the latter form a *bigraduation* of types Δ, Δ on this **Z**-module. Consider on $M \otimes_{\mathbf{Z}} N$ the *total graduation* of type Δ associated with this bigraduation (no. 1, *Example* 4); it consists of the sub-**Z**-modules $P_\lambda = \sum_{\mu+\nu=\lambda} (M_\mu \otimes_{\mathbf{Z}} N_\nu)$. It is known that the **Z**-module $M \otimes_A N$ is the quotient of $M \otimes_{\mathbf{Z}} N$ by the sub-**Z**-module Q generated by the elements $(xa) \otimes y - x \otimes (ay)$, where $x \in M$, $y \in N$ and $a \in A$ (§ 3, no. 1); if, for all $\lambda \in \Delta$, $x_\lambda, y_\lambda, a_\lambda$ are the homogeneous components of degree λ of x, y, a respectively, clearly $(xa) \otimes y - x \otimes (zy)$ is the sum of the homogeneous elements $(x_\lambda a_\nu) \otimes y_\mu - x_\lambda \otimes (a_\nu y_\mu)$, in other words Q is a *graded* sub-**Z**-module of $M \otimes_{\mathbf{Z}} N$ (no. 3, Proposition 2) and the quotient

$$M \otimes_A N = (M \otimes_{\mathbf{Z}} N)/Q$$

therefore has canonically a graded **Z**-module structure of type Δ (no. 3). Moreover (no. 3, Proposition 5), the *centre* C of A is a graded subring of A; the graduation which we have just defined on $M \otimes_A N$ is *compatible with its module structure over the graded ring* C. For $M \otimes_{\mathbf{Z}} N$ has canonically *two* C-module structures, for which respectively $c(x \otimes y) = (xc) \otimes y$ and $(x \otimes y)c = x \otimes (cy)$ for $x \in M$, $y \in N$, $c \in C$ (§ 3, no. 3); if $x \in M_\lambda$, $y \in N_\mu$, $c \in C \cap A_\nu$, the two elements $c(x \otimes y)$ and $(x \otimes y)c$ belong to $(M \otimes_{\mathbf{Z}} N)_{\lambda+\mu+\nu}$ and their difference belongs to Q and hence their common image in $M \otimes_A N$ belongs to $(M \otimes_A N)_{\lambda+\mu+\nu}$, which establishes our assertion. When we speak of $M \otimes_A N$ as a *graded* C-*module*, we always mean with the structure thus defined, unless otherwise mentioned. Note that $(M \otimes_A N)_\lambda$ can be defined as the additive group of $M \otimes_A N$ generated by the $x_\mu \otimes y_\nu$, where $x_\mu \in M_\mu$, $y_\nu \in N_\nu$ and $\mu + \nu = \lambda$.

Let M′ (resp. N′) be another graded right (resp. left) A-module and $u: M \to M'$, $v: N \to N'$ graded homomorphisms of respective degrees α and β. Then it follows immediately from the above remark that $u \otimes v$ is a *graded* (C-module) homomorphism of degree $\alpha + \beta$.

When A is commutative, a graduation (compatible with the A-module structure) is similarly defined on the tensor product of any finite number of graded A-modules; it is moreover immediate that the associativity isomorphisms such as $(M \otimes N) \otimes P \to M \otimes (N \otimes P)$ (§ 3, no. 8, Proposition 8) are isomorphisms of *graded* modules.

Remark. When A has the *trivial* graduation (no. 1, *Example* 1), $(M \otimes_A N)_\lambda$ is then simply the direct sum of the sub-Y-modules $M_\mu \otimes_A N_\nu$ of $M \otimes_A N$ such that $\mu + \nu = \lambda$.

Let M (resp. N) be a graded right (resp. left) A-module of type Δ, P a graded **Z**-module of type Δ and let f be a **Z**-bilinear mapping of $M \times N$ into P satisfying condition (1) of § 3, no. 1, and such moreover that

$$f(x_\lambda, y_\mu) \in P_{\lambda+\mu} \quad \text{for } x_\lambda \in M_\lambda, y_\mu \in N_\mu, \lambda, \mu \text{ in } \Delta.$$

Then $f(x, y) = g(x \otimes y)$, where $g: M \otimes_A N \to P$ is a **Z**-linear mapping (§ 3, no. 1, Proposition 1) and it follows from the above condition that g is a *graded* **Z**-module homomorphism of degree 0.

Let B be another graded ring of type Δ and $\rho: A \to B$ a homomorphism of graded rings (no. 2); then $\rho_*(B_d)$ is a graded right A-module of type Δ. If E is a graded left A-module of type Δ and $\rho^*(B_d) \otimes_A E$ is given the graded **Z**-module structure of type Δ defined above, the canonical left B-module structure (§ 5, no. 1) is compatible with the graduation of

$$E_{(B)} = \rho^*(E) = \rho_*(B_d) \otimes_A E.$$

The graded B-module thus obtained is said to be obtained by extending the ring of scalars to B by means of ρ and when we speak of $E_{(B)}$ or $\rho^*(E)$ as a graded B-module, we always mean this structure, unless otherwise mentioned.

6. GRADED MODULES OF GRADED HOMOMORPHISMS

We suppose in this no. that the monoid Δ is a *group*. Let A be a graded ring of type Δ and M, N two graded left (for example) A-modules of type Δ. Let H_λ denote the additive group of *graded homomorphisms of degree* λ of M into N (no. 2); in the additive group $\operatorname{Hom}_A(M, N)$ of *all* homomorphisms of M into N (with the *non-graded* A-module structures) the sum (for $\lambda \in \Delta$) of the H_λ is *direct*. For, if there is a relation $\sum_\lambda u_\lambda = 0$ with $u_\lambda \in H_\lambda$ for all λ, it follows that $\sum_\lambda u_\lambda(x_\mu) = 0$ for all μ and all $x_\mu \in M_\mu$. As the elements of Δ are cancellable, $u_\lambda(x_\mu)$ is the homogeneous component of $\sum_\lambda u_\lambda(x_\mu)$ of degree $\lambda + \mu$; hence $u_\lambda(x_\mu) = 0$ for every ordered pair (μ, λ) and every $x_\mu \in M_\mu$, which implies $u_\lambda = 0$ for all $\lambda \in \Delta$. We shall denote (in this paragraph) by $\operatorname{Homgr}_A(M, N)$ the additive subgroup of $\operatorname{Hom}_A(M, N)$ the sum of the H_λ and we shall call it the additive group of *graded A-module homomorphisms* of M into N. Let C be the centre of A, which is a graded subring (no. 3, Corollary to Proposition 5); for the canonical C-module structure on $\mathrm{Hom}_A(M, N)$ (§ 1, no. 14, *Remark* 1), $\mathrm{Homgr}_A(M, N)$ is a *submodule* and the graduation (H_λ) is *compatible* with the C-module structure: for, if $c_\nu \in C \cap A_\nu$, $x_\mu \in N_\mu$ and $u_\lambda \in H_\lambda$, then by definition $(c_\nu u_\lambda)(x_\mu) = c_\nu . u_\lambda(x_\mu) \subset N_{\lambda+\mu+\nu}$ and hence $c_\nu u_\lambda \in H_{\lambda+\nu}$.

Let M′ and N′ be two graded left A-modules of type Δ and $u' : M' \to M$, $v' : N \to N'$ graded homomorphisms of respective degrees α and β. Then it is immediate that $\mathrm{Hom}(u', v') : w \mapsto v' \circ w \circ u'$ maps $\mathrm{Homgr}_A(M, N)$ into $\mathrm{Homgr}_A(M', N')$ and that its restriction to $\mathrm{Homgr}_A(M, N)$ is a *graded* homomorphism into $\mathrm{Homgr}_A(M', N')$ *of degree* $\alpha + \beta$.

In particular $\mathrm{Homgr}_A(M, M)$ is a *graded subring* of $\mathrm{End}_A(M)$, which is denoted by $\mathrm{Endgr}_A(M)$.

Remark. If M and N are graded left A-modules, $\mathrm{Homgr}_A(M, N)$ is in general distinct from $\mathrm{Hom}_A(M, N)$. However these two sets are equal when M is a *finitely generated* A-module. For M is then generated by a finite number of homogeneous elements x_i $(1 \leqslant i \leqslant n)$; let $d(i)$ be the degree of x_i; let $u \in \mathrm{Hom}_A(M, N)$ and for all $\lambda \in \Delta$ let $z_{i,\lambda}$ denote the homogeneous component of $u(x_i)$ of degree $\lambda + d(i)$. We show that there exists a homomorphism $u_\lambda : M \to N$ such that $u_\lambda(x_i) = z_{i,\lambda}$ for all i. It suffices to prove that if $\sum_i a_i x_i = 0$ with $a_i \in A$ for $1 \leqslant i \leqslant n$, then $\sum_i a_i z_{i,\lambda} = 0$ for all $\lambda \in \Delta$ (§ 1, no. 7, *Remark*). It can be assumed that each a_i is homogeneous of degree $d'(i)$ such that $d(i) + d'(i) = \mu$ for all i (no. 3, *Remark* 1); then $\sum_i a_i u(x_i) = 0$; taking the homogeneous component of degree $\lambda + \mu$ on the left-hand side, we obtain $\sum_i a_i z_{i,\lambda} = 0$, whence the existence of the homomorphism u_λ; clearly moreover u_λ is *graded* of degree λ. Finally, $u_\lambda = 0$ except for a finite number of values of λ, and $u = \sum_\lambda u_\lambda$ by definition, which proves our assertion.

In particular, $\mathrm{Homgr}_A(A_s, M) = \mathrm{Hom}_A(A_s, M)$ for every graded left A-module M; moreover $\mathrm{Hom}_A(A_s, M)$ has a *graded left A-module* structure (and not just a graded C-module structure), and it is immediate that with this structure the canonical mapping of M into $\mathrm{Hom}_A(A_s, M)$ (§ 1, no. 14, *Remark* 2) is a *graded A-module isomorphism.*

Similarly, $\mathrm{Homgr}_A(M, A_s)$ has a *graded right A-module* structure (and not only a graded C-module structure); it is called the *graded dual* of the graded A-module M and is denoted by M^{*gr}, or simply M^* when no confusion results. If $u : M \to N$ is a graded homomorphism of degree δ, it follows from the above that the restriction to N^{*gr} of ${}^t u = \mathrm{Hom}(u, 1_{A_s})$ is a graded homomorphism of the graded dual N^{*gr} into the graded dual M^{*gr}, of degree δ, called the *graded transpose* of u.

We sometimes consider on the graded dual M^{*gr} the graduation derived from the above with the aid of the isomorphism $\lambda \mapsto -\lambda$ of Δ (no. 1, *Example* 2) so that the homogeneous elements of degree λ in M^{*gr} are the graded linear forms *of degree* $-\lambda$ on M (when A has the trivial graduation, these are the zero linear forms of the M_μ of index $\mu \neq \lambda$). Then, if $u\colon M \to N$ is a graded homomorphism of degree δ, u' becomes a graded homomorphism of degree $-\delta$.

Suppose A is *commutative* and graded of type Δ and let M, N, P, Q be four *graded* A-modules of type Δ. Then there are *canonical graded homomorphisms of degree* 0

$$\mathrm{Homgr}_A(M, \mathrm{Homgr}_A(N, P)) \to \mathrm{Homgr}_A(M \otimes_A N, P) \tag{3}$$

$$\mathrm{Homgr}_A(M, N) \otimes_A P \to \mathrm{Homgr}_A(M, N \otimes_A P) \tag{4}$$

$$\mathrm{Homgr}_A(M, P) \otimes_A \mathrm{Homgr}_A(N, Q) \to \mathrm{Homgr}_A(M \otimes_A N, P \otimes_A Q) \tag{5}$$

(the tensor products being given the graduations defined in no. 5) obtained by restricting the canonical homomorphisms defined in § 4, nos. 1, 2 and 4; for, if $u\colon M \to \mathrm{Homgr}_A(N, P)$ is graded of degree δ, then, for all $x \in M_\lambda$, $u(x)$ is a graded homomorphism $N \to P$ of degree $\delta + \lambda$ and hence, for $y \in N_\mu$, $u(x)(y) \in P_{\delta+\lambda+\mu}$; if $v\colon M \otimes_A N \to P$ corresponds canonically to u, it is then seen that v is a graded homomorphisms of degree δ, whence our assertion concerning (3); moreover it is seen that this homomorphism is *bijective*. The argument is similar for (4) and (5).

If in particular $P = Q = A$ in (5), then there is a canonical *graded* homomorphism of degree 0

$$M^{*gr} \otimes_A N^{*gr} \to (M \otimes_A N)^{*gr}. \tag{6}$$

APPENDIX

PSEUDOMODULES

1. ADJUNCTION OF A UNIT ELEMENT TO A PSEUDO-RING

Let A be a pseudo-ring (I, § 8, no. 1). On the set $A' = \mathbf{Z} \times A$ we define the following laws of composition:

$$(1) \qquad \begin{cases} (m, a) + (n, b) = (m + n, a + b) \\ (m, a)(n, b) = (mn, mb + na + ab). \end{cases}$$

It is immediately verified that A′ with these two laws of composition is a *ring* in which the element (1, 0) is the *unit element*. The set $\{0\} \times A$ is a two-sided ideal of A′ and $\iota: x \mapsto (0, x)$ is an isomorphism of the pseudo-ring A onto the sub-pseudo-ring $\{0\} \times A$ by means of which A and $\{0\} \times A$ are identified. A′ is called the *ring derived from the pseudo-ring* A *by adjoining a unit element.*

If A already has an identity element ε, the element $e = (0, \varepsilon)$ of A′ is an *idempotent* belonging to the centre of A′ and such that

$$A = eA' = A'e.$$

Then $(eA', (1-e)A')$ is a direct decomposition (I, § 8, no. 11) of A′ and the ring $(1 - e)A'$ is *isomorphic* to $\mathbf{Z}$.

2. PSEUDOMODULES

Given a pseudo-ring A with or without a unit element, a *left pseudomodule* over A is a commutative group E (written additively) admitting A as set of operators and satisfying axioms (M_I), (M_{II}) and (M_{III}) of § 1, no. 1, Definition 1. Right pseudomodules over A are defined similarly.

Let A′ be the ring obtained by adjoining a unit element to A. If E is a left pseudomodule over A, a *left* A′*-module* structure on E is associated with it by writing, for all $x \in E$ and every element $(n, a) \in A'$,

$$(2) \qquad (n, a).x = nx + ax.$$

Axioms (M_I) to (M_{IV}) of § 1, no. 1, Definition 1 are immediately verified;

moreover, by restricting the set of operators of this module structure to $\{0\} \times A$ (identified with A), we obtain on E the pseudomodule structure given initially.

For a subset M of E to be a subgroup with operators of the pseudomodule E (in which case the induced structure is obviously also a left pseudomodule structure over A), it is necessary and sufficient that M be a *submodule* of the associated A′-module E and this sub-A′-module is associated with the pseudomodule M. Moreover, the quotient A′-module E/M is then associated with the quotient group with operators E/M, which is obviously a pseudomodule over A.

If E, F are two pseudomodules over A, the homomorphisms $E \to F$ of groups with operators are identical with the A′-linear mappings $E \to F$ of the A′-modules associated respectively with the pseudomodules E and F. If $(E_\iota)_{\iota \in I}$ is a family of pseudomodules over A, the groups with operators $\prod_{\iota \in I} E$ and $\bigoplus_{\iota \in I} E_\iota$ are pseudomodules over A and the associated A′-modules are respectively the product and direct sum of the associated A′-modules E_ι. There are analogous results for inverse and direct limits of pseudomodules. The theory of pseudomodules over A can thus be entirely reduced to that of A′-modules.

EXERCISES

§ 1

1. Let E, F be two A-modules and $u: E \to F$ a linear mapping. Let M be a submodule of E, N a submodule of F, $i: M \to E$ the canonical injection and $q: F \to F/N$ the canonical surjection. Show that $u(M) = \operatorname{Im}(u \circ i)$ and $\overset{-1}{u}(N) = \operatorname{Ker}(q \circ u)$.

2. Let E be a left A-module; for every left ideal $\mathfrak{a}$ of A let $\mathfrak{a}E$ denote the sum of the $\mathfrak{a}x$ where $x \in E$, which is therefore a submodule of E.

(*a*) Show that for every family $(E_\lambda)_{\lambda \in L}$ of A-modules, $\mathfrak{a}.\bigoplus_{\lambda \in L} E_\lambda$ is canonically identified with $\bigoplus_{\lambda \in L} \mathfrak{a}E_\lambda$.

(*b*) If $\mathfrak{a}$ is a finitely generated left ideal of A, show that for every family $(E_\lambda)_{\lambda \in L}$ of A-modules, $\mathfrak{a}.\prod_{\lambda \in L} E_\lambda$ is canonically identified with $\prod_{\lambda \in L} \mathfrak{a}E_\lambda$. Give an example where $\mathfrak{a}$ is not finitely generated and the above property fails to hold (take A commutative and all the E_λ equal to A).

*(*c*) Let K be a commutative field, A the ring K[X, Y] of polynomials in two indeterminates over K, $\mathfrak{m} = AX$ and $\mathfrak{n} = AY$. If E is the quotient A-module of A^2 by the submodule of A^2 generated by $Xe_1 - Ye_2$ (where (e_1, e_2) is the canonical basis of A^2), show that $(\mathfrak{m} \cap \mathfrak{n})E \neq (\mathfrak{m}E) \cap (\mathfrak{n}E)$.$_*$

3. Let E be an A-module.

(*a*) Let $(L_\mu)_{\mu \in M}$ be a family of right directed ordered sets and, for all $\mu \in M$, let $(F_{\lambda\mu})_{\lambda \in L_\mu}$ be an increasing family of submodules of E. Show that

$$\bigcap_{\mu \in M}\Big(\sum_{\lambda \in L_\mu} F_{\lambda\mu}\Big) = \sum_{\rho \in \prod_\mu L_\mu}\Big(\bigcap_{\mu \in M} F_{\rho(\mu),\mu}\Big).$$

(*b*) Take A to be a field and E the A-module $A^{\mathbf{N}}$; let G be the submodule $A^{(\mathbf{N})}$ of E; on the other hand, for every finite subset H of $\mathbf{N}$, let F_H be the submodule $\prod_{n\in\mathbf{N}} M_n$ of E, where $M_n = A$ for $n \notin H$ and $M_n = 0$ for $n \in H$. Show that the family (F_H), where H runs throughout the set Φ of finite subsets of $\mathbf{N}$ is right directed and that

$$\left(\bigcap_{H\in\Phi} F_H\right) + G \neq \bigcap_{H\in\Phi} (F_H + G).$$

4. Let E, F_1, F_2 be three A-modules and $f_1: F_1 \to E$, $f_2: F_2 \to E$ two A-linear mappings. The *fibre product* of F_1 and F_2 relative to f_1 and f_2 is the submodule of the product $F_1 \times F_2$ consisting of the ordered pairs (x_1, x_2) such that $f_1(x_1) = f_2(x_2)$; it is denoted by $F_1 \times_E F_2$.

(*a*) For every ordered pair of A-linear mappings $u_1: G \to F_1$, $u_2: G \to F_2$ such that $f_1 \circ u_1 = f_2 \circ u_2$, there exists one and only one A-linear mapping $u: G \to F_1 \times_E F_2$ such that $p_1 \circ u = u_1$, $p_2 \circ u = u_2$, where p_1 and p_2 denote the restrictions to $F_1 \times_E F_2$ of the projections pr_1 and pr_2.

(*b*) Let E', F'_1, F'_2 be three A-modules, $f'_1: F'_1 \to E'$, $f'_2: F'_2 \to E'$ A-linear mappings and $F'_1 \times_{E'} F'_2$ the fibre product relative to these mappings. For every system of A-linear mappings $v_1: F_1 \to F'_1$, $v_2: F_2 \to F'_2$, $w: E \to E'$ such that $f'_1 \circ v_1 = w \circ f_1$, $f'_2 \circ v_2 = w \circ f_2$, let v be the A-linear mapping of $F_1 \times_E F_2$ into $F'_1 \times_{E'} F'_2$ corresponding to the two linear mappings $v_1 \circ p_1$ and $v_2 \circ p_2$. Show that if v_1 and v_2 are injective, so is v; give an example where v_1 and v_2 are surjective but v is not surjective (take $E' = \{0\}$). If w is injective and v_1 and v_2 surjective, show that v is surjective.

5. Let E, F_1, F_2 be three A-modules and $f_1: E \to F_1$, $f_1: E \to F_2$ two linear mappings. The *amalgamated sum* of F_1 and F_2 relative to f_1 and f_2 is the quotient module of $F_1 \times F_2$ by the submodule the image of E under the mapping $z \mapsto (f_1(z), -f_2(z))$; it is denoted by $F_1 \oplus_E F_2$.

(*a*) For every ordered pair of A-linear mappings $u_1: F_1 \to G$, $u_2: F_2 \to G$ such that $u_1 \circ f_1 = u_2 \circ f_2$, show that there exists one and only one A-linear mapping $u: F_1 \oplus_E F_2 \to G$ such that $u \circ j_1 = u_1$, $u \circ j_2 = u_2$, where j_1 and j_2 denote the respective compositions of the canonical mapping

$$F_1 \times F_2 \to F_1 \oplus_E F_2$$

with the canonical injections $F_1 \to F_1 \times F_2$, $F_2 \to F_1 \times F_2$.

(*b*) Let E', F'_1, F'_2 be three A-modules, $f'_1: E' \to F'_1$, $f'_2: E' \to F'_2$ two A-linear mappings and $F'_1 \oplus_{E'} F'_2$ the amalgamated sum relative to these mappings. For every system of A-linear mappings $v_1: F'_1 \to F_1$, $v_2: F'_2 \to F_2$, $w: E' \to E$ such that $v_1 \circ f'_1 = f_1 \circ w$, $v_2 \circ f'_2 = f_2 \circ w$, let v be the A-linear mapping of $F'_1 \oplus_{E'} F'_2$ into $F_1 \oplus_E F_2$ corresponding to the two linear mappings $j_1 \circ v_1$ and $j_2 \circ v_2$. Show that if v_1 and v_2 are surjective, so is v; give an example where v_1 and v_2 are injective but v is not injective. Show that if w is surjective and v_1 and v_2 injective, then v is injective.

6. Given an A-module E, let $\gamma(E)$ denote the *least* of the cardinals of the generating systems of E.

(*a*) If $0 \to E \to F \to G \to 0$ is an exact sequence of A-modules, then

$$\gamma(G) \leqslant \gamma(F) \leqslant \gamma(E) + \gamma(G).$$

Give an example of a product $F = E \times G$ such that $\gamma(E) = \gamma(F) = \gamma(G) = 1$ (take E and G to be quotients of $\mathbf{Z}$).

(*b*) If E is the sum of a family $(F_\lambda)_{\lambda \in L}$ of submodules, then $\gamma(E) \leqslant \sum_{\lambda \in L} \gamma(F_\lambda)$.

(*c*) If M and N are submodules of a module E, show that

$$\sup(\gamma(M), \gamma(N)) \leqslant \gamma(M \cap N) + \gamma(M + N).$$

7. (*a*) Let A be a ring and E an (A, A)-bimodule; on the product $B = A \times E$ a ring structure is defined by setting

$$(a, x)(a', x') = (aa', ax' + xa');$$

E (canonically identified with $\{0\} \times E$) is then a two-sided ideal of B such that $E^2 = \{0\}$.

(*b*) By suitably choosing A and E, show that E can be a left B-module with $\gamma(E)$ (Exercise 6) an arbitrary cardinal, although $\gamma(B_s) = 1$.

8. Let A be a commutative ring, $B = A[X_n]_{n \geqslant 1}$ a polynomial ring over A in an infinity of indeterminates and C the quotient ring of B by the ideal generated by the polynomials $(X_1 - X_2)X_i$ for $i \geqslant 3$. If ξ_1 and ξ_2 are the classes in C of X_1 and X_2 respectively, show that the intersection in C of the monogenous ideals $C\xi_1$ and $C\xi_2$ is not a finitely generated ideal.

9. Show that in an A-module E, if $(F_n)_{n \geqslant 1}$ is a *strictly increasing* sequence of finitely generated submodules, the submodule F of E the union of the F_n is not finitely generated. Deduce that if E is an A-module which is not finitely generated, there exists a submodule F of E such that $\gamma(F) = \aleph_0$ $(= \mathrm{Card}(\mathbf{N}))$.

10. Let E be a $\mathbf{Z}$-module the direct sum of two submodules M, N respectively isomorphic to $\mathbf{Z}/2\mathbf{Z}$ and $\mathbf{Z}/3\mathbf{Z}$; show that there exists no other decomposition of E as a direct sum of two non-zero submodules.

11. Let A be a ring admitting a field of left fractions (I, § 9, Exercise 15). Show that in the A-module A_s, a submodule distinct from A_s and $\{0\}$ admits no supplementary submodule (cf. I, § 9, Exercise 16).

12. Let G be a module and E, F two submodules such that $E \subset F$.

(*a*) If F is a direct factor of G, F/E is a direct factor of G/E. If also E is a direct factor of F, E is a direct factor of G.

(*b*) If E is a direct factor of G, then E is a direct factor of F. If also F/E is a direct factor of G/E, then F is a direct factor in G.

(*c*) Give an example of two submodules M, N of the **Z**-module $E = \mathbf{Z}^2$ such that M and N are direct factors of E but $M + N$ is not a direct factor of E.

(*d*) Let p be a prime number and U_p the submodule of the **Z**-module $\mathbf{Q}/\mathbf{Z}$ consisting of the classes mod **Z** of rational numbers of the form k/p^n ($k \in \mathbf{Z}$, $n \in \mathbf{N}$). Let E be the product **Z**-module $M \times N$, where M and N are both isomorphic to U_p, M and N being canonically identified with submodules of E. Consider the endomorphism $u: (x, y) \mapsto (x, y + px)$ of E; show that u is bijective. If $M' = u(M)$, show that M and M′ are both direct factors of E but that $M \cap M'$ is not a direct factor of E (use (*b*), showing that the **Z**-module U_p has no direct factor other than itself and $\{0\}$).

¶ 13. Let E be an A-module and B the ring $\operatorname{End}_A E$. Show that for an element $u \in B$ the following three conditions are equivalent: (1) Bu is a direct factor of the left B-module B_s; (2) uB is a direct factor of the right B-module B_d; (3) $\operatorname{Ker}(u)$ and $\operatorname{Im}(u)$ are direct factors of the A-module E. (Observe that every left ideal which is a direct factor of B_s is of the form Bp, where p is a projector in E; cf. I, § 8, Exercise 12 and use II, § 1, no. 9, Corollaries 1 and 2 to Proposition 15.)

14. Let L be a well-ordered set, E an A-module and $(E_\lambda)_{\lambda \in L}$ an increasing family of submodules of E such that: (1) there exists a $\lambda \in L$ such that $E = \{0\}$; (2) E is the union of the E_λ for $\lambda \in L$; (3) if $\lambda \in L$ is such that the set of $\mu < \lambda$ admits a greatest element λ', E_λ is the direct sum $E_{\lambda'}$ and a submodule F_λ; (4) if $\lambda \in L$ is the least upper bound of the set of $\mu < \lambda$, then E_λ is the union of the E_μ for $\mu < \lambda$. Under these conditions, show that E is the *direct sum* of the F_λ. (Prove by transfinite induction that each E_λ is the direct sum of the F_μ such that $\mu \leqslant \lambda$.)

¶ 15. Let $\mathfrak{c}$ be an infinite cardinal.

(*a*) Let I be a set and $R\{a, b\}$ a relation between elements of I such that for all $\alpha \in I$ the set of $\beta \in I$ for which $R\{\alpha, \beta\}$ holds has cardinal $\leqslant \mathfrak{c}$. Show that there exists a well-ordered set L and an increasing family $(I_\lambda)_{\lambda \in L}$ of subsets of I such that: (1) there exists a $\lambda \in L$ such that $I_\lambda = \varnothing$; (2) I is the union of the I_λ for $\lambda \in L$; (3) if $\lambda \in L$ is such that the set of $\mu < \lambda$ has a greatest element λ', $\operatorname{Card}(I_\lambda - I_{\lambda'}) \leqslant \mathfrak{c}$; (4) if $\lambda \in L$ is the least upper bound of the set of $\mu < \lambda$, I_λ is the union of the I_μ for $\mu < \lambda$; (5) if $\alpha \in I_\lambda$ and $R\{\alpha, \beta\}$ holds, then $\beta \in I_\lambda$. (Note first that for all $\alpha \in I$ the set of $\beta \in I$ for which there exists an integer n and a sequence $(\gamma_j)_{1 \leqslant j \leqslant n}$ of elements of I such that $\gamma_1 = \alpha$, $\gamma_n = \beta$ and $R\{\gamma_j, \gamma_{j+1}\}$ holds for $1 \leqslant j \leqslant n - 1$, has cardinal $\leqslant \mathfrak{c}$. Then construct the I_λ by transfinite induction.)

(*b*) Let E be an A-module the direct sum of a family $(M_\alpha)_{\alpha \in I}$ of submodules such that $\gamma(M_\alpha) \leqslant \mathfrak{c}$ for all $\alpha \in I$ (cf. Exercise 6). Let f be an endomorphism of E. Show that there exists a well-ordered set L and an increasing family $(I_\lambda)_{\lambda \in L}$ of subsets of I satisfying properties (1), (2), (3) and (4) of (*a*) and such moreover that if we write $E_\lambda = \bigoplus_{\alpha \in I_\lambda} M_\alpha$, then $f(E_\lambda) \subset E_\lambda$. (Apply (*a*) to the relation $R\{\alpha, \beta\}$: "there exists $x \in M_\alpha$ such that the component of $f(x)$ in M_β is $\neq 0$".) For all $\lambda \in L$ such that the set of $\mu < \lambda$ has a greatest element λ', write $F_\lambda = \bigoplus_{\alpha \in I_\lambda - I_{\lambda'}} M_\alpha$. Show that E is the direct sum of the F_λ (apply Exercise 14).

(*c*) With the hypotheses and notation of (*b*), suppose further that f is a *projector* and write $P = f(E)$. Let $P_\lambda = P \cap E_\lambda$; show that if $\lambda \in L$ is such that the set of $\mu < \lambda$ has a greatest element λ', $P_{\lambda'}$ is a direct factor of P_λ (cf. Exercise 12), that a supplementary submodule P'_λ of $P_{\lambda'}$ in P_λ is isomorphic to a direct factor of F_λ and that $\gamma(P'_\lambda) \leqslant c$. Show that P is the direct sum of the P'_λ (apply Exercise 14).

(*d*) Deduce from (*c*) that if an A-module E is the direct sum of a family $(M_\alpha)_{\alpha \in I}$ of submodules such that $\gamma(M_\alpha) \leqslant c$, then every *direct factor* P of E is also the direct sum of a family $(N_\lambda)_{\lambda \in L}$ of submodules such that $\gamma(N_\lambda) \leqslant c$.

16. Let A be a ring such that there exists an A-module M with a generating system of n elements but containing a free system of $n + 1$ elements.

(*a*) Show that there exists in A_s^n a free system of $n + 1$ elements.

(*b*) Deduce that there exists in M an *infinite* free system (construct such a system by induction using (*a*)).

(*c*) Let C be a ring, E the free C-module $C_s^{(N)}$, (e_n) its canonical basis and A the endomorphism ring of E. Let u_1 and u_2 denote the endomorphisms of E defined by the conditions $u_1(e_{2n}) = e_n$, $u_1(e_{2n+1}) = 0$, $u_2(e_{2n+1}) = e_n$, $u_2(e_{2n}) = 0$ for all $n \geqslant 0$. Show that u_1 and u_2 form a *basis* of the A-module A_s and deduce that in the A-module A_s there exist infinite free systems.

*(*d*) Let A be the tensor algebra of a vector space E of dimension $\geqslant 2$ over a commutative field (cf. III, § 5, no. 1). Show that in the A-module A_s there exist infinite free systems (observe that two linearly independent vectors in E are linearly independent in A_s and use (*b*)). On the other hand, for every integer $n > 0$, every basis of the A-module A_s^n has n elements (cf. § 7, no. 2, Proposition 3).*

17. Let A be a ring.

(*a*) Show that if the A-module A_s^n is isomorphic to no A_s^m for $m > n$, then, for all $p < n$, A_s^p is isomorphic to no A_s^q for $q > p$.

(*b*) Show that if the A-module A_s^n contains no free system of $n + 1$ elements, the A-module A_s^p contains no free system of $p + 1$ elements for $p < n$.

18. Let A be a ring for which there exists an integer p with the following property: for every family $(a_i)_{1 \leqslant i \leqslant p}$ of p elements of A, there exists a family $(c_i)_{1 \leqslant i \leqslant p}$ of elements of A, one at least of which is not a right divisor of 0, such that $\sum_{i=1}^{p} c_i a_i = 0$.

(*a*) Show, by induction on n, that for every family (x_i) of p^n elements of A_s^n, there exists a family (c_j) of p^n elements of A, one at least of which is not a right divisor of 0, such that $\sum_j c_j x_j = 0$.

(*b*) Deduce from (*a*) and Exercise 16 that for all $n > 0$, the A-module A_s^n contains no free system of $n + 1$ elements.

19. Let A be a non-zero ring with no divisors of zero.

(*a*) Show that for $n > 1$, the A-module A_s^n is not monogenous.

(*b*) Given an integer $n \geqslant 1$, for the A-module A_s^n to contain no free system of $n + 1$ elements, it is necessary that A admit a field of left fractions (I, § 9, Exercise 15); conversely, if this condition is satisfied, A_s^n contains a free system of $n + 1$ elements for *no* integer $n > 0$. (Use Exercises 17 and 18 and I, § 9, Exercise 16.)

(*c*) Show that if every left ideal of A is monogenous, A admits a field of left fractions (use (*a*) and I, § 9, Exercise 16).

20. (*a*) Let A be a ring admitting a field of left fractions (I, § 9, Exercise 15). Let E be an A-module; show that, if in E $(x_i)_{1 \leqslant i \leqslant r}$ is a free system and y, z two elements such that $x_1, \ldots, x_r, y$ on the one hand and $x_1, \ldots, x_r, z$ on the other are two related systems, then $x_2, \ldots, x_r, y, z$ is a related system.

(*b*) Let A be the quotient ring $\mathbf{Z}/6\mathbf{Z}$ and let (e_1, e_2) be the canonical basis of A^2; show that if $a = 2e_1 + 3e_2$, a and e_1 form a related system, as do a and e_2, although e_1 and e_2 form a free system.

21. Let A be a ring, b, c two elements of A which are not right divisors of 0, M the A-module A/Abc and N the submodule Ac/Abc of M. For N to be a direct factor in M, it is necessary and sufficient that there exist two elements x, y of A such that $xb + cy = 1$. (If ε is the class of 1 in M, show that a supplementary subspace of N in M is generated by an element of the form $(1 - yc)\varepsilon$ whose annihilator is the ideal Ac.)

22. If an A-module E admits a basis whose indexing set is I and one of the two sets A, I is infinite, show that E is equipotent to $A \times I$.

23. (*a*) Let M, N be two subsets of an A-module E and $\mathfrak{m}$ and $\mathfrak{n}$ their annihilators; show that the annihilator of $M \cap N$ contains $\mathfrak{m} + \mathfrak{n}$ and give an example where it is distinct from $\mathfrak{m} + \mathfrak{n}$.

(*b*) In a product module $\prod_\iota E_\iota$, the annihilator of a subset F is the intersection of the annihilators of its projections.

(*c*) In a free A-module, the annihilator of an element $\neq 0$ contains only left divisors of 0 in A; in particular, if A is a ring with no divisors of 0, every element $\neq 0$ of a free module is free.

24. Let E be an A-module and M, N two submodules of E. The *transporter* of M into N, denoted by N:M, is the set of $a \in A$ such that $aM \subset N$; it is a two-sided ideal of A, equal to $\mathrm{Ann}((M + N)/N)$.

(*a*) Let A be a commutative ring, M an A-module, N a sub-module of M and x an element of M; show that $N \cap Ax = (N:Ax)x$.

(*b*) Let A be a commutative ring and $\mathfrak{a}$, $\mathfrak{b}$ two ideals of A. Show that $\mathfrak{a}:\mathfrak{b} \supset \mathfrak{a}$ and that $(\mathfrak{a}:\mathfrak{b})/\mathfrak{a}$ is an A-module isomorphic to $\mathrm{Hom}(A/\mathfrak{b}, A/\mathfrak{a})$.

25. Let E, F be two A-modules and $u: E \rightarrow F$ a linear mapping. Show that the mapping $(x, y) \mapsto (x, y - u(x))$ of the product module $E \times F$ to itself is an automorphism of $E \times F$. Deduce that if there exists a linear mapping $v: F \rightarrow E$ and an $a \in E$ such that $v(u(a)) = a$, there exists an automorphism w of $E \times F$ such that $w(a, 0) = (0, u(a))$.

26. Let E be a free A-module with a basis containing at least two elements.

(*a*) Show that every semi-linear mapping of E into itself (relative to an automorphism of A) which permutes with all the automorphisms of E, is necessarily a homothety $x \mapsto \alpha x$ $(\alpha \in A)$.

(*b*) Deduce from (*a*) that the centre of $\mathrm{End}_A(E)$ is the ring of central homotheties, isomorphic to the centre of A.

(*c*) Deduce from (*a*) that the centre of the group $\mathbf{GL}(E)$ is the group of invertible central homotheties of E, which is isomorphic to the multiplicative group of invertible elements of the centre of A.

27. Let A be a commutative ring. Show that if an ideal $\mathfrak{J}$ of A is a free A-module, then $\mathfrak{J}$ is a monogenous A-module (in other words a principal ideal). Give an example showing that the proposition does not extend to the left ideals in a non-commutative ring (Exercise 16).

§ 2

1. (*a*) If A is a ring with no divisors of zero, show that every element $\neq 0$ of a projective A-module is free.

(*b*) Give an example of a projective A-module whose annihilator is non-zero (cf. I, § 8, no. 11).

(*c*) Show that the $\mathbf{Z}$-module $\mathbf{Q}$ is not a projective $\mathbf{Z}$-module. * (See *Commutative Algebra*, II, § 5, Exercise 11 for an example of a finitely generated projective module over an integral domain, which is not free.) *

2. Show that every projective A-module is the direct sum of a family of projective submodules each of which admits a countable number of generators ("*Kaplansky's Theorem*"; apply Exercise 15 of § 1, to the case where E is a free module).

3. Let $\mathfrak{c}$ be a cardinal and P a projective A-module such that $\gamma(P) \leqslant \mathfrak{c}$ (§ 1,

Exercise 6). Show that if L is a free A-module with an infinite basis of cardinal $\geqslant \mathfrak{c}$, $P \oplus L$ is isomorphic to L. (Observe that there exists a free A-module M, whose basis has cardinal $\leqslant \mathfrak{c}$, isomorphic to the direct sum of P and an A-module Q. Note that $M^{(\mathbf{N})}$ and $P \oplus M^{(\mathbf{N})}$ are isomorphic.)

¶ 4. (*a*) Let E, E′ be two A-modules, N a submodule of E and N′ a submodule of E′ such that E is projective and E/N and E′/N′ are isomorphic. Show that there exists an exact sequence

$$0 \longrightarrow N \xrightarrow{u} E \oplus N' \xrightarrow{v} E' \longrightarrow 0.$$

(Observe that there exists a homomorphism $f: E \to E'$ such that $f(N) \subset N'$, which gives when passing to the quotients an isomorphism $E/N \to E'/N'$; define u and v in an analogous way to that used for the exact sequence (29) of § 1, no. 7.)

(*b*) Deduce from (*a*) that if E′ is projective, $E \oplus N'$ and $E' \oplus N$ are isomorphic.

(*c*) Let $0 \to E_m \to E_{m-1} \to \ldots \to E_1 \to E_0 \to 0$ be an exact sequence of A-modules such that $E_0, E_1, \ldots, E_{m-2}$ are projective. Show that the A-modules $\bigoplus_{h \geqslant 0} E_{m-2h}$ and $\bigoplus_{h \geqslant 0} E_{m-2h+1}$ are isomorphic (with the convention that $E_i = \{0\}$ for $i < 0$).

5. Let u be a linear mapping from an A-module E to an A-module F and $v = {}^t u$ its transpose. For every submodule N′ of F*, the submodule of E orthogonal to $v(N')$ is $\overset{-1}{u}(N)$, where N is the submodule of F orthogonal to N′.

6. (*a*) If E is an A-module, every injective endomorphism u of E is not a left divisor of zero in the ring $\mathrm{End}_A(E)$. Conversely, if for every sub-A-module $F \neq \{0\}$ of E there exists an endomorphism $v \neq 0$ of E such that $v(E) \subset F$, an element which is not a left divisor of zero in $\mathrm{End}_A(E)$ is an injective endomorphism. The above condition is satisfied if there exists a linear form $x' \in E^*$ and an $x \in E$ such that $\langle x, x' \rangle$ is invertible in A and in particular if E is free.

(*b*) Every endomorphism $\neq 0$ of the **Z**-module **Q** is bijective, although there exists no linear form $\neq 0$ on **Q**.

(*c*) Let U_p be the **Z**-module defined in § 1, Exercise 12(*d*). Show that the endomorphism $x \mapsto px$ of U_p is not injective and is not a left divisor of zero in $\mathrm{End}_{\mathbf{Z}}(U_p)$.

7. (*a*) If u is a surjective endomorphism of an A-module E, u is not a right divisor of zero in $\mathrm{End}_A(E)$. Conversely, if for every submodule $F \neq E$ of E there exists a linear form $x^* \in E^*$ which is zero on F and surjective, every element of $\mathrm{End}_A(E)$ which is not a right divisor of zero is a surjective endomorphism.

(*b*) If E is the **Z**-module defined in Exercise 6(*c*), show that every endomorphism $\neq 0$ of E is surjective, although there exists no linear form $\neq 0$ on E.

(*c*) Show that if E is a free **Z**-module $\neq 0$, there exist non-surjective endomorphisms of E which are not right divisors of 0 in $\mathrm{End}_{\mathbf{Z}}(E)$.

8. (*a*) Let E be a free A-module, F, G two A-modules and $u: E \rightarrow G$, $v: F \rightarrow G$ two linear mappings. Show that if $u(E) \subset v(F)$, there exists a linear mapping $w: E \rightarrow F$ such that $u = v \circ w$.

(*b*) Give an example of two non-zero endomorphisms u, v of the **Z**-module E defined in Exercise 6(*c*) such that there exists no endomorphism w of E for which $u = v \circ w$ (although $u(E) = v(E) = E$).

(*c*) Give an example of two endomorphisms u, v of the **Z**-module **Z** such that $\overset{-1}{u}(0) = \overset{-1}{v}(0) = \{0\}$, but such that there exists no endomorphism w of **Z** for which $u = w \circ v$ (cf. § 7, Exercise 14).

9. Given an A-module E, for every submodule M of E (resp. every submodule M′ of E*), let M° (resp. M′°) denote the orthogonal of M in E* (resp. the orthogonal of M′ in E). Consider the following four properties:

(A) The canonical homomorphism $c_E: E \rightarrow E^{**}$ is bijective.

(B) For every submodule M of E, the canonical homomorphism $E^*/M^\circ \rightarrow M^*$ is bijective.

(C) For every submodule M of E, $M^{\circ\circ} = M$.

(D) For every ordered pair of submodules M, N of E,

$$(M \cap N)^\circ = M^\circ + N^\circ.$$

(*a*) Show that condition (B) implies (D) (prove that for $x^* \in (M \cap N)^\circ$, there exists $y^* \in E^*$ such that $\langle x + y, y^* \rangle = \langle x, x^* \rangle$ for $x \in M$ and $y \in N$).

(*b*) Let A be a ring with no divisors of zero, but having no field of left fractions *(for example the tensor algebra of a vector space of dimension > 1, cf. III, § 5, Exercise 5)*. If we take $E = A_s$, condition (A) is satisfied, but none of conditions (B), (C), (D) (for an example where (B), (C), (D) hold, but not (A), see § 7, no. 5, Theorem 6).

(*c*) Take A to be a product ring K^I, where K is a commutative field and I an infinite set. Show that the A-module $E = A$ satisfies conditions (A) and (B), but not (C) (observe that the annihilators of the ideals of A are all of the form K^J, where $J \subset I$).

(*d*) Suppose that for two submodules M, N of E such that $N \subset M$ and $N \neq M$, the dual $(M/N)^*$ is non-zero; then show that condition (B) implies (C).

(*e*) The kernel of c_E is the orthogonal $(E^*)^\circ$ of E* in E. Give an example where neither E* nor $(E^*)^\circ$ is 0 (consider a module containing an element whose annihilator contains an element which is not a divisor of 0).

(*f*) Let M be a direct factor of E; show that $M^{\circ\circ} = M + (E^*)^\circ$.

10. Give an example of an A-linear mapping $u: E \rightarrow F$ such that ${}^t u$ is bijective but u is neither injective nor surjective (cf. Exercise 6(*b*) and (*c*)).

Deduce an example where $y_0 \in F$ is orthogonal to the kernel of ${}^t u$ but where the equation $u(x) = y_0$ has no solution.

¶ 11. Let A be a ring and I an A-module. I is called *injective* if, for every exact sequence $E' \to E \to E''$ of A-linear mappings, the sequence

$$\mathrm{Hom}(E'', I) \to \mathrm{Hom}(E, I) \to \mathrm{Hom}(E', I)$$

is exact.

(*a*) Show that the following properties are equivalent:

(α) I is injective.

(β) For every exact sequence $0 \to E' \to E \to E'' \to 0$ of A-linear mappings, the sequence

$$0 \to \mathrm{Hom}(E'', I) \to \mathrm{Hom}(E, I) \to \mathrm{Hom}(E', I) \to 0$$

is exact.

(γ) For every A-module M and every submodule N of M, every A-linear mapping of N into I can be extended to an A-linear mapping of M into I.

(δ) For every left ideal $\mathfrak{a}$ of A and every A-linear mapping $f: \mathfrak{a} \to I$, there exists an element $b \in I$ such that $f(a) = ab$ for all $a \in \mathfrak{a}$.

(ζ) I is a direct factor of every A-module containing it.

(θ) For every A-module E which is the sum of I and a monogenous submodule, I is a direct factor of E.

(To prove that (δ) implies (γ), show that property (δ) implies that if E is an A-module and F a submodule of E such that E/F is monogenous, then every linear mapping of F into I can be extended to a linear mapping of E into I. Then use Zorn's Lemma. To prove that (θ) implies (δ), consider the A-module $A_s \times I = M$, the submodule N of M consisting of the elements $(a, -f(a))$ for $a \in \mathfrak{a}$, and apply (θ) to the quotient module M/N.)

(*b*) For a **Z**-module E to be injective, it is necessary and sufficient that for all $x \in E$ and every integer $n \neq 0$, there exists $y \in E$ such that $ny = x$. In particular, the **Z**-modules **Q** and **Q**/**Z** are injective.

12. (*a*) For a product $\prod_{\lambda \in L} E_\lambda$ of A-modules to be injective (Exercise 11), it is necessary and sufficient that each of the E_λ be injective.

(*b*) Let K be a commutative field, L an infinite set and A the product ring K^L. Show that A is an injective A-module (use (*a*)) but that the ideal $\mathfrak{a}$ of A the direct sum of the factors of K^L (canonically identified with ideals of A) is not an injective A-module.

¶ 13. (*a*) Let A be a ring and I an injective A-module (Exercise 11) such that for every non-zero monogenous A-module E, there exists a non-zero A-homomorphism of E into I. Let M be an A-module, N a submodule of M and $\tilde{N}$ the set of $u \in \mathrm{Hom}_A(M, I)$ such that $u(x) = 0$ in N. Show that for all $y \notin N$, there exists a $u \in \tilde{N}$ such that $u(y) \neq 0$.

(*b*) Let A be a ring. For every (A, A)-bimodule T and every left (resp. right) A-module M, $\mathrm{Hom}_A(M, T)$ has canonically a right (resp. left) A-module structure deriving from that on T, and there is a canonical A-homomorphism $c_{M,T}: M \to \mathrm{Hom}_A(\mathrm{Hom}_A(M, T), T)$ such that $c_{M,T}(x)$ is the A-homomorphism $u \mapsto u(x)$. Show that if I is an (A, A)-bimodule which, as a left A-module, satisfies the conditions of (*a*), then the canonical homomorphism $c_{M,I}$ is injective for every left A-module M.

(*c*) Suppose that I is an (A, A)-bimodule which is injective as a left A-module and a right A-module and that for every (left or right) monogenous A-module $E \neq 0$, there exists a non-zero A-homomorphism of E into I. Show that under these conditions, if P is a left (resp. right) projective A-module, $\mathrm{Hom}_A(P, I)$ is a right (resp. left) injective A-module. (Suppose that P is a left projective A-module. Observe that for every submodule N′ of a right A-module N, $\mathrm{Hom}_A(N', I)$ is isomorphic to a quotient module of $\mathrm{Hom}_A(N, I)$ and apply (*b*) to the right module N and the left module P.)

(*d*) Let C be a commutative ring, A a C-algebra and I an injective C-module such that for every monogenous C-module $E \neq \{0\}$ there exists a non-zero C-homomorphism of E into I. For every A-module M, $\mathrm{Hom}_C(M, I)$ is an A-module; show that for every projective A-module P, $\mathrm{Hom}_C(P, I)$ is an injective A-module (same argument as in (*c*)).

14. Let A be a ring. Show that for every A-module M there exists an injective A-module I such that M is isomorphic to a submodule of I. (Apply Exercise 13(*d*) with $C = \mathbf{Z}$, using Exercise 11(*b*) and the fact that every A-module is the quotient of a free A-module.)

¶ 15. An A-module homomorphism $u: E \to F$ is called *essential* if it is injective and if, for every submodule $P \neq \{0\}$ of F, $\overset{-1}{u}(P) \neq \{0\}$; it suffices that this condition holds for every *momogenous* submodule $P \neq \{0\}$ of F. If E is a submodule of F, F is called an *essential extension* of E if the canonical injection $E \to F$ is essential.

(*a*) Let $u: E \to F$, $v: F \to G$ be two A-homomorphisms of A-modules. If u and v are essential, so is $v \circ u$. Conversely, if $v \circ u$ is essential and v is injective, then u and v are essential.

(*b*) Let E be a submodule of an A-module F and let $(F_\lambda)_{\lambda \in L}$ be a family of submodules of F containing E and whose union is F. Show that if each of the F_λ is an essential extension of E, so is F.

(*c*) Let $(u_\lambda)_{\lambda \in L}$ be a family of essential homomorphisms $u_\lambda: E_\lambda \to F_\lambda$. Show that the homomorphism $\bigoplus_{\lambda \in L} u_\lambda: \bigoplus_{\lambda \in L} E_\lambda \to \bigoplus_{\lambda \in L} F_\lambda$ is essential. (Prove it first when L has two elements, then use (*b*).)

(*d*) The **Z**-module **Q** is an essential extension of **Z** but the product $\mathbf{Q}^{\mathbf{N}}$ is not an essential extension of $\mathbf{Z}^{\mathbf{N}}$.

(*e*) Let E be a submodule of a module F. Show that there exists a submodule

Q of F such that $Q \cap E = \{0\}$ and that the restriction to E of the canonical homomorphism $F \to F/Q$ is essential.

16. A submodule E of an A-module F is called *irreducible with respect to* F (or *in* F) if $E \neq F$ and E is not the intersection of two submodules of F distinct from E.

(*a*) For an A-module F to be an essential extension of each of its submodules $\neq \{0\}$ it is necessary and sufficient that $\{0\}$ be irreducible with respect to F.

(*b*) Let $(E_\lambda)_{\lambda \in L}$ be a family of submodules of an A-module F. $(E_\lambda)_{\lambda \in L}$ is called a *reduced irreducible decomposition* of $\{0\}$ in F if each of the F_λ is irreducible in F, if the intersection of the E_λ reduces to 0 and if none of the E_λ contains the intersection of the E_μ for $\mu \neq \lambda$. When this is so, show that the canonical homomorphism of F into $\bigoplus_{\lambda \in L} (F/E_\lambda)$ is essential (if $E'_\lambda = \bigcap_{\mu \neq \lambda} E_\mu$, observe that the image F'_λ of E'_λ in F/E_λ is non-zero and that the image of F in $\bigoplus_{\lambda \in L} (F/E_\lambda)$ contains $\bigoplus_{\lambda \in L} F'_\lambda$; then apply Exercise 15(*c*)). Conversely, if $(E_\lambda)_{\lambda \in L}$ is a family of submodules of F irreducible in F and the canonical homomorphism $F \to \bigoplus_{\lambda \in L} (F/E_\lambda)$ is essential, the family (E_λ) is a reduced irreducible decomposition of $\{0\}$ in F.

¶ 17. (*a*) Let M, N, M′, N′ be four submodules of a module E such that $M \cap N = M' \cap N' = P$. Show that

$$M = (M + (N \cap M')) \cap (M + (N \cap N')).$$

Deduce that if M is irreducible in E, then necessarily $P = M' \cap N$ or $P = N' \cap N$.

(*b*) Let $(N_i)_{1 \leqslant i \leqslant n}$ be a finite family of submodules of E irreducible in E and M its intersection. Show that if $(P_j)_{1 \leqslant j \leqslant m}$ is a finite family of submodules of E such that $M = \bigcap_j P_j$, then for every index i such that $1 \leqslant i \leqslant n$ there exists an index $\phi(i)$ such that $1 \leqslant \phi(i) \leqslant m$ and such that, if we write $N'_i = \bigcap_{k \neq i} N_k$, then $M = P_{\phi(i)} \cap N'_i$ (use (*a*)). Deduce that if none of the P_j contains the intersection of the P_k of index $\neq j$, then $m \leqslant n$ (successively replace the N_i by suitable P_j, using the above result). Conclude that two reduced irreducible decompositions of M in E, one of which is finite, necessarily have the same number of terms (cf. Exercise 23).

¶ 18. (*a*) For an A-module I to be injective, show that it is necessary and sufficient that I admit no essential extension distinct from itself (show that this condition implies condition (δ) of Exercise 11(*a*), arguing as for the proof that (θ) implies (δ) in that exercise).

(*b*) Let I be an injective A-module. For a submodule E of I to be injective, it is necessary and sufficient that E admit no essential extension distinct from itself and contained in I (using Exercise 15(*e*)), show that this condition implies that E is a direct factor of I).

(*c*) Let I be an injective A-module, E a submodule of I and $h: E \to F$ an essential homomorphism. Show that there exists an *injective* homomorphism $j: F \to I$ such that $j \circ h$ is the canonical injection of E into I.

(*d*) Let I be an A-module and E a submodule of I. Show that the following properties are equivalent:

(α) I is injective and is an essential extension of E.

(β) I is injective and for every injective homomorphism j of E into an injective A-module J, there exists an injective homomorphism of I into J extending j.

(γ) I is injective and is the smallest injective submodule of I containing E.

(δ) I is an essential extension of E and for every essential homomorphism $h: E \to F$, there exists an injective homomorphism $j: F \to I$ such that $j \circ h$ is the canonical injection of E into I.

(Use (*c*) and (*a*).) When these equivalent conditions hold, I is called an *injective envelope* of E; I is then also an injective envelope of every submodule of I containing E.

(*e*) Let I be an injective A-module and E a submodule of I. Show that every maximal element of the set of essential extensions of E contained in I is an injective envelope of E (use (*b*)). In particular, every A-module admits an injective envelope (cf. Exercise 14).

(*f*) If I, I′ are two injective envelopes of the same A-module E, show that there exists an isomorphism of I onto I′ leaving invariant the elements of E.

¶ 19. (*a*) Let E, F be two A-modules and M an injective submodule of $E \oplus F$. Let I be an injective envelope of $M \cap E$ in M and J a supplementary submodule of I in M. Show that the restriction to I (resp. J) of the canonical projection of $E \oplus F$ onto E (resp. F) is injective (compose with the projection of I into E an extension $E \to I$ of the injection $M \cap E \to I$).

(*b*) Let E be an A-module and M a submodule of E, maximal in the set of injective submodules of E; M admits in E a supplementary submodule N which contains no injective submodule $\neq \{0\}$. Show that for every injective submodule P of E, the image of P under the projection of E onto M (under the decomposition of E as a direct sum $M \oplus N$) is an injective envelope of $P \cap M$ in M (use (*a*)). If M′ is another submodule of E which is maximal in the set of injective submodules of E, show that there exists an automorphism of E transforming M into M′ and leaving invariant the elements of N.

¶ 20. (*a*) Let A be a ring admitting a field of left fractions K (I, § 9, Exercise 15). Show that K, considered as a left A-module, is an injective envelope of A_s. (Apply criterion (α) of Exercise 18(*d*) and criterion (δ) of Exercise 11(*a*) noting that if $f: \mathfrak{a} \to K$ is an A-linear mapping of a left ideal $\mathfrak{a}$ of A into K, the element $x^{-1}f(x)$ for $x \in \mathfrak{a}$, $x \neq 0$ (the inverse being taken in K) does not depend on $x \in \mathfrak{a}$.)

(*b*) Let U_p be the sub-**Z**-module of $\mathbf{Q}/\mathbf{Z}$ consisting of the canonical images of the rational numbers of the form k/p^n, where p is a given prime number, $k \in \mathbf{Z}$ and $n \in \mathbf{N}$ (II, § 1, Exercise 12(*d*)). Show that U_p is an injective envelope of each of the **Z**-modules $p^{-n}\mathbf{Z}/\mathbf{Z}$ isomorphic to $\mathbf{Z}/p^n\mathbf{Z}$. Show that there exist automorphisms of U_p distinct from the identity automorphism leaving invariant the elements of a submodule $p^{-n}\mathbf{Z}/\mathbf{Z}$.

¶ 21. (*a*) Let $(E_j)_{j \in J}$ be a finite family of A-modules and for all $j \in J$ let I_j be an injective envelope of E_j; show that $\bigoplus_{j \in J} I_j$ is an injective envelope of $\bigoplus_{j \in J} E_j$.

(*b*) An A-module M is called *indecomposable* if it is distinct from $\{0\}$ and it is not the direct sum of two submodules distinct from $\{0\}$. Show that if I is an injective A-module, the following properties are equivalent:

(α) $\{0\}$ is an irreducible submodule in I.

(β) I is indecomposable.

(γ) I is an injective envelope of each of its submodules $\neq 0$. (Use (*a*) and Exercise 18 (*e*).)

(δ) I is isomorphic to the injective envelope $I(A/\mathfrak{q})$, where $\mathfrak{q}$ is an irreducible left ideal of A.

Moreover, when this is so, for all $x \neq 0$ in I, $\mathrm{Ann}(x)$ is an irreducible left ideal of A and I is isomorphic to $I(A/\mathrm{Ann}(x))$.

Deduce that for the injective envelope of an A-module E to be indecomposable, it is necessary and sufficient that $\{0\}$ be an irreducible submodule of E.

(*c*) Give an example of an indecomposable **Z**-module E such that its injective envelope $I(E)$ is decomposable (cf. VII, § 3, Exercise 5).

(*d*) Let E be an A-module, I an injective envelope of E and $I = \bigoplus_{\lambda \in L} I_\lambda$ a decomposition of I as a direct sum of a *finite* family of indecomposable injective submodules; for all $\lambda \in L$ let $J_\lambda = \bigoplus_{\mu \neq \lambda} I_\mu$ and let $N_\lambda = J_\lambda \cap E$. Show that $(N_\lambda)_{\lambda \in L}$ is a reduced irreducible decomposition of $\{0\}$ in E (Exercise 16 (*b*)) and that I_λ is an injective envelope of E/N_λ. Conversely, every reduced irreducible decomposition of $\{0\}$ in E can be obtained by the above procedure uniquely up to isomorphism (use Exercise 16 (*b*)).

22. Let I be an injective A-module. If I is indecomposable, every injective endomorphism of I is an automorphism of I. Deduce that, for I to be indecomposable, it is necessary and sufficient that the non-invertible elements of the ring $\mathrm{End}(I)$ form a two-sided ideal of that ring.

¶ 23. Let M be an A-module, the direct sum of a family (finite or otherwise) $(M_\lambda)_{\lambda \in L}$ of submodules such that for all $\lambda \in L$ the non-invertible elements of $\mathrm{End}(M_\lambda)$ form a two-sided ideal of this ring (which implies that M_λ is *indecomposable*).

(*a*) Let f, g be two endomorphisms of M such that $f + g = 1_M$. Show that for every finite sequence $(\lambda_k)_{1 \leqslant k \leqslant s}$ of distinct indices in L there exists a family of submodules N_k $(1 \leqslant k \leqslant s)$ of M such that for every k one of the endomorphisms f, g restricted to $M_{\lambda k}$ is an isomorphism of this submodule onto N_k and that M is the direct sum of the N_k $(1 \leqslant k \leqslant s)$ and the M_λ for λ distinct from λ_k. (Reduce it to the case $s = 1$; if π_λ is the canonical projection of M onto M_λ corresponding to the decomposition as a direct sum of M_λ and $\bigoplus_{\mu \neq \lambda} M_\mu$, note that either $\pi_\lambda \circ f$ or $\pi_\lambda \circ g$, restricted to M_λ, is an automorphism of M_λ.)

(*b*) Let f be an *idempotent* endomorphism of M. Show that there exists at least one index $\lambda \in L$ such that the restriction of f to M_λ is an isomorphism of M_λ onto $f(M_\lambda)$; moreover $f(M_\lambda)$ is a direct factor of M. (If $g = 1_M - f$, note that there exists a finite sequence of indices $(\lambda_k)_{1 \leqslant k \leqslant s}$ such that the intersection of $\mathrm{Ker}(g)$ and the direct sum of the M_{λ_k} is $\neq \{0\}$ and use (*a*).)

(*c*) Deduce from (*b*) that every indecomposable direct factor of M is isomorphic to one of the M_λ.

(*d*) Let $(N_\kappa)_{\kappa \in K}$ be a family of indecomposable submodules of M of which M is the direct sum; every N_κ is therefore isomorphic to an M_λ and vice versa, by virtue of (*c*). Let T be the set of classes of indecomposable submodules of M (under the relation of isomorphism) such that an M_λ (or an N_κ) belongs to one of these classes. For all $t \in T$, let $R(t)$ (resp. $S(t)$) be the set of $\lambda \in L$ (resp. $\kappa \in K$) such that $M_\lambda \in t$ (resp. $N_\kappa \in t$). Show that, for all $t \in T$, $\mathrm{Card}(R(t)) = \mathrm{Card}(S(t))$. (In the notation of (*a*), let $J(\kappa)$ be the set $\lambda \in L$ such that the restriction of π_λ to N_κ is an isomorphism of N_κ onto M_λ; show that $J(\kappa)$ is finite and that the $J(\kappa)$ cover $R(t)$ when κ runs through $S(t)$. Deduce that $\mathrm{Card}(S(t)) \leqslant \mathrm{Card}(R(t))$ when $R(t)$ is infinite. When $R(t)$ is finite, show with the aid of (*a*) that M is the direct sum of the M_λ for $\lambda \notin R(t)$ and a subfamily of $(N_\kappa)_{\kappa \in S(t)}$ of cardinal equal to that of $R(t)$.)

(*e*) Deduce from (*d*) and Exercises 22 and 21 (*d*) that if $(E_\lambda)_{\lambda \in L}$ and $(E'_\kappa)_{\kappa \in K}$ are two reduced irreducible decompositions of $\{0\}$ in a module F, L and K are equipotent.

24. Let M be an A-module, I its injective envelope (Exercise 18), $\mathfrak{a}$ a two-sided ideal of A and Q the submodule of I consisting of the $z \in I$ such that $\mathfrak{a}z = \{0\}$; Q has a natural $(A/\mathfrak{a})$-module structure. Let N be the submodule of M consisting of the $x \in M$ such that $\mathfrak{a}x = \{0\}$; if N is considered as an $(A/\mathfrak{a})$-module, show that Q is isomorphic to an injective envelope of N.

25. (*a*) For an A-module Q to be injective, it is sufficient that, for every

projective A-module P and every submodule P′ of P, every linear mapping of P′ into Q can be extended to a linear mapping of P into Q (use the fact that every A-module is a quotient of a projective A-module).

(*b*) For an A-module P to be projective, it is sufficient that, for every *injective* A-module Q and every quotient module Q″ of Q, every homomorphism of P into Q″ be of the form $\phi \circ u$, where $\phi : Q \to Q''$ is the canonical homomorphism and u is a homomorphism of P into Q (use the fact that every A-module is a submodule of an injective A-module).

26. For every **Z**-module G and every integer $n > 0$, let ${}_nG$ denote the kernel of the endomorphism $x \mapsto nx$ of G. If $0 \to G' \to G \to G'' \to 0$ is an exact sequence, define a canonical homomorphism $d : {}_nG'' \to G'/nG'$ such that the sequence

$$0 \longrightarrow {}_nG' \longrightarrow {}_nG'' \xrightarrow{d} G'/nG' \longrightarrow G/nG \longrightarrow G''/nG'' \longrightarrow 0$$

is exact. If ${}_nG$ and G/nG are finite, we write

$$h_n(G) = \mathrm{Card}(G/nG) - \mathrm{Card}({}_nG);$$

then show that ${}_nG'$, ${}_nG''$, G'/nG', G''/nG'' are finite and that

$$h_n(G) = h_n(G') + h_n(G'').$$

§ 3

1. Let A be a ring, E a right A-module and F a left A-module. Let f be a function defined on the set S of all finite sequences

$$((x_1, y_1), (x_2, y_2), \ldots, (x_n, y_n)) \quad (n \text{ arbitrary})$$

of elements of E × F with values in a set G and such that:

(1) $$f((x_1, y_1), \ldots, (x_n, y_n)) = f((x_{\sigma(1)}, y_{\sigma(1)}), \ldots, (x_{\sigma(n)}, y_{\sigma(n)}))$$

for every permutation $\sigma \in \mathfrak{S}_n$;

(2) $$f((x_1 + x'_1, y_1), (x_2, y_2), \ldots, (x_n, y_n)) = f((x_1, y_1), (x'_1, y_1), (x_2, y_2), \ldots, (x_n, y_n));$$

(3) $$f((x_1, y_1 + y'_1), (x_2, y_2), \ldots, (x_n, y_n)) = f((x_1, y_1), (x_1, y'_1), (x_2, y_2), \ldots, (x_n, y_n));$$

(4) $$f((x_1\lambda, y_1), \ldots, (x_n, y_n)) = f((x_1, \lambda y_1), \ldots, (x_n, y_n))$$

for all $\lambda \in A$.

Show that there exists one and only one mapping g of $E \otimes_A F$ into G such that $f((x_1, y_1), \ldots, (x_n, y_n)) = g\left(\sum_{i=1}^{n} x_i \otimes y_i\right)$. (Note that if

$$\sum_i x_i \otimes y_i = \sum_j x'_j \otimes y'_j,$$

the difference $\sum_i (x_i, y_i) - \sum_j (x'_j, y'_j)$ in the **Z**-module $\mathbf{Z}^{(E \times F)}$ is a linear combination with integer coefficients of elements of the form (2) of no. 1.)

*2. Consider the field **C** of complex numbers as a vector space over the field **R** of real numbers.

(*a*) Show that the canonical mapping $\mathbf{C} \otimes_{\mathbf{R}} \mathbf{C} \to \mathbf{C} \otimes_{\mathbf{C}} \mathbf{C}$ is not injective.

(*b*) Show that the two **C**-module structures on $\mathbf{C} \otimes_{\mathbf{R}} \mathbf{C}$ arising from each of the factors are distinct$_*$.

3. (*a*) Let $(E_\lambda)_{\lambda \in L}$ be a family of right A-modules and F a left A-module. Show that if F is finitely generated, the canonical mapping

$$\left(\prod_{\lambda \in L} E_\lambda\right) \otimes_A F \to \prod_{\lambda \in L} (E_\lambda \otimes_A F)$$

is surjective.

(*b*) Give an example of a commutative ring A and an ideal $\mathfrak{m}$ of A such that the canonical mapping $A^{\mathbf{N}} \otimes_A (A/\mathfrak{m}) \to (A/\mathfrak{m})^{\mathbf{N}}$ is not injective (cf. § 1, Exercise 2 (*b*)).

(*c*) Give an example of a commutative ring A such that the canonical mapping $A^{\mathbf{N}} \otimes_A A^{\mathbf{N}} \to A^{\mathbf{N} \times \mathbf{N}}$ is not surjective (observe that for given n, the image of **N** under a mapping of the form $m \mapsto \sum_{i=1}^{r} u_i(m) v_i(n)$, where the u_i and v_i belong to $A^{\mathbf{N}}$, generate a finitely generated ideal of A).

(*d*) Deduce from (*b*) and (*c*) an example where the canonical homomorphism (22) (no. 7) is neither injective nor surjective.

4. Let E be a right A-module and F a *free* left A-module. Show that if x is a free element of E and y an element $\neq 0$ of F, then $x \otimes y \neq 0$; if also A is commutative and y is a free element of F, then $x \otimes y$ is free in the A-module $E \otimes_A F$.

§ 4

1. Let A, B be two rings, E a left A-module, F a right A-module and G an (A, B)-bimodule; $\mathrm{Hom}_B(F, G)$ then has a left A-module structure and $\mathrm{Hom}_A(E, G)$ a right B-module structure. Show that the **Z**-modules

$$\mathrm{Hom}_A(E, \mathrm{Hom}_B(F, G)) \quad \text{and} \quad \mathrm{Hom}_B(F, \mathrm{Hom}_A(E, G))$$

are both canonically isomorphic to the **Z**-module of **Z**-bilinear mappings f of $E \times F$ into G such that $f(\alpha x, y) = \alpha f(x, y)$ and $f(x, y\beta) = f(x, y)\beta$ for $\alpha \in A$, $\beta \in B$, $x \in E$, $y \in F$.

2. Let A be the ring $\mathbf{Z}/4\mathbf{Z}$, $\mathfrak{m}$ the ideal $2\mathbf{Z}/4\mathbf{Z}$ of A and E the A-module

A/$\mathfrak{m}$. Show that the canonical mapping $E^* \otimes_A E \to \mathrm{Hom}_A(E, E)$ is neither injective nor surjective. The same is true of the canonical mapping

$$E \otimes_A E \to \mathrm{Hom}_A(E^*, E)$$

and the canonical mapping $E^* \otimes_A E^* \to (E \otimes_A E)^*$.

3. Show that under the conditions of no. 2, if E is assumed to be a finitely generated left A-module and F a projective right B-module, the canonical homomorphism (7) (no. 2) is bijective.

4. Given an example where the canonical homomorphism (21) (no. 4) is neither injective nor surjective, although $F_2 = C$, $E_1 = C$ and E_2 and F_1 are finitely generated C-modules (cf. Exercise 2).

5. Let A, B be two rings, E a right A-module, F a left A-module and G a (B, A)-bimodule.

(*a*) Show that there exists one and only one **Z**-linear mapping

$$\eta: E \otimes_A F \to \mathrm{Hom}_B(\mathrm{Hom}_A(E, G), G \otimes_A F)$$

which maps the tensor product $x \otimes y$, where $x \in E$ and $y \in F$, to the mapping $v_{x,y}: u \mapsto u(x) \otimes y$ of $\mathrm{Hom}_A(E, G)$ into $G \otimes_A F$. When $B = A$ and $G = {}_sA_d$, the homomorphism reduces to the canonical homomorphism (15) (no. 2).

(*b*) Show that if F is a projective A-module and E and G are such that for $x \neq 0$ in E there exists $u \in \mathrm{Hom}_A(E, G)$ such that $u(x) \neq 0$, then the homomorphism η is injective.

¶ 6. Let A, B be two rings, E a left A-module, F an (A, B)-bimodule and G a right B-module.

(*a*) Show that there exists one and only one **Z**-linear mapping

$$\sigma: \mathrm{Hom}_B(F, G) \otimes_A E \to \mathrm{Hom}_B(\mathrm{Hom}_A(E, F), G)$$

such that for $x \in E$ and $u \in \mathrm{Hom}_B(F, G)$, $\sigma(u \otimes x)$ is the mapping $v \mapsto u(v(x))$ of $\mathrm{Hom}_A(E, F)$ into G.

(*b*) If E is a finitely generated projective A-module, show that σ is bijective.

(*c*) Suppose that G is an *injective* B-module (§ 2, Exercise 11) and that E is the cokernel of an A-linear mapping $A_s^m \to A_s^n$. Show that σ is bijective (start from the exact sequence $A_s^m \to A_s^n \to E \to 0$ and use (*b*), the definition of injective modules and Theorem 1 of § 2, no. 1, and Proposition 5 of § 3, no. 6).

7. Show that, if E_1, E_2, F_1, F_2 are four modules over a commutative ring

C, the canonical homomorphism (21) (no. 4) is composed of the homomorphisms

$$\begin{aligned}\mathrm{Hom}(E_1, F_1) \otimes \mathrm{Hom}(E_2, F_2) &\to \mathrm{Hom}(E_1, F_2 \otimes \mathrm{Hom}(E_2, F_2))\\ &\to \mathrm{Hom}(E_1, \mathrm{Hom}(E_2, F_1 \otimes F_2))\\ &\to \mathrm{Hom}(E_1 \otimes E_2, F_1 \otimes F_2)\end{aligned}$$

where the first two homomorphisms come from the canonical homomorphism (7) (no. 2) and the last is the isomorphism of no. 1, Proposition 1.

8. Let E, F be two finitely generated projective modules over a commutative ring C. Show that if u is an endomorphism of E and v an endomorphism of F then $\mathrm{Tr}(u \otimes v) = \mathrm{Tr}(u)\mathrm{Tr}(v)$.

9. Let A be a ring, C its centre, E a right A-module, F a left A-module and E^*, F^* the respective duals of E and F. An additive mapping f of $E^* \otimes_C F^*$ into A is called *doubly linear* if $f(aw) = af(w)$ and $f(wa) = f(w)a$ for all $w \in E^* \otimes_C F^*$ and all $a \in A$. Show that there exists one and only one C-linear mapping ϕ of $E \otimes_A F$ into the C-module L of doubly linear mappings of $E^* \otimes_C F^*$ into A such that

$$(\phi(x \otimes y))(x^* \otimes y^*) = \langle x^*, x\rangle\langle y, y^*\rangle$$

for all $x \in E$, $y \in F$, $x^* \in E^*$, $y^* \in F^*$. If E and F are finitely generated projective modules, ϕ is bijective.

§ 5

1. Give an example of a homomorphism $\rho: A \to B$ of commutative rings and two A-modules E, F such that the canonical homomorphism (17) (no. 3) is neither injective nor surjective (cf. § 4, Exercise 2).

2. Give an example of a free A-module and a ring B such that the homomorphism (20) (no. 4) is not injective (§ 3, Exercise 3).

3. Give an example of a monogenous A-module E and a ring B such that the homomorphism (20) (no. 4) is not surjective (cf. § 4, Exercise 2).

4. Let $\rho: A \to B$ be a ring homomorphism. For every A-module E, show that the diagram

$$\begin{array}{ccccc} E & \xrightarrow{c_E} & E^{**} & \longrightarrow & (E^{**})_{(B)} \\ \downarrow & & \downarrow & & \downarrow{\scriptstyle \upsilon_{E^*}} \\ E_{(B)} & \xrightarrow[c_{E_{(B)}}]{} & (E_{(B)})^{**} & \xrightarrow[{}^t\upsilon_E]{} & ((E^*)_{(B)})^* \end{array}$$

is commutative.

5. Let K be a field, A the product ring K^N, $\mathfrak{a}$ the ideal $K^{(N)}$ of A and B the ring $A/\mathfrak{a}$. Show that $\mathfrak{a}$ is a projective A-module which is not finitely generated, although $\mathfrak{a}_{(B)} = \{0\}$.

6. With the rings A and B defined as in § 1, Exercise 7, let M be an A-module; show that for M to be finitely generated (resp. free, projective), it is necessary and sufficient that $M \otimes_A B$ be finitely generated (resp. free, projective).

7. Let $\rho: A \to B$ be a ring homomorphism. For every left B-module F, define a canonical B-module homomorphism

$$F \to \tilde{\rho}(\rho_*(F))$$

and show that if E is a left A-module, the composite mappings

$$\tilde{\rho}(E) \to \tilde{\rho}(\rho_*(\tilde{\rho}(E))) \to \tilde{\rho}(E)$$

$$\rho_*(F) \to \rho_*(\tilde{\rho}(\rho_*(F))) \to \rho_*(F)$$

are the identity mappings.

§ 6

1. Let (E_n, f_{nm}) be the inverse system of **Z**-modules whose indexing set is **N**, such that $E_n = \mathbf{Z}$ for all n and, for $n \leqslant m$, f_{nm} is the mapping $x \mapsto 3^{m-n}x$. For all n, let u_n be the canonical mapping $E_n \to \mathbf{Z}/2\mathbf{Z}$. Show that (u_n) is an inverse system of surjective linear mappings but that $\varprojlim u_n$ is not surjective.

2. Let $(E_\alpha, f_{\beta\alpha})$ be a direct system of A-modules. Show that the A-module $\varinjlim E_\alpha$ is canonically isomorphic to the quotient of the direct sum $\bigoplus_\alpha E_\alpha = F$ by the submodule N generated by the elements of the form $j_\beta(f_{\beta\alpha}(x_\alpha)) - j_\alpha(x_\alpha)$, for every ordered pair (α, β) such that $\alpha \leqslant \beta$ and all $x_\alpha \in E_\alpha$, the $j_\alpha: E_\alpha \to F$ being the canonical injections.

3. Let (F_n, f_{mn}) be the direct system of **Z**-modules such that F_n is equal to U_p (§ 1, Exercise 12 (*d*)) for all $n \geqslant 0$ and f_{mn} is the endomorphism $x \mapsto p^{m-n}x$ of U_p for $n \leqslant m$. Show that $\varinjlim F_n = \{0\}$ although the f_{mn} are surjective.

¶ 4. Let $(F_\alpha, f_{\beta\alpha})$ be a direct system of A-modules.

(*a*) For every A-module E, define a canonical **Z**-module homomorphism

$$\varepsilon: \varinjlim \mathrm{Hom}_A(E, F_\alpha) \to \mathrm{Hom}_A(E, \varinjlim F_\alpha).$$

(*b*) Show that if E is finitely generated, ε is injective. Give an example where ε is not injective (cf. Exercise 3, taking $E = \varinjlim F_\alpha$).

(*c*) Show that if E is finitely generated and the $f_{\beta\alpha}$ are injective, ε is surjective. Give an example where $E = \varinjlim F_\alpha$ is free, the $f_{\beta\alpha}$ are injective and ε is not surjective.

(*d*) Show that if E is projective and finitely generated, ε is bijective.

*(*e*) Let A be an integral domain and $\mathfrak{a}$ an ideal of A which is not finitely generated. If $(\mathfrak{a}_\alpha)$ is the family of finitely generated ideals contained in $\mathfrak{a}$, show that $E = A/\mathfrak{a}$ is canonically isomorphic to the direct limit of the $F_\alpha = A/\mathfrak{a}_\alpha$, but that the homomorphism ε is not surjective.*

§ 7

1. Show that if E is a vector space $\neq \{0\}$ over a field K with an infinity of elements, the set of generating systems of E is not inductive for the order relation $\supset$ (form a decreasing sequence (S_n) of generating systems of E such that the intersection of the S_n is empty).

¶ 2. Let K be a field, L a subfield of K, I a set and $(x_i)_{1 \leqslant i \leqslant m}$ a free family of vectors in K_s^I such that the coordinates of each of the x_i belong to L. Let V be the vector subspace of K_s^I generated by the x_i; show that there exists a finite subset J of I, with m elements, such that for every vector $z = (\zeta_\alpha)_{\alpha \in I}$ of V, all the ζ_α belong to the subfield of K generated by L and the m elements ζ_β where $\beta \in J$. (By applying Corollary 3 of no. 5, to the space L_s^I, show that it can be reduced to the case where there exist m indices $\beta_i \in I$ $(1 \leqslant i \leqslant m)$ such that $\mathrm{pr}_{\beta i}(x_j) = \delta_{ij}$).

¶ 3. (*a*) Let G be a group and M an infinite subset of G. Show that if G′ is the subgroup of G generated by M, then $\mathrm{Card}(G') = \mathrm{Card}(M)$.

(*b*) Let K be a field and M an infinite subset of K. Show that if K′ is the subfield of K generated by M, then $\mathrm{Card}(K') = \mathrm{Card}(M)$. (Consider two sequences $(A_n)_{n \geqslant 0}$, $(P_n)_{n \geqslant 0}$ of subsets of K such that $A_0 = M$, P_n is the multiplicative subgroup of K^* generated by $A_n \cap K^*$ and A_{n+1} the additive subgroup of K generated by P_n; then apply (*a*).)

(*c*) Let K be a field and I an infinite set; show that $\mathrm{Card}(K) \leqslant \dim(K_s^I)$. (Reduce it to the case $I = \mathbf{N}$ and argue by *reductio ad absurdum*. Let B be a basis of $K_s^{\mathbf{N}}$ such that $\mathrm{Card}(B) < \mathrm{Card}(K)$ and let L be the subfield of K generated by the coordinates of all the elements of B; then $\mathrm{Card}(L) < \mathrm{Card}(K)$. Form a sequence $(\xi_n)_{n \geqslant 1}$ of elements of K such that for all n, ξ_n does not belong to the subfield of K generated by L and the ξ_i of index $i < n$; then apply Exercise 2 to obtain a contradiction, by considering the point $x = (\xi_n)$ of $K_s^{\mathbf{N}}$.)

(*d*) Deduce from (*c*) that for every field K and every infinite set I,

$$\dim(K_s^I) = (\mathrm{Card}(K))^{\mathrm{Card}(I)}$$

("*Erdös–Kaplansky theorem*").

4. Give an example of an infinite sequence (F_n) of vector subspaces of a vector space E such that $\operatorname{codim}\left(\bigcap_n F_n\right) > \sum_n \operatorname{codim}(F_n)$. (Take $\operatorname{codim}(F_n) = 1$ for all n and use Exercise 3 (*d*).)

5. Let $(H_\lambda)_{\lambda \in L}$ be a family of hyperplanes (passing through 0) of a vector space E over a field K, which is a *covering* of E.

(*a*) Show that, if K is finite, then $\operatorname{Card}(L) \geqslant 1 + \operatorname{Card}(K)$ and, if K is infinite, $\operatorname{Card}(L) \geqslant \aleph_0 = \operatorname{Card}(\mathbf{N})$. (Prove by induction on r that if K has at least r elements, then E cannot be the union of r hyperplanes.) Show by examples that these inequalities cannot be improved without an additional hypothesis (when K is infinite, consider the space $E = K_s^{(\mathbf{N})}$).

(*b*) If E is finite-dimensional, show that $\operatorname{Card}(L) \geqslant 1 + \operatorname{Card}(K)$. (Argue by induction on $\dim(E)$.)

¶ 6. Let S be a non-empty finite subset of a vector K-space E, not containing 0. Suppose that there exists an integer $k \geqslant 1$ with the following property:

(*) for all $x \in S$, there is a partition of $S - \{x\}$ into h free subsets with $h \leqslant k$.

For such a partition $(N_i)_{1 \leqslant i \leqslant h}$ and every sequence $(r_j)_{1 \leqslant j \leqslant m}$ of integers of $(1, h)$ a subspace $F_{r_1 r_2 \ldots r_m}$ of E is defined by induction on m as follows: F_{r_1} is the subspace generated by N_{r_1}, $F_{r_1 \ldots r_p}$ the subspace generated by the intersection of $F_{r_1 \ldots r_{p-1}}$ and N_{r_p}. Suppose that:

(**) for all $x \in S$ and every partition $(N_i)_{1 \leqslant i \leqslant h}$ of $S - \{x\}$ into at most k free subsets, there exists at least one sequence $(r_j)_{1 \leqslant j \leqslant m}$ such that

$$x \notin F_{r_1 r_2 \ldots r_m}.$$

(*a*) Let n be the least of the integers $m \geqslant 1$ for all possible choices of $x \in S$ and let $(N_i)_{1 \leqslant i \leqslant h}$ and $(r_j)_{1 \leqslant j \leqslant m}$ satisfy conditions (*) and (**). Show that of necessity $n = 1$. (Argue by *reductio ad absurdum* by assuming $x \notin F_{r_1 r_2 \ldots r_n}$ and $n > 1$. Then necessarily $x \in F_{r_n}$. For simplicity write $E_p = F_{r_1 r_2 \ldots r_p}$ for $1 \leqslant p \leqslant n$ and $x = \sum_i \alpha_i y_i$ with $\alpha_i \in K$ and $y_i \in N_{r_n}$; there is at least one y_i not belonging to E_{n-1}, denote it by y; then write $N = N_{r_n} - \{y\}$, $N'_i = N_i$ for $1 \leqslant i \leqslant h$ and $i \neq r_n$, $N'_{r_n} = N \cup \{x\}$, so that $(N'_i)_{1 \leqslant i \leqslant h}$ is a partition of $S - \{y\}$ consisting of free subsets. Define $E'_0 = E$ and for $1 \leqslant p \leqslant n$, E'_p as the subspace generated by the intersection of E'_{p-1} and N'_{r_p}. Now $y \notin E_{n-1}$; let q be the least integer such that $y \notin E_q$; show that $E'_q \not\subset E_q$. Let s be the least integer such that $E'_s \not\subset E_s$; show that $r_s = r_n$; deduce finally a contradiction from the relations $y \in E_s$, $E'_{s-1} \subset E_{s-1}$ and $E'_s \not\subset E_s$.)

(*b*) Conclude from (*a*) that with the given hypotheses there exists a partition of S into h free subsets for an integer $h \leqslant k$.

¶ 7. Let S be a non-empty finite subset of a vector K-space E, not containing 0. Suppose that there exists an integer $k \geqslant 1$ such that for every subset

T of S, $\mathrm{Card}(T) \leqslant k.\mathrm{rg}(T)$. Show that there exists a partition of S into h free subsets for some integer $h \leqslant k$. (Prove that S satisfies condition (**) of Exercise 6, arguing by induction on Card(S) and taking amongst all the subspaces $F_{r_1 \ldots r_m}$ one of those with the smallest possible dimension; if G is such a subspace and j an index for which $\mathrm{Card}(G \cap N_j)$ is the least possible, prove that

$$k.\mathrm{Card}(G \cap N_j) \leqslant \mathrm{Card}(G \cap (S - \{x\})) \leqslant \mathrm{Card}(G \cap S) \leqslant k.\mathrm{Card}(G \cap N_j).)$$

8. Let E be a vector space. Show that every endomorphism of E which is not a right divisor of 0 in End(E) is surjective (§ 2, Exercise 7).

9. Let K be a field.

(*a*) In the vector space $E = K_s^{(\mathbf{Z})}$, let $(e_n)_{n \in \mathbf{Z}}$ denote the canonical basis; let v be the automorphism of E defined by the relations $v(e_n) = e_{n+1}$ for all $n \in \mathbf{Z}$. If $u = 1_E - v$, show that u is an injective endomorphism of E, but that $\dim(\mathrm{Coker}(u)) = 1$.

(*b*) In the vector space $E = K_s^{(\mathbf{N})}$, let $(e_n)_{n \in \mathbf{N}}$ denote the canonical basis; let v be the endomorphism of E defined by $v(e_0) = e_0$, $v(e_n) = e_{n-1} + e_n$ for $n \geqslant 1$. Show that v is an automorphism of E and that, if $u = 1_E - v$, u is a surjective endomorphism of E such that $\dim(\mathrm{Ker}(u)) = 1$.

(*c*) Let I be a non-empty set. Show that every automorphism u of the vector space $E = K_s^{(I)}$ can be written as a difference $u = v - w$ of two automorphisms of E, except when K has only two elements and I consists of a single element. (Note that it suffices to prove that *one* particular automorphism u is of the desired form; when K is a field with two elements, consider separately the cases where $\mathrm{Card}(I) = 2$ and $\mathrm{Card}(I) = 3$.)

¶ 10. Let E be a vector space over a field K. Show that every endomorphism u of E is an automorphism or can be written as $u = v - w$, where v and w are automorphisms of E. (Observe that u can always be replaced by $s_1 \circ u \circ s_2$, where s_1 and s_2 are two automorphisms of E. Distinguish two cases, according to whether $\dim(\mathrm{Ker}(u)) \leqslant \dim(\mathrm{Coker}(u))$ or $\dim(\mathrm{Ker}(u)) > \dim(\mathrm{Coker}(u))$; when $\mathrm{rg}(u)$ is finite, then the first case always holds. When

$$\dim(\mathrm{Ker}(u)) \leqslant \dim(\mathrm{Coker}(u)),$$

it can be reduced to the case where $\mathrm{Ker}(u) \cap \mathrm{Im}(u) = \{0\}$; if

$$\dim(\mathrm{Ker}(u)) = \dim(\mathrm{Coker}(u)),$$

it can also be assumed that $\mathrm{Im}(u) + \mathrm{Ker}(u) = E$ and then apply Exercise 9 (*c*). If $\dim(\mathrm{Ker}(u)) < \dim(\mathrm{Coker}(u))$, there is a supplementary subspace in E of $\mathrm{Ker}(u)$ of the form $W \oplus \mathrm{Im}(u)$ with $\dim(W) = \dim(\mathrm{Im}(u)) = \dim(E)$; apply the remark at the beginning with $s_1(u(\mathrm{Im}(u))) = \mathrm{Im}(u)$ and use Exercise 9 (*a*) and (*c*). When $\dim(\mathrm{Ker}(u)) > \dim(\mathrm{Coker}(u))$, it can be reduced to the case where $\mathrm{Ker}(u) \subset \mathrm{Im}(u)$; this time take $s_1(\mathrm{Ker}(u)) = \overset{-1}{u}(\mathrm{Ker}(u))$ and use the results of the preceding cases and Exercise 9.)

11. Let E, F be two vector spaces over a field K and $u: \mathrm{E} \to \mathrm{F}$ a linear mapping. Show that for every vector subspace V of F,

$$\dim(\overset{-1}{u}(\mathrm{V})) = \dim(\mathrm{V} \cap \mathrm{Im}(u)) + \dim(\mathrm{Ker}(u)).$$

12. Let E, F be two vector spaces and u, v two linear mappings of E into F.

(*a*) Show that the following inequalities hold:

$$\mathrm{rg}(v) \leqslant \mathrm{rg}(u + v) + \mathrm{rg}(u)$$
$$\mathrm{rg}(u + v) \leqslant \inf(\dim(\mathrm{E}), \dim(\mathrm{F}), \mathrm{rg}(u) + \mathrm{rg}(v)).$$

If E and F are finite-dimensional, show that $\mathrm{rg}(u + v)$ can take any integer value satisfying the inequalities

$$|\mathrm{rg}(u) - \mathrm{rg}(v)| \leqslant \mathrm{rg}(u + v) \leqslant \inf(\dim(\mathrm{E}), \dim(\mathrm{F}), \mathrm{rg}(u) + \mathrm{rg}(v)).$$

(*b*) Show that

$$\dim(\mathrm{Ker}(u + v)) \leqslant \dim(\mathrm{Ker}(u) \cap \mathrm{Ker}(v)) + \dim(\mathrm{Im}(u) \cap \mathrm{Im}(v)).$$

(*c*) Show that

$$\dim(\mathrm{Coker}(u + v)) \leqslant \dim(\mathrm{Coker}(u)) + \mathrm{rg}(v).$$

(If V is a supplementary subspace of $\mathrm{Ker}(v)$ in E, note that

$$u(\mathrm{Ker}(v)) \subset \mathrm{Im}(u + v) \quad \text{and} \quad \dim(u(\mathrm{V})) \leqslant \mathrm{rg}(v).)$$

13. Let E, F, G be three vector spaces and $u: \mathrm{E} \to \mathrm{F}$, $v: \mathrm{F} \to \mathrm{G}$ two linear mappings.

(*a*) Show that there exists a decomposition of E as a direct sum of $\mathrm{Ker}(u)$ and two subspaces M, N such that $\mathrm{Ker}(v \circ u) = \mathrm{M} \oplus \mathrm{Ker}(u)$ and

$$\mathrm{Im}(v \circ u) = v(u(\mathrm{N})).$$

(*b*) $\dim(\mathrm{Ker}(u)) \leqslant \dim(\mathrm{Ker}(v \circ u)) \leqslant \dim(\mathrm{Ker}(u)) + \dim(\mathrm{Ker}(v))$; if $\mathrm{Ker}(u)$ and $\mathrm{Ker}(v)$ are finite-dimensional, show that $\dim(\mathrm{Ker}(v \circ u))$ can take every integer value satisfying the above inequalities.

(*c*) The following equalities hold:

$$\mathrm{rg}(u) = \mathrm{rg}(v \circ u) + \dim(\mathrm{Im}(u) \cap \mathrm{Ker}(v))$$
$$\mathrm{rg}(v) = \mathrm{rg}(v \circ u) + \mathrm{codim}_{\mathrm{F}}(\mathrm{Im}(u) + \mathrm{Ker}(v)).$$

If F is finite-dimensional, show that $\mathrm{rg}(v \circ u)$ can take any integer value satisfying the inequality

$$\sup(0, \mathrm{rg}(u) + \mathrm{rg}(v) - n) \leqslant \mathrm{rg}(v \circ u) \leqslant \inf(\mathrm{rg}(u), \mathrm{rg}(v)).$$

14. Let E, F, G be three vector spaces and $u: \mathrm{E} \to \mathrm{F}$, $w: \mathrm{E} \to \mathrm{G}$ two linear mappings. Show that if $\overset{-1}{u}(0) \subset \overset{-1}{w}(0)$, there exists a linear mapping v of F into G such that $w = v \circ u$ (cf. § 2, Exercise 8).

15. Let E, F be two vector spaces over a field K and $u: E \to F$ a linear mapping.

(a) The following conditions are equivalent: (1) $\mathrm{Ker}(u)$ is finite-dimensional; (2) there exists a linear mapping v of F into E such that $v \circ u = 1_E + w$, where w is an endomorphism of E of finite rank; (3) for every vector space G over K and every linear mapping $f: G \to E$, the relation $\mathrm{rg}(u \circ f) < +\infty$ implies $\mathrm{rg}(f) < +\infty$.

(b) The following conditions are equivalent: (1) $\mathrm{Coker}(u)$ is finite-dimensional; (2) there exists a linear mapping v of F into E such that $u \circ v = 1_F + w$, where w is an endomorphism of F of finite rank; (3) for every vector space G over K and every linear mapping $g: F \to G$, the relation $\mathrm{rg}(g \circ u) < +\infty$ implies $\mathrm{rg}(g) < +\infty$.

16. Let E, F be two vector spaces over K and $u: E \to F$ a linear mapping. u is said to be of *finite index* if $\mathrm{Ker}(u)$ and $\mathrm{Coker}(u)$ are finite-dimensional and the number $d(u) = \dim(\mathrm{Ker}(u)) - \dim(\mathrm{Coker}(u))$ is then called the *index* of u.

Let G be a third vector space over K and $v: F \to G$ a linear mapping. Show that if two of the three linear mappings u, v, $v \circ u$ are of finite index, so is the third and $d(v \circ u) = d(u) + d(v)$ (use Exercise 13 (a)).

17. Let E, F, G, H be four vector spaces over K and $u: E \to F$, $v: F \to G$, $w: G \to H$ three linear mappings. Show that

$$\mathrm{rg}(v \circ u) + \mathrm{rg}(w \circ v) \leqslant \mathrm{rg}(v) + \mathrm{rg}(w \circ v \circ u).$$

18. Let E, F be two vector spaces over a field K and u, v two linear mappings of E into F. For there to exist an automorphism f of E and an automorphism g of F such that $v = g \circ u \circ f$, it is necessary and sufficient that $\mathrm{rg}(u) = \mathrm{rg}(v)$, $\dim(\mathrm{Ker}(u)) = \dim(\mathrm{Ker}(v))$ and $\dim(\mathrm{Coker}(u)) = \dim(\mathrm{Coker}(v))$.

19. Let E be a vector space over a field K.

(a) Show that every mapping f of E into E which is permutable with every automorphism of E is of the form $x \mapsto \alpha x$, where $\alpha \in K$ (show first that for all $x \in E$ there exists $\rho(x) \in K$ such that $f(x) = \rho(x)x$, using the fact that f commutes with every automorphism of E leaving x invariant).

(b) Let g be a mapping of $E \times E$ into E such that, for every automorphism u of E, $g(u(x), u(y)) = u(g(x, y))$ for all x, y in E. Show that there exist two elements α, β of K such that $g(x, y) = \alpha x + \beta y$ on the set of ordered pairs (x, y) of linearly independent elements of E; moreover there exists a mapping ϕ of $K \times K$ into K such that $g(\lambda x, \mu x) = \phi(\lambda, \mu)x$ for all $x \in E$ (same method). If further $g(u(x), u(y)) = u(g(x, y))$ for every endomorphism u of E, then $g(x, y) = \alpha x + \beta y$ for all x, y in E. Generalize to mappings of E^n into E.

20. Let E be a vector space.

(a) Show that if M and N are two vector subspaces of E and M′, N′ the orthogonals of M and N respectively in E^*, the orthogonal of $M \cap N$ is $M' + N'$ (cf. § 2, Exercise 9 (*a*)).

(*b*) If E is infinite-dimensional, show that there exist hyperplanes H' of E^* such that the subspace of E orthogonal to H' reduces to 0 (consider a hyperplane containing the coordinate forms corresponding to a basis of E).

(*c*) Show that if E is infinite-dimensional there exists an infinite family (V_ι) of subspaces of E such that, if V'_ι is the subspace of E^* orthogonal to V_ι, the subspace of E^* orthogonal to $\bigcap_\iota V_\iota$ is distinct from $\sum_\iota V'_\iota$.

(*d*) Deduce from (*b*) that if E is infinite-dimensional, there exists a decomposition $V' \oplus W'$ of E^* as a direct sum, W' being finite-dimensional, for which the sum $V + W$ of subspaces V, W of E orthogonal respectively to V' and W' is distinct from E.

(*e*) For a subspace of E to be finite-dimensional, it is necessary and sufficient that its orthogonal in E^* be of finite codimension.

21. With the notation of Theorem 8 of no. 6, let u denote the linear mapping $x \mapsto (\langle x, x_\iota^* \rangle)$ of E into K_s^I and $y_0 = (\eta_\iota)$. $K_d^{(I)}$ is identified with a subspace of the dual of K_s^I under the canonical mapping of $K_d^{(I)}$ into its bidual; then show that if N' is the kernel of ${}^t u$, the condition of Theorem 3 (no. 5) expresses the fact that y_0 is orthogonal to the intersection of N' and $K_d^{(I)}$. When I is infinite and the system (21) (no. 5) is of finite rank, show that this intersection is distinct from N' (note that N' is then of finite codimension).

22. Let E and F be two vector spaces over a field K and u a linear mapping of E into F. If V is a vector subspace of E and V' the orthogonal of V in E^*, show that the dual of $u(V)$ is isomorphic to ${}^t u(F^*)/(V' \cap {}^t u(F^*))$. If W' is a subspace of F^* such that ${}^t u(W')$ is finite-dimensional and W is the orthogonal of W' in F, ${}^t u(W')$ is isomorphic to the dual of the space $u(E)/(W \cap u(E))$.

23. Show that for a linear mapping u of a vector space E into a vector space F to be such that $\operatorname{rg}({}^t u) = \operatorname{rg}(u)$, it is necessary and sufficient that $\operatorname{rg}(u)$ be finite (cf. Exercise 3 (*d*)).

24. Let E be a vector space of finite dimension $n > 1$ over a commutative field K; show that, unless $n = 2$ and K is a field with two elements, there exists no isomorphism ϕ of E onto E^* depending only on the vector space structure of E. (Note that if ϕ is such an isomorphism, then necessarily $\langle x, \phi(y) \rangle = \langle u(x), \phi(u(y)) \rangle$ for x, y in E and for every automorphism u of E (*Set Theory*, IV, § 1, no. 5).)

25. (*a*) Let K be a field and L, M two sets; there are canonical inclusions $(K_s^L)^{(M)} \subset (K_s^{(M)})^L \subset K_s^{L \times M}$. Show that for two of these vector spaces to be equal, it is necessary and sufficient that one of the sets L, M be finite.

(*b*) Let $(E_\lambda)_{\lambda \in L}$ be a family of right vector spaces over K and $(F_\mu)_{\mu \in M}$ a family of left vector spaces over K. For the canonical mapping (26) (no. 7) to be bijective, it is necessary and sufficient that one of the vector spaces $\prod_{\lambda \in L} E_\lambda$, $\prod_{\mu \in M} F_\mu$ be finite-dimensional (use (*a*)).

26. (*a*) Let E, F be two left vector spaces over a field K; for the canonical homomorphism $E^* \otimes_K F \to \mathrm{Hom}_K(E, F)$ to be bijective, it is necessary and sufficient that one of the spaces E, F be finite-dimensional.

(*b*) Let E be a right vector space and F a left vector space over a field K. For the canonical homomorphism $E \otimes_K F \to \mathrm{Hom}_K(E^*, F)$ to be bijective, it is necessary and sufficient that E be finite-dimensional (use Exercise 3 (*d*)).

27. (*a*) Under the hypotheses of Proposition 16 (i) (no. 7), for the canonical homomorphism (27) (no. 7) to be bijective, it is necessary and sufficient that E or F be finite-dimensional (note that the image under (27) (no. 7) of an element of $\mathrm{Hom}_K(E, G) \otimes_L F$ maps E to a subspace of $G \otimes_L F$ contained in a subspace of the form $G \otimes_L F'$, where F′ is a finite-dimensional subspace of F).

(*b*) Under the hypotheses of Proposition 16 (ii) (no. 7), for the canonical homomorphism (28) (no. 7) to be bijective, it is necessary and sufficient that one of the ordered pairs (E_1, E_2), (E_1, F_1), (E_2, F_2) consist of finite-dimensional vector spaces (use (*a*) and Exercise 7 of § 4).

28. Let $\rho: K \to A$ be an injective homomorphism of a field K into a ring A and let E be a left vector space over K. For the canonical mapping $(E^*)_{(A)} \to (E_{(A)})^*$ to be bijective, it is necessary and sufficient that E be finite-dimensional or that A, considered as a right vector K-space, be finite-dimensional (use Exercise 25 (*b*)). When one of these two cases holds, there exists a canonical bijection $(E^{**})_{(A)} \to (E_{(A)})^{**}$.

29. Let A be an integral domain, K its field of fractions, E an A-module and T(E) its torsion module.

(*a*) Show that every linear form on E is zero on T(E), so that the duals E^* and $(E/T(E))^*$ are canonically isomorphic.

(*b*) Show that the canonical mapping $(E^*)_{(K)} \to (E_{(K)})^*$ is injective and that it is bijective when E is finitely generated.

(*c*) Give an example of a torsion-free A-module E such that $E^* = \{0\}$ and $E_{(K)} \neq \{0\}$.

¶ 30. Let A be an integral domain and E, F two A-modules. Show that if x is a free element of E and y a free element of F, then $x \otimes y \neq 0$ in $E \otimes_A F$. (Reduce it to the case where E and F are torsion-free, then to the case where E and F are finitely generated and use Exercise 29 (*b*) to show that there is an A-bilinear mapping f of $E \times F$ into K such that $f(x, y) \neq 0$.)

31. Let K_0 be a commutative field, A the polynomial ring $K_0[X, Y]$ in two indeterminates over K_0, which is an integral domain, and E the ideal $AX + AY$ of A. Show that in the tensor product $E \otimes_A E$ the element $X \otimes Y - Y \otimes X$ is $\neq 0$ and that $XY(X \otimes Y - Y \otimes X) = 0$ (consider the A-bilinear mappings of $E \times E$ into the quotient module A/E).

32. Extend the results of no. 10 to the case where A is a non-commutative ring admitting a *field of left fractions* K (I, § 9, Exercise 15). (Note that if ξ_i $(1 \leqslant i \leqslant n)$ are elements of K, there exists an $\alpha \neq 0$ in A such that all the $\alpha\xi_i$ belong to A.)

33. Let A be an integral domain. An A-module E is called *divisible* if for all $x \in E$ and all $\alpha \neq 0$ in A, there exists $y \in E$ such that $\alpha y = x$.

(*a*) Let E, F be two A-modules. Show that if E is divisible, so is $E \otimes_A F$. If E is divisible and F is torsion-free, $\mathrm{Hom}_A(E, F)$ is torsion-free. If E is a torsion A-module and F is divisible, then $E \otimes_A F = \{0\}$. If E is a torsion A-module and F is torsion-free, then $\mathrm{Hom}_A(E, F) = \{0\}$.

(*b*) Show that every injective A-module (§ 2, Exercise 11) is divisible and conversely that every divisible torsion-free A-module is injective (use criterion (δ) of 2, Exercise 11).

34. Let A be an integral domain, P a projective A-module, P′ a projective submodule of P and $j: P' \to P$ the canonical injection. Show that for every torsion-free A-module E, the homomorphism $j \otimes 1: P' \otimes_A E \to P \otimes_A E$ is injective (reduce it to the case where E is finitely generated and embed E in a free A-module).

¶ 35. (*a*) Let K be a field and E a (K, K)-bimodule. Suppose that the dimensions of E, as a left and right vector space over K, are equal to the same finite number n. Show that there exists a family (e_i) of n elements of E, which is a basis for each of the two vector space structures on E. (Note that if $(b_j)_{1 \leqslant j \leqslant m}$ is a family of $m < n$ elements of E, which is free for each of the two vector space structures on E and V is the left vector subspace and W the right vector subspace generated by (b_j), either $V + W \neq E$, or $V \cap \complement W$ and $W \cap \complement V$ are non-empty; in the latter case, if $y \in V \cap \complement W$, $z \in W \cap \complement V$, $y + z$ forms with the b_j a family of $m + 1$ elements which is free for each of the two vector space structures on E.)

(*b*) Let F be a subbimodule of E, whose two vector space structures over K have the same dimension $p < n$. Show that there exists a family $(e_i)_{1 \leqslant i \leqslant n}$ of elements of E which is a basis of E for each of its vector space structures and such that $(e_i)_{1 \leqslant i \leqslant p}$ is a basis of F for each of its two vector space structures (same method).

(*c*) Let $(b_j)_{1 \leqslant j \leqslant n-1}$ be a family of $n - 1$ elements of E, which is free for each of the two vector space structures on E. Let V (resp. W) be the hyperplane generated by the (b_j) in E considered as a left (resp. right) vector space.

Show that if $V \subset W$ (resp. $W \subset V$), then necessarily $V = W$ (note that if $a \notin W$, the set of $\lambda \in K$ such that $\lambda a \in W$ is a left ideal).

*36. (*a*) Let K_0 be a commutative field and $K = K_0(X)$ the field of rational functions in one indeterminate over K_0 (Chapter IV). A (K, K)-bimodule structure is defined on K as follows: the product $t.u$ of an element $u \in K$ by a left operator $t \in K$ is the rational function $t(X)u(X)$; the product $u.t$ of u by a right operator $t \in K$ is the rational function $u(X)t(X^2)$; the left (resp. right) vector K-space structure thus defined on K is then 1-dimensional (resp. 2-dimensional). Deduce examples of (K, K)-bimodules such that dimensions of the two vector space structures of such a bimodule are arbitrary integers.

(*b*) Deduce from (*a*) an example of a (K, K)-bimodule E whose two vector space structures have the same dimension and such that there exists a sub-bimodule F of E whose two vector space structures do not have the same dimension.*

¶ 37. (*a*) Let E be a left vector space (finite-dimensional or otherwise) and A_i $(1 \leqslant i \leqslant n)$ vector subspaces of E. Suppose that, for every sequence $(a_i)_{1 \leqslant i \leqslant n}$ of points of E such that $a_i \in A_i$ for all i, the vector subspace of E generated by the a_i is of dimension $\leqslant m$ (m an integer $\leqslant n$). Then prove that there exists a subspace W of E of dimension $h \leqslant m$, containing $h + (n - m)$ of the subspaces A_i. (Argue by induction on n. Prove first that (for fixed n) attention may be confined to the case where $m < n$ and $\dim(A_i) \leqslant m$ for all i; it may also be assumed that $A_i \neq \{0\}$ for all i and that the A_i are not all of dimension 1. Then argue (for fixed n) by induction on $d = \sum_{i=1}^{m} \dim(A_i)$. If for example $\dim(A_n) \geqslant 2$, consider a 1-dimensional subspace B_n of A_n of dimension 1 and apply the induction hypothesis to $A_1, \ldots, A_{n-1}, B_n$; conclude that there exists a number k such that $1 \leqslant k \leqslant m$ and a k-dimensional subspace U of E containing k of the A_i, for example $A_1, \ldots, A_k$; replacing, if need be, k by an integer k' such that $1 \leqslant k' \leqslant k$, show also that it can be assumed that there exist vectors $b_i \in A_i$ for $1 \leqslant i \leqslant k$ such that $b_1, \ldots, b_k$ form a basis of U; then project onto a supplementary subspace of U in E and use the induction hypothesis.)

(*b*) Let K be a finite field and let E be an N-dimensional vector space over K. Let U_i $(1 \leqslant i \leqslant n)$ be vector subspaces of E such that, for every subset H of $[1, n]$, the intersection of the U_i such that $i \in H$ has dimension $\leqslant N - \mathrm{Card}(H)$. Show that the union of the U_i cannot be equal to E (argue by duality using (*a*)).

38. Let K be a commutative field of characteristic 0 and E a finite-dimensional vector space over K. Let $p_1, \ldots, p_m$ be projectors of E onto vector subspaces of E such that $1_E = p_1 + p_2 + \cdots + p_m$. Show that E is the direct

sum of the $p_i(E)$ and that the p_i are pairwise permutable (take the traces of the two sides).

39. Let K be a field, L a subfield of K and $(K_\alpha)_{\alpha \in I}$ a left directed family of subfields of K such that $L = \bigcap_{\alpha \in I} K_\alpha$. Let E be a vector space over K and $(a_i)_{1 \leqslant i \leqslant m}$ a finite family of elements of E; show that if the family (a_i) is free over L, there exists an index α such that (a_i) is free over K_α. (Argue by induction on $m - r$, where r is the rank of the family (a_i) over K.)

§ 8

1. Let K be a field, K′ a subfield of K and V a non-zero right vector space over K.

(*a*) Let V′ be a K′-structure on V. Show that K′ is equal to the set of $\mu \in K$ such that, for all $x \in V'$, $x\mu \in V'$.

(*b*) Let Γ' be the group of automorphisms of K leaving invariant the elements of K′. Show that if there exists no element of K not belonging to K′ and invariant under all the automorphisms $\sigma \in \Gamma'$, there exists no element of V not belonging to V′ and invariant under every bijective dimorphism of V (relative to an automorphism of K) which leaves invariant all the elements of V′.

(*c*) Conversely, let E be a subset of V and G the group of bijective dimorphisms of V leaving invariant all the elements of E; for all $u \in G$, let σ_u be the corresponding automorphism of K and let Γ be the group of automorphisms of K the image of G under the homomorphism $u \mapsto \sigma_u$. Suppose that G contains no automorphism of V distinct from the identity automorphism and that there exists no element of V not belonging to E and invariant under all the dimorphisms $u \in G$. Show that if K′ is the subfield of elements of K invariant under all the automorphism $\sigma \in \Gamma$, E is a K′-structure on V. (Prove first that E is an additive subgroup of V containing a basis of V; let K″ be the set of elements $\mu \in K$ such that the relation $x \in E$ implies $x\mu \in E$. Show that the elements of K″ are invariant under all $\sigma \in \Gamma$ and that K″ is a subfield of K; deduce that E is a K″-structure on V.)

2. Let K be a field, K′ a subfield of K, V a right vector space over K, V′ a K′-structure on V and R′ the canonical image of the dual V'^* in the dual V^* of V (no. 4). For R′ to be a generating system of V^*, it is necessary and sufficient that V be finite-dimensional or that K be a finite-dimensional right vector K′-space (use Exercise 25 of § 7).

¶ 3. Let K be a field, L a subfield of K and let K_L denote the field K considered as a left vector space over L; $E = \mathrm{End}_L(K_L)$ has canonically a (K, K)-bimodule structure. The dual $(K_L)^*$ is contained in E and is a (K, L)-subbimodule of E. When L is contained in the centre of K, the two vector L-space structures on E, obtained by restricting to L the field of scalars of the two vector K-space structures on E, are identical.

Suppose henceforth that L is contained in the *centre* of K and that the dimension of K_L is finite and equal to n.

(*a*) Show that $(K_L)^*$ is 1-dimensional under its left vector space structure over K (calculate the dimension of $(K_L)^*$ over L in two ways).

(*b*) Show that E is an n-dimensional left vector space over K (same method).

(*c*) Let F be a finite-dimensional left vector space over K and $F_{[L]}$ the corresponding vector space over L. If $u_0 \neq 0$ is a linear form on K_L, show that the mapping $x' \mapsto u_0 \circ x'$ of the dual F^* of F (considered as a vector space over L) onto the dual $(F_{[L]})^*$ of $F_{[L]}$ is bijective (use (*a*)).

¶ 4. Let K be a field of finite rank over its centre Z and L a subfield of K containing Z.

(*a*) Show that the left and right vector space structures of K with respect to L have the same dimension.

(*b*) Show that properties (*a*), (*b*), (*c*) of Exercise 3 also hold in this case (if u_0 is a linear form $\neq 0$ on L_Z and v_0 a linear form $\neq 0$ on K_L, note that $\xi.(u_0 \circ v_0) = u_0 \circ (\xi v_0)$ for $\xi \in K$, that $\xi.(u_0 \circ v_0)$ runs through the dual $(K_Z)^*$ when ξ runs through K and use Exercise 3 (*c*)).

5. Let K_0 be a subfield of a field K such that K is of dimension 2 as a right vector space over K_0. Let E be a left vector space over K.

(*a*) Let E_0 be a subset of E which is a vector space over K_0 and let V be the largest vector sub-K-space of E_0; if W_0 is a vector sub-K_0-space of E_0 supplementary to V, show that the vector sub-K-space W of E generated by W_0 is such that $V \cap W = \{0\}$ (note that if an element $\mu \in K$ does not belong to K_0, the relation $\mu x \in E_0$ for an $x \in E_0$ implies $x \in V$).

(*b*) Let E'_0 be another vector sub-K_0-space of E and V′ the largest vector sub-K-space of E'_0. For there to exist a K-automorphism of E mapping E_0 to E'_0, it is necessary and sufficient that V and V′ have the same dimension with respect to K, that the codimensions of V in E_0 and V′ in E'_0 (with respect to K_0) be equal and that the codimensions of E_0 and E'_0 in E (with respect to K_0) be equal (use (*a*)).

§ 9

*1. In an affine space E over a field K, a quadruple (a, b, c, d) of points of E is called a *parallelogram* if $b - a = c - d$, in which case (a, d, c, b) is also a parallelogram. Show that if K is of characteristic $\neq 2$ the midpoints of the ordered pairs (a, c) and (b, d) are then equal; what can be said when K is of characteristic 2?$_*$

*2. Let E be an affine space over a field of characteristic $\neq 2$ and a, b, c, d

any four points of E. Show that if x, y, z, t are the respective midpoints of the ordered pairs (a, b), (b, c), (c, d), (d, a), the quadruple (x, y, z, t) is a parallelogram (Exercise 1).$_*$

*3. Let K be a commutative field of characteristic $\neq 2$, E an affine plane over K and a, b, c, d four points of E any three of which do not lie on a straight line. Let D_{xy} denote the line passing through two distinct points x, y of E. Suppose that the lines D_{ab} and D_{cd} have a common point e and that D_{ad} and D_{bc} have a common point f. Show that the midpoints of the three ordered pairs (a, c), (b, d), (e, f) lie on a straight line. What becomes of this property when D_{ab} and D_{cd} are parallel or when D_{ad} and D_{bc} are parallel? Consider the case where K is a field with 3 elements.$_*$

*4. Let K be a field whose characteristic is different from 2 and 3, E an affine space over K, a, b, c three points of E not on a straight line and a', b', c' the respective midpoints of the ordered pairs (b, c), (c, a) and (a, b). Show that (in the notation of Exercise 3) the lines $D_{aa'}$, $D_{bb'}$ and $D_{cc'}$ pass through the barycentre of the three points a, b, c. What becomes of this property when K is of characteristic 2 or of characteristic 3? Generalize to a system of n affinely independent points.$_*$

5. For a non-empty subset V of an affine space E over a field K with at least three elements to be a linear variety, it is necessary and sufficient that for every ordered pair (x, y) of distinct points of V the line D_{xy} passing through x and y be entirely contained in V. If K has two elements, for V to be a linear variety, it is necessary and sufficient that the barycentre of any three points of V belong to V.

6. (a) Let E be an affine space of dimension $\geqslant 2$ over a field K. For an affine mapping u of E into itself to transform every line of E into a parallel line, it is necessary and sufficient that the linear mapping v associated with u be a homothety $t \mapsto \gamma t$ of ratio $\gamma \neq 0$ belonging to the centre of K. If $\gamma = 1$, u is a translation; show that if $\gamma \neq 1$, there exists one and only one point $a \in E$ such that $u(a) = a$. If a is taken as origin of E, u is then identified with a central homothety for the vector space structure thus determined on E; u is called a *central homothety* of the affine space E of centre a and ratio γ.

(b) Let u_1, u_2 be two affine mappings of E into E each of which is either a translation or a central homothety. Show that $u_1 \circ u_2$ is a translation or a central homothety of E; if u_1, u_2 and $u_1 \circ u_2$ are all three central homotheties show that their centres lie on a straight line. What can be said when u_1 and u_2 are central homotheties and $u_1 \circ u_2$ is a translation?

(c) Show that the set of translations and central homotheties is a normal subgroup H of the affine group of E and that H/T is isomorphic to the multiplicative group of the centre of K: show that H can only be commutative if H = T, in other words if the centre of K has only two elements.

¶ 7. Let E (resp. E′) be an affine space of finite dimension $n \geqslant 2$ over field K with at least 3 elements (resp. over a field K′) and let u be an injective mapping of E into E′ mapping any three points in a straight line in E to three points in a straight line and such that the affine linear variety generated by $u(E)$ in E′ is equal to E′.

(*a*) Show that u maps every system of affinely independent points of E to a system of affinely independent points (use Exercise 5).

(*b*) Let D_1, D_2 be two parallel lines in E and D'_1, D'_2 the lines of E′ containing respectively $u(D_1)$ and $u(D_2)$. Show that D'_1 and D'_2 are in the same plane; if further u is surjective, show that D'_1 and D'_2 are parallel (in the contrary case, show that there would be 3 points not in a straight line in E whose images under u would be in a straight line) (cf. Exercise 17).

(*c*) Suppose henceforth that if D_1 and D_2 are parallel lines in E, the lines of E′ containing respectively $u(D_1)$ and $u(D_2)$ are parallel. Show that if an origin a in E and the origin $a' = u(a)$ in E′ are taken, there exists an isomorphism σ of K onto a subfield K_1 of K′ such that if E is considered as a vector space over K and E′ as a vector space over K_1, u is an injective semi-linear mapping (relative to σ) of E into E′ (§ 1, no. 13). (Consider first the case $n = 2$; given a basis (e_1, e_2) of E, show that for any two elements α, β of K, the points $(\alpha + \beta)e_1$ and $(\alpha\beta)e_1$ of E can be constructed starting with the points $0, e_1, e_2, \alpha e_1, \beta e_1$ by constructing parallels to given lines and intersections of given lines; deduce that $u(\lambda e_1) = \lambda^\sigma u(e_1)$, where σ is an isomorphism of K onto a subfield of K′, then show that also $u(\lambda e_2) = \lambda^\sigma u(e_2)$ by considering the line joining the points λe_1 and λe_2. Pass finally from here to the case where n is arbitrary, arguing by induction on n.) If u is bijective, show that $K_1 = K'$.

(*d*) Extend the result of (*c*) to the case where K is a field with 2 elements assuming also that u maps every system of affinely independent points of E to a system of affinely independent points of E′.

8. Let E be a left affine space over a field K and T its translation space.

(*a*) For a mapping f of E into a left vector space L over K to be affine, it is necessary and sufficient that

$$f(\mathbf{t} + x) - f(x) = f(\mathbf{t} + y) - f(y)$$
$$f(\lambda\mathbf{t} + x) - f(x) = \lambda f(\mathbf{t} + x) - f(x))$$

for all $\lambda \in K$, $\mathbf{t} \in T$, x, y in E. Let $x \mapsto [x]$ be the canonical injection of E into the vector space $K_s^{(E)}$ of formal linear combinations of elements of E and let N be the subspace of $K_s^{(E)}$ generated by the elements

$$[\mathbf{t} + x] - [x] - [\mathbf{t} + y] + [y]$$
$$[\lambda\mathbf{t} + x] - [x] - \lambda[\mathbf{t} + x] + \lambda[x]$$

for $\lambda \in K$, $\mathbf{t} \in T$, x, y in E; finally, let V be the quotient space of $K_s^{(E)}$ by N and ϕ the mapping of E into V which associates with every $x \in E$ the class of $[x]$ modulo N. Then ϕ is an affine mapping and for every affine mapping f of E into a left vector space L over K, there exists one and only one linear mapping g of V into L such that $f = g \circ \phi$.

(*b*) Let $\phi_0 : T \to V$ be the linear mapping associated with ϕ, so that $\phi(\mathbf{t} + x) - \phi(x) = \phi_0(\mathbf{t})$. Show that ϕ (and therefore also ϕ_0) is injective (consider the mapping $x \mapsto x - a$ of E into T for some $a \in E$); for every family $(\lambda_\iota)_{\iota \in I}$ of elements of K with finite support and every family $(x_\iota)_{\iota \in I}$ of elements of E,

$$\sum_\iota \lambda_\iota \phi(x_\iota) = \phi_0\Big(\sum_\iota \lambda_\iota x_\iota\Big) \qquad \text{if } \sum_\iota \lambda_\iota = 0$$

$$\sum_\iota \lambda_\iota \phi(x_\iota) = \mu\phi_0\Big(\sum_\iota \mu^{-1}\lambda_\iota x_\iota\Big) \qquad \text{if } \sum_\iota \lambda_\iota = \mu \neq 0.$$

Deduce that $\phi_0(T)$ is a hyperplane passing through 0 in V and $\phi(E)$ a hyperplane parallel to $\phi_0(T)$.

9. Let K be a finite field with q elements and V an n-dimensional vector space over K.

(*a*) Show that the set of sequences $(x_1, x_2, \dots, x_m)$ of $m \leqslant n$ vectors of V forming a free system has cardinal equal to

$$(q^n - 1)(q^n - q)\dots(q^n - q^{m-1})$$

(argue by induction on n).

(*b*) Deduce from (*a*) that the cardinal of the set of m-dimensional linear varieties in an n-dimensional projective space over K is equal to

$$\frac{(q^{n+1} - 1)(q^{n+1} - q)\dots(q^{n+1} - q^m)}{(q^{m+1} - 1)(q^{m+1} - q)\dots(q^{m+1} - q^m)}.$$

10. In an n-dimensional projective space $\mathbf{P}(V)$ over a field K, a *projective basis* is a set S of $n + 2$ points any two of which form a projectively free system. If $S = (a_i)_{0 \leqslant i \leqslant n+1}$ and $S' = (a'_i)_{0 \leqslant i \leqslant n+1}$ are any two projective bases of $\mathbf{P}(V)$, show that there exists a transformation $f \in \mathbf{PGL}(V)$ such that $f(a_i) = a'_i$ for $0 \leqslant i \leqslant n + 1$. For this transformation to be unique, it is necessary and sufficient that K be commutative. (Reduce it to the case where $a'_i = a_i$ for all i and note that it is always possible to write (in the notation of no. 6) $a_i = \pi(b_i)$, where $(b_i)_{1 \leqslant i \leqslant n+1}$ is a basis of V and $b_0 = b_1 + b_2 + \dots + b_{n+1}$.) *Give an example where K is a field of quaternions and there exists an infinite set T of points of $\mathbf{P}(V)$ any $n + 1$ of which form a projectively free system and a transformation $f \in \mathbf{PGL}(V)$, distinct from the identity, leaving invariant all the points of T.*

11. Let V be a 2-dimensional vector space over a field K and a, b, c, d four

distinct points of the projective line $\mathbf{P}(V)$. The *cross-ratio* of the quadruple (a, b, c, d), denoted by $\begin{bmatrix} a & b \\ d & c \end{bmatrix}$, is the set of elements $\xi \in K$ such that there exist two vectors u, v in V for which (in the notation of no. 6) $a = \pi(u)$, $b = \pi(v)$, $c = \pi(u + v)$, $d = \pi(u + \xi v)$. This definition extends immediately to any quadruple of distinct points of a set with a projective line structure (no. 11).

(*a*) Show that $\begin{bmatrix} a & b \\ d & c \end{bmatrix}$ is the set of conjugates of an element $\neq 1$ of the multiplicative group K^* and conversely that, when a, b, c, are distinct points of $\mathbf{P}(V)$ and ρ the set of conjugates of an element $\neq 1$ of K^*, there exists a point $d \in \mathbf{P}(V)$ such that $\begin{bmatrix} a & b \\ d & c \end{bmatrix} = \rho$. For d to be unique, it is necessary and sufficient that ρ consist of a single point.

(*b*) Show that

$$\begin{bmatrix} a & b \\ c & d \end{bmatrix} = \begin{bmatrix} b & a \\ d & c \end{bmatrix} = \begin{bmatrix} a & b \\ d & c \end{bmatrix}^{-1}$$

and

$$\begin{bmatrix} d & a \\ c & b \end{bmatrix} = 1 - \begin{bmatrix} a & b \\ d & c \end{bmatrix}$$

(denoting by ρ^{-1} (resp. $1 - \rho$) the set of conjugates $\lambda\xi^{-1}\lambda^{-1} = (\lambda\xi\lambda^{-1})^{-1}$ (resp. $1 - \lambda\xi\lambda^{-1} = \lambda(1 - \xi)\lambda^{-1}$), where ξ is an element of ρ).

(*c*) Let (a, b, c, d), (a', b', c', d') be two quadruples of distinct points of $\mathbf{P}(V)$. For there to exist a bijective semi-linear mapping of V onto itself such that the bijective mapping f of $\mathbf{P}(V)$ onto itself, obtained when passing to the quotients, satisfies the conditions $f(a) = a'$, $f(b) = b'$, $f(c) = c'$, $f(d) = d'$, it is necessary and sufficient that there exist an automorphism σ of K such that

$$\begin{bmatrix} a' & b' \\ d' & c' \end{bmatrix} = \begin{bmatrix} a & b \\ d & c \end{bmatrix}^{\sigma}.$$

For there to exist a transformation f of the projective group $\mathbf{PGL}(V)$ satisfying the above conditions, it is necessary and sufficient that

$$\begin{bmatrix} a' & b' \\ d' & c' \end{bmatrix} = \begin{bmatrix} a & b \\ d & c \end{bmatrix}.$$

12. Let $\mathbf{P}(V)$ be a (left) projective space of finite dimension n over a field K. Show that there exists on the set of projective hyperplanes of $\mathbf{P}(V)$ an n-dimensional (right) projective space structure over K canonically isomorphic to that of $\mathbf{P}(V^*)$ (V^* being the dual of V). If M is a linear variety of dimension $r < n$ in $\mathbf{P}(V)$, derive an $(n - r - 1)$-dimensional projective space structure on the set of projective hyperplanes containing M. In particular if M is of

dimension $n-2$, the cross-ratio $\begin{bmatrix} H_1 & H_2 \\ H_4 & H_3 \end{bmatrix}$ of a quadruple (H_1, H_2, H_3, H_4) of distinct hyperplanes containing M can be defined. Show that if $D \subset \mathbf{P}(V)$ is a line not meeting M and a_i is the intersection of D with H_i $(1 \leqslant i \leqslant 4)$, then

$$\begin{bmatrix} a_1 & a_2 \\ a_4 & a_3 \end{bmatrix} = \begin{bmatrix} H_1 & H_2 \\ H_4 & H_3 \end{bmatrix}.$$

*13. In a projective plane $\mathbf{P}(V)$ over a field K of characteristic $\neq 2$, let a, b, c, d be four points forming a projective basis (Exercise 10); denoting by D_{xy} the line passing through two distinct points x, y of $\mathbf{P}(V)$, let e, f, g be the points of intersection of the lines D_{ab} and D_{cd}, D_{ac} and D_{bd}, D_{ad} and D_{bc} respectively; let h be the point of intersection of D_{bc} and D_{ef}; show that $\begin{bmatrix} b & c \\ h & g \end{bmatrix} = \{-1\}$ ("*theorem of the complete quadrilateral*"; reduce it to the case where D_{ad} is the line at infinity of an affine plane). What is the corresponding result when K is of characteristic 2?*

14. In a projective plane $\mathbf{P}(V)$ over a field K with at least three elements, let D, D′ be two distinct lines. In order that, for any distinct points a, b, c of D, a', b', c' of D′, the intersection points r of $D_{ab'}$ and $D_{ba'}$, q of $D_{ac'}$ and $D_{ca'}$, p of $D_{bc'}$ and $D_{cb'}$ lie on a straight line, it is necessary and sufficient that K be commutative ("*Pappus's theorem*"; reduce it to the case where q and r are on the line at infinity of an affine plane). Apply this theorem to the projective space of lines of $\mathbf{P}(V)$ (Exercise 12).

15. In a projective plane over a field K with at least three elements, let s, a, b, c, a', b', c' be seven distinct points such that $\{s, a, b, c\}$ and $\{s, a', b', c'\}$ are projective bases (Exercise 10) and the lines D_{sa}, D_{sb}, D_{sc} pass respectively through a', b', c'. Show that the intersection points r of D_{ab} and $D_{a'b'}$, p of D_{bc} and $D_{b'c'}$, q of D_{ca} and $D_{c'a'}$ lie on a straight line ("*Desargues's theorem*"; method analogous to that of Exercise 14).

16. (*a*) Let $E = \mathbf{P}(V)$ and $E' = \mathbf{P}(V')$ be two projective spaces of the same dimension $n \geqslant 2$ over two fields K, K′ respectively and let u be a *bijective* mapping of E onto E′, mapping any three points on a straight line to three points on a straight line. Show that there exist an isomorphism σ of K onto K′ and a bijective semi-linear mapping v of V onto V′ (relative to σ) such that u is the mapping obtained from v when passing to the quotients ("*fundamental theorem of projective geometry*"; use Exercise 7). Suppose also that $V' = V$ and K is commutative; for u to be a projective mapping, it is necessary and sufficient that also $\begin{bmatrix} u(a) & u(b) \\ u(d) & u(c) \end{bmatrix} = \begin{bmatrix} a & b \\ d & c \end{bmatrix}$ for every quadruple (a, b, c, d) of distinct points on a straight line in $\mathbf{P}(V)$.

(*b*) Let p be an integer such that $1 \leqslant p \leqslant n - 1$. Show that the first conclusion of (*a*) holds when the image under u of every p-dimensional projective linear variety is contained in a p-dimensional projective linear variety.

17. Let V be a vector space of finite dimension n over a field K, $(e_i)_{1 \leqslant i \leqslant n}$ a basis of V, K′ a subfield of K and V′ the n-dimensional vector space over K′ generated by the e_i. Give an example of an injective mapping of V′ into V, mapping any three points on a straight line of the affine space V′ to three points on a straight line in the affine space V, but not necessarily mapping two parallel lines to sets contained in two parallel lines. (Embed V in the projective space E canonically associated with it and consider a projective transformation u of E into itself such that the inverse image under u of the hyperplane at infinity is distinct from this hyperplane and contains no point of V′; for example K can be taken to be infinite and K′ finite.)

¶ 18. Let $E = \mathbf{P}(V)$ be a projective plane over a field K and u a bijective mapping of E onto itself, arising when passing to the quotients from a bijective semi-linear mapping v of V onto itself, relative to an automorphism σ of K.

(*a*) Show that the four following properties are equivalent: (α) for all $x \in E$, x, $u(x)$ and $u^2(x)$ lie on a straight line; (β) every line of E contains a point invariant under u; (γ) through every point of E there passes a line invariant under u; (δ) for every line D of E, the three line D, $u(D)$ and $u^2(D)$ have a common point. (Show first that (α) and (γ) are equivalent; deduce by duality (Exercise 12) that (β) and (δ) are equivalent; prove finally that (γ) implies (β) and deduce by duality that (β) implies (γ).)

(*b*) Suppose that u has the properties stated in (*a*). Show that if there exists in E a line D invariant under u and containing only a single point invariant under u, u arises from a transvection v of V when passing to the quotients (§ 10, Exercise 11). (Show that every line invariant under u contains a; by considering a line not passing through a, show that there exists a line D_0 passing through a and containing at least two points invariant under u; conclude that all the points of D_0 are necessarily invariant under u.)

(*c*) Suppose that u has the properties stated in (*a*). Show that if there exists in E a line E invariant under u and containing only two points invariant under u, u arises from a dilatation v of V when passing to the quotients (§ 10, Exercise 11). (If a, b are the two points of D invariant under u, show that every line invariant under u passes through a or through b; then note that there exist at least two other points c, d distinct from a, b and invariant under u and therefore the line D_{cd} passes through a or b; conclude by proving that all the points of D_{cd} are invariant under u.)

(*d*) Suppose that u has the properties stated in (*a*) and that every line of E invariant under u contains at least three distinct points invariant under u; then there exists in E a projective basis (Exercise 10) each point of which is invariant under u; conclude that there exists a basis $(e_i)_{1 \leqslant i \leqslant 3}$ of V such that u

arises from a semi-linear mapping v of V into itself such that $v(e_i) = e_i$ for $1 \leqslant i \leqslant 3$, when passing to the quotients. The set of points of E invariant under u is then the projective plane $\mathbf{P}(V_0)$, where V_0 is the vector space over the field K_0 of invariants of σ, generated by e_1, e_2, e_3.

(*e*) Suppose henceforth that u satisfies the conditions of (*a*) and (*d*) and that neither u nor u^2 is the identity. Show that there exists $\gamma \in K$ such that $\gamma^\sigma = \gamma$ and

$$(\xi^\sigma - \xi)^\sigma = \gamma(\xi^\sigma - \xi) \tag{1}$$

for all $\xi \in K$ (use conditions (α) of (*a*) and the existence of $\zeta \in K$ such that $\zeta^\sigma \neq \zeta$). Show that $\gamma \neq -1$ and that, for all $\xi \in K$ such that $\xi^\sigma \neq \xi$,

$$(1 + \gamma)\xi^\sigma\gamma = \gamma\xi(1 + \gamma) \tag{2}$$

(apply (1) replacing ξ by ξ^2); extend (2) to all $\xi \in K$ by noting that $\xi = \eta - \zeta$, where $\eta^\sigma \neq \eta$ and $\xi^\sigma \neq \zeta$. Conclude that $\gamma \neq 1$, then deduce from (1) and (2) that

$$\xi^\sigma = (1 + \gamma)\xi(1 + \gamma)^{-1} \tag{3}$$

for all $\xi \in K$ and that $\gamma + \gamma^{-1}$ belongs to the centre of K. Obtain the converse.

Give an example where γ does not belong to the centre of K, but where $\gamma + \gamma^{-1}$ is in this centre.

19. Let K be a commutative field, $\tilde{K}$ the projective field obtained by adjoining to K a point at infinity (no. 9) and f and g two elements of K(X). Show that, writing $h(X) = f(g(X))$, if $\tilde{f}$, $\tilde{g}$, $\tilde{h}$ are the canonical extensions of f, g, h to $\tilde{K}$, then $\tilde{h} = \tilde{f} \circ \tilde{g}$.

§ 10

1. (*a*) Let E be a right A-module. On the additive group E^r (with $r \geqslant 1$) an external law of composition is defined with $\mathbf{M}_r(A)$ as set of operators, denoting by $x.P$, for every element $x = (x_i)_{1 \leqslant i \leqslant r}$ of E^r and every square matrix $P = (\alpha_{ij}) \in \mathbf{M}_r(A)$, the element $y = (y_i)$ of E^r such that

$$y_i = \sum_{j=1}^{r} x_j \alpha_{ji} \qquad (1 \leqslant i \leqslant r).$$

This external law defines with the additive law on E^r a right $\mathbf{M}_r(A)$-module structure on E^r; by restricting the ring of operators of the ring of scalar matrices $I.\alpha$, the product A-module structure on E^r is recovered. Show that for the A-module E to have a generating system of r elements, it is necessary and sufficient that the $\mathbf{M}_r(A)$-module E^r be monogenous.

(*b*) Let $(x_i)_{1 \leqslant i \leqslant n}$ be a generating system of E and $(y_j)_{1 \leqslant j \leqslant m}$ a family of elements of E (resp. a generating system of E). Let z, z', z'' be three elements of E^{m+n} such that $z_i = x_i$ for $1 \leqslant i \leqslant n$, $z_{n+j} = 0$ for $1 \leqslant j \leqslant m$, $z'_i = x_i$ for

$1 \leqslant i \leqslant n$, $z'_{n+j} = y_j$ for $1 \leqslant j \leqslant n$, $z''_i = 0$ for $1 \leqslant i \leqslant n$, $z''_{n+j} = y_j$ for $1 \leqslant j \leqslant n$. Show that there exist two invertible matrices P, Q of $\mathbf{M}_{n+m}(A)$ such that $z' = z.P$ (resp. $z'' = z.Q$).

(*c*) If A is commutative, $u_P : x \mapsto x.P$ is an endomorphism of the A-module E^r and the mapping $P \mapsto u_P$ is a homomorphism of the ring $\mathbf{M}_r(A)$ into the ring $\mathrm{End}_A(E^r)$. If E is a faithful A-module this homomorphism is injective.

2. (*a*) Let X be a square matrix over a ring A, which can be written as an upper triangular matrix (X_{ij}) of matrices ($1 \leqslant i \leqslant p$, $1 \leqslant j \leqslant p$). Show that if each of the square matrices X_{ii} ($1 \leqslant i \leqslant p$) is invertible, so is X and X^{-1} can be written as an upper triangular matrix (Y_{ij}) ($1 \leqslant i \leqslant p$, $1 \leqslant j \leqslant p$) corresponding to the same partition of the indexing set. When A is a field, prove that this sufficient condition for X to be invertible is also necessary.

(*b*) Give an example of a ring A and a matrix $\begin{pmatrix} a & 1 \\ 0 & b \end{pmatrix}$ (with $a \in A$, $b \in A$) which is invertible without either a or b being invertible in A and whose inverse $\begin{pmatrix} a' & b' \\ c' & d' \end{pmatrix}$ is such that $b' \neq 0$ and $c' \neq 0$ (take A to be the endomorphism ring of an infinite-dimensional vector space).

3. Let A be the quotient ring $\mathbf{Z}/30\mathbf{Z}$. Show that in the matrix

$$\begin{pmatrix} 1 & 1 & -1 \\ 0 & 2 & 3 \end{pmatrix}$$

over the ring A, the two rows are linearly independent but any two columns are linearly dependent.

4. Let A be a ring, C its centre, B the ring of matrices $\mathbf{M}_n(A)$, Δ the additive subgroup of A generated by the elements $\alpha\beta - \beta\alpha$ for $\alpha \in A$, $\beta \in A$, and D the additive subgroup of B generated by the matrices $XY - YX$ for $X \in B$, $Y \in B$; Δ and D are C-modules. For every matrix $X = (\xi_{ij}) \in B$, let $\theta(X)$ be the element $\sum_{i=1}^{n} \bar{\xi}_{ii}$ of A/Δ, where, for all $\alpha \in A$, $\bar{\alpha}$ denotes the class of α mod. Δ. Show that θ is a surjective homomorphism of B onto A/Δ, whose kernel is equal to D, so that B/D is isomorphic to A/Δ as a C-module. (Observe that D contains the matrices αE_{ij} and the matrices $\alpha(E_{ii} - E_{jj})$ for $\alpha \in A$ and $i \neq j$.)

¶ 5. Let A be a ring, L, M, N any three indexing sets $U = (a_{\lambda\mu})$ a matrix of $A^{L \times M}$ and $V = (b_{\mu\nu})$ a matrix of $A^{M \times N}$; if, for every ordered pair $(\lambda, \nu) \in L \times N$, the family $(a_{\lambda\mu}b_{\mu\nu})$ $(\mu \in M)$ has *finite* support, the element $c_{\lambda\nu} = \sum_{\mu} a_{\lambda\mu}b_{\mu\nu}$ is defined and it is also said that the matrix $(c_{\lambda\nu})$ is the *product* UV of U by V. When the products UV' and UV'' are defined, so is $U(V' + V'')$

and $UV' + UV'' = U(V' + V'')$; is the converse true? Give an example of three infinite matrices, U, V, W such that the products UV, VW, $U(VW)$ and $(UV)W$ are defined but $U(VW) \neq (UV)W$ (take U to be a matrix with one row all of whose elements are equal to 1, W its transpose and V a matrix all of whose elements are 0, 1 or -1 and which has only a finite number of elements $\neq 0$ in each row and each column. Determine by induction the elements of V such that $UV = 0$ but VW has a single element $\neq 0$.)

6. Let X be a matrix with m rows and n columns over a field K; show that the rank rg(X) is equal to the greatest of the ranks of the submatrices of X with an equal number of rows and columns. (Let $\text{rg}(X) = r$; if $a_1, \ldots, a_r$ are r columns of X forming a free system in K_d^m and a basis of K_d^m is formed with these r vectors and $m - r$ vectors of the canonical basis, show that the components of $a_1, \ldots, a_r$ over the r other vectors of the canonical basis form a matrix of rank r.)

7. Let X be a matrix with m rows and n columns over a field K; if r is the rank of X, show that the rank of a submatrix with m rows and s columns, obtained by suppressing $n - s$ columns of X, is $\geqslant r + s - n$.

8. Let $X = (\alpha_{ij})$ be a matrix with m rows and n columns over a field K. For X to be of rank 1, it is necessary and sufficient that there exist in K a family $(\lambda_i)_{1 \leqslant i \leqslant m}$ of m elements not all zero and a family $(\mu_j)_{1 \leqslant j \leqslant n}$ of n elements not all zero such that $\alpha_{ij} = \lambda_i \mu_j$ for every ordered pair of indices.

9. Let X, Y be two matrices with m rows and n columns over a field K; if there exist two square matrices P, P_1 of order m and two square matrices Q, Q_1 of order n such that $Y = PXQ$ and $X = P_1XQ_1$, show that X and Y are equivalent.

10. Let X, X', Y, Y' be four square matrices of order n over a ring A, such that X is invertible. For there to exist two invertible square matrices P, Q of order n such that $X' = PXQ$ and $Y' = PYQ$, it is necessary and sufficient that X' be invertible and that the matrices YX^{-1} and $Y'X'^{-1}$ be similar.

11. Let E be a right vector space over a field K, of dimension $\geqslant 1$, and H a hyperplane of E. Every endomorphism u of E leaving invariant each of the elements of H gives, on passing to the quotients, an endomorphism of the 1-dimensional quotient space E/H, an endomorphism which is therefore of the form $\dot{x} \mapsto \dot{x}\mu(\dot{x})$, where $\mu(\dot{x}) \in K$ is such that $\mu(\dot{x}\lambda) = \lambda^{-1}\mu(\dot{x})\lambda$ for $\lambda \in K^*$. An *automorphism* of E leaving invariant the elements of H is called a *transvection of hyperplane* H if the corresponding automorphism of E/H is the identity and a *dilatation of hyperplane* H otherwise; when u is a dilatation, the set of elements $\mu(\dot{x})$ for $\dot{x} \in E/H$ which is a class of conjugate elements in the multiplicative group K^*, is called the *class* of the dilatation u.

(*a*) Show that for every dilatation there exists one and only one supplementary line of H, invariant under the dilatation.

(*b*) Let ϕ be a linear form on E such that $H = \overset{-1}{\phi}(0)$; show that for every transvection u of hyperplane H there exists a unique vector $a \in H$ such that $u(x) = x + a\phi(x)$ for all $x \in E$. Let $\Gamma(E, H)$ be the subgroup of $\mathbf{GL}(E)$ consisting of the automorphisms leaving invariant each element of H; show that the transvections of hyperplane H form a normal commutative subgroup $\Theta(E, H)$ of $\Gamma(E, H)$ isomorphic to the additive group H; the quotient group $\Gamma(E, H)/\Theta(E, H)$ is isomorphic to the multiplicative group K^*. For $\Gamma(E, H)$ to be commutative, it is necessary and sufficient that one of the two following conditions be satisfied: (α) K is a field with two elements; (β) K is commutative and $\dim E = 1$.

(*c*) Suppose that E is finite-dimensional; show that for every transvection u there exists a basis of E such that the matrix of u with respect to this basis has all its diagonal elements equal to 1 and at most one other element $\neq 0$.

(*d*) Show that the *centralizer* (I, § 5, no. 3) of the group $\Theta(E, H)$ in the group $\mathbf{GL}(E)$ is the composition $Z(E)\Theta(E, H) = \Theta(E, H)Z(E)$ of $\Theta(E, H)$ and the centre $Z(E)$ of $\mathbf{GL}(E)$ (cf. § 1, Exercise 26). The only automorphisms belonging to this centralizer and leaving invariant at least one element $\neq 0$ of E are the transvections of the group $\Theta(E, H)$. If K has at least 3 elements, the centralizer of $\Gamma(E, H)$ in $\mathbf{GL}(E)$ is equal to $Z(E)$.

(*e*) Show that the *normalizers* (I, § 5, no. 3) of $\Theta(E, H)$ and $\Gamma(E, H)$ in $\mathbf{GL}(E)$ are both equal to the subgroup consisting of the automorphisms leaving H invariant.

¶ 12. Let E be a right vector space of dimension $\geqslant 1$ over a field K. Let F(E) denote the normal subgroup of $\mathbf{GL}(E)$ consisting of the automorphisms u such that the kernel of $1_E - u$ is of *finite* codimension (hence $F(E) = \mathbf{GL}(E)$ if E is finite-dimensional). Let $\mathbf{SL}(E)$ denote the normal subgroup of $\mathbf{GL}(E)$ generated by all the transvections (Exercise 11); it is contained in F(E).

(*a*) If $\dim E \geqslant 2$, show that for every ordered pair of non-zero vectors x, y of E, there exists a transvection or product of two transvections, which maps x to y (in other words $\mathbf{SL}(E)$ operates *transitively* on $E - \{0\}$).

(*b*) Let V, W be two hyperplanes of E, $\dot{x}_0 = x_0 + V$ a class mod. V distinct from V and $\dot{y}_0 = y_0 + W$ a class mod. W distinct from W. Show that if $\dim E \geqslant 2$, there exists a transvection or product of two transvections, which maps V to W and $\dot{x}_0$ to $\dot{y}_0$ (consider first the case where V and W are distinct).

(*c*) If $\dim E \geqslant 2$, show that any two transvections distinct from the identity are *conjugate in the group* F(E).

(*d*) If $\dim E \geqslant 3$, show that any two transvections distinct from the identity are *conjugate in the group* $\mathbf{SL}(E)$ (reduce it with the aid of (*b*) to the case where the hyperplanes of the two transvections are identical, then use (*a*)).

(*e*) Suppose that $\dim E = 2$. For any two transvections to be conjugate in **SL**(E), it is necessary and sufficient that the subgroup Q of K^* generated by the *squares* of elements of K^* be identical with K^*. (If u is a transvection distinct from the identity, a a vector of E which is not invariant under u and $b = u(a) - a$, show that, for every transvection u' conjugate to u in **SL**(E), $u'(a) - a = a\lambda + b\mu$, with $\mu = 0$ or $\mu \in Q$; for this use the fact that in a group G every product $sts^{-1}t$ is a product of squares.)

(*f*) Suppose that $\dim E \geqslant 2$. For two dilatations u, u' to be such that there exists a $v \in \mathbf{SL}(E)$ such that $u' = vuv^{-1}$, it is necessary and sufficient that the classes (Exercise 11) of u and u' be the same (use (*b*)). Deduce that if the class of a dilatation is contained in the commutator group of K^*, this dilatation belongs to the group **SL**(E) (observe that $(vuv^{-1})u^{-1} = v(uv^{-1}u^{-1})$).

¶ 13. Let E be a right vector space of dimension $\geqslant 2$ and H_0 a hyperplane of E.

(*a*) Show that every automorphism u belonging to F(E) (Exercise 12) is the product of an automorphism of **SL**(E) and possibly a dilatation of the hyperplane H_0 (proceed by induction on the codimension of the kernel of $1_E - u$, using Exercises 12(*a*) and 12(*f*)).

(*b*) Show that **SL**(E) contains the commutator group of F(E) and is identical with this group except when E is a space of dimension 2 over the field with 2 elements. (To show that **SL**(E) contains the commutator group of F(E), use (*a*) and Exercise 12(*f*). To see that **SL**(E) is contained in the commutator group of F(E), except in the case indicated, show that for every homomorphism of F(E) into a commutative group, the image of a transvection is the identity element, using Exercises 11(*b*) and 12(*c*).)

(*c*) Show that **SL**(E) is equal to its commutator group except when $\dim E = 2$ and K has 2 or 3 elements. (If $\dim E \geqslant 3$, note that every transvection u can be written as $vwv^{-1}w^{-1}$, where v is a transvection and $w \in \mathbf{SL}(E)$, using Exercise 12(*d*). If $\dim E = 2$, note first that every automorphism u whose matrix with respect to a basis of E is of the form $\begin{pmatrix}\lambda & 0\\ 0 & \lambda^{-1}\end{pmatrix}$ is a product of transvections, arguing as in (*a*); then consider the commutator $uvu^{-1}v^{-1}$, where v is the transvection whose matrix with respect to the same basis is $\begin{pmatrix}1 & 1\\ 0 & 1\end{pmatrix}$.)

¶ 14. Let E be a vector space of dimension $\geqslant 2$ over a field K.

(*a*) Show that the projective group **PGL**(E) is canonically isomorphic to the quotient of the linear group **GL**(E) by its centre (isomorphic to the multiplicative group of the centre of K, cf. § 1, Exercise 26).

(*b*) Let **PSL**(E) denote the canonical image of the group **SL**(E) in **PGL**(E); it is a normal subgroup of **PGL**(E), containing the commutator group of **PGL**(E) when E is finite-dimensional. Show that **PSL**(E) is a *doubly transitive* (cf. I, § 5, Exercise 14) group of permutations of **P**(E).

(*c*) Let $a \neq 0$ be an element of E and let $e = \pi(a) \in \mathbf{P}(E)$ (in the notation of § 9, no. 6). Show that the transvections of the form $x \mapsto x + a\phi(x)$ (where $\phi \in E^*$ is such that $\phi(a) = 0$) constitute a commutative subgroup of **SL**(E); if Γ_e is the image of this subgroup in **PSL**(E), Γ_e is a normal subgroup of the subgroup Φ_e of **PSL**(E) leaving e invariant. Show also that the union of the subgroups conjugate to Γ_e in **PSL**(E) generates **PSL**(E).

(*d*) Let Σ be a primitive group (I, § 5, Exercise 13) of permutations of a set F which is equal to its commutator group. Suppose that there exists an element $c \in F$ such that the subgroup Φ_c of Σ leaving c invariant contains a commutative normal subgroup Γ_c such that the union of the subgroups conjugate to Γ_c in Σ generates Σ. Show that under these conditions Σ is *simple*. (Let Δ be a normal subgroup of Σ distinct from the identity; using I, § 5, Exercise 17, show that $\Sigma = \Delta.\Phi_c = \Delta.\Gamma_c$ and, using the fact that Γ_c is commutative, deduce that every commutator in Σ is contained in Δ.)

(*e*) Deduce from (*c*), (*d*) and Exercise 13(*c*) that the group **PSL**(E) is *simple* except when dim E = 2 and K has 2 or 3 elements.

*(*f*) If $\dim E = n + 1$ and K is a field with q elements, show that the order of **PGL**(E) is

$$(q^{n+1} - 1)(q^{n+1} - q)\dots(q^{n+1} - q^{n-1})q^n$$

(use (*a*), Exercise 9(*a*) of § 9, and the fact that K is necessarily commutative (V, § 11, Exercise 14 and VIII, § 11, no. 1, Theorem 1)).*

(*g*) Show that if dim E = 2 and K is a field with 2 (resp. 3) elements, **PSL**(E) is isomorphic to the symmetric group $\mathfrak{S}_3$ (resp. the alternating group $\mathfrak{A}_4$). (Considering **PSL**(E) as a group of permutations of **P**(E), note that it contains every transposition (resp. every cycle $(a\,b\,c)$ and use *f*).)

(*h*) Show that, except in the cases considered in (*g*), every normal subgroup of **SL**(E) distinct from **SL**(E) is contained in the centre of **SL**(E) (use (*d*) and the fact that every transvection is a commutator in **SL**(E) (Exercise 13(*c*))).

¶ 15. Let A be a ring; for every integer $n \geqslant 1$, A^n is canonically identified with the submodule of $A^{(N)}$ consisting of the elements whose coordinates of index $\geqslant n$ are 0 and A'_n denotes the complementary submodule consisting of the elements whose coordinates of index $<n$ are 0. Every endomorphism $u \in \mathbf{GL}_n(A)$ is identified with the automorphism of $A^{(N)}$ whose restriction to A^n is u and whose restriction to A'_n is the identity, so that $\mathbf{GL}_n(A)$ is identified with a subgroup of $\mathbf{GL}(A^{(N)})$; let F denote the subgroup of $\mathbf{GL}(A^{(N)})$ the union of the $\mathbf{GL}_n(A)$, in other words the subgroup leaving invariant the elements of the canonical basis of $A^{(N)}$ except for a finite number of them (cf. Exercise 12).

Let T_n be the subgroup of $\mathbf{GL}_n(\mathrm{A})$ generated by the $\mathrm{B}_{ij}(\lambda)$ for $0 \leqslant i < n$, $0 \leqslant j < n$, $i \neq j$, $\lambda \in \mathrm{A}$ (no. 13) and let T be the subgroup of F the union of the T_n.

(*a*) Show that for $n \geqslant 3$, T_n is equal to its commutator group.

(*b*) Let a, b be two *invertible* elements of A. Show that

$$\begin{pmatrix} ab & 0 \\ 0 & 1 \end{pmatrix} = P \begin{pmatrix} a & 0 \\ 0 & b \end{pmatrix} Q \quad \text{and} \quad \begin{pmatrix} a & 0 \\ 0 & b \end{pmatrix} = P' \begin{pmatrix} b & 0 \\ 0 & a \end{pmatrix} Q'$$

where P, Q, P', Q' belong to T_2. Deduce that the matrix $\begin{pmatrix} aba^{-1}b^{-1} & 0 \\ 0 & 1 \end{pmatrix}$ belongs to T_2.

(*c*) Deduce from (*b*) that the commutator group of $\mathbf{GL}_n(\mathrm{A})$ is contained in T_{2n} (replace A by $\mathbf{M}_n(\mathrm{A})$ in (*b*)). Conclude (using (*a*)) that the commutator group of F is equal to T.

16. (*a*) Let U be a square matrix of order n over the field $\mathbf{Q}$ all of whose non-diagonal elements are equal to the same rational number $r > 0$ and whose diagonal $(d_1, \ldots, d_n)$ is such that $d_i \geqslant r$ for all i and $d_i > r$ except perhaps for one value of i. Show that U is invertible (if $x = (x_i)_{1 \leqslant i \leqslant n}$ is a solution of the equation $U.x = 0$, show that the x_i all have the same sign and deduce that they are all zero).

(*b*) Let E be a finite set with m elements denoted by a_i $(1 \leqslant i \leqslant m)$ and $(\mathrm{A}_j)_{1 \leqslant j \leqslant n}$ a family of n subsets of E, distinct from one another; suppose that $\mathrm{Card}(\mathrm{A}_i \cap \mathrm{A}_j) = r \geqslant 1$ for every ordered pair (i, j) of distinct elements of $(1, n)$. Show that necessarily $n \leqslant m$. (Let $V = (c_{ij})$ be the matrix of type (m, n) such that $c_{ij} = 1$ when $a_i \in \mathrm{A}_j$, $c_{ij} = 0$ otherwise. Apply (*a*) to the matrix ${}^tV.V$ of order n.)

¶ 17. Let K be a field (commutative or otherwise), E an n-dimensional right vector space over K and u an endomorphism of E.

(*a*) Show that if $n > 1$ and u is not a central homothety, there exists a basis of E such that the matrix of u with respect to this basis is of the form (α_{ij}) with $\alpha_{ij} = 0$ for $j > i + 1$ and $\sum_{i=1}^{n-1} \alpha_{i,i+1} = 1$ (argue by induction on n).

(*b*) Show that if $n > 2$, there exists a basis of E such that the matrix of u with respect to this basis is of the form (α_{ij}) with $\alpha_{ij} = 0$ for $j > i + 1$ and $\sum_{i=1}^{n-1} \alpha_{i,i+1} = 0$. (Argue by induction on n using (*a*), noting that it amounts to the same to prove the proposition for u or for $u + \gamma 1_{\mathrm{E}}$, where γ belongs to the centre of K, and studying separately the case where u is of rank 1 or 2.)

(*c*) Let $A = (\alpha_{ij})$ be a matrix with the properties described in (*b*). Also let S be the matrix of order n (ε_{ij}) such that $\varepsilon_{ij} = 0$ except when $i = j + 1$, in which case $\varepsilon_{j+1,j} = 1$. Show that there exists a matrix $B = (\beta_{ij})$ of order n such that $A = \lambda E_{nn} + (BS - SB)$ with $\lambda \in K$. (Take B such that $\beta_{i1} = 0$ for $1 \leqslant i \leqslant n$ and $\beta_{ij} = 0$ for $j > i + 2$.)

(*d*) Suppose that K is commutative. Deduce from (*c*) that every endomorphism u of E such that $\mathrm{Tr}(u) = 0$ can be written as $vw - wv$, where v and w are endomorphisms of E.

§ 11

1. Let Δ be a commutative monoid, written additively, with its identity element denoted by 0. Let $\mathbf{Z}^{(\Delta)}$ be the algebra of Δ over $\mathbf{Z}$ (III, § 2, no. 6), that is a commutative ring, which is a free $\mathbf{Z}$-module with a basis $(X^\sigma)_{\sigma\in\Delta}$ and the multiplication $(aX^\sigma)(bX^\tau) = abX^{\sigma+\tau}$. For every $\mathbf{Z}$-module, we write $E^{(\Delta)} = E \otimes_{\mathbf{Z}} \mathbf{Z}^{(\Delta)}$, every element of $E^{(\Delta)}$ hence being written uniquely in the form $\sum_{\sigma\in\Delta} z_\sigma \otimes X^\sigma$, with $z_\sigma \in E$, the family (z_σ) having finite support. For every $\mathbf{Z}$-module homomorphism $f: E \to E'$, let $f^{(\Delta)}$ denote the $\mathbf{Z}^{(\Delta)}$-module homomorphism $f \otimes 1: E^{(\Delta)} \to E'^{(\Delta)}$. Similarly, if Δ' is another additive monoid with its identity element denoted by 0 and if $\alpha: \Delta \to \Delta'$ is a monoid homomorphism, $\alpha(E)$ denotes the homomorphism $E^{(\Delta)} \to E^{(\Delta')}$ such that $\alpha(E)(z_\sigma \otimes X^\sigma) = z_\sigma \otimes X^{\alpha(\sigma)}$. $(E^{(\Delta)})^{(\Delta')}$ is identified with $E^{(\Delta\times\Delta')}$ by associating with $\sum_{\sigma'\in\Delta'}\left(\sum_{\sigma\in\Delta} z_{\sigma,\sigma'} \otimes X^\sigma\right) \otimes X^{\sigma'}$ the element $\sum_{\sigma,\sigma'} z_{\sigma,\sigma'} \otimes X^{(\sigma,\sigma')}$. Finally, let $\varepsilon(E)$ denote the homomorphism $E^{(\Delta)} \to E$ mapping $\sum_\sigma z_\sigma \otimes X^\sigma$ to $\sum_\sigma z_\sigma$.

(*a*) Let E be a $\mathbf{Z}$-module, $(E_\sigma)_{\sigma\in\Delta}$ a graduation on E of type Δ and, for all $\sigma \in \Delta$, let p_σ be the projector $E \to E_\sigma$ corresponding to the decomposition of E as the direct sum of the E_σ. For all $x \in E$, let $\phi_E(x) = \sum_{\sigma\in\Delta} p_\sigma(x) \otimes X^\sigma$. Show that the homomorphism $\phi_E: E \to E^{(\Delta)}$ thus defined has the following properties:

$$\varepsilon(E) \circ \phi_E = 1_E \tag{1}$$

$$\delta(E) \circ \phi_E = \phi_E^{(\Delta)} \circ \phi_E, \tag{2}$$

where $\delta: \Delta \to \Delta \times \Delta$ is the diagonal mapping.

Show that a *bijection* of the set of graduations on E of type Δ is thus defined on to the set of homomorphisms of E into $E^{(\Delta)}$ satisfying conditions (1) and (2).

(*b*) Let E, F be two graded $\mathbf{Z}$-modules of type Δ. For a homomorphism $f: E \to F$ to be graded of degree 0, it is necessary and sufficient that $\phi_F \circ f = f^{(\Delta)} \circ \phi_E$. For a submodule E′ of E to be graded, it is necessary and sufficient that $\phi_E(E') \subset E'^{(\Delta)}$.

(*c*) Let A be a ring; $A^{(\Delta)} = A \otimes_{\mathbf{Z}} \mathbf{Z}^{(\Delta)}$ is given a ring structure such that $(a \otimes X^\sigma)(b \otimes X^\tau) = (ab) \otimes X^{\sigma+\tau}$. Let $(A_\sigma)_{\sigma\in\Delta}$ be a graduation of the additive group A; for this graduation to be compatible with the ring structure on A and $1 \in A_0$, it is necessary and sufficient that $\phi_A : A \to A^{(\Delta)}$ be a ring homomorphism. State and prove an analogous result for graded A-modules.

APPENDIX

1. Let A be a pseudo-ring and M a left pseudomodule over A such that for all $z \neq 0$ in M, $z \in Az$. Show that if x, y are two elements of M such that $\mathrm{Ann}(x) \subset \mathrm{Ann}(y)$ in A, there exists an endomorphism u of M such that $u(x) = y$ (prove that the relation $bx = ax$ for a, b in A implies $by = ay$).

CHAPTER III

Tensor Algebras, Exterior Algebras, Symmetric Algebras

Recall the exponential notation introduced in Chapter I, *of which we shall make frequent use* (I, § 7, no. 8):

Let $(x_\lambda)_{\lambda \in L}$ *be a family of pairwise permutable elements of a ring* A; *for every mapping* $\alpha : L \to \mathbf{N}$ *of finite support we shall write*

$$x^\alpha = \prod_{\lambda \in L} x_\lambda^{\alpha(\lambda)}.$$

If β *is another mapping of* L *into* **N** *of finite support,* $\alpha + \beta$ *denotes the mapping*

$$\lambda \mapsto \alpha(\lambda) + \beta(\lambda)$$

of L *into* **N**; *with this law of composition the set* $\mathbf{N}^{(L)}$ *of mappings of* L *into* **N** *of finite support is the free commutative monoid derived from* L *and*

$$x^\alpha x^\beta = x^{\alpha + \beta}.$$

For all $\alpha \in \mathbf{N}^{(L)}$, *we write* $|\alpha| = \sum_{\lambda \in L} \alpha(\lambda) \in \mathbf{N}$; *then* $|\alpha + \beta| = |\alpha| + |\beta|$; $|\alpha|$ *is called the order of the "multiindex"* α. *For all* $\lambda \in L$, *let* δ_λ *denote the element of* $\mathbf{N}^{(L)}$ *such that* $\delta_\lambda(\lambda) = 1$, $\delta_\lambda(\mu) = 0$ *for* $\mu \neq \lambda$ *(Kronecker index); the* δ_λ *for* $\lambda \in L$ *are the only elements of* $\mathbf{N}^{(L)}$ *of order* 1. $\mathbf{N}^{(L)}$ *is given the ordering induced by the product ordering on* $\mathbf{N}^L$, *so that the relation* $\alpha \leqslant \beta$ *is equivalent to "*$\alpha(\lambda) \leqslant \beta(\lambda)$ *for all* $\lambda \in L$*"; then the multiindex* $\lambda \mapsto \beta(\lambda) - \alpha(\lambda)$ *is denoted by* $\beta - \alpha$, *so that it is the unique multiindex such that* $\alpha + (\beta - \alpha) = \beta$. *For all* $\alpha \in \mathbf{N}^{(L)}$, *there are only a finite number of multiindices* $\beta \leqslant \alpha$; *the* δ_λ *are the minimal elements of the set* $\mathbf{N}^{(L)} - \{0\}$; *the relation* $\alpha \leqslant \beta$ *implies* $|\alpha| \leqslant |\beta|$ *and if both* $\alpha \leqslant \beta$ *and* $|\alpha| = |\beta|$, *then* $\alpha = \beta$. *Finally, we write* $\alpha! = \prod_{\lambda \in L} (\alpha(\lambda))!$, *which is meaningful since* $0! = 1$.

From § 4 *to* § 8 *inclusive,* A *denotes a commutative ring and, unless otherwise stated, the algebras considered are assumed to be associative and unital and the algebra homomorphisms are assumed to be unital.*

§ 1. ALGEBRAS

1. DEFINITION OF AN ALGEBRA

DEFINITION 1. *Let* A *be a* commutative *ring. An algebra over* A *(or an* A*-algebra, or simply an algebra, when no confusion is to be feared) is a set* E *with a structure defined by giving the following:*

(1) *an* A*-module structure on* E*;*

(2) *an* A*-bilinear mapping* (II, § 3, no. 5) *of* E × E *into* E.

The A-bilinear mapping of E × E into E occurring in this definition is called the *multiplication* of the algebra E; it is usually denoted by $(x, y) \mapsto x.y$, or simply $(x, y) \mapsto xy$.

Let $(\alpha_i)_{i \in I}$ and $(\beta_j)_{j \in J}$ be two families of elements of A, *of finite support* (I, § 2, no. 1). Then, for all families $(x_i)_{i \in I}$ and $(y_j)_{j \in J}$ of elements of E, the general distributivity formula (I, § 3, no. 4)

$$(1) \qquad \Big(\sum_{i \in I} \alpha_i x_i\Big)\Big(\sum_{j \in J} \beta_j y_j\Big) = \sum_{(i, j) \in I \times J} (\alpha_i \beta_j)(x_i y_j)$$

holds; in particular

$$(2) \qquad (\alpha x) y = x(\alpha y) = \alpha(xy) \quad \text{for } \alpha \in A, x \in E \text{ and } y \in E.$$

The bilinear mapping $(x, y) \mapsto yx$ of E × E into E and the A-module structure on E define an A-algebra structure on E, called *opposite* to the given algebra structure. The set E with this new structure is called the *opposite algebra* to the algebra E; it is often denoted by E^0. The A-algebra E is called *commutative* if it is identical with its opposite, in other words if multiplication in E is commutative. An isomorphism of E onto E^0 is also called an *anti-automorphism* of the algebra E.

When multiplication in the algebra E is associative, E is called an *associative* A-algebra. When multiplication in E admits an identity element (necessarily unique (I, § 2, no. 1)), this element is called the *unit element* of E and E is called a *unital* algebra.

Examples. (1) Every commutative ring A can be considered as an (associative and commutative) A-algebra.

(2) Let E be a pseudo-ring (I, § 8, no. 1). Multiplication on E and the unique **Z**-module structure on E define on E an associative **Z**-algebra structure.

(3) Let F be a set and A a commutative ring. The set A^F of all mappings of F into A, with the product ring structure (I, § 8, no. 10) and the product

A-module structure (II, § 1, no. 5) is an associative and commutative A-algebra.

(4) Let E be an A-algebra; the internal laws $(x, y) \mapsto xy + yx$ and $(x, y) \mapsto xy - yx$ define (with the A-module structure on E) two A-algebra structures on E, which are not in general associative; the first law

$$(x, y) \mapsto xy + yx$$

is always commutative.

DEFINITION 2. *Given two algebras* E, E' *over a commutative ring* A, *a homomorphism of* E *into* E' *is a mapping* $f: E \to E'$ *such that*

(1) f *is an* A-*module homomorphism;*

(2) $f(xy) = f(x)f(y)$ *for all* $x \in E$ *and* $y \in E$.

Clearly the composition of two A-algebra homomorphisms is an A-algebra homomorphism. Every bijective algebra homomorphism is an isomorphism. Therefore A-algebra homomorphisms may be taken as *morphisms* of the species of A-algebra structure (*Set Theory*, IV, 2, no. 1). We shall always suppose in what follows that this choice of morphisms has been made. If E, E' are two A-algebras, let $\mathrm{Hom}_{A\text{-alg}}(E, E')$ denote the set of A-algebra homomorphisms of E into E'.

Let E, E' be two algebras each with a unit element. A homomorphism of E into E' mapping the unit element of E to the unit element of E' is called a *unital homomorphism* (or *unital algebra morphism*).

2. SUBALGEBRAS. IDEALS. QUOTIENT ALGEBRAS

Let A be a commutative ring and E an A-algebra. If F is a sub-A-module of W which is stable under the multiplication on E, the restriction to $F \times F$ of the multiplication of E defines (with the A-module structure on F) an A-algebra structure on F. F, with this structure, is called a *subalgebra* of the A-algebra E. Every intersection of subalgebras of E is a subalgebra of E. For every family $(x_\iota)_{\iota \in I}$ of elements of E, the intersection of the subalgebras of E containing all the x_ι is called the subalgebra of E *generated* by the family $(x_\iota)_{\iota \in I}$ and $(x_\iota)_{\iota \in I}$ is called a *generating system* (or *generating family*) of this subalgebra. If $u: E \to E'$ is an A-algebra homomorphism, the image $u(F)$ of every subalgebra F of E is a subalgebra of E'.

Let E be an *associative* algebra. For every subset M of E, the set M' of elements of E which are permutable with all the elements of M is a subalgebra of E called the *centralizer* subalgebra of M in E (I, § 1, no. 5). The centralizer M'' of M' in E is also called the *bicentralizer* of M; clearly $M \subset M''$. It follows that M' is contained in its bicentralizer M''', which is just the centralizer of M''; but the relation $M \subset M''$ implies $M''' \subset M'$, so that $M' = M'''$ (cf. *Set Theory*, III, § 1, no. 7, Proposition 2). If F is a subalgebra of E, the *centre* of F is the intersection $F \cap F'$ of F and its centralizer F' in E. Note that if

F is *commutative*, then $F \subset F'$ and hence $F' \supset F''$; the bicentralizer F'' of F is in this case the *centre* of F.

For certain non-associative algebras (for example Lie algebras) the notions of centralizer of a subalgebra and centre are defined differently (*Lie Groups and Lie algebras*, I, § 1, no. 6).

A subset $\mathfrak{a}$ of an A-algebra E is called a *left ideal* (resp. *right ideal*) of E when $\mathfrak{a}$ is a sub-A-module of E and the relations $x \in \mathfrak{a}$, $y \in E$ imply $yx \in \mathfrak{a}$ (resp. $xy \in \mathfrak{a}$). It amounts to the same to say that $\mathfrak{a}$ is a left ideal of E or a right ideal of the opposite algebra E^0. A *two-sided ideal* of E is a subset $\mathfrak{a}$ of E which is both a left ideal and a right ideal. When E is associative and admits a unit element e, then, for $\alpha \in A$ and $x \in E$, $\alpha x = (\alpha e)x = x(\alpha e)$ by virtue of (2) (no. 1) and hence the (right, left, two-sided) ideals of the *ring* E (I, § 8, no. 6) are identical with the (right, left, two-sided) ideals of the *algebra* E. Every sum and every intersection of left (resp. right, two-sided) ideals of the algebra E is a left (resp. right, two-sided) ideal. The intersection of the left (resp. right, resp. two-sided) ideals containing a subset X of E is called the left (resp. right, resp. two-sided) ideal of E *generated* by X.

Let $\mathfrak{b}$ be a *two-sided* ideal of an A-algebra E. If $x \equiv x' \pmod{\mathfrak{b}}$ and $y \equiv y' \pmod{\mathfrak{b}}$, then

$$x(y - y') \in \mathfrak{b} \quad \text{and} \quad (x - x')y' \in \mathfrak{b}$$

and hence $xy \equiv x'y' \pmod{\mathfrak{b}}$. Hence an internal law can be defined on the quotient A-module $E/\mathfrak{b}$, which is the quotient of the multiplication law $(x, y) \mapsto xy$ of E by the equivalence relation $x \equiv x' \pmod{\mathfrak{b}}$ (I, § 1, no. 6). It is immediately verified that this quotient law is an A-bilinear mapping of $(E/\mathfrak{b}) \times (E/\mathfrak{b})$ into $E/\mathfrak{b}$; it therefore defines with the A-module structure on $E/\mathfrak{b}$ an A-algebra structure on $E/\mathfrak{b}$. $E/\mathfrak{b}$, with this algebra structure, is called the *quotient algebra* of the algebra E by the two-sided ideal $\mathfrak{b}$. The canonical mapping $p: E \to E/\mathfrak{b}$ is an algebra homomorphism.

Let E, E′ be two A-algebras and $u: E \to E'$ an algebra homomorphism. The image $u(E)$ is a subalgebra of E′ and the kernel $\mathfrak{b} = \overset{-1}{u}(0)$ is a two-sided ideal of E; further, in the canonical decomposition of u:

$$E \xrightarrow{p} E/\mathfrak{b} \xrightarrow{v} u(E) \xrightarrow{j} E'$$

v is an *algebra isomorphism*. More generally, all the results of Chapter I, § 8, no. 9 are still valid (and also their proofs) when the word "ring" is replaced by "algebra".

Let A be a commutative ring and E an A-algebra. On the set $\tilde{E} = A \times E$, we define the following laws of composition:

$$(\lambda, x) + (\mu, y) = (\lambda + \mu, x + y)$$
$$(\lambda, x)(\mu, y) = (\lambda\mu, xy + \mu x + \lambda y)$$
$$\lambda(\mu, x) = (\lambda\mu, \lambda x).$$

It is immediately verified that $\tilde{E}$, with these laws of composition, is an *algebra over* A and $(1, 0)$ is a unit element of this algebra. The set $\{0\} \times E$ is a two-sided ideal of $\tilde{E}$ and $x \mapsto (0, x)$ is an isomorphism of the algebra E onto the subalgebra $\{0\} \times E$, by means of which E and $\{0\} \times E$ are identified. $\tilde{E}$ is called the *algebra derived from* E *by adjoining a unit element*; it is associative (resp. commutative) if and only if E is.

3. DIAGRAMS EXPRESSING ASSOCIATIVITY AND COMMUTATIVITY

Let A be a *commutative* ring and E an A-module; being given a bilinear mapping of $E \times E$ into E is equivalent to being given an A-*linear* mapping:

$$m: E \otimes_A E \to E$$

(II, § 3, no. 5). An A-algebra structure on E is therefore defined by giving an A-module structure on E and an A-linear mapping of $E \otimes_A E$ into E.

Let E′ be another A-algebra and $m': E' \otimes_A E' \to E'$ the A-linear mapping defining the multiplication of E′. A mapping $f: E \to E'$ is an A-algebra homomorphism if and only if f is a mapping rendering commutative the diagram

$$\begin{array}{ccc} E \otimes_A E & \xrightarrow{f \otimes f} & E' \otimes_A E' \\ \downarrow{\scriptstyle m} & & \downarrow{\scriptstyle m'} \\ E & \xrightarrow[f]{} & E' \end{array}$$

For an A-algebra E to be *associative*, it is necessary and sufficient (taking account of the associativity of tensor products, cf. II, § 3, no. 8) that the diagram of A-linear mappings

$$\begin{array}{ccc} E \otimes_A E \otimes_A E & \xrightarrow{m \otimes 1_E} & E \otimes_A E \\ \downarrow{\scriptstyle 1_E \otimes m} & & \downarrow{\scriptstyle m} \\ E \otimes_A E & \xrightarrow[m]{} & E \end{array}$$

be commutative. Similarly, for the A-algebra E to be *commutative*, it is necessary and sufficient that the diagram of A-linear mappings

$$\begin{array}{ccc} E \otimes_A E & \xrightarrow{\sigma} & E \otimes_A E \\ & {\scriptstyle m} \searrow \quad \swarrow {\scriptstyle m} & \\ & E & \end{array}$$

be commutative, where σ denotes the canonical A-linear mapping defined by

$\sigma(x \otimes y) = y \otimes x$ for $x \in \mathrm{E}$, $y \in \mathrm{E}$ (II, § 3, no. 1, Corollary 2 to Proposition 1).

For all $c \in \mathrm{E}$, let η_c denote the A-linear mapping of A into E defined by the condition $\eta_c(1) = c$. For c to be a *unit element* of E, it is necessary and sufficient that the two diagrams

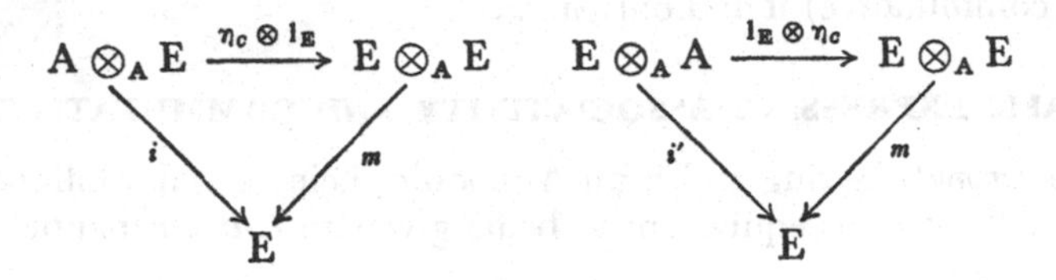

be commutative (i and i' denoting the canonical isomorphisms (II, § 3, no. 4, Proposition 4)).

Let E be an A-algebra with unit element e and let $\eta = \eta_e$ (also denoted by η_E); then $\eta(\alpha\beta) = \eta(\alpha)\eta(\beta) = \alpha\eta(\beta)$, for by (2) (no. 1),

$$(\alpha e)(\beta e) = (\alpha\beta)e = \alpha(\beta e);$$

hence η is an A-*algebra* homomorphism. Observe that the A-module structure on E can be defined using η, for

$$\alpha x = \eta(\alpha).x \quad \text{for } \alpha \in \mathrm{A},\ x \in \mathrm{E} \tag{3}$$

(where, on the right hand side, multiplication is in E). The image of the homomorphism η is a *subalgebra* of E whose elements commute with all those of E. The kernel of the homomorphism η is the *annihilator* of the element e of the A-module E; by (3), it is also the annihilator of the A-module E (II, § 1, no. 12).

When the algebra E is *unital* and *associative*, η is a ring homomorphism. Conversely, let $\rho : \mathrm{A} \to \mathrm{B}$ be a *ring* homomorphism such that the image $\rho(\mathrm{A})$ is *contained in the centre* of B, assuming also that the ring A is commutative; then an A-*algebra* structure is defined on B which is associative and unital, by writing (cf. (3))

$$\lambda x = \rho(\lambda).x \quad \text{for } \lambda \in \mathrm{A},\ x \in \mathrm{E}.$$

4. PRODUCTS OF ALGEBRAS

Let $(\mathrm{E}_\iota)_{\iota \in \mathrm{I}}$ be a family of algebras over the same commutative ring A. It is immediately verified that on the product set $\mathrm{E} = \prod_{\iota \in \mathrm{I}} \mathrm{E}_\iota$, the product A-module structure (II, § 1, no. 5) and the multiplication

$$((x_\iota), (y_\iota)) \mapsto (x_\iota y_\iota) \tag{4}$$

define an A-algebra structure; with this structure, the set E is called the *product algebra* of the family of algebras $(E_\iota)_{\iota \in I}$.

When all the algebras E_ι are associative (resp. commutative, resp. unital), so is their product. Moreover, all the properties stated in I, § 8, no. 10, extend without modification to arbitrary products of algebras.

5. RESTRICTION AND EXTENSION OF SCALARS

Let A_0 and A be two commutative rings and $\rho: A_0 \to A$ a ring homomorphism. If E is an A-algebra, we denote (conforming with II, § 1, no. 13) by $\rho_*(E)$ the A_0-module defined by addition on E and the external law

$$\lambda . x = \rho(\lambda) x \quad \text{for all } \lambda \in A_0 \text{ and all } x \in E.$$

Multiplication in E and the A_0-module structure on $\rho_*(E)$ define an A_0-*algebra* structure on $\rho_*(E)$. When A_0 is a subring of A and ρ the canonical injection, the algebra $\rho_*(E)$ is said to be obtained from E by *restricting* the ring A of scalars to A_0. By an abuse of language, this is also sometimes said when ρ is arbitrary.

Let F be an A_0-algebra. An A_0-algebra homomorphism $F \to \rho_*(E)$ is called a *semi-homomorphism* (relative to ρ) or a ρ-*homomorphism* of F into the A-algebra E; it is also called an A_0-homomorphism if no confusion arises. If E, E′ are two A-algebras, every A-algebra homomorphism $E \to E'$ is also an A_0-algebra homomorphism $\rho_*(E) \to \rho_*(E')$.

Consider now two commutative rings A and B and a ring homomorphism $\rho: A \to B$. For every A-module E, the B-module $\rho^*(E) = E \otimes_A B$, obtained from E by *extending* the ring A of scalars to B, has been defined (II, § 5, no. 1). If E is also an A-algebra, we shall define on $\rho^*(E)$ a B-*algebra* structure. For this, observe that $(E \otimes_A B) \otimes_B (E \otimes_A B)$ is canonically isomorphic to $(E \otimes_A E) \otimes_A B$ (II, § 5, no. 1, Proposition 3). If $m: E \otimes_A E \to E$ defines the multiplication on E, the mapping $m \otimes 1_B: (E \otimes_A E) \otimes_A B \to E \otimes_A B$ is therefore canonically identified with a B-linear mapping

$$m': \rho^*(E) \otimes_B \rho^*(E) \to \rho^*(E)$$

which defines the desired B-algebra structure on $\rho^*(E)$. Hence

$$(x \otimes \beta)(x' \otimes \beta') = (xx') \otimes (\beta\beta') \tag{5}$$

for x, x' in E, β and β' in B. The B-algebra $\rho^*(E)$ is said to be derived from the A-algebra E by *extending* the ring A of scalars to B (by means of ρ). It is also denoted by $E_{(B)}$ or $E \otimes_A B$. When E is associative (resp. commutative, resp. unital), so is $\rho^*(E)$.

PROPOSITION 1. *For every* A-*algebra* E, *the canonical mapping* $\phi_E: x \mapsto x \otimes 1$ *of* E *into* $E_{(B)}$ *is an* A-*homomorphism of algebras. Moreover, for every* B-*algebra* F *and every* A-*homomorphism* $f: E \to F$, *there exists one and only one* B-*homomorphism* $\bar{f}: E_{(B)} \to F$ *such that* $\bar{f}(x \otimes 1) = f(x)$ *for all* $x \in E$.

The first assertion follows immediately from the definition of multiplication in $E_{(B)}$, which gives $(x \otimes 1)(x' \otimes 1) = (xx') \otimes 1$ for $x \in E$ and $x' \in E$. The existence and uniqueness of the B-*linear* mapping $\bar{f}$ of $E_{(B)}$ into F satisfying the relation $\bar{f}(x \otimes 1) = f(x)$ for all $x \in E$ follow from II, § 5, no. 1, Proposition 1; here it all amounts to verifying that $\bar{f}(yy') = \bar{f}(y)\bar{f}(y')$ for y and y' in $E_{(B)}$; as the elements of the form $x \otimes 1$ (with $x \in E$) generate the B-module $E_{(B)}$, attention may be confined to the case where $y = x \otimes 1$, $y' = x' \otimes 1$ with $x \in E$, $x' \in E$; as $yy' = (xx') \otimes 1$, the relation $\bar{f}(yy') = \bar{f}(y)\bar{f}(y')$ then follows from $f(xx') = f(x)f(x')$.

It can also be said that $f \mapsto \bar{f}$ is a *canonical bijection*

$$\mathrm{Hom}_{A\text{-alg.}}(E, \rho_*(F)) \to \mathrm{Hom}_{B\text{-alg.}}(\rho^*(E), F). \tag{6}$$

The ordered pair consisting of $E_{(B)}$ and ϕ_E is therefore a solution of the *universal mapping problem* (*Set Theory*, IV, § 3, no. 1) where Σ is the species of B-algebra structure and the α-mappings the A-homomorphisms from E to a B-algebra.

Corollary. *Let* E, E′ *be two* A-*algebras; for every* A-*homomorphism of algebras* $u: E \to E'$, $u \otimes 1_B$ *is the unique* B-*homomorphism of algebras* $v: E \otimes_A B \to E' \otimes_A B$ *rendering commutative the diagram*

$$\begin{array}{ccc} E & \xrightarrow{\phi_E} & E \otimes_A B \\ {\scriptstyle u}\downarrow & & \downarrow{\scriptstyle v} \\ E' & \xrightarrow[\phi_{E'}]{} & E' \otimes_A B \end{array}$$

Let C be a third commutative ring and $\sigma: B \to C$ a ring homomorphism; it is immediate that the canonical C-homomorphism

$$\sigma^*(\rho^*(E)) \to (\sigma \circ \rho)^*(E)$$

mapping $(x \otimes 1) \otimes 1$ to $x \otimes 1$ for all $x \in E$ (II, § 5, no. 1, Proposition 2) is an *algebra* isomorphism.

6. INVERSE AND DIRECT LIMITS OF ALGEBRAS

Let I be a preordered set and (A_i, ϕ_{ij}) an inverse system of *commutative rings* with I as indexing set. Let (E_i, f_{ij}) be an inverse system of A_i-modules with I as indexing set (II, § 6, no. 1) and suppose further that each E_i has an A_i-algebra structure and that, for $i \leqslant j$, f_{ij} is an A_j-homomorphism of algebras (relative to ϕ_{ij}) (no. 5). Let $A = \varprojlim A_i$ and $E = \varprojlim E_i$, which has an A-module structure, the inverse limit of the structure of the A_i-modules E_i (II, § 6, no. 1); it is immediately verified that the law of composition on E, considered as the inverse limit of the E_i considered as magmas under multiplication (I, § 10, no. 1), with the A-module structure on E, defines on E an A-*algebra* structure; (E_i, f_{ij}) is called an *inverse system* of A_i-algebras and the A-algebra E is called its *inverse limit*. If $f_i: E \to E_i$, $\phi_i: A \to A_i$ are the canonical mappings, f_i is an A-*homomorphism* of algebras (relative to ϕ_i). If the E_i are associative (resp. commutative), so is E; if each E_i admits a unit element e_i and $f_{ij}(e_j) = e_i$ for $i \leqslant j$, $e = (e_i)$ is a unit element of the algebra E.

Let (E'_i, f'_{ij}) be another inverse system of A_i-algebras and for all i let $u_i: E_i \to E'_i$ be an A_i-algebra homomorphism, these mappings forming an *inverse system*; then $u = \varprojlim u_i$ is an A-*algebra homomorphism*.

Suppose now that all the A_i are equal to the *same* commutative *ring* A and the ϕ_{ij} to Id_A, so that $E = \varprojlim E_i$ is an A-algebra. Let F be an A-algebra and, for all $i \in I$, let $u_i: F \to E_i$ be an A-algebra homomorphism such that (u_i) is an inverse system of mappings; then $u = \lim u_i$ is a homomorphism of the algebra F into the algebra E. *Conversely*, for every A-algebra homomorphism $v: F \to E$, the family of $v_i = f_i \circ v$ is an inverse system of A-algebra homomorphisms such that $v = \varprojlim v_i$. As moreover, writing $\bar{f}_{ij} = \mathrm{Hom}(1_F, f_{ij})$, $(\mathrm{Hom}_{A\text{-alg.}}(F, E_i), \bar{f}_{ij})$ is clearly an inverse system of sets, it is seen that the above remarks can also be expressed by saying that the canonical mapping $v \mapsto (f_i \circ v)$ is a *bijection*

$$l_F: \mathrm{Hom}_{A\text{-alg.}}(F, \varprojlim E_i) \to \varprojlim \mathrm{Hom}_{A\text{-alg.}}(F, E_i).$$

Moreover, for every A-algebra homomorphism $w: F \to F'$, the

$$\bar{w}_i = \mathrm{Hom}(w, 1_{E_i}): \mathrm{Hom}_{A\text{-alg.}}(F', E_i) \to \mathrm{Hom}_{A\text{-alg.}}(F, E_i)$$

form an inverse system of mappings and the diagram

$$\begin{array}{ccc}
\mathrm{Hom}_{A\text{-alg.}}(F', \varprojlim E_i) & \xrightarrow{l_{F'}} & \varprojlim \mathrm{Hom}_{A\text{-alg.}}(F', E_i) \\
\Big\downarrow{\scriptstyle \mathrm{Hom}(w, 1_E)} & & \Big\downarrow{\scriptstyle \varprojlim \bar{w}_i} \\
\mathrm{Hom}_{A\text{-alg.}}(F, \varprojlim E_i) & \xrightarrow[l_F]{} & \varprojlim \mathrm{Hom}_{A\text{-alg.}}(F, E_i)
\end{array}$$

is commutative.

Suppose now that I is right *directed*. Consider a direct system of commutative rings (A_i, ϕ_{ji}) and a direct system (E_i, f_{ji}) of A_i-modules, with I as indexing set; suppose that each E_i has an A_i-algebra structure and that, for $i \leqslant j$, f_{ji} is an A_i-homomorphism of algebras (relative to ϕ_{ji}) (no. 5). Let $A = \varinjlim A_i$, $E = \varinjlim E_i$; E has an A-module structure, the direct limit of the structures of the A_i-modules E_i (II, § 6, no. 2); moreover, the law of composition on E considered as the direct limit of the E_ι, considered as magmas under multiplication (I, § 10, no. 3), with the A-module structure on E, defines an A-*algebra* structure on E; $(E_\iota, f_{j\iota})$ is called a *direct system* of A_ι-*algebras* and the A-algebra E is called its *direct limit*. If $f_\iota: E_\iota \to E$, $\phi_\iota: A_\iota \to A$ are the canonical mappings, f_ι is an A_ι-*homomorphism* of algebras (relative to ϕ_ι). If the E_ι are associative (resp. commutative), so is E; if each E_ι admits a unit element e_ι and $f_{j\iota}(e_\iota) = e_j$ for $i \leqslant j$, E admits a unit element e such that $f_\iota(e_\iota) = e$ for all $i \in I$.

Let $(E'_\iota, f'_{j\iota})$ be another direct system of A_ι-algebras and for all i let $u_\iota: E_\iota \to E'_\iota$ be an A_ι-algebra homomorphism, these mappings forming a *direct system*; then $u = \varinjlim u_\iota$ is an A-*algebra homomorphism*.

Suppose now that all the rings A_ι are equal to the *same ring* A and the $\phi_{j\iota}$ to Id_A, so that $E = \varinjlim E_\iota$ is an A-algebra. Let F be an A-algebra and for all i let $u_\iota: E_\iota \to F$ an A-algebra homomorphism such that (u_ι) is a direct system of mappings; then $u = \varinjlim u_\iota$ is a homomorphism of the algebra E into the algebra F. *Conversely*, for every A-algebra homomorphism $v: E \to F$, the family of $v_\iota = v \circ f_\iota$ is a direct system of A-algebra homomorphisms such that $v = \varinjlim v_\iota$. As moreover, writing $\bar{f}_{\iota j} = \mathrm{Hom}(f_{\iota j}, 1_F)$, $(\mathrm{Hom}_{A\text{-alg.}}(E_\iota, F), \bar{f}_{\iota j})$ is clearly an inverse system of sets, it is seen that the above remarks can also be expressed by saying that the canonical mapping $v \mapsto (v \circ f_\iota)$ is a *bijection*

$$d_F: \mathrm{Hom}_{A\text{-alg.}}(\varinjlim E_\iota, F) \to \varprojlim \mathrm{Hom}_{A\text{-alg.}}(E_\iota, F).$$

Further, for every A-algebra homomorphism $w: F \to F'$, the

$$\bar{w}_\iota = \mathrm{Hom}(1_{E_\iota}, w): \mathrm{Hom}_{A\text{-alg.}}(E_\iota, F) \to \mathrm{Hom}_{A\text{-alg.}}(E_\iota, F')$$

form an inverse system of mappings and the diagram

$$\begin{array}{ccc}
\mathrm{Hom}_{A\text{-alg.}}(\varinjlim E_\iota, F) & \xrightarrow{d_F} & \varprojlim \mathrm{Hom}_{A\text{-alg.}}(E_\iota, F) \\
\Big\downarrow{\scriptstyle \mathrm{Hom}(1_E, w)} & & \Big\downarrow{\scriptstyle \varprojlim \bar{w}_\iota} \\
\mathrm{Hom}_{A\text{-alg.}}(\varinjlim E_\iota, F') & \xrightarrow[d_{F'}]{} & \varprojlim \mathrm{Hom}_{A\text{-alg.}}(E_\iota, F')
\end{array}$$

is commutative.

7. BASES OF AN ALGEBRA. MULTIPLICATION TABLE

By definition, a *basis* of an A-algebra E is a basis of E for its A-module structure. Let $(a_\iota)_{\iota \in I}$ be a basis of E; there exists a unique family $(\gamma_{ij}^k)_{(i,j,k) \in I \times I \times I}$ of elements of the ring A such that for every ordered pair $(i, j) \in I \times I$, the set of $k \in I$ such that $\gamma_{ij}^k \neq 0$ is *finite* and

$$(7) \qquad a_i a_j = \sum_{k \in I} \gamma_{ij}^k a_k.$$

The γ_{ij}^k are called the *constants of structure* of the algebra E with respect to the basis (a_i) and relations (7) constitute the *multiplication table* of the algebra E (relative to the basis (a_i)).

Relations (7) can be imagined written down by setting out the right hand sides of these relations in a square table

		...	a_j	...
⋮				
a_i			$\sum_k \gamma_{ij}^k a_k$	
⋮				

it being understood that the element appearing in the row of index i and the column of index j is equal to the product $a_i a_j$.

Conversely, given an A-*module* E, a basis $(a_i)_{i \in I}$ of E and a family (γ_{ij}^k) of elements of A such that for every ordered pair $(i, j) \in I \times I$ the set of $k \in I$ such that $\gamma_{ij}^k \neq 0$ is finite, then there is on E one and only one A-algebra structure under which relations (7) hold, since the A-module $E \otimes_A E$ is free and admits as basis $(a_i \otimes a_j)_{(i,j) \in I \times I}$ (cf. II, § 3, no. 7, Corollary 2 to Proposition 7).

Let E be an A-algebra and $(a_i)_{i \in I}$ a generating system of the A-module E (for example a basis). For E to be *associative*, it is necessary and sufficient that the a_i satisfy the *associativity relations*

$$(8) \qquad (a_i a_j) a_k = a_i (a_j a_k) \quad \text{for all } i, j, k$$

The mapping $(x, y, z) \mapsto (xy)z - x(yz)$ is an A-trilinear mapping

$$E \times E \times E \to E$$

and hence defines an A-linear mapping $E \otimes_A E \otimes_A E \to E$; if the latter mapping is zero for all the elements $a_i \otimes a_j \otimes a_k$, which form a generating system of the A-module $E \otimes_A E \otimes_A E$, it is identically zero.

Similarly, for E to be *commutative*, it is necessary and sufficient that the a_i satisfy the *commutativity relations*

$$(9) \qquad a_i a_j = a_j a_i \quad \text{for all } i, j.$$

The proof is analogous, this time considering the A-bilinear mapping $(x, y) \mapsto xy - yx$. Finally, for an element $e \in E$ to be a unit element, it is necessary and sufficient that the a_i satisfy the relations

$$(10) \qquad a_i = e a_i = a_i e \quad \text{for all } i,$$

as is seen this time by considering the A-linear mappings $x \mapsto x - xe$ and $x \mapsto x - ex$.

When $(a_i)_{i \in I}$ is a basis of E and (γ_{ij}^k) the corresponding family of constants of structure, relations (8) are equivalent to the relations

$$\sum_r \gamma_{ij}^r \gamma_{rk}^s = \sum_r \gamma_{ir}^s \gamma_{jk}^r$$

for all i, j, k, s. Similarly relations (9) are equivalent to $\gamma_{ij}^k = \gamma_{ji}^k$ for all i, j, k.

Let $(a_i)_{i \in I}$ be a basis of the A-algebra E; if $\rho : A \to B$ is a ring homomorphism, $(a_i \otimes 1)$ is a basis of the B-algebra $\rho^*(E) = E \otimes_A B$ (II, § 5, no. 1, Proposition 4). If (γ_{ij}^k) is the family of constants of structure relative to the basis (a_i), the family $(\rho(\gamma_{ij}^k))$ is the family of constants of structure of $\rho^*(E)$ relative to the basis $(a_i \otimes 1)$.

§ 2. EXAMPLES OF ALGEBRAS

Throughout this paragraph, A *denotes a commutative ring.*

1. ENDOMORPHISM ALGEBRAS

Let B be an associative A-algebra with a unit element denoted by 1 and let M be a right B-module. We know that the ring $E = \text{End}_B(M)$ also has a module structure over the centre of B. Now the image of the homomorphism $h : \alpha \mapsto \alpha . 1$ of A into B is contained in the centre of B (§ 1, no. 1); hence h gives E an A-*module* structure. Further, for $\alpha \in A$ and f, g in E,

$$\alpha(f \circ g) = f \circ (\alpha g) = (\alpha f) \circ g;$$

hence multiplication in E and the A-module structure on E define an *associative* A-*algebra* structure on E; the identity mapping of M is a unit element of this algebra.

2. MATRIX ELEMENTS

Let B be a unital associative A-algebra and $\mathbf{M}_n(B)$ the set of *square matrice of order n* over B (II, § 10, no. 7). Then $\mathbf{M}_n(B)$ has an A-module structure defined by $\alpha . (b_{ij}) = (\alpha b_{ij})$ ($\alpha \in A$, $b_{ij} \in B$, $1 \leqslant i \leqslant n$, $1 \leqslant j \leqslant n$); this structure and matrix multiplication define a *unital associative* A-*algebra* structure on $\mathbf{M}_n(B)$. The canonical bijection of $\mathbf{M}_n(B)$ onto $\text{End}_B(B_d^n)$ (II, § 10, no. 7) is an A-algebra isomorphism.

When B = A, the A-algebra $\mathbf{M}_n(A)$ admits a *canonical basis* (E_{ij}) consisting of the matrix units (II, § 10, no. 3); the corresponding multiplication table is

(1) $$E_{ij}E_{hk} = \delta_{jh}E_{ik}.$$

The unit element I_n is equal to $\sum_{i=1}^{n} E_{ii}$.

3. QUADRATIC ALGEBRAS

Let α, β be two elements of A and (e_1, e_2) the canonical basis of A^2. The *quadratic algebra of type* (α, β) *over* A is the A-module A^2 with the algebra structure defined by the multiplication table (§ 1, no. 7)

(2) $$e_1^2 = e_1, \quad e_1e_2 = e_2e_1 = e_2, \quad e_2^2 = \alpha e_1 + \beta e_2.$$

An A-algebra E isomorphic to a quadratic algebra is also called a *quadratic algebra.* It amounts to the same to say that E admits a basis of two elements one of which is the unit element.

It can be shown that every unital A-algebra which admits a basis of two elements is a quadratic algebra (Exercise 1).

If a basis (e_1, e_2) of an A-algebra has multiplication table (2), it is called a *basis of type* (α, β). By an abuse of language, a quadratic algebra is said to be *of type* (α, β) when it has a basis of type (α, β).

PROPOSITION 1. *A quadratic algebra* E *is associative and commutative.*

The fact that E is commutative follows from the equation $e_1e_2 = e_2e_1$ in (2); similarly, to verify associativity, it suffices to see that $x(yz) = (xy)z$ when x, y, z are each equal to e_1 or e_2. Now, this relation is obvious if at least one of the elements x, y, z is equal to e_1; it is also true for $x = y = z = e_2$ since E is commutative; whence the proposition.

Let e denote the unit element of a quadratic algebra E and let (e, i) be a basis of E of type (α, β); every other basis of E containing e is therefore of the form (e, j) with $j = \gamma e + \delta i$ (II, § 7, no. 2, Corollary to Proposition 3); moreover, for (e, j) to be a basis of E, it is necessary and sufficient that δ be *invertible* in A; the condition is obviously sufficient; conversely, if $\bar{i}$ is the canonical image of i in E/Ae, $\bar{i}$ and $\bar{j} = \delta\bar{i}$ must each form a basis of E/Ae, whence the necessity of the condition. Then

$$j^2 = (\gamma^2 + \alpha\delta^2)e + (2\gamma\delta + \beta\delta^2)i = (\alpha\delta^2 - \gamma^2 - \beta\gamma\delta)e + (2\gamma + \beta\delta)j;$$

thus it is seen that E is of type

(3) $$(\alpha\delta^2 - \gamma^2 - \beta\gamma\delta, 2\gamma + \beta\delta)$$

for all invertible $\delta \in A$ and all $\gamma \in A$. In particular, if E is of type $(\alpha, 2\beta')$, i is also of type $(\alpha + \beta'^2, 0)$ as is seen by taking $\gamma = -\beta'$ and $\delta = 1$.

PROPOSITION 2. *Let E be a quadratic A-algebra and e its unit element. For all $u \in E$, let $T(u)$ be the trace of the endomorphism $m_u : x \mapsto ux$ of the free A-module E (II, § 4, no. 3). Then the mapping s defined by $s(u) = T(u).e - u$ is an automorphism of the algebra E and $s^2(u) = u$ for all $u \in E$.*

Let (e, i) be a basis of E of type (α, β); then $T(e) = 2$, when $s(e) = e$, and $T(i) = \beta$, whence $s(i) = \beta e - i$. Hence $(e, s(i))$ is a basis of E, whose type is given by (3) with $\gamma = \beta$ and $\delta = -1$, which again gives (α, β); it follows that s is an automorphism of the algebra E. As $m_{s(u)} = sm_u s^{-1}$, the endomorphisms m_u and $m_{s(u)}$ of the A-module E have the same trace (II, § 4, no. 3, Proposition 3), whence

$$s^2(u) = T(u).e - s(u) = T(u).e - (T(u).e - u) = u$$

for all $u \in E$.

The automorphism s is called *conjugation* of the A-algebra E and $s(u)$ the *conjugate* of u.

If $u = \xi e + \eta i$, with ξ, η in A, then $s(u) = (\xi + \beta\eta)e - \eta i$, whence

$$T(u)e = u + s(u) = (2\xi + \beta\eta)e \tag{4}$$

$$u.s(u) = (\xi^2 + \beta\xi\eta - \alpha\eta^2)e = N(u)e \tag{5}$$

where we have written $N(u) = \xi^2 + \beta\xi\eta - \alpha\eta^2$. The elements $T(u)$ and $N(u)$ (or, when A and Ae are canonically identified, the elements $T(u)e$ and $N(u)e$) are called respectively the *trace* and *norm* of u.

When $\beta = 0$, the above formulae are simplified to

$$s(\xi e + \eta i) = \xi e - \eta i, \quad T(\xi e + \eta i) = 2\xi, \quad N(\xi e + \eta i) = \xi^2 - \alpha\eta^2. \tag{6}$$

Clearly T is a *linear form* on E *and N is a *quadratic form* on E (IX, § 3, no. 4)*. As E is commutative and associative, it follows from (5) that

$$N(uv) = N(u))N(v). \tag{7}$$

For u to be *invertible in* E, it is necessary and sufficient that $N(u)$ be *invertible in* A. For, as $N(e) = 1$, the necessity of the condition follows from (7), writing $v = u^{-1}$. Conversely, if $N(u)$ is invertible in A, it follows from (5) that u is invertible and that

$$u^{-1} = (N(u))^{-1}s(u). \tag{8}$$

*It can be proved that $N(u)$ is the determinant (§ 8, no. 1) of the endomorphism m_u (cf. § 9 no. 3, *Example* 1).*

The following proposition gives the structure of quadratic algebras over a commutative *field*:

PROPOSITION 3. *Let* E *be a quadratic* A*-algebra of type* (α, β).

(i) *If* A *is a field and contains no element* ζ *such that* $\zeta^2 = \alpha + \beta\zeta$, E *is a (commutative) field* (cf. V, § 3).

(ii) *If the ring* A *contains an element* ζ *such that* $\zeta^2 = \alpha + \beta\zeta$ *and* $\beta - 2\zeta$ *is invertible* (resp. *zero*), E *is isomorphic to* A × A (resp. *is of type* (0, 0)).

We prove (i). Let ξ, η be two elements of A and $u = \xi e + \eta i$. If $\eta \neq 0$ and we write $\theta = -\xi\eta^{-1}$, then $N(u) = \eta^2(\theta^2 - \beta\theta - \alpha)$ by (5), whence $N(u) \neq 0$ by virtue of the hypothesis on A; if $\eta = 0$, then $N(u) = \xi^2$. In any case, if $u \neq 0$, then $N(u) \neq 0$, hence $N(u)$ is invertible in A and therefore u is invertible in E.

We now prove (ii). The canonical basis (e_1, e_2) of the algebra A × A is of type (0, 1). We have seen (formula (3)) that E is of type

$$(\alpha\delta^2 - \gamma^2 - \beta\gamma\delta, 2\gamma + \beta\delta)$$

for all $\gamma \in A$ and all δ invertible in A. If $\beta - 2\zeta$ is invertible, take $\delta = (\beta - 2\zeta)^{-1}$ and $\gamma = -\zeta(\beta - 2\zeta)^{-1}$; then $2\gamma + \beta\delta = 1$ and

$$\alpha\delta^2 - \gamma^2 - \beta\gamma\delta = \delta^2(\alpha - \zeta^2 + \beta\zeta) = 0;$$

thus E is of type (0, 1) and hence isomorphic to A × A. If $\beta - 2\zeta = 0$, it has already been remarked that E is of type $(\alpha + \zeta^2, 0)$ and hence of type (0, 0) since $\alpha + \zeta^2 = 2\zeta^2 - \beta\zeta = 0$.

A quadratic A-algebra of type (0, 0) is also called an *algebra of dual numbers* over A.

4. CAYLEY ALGEBRAS

DEFINITION 1. *A Cayley algebra over* A *is an ordered pair* (E, *s*), *where* E *is an algebra over* A *with a unit element* e *and* s *is an antiautomorphism of* E *such that*

$$u + s(u) \in Ae \quad \text{and} \quad u.s(u) \in Ae$$

for all $u \in E$.

s is called the *conjugation* of the Cayley algebra (E, s) and $s(u)$ the *conjugate* of u. The condition $u + s(u) \in Ae$ implies that u and $s(u)$ are *permutable*. We write

$$T(u) = u + s(u) \tag{9}$$

$$N(u) = u.s(u) = s(u).u \tag{10}$$

and these elements of the subalgebra Ae are called respectively the *Cayley trace* and *norm* of u.

The ordered pair consisting of a quadratic algebra E and its conjugation s (which is an antiautomorphism since E is commutative) (no. 3) is a Cayley algebra.

Let (E, s) be a Cayley algebra; as $s(e) = e$, $s(u + s(u)) = u + s(u)$, in other words $s(u) + s^2(u) = u + s(u)$ or also

$$s^2(u) = u \tag{11}$$

so that s^2 is the identity mapping of E. It follows that

$$T(s(u)) = T(u), \qquad N(s(u)) = N(u). \tag{12}$$

Finally, the relation $(u - u)(u - s(u)) = 0$ gives

$$u^2 - T(u).u + N(u) = 0 \tag{13}$$

for all $u \in E$.

PROPOSITION 4. *Let* E *be an* A-*algebra and* s *and* s' *antiautomorphisms of* E *such that* (E, s) *and* (E, s') *are Cayley algebras. If* E *admits a basis containing the unit element* e, *then* $s' = s$.

Clearly $s'(u) = s(u) = u$ for all $u \in Ae$. If T, N (resp. T', N') are the trace and norm functions for (E, s) (resp. (E, s')), it follows from (13) that

$$(T(u) - T'(u)).u - (N(u) - N'(u)) = 0.$$

Let B be a basis of E containing e and u an element of B distinct from e; then $T(u) - T'(u) = 0$, whence $s(u) = s'(u)$. As s and s' coincide on B, they are equal.

In what follows, we shall write $\bar{u} = s(u)$, so that

$$\begin{cases} u + \bar{u} = T(u), & u\bar{u} = \bar{u}u = N(u), \quad \bar{\bar{u}} = u \\ \overline{u + v} = \bar{u} + \bar{v}, & \overline{\alpha u} = \alpha\bar{u}, \quad \overline{uv} = \bar{v}.\bar{u} \end{cases} \tag{14}$$

for u, v in E, $\alpha \in A$; moreover

$$T(e) = 2e, \qquad N(e) = e. \tag{15}$$

From the formula

$$T(uv) = uv + \overline{uv} = uv + \bar{v}.\bar{u} = uv + (T(v) - v)(T(u) - u),$$

we deduce that

$$uv + vu = T(u)v + T(v)u + (T(uv) - T(u)T(v)) \tag{16}$$

whence, exchanging u and v,

$$T(vu) = T(uv). \tag{17}$$

On the other hand, $N(u+v) = (u+v)(\bar{u}+\bar{v}) = N(u) + N(v) + T(u\bar{v})$, whence

$$(18) \qquad T(v\bar{u}) = T(u\bar{v}) = N(u+v) - N(u) - N(v).$$

Now, (16) applied with u replaced by $\bar{u}$ gives

$$T(u\bar{v}) = T(u)T(v) + \bar{u}v + v\bar{u} - T(u)v - T(v)\bar{u} = T(u)T(v) - uv - \bar{v}.\bar{u};$$

whence

$$(19) \quad T(v\bar{u}) = T(u\bar{v}) = N(u+v) - N(u) - N(v) = T(u)T(v) - T(uv).$$

Finally, clearly for all $\alpha \in A$,

$$(20) \qquad N(\alpha u) = \alpha^2 N(u);$$

in particular $N(2u) = 4N(u)$, so that formula (19) gives

$$(21) \qquad (T(u))^2 - T(u^2) = 2N(u).$$

Clearly T is a linear form on the (Ae)-module E. As $(u, v) \mapsto T(v\bar{u})$ is a bilinear form on this module, *it follows from (18) and (20) that N is a quadratic form (cf. IX, § 3, no. 4).*

5. CONSTRUCTION OF CAYLEY ALGEBRAS. QUATERNIONS

Let (E, s) be a Cayley algebra over A, for which we shall use the notation of no. 4, and let $\gamma \in A$. Let F be the algebra over A whose underlying module is $E \times E$ and whose multiplication is defined by

$$(22) \qquad (x, y)(x', y') = (xx' + \gamma\bar{y}'y, y\bar{x}' + y'x);$$

clearly $(e, 0)$ is unit element of F and $E \times \{0\}$ is a subalgebra of F isomorphic to E; we shall *identify* it with E in what follows, so that $x \in E$ is identified with $(x, 0)$ and in particular e is identified with the unit element of F.

Let t be the permutation of F defined by

$$(23) \qquad t((x, y)) = (\bar{x}, -y) \qquad (x \in E, y \in E).$$

PROPOSITION 5. (i) *The ordered pair* (F, t) *is a Cayley algebra over* A.

(ii) *Let* $j = (0, e)$, *so that* $(x, y) = xe + yj$ *for* $x \in E$, $y \in E$. *The Cayley trace and norm* T_F *and* N_F *of* F *are given by the formulae*

$$(24) \qquad T_F(xe + yj) = T(x), \qquad N_F(xe + yj) = N(x) - \gamma N(y).$$

(iii) *For* F *to be associative, it is necessary and sufficient that* E *be associative and commutative.*

For $(x, y) \in F$,

$$(x, y) + t((x, y)) = (x + \bar{x}, 0) = T(x)e \tag{25}$$

$$\begin{aligned}(x, y)t((x, y)) &= (x, y)(\bar{x}, -y) = (x\bar{x} - \gamma\bar{y}y, y\bar{\bar{x}} - yx) \\ &= (N(x) - \gamma N(y), 0) = (N(x) - \gamma N(y))e.\end{aligned} \tag{26}$$

To prove both (i) and (ii), it therefore suffices to show that t is an antiautomorphism of F. Clearly t is an A-linear bijection. On the other hand,

$$\begin{aligned}t((x, y) . (x', y')) &= t((xx' + \gamma\bar{y}'y, y\bar{x}' + y'x)) = (\bar{x}'\bar{x} + \gamma\bar{y}y', -y\bar{x} - y'x) \\ &= (\bar{x}', -y')(\bar{x}, -y) = t((x', y'))t((x, y))\end{aligned}$$

and hence t is an antiautomorphism.

It remains to prove (iii). As E is identified with a subalgebra of F, E may be assumed to be associative. Let $u = (x, y)$, $u' = (x', y')$, $u'' = (x'', y'')$ be elements of F. Then

$$\begin{cases} (uu')u'' = ((xx' + \gamma\bar{y}'y)x'' + \gamma\bar{y}''(\bar{y}x' + y'x), \\ \qquad\qquad (y\bar{x}' + y'x)\bar{x}'' + y''(xx' + \gamma\bar{y}'y)) \\ u(u'u'') = (x(x'x'' + \gamma\bar{y}''y') + (\gamma(x''\bar{y}' + \bar{x}'\bar{y}'')y, \\ \qquad\qquad y(\bar{x}''\bar{x}' + \gamma\bar{y}'y'') + (y'\bar{x}'' + y''x')x). \end{cases} \tag{27}$$

Examining these formulae shows that the commutativity of E implies the associativity of F. Conversely, if F is associative, formulae (27) applied with $y = y' = 0$, $x'' = 0$ and $y'' = e$ give $(0, x'x) = (0, xx')$, that is $x'x = xx'$ for all x, x' in E; thus E is then commutative.

Note also that, in the above notation, for x, y in E,

$$yj = j\bar{y}, \qquad x(yj) = (yx)j, \qquad (xj)y = (x\bar{y})j, \qquad (xj)(yj) = \bar{y}xe \tag{28}$$

$$j^2 = e. \tag{29}$$

The Cayley algebra (F, t) is called the *Cayley extension of* (E, s) *defined by* γ.

Examples. (1) If we take $E = A$ (and hence $s = 1_A$), the algebra F is a *quadratic* A-*algebra* with basis (e, j) where $j^2 = \gamma e$.

(2) Take E to be a *quadratic algebra of type* (α, β), so that the underlying module of E is A^2, with multiplication table (2) (no. 3) for the canonical basis. Take s to be conjugation of E (no. 3, Proposition 2). Then, for all $\gamma \in A$, the Cayley extension F of (E, s) defined by γ is called the *quaternion algebra of type* (α, β, γ), which is *associative* by no. 3, Proposition 1 and Proposition 5 above; its underlying module is A^4 and, if (e, i, j, k) denotes the canonical basis of A^4, the corresponding multiplication table is given by

$$\begin{cases} i^2 = \alpha e + \beta i, & ij = k, & ik = \alpha j + \beta k, \\ ji = \beta j - k, & j^2 = \gamma e, & jk = \beta\gamma e - \gamma i, \\ ki = -\alpha j, & kj = \gamma i, & k^2 = -\alpha\gamma e. \end{cases} \tag{30}$$

Further, for $u = \rho e + \xi i + \eta j + \eta k$ (with ρ, ξ, η, ζ in A), we have (writing $\bar{u}$ instead of $t(u)$ and identifying A with Ae):

$$(31) \qquad \begin{cases} u = (\rho + \beta\xi)e - \xi i - \eta j - \zeta k \\ T_F(u) = 2\rho + \beta\xi \\ N_F(u) = \rho^2 + \beta\rho\xi - \alpha\xi^2 - \gamma(\eta^2 + \beta\eta\zeta - \alpha\zeta^2). \end{cases}$$

Formulae (30) follow from (28) and (29) and formulae (31) from (23) and (24), taking account of the formulae for the quadratic algebra E.

Then, for u, v in F,

$$(32) \qquad N_F(uv) = N_F(u)N_F(v)$$

for $N_F(uv) = uv.\overline{uv} = uv(\bar{v}.\bar{u}) = u(v\bar{v})\bar{u} = (u\bar{u})(v\bar{v})$ by virtue of the associativity and the fact that $N_F(u)$ belongs to the centre of F.

An A-algebra isomorphic to a quaternion algebra is also called a *quaternion algebra*; if a basis of such an algebra has multiplication table (30), it is called a *basis of type* (α, β, γ). By an abuse of language, a quaternion algebra is said to be *of type* (α, β, γ) when it has a basis of type (α, β, γ).

When $\beta = 0$, formulae (30) and (31) simplify to

$$(33) \qquad \begin{cases} i^2 = \alpha e, & ij = k, & ik = \alpha j, \\ ji = -k, & j^2 = \gamma e, & jk = -\gamma i, \\ ki = -\alpha i, & kj = \gamma i, & k^2 = -\alpha\gamma e, \end{cases}$$

and

$$(34) \qquad \begin{cases} \bar{u} = \rho e - \xi i - \eta j - \zeta k \\ T_F(u) = 2\rho \\ N_F(u) = \rho^2 - \alpha\xi^2 - \gamma\eta^2 + \alpha\gamma\zeta^2. \end{cases}$$

Then (α, β, γ) is replaced throughout by (α, γ) in the above expressions. It is immediate that the quaternion algebras of types (α, γ) and (γ, α) are *isomorphic*.

Note that formulae (32) show that F is not commutative when $-1 \neq 1$ in A.

*Taking A to be the field **R** of real numbers and $\alpha = \gamma = -1$, $\beta = 0$, the corresponding algebra F is called the *algebra of Hamiltonian quaternions* and is denoted by **H**. If $u = \rho e + \xi i + \eta j + \zeta k$ (ρ, ξ, η, ζ in **R**) is an element $\neq 0$ in **H**, the formula $u\bar{u} = \bar{u}u = \rho^2 + \xi^2 + \eta^2 + \zeta^2$ (cf. (34)) shows that $N(u) \neq 0$ in **R**, so that u admits an *inverse* $u^{-1} = N(u)^{-1}\bar{u}$ in **H** and that **H** is therefore a *non-commutative field*.

(3) If E is taken to be a quaternion algebra (cf. *Example* 2), the Cayley extension of E defined by an element $\delta \in A$ is in general non-associative

(Proposition 5); it is called an *octonion algebra* over A (cf. Appendix, no. 3).

6. ALGEBRA OF A MAGMA, A MONOID, A GROUP

Recall that a *magma* is a set with a law of composition (I, § 1, no. 1). Let S be a magma written multiplicatively and let $E = A^{(S)}$ be the A-module of formal linear combinations of elements of S (II, § 1, no. 11); we know that a canonical mapping $s \mapsto e_s$ is defined of S into $A^{(S)}$ such that the family $(e_s)_{s \in S}$ is a basis (called *canonical*) of $A^{(S)}$, every element of $A^{(S)}$ being then written uniquely in the form $\sum_{s \in S} \alpha_s e_s$, where (α_s) is a family of elements of A of finite support. Then an A-*algebra* structure is defined on E by taking as multiplication table of the canonical basis

$$e_s e_t = e_{st}. \tag{35}$$

The algebra E thus defined is called the *algebra of the magma* S over A. If $x = \sum_{s \in S} \xi_s e_s$ and $y = \sum_{s \in S} \eta_s e_s$ are two elements of E, then

$$xy = \sum_{s \in S} \Big(\sum_{tu=s} \xi_t \eta_u \Big) e_s.$$

When S is a monoid (resp. group), E is called the *algebra of the monoid* (resp. *group*) S over A; it is then an *associative* algebra (§ 1, no. 7); similarly, when S is a commutative monoid, its algebra is *associative and commutative*. Finally if the magma S admits an identity element u, e_u is unit element of the algebra E; as the element e_u is free, A is then identified with the subalgebra Ae_u of E.

When $A \neq \{0\}$, S is sometimes identified with its image under the injection $s \mapsto e_s$, so that an element of E is written as $\sum_{s \in S} \alpha_s s$; but this identification is not possible (without causing confusion) when S is written additively. Then e^s is also often written instead of e_s.

Let B be another commutative ring and $\rho: A \to B$ a ring homomorphism; consider the algebras $E = A^{(S)}$ and $E' = B^{(S)}$ of the same magma S over A and B and let $(e_s)_{s \in S}$ and $(e'_s)_{s \in S}$ be their respective canonical bases. The algebra $B^{(S)}$ is canonically identified, under the A-linear mapping j such that $j(e_s \otimes 1) = e'_s$ for all $s \in S$, with the algebra $A^{(S)} \otimes_A B$ obtained from $A^{(S)}$ by extending the ring of scalars to B (II, § 1, no. 11, Corollary 3 to Proposition 17).

PROPOSITION 6. *Let* S *be a magma,* F *an* A-*algebra and* f *a homomorphism of*

S *into* F *with only its multiplicative structure. Then there exists one and only one* A-*algebra homomorphism* $\bar{f}:A^{(S)} \to F$ *rendering commutative the diagram*

$$\begin{array}{ccc} S & \xrightarrow{f} & F \\ \downarrow & & \downarrow 1_F \\ A^{(S)} & \xrightarrow[\bar{f}]{} & F \end{array} \tag{36}$$

(*where the vertical arrow on the left is the canonical mapping* $s \mapsto e_s$).

Let $\bar{f}:A^{(S)} \to F$ be the unique A-module homomorphism such that $\bar{f}(e_s) = \bar{f}(s)$ (II, § 1, no. 11, Corollary 3 to Proposition 17); it suffices to verify that $\bar{f}$ is an algebra homomorphism and for this it suffices to prove that $\bar{f}(e_s e_t) = \bar{f}(e_s)\bar{f}(e_t)$, which follows immediately from the definition and the hypothesis $f(st) = f(s)f(t)$.

Proposition 6 expresses the fact that the ordered pair consisting of $A^{(S)}$ and the canonical mapping $s \mapsto e_s$ is a solution of the universal mapping problem (*Set Theory*, IV, § 3, no. 1) where Σ is the species of A-algebra structure and the α-mappings the homomorphisms of S into an A-algebra with only its multiplicative law.

Corollary. *Let* S, S′ *be two magmas and* $g:S \to S'$ *a homomorphism. Then there exists one and only one* A-*algebra homomorphism* $u:A^{(S)} \to A^{(S')}$ *rendering commutative the diagram*

$$\begin{array}{ccc} S & \xrightarrow{g} & S' \\ \downarrow & & \downarrow \\ A^{(S)} & \xrightarrow[u]{} & A^{(S')} \end{array}$$

(*where the vertical arrows are the canonical mappings*).

It suffices to apply Proposition 6 taking f to be the composite mapping $S \xrightarrow{g} S' \to A^{(S')}$.

In particular, if T is a *stable subset* of the magma S (I, § 1, no. 4), the set of elements $\sum_{s \in T} \alpha_s e_s$ of $A^{(S)}$ is a *subalgebra* of $A^{(S)}$ canonically isomorphic to the algebra $A^{(T)}$ and sometimes identified with the latter.

Example. Let V be an A-module and S a monoid which *operates* on V *on the left*; this means (I, § 5, no. 1) that there is given a mapping $(s, x) \mapsto s.x$ of S into V such that $s.(x + y) = s.x + s.y$, $s.(\alpha x) = \alpha(s.x)$ and $s.(t.x) = (st).x$ for s, t in S, x, y in V and $\alpha \in A$ and, denoting by e the identity element of S,

$e.x = x$ for $x \in V$. Writing $f(s)(x) = s.x$, f is a *homomorphism* of S into the algebra $\mathrm{End}_A(V)$ (with only its multiplicative law), mapping the identity element e to the unit element 1_V. Applying Proposition 6, an A-algebra homomorphism $\bar{f}: A^{(S)} \to \mathrm{End}_A(V)$ is obtained, which gives the underlying group of V a *left module* structure over $A^{(S)}$.

This allows us to reduce the study of commutative groups with operators to that of modules. For let M be a commutative group with operators written additively, all of whose external laws are written multiplicatively. Let Ω be the sum set (*Set Theory*, II, § 4, no. 8) of the domains of operators of the various external laws on M, each of these domains being canonically identified with a subset of Ω. Let $\mathrm{Mo}(\Omega)$ be the *free monoid* (I § 7, no. 2) constructed on Ω; a law of action

$$(s, x) \mapsto s.x$$

is defined on M with $\mathrm{Mo}(\Omega)$ as domain of operators, by induction on the length of the *word* s in $\mathrm{Mo}(\Omega)$; if s is of length 0, it is the empty word e and we write $e.x = x$ for all $x \in M$. If x is of length $n \geqslant 1$, it can be written uniquely as $s = tu$, where u is of length $n - 1$ and t of length 1, so that $t \in \Omega$; we then write $s.x = t.(u.x)$. For any two words s, s' in $\mathrm{Mo}(\Omega)$, the relation $s.(s'.x) = (ss').x$ is verified by induction on the length of s.

Then applying the method described above, a left $\mathbf{Z}^{(\mathrm{Mo}(\Omega))}$-module structure is obtained on M and it is verified without difficulty that the usual notions in the theory of groups with operators (stable subgroups, homomorphisms) are the same for commutative groups with operators and the modules thus associated with them.

7. FREE ALGEBRAS

DEFINITION 2. *Let* I *be a set; let* M(I) (resp. Mo(I), resp. $\mathbf{N}^{(I)}$) *denote the free magma* (resp. *free monoid*, resp. *free commutative monoid*) *derived from* I. *The algebra of* M(I) (resp. Mo(I), resp. $\mathbf{N}^{(I)}$) *over* A *is called the free algebra* (resp. *free associative algebra*, resp. *free commutative associative algebra* (or, by an abuse of language, *free commutative algebra*)) *of the set* I *over the ring* A.

We shall denote the free algebra (resp. free associative algebra, resp. free commutative algebra) of I over A by $\mathrm{Lib}_A(I)$ (resp. $\mathrm{Libas}_A(I)$, resp. $\mathrm{Libasc}_A(I)$). By composing the canonical mapping of I into M(I) (resp. Mo(I), resp. $\mathbf{N}^{(I)}$) with the canonical mapping of M(I) (resp. Mo(I), resp. $\mathbf{N}^{(I)}$) into $\mathrm{Lib}_A(I)$ (resp. $\mathrm{Libas}_A(I)$, resp. $\mathrm{Libasc}_A(I)$), a canonical mapping is obtained of I into $\mathrm{Lib}_A(I)$ (resp. $\mathrm{Libas}_A(I)$, resp. $\mathrm{Libasc}_A(I)$), which is injective if $A \neq \{0\}$. We shall denote the image of an element $i \in I$ under this canonical mapping by X_i and we shall say that X_i is the *indeterminate of index* i of $\mathrm{Lib}_A(I)$ (resp. $\mathrm{Libas}_A(I)$, resp. $\mathrm{Libasc}_A(I)$).

As Mo(I) and $\mathbf{N}^{(I)}$ each have an identity element, $\mathrm{Libas}_A(I)$ and $\mathrm{Libasc}_A(I)$

are unital associative algebras and further $\text{Libasc}_A(I)$ is commutative. If e is the unit element of $\text{Libas}_A(I)$ (resp. $\text{Libasc}_A(I)$), the mapping $\alpha \mapsto \alpha e$ is an isomorphism of A onto a subring of the centre of $\text{Libas}_A(I)$ (resp. $\text{Libasc}_A(I)$), which is identified with A (no. 1).

PROPOSITION 7. *Let* I *be a set and* F *an algebra* (resp. *unital associative algebra*, resp. *unital commutative associative algebra*) *over* A. *For every mapping* $f: I \rightarrow F$, *there exists one and only one homomorphism* (resp. *unital homomorphism*) $\bar{f}$ *of* $\text{Lib}_A(I)$ (resp. $\text{Libas}_A(I)$, resp. $\text{Libasc}_A(I)$) *into* F *such that* $\bar{f}(X_i) = f(i)$ *for all* $i \in I$.

Let F_m be the magma (resp. monoid) obtained by giving the set F its multiplicative law of composition. There is one and only one homomorphism (resp. unital homomorphism) g of M(I) (resp. Mo(I), resp. $\mathbf{N}^{(I)}$) into F_m such that $g(i) = f(i)$ for all $i \in I$ (I, § 7, nos. 1, 2 and 7); Proposition 7 then follows from no. 6, Proposition 6.

Remarks. (1) We shall later define an isomorphism of $\text{Libas}_A(I)$ onto the tensor algebra of the free module $A^{(I)}$ (§ 5, no. 5) and also an isomorphism of $\text{Libasc}_A(I)$ onto the symmetric algebra of $A^{(I)}$ (§ 6, no. 6).

(2) Let ρ be a unital homomorphism of A into a commutative ring B. As has been seen (§ 2, no. 6), an isomorphism σ is derived from ρ of $\text{Lib}_B(I)$ (resp. $\text{Libas}_B(I)$, resp. $\text{Libasc}_B(I)$) onto the algebra $(\text{Lib}_A(I))_{(B)}$ (resp. $(\text{Libas}_A(I))_{(B)}$, resp. $(\text{Libasc}_A(I))_{(B)}$) obtained by extending the scalars to B by means of ρ; if X_i^A, X_i^B are the indeterminates of index i corresponding respectively to A and B, then $\sigma(X_i^B) = X_i^A \otimes 1$.

(3) Let J be a subset of I; we know that M(J) is identified with a stable subset of the magma M(I) and hence (no. 6) $\text{Lib}_A(J)$ is canonically identified with a subalgebra of $\text{Lib}_A(I)$, generated by the X_i such that $i \in J$; it is said that only the indeterminates of indices belonging to J occur in an element of $\text{Lib}_A(J)$. The definition given in no. 6 of the algebra of a magma shows that $\text{Lib}_A(I)$ is the union of the directed family of subalgebras $\text{Lib}_A(J)$ when J runs through the set of *finite* subsets of I. There are analogous results for $\text{Libas}_A(I)$ and $\text{Libasc}_A(I)$.

(4) With element s of M(I) (resp. Mo(I), resp. $\mathbf{N}^{(I)}$) is associated its *length* $l(s)$, which is an integer $\geqslant 1$ (resp. $\geqslant 0$) such that $l(ss') = l(s) + l(s')$ (I, § 7, nos. 1, 2 and 7). If e_s is the element of $\text{Lib}_A(I)$ (resp. $\text{Libas}_A(I)$, resp. $\text{Libasc}_A(I)$) corresponding to s, the *total degree* (or simply the *degree* of an element $x = \sum_s \alpha_s e_s \neq 0$ of $\text{Lib}_A(I)$ (resp. $\text{Libas}_A(A)$, resp. $\text{Libasc}_A(I)$) is the greatest of the numbers $l(s)$ when s runs through the (non-empty by hypothesis) set of elements such that $\alpha_s \neq 0$. For example, if i, j, k are three distinct elements of I, the element $(X_i(X_jX_k))X_i - (X_iX_j)(X_kX_i)$ is an element $\neq 0$ of total degree 4 in $\text{Lib}_A(I)$.

8. DEFINITION OF AN ALGEBRA BY GENERATORS AND RELATIONS

Let F be an algebra over A and $(x_i)_{i \in I}$ a family of elements of F. By no. 7, Proposition 7, there exists a unique homomorphism $f: \mathrm{Lib}_A(I) \to F$ such that $f(X_i) = x_i$ for all $i \in I$. For f to be surjective, it is necessary and sufficient that $(x_i)_{i \in I}$ be a generating system of F.

If $U \in \mathrm{Lib}_A(I)$, $f(U)$ is called the *element of* F *derived from* U *by substituting the elements* x_i *for the indeterminates* X_i, or also the *value* of U for the values x_i of the indeterminates X_i; it is usually denoted by $U((x_i)_{i \in I})$; in particular $U((X_i)_{i \in I}) = U$. If λ is a homomorphism of F into an algebra F′ over A, then

$$\lambda(U((x_i)_{i \in I})) = U((\lambda(x_i))_{i \in I}).$$

Consider in particular the case where $F = \mathrm{Lib}_A(J)$, where J is another set; for every family $(H_i)_{i \in I}$ of elements of $\mathrm{Lib}_A(J)$ and every family $(y'_j)_{j \in J}$ of elements of an A-algebra F′,

$$U((H_i)_{i \in I})((y'_j)_{j \in J}) = U((H_i((y'_j)_{j \in J}))_{i \in I}). \tag{37}$$

In the above notation, every element U of $\mathrm{Lib}_A(I)$ such that $U((x_i)_{i \in I}) = 0$, or also such that $f(U) = 0$, is called a *relator* of the family $(x_i)_{i \in I}$ in F. The two-sided ideal $\mathrm{Ker}(f)$ consisting of these elements is called the *ideal of relators* of (x_i).

Let $(R_j)_{j \in J}$ be a family of elements of $\mathrm{Lib}_A(I)$. $((x_i)_{i \in I}, (R_j)_{j \in J})$ is called a *presentation* of the algebra F if $(x_i)_{i \in I}$ is a generating system of F and the two-sided ideal of $\mathrm{Lib}_A(I)$ generated by the R_j is equal to the ideal of relators of the family $(x_i)_{i \in I}$; the x_i are called the *generators* and the R_j the *relators* of the presentation.

Consider now any set I and a family $(R_j)_{j \in J}$ of elements of $\mathrm{Lib}_A(I)$. The quotient algebra E of $\mathrm{Lib}_A(I)$ by the two-sided ideal generated by the family (R_j) is called the *universal algebra defined by the generating system* I *related by the family of relators* $(R_j)_{j \in J}$. Clearly, if $\overline{X}_i$ denotes the image of X_i in E,

$$((\overline{X}_i)_{i \in I}, (R_j)_{j \in J})$$

is a *presentation* of E. Moreover, if $(x_i)_{i \in I}$ is a family of elements of an algebra F and $R_j((x_i)_{i \in I}) = 0$ for all $j \in J$, there exists a unique homomorphism $g: E \to F$ such that $g(\overline{X}_i) = x_i$ for all $i \in I$; for $((x_i)_{i \in I}, (R_j)_{j \in J})$ to be a presentation of F, it is necessary and sufficient that g be *bijective*.

These remarks justify the following abuse of language: instead of saying "$((x_i)_{i \in I}, (R_j)_{j \in J})$ is a presentation of F", it is also said that "F is the algebra generated by the generators x_i subject to the relations $R_j((x_i)_{i \in I}) = 0$". When the R_j are of the form $P_j - Q_j$, it is also said that "F is the algebra generated by the x_i subject to the relations $P_j((x_i)) = Q_j((x_i))$".

Let H be a set; we shall say that an element S of $\mathrm{Lib}_A(H)$ is a *universal relator* for an A-algebra F is $S((x_h)_{h \in H}) = 0$ for *every* family $(x_h)_{h \in H}$ of elements of F with H as indexing set.

Examples. (1) Take $H = \{1, 2, 3\}$; the algebras which admit

$$(X_1X_2)X_3 - X_1(X_2X_3)$$

as universal relator are the associative algebras. The algebras which admit $X_1X_2 - X_2X_1$ as universal relator are the commutative algebras. *The algebras which admit the universal relators X_1X_1 and

$$(X_1X_2)X_3 + (X_2X_3)X_1 + (X_3X_1)X_2$$

as universal relator are the Lie algebras.*

Let I be a set; let a family $(S_k)_{k \in K}$ of elements of $\text{Lib}_A(H)$ be given and consider the set T of elements of $\text{Lib}_A(I)$ of the form $S_k((U_h))_{h \in H})$, where k runs through K and, for each k, $(U_h)_{h \in H}$ runs through the set of families of elements of $\text{Lib}_A(I)$ with H as indexing set; consider a family $(R_j)_{j \in J}$ with T as set of elements. Then let F be the universal algebra defined by the generating system I related by the family $(R_j)_{j \in J}$ and let $u: \text{Lib}_A(I) \to F$ be the canonical homomorphism, so that $\text{Ker}(u)$ is generated by the elements $S_k((U_h)_{h \in H})$ for all $k \in K$ and all families $(U_h)_{h \in H}$ of elements of $\text{Lib}_A(I)$; clearly each of the S_k $(k \in K)$ is a *universal relator* for F. Now let F′ be an algebra admitting a generating system $(x_i)_{i \in I}$, for which each of the S_k is a universal relator, and let $u': \text{Lib}_A(I) \to F'$ be the homomorphism such that $u'(X_i) = x_i$ for all $i \in I$; clearly $\text{Ker}(u) \subset \text{Ker}(u')$ and hence u' can be written uniquely in the form $u' = h \circ u$, where $h: F \to F'$ is a homomorphism such that $h(\overline{X}_i) = x_i$ for all $i \in I$. For this reason F is called *the universal algebra defined by the generating system* I, *corresponding to the family of universal relators* $(S_k)_{k \in K}$. By an abuse of language, F is sometimes called the universal algebra generated by I and subject to the *identities* $S_k((u_h)) = 0$ for every family $(u_h)_{h \in H}$ of elements of F.

Example 2. Let L′ be the universal algebra generated by I and subject to the identities $(uv)w - u(vw) = 0$ for every family of three elements of L′ and let L″ be the algebra obtained by adjoining a unit element to L′; then there exists a unique unital isomorphism g of L″ onto $\text{Libas}_A(I)$ such that $g(\overline{X}_i) = X_i$ for all $i \in I$. For clearly L″ is associative and the existence of the homomorphism g follows from the definition of L′ and the remarks preceding it; but then clearly L″ satisfies the universal property (no. 7, Proposition 7) which characterizes $\text{Libas}_A(I)$, whence the conclusion.

Considerations analogous to the above can be applied to associative algebras (resp. commutative associative algebras), taking account of the following remarks. When the context gives sufficient indication that the algebras considered are unital algebras, by an abuse of language, a family of elements $(x_\iota)_{\iota \in I}$ of an algebra F, such that the subalgebra generated by the x_ι ($\iota \in I$) *and the unit element* is equal to F, is often called a *generating system* of F. Then let F be a unital associative algebra over A and let $(x_\iota)_{\iota \in I}$ be a family of elements of F; by virtue of no. 7, Proposition 7, there exists a unique *unital* homomorphism $f: \mathrm{Libas}_A(I) \to F$ such that $f(X_\iota) = x_\iota$ for all $\iota \in I$; if $U \in \mathrm{Libas}_A(I)$, $f(U)$ is also called *the element of* F *derived from* U *by substituting the elements* x_ι *for the indeterminates* X_ι and it is also denoted by $U((x_\iota)_{\iota \in I})$. Then the notions of *relator, presentation* and *universal relator* go over immediately to associative algebras; it suffices simply to replace $\mathrm{Lib}_A(I)$ throughout by $\mathrm{Libas}_A(I)$. *The universal unital associative algebra defined by the generating system* I *related by the family of relators* $(R_j)_{j \in J}$ is the quotient algebra of $\mathrm{Libas}_A(I)$ by the two-sided ideal generated by the family (R_j). The universal unital associative algebra generated by the generating system I, corresponding to a family of universal relators is defined similarly. We leave to the reader the task of stating the analogous definitions relative to commutative associative algebras with $\mathrm{Libasc}_A(I)$ substituted for $\mathrm{Libas}_A(I)$.

Example 3. Let L′ be the universal unital associative algebra generated by I and subject to the identities $uv - vu = 0$ for every family of two elements of L′. It is seen, as in *Example* 2, that L′ is canonically isomorphic to $\mathrm{Libasc}_A(I)$.

9. POLYNOMIAL ALGEBRAS

Let B be a unital *commutative* associative A-algebra and let $(x_\iota)_{\iota \in I}$ be a family of elements of B; the subalgebra of B generated by the x_ι ($\iota \in I$) and the unit element is denoted by $A[(x_\iota)_{\iota \in I}]_B$ or simply $A[(x_\iota)_{\iota \in I}]$ when no confusion can arise. For every set I, the algebra $\mathrm{Libasc}_A(I)$ is therefore equal to $A[(X_\iota)_{\iota \in I}]$ (also denoted by $A[X_\iota]_{\iota \in I}$); the latter notation, which has the advantage of indicating the notation chosen to denote the indeterminates, is the one we shall generally use in the rest of this Treatise. The elements of $A[(X_\iota)_{\iota \in I}]$ are called *polynomials in the indeterminates* X_ι ($\iota \in I$) *with coefficients in* A; it is a convention that, when it is said "let $A[(X_\iota)_{\iota \in I}]$ be a polynomial algebra", the X_ι are always understood to be the indeterminates. For every subset J of I, the use of the above notation amounts to identifying $\mathrm{Libasc}_A(J)$ with the algebra of $\mathrm{Libasc}_A(I)$ generated by the X_ι of index $\iota \in J$ and the unit element (cf. no. 7, *Remark* (3)). For $I = \{1, 2, \dots, n\}$, we write $A[X_1, X_2, \dots, X_n]$ instead of $A[(X_\iota)_{\iota \in I}]$.

If I and I′ are two equipotent sets, the algebras $\mathrm{Libasc}_A(I)$ and $\mathrm{Libasc}_A(I')$ are isomorphic. A[X] is often used to denote the polynomial algebra corresponding to an unspecified indexing set with *a single* element, X denoting the unique indeterminate; similarly, A[X, Y], A[X, Y, Z], ... are used to denote the polynomial algebras corresponding to unspecified indexing sets with

2, 3, ... elements. Note that, by virtue of the conventions made above, A[X] and A[Y] are for example (distinct) subalgebras of A[X, Y, Z] if $A \neq \{0\}$.

The elements

$$X^{\nu} = \prod_{\iota \in I} X_{\iota}^{\nu(\iota)},$$

where ν runs through $\mathbf{N}^{(I)}$ form a basis of the polynomial algebra $A[(X_{\iota})_{\iota \in I}]$. These elements are called *monomials* in the indeterminates X_{ι} and the number $|\nu| = \sum_{\iota \in I} \nu(\iota)$ is called the *degree* (or *total degree*) of the monomial X^{ν}. The unique monomial of degree 0 is the unit element of $A[(X_{\iota})_{\iota \in I}]$; it is often identified with the unit element 1 of A. Every polynomial u of $A[(X_{\iota})_{\iota \in I}]$ can be written uniquely as

$$u = \sum_{\nu \in \mathbf{N}^{(I)}} \alpha_{\nu} X^{\nu}$$

with $\alpha_{\nu} \in A$; the elements α_{ν}, zero except for a finite number of indices $\nu \in \mathbf{N}^{(I)}$, are called the *coefficients* of u; the elements $\alpha_{\nu}X^{\nu}$ are called the *terms* of u (the element $\alpha_{\nu}X^{\nu}$ often being called the "term in X^{ν}"); in particular the term $\alpha_0 X^0$ (identified with $\alpha_0 \in A$) is called the *constant term* of u. If J is a subset of I, u belongs to $A[(X_{\iota})_{\iota \in J}]$ if and only if $\alpha_{\nu} = 0$ for $\nu \notin \mathbf{N}^{(J)}$. It follows that $A[(X_{\iota})_{\iota \in I}]$ is the union of the subalgebras $A[(X_{\iota})_{\iota \in J}]$, where J runs through the set of finite subsets of I. If $\alpha_{\nu} = 0$ for $|\nu| > n$, u is said to be a *polynomial of degree* $\leqslant n$. When $\alpha_{\nu} = 0$, (by an abuse of language) u is said to *contain no term in* X^{ν}; in particular, when $\alpha_0 = 0$, u is said to be a polynomial *with no constant term.*

For every *non-zero* polynomial $u = \sum_{\nu} \alpha_{\nu} X^{\nu}$, the *degree* (or *total degree*) of u is the greatest of the integers $|\nu|$ of the multiindices ν such that $\alpha_{\nu} \neq 0$.

Let F be a unital associative A-algebra and let $(x_{\iota})_{\iota \in I}$ be a family of elements of F, which are *pairwise permutable.* The subalgebra F′ of F generated by the x_{ι} and the unit element is commutative (§ 1, no. 7), which allows us to define substituting the x_{ι} for the X_{ι} in the polynomial $u \in A[(X_{\iota})_{\iota \in I}]$ (although F is not necessarily commutative): $u((x_{\iota})_{\iota \in I})$ is an element of F′ and therefore of F and $h: u \to u((x_{\iota})_{\iota \in I})$ is a homomorphism of $A[(X_{\iota})_{\iota \in I}]$ into F. The elements of the kernel of h are the relators of the family (x_{ι}) in $A[(X_{\iota})_{\iota \in I}]$, also called *polynomial relators* (with coefficients in A) between the x_{ι}. The image of the homomorphism h is the subalgebra F′, also denoted by $A[(x_{\iota})_{\iota \in I}]$ (even when F is not commutative); if $\mathfrak{a}$ is the ideal of polynomial relators between the x_{ι}, then there is an exact sequence of A-modules

$$0 \longrightarrow \mathfrak{a} \longrightarrow A[(X_{\iota})_{\iota \in I}] \xrightarrow{h} A[(x_{\iota})_{\iota \in I}] \longrightarrow 0.$$

PROPOSITION 8. *Let* $A[(X_{\iota})_{\iota \in I}]$ *be a polynomial algebra,* J *a subset of* I *and* K *the complement of* J *in* I. *Writing* $A' = A[(X_j)_{j \in J}]$ *and denoting by* X'_k $(k \in K)$ *the indeterminates in the polynomial algebra* $\mathrm{Libasc}_{A'}(K) = A'[(X'_k)_{k \in K}]$, *there exists a unique ring isomorphism of* $A'[(X'_k)_{k\in K}]$ *onto* $A[(X_i)_{i\in I}]$ *which coincides with the identity on* A' *and maps* X'_k *to* X_k *for all* $k \in K$.

Clearly $A[(X_i)_{i\in I}]$ is an A'-algebra generated by the X_k for $k \in K$. On the other hand, as a polynomial relator between the X_k $(k \in K)$ with coefficients in A' can be written uniquely as $\sum_{\nu} h_\nu((X_j)_{j\in J})X^\nu$ where ν runs through a finite subset of $\mathbf{N}^{(K)}$ and where the h_ν are elements of $A[(X_j)_{j\in J}]$, the h_ν must be polynomial relators between the X_j with coefficients in A and hence are all zero, which proves the proposition.

The isomorphism described in Proposition 8 is often used to identify the elements of $A[(X_i)_{i\in I}]$ with polynomials with coefficients in $A' = A[(X_j)_{j\in J}]$. If u is an element $\neq 0$ of $A[(X_i)_{i\in I}]$, its total degree considered as an element of $A'[(X_k)_{k\in K}]$ is also called its *degree with respect to the* X_i *of index* $i \in K$.

Remark. Let I and J be two sets and $(P_j)_{j\in J}$ a family of elements of $\mathbf{Z}[(X_i)_{i\in I}]$; if Q is an element of $\mathbf{Z}[(X_j)_{j\in J}]$ such that $Q((P_j)_{j\in J}) = 0$, then, for every family $(b_i)_{i\in I}$ of pairwise permutable elements of a ring B,

$$Q((P_j((b_i)_{i\in I}))_{j\in J}) = 0.$$

The relations which hold of the form $Q((P_j)_{j\in J}) = 0$ are sometimes called *polynomial identities*. For example

$$(X_1 + X_2)^2 - X_1^2 - X_2^2 - 2X_1X_2 = 0$$

with

$$Q = Y_1^2 - Y_2, \qquad P_1 = X_1 + X_2, \qquad P_2 = X_1^2 + X_2^2 + 2X_1X_2$$

$$X_1^n - X_2^n - (X_1 - X_2)(X_1^{n-1} + X_1^{n-2}X_2 + \cdots + X_2^{n-1}) = 0$$

with

$$Q = Y_1 - Y_2Y_3, \qquad P_1 = X_1^n - X_2^n, \qquad P_2 = X_1 - X_2,$$
$$P_3 = X_1^{n-1} + X_1^{n-2}X_2 + \cdots + X_2^{n-1}$$

are polynomial identities.

10. TOTAL ALGEBRA OF A MONOID

The algebra of a monoid S over A is (as an A-module) the submodule of the product A^S consisting of the families $(\alpha_s)_{s\in S}$ of finite support; the multiplication in this algebra is defined by the relations $(\alpha_s)(\beta_s) = (\gamma_s)$, where, for all $s \in S$,

$$\gamma_s = \sum_{tu=s} \alpha_t\beta_u \tag{38}$$

(cf. no. 6, formula (35)). The sum on the right hand side of (38) is meaningful because (α_s) and (β_s) are families of finite support and so therefore is the double family $(\alpha_t\beta_u)_{(t,u)\in S\times S}$. But the right hand side of (38) is also meaningful for *arbitrary* elements (α_s), $\beta_s)$ of A^S when the monoid S satisfies the following condition:

(D) *For all* $s \in S$, *there exists only a finite number of ordered pairs* (t, u) *in* $S \times S$ *such that* $tu = s$.

Suppose then that S satisfies condition (D); a multiplication law can then be defined on the product A-module A^S by formula (38). It is immediate that the multiplication thus defined on A^S is A-bilinear; also it is *associative*, since, for α, β, γ in A^S,

$$\sum_{uvw=t} \alpha_u\beta_v\gamma_w = \sum_{rw=t}\Big(\Big(\sum_{uv=r} \alpha_u\beta_v\Big)\gamma_w\Big) = \sum_{us=t}\Big(\alpha_u\Big(\sum_{vw=s} \beta_v\gamma_w\Big)\Big).$$

This multiplication and the A-module structure on A^S therefore define on A^S a unital *associative algebra* structure over A; we shall say that the set A^S, with this structure, is the *total algebra* of the monoid S over A.

It is immediate that the *algebra* $A^{(S)}$ of the monoid S over A (also called the *restricted algebra* of S when necessary to avoid confusion) is a *subalgebra* of the total algebra of S over A (and is identical with the latter when S is finite). *By an abuse of language*, every element $(\xi_s)_{s\in S}$ of the total algebra of S over A is also denoted by the same notation $\sum_{s\in S} \xi_s e_s$ $\Big($or even $\sum_{s\in S} \xi_s . s\Big)$ as the elements of the restricted algebra of S; of course the summation symbol appearing in this notation corresponds to no algebraic operation because it is taken over an *infinity* of terms $\neq 0$ in general. With this notation, multiplication in the total algebra of S is also given by formula (35) of no. 6.

If S is commutative, so is its total algebra A^S. If T is a submonoid of S, the total algebra A^T of the monoid is canonically identified with a subalgebra of the total algebra of S. If $\rho: A \to B$ is a ring homomorphism, the canonical extension $\rho^S: A^S \to B^S$ is an A-homomorphism of the total algebra of S over A into the total algebra of S over B, which extends the canonical homomorphism $A^{(S)} \to B^{(S)}$.

11. FORMAL POWER SERIES OVER A COMMUTATIVE RING

For every set I, the additive monoid $\mathbf{N}^{(I)}$ satisfies condition (D) of no. 10; for, if $s = (n_\iota)_{\iota\in I}$ with $n_\iota = 0$ except for the indices ι in a finite subset H of I, the relation $s = t + u$ with $t = (p_\iota)_{\iota\in I}$ and $u = (q_\iota)_{\iota\in I}$ is equivalent to $p_\iota + q_\iota = n_\iota$ for all ι; but this implies $p_\iota = q_\iota = 0$ for $\iota \notin H$ and $p_\iota \leqslant n_\iota$, $q_\iota \leqslant n_\iota$ for $\iota \in H$; there are therefore $\prod_{\iota\in H} (n_\iota + 1)$ ordered pairs (t, u) in $\mathbf{N}^{(I)}$ such that $t + u = s$.

We can therefore consider the *total algebra* of the monoid $\mathbf{N}^{(I)}$ over A, which contains the (restricted) algebra $A[X_i]_{i \in I}$ of this monoid. It is a unital commutative associative algebra called the *algebra of formal power series in the indeterminates* X_i $(i \in I)$ *with coefficients in* A and denoted by $A[[X_i]]_{i \in I}$; its elements are called *formal power series* in the indeterminates X_i $(i \in I)$ with coefficients in A. Such an element $(\alpha_\nu)_{\nu \in \mathbf{N}^{(I)}}$ is also denoted, following the convention made in no. 10, by $\sum_{\nu \in \mathbf{N}^{(I)}} \alpha_\nu X^\nu$; the α_ν are the *coefficients* of the formal power series and the $\alpha_\nu X^\nu$ its *terms*; a polynomial in the X_i is therefore a formal power series with only a *finite* number of terms $\neq 0$.

Clearly an algebra isomorphism $A[[X_i]]_{i \in I_1} \to A[[X_i]]_{i \in I_2}$ is canonically derived from every bijection $\sigma \colon I_1 \to I_2$ by mapping the formal power series $\sum_{(n_i)} \alpha_{(n_i)} \cdot \prod_{i \in I_1} X_i^{n_i}$ to the formal power series $\sum_{(n_i)} \alpha_{(n_i)} \cdot \prod_{i \in I_1} X_{\sigma(i)}^{n_i}$.

Let J be a subset of I; the algebra $A[[X_i]]_{i \in J}$ can be identified with a subalgebra of $A[[X_i]]_{i \in I}$ consisting of the formal power series $\sum_{(n_i)} \alpha_{(n_i)} \cdot \prod_{i \in I} X_i^{n_i}$ where $\alpha_{(n_i)} = 0$ for every element $(n_i) \in \mathbf{N}^{(I)}$ such that $n_i \neq 0$ for at least one index $i \in I - J$. Further, if $K = I - J$, $A[[X_i]]_{i \in I}$ is canonically identified with $(A[[X_j]]_{j \in J})[[X_k]]_{k \in K}$, by identifying the formal power series $\sum_{(n_i)} \alpha_{(n_i)} \cdot \prod_{i \in I} X_i^{n_i}$ with the formal power series $\sum_{(m_k)} \beta_{(m_k)} \cdot \prod_{k \in K} X_k^{m_k}$, where

$$\beta_{(m_k)} = \sum_{(p_j)} \gamma_{(p_j)} \cdot \prod_{j \in J} X_j^{p_j}$$

with $\gamma_{(p_j)} = \alpha_{(n_i)}$ for the sequence (n_i) such that $n_i = p_i$ for $i \in J$ and $n_i = m_i$ for $i \in K$.

Given a formal power series $u = \sum_\nu \alpha_\nu X^\nu$, the terms $\alpha_\nu X^\nu$ such that $|\nu| = p$ are called the terms in u *of total degree* p. The formal power series u_p whose terms of total degree p are those of u and whose other terms are zero, is called the *homogenous part* of u *of degree* p; when I is *finite*, u_p is a *polynomial* for all p; u_0 is identified with an element of A (also called the *constant term* of u). If u and v are two formal power series and $w = uv$, then

$$w_p = \sum_{r=0}^{p} u_r v_{p-r} \tag{39}$$

for every integer $p \geqslant 0$.

For every formal power series $u \neq 0$, the least integer $p \geqslant 0$ such that $u_p \neq 0$ is called the *total order* (or simply the *order*) of u. If this order is denoted by $\omega(u)$ and u and v are two formal power series $\neq 0$, then

$$\omega(u + v) \geqslant \inf(\omega(u), \omega(v)) \quad \text{if } u + v \neq 0 \tag{40}$$

$$\omega(uv) \geqslant \omega(u) + \omega(v) \quad \text{if } uv \neq 0. \tag{41}$$

Further, if $\omega(u) \neq \omega(v)$, then necessarily $u + v \neq 0$ and the two sides of (40) are *equal*.

Note that the order of 0 *is not defined*. By an abuse of language, it is a convention to say that "f is a formal power series of order $\geqslant p$ (resp. $> p$)" if the homogeneous part of f of degree n is zero for all $n < p$ (resp. $n \leqslant p$); 0 is therefore a "formal power series of order $> p$" for *every* integer $p \geqslant 0$.

Let J be a subset of I and let $A[[X_\iota]]_{\iota \in I}$ be identified as above with $B[[X_k]]_{k \in K}$, where $K = I - J$ and $B = A[[X_j]]_{j \in J}$; corresponding to the above definitions applied to $B[[X_k]]_{k \in K}$ there are new definitions for the formal power series $u \in A[[X_\iota]]_{\iota \in I}$; a term $\alpha_{(n_\iota)} \cdot \prod_{\iota \in I} X_\iota^{n_\iota}$ is said to be *of degree p in the* X_ι *of index $i \in K$* if $\sum_{\iota \in K} n_\iota = p$ and the formal power series of $B[[X_k]]_{k \in K}$ with the same terms of degree p as u and the others zero is *called the homogenous part of degree p in the* X_ι *of index $i \in K$*. If $u \neq 0$, the *order* $\omega_K(u)$ with respect to the X_ι of index $i \in K$ is the smallest of the integers $p \geqslant 0$ such that the homogeneous part of u of degree p in the X_ι of index $i \in K$ is $\neq 0$. Inequalities (40) and (41) still hold when ω is replaced by ω_K.

§ 3. GRADED ALGEBRAS

The graduations considered in this paragraph will have as set of degrees a *commutative monoid written additively whose identity element is denoted by* 0.

1. GRADED ALGEBRAS

DEFINITION 1. *Let Δ be a commutative monoid*, A *a graded ring of type* Δ (II, § 11, no. 2), $(A_\lambda)_{\lambda \in \Delta}$ *its graduation and* E *an* A*-algebra. A graduation* $(E_\lambda)_{\lambda \in \Delta}$ *of type* Δ *on the additive group* E *is said to be compatible with the* A*-algebra structure on* E *if it is compatible both with the* A*-module and with the ring structure on* E, *in other words, if, for all* λ, μ *in* Δ,

$$A_\lambda E_\mu \subset E_{\lambda+\mu} \tag{1}$$

$$E_\lambda E_\mu \subset E_{\lambda+\mu}. \tag{2}$$

The A*-algebra* E, *with this graduation, is then called a graded algebra of type* Δ *over the graded ring* A.

When the graduation on A is *trivial* (that is (II, § 11, no. 1) $A_0 = A$, $A_\lambda = \{0\}$ for $\lambda \neq 0$), condition (1) means that the E_λ are *sub-*A*-modules* of E. This leads to the definition of the notion of graded algebra of type Δ over a *non-graded* commutative ring A: A is given the trivial graduation of type Δ and the above definition is applied.

When we consider graded A-algebras E with a *unit element* e, it will always be understood that e is *of degree* 0 (cf. Exercise 1).

It follows that if an invertible element $x \in E$ is *homogeneous* and of degree p, its inverse x^{-1} is *homogeneous* and of degree $-p$: it suffices to decompose x^{-1} as a sum of homogeneous elements in the relations $x^{-1}x = xx^{-1} = e$.

Let E and E' be two graded algebras of type Δ over a graded ring A of type Δ. An A-algebra homomorphism $u: E \to E'$ is called a *graded algebra homomorphism* if $u(E_\lambda) \subset E'_\lambda$ for all $\lambda \in \Delta$ (where (E_λ) and (E'_λ) denote the respective graduations of E and E'); where E and E' are associative and unital and u is unital, this condition means that u is a graded ring homomorphism (II, § 11, no. 2).

Let E be a graded A-algebra of type **N**. E is identified with a graded A-algebra of type **Z** by writing $E_n = \{0\}$ for $n < 0$.

Remark. Definition 1 can also be interpreted by saying that E is a graded A-module and that the A-linear mapping

$$m: E \otimes_A E \to E$$

defining the multiplication on E (§ 1, no. 3), is *homogeneous of degree* 0 when $E \otimes_A E$ is given its graduation of type Δ (II, § 11, no. 5).

To define a graded A-algebra structure of type Δ on the graded ring A, with E as underlying graded A-module, therefore amounts to defining for each ordered pair (λ, μ) of elements of Δ a **Z**-bilinear mapping

$$m_{\lambda\mu}: E_\lambda \times E_\mu \to E_{\lambda+\mu}$$

such that for every triple of indices (λ, μ, ν) and for $\alpha \in A_\lambda$, $x \in E_\mu$, $y \in E_\nu$, $\alpha . m_{\mu\nu}(x, y) = m_{\lambda+\mu,\nu}(\alpha x, y) = m_{\mu,\lambda+\nu}(x, \alpha y)$.

Examples. (1) Let B be a *graded ring* of type Δ; if B is given its canonical **Z**-algebra structure (§ 1, no. 1, *Example* 2), B is a graded A-algebra (**Z** being given the trivial graduation).

(2) Let A be a graded commutative ring of type Δ and M a graded A-module of type Δ. Suppose that all the elements of the monoid Δ are *cancellable*, which allows (II, § 11, no. 6) us to define on $\mathrm{Homgr}_A(M, M) = \mathrm{Endgr}_A(M)$ a graded A-module structure of type Δ; as this graduation is compatible with the ring structure on $\mathrm{Endgr}_A(M)$ (II, § 11, no. 6), it defines a *unital graded A-algebra* structure on the A-algebra $\mathrm{Endgr}_A(M)$.

(3) *Algebra of a magma*. Let S be a magma and $\phi: S \to \Delta$ a homomorphism. For all $\lambda \in \Delta$, we write $S_\lambda = \overset{-1}{\phi}(\lambda)$; then $S_\lambda S_\mu \subset S_{\lambda+\mu}$. Let A be a graded commutative ring of type Δ and $(A_\lambda)_{\lambda\in\Delta}$ its graduation; we shall define a graded A-algebra structure on the algebra $E = A^{(S)}$ of the magma S (§ 2,

no. 6). To this end, let E_λ denote the additive subgroup of E generated by the elements of the form $\alpha . s$ such that $\alpha \in A_\mu$, $s \in S_\nu$ and $\mu + \nu = \lambda$. As the S_λ are pairwise disjoint, E is the direct sum of the $A_\mu S_\nu$ and hence also the direct sum of the E_λ and it is immediate that the E_λ satisfy conditions (1) and (2) and therefore define on E the desired graded A-algebra structure. If S admits an identity element e, it may also be supposed that $\phi(e) = 0$. A particular case is the one where the graduation of the ring A is trivial; then E_λ is the sub-A-module of E generated by S_λ. More particularly, if we take $S = \mathbf{N}^{(I)}$, $\Delta = \mathbf{N}$ and ϕ the mapping such that $\phi((n_\iota)) = \sum_{\iota \in I} n_\iota$, the ring A having the trivial graduation, a graduation is thus obtained on the polynomial algebra $A[X_\iota]_{\iota \in I}$, for which the degree of a homogeneous polynomial $\neq 0$ is the *total degree* defined in § 2, no. 9 (cf. § 6, no. 6).

We now take S to be the *free monoid* Mo(B) of a set B (I, § 7, no. 2) and ϕ the homomorphism $\mathrm{Mo(B)} \to \mathbf{N}$ which associates with each word its *length*. Thus a graded A-algebra structure is obtained on the *free associative algebra* of the set B (§ 2, no. 7; cf. § 5, no. 5).

2. GRADED SUBALGEBRAS, GRADED IDEALS OF A GRADED ALGEBRA

Let E be a graded algebra of type Δ over a graded ring A of type Δ. If F is a *sub-A-algebra* of E which is a *graded sub-A-module*, then the graduation (F_λ) on F is compatible with its A-algebra structure, since $F_\lambda = F \cap E_\lambda$; in this case F is called a *graded subalgebra* of E and the canonical injection $F \to E$ is a graded algebra homomorphism.

Similarly, if $\mathfrak{a}$ is a left (resp. right) *ideal* of E which is a *graded sub-A-module*, then $E_\lambda \mathfrak{a}_\mu \subset \mathfrak{a}_{\lambda+\mu}$ (resp. $\mathfrak{a}_\lambda E_\mu \subset \mathfrak{a}_{\lambda+\mu}$), since $\mathfrak{a}_\lambda = \mathfrak{a} \cap E_\lambda$; then $\mathfrak{a}$ is called a *graded ideal* of the algebra E. If $\mathfrak{b}$ is a graded two-sided ideal of E the quotient graduation on the module $E/\mathfrak{b}$ is compatible with the algebra structure on $E/\mathfrak{b}$ and the canonical homomorphism $E \to E/\mathfrak{b}$ is a graded algebra homomorphism.

If $u: E \to E'$ is a graded algebra homomorphism, $\mathrm{Im}(u)$ is a graded subalgebra of E', $\mathrm{Ker}(u)$ is a graded two-sided ideal of E and the bijection $E/\mathrm{Ker}(u) \to \mathrm{Im}(u)$ canonically associated with u is a graded algebra isomorphism.

PROPOSITION 1. *Let* A *be a graded commutative ring of type* Δ, E *a graded* A*-algebra of type* Δ *and* S *a set of homogeneous elements of* E. *Then the sub-A-algebra* (resp. *left ideal, right ideal, two-sided ideal*) *generated by* S *is a graded subalgebra* (resp. *graded ideal*).

The subalgebra of E generated by S is the sub-A-module generated by the finite products of elements of S, which are homogeneous; similarly, the left (resp. right) ideal generated by S is the sub-A-module generated by the elements of the form $u_1(u_2(\ldots(u_n s))\ldots)$ (resp. $(\ldots((s u_n)u_{n-1})\ldots)u_2)u_1)$ where

$s \in S$ and the $u_j \in E$ are homogeneous (n arbitrary) and these products are homogeneous, whence in this case the conclusion by virtue of II, § 11, no. 3, Proposition 2); finally the two-sided ideal generated by S is the union of the sequence $(\mathfrak{J}_n)_{n \geqslant 1}$, where $\mathfrak{J}_1$ is the left ideal generated by S and $\mathfrak{J}_{2n}$ (resp. $\mathfrak{J}_{2n+1}$) the right (resp. left) ideal generated by $\mathfrak{J}_{2n-1}$ (resp. $\mathfrak{J}_{2n}$), which completes the proof.

3. DIRECT LIMITS OF GRADED ALGEBRAS

Let $(A_\alpha, \phi_{\beta\alpha})$ be a directed direct system of graded commutative rings of type Δ (II, § 11, no. 3, *Remark* 3) and for each α let E_α be a graded A_α-algebra of type Δ; for $\alpha \leqslant \beta$ let $f_{\beta\alpha}: E_\alpha \to E_\beta$ be an A_α-homomorphism of *graded algebras* and suppose that $f_{\gamma\alpha} = f_{\gamma\beta} \circ f_{\beta\alpha}$ for $\alpha \leqslant \beta \leqslant \gamma$; then we shall call $(E_\alpha, f_{\beta\alpha})$ a *directed direct system of graded algebras of type* Δ over the directed direct system $(A_\alpha, \phi_{\beta\alpha})$ of graded commutative rings of type Δ. Then we know (II, § 11, no. 3) that $E = \varinjlim E_\alpha$ has canonically a graded module structure of type Δ over the graded ring $A = \varinjlim A_\alpha$ and a multiplication such that $E^\lambda E^\mu \subset E^{\lambda+\mu}$ (where (E^λ) denotes the graduation on E); then this multiplication and the graded A-module structure on E define on E a *graded* A-*algebra* structure *of type* Δ; E, with this structure, is called the *direct limit* of the direct system $(E_\alpha, f_{\beta\alpha})$ of graded algebras. The canonical homomorphisms $E_\alpha \to E$ are taken A_α-homomorphisms of graded algebras. Moreover, if F is a graded A-algebra of type Δ and (u_α) a direct system of A_α-homomorphisms $u_\alpha: E_\alpha \to F$, $u = \varinjlim u_\alpha$ is an A-homomorphism of graded algebras.

§ 4. TENSOR PRODUCTS OF ALGEBRAS

From § 4 to § 8 inclusive, A *denotes a commutative ring and, unless otherwise mentioned, the algebras considered are assumed to be associative and unital and the algebra homomorphisms are assumed to be unital.*

1. TENSOR PRODUCT OF A FINITE FAMILY OF ALGEBRAS

A always denotes a commutative ring with unit element. Let $(E_\iota)_{\iota \in I}$ be a *finite* family of A-algebras and let $E = \bigotimes_{\iota \in I} E_\iota$ be the tensor product A-module of the A-modules E_ι (II, § 3, no. 9). We shall define an A-*algebra* structure on E. Let $m_\iota: E_\iota \otimes_A E_\iota \to E_\iota$ be the A-linear mapping defining the multiplication on E_ι (§ 1, no. 3). Consider the A-linear mapping

$$m' = \bigotimes_{\iota \in I} m_\iota : \bigotimes_{\iota \in I} (E_\iota \otimes_A E_\iota) \to \bigotimes_{\iota \in I} E_\iota = E;$$

the composite mapping

$$\Big(\bigotimes_{\iota\in I} E_\iota\Big) \otimes_A \Big(\bigotimes_{\iota\in I} E_\iota\Big) \xrightarrow{\tau} \bigotimes_{\iota\in I} (E_\iota \otimes_A E_\iota) \xrightarrow{m'} \bigotimes_{\iota\in I} E_\iota$$

where τ is the associativity isomorphism (II, § 3, no. 9) is an A-linear mapping $m\colon E \otimes_A E \to E$; we shall see that m defines an (associative and unital) algebra structure on E. For, on explicitly performing the multiplication defined by m, we obtain the formula

$$\text{(1)} \qquad \Big(\bigotimes_{\iota\in I} x_\iota\Big)\Big(\bigotimes_{\iota\in I} y_\iota\Big) = \bigotimes_{\iota\in I} (x_\iota y_\iota) \quad \text{for } x_\iota, y_\iota \text{ in } E_\iota \text{ and } \iota \in I.$$

It is therefore seen already, by linearity, that if e_ι is the unit element of E_ι, $e = \bigotimes_{\iota\in I} e_\iota$ is unit element of E. On the other hand, the associativity of each of the E_ι implies the relation

$$\Big(\Big(\bigotimes_{\iota\in I} x_\iota\Big)\Big(\bigotimes_{\iota\in I} y_\iota\Big)\Big)\Big(\bigotimes_{\iota\in I} z_\iota\Big) = \bigotimes_{\iota\in I} (x_\iota y_\iota z_\iota) = \Big(\bigotimes_{\iota\in I} x_\iota\Big)\Big(\Big(\bigotimes_{\iota\in I} y_\iota\Big)\Big(\bigotimes_{\iota\in I} z_\iota\Big)\Big)$$

whence, by linearity, the relation $x(yz) = (xy)z$ for all x, y, z in E.

DEFINITION 1. *Given a family* $(E_\iota)_{\iota\in I}$ *of algebras over* A, *the tensor product of this family, denoted by* $\bigotimes_{\iota\in I} E_\iota$ (or, when I is the interval $[1, n]$ of $\mathbf{N}$, $E_1 \otimes_A E_2 \otimes \cdots \otimes_A E_n$, or simply $E_1 \otimes E_2 \otimes \cdots \otimes E_n$) *is the algebra obtained by giving the tensor product of the* A-*modules* E_ι *the multiplication defined by* (1).

Relation (1) shows that the tensor product $\bigotimes_{\iota\in I} E_\iota^0$ of the *opposite* algebras to the E_ι is the opposite algebra to $\bigotimes_{\iota\in I} E_\iota$; in particular, if the E_ι are *commutative*, so is $\bigotimes_{\iota\in I} E_\iota$.

Let $(E_\iota)_{\iota\in I}$ and $(F_\iota)_{\iota\in I}$ be two families of A-algebras with the same finite indexing set I. For each $\iota \in I$, let $f_\iota\colon E_\iota \to F_\iota$ be an A-algebra homomorphism. Then the A-linear mapping

$$f = \bigotimes_{\iota\in I} f_\iota \colon \bigotimes_{\iota\in I} E_\iota \to \bigotimes_{\iota\in I} F_\iota$$

is an A-*algebra homomorphism*, as follows from (1).

For every partition $(I_j)_{j\in J}$ of I, the associativity isomorphisms

$$\bigotimes_{j\in J}\Big(\bigotimes_{\iota\in I_j} E_\iota\Big) \to \bigotimes_{\iota\in I} E_\iota$$

(II, § 3, no. 9) are also *algebra* isomorphisms, as follows from (1) and their definitions.

When I is the interval $(1, n)$ of $\mathbf{N}$ and all the algebras E_ι are equal to the same algebra E, the tensor product algebra $\bigotimes_{\iota \in I} E_\iota$ is also denoted by $E^{\otimes n}$.

We shall restrict our attention in the remainder of this no. to the properties of tensor products of two algebras, leaving to the reader the task of extending them to tensor products of arbitrary finite families.

Let E, F be two A-algebras; if $\mathfrak{a}$ (resp. $\mathfrak{b}$) is a left ideal of E (resp. F), the canonical image $\operatorname{Im}(\mathfrak{a} \otimes \mathfrak{b})$ of $\mathfrak{a} \otimes_A \mathfrak{b}$ in $E \otimes_A F$ is a left ideal of $E \otimes_A F$; there are analogous results when "left ideal" is replaced by "right ideal" or "two-sided ideal". Moreover:

Proposition 1. *Let* E, F *be two* A*-algebras and* $\mathfrak{a}$ (resp. $\mathfrak{b}$) *a two-sided ideal of* E (resp. F). *Then the canonical* A*-module isomorphism*

$$(E/\mathfrak{a}) \otimes (E/\mathfrak{b}) \to (E \otimes F)/(\operatorname{Im}(\mathfrak{a} \otimes F) + \operatorname{Im}(E \otimes \mathfrak{b}))$$

(II, § 3, no. 6, Corollary 1 to Proposition 6) *is an algebra isomorphism.*

This follows from (1) and the definition given *loc. cit.*

Corollary 1. *Let* E *be an* A*-algebra and* $\mathfrak{a}$ *an ideal of* A. *Then the* A*-module* $\mathfrak{a}E$ *is a two-sided ideal of* E *and the canonical* $(A/\mathfrak{a})$*-module isomorphism*

$$(A/\mathfrak{a}) \otimes_A E \to E/\mathfrak{a}E$$

is an $(A/\mathfrak{a})$*-algebra isomorphism.*

Corollary 2. *If* $\mathfrak{a}, \mathfrak{b}$ *are two ideals of* A, $(A/\mathfrak{a}) \otimes_A (A/\mathfrak{b})$ *is canonically isomorphic to* $A/(\mathfrak{a} + \mathfrak{b})$.

Corollary 3. *Let* E, F *be two* A*-algebras and* $\mathfrak{a}$ *an ideal of* A *contained in the annihilator of* F. *Then the* $(A/\mathfrak{a})$*-algebra* $E \otimes_A F$ *is canonically isomorphic to* $(E/\mathfrak{a}E) \otimes_{A/\mathfrak{a}} F$.

Proposition 2. *Let* $(E_\lambda)_{\lambda \in L}$ *and* $(F_\mu)_{\mu \in M}$ *be two families of* A*-algebras. The canonical mapping* (II, § 3, no. 7)

$$\Big(\bigoplus_{\lambda \in L} E_\lambda\Big) \otimes_A \Big(\bigoplus_{\mu \in M} F_\mu\Big) \to \bigoplus_{(\lambda, \mu) \in L \times M} (E_\lambda \otimes_A F_\mu)$$

is an algebra isomorphism.

This follows immediately from II, § 3, no. 7, Proposition 7 and the definition of multiplication on $E \otimes F$.

Proposition 3. *Let* A, B *be two commutative rings,* $\rho: A \to B$ *a ring homomorphism and* E, F *two* A*-algebras. Then the canonical* B*-module isomorphism*

$$\rho^*(E) \otimes_B \rho^*(F) \to \rho^*(E \otimes_A F)$$

(II, § 5, no. 1, Proposition 3) *is a* B*-algebra isomorphism.*

PROPOSITION 4. *Let* A, B *be two commutative rings*, $\rho: A \to B$ *a ring homomorphism*, E *an* A-*algebra and* F *a* B-*algebra. Then the canonical* A-*module isomorphism*

$$\rho_*(F) \otimes_A E \to \rho_*(F \otimes_B \rho^*(E))$$

(II, § 5, no. 2, Proposition 6) *is an* A-*algebra isomorphism.*

The verifications are trivial on account of § 1, no. 5.

In particular, the A-algebra structure on $B \otimes_A E$, obtained by restricting the ring B of scalars to A, is identical with the structure of the algebra $B \otimes_A E$, the tensor product of the A-algebras B and E.

Finally, if (A_i, ϕ_{ji}) is a direct system of commutative rings, (E_i, f_{ji}) and (F_i, g_{ji}) two direct systems of A_i-algebras (§ 1, no. 6) and $A = \varinjlim A_i$, the canonical A-module isomorphism

$$\varinjlim (E_i \otimes_{A_i} F_i) \to (\varinjlim E_i) \otimes_A (\varinjlim F_i)$$

(II, § 6, no. 3, Proposition 7) is also an A-*algebra isomorphism*, as follows from the definitions.

Examples of tensor products of algebras. (1) Let A be a commutative ring and M, N two A-modules; the canonical mapping

$$\text{(2)} \qquad \mathrm{End}_A(M) \otimes_A \mathrm{End}_A(N) \to \mathrm{End}_A(M \otimes_A N)$$

(II, § 4, no. 4) is an A-*algebra* homomorphism, as follows from II, § 3, no. 2, formula (5). When M or N is a *finitely generated projective* A-module, we know that this homomorphism is *bijective* (II, § 4, no. 4, Proposition 4). In particular we recover the definition of the tensor product of two square matrices.

(2) Let S, T be two monoids and $A^{(S)}$ and $A^{(T)}$ the algebras of the monoids S and T over the ring A (III, § 2, no. 6); then there is a canonical A-algebra isomorphism

$$\text{(3)} \qquad A^{(S)} \otimes_A A^{(T)} \to A^{(S \times T)}.$$

The elements $e_s \otimes e_t$ (resp. $e_{(s,t)}$), where s runs through S and t runs through T, form a basis of $A^{(S)} \otimes_A A^{(T)}$ by virtue of II, § 3, no. 7, Corollary 2 to Proposition 7 (resp. of $A^{(S \times T)}$; the desired isomorphism is obtained by mapping $e_s \otimes e_t$ to $e_{(s,t)}$ and it follows from the definitions that this is indeed an *algebra* isomorphism.

2. UNIVERSAL CHARACTERIZATION OF TENSOR PRODUCTS OF ALGEBRAS

PROPOSITION 5. *Let* $(E_i)_{i \in I}$ *be a finite family of* A-*algebras and, for each* $i \in I$, *let* e_i *be the unit element of* E_i. *For each* $i \in I$, *let* $u_i: E_i \to E = \bigotimes_{i \in I} E_i$ *be the* A-*linear mapping defined by*

$$u_i(x_i) = \bigotimes_j x'_j \quad \textit{with } x'_i = x_i \textit{ and } x'_j = e_j \textit{ for } j \neq i.$$

(i) *The u_i are* A-*algebra isomorphisms; further, for $i \neq j$, the elements $u_i(x_i)$ and $u_j(x_j)$ are permutable in* E *for all $x_i \in E_i$ and $x_j \in E_j$ and* E *is generated by the union of the subalgebras $u_i(E_i)$.*

(ii) *Let* F *be an* A-*algebra and, for all $i \in I$, let $v_i: E_i \to F$ be an* A-*algebra homomorphism, where the v_i are such that, for $i \neq j$, $v_i(x_i)$ and $v_j(x_j)$ are permutable in* F *for all $x_i \in E_i$ and $x_j \in E_j$. Then there exists one and only one* A-*algebra homomorphism $w: E \to F$ such that*

$$v_i = w \circ u_i \quad \textit{for all } i \in I. \tag{4}$$

(i) The mapping u_i is an algebra homomorphism by definition of the multiplication on E. If $i \neq j$, $x_i \in E_i$, $x_j \in E_j$, then

$$u_i(x_i) = \bigotimes_k x'_k \quad \text{with } x'_i = x_i,\ x'_k = e_k \text{ for } k \neq i,$$

$$u_j(x_j) = \bigotimes_k x''_k \quad \text{with } x''_j = x_j,\ x''_k = e_k \text{ for } k \neq j.$$

Clearly $x'_k x''_k = x''_k x'_k$ for all $k \in I$ and hence $u_i(x_i)$ and $u_j(x_j)$ commute in E by formula (1) (no. 1) defining the multiplication in E. The last assertion follows from the relation $\bigotimes_i x_i = \prod_{i \in I} u_i(x_i)$.

(ii) For each $i \in I$, let x_i be an element of E_i. The product $\prod_{i \in I} v_i(x_i)$ is then defined in F independently of any ordering on I since the algebra F is associative and the elements $v_i(x_i)$ are pairwise permutable. The mapping $(x_i)_{i \in I} \to \prod_{i \in I} v_i(x_i)$ of $\prod_{i \in I} E_i$ into F is obviously A-multilinear and there therefore exists one and only one A-linear mapping $w: E \to F$ such that

$$w\Big(\bigotimes_i x_i\Big) = \prod_i v_i(x_i). \tag{5}$$

Now, the desired A-algebra homomorphism $w: E \to F$ must satisfy (5), which follows from (4) and the fact that $\bigotimes_i x_i = \prod_{i \in I} u_i(x_i)$. This proves the uniqueness of w; it remains to show that the A-linear mapping w defined by (5) is an A-algebra homomorphism and satisfies (4). The fact that w satisfies (4) is obvious: it suffices to apply (5) to the case where $x_j = e_j$ for $j \neq i$ and we obtain $w(u_i(x_i)) = v_i(x_i)$. Finally, w is an algebra homomorphism, for

$$\begin{aligned} w\Big(\Big(\bigotimes_i x_i\Big)\Big(\bigotimes_i y_i\Big)\Big) &= w\Big(\bigotimes_i (x_i y_i)\Big) = \prod_i v_i(x_i y_i) \\ &= \prod_i (v_i(x_i) v_i(y_i)) = \Big(\prod_i v_i(x_i)\Big) . \Big(\prod_i v_i(y_i)\Big) \end{aligned}$$

since $v_i(x_i)$ commutes with $v_j(y_j)$ for $j \neq i$; hence

$$ww\Big(\Big(\bigotimes_i x_i\Big)\Big(\bigotimes_i y_i\Big)\Big) = w\Big(\bigotimes_i x_i\Big).w\Big(\bigotimes_i y_i\Big)$$

which, by linearity, completes the proof.

The ordered pair consisting of E and the canonical mapping $\phi: (x_i) \mapsto \bigotimes_i x_i$ of $\prod_i E_i$ into E is a solution of the *universal mapping problem* (*Set Theory*, IV, § 3, no. 1) where Σ is the species of A-algebra structure, the morphisms being the A-algebra homomorphisms and the α-mappings the mappings $\prod_i u_i$ of $\prod_i E_i$ into an A-algebra such that the u_i are A-algebra homomorphisms and $u_i(x_i)$ and $u_j(x_j)$ commute for $i \neq j$, for all $x_i \in E_i$ and $x_j \in E_j$.

Corollary. *Let* $(E_i)_{i \in I}$, $(F_i)_{i \in I}$ *be two finite families of* A-*algebras and, for all* $i \in I$, *let* $f_i: E_i \to F_i$ *be an algebra homomorphism. If* $u_i: E_i \to \bigotimes_{j \in I} E_j$, $v_i: F_i \to \bigotimes_{j \in I} F_j$ *are the canonical homomorphisms, the mapping* $f = \bigotimes_i f_i$ (cf. no. 1) *is the unique* A-*algebra homomorphism such that* $f \circ u_i = v_i \circ f_i$ *for all* $i \in I$.

It suffices to note that the homomorphisms $g_i = v_i \circ f_i$ are such that $g_i(x_i) = v_i(f_i(x_i))$ and $g_j(x_j) = v_j(f_j(x_j))$ commute for $i \neq j$, $x_i \in E_i$ and $x_j \in E_j$; then apply Proposition 5.

When, in Proposition 5, the algebra F is assumed to be *commutative*, the hypothesis that $v_i(x_i)$ and $v_j(x_j)$ are permutable for $i \neq j$ is automatically satisfied. Hence, *when* F *is commutative*, there is a canonical bijection

$$\text{(6)} \qquad \mathrm{Hom}_{A\text{-alg.}}\Big(\bigotimes_i E_i, F\Big) \to \prod_i \mathrm{Hom}_{A\text{-alg.}}(E_i, F),$$

namely the one which associates with every homomorphism w of $\bigotimes_i E_i$ into F the family of $w \circ u_i$.

Note that if E is a commutative A-algebra, the ring structure of $E \otimes_A F$ is the same as that of $F_{(E)}$ (§ 1, no. 5).

3. MODULES AND MULTIMODULES OVER TENSOR PRODUCTS OF ALGEBRAS

Definition 2. *Let* E *be a* (*unital*) A-*algebra. A left* (resp. *right*) E-*module is a left* (resp. *right*) *module over the underlying ring of* E.

Unless otherwise mentioned, all the modules and multimodules considered in this no. are left modules and multimodules.

If M is an E-module, the homomorphism $\eta: A \to E$ (§ 1, no. 4) then defines on M an A-module structure, said to be *underlying* the E-module structure on M; for $\alpha \in A$, $s \in E$, $x \in M$,

$$\alpha(sx) = s(\alpha x) = (\alpha s)x, \tag{7}$$

so that for all $s \in E$, the homothety $h_s\colon x \mapsto sx$ of M is an *endomorphism* of the underlying A-module structure. Conversely, being given an E-module structure on M is equivalent to being given an A-*module* structure on M and an A-*algebra homomorphism* $s \mapsto h_s$ of E into $\mathrm{End}_A(M)$.

DEFINITION 3. *Let* E *and* F *be two (unital)* A-*algebras and* M *a set with an* E-*module structure and an* F-*module structure.* M *is called a (left) bimodule over the algebras* E *and* F *if:*

(1) M *is a bimodule over the underlying rings of* E *and* F (II, § 1, no. 14);

(2) *the two* A-*module structures underlying the* E-*module and* F-*module structures on* M *are identical.*

The latter condition says that if e and e' are the unit elements of E and F respectively, then

$$(\alpha e)x = (\alpha e')x \quad \text{for } \alpha \in A,\ x \in M; \tag{8}$$

then αx is used to denote the common value of the two sides.

It can also be said that being given on M a bimodule structure over E and F is equivalent to being given an A-*module structure* on M and also two A-algebra homomorphisms $s \mapsto h'_s$ of E into $\mathrm{End}_A(M)$ and $t \mapsto h''_t$ of F into $\mathrm{End}_A(M)$ such that $h'_s h''_t = h''_t h'_s$ for all $s \in E$ and $t \in F$. Consequently (no. 2, Proposition 5) an A-algebra homomorphism $u \mapsto h_u$ of $E \otimes_A F$ into $\mathrm{End}_A(M)$ is canonically derived such that $h_{s \otimes t} = h'_s h''_t = h''_t h'_s$ for $s \in E$ and $t \in F$. In other words, an $(E \otimes_A F)$-*module* structure is thus defined on M, which is said to be *associated* with the given bimodule structure over E and F and under which

$$(s \otimes t).x = s(tx) = t(sx) \quad \text{for } s \in E,\ t \in F \text{ and } x \in M.$$

The given E-module and F-module structures on M can be derived from this $(E \otimes_A F)$-module structure by restrictions of the ring of scalars, corresponding to the two canonical homomorphisms $E \to E \otimes_A F$ and $F \to E \otimes_A F$.

Conversely, if an $(E \otimes_A F)$-module structure is given on M, by means of the canonical homomorphisms $E \to E \otimes_A F$ and $F \to E \otimes_A F$ an E-module structure and an F-module structure on M and it is immediate that M is a *bimodule* over the algebras E and F with these two structures and that the given $(E \otimes_A F)$-module structure is associated with this bimodule structure.

Thus a one-to-one correspondence has been established between $(E \otimes_A F)$-module and bimodules over the algebras E and F. Clearly every sub-bimodule of M is a submodule for the associated $(E \otimes_A F)$-module structure and conversely. There are analogous results for quotients, products, direct sums and inverse and direct limits. Finally, if M′ is another bimodule over the algebras E and F and $f: M \to M'$ is a bimodule homomorphism, f is also an $(E \otimes_A F)$-module homomorphism and conversely.

There are obviously corresponding statements for right bimodule structures, or when for example there is a left E-module structure and a right F-module structure; in this case we speak of an (E, F)-*bimodule* and being given such a structure amounts to being given a *left* bimodule structure over E and F^o.

Examples. (1) Let B be an A-algebra; the ring B has canonically a (B, B)-*bimodule* structure (II, § 1, no. 14, *Example* 1) and, if e is the unit element of B, then $(\alpha e)x = x(\alpha e) = \alpha x$ for all $x \in B$ and all $\alpha \in A$; B can therefore be considered as a *left bimodule* over the algebras B and B^o (opposite to B); there is therefore associated with the (B, B)-bimodule structure on B a $(B \otimes_A B^o)$-*module* structure such that, for b, x and b' in B,

$$(b \otimes b').x = bxb' \tag{9}$$

the right-hand side being the product in the ring B.

(2) Let E and F be two A-algebras, e, e' their respective unit elements, M an E-module and N an F-module; these module structures define on M a bimodule structure over the rings A and E and on N a bimodule structure over the rings A and F; from these is therefore derived a bimodule structure over the rings E and F on the tensor product $M \otimes_A N$, defined by

$$x.(m \otimes n) = (x.m) \otimes n, \qquad y.(m \otimes n) = m \otimes (y.n)$$

for $x \in E$, $y \in F$, $m \in M$, $n \in N$ (II, § 3, no. 4); it is also seen that conditions (8) hold and hence the above bimodule structure is associated with an $(E \otimes_A F)$-*module* structure on $M \otimes_A N$, such that

$$(x \otimes y).(m \otimes n) = (x.m) \otimes (y.n) \tag{10}$$

for $x \in E$, $y \in F$, $m \in M$, $n \in N$.

In particular, taking $M = E_s$, $E_s \otimes_A N$ has canonically an $(E \otimes_A F)$-module structure; on the other hand, $E \otimes_A N$ is canonically identified with $E \otimes_A (F_d \otimes_F N) = (E \otimes_A F) \otimes_F N$, where $E \otimes_A F$ is considered as having its right F-module structure defined by the canonical homomorphism $v: F \to E \otimes_A F$; for x, x' in E, $y \in F$, $n \in N$, $x' \otimes n$ is thus identified with $(x' \otimes e') \otimes n$ and $(x \otimes y).(x' \otimes n') = (xx') \otimes (y.n)$ with $((xx') \otimes y) \otimes n$. The $(E \otimes_A F)$-module $E_s \otimes_A N$ is thus identified with the $(E \otimes_A F)$-module derived from N by extending the scalars to $E \otimes_A F$ by means of the homomorphism v (II, § 5, no. 1). The canonical mapping $n \mapsto e \otimes n$ of N into $E_s \otimes_A N$ is identified with the canonical mapping $n \mapsto (e \otimes e') \otimes n$ of N into $(E \otimes_A F) \otimes_F N$; this is known to be an F-homomorphism.

With the same notation, let P be a right $(E \otimes_A F)$-module; then there is a canonical **Z**-module isomorphism

$$(11) \qquad P \otimes_{E \otimes_A F} (E_s \otimes_A N) \to P \otimes_F N$$

where on the right-hand side P is considered as a right F-module by means of the canonical homomorphism v. For P is canonically identified with $P \otimes_{E \otimes_A F} (E \otimes_A F)$ and $(E \otimes_A F) \otimes_F N$ with $E \otimes_A (F \otimes_F N)$ and hence with $E \otimes_A N$, which establishes the stated isomorphism (II, § 3, no. 8, Proposition 8 and II, § 3, no. 4, proposition 4).

All the above extends to *multimodules* (II, § 1, no. 14).

4. TENSOR PRODUCT OF ALGEBRAS OVER A FIELD

Let K be a commutative *field* and E, F two algebras over K whose respective unit elements e, e' are *non-zero*. Then the homomorphisms $\eta_E: K \to E$ and $\eta_F: K \to F$ (§ 1, no. 3) are injections which allow us to identify K with a subfield of E (resp. F). The canonical homomorphisms $u: E \to E \otimes_K F$ and $v: F \to E \otimes_K F$, defined by $u(x) = x \otimes e'$ and $v(y) = e \otimes y$ are *injective* (II, § 7, no. 9, Proposition 19) and allow us to identify E and F with *subalgebras* of $E \otimes_K F$, both having as unit element the unit element $e \otimes e'$ of $E \otimes_K F$. In $E \otimes_K F$, $E \cap F = K$ (II, § 7, no. 9, Proposition 19).

If E′ and F′ are subalgebras of E and F respectively, the canonical homomorphism $E' \otimes_K F' \to E \otimes_K F$ is injective and allows us to identify $E' \otimes_K F'$ with the subalgebra of $E \otimes_K F$ generated by $E' \cup F'$ (II, § 7, no. 7, Proposition 14).

PROPOSITION 6. *Let* E, F *be two non-zero algebras over a commutative field,* K, C (resp. D) *a subalgebra of* E (resp. F) *and* C′ (resp. D′) *the centralizer of* C *in* E (resp. D *in* F). *Then the centralizer of* $C \otimes_K D$ in $E \otimes_K F$ *is* $C' \otimes_K D'$.

It all reduces to verifying that an element $z = \sum_i x_i \otimes y_i$ of the centralizer of $C \otimes_K D$ ($x_i \in F, y_i \in F$) belongs to $C' \otimes_K D'$; we know that

$$C' \otimes_K D' = (C' \otimes_K F) \cap (E \otimes_K D')$$

(II, § 7, no. 7, Corollary to Proposition 14). The y_i may be assumed to be linearly independent over K; for all $x \in C$, of necessity $(x \otimes e')z = z(x \otimes e')$, that is $\sum_i (xx_i - x_i x) \otimes y_i = 0$, whence $xx_i = x_i x$ for all i (II, § 3, no. 7, Corollary 1 to Proposition 7); hence of necessity $x_i \in C'$ for all i and therefore $z \in C' \otimes_K F$; it can similarly be shown that $z \in E \otimes_K D'$, whence the proposition.

COROLLARY. *If* Z *and* Z′ *are the respective centres of* E *and* F, *the centre of* $E \otimes_K F$ *is* $Z \otimes_K Z'$.

Let E and F be two subalgebras of an algebra G over a commutative field K; suppose that every element of E *commutes* with every element of F. Then the canonical injections $i: E \to G$, $j: F \to G$ define a canonical homomorphism $h = i \otimes j: E \otimes_K F \to G$ (no. 2, Proposition 5) such that

$$(i \otimes j)(x \otimes y) = xy \quad \text{for } x \in E, y \in F.$$

DEFINITION 4. *Given an algebra* G *over a commutative field* K, *two subalgebras* E, F *of* G *are said to be linearly disjoint over* K *if they satisfy the following conditions:*

(1) *every element of* E *commutes with every element of* F;

(2) *the canonical homomorphism of* $E \otimes_K F$ *into* G *is injective.*

PROPOSITION 7. *Let* G *be an algebra over a commutative field* K *and* E, F *two subalgebras of* G *such that every element of* E *commutes with every element of* F. *For* E *and* F *to be linearly disjoint over* K, *it is necessary and sufficient that there exist a basis of* E *over* K *which is a free subset of* G *for the right* F*-module structure on* G. *When this is so:*

(i) *the canonical homomorphism* $h: E \otimes_K F \to G$ *is an isomorphism of* $E \otimes_K F$ *onto the subalgebra of* G *generated by* $E \cup F$;

(ii) $E \cap F = K$;

(iii) *every free subset of* E (resp. F) *over* K *is a free subset of* G *with its right or left* F*-module* (resp. E*-module*) *structure.*

The condition of the statement is obviously necessary, since every basis of E over K is a basis of $E \otimes_K F$ with its right F-module structure (II, § 3, no. 7, Corollary 1 to Proposition 7). To see that the condition is sufficient, note that the image H of $E \otimes_K F$ under h is the set of sums $\sum_i x_i y_i = \sum_i y_i x_i$ in G, with $x_i \in E$ and $y_i \in F$; if (a_λ) is a basis of E over K, H is therefore also the *submodule* of the (right or left) F-module G, generated by (a_λ). The condition of the statement therefore means that there exists a basis (a_λ) of E which is also a basis of the F-module H; it follows that h is injective. Assertion (iii) follows from the fact that every free subset of E is contained in a basis of E (II, § 7, no. 1, Theorem 2).

COROLLARY 1. *For the canonical homomorphism of* $E \otimes_K F$ *into* G *to be bijective, it is necessary and sufficient that there exist a basis of* E *over* K *which is a basis of the* (*right or left*) F*-module* G.

COROLLARY 2. *Let* E, F *be two subalgebras of* G, *of finite rank over* K *and such that every element of* E *commutes with every element of* F. *For* E *and* F *to be linearly disjoint over* K, *it is necessary and sufficient that the subalgebra* H *of* G *generated by* $E \cup F$ *be such that*

$$[H:K] = [E:K].[F:K]. \tag{12}$$

This says that the rank over K of the surjective canonical homomorphism

$h: E \otimes_K F \to H$ is equal to the rank of $E \otimes_K F$ over K, which is equivalent to saying that this homomorphism is bijective (II, § 7, No. 4, Proposition 9).

5. TENSOR PRODUCT OF AN INFINITE FAMILY OF ALGEBRAS

Let A be a commutative ring and $(E_\iota)_{\iota \in I}$ an arbitrary family of (unital) A-algebras. For every finite subset J of I, let E_J denote the tensor product $\bigotimes_{\iota \in J} E_\iota$ of the algebras E_ι of index $i \in J$; let e_ι denote the unit element of E_ι and $e_J = \bigotimes_{\iota \in J} e_\iota$ the unit element of E_J; let $f_{J,\iota}$ denote the canonical homomorphism $E_\iota \to E_J$ for $i \in J$ (no. 2, Proposition 5). If J, J′ are two finite subsets of I such that $J \subset J'$, a homomorphism $f_{J'J}: E_J \to E_{J'}$ is canonically derived (no. 2, Proposition 5), by the condition $f_{J'J} \circ f_{J,\iota} = f_{J',\iota}$ for all $i \in J$. Moreover the uniqueness of $f_{J'J}$ implies that if J, J′, J″ are three finite subsets of I such that $J \subset J' \subset J''$, then $f_{J''J} = f_{J''J'} \circ f_{J'J}$. In other words, $(E_J, f_{J'J})$ is a *direct system of* A-*algebras* whose indexing set is the right directed set $\mathfrak{F}(I)$ of finite subsets of I.

DEFINITION 5. *The direct limit* E *of the direct system* $(E_J, f_{J'J})$ *is called the tensor product of the family of* A-*algebras* $(E_\iota)_{\iota \in I}$.

If I is finite, E is identified with $\bigotimes_{\iota \in I} E_\iota$. By an abuse of notation, E is also denoted by $\bigotimes_{\iota \in I} E_\iota$ even if I is infinite.

For every finite subset J of I, let f_J denote the canonical homomorphism $\bigotimes_{\iota \in J} E_\iota \to \bigotimes_{\iota \in I} E_\iota$ (writing f_ι instead of $f_{\{\iota\}}$); if e is the unit element of $\bigotimes_{\iota \in I} E_\iota$, then $f_J(e_J) = e$ for all $J \in \mathfrak{F}(I)$. It is immediate that if all the algebras E_ι are commutative, so is $\bigotimes_{\iota \in I} E_\iota$.

PROPOSITION 8. (i) *The homomorphisms* $f_\iota: E_\iota \to E = \bigotimes_{k \in I} E_k$ *are such that for two indices* i, j *such that* $i \neq j$, $f_i(x_i)$ *and* $f_j(x_j)$ *commute in* E *for all* $x_i \in E_i$ *and* $x_j \in E_j$; *further,* E *is generated by the union of the subalgebras* $f_\iota(E_\iota)$.

(ii) *Let* F *be an* A-*algebra and, for all* $i \in I$, *let* $u_\iota: E_\iota \to F$ *be an* A-*algebra homomorphism such that, for* $i \neq j$, $u_i(x_i)$ *and* $u_j(x_j)$ *commute in* F *for all* $x_i \in E_i$ *and* $x_j \in E_j$. *Then there exists one and only one* A-*algebra homomorphism* $u: E \to F$ *such that* $u_\iota = u \circ f_\iota$ *for all* $i \in I$.

(i) As, for every finite subset J of I, $f_\iota = f_J \circ f_{J,\iota}$, the first assertion in (i) follows from no. 2, Proposition 5, taking J containing i and j; the second also follows from no. 2, Proposition 5, taking account of the fact that E is the union of the $f_J(E_J)$ when J runs through $\mathfrak{F}(I)$.

(ii) For every finite subset J of I, it follows from no. 2, Proposition 5 that there exists a unique homomorphism $u_J: E_J \to F$ such that $u_J \circ f_{J,\iota} = u_\iota$

for all $i \in J$; it immediately follows from this uniqueness property that, for $J \subset J'$, $u_J = u_{J'} \circ f_{J'J}$; in other words, the u_J form a *direct system* of homomorphisms. Let $u = \varinjlim u_J : E \to F$; then by definition $u_J = u \circ f_J$ for every finite subset J of I and in particular $u_i = u \circ f_i$ for all $i \in I$; the uniqueness of u follows from these relations and the fact that the $f_i(E_i)$ generate the algebra E.

COROLLARY. *Let* $(E_i)_{i \in I}$, $(E'_i)_{i \in I}$ *be two families of A-algebras with the same indexing set and, for all* $i \in I$, *let* $u_i : E_i \to E'_i$ *be an algebra homomorphism. Then there exists one and only one A-algebra homomorphism* $u : \bigotimes_{i \in I} E_i \to \bigotimes_{i \in I} E'_i$ *such that, for all* $i \in I$, *the diagram*

$$\begin{array}{ccc} E_i & \xrightarrow{u_i} & E'_i \\ \downarrow f_i & & \downarrow f'_i \\ \bigotimes_{i \in I} E_i & \xrightarrow[u]{} & \bigotimes_{i \in I} E'_i \end{array}$$

is commutative, f_i *and* f'_i *denoting the canonical homomorphisms.*

It suffices to apply Proposition 8 to the homomorphism $f'_i \circ u_i$.

The homomorphism u defined in the Corollary to Proposition 8 is denoted by $\bigotimes_{i \in I} u_i$. If J is any subset of I, Proposition 8 can be applied to the family $(f_i)_{i \in J}$ of canonical homomorphisms $f_i : E_i \to \bigotimes_{i \in I} E_i = E$; a canonical homomorphism $E_J \to E$ is derived which is also denoted by f_J and which, when J is *finite*, coincides with the homomorphism denoted thus above.

Now let $(x_i)_{i \in I}$ be an element of $\prod_{i \in I} E_i$ such that the family $(x_i - e_i)_{i \in I}$ has *finite support* H. It is immediate that, if J and J′ are two finite subsets of I containing H, then

$$f_J((x_i)_{i \in J}) = f_{J'}((x_i)_{i \in J'}).$$

The common value of the $f_J((x_i)_{i \in J})$ for the finite subsets $J \supset H$ of I is denoted by $\bigotimes_{i \in I} x_i$.

PROPOSITION 9. *Let* $(E_i)_{i \in I}$ *be a family of A-algebras and for each* $i \in I$ *let* B_i *be a basis of* E_i *such that the unit element* e_i *belongs to* B_i. *Let* B *be the set of elements of the form* $\bigotimes_{i \in I} x_i$, *where* (x_i) *runs through the set of elements of* $\prod_{i \in I} B_i$ *such that the family* $(x_i - e_i)$ *has finite support. Then* B *is a basis of the algebra* $\bigotimes_{i \in I} E_i$ *and this basis contains the unit element* e.

For every finite subset J of I, let B_J be the basis of $E_J = \bigotimes_{i \in J} E_i$ the tensor product of the bases B_i for $i \in J$ (II, § 3, no. 9). It follows immediately from the definitions that B is the union of the $f_J(B_J)$ when J runs through $\mathfrak{F}(I)$ and that $f_{J'J}(B_J) \subset B_{J'}$ when $J \subset J'$; hence (B_J) is a direct system of subsets of the E_J and $B = \varinjlim B_J$; the conclusion then follows from II, § 6, no. 2, Corollary to Proposition 5.

The basis B is also called the *tensor product* of the bases B_i for $i \in I$; when the conditions of Proposition 9 are fulfilled, the canonical homomorphisms $f_J: E_J \to E = \bigotimes_{i \in I} E_i$ are *injective* for every subset J of I, for if B_J is the basis of E_J the tensor product of the B_i for $i \in J$, it is immediately verified that the restriction of f_J to B_J is injective and maps B_J onto a subset of B.

6. COMMUTATION LEMMAS

Lemma 1. *Let* A *be a commutative ring,* E *an* A*-algebra,* $(x_i)_{1 \leqslant i \leqslant n}$ *a finite sequence of elements of* E, $(\lambda_i)_{1 \leqslant i \leqslant n}$ *a finite sequence of elements of* A *and* y *an element of* E; *suppose that*

$$x_i y = \lambda_i y x_i \quad \textit{for } 1 \leqslant i \leqslant n. \tag{13}$$

Then

$$(x_1 x_2 \ldots x_n) y = (\lambda_1 \lambda_2 \ldots \lambda_n) y (x_1 x_2 \ldots x_n). \tag{14}$$

The lemma being trivial for $n = 1$, we argue by induction on $n \geqslant 2$. Now

$$\begin{aligned}(x_1 x_2 \ldots x_n) y &= (x_1 \ldots x_{n-1})(x_n y) \\ &= (x_1 \ldots x_{n-1})(\lambda_n y x_n) = \lambda_n((x_1 \ldots x_{n-1}) y) x_n,\end{aligned}$$

which, by the induction hypothesis, is equal to

$$\lambda_n(\lambda_1 \ldots \lambda_{n-1}) y (x_1 \ldots x_{n-1}) x_n = (\lambda_1 \ldots \lambda_{n-1} \lambda_n) y (x_1 \ldots x_{n-1} x_n),$$

whence the lemma.

Lemma 2. *Let* A *be a commutative ring,* E *an* A*-algebra and* $(x_i)_{1 \leqslant i \leqslant n}$ *and* $(y_i)_{1 \leqslant i \leqslant n}$ *two finite sequences of* n *elements of* E; *suppose that for* $1 \leqslant j \leqslant i \leqslant n$,

$$x_i y_j = \lambda_{ij} y_j x_i \quad \textit{with } \lambda_{ij} \in A. \tag{15}$$

Then

$$(x_1 x_2 \ldots x_n)(y_1 y_2 \ldots y_n) = \Big(\prod_{i > j} \lambda_{ij}\Big)(x_1 y_1)(x_2 y_2) \ldots (x_n y_n). \tag{16}$$

The lemma being trivial for $n = 1$, we again argue by induction on n for $n \geqslant 2$. By virtue of Lemma 1,

$$\begin{aligned}(x_1 \ldots x_n)(y_1 \ldots y_n) &= x_1 (x_2 \ldots x_n) y_1 (y_2 \ldots y_n) \\ &= \Big(\prod_{i > 1} \lambda_{i1}\Big)(x_1 y_1)(x_2 \ldots x_n)(y_2 \ldots y_n)\end{aligned}$$

and it then suffices to apply the induction hypothesis to obtain (16).

For every family $\lambda = (\lambda_{ij})$ of elements of A, with $1 \leqslant j < i \leqslant n$, and for every permutation $\sigma \in \mathfrak{S}_n$, we write

(17) $$\varepsilon_\sigma(\lambda) = \prod_{i>j,\, \sigma^{-1}(i)<\sigma^{-1}(j)} \lambda_{ij} = \prod_{i<j,\, \sigma(i)>\sigma(j)} \lambda_{\sigma(i),\sigma(j)}.$$

Observe that, when $A = \mathbf{Z}$ and $\lambda_{ij} = -1$ for every ordered pair (i, j) such that $1 \leqslant j < i \leqslant n$, $\varepsilon_\sigma(\lambda)$ is just the signature ε_σ of the permutation σ (I, § 5, no. 7).

Lemma 3. Let A *be a commutative ring,* E *an* A*-algebra,* $(x_i)_{1 \leqslant i \leqslant n}$ *a finite sequence of elements of* E *and suppose that, for every ordered pair* (i, j) *of integers such that* $1 \leqslant j < i \leqslant n$,

(18) $$x_i x_j = \lambda_{ij} x_j x_i \quad \textit{with } \lambda_{ij} \in A.$$

Then, for every permutation $\sigma \in \mathfrak{S}_n$,

(19) $$x_{\sigma(1)} x_{\sigma(2)} \ldots x_{\sigma(n)} = \varepsilon_\sigma(\lambda) x_1 x_2 \ldots x_n.$$

The lemma is trivial for $n = 1$ and $n = 2$; we proceed by induction on n for $n \geqslant 3$. If $\sigma(n) = n$, relation (19) follows from the induction hypothesis. Suppose therefore that $\sigma(n) = k$, $k \neq n$, and let τ be the permutation of $[1, n]$ defined by

$$\begin{cases} \tau(i) = i & \text{for } i < k \\ \tau(i) = i + 1 & \text{for } k \leqslant i < n \\ \tau(n) = k. \end{cases}$$

Let $\pi = \tau^{-1} \circ \sigma$; the permutation π leaves n fixed; now $\sigma = \tau \circ \pi$ and therefore, writing $y_i = x_{\tau(i)}$, $y_{\pi(i)} = x_{\sigma(i)}$. If $i \neq n$ and $j \neq n$, the relations $\pi(i) > \pi(j)$ and $\sigma(i) > \sigma(j)$ are equivalent (since τ is a strictly increasing mapping of $[1, n-1]$ into $[1, n]$. For $i \neq n, j \neq n$ and $\sigma(i) > \sigma(j)$,

$$y_{\pi(i)} y_{\pi(j)} = x_{\sigma(i)} x_{\sigma(j)} = \lambda_{\sigma(i),\sigma(j)} x_{\sigma(j)} x_{\sigma(i)} = \lambda_{\sigma(i),\sigma(j)} y_{\pi(j)} y_{\pi(i)}$$

whence, by the induction hypothesis (using the fact that $\pi(n) = n$):

$$y_{\pi(1)} y_{\pi(2)} \ldots y_{\pi(n)} = \Big(\prod_{i<j<n,\, \sigma(i)>\sigma(j)} \lambda_{\sigma(i),\sigma(j)}\Big) y_1 y_2 \ldots y_n$$

that is

(20) $$x_{\sigma(1)} x_{\sigma(2)} \ldots x_{\sigma(n)} = \Big(\prod_{i<j<n,\, \sigma(i)>\sigma(j)} \lambda_{\sigma(i),\sigma(j)}\Big) x_{\tau(1)} \ldots x_{\tau(n)}.$$

Now

$$x_{\tau(1)} \ldots x_{\tau(n)} = x_1 \ldots x_{k-1} x_{k+1} \ldots x_n x_k$$

and this, by Lemma 1, is equal to

(21) $$\Big(\prod_{j>k} \lambda_{jk}\Big) x_1 \ldots x_n = \Big(\prod_{\sigma(i)>\sigma(n)} \lambda_{\sigma(i),\sigma(n)}\Big) x_1 \ldots x_n.$$

Finally, (20) and (21) give

$$x_{\sigma(1)} \ldots x_{\sigma(n)} = \alpha . x_1 \ldots x_n$$

with

$$\alpha = \Big(\prod_{i<j<n,\ \sigma(i)>\sigma(j)} \lambda_{\sigma(i),\sigma(j)}\Big) . \Big(\prod_{i<n,\ \sigma(i)>\sigma(n)} \lambda_{\sigma(i),\sigma(n)}\Big)$$
$$= \prod_{i<j,\ \sigma(i)>\sigma(j)} \lambda_{\sigma(i),\sigma(j)} = \varepsilon_\sigma(\lambda)$$

which completes the proof of Lemma 3.

7. TENSOR PRODUCT OF GRADED ALGEBRAS RELATIVE TO COMMUTATION FACTORS

DEFINITION 6. *Let $(\Delta_i)_{i \in I}$ be a finite family of commutative monoids written additively. A system of commutation factors over the Δ_i with values in a commutative ring* A *is a system of mappings $\varepsilon_{ij}: \Delta_i \times \Delta_j \to A$, where $i \in I$, $j \in I$, $i \neq j$ satisfying the following conditions:*

(22) $$\varepsilon_{ij}(\alpha_i + \alpha'_i, \beta_j) = \varepsilon_{ij}(\alpha_i, \beta_j)\varepsilon_{ij}(\alpha'_i, \beta_j)$$

(23) $$\varepsilon_{ij}(\alpha_i, \beta_j + \beta'_j) = \varepsilon_{ij}(\alpha_i, \beta_j)\varepsilon_{ij}(\alpha_i, \beta'_j)$$

(24) $$\varepsilon_{ij}(\alpha_i, \beta_j)\varepsilon_{ji}(\beta_j, \alpha_i) = 1,$$

for all α_i, α'_i in Δ_i, β_j, β'_j in Δ_j.

If I is given a total ordering and the Δ_i are groups, a system of commutation factors is defined over the Δ_i by taking for every ordered pair (i, j) such that $i < j$ an arbitrary **Z**-*bilinear* mapping of $\Delta_i \times \Delta_j$ into the (*multiplicative*) **Z**-module A^* of *invertible* elements of the ring A and then writing

$$\varepsilon_{ji}(\beta_j, \alpha_i) = (\varepsilon_{ij}(\alpha_i, \beta_j))^{-1}$$

for $i < j$.

Note that, since the $\varepsilon_{ij}(\alpha_i, \beta_j)$ are invertible,

$$\varepsilon_{ij}(0, \beta_j) = \varepsilon_{ij}(\alpha_i, 0) = 1,$$

by virtue of (22) and (23).

Examples. (1) The *trivial* system of commutation factors consists of the ε_{ij} such that $\varepsilon_{ij}(\alpha_i, \beta_j) = 1$ for all i, j, $\alpha_i \in \Delta_i$, $\beta_j \in \Delta_j$.

(2) If we take $A = \mathbf{Z}$ and $\Delta_i = \mathbf{Z}$ for all $i \in I$, a system of commutation factors is obtained by taking $\varepsilon_{ij}(\alpha_i, \beta_j) = (-1)^{\alpha_i\beta_j}$. Note that this number depends only on the parities of α_i and β_j and the ε_{ij} can therefore be considered as commutation factors when certain of the Δ_i are equal to $\mathbf{Z}/2\mathbf{Z}$ and the others to **Z**.

These two examples are the most frequent cases encountered in applications.

PROPOSITION 10. *Let* A *be a commutative ring and* $(\Delta_\iota)_{\iota\in I}$ *a finite family of commutative monoids written additively; for each* $\iota \in I$ *let* E_ι *be a graded* A*-algebra of type* Δ_ι. *Finally, let* $(\varepsilon_{\iota j})$ *be a system of commutation factors over the* Δ_ι *with values in* A. *Then there exist a graded* A*-algebra* E *of type* $\Delta = \prod_{\iota\in I} \Delta_\iota$ *and for each* $\iota \in I$ *an algebra homomorphism* $h_\iota : E_\iota \to E$, *with the following properties:*

(i) *If* $\phi_\iota : \Delta_\iota \to \Delta$ *is the canonical homomorphism, then* h_ι *is a graded homomorphism* (II, § 11, no. 2), *in other words,* $h_\iota(E_\iota^{\alpha_\iota}) \subset E^{\phi_\iota(\alpha_\iota)}$, *where* $(E_\iota^{\alpha_\iota})$ *and* (E^α) *denote the respective graduations on* E_ι *and* E.

(ii) *If* $\iota \neq j$ *and* x_ι (resp. x_j) *is a homogeneous element of* E_ι (resp. E_j) *of degree* $\alpha_\iota \in \Delta_\iota$ (resp. $\beta_j \in \Delta_j$), *then*

$$h_\iota(x_\iota)h_j(x_j) = \varepsilon_{\iota j}(\alpha_\iota, \beta_j)h_j(x_j)h_\iota(x_\iota). \tag{25}$$

(iii) *For every* A*-algebra* F *and every system of homomorphisms* $f_\iota : E_\iota \to F$ *satisfying the conditions*

$$f_\iota(x_\iota)f_j(x_j) = \varepsilon_{\iota j}(\alpha_\iota, \beta_j)f_j(x_j)f_\iota(x_\iota), \tag{26}$$

where $\iota, j, x_\iota, x_j, \alpha_\iota, \beta_j$ *are as in* (ii), *then there exists one and only one algebra homomorphism* $f : E \to F$ *such that* $f_\iota = f \circ h_\iota$ *for all* $\iota \in I$. *Moreover the underlying* A*-module of* E *is the tensor product* $\bigotimes_{\iota\in I} E_\iota$.

Consider the A*-module* $E = \bigotimes_{\iota\in I} E_\iota$; it is identified with the direct sum of the submodules E^α, where, for each $\alpha = (\alpha_\iota) \in \Delta$, we write $E^\alpha = \bigotimes_{\iota\in I} E_\iota^{\alpha_\iota}$; the E^α therefore form a graduation of type Δ on the A-module E. We shall define on E a *graded* A*-algebra* structure of type Δ. For this let I be given a total ordering; for every ordered pair of elements $\alpha = (\alpha_\iota)$, $\beta = (\beta_\iota)$ of Δ, we must first define an A-bilinear mapping of $E^\alpha \times E^\beta$ into $E^{\alpha+\beta}$, or alternatively an A-linear mapping $m_{\alpha\beta}$ of $E^\alpha \otimes_A E^\beta$ into $E^{\alpha+\beta}$. We shall define $m_{\alpha\beta}$ by the condition

$$m_{\alpha\beta}\Big(\Big(\bigotimes_{\iota\in I} x_\iota\Big) \otimes \Big(\bigotimes_{\iota\in I} y_\iota\Big)\Big) = \varepsilon(\alpha, \beta) \bigotimes_{\iota\in I} (x_\iota y_\iota) \tag{27}$$

for $x_\iota \in E_\iota^{\alpha_\iota}$, $y_\iota \in E_\iota^{\alpha_\iota}$, where

$$\varepsilon(\alpha, \beta) = \prod_{\iota > j} \varepsilon_{\iota j}(\alpha_\iota, \beta_j). \tag{28}$$

The right hand side of (27) obviously belongs to $E^{\alpha+\beta}$ and the mapping $(x_1, \dots, x_n, y_1, \dots, y_n) \mapsto \varepsilon(\alpha, \beta) \bigotimes_{\iota\in I} (x_\iota y_\iota)$ is A-multilinear in the product of the $E_\iota^{\alpha_\iota}$ and the $E_\iota^{\beta_\iota}$ $(1 \leqslant i \leqslant n)$. Then it must be proved that the multiplication thus defined on E is *associative*; now, if $\gamma = (\gamma_i)$ is a third element of Δ and $z_i \in E_i^{\gamma_i}$ for $1 \leqslant i \leqslant n$, then

$$\Big(\Big(\bigotimes_i x_i\Big)\Big(\bigotimes_i y_i\Big)\Big)\Big(\bigotimes_i z_i\Big) = \varepsilon(\alpha + \beta, \gamma)\varepsilon(\alpha, \beta) \bigotimes_i (x_i y_i z_i)$$

$$\Big(\bigotimes_i x_i\Big)\Big(\Big(\bigotimes_i y_i\Big)\Big(\bigotimes_i z_i\Big)\Big) = \varepsilon(\alpha, \beta + \gamma)\varepsilon(\beta, \gamma) \bigotimes_i (x_i y_i z_i)$$

and it reduces to verifying the identity

$$\varepsilon(\alpha + \beta, \gamma)\varepsilon(\alpha, \beta) = \varepsilon(\alpha, \beta + \gamma)\varepsilon(\beta, \gamma).$$

But the latter follows immediately from the relations

$$\varepsilon(\alpha + \beta, \gamma) = \varepsilon(\alpha, \beta)\varepsilon(\beta, \gamma)$$

$$\varepsilon(\alpha, \beta + \gamma) = \varepsilon(\alpha, \beta)\varepsilon(\alpha, \gamma)$$

themselves immediate consequences of the definition (28) and (22) and (23).

If, for all $i \in I$, e_i denotes the unit element of E_i, we know that e_i is homogeneous of degree 0 (§ 3, no. 1), hence $e = \bigotimes_{i \in I} e_i$ is homogeneous of degree 0 and it follows from (27), (28) and the relations

$$\varepsilon_{ij}(\alpha_i, 0) = \varepsilon_{ij}(0, \beta_j) = 1$$

that e is unit element of E, which completes the definition on E of the desired graded A-algebra structure. Then take $h_i(x_i) = x_i \otimes \bigotimes_{j \neq i} e_j$; to verify that $h_i(x_i x_i') = h_i(x_i)h_i(x_i')$ for x_i, x_i' in E_i, attention may be confined to the case where x_i and x_i' are homogeneous and then this relation follows immediately from (27) and the relations $\varepsilon_{ij}(\alpha_i, 0) = \varepsilon_{ij}(0, \beta_j) = 1$; the same relations and (24) prove also that the h_i satisfy conditions (i) and (ii) of the statement and that

$$\bigotimes_{i \in I} x_i = \prod_{i \in I} h_i(x_i) \tag{29}$$

where the right hand side is the product of the *ordered sequence* $(h_i(x_i))_{i \in I}$ in E with the given total ordering on I (I, § 1, no. 2) (it suffices to argue by induction on the number of x_i (assumed homogeneous) distinct from the e_i).

It remains to prove condition (iii); note that the mapping

$$(x_i)_{i \in I} \mapsto \prod_{i \in I} f_i(x_i),$$

where the right hand side is the product of the *ordered sequence* $(f_i(x_i))_{i \in I}$ with the given total ordering on I, is A-multilinear. Then there exists one and only one A-linear mapping $f : E \to F$ such that

$$f\Big(\bigotimes_{i \in I} x_i\Big) = \prod_{i \in I} f_i(x_i). \tag{30}$$

Clearly $f(e)$ is the unit element of F and $f \circ h_i = f_i$; to verify that f is an algebra homomorphism, in other words that $f(x)f(y) = f(xy)$ for x, y in E, attention may be confined, by linearity, to the case where $x = \bigotimes_{i \in I} x_i$ and $y = \bigotimes_{i \in I} y_i$, x_i (resp. y_i) being homogeneous of degree α_i (resp. β_i) for all $i \in I$. The relation to be verified then reduces, by (27), to

$$\Big(\prod_{i \in I} f_i(x_i)\Big)\Big(\prod_{i \in I} f_i(y_i)\Big) = \varepsilon(\alpha, \beta) \prod_{i \in I} (f_i(x_i) f_i(y_i)).$$

But, taking account of relations (26), this is a consequence of Lemma 2 of no. 6.

Clearly the algebra E and the canonical mapping $\psi : \bigotimes_{i \in I} E_i \to E$ constitute a solution of the *universal mapping problem* (*Set Theory*, IV, § 3, no. 1), where Σ is the species of A-algebra structure and the α-mappings $\prod_i f_i$ from $\prod_i E_i$ to an A-algebra, satisfying conditions (26).

For a fixed total ordering on I, the graded algebra E defined in the proof of Proposition 10 will be called a *graded tensor* ε*-product of type* Δ of the family $(E_i)_{i \in I}$ of graded algebras of type Δ_i and will be denoted by ${}^{\varepsilon}\bigotimes_{i \in I} E_i$ (if no confusion can arise over the ordering on I); similarly, the homomorphism $f : E \to F$ defined in the proof of Proposition 10 will be denoted by ${}^{\varepsilon}\bigotimes_{i \in I} f_i$. The homomorphisms h_i are called *canonical*. We also write ${}^{\varepsilon}G^{\otimes n}$ when $I = [1, n]$ and all the E_i are equal to the same algebra G.

Remarks. (1) We recover the tensor product of algebras defined in no. 1 (with moreover the graduation the tensor product of those of its factors) taking $\varepsilon_{ij}(\alpha_i, \beta_j) = 1$ for all i, j, α_i and β_j.

(2) Suppose that all the Δ_i are equal to **Z** and write $\varepsilon_{ij}(\alpha_i, \beta_j) = (-1)^{\alpha_i \beta_j}$; the tensor ε-product ${}^{\varepsilon}\bigotimes_{i \in I} E_i$ corresponding to this system of commutation factors is then called the *skew* tensor product of the graded algebras E_i of type **Z** and denoted by ${}^{g}\bigotimes_{i \in I} E_i$ (or $E\ {}^{g}\!\otimes_A F$ for two algebras, or ${}^{g}G^{\otimes n}$ instead of ${}^{\varepsilon}G^{\otimes n}$).

COROLLARY 1. *In the notation of Proposition* 10, *suppose further that* F *is a graded* A-*algebra of type* Δ *and that each* f_i *is a graded algebra homomorphism relative to* $\phi_i:\Delta_i \to \Delta$; *then* $f = {}^{\varepsilon}\bigotimes_i f_i$ *is a graded algebra homomorphism.*

This follows immediately from the definition of f and the fact that

$$\sum_{i \in I} \phi_i(\alpha_i) = (\alpha_i)$$

by definition of the ϕ_i.

It is therefore seen that (E, ψ) is *also* a solution of another universal mapping problem, where this time Σ is the species of *graded* A-*algebra* structure *of type* Δ, the morphisms being graded algebra homomorphisms of type Δ and the α-mappings the mappings $\prod_i f_i$, where, in addition to conditions (26), it is assumed that f_i is a graded algebra homomorphism relative to ϕ_i.

COROLLARY 2. *Let* $(E_i)_{i \in I}$, $(F_i)_{i \in I}$ *be two finite families of* A-*algebras, with* E_i *and* F_i *graded of type* Δ_i *for all* $i \in I$. *For each* $i \in I$, *let* $g_i: E_i \to F_i$ *be a graded algebra homomorphism of type* Δ_i. *Then, if* $h_i: E_i \to {}^{\varepsilon}\bigotimes_{i \in I} E_i$ *and* $h'_i: F_i \to {}^{\varepsilon}\bigotimes_{i \in I} F_i$ *are the canonical homomorphisms, there exists one and only one homomorphism of graded* A-*algebras of type* Δ, $g: {}^{\varepsilon}\bigotimes_{i \in I} E_i \to {}^{\varepsilon}\bigotimes_{i \in I} F_i$ *such that* $g \circ h_i = h'_i \circ g_i$ *for all* $i \in I$. *Also, if each* g_i *is bijective, so is* g.

It suffices to apply Corollary 1 to $f_i = h'_i \circ g_i$, noting that conditions (26) then follow from relations (25) applied to the h'_i.

The homomorphism defined in Corollary 2 is also denoted by ${}^{\varepsilon}\bigotimes_i g_i$ (if no confusion can arise); if, for each $i \in I$, G_i is a third graded A-algebra of type Δ_i and $g'_i: F_i \to G_i$ a graded algebra homomorphism, then

$$(31) \qquad \left({}^{\varepsilon}\bigotimes_i g'_i\right) \circ \left({}^{\varepsilon}\bigotimes_i g_i\right) = {}^{\varepsilon}\bigotimes_i (g'_i \circ g_i),$$

as follows immediately from (30).

In the case of a *skew* tensor product of graded algebras of type $\mathbf{Z}$, we write ${}^{g}\bigotimes_i f_i$ instead of ${}^{\varepsilon}\bigotimes_i f_i$ for homomorphisms $f_i: E_i \to F_i$ of graded algebras of type $\mathbf{Z}$; when $I = \{1, 2\}$, this homomorphism is also denoted by $f_1 \,{}^{g}\!\otimes f_2$; when $I = [1, n]$ and all the E_i (resp. F_i) are equal and all the f_i equal to the same homomorphism f, we write ${}^{g}f^{\otimes n}$.

Remark. In the proof of Proposition 10, a total ordering on I was used to define an *algebra* structure on the tensor product $\bigotimes_{i \in I} E_i$ of the A-modules

E_ι. If the ordering on I is changed, another multiplicative structure arises on $\bigotimes_{\iota \in I} E_\iota$, but the new algebra thus obtained is *canonically isomorphic* to the above, since both are solutions of the same universal mapping problem. For example, when $I = \{1, 2\}$, the canonical isomorphism of the algebra $E_1 \,{}^{\varepsilon}\!\otimes_A E_2$ onto the algebra $E_2 \,{}^{\varepsilon}\!\otimes_A E_1$ maps $x_1 \otimes x_2$ to $\varepsilon_{2,1}(\alpha, \beta) x_2 \otimes x_1$, where x_1 is homogeneous of degree α and x_2 homogeneous of degree β.

Let J be a subset of I and, for each $i \in J$, consider the canonical homomorphism $h_i : E_i \to {}^{\varepsilon}\!\bigotimes_{\iota \in I} E_\iota = E$. By virtue of relations (25) a canonical homomorphism $h : E' = {}^{\varepsilon}\!\bigotimes_{\iota \in J} E_\iota \to E$ is derived canonically (by Proposition 10) from these homomorphisms, such that, for all $i \in J$, $h'_i = h \circ h_i$, h'_i being the canonical homomorphism $E_i \to E'$. Taking the total ordering on J induced by that chosen on I, we obtain

$$h\Big(\bigotimes_{\iota \in J} x_\iota\Big) = \prod_{\iota \in I} h_\iota(x_\iota) = \bigotimes_{\iota \in I} x'_\iota$$

where the middle term is the product of the *ordered sequence* $(h_\iota(x_\iota))_{\iota \in J}$ and where, in the right hand term, $x'_\iota = x_\iota$ for $\iota \in J$, $x'_\iota = e_\iota$ for $\iota \notin J$.

PROPOSITION 11. ("associativity" of the tensor ε-product). *In the notation of Proposition* 10, *let* $(J_\lambda)_{\lambda \in L}$ *be a partition of* I *and write* $\Delta'_\lambda = \prod_{\iota \in J_\lambda} \Delta_\iota$ *for all* $\lambda \in L$. *Let* E'_λ *be a graded tensor* ε*-product of type* Δ'_λ *of the family* $(E_\iota)_{\iota \in J_\lambda}$ *(for some total ordering chosen on* J_λ*). On the other hand, for* λ, μ *in* L *and* $\lambda \neq \mu$, *we write, for* $\alpha'_\lambda = (\alpha_\iota)_{\iota \in J_\lambda}$, $\beta'_\mu = (\beta_j)_{j \in J_\mu}$,

$$\varepsilon'_{\lambda\mu}(\alpha'_\lambda, \beta'_\mu) = \prod_{\iota \in J_\lambda,\, j \in J_\mu} \varepsilon_{\iota j}(\alpha_\iota, \beta_j). \tag{32}$$

Then $(\varepsilon'_{\lambda\mu})$ *is a system of commutation factors over the* Δ'_λ *with values in* A *and there exists one and only one homomorphism of graded algebras of type* Δ, $v : {}^{\varepsilon'}\!\bigotimes_{\lambda \in L} E'_\lambda \to {}^{\varepsilon}\!\bigotimes_{\iota \in I} E_\iota$, *such that*

$$v\Big(\bigotimes_{\lambda \in L}\Big(\bigotimes_{\iota \in J_\lambda} x_\iota\Big)\Big) = \bigotimes_{\iota \in I} x_\iota \tag{33}$$

for all $(x_\iota) \in \prod_{\iota \in I} E_\iota$, *provided that the total ordering is taken on* I *which induces on each* J_λ *the chosen total ordering, and which is such that, for* $\lambda < \mu$ *in* L, $i \in J_\lambda$ *and* $j \in J_\mu$, $i < j$.

The fact that the $\varepsilon'_{\lambda\mu}$ form a system of commutation factors is trivial. Let $h_{\iota,\lambda} : E_\iota \to E'_\lambda$, $h'_\lambda : E'_\lambda \to {}^{\varepsilon'}\!\bigotimes_{\lambda \in L} E'_\lambda$ be the canonical homomorphisms (for $\lambda \in L$, $i \in J_\lambda$) and write $h''_\iota = h'_\lambda \circ h_{\iota,\lambda}$; it will suffice by virtue of the uniqueness of the solution of a universal mapping problem, to show that $\overset{\varepsilon'}{\bigotimes}_{\lambda \in L} E'_\lambda$ and the h''_i satisfy the conditions of Proposition 10. Now, for all $\lambda \in L$, let $f'_\lambda : E'_\lambda \to F$ be the unique algebra homomorphism such that $f'_\lambda \circ h_{i,\lambda} = f_i$ for all $i \in J_\lambda$. We show that, for $\lambda \neq \mu$, $\alpha'_\lambda = (\alpha_i)_{i \in J_\lambda}$, $\beta'_\mu = (\beta_j)_{j \in J_\mu}$,

$$f'_\lambda(x'_\lambda) f'_\mu(x'_\mu) = \varepsilon'_{\lambda\mu}(\alpha'_\lambda, \beta'_\mu) f'_\mu(x'_\mu) f'_\lambda(x'_\lambda) \tag{34}$$

for $x'_\lambda \in E'_\lambda$ (resp. $x'_\mu \in E'_\mu$) homogeneous of degree α'_λ (resp. β'_μ); it suffices, by linearity, to verify it when $x'_\mu = \bigotimes_{i \in J_\lambda} x_i$, $x'_\mu = \bigotimes_{j \in J_\mu} x_j$, x_i (resp. x_j) being homogeneous of degree α_i (resp. β_j) in E_i (resp. E_j) for $i \in J_\lambda$, $j \in J_\mu$. But this follows from formula (30) which defines the f'_λ and Lemma 3 of no. 6, taking account of hypothesis (26) and definition (32). There is therefore one and only one algebra homomorphism $f : \overset{\varepsilon'}{\bigotimes}_{\lambda \in L} E'_\lambda \to F$ such that $f \circ h'_\lambda = f'_\lambda$ for all $\lambda \in L$; whence $f \circ h''_i = f_i$ for all $i \in I$ and the uniqueness of f is trivial.

8. TENSOR PRODUCT OF GRADED ALGEBRAS OF THE SAME TYPES

Assuming the hypotheses of no. 7, Proposition 10 hold, suppose further that all the Δ_i are equal to the *same commutative monoid* Δ_0; we can then consider on the tensor ε-product $\overset{\varepsilon}{\bigotimes}_{i \in I} E_i$ the *total graduation* of type Δ_0, associated with the graduation of type $\Delta = \Delta_0^{(I)}$ on this algebra (II, § 11, no. 1); we shall call $\overset{\varepsilon}{\bigotimes}_{i \in I} E_i$, with this graduation, a *graded tensor ε-product of type* Δ_0 of the family $(E_i)_{i \in I}$ of graded algebras of type Δ_0.

Always preserving the notation of Proposition 10 of no. 7, suppose that F is also a *graded* A-*algebra of type* Δ_0 and that the f_i are *homomorphisms of graded algebras of type* Δ_0. Then $f : \overset{\varepsilon}{\bigotimes}_{i \in I} E_i \to F$ is also a *homomorphism of graded algebras of type* Δ_0: for it follows from formula (30) (no. 7) that if x_i is homogeneous and of degree $\alpha_i \in \Delta_0$, $\bigotimes_{i \in I} x_i$ and $\prod_{i \in I} f_i(x_i)$ are both homogeneous of degree $\sum_{i \in I} \alpha_i \in \Delta_0$.

It can therefore be said that $\overset{\varepsilon}{\bigotimes}_{i \in I} E_i$, with the total graduation of type Δ_0, constitutes, together with the canonical mapping ψ, a solution of a third universal mapping problem, where Σ is the species of *graded* A-*algebra of type* Δ_0, the morphisms are homomorphisms of graded algebras of type Δ_0 and the α-mappings are the mappings $\prod_i f_i$, where, in addition to conditions (26) (of no. 7), it is assumed that each f_i is a homomorphism of graded algebras of type Δ_0.

For every subset J of I, the canonical homomorphism $\overset{\varepsilon}{\bigotimes}_{i \in J} E_i \to \overset{\varepsilon}{\bigotimes}_{i \in I} E_i$ (no. 7) is, in fact, a homomorphism of graded algebras of type Δ_0, as follows immediately from the above.

PROPOSITION 12. ("associativity" of the tensor ε-product of graded algebras of the same types). *With the notation of Proposition* 10 *of* no. 7, *suppose that all the Δ_i are equal to the same monoid Δ_0; let $(J_\lambda)_{\lambda \in L}$ be a partition of* I. *With the notation of Proposition* 11 *of* no. 7, *suppose that the right hand side of formula* (32) (no. 7) *depends only on the sums* $\alpha''_\lambda = \sum_{i \in J_\lambda} \alpha_i$, $\beta''_\mu = \sum_{j \in J_\mu} \beta_j$, *for every ordered pair* (λ, μ) *of distinct indices, all* $\alpha'_\lambda \in \Delta'_\lambda$ *and all* $\beta'_\mu \in \Delta'_\mu$; *let* $\varepsilon''_{\lambda\mu}(\alpha''_\lambda, \beta''_\mu)$ *denote the right hand side of* (32). *Then* $(\varepsilon''_{\lambda\mu})$ *is a system of commutation factors over the family* $(\Delta''_\lambda)_{\lambda \in L}$, *where* $\Delta''_\lambda = \Delta_0$ *for all* $\lambda \in L$. *If* E''_λ *is the graded tensor* ε-*product of type* Δ_0 *of the family* $(E_i)_{i \in J_\lambda}$, *there exists one and only one isomorphism of graded algebras of type* Δ_0, $w\colon \overset{\varepsilon''}{\bigotimes}_{\lambda \in L} E''_\lambda \to \overset{\varepsilon}{\bigotimes}_{i \in I} E_i$, *such that*

$$w\Big(\bigotimes_{\lambda \in L}\Big(\bigotimes_{i \in J_\lambda} x_i\Big)\Big) = \bigotimes_{i \in I} x_i \tag{35}$$

provided that total orderings are chosen on the J_λ *and on* I *as described in* no. 7, *Proposition* 11.

By the hypothesis, for γ, δ in Δ_0, $\varepsilon''_{\lambda\mu}(\gamma, \delta) = \varepsilon_{i_0 j_0}(\gamma, \delta)$ for some $i_0 \in J_\lambda$ and some $j_0 \in J_\mu$, as is seen by considering the elements $\alpha'_\lambda = (\alpha_i)_{i \in J_\lambda}$ and $\beta'_\mu = (\beta_j)_{j \in J_\mu}$ such that $\alpha_{i_0} = \gamma$, $\alpha_i = 0$ for $i \neq i_0$, $\beta_{j_0} = \delta$, $\beta_j = 0$ for $j \neq j_0$; it follows immediately that the $\varepsilon''_{\lambda\mu}$ form a system of commutation factors. The rest of the proof is then analogous to that of Proposition 11 (no. 7) and is left to the reader.

Note that the additional hypotheses of Proposition 12 are fulfilled when $\Delta_0 = \mathbf{Z}$ and that (ε_{ij}) is, either the trivial system of factors ($\varepsilon_{ij}(\alpha_i, \beta_j) = 1$ for all i, j), or the system of factors defined by $\varepsilon_{ij}(\alpha_i, \beta_j) = (-1)^{\alpha_i \beta_j}$; in the latter case, the right hand side of formula (32) is equal to $(-1)^\gamma$, where $\gamma = \sum_{i \in J_\lambda, j \in J_\mu} \alpha_i \beta_j = \Big(\sum_{i \in J_\lambda} \alpha_i\Big)\Big(\sum_{j \in J_\mu} \beta_j\Big)$.

Remarks. (1) Let I be an *infinite* indexing set and Δ_0 a commutative monoid; let $(\Delta_i)_{i \in I}$ denote the family such that $\Delta_i = \Delta_0$ for all i and suppose given for every ordered pair of distinct indices (i, j) of I a mapping $\varepsilon_{ij}\colon \Delta_i \times \Delta_j \to A$ satisfying conditions (22), (23) and (24) (no. 7); this will also be called a *system of commutation factors over the family* (Δ_i). Consider a family $(E_i)_{i \in I}$ of graded A-algebras of type Δ_0; for each finite subset J of I, let E_J denote a *graded tensor* ε-*product of type* Δ_0 of the subfamily $(E_i)_{i \in J}$ (with an arbitrary choice of a total ordering on J). If J, J′ are two finite subsets of I such that $J \subset J'$, a canonical homomorphism of graded algebras of type Δ_0,

$h_{J'J}: E_J \to E_{J'}$, has been defined above and the uniqueness properties of these homomorphisms show immediately that if $J \subset J' \subset J''$ are three finite subsets of I, then $h_{J''J} = h_{J''J'} \circ h_{J'J}$. Thus there is a direct system $(E_J, h_{J'J})$ of graded algebras of type Δ_0 (§ 3, no. 3), whose indexing set is the right directed set $\mathfrak{F}(I)$ of finite subsets of I. The graded algebra of type Δ_0, the *direct limit* of this direct system (§ 3, no. 3), is called a *graded tensor ε-product of type* Δ_0 of the family $(E_\iota)_{\iota \in I}$; it is also denoted by $\overset{\varepsilon}{\bigotimes}_{\iota \in I} E_\iota$. When all the Δ_ι are equal to **Z** and $\varepsilon_{\iota\jmath}(\alpha_\iota, \beta_\jmath) = (-1)^{\alpha_\iota \beta_\jmath}$, the tensor product $\overset{\varepsilon}{\bigotimes}_{\iota \in I} E_\iota$ is also called the *skew* tensor product of the family $(E_\iota)_{\iota \in I}$ and is denoted by $\overset{g}{\bigotimes}_{\iota \in I} E_\iota$. We leave to the reader the task of formulating and proving the proposition which generalizes Proposition 10 of no. 7 to the case where I is infinite, as Proposition 8 of no. 5 generalizes Proposition 5 of no. 2 to the case where I is infinite. Note that the underlying A-module of $\overset{\varepsilon}{\bigotimes}_{\iota \in I} E_\iota$ is the same as that underlying the (non-graded) tensor product of the family $(E_\iota)_{\iota \in I}$ of non-graded algebras defined in no. 5.

(2) Let E be a graded A-algebra of type Δ_0 (where Δ_0 is a commutative monoid) and $\rho: A \to B$ a ring homomorphism; the graduation on $\rho^*(E)$ (II, § 11, no. 5) is identical with the graduation on the graded tensor product $B \otimes_A E$, where B has the trivial graduation.

9. ANTICOMMUTATIVE ALGEBRAS AND ALTERNATING ALGEBRAS

DEFINITION 7. *A graded* A-*algebra* E *of type* **Z** *is called anticommutative if for all non-zero homogeneous elements x, y of* E

$$xy = (-1)^{\deg(x)\deg(y)} yx. \tag{36}$$

The algebra E *is called alternating if it is anticommutative and also $x^2 = 0$ for every homogeneous element $x \in E$ of odd degree.*

Remarks. (1) Let E^+ be the graded subalgebra of E the direct sum of the E_{2n} ($n \in \mathbf{Z}$); it follows from Definition 7 that if E is anticommutative, E^+ is a *subalgebra contained in the centre of* E (and hence commutative).

(2) Suppose that 2 is not a divisor of 0 in E; then if E is anticommutative E is alternating, since for $x \in E$ homogeneous and of odd degree, $x^2 = -x^2$ by (36), whence $2x^2 = 0$ and $x^2 = 0$ by virtue of the hypothesis.

(3) We shall study in detail in § 7 important examples of alternating algebras.

Lemma 4. *Let* E *be a graded algebra of type* **Z** *and* S *a set of homogeneous elements* $\neq 0$; *the set* F *of elements of* E *all of whose homogeneous components* $x \neq 0$ *satisfy relation* (36) *for all* $y \in S$ *is a graded subalgebra of* E.

It suffices to note that: (1) if x', x'' are two homogeneous elements of the same degree p, y a homogeneous element of degree q and $x'y = (-1)^{pq}yx'$, $x''y = (-1)^{pq}yx''$, then also $(x' + x'')y = (-1)^{pq}y(x' + x'')$; (2) if x', x'' are two homogeneous elements of respective degrees p', p'', y a homogeneous element of degree q and $x'y = (-1)^{p'q}yx'$, $x''y = (-1)^{p''q}yx''$, then

$$(x'x'')y = (-1)^{(p'+p'')q}y(x'x'').$$

PROPOSITION 13. *Let* E *be a graded* A*-algebra of type* **Z** *and* S *a generating system of the algebra* E *consisting of homogeneous elements* $\neq 0$*; for* E *to be anticommutative* (resp. *alternating*), *it is necessary and sufficient that* (36) *hold for all* $x \in S$ and $y \in S$ (resp. *that this condition hold and further that* $x^2 = 0$ *for all* x *homogeneous of odd degree belonging to* S).

We consider first the case of anticommutative algebras. By Lemma 4, the subalgebra F consisting of the elements all of whose homogeneous components $x \neq 0$ satisfy (36) for all $y \in S$, contains all the elements of S and hence F = E. If now F′ is similarly the subalgebra of E consisting of the elements all of whose homogeneous components $x \neq 0$ satisfy (36) for every homogeneous element $y \neq 0$, it follows from the above that F′ contains all the elements of S and hence F′ = E, which completes the proof of the proposition in this case.

To prove the proposition in the case of alternating algebras, it can be assumed that E is already anticommutative; it is then immediate that every homogeneous element of odd degree in E is of the form $\sum_i z_i x_i$, where $z_i \in E^+$ and $x_i \in S$ is of odd degree (using the fact that E^+ is contained in the centre of E); it follows that $\left(\sum_i z_i x_i\right)^2 = \sum_i z_i^2 x_i^2 + \sum_{i<j} z_i z_j (x_i x_j + x_j x_i) = 0$ since $x_i^2 = 0$ by hypothesis and $x_i x_j + x_j x_i = 0$ by (36).

PROPOSITION 14. *Let* E *and* F *be two graded* A*-algebras of type* **Z**, *both anticommutative* (resp. *alternating*). *Then the skew tensor product* $E\ {}^g\!\otimes_A F$ (no. 7) *is an anticommutative* (resp. *alternating*) *algebra.*

A generating system of $E\ {}^g\!\otimes_A F$ consists of the $x \otimes y$, where x (resp. y) is a homogeneous element $\neq 0$ of E (resp. F). Consider two such elements $x \otimes y$, $x' \otimes y'$, with $\deg(x) = p$, $\deg(y) = q$, $\deg(x') = p'$, $\deg(y') = q'$, so that $x \otimes y$ is of degree $p + q$ and $x' \otimes y'$ of degree $p' + q'$. Then by definition (no. 7, formula (27)) and by virtue of (36),

$$\begin{aligned}(x \otimes y)(x' \otimes y') &= (-1)^{qp'}(xx') \otimes (yy')\\ (x' \otimes y')(x \otimes y) &= (-1)^{pq'}(x'x) \otimes (y'y)\\ &= (-1)^{pq'+pp'+qq'}(xx') \otimes (yy')\end{aligned}$$

and the criterion of Proposition 13 shows that $E\ {}^g\!\otimes_A F$ is anticommutative since $pq' + pp' + qq' - qp' \equiv (p + q)(p' + q')$ (mod. 2). If further E and F are alternating and $p + q$ is odd, one of the numbers p, q is necessarily odd, hence $(x \otimes y)^2 = \pm (x^2) \otimes (y^2) = 0$ and Proposition 13 shows that $E \,{}^g\!\otimes_A F$ is alternating.

COROLLARY. *Let* E *be an anticommutative* (resp. *alternating*) *graded* A-*algebra of type* **Z**. *Then for every ring homomorphism* $\rho: A \to B$, *the graded* B-*algebra* $\rho^*(E)$ (no. 8, *Remark* 2) *is anticommutative* (resp. *alternating*).

The ring B with the trivial graduation can be considered as an alternating A-algebra and $\rho^*(E) = E \,{}^g\!\otimes_A B$, hence Proposition 14 can be applied.

Remark. Let E be an anticommutative graded A-algebra of type **Z**. Then the A-linear mapping of $E \otimes_A E$ into E defined by multiplication of E (§ 1, no. 3) is a homomorphism of the graded A-algebra $E \,{}^g\!\otimes_A E$ into E, for in the notation of Proposition 14, in the algebra E,

$$(xy)(x'y') = (-1)^{qp'}(xx')(yy').$$

§ 5. TENSOR ALGEBRA, TENSORS

1. DEFINITION OF THE TENSOR ALGEBRA OF A MODULE

Let A be a commutative ring and M an A-module. For every integer $n \geqslant 0$, the A-module the tensor product of n modules equal to M (also called the *n-th tensor power* of M) is denoted by $\bigotimes^n M$, or $M^{\otimes n}$, or $T^n(M)$, or $T^n_A(M)$, or $\mathrm{Tens}^n(M)$; then $T^1(M) = M$; also we write $T^0(M) = A$. The A-module the *direct sum* $\bigoplus_{n \geqslant 0} T^n(M)$ is denoted by $T(M)$ or $\mathrm{Tens}(M)$. We shall define a graded A-algebra structure of type **N** on $T(M)$, by defining for every ordered pair of integers $p \geqslant 0$, $q \geqslant 0$, an A-linear mapping

$$m_{pq}: T^p(M) \otimes_A T^q(M) \to T^{p+q}(M)$$

(§ 3, no. 1, *Remark*). For $p > 0$ and $q > 0$, m_{pq} is the associativity isomorphism (II, § 3, no. 9) and, when $p = 0$ (resp. $q = 0$), $m_{0,q}$ is the canonical isomorphism of $A \otimes_A T^q(M)$ onto $T^q(M)$ (resp. $m_{p,0}$ is the canonical isomorphism of $T^p(M) \otimes_A A$ onto $T^p(M)$ (II, § 3, no. 4, Proposition 4). Then, for $x_i \in M$, $\alpha \in A$,

$$(1) \quad \begin{cases} (x_1 \otimes \cdots \otimes x_p).(x_{p+1} \otimes \cdots \otimes x_{p+q}) \\ \qquad = x_1 \otimes \cdots \otimes x_p \otimes x_{p+1} \otimes \cdots \otimes x_{p+q} \\ \alpha.(x_1 \otimes \cdots \otimes x_p) = \alpha(x_1 \otimes \cdots \otimes x_p). \end{cases}$$

It is immediate that the multiplication thus defined on $T(M)$ is *associative* and admits as unit element the unit element 1 of $A = T^0(M)$.

DEFINITION 1. *For every module* M *over a commutative ring* A, *the tensor algebra of* M, *denoted by* T(M), *or* Tens(M), *or* $T_A(M)$, *is the algebra* $\bigoplus_{n \geq 0} T^n(M)$ *with the multiplication defined in* (1). *The canonical injection* $\phi: T^1(M) \to T(M)$ (II, § 1, no. 12) (also denoted by ϕ_M) *is called the canonical injection of* M *into* T(M).

PROPOSITION 1. *Let* E *be a (unital)* A-*algebra and* $f: M \to E$ *an* A-*linear mapping. There exists one and only one* A-*algebra homomorphism* $g: T(M) \to E$ *such that* $f = g \circ \phi$.

In other words, (T(M), ϕ) is a solution of the *universal mapping problem* (*Set Theory*, IV, § 3, no. 1), where Σ is the species of A-algebra structure, the α-mappings being the A-linear mappings from the module M to an A-algebra. Observe that here there is no question of a graduation on T(M).

For every finite family $(x_i)_{1 \leq i \leq n}$ of n elements of M, by definition of the product in T(M), $x_1 \otimes x_2 \otimes \cdots \otimes x_n = \phi(x_1)\phi(x_2)\ldots\phi(x_n)$; then necessarily $g(x_1 \otimes x_2 \otimes \cdots \otimes x_n) = f(x_1)f(x_2)\ldots f(x_n)$ for $n \geq 1$ and $g(\alpha) = \alpha e$ (if e is the unit element of E) for $\alpha \in A$, which proves the uniqueness of g. Conversely, note that, for all $n > 0$, the mapping

$$(x_1, \ldots, x_n) \mapsto f(x_1)f(x_2)\ldots f(x_n)$$

of M^n into E is A-multilinear; hence there corresponds to it an A-linear mapping $g_n: T^n(M) \to E$ such that

$$g(x_1 \otimes x_2 \otimes \cdots \otimes x_n) = f(x_1)f(x_2)\ldots f(x_n). \tag{2}$$

Also we define the mapping $g_0: T^0(M) \to E$ as equal to η_E (§ 1, no. 3), in other words $g_0(\alpha) = \alpha e$ for $\alpha \in A$. Let g be the unique A-linear mapping of T(M) into E whose restriction to $T^n(M)$ is g_n ($n \geq 0$); it is immediate that $g \circ \phi = g_1 = f$ and it remains to verify that g is an A-algebra homomorphism. By construction $g(1) = e$ and it suffices by linearity to show that $g(uv) = g(u)g(v)$ for $u \in T^p(M)$ and $v \in T^q(M)$ ($p > 0$, $q > 0$); now it follows from formulae (1) and (2) that this relation is true when $u = x_1 \otimes x_2 \otimes \cdots \otimes x_p$ and $v \in x_{p+1} \otimes \cdots \otimes x_{p+q}$ (where the x_i belong to E). It is therefore true for $u \in T^p(M)$ and $v \in T^q(M)$ by linearity.

Remark. Suppose that E is a *graded* A-algebra of type **Z**, with graduation (E_n), and suppose also that

$$f(M) \subset E_1. \tag{3}$$

Then it follows from (2) that $g(T^p(M)) \subset E_p$ for all $p \geq 0$ and hence g is a *graded algebra* homomorphism.

2. FUNCTORIAL PROPERTIES OF THE TENSOR ALGEBRA

PROPOSITION 2. *Let* A *be a commutative ring,* M *and* N *two* A-*modules and*

$$u: M \to N$$

an A*-linear mapping. There exists one and only one* A*-algebra homomorphism*

$$u' : \mathrm{T}(\mathrm{M}) \to \mathrm{T}(\mathrm{N})$$

such that the diagram

$$\begin{array}{ccc} \mathrm{M} & \xrightarrow{u} & \mathrm{N} \\ \downarrow{\scriptstyle \phi_\mathrm{M}} & & \downarrow{\scriptstyle \phi_\mathrm{N}} \\ \mathrm{T}(\mathrm{M}) & \xrightarrow[u']{} & \mathrm{T}(\mathrm{N}) \end{array}$$

is commutative. Further, u' is a graded algebra homomorphism.

The existence and uniqueness of u' follow from no. 1, Proposition 1, applied to the algebra $\mathrm{T}(\mathrm{N})$ and the linear mapping $\phi_\mathrm{N} \circ u : \mathrm{M} \to \mathrm{T}(\mathrm{N})$; as

$$u(\mathrm{M}) \subset \mathrm{T}^1(\mathrm{N}) = \mathrm{N},$$

the fact that u' is a graded algebra homomorphism follows from the *Remark* of no. 1.

The homomorphism u' of Proposition 2 will henceforth be denoted by $\mathrm{T}(u)$. If P is an A-module and $v : \mathrm{N} \to \mathrm{P}$ an A-linear mapping, then

$$\mathrm{T}(v \circ u) = \mathrm{T}(v) \circ \mathrm{T}(u) \tag{4}$$

for $\mathrm{T}(v) \circ \mathrm{T}(u)$ is an algebra homomorphism rendering commutative the diagram

$$\begin{array}{ccc} \mathrm{M} & \xrightarrow{v \circ u} & \mathrm{P} \\ \downarrow{\scriptstyle \phi_\mathrm{M}} & & \downarrow{\scriptstyle \phi_\mathrm{P}} \\ \mathrm{T}(\mathrm{M}) & \xrightarrow[\mathrm{T}(v) \circ \mathrm{T}(u)]{} & \mathrm{T}(\mathrm{P}) \end{array}$$

$\mathrm{T}(u)$ is sometimes called the *canonical extension* of u to $\mathrm{T}(\mathrm{M})$ (which contains $\mathrm{M} = \mathrm{T}^1(\mathrm{M})$). Note that the restriction $\mathrm{T}^n(u) : \mathrm{T}^n(\mathrm{M}) \to \mathrm{T}^n(\mathrm{N})$ is just the linear mapping $u^{\otimes n} = u \otimes u \otimes \cdots \otimes u$ (n times), for

$$\mathrm{T}^n(u)(x_1 \otimes \cdots \otimes x_n) = u(x_1) \otimes \cdots \otimes u(x_n)$$

since $\mathrm{T}(u)$ is an algebra homomorphism and $\mathrm{T}^1(u) = u$; the restriction $\mathrm{T}^0(u)$ to A is the identity mapping. $\mathrm{T}^n(u)$ is called the *n-th tensor power* of u.

PROPOSITION 3. *If $u : \mathrm{M} \to \mathrm{N}$ is a surjective* A*-linear mapping, the homomorphism $\mathrm{T}(u) : \mathrm{T}(\mathrm{M}) \to \mathrm{T}(\mathrm{N})$ is surjective and its kernel is the two-sided ideal of $\mathrm{T}(\mathrm{M})$ generated by the kernel $\mathrm{P} \subset \mathrm{M} \subset \mathrm{T}(\mathrm{M})$ of u.*

$\mathrm{T}^0(u) : \mathrm{T}^0(\mathrm{M}) \to \mathrm{T}^0(\mathrm{N})$ is bijective and for every integer $n > 0$,

$$\mathrm{T}^n(u) : \mathrm{T}^n(\mathrm{M}) \to \mathrm{T}^n(\mathrm{N})$$

is surjective, as is seen by induction on n using II, § 3, no. 6, Proposition 6; the latter proposition also shows, by induction on n, that the kernel $\mathfrak{J}_n$ of $\mathsf{T}^n(u)$ is the submodule of $\mathsf{T}^n(\mathrm{M})$ generated by the products

$$x_1 \otimes x_2 \otimes \cdots \otimes x_n$$

where at least one of the x_i belongs to P. This shows that the kernel $\mathfrak{J} = \bigoplus_{n \geqslant 1} \mathfrak{J}_n$ of $\mathsf{T}(u)$ is the two-sided ideal generated by P in $\mathsf{T}(\mathrm{M})$.

If $u : \mathrm{M} \to \mathrm{N}$ is an *injective* linear mapping, it is not always true that $\mathsf{T}(u)$ is an injective mapping (Exercise 1). However, this is true when u is an injection such that $u(\mathrm{M})$ is a *direct factor* of N, for then there exists a linear mapping $v : \mathrm{N} \to \mathrm{M}$ such that $v \circ u$ is the identity mapping on M and therefore

$$\mathsf{T}(v \circ u) = \mathsf{T}(v) \circ \mathsf{T}(u)$$

is the identity mapping on $\mathsf{T}(\mathrm{M})$, hence $\mathsf{T}(u)$ is injective and its image (isomorphic to $\mathsf{T}(\mathrm{M})$) is a *direct factor* of $\mathsf{T}(\mathrm{N})$ (II, § 1, no. 9, Proposition 15). More precisely:

PROPOSITION 4. *Let* N *and* P *be two submodules of an* A*-module* M *such that their sum* $\mathrm{N} + \mathrm{P}$ *is a direct factor in* M *and their intersection* $\mathrm{N} \cap \mathrm{P}$ *is a direct factor in* N *and in* P. *Then the homomorphisms* $\mathsf{T}(\mathrm{N}) \to \mathsf{T}(\mathrm{M})$, $\mathsf{T}(\mathrm{P}) \to \mathsf{T}(\mathrm{M})$ *and*

$$\mathsf{T}(\mathrm{N} \cap \mathrm{P}) \to \mathsf{T}(\mathrm{M}),$$

canonical extensions of the canonical injections, are injective; if $\mathsf{T}(\mathrm{N})$, $\mathsf{T}(\mathrm{P})$ *and* $\mathsf{T}(\mathrm{N} \cap \mathrm{P})$ *are identified with subalgebras of* $\mathsf{T}(\mathrm{M})$ *by means of these homomorphisms, then*

$$\mathsf{T}(\mathrm{N} \cap \mathrm{P}) = \mathsf{T}(\mathrm{N}) \cap \mathsf{T}(\mathrm{P}). \tag{5}$$

By hypothesis, there exist submodules $\mathrm{N}' \subset \mathrm{N}$ and $\mathrm{P}' \subset \mathrm{P}$ such that $\mathrm{N} = \mathrm{N}' \oplus (\mathrm{N} \cap \mathrm{P})$, $\mathrm{P} = \mathrm{P}' \oplus (\mathrm{N} \cap \mathrm{P})$; then

$$\mathrm{N} + \mathrm{P} = \mathrm{N}' \oplus \mathrm{P}' \oplus (\mathrm{N} \cap \mathrm{P})$$

and there exists by hypothesis a submodule M′ of M such that

$$\begin{aligned} \mathrm{M} = \mathrm{M}' \oplus (\mathrm{N} + \mathrm{P}) &= \mathrm{M}' \oplus \mathrm{N}' \oplus \mathrm{P}' \oplus (\mathrm{N} \cap \mathrm{P}) \\ &= \mathrm{M}' \oplus \mathrm{P}' \oplus \mathrm{N} = \mathrm{M}' \oplus \mathrm{N}' \oplus \mathrm{P}. \end{aligned}$$

In particular, $\mathrm{N} + \mathrm{P}$, N, P and $\mathrm{N} \cap \mathrm{P}$ are direct factors in M, which implies, as has been seen above, that the canonical homomorphisms

$$\mathsf{T}(\mathrm{N} + \mathrm{P}) \to \mathsf{T}(\mathrm{M}), \quad \mathsf{T}(\mathrm{N}) \to \mathsf{T}(\mathrm{M}),$$
$$\mathsf{T}(\mathrm{P}) \to \mathsf{T}(\mathrm{M}), \quad \mathsf{T}(\mathrm{N} \cap \mathrm{P}) \to \mathsf{T}(\mathrm{M})$$

are injective. The three algebras $\mathsf{T}(\mathrm{N})$, $\mathsf{T}(\mathrm{P})$ and $\mathsf{T}(\mathrm{N} \cap \mathrm{P})$ are thus identified with subalgebras of $\mathsf{T}(\mathrm{N} + \mathrm{P})$ and the latter with a subalgebra of $\mathsf{T}(\mathrm{M})$;

writing $Q = N \cap P$, it remains to show that, if $T(Q)$, $T(N' \oplus Q)$ and $T(P' \oplus Q)$ are identified with subalgebras of $T(N' \oplus P' \oplus Q)$, then

$$T(N' \oplus Q) \cap T(P' \oplus Q) = T(Q). \tag{6}$$

Now, consider the commutative diagram

$$\begin{array}{ccc} N' \oplus Q & \longrightarrow & N' \oplus P' \oplus Q \\ \downarrow & & \downarrow \\ Q & \longrightarrow & P' \oplus Q \end{array}$$

where the horizontal arrows are the canonical injections and the vertical arrows the canonical projections. We derive a commutative diagram

$$\begin{array}{ccc} T(N' \oplus Q) & \xrightarrow{u} & T(N' \oplus P' \oplus Q) \\ {\scriptstyle r}\downarrow & & \downarrow{\scriptstyle s} \\ T(Q) & \xrightarrow[v]{} & T(P' \oplus Q) \end{array} \tag{7}$$

where r and s are surjective homomorphisms (Proposition 3) and u and v injective homomorphisms. Hence, to prove (6), note that the right hand side is obviously contained in the left; it therefore suffices to verify that if

$$x \in T(N' \oplus Q) \cap T(P' \oplus Q),$$

then $x \in T(Q)$. Now the definition of the homomorphism s shows that its restriction to $T(P' \oplus Q)$ (identified with a subalgebra of $T(N' \oplus P' \oplus Q)$ is the identity mapping; the hypothesis on x therefore implies that $s(u(x)) = x$. But then also $v(r(x)) = x$, in other words x belongs to the image of $T(Q)$ in $T(P' \oplus Q)$, which was to be proved.

Remark. Note in particular that the hypotheses of Proposition 4 always hold for arbitrary submodules N, P of M when A is a *field* (II, § 7, no. 3, Proposition 4). Moreover, if $N \subset P$ and $N \neq P$, then $T^n(N) \neq T^n(P)$ for all $n \geqslant 1$, since if R is a complement of P in N, then $T^n(P) \cap T^n(R) = \{0\}$ by (4) and $T^n(R) \neq \{0\}$.

COROLLARY. *Let* K *be a commutative field and* M *a vector space over* K. *For every element* $z \in T(M)$, *there exists a smallest vector space* N *of* M *such that* $z \in T(N)$ *and* N *is of finite rank over* K.

It is understood in this statement that for every vector subspace P of M, $T(P)$ is canonically identified with a subalgebra of $T(M)$. Let $z \in T(M)$; z can be expressed as a linear combination of elements each of which is a finite product of elements of $M = T^1(M)$; all the elements of M which occur in these products generate a vector subspace Q of finite rank and $z \in T(Q)$.

Let $\mathfrak{F}$ be the (non-empty) set of vector subspaces P of finite rank such that $z \in \mathsf{T}(\mathrm{P})$. Every decreasing sequence of elements of $\mathfrak{F}$ is stationary, since they are vector spaces of finite rank. Hence $\mathfrak{F}$ has a minimal element N (*Set Theory*, III, § 6, no. 5). It remains to verify that every $\mathrm{P} \in \mathfrak{F}$ contains N; now, $z \in \mathsf{T}(\mathrm{P}) \cap \mathsf{T}(\mathrm{N}) = \mathsf{T}(\mathrm{P} \cap \mathrm{N})$ (Proposition 4); in view of the definition of N, this implies $\mathrm{N} \cap \mathrm{P} = \mathrm{N}$, that is $\mathrm{P} \supset \mathrm{N}$.

The subspace N of M is said to be *associated* with z.

3. EXTENSION OF THE RING OF SCALARS

Let A, A′ be two commutative rings and $\rho : \mathrm{A} \to \mathrm{A}'$ a ring homomorphism. Let M be an A-module, M′ an A′-module and $u : \mathrm{M} \to \mathrm{M}'$ an A-homomorphism; as the canonical injection $\phi_{\mathrm{M}'} : \mathrm{M}' \to \mathsf{T}_{\mathrm{A}'}(\mathrm{M}')$ is also an A-homomorphism (by restriction of scalars), so is the composition $\mathrm{M} \xrightarrow{u} \mathrm{M}' \xrightarrow{\phi_{\mathrm{M}'}} \mathsf{T}_{\mathrm{A}'}(\mathrm{M}')$. An A-algebra homomorphism $\mathsf{T}_{\mathrm{A}}(\mathrm{M}) \to \rho^*(\mathsf{T}_{\mathrm{A}'}(\mathrm{M}'))$ is derived (no. 2), also denoted by $\mathsf{T}(u) : \mathsf{T}_{\mathrm{A}}(\mathrm{M}) \to \mathsf{T}_{\mathrm{A}'}(\mathrm{M}')$, which is the unique A-homomorphism rendering commutative the diagram

$$\begin{array}{ccc} \mathrm{M} & \xrightarrow{u} & \mathrm{M}' \\ \downarrow{\scriptstyle \phi_{\mathrm{M}}} & & \downarrow{\scriptstyle \phi_{\mathrm{M}'}} \\ \mathsf{T}_{\mathrm{A}}(\mathrm{M}) & \xrightarrow[\mathsf{T}(u)]{} & \mathsf{T}_{\mathrm{A}'}(\mathrm{M}') \end{array} \tag{8}$$

If $\sigma : \mathrm{A}' \to \mathrm{A}''$ is a commutative ring homomorphism, M″ an A″-module and $v : \mathrm{M}' \to \mathrm{M}''$ an A′-homomorphism, the above uniqueness property shows that

$$\mathsf{T}(v \circ u) = \mathsf{T}(v) \circ \mathsf{T}(u). \tag{9}$$

PROPOSITION 5. *Let* A, B *be two commutative rings,* $\rho : \mathrm{A} \to \mathrm{B}$ *a ring homomorphism and* M *an* A-*module. The canonical extension*

$$\psi : \mathsf{T}_{\mathrm{B}}(\mathrm{B} \otimes_{\mathrm{A}} \mathrm{M}) \to \mathrm{B} \otimes_{\mathrm{A}} \mathsf{T}_{\mathrm{A}}(\mathrm{M})$$

of the B-*linear mapping* $1_{\mathrm{B}} \otimes \phi_{\mathrm{M}} : \mathrm{B} \otimes_{\mathrm{A}} \mathrm{M} \to \mathrm{B} \otimes_{\mathrm{A}} \mathsf{T}_{\mathrm{A}}(\mathrm{M})$ *is an isomorphism of graded* B-*algebras.*

Consider the two A-algebra homomorphisms: the canonical injection $j : \mathrm{B} = \mathsf{T}^0(\mathrm{B} \otimes_{\mathrm{A}} \mathrm{M}) \to \mathsf{T}(\mathrm{B} \otimes_{\mathrm{A}} \mathrm{M})$ and the homomorphism

$$h = \mathsf{T}(i) : \mathsf{T}(\mathrm{M}) \to \mathsf{T}(\mathrm{B} \otimes_{\mathrm{A}} \mathrm{M})$$

derived (cf. formula (8)) from the canonical A-linear mapping

$$i : \mathrm{M} \to \mathrm{B} \otimes_{\mathrm{A}} \mathrm{M}.$$

As $T^0(B \otimes_A M)$ is contained in the centre of $T(B \otimes_A M)$, Proposition 5 of § 4, no. 2 can be applied and an A-algebra homomorphism

$$\psi' : B \otimes_A T(M) \to T(B \otimes_A M)$$

is obtained such that, for $\beta \in B$, $x_i \in M$ for $1 \leqslant i \leqslant n$,

$$\psi'(\beta \otimes (x_1 \otimes x_2 \otimes \cdots \otimes x_n)) = \beta((1 \otimes x_1) \otimes (1 \otimes x_2) \otimes \cdots \otimes (1 \otimes x_n)),$$

which shows immediately that ψ' is also a B-algebra homomorphism. It suffices to prove that $\psi \circ \psi'$ and $\psi' \circ \psi$ are the identity mappings on $B \otimes_A T(M)$ and $T(B \otimes_A M)$ respectively. Now, these two algebras are generated by $B \otimes_A M$ and clearly $\psi \circ \psi'$ and $\psi' \circ \psi$ coincide with the identity mapping on $B \otimes_A M$, whence the conclusion.

4. DIRECT LIMIT OF TENSOR ALGEBRAS

Let $(A_\alpha, \phi_{\beta\alpha})$ be a directed direct system of commutative rings and $(M_\alpha, f_{\beta\alpha})$ a direct system of A_α-modules (II, § 6, no. 2); let $A = \varinjlim A_\alpha$ and $M = \varinjlim M_\alpha$, which is an A-module. For $\alpha \leqslant \beta$ an A_α-algebra homomorphism (no. 3, formula (8)) $f'_{\beta\alpha} = T(f_{\beta\alpha}) : T_{A\alpha}(M_\alpha) \to T_{A\beta}(M_\beta)$ is derived canonically from the A_α-homomorphism $f_{\beta\alpha} : M_\alpha \to M_\beta$ and it follows from (9) (no. 3) that $(T_{A\alpha}(M_\alpha), f'_{\beta\alpha})$ is a *direct system of* A_α-*algebras*. On the other hand let $f_\alpha : M_\alpha \to M$ be the canonical A_α-homomorphism; an A_α-algebra homomorphism $f'_\alpha : T_{A\alpha}(M_\alpha) \to T_A(M)$ is derived (no. 3, formula (8)) and it also follows from (9) (no. 3) that the f'_α constitute a direct system of A_α-homomorphisms.

PROPOSITION 6. *The* A-*homomorphism* $f' = \varinjlim f'_\alpha : \varinjlim T_{A_\alpha}(M_\alpha) \to T_A(M)$ *is a graded algebra isomorphism.*

To simplify we write $E = T_A(M)$ and $E' = \varinjlim T_{A_\alpha}(M_\alpha)$ and let

$$g_\alpha : T_{A_\alpha}(M_\alpha) \to E'$$

be the canonical A_α-homomorphism. Clearly the composite A_α-linear mappings $M_\alpha \xrightarrow{\phi_{M_\alpha}} T_{A_\alpha}(M_\alpha) \xrightarrow{g_\alpha} E'$ form a direct system and there is therefore one and only one A-linear mapping $u = \varinjlim(g_\alpha \circ \phi_{M_\alpha}) : M \to E'$ such that

$$u \circ f_\alpha = g_\alpha \circ \phi_{M_\alpha}$$

for all α. This mapping itself factorizes uniquely (no. 1, Proposition 1) into $M \xrightarrow{\phi_M} E \to E'$, where h is an A-algebra homomorphism. It will suffice to prove that $h \circ f' = 1_{E'}$ and $f' \circ h = 1_E$.

To this end note that, for every index α, (no. 3, formula (8))

$$h \circ f'_\alpha \circ \phi_{M_\alpha} = h \circ \phi_M \circ f_\alpha = u \circ f_\alpha = g_\alpha \circ \phi_{M_\alpha}$$

whence, by the uniqueness assertion of no. 1, Proposition 1,

$$h \circ f'_\alpha = g_\alpha$$

for all α; it follows that $(h \circ f') \circ g_\alpha = g_\alpha$ for all α and hence $h \circ f' = 1_E$, by definition of a direct limit.

On the other hand, by virtue of no. 3, formula (8),

$$f' \circ u \circ f_\alpha = f' \circ g_\alpha \circ \phi_{M_\alpha} = f'_\alpha \circ \phi_{M_\alpha} = \phi_M \circ f_\alpha,$$

whence again $f' \circ u = \phi_M$ by definition of a direct limit; we conclude that $f' \circ h \circ \phi_M = \phi_M$ and the uniqueness property of no. 1, Proposition 1 gives $f' \circ h = 1_E$.

Proposition 6 can also be shown by observing that, for every integer $n \geqslant 1$, there is a canonical A-module isomorphism $\varinjlim \mathrm{T}^n_{A_\alpha}(M_\alpha) \to \mathrm{T}^n_A(M)$, as follows by induction on n from II, § 6, no. 3, Proposition 7. It is immediately verified that these isomorphisms are the restrictions of f'.

5. TENSOR ALGEBRA OF A DIRECT SUM. TENSOR ALGEBRA OF A FREE MODULE. TENSOR ALGEBRA OF A GRADED MODULE

Let A be a commutative ring and $M = \bigoplus_{\lambda \in L} M_\lambda$ the direct sum of a family of A-modules. It follows from II, § 3, no. 7, Proposition 7, by induction on n, that $\mathrm{T}^n(M)$ is the direct sum of the submodules the images of the canonical injections

$$M_{\lambda_1} \otimes M_{\lambda_2} \otimes \cdots \otimes M_{\lambda_n} \to \mathrm{T}^n(M) = M^{\otimes n}$$

relative to *all* the sequences $(\lambda_i) \in L^n$. Identifying $M_{\lambda_1} \otimes M_{\lambda_2} \otimes \cdots \otimes M_{\lambda_n}$ with this image, it is seen that T(M) is the *direct sum of all the modules*

$$M_{\lambda_1} \otimes M_{\lambda_2} \otimes \cdots \otimes M_{\lambda_n}$$

where n runs through **N** and, for each n, (λ_i) runs through L^n.

We first deduce the following consequence:

THEOREM 1. *Let* A *be a commutative ring,* M *a free* A-*module and* $(e_\lambda)_{\lambda \in L}$ *a basis of* M. *Then the elements* $e_s = e_{\lambda_1} \otimes e_{\lambda_2} \otimes \cdots \otimes e_{\lambda_n}$, *where* $s = (\lambda_1, \ldots, \lambda_n)$ *runs through the set of all finite sequences of elements of* L *and* $e_\varnothing$ *is used to denote the unit element of* T(M), *form a basis of the* A-*module* T(M).

The elements of this basis are obviously homogeneous and the multiplication table is given by

$$e_s e_t = e_{st} \tag{10}$$

where st denotes the sequence of elements of L obtained by *juxtaposing* the sequences s and t (I, § 7, no. 2).

It is seen that the basis (e_s) of $\mathsf{T}(M)$, with the multiplicative law (10), is canonically isomorphic to the *free monoid* of the set L (I, § 7, no. 2), the isomorphism being obtained by mapping each word s of this monoid to the element e_s. It follows (§ 2, no. 7) that *the tensor algebra* $\mathsf{T}(M)$ *of a free module* M, *with a basis of indexing set* L, *is canonically isomorphic to the free associative algebra of* L *over* A. In particular (§ 2, no. 7, Proposition 7), for every mapping $f: L \to E$ from L to an A-algebra E, there exists one and only one A-algebra homomorphism $\bar{f}: \mathsf{T}(M) \to E$ such that $\bar{f}(e_\lambda) = f(\lambda)$.

Remark. The above results can equally be obtained as a consequence of the universal properties of the free associative algebra of the tensor algebra, using II, § 3, no. 7, Corollary 2 to Proposition 7.

COROLLARY. *If* M *is a projective* A*-module,* $\mathsf{T}(M)$ *is a projective* A*-module.*

M is a direct factor of a free A-module N (II, § 2, no. 2, Proposition 4) and hence $\mathsf{T}(M)$ is a direct factor of $\mathsf{T}(N)$ (no. 2); as $\mathsf{T}(N)$ is free (Theorem 1), this shows that $\mathsf{T}(M)$ is projective (II, § 2, no. 2).

PROPOSITION 7. *Let* Δ *be a commutative monoid,* M *a graded* A*-module of type* Δ *and* $(M_\alpha)_{\alpha \in \Delta}$ *its graduation. For every ordered pair* $(\alpha, n) \in \Delta \times \mathbf{N}$, *let* $\mathsf{T}^{\alpha, n}(M)$ *be the (direct) sum of the submodules* $M_{\alpha_1} \otimes M_{\alpha_2} \otimes \cdots \otimes M_{\alpha_n}$ *of* $\mathsf{T}^n(M)$ *such that* $\sum_{i=1}^{n} \alpha_i = \alpha$; *then* $(\mathsf{T}^{\alpha, n}(M))_{(\alpha, n) \in \Delta \times \mathbf{N}}$ *is the only graduation of type* $\Delta \times \mathbf{N}$ *compatible with the algebra structure on* $\mathsf{T}(M)$ *and inducing on* $M = \mathsf{T}^1(M)$ *the given gradation.*

It has been seen at the beginning of this no. that $\mathsf{T}(M)$ is the direct sum of the $\mathsf{T}^{\alpha, n}(M)$ and the fact that this is a graduation compatible with the algebra structure follows immediately from the definitions. If $(\mathsf{T}'^{\alpha, n})$ is another graduation of type $\Delta \times \mathbf{N}$ on $\mathsf{T}(M)$ compatible with the algebra structure and such that $\mathsf{T}^{\alpha, 1}(M) = \mathsf{T}'^{\alpha, 1}$ for $\alpha \in \Delta$, it follows immediately from the definitions that, for all $n \geqslant 1$ and all $\alpha \in \Delta$, $\mathsf{T}^{\alpha, n}(M) \subset \mathsf{T}'^{\alpha, n}$; but since $\mathsf{T}(M)$ is also the direct sum of the $\mathsf{T}^{\alpha, n}(M)$, this implies that $\mathsf{T}'^{\alpha, n} = \mathsf{T}^{\alpha, n}(M)$ (II, § 1, no. 8, *Remark*).

6. TENSORS AND TENSOR NOTATION

Let A be a commutative ring, M an A-module, M^* the *dual* of M (II, § 2, no. 3) and I and J two *disjoint finite* sets; the A-module $\bigotimes_{\iota \in I \cup J} E_\iota$, where $E_\iota = M$ if $\iota \in I$, $E_\iota = M^*$ if $\iota \in J$, is denoted by $\mathsf{T}^I_J(M)$; the elements of $\mathsf{T}^I_J(M)$ are called *tensors of type* (I, J) over M. They are called *contravariant* if $J = \varnothing$, *covariant* if $I = \varnothing$ and *mixed* otherwise.

Let I′, I″ be two subsets of I and J′, J″ two subsets of J such that $I' \cup I'' = I$,

$I' \cap I'' = \varnothing$, $J' \cup J'' = J$, $J' \cap J'' = \varnothing$; then there is a canonical associativity isomorphism (II, § 3, no. 9)

$$\mathrm{T}^{\mathrm{I}}_{\mathrm{J}}(\mathrm{M}) \to \mathrm{T}^{\mathrm{I}'}_{\mathrm{J}'}(\mathrm{M}) \otimes_{\mathrm{A}} \mathrm{T}^{\mathrm{I}''}_{\mathrm{J}''}(\mathrm{M}). \tag{11}$$

Considering the tensor algebra $\mathrm{T}(\mathrm{M} \oplus \mathrm{M}^*)$, it follows from no. 5 that $\mathrm{T}^n(\mathrm{M} \oplus \mathrm{M}^*)$ is canonically identified with the direct sum of the $\mathrm{T}^{\mathrm{I}}_{\mathrm{J}}(\mathrm{M})$ where I runs through the set of subsets of the interval $(1, n)$ of **N** and J is the complement of I in $(1, n)$.

When $\mathrm{I} = (1, p)$ and $\mathrm{J} = (p + 1, p + q)$ with integers $p \geqslant 0$, $q \geqslant 0$ (where we replace I (resp. J) by $\varnothing$ when $p = 0$ (resp. $q = 0$)), the A-module $\mathrm{T}^{\mathrm{I}}_{\mathrm{J}}(\mathrm{M})$ is also denoted by $\mathrm{T}^p_q(\mathrm{M})$; the A-modules $\mathrm{T}^n_0(\mathrm{M})$ and $\mathrm{T}^0_n(\mathrm{M})$ are therefore by definition the A-modules $\mathrm{T}^n(\mathrm{M})$ and $\mathrm{T}^n(\mathrm{M}^*)$ respectively. When I and J are arbitrary finite sets of cardinals $p = \mathrm{Card}(\mathrm{I})$ and $q = \mathrm{Card}(\mathrm{J})$, we give each of them a total ordering; then there exists an increasing bijection of I (resp. J) onto $(1, p)$ (resp. $(p + 1, p + q)$) and these bijections therefore define an isomorphism

$$\mathrm{T}^{\mathrm{I}}_{\mathrm{J}}(\mathrm{M}) \to \mathrm{T}^p_q(\mathrm{M}).$$

When M is a *finitely generated projective* A-module, it follows from II, § 2, no. 7, Corollary 4 to Proposition 13 and II, § 4, no. 4, Corollary 1 to Proposition 4 that there is a canonical isomorphism

$$(\mathrm{T}^{\mathrm{I}}_{\mathrm{J}}(\mathrm{M}))^* \to \mathrm{T}^{\mathrm{J}}_{\mathrm{I}}(\mathrm{M}).$$

Suppose now that M is a *finitely generated free* A-module and let $(e_\lambda)_{\lambda \in \mathrm{L}}$ be a basis of M (L therefore being a *finite* set). The basis of M^* *dual* to (e_λ) (II, § 2, no. 6) is denoted by $(e^\lambda)_{\lambda \in \mathrm{L}}$. The bases (e_λ) and (e^λ) of M and M^* respectively define (no. 5) a basis of $\mathrm{T}^{\mathrm{I}}_{\mathrm{J}}(\mathrm{M})$, which we give explicitly as follows: given two mappings $f: \mathrm{I} \to \mathrm{L}$ and $g: \mathrm{J} \to \mathrm{L}$, let e^g_f be the element $\bigotimes_{i \in \mathrm{I} \cup \mathrm{J}} x_i$ of $\mathrm{T}^{\mathrm{I}}_{\mathrm{J}}(\mathrm{M})$ defined by

$$x_i = e_{f(i)} \quad \text{if } i \in \mathrm{I}, \qquad x_i = e^{g(i)} \quad \text{if } i \in \mathrm{J}.$$

When (f, g) runs through the set of ordered pairs of mappings $f: \mathrm{I} \to \mathrm{L}$ and $g: \mathrm{J} \to \mathrm{L}$, the e^g_f form a *basis* of the A-module $\mathrm{T}^{\mathrm{I}}_{\mathrm{J}}(\mathrm{M})$, said to be *associated* with the given basis (e_λ) of M. For $z \in \mathrm{T}^{\mathrm{I}}_{\mathrm{J}}(\mathrm{M})$, we can therefore write

$$z = \sum_{(f, g)} \alpha^f_g(z) . e^g_f \tag{12}$$

where the α^f_g are the coordinate forms relative to the basis (e^g_f); by an abuse of language, the $\alpha^f_g(z)$ are called the coordinates of the tensor z *with respect to the basis* (e_λ) of the module M. The α^f_g constitute the dual basis of (e^g_f), in other words they are identified with the elements of the basis of $\mathrm{T}^{\mathrm{J}}_{\mathrm{I}}(\mathrm{M})$, *associated* with (e_λ). When I and J are complementary subsets of an interval $(1, n)$ of **N**, α^f_g (or $\alpha^f_g(z)$) is denoted as follows: each element $f(i)$ for $i \in \mathrm{I}$ is written as an upper index in the i-th place with a dot in the i-th place for $i \in J$; similarly, $g(i)$ for $i \in J$ is written as a lower index in the i-th place with a dot in the i-th place for $i \in I$. For example, for $I = \{1, 4\}$, $J = \{2, 3\}$, we write $\alpha^{\lambda\cdot\cdot\mu}_{\cdot\nu\rho\cdot}$ if $f(1) = \lambda$, $f(4) = \mu$, $g(2) = \nu$, $g(3) = \rho$.

Let $(\bar{e}_\lambda)_{\lambda \in L}$ be another basis of M and P the matrix of passing from (e_λ) to $(\bar{e}_\lambda)$ (II, § 10, no. 8). Then the matrix of passing from (e^λ) to $(\bar{e}^\lambda)$ (dual basis of $(\bar{e}_\lambda)$) is the *contragredient* ${}^tP^{-1}$ of P (II, § 10, no. 8, Proposition 5). It follows (II, § 10, no. 10) that the matrix of passing from the basis (e^g_f) of $T^I_J(M)$ to the basis $(\bar{e}^g_f)$ (where f (resp. g) runs through the set of mappings of I into L (resp. of J into L)) is the matrix

$$(13) \qquad \bigotimes_{\iota \in I \cup J} Q_\iota, \quad \text{where } Q_\iota = P \text{ if } \iota \in I,\ Q_\iota = {}^tP^{-1} \text{ if } \iota \in J.$$

The transpose of this matrix is therefore identified with

$$(14) \qquad \bigotimes_{\iota \in I \cup J} R_\iota, \quad \text{where } R_\iota = {}^tP^{-1} \text{ if } \iota \in I,\ R_\iota = P \text{ if } \iota \in J.$$

Suppose now that M is an arbitrary module. Let $i \in I$, $j \in J$ and write $I' = I - \{i\}$, $J' = J - \{j\}$; we shall define a canonical A-linear mapping

$$c^i_j : T^I_J(M) \to T^{I'}_{J'}(M),$$

called *contraction of the index i and the index j*. For this, note that the mapping of $M^I \times (M^*)^J$, which associates with every family $(x_\iota)_{\iota \in I \cup J}$, where $x_\iota \in M$ if $\iota \in I$ and $x_\iota \in M^*$ if $\iota \in J$, the element

$$(15) \qquad \langle x_i, x_j \rangle \bigotimes_{k \in (I \cup J) - \{i, j\}} x_k$$

of $T^{I'}_{J'}(M)$, is A-*multilinear*; c^i_j is the corresponding A-linear mapping.

Suppose now that M is free and finitely generated and let $(e_\lambda)_{\lambda \in L}$ be a basis of M. Given two mappings $f: I \to L$, $g: J \to L$, let f_i denote the restriction of f to $I' = I - \{i\}$ and g_j the restriction of g to $J' = J - \{j\}$; then by virtue of (12)

$$(16) \qquad c^i_j(e^g_f) = \begin{cases} 0 & \text{if } f(i) \neq g(j) \\ e^{g_j}_{f_i} & \text{if } f(i) = g(j) \end{cases}$$

The expression for the coordinates of $c^i_j(z)$ as a function of those of z is obtained; for every mapping f' (resp. g') of I' into L (resp. of J' into L) and every $\lambda \in L$, let (f', λ) (resp. (g', λ)) denote the mapping of I into L (resp. of J into L) whose restriction to I' (resp. J') is f' (resp. g') and which takes the value λ at the element i (resp. j). Then, if the coordinate forms relative to the basis $(e^{g'}_{f'})$ of $T^{I'}_{J'}(M)$ are denoted by $\beta^{f'}_{g'}$,

$$(17) \qquad \beta^{f'}_{g'}(c^i_j(z)) = \sum_{\lambda \in L} \alpha^{(f', \lambda)}_{(g', \lambda)}(z).$$

Examples of tensors. (1) Let M be a *finitely generated projective* A-module. We know (II, § 4, no. 2, Corollary to Proposition 2), that there is a canonical A-module isomorphism

$$\theta_M : M^* \otimes_A M \to \mathrm{End}_A(M)$$

such that $\theta_M(x^* \otimes x)$ (for $x \in M$, $x^* \in M^*$) is the endomorphism

$$y \mapsto \langle y, x^* \rangle x.$$

Hence, by means of θ_M, $T^{\{2\}}_{\{1\}}(M)$ (isomorphic to $T^1_1(M)$) can be identified with the A-module $\mathrm{End}_A(M)$. Suppose that M is a free module and let $(e_\lambda)_{\lambda \in L}$ be a basis of M; then the coordinates of a tensor $z \in M^* \otimes M$ relative to the basis $(e^\mu \otimes e_\lambda)$ of this module are denoted by $\zeta_{\mu\cdot}^{\cdot\lambda}$. As $\theta_M(e^\mu \otimes e_\lambda)$ is the endomorphism $y \mapsto \langle y, e^\mu \rangle e_\lambda$, the endomorphism $u = \theta_M(z) = \theta_M\Big(\sum_{\lambda, \mu} \zeta_{\mu\cdot}^{\cdot\lambda} e^\mu \otimes e_\lambda\Big)$ maps y to $\sum_{\lambda, \mu} \zeta_{\mu\cdot}^{\cdot\lambda} \langle y, e^\mu \rangle e_\lambda$; writing $y = e_\lambda$, we obtain the relation

$$u(e_\lambda) = \sum_{\mu \in L} \zeta_{\lambda\cdot}^{\cdot\mu} e_\mu \tag{18}$$

in other words, *the matrix of the linear mapping* $u = \theta_M(z)$ *is that whose element in the row of index* μ *and column of index* λ *is* $\zeta_{\lambda\cdot}^{\cdot\mu}$.

The definition of the *trace* of u (II, § 4, no. 3) shows immediately that

$$\mathrm{Tr}(\theta_M(z)) = c^2_1(z).$$

Therefore the element $z_0 = \sum_{\lambda \in L} e^\lambda \otimes e_\lambda$ (whose coordinates $\zeta_{\mu\cdot}^{\cdot\lambda}$ are zero for $\lambda \neq \mu$ and equal to 1 for $\lambda = \mu$), which is such that $\theta_M(z_0) = 1_M$, is the image of the element $1 \in A = T^0_0(M)$ under the mapping the *transpose of the contraction*

$$c^2_1 : T^{\{2\}}_{\{1\}}(M) \to A.$$

(2) Suppose always that M is a *finitely generated projective* A-module; there is a canonical A-module isomorphism

$$\mu : M^* \otimes_A M^* \to (M \otimes_A M)^*$$

(II, § 4, no. 4, Corollary 1 to Proposition 2) and a canonical isomorphism

$$\theta : (M \otimes_A M)^* \otimes_A M \to \mathrm{Hom}_A(M \otimes_A M, M)$$

(II, § 4, no. 2, Corollary to Proposition 2); also $\mathrm{Hom}_A(M \otimes_A M, M)$ is canonically isomorphic to the A-module $\mathscr{L}_2(M, M; M)$ of A-*bilinear* mappings of $M \times M$ into M (II, § 3, no. 9). Composing these isomorphisms, a canonical isomorphism is obtained

$$\chi_M : T^{\{3\}}_{\{1,2\}}(M) = M^* \otimes M^* \otimes M \to \mathscr{L}_2(M, M; M)$$

such that, for x^*, y^* in M^*, $z \in M$, $\chi_M(x^* \otimes y^* \otimes z)$ is the bilinear mapping

$$(u, v) \mapsto \langle u, x^*\rangle\langle v, y^*\rangle z.$$

Hence, by means of χ_M, $T^{(3)}_{(1,2)}(M)$ (isomorphic to $T^1_2(M)$) can be identified with the A-module $\mathscr{L}_2(M, M; M)$. Suppose that M is a free A-module and let $(e_\lambda)_{\lambda \in L}$ be a basis of M; then the coordinates of a tensor $z \in M^* \otimes M^* \otimes M$ relative to the basis $(e^\lambda \otimes e^\mu \otimes e_\nu)$ of this module are denoted by $\zeta_{\lambda\mu\cdot}^{\cdot\cdot\nu}$. The bilinear mapping $\chi_M(z)$ maps the ordered pair (e_λ, e_μ) to

$$\sum_{\nu \in L} \zeta_{\lambda\mu\cdot}^{\cdot\cdot\nu} e_\nu$$

and therefore the $\zeta_{\lambda\mu\cdot}^{\cdot\cdot\nu}$ are just the *constants of structure* of the (in general non-associative) algebra defined on M by the bilinear mapping $\chi_M(z)$, with respect to the basis (e_λ) (§ 1, no. 7).

Remark 2. Let $(e_\lambda)_{\lambda \in L}$, $(\bar{e}_\lambda)_{\lambda \in L}$ be two bases of M and P the matrix of passage from (e_λ) to $(\bar{e}_\lambda)$. On account of what was seen in *Example* 1, the element of P appearing in the row of index λ and the column of index μ is denoted by α_μ^λ and the element of the contragredient ${}^tP^{-1}$ appearing in the row of index λ and the column of index μ is denoted by β_λ^μ. Then (in the notation introduced above)

$$\text{(19)} \qquad \begin{cases} \bar{e}_\mu = \sum_\lambda \alpha_\mu^\lambda e_\lambda \\ \bar{e}^\mu = \sum_\lambda \beta_\lambda^\mu e^\lambda \end{cases}$$

$$\text{(20)} \qquad \bar{e}_{f'}^{g'} = \sum_{(f,g)} \Big(\prod_{i \in I} \alpha_{f'(i)}^{f(i)}\Big)\Big(\prod_{j \in J} \beta_{g(j)}^{g'(j)}\Big) e_f^g$$

for all mappings $f' : I \to L$ and $g' : J \to L$. The coordinates ζ_g^f of a tensor $z \in T_J^I(M)$ with respect to the basis (e_λ) can therefore be expressed in terms of the coordinates $\bar{\zeta}_{g'}^{f'}$ of z with respect to the basis $(\bar{e}_\lambda)$ using the formulae

$$\text{(21)} \qquad \zeta_g^f = \sum_{(f',g')} \Big(\prod_{i \in I} \alpha_{f'(i)}^{f(i)}\Big)\Big(\prod_{j \in J} \beta_{g(j)}^{g'(j)}\Big) \bar{\zeta}_{g'}^{f'}.$$

The matrix P^{-1} of passage from the basis $(\bar{e}_\lambda)$ to the basis (e_λ) is the transpose of ${}^tP^{-1}$, so that β_λ^μ is the element which appears in the column of index λ and the row of index μ of P^{-1}. The calculation of e_f^g in terms of the $\bar{e}_{f'}^{g'}$ and that of the $\bar{\zeta}_{g'}^{f'}$ in terms of the ζ_g^f are therefore made by replacing α_μ^λ by β_λ^μ and β_λ^μ by α_μ^λ in the above calculations and exchanging the roles of f and f' and those of g and g'. Then

$$\text{(22)} \qquad e_f^g = \sum_{(f',g')} \Big(\prod_{j \in J} \alpha_{g'(j)}^{g(j)}\Big)\Big(\prod_{i \in I} \beta_{f(i)}^{f'(i)}\Big) \bar{e}_{f'}^{g'}$$

$$(23) \qquad \bar{\zeta}^{f'}_{g'} = \sum_{(f,g)} \left(\prod_{j \in J} \alpha^{g(j)}_{g'(j)}\right)\left(\prod_{i \in I} \beta^{f'(i)}_{f(i)}\right)\zeta^{f}_{g}.$$

The above formulae are such that the summation is over indices which appear once as a lower index and once as an upper index. Certain authors allow themselves because of this circumstance to suppress the summation signs.

§ 6. SYMMETRIC ALGEBRAS

1. DEFINITION OF THE SYMMETRIC ALGEBRA OF A MODULE

DEFINITION 1. *Let* A *be a commutative ring and* M *an* A*-module. The symmetric algebra of* M, *denoted by* S(M), *or* Sym(M), *or* $S_A(M)$, *is the quotient algebra over* A *of the tensor algebra* T(M) *by the two-sided ideal* $\mathfrak{J}'$ (*also denoted by* $\mathfrak{J}'_M$) *generated by the elements* $xy - yx = x \otimes y - y \otimes x$ *of* T(M), *where* x *and* y *run through* M.

Since the ideal $\mathfrak{J}'$ is generated by homogeneous elements of degree 2, it is a *graded ideal* (II, § 11, no. 3, Proposition 2); we write $\mathfrak{J}'_n = \mathfrak{J}' \cap T^n(M)$; the algebra S(M) is then graded by the graduation (called *canonical*) consisting of the $S^n(M) = T^n(M)/\mathfrak{J}'_n$. Now $\mathfrak{J}'_0 = \mathfrak{J}'_1 = \{0\}$ and hence $S^0(M)$ is canonically identified with A and $S^1(M)$ with $T^1(M) = M$; we shall always make these identifications in what follows and denote by ϕ' or ϕ'_M the canonical injection $M \to S(M)$.

PROPOSITION 1. *The algebra* S(M) *is commutative.*

By definition $\phi'(x)\phi'(y) = \phi'(y)\phi'(x)$ for x, y in M and, as the elements $\phi'(x)$, where x runs through M, generate S(M), the conclusion follows from § 1, no. 7.

PROPOSITION 2. *Let* E *be an* A*-algebra and* $f\colon M \to E$ *an* A*-linear mapping such that*

$$(1) \qquad f(x)f(y) = f(y)f(x) \text{ for all } x, y \text{ in } M.$$

There exists one and only one A*-algebra homomorphism* $g\colon S(M) \to E$ *such that* $f = g \circ \phi'$.

In other words, (S(M), ϕ') is a solution of the *universal mapping problem* (*Set Theory*, IV, § 3, no. 1), where Σ is the species of A-algebra structure, the α-mappings being the linear mappings of the A-module M to an A-algebra satisfying (1).

The uniqueness of g follows from the fact that $\phi'(M) = M$ generates S(M). To prove the existence of g, note that by virtue of § 5, No. 1, Proposition 1 there exists an A-algebra homomorphism $g_1\colon T(M) \to E$ such that $f = g_1 \circ \phi$;

all that needs to be proved is that g_1 is zero on the ideal $\mathfrak{J}'$, for then, if $p\colon \mathsf{T}(M) \to \mathsf{S}(M) = \mathsf{T}(M)/\mathfrak{J}'$ is the canonical homomorphism, we can write $g_1 = g \circ p$, where $g\colon \mathsf{S}(M) \to E$ is an algebra homomorphism, and the conclusion will follow from the fact that $p \circ \phi = \phi'$. Now the kernel of g_1 is a two-sided ideal which, by virtue of (1) and the relation $g_1 \circ \phi = f$, contains the elements $x \otimes y - y \otimes x$ for x, y in M. This completes the proof.

Remarks. (1) Suppose that E is a *graded* A-algebra of type **Z**, with graduation (E_n), and suppose also that the linear mapping f (assumed to satisfy (1)) is such that

$$f(M) \subset E_1. \tag{2}$$

Then the relation $g(x_1x_2\ldots x_p) = f(x_1)f(x_2)\ldots f(x_p)$ with the $x_i \in M$ shows that $g(\mathsf{S}^p(M)) \subset E_p$ for all $p \geqslant 0$ and hence g is a *graded algebra* homomorphism.

(2) Every element of $\mathsf{S}(M)$ is a sum of products of the form $x_1x_2\ldots x_n$, where the x_i belong to M; care should be taken not to confuse such products taken *in* $\mathsf{S}(M)$ with the analogous products taken *in* $\mathsf{T}(M)$.

(3) If $n!.1$ is invertible in A, the A-module $\mathsf{S}^n(M)$ is generated by the elements of the form x^n, where $x \in M$; this follows from the above remark and I, § 8, no. 2, Proposition 2.

2. FUNCTORIAL PROPERTIES OF THE SYMMETRIC ALGEBRA

PROPOSITION 3. *Let* A *be a commutative ring,* M *and* N *two* A*-modules and* $u\colon M \to N$ *an* A*-linear mapping. There exists one and only one* A*-algebra homomorphism* $u'\colon \mathsf{S}(M) \to \mathsf{S}(N)$ *such that the diagram*

$$\begin{array}{ccc} M & \xrightarrow{u} & N \\ {\scriptstyle \phi'_M}\downarrow & & \downarrow{\scriptstyle \phi'_N} \\ \mathsf{S}(M) & \xrightarrow[u']{} & \mathsf{S}(N) \end{array}$$

is commutative. Moreover, u' *is a graded algebra homomorphism.*

The existence and uniqueness of u' follow from no. 1, Proposition 2 applied to the commutative algebra $\mathsf{S}(N)$ and $f = \phi'_N \circ u\colon M \to \mathsf{S}(N)$; as

$$f(M) \subset \mathsf{S}^1(N) = N,$$

the fact that u' is a graded algebra homomorphism follows from no. 1, *Remark* 1.

The homomorphism u' of Proposition 3 will henceforth be denoted by $\mathsf{S}(u)$. If P is an A-module and $v\colon N \to P$ an A-linear mapping, then

$$\mathsf{S}(v \circ u) = \mathsf{S}(v) \circ \mathsf{S}(u) \tag{3}$$

for $\mathsf{S}(v) \circ \mathsf{S}(u)$ is an algebra homomorphism rendering commutative the diagram

$$\begin{array}{ccc} M & \xrightarrow{v \circ u} & P \\ \downarrow{\scriptstyle \phi'_M} & & \downarrow{\scriptstyle \phi'_P} \\ \mathsf{S}(M) & \xrightarrow[\mathsf{S}(v) \circ \mathsf{S}(u)]{} & \mathsf{S}(P) \end{array}$$

As $\mathsf{S}(M)$ contains $M = \mathsf{S}^1(M)$, $\mathsf{S}(u)$ is sometimes called the *canonical extension* of u to $\mathsf{S}(M)$. The restriction $\mathsf{S}^n(u) : \mathsf{S}^n(M) \rightarrow \mathsf{S}^n(N)$ is such that

$$\mathsf{S}^n(u)(x_1x_2\ldots x_n) = u(x_1)u(x_2)\ldots u(x_n)$$

where the $x_i \in M$, since $\mathsf{S}(u)$ is an algebra homomorphism and $\mathsf{S}^1(u) = u$; the restriction $\mathsf{S}^0(u)$ to A is the identity mapping. Note that $\mathsf{S}^n(u)$ can be obtained from $\mathsf{T}^n(u) : \mathsf{T}^n(M) \rightarrow \mathsf{T}^n(N)$ by passing to the quotients.

PROPOSITION 4. *If $u : M \rightarrow N$ is a surjective A-linear mapping, the homomorphism $\mathsf{S}(u) : \mathsf{S}(M) \rightarrow \mathsf{S}(N)$ is surjective and its kernel is the ideal of $\mathsf{S}(M)$ generated by the kernel $P \subset M \subset \mathsf{S}(M)$ of u.*

We write $v = \mathsf{T}(u) : \mathsf{T}(M) \rightarrow \mathsf{T}(N)$; we know (§ 5, no. 2, Proposition 3) that v is surjective and hence it follows from the definitions that $v(\mathfrak{J}'_M) = \mathfrak{J}'_N$; if $\mathfrak{K}$ is the kernel of v, then $\overset{-1}{v}(\mathfrak{J}'_N) = \mathfrak{K} + \mathfrak{J}'_M$. As $\mathsf{S}(u) : \mathsf{T}(M)/\mathfrak{J}'_M \rightarrow \mathsf{T}(N)/\mathfrak{J}'_N$ is derived from v by passing to the quotients, it is a surjective homomorphism whose kernel is $\mathfrak{K}' = (\mathfrak{K} + \mathfrak{J}'_M)/\mathfrak{J}'_M$. As $\mathfrak{K}$ is generated by the kernel P of u (§ 5, no. 2), so is $\mathfrak{K}'$.

If $u : M \rightarrow N$ is an *injective* linear mapping, it is not always true that $\mathsf{S}(u)$ is an injective mapping (Exercise 1). However it is so when u is an injection such that $u(M)$ is a *direct factor* in N and then the image of $\mathsf{S}(u)$ (isomorphic to $\mathsf{S}(M)$) is a *direct factor* of $\mathsf{S}(N)$; the proof is the same as that for the analogous assertions for $\mathsf{T}(u)$ (§ 5, no. 2) replacing T by S.

PROPOSITION 5. *Let N and P be two submodules of an A-module M such that their sum $N + P$ is a direct factor in M and their intersection $N \cap P$ is a direct factor in N and in P. Then the homomorphisms $\mathsf{S}(N) \rightarrow \mathsf{S}(M)$, $\mathsf{S}(P) \rightarrow \mathsf{S}(M)$ and*

$$\mathsf{S}(N \cap P) \rightarrow \mathsf{S}(M),$$

canonical extensions of the canonical injections, are injective; if $\mathsf{S}(N)$, $\mathsf{S}(P)$ and $\mathsf{S}(N \cap P)$ are identified with subalgebras of $\mathsf{S}(M)$ by means of these homomorphisms, then

$$\mathsf{S}(N \cap P) = \mathsf{S}(N) \cap \mathsf{S}(P). \tag{4}$$

The proof reduces to that of § 5, no. 2, Proposition 4 replacing T by S throughout. The hypotheses of Proposition 5 always hold for *arbitrary* submodules N, P of M when A is a field.

COROLLARY. *Let* K *be a commutative field and* M *a vector space over* K. *For every element* $z \in \mathrm{S}(\mathrm{M})$ *there exists a smallest vector space* N *of* M *such that* $z \in \mathrm{S}(\mathrm{N})$ *and* N *is finite-dimensional.*

The proof is derived from that of § 5, no. 2, Corollary to Proposition 4 replacing T by S throughout.

N is called the vector subspace of M *associated* with z.

3. n-th SYMMETRIC POWER OF A MODULE AND SYMMETRIC MULTILINEAR MAPPINGS

Let X, Y be two sets and n an integer $\geqslant 1$. A *symmetric mapping* of X^n into Y is any mapping $f: \mathrm{X}^n \rightarrow \mathrm{Y}$ such that, for every permutation $\sigma \in \mathfrak{S}_n$ and every element $(x_i) \in \mathrm{X}^n$,

$$f(x_{\sigma(1)}, x_{\sigma(2)}, \ldots, x_{\sigma(n)}) = f(x_1, x_2, \ldots, x_n). \tag{5}$$

As the transpositions which exchange two consecutive integers generate the group $\mathfrak{S}_n$ (I, § 5, no. 7), it suffices that condition (5) hold when σ is such a transposition.

When Y is a *module* over a commutative ring A, clearly the set of symmetric mappings of X^n into Y is a *submodule* of the A-module $\mathrm{Y}^{\mathrm{X}^n}$ of all mappings of X^n into Y.

PROPOSITION 6. *Let* A *be a commutative ring and* M *and* N *two* A*-modules. If with every* A*-linear mapping* $g: \mathrm{S}^n(\mathrm{M}) \rightarrow \mathrm{N}$ $(n \geqslant 1)$ *is associated the n-linear mapping*

$$(x_1, x_2, \ldots, x_n) \mapsto g(x_1 x_2 \ldots x_n) \tag{6}$$

(*where on the right-hand side the product is taken in the algebra* $\mathrm{S}(\mathrm{M})$), *a bijective* A*-linear mapping is obtained of the* A*-module* $\mathrm{Hom}_{\mathrm{A}}(\mathrm{S}^n(\mathrm{M}), \mathrm{N})$ *onto the* A*-module of* symmetric *n*-linear *mappings of* M^n *into* N.

Recall (II, § 3, no. 9) that there is a canonical bijection of the A-module $\mathrm{Hom}_{\mathrm{A}}(\mathrm{T}^n(\mathrm{M}), \mathrm{N})$ onto the A-module $\mathscr{L}_n(\mathrm{M}, \ldots, \mathrm{M}; \mathrm{N})$ of *all* n-linear mappings of M^n into N obtained by associating with every A-linear mapping f: $\mathrm{T}^n(\mathrm{M}) \rightarrow \mathrm{N}$ the n-linear mapping

$$\bar{f}: (x_1, x_2, \ldots, x_n) \mapsto f(x_1 \otimes x_2 \otimes \cdots \otimes x_n). \tag{7}$$

On the other hand, the A-linear mappings $g: \mathrm{S}^n(\mathrm{M}) \rightarrow \mathrm{N}$ are in one-to-one correspondence with the A-linear mappings $f: \mathrm{T}^n(\mathrm{M}) \rightarrow \mathrm{N}$ such that f *is zero on* $\mathfrak{J}'_n$, by associating with g the mapping $f = g \circ p_n$, where

$$p_n: \mathrm{T}^n(\mathrm{M}) \rightarrow \mathrm{S}^n(\mathrm{M}) = \mathrm{T}^n(\mathrm{M})/\mathfrak{J}'_n$$

is the canonical homomorphism (II, § 2, no. 1, Theorem 1). But as $\mathfrak{J}'_n$ is a linear combination of elements of the form

$$(u_1 \otimes u_2 \otimes \cdots \otimes u_p) \otimes (x \otimes y - y \otimes x) \otimes (v_1 \otimes \cdots \otimes v_{n-p-2})$$

(x, y, u_i, v_j in M), to say that the function f is of the form $g \circ p_n$ means that the corresponding n-linear function $\bar{f}$ satisfies the relation

$$\bar{f}(u_1, \ldots, u_p, x, y, v_1, \ldots, v_{n-p-2}) = \bar{f}(u_1, \ldots, u_p, y, x, v_1, \ldots, v_{n-p-2});$$

in other words, by what has been seen above, this means that $\bar{f}$ is *symmetric*; whence the proposition, taking account of the fact that

$$p_n(x_1 \otimes x_2 \otimes \cdots \otimes x_n) = x_1 x_2 \ldots x_n$$

with the $x_i \in \mathrm{M}$.

The A-module $\mathrm{S}^n(\mathrm{M})$ is called the *n-th symmetric power* of M. For every A-module homomorphism $u: \mathrm{M} \to \mathrm{N}$, the mapping $\mathrm{S}^n(u): \mathrm{S}^n(\mathrm{M}) \to \mathrm{S}^n(\mathrm{N})$ which coincides with $\mathrm{S}(u)$ on $\mathrm{S}^n(\mathrm{M})$ is called the *n-th symmetric power of u*.

Remark. Let σ be a permutation in $\mathfrak{S}_n$; as the mapping

$$(x_1, x_2, \ldots, x_n) \mapsto x_{\sigma^{-1}(1)} \otimes x_{\sigma^{-1}(2)} \otimes \cdots \otimes x_{\sigma^{-1}(n)}$$

of M^n into $\mathrm{T}^n(\mathrm{M})$ is A-multilinear, it may be written uniquely as

$$(x_1, \ldots, x_n) \mapsto u_\sigma(x_1 \otimes x_2 \otimes \cdots \otimes x_n),$$

where u_σ is an *endomorphism* of the A-module $\mathrm{T}^n(\mathrm{M})$, also denoted by $z \mapsto \sigma.z$. Clearly, if σ is the identity element of $\mathfrak{S}_n$, u_σ is the identity; on the other hand, writing $y_i = x_{\sigma^{-1}(i)}$, we obtain, for every permutation $\tau \in \mathfrak{S}_n$, $y_{\tau^{-1}(i)} = x_{\sigma^{-1}(\tau^{-1}(i))}$ and hence $\tau.(\sigma.z) = (\tau\sigma).z$; in other words, the A-module $\mathrm{T}^n(\mathrm{M})$ is a left $\mathfrak{S}_n$*-set* under the operation $(\sigma, z) \mapsto \sigma.z$ (I, § 5, no. 1). The elements of $\mathrm{T}^n(\mathrm{M})$ such that $\sigma.z = z$ for *all* $\sigma \in \mathfrak{S}_n$ are called (contravariant) *symmetric tensors of order n*; they form a sub-A-module $\mathrm{S}'_n(\mathrm{M})$ of $\mathrm{T}^n(\mathrm{M})$.

For all $z \in \mathrm{T}^n(\mathrm{M})$, we write $s.z = \sum_{\sigma \in \mathfrak{S}_n} \sigma.z$ and call $s.z$ the *symmetrization* of the tensor z; clearly $s.z$ is a symmetric tensor and therefore $z \mapsto s.z$ is an endomorphism of $\mathrm{T}^n(\mathrm{M})$ whose image $\mathrm{S}''_n(\mathrm{M})$ is contained in $\mathrm{S}'_n(\mathrm{M})$; in general, $\mathrm{S}''_n(\mathrm{M}) \neq \mathrm{S}'_n(\mathrm{M})$ (Exercise 5). If z is a symmetric tensor, then $s.z = n!z$; it follows that *when $n!$ is invertible in* A, *the endomorphism* $z \mapsto (n!)^{-1}s.z$ is a *projector* of $\mathrm{T}^n(\mathrm{M})$ (II, § 1, no. 8), whose image is $\mathrm{S}'_n(\mathrm{M}) = \mathrm{S}''_n(\mathrm{M})$; moreover the *kernel* on this projector is just $\mathfrak{J}'_n$. For, obviously $\sigma(\mathfrak{J}'_n) \subset \mathfrak{J}'_n$ for all $\sigma \in \mathfrak{S}_n$ and $\mathfrak{J}'_n$ is by definition generated by the tensors $z - \rho.z$, where ρ is a transposition exchanging two consecutive numbers in $[1, n]$; also, if σ, τ are two permutations in $\mathfrak{S}_n$, then $z - (\sigma\tau).z = z - \sigma.z + \sigma.(z - \tau.z)$, whence it follows (since every permutation in $\mathfrak{S}_n$ is a product of transpositions exchanging two consecutive numbers) that $z - \sigma.z \in \mathfrak{J}'_n$ for all $z \in \mathrm{T}^n(\mathrm{M})$ and $\sigma \in \mathfrak{S}_n$. Therefore (always supposing that $n!$ is invertible in A), it is seen that $z - (n!)^{-1}s.z = \sum_{\sigma \in \mathfrak{S}_n} (n!)^{-1}(z - \sigma.z) \in \mathfrak{J}'_n$ for all $z \in \mathrm{T}^n(\mathrm{M})$, which proves our assertion.

When $n!$ is invertible in A, the submodules $\mathrm{S}'_n(\mathrm{M})$ and $\mathfrak{I}'_n$ of $\mathrm{T}^n(\mathrm{M})$ are therefore *supplementary* and the restriction to $\mathrm{S}'_n(\mathrm{M})$ of the canonical homomorphism $\mathrm{T}^n(\mathrm{M}) \to \mathrm{S}^n(\mathrm{M}) = \mathrm{T}^n(\mathrm{M})/\mathfrak{I}'_n$ is an *A-module isomorphism,* which allows us in the case envisaged to identify symmetric tensors of order n with the elements of the n-th symmetric power of M. Note however that this identification is not compatible with multiplication, the product (in T(M)) of two symmetric tensors not being symmetric in general and not therefore having as image in S(M) the product of the images of the symmetric tensors considered.

4. EXTENSION OF THE RING OF SCALARS

Let A, A′ be two commutative rings, $\rho: \mathrm{A} \to \mathrm{A}'$ a ring homomorphism, M an A-module, M′ an A′-module and $f: \mathrm{M} \to \mathrm{M}'$ an *A-homomorphism* (relative to ρ) of M into M′. The composite mapping $\mathrm{M} \xrightarrow{f} \mathrm{M}' \xrightarrow{\phi'_{\mathrm{M}'}} \mathrm{S}_{\mathrm{A}'}(\mathrm{M}')$ is an A-linear mapping of M into the commutative algebra $\rho_*(\mathrm{S}_{\mathrm{A}}(\mathrm{M}'))$; then there exists no. 1, Proposition 2) one and only one *A-homomorphism* of algebras $f': \mathrm{S}_{\mathrm{A}}(\mathrm{M}) \to \mathrm{S}_{\mathrm{A}'}(\mathrm{M}')$ rendering commutative the diagram

$$\begin{array}{ccc} \mathrm{M} & \xrightarrow{f} & \mathrm{M}' \\ \downarrow{\scriptstyle \phi'_{\mathrm{M}}} & & \downarrow{\scriptstyle \phi'_{\mathrm{M}'}} \\ \mathrm{S}_{\mathrm{A}}(\mathrm{M}) & \xrightarrow[f']{} & \mathrm{S}_{\mathrm{A}'}(\mathrm{M}') \end{array} \tag{8}$$

It follows immediately that if $\sigma: \mathrm{A}' \to \mathrm{A}''$ is another ring homomorphism, M″ an A″-module, $g: \mathrm{M}' \to \mathrm{M}''$ an A′-homomorphism (relative to σ) and $g': \mathrm{S}_{\mathrm{A}'}(\mathrm{M}') \to \mathrm{S}_{\mathrm{A}''}(\mathrm{M}'')$ the corresponding A′-homomorphism of algebras, then the composite A-homomorphism of algebras

$$\mathrm{S}_{\mathrm{A}}(\mathrm{M}) \xrightarrow{f'} \mathrm{S}_{\mathrm{A}'}(\mathrm{M}') \xrightarrow{g'} \mathrm{S}_{\mathrm{A}''}(\mathrm{M}'') \tag{9}$$

corresponds to the composite A-homomorphism $g \circ f: \mathrm{M} \to \mathrm{M}''$ (relative to $\sigma \circ \rho$).

PROPOSITION 7. *Let* A, B *be two commutative rings,* $\rho: \mathrm{A} \to \mathrm{B}$ *a ring homomorphism and* M *an A-module. The canonical extension*

$$\psi: \mathrm{S}_{\mathrm{B}}(\mathrm{B} \otimes_{\mathrm{A}} \mathrm{M}) \to \mathrm{B} \otimes_{\mathrm{A}} \mathrm{S}_{\mathrm{A}}(\mathrm{M})$$

of the B*-linear mapping* $1_{\mathrm{B}} \otimes \phi'_{\mathrm{M}}: \mathrm{B} \otimes_{\mathrm{A}} \mathrm{M} \to \mathrm{B} \otimes_{\mathrm{A}} \mathrm{S}_{\mathrm{A}}(\mathrm{M})$ *is a graded* B*-algebra isomorphism.*

The proof is derived from that of § 5, no. 3, Proposition 5 replacing T by S and ϕ_{M} by ϕ'_{M}.

5. DIRECT LIMIT OF SYMMETRIC ALGEBRAS

Let $(A_\alpha, \phi_{\beta\alpha})$ be a directed direct system of commutative rings, $(M_\alpha, f_{\beta\alpha})$ a direct system of A_α-modules, $A = \varinjlim A_\alpha$ and $M = \varinjlim M_\alpha$. For $\alpha \leqslant \beta$, we derive canonically from the A_α-homomorphism $f_{\beta\alpha}: M_\alpha \to M_\beta$ an A_α-algebra homomorphism (no. 4, formula (8)) $f'_{\beta\alpha}: S_{A_\alpha}(M_\alpha) \to S_{A_\beta}(M_\beta)$ and it follows from (9) (no. 4) that $(S_{A_\alpha}(M_\alpha), f'_{\beta\alpha})$ is a *direct system of* A_α*-algebras*. On the other hand, $f_\alpha: M_\alpha \to M$ be the canonical A-homomorphism; we derive (no. 4, formula (8)) an A_α-algebra homomorphism

$$f'_\alpha: S_{A_\alpha}(M) \to S_A(M)$$

and it also follows from (9) that the f'_α constitute a direct system of A_α-homomorphisms.

PROPOSITION 8. *The* A*-homomorphism* $f' = \varinjlim f'_\alpha: \varinjlim S_{A_\alpha}(M_\alpha) \to S_A(M)$ *is a graded algebra isomorphism.*

The proof is the same as that of § 5, no. 5, Proposition 6 replacing throughout T by S and ϕ by ϕ' and taking account of the fact that a direct limit of commutative algebras is commutative.

6. SYMMETRIC ALGEBRA OF A DIRECT SUM. SYMMETRIC ALGEBRA OF A FREE MODULE. SYMMETRIC ALGEBRA OF A GRADED MODULE

Let A be a commutative ring, $M = \bigoplus_{\lambda \in L} M_\lambda$ the direct sum of a family of A-modules and $j_\lambda: M_\lambda \to M$ the canonical injection; we derive an A-homomorphism of algebras $S(j_\lambda): S(M_\lambda) \to S(M)$. Since $S(M)$ is commutative, Proposition 8 of § 4, no. 5, can be applied to the homomorphisms $S(j_\lambda)$ and there therefore exists a unique algebra homomorphism

$$g: \bigotimes_{\lambda \in L} S(M_\lambda) \to S(M) \tag{10}$$

(also denoted by g_M) such that $S(j_\lambda) = g \circ f_\lambda$ for all $\lambda \in L$, where

$$f_\lambda: S(M_\lambda) \to \bigotimes_{\lambda \in L} S(M_\lambda)$$

denotes the canonical homomorphism.

PROPOSITION 9. *The canonical homomorphism g (formula* (10)) *is a graded algebra isomorphism* (cf. § 4, no. 8, *Remark* 1).

To prove that g is bijective, we define an algebra homomorphism

$$h: S(M) \to \bigotimes_{\lambda \in L} S(M_\lambda) \tag{11}$$

such that $g \circ h$ and $h \circ g$ are respectively the identity mappings on $S(M)$ and $\bigotimes_{\lambda \in L} S(M_\lambda)$. For each $\lambda \in L$, let u_λ be the composite linear mapping

$$M_\lambda \xrightarrow{\phi'_{M_\lambda}} S(M_\lambda) \xrightarrow{f_\lambda} \bigotimes_{\lambda \in L} S(M_\lambda).$$

There exists one and only one A-linear mapping $u: M \to \bigotimes_{\lambda \in L} S(M_\lambda)$ such that $u \circ j_\lambda = u_\lambda$ for all $\lambda \in L$. As the $S(M_\lambda)$ are commutative, so is their tensor product (§ 4, no. 5) and hence (no. 1, Proposition 2) there exists a unique algebra homomorphism $h: S(M) \to \bigotimes_{\lambda \in L} S(M_\lambda)$ such that $h \circ \phi'_M = u$; on the other hand, it is immediate that $u(M)$ is contained in the submodule of elements of degree 1 of the graded algebra $\bigotimes_{\lambda \in L} S(M_\lambda)$ and hence h is a graded algebra homomorphism. For $x_\lambda \in M_\lambda$,

$$h(g(u_\lambda(x_\lambda))) = h(g(f_\lambda(\phi'_{M_\lambda}(x_\lambda)))) = h(S(j_\lambda)(\phi'_{M_\lambda}(x_\lambda))) = h(\phi'_M(j_\lambda(x_\lambda))) = u_\lambda(x_\lambda);$$

as the $u_\lambda(x_\lambda)$ generate the algebra $\bigotimes_{\lambda \in L} S(M_\lambda)$ (§ 4, no. 5, Proposition 8), $h \circ g$ is certainly the identity mapping. Similarly,

$$g(h(\phi'_M(j_\lambda(x_\lambda)))) = g(u_\lambda(x_\lambda)) = g(f_\lambda(\phi'_{M_\lambda}(x_\lambda))) = S(j_\lambda)(\phi'_{M_\lambda}(x_\lambda)) = \phi'_M(j_\lambda(x_\lambda))$$

and, as the elements $\phi'_M(j_\lambda(x_\lambda))$ generate the algebra $S(M)$, $g \circ h$ is certainly the identity mapping.

Remark (1) Let $N = \bigoplus_{\lambda \in L} N_\lambda$ be the direct sum of another family of A-modules with L as indexing set and, for all $\lambda \in L$, let $v_\lambda: M_\lambda \to N_\lambda$ be an A-linear mapping, whence there is an A-linear mapping $v = \bigoplus_\lambda v_\lambda: M \to N$ (II, § 1, no. 6, Proposition 6). Then the diagram

$$\begin{array}{ccc} \bigotimes_{\lambda \in L} S(M_\lambda) & \xrightarrow{g_M} & S(M) \\ {\scriptstyle \bigoplus_{\lambda \in L} S(v_\lambda)} \downarrow & & \downarrow {\scriptstyle S(v)} \\ \bigotimes_{\lambda \in L} S(N_\lambda) & \xrightarrow[g_N]{} & S(N) \end{array}$$

is commutative, as follows from the definitions (§ 4, no. 5, Corollary to Proposition 8).

The sub-A-module of $\bigotimes_{\lambda \in L} S(M_\lambda)$ with which $S^n(M)$ is identified by means of the isomorphism g can be described more precisely. For every finite subset J of L, we write $E_J = \bigotimes_{\lambda \in J} S(M_\lambda)$, so that $\bigotimes_{\lambda \in L} S(M_\lambda) = \varinjlim E_J$ relative to the directed set $\mathfrak{F}(L)$ of finite subsets of L, by definition (§ 4, no. 5). For every family $\nu = (n_\lambda) \in \mathbf{N}^{(L)}$ (thus having *finite* support) such that $\sum_{\lambda \in L} n_\lambda = n$ and every finite subset J of L containing the support of the family ν, we write

$$S^{J,\nu}(M) = \bigotimes_{\lambda \in J} S^{n_\lambda}(M_\lambda) \tag{12}$$

so that the submodule $E_{J,n}$ of elements of degree n in E_J is the *direct sum* of the $S^{J,\nu}(M)$ over all the families ν of support contained in J and such that $\sum_{\lambda \in L} n_\lambda = n$ (§ 4, no. 7, Proposition 10 and § 4, no. 8). As a convention we write $S^{J,\nu}(M) = \{0\}$ for the families ν whose support is not contained in J; then $E_{J,n}$ can also be called the *direct sum* of *all* the $S^{J,\nu}(M)$, where ν runs through the set H_n of *all* families $\nu = (n_\lambda)_{\lambda \in L}$ such that $\sum_{\lambda \in L} n_\lambda = n$. Since $S^0(M_\lambda)$ is identified with A, clearly also, for two finite subsets $J \subset J'$ of L and a family ν of support contained in J, the canonical mapping $S^{J,\nu}(M) \to S^{J',\nu}(M)$ (restriction of the canonical mapping $E_J \to E_{J'}$ to $S^{J,\nu}(M)$) is *bijective*. If we write, for all $\nu \in H_n$,

$$S^\nu(M) = \varinjlim S^{J,\nu}(M) \tag{13}$$

it is seen that, taking account of II, § 6, no. 2, Proposition 5:

COROLLARY. *The* A-*module* $S^n(M)$ *is the image under isomorphism* (10) *of the submodule of* $\bigotimes_{\lambda \in L} S(M_\lambda)$ *the direct sum of the submodules* $S^\nu(M)$ *for all the families* $\nu = (n_\lambda) \in \mathbf{N}^{(L)}$ *such that* $\sum_{\lambda \in L} n_\lambda = n$; *if* J *is the support of* ν, $S^\nu(M)$ *is canonically isomorphic to* $\bigotimes_{\lambda \in J} S^{n_\lambda}(M_\lambda)$.

In general $S^\nu(M)$, $\bigotimes_{\lambda \in J} S^{n_\lambda}(M_\lambda)$ and their image in $S^n(M)$ are identified.

THEOREM 1. *Let* A *be a commutative ring and* M *a free* A-*module with basis* $(e_\lambda)_{\lambda \in L}$. *For every mapping* $\alpha : L \to \mathbf{N}$ *of finite support, we write*

$$e^\alpha = \prod_{\lambda \in L} e_\lambda^{\alpha(\lambda)} \tag{14}$$

(*product in the commutative algebra* S(M)). *Then, when* α *runs through the set* $\mathbf{N}^{(L)}$ *of mappings of* L *into* $\mathbf{N}$, *of finite support, the* e^α *form a basis of the* A-*module* S(M).

As M is the direct sum of the $M_\lambda = Ae_\lambda$, it suffices to prove the theorem when L is reduced to a single element and then apply Proposition 9. But when $M = Ae$ (e a free element), then $x \otimes y - y \otimes x = 0$ for all x, y in M; the ideal $\mathfrak{I}'$ is therefore zero, whence $T(Ae) = S(Ae)$ and the theorem then follows from § 5, no. 5, Theorem 1.

The multiplication table of the basis (14) is given by

(15) $$e^{\alpha}e^{\beta} = e^{\alpha+\beta}$$

where $\alpha + \beta$ is the mapping $\lambda \mapsto \alpha(\lambda) + \beta(\lambda)$ of L into **N**. In other words, the basis (e^{α}) of S(M), with the multiplicative law (15), is canonically isomorphic to the free commutative monoid $\mathbf{N}^{(L)}$ derived from L; it follows (§ 2, no. 9) that the *symmetric algebra* S(M) *of a free module* M *with basis whose indexing set is* L, *is canonically isomorphic to the polynomial algebra* $A[(X_{\lambda})_{\lambda \in L}]$, the canonical isomorphism being obtained by mapping e_{λ} to X_{λ}. In particular (§ 2, no. 7, Proposition 7), for every mapping f: L → E of L into a *commutative* A-algebra E, there exists one and only one A-algebra homomorphism $\bar{f}$: S(M) → E such that $\bar{f}(e_{\lambda}) = f(\lambda)$.

Remark (2). The above results can equally well be obtained as a consequence of the universal properties of polynomial algebras and symmetric algebras, taking account of II, § 1, no. 11, Corollary 3 to Proposition 17.

Corollary. *If* M *is a projective* A-*module*, S(M) *is a projective* A-*module.*

The proof is the same as that for § 5, no. 5, Corollary to Theorem 1, replacing **T** by **S**.

Proposition 10. *Let* Δ *be a commutative monoid,* M *a graded* A-*module of type* Δ *and* $(M_{\alpha})_{\alpha \in \Delta}$ *its graduation. For every ordered pair* $(\alpha, n) \in \Delta \times \mathbf{N}$, *let* $S^{\alpha, n}(M)$ *be the submodule of* $S^n(M)$ *the direct sum of the submodules* $\bigotimes_{\lambda \in J} S^{n_{\lambda}}(M_{\alpha_{\lambda}})$, *where* $(n_{\lambda})_{\lambda \in L}$ *runs through the set of families of integers* $\geqslant 0$ *such that* $\sum_{\lambda \in L} n_{\lambda} = n$, J *is its support and, for each* (n_{λ}), $(\alpha_{\lambda})_{\lambda \in J}$ *runs through the set of families of* Δ^J *such that* $\sum_{\lambda \in J} \alpha_{\lambda} = \alpha$. *Then* $(S^{\alpha, n}(M))_{(\alpha, n) \in \Delta \times \mathbf{N}}$ *is the only graduation of type* $\Delta \times \mathbf{N}$ *compatible with the algebra structure on* S(M) *and which induces on* M = $S^1(M)$ *the given graduation.*

The fact that S(M) is the direct sum of the $S^{\alpha, n}(M)$ follows from the Corollary to Proposition 9; the rest of the proof is identical with the end of the proof of § 5, no. 5, Proposition 7.

Suppose more particularly that $\Delta = \mathbf{Z}$ and let S(M) be given the *total* graduation (of type **Z**) (II, § 11, no. 1) corresponding to the graduation of type $\mathbf{Z} \times \mathbf{N}$ (and hence also of type $\mathbf{Z} \times \mathbf{Z}$) defined above; the homogeneous elements of degree $n \in \mathbf{Z}$ under this graduation are therefore those of the direct sum of the $S^{q, m}(M)$ for $q + m = n$. Let f be a *homogeneous* linear mapping *of degree* 0 of the graded A-module M into a *commutative* graded A-algebra F of type **Z**; then the algebra homomorphism g: S(M) → F such that $f = g \circ \phi'_M$ is a *homomorphism of graded algebras of type* **Z**, as follows from the formula $g(x_1 x_2 \ldots x_n) = f(x_1) f(x_2) \ldots f(x_n)$ for homogeneous x_i in M, from the hypothesis on f and from the definition of the graduation of type **Z** on S(M).

§ 7. EXTERIOR ALGEBRAS

1. DEFINITION OF THE EXTERIOR ALGEBRA OF A MODULE

DEFINITION 1. *Let* A *be a commutative ring and* M *an* A*-module. The exterior algebra of* M, *denoted by* $\bigwedge(M)$ *or* $\mathrm{Alt}(M)$ *or* $\bigwedge_A(M)$, *is the algebra over* A *the quotient of the tensor algebra* $\mathsf{T}(M)$ *by the two-sided ideal* $\mathfrak{J}''$ *(also denoted by* $\mathfrak{J}''_M$*) generated by the elements* $x \otimes x$, *where* x *runs through* M.

Since the ideal $\mathfrak{J}''$ is generated by homogeneous elements of degree 2, it is a *graded ideal* (II, § 11, no. 3, Proposition 2); we write $\mathfrak{J}''_n = \mathfrak{J}'' \cap \mathsf{T}^n(M)$; the algebra $\bigwedge(M)$ is therefore graded by the graduation (called *canonical*) consisting of the $\bigwedge^n(M) = \mathsf{T}^n(M)/\mathfrak{J}''_n$. Then $\mathfrak{J}''_0 = \mathfrak{J}''_1 = \{0\}$ and hence $\bigwedge^0(M)$ is identified with A and $\bigwedge^1(M)$ with $\mathsf{T}^1(M) = M$; in what follows we shall always make these identification and the canonical injection $M \to \bigwedge(M)$ will be denoted by ϕ'' or ϕ''_M.

PROPOSITION 1. *Let* E *be an* A*-algebra and* $f: M \to E$ *an* A*-linear mapping such that*

$$(f(x))^2 = 0 \quad \textit{for all } x \in M. \tag{1}$$

There exists one and only one A*-algebra homomorphism* $g: \bigwedge(M) \to E$ *such that* $f = g \circ \phi''$.

In other words, $(\bigwedge(M), \phi'')$ is a solution of the *universal mapping problem* (*Set Theory*, IV, § 3, no. 1), where Σ is the species of A-algebra structure, the α-mappings being the linear mappings from the A-module M to an A-algebra satisfying (1).

The uniqueness of g follows from the fact that $\phi''(M) = M$ generates $\bigwedge(M)$. To prove the existence of g, we note that by § 5, no. 1, Proposition 1 there exists an A-algebra homomorphism $g_1: \mathsf{T}(M) \to E$ such that $f = g_1 \circ \phi$; we need to prove that g_1 is zero on the ideal $\mathfrak{J}''$, for then if

$$p: \mathsf{T}(M) \to \bigwedge(M) = \mathsf{T}(M)/\mathfrak{J}''$$

is the canonical homomorphism, we can write $g_1 = g \circ p$, where $g: \bigwedge(M) \to E$ is an algebra homomorphism and the conclusion will follow from the fact that $p \circ \phi = \phi''$. Now, the kernel of g_1 is a two-sided ideal which, by virtue of (1) and the relation $g_1 \circ \phi = f$, contains the elements $x \otimes x$ for $x \in M$. This completes the proof.

Remarks. (1) Suppose that E is a *graded* A-algebra of type **Z**, of graduation (E_n), and suppose also that the linear mapping f (assumed to satisfy (1)) is such that

$$f(M) \subset E_1. \tag{2}$$

Then the relation $g(x_1x_2\ldots x_p) = f(x_1)f(x_2)\ldots f(x_p)$ with the $x_i \in M$ shows that $g(\bigwedge^p(M)) \subset E_p$ for all $p \geqslant 0$ and hence g is a *graded algebra* homomorphism.

(2) To avoid confusion, the product of two elements u, v of the exterior algebra $\bigwedge(M)$ is usually denoted by $u \wedge v$ and is called the *exterior product* of u by v. The elements of $\bigwedge^n(M)$ are therefore the sums of elements of the form $x_1 \wedge x_2 \wedge \cdots \wedge n_n$ with $x_i \in M$ for $1 \leqslant i \leqslant n$ and are often called *n-vectors.*

2. FUNCTORIAL PROPERTIES OF THE EXTERIOR ALGEBRA

PROPOSITION 2. *Let* A *be a commutative ring,* M *and* N *two* A*-modules and* u: M $\rightarrow$ N *an* A*-linear mapping. There exists one and only one* A*-algebra homomorphism*

$$u'' : \bigwedge(M) \rightarrow \bigwedge(N)$$

such that the diagram

$$\begin{array}{ccc} M & \xrightarrow{u} & N \\ \downarrow{\scriptstyle \phi''_M} & & \downarrow{\scriptstyle \phi''_N} \\ \bigwedge(M) & \xrightarrow[u'']{} & \bigwedge(N) \end{array}$$

is commutative. Moreover, u'' *is a graded algebra homomorphism.*

The existence and uniqueness of u' follow from no. 1, Proposition 1 applied to the algebra $\bigwedge(N)$ and $f = \phi''_N \circ u$: M $\rightarrow \bigwedge(N)$; for $f(M) \subset N$ and hence f satisfies condition (1) by definition of $\mathfrak{J}''_N$: as $f(M) \subset \bigwedge^1(N) = N$, the fact that u'' is a graded algebra homomorphism follows from *Remark* 1 of no. 1.

The homomorphism u'' of Proposition 2 will henceforth be denoted by $\bigwedge(u)$. If P is an A-module and v: N $\rightarrow$ P is an A-linear mapping, then

$$\bigwedge(v \circ u) = \bigwedge(v) \circ \bigwedge(u) \tag{3}$$

for $\bigwedge(v) \circ \bigwedge(u)$ is an algebra homomorphism which renders commutative the diagram

$$\begin{array}{ccc} M & \xrightarrow{v\ u} & P \\ \downarrow{\scriptstyle \phi''_M} & & \downarrow{\scriptstyle \phi''_P} \\ \bigwedge(M) & \xrightarrow[\bigwedge(v) \circ \bigwedge(u)]{} & \bigwedge(P) \end{array}$$

Since $\bigwedge(M)$ contains $M = \bigwedge^1(M)$, $\bigwedge(u)$ is sometimes called the *canonical extension* of u to $\bigwedge(M)$. The restriction $\bigwedge^n(u): \bigwedge^n(M) \to \bigwedge^n(N)$ is such that

(4) $$\bigwedge^n(u)(x_1 \wedge x_2 \wedge \cdots \wedge x_n) = u(x_1) \wedge u(x_2) \wedge \cdots \wedge u(x_n)$$

with the $x_i \in M$, since $\bigwedge(u)$ is an algebra homomorphism and $\bigwedge^1(u) = u$; the restriction $\bigwedge^0(u)$ to A is the identity mapping. Note that $\bigwedge^n(u)$ is obtained from $T^n(u): T^n(M) \to T^n(N)$ by passing to the quotients.

PROPOSITION 3. *If* $u: M \to N$ *is a surjective A-linear mapping, the homomorphism* $\bigwedge(u): \bigwedge(M) \to \bigwedge(N)$ *is surjective and its kernel is the two-sided ideal of* $\bigwedge(M)$ *generated by the kernel* $P \subset M \subset \bigwedge(M)$ *of* u.

The proof is derived from that of § 6, no. 2, Proposition 4, replacing S by $\bigwedge$ and $\mathfrak{J}'$ by $\mathfrak{J}''$.

If $u: M \to N$ is an *injective* linear mapping, it is not always true that $\bigwedge(u)$ is an injective mapping (§ 6, Exercise 3) (see however below no. 9, Corollary to Proposition 12). However this is so when u is an injection such that $u(M)$ is *a direct factor* of N and then the image of $\bigwedge(u)$ (isomorphic to $\bigwedge(M)$) is a *direct factor* of $\bigwedge(N)$; the proof is the same as that for the analogous assertions for $T(u)$ (§ 5, no. 2) replacing T by $\bigwedge$.

PROPOSITION 4. *Let* N *and* P *be two submodules of an A-module* M *such that their sum* $N + P$ *is a direct factor in* M *and their intersection* $N \cap P$ *is a direct factor in* N *and in* P. *Then the homomorphisms* $\bigwedge(N) \to \bigwedge(M)$, $\bigwedge(P) \to \bigwedge(M)$ *and* $\bigwedge(N \cap P) \to \bigwedge(M)$, *canonical extensions of the canonical injections, are injective; if* $\bigwedge(N)$, $\bigwedge(P)$ *and* $\bigwedge(N \cap P)$ *are identified with subalgebras of* $\bigwedge(M)$ *by means of these homomorphisms, then*

(5) $$\bigwedge(N \cap P) = \bigwedge(N) \cap \bigwedge(P).$$

The proof is derived from that of § 5, no. 2, Proposition 4 replacing T by $\bigwedge$ throughout. The hypotheses of Proposition 4 are always satisfied by *arbitrary* submodules N, P of M when A is a *field*.

COROLLARY. *Let* K *be a commutative field and* M *a vector space over* K. *For every element* $z \in \bigwedge(M)$, *there exists a smallest vector subspace* N *of* M *such that* $z \in \bigwedge(N)$ *and* N *is finite-dimensional.*

The proof is deduced from that of § 5, no. 2, Corollary to Proposition 4 replacing T by $\bigwedge$ throughout.

N is called the vector subspace of M *associated* with the element z of $\bigwedge(M)$.

3. ANTICOMMUTATIVITY OF THE EXTERIOR ALGEBRA

PROPOSITION 5. (i) *Let $(x_i)_{1 \leqslant i \leqslant n}$ be a finite sequence of elements of the module* M; *for every permutation σ in the symmetric group $\mathfrak{S}_n$,*

$$x_{\sigma(1)} \wedge x_{\sigma(2)} \wedge \cdots \wedge x_{\sigma(n)} = \varepsilon_\sigma . x_1 \wedge x_2 \wedge \cdots \wedge x_n \tag{6}$$

where ε_σ denotes the signature of the permutation σ.

(ii) *If there exist two distinct indices i, j such that $x_i = x_j$, the product*

$$x_1 \wedge x_2 \wedge \cdots \wedge x_n$$

is zero.

(i) First of all, since $x \wedge x = 0$ for all $x \in$ M by definition of the ideal $\mathfrak{J}''$, also, for x, y in M,

$$x \wedge y + y \wedge x = (x + y) \wedge (x + y) - x \wedge x - y \wedge y = 0.$$

This establishes (6) in the case $n = 2$. The general case then follows from § 4, no. 6, Lemma 3.

(ii) Under the hypothesis of (ii), there exists a permutation $\sigma \in \mathfrak{S}_n$ such that $\sigma(1) = i$ and $\sigma(2) = j$; then the left hand side of (6) is zero for this permutation and hence so is the right hand side.

COROLLARY 1. *Let* H, K *be two complementary subsets of the interval* $[1, n]$ *of* **N** *and let $(i_h)_{1 \leqslant h \leqslant p}$, $(j_k)_{1 \leqslant k \leqslant n-p}$ be the sequences of elements of* H *and* K *respectively, arranged in increasing order; we write*

$$x_{\mathrm{H}} = x_{i_1} \wedge x_{i_2} \wedge \cdots \wedge x_{i_p}, \qquad x_{\mathrm{K}} = x_{j_1} \wedge x_{j_2} \wedge \cdots \wedge x_{j_{n-p}};$$

then

$$x_{\mathrm{H}} \wedge x_{\mathrm{K}} = (-1)^{\nu} x_1 \wedge x_2 \wedge \cdots \wedge x_n \tag{7}$$

where ν is the number of ordered pairs $(i, j) \in \mathrm{H} \times \mathrm{K}$ such that $i > j$.

By Proposition 5 this reduces to proving

Lemma 1. *If $\sigma \in \mathfrak{S}_n$ is the permutation such that $\sigma(h) = i_h$ for $1 \leqslant h \leqslant p$, $\sigma(h) = j_{h-p}$ for $p + 1 \leqslant h \leqslant n$, then $\varepsilon_\sigma = (-1)^\nu$.*

For $1 \leqslant h < h' \leqslant p$ or $p + 1 \leqslant h < h' \leqslant n$, $\sigma(h') > \sigma(h)$ and the number of ordered pairs (h, h') such that $1 \leqslant h \leqslant p < h' \leqslant n$ and $\sigma(h) > \sigma(h')$ is equal to ν.

COROLLARY 2. *The graded algebra* $\bigwedge$(M) *is alternating* (§ 4, no. 9).

It suffices to apply Proposition 13 of § 4, no. 9 to $\bigwedge$(M), taking as generating system the set M and using Proposition 5.

PROPOSITION 6. *If* M *is a finitely generated* A-*module*, $\bigwedge(\mathrm{M})$ *is a finitely generated* A-*module; also, if* M *admits a generating system with n elements, then* $\bigwedge^p(\mathrm{M}) = \{0\}$ *for* $p > n$.

Let $(x_i)_{1 \leqslant i \leqslant n}$ be a generating system of M. Every element of $\bigwedge^p(\mathrm{M})$ is a linear combination of elements of the form

$$x_{i_1} \wedge x_{i_2} \wedge \cdots \wedge x_{i_p}$$

where the indices i_k belong to $[1, n]$; by Proposition 5, these indices can be assumed to be distinct (otherwise the corresponding element is zero). If $p > n$, there is no such sequence of indices and hence $\bigwedge^p(\mathrm{M}) = \{0\}$. If $p \leqslant n$, these sequences are finite in number, which completes the proof.

4. n-th EXTERIOR POWER OF A MODULE AND ALTERNATING MULTILINEAR MAPPINGS

Given two modules M, N over a commutative ring A, an *alternating n-linear* mapping of M^n into N is any n-linear mapping $f\colon \mathrm{M}^n \to \mathrm{N}$ such that, for all $p \leqslant n - 2$,

$$f(u_1, \ldots, u_p, x, x, v_1, \ldots, v_{n-p-2}) = 0 \tag{8}$$

for all x, the u_i $(1 \leqslant i \leqslant p)$ and the v_j $(1 \leqslant j \leqslant n - p - 2)$ in M.

PROPOSITION 7. *Let* A *be a commutative ring and* M *and* N *two* A-*modules. If with every* A-*linear mapping* $g\colon \bigwedge^n(\mathrm{M}) \to \mathrm{N}$ $(n \geqslant 2)$ *is associated the n-linear mapping*

$$(x_1, x_2, \ldots, x_n) \mapsto g(x_1 \wedge x_2 \wedge \cdots \wedge x_n) \tag{9}$$

a bijective A-*linear mapping is obtained of the* A-*module* $\mathrm{Hom}_{\mathrm{A}}(\bigwedge^n(\mathrm{M}), \mathrm{N})$ *onto the* A-*module of* alternating n-linear *mappings of* M^n *into* N.

We consider the canonical bijection of the A-module $\mathrm{Hom}_{\mathrm{A}}(\mathrm{T}^n(\mathrm{M}), \mathrm{N})$ onto the A-module $\mathscr{L}_n(\mathrm{M}, \ldots, \mathrm{M}; \mathrm{N})$ of *all* n-linear mappings of M^n into N, obtained by associating with every A-linear mapping $f\colon \mathrm{T}^n(\mathrm{M}) \to \mathrm{N}$ the n-linear mapping

$$\bar{f}\colon (x_1, \ldots, x_n) \mapsto f(x_1 \otimes x_2 \otimes \cdots \otimes x_n)$$

(II, § 3, no. 9). On the other hand, the A-linear mappings $g\colon \bigwedge^n(\mathrm{M}) \to \mathrm{N}$ are in one-to-one correspondence with the A-linear mappings $f\colon \mathrm{T}^n(\mathrm{M}) \to \mathrm{N}$ such that f *is zero on* $\mathfrak{J}''_n$, by associating with g the mapping $f = g \circ p_n$, where

$$p_n\colon \mathrm{T}^n(\mathrm{M}) \to \bigwedge^n(\mathrm{M}) = \mathrm{T}^n(\mathrm{M})/\mathfrak{J}''_n$$

is the canonical homomorphism (II, § 2, no. 1, Theorem 1). But as $\mathfrak{J}''_n$ is a linear combination of elements of the form

$$(u_1 \otimes u_2 \otimes \cdots \otimes u_p) \otimes (x \otimes x) \otimes (v_1 \otimes \cdots \otimes v_{n-p-2})$$

(x, u_i, v_j in M), to say that f is of the form $g \circ p_n$ means that the corresponding n-linear function $\bar{f}$ satisfies (8), in other words it is *alternating*.

The A-module $\bigwedge^n(\mathrm{M})$ is called the *n-th exterior power* of M. For every A-module homomorphism $u\colon \mathrm{M} \to \mathrm{N}$, the mapping

$$\bigwedge^n(u) : \bigwedge^n(\mathrm{M}) \to \bigwedge^n(\mathrm{N})$$

which coincides with $\bigwedge(u)$ on $\bigwedge^n(\mathrm{M})$ is called the *n-th exterior power of u*.

COROLLARY 1. *For every alternating n-linear mapping* $g\colon \mathrm{M}^n \to \mathrm{N}$, *for every permutation* $\sigma \in \mathfrak{S}_n$,

$$g(x_{\sigma(1)}, x_{\sigma(2)}, \ldots, x_{\sigma(n)}) = \varepsilon_\sigma \cdot g(x_1, x_2, \ldots, x_n) \tag{10}$$

for all $x_i \in \mathrm{M}$; *moreover if* $x_i = x_j$ *for two distinct indices i, j, then*

$$g(x_1, x_2, \ldots, x_n) = 0.$$

This is an obvious consequence of Proposition 7 and no. 3, Proposition 5.

COROLLARY 2. *Let* $(x_i)_{1 \leqslant i \leqslant n}$ *be a sequence of n elements of* M *such that*

$$x_1 \wedge x_2 \wedge \cdots \wedge x_n = 0;$$

then, for every alternating n-linear mapping $g\colon \mathrm{M}^n \to \mathrm{N}$, $g(x_1, \ldots, x_n) = 0$.

COROLLARY 3. *Let* $f\colon \mathrm{M}^{n-1} \to \mathrm{A}$ *be an alternating* $(n-1)$*-linear form. If* $(x_i)_{1 \leqslant i \leqslant n}$ *is a sequence of n elements of* M *such that* $x_1 \wedge x_2 \wedge \cdots \wedge x_n = 0$, *then*

$$\sum_{i=1}^{n} (-1)^i f(x_1, \ldots, \hat{x}_i, \ldots, x_n) . x_i = 0 \tag{11}$$

(*where we write* $f(x_1, \ldots, \hat{x}_i, \ldots, x_n) = f(x_1, \ldots, x_{i-1}, x_{i+1}, \ldots, x_n)$ *for* $1 \leqslant i \leqslant n$.

It suffices to prove that the n-linear mapping

$$(x_1, \ldots, x_n) \mapsto \sum_{i=1}^{n} (-1)^i f(x_1, \ldots, \hat{x}_i, \ldots, x_n) . x_i$$

of M^n to M is *alternating*. Now, if $x_i = x_{i+1}$, all the terms in the sum on the right hand side have zero coefficients except x_i and x_{i+1}, since f is alternating; on the other hand, the coefficient of x_i is

$$(-1)^i f(x_1, \ldots, x_{i-1}, x_{i+1}, x_{i+2}, \ldots, x_n)$$

and that of x_{i+1} is $(-1)^{i+1}f(x_1, \dots, x_i, x_{i+2}, \dots, x_n)$ and they are inverses by hypothesis.

Remark. An element z of $T^n(M)$ is called a (contravariant) *skew-symmetric tensor of order n* if $\sigma . z = \varepsilon_\sigma z$ for every permutation $\sigma \in \mathfrak{S}_n$ (cf. § 6, no. 3, *Remark*); these elements form a sub-A-module $A'_n(M)$ of $T^n(M)$. For all $z \in T^n(M)$, we write $\boldsymbol{a}.z = \sum_{\sigma \in \mathfrak{S}_n} \varepsilon_\sigma(\sigma . z)$ and call $\boldsymbol{a}.z$ the *skew-symmetrization* of z; as $\varepsilon_{\sigma\tau} = \varepsilon_\sigma \varepsilon_\tau$, it is immediately seen that $\boldsymbol{a}.z$ is a skew-symmetric tensor and therefore $z \mapsto \boldsymbol{a}.z$ is an endomorphism of $T^n(M)$ whose image $A''_n(M)$ is contained in $A'_n(M)$; in general $A''_n(M) \neq A'_n(M)$ (Exercise 8). If z is a skew-symmetric tensor, then $\boldsymbol{a}.z = n!z$; hence, *when $n!$ is invertible in* A, the endomorphism $z \mapsto (n!)^{-1}\boldsymbol{a}.z$ is a *projector* of $T^n(M)$ whose image is

$$A'_n(M) = A''_n(M).$$

Moreover the *kernel* of this projector is just $\mathfrak{J}''_n$; for attention may obviously be confined to the case where $n \geqslant 2$, hence 2 (a divisor of $n!$) is invertible in A and $x \otimes x = 2^{-1}(x \otimes x + x \otimes x)$; therefore $\mathfrak{J}''_n$ is generated by the elements $z + \rho . z$, where ρ is a transposition exchanging two consecutive numbers in $[1, n]$; moreover, for two permutations σ, τ in $\mathfrak{S}_n$, we can write

$$z - \varepsilon_{\sigma\tau}((\sigma\tau).z) = z - \varepsilon_\tau(\tau . z) + \varepsilon_\tau(\tau . z - \varepsilon_\sigma \sigma . (\tau . z))$$

whence it follows that $z - \varepsilon_\sigma(\sigma . z) \in \mathfrak{J}''_n$ for all $z \in T^n(M)$ and $\sigma \in \mathfrak{S}_n$. Therefore (always assuming $n!$ is invertible in A), it is seen that

$$z - (n!)^{-1}\boldsymbol{a}.z = \sum_{\sigma \in \mathfrak{S}_n} (n!)^{-1}(z - \varepsilon_\sigma(\sigma . z)) \in \mathfrak{J}''_n$$

for all $z \in T^n(M)$, which establishes our assertion.

When $n!$ is invertible in A, the submodules $A'_n(M)$ and $\mathfrak{J}''_n$ of $T^n(M)$ are therefore *supplementary* and the restriction to $A'_n(M)$ of the canonical homomorphism $T^n(M) \to \bigwedge^n(M) = T^n(M)/\mathfrak{J}''_n$ is an A-module *isomorphism*, which allows us in the case in question to identify skew-symmetric tensors of order n with the elements of the n-th exterior power of M. Note here also that this identification is not compatible with multiplication, the product in T(M) of two skew-symmetric tensors not being skew-symmetric in general.

5. EXTENSION OF THE RING OF SCALARS

Let A, A′ be two commutative rings, $\rho : A \to A'$ a ring homomorphism, M an A-module, M′ an A′-module and $f : M \to M'$ an A-*homomorphism* (relative to ρ) of M into M′. The composite mapping $M \xrightarrow{f} M' \xrightarrow{\phi_{M'}} \bigwedge_{A'}(M')$ is an A-linear mapping of M into the A-algebra $\bigwedge_{A'}(M')$ and, as the elements of

$f(M) \subset M'$ are of zero square in $\bigwedge_{A'}(M')$, there exists (no. 1, Proposition 1) one and only one A-*homomorphism* of algebras $f'': \bigwedge_A(M) \to \bigwedge_{A'}(M')$ rendering commutative the diagram

$$\begin{array}{ccc} M & \xrightarrow{f} & M' \\ \phi''_M \downarrow & & \downarrow \phi''_{M'} \\ \bigwedge_A(M) & \xrightarrow[f'']{} & \bigwedge_{A'}(M') \end{array} \tag{12}$$

and f'' is a graded algebra homomorphism. It is immediately deduced that if $\sigma: A' \to A''$ is another ring homomorphism, M'' an A''-module, $g: M' \to M''$ an A'-homomorphism (relative to σ) and $g'': \bigwedge_{A'}(M') \to \bigwedge_{A''}(M'')$ the corresponding A''-homomorphism of algebras, then the composite A-homomorphism of algebras

$$\bigwedge_A(M) \xrightarrow{f''} \bigwedge_{A'}(M') \xrightarrow{g''} \bigwedge_{A''}(M'') \tag{13}$$

corresponds to the composite A-homomorphism $g \circ f: M \to M''$ (relative to $\sigma \circ \rho$).

PROPOSITION 8. *Let* A, B *be two commutative rings,* $\rho: A \to B$ *a ring homomorphism and* M *an* A-*module. The canonical extension*

$$\psi: \bigwedge_B(B \otimes_A M) \to B \otimes_A \bigwedge_A(M)$$

of the B-*linear mapping* $1_B \otimes \phi''_M: B \otimes_A M \to B \otimes_A \bigwedge_A(M)$ *is a graded* B-*algebra isomorphism.*

The proof is derived from that of § 5, no. 4, Proposition 5 replacing T by $\bigwedge$ and ϕ_M by ϕ''_M.

6. DIRECT LIMITS OF EXTERIOR ALGEBRAS

Let $(A_\alpha, \phi_{\beta\alpha})$ be a directed direct system of commutative rings, $(M_\alpha, f_{\beta\alpha})$ a direct system of A_α-modules, $A = \varinjlim A_\alpha$ and $M = \varinjlim M_\alpha$. For $\alpha \leqslant \beta$, we derive canonically from the A_α-homomorphism $f_{\beta\alpha}: M_\alpha \to M_\beta$ an A_α-algebra homomorphism (no. 5, formula (12)) $f''_{\beta\alpha}: \bigwedge_{A_\alpha}(M_\alpha) \to \bigwedge_{A_\beta}(M_\beta)$ and it follows from (13) that $(\bigwedge_{A_\alpha}(M_\alpha), f''_{\beta\alpha})$ is a *direct system of graded* A_α-*modules.* On the other hand let $f_\alpha: M_\alpha \to M$ be the canonical A_α-homomorphism; we derive (no. 5, formula (12)) a graded A_α-algebra homomorphism $f''_\alpha: \bigwedge_{A_\alpha}(M_\alpha) \to \bigwedge_A(M)$ and it also follows from (13) that the f''_α constitute a direct system of A_α-homomorphisms.

PROPOSITION 9. *The A-homomorphism* $f'' = \varinjlim f''_\alpha : \varinjlim \bigwedge_{A_\alpha}(M_\alpha) \to \bigwedge_A(M)$ *is a graded algebra isomorphism.*

The proof is the same as that of § 5, no. 4, Proposition 6 replacing throughout T by $\bigwedge$ and ϕ by ϕ'' and taking account of the fact that a direct limit of alternating algebras is alternating.

7. EXTERIOR ALGEBRA OF A DIRECT SUM. EXTERIOR ALGEBRA OF A GRADED MODULE

Let A be a commutative ring, $M = \bigoplus_{\lambda \in L} M_\lambda$ the direct sum of a family of a family of A-modules and $j_\lambda : M_\lambda \to M$ the canonical injection; an A-homomorphism of graded algebras $\bigwedge(j_\lambda) : \bigwedge(M_\lambda) \to \bigwedge(M)$ is derived. Since $\bigwedge(M)$ is anticommutative, Proposition 10 of § 4, no. 7 (or if need be generalized to the case where L is infinite, cf. § 4, no. 8, *Reamrk* 1) can be applied to the homomorphisms $\bigwedge(j_\lambda)$; then there exists a unique algebra homomorphism

$$g : {}^g\bigotimes_{\lambda \in L} \bigwedge(M_\lambda) \to \bigwedge(M) \tag{14}$$

(also denoted by g_M) such that $\bigwedge(j_\lambda) = g \circ f_\lambda$, where

$$f_\lambda : \bigwedge(M_\lambda) \to {}^g\bigotimes_{\lambda \in L} \bigwedge(M_\lambda)$$

denotes the canonical homomorphism.

PROPOSITION 10. *The canonical homomorphism g (formula* (14)) *is a graded algebra isomorphism.*

To prove that g is bijective, we define a graded algebra homomorphism

$$h : \bigwedge(M) \to {}^g\bigotimes_{\lambda \in L} \bigwedge(M_\lambda) \tag{15}$$

such that $g \circ h$ and $h \circ g$ are respectively the identity mappings on $\bigwedge(M)$ and ${}^g\bigotimes_{\lambda \in L} \bigwedge(M_\lambda)$. For each $\lambda \in L$, consider the composite linear mapping

$$u_\lambda : M_\lambda \xrightarrow{\phi''_{M_\lambda}} \bigwedge(M_\lambda) \xrightarrow{f_\lambda} {}^g\bigotimes_{\lambda \in L} \bigwedge(M_\lambda).$$

There exists one and only one A-linear mapping $u : M \to {}^g\bigotimes_{\lambda \in L} \bigwedge(M_\lambda)$ such that $u \circ j_\lambda = u_\lambda$ for all $\lambda \in L$. The skew tensor product ${}^g\bigotimes_{\lambda \in L} \bigwedge(M_\lambda)$ is an *alternating* algebra: for, for L finite, this follows from § 4, no. 9, Proposition 14 and, for L arbitrary, this follows from the definition of this product given in § 4, no. 8, *Remark* 1 and the fact that a direct limit of alternating graded algebras is alternating. As also $u(M)$ is contained in the submodule of elements of degree 1 in the graded algebra ${}^g\bigotimes_{\lambda \in L} \bigwedge(M_\lambda)$, it follows from no. 1, Proposition 1 and *Remark* 1 that there exists a unique graded algebra homomorphism.

$$h : \bigwedge(M) \to {}^g\bigotimes_{\lambda \in L} \bigwedge(M_\lambda)$$

such that $h \circ \phi''_M = u$. The verification of the fact that $g \circ h$ and $h \circ g$ are the identity mappings is then performed as in § 6, no. 6, Proposition 9 replacing S by $\bigwedge$ and ϕ' by ϕ''.

Remark. Let $N = \bigoplus_{\lambda \in L} N_\lambda$ be the direct sum of another family of A-modules with L as indexing set and, for all $\lambda \in L$, let $v_\lambda : M_\lambda \to N_\lambda$ be an A-linear mapping, whence there is an A-linear mapping $v = \bigoplus_\lambda v_\lambda : M \to N$ (II, § 1, no. 6, Proposition 7). Then the diagram

$$\begin{array}{ccc} {}^g\bigotimes_{\lambda \in L} \bigwedge(M_\lambda) & \xrightarrow{g_M} & \bigwedge(M) \\ {\scriptstyle \bigotimes_{\lambda \in L} \bigwedge(v_\lambda)} \downarrow & & \downarrow {\scriptstyle \bigwedge(v)} \\ {}^g\bigotimes_{\lambda \in L} \bigwedge(N_\lambda) & \xrightarrow[g_N]{} & \bigwedge(N) \end{array}$$

is commutative (cf. § 4, no. 5, Corollary to Proposition 8).

The sub-A-module of ${}^g\bigotimes_{\lambda \in L} \bigwedge(M_\lambda)$ with which $\bigwedge^n(M)$ is identified by means of the isomorphism g can be described more precisely. For every finite subset J of L, we write $E_J = {}^g\bigotimes_{\lambda \in J} \bigwedge(M_\lambda)$, so that ${}^g\bigotimes_{\lambda \in L} \bigwedge(M_\lambda) = \varinjlim E_J$ relative to the directed set $\mathfrak{F}(L)$ of finite subsets of L, by definition (§ 4, no. 8, *Remark* 1). For every family $\nu = (n_\lambda) \in \mathbf{N}^{(L)}$ (therefore with *finite* support) such that $\sum_{\lambda \in L} n_\lambda = n$ and every finite subset J of L containing the support of the family ν, we write

$$\bigwedge^{J,\nu}(M) = \bigotimes_{\lambda \in J} \bigwedge^{n_\lambda}(M_\lambda) \tag{16}$$

so that the submodule $E_{J,n}$ of elements of degree n in E_J is the *direct sum* of the $\bigwedge^{J,\nu}(M)$ for all families ν of support contained in J and such that

$$\sum_{\lambda \in L} n_\lambda = n$$

(§ 4, no. 7, Proposition 10 and no. 8). By way of convention we write $\bigwedge^{J,\nu}(M) = \{0\}$ for the families ν whose support is not contained in J; then

$E_{J,n}$ can also be called the *direct sum* of *all* the $\bigwedge^{J,\nu}(M)$, where ν runs through the set H_n of *all* families $\nu = (n_\lambda)_{\lambda \in L}$ such that $\sum_{\lambda \in L} n_\lambda = n$. Since $\bigwedge^0(M_\lambda)$ is identified with A, clearly also for two finite subsets $J \subset J'$ of L and a family ν of support contained in J, the canonical mapping $\bigwedge^{J,\nu}(M) \to \bigwedge^{J',\nu}(M)$ (restriction of the canonical mapping $E_J \to E_{J'}$ to $\bigwedge^{J,\nu}(M)$) is *bijective*. If we write, for all $\nu \in H_n$,

$$\bigwedge^{\nu}(M) = \varinjlim \bigwedge^{J,\nu}(M) \tag{17}$$

it is then seen that, taking account of II, § 6, no. 2, Proposition 5, we have:

Corollary. *The A-module* $\bigwedge^n(M)$ *is the image under isomorphism* (14) *of the submodule of* ${}^g\bigotimes_{\lambda \in L} \bigwedge(M_\lambda)$ *the direct sum of the submodules* $\bigwedge^\nu(M)$ *for all families* $\nu = (n_\lambda) \in \mathbf{N}^{(L)}$ *such that* $\sum_{\lambda \in L} n_\lambda = n$; *if* J *is the support of* ν, $\bigwedge^\nu(M)$ *is canonically isomorphic to* $\bigotimes_{\lambda \in J} \bigwedge^{n_\lambda}(M_\lambda)$.

In general $\bigwedge^\nu(M)$, $\bigotimes_{\lambda \in J} \bigwedge^{n_\lambda}(M_\lambda)$ and their image in $\bigwedge^n(M)$ are identified. With this convention:

Proposition 11. *Let* Δ *be a commutative monoid,* M *a graded A-module of type* Δ *and* $(M_\alpha)_{\alpha \in \Delta}$ *its graduation. For every ordered pair* $(\alpha, n) \in \Delta \times \mathbf{N}$, *let* $\bigwedge^{\alpha,n}(M)$ *be the submodule of* $\bigwedge^n(M)$ *the direct sum of the submodules* $\bigotimes_{\lambda \in J} \bigwedge^{n_\lambda}(M_{\alpha_\lambda})$, *where* $(n_\lambda)_{\lambda \in L}$ *runs through the set of families of integers* $\geqslant 0$ *such that* $\sum_{\lambda \in L} n_\lambda = n$, J *is its support and, for each* (n_λ), $(\alpha_\lambda)_{\lambda \in J}$ *runs through the set of families of* Δ^J *such that* $\sum_{\lambda \in J} \alpha_\lambda = \alpha$. *Then* $(\bigwedge^{\alpha,n}(M))_{(\alpha,n) \in \Delta \times \mathbf{N}}$ *is the only graduation of type* $\Delta \times \mathbf{N}$ *compatible with the algebra structure on* $\bigwedge(M)$ *and which induces on* $M = \bigwedge^1(M)$ *the given graduation.*

The fact that $\bigwedge(M)$ is the direct sum of the $\bigwedge^{\alpha,n}(M)$ follows from the Corollary to Proposition 10; the rest of the proof is identical with the end of the proof of § 5, no. 5, Proposition 7.

8. EXTERIOR ALGEBRA OF A FREE MODULE

Theorem 1. *Let* M *be an A-module with a basis* $(e_\lambda)_{\lambda \in L}$. *Let* L *be given the structure of a totally ordered set* (*Set Theory*, III, § 2, no. 3, Theorem 1) *and for every finite subset* J *of* L *we write*

$$e_J = e_{\lambda_1} \wedge e_{\lambda_2} \wedge \cdots \wedge e_{\lambda_n} \tag{18}$$

where $(\lambda_k)_{1 \leqslant k \leqslant n}$ *is the sequence of elements of* J *arranged in increasing order* (*Set Theory*, III, § 5, no. 3, Proposition 6); *we write* $e_\varnothing = 1$, *the unit element of* A. *Then the* e_J, *where* J *runs through the set* $\mathfrak{F}(L)$ *of finite subsets of* L, *form a basis of the exterior algebra* $\bigwedge(M)$.

Since the e_λ generate the A-module M, every element of $\bigwedge(M)$ is a linear combination of a finite number of products of the elements e_λ and hence (taking account of no. 3, Proposition 5) is a linear combination of a finite number of elements e_J for $J \in \mathfrak{F}(L)$. It reduces to proving that the e_J are linearly independent over A. Otherwise, there would exist between these elements a linear relation with coefficients not all zero; the union of the subsets J which correspond to the e_J whose coefficients in this relation are $\neq 0$ is a *finite* subset K of L (since there is only a finite number of coefficients $\neq 0$). Let N be the submodule of M generated by the e_λ such that $\lambda \in K$; N is a direct factor in M, hence (no. 2) $\bigwedge(N)$ is identified with a subalgebra of $\bigwedge(M)$ and, if we show that the e_J with $J \subset K$ form a basis of $\bigwedge(N)$, we shall obtain the desired contradiction.

It therefore reduces to providing Theorem 1 when the basis of M is *finite*; we may therefore suppose that $L = [1, m] \subset \mathbf{N}$. For each $i \in L$, let M_i be the free submodule Ae_i of M; M is the direct sum of the M_i and $\bigwedge(M_i)$ is the direct sum of $\bigwedge^0(M_i) = A$ and $\bigwedge^1(M_i) = M_i$ (no. 3, Proposition 6). Let $\bigwedge(M)$ be canonically identified with the A-module the tensor product of the $\bigwedge(M_i)$ (no. 7, Proposition 10); the latter has as basis the tensor product of the bases $(1, e_i)$ of the $\bigwedge(M_i)$ (II, § 3, no. 7, Corollary 2 to Proposition 7); thus we obtain all the elements

$$u_1 \otimes u_2 \otimes \cdots \otimes u_m$$

where either $u_i = 1$ or $u_i = e_i$; if J is the set of indices i such that $u_i = e_i$, $u_1 \otimes u_2 \otimes \cdots \otimes u_m$ is identical with e_J, which completes the proof.

COROLLARY 1. *Suppose that* $L = [1, m]$; *then the basis* $(e_J)_{J \in \mathfrak{P}(L)}$ *of* $\bigwedge(M)$ *has* 2^m *elements. For* $p > m$, $\bigwedge^p(M) = \{0\}$; $\bigwedge^m(M)$ *has a basis consisting of a single element* e_L; *for* $0 \leqslant p \leqslant m$ *the number of elements in the basis* (e_J) *of* $\bigwedge^p(M)$ *consisting of the* e_J *such that* Card(J) $= p$ *is*

$$\binom{m}{p} = \frac{m!}{p!(m-p)!}.$$

This follows from *Set Theory*, III, § 3, no. 5, Proposition 12 and *Set Theory*, III, § 5, no. 8, Corollary 1 to Proposition 11.

We return to the case where the set L in Theorem 1 is arbitrary and give explicitly the *multiplication table* (§ 1, no. 7) of the basis (e_J). Given two finite subsets J, K of the totally ordered set L, we write

$$\begin{cases} \rho_{J,K} = 0 & \text{if } J \cap K \neq \varnothing \\ \rho_{J,K} = (-1)^{\nu} & \text{if } J \cap K = \varnothing \end{cases} \tag{19}$$

where ν denotes the number of ordered pairs $(\lambda, \mu) \in J \times K$ such that $\lambda > \mu$. Then Corollary 1 to Proposition 4 of no. 2 immediately implies the relation

$$e_J \wedge e_K = \rho_{J,K} e_{J \cup K}. \tag{20}$$

Note the formula

$$\rho_{J,K}\rho_{K,J} = (-1)^{jk} \tag{21}$$

when $J \cap K = \varnothing$, $j = \text{Card}(J)$, $k = \text{Card}(K)$ (no. 3, Corollary 2 to Proposition 5.)

COROLLARY 2. *If* M *is a projective* A*-module,* $\bigwedge(M)$ *is a projective* A*-module.*

The proof is the same as that of § 5, no. 5, Corollary to Theorem 1 replacing T by $\bigwedge$.

COROLLARY 3. *Let* M *be a projective* A*-module and* $(x_i)_{1 \leq i \leq n}$ *a finite sequence of elements of* M. *For there to exist on* M *an alternating n-linear form f such that*

$$f(x_1, x_2, \ldots, x_n) \neq 0,$$

it is necessary and sufficient that $x_1 \wedge x_2 \wedge \cdots \wedge x_n \neq 0$.

We know already (with no hypothesis on M) that the condition is necessary (no. 4, Proposition 7). Suppose now that M is projective and that

$$x_1 \wedge x_2 \wedge \cdots \wedge x_n \neq 0.$$

Then $\bigwedge^n(M)$ is projective (Corollary 2) and hence the canonical mapping

$$\textstyle\bigwedge^n(M) \to (\bigwedge^n(M))^{**}$$

is injective (II, § 2, no. 7, Corollary 4 to Proposition 13); we conclude that there exists a linear form $g: \bigwedge^n(M) \to A$ such that $g(x_1 \wedge x_2 \wedge \cdots \wedge x_n) \neq 0$. If f is the alternating n-linear form corresponding to g (no. 4, Proposition 7), then $f(x_1, \ldots, x_n) \neq 0$.

9. CRITERIA FOR LINEAR INDEPENDENCE

PROPOSITION 12. *Let* M *be a projective* A*-module. For elements* $x_1, x_2, \ldots, x_n$ *of* M *to be linearly dependent, it is necessary and sufficient that there exist* $\lambda \neq 0$ *in* A *such that*

$$\lambda x_1 \wedge x_2 \wedge \cdots \wedge x_n = 0. \tag{22}$$

The condition is necessary (with no hypothesis on M), for if for example λx_1 (with $\lambda \neq 0$) is a linear combination of $x_2, \ldots, x_n$, relation (22) holds (no. 3, Proposition 5). We show that the condition is sufficient by induction on n; for $n = 1$, it is a trivial consequence of the definition. Suppose therefore that $n > 1$ and that condition (22) holds for some $\lambda \neq 0$. If

$$\lambda x_2 \wedge x_3 \wedge \cdots \wedge x_n = 0,$$

then the induction hypothesis implies that $x_2, \ldots, x_n$ are linearly dependent and hence *a fortiori* $x_1, x_2, \ldots, x_n$ are. If $\lambda x_2 \wedge x_3 \wedge \cdots \wedge x_n \neq 0$, it follows from no. 8, Corollary 3 to Theorem 1 that there exists an alternating $(n-1)$-linear form f such that $f(\lambda x_2 \wedge x_3 \wedge \cdots \wedge x_n) = \mu \neq 0$. Since

$$x_1 \wedge (\lambda x_2) \wedge \cdots \wedge x_n = 0,$$

it follows from no. 8, Corollary 3 to Theorem 1 that μx_1 is a linear combination of $\lambda x_2, x_3, \ldots, x_n$; hence $x_1, x_2, \ldots, x_n$ are linearly dependent.

Corollary. *Let* M *and* N *be two projective* A*-modules and* $f: \mathrm{M} \to \mathrm{N}$ *an* A*-linear mapping. If* f *is injective, then* $\bigwedge(f): \bigwedge(\mathrm{M}) \to \bigwedge(\mathrm{N})$ *is injective.*

We prove this first under the assumption that M is *free*; let $(e_\lambda)_{\lambda \in \mathrm{L}}$ be a basis of M, so that $(e_\mathrm{J})_{\mathrm{J} \in \mathfrak{F}(\mathrm{L})}$ (formula (18)) is a basis of $\bigwedge(\mathrm{M})$. Suppose that the kernel of $\bigwedge(f)$ contains an element $u = \sum_\mathrm{J} \alpha_\mathrm{J} e_\mathrm{J} \neq 0$. Let K be an element minimal among the finite subsets J such that $\alpha_\mathrm{J} \neq 0$ and let H be a finite subset of L disjoint from K and such that $\mathrm{K} \cup \mathrm{H}$ contains all the J (finite in number) such that $\alpha_\mathrm{J} \neq 0$; for all $\mathrm{J} \neq \mathrm{K}$ such that $\alpha_\mathrm{J} \neq 0$, we have therefore by definition $\mathrm{J} \cap \mathrm{H} \neq 0$ and consequently (no. 8, formula (20)).

$$u \wedge e_\mathrm{H} = +\alpha_\mathrm{K} e_{\mathrm{K} \cup \mathrm{H}}$$

belongs to the two-sided ideal of $\bigwedge(\mathrm{M})$, the kernel of $\bigwedge(f)$. We write $e_{\mathrm{K} \cup \mathrm{H}} = e_{\lambda_1} \wedge e_{\lambda_2} \wedge \cdots \wedge e_{\lambda_n}$; then $\alpha_\mathrm{K} f(e_{\lambda_1}) \wedge f(e_{\lambda_2}) \wedge \cdots \wedge f(e_{\lambda_n}) = 0$; by virtue of Proposition 12, the elements $f(e_{\lambda_i})$ $(1 \leqslant i \leqslant n)$ of N are linearly dependent. But this contradicts the hypothesis that f is injective (II, § 1, no. 11, Corollary 3 to Proposition 17).

We now consider the general case; let M′ be an A-module such that $\mathrm{M} \oplus \mathrm{M}' = \mathrm{P}$ is free (II, § 2, no. 2, Proposition 4). Consider the linear mapping $g: \mathrm{M} \oplus \mathrm{M}' \to \mathrm{N} \oplus \mathrm{M} \oplus \mathrm{M}'$ such that $g(x, y) = (f(x), 0, y)$, so that there is a commutative diagram

$$\begin{array}{ccc} \mathrm{M} & \xrightarrow{f} & \mathrm{N} \\ {\scriptstyle j}\downarrow & & \downarrow{\scriptstyle j'} \\ \mathrm{P} & \xrightarrow[g]{} & \mathrm{N} \oplus \mathrm{P} \end{array}$$

where the vertical arrows are the canonical injections. Since g is injective and P is free, $\bigwedge(g)$ is an injective homomorphism as has been seen above. Now, $\bigwedge(j):\bigwedge(\mathrm{M}) \to \bigwedge(\mathrm{P})$ is an injective homomorphism since M is a direct factor in P (no. 2). The composite homomorphism

$$\bigwedge(\mathrm{M}) \xrightarrow{\bigwedge(j)} \bigwedge(\mathrm{P}) \xrightarrow{\bigwedge(g)} \bigwedge(\mathrm{N} \oplus \mathrm{P})$$

is therefore injective and, as it is also equal to the composite homomorphism

$$\bigwedge(\mathrm{M}) \xrightarrow{\bigwedge(f)} \bigwedge(\mathrm{N}) \xrightarrow{\bigwedge(j')} \bigwedge(\mathrm{N} \oplus \mathrm{P})$$

we conclude that $\bigwedge(f)$ is injective.

PROPOSITION 13. *Let* M *be an* A-*module,* N *a direct factor submodule of* M *which is free of dimension* p *and* $\{u\}$ *a basis of* $\bigwedge^p(\mathrm{N})$. *For an element* $x \in \mathrm{M}$ *to belong to* N, *it is necessary and sufficient that* $u \wedge x = 0$.

Let P be a submodule of M complementary to N and let $y \in \mathrm{N}$, $z \in \mathrm{P}$ be such that $x = y + z$. Then $u \wedge x = u \wedge z$. As $\bigwedge^p(\mathrm{N})$ is free of dimension 1, the mapping $\phi:\mathrm{P} \to \mathrm{P} \otimes \bigwedge^p(\mathrm{N})$ such that $\phi(p) = p \otimes u$ is bijective (II, § 3, no. 4, Proposition 4). On the other hand (no. 7, Proposition 10), the composition of the canonical homomorphisms

$$\psi:\mathrm{P} \otimes \bigwedge^p(\mathrm{N}) \to \bigwedge(\mathrm{P}) \otimes \bigwedge(\mathrm{N}) \to \bigwedge(\mathrm{M})$$

is injective. The mapping $\psi \circ \phi$ is therefore injective, whence the proposition.

THEOREM 2. *Let* M *be an* A-*module with a finite basis* $(e_i)_{1 \leqslant i \leqslant n}$. *For a sequence* $(x_i)_{1 \leqslant i \leqslant n}$ *of* n *elements of* M *to form a basis of* M, *it is necessary and sufficient that the element* $\lambda \in \mathrm{A}$ *such that*

$$x_1 \wedge x_2 \wedge \ldots \wedge x_n = \lambda . e_1 \wedge e_2 \wedge \cdots \wedge e_n \tag{23}$$

be invertible in A.

Recall that $e_1 \wedge e_2 \wedge \cdots \wedge e_n$ is the unique element of a basis of $\bigwedge^n(\mathrm{M})$ (no. 8, Corollary 1 to Theorem 1) so that the element $\lambda \in \mathrm{A}$ satisfying (23) is determined uniquely. If $(x_i)_{1 \leqslant i \leqslant n}$ is a basis of M, $x_1 \wedge x_2 \wedge \cdots \wedge x_n$ is the unique element in a basis of $\bigwedge^n(\mathrm{M})$ (no. 8), then λ is invertible. Conversely, suppose λ is invertible; then the alternating n-linear form f corresponding to the linear mapping $g:\bigwedge^n(\mathrm{M}) \to \mathrm{A}$ such that

$$g(e_1 \wedge e_2 \wedge \cdots \wedge e_n) = \lambda^{-1}$$

is such that $f(x_1, x_2, \ldots, x_n) = 1$. For all $x \in \mathrm{M}$, obviously

$$x \wedge x_1 \wedge \cdots \wedge x_n = 0$$

(no. 3, Proposition 6); applying no. 8, Corollary 3 to Theorem 1, we obtain

$$f(x_1, x_2, \dots, x_n).x = \sum_{i=1} (-1)^{i-1} f(x, x_1, \dots, \hat{x}_i, \dots, x_n).x_i.$$

As $f(x_1, \dots, x_n) = 1$, this shows that every $x \in M$ is a linear combination of $x_1, x_2, \dots, x_n$ and, as the latter are linearly independent (since

$$x_1 \wedge x_2 \wedge \cdots \wedge x_n \neq 0),$$

they form a basis of M.

§ 8. DETERMINANTS

1. DETERMINANTS OF AN ENDOMORPHISM

Let M be an A-module with a *finite basis* of n elements and u an endomorphism of M. The A-module $\bigwedge^n(M)$ is a monogenous free module, that is isomorphic to A (no. 8, Corollary 1 to Theorem 1); $\bigwedge^n(u)$ is an endomorphism of this module and is therefore a homothety $z \mapsto \lambda z$ of ratio $\lambda \in A$ determined uniquely (II, § 2, no. 3, Proposition 5).

DEFINITION 1. *The determinant of an endomorphism u of a free* A*-module* M *of finite dimension n* (II, § 7, no. 2, Corollary to Proposition 3 and *Remark* 1), *denoted by* det u, *is the scalar λ such that $\bigwedge^n(u)$ is the homothety of ratio λ.*

By formula (4) of § 7, no. 2, det u is the unique scalar such that

$$u(x_1) \wedge u(x_2) \wedge \cdots \wedge u(x_n) = (\det u).x_1 \wedge x_2 \wedge \cdots \wedge x_n \tag{1}$$

for every sequence $(x_i)_{1 \leqslant i \leqslant n}$ of n elements of M. If $\det(u) = 1$, u is said to be *unimodular.*

PROPOSITION 1. (i) *If u and v are two endomorphisms of a finite-dimensional free* A*-module* M, *then*

$$\det(u \circ v) = (\det u)(\det v). \tag{2}$$

(ii) $\det(1_M) = 1$; *for every automorphism u of* M, det u *is invertible in* A *and*

$$\det(u^{-1}) = (\det u)^{-1}. \tag{3}$$

If n is the dimension of M, this follows immediately from the relation $\bigwedge^n(u \circ v) = (\bigwedge^n(u)) \circ (\bigwedge^n(v))$ § 7, no. 2, formula (3)).

Let M be a free A-module with a finite basis $(e_i)_{1 \leqslant i \leqslant n}$; given a sequence $(x_i)_{1 \leqslant i \leqslant n}$ of n elements of M, the determinant of this sequence *with respect to the given basis* (e_i), denoted by $\det(x_1, x_2, \ldots, x_n)$ when no confusion can arise over the basis, is the determinant of the endomorphism u of M such that $u(e_i) = x_i$ for $1 \leqslant i \leqslant n$. Then by formula (1)

$$(4) \qquad x_1 \wedge x_2 \wedge \cdots \wedge x_n = \det(x_1, x_2, \ldots, x_n)\, e_1 \wedge e_2 \wedge \cdots \wedge e_n$$

and this relation characterizes the mapping $(x_i) \mapsto \det(x_1, x_2, \ldots, x_n)$ of M^n into A. It shows that this mapping is an *alternating n-linear form*. As, by virtue of § 7, no. 4, Proposition 7, the A-module of alternating n-linear forms is canonically isomorphic to the dual of $\bigwedge^n(M)$ and $\bigwedge^n(M)$ is isomorphic to A, it is seen that *every alternating n-linear form on* M^n *can be written*

$$(x_1, \ldots, x_n) \mapsto \alpha \det(x_1, x_2, \ldots, x_n)$$

for some $\alpha \in A$.

PROPOSITION 2. *Let* M *be a free* A*-module with a finite basis* $(e_i)_{1 \leqslant i \leqslant n}$ *and* v *an endomorphism of* M. *For every sequence* $(x_i)_{1 \leqslant i \leqslant n}$ *of* n *elements of* M,

$$(5) \qquad \det(v(x_1), \ldots, v(x_n)) = (\det v)\det(x_1, \ldots, x_n).$$

If u is the endomorphism of M such that $u(e_i) = x_i$ for all i, then

$$v(x_i) = (v \circ u)(e_i)$$

and (5) therefore follows from (2).

2. CHARACTERIZATION OF AUTOMORPHISMS OF A FINITE-DIMENSIONAL FREE MODULE

THEOREM 1. *Let* M *be a finite-dimensional free* A*-module and* u *an endomorphism of* M. *The following conditions are equivalent:*

(a) u *is bijective;*
(b) u *is right invertible* (II, § 1, no. 9, Corollary 1 to Proposition 15)*;*
(c) u *is left invertible* (II, § 1, no. 9, Corollary 2 to Proposition 15)*;*
(d) u *is surjective;*
(e) $\det u$ *is invertible in* A.

Let $(e_i)_{1 \leqslant i \leqslant n}$ be a basis of M. If $x_i = u(e_i)$ for $1 \leqslant i \leqslant n$, then

$$x_1 \wedge x_2 \wedge \cdots \wedge x_n = (\det u) e_1 \wedge e_2 \wedge \cdots \wedge e_n.$$

By § 7, no. 9, Theorem 2 a necessary and sufficient condition for the x_i to form a basis of M is that $\det u$ be an invertible element of A; this proves the equivalence of (a) and (e). Observe that (a) obviously implies each of conditions (b), (c) and (d); it remains to prove that each of conditions (b), (c) and (d) implies (e). Now, if there exists an endomorphism v of M such that $v \circ u = 1_M$ or $u \circ v = 1_M$, then $(\det v)(\det u) = 1$ and hence $\det u$ is invertible in A. If u is surjective, so is $\bigwedge^n(u)$ (§ 7, no. 2, Proposition 3), in other words the homothety of ratio det u in A is surjective, which immediately implies that det u is invertible.

PROPOSITION 3. *Let* M *be a finite-dimensional free* A-*module. For every endomorphism* u *of* M, *the following conditions are equivalent:*

(f) u *is injective;*

(g) det u *is not a divisor of zero in* A.

With the same notation as in the proof of Theorem 1, for u to be injective, it is necessary and sufficient that the x_i be linearly independent. By § 7, no. 9, Proposition 12, it is necessary and sufficient for this that the relation $\lambda x_1 \wedge x_2 \wedge \cdots \wedge x_n = 0$ (with $\lambda \in A$) imply $\lambda = 0$. But this is equivalent to $\lambda(\det u) = 0$ since $e_1 \wedge \cdots \wedge e_n$ is a basis of $\bigwedge^n(M)$; whence the proposition.

Remark. When A is a field, condition (e) of Theorem 2 is equivalent to condition (g) of Proposition 3, since they both mean that det $u \neq 0$. In this case therefore all the conditions of Theorem 1 and Proposition 3 are equivalent (cf. II, § 7, no. 4, Corollary to Proposition 9).

3. DETERMINANT OF A SQUARE MATRIX

DEFINITION 2. *Let* I *be a finite set,* A *a commutative ring and* X *a square matrix of type* (I, I) *over the ring* A (II, § 10, no. 7). *The determinant of the endomorphism* u *of the* A-*module* A^I, *whose matrix with respect to the canonical basis of* A^I *is* X, *is called the determinant of* X *and denoted by* det X.

If $X = (\xi_{ij})_{(i,j) \in I \times I}$ and $(e_i)_{i \in I}$ is the canonical basis of A^I, the endomorphism u is then given by

$$u(e_i) = \sum_{j \in J} \xi_{ji} e_j.$$

When $I = [1, n] \subset \mathbf{N}$, if we write $x_i = u(e_i)$ for $i \in I$, the determinant of X is then defined in the relation

$$x_1 \wedge x_2 \wedge \cdots \wedge x_n = (\det X) e_1 \wedge e_2 \wedge \cdots \wedge e_n \tag{6}$$

in other words, det X is equal to the determinant $\det(x_1, x_2, \ldots, x_n)$ with respect to the canonical basis of A^n. Consequently:

PROPOSITION 4. *For* n *vectors* $x_1, \ldots, x_n$ *of* A^n, *let* $X(x_1, \ldots, x_n)$ *denote the square matrix of order* n *whose* i-*th column is* x_i *for* $1 \leqslant i \leqslant n$. *Then the mapping*

$$(x_1, \ldots, x_n) \mapsto \det(X(x_1, \ldots, x_n)$$

of $(A^n)^n$ *into* A *is alternating and* n-*linear.*

In particular, the determinant of a matrix two of whose columns are equal is zero. If a permutation is performed on the columns of a matrix, the determinant of the new matrix is equal to that of the old multiplied by ε_σ. If to one column of a matrix is added a scalar multiple of a column of a different index, the determinant of the new matrix is equal to that of the old.

More generally, let M be a free A-module of finite dimension n and let $(e_\iota)_{\iota \in \mathrm{I}}$ be a basis of M; for every automorphism u of M, if X is the matrix of u with respect to the basis (e_ι), then

$$\det(u) = \det(X) \tag{7}$$

as follows immediately from the definitions.

When $\mathrm{I} = [1, n]$, the determinant of X is also denoted by

$$\det(\xi_{ij})_{1 \leqslant i \leqslant n, 1 \leqslant j \leqslant n}$$

or simply $\det(\xi_{ij})$ if this causes no confusion, or also

$$\begin{vmatrix} \xi_{11} & \xi_{12} & \cdots & \xi_{1n} \\ \xi_{21} & \xi_{22} & \cdots & \xi_{2n} \\ \cdot & \cdot & \cdot & \cdot \\ \xi_{n1} & \xi_{n2} & \cdots & \xi_{nn} \end{vmatrix}$$

When $X = 1$, the matrix X is called *unimodular*.

Examples. (1) The determinant of the empty matrix is equal to 1; the determinant of a square matrix of order 1 is equal to the unique element of this matrix. For a matrix of order 2

$$\begin{pmatrix} \xi_{11} & \xi_{12} \\ \xi_{21} & \xi_{22} \end{pmatrix}$$

then, in the above notation,

$$x_1 \wedge x_2 = (\xi_{11}e_1 + \xi_{21}e_2) \wedge (\xi_{12}e_1 + \xi_{22}e_2) = \xi_{11}\xi_{22}e_1 \wedge e_2 + \xi_{21}\xi_{12}e_2 \wedge e_1$$

whence

$$\begin{vmatrix} \xi_{11} & \xi_{12} \\ \xi_{21} & \xi_{22} \end{vmatrix} = \xi_{11}\xi_{22} - \xi_{12}\xi_{21}.$$

We translate into the language of matrices some of the results of nos. 1 and 2:

Proposition 5. *If X and Y are two square matrices over a commutative ring* A *with the same finite indexing set, then*

$$\det(XY) = (\det X)(\det Y). \tag{8}$$

For X to be invertible, it is necessary and sufficient that $\det X$ *be an invertible element of* A, *and then*

$$\det(X^{-1}) = (\det X)^{-1}. \tag{9}$$

This follows immediately from no. 1, Proposition 1 and no. 2, Theorem 1.

COROLLARY. *Two similar square matrices have equal determinants.*

If P is an invertible square matrix, then $\det(PXP^{-1}) = \det X$ by (8) and (9).

PROPOSITION 6. *For the columns of a square matrix X of finite order to be linearly independent, it is necessary and sufficient that* $\det X$ *be not a divisor of zero in* A.

This follows from no. 2, Proposition 3.

4. CALCULATION OF A DETERMINANT

Lemma 1. *Let* A *be a commutative ring and* M *a free* A*-module with a basis* $(e_j)_{j\in J}$, *where the indexing set* J *is totally ordered. For every integer $p \leqslant$* Card(J), *every alternating p-linear function $f: M^p \to N$ (where* N *is an* A*-module) and every family of p elements $x_i = \sum_{j\in J} \xi_{ji} e_j$ of* M $(1 \leqslant i \leqslant p)$,

$$\begin{aligned} (10)\quad & f(x_1, x_2, \ldots, x_p) \\ &= \sum_{j_1<j_2<\cdots<j_p} \Big(\sum_{\sigma\in\mathfrak{S}_p} \varepsilon_\sigma \cdot \xi_{j_{\sigma(1)},1}\xi_{j_{\sigma(2)},2}\ldots\xi_{j_{\sigma(p)},p}\Big) f(e_{j_1}, \ldots, e_{j_p}) \end{aligned}$$

where $(j_k)_{1\leqslant k\leqslant p}$ runs through the set of strictly increasing sequences of p elements of J.

Now

$$f(x_1, \ldots, x_p) = \sum_{(j_k)} \xi_{j_1,1}\xi_{j_2,2}\ldots\xi_{j_p,p} f(e_{j_1}, e_{j_2}, \ldots, e_{j_p})$$

where $(j_k)_{1\leqslant k\leqslant p}$ runs through *all* the sequences of p elements of J; it then suffices to apply Corollary 1 to Proposition 7 of § 7, no. 4 to f.

In particular, if J is finite and has n elements and $x_i = \sum_{j\in J} \xi_{ji} e_j$ $(1 \leqslant i \leqslant n)$ are n elements of M, then

$$\begin{aligned} (11)\quad & x_1 \wedge x_2 \wedge \cdots \wedge x_n \\ &= \Big(\sum_{\sigma\in\mathfrak{S}_n} \varepsilon_\sigma \xi_{j_{\sigma(1)},1}\xi_{j_{\sigma(2)},2}\ldots\xi_{j_{\sigma(n)},n}\Big) e_{j_1} \wedge e_{j_2} \wedge \cdots \wedge e_{j_n} \end{aligned}$$

where $(j_k)_{1\leqslant k\leqslant n}$ is the unique sequence of the n elements of J arranged in increasing order, whence

$$\det(x_1, x_2, \ldots, x_n) = \sum_{\sigma\in\mathfrak{S}} \varepsilon_\sigma \cdot \xi_{j_{\sigma(1)},1}\xi_{j_{\sigma(2)},2}\ldots\xi_{j_{\sigma(n)},n}. \tag{12}$$

With the notation of Lemma 1, comparing formulae (10) and (12) gives

$$(13)\qquad x_1 \wedge x_2 \wedge \cdots \wedge x_p = \sum_{H \in \mathfrak{F}_p(J)} \det(x_{H,1}, x_{H,2}, \ldots, x_{H,p}) e_H$$

where $\mathfrak{F}_p(J)$ is the set of subsets of J with p elements and, for every subset $H \in \mathfrak{F}_p(J)$, we write $x_{H,i} = \sum_{j \in H} \xi_{ji} e_j$ and $e_H = e_{j_1} \wedge e_{j_2} \wedge \cdots \wedge e_{j_p}$, $(j_k)_{1 \leqslant k \leqslant p}$ being the sequence of elements of H arranged in increasing order, it being understood that $\det(x_{H,1}, \ldots, x_{H,p})$ is taken with respect to the basis $(e_{j_k})_{1 \leqslant k \leqslant p}$.

PROPOSITION 7. *Let* I *be a finite set and* $X = (\xi_{ji})_{(j,i) \in I \times I}$ *a square matrix of type* (I, I) *over a commutative ring* A. *Then*

$$(14)\qquad \det X = \sum_{\sigma \in \mathfrak{S}_I} \varepsilon_\sigma \Big(\prod_{i \in I} \xi_{\sigma(i), i} \Big)$$

where σ *runs through the group* $\mathfrak{S}_I$ *of permutations of* I *and* ε_σ *is the signature of* σ (I, § 5, no. 7).

Attention may be confined to the case where $I = [1, n] \subset \mathbf{N}$ and it then suffices to apply formula (12), where $(e_i)_{1 \leqslant i \leqslant n}$ is the canonical basis of A^n and the x_i are the columns of X (cf. no. 3, formula (6)).

In particular, for the determinant of a matrix of order 3

$$X = \begin{pmatrix} \xi_{11} & \xi_{12} & \xi_{13} \\ \xi_{21} & \xi_{22} & \xi_{23} \\ \xi_{31} & \xi_{32} & \xi_{33} \end{pmatrix}$$

we have

$$\det(X) = \xi_{11}\xi_{22}\xi_{33} + \xi_{12}\xi_{23}\xi_{31} + \xi_{21}\xi_{32}\xi_{13} - \xi_{13}\xi_{22}\xi_{31} - \xi_{12}\xi_{21}\xi_{33} - \xi_{11}\xi_{23}\xi_{32}.$$

PROPOSITION 8. *For every square matrix* X *over a commutative ring, the determinant of the transpose matrix* ${}^tX$ *is equal to the determinant of* X.

Suppose that X is of type (I, I). For every ordered pair of permutations σ, τ of $\mathfrak{S}_I$, (since multiplication is commutative)

$$\prod_{i \in I} \xi_{\sigma(i), i} = \prod_{j \in I} \xi_{\sigma(\tau(j)), \tau(j)}.$$

In particular take $\tau = \sigma^{-1}$; using the fact that $\varepsilon_{\sigma^{-1}} = \varepsilon_\sigma$, it is seen that

$$(15)\qquad \det X = \sum_{\sigma \in \mathfrak{S}_I} \varepsilon_\sigma \cdot \Big(\prod_{i \in I} \xi_{i, \sigma(i)} \Big),$$

which proves the proposition.

COROLLARY 1. *For n vectors* $x_1, \ldots, x_n$ *of* A^n, *let* $Y(x_1, \ldots, x_n)$ *denote the square matrix of order n whose i-th row is* x_i, *for* $1 \leqslant i \leqslant n$. *Then the mapping*

$$(x_1, \ldots, x_n) \mapsto \det(Y(x_1, \ldots, x_n))$$

from $(A^n)^n$ *to* A *is alternating and n-linear.*

COROLLARY 2. *For a square matrix X of finite order over a commutative ring* A *the following conditions are equivalent:*

(i) *the rows of X are linearly independent;*
(ii) *the columns of X are linearly independent;*
(iii) $\det X$ *is not a divisor of zero in* A.

This follows immediately from no. 3, Proposition 6 and Proposition 8 above.

COROLLARY 3. *Let u be an endomorphism of a finite-dimensional free* A-*module* M *and* ${}^t u$ *the transpose endomorphism of the dual* M* (II, § 2, no. 5, Definition 5); *then*

$$\det({}^t u) = \det(u). \tag{16}$$

If X is the matrix of u with respect to a basis of M, ${}^t X$ is the matrix of ${}^t u$ with respect to the dual basis (II, § 10, no. 4, Proposition 3); as $\det(u) = \det(X)$ and $\det({}^t u) = \det({}^t X)$, the conclusion follows from Proposition 8.

5. MINORS OF A MATRIX

Let X be a rectangular matrix $(\xi_{ij})_{(i,j)\in I\times J}$ of type (I, J) whose indexing sets I and J are *totally ordered*. If $H \subset I$ and $K \subset J$ are finite subsets with the *same number p* of elements, there exists a unique *increasing* bijection $\phi : H \to K$ (*Set Theory*, III, § 5, no. 3, Proposition 6); we shall denote by $X_{H,K}$ the *square matrix of type* (H, H) equal to $(\xi_{i,\phi(j)})_{(i,j)\in H\times H}$. If the elements of X belong to a commutative ring A, the determinant $\det(X_{H,K})$ is called the *minor* of the matrix X of indices H, K; these determinants (for all ordered pairs (H, K) of subsets of I and J respectively with p elements) are also called the *minors* of X *of order p*. With this notation:

PROPOSITION 9. *Let* M *be an* A-*module with a basis* $(e_i)_{i\in J}$ (*finite or otherwise*) *whose indexing set* J *is totally ordered. For every integer* $p > 0$, *let* $(e_H)_{H\in \mathfrak{F}_p(J)}$ *be the corresponding basis of* $\bigwedge^p(M)$ (§ 7, no. 8). *Let* $(x_i)_{1\leqslant i\leqslant p}$ *be a sequence of p elements of* M; *let*

$$x_i = \sum_{j\in J} \xi_{ji} e_j \quad \textit{for } i \in I = (1, p)$$

and let X denote the matrix (ξ_{ji}) of type (J, I). *Then*

$$(17) \qquad x_1 \wedge x_2 \wedge \cdots \wedge x_p = \sum_{\mathrm{H} \in \mathfrak{F}_p(\mathrm{J})} (\det X_{\mathrm{H},\mathrm{I}}) e_{\mathrm{H}}.$$

where H *runs through the set* $\mathfrak{F}_p(\mathrm{J})$ *of subsets of* J *with p elements.*

This follows immediately from formula (12) of no. 4 and formula (6) of no. 3.

PROPOSITION 10. *Let* M *and* N *be two free* A*-modules of respective dimensions m and n, $u : \mathrm{M} \to \mathrm{N}$ a linear mapping and X the matrix of u with respect to a basis $(e_i)_{1 \leqslant i \leqslant m}$ of* M *and a basis $(f_j)_{1 \leqslant j \leqslant n}$ of* N. *Then, for every integer $p \leqslant \inf(m, n)$, the matrix of $\bigwedge^p(u)$ with respect to the basis $(e_{\mathrm{K}})_{\mathrm{K} \in \mathfrak{F}_p(\mathrm{I})}$ of $\bigwedge^p(\mathrm{M})$ and the basis $(f_{\mathrm{H}})_{\mathrm{H} \in \mathfrak{F}_p(\mathrm{J})}$ of $\bigwedge^p(\mathrm{N})$ (where we have written $\mathrm{I} = (1, m)$ and $\mathrm{J} = (1, n)$ is the matrix $(\det(X_{\mathrm{H},\mathrm{K}}))$ of type $(\mathfrak{F}_p(\mathrm{J}), \mathfrak{F}_p(\mathrm{I}))$* (and hence with $\binom{n}{p}$ rows and $\binom{m}{p}$ columns).

For a subset $\mathrm{K} \subset \mathrm{J}$ with p elements, let $(j_k)_{1 \leqslant k \leqslant p}$ be the sequence of elements of K arranged in increasing order; by definition of $\bigwedge^p(u)$, by § 7, no. 2, formula (4)

$$\textstyle\bigwedge^p(u)(e_{\mathrm{K}}) = u(e_{j_1}) \wedge u(e_{j_2}) \wedge \cdots \wedge u(e_{j_p}).$$

Hence the element of the matrix of $\bigwedge^p(u)$ which is in the row of index H and the column of index K is the component of index H of the element $u(e_{j_1}) \wedge \cdots \wedge u(e_{j_p})$; it is therefore equal to $\det(X_{\mathrm{H},\mathrm{K}})$ by Proposition 9.

The matrix $(\det(X_{\mathrm{H},\mathrm{K}}))$ is called the *p-th exterior power* of the matrix X and is denoted by $\bigwedge^p(X)$. When $p = m = n$, $\bigwedge^n(X)$ is the matrix with the single element $\det(X)$.

PROPOSITION 11. *Let* M *be a free* A*-module of finite dimension n; for every endomorphism u of* M *and every ordered pair of elements ξ, η of* A,

$$(18) \qquad \det(\xi 1_{\mathrm{M}} + \eta u) = \sum_{k \geqslant 0} \mathrm{Tr}(\textstyle\bigwedge^k(u)) \xi^{n-k} \eta^k.$$

Let $(e_i)_{1 \leqslant i \leqslant n}$ be a basis of M and let $\mathrm{I} = (1, n)$; to calculate the left hand side of (18), we must form the product

$$(\xi e_1 + \eta u(e_1)) \wedge (\xi e_2 + \eta u(e_2)) \wedge \cdots \wedge (\xi e_n + \eta u(e_n))$$

which is equal to the sum of the terms $\xi^{n-p}\eta^p z_{\mathrm{K}}$, where

$$z_{\mathrm{K}} = x_1 \wedge x_2 \wedge \cdots \wedge x_n$$

with $x_i = u(e_i)$ for $i \in \mathrm{K}$, $x_i = e_i$ for $i \in \mathrm{H} = \mathrm{I} - \mathrm{K}$, where the integer p runs through the interval $(0, n)$ and, for each p, K runs through the set of subsets of I with p elements. If $i_1 < i_2 < \cdots < i_{n-p}$ (resp. $j_1 < j_2 < \cdots < j_p$) are the elements of H (resp. K) arranged in increasing order, we can write (§ 7, no. 8, Corollary 1 to Theorem 1 and formula (19))

$$z_{\mathrm{K}} = \rho_{\mathrm{H,K}} e_{i_1} \wedge e_{i_2} \wedge \cdots \wedge e_{i_{n-p}} \wedge u(e_{j_1}) \wedge \cdots \wedge u(e_{j_p}).$$

But if X is the matrix of u with respect to the basis (e_i), then by Proposition 10

$$u(e_{j_1}) \wedge \cdots \wedge u(e_{j_p}) = \sum_{\mathrm{L} \in \mathfrak{F}_p(\mathrm{I})} (\det X_{\mathrm{L,K}}) e_{\mathrm{L}}$$

and hence

$$z_{\mathrm{K}} = \rho_{\mathrm{H,K}} \sum_{\mathrm{L} \in \mathfrak{F}_p(\mathrm{I})} (\det X_{\mathrm{L,K}}) e_{\mathrm{H}} \wedge e_{\mathrm{L}}.$$

Now $\mathrm{H} \cap \mathrm{L} \neq \varnothing$ except for $\mathrm{L} = \mathrm{K}$; it therefore follows from § 7, no. 8, formula (20) that $z_{\mathrm{K}} = (\det X_{\mathrm{K,K}}) e_1 \wedge e_2 \wedge \cdots \wedge e_n$ and formula (18) then follows from Proposition 10 and the definition of the trace of a matrix (II, § 10, no. 11, formulae (49) and (50)).

Corollary. *Under the same hypotheses as in Proposition* 11, *for the endomorphism* $\bigwedge(u)$ *of the* A*-module* $\bigwedge(\mathrm{M})$

$$\mathrm{Tr}(\textstyle\bigwedge(u)) = \det(1_{\mathrm{M}} + u). \tag{19}$$

It suffices to replace ξ and η by 1 in (18) and observe that the matrix of $\bigwedge(u)$ with respect to the basis of the e_{H} ($\mathrm{H} \in \mathfrak{F}(\mathrm{I})$) is the diagonal matrix of the matrices of the $\bigwedge^k(u)$ for $\geqslant k$ 0 (II, § 10, no. 7, *Example* IV).

6. EXPANSIONS OF A DETERMINANT

Let I be a totally ordered finite indexing set. For every subset H of I let H′ denote the complement I − H. Let $X = (\xi_{ji})$ be a square matrix of type (I, I), which can be considered as the matrix of an endomorphism u of $\mathrm{M} = \mathrm{A}^{\mathrm{I}}$ with respect to the canonical basis $(e_i)_{i \in \mathrm{I}}$ of M. Let $n = \mathrm{Card(I)}$ and let H be a subset of I with $q \leqslant n$ elements and K a subset of I with $n - q$ elements; then we can write (no. 5, Proposition 10)

$$(\textstyle\bigwedge^q(u))(e_{\mathrm{H}}) = \sum_{\mathrm{R}} \det(X_{\mathrm{R,H}}) e_{\mathrm{R}}$$

$$(\textstyle\bigwedge^{n-q}(u))(e_{\mathrm{K}}) = \sum_{\mathrm{S}} \det(X_{\mathrm{S,K}}) e_{\mathrm{S}}$$

where R (resp. S) runs through the set of subsets of I with q (resp. $n - q$) elements. It follows from § 7, no. 8, formulae (19) and (20) that

$$e_{\mathrm{R}} \wedge e_{\mathrm{S}} = 0$$

except when $S = R'$, whence the formula

$$(20)\quad (\textstyle\bigwedge^q(u)(e_H)) \wedge (\bigwedge^{n-q}(u)(e_K)) = \sum_R \rho_{R,R'} \det(X_{R,H})\det(X_{R',K})e_I$$

where R runs through the set $\mathfrak{F}_q(I)$ of subsets of I with q elements.

If we take $K = H'$, it follows from the definition of $\bigwedge^n(u)$ (§ 7, no. 2, formula (4)) and § 7, no. 3, Corollary 1 to Proposition 5 that the right hand side of (20) is $\rho_{H,H'}\bigwedge^n(u)(e_I)$. Hence (no. 1, formula (1) and § 7, no. 2, formula (4))

$$(21)\qquad \det(X) = \rho_{H,H'} \sum_{R\in\mathfrak{F}_q} \rho_{R,R'} \det(X_{R,H})\det(X_{R',H'}).$$

If on the other hand $K \neq H'$, then $H \cap K \neq \varnothing$; as the left hand side of (20) is $\pm\bigwedge^n(u)(e_H \wedge e_K)$, it is *zero*, whence

$$(22)\qquad \sum_R \rho_{R,R'} \det(X_{R,H})\det(X_{R',K}) = 0 \quad \text{for } K \neq H'.$$

The right hand side of (21) is called the *Laplace expansion* of the determinant of the matrix X *by the* q *columns whose indices belong to* H *and the* $n - q$ *columns whose indices belong to the complement* H' *of* H. The minors $\det(X_{R,H})$ and $\det(X_{R',H'})$ are sometimes called *complementary*.

An important simple case of the Laplace expansion is that where $I = [1, n]$ and $q = 1$, hence $H = \{i\}$; for every subset $R = \{j\}$ of I with one element then $\det X_{R,H} = \xi_{ji}$. The minor of $\det X_{R',H'}$ is the determinant of the square matrix derived canonically (no. 5) from the matrix obtained by suppressing in X the row of index j and the column of index i. We denote this square matrix by X^{ji}. Obviously $\rho_{H,H'} = (-1)^{i-1}$ and $\rho_{R,R'} = (-1)^{j-1}$; therefore (21) becomes in this case

$$(23)\qquad \det X = \sum_{j=1}^{n} (-1)^{i+j}\xi_{ji} \det(X^{ji})$$

and we obtain similarly from (22)

$$(24)\qquad \sum_{j=1}^{n} (-1)^{j}\xi_{ji} \det(X^{jk}) = 0 \quad \text{for } k \neq i.$$

Formula (23) is known as the *expansion of the determinant of* X *by the column of index* i. The scalar $(-1)^{i+j}\det(X^{ji})$ is called the *cofactor* of indices j and i (or, by an abuse of language, the cofactor of ξ^{ji}) in X.

The *matrix of cofactors* of X is the matrix

$$(25)\qquad Y = ((-1)^{i+j}\det(X^{ji}))$$

whose element in the j-th row and i-th column is the cofactor of indices j and i. Formulae (23) and (24) are equivalent to the formula

$$ {}^tY.X = (\det X)I_n. \tag{26} $$

Therefore:

PROPOSITION 12. *For every invertible square matrix X of type (n, n), the inverse of X is given by the formula*

$$ X^{-1} = (\det X)^{-1}.{}^tY \tag{27} $$

where Y is the cofactor matrix of X.

By considering the *transpose* of X and using Proposition 8 of no. 5, the Laplace expansions could be obtained relative to two complementary sets of rows and, in particular, the expansion of $\det X$ by a row, thus there are formulae equivalent to

$$ X.{}^tY = (\det X)I_n, \tag{28} $$

in the above notation.

It is easily verified that if X is the matrix of an endomorphism u of a free A-module M of dimension n with respect to a basis $(e_i)_{1 \leqslant i \leqslant n}$, ${}^tY$ is the matrix of the endomorphism $\tilde{u}$ of M defined by the following condition: for every set of n elements $x, y_2, \ldots, y_n$ of M,

$$ \tilde{u}(x) \wedge y_2 \wedge \cdots \wedge y_n = x \wedge u(y_2) \wedge \cdots \wedge u(y_n). $$

$\tilde{u}$ is called the *cotranspose* of u (cf. § 11, no. 11, Corollary to Proposition 13).

Examples. (1) *Vandermonde determinant.* Given a sequence $(\zeta_i)_{1 \leqslant i \leqslant n}$ of n elements of A, the *Vandermonde determinant* of this sequence is the determinant

$$ V(\zeta_1, \zeta_2, \ldots, \zeta_n) = \begin{vmatrix} 1 & 1 & \ldots & 1 \\ \zeta_1 & \zeta_2 & \ldots & \zeta_n \\ \zeta_1^2 & \zeta_2^2 & \ldots & \zeta_n^2 \\ \cdot & \cdot & \cdot & \cdot \\ \zeta_1^{n-1} & \zeta_2^{n-1} & \ldots & \zeta_n^{n-1} \end{vmatrix} $$

We shall show that

$$ V(\zeta_1, \zeta_2, \ldots, \zeta_n) = \prod_{i<j} (\zeta_j - \zeta_i). \tag{29} $$

Since the proposition is immediate for $n = 1$, we argue by induction on n.

For each index $k \geqslant 2$, we subtract from the row of index k the row of index $k-1$ multiplied by ζ_1; the value of the determinant is unaltered and hence

$$V(\zeta_1, \zeta_2, \ldots, \zeta_n) = \begin{vmatrix} 1 & 1 & \ldots & 1 \\ 0 & \zeta_2 - \zeta_1 & \ldots & \zeta_n - \zeta_1 \\ 0 & \zeta_2(\zeta_2 - \zeta_1) & \ldots & \zeta_n(\zeta_n - \zeta_1) \\ \cdot & \cdot & \cdot & \cdot \\ 0 & \zeta_2^{n-2}(\zeta_2 - \zeta_1) & \ldots & \zeta_n^{n-2}(\zeta_n - \zeta_1) \end{vmatrix}$$

whence, expanding by the first column and then taking out the factor $\zeta_k - \zeta_1$ from the column of index $k-1$ from the minor thus obtained $(2 \leqslant k \leqslant n)$

$$V(\zeta_1, \ldots, \zeta_n) = (\zeta_2 - \zeta_1)(\zeta_3 - \zeta_1) \ldots (\zeta_n - \zeta_1) V(\zeta_2, \ldots, \zeta_n)$$

which establishes (29) by induction.

(2) Consider a square matrix of order n which is presented in the form of an "upper triangular matrix of matrices" (II, § 10, no. 7, *Example* IV)

$$X = \begin{pmatrix} Y & T \\ 0 & Z \end{pmatrix}$$

We show that

$$\det X = (\det Y)(\det Z). \tag{30}$$

Let n be the order of the matrix X, h that of Y, $(e_i)_{1 \leqslant i \leqslant n}$ the canonical basis of A^n and x_i $(1 \leqslant i \leqslant n)$ the columns of X; the hypothesis implies that the columns $x_1, \ldots, x_h$ belong to the submodule of A^n with basis $e_1, \ldots, e_h$ and then by definition (no. 3, formula (6))

$$x_1 \wedge x_2 \wedge \cdots \wedge x_h = (\det Y) e_1 \wedge e_2 \wedge \cdots \wedge e_h.$$

On the other hand, for every index $i > h$, we can write $x_i = y_i + z_i$, where y_i is a linear combination of $e_1, e_2, \ldots, e_h$ and z_i is a linear combination of $e_{h+1}, \ldots, e_n$. By (30), $x_1 \wedge x_2 \wedge \cdots \wedge x_h \wedge y_i = 0$ for all $i > h$, therefore

$$x_1 \wedge x_2 \wedge \cdots \wedge x_n = (\det Y) e_1 \wedge e_2 \wedge \cdots \wedge e_h \wedge z_{h+1} \wedge \cdots \wedge z_n.$$

But by definition

$$z_{h+1} \wedge z_{h+2} \wedge \cdots \wedge z_n = (\det Z) e_{h+1} \wedge e_{h+2} \wedge \cdots \wedge e_n$$

whence formula (30).

By induction on p, it follows that if X is of the form of an upper triangular matrix of matrices:

$$X = \begin{pmatrix} X_{11} & X_{12} & \dots & X_{1p} \\ 0 & X_{22} & \dots & X_{2p} \\ \cdot & \cdot & \cdot & \cdot \\ 0 & 0 & \dots & X_{pp} \end{pmatrix}$$

$$\det X = (\det X_{11})(\det X_{22}) \dots (\det X_{pp}). \tag{31}$$

This can be applied in particular to a triangular matrix (where all the X_{ii} are of order 1) and more particularly to a diagonal matrix:

$$\det(\text{diag}(\alpha_1, \alpha_2, \dots, \alpha_n)) = \alpha_1\alpha_2 \dots \alpha_n. \tag{32}$$

(3) Let M, M′ be two free A-modules of respective dimensions n, n', u an endomorphism of M and u' an endomorphism of M′. Then

$$\det(u \otimes u') = (\det u)^{n'}(\det u')^{n}. \tag{33}$$

For we can write $u \otimes u' = (u \otimes 1_{M'}) \circ (1_M \otimes u')$ and are then led to the case where one of the two endomorphisms u, u' is the identity. For example if $u' = 1_{M'}$ and X is the matrix of u with respect to a basis (e_i) of M, then the matrix of $u \otimes 1_{M'}$ with respect to the tensor product of (e_i) and a basis of M′ can be written as a matrix (with n' rows and n' columns) of matrices with n rows and n columns

$$\begin{pmatrix} X & 0 & \dots & 0 \\ 0 & X & \dots & 0 \\ \cdot & \cdot & \cdot & \cdot \\ 0 & 0 & \dots & X \end{pmatrix}$$

whence, by virtue of *Example* 2,

$$\det(u \otimes 1_{M'}) = (\det X)^{n'} = (\det u)^{n'}$$

which immediately gives formula (33).

7. APPLICATION TO LINEAR EQUATIONS

Consider a system of n scalar linear equations in n unknowns over a (*commutative*) ring A (II, § 2, no. 8):

$$\sum_{j=1}^{n} \lambda_{ij}\xi_j = \eta_i \qquad (1 \leqslant i \leqslant n). \tag{34}$$

Let L be the square matrix (λ_{ij}) of order n; by identifying as usual the matrix with one column consisting of the ξ_i (resp. the η_i) with an element $x = (\xi_i)$ of A^n (resp. the element $y = (\eta_i)$ of A^n), system (34) may also be written (II, § 10, no. 3, Proposition 2)

$$L.x = y. \tag{35}$$

Let u be the endomorphism $x \mapsto L.x$ of A^n, with L as matrix with respect to the canonical basis; to say that equation (35) has (at least) one solution for *all* $y \in A^n$ means that u is *surjective*; Theorem 1 of no. 2 then implies the following proposition:

PROPOSITION 13. *For a system of n linear equations in n unknowns over a commutative ring to admit at least one solution regardless of the right hand sides, it is necessary and sufficient that the determinant of the matrix of the system be invertible; in this case the system admits a single solution.*

If $\det L$ is not a divisor of zero in A, equation (34) is equivalent to the equation

$$(\det L)L.x = (\det L)y.$$

If M is the cofactor matrix of L, we derive from (34) and formula (26) of no. 6 the relation

$$(\det L)x = {}^tM.y \tag{36}$$

which may also be written

$$(\det L)\xi_i = \sum_{j=1}^{n} (-1)^{i+j}(\det L^{ij})\eta_j = \det L_i \qquad (1 \leqslant i \leqslant n) \tag{37}$$

where L_i denotes the matrix obtained by replacing the column of L of index i by y. Formulae (37) are called *Cramer's formulae* for system (34); every solution of (34) is also a solution of (37). Conversely, we derive from (36), taking account of formula (28) of no. 6,

$$(\det L)(L.x - y) = 0 \tag{38}$$

and hence, if $\det L$ is not a divisor of zero in A, systems (34) and (37) are *equivalent*; if $\det L$ is invertible, the unique solution of (34) is given by

$$\xi_i = (\det L)^{-1}(\det L_i) \qquad (1 \leqslant i \leqslant n). \tag{39}$$

A system (34) such that $\det L$ is invertible is also called a *Cramer system*.

In particular let $y = 0$; it then follows from Proposition 3 of no. 2 that:

PROPOSITION 14. *For a* homogeneous *linear system of n equations in n unknowns over a commutative ring to admit a non-zero solution, it is necessary and sufficient that the determinant of its matrix be a divisor of zero.*

8. CASE OF A COMMUTATIVE FIELD

All the above applies when the ring A is a commutative field; but there are simplifications and additional results.

Thus Proposition 12 of § 7, no. 9 can be formulated in this case as follows:

PROPOSITION 15. *Let* E *be a vector space over a commutative field; for p vectors $x_i \in E$ $(1 \leqslant i \leqslant p)$ to be linearly independent, it is necessary and sufficient that* $x_1 \wedge x_2 \wedge \cdots \wedge x_p \neq 0$.

COROLLARY. *Let X be a matrix of type (m, n) over a commutative field. The rank of X is equal to the greatest integer p such that there exists at least one minor of X of order p which is $\neq 0$.*

The rank of X is the maximum number of linearly independent columns of X (II, § 10, no. 12, Definition 7). The corollary then follows from Proposition 15 and formula (17) of no. 5.

Consider now the case of a *system of m linear equations in n unknowns over a commutative field* K:

$$\sum_{j=1}^{n} \lambda_{ij}\xi_j = \eta_i \qquad (1 \leqslant i \leqslant m). \tag{41}$$

PROPOSITION 16. *Let $L = (\lambda_{ij})$ be the matrix (of type (m, n)) of system* (41). *Let M be the matrix of type $(m, n + 1)$ obtained by adding to L the $(n + 1)$-th column* (η_i) (II, § 10, no. 1). *Let p be the rank of L (calculated by applying the Corollary to Proposition* 15). *Suppose that the minor Δ of L, the determinant of the matrix by suppressing the rows and columns of index $\geqslant p + 1$ in L, is $\neq 0$ (which is always possible by means of a suitable permutation on the rows of L and a suitable permutation on the columns of L). Then, for system* (41) *to have at least one solution it is necessary and sufficient that all the minors of order $p + 1$ of M, the determinants of the submatrices of order $p + 1$ of M whose columns have indices $1, 2, \ldots, p$ and $n + 1$, be zero. If this is so, the solutions of system* (41) *are those of the system consisting of the first p equations; if they are written*

$$\sum_{j=1}^{p} \lambda_{ij}\xi_j = \eta_i - \sum_{k=p+1} \lambda_{ik}\xi_k \qquad (1 \leqslant i \leqslant p) \tag{42}$$

all the solutions of this system are obtained by taking for the ξ_k of index $k > p$ arbitrary values and applying Cramer's formulae (no. 7, *formulae* (37)) *to calculate the ξ_j of index $j \leqslant p$.*

We know (II, § 10, no. 12, Proposition 12) that for the system (41) to have at least one solution, it is necessary and sufficient that the matrices L and M have the *same rank.* With the rows and columns of L permuted to satisfy the condition of the statement, let a_i ($1 \leqslant i \leqslant p$) denote the first p columns of L and $y = (\eta_i)$ the $(n+1)$-*th* column of M; since all the columns of L are by hypothesis linear combinations of the a_i, to say that M has the same rank p as L means that y is a linear combination of the a_i, or also (Proposition 15) that $a_1 \wedge \cdots \wedge a_p \wedge y = 0$. The possibility condition in the statement is the translation of the latter relation, taking account of formula (17) of no. 5. Moreover, since the first p rows of M are linearly independent, the rows of index $>p$ are linear combinations of them and hence every solution of (42) is also a solution of (41). The last assertion is then an immediate consequence of Proposition 13 of no. 7.

9. THE UNIMODULAR GROUP SL(n, A)

Let $\mathbf{M}_n(\mathrm{A})$ denote the ring of square matrices of order n over A. Consider the mapping $\det : \mathbf{M}_n(\mathrm{A}) \rightarrow \mathrm{A}$. The group $\mathbf{GL}(n, \mathrm{A})$ of invertible elements of $\mathbf{M}_n(\mathrm{A})$ (isomorphic to the group of automorphisms of the A-module A^n (II, § 10, no. 7)) is just the inverse image under this mapping of the multiplicative group A^* of invertible elements of A (no. 3, Proposition 5). Note on the other hand that the mapping $\det : \mathbf{GL}(n, \mathrm{A}) \rightarrow \mathrm{A}^*$ is a group homomorphism (no. 3, Proposition 5).

The mapping $\det : \mathbf{M}_n(\mathrm{A}) \rightarrow \mathrm{A}$ is moreover *surjective* (and therefore so is the homomorphism $\det : \mathbf{GL}(n, \mathrm{A}) \rightarrow \mathrm{A}^*$); since, for all $\lambda \in \mathrm{A}$,

$$\det(\operatorname{diag}(\lambda, 1, \ldots, 1)) = \lambda$$

by virtue of formula (32) of no. 6.

The *kernel* of the surjective homomorphism $\det : \mathbf{GL}(n, \mathrm{A}) \rightarrow \mathrm{A}^*$ is a normal subgroup of $\mathbf{GL}(n, \mathrm{A})$, which is composed of *unimodular* matrices; it is denoted by $\mathbf{SL}_n(\mathrm{A})$ or $\mathbf{SL}(n, \mathrm{A})$ and is often called the *unimodular group* or *special linear group* of square matrices of order n over A.

In this no. we shall examine the case where A is a *field*. Recall that for $1 \leqslant i \leqslant n$, $1 \leqslant j \leqslant n$, E_{ij} denotes the square matrix of order n all of whose elements are zero except the one in the row of index i and the column of index j, which is equal to 1; with I_n denoting the unit matrix of order n, we write $B_{ij}(\lambda) = I_n + \lambda E_{ij}$ for every ordered pair of *distinct* indices i, j and all $\lambda \in \mathrm{A}$ (II, § 10, no. 13).

PROPOSITION 17. *Let* K *be a commutative field. The unimodular group* $\mathbf{SL}(n, \mathrm{K})$ *is generated by the matrices* $B_{ij}(\lambda)$ *with* $i \neq j$ *and* $\lambda \in \mathrm{K}$.

By II, § 10, no. 13, Proposition 14, we know that every matrix in $\mathbf{GL}(n, \mathrm{K})$ is a product of matrices of the form $B_{ij}(\lambda)$ and a matrix of the form $\operatorname{diag}(1, 1, \ldots, 1, \alpha)$ with $\alpha \in \mathrm{K}^*$. Now it is immediate that $\det(B_{ij}(\lambda)) = 1$ and $\det(\operatorname{diag}(1, \ldots, 1, \alpha)) = \alpha$ (no. 6, *Example* 2); whence the proposition.

COROLLARY. *The group* $\mathbf{SL}(n, K)$ *is the group of commutators of* $\mathbf{GL}(n, K)$, *except in the case where* $n = 2$ *and where* K *is a field with 2 elements.*

As $\mathbf{SL}(n, K)$ is the kernel of the homomorphism det of $\mathbf{GL}(n, K)$ to a commutative group K^*, $\mathbf{SL}(n, K)$ contains the commutator group Γ of $\mathbf{GL}(n, K)$ (I, § 6, no. 2). To prove that $\mathbf{SL}(n, K) = \Gamma$, it will suffice, by Proposition 17, to show that, for all $\lambda \in K^*$, $B_{ij}(\lambda)$ belongs to Γ. Now, $B_{ij}(\lambda)$ is a conjugate of $B_{ij}(1)$ in $\mathbf{GL}(n, K)$ since $B_{ij}(\lambda) = Q.B_{ij}(1).Q^{-1}$, where Q denotes the matrix with respect to the canonical basis (e_i) of the automorphism v of K^n such that $v(e_i) = \lambda e_i$, $v(e_k) = e_k$ for $k \neq i$. On the other hand, let u_{ij} (for $i \neq j$) be the automorphism of K^n such that $u_{ij}(e_i) = -e_j$, $u_{ij}(e_j) = e_i$, $u_{ij}(e_k) = e_k$ for $k \notin \{i, j\}$, which belongs to $\mathbf{SL}(n, K)$; then $B_{ji}(\lambda) = U_{ij}B_{ij}(-\lambda)U_{ij}^{-1}$, where U_{ij} is the matrix of u_{ij} with respect to the canonical basis. Similarly, if $1 < i < j$, then $B_{1j}(\lambda) = U_{1i}B_{ij}(\lambda)U_{1i}^{-1}$ and finally, for $2 < j$, $B_{12}(\lambda) = U_{2j}B_{1j}(\lambda)U_{2j}^{-1}$. This proves that all the $B_{ij}(\lambda)$ have the same image s in $\mathbf{GL}(n, K)/\Gamma$ and it remains to show that s is the identity element.

Suppose first that K contains an element λ distinct from 0 and 1; then $1 = \lambda + (1 - \lambda)$, the two terms on the right hand side being $\neq 0$; the relation $B_{12}(1) = B_{12}(\lambda)B_{12}(1 - \lambda)$ shows that $s^2 = s$ and hence s is the identity element.

Suppose now that $n \geqslant 3$. The product $B_{21}(1)B_{31}(1)$ is the matrix of an automorphism u of K^n such that $u(e_1) = e_1 + e_2 + e_3$, $u(e_i) = e_i$ for $i \neq 1$. If S is the matrix of the automorphism u' of K^n such that $u'(e_2) = e_2 + e_3$, $u'(e_i) = e_i$ for $i \neq 2$, then $S.B_{21}(1)B_{31}(1).S^{-1} = B_{21}(1)$; we also deduce that $s^2 = s$, which completes the proof.

Remarks. (1) $\mathbf{GL}(2, \mathbf{F}_2) = \mathbf{SL}(2, \mathbf{F}_2)$; this is a solvable group of order 6, whose commutator group is of index 2 (II, § 10, Exercise 14).

(2) With the same notation as above, it can be proved as in I, § 5, no. 7, Proposition 9 that, for $i < j$, $j - i > 1$, $u_{ij} = u_{j-1,j}u_{i,j-1}u_{j-1,j}^{-1}$; hence *the group* $\mathbf{SL}(n, K)$ *is generated by the matrices* $B_{12}(\lambda)$ *and* $U_{i,i+1}$ *for* $1 \leqslant i \leqslant n - 1$.

10. THE A[X]-MODULE ASSOCIATED WITH AN A-MODULE ENDOMORPHISM

Let M be an A-module and u an endomorphism of M. Consider the polynomial ring A[X] in one indeterminate X over A. For every polynomial $p \in A[X]$ and all $x \in M$, we write

$$p.x = p(u)(x). \tag{43}$$

As $(pq)(u) = p(y) \circ q(u)$ for two polynomials p, q of A[X], and A[X]-module structure is thus defined on M; the set M, with this structure, will be denoted by M_u; the A-module structure given on M is obtained by restricting the ring of operators of M_u to A. Note that the submodules of M_u are just the submodules of M which are *stable* under u.

As the mapping $(p, x) \mapsto p.x$ of $A[X] \times M$ into M is A-bilinear, it defines canonically an A-linear mapping $\phi: A[X] \otimes_A M \to M$ such that

$$\phi(p \otimes x) = p.x = p(u)(x). \tag{44}$$

On the other hand, $A[X] \otimes_A M$ has canonically an A[X]-module structure (II, § 5, no. 1); we shall denote this A[X]-module by M[X]; the mapping $\phi: M[X] \to M_u$ is A[X]-*linear* since, for p, q in A[X] and $x \in M$,

$$\phi(q(p \otimes x)) = \phi((qp) \otimes x) = (qp).x = q(u)(p(u)(x)) = q.\phi(p \otimes x).$$

Moreover, u is an A[X]-endomorphism of M_u, for

$$u(p.x) = u(p(u)(x)) = (up(x))(x) = p.u(x).$$

Finally, an A[X]-endomorphism $\bar{u}$ of M[X] is defined by writing (II, § 5, no. 1)

$$\bar{u}(p \otimes x) = p \otimes u(x). \tag{45}$$

Moreover it follows from formulae (44) and (45) that that A[X]-linear mappings u, $\bar{u}$ and ϕ are related by the relation

$$\phi \circ \bar{u} = u \circ \phi. \tag{46}$$

Let ψ denote the A[X]-endomorphism $X - \bar{u}$ of M[X], so that $\psi(p \otimes x) = (Xp) \otimes x - p \otimes u(x)$. We have the following proposition:

PROPOSITION 18. *The sequence of* A[X]-*homomorphisms*

$$M[X] \xrightarrow{\psi} M[X] \xrightarrow{\phi} M_u \longrightarrow 0 \tag{47}$$

is exact.

As $\phi(1 \otimes x) = x$ for all $x \in M$, clearly ϕ is surjective; on the other hand,

$$\phi(X(p \otimes x)) = X.\phi(p \otimes x) = u(\phi(p \otimes x)),$$

in other words, $\phi \circ X = u \circ \phi = \phi \circ \bar{u}$ by (46); this proves that $\phi \circ \psi = 0$. It remains to verify that $\operatorname{Ker} \phi \subset \operatorname{Im} \psi$. For this note that, since the monomials X^k $(k \geqslant 0)$ form a basis of the A-module A[X], every element $z \in M[X]$ can be written uniquely in the form $z = \sum_k X^k \otimes x_k$, where (x_k) is a family of elements of M, of finite support. If $z \in \operatorname{Ker} \phi$, then $\phi(z) = \sum_k u^k(x_k) = 0$ and we can write

$$z = \sum_k (X^k \otimes x_k - 1 \otimes u^k(x_k)) = \sum_k (X^k - \bar{u}^k)(1 \otimes x_k).$$

But as the A[X]-emdomorphisms X and $\bar{u}$ of M[X] are permutable, then

$X^k - \bar{u}^k = (X - \bar{u}) \circ \left(\sum_{j=0}^{k-1} X^j \bar{u}^{k-j-1}\right)$ which proves that there exists a $y \in M[X]$ such that $z = \psi(y)$.

Now let M′ be another A-module and u' an endomorphism of M′; let $M'_{u'}$, ϕ', $\bar{u}'$, ψ' be the module and mapping obtained from M′ and u' as M_u, ϕ, $\bar{u}$, ψ are obtained from M and u. Then:

PROPOSITION 19. *For a mapping g of M into M′ to be an A[X]-homomorphism of M_u into $M'_{u'}$, it is necessary and sufficient that g be an A-homomorphism of M into M′ such that $g \circ u = u' \circ g$. When this is so, if $\bar{g}$ is the A[X]-homomorphism of M[X] into M′[X] equal to $1_{A[X]} \otimes g$ (II, § 5, no. 1), the diagram*

$$\text{(48)} \qquad \begin{array}{ccccccc} M[X] & \xrightarrow{\psi} & M[X] & \xrightarrow{\phi} & M_u & \longrightarrow & 0 \\ \downarrow{\scriptstyle \bar{g}} & & \downarrow{\scriptstyle \bar{g}} & & \downarrow{\scriptstyle g} & & \\ M'[X] & \xrightarrow[\psi']{} & M'[X] & \xrightarrow[\phi']{} & M'_{u'} & \longrightarrow & 0 \end{array}$$

is commutative.

The condition $g \circ u = u' \circ g$ is obviously necessary by (43) for g to be an A[X]-homomorphism; it is sufficient, for it implies by induction that $g \circ u^n = u'^n \circ g$ for every integer $n > 0$. On the other hand, for all $x \in M$ and all $p \in A[X]$,

$$\phi'(\bar{g}(p \otimes x)) = \phi'(p \otimes g(x)) = p(u')(g(x)) = g(p(u)(x)) = g(\phi(p \otimes x))$$

and

$$\bar{u}'(g(p \otimes x)) = \bar{u}'(p \otimes g(x)) = p \otimes u'(g(x)) = p \otimes g(u(x)) = \bar{g}(\bar{u}(p \otimes x))$$

which proves the commutativity of diagram (48).

11. CHARACTERISTIC POLYNOMIAL OF AN ENDOMORPHISM

Let M be a free A-module of dimension n and u an endomorphism of M. Consider the polynomial ring in two indeterminates A[X, Y] and the A[X, Y]-module $M[X, Y] = A[X, Y] \otimes_A M$; let $\bar{u}$ be the endomorphism of the A[X, Y]-module M[X, Y] canonically derived from u (II, § 5, no. 1). It follows from no. 5, Proposition 11 that

$$\text{(49)} \qquad \det(X - Y\bar{u}) = \sum_{j=0}^{n} (-1)^j \operatorname{Tr}(\textstyle\bigwedge^j(u)) X^{n-j} Y^j$$

for if U is the matrix of u with respect to a basis $(e_i)_{1 \leqslant i \leqslant n}$ of M, U is the matrix of $\bar{u}$ with respect to the basis $(1 \otimes e_i)_{1 \leqslant i \leqslant n}$ of M[X, Y], hence

$$\operatorname{Tr}(\textstyle\bigwedge^j(\bar{u})) = \operatorname{Tr}(\textstyle\bigwedge^j(u)).$$

DEFINITION 3. *Let* M *be a finite-dimensional free* A*-module and* u *an endomorphism of* M. *The determinant of the endomorphism* $X - \bar{u}$ *of the free* A[X]*-module* M[X] *is called the characteristic polynomial of* u *and is denoted by* $\chi_u(X)$.

If M is of rank n, it follows from (49) that

$$\chi_u(X) = \sum_{j=0}^{n} (-1)^j \operatorname{Tr}(\textstyle\bigwedge^j(u)) X^{n-j} \tag{50}$$

for $\det(X - Y\bar{u}) = \det(X.I_n + YU)$ and $\det(X - \bar{u}) = \det(X.I_n - U)$. It is therefore seen that $\chi_u(X)$ is a monic polynomial of degree n, in which the coefficient of X^{n-1} is $-\operatorname{Tr}(u)$ and the constant term is $(-1)^n \det(u)$.

PROPOSITION 20 ("Cayley–Hamilton theorem"). *For every endomorphism* u *of a finite-dimensional free* A*-module,* $\chi_u(u) = 0$.

In the notation of Proposition 18 (§ 3, no. 10), for all $x \in M$, $\chi_u(u)(x)$ is the image under ϕ of $\chi_u(X) \otimes x$. But if v is the endomorphism of M[X], the *co-transpose* of $X - \bar{u}$ (no. 6), then

$$\chi_u(X) \otimes x = \chi_u(X)(1 \otimes x) = (X - \bar{u})(v(1 \otimes x))$$

and the conclusion follows from Proposition 18 of no. 10.

§ 9. NORMS AND TRACES

Throughout this paragraph, K *will denote a commutative ring and* A *a unital associative* K*-algebra. Every* A*-module will be assumed to be given the* K*-module structure obtained by restricting the scalars to* K.

1. NORMS AND TRACES RELATIVE TO A MODULE

DEFINITION 1. *Let* M *be an* A*-module admitting a finite basis as a* K*-module. For all* $a \in A$, *let* a_M *be the endomorphism* $x \mapsto ax$ *of the* K*-module* M. *The trace, determinant and characteristic polynomial of* a_M *are called respectively the trace, norm and characteristic polynomial of* z *relative to* M.

The trace and norm of a are therefore elements of K, denoted respectively by $\operatorname{Tr}_{M/K}(a)$ and $N_{M/K}(a)$; the characteristic polynomial of a is an element of K[X], denoted by $\operatorname{Pc}_{M/K}(a; X)$. We omit K in the above notation when there is no risk of confusion.

From the properties of the trace and determinant of an endomorphism (II, § 4, no. 3 and § 8, no. 1) we obtain the relations

$$\operatorname{Tr}_M(a + a') = \operatorname{Tr}_M(a) + \operatorname{Tr}_M(a') \tag{1}$$

$$(2)\qquad \operatorname{Tr}_{M}(aa') = \operatorname{Tr}_{M}(a'a)$$

$$(3)\qquad N_{M}(aa') = N_{M}(a)N_{M}(a')$$

for all a, a' in A.

Let $(e_i)_{1 \leqslant i \leqslant n}$ be a basis of the K-module M and $(m_{ij}(a))$ the matrix of the endomorphism a_M with respect to this basis. The functions m_{ij} are linear forms on the K-module A and

$$(4)\qquad \operatorname{Tr}_{M}(a) = \sum_{i=1}^{n} m_{ii}(a)$$

$$(5)\qquad N_{M}(a) = \det(m_{ij}(a))$$

$$(6)\qquad \operatorname{Pc}(a; X) = \det(\delta_{ij}X - m_{ij}(a)).$$

It follows from the method of calculating a determinant (§ 8, no. 11, formula (50)) that

$$(7)\qquad \operatorname{Pc}_{M}(a; X) = X^n + c_1X^{n-1} + \cdots + c_n$$

where

$$(8)\qquad c_1 = -\operatorname{Tr}_{M}(a), \qquad c_n = (-1)^n N_{M}(a).$$

For $\lambda \in K$,

$$(9)\qquad \operatorname{Tr}_{M}(\lambda) = n.\lambda, \qquad N_{M}(\lambda) = \lambda^n, \qquad \operatorname{Pc}_{M}(\lambda; X) = (X - \lambda)^n.$$

Let K' be a commutative K-algebra. Let $M' = K' \otimes_K M$ and $A' = K' \otimes_K A$, so that M' has an A'-module structure (§ 4, *Example* 2). As a K'-module M' admits the basis consisting of the $1 \otimes e_i$ $(1 \leqslant i \leqslant n)$ and the matrix of a_M with respect to (e_i) is equal to the matrix of $(1 \otimes a)_{M'}$ with respect to (e'_i). Then

$$(12)\qquad \begin{gathered} \operatorname{Tr}_{M'}(1 \otimes a) = \operatorname{Tr}_{M}(a).1, \qquad N_{M'}(1 \otimes a) = N_{M}(a).1 \\ \operatorname{Pc}_{M'}(1 \otimes a; X) = \operatorname{Pc}_{M}(a; X).1 \end{gathered}$$

for all $a \in A$, where 1 denotes the unit element of K'. If in particular we take $K' = K[X]$, then

$$(13)\qquad \operatorname{Pc}_{M/K}(a; X) = N_{M[X]/K[X]}(X - a).$$

2. PROPERTIES OF NORMS AND TRACES RELATIVE TO A MODULE

If M and M' are two *isomorphic* A-modules with finite bases over K, then, for all $a \in A$,

$$(14)\qquad \operatorname{Tr}_{M'}(a) = \operatorname{Tr}_{M}(a), \qquad N_{M'}(a) = N_{M}(a), \qquad \operatorname{Pc}_{M'}(a; X) = \operatorname{Pc}_{M}(a; X)$$

for if f is an isomorphism of M onto M', the matrix of a_M with respect to a basis B of M over K is the same as the matrix of $a_{M'}$ with respect to the basis $f(B)$ of M'.

PROPOSITION 1. *Let* $M = M_0 \supset M_1 \supset \cdots \supset M_r = \{0\}$ *be a decreasing sequence of submodules of an* A*-module* M *such that each of the* K*-modules* $P_i = M_{i-1}/M_i$ $(1 \leqslant i \leqslant r)$ *admits a finite basis. Then the* K*-module* M *admits a finite basis and*

$$\mathrm{Tr}_{M}(a) = \sum_{i=1}^{r} \mathrm{Tr}_{P_i}(a), \qquad N_M(a) = \prod_{i=1}^{r} N_{P_i}(a) \tag{15}$$

$$\mathrm{Pc}_M(a; X) = \prod_{i=1}^{r} \mathrm{Pc}_{P_i}(a; X).$$

Let B'_i be a basis of P_i over K; then a system of representatives B_i of B'_i (mod. M_i) is a basis of a supplementary submodule of the K-module M_i in the K-module M_{i-1} (II, § 1, no. 11, Proposition 21). The union B of the B_i $(1 \leqslant i \leqslant r)$ is a basis of M over K. Let X_{ii} be the matrix of the endomorphism a_{P_i} with respect to the basis B'_i. It is immediate that the matrix of a_M with respect to B is of the form

$$\begin{pmatrix} X_{rr} & X_{r,r-1} & \dots & X_{r,1} \\ 0 & X_{r-1,r-1} & \dots & X_{r-1,1} \\ \cdot & \cdot & \cdot & \cdot \\ 0 & 0 & \dots & X_{11} \end{pmatrix}$$

and the proposition follows from formulae (4), (5) and (6) of no. 1 and formula (31) of § 8, no. 6.

PROPOSITION 2. *Let* A, A′ *be two* K*-algebras,* M *an* A*-module and* M′ *an* A′*-module. Suppose that* M *and* M′ *are free* K*-modules of respective dimensions* n *and* n' *and consider* $M \otimes_K M'$ *as an* $(A \otimes_K A')$*-module* (§ 4, no. 3, *Example* 2). *Then, for* $a \in A$ *and* $a' \in A'$,

$$\mathrm{Tr}_{M\otimes M'}(a \otimes a') = \mathrm{Tr}_M(a)\mathrm{Tr}_{M'}(a') \tag{16}$$

$$N_{M\otimes M'}(a \otimes a') = (N_M(a))^{n'}(N_{M'}(a'))^{n}. \tag{17}$$

Formula (16) follows from II, § 4, no. 4, formula (26) and formula (17) from § 8, no. 6, formula (33).

3. NORM AND TRACE IN AN ALGEBRA

DEFINITION 2. *Let* A *be a* K*-algebra which is a finite-dimensional free* K*-module. For every element* $a \in A$, *the trace* (resp. *norm,*† resp. *characteristic polynomial*) *of* a *relative to* A *and* K *is the trace* (resp. *determinant*, resp. *characteristic polynomial*) *of the endomorphism* $x \mapsto ax$ *of the* K*-module* A.

† This notion should not be confused with the notion for norm in an algebra over a valued field (*General Topology*), IX, § 3, no. 7).

The trace, norm and characteristic polynomial of $a \in A$ relative to A and K are denoted by $\mathrm{Tr}_{A/K}(a)$, $N_{A/K}(a)$ and $\mathrm{Pc}_{A/K}(a; X)$; we omit K and even A from this notation when there is no risk of confusion. Note that the trace (resp. norm, characteristic polynomial) of $a \in A$ is just the trace (resp. norm, characteristic polynomial) of a relative to the A-module A_s.

Suppose that A is the product $A_1 \times A_2 \times \cdots \times A_m$ of a finite number finite-dimensional algebras over K which are free K-modules. Using the above remark and Proposition 1 of no. 2, we have, for every element

$$a = (a_1, \ldots, a_m) \in A,$$

$$\mathrm{Tr}_{A/K}(a) = \sum_{i=1}^{m} \mathrm{Tr}_{A_i/K}(a_i), \qquad N_{A/K}(a) = \prod_{i=1}^{m} N_{A_i/K}(a_i)$$

(18)

$$\mathrm{Pc}_{A/K}(a; X) = \prod_{i=1}^{m} \mathrm{Pc}_{A_i/K}(a_i; X).$$

Similarly, Proposition 2 of no. 2 shows that if A and A′ are two algebras, which are free K-modules, of finite dimensions n, n' respectively over K, then, for $a \in A$, $a' \in A'$.

$$\mathrm{Tr}_{A \otimes A'}(a \otimes a') = \mathrm{Tr}_A(a)\,\mathrm{Tr}_{A'}(a') \tag{19}$$

$$N_{A \otimes A'}(a \otimes a') = (N_A(a))^{n'}(N_{A'}(a'))^{n}. \tag{20}$$

Finally let A be a finite-dimensional algebra over K which is a free K-module, h a homomorphism of K into a commutative ring K′ and $A' = A_{(K')}$ the K′-algebra derived from A by extending the scalars by means of h. It follows from formula (12) of no. 1 that, for all $a \in A$,

(21)

$$\mathrm{Tr}_{A'/K'}(1 \otimes a) = h(\mathrm{Tr}_{A/K}(a)), \qquad N_{A'/K'}(1 \otimes a) = h(N_{A/K}(a))$$

$$\mathrm{Pc}_{A'/K'}(1 \otimes a; X) = \bar{h}(\mathrm{Pc}_{A/K}(a; X))$$

where $\bar{h}$ is the homomorphism $K[X] \to K'[X]$ derived from h. In particular, for $K' = K[X]$, we obtain, writing $A[X] = A \otimes_K K[X]$,

$$\mathrm{Pc}_{A/K}(a; X) = N_{A[X]/K[X]}(X - a). \tag{22}$$

More generally, if K′ is a commutative K-algebra and $A' = A \otimes_K K'$, then, for all $x \in A'$,

$$\mathrm{Pc}_{A/K}(a; x) = N_{A'/K'}(x - a).$$

Examples. (1) Let A be a quadratic algebra over K of type (α, β) and (e_1, e_2) a basis of type (α, β) (§ 2, no. 3). For $x = \xi e_1 + \eta e_2$, $\mathrm{Tr}_{A/K}(x) = 2\xi + \beta\eta$ and $N_{A/K}(x) = \xi^2 + \beta\xi\eta - \alpha\eta^2$; these functions are therefore identical with the Cayley trace and norm of x (§ 2, no. 24).

(2) Let A be a quaternion algebra over K. A direct calculation allows us to

verify that $\mathrm{Tr}_{A/K}(x) = 2T(x)$ and $N_{A/K}(x) = (N(x))^2$, where T and N are the Cayley trace and norm (§ 2, no. 4).

(3) Let $A = \mathbf{M}_n(K)$ and let the canonical basis (E_{ij}) of A (II, § 10, no. 3) be arranged in lexicographic order. Then it is immediately seen that for every matrix $X = \sum_{i,j} \xi_{ij} E_{ij}$ the matrix (of order n^2) of the endomorphism $Y \mapsto XY$ is of the form

$$\begin{pmatrix} X & 0 & \dots & 0 \\ 0 & X & \dots & 0 \\ . & . & . & . \\ 0 & 0 & \dots & X \end{pmatrix}$$

whence $\mathrm{Tr}_{A/K}(X) = n.\mathrm{Tr}(X)$ and $N_{A/K}(X) = (\det(X))^n$.

4. PROPERTIES OF NORMS AND TRACES IN AN ALGEBRA

PROPOSITION 3. *Let A be a K-algebra admitting a finite basis. For an element $a \in A$ to be invertible, it is necessary and sufficient that its $N_{A/K}(a)$ be invertible in K.*

If a admits an inverse a' in A, then

$$N_{A/K}(a)N_{A/K}(a') = N_{A/K}(aa') = N_{A/K}(1) = 1$$

by formula (3) of no. 1. Conversely, if $N_{A/K}(a)$ is invertible, the endomorphism $h: x \mapsto ax$ is bijective (§ 8, no. 2, Theorem 1). Then there exists $a' \in A$ such that $aa' = 1$; then $h(a'a - 1) = aa'a - a = (aa' - 1)a = 0$, whence $aa' = 1$ since h is injective. Hence a' is the inverse of a.

PROPOSITION 4. *Let A be a K-algebra admitting a finite basis. For all $a \in A$, $\mathrm{Pc}_{A/K}(a; a) = 0$.*

This follows immediately from the Cayley-Hamilton theorem (§ 8, no. 11, Proposition 20).

PROPOSITION 5. *Let A be a K-algebra and $\mathfrak{m}$ a two-sided ideal of A. Suppose that $A_0 = A/\mathfrak{m}$ is a free K-module of finite dimension n, that there exists an integer $r > 0$ such that $\mathfrak{m}^r = \{0\}$ and that $\mathfrak{m}^{i-1}/\mathfrak{m}^i$ is a free A_0-module of finite dimension s_i for $1 \leqslant i \leqslant r$. Let $s = s_1 + \dots + s_r$ and for all $a \in A$ let a_0 denote the class of a* mod. $\mathfrak{m}$. *Then A is a free K-module of dimension ns and, for all $a \in A$,*

$$\mathrm{Tr}_A(a) = s.\mathrm{Tr}_{A_0}(a_0), \qquad N_A(a) = (N_{A_0}(a_0))^s$$
$$\mathrm{Pc}_A(a; X) = (\mathrm{Pc}_{A_0}(a_0; X))^s. \tag{23}$$

By virtue of II, § 1, no. 13, Proposition 25, $\mathfrak{m}^{i-1}/\mathfrak{m}^i$ is a free K-module of dimension ns_i. Hence Proposition 1 of no. 2 can be applied with $P_i = \mathfrak{m}^{i-1}/\mathfrak{m}^i$;

this shows in the first place that A is a free K-module of dimension $n(s_1 + \cdots + s_r) = ns$. Moreover, the hypothesis implies that the A-module P_i is isomorphic to a direct sum of s_i submodules isomorphic to the A-module A_0; by Proposition 1 of no. 2, therefore $N_{P_i}(a) = N_{A_0}(a)^{s_i}$; finally therefore

$$N_A(a) = N_{A_0}(a)^s.$$

In this formula $N_{A_0}(a)$ is defined by considering A_0 as a left A-module and it is equal to the determinant of the K-linear mapping $x \mapsto ax$ of A_0 into itself; but, as $ax = a_0x$ for $x \in A_0$, $N_{A_0}(a) = N_{A_0}(a_0)$, which completes the proof of formula (23) for the norm. The two others are shown analogously.

PROPOSITION 6. *Let* A *be a* commutative K-*algebra admitting a finite basis over* K *and* V *an* A-*module admitting a finite basis over* A. *Then* V *admits a finite basis over* K *and for every* A-*endomorphism* u *of* V, *if* u_K *is the mapping* u *considered as a* K-*endomorphism of* V,

$$(24)\qquad \begin{gathered} \mathrm{Tr}(u_K) = \mathrm{Tr}_{A/K}(\mathrm{Tr}(u)), \qquad \det(u_K) = \mathrm{N}_{A/K}(\det(u)) \\ \mathrm{Pc}(u_K; X) = \mathrm{N}_{A[X]/K[X]}(\mathrm{Pc}(u; X)). \end{gathered}$$

Let $(a_i)_{1 \leqslant i \leqslant m}$ be a basis of A over K and $(e_j)_{1 \leqslant j \leqslant n}$ a basis of V over A; then (a_ie_j) is a basis of V over K (II, § 1, no. 13, Proposition 25). On the other hand the third of formulae (24) can be deduced from the second applied to the endomorphism $X - \bar{u}$ of the A[X]-module $A[X] \otimes_A V$ (§ 8, no. 10). It will therefore suffice to show the first two formulae in (24). We shall first establish the following lemma:

Lemma 1. *Let* X_{ij} $(1 \leqslant i \leqslant n,\ 1 \leqslant j \leqslant n)$ *be* n^2 *indeterminates,* X *the square matrix* (X_{ij}) *of order* n *and* $D(X_{11}, \ldots, X_{nn}) \in \mathbf{Z}[X_{11}, \ldots, X_{nn}]$ *the determinant* $\det(X)$. *On the other hand let* A *be a commutative ring,* M_{ij} $(1 \leqslant i \leqslant n, 1 \leqslant j \leqslant n)$ n^2 *matrices of order* m *over* A, *which are pairwise permutable, and* M *the square matrix of order* mn *over* A *which can be expressed as a square matrix of matrices* (II, § 10, no. 7)

$$M = \begin{pmatrix} M_{11} & M_{12} & \ldots & M_{1n} \\ M_{21} & M_{22} & \ldots & M_{2n} \\ \cdot & \cdot & \cdot & \cdot \\ M_{n1} & M_{n2} & \ldots & M_{nn} \end{pmatrix}$$

Then the determinant of M *is equal to the determinant of the square matrix*

$$D(M_{11}, \ldots, M_{nn})$$

of order m.

We proceed by induction on n, the cases $n = 0$ and $n = 1$ being trivial.

Let Z be a new indeterminate and N_{ij} the matrix $M_{ij} + \delta_{ij}ZI_m$ (δ_{ij} the Kronecker index). If $D^{ij}(X_{11}, \ldots, X_{nn})$ is the *cofactor* of X_{ij} in the matrix X (§ 8, no. 6), then

$$X_{ji}D^{kl}(X_{11}, \ldots, X_{nn}) = \delta_{jk}D(X_{11}, \ldots, X_{nn}) \tag{25}$$

(§ 8, no. 6, formula (28)). We write $N'_{ij} = D^{ij}(N_{11}, \ldots, N_{nn})$, which is a square matrix of order m over A[Z] and consider the product $N.U$, where

$$U = \begin{pmatrix} N'_{11} & 0 & \ldots & 0 \\ N'_{12} & I_m & \ldots & 0 \\ \cdot & \cdot & \cdot & \cdot \\ N'_{1n} & 0 & \ldots & I_m \end{pmatrix},$$

$$N = \begin{pmatrix} N_{11} & N_{12} & \ldots & N_{1n} \\ N_{21} & N_{22} & \ldots & N_{2n} \\ \cdot & \cdot & \cdot & \cdot \\ N_{n1} & N_{n2} & \ldots & N_{nn} \end{pmatrix}$$

Performing this product in blocks (II, § 10, no. 5) and using formulae (25), we obtain

$$N.U = \begin{pmatrix} P & N_{12} & \ldots & N_{1n} \\ 0 & N_{22} & \ldots & N_{2n} \\ \cdot & \cdot & \cdot & \cdot \\ 0 & N_{n2} & \ldots & N_{nn} \end{pmatrix}$$

where we have written $P = D(N_{11}, \ldots, N_{nn})$. Let

$$Q = \begin{pmatrix} N_{22} & \ldots & N_{2n} \\ \cdot & \cdot & \cdot \\ N_{n2} & \ldots & N_{nn} \end{pmatrix}$$

which is a matrix of order $m(n-1)$; then (§ 8, no. 6, formula (31)) $(\det N)(\det U) = (\det P)(\det Q)$ and $\det U = \det N'_{11}$. But by the induction hypothesis, $\det Q = \det(D^{11}(N_{11}, \ldots, N_{nn})) = \det N'_{11}$ and by virtue of the definition of the N_{ij}, clearly $\det Q$ is a polynomial in A[Z] of degree $m(n-1)$, whose term in $Z^{m(n-1)}$ has coefficient 1; it follows immediately that $\det Q$ is not a divisor of zero in the graded algebra A[Z]. We therefore conclude that $\det N = \det(D(N_{11}, \ldots, N_{nn}))$ in A[Z]; if we substitute 0 for Z in these polynomials, then $\det M = \det(D(M_{11}, \ldots, M_{nn}))$.

Having shown this lemma, the K-module V is the direct sum of the K-modules Ae_j $(1 \leqslant j \leqslant n)$; we write $u(e_j) = \sum_{k=1}^{n} c_{jk}e_k$. For every element $xe_j \in Ae_j$, with $x \in A$, the component of $u(xe_j)$ in Ae_k is $xc_{jk}e_k$; it follows that the matrix of u_K with respect to the basis (a_ie_j) of the K-module V can be expressed in the form of a square matrix of matrices (M_{jk}), where M_{jk} is the matrix of the K-linear mapping $xe_j \mapsto xc_{jk}e_k$ of Ae_j into Ae_k with respect to the bases $(a_ie_j)_{1\leqslant i\leqslant m}$ and $(a_ie_k)_{1\leqslant i\leqslant m}$ of these two K-modules (II, § 10, no. 5). If for all $t \in A$, $M(t)$ denotes the matrix, with respect to the basis $(a_i)_{1\leqslant i\leqslant m}$ of A over K, of the endomorphism $x \mapsto xt$ of A, then $M_{jk} = M(c_{jk})$; as $t \mapsto M(t)$ is a ring homomorphism the matrices M_{jk} are *permutable with one another*. Then

$$\det u_K = \det(D(M_{11}, \ldots, M_{nn}))$$

by Lemma 1. But as $t \mapsto M(t)$ is a ring homomorphism, $D(M_{11}, \ldots, M_{nn})$ is the matrix of the K-endomorphism $x \mapsto x.\det(c_{jk})$ of A with respect to basis (a_i); by definition its determinant is therefore $N_{A/K}(\det(u))$, which proves the second formula of (24). On the other hand,

$$\mathrm{Tr}(u_K) = \sum_{j=1}^{n} \mathrm{Tr}(M_{jj}) = \sum_{j=1}^{n} \mathrm{Tr}_{A/K}(c_{jj}) = \mathrm{Tr}_{A/K}\left(\sum_{j=1}^{n} c_{jj}\right) = \mathrm{Tr}_{A/K}(\mathrm{Tr}(u))$$

and the proof of Proposition 6 is complete.

COROLLARY. *Let* A *be a commutative* K*-algebra admitting a finite basis over* K *and* B *an* A*-algebra admitting a finite basis over* A. *Then* B *admits a finite basis over* K, *and, for all* $b \in B$ ("transitivity formulae")

$$\mathrm{Tr}_{B/K}(b) = \mathrm{Tr}_{A/K}(\mathrm{Tr}_{B/A}(b)), \qquad N_{B/K}(b) = N_{A/K}(N_{B/A}(b))$$
$$\mathrm{Pc}_{B/K}(b; X) = N_{A[X]/K[X]}(\mathrm{Pc}_{B/A}(b; X)). \tag{26}$$

This follows immediately from Proposition 6, setting $V = B$ and $u(x) = bx$.

Remark. Suppose that the homomorphism $\lambda \mapsto \lambda.1$ of K into A is injective and let K be identified with its image in A; suppose that A admits a finite basis $(e_i)_{1\leqslant i\leqslant n}$ as a K-module. Let s be an automorphism of A such that $s(K) = K$. Let a be an element of A; then, by transporting the structure

$$\mathrm{Tr}_{A/K}(s(a)) = s(\mathrm{Tr}_{A/K}(a)) \tag{27}$$

$$N_{A/K}(s(a)) = s(N_{A/K}(a)). \tag{28}$$

*Consider also a derivation D of A (§ 10, no. 2) such that $D(K) \subset K$ and write $D(e_i) = \sum_{j=1}^{n} e_j\mu_{ji}$ where $\mu_{ji} \in K$; write

$$ae_i = \sum_{j=1}^{n} e_j\lambda_{ji} \quad \text{with} \quad \lambda_{ji} \in K.$$

Then

$$\mathrm{D}(a)e_i + a\mathrm{D}(e_i) = \mathrm{D}(ae_i) = \sum_{j=1}^{n} (\mathrm{D}(e_j)\lambda_{ji} + e_j\mathrm{D}(\lambda_{ji})).$$

It follows that

$$\mathrm{D}(a)e_i = \sum_{j=1}^{n} e_j\nu_{ji}$$

with $\nu_{ji} = \mathrm{D}(\lambda_{ji}) + \sum_{k=1}^{n} (\mu_{jk}\lambda_{ki} - \lambda_{jk}\mu_{ki})$. As $\sum_{i,k} (\mu_{ik}\lambda_{ki} - \lambda_{ik}\mu_{ki}) = 0$, therefore $\mathrm{Tr}_{\mathrm{A/K}}(\mathrm{D}(a)) = \sum_{i=1}^{n} \mathrm{D}(\lambda_{ii})$, in other words

(29) $$\mathrm{Tr}_{\mathrm{A/K}}(\mathrm{D}(a)) = \mathrm{D}(\mathrm{Tr}_{\mathrm{A/K}}(a)).*$$

5. DISCRIMINANT OF AN ALGEBRA

DEFINITION 3. *Let* A *be a* K*-algebra admitting a finite basis of n elements. The discriminant of a sequence* $(x_1, \dots, x_n)$ *of n elements of* A*, with respect to* K*, denoted by* $\mathrm{D}_{\mathrm{A/K}}(x_1, \dots, x_n)$*, is the discriminant of the square matrix*

$$(\mathrm{Tr}_{\mathrm{A/K}}(x_ix_j))_{1 \leqslant i \leqslant n, 1 \leqslant j \leqslant n}.$$

Consider first a basis $(e_i)_{1 \leqslant i \leqslant n}$ of A over K and write

(30) $$e_ie_j = \sum_{k=1}^{n} c_{ijk}e_k \quad \text{with} \quad c_{ijk} \in \mathrm{K}.$$

Then $\mathrm{Tr}_{\mathrm{A/K}}(e_i) = \sum_{s=1}^{n} c_{iss}$, whence $\mathrm{Tr}_{\mathrm{A/K}}(e_ie_j) = \sum_{k,s} c_{ijk}c_{kss}$ and therefore

(31) $$\mathrm{D}_{\mathrm{A/K}}(e_1, \dots, e_n) = \det\Big(\Big(\sum_{k,s} c_{ijk}c_{kss}\Big)_{1 \leqslant i \leqslant n, 1 \leqslant j \leqslant n}\Big).$$

Now let $(x_i)_{1 \leqslant i \leqslant n}$, $(x'_i)_{1 \leqslant i \leqslant n}$ be two sequences of n elements of A and suppose that there exists a square matrix of order n, $M = (m_{ij})$, with coefficients in K, such that $x_i = \sum_{j=1}^{n} m_{ij}x'_j$ for $1 \leqslant i \leqslant n$. We write

$$T = (\mathrm{Tr}_{\mathrm{A/K}}(x_ix_j))_{1 \leqslant i \leqslant n, 1 \leqslant j \leqslant n}, \qquad T' = (\mathrm{Tr}_{\mathrm{A/K}}(x'_ix'_j))_{1 \leqslant i \leqslant n, 1 \leqslant j \leqslant n}.$$

Then $\mathrm{Tr}_{\mathrm{A/K}}(x_ix_j) = \sum_{p,q} m_{ip}m_{jq}\,\mathrm{Tr}_{\mathrm{A/K}}(x'_px'_q)$, whence $T = M.T'.{}^tM$; the rule of multiplication of determinants therefore gives

$$\det T = \det M.\det T'.\det {}^tM = (\det M)^2 \det T'$$

whence finally

(32) $$\mathrm{D}_{\mathrm{A/K}}(x_1, \dots, x_n) = (\det M)^2\mathrm{D}_{\mathrm{A/K}}(x'_1, \dots, x'_n).$$

The above formula shows in particular that the discriminants of two *bases* of A over K differ by the square of an *invertible* element of K and therefore generate the same (*principal*) ideal of K. This ideal $\Delta_{A/K}$ is called the *discriminant ideal* of A over K; by formula (32) the discriminant of every sequence of n elements of A which differ only in the order of the terms have the same discriminant, for the determinant of a permutation matrix is equal to ± 1.

Examples. (1) If A is a quadratic algebra of type (α, β) over K, then (in the notation of § 2, no. 3) $\mathrm{Tr}(e_1) = 2$, $\mathrm{Tr}(e_2) = \beta$,

$$\mathrm{Tr}(e_2^2) = \alpha\,\mathrm{Tr}(e_1) + \beta\,\mathrm{Tr}(e_2) = 2\alpha + \beta^2,$$

whence $D_{A/K}(e_1, e_2) = \beta^2 + 4\alpha$.

(2) Let $A = K[X]/K[X]P$, where $P(X) = X^3 + pX + q$, so that if x is the image of X in A, $1, x, x^2$ form a basis of A over K and $x^3 = -px - q$. It is seen immediately that $\mathrm{Tr}(1) = 3$, $\mathrm{Tr}(x) = 0$, $\mathrm{Tr}(x^2) = -2p$, taking account of the relation $x^3 = -px - q$, $\mathrm{Tr}(x^3) = -3q$ and $\mathrm{Tr}(x^4) = 2p^2$, whence easily $D_{A/K}(1, x, x^2) = -4p^3 - 27q^2$.

(3) Let A be a quaternion algebra of type (α, β, γ) over K and $(1, i, j, k)$ a basis of A of type (α, β, γ); taking account of § 3, no. 5, formula (30), it is easily found that $\mathrm{Tr}(1) = 4$, $\mathrm{Tr}(i) = 2\beta$, $\mathrm{Tr}(j) = \mathrm{Tr}(k) = 0$, then

$$D_{A/K}(1, i, j, k) = -16\gamma^2(\beta^2 + 4\alpha)^2.$$

(4) Let $A = \mathbf{M}_n(K)$ and consider the canonical basis $(E_{ij})_{1 \leqslant i \leqslant n, 1 \leqslant j \leqslant n}$ of A over K (II, § 10, no. 3). It is immediate that $\mathrm{Tr}_{A/K}(E_{ij}) = 0$ if $j \neq i$ and $\mathrm{Tr}_{A/K}(E_{ii}) = n$ for all i; it follows without difficulty that the matrix $(\mathrm{Tr}(E_{ij}E_{hk}))$ of order n^2 is of the form $n.P$, where P is a permutation matrix, whence $D_{A/K}((E_{ij})) = \pm n^{n^2}$.

§ 10. DERIVATIONS

In this paragraph, and unless otherwise mentioned, the algebras considered are not assumed to be associative nor necessarily to possess a unit element; K *denotes a commutative ring.*

1. COMMUTATION FACTORS

When in this paragraph we speak of *graduations* without specifying them, we shall always mean graduations of type Δ, where Δ is a *commutative group* written additively. In this paragraph, a commutation factor over Δ with values in $\mathbf{Z}$ is called a *commutation factor* over Δ (§ 4, no. 7, Definition 6). A commutation factor over Δ is therefore identified with a mapping $\varepsilon: (\alpha, \beta) \mapsto \varepsilon_{\alpha\beta} = \varepsilon(\alpha,\beta)$ of $\Delta \times \Delta$ into the multiplicative group $\{-1, 1\}$ such that for $\alpha, \alpha', \beta, \beta'$ in Δ,

$$(1) \qquad \begin{cases} \varepsilon(\alpha + \alpha', \beta) = \varepsilon(\alpha, \beta)\varepsilon(\alpha', \beta) \\ \varepsilon(\alpha, \beta + \beta') = \varepsilon(\alpha, \beta)\varepsilon(\alpha, \beta') \\ \varepsilon(\beta, \alpha) = \varepsilon(\alpha, \beta). \end{cases}$$

It follows that $\varepsilon(2\alpha, \beta) = \varepsilon(\alpha, 2\beta) = 1$.

When $\Delta = \mathbf{Z}$, every commutation factor ε is determined by giving $\varepsilon(1, 1)$; there are therefore only *two* such factors, the first defined by

$$(2) \qquad \varepsilon(p, q) = 1 \quad \text{for} \quad p, q \text{ in } \mathbf{Z}$$

and the second by

$$(3) \qquad \varepsilon(p, q) = (-1)^{pq} \quad \text{for} \quad p, q \text{ in } \mathbf{Z}.$$

2. GENERAL DEFINITION OF DERIVATIONS

Consider a commutative ring K, six graded K-modules of type Δ: A, A′, A″, B, B′, B″, and three K-linear mappings

$$\mu: A \times A' \to A'', \qquad \lambda_1: B \times A' \to B'', \qquad \lambda_2: A \times B' \to B''$$

such that the corresponding K-linear mappings

$$A \otimes_K A' \to A'', \qquad B \otimes_K A' \to B'', \qquad A \otimes_K B' \to B''$$

are *graded of degree* 0. The image $\mu(a, a')$ for $a \in A$, $a' \in A'$ is simply denoted by $a.a'$ or even aa', and similarly for the two other bilinear mappings. The *degree* of $a.a'$ is therefore the *sum* of the degrees of a and a'.

DEFINITION 1. *Given the above and a commutation factor ε on $\Delta \times \Delta$, an ε-derivation (or (K, ε)-derivation) of degree $\delta \in \Delta$ of (A, A′, A″) into (B, B′, B″) is a triple of graded K-linear mappings of degree δ:*

$$d: A \to B, \qquad d': A' \to B', \qquad d'': A'' \to B''$$

such that, for every homogeneous *element $a \in A$ and every element $a' \in A'$*

$$(4) \qquad d''(a.a') = (da).a' + \varepsilon_{\delta, \deg(a)} a.(d'a').$$

It obviously suffices by linearity to verify relation (4) when a and a' run through respective generating systems of A and A′.

It is often convenient to denote the three mappings d, d', d'' by the same letter d (which can be justified by denoting equally by d the graded K-linear mapping of degree δ

$$(a, a', a'') \mapsto (da, d'a', d''a'')$$

of $A \oplus A' \oplus A''$ into $B \oplus B' \oplus B''$). Relation (4) can then be written more simply

$$d(a.a') = (da).a' + \varepsilon_{\delta,\deg(a)}a.(da'). \tag{5}$$

The ε-derivations of (A, A', A'') into (B, B', B'') of *given* degree form a sub-K-module of the K-module of graded linear mappings

$$\mathrm{Homgr}_K(A \oplus A' \oplus A'', B \oplus B' \oplus B'').$$

When $\varepsilon(\alpha, \beta) = 1$ for all α, β in Δ, we say simply *derivation* (or K-*derivation*) instead of ε-derivation. Derivations form a sub-K-module of

$$\mathrm{Hom}_K(A \oplus A' \oplus A'', B \oplus B' \oplus B'').$$

When $\Delta = \mathbf{Z}$ and $\varepsilon(p.q) = (-1)^{pq}$, every ε-derivation of *even* degree is a derivation; every ε-derivation of *odd* degree is often called an *antiderivation* (or K-*antiderivation*); an antiderivation d therefore satisfies

$$d(a.a') = (da).a' + (-1)^{\deg(a)}a.(da') \tag{6}$$

for a *homogeneous* element $a \in A$.

Remarks. (1) The notion of *derivation* can be defined for non-graded modules by agreeing to give these modules the trivial graduation.

(2) If only ε-derivations of given degree δ are considered, the commutation factor ε may be disposed of as follows: the bilinear mapping $\lambda_2: A \times B' \to B''$ is modified by replacing it by the bilinear mapping $\lambda'_2: A \times B' \to B''$ such that, for every *homogeneous* a in A and all $b' \in B'$,

$$\lambda'_2(a, b') = \varepsilon_{\delta,\deg(a)}\lambda_2(a, b').$$

Then d is a derivation relative to the bilinear mappings μ, λ_1, λ'_2.

The general definition of ε-derivations given above is especially used in two cases:

Case (I): $A = B$, $A' = B'$, $A'' = B''$ and the three bilinear mappings μ, λ_1, λ_2 are equal to the *same* mapping.

Case (II): $A = A' = A''$, $B = B' = B''$, so that (for μ) A is a *graded algebra* and the two K-bilinear mappings.

$$\lambda_1: B \times A \to B, \qquad \lambda_2: A \times B \to B \tag{7}$$

are such that the corresponding K-linear mappings $B \otimes_K A \to B$, $A \otimes_K B \to B$ are graded of degree 0. An ε-derivation of degree δ of A into B is then a graded K-linear mapping $d: A \to B$ of degree δ, such that for every homogeneous x in A and every $y \in A$, we have the relation

$$d(xy) = (dx)y + \varepsilon_{\delta,\deg(a)}x(dy). \tag{8}$$

Consider in particular in case (II) the case where A is a unital *associative* K-algebra and λ_1 and λ_2 are the external laws of an (A, A)-*bimodule* (§ 4, no. 3, Definition 3). This holds notably when A and B are two unital *associative* K-*algebras*, a unital homomorphism of graded K-algebras $\rho: A \to B$ is given and an (A, A)-bimodule structure is considered on B defined by the two external laws

$$\lambda_1: (b, a) \mapsto b\rho(a), \qquad \lambda_2: (a, b) \mapsto \rho(a)b$$

for $a \in A$, $b \in B$.

Cases (I) and (II) have the following case in common: consider a graded K-algebra A, take B = A, mappings (7) both being multiplication on A. We then speak of an ε-*derivation* (or (K, ε)-*derivation*) *of the graded algebra* A: it is a graded K-linear mapping of A into itself, of degree δ, satisfying (8) for every homogeneous x in A and all $y \in A$. In particular if A is a *graded ring*, considered as an (associative) **Z**-algebra, we speak of the ε-*derivation of the ring* A.

Let A be a *unital commutative associative* K-algebra and B an A-*module*; when we speak of a *derivation of* A *into* B, it will always be understood that we mean with the A-bimodule structure on B derived from its A-module structure; then the formula

$$d(xy) = x(dy) + y(dx) \quad \text{for} \quad x \in A, y \in A \tag{9}$$

holds for such a derivation $d: A \to B$.

3. EXAMPLES OF DERIVATIONS

Example 1. *Let A be an **R**-algebra of differentiable mappings of **R** into **R** and let x_0 be a point of **R**; **R** can be considered as an A-module with the external law $(f, a) \mapsto f(x_0)a$. Then the mapping $f \mapsto Df(x_0)$ is a derivation, since (*Functions of a Real Variable*, I, § 1, no. 3)

$$(D(fg))(x_0) = (Df(x_0))g(x_0) + f(x_0)(Dg(x_0)).*$$

Example 2. *Let X be a differentiable manifold of class C^∞ and let A be the graded **R**-algebra of differential forms on X. The mapping which associates with every differential form ω on X its exterior differential $d\omega$ is an anti-derivation of degree +1 (*Differentiable and Analytic Manifolds*, R, § 8).*

Example 3. Let A be an associative K-algebra. For all $a \in A$, the mapping $x \mapsto ax - xa$ is a *derivation* of the algebra A (cf. no. 6).

Example 4. Let M be a K-module and A the exterior algebra $\bigwedge(M^*)$ with its usual graduation (§ 7, no. 1). *It will be seen in § 11, no. 9 that, for all $x \in M$, the right interior product $i(x)$ is an *antiderivation* of A of degree −1.*

Example 5. Returning to the general situation of Definition 1 of no. 2, let $\bar{\mathrm{K}}$ be another commutative ring and $\rho: \mathrm{K} \to \bar{\mathrm{K}}$ a ring homomorphism; let $\bar{\mathrm{A}}, \bar{\mathrm{A}}', \bar{\mathrm{A}}'', \bar{\mathrm{B}}, \bar{\mathrm{B}}', \bar{\mathrm{B}}''$ denote the graded $\bar{\mathrm{K}}$-modules obtained respectively from A, A′, A″, B, B′, B″ by extending the ring of scalars to $\bar{\mathrm{K}}$ (II, § 11, no. 5); we derive from μ, λ_1 and λ_2 $\bar{\mathrm{K}}$-bilinear mappings

$$\bar{\mu}: \bar{\mathrm{A}} \times \bar{\mathrm{A}}' \to \bar{\mathrm{A}}'', \qquad \bar{\lambda}_1: \bar{\mathrm{B}} \times \bar{\mathrm{A}}' \to \bar{\mathrm{B}}'', \qquad \bar{\lambda}_2: \bar{\mathrm{A}} \times \bar{\mathrm{B}}' \to \bar{\mathrm{B}}''$$

by considering the tensor products by $1_{\bar{\mathrm{K}}}$ of the corresponding K-linear mappings to μ, λ_1 and λ_2 (II, § 5, no. 1). Then, if d is an ε-derivation of degree δ of (A, A′, A″) into (B, B′, B″), the mapping $\bar{d} = d \otimes 1_{\bar{\mathrm{K}}}$ of $\bar{\mathrm{A}} \oplus \bar{\mathrm{A}}' \oplus \bar{\mathrm{A}}''$ into $\bar{\mathrm{B}} \oplus \bar{\mathrm{B}}' \oplus \bar{\mathrm{B}}''$ is an ε-derivation of degree δ of $(\bar{\mathrm{A}}, \bar{\mathrm{A}}', \bar{\mathrm{A}}'')$ into $(\bar{\mathrm{B}}, \bar{\mathrm{B}}', \bar{\mathrm{B}}'')$.

Example 6. Let A be a graded K-algebra of type **Z**; a graded linear K-mapping of degree 0, $d: \mathrm{A} \to \mathrm{A}$, is defined by taking, for $x_n \in \mathrm{A}_n (n \in \mathbf{Z})$, $d(x_n) = nx_n$. This mapping is a *derivation* of A, since, for $x_p \in \mathrm{A}_p$, $x_q \in \mathrm{A}_q$,

$$d(x_p x_q) = (p + q)x_p x_q = d(x_p)x_q + x_p d(x_q).$$

4. COMPOSITION OF DERIVATIONS

We suppose in this no. that case (I) of no. 2 holds, that is that A, A′, A″ are three graded K-modules of type Δ and that we are given a K-bilinear mapping $\mu: \mathrm{A} \times \mathrm{A}' \to \mathrm{A}''$ corresponding to a graded K-linear mapping of degree 0, $\mathrm{A} \otimes_{\mathrm{K}} \mathrm{A}' \to \mathrm{A}''$. The graded endomorphisms f of $\mathrm{A} \oplus \mathrm{A}' \oplus \mathrm{A}''$ such that $f(\mathrm{A}) \subset \mathrm{A}$, $f(\mathrm{A}') \subset \mathrm{A}'$ and $f(\mathrm{A}'') \subset \mathrm{A}''$ form a *graded subalgebra* of the graded associative algebra $\mathrm{Endgr}_{\mathrm{K}}(\mathrm{A} \oplus \mathrm{A}' \oplus \mathrm{A}'')$ (§ 3, no. 1, *Example* 2). In particular two ε-derivations of (A, A′, A″) can be composed, but it should not be thought that the composition of two ε-derivations is an ε-derivation.

On every graded algebra B of type Δ is defined the *ε-bracket* (or simply *bracket* when $\varepsilon = 1$) of two homogeneous elements u, v, by the formula (10)

$$[u, v]_\varepsilon = uv - \varepsilon_{\deg u, \deg v} vu \text{ (denoted simply by } [u, v] \text{ if } \varepsilon = 1).$$

By extending this mapping by linearity, a K-bilinear mapping $(u, v) \mapsto [u, v]_\varepsilon$ of $\mathrm{B} \times \mathrm{B}$ into B is obtained. Then, for homogeneous u and v in B

$$[v, u]_\varepsilon = -\varepsilon_{\deg u, \deg v}[u, v]_\varepsilon.$$

Applying this definition to the graded algebra $\mathrm{Endgr}_{\mathrm{K}}(\mathrm{A} \oplus \mathrm{A}' \oplus \mathrm{A}'')$, the ε-bracket of two graded endomorphisms is thus defined.

PROPOSITION 1. *Let d_1, d_2 be two ε-derivations of* (A, A′, A″) *of respective degrees δ_1, δ_2. Then the ε-bracket*

$$[d_1, d_2]_\varepsilon = d_1 \circ d_2 - \varepsilon_{\delta_1, \delta_2} d_2 \circ d_1$$

is an ε-derivation of degree $\delta_1 + \delta_2$. Moreover, if d is an ε-derivation of (A, A′, A″) *of degree δ and if $\varepsilon_{\delta, \delta} = -1$, then $d^2 = d \circ d$ is a derivation.*

Suppose $x \in A$ is homogeneous of degree ξ; for all $y \in A'$,

$$(11) \quad d_1(d_2(xy)) = ((d_1d_2)(x))y + \varepsilon_{\delta_1,\delta_2+\xi}(d_2x)(d_1y) + \varepsilon_{\delta_2,\xi}(d_1x)(d_2y) + \varepsilon_{\delta_1+\delta_2,\xi}x((d_1d_2)(y))$$

taking account of formulae (1) of no. 1. Exchanging the roles of d_1 and d_2, we obtain, after simplifications again using (1) (no. 1),

$$(d_1d_2)(xy) - \varepsilon_{\delta_1,\delta_2}(d_2d_1)(xy) = ((d_1d_2)(x))y - \varepsilon_{\delta_1,\delta_2}((d_2d_1)(x))y + \varepsilon_{\delta_1+\delta_2,\xi}x((d_1d_2)(y)) - \varepsilon_{\delta_1,\delta_2}\varepsilon_{\delta_1+\delta_2,\xi}x((d_2d_1)(y))$$

that is, writing $d = [d_1, d_2]_\varepsilon$ and $\delta = \delta_1 + \delta_2$,

$$d(xy) = (dx)y + \varepsilon_{\delta,\xi}x(dy)$$

which proves that d is an ε-derivation.

On the other hand, if, in (11), we let $d_1 = d_2 = d$, $\delta_1 = \delta_2 = \delta$ and $\varepsilon_{\delta,\delta} = -1$, we obtain, since then $\varepsilon_{\delta,\delta+\xi} = -\varepsilon_{\delta,\xi}$ by (1),

$$d^2(xy) = (d^2x)y + \varepsilon_{2\delta,\xi}x(d^2y)$$

and as $\varepsilon_{2\delta,\xi} = 1$ it is seen that d^2 is a derivation.

COROLLARY. *Suppose that* $\Delta = \mathbf{Z}$. *Then:*

(i) *The square of an antiderivation is a derivation.*

(ii) *The bracket of two derivations is a derivation.*

(iii) *The bracket of an antiderivation and a derivation of even degree is an antiderivation.*

(iv) *If* d_1 *and* d_2 *are antiderivations,* $d_1d_2 + d_2d_1$ *is a derivation.*

Under the hypotheses of the beginning of this no., consider now a finite sequence $D = (d_i)_{1 \leqslant i \leqslant n}$ of *pairwise permutable derivations* of (A, A', A''). For every polynomial $P(X_1, \ldots, X_n)$ in the algebra $K[X_1, \ldots, X_n]$, the element $P(d_1, \ldots, d_n)$ of $\mathrm{Endgr}_K(A \oplus A' \oplus A'')$ is then defined (§ 2, no. 9); its abbreviated notation is $P(D)$.

PROPOSITION 2. *With the above hypotheses and notation, consider* $2n$ *indeterminates* $T_1, \ldots, T_n$, $T'_1, \ldots, T'_n$ *and for every polynomial* $F \in K[X_1, \ldots, X_n]$ *write* $F(\mathbf{T}) = F(T_1, \ldots, T_n)$, $F(\mathbf{T}') = F(T'_1, \ldots, T'_n)$ *and*

$$F(\mathbf{T} + \mathbf{T}') = F(T_1 + T'_1, \ldots, T_n + T'_n).$$

Suppose that

$$P(\mathbf{T} + \mathbf{T}') = \sum_i Q_i(\mathbf{T})R_i(\mathbf{T}')$$

where the Q_i *and* R_i *belong to* $K[X_1, \ldots, X_n]$. *Then, for all* $x \in A$ *and* $y \in A'$,

$$(12) \quad P(D)(xy) = \sum_i (Q_i(D)x)(R_i(D)y).$$

We introduce n other indeterminates $T''_1, \ldots, T''_n$ and consider the polynomial algebra $K[T_1, \ldots, T_n, T'_1, \ldots, T'_n, T''_1, \ldots, T''_n] = B$; on the other hand we consider the K-module M of bilinear mappings of $A \times A'$ into A''; a B-*module* structure is defined on M by writing, for every K-bilinear mapping $f \in M$ and $1 \leqslant i \leqslant n$,

$$\text{(13)} \qquad \begin{cases} (T_i f)(a, a') = f(d_i a, a') \\ (T'_i f)(a, a') = f(a, d_i a') \\ (T''_i f)(a, a') = d_i(f(a, a')). \end{cases}$$

Since the d_i are permutable with one another, it is seen that, for every polynomial $F \in K[X_1, \ldots, X_n]$, $(F(\mathbf{T})f)(a, a') = f(F(D)a, a')$,

$$(F(\mathbf{T}')f)(a, a') = f(a, F(D)a')$$

and $(F(\mathbf{T}'')f) = F(D)(f(a, a'))$. Hence, to prove (12) it suffices to show that

$$\text{(14)} \qquad (P(\mathbf{T}'') - \sum_i Q_i(\mathbf{T})R_i(\mathbf{T}')).\mu = 0$$

or also $(P(\mathbf{T}'') - P(\mathbf{T} + \mathbf{T}')).\mu = 0$ in the B-module M. Now, the hypothesis that the d_i are derivations can also be expressed by saying that, for $1 \leqslant i \leqslant n$,

$$\text{(15)} \qquad (T''_i - T_i - T'_i).\mu = 0$$

in the B-module M. By considering successively the polynomials

$$P(T''_1, T''_2, \ldots, T''_n) - P(T_1 + T'_1, T''_2, \ldots, T''_n)$$
$$P(T_1 + T'_1, T''_2, \ldots, T''_n) - P(T_1 + T'_1, T_2 + T'_2, \ldots, T''_n)$$
$$\cdots\cdots\cdots$$
$$P(T_1 + T'_1, \ldots, T_{n-1} + T'_{n-1}, T''_n) - P(T_1 + T'_1, \ldots, T_{n-1} + T'_{n-1}, T_n + T'_n)$$

it is seen that the difference $P(\mathbf{T}'') - P(\mathbf{T} + \mathbf{T}')$ can be written in the form

$$\sum_{i=1}^{n} (\mathbf{T}''_i - \mathbf{T}_i - \mathbf{T}'_i) G_i(\mathbf{T}, \mathbf{T}', \mathbf{T}'')$$

where the G_i are elements of B. Relation (14) is therefore an immediate consequence of relations (15).

COROLLARY (Leibniz's formula). *Let* d_i $(1 \leqslant i \leqslant n)$ *be* n *derivations of* (A, A', A'') *which are permutable with one another. For all* $\alpha = (\alpha_1, \ldots, \alpha_n) \in \mathbf{N}^n$, *we write*

$$\text{(16)} \qquad d^\alpha = d_1^{\alpha_1} d_2^{\alpha_2} \ldots d_n^{\alpha_n}.$$

Then, for $x \in \mathrm{A}$ *and* $y \in \mathrm{A}'$,

$$d^{\alpha}(xy) = \sum_{\beta+\gamma=\alpha} ((\beta, \gamma))d^{\beta}(x)d^{\gamma}(y) \tag{17}$$

where we have written (in the notation introduced at the beginning of the chapter)

$$((\beta, \gamma)) = (\beta + \gamma)!/(\beta!\gamma!). \tag{18}$$

This follows immediately from the multinomial formula (I, § 8, no. 2)

$$(\mathbf{T} + \mathbf{T}')^{\alpha} = \sum_{\beta+\gamma=\alpha} ((\beta, \gamma))\mathbf{T}^{\beta}\mathbf{T}'^{\gamma}$$

and Proposition 2.

5. DERIVATIONS OF AN ALGEBRA A INTO AN A-MODULE

We suppose in this no. that *Case* (II) of no. 2 holds. Then there is a graded K-algebra A and a graded K-module E and also two K-linear mappings of degree 0

$$\mathrm{E} \otimes_{\mathrm{K}} \mathrm{A} \to \mathrm{E}, \quad \mathrm{A} \otimes_{\mathrm{K}} \mathrm{E} \to \mathrm{E}$$

denoted by

$$x \otimes a \mapsto x.a \quad \text{and} \quad a \otimes x \mapsto a.x \quad \text{for } a \in \mathrm{A} \text{ and } x \in \mathrm{E}.$$

PROPOSITION 3. *Let* $d : \mathrm{A} \to \mathrm{E}$ *be an* ε*-derivation of degree* δ. *Then* Ker(d) *is a graded subalgebra of* A; *if* A *admits a unit element, it belongs to* Ker(d).

Clearly Ker(d) is a graded sub-K-module of A; further, relation (8) of no. 2 shows that, if x and y are two homogeneous elements belonging to Ker(d), then $d(xy) = 0$ and hence $xy \in \mathrm{Ker}(d)$. Finally, if A admits a unit element 1 (of degree 0, cf. § 3, no. 1), relation (8) of no. 2, where x and y are replaced by 1, gives $d(1) = d(1) + d(1)$ and hence $d(1) = 0$.

COROLLARY. *Let* d_1 *and* d_2 *be two* ε*-derivations from* A *to* E *of the same degree* δ. *If* d_1 *and* d_2 *take the same values at each element of a generating system of the algebra* A, *then* $d_1 = d_2$.

$d_1 - d_2$ is an ε-derivation of degree δ, hence Ker$(d_1 - d_2)$ is a subalgebra of A which contains a generating system of A and hence is equal to A.

PROPOSITION 4. *Let* $d : \mathrm{A} \to \mathrm{E}$ *be an* ε*-derivation of degree* δ. *Suppose that* A *has a unit element* 1 *and let* x *be a homogeneous element of* A *with an inverse* x^{-1} *in* A. *Then*

$$d(x^{-1}) = -\varepsilon_{\delta, \deg(x)}x^{-1}((dx)x^{-1}) = -\varepsilon_{\delta, \deg(x)}(x^{-1}(dx))x^{-1}. \tag{19}$$

We have $d(xx^{-1}) = d(1) = 0$ (Proposition 3), whence

$$(dx)x^{-1} + \varepsilon_{\delta, \deg(x)}x(d(x^{-1})) = 0$$

which proves the first formula of (19). On the other hand, x^{-1} is homogeneous of degree $-\deg(x)$ and $\varepsilon_{\delta, \deg(x)} = \varepsilon_{\delta, -\deg(x)}$ by formulae (1) of no. 1; writing $d(x^{-1}x) = 0$, the second formula of (19) is obtained similarly.

PROPOSITION 5. *Suppose that* A *is an integral domain and let* L *be its field of fractions. Every derivation of* A *into a vector space* E *over* L *(considered as an* A*-module) can be extended uniquely to a derivation of* L *into* E.

Let d be a derivation of A into E and $\bar{d}$ a derivation of L into E extending d; then, for $u \in A$, $v \in A$, $v \neq 0$, of necessity, by virtue of (19),

$$\bar{d}(u/v) = v^{-1}du - uv^{-2}dv \tag{20}$$

which proves the uniqueness of $\bar{d}$. Conversely, we show that $\bar{d}$ can be defined by formula (20); it must first be verified that if $u/v = u'/v'$ the value of the right hand side of (20) does not change when u is replaced by u' and v by v'. Now, $uv' = vu'$, hence $v'(du) + u(dv') = v(du') + u'(dv)$ and therefore $v'(du - uv^{-1}dv) = v(du' - u'v'^{-1}dv')$, since $uv'v^{-1} = u'$ and $u'v'^{-1}v = u$. Thus a mapping $\bar{d}: L \rightarrow E$ has been defined which extends d; it is immediately verified that it is K-linear and a derivation.

PROPOSITION 6. *Suppose that* A *is a unital associative graded* K*-algebra and* E *a graded* (A, A)*-bimodule. If* $d: A \rightarrow E$ *is an* ε*-derivation of degree* δ*, then, for every finite sequence* $(x_i)_{1 \leqslant i \leqslant n}$ *of homogeneous elements of* A*, of respective degrees* ξ_i $(1 \leqslant i \leqslant n)$,

$$d(x_1x_2 \ldots x_n) = \sum_{i=1}^{n} \varepsilon_{\delta, \xi_1 + \cdots + \xi_{i-1}} x_1 \ldots x_{i-1}(dx_i)x_{i+1} \ldots x_n. \tag{21}$$

Formula (21) is trivial for $n = 0$ and is proved by induction on n, taking account of (4) (no. 2).

COROLLARY. *Suppose that* A *is a unital commutative associative algebra and* E *an* A*-module. If* $d: A \rightarrow E$ *is a derivation, then, for every integer* $n \geqslant 0$,

$$d(x^n) = nx^{n-1}(dx) \quad \textit{for all } x \in A. \tag{22}$$

It suffices to give A the trivial graduation and apply (21) with all the x_i equal to x.

We return to the general case of an ε-derivation $d: A \rightarrow E$ of degree δ. Let Z_ε be the set of $a \in A$ such that for every homogeneous component a_α of a of degree α, for every homogeneous x in E,

$$xa_\alpha = \varepsilon_{\alpha, \deg(x)}a_\alpha x. \tag{23}$$

If A is a unital associative graded algebra and E a graded (A, A)-bimodule it follows immediately from this definition that Z_ε is a *graded subalgebra* of A containing the unit element.

PROPOSITION 7. *Suppose that* A *is a unital associative graded algebra and* E *a graded* (A, A)*-bimodule. Let* $d: A \rightarrow E$ *be an* ε*-derivation of degree* δ *and* a *a homogeneous element of* Z_ε *of degree* α. *Then the mapping* $x \mapsto a(dx)$ *is an* ε*-derivation of degree* $\delta + \alpha$.

We write $d'(x) = a(dx)$; for x homogeneous of degree ξ in A and $y \in A$, by virtue of (23) and (1) (no. 1),

$$\begin{aligned} d'(xy) &= a((dx)y) + \varepsilon_{\delta,\xi}a(x(dy)) = (a(dx))y + \varepsilon_{\delta+\alpha,\xi}(xa)(dy) \\ &= (d'x)y + \varepsilon_{\delta+\alpha,\xi}x(d'y). \end{aligned}$$

Proposition 7 says that the K-module of ε-derivations of A into E is a *graded* Z_ε*-module of type* Δ.

6. DERIVATIONS OF AN ALGEBRA

Let A be a graded K-algebra; for every *homogeneous* element $a \in A$, let $\mathrm{ad}_\varepsilon(a)$, or simply $\mathrm{ad}(a)$ if no confusion can arise, denote the K-linear mapping of A into A

$$x \mapsto [a, x]_\varepsilon$$

(no. 4, formula (10)) which is *graded of degree* deg a.

PROPOSITION 8. *Let* A *be a graded* K*-algebra.*

(i) *For every* ε*-derivation* $d: A \rightarrow A$ *and every homogeneous element* a *of* A,

$$[d, \mathrm{ad}_\varepsilon(a)]_\varepsilon = \mathrm{ad}_\varepsilon(da). \tag{24}$$

(ii) *If the algebra* A *is associative,* $\mathrm{ad}_\varepsilon(a)$ *is an* ε*-derivation of* A *of degree* $\deg(a)$.

(i) Suppose that d is of degree δ, let $\alpha = \deg a$ and write $f = [d, \mathrm{ad}_\varepsilon(a)]_\varepsilon$. For every homogeneous element $x \in A$ of degree ξ, we have, by (1) (no. 1),

$$\begin{aligned} f(x) &= d(ax - \varepsilon(\alpha, \xi)xa) - \varepsilon_{\delta,\alpha}(a(dx) - \varepsilon_{\alpha,\delta+\xi}(dx)a) \\ &= (da)x + \varepsilon_{\delta,\alpha}a(dx) - \varepsilon_{\alpha,\xi}(dx)a - \varepsilon_{\delta+\alpha,\xi}x(da) \\ &\quad - \varepsilon_{\delta,\alpha}a(dx) + \varepsilon_{\alpha,\xi}(dx)a \\ &= (da)x - \varepsilon_{\delta+\alpha,\xi}x(da) = [da, x]_\varepsilon. \end{aligned}$$

(ii) For all x homogeneous of degree ξ and all y homogeneous of degree η in A,

$$\begin{aligned} \mathrm{ad}_\varepsilon(a)(xy) &= a(xy) - \varepsilon_{\alpha,\xi+\eta}(xy)a \\ &= (ax - \varepsilon_{\alpha,\xi}xa)y + \varepsilon_{\alpha,\xi}x(ay - \varepsilon_{\alpha,\eta}ya) \\ &= \mathrm{ad}_\varepsilon(a)(x)\cdot y + \varepsilon_{\alpha,\xi}x\cdot\mathrm{ad}_\varepsilon(a)(y) \end{aligned}$$

taking account of (1) and the associativity of A.

When A is associative, $\mathrm{ad}_\varepsilon(a)$ is called the *inner ε-derivation* of A defined by *a*.

COROLLARY. *Let* A *be an associative graded algebra. For two homogeneous elements, a, b of* A,

$$[\mathrm{ad}_\varepsilon(a), \mathrm{ad}_\varepsilon(b)]_\varepsilon = \mathrm{ad}_\varepsilon([a, b]_\varepsilon). \tag{25}$$

It suffices to replace d by $\mathrm{ad}_\varepsilon(a)$ and $\mathrm{ad}_\varepsilon(a)$ by $\mathrm{ad}_\varepsilon(b)$ in (24).

If $\deg a = \alpha$, $\deg b = \beta$, formula (25) is equivalent to the following relation for every homogeneous element $c \in \mathrm{A}$ of degree γ

$$\varepsilon_{\alpha,\gamma}[a, [b, c]_\varepsilon]_\varepsilon + \varepsilon_{\beta,\alpha}[b, [c, a]_\varepsilon]_\varepsilon + \varepsilon_{\gamma,\beta}[c, [a, b]_\varepsilon]_\varepsilon = 0 \tag{26}$$

called the *Jacobi identity*.

7. FUNCTORIAL PROPERTIES

In this no., all algebras are assumed to be associative and unital and every algebra homomorphism is assumed to be unital.

PROPOSITION 9. *Let* A, B *be two graded* K*-algebras,* E *an* (A, A)*-bimodule and* F *a graded* (B, B)*-bimodule; let* $\rho : \mathrm{A} \to \mathrm{B}$ *be a graded algebra homomorphism and* $\theta : \mathrm{E} \to \mathrm{F}$ *a graded* A*-homomorphism of* A*-bimodules (relative to* ρ*), of degree* 0. *Then:*

(i) *For every ε-derivation* $d' : \mathrm{B} \to \mathrm{F}$, $d' \circ \rho : \mathrm{A} \to \rho^*(\mathrm{F})$ *is an ε-derivation of the same degree.*

(ii) *For every ε-derivation* $d : \mathrm{A} \to \mathrm{E}$, $\theta \circ d : \mathrm{A} \to \rho^*(\mathrm{F})$ *is an ε-derivation of the the same degree.*

The two assertions follow immediately from the relations

$$d'(\rho(xy)) = d'(\rho(x)\rho(y)) = d'(\rho(x))\rho(y) + \varepsilon_{\delta',\xi}\rho(x)d'(\rho(y))$$
$$\theta(d(xy)) = \theta((dx)y + \varepsilon_{\delta,\xi}x(dy)) = \theta(dx)\rho(y) + \varepsilon_{\delta,\xi}\rho(x)\theta(dy)$$

for $x \in \mathrm{A}$ homogeneous of degree ξ and $y \in \mathrm{A}$, δ and δ' denoting the respective degrees of d and d'.

COROLLARY. *Let* S *be a generating system of the algebra* A. *In order that* $d' \circ \rho = \theta \circ d$, *it is necessary and sufficient that* $d'(\rho(x)) = \theta(d(x))$ *for all* $x \in \mathrm{S}$.

This is an immediate consequence of Proposition 9 and no. 5, Corollary to Proposition 3.

Under the conditions of Proposition 9, we know that B has (by means of ρ) an (A, A)-bimodule structure (II, § 1, no. 14, *Example* 1).

PROPOSITION 10. *Under the conditions of Proposition* 9, *for an ε-derivation* $d' : \mathrm{B} \to \mathrm{F}$ *to be* A*-linear for the left* (resp. *right*) A*-module structures on* B *and* $\rho_*(\mathrm{F})$, *it is necessary and sufficient that* d' *be zero on the subalgebra* $\rho(\mathrm{A})$ *of* B.

We perform the proof for left A-module structures. For $a \in A$, $b \in B$,

$$d'(\rho(a)b) = d'(\rho(a))b + \rho(a)d'b$$

and hence if $d' \circ \rho = 0$, d' is linear for the left A-module structures on B and $\rho^*(F)$. Conversely, if this is so, in particular

$$d'(\rho(a)) = d'(\rho(a).1) = \rho(a)d'(1) = 0$$

(no. 5, Proposition 3).

In particular let $D_K(B, F)$ denote the K-module of *derivations* of B into F (no. 2); those among these derivations which are A-*linear*, in other words those which are zero on $\rho(A)$, form a *sub*-K-*module* of $D_K(B, F)$, denoted by $D_{A,\rho}(B, F)$ or simply $D_A(B, F)$ (obviously $D_K(B, F) = D_{K,\phi}(B, F)$, where $\phi: K \to B$ is the homomorphism defining the K-algebra structure on B).

Now let A, B, C be three graded K-algebras, $\rho: A \to B$, $\sigma: B \to C$ two graded algebra homomorphisms and G a graded (C, C)-bimodule; if $D_A(B, G)$, $D_B(C, G)$ and $D_A(C, G)$ denote the respective K-modules $D_{A,\rho}(B, \sigma_*(G))$, $D_{B,\sigma}(C, G)$ and $D_{A,\sigma\circ\rho}(C, G)$, $D_B(C, G)$ is clearly a *sub*-K-*module* of $D_A(C, G)$ since $\sigma(\rho(A)) \subset \sigma(B)$.

PROPOSITION 11. *Under the above conditions, there is an exact sequence of* K-*homomorphisms*

$$0 \to D_B(C, G) \xrightarrow{u} D_A(C, G) \xrightarrow{v} D_A(B, G) \tag{27}$$

where u is the canonical injection and v the homomorphism $d \mapsto d \circ \sigma$ (Proposition 9).

The kernel of v is the set of derivations $d: C \to G$ such that $d(\sigma(b)) = 0$ for all $b \in B$, which is precisely the image of u.

8. RELATIONS BETWEEN DERIVATIONS AND ALGEBRA HOMOMORPHISMS

We suppose again in this no. that *Case* (II) of no. 2 holds and the graded K-algebra A is not assumed to be associative. Given an element $\delta \in \Delta$, consider the graded K-module $E(\delta)$ (II, § 11, no. 2) such that

$$(E(\delta))_\mu = E_{\mu+\delta}$$

for all $\mu \in \Delta$. We define on the graded K-module $A \oplus E(\delta)$ a *graded* K-*algebra* structure by setting, for every homogeneous element $a \in A$ and arbitrary elements $a' \in A$, x, x' in $E(\delta)$

$$(a, x)(a', x') = (aa', x.a' + \varepsilon_{\delta,\deg(a)}a.x'); \tag{28}$$

the verification of the fact that this multiplication defines a graded ring structure is immediate.

The projection $p:(a, x) \mapsto a$ is called the *augmentation* of the algebra $A \oplus E(\delta)$ and is a graded algebra homomorphism. The graded K-linear mappings $g: A \to A \oplus E(\delta)$ *of degree* 0 such that the composition

$$A \xrightarrow{g} A \oplus E(\delta) \xrightarrow{p} A$$

is the identity 1_A are the mappings of the form $x \mapsto (x, f(x))$, where $f: A \to E$ is a graded K-linear mapping *of degree* δ.

PROPOSITION 12. *For a graded* K*-linear mapping* $f: A \to E$ *of degree* δ *to be an* ε*-derivation, it is necessary and sufficient that the mapping* $x \mapsto (x, f(x))$ *of* A *into* $A \oplus E(\delta)$ *be a graded* K*-algebra homomorphism.*

Using the fact that for x homogeneous in A and $y \in A$

$$(xy, f(xy)) = (x, f(x)) . (y, f(y)),$$

we obtain, taking account of (28), the equivalent relation

$$f(xy) = f(x) . y + \varepsilon_{\delta, \deg(x)} x . f(y),$$

whence the proposition.

PROPOSITION 13. *For the algebra* $A \oplus E(\delta)$ *to be associative and unital, it is necessary and sufficient that* A *be associative and unital, and that the mappings* $(a, x) \mapsto a.x$ *and* $(a, x) \mapsto x.a$ *define on* E *an* (A, A)*-bimodule structure; the unit element of* $A \oplus E(\delta)$ *is then* (1, 0).

If an element $(u, m) \in A \oplus E(\delta)$ is written as unit element of this algebra, it is immediately found that u must be the unit element of A; writing $(u, m).(0, x) = (0, x).(u, m) = (0, x)$, we obtain $u.x = x.u = x$ for $x \in E$ and, writing $(u, m).(u, 0) = (u, 0).(u, m) = (u, 0)$, we obtain $m = 0$. The fact that A is associative when $A \oplus E(\delta)$ is follows from the fact that the augmentation is a surjective homomorphism. The condition $(x.a').a'' = x.(a'.a'')$ is then equivalent to $((0, x)(a', 0))(a'', 0) = (0, x)((a', 0)(a'', 0))$ and similarly the condition $a.(a'.x) = (a.a').x$ is equivalent to

$$(a, 0)((a', 0)(0, x)) = ((a, 0)(a', 0))(0, x);$$

finally the condition $a.(x.a') = (a.x).a'$ is equivalent to

$$(a, 0)((0, x)(a', 0)) = ((a, 0)(0, x))(a', 0).$$

9. EXTENSION OF DERIVATIONS

PROPOSITION 14. *Let* A *be a commutative ring,* M *an* A*-module,* B *the* A*-algebra* T(M) (resp. S(M), resp. $\bigwedge$(M)) *and* E *a* (B, B)*-bimodule. Let* $d_0: A \to E$ *be a* *derivation of the ring* A *into the* A*-module* E *and* $d_1: M \to E$ *an additive group homomorphism such that, for all* $a \in A$ *and all* $x \in M$,

$$d_1(ax) = ad_1(x) + d_0(a).x, \tag{29}$$

and further, when $B = S(M)$,

$$x.d_1(y) + d_1(x).y = y.d_1(x) + d_1(y).x \tag{30}$$

for all x, y *in* M, *and, when* $B = \bigwedge(M)$,

$$x.d_1(x) + d_1(x).x = 0 \tag{31}$$

for all $x \in M$. *Then there exists one and only one derivation* d *of* B *(considered as a* $\mathbf{Z}$*-algebra) into the* (B, B)*-bimodule* E *such that* $d|A = d_0$ *and* $d|M = d_1$.

We take on the $\mathbf{Z}$-module $B \oplus E$ the associative $\mathbf{Z}$-algebra structure defined by

$$(b, t)(b', t') = (bb', bt' + tb')$$

which has (1, 0) as unit element (no. 8, Proposition 13). Under the canonical injection $t \mapsto (0, t)$, E is identified with a two-sided ideal of $B \oplus E$ such that $E^2 = \{0\}$. On the other hand, the mapping h_0: $B \oplus E$ defined by $h_0(a) = (a, d_0(a))$ is a unital ring homomorphism (no. 8, Proposition 12); under this mapping, $B \oplus E$ then becomes an A-*algebra*. Moreover, if, for all $x \in M$, we write $h_1(x) = (x, d_1(x))$, it follows from the definition of h_0 and (29) that $h_1(ax) = h_0(a)h_1(x)$; in other words h_1 is an A-*linear* mapping of M into $B \oplus E$. Then there exists one and only one A-*algebra homomorphism*, $h: B \to B \oplus E$ such that $h | M = h_1$ (and necessarily $h | A = h_0$): for, if $B = T(M)$, this follows from § 5, no. 1, Proposition 1; if $B = S(M)$, condition (30) shows that $h(x)h(y) = h(y)h(x)$ for all x, y in M and the conclusion follows from § 6, no. 1, Proposition 2; finally if $B = \bigwedge(M)$, condition (31) shows that $(h(x))^2 = 0$ for all $x \in M$, since $x \wedge x = 0$ and the conclusion follows from § 7, no. 1, Proposition 1. The homomorphism h is such that the composition $p \circ h: B \to B$ with the augmentation $\rho: B \oplus E \to B$ is the identity 1_B, for $p \circ h$ and 1_B coincide by definition for the elements of A and those of M and the set of these elements is a generating system of B. We can therefore write $h(b) = (b, d(b))$ for all $b \in B$ and the mapping $b \mapsto d(b)$ of B into E is a derivation with the required properties, by virtue of Proposition 12 of no. 8.

COROLLARY. *Let* M *be a graded* K*-module of type* Δ; *the* K*-algebras* T(M), S(M) *and* $\bigwedge(M)$ *are given the corresponding graduations of type* $\Delta' = \Delta \times \mathbf{Z}$ (§ 5, no. 5, Proposition 7, § 6, no. 6, Proposition 10 and § 7, no. 7, Proposition 11). *On the other hand* M *is given the graduation of type* Δ' *such that* $M_{\alpha, 1} = M_\alpha$ *for all* $\alpha \in \Delta$ *and* $M_{\alpha, n} = \{0\}$ *for* $\alpha \in \Delta$ *and* $n \neq 1$. *Let* ε' *be a commutation factor over* Δ'.

(i) *Let* E *be a graded* (*left and right*) T(M)*-bimodule of type* Δ'; *for all* $\delta \in \Delta$ *and every integer* $n \in \mathbf{Z}$, *every graded* K*-linear mapping* $f: M \to E$ *of degree* $\delta_1' = (\delta, n)$ *can be extended uniquely to an* ε'*-derivation* $d: T(M) \to E$ *of degree* δ'.

(ii) *Let* E *be a graded* S(M)*-module of type* Δ'; *for a graded* K*-linear mapping* $f: M \to E$ *of degree* δ' *to be extendable to an* ε'*-derivation* $d: S(M) \to E$ *of degree* δ', *it is necessary and sufficient that, for every ordered pair* (x, y) *of homogeneous elements of* M,

$$(33) \qquad x.f(y) + \varepsilon_{\delta', (\deg(y), 1)} y.f(x) = y.f(x) + \varepsilon'_{\delta', (\deg(x), 1)} x.f(y).$$

The ε'*-derivation* d *is then unique.*

(iii) *Let* E *be a graded* (*left and right*) $\bigwedge$(M)*-bimodule of type* Δ'; *for a graded* K*-linear mapping* $f: M \to E$ *of degree* δ' *to be extendable to an* ε'*-derivation*

$$d: \bigwedge(M) \to E$$

of degree δ', *it is necessary and sufficient that, for every homogeneous element* x *of* M,

$$(34) \qquad x.f(x) + \varepsilon'_{\delta', (\deg(x), 1)} f(x).x = 0.$$

The ε'*-derivation* d *is then unique.*

Remark 2 of no. 2 is applied with one of the external B-module laws on E (with B equal to T(M), S(M) or $\bigwedge$(M)) modified; the external law thus modified is still, by (1) (no. 1), a B-module law and the B-module law thus obtained on E is still compatible with the other B-module structure. It then suffices to apply Proposition 14 with A = K and $d_0 = 0$.

Example (1). In the application of Proposition 14 note that if $d_0 = 0$ condition (29) means simply that d_1 is A-*linear*. If we take in particular E = B and the (B, B)-bimodule structure derived from the ring structure on B, conditions (30) and (31) are automatically satisfied when d_1 is taken to be the composition of an *endomorphism* s of M and the canonical injection $M \to B$; this is obvious for (30) since S(M) is commutative and for (31) this follows from the fact that x and $s(x)$ are of degree 1 in $\bigwedge$(M). It is therefore seen that *every endomorphism* s *of* M *can be extended uniquely to a derivation* D_s of T(M) (resp. S(M), resp. $\bigwedge$(M)), which is of degree 0. Moreover, for two endomorphisms s, t of M,

$$(35) \qquad [D_s, D_t] = D_{[s, t]}$$

for both sides are derivations of T(M) (resp. S(M), resp. $\bigwedge$(M)) which are equal to $[s, t]$ on M.

The expression for D_s is obtained using formula (21) of no. 5, which gives respectively, for $x_1, x_2, \ldots, x_n$ in M,

$$(36)\quad \begin{cases} D_s(x_1 \otimes x_2 \otimes \cdots \otimes x_n) \\ \qquad = \sum_{i=1}^{n} x_1 \otimes \cdots \otimes x_{i-1} \otimes s(x_i) \otimes x_{i+1} \otimes \cdots \otimes x_n \\ D_s(x_1 x_2 \ldots x_n) = \sum_{i=1}^{n} x_1 \ldots x_{i-1} s(x_i) x_{i+1} \ldots x_n \\ D_s(x_1 \wedge x_2 \wedge \cdots \wedge x_n) \\ \qquad = \sum_{i=1}^{n} x_1 \wedge \cdots \wedge x_{i-1} \wedge s(x_i) \wedge x_{i+1} \wedge \cdots \wedge x_n. \end{cases}$$

In the case of the algebra $\bigwedge(M)$, there is the following interpretation of D_s:

PROPOSITION 15. *If* M *is a free* K*-module of finite rank n, then, for every endomorphism s of* M, *the restriction to* $\bigwedge^n(M)$ *of the derivation* D_s *is the homothety of ratio* $Tr(s)$.

Let $(e_j)_{1 \leqslant j \leqslant n}$ be a basis of M and write $e = e_1 \wedge e_2 \wedge \cdots \wedge e_n$. If

$$s(e_j) = \sum_{k=1}^{n} \alpha_{jk} e_k,$$

the third formula in (36) gives

$$D_s(e) = \sum_{i=1}^{n} e_1 \wedge \cdots \wedge e_{i-1} \wedge s(e_i) \wedge e_{i+1} \wedge \cdots \wedge e_n = \left(\sum_{j=1}^{n} \alpha_{jj}\right) e.$$

Example (2). In the Corollary to Proposition 14, part (iii), let $\Delta = \{0\}$, the graduation on $\bigwedge(M)$ then being the usual graduation of type **Z**; on the other hand take $\varepsilon(p, q) = (-1)^{pq}$. Then, for every linear form $x^* \in M^*$ on M, $x \mapsto \langle x, x^* \rangle$ is a graded K-linear mapping of degree -1 of M into $\bigwedge(M)$ satisfying relation (34); then there exists an *antiderivation* $i(x^*)$ of $\bigwedge(M)$, of degree -1, such that (by virtue of formula (21) of no. 5)

$$i(x^*)(x_1 \wedge \cdots \wedge x_n) \\ = \sum_{i=1}^{n} (-1)^{i-1} \langle x_i, x^* \rangle x_1 \wedge \cdots \wedge x_{i-1} \wedge x_{i+1} \wedge \cdots \wedge x_n$$

and which is a special case of the inner product to be defined in § 11, no. 9, formula (68).

PROPOSITION 16. *Let* A *be a commutative* K*-algebra*, M_i $(1 \leqslant i \leqslant n)$ *and* P A*-modules and* H *the* A*-module of* A*-multilinear mappings of* $M_1 \times M_2 \times \cdots \times M_n$

into P. *Suppose that there is given a* K-*derivation* $d_0: A \to A$ *of the algebra* A, *for each* i, *a* K-*linear mapping* $d_i: M_i \to M_i$ *and a* K-*linear mapping* $D: P \to P$, *so that, for* $1 \leqslant i \leqslant n$, (d_0, d_i, d_i) *is a* K-*derivation of* (A, M_i, M_i) *into itself and* (d_0, D, D) *is a* K-*derivation of* (A, P, P) *into itself. Then there exists a* K-*linear mapping* $D': H \to H$ *such that* (d_0, D', D') *is a* K-*derivation of* (A, H, H) *into itself and*

$$(37)\quad D(f(x_1, \ldots, x_n)) = (D'f)(x_1, \ldots, x_n) + \sum_{i=1}^{n} f(x_1, \ldots, x_{i-1}, d_i x_i, x_{i+1}, \ldots, x_n)$$

for all $x_i \in M_i$ *for* $1 \leqslant i \leqslant n$ *and* $f \in H$.

We show that for $f \in H$, the mapping $D'f$ of $M_1 \times M_2 \times \cdots \times M_n$ into P defined by (37) is A-multilinear. For $a \in A$,

$$(D'f)(ax_1, x_2, \ldots, x_n) = D(af(x_1, \ldots, x_n)) - f(d_1(ax_1), x_2, \ldots, x_n) - a\sum_{i=2}^{n} f(x_1, \ldots, x_{i-1}, d_i x_i, x_{i+1}, \ldots, x_n)$$

and by hypothesis

$$D(af(x_1, \ldots, x_n)) = (d_0 a) f(x_1, \ldots, x_n) + aD(f(x_1, \ldots, x_n))$$

and $d_1(ax_1) = (d_0 a)x_1 + a.d_1 x_1$, which gives

$$(D'f)(ax_1, x_2, \ldots, x_n) = a.(D'f)(x_1, \ldots, x_n)$$

and linearity in each of the x_i is proved similarly. On the other hand,

$$\begin{aligned}(D'(af))(x_1, \ldots, x_n) &= D(af(x_1, \ldots, x_n)) - \sum_{i=1}^{n} af(x_1, \ldots, x_{i-1}, d_i x_i, x_{i+1}, \ldots, x_n) \\ &= (d_0 a) f(x_1, \ldots, x_n)) + aD(f(x_1, \ldots, x_n)) - \sum_{i=1}^{n} af(x_1, \ldots, x_{i-1}, d_i x_i, x_{i+1}, \ldots, x_n) \\ &= (d_0 a) f(x_1, \ldots, x_n) + a(D'f)(x_1, \ldots, x_n)\end{aligned}$$

in other words

$$D'(af) = (d_0 a) f + a(D'f)$$

which establishes the proposition.

Examples. (3) Applying Proposition 16 to the case $n = 1$, $M_1 = M$, $P = A$, then $H = M^*$, the *dual* of M, and it is seen that for a K-derivation (d_0, d, d) of (A, M, M) is derived a K-derivation (d_0, d^*, d^*) of (A, M^*, M^*) such that

$$(38)\quad d_0\langle m, m^*\rangle = \langle dm, m^*\rangle + \langle m, d^* m^*\rangle$$

for $m \in M$ and $m^* \in M^*$. The K-linear mapping of $M \oplus M^*$ into itself which is equal to d on M and to d^* on M* then satisfies condition (29) and there is therefore a K-*derivation* D of the A-algebra $T(M \oplus M^*)$, which reduces to d_0 on A, to d on M and to d^* on M*. The restriction d_J^I of D to the sub-A-module $T_J^I(M)$ of $T(M \oplus M^*)$ (§ 5, no. 6) is a K-endomorphism of $T_J^I(M)$ such that (d_0, d_J^I, d_J^I) is a K-*derivation* of $(A, T_J^I(M), T_J^I(M))$. Moreover, for $i \in I$, $j \in J$, if we write $I' = I - \{i\}$, $J' = J - \{j\}$, it is immediately verified that for the contraction c_j^i (§ 5, no. 6)

$$c_j^i(d_J^I(z)) = d_{J'}^{I'}(c_j^i(z)) \quad \text{for all } z \in T_J^I(M).$$

(4) Let M_i $(1 \leqslant i \leqslant 3)$ be three A-modules and, for each i, suppose that (d_0, d_i, d_i) is a derivation of (A, M_i, M_i); applying Proposition 16 again for $n = 1$, a derivation (d_0, d_{ij}, d_{ij}) of (A, H_{ij}, H_{ij}), where $H_{ij} = \mathrm{Hom}_A(M_i, M_j)$, is derived for each ordered pair (i, j). With this notation, for $u \in \mathrm{Hom}_A(M_1, M_2)$ and $v \in \mathrm{Hom}_A(M_2, M_3)$,

$$d_{13}(v \circ u) = (d_{23}v) \circ u + v \circ (d_{12}u) \tag{39}$$

as is immediately verified from the definitions.

10. UNIVERSAL PROBLEM FOR DERIVATIONS; NON-COMMUTATIVE CASE

Throughout the rest of § 10 all the algebras are assumed to be associative and unital and all the algebra homomorphisms are assumed to be unital.

Let A be a K-algebra; the tensor product $A \otimes_K A$ has canonically an (A, A)-bimodule structure under which

$$x.(u \otimes v).y = (xu) \otimes (vy) \tag{40}$$

for all x, y, u, v in A (§ 4, no. 3, *Example* 2). The K-linear mapping $m: A \otimes_K A \to A$ corresponding to multiplication in A (and hence such that $m(x \otimes y) = xy$) is an (A, A)-*bimodule* homomorphism; its *kernel* I is therefore a sub-bimodule of $A \otimes_K A$.

Lemma 1. *The mapping* $\delta_A: x \mapsto x \otimes 1 - 1 \otimes x$ *is a* K-*derivation of* A *into* I *and* I *is generated, as a left* A-*module, by the image of* δ_A.

The first assertion follows from the fact that

$$(xy) \otimes 1 - 1 \otimes (xy) = (x \otimes 1 - 1 \otimes x).y + x.(y \otimes 1 - 1 \otimes y)$$

by (40). On the other hand, if the element $\sum_i x_i \otimes y_i$ (for x_i, y_i in A) belongs to I, by definition $\sum_i x_i y_i = 0$ and hence

$$\sum_i (x_i \otimes y_i) = \sum_i x_i(1 \otimes y_i - y_i \otimes 1)$$

by (40), which completes the proof of the lemma.

PROPOSITION 17. *The derivation* δ_A *has the following universal property: for every* (A, A)*-bimodule* E *and every* K*-derivation* $d: A \to E$, *there exists one and only one* (A, A)*-bimodule homomorphism* $f: I \to E$ *such that* $d = f \circ \delta_A$.

Note first that, for every (A, A)-bimodule homomorphism $f: I \to E$, $f \circ \delta_A$ is a derivation (no. 7, Proposition 9). Conversely, let $d: A \to E$ be a K-derivation; then we prove first that if there exists an (A, A)-bimodule homomorphism $f: I \to E$ such that $d = f \circ \delta_A$, f *is uniquely determined* by this condition for the definition of δ_A gives

$$f(x \otimes 1 - 1 \otimes x) = dx$$

and our assertion follows from the fact that the image of δ_A already generates I as a left A-module (Lemma 1): hence of necessity

$$f\Big(\sum_i x_i \otimes y_i\Big) = \sum_i x_i . f(1 \otimes y_i - y_i \otimes 1) = -\sum_i x_i . dy_i.$$

Conversely, as the mapping $(x, y) \mapsto -x.dy$ of $A \times A$ into E is K-bilinear, there exists one and only one K-linear mapping $g: A \otimes_K A \to E$ such that $g(x \otimes y) = -x.dy$; it suffices to verify that the restriction f of g to I is A-linear for the left and right A-module structures. The first assertion is obvious since $(xx').dy = x.(x'.dy)$; to prove the second, note that, if $\sum_i x_i \otimes y_i \in I$ and $x \in A$, then

$$\sum_i x_i . d(y_i x) = \sum_i x_i . dy_i . x + \sum_i (x_i y_i) . dx$$

but since $\sum_i x_i y_i = 0$ by definition of I, this completes the proof.

We have thus defined a *canonical* K-module *isomorphism* $f \mapsto f \circ \delta_A$

$$\mathrm{Hom}_{(A, A)}(I, E) \to D_K(A, E)$$

the left hand side being the K-module of (A, A)-bimodule homomorphisms of A into E.

11. UNIVERSAL PROBLEM FOR DERIVATIONS; COMMUTATIVE CASE

Suppose now that A is a *commutative* K-algebra and E an A-*module*; E can be considered as an (A, A)-bimodule whose two external laws are identical with the given A-module law. On the other hand the (A, A)-bimodule structure on $A \otimes_K A$ is identical with its $(A \otimes_K A)$-module structure arising from the *commutative ring* structure on $A \otimes_K A$, since in this case, for x, y, u, v in A,

$$x.(u \otimes v).y = (xu) \otimes (vy) = (xu) \otimes (yv) = (x \otimes y)(u \otimes v).$$

The kernel $\mathfrak{I}$ of m is therefore in this case an *ideal* of the ring $A \otimes_K A$ and, as $m: A \otimes_K A \to A$ is surjective, $(A \otimes_K A)/\mathfrak{I}$ is isomorphic to A; if also E is considered as an $(A \otimes_K A)$-module by means of m (in other words the $(A \otimes_K A)$-module $m_*(E)$), the (A, A)-*bimodule* homomorphisms $\mathfrak{I} \to E$ are identified with the $(A \otimes_K A)$-*module* homomorphisms $\mathfrak{I} \to E$ (§ 4, no. 3), in other words there is a canonical K-module isomorphism.

$$\mathrm{Hom}_{(A, A)}(\mathfrak{I}, E) \to \mathrm{Hom}_{A \otimes_K A}(\mathfrak{I}, E).$$

On the other hand, $\mathfrak{I}E = \{0\}$, for the elements $1 \otimes x - x \otimes 1$ generate $\mathfrak{I}$ as an $(A \otimes_K A)$-module (no. 10, Lemma 1) and, for all $z \in E$,

$$(1 \otimes x - x \otimes 1)z = 0$$

by virtue of the definition of the $(A \otimes_K A)$-module structure on E. Since $\mathfrak{I}$ is contained in the annihilator of the $(A \otimes_K A)$-module E and the $((A \otimes_K A)/\mathfrak{I})$-module structure on E is by definition just the initial A-module structure given on E, there is, taking account of the canonical isomorphism of

$$\mathfrak{I} \otimes_K ((A \otimes_K A)/\mathfrak{I})$$

onto $\mathfrak{I}/\mathfrak{I}^2$ (§ 4, no. 1, Corollary 1 to Proposition 1), a canonical K-module isomorphism

$$\mathrm{Hom}_{A \otimes_K A}(\mathfrak{I}, E) \to \mathrm{Hom}_A(\mathfrak{I}/\mathfrak{I}^2, E).$$

Taking account of Proposition 17 of no. 10, it is seen that we have proved the following proposition:

PROPOSITION 18. *Let* A *be a commutative* K-*algebra and* $\mathfrak{I}$ *the ideal the kernel of the surjective canonical homomorphism* $m: A \otimes_K A \to A$, *so that* A *is isomorphic to* $(A \otimes_K A)/\mathfrak{I}$ *and* $\mathfrak{I}/\mathfrak{I}^2$ *has canonically an* A-*module structure. Let* $d_{A/K}: A \to I/I^2$ *be the* K-*linear mapping which associates with every* $x \in A$ *the class of* $x \otimes 1 - 1 \otimes x$ *modulo* $\mathfrak{I}^2$. *The mapping* $d_{A/K}$ *is a* K-*derivation and, for every* A-*module* E *and every* K-*derivation* $D: A \to E$, *there exists one and only one* A-*linear mapping* $g: \mathfrak{I}/\mathfrak{I}^2 \to E$ *such that* $D = g \circ d_{A/K}$.

The A-module $\mathfrak{I}/\mathfrak{I}^2$ is called the A-*module of* K-*differentials* of A and is denoted by $\Omega_K(A)$; for all $x \in A$, $d_{A/K}(x)$ (also denoted by dx) is called the *differential* of x; it has been seen (no. 10, lemma 1) that the elements $d_{A/K}(x)$, for $x \in A$, form a *generating system* of the A-module $\Omega_K(A)$. Proposition 18 shows that the mapping $g \mapsto g \circ d_{A/K}$ is a *canonical* A-*module isomorphism*

$$\phi_A: \mathrm{Hom}_A(\Omega_K(A), E) \to D_K(A, E)$$

(the A-module structure on $D_K(A, E)$ being defined by Proposition 7 of no. 5).

The ordered pair $(\Omega_K(A), d_{A/K})$ is therefore the solution of the universal mapping problem where Σ is the species of A-module structure and the α-mappings the K-derivations from A to an A-module (*Set Theory,* IV, § 3, no. 1).

Example. Let M be a K-module; it follows from Proposition 14 of no. 9 that for every S(M)-module E, the mapping $D \mapsto D \mid M$ defines an S(M)-module isomorphism of $D_K(S(M), E)$ onto $\mathrm{Hom}_K(M, E)$; on the other hand, since E is an S(M)-module, $\mathrm{Hom}_K(M, E)$ is canonically isomorphic to

$$\mathrm{Hom}_{S(M)}(M \otimes_K S(M), E),$$

every K-homomorphism of M into E being uniquely expressible in the form $x \mapsto h(x \otimes 1)$, where

$$h \in \mathrm{Hom}_{S(M)}(M \otimes_K S(M), E)$$

(II, § 5, no. 1). Let D_0 be the K-derivation $S(M) \to M \otimes_K S(M)$ whose restriction to M is the canonical homomorphism $x \mapsto x \otimes 1$; every K-derivation $D : S(M) \to E$ can therefore be written uniquely as $h \circ D_0$ with

$$h \in \mathrm{Hom}_{S(M)}(M \otimes_K S(M), E).$$

By the uniqueness of a solution of a universal mapping problem, it is seen that there exists a *unique* S(M)-module isomorphism

$$\omega : M \otimes_K S(M) \to \Omega_K(S(M))$$

such that $D_0 \circ \omega = d_{S(M)/K}$; in other words, for all $x \in M$, $\omega(x \otimes 1) = dx$.

In particular, *if* M *is a free* K-*module with basis* $(e_\lambda)_{\lambda \in L}$, $\Omega_K(S(M))$ *is a free* S(M)-*module with basis the set of differentials* de_λ. Consider in particular the case where $L = [1, n]$, so that S(M) is identified with the polynomial algebra $K[X_1, \ldots, X_n]$ (§ 6, no. 6); for every polynomial $P \in K[X_1, \ldots, X_n]$, we can write uniquely

$$dP = \sum_{i=1}^{n} D_i P . dX_i$$

with $D_i P \in K[X_1, \ldots, X_n]$ and, by virtue of the above, the mappings $P \mapsto D_i P$ are the K-*derivations* of $K[X_1, \ldots, X_n]$ corresponding to the coordinate forms on $\Omega_K(S(M))$ for the basis (dX_i); we also write $\frac{\partial P}{\partial X_i}$ instead of $D_i P$ and this is called the *partial derivative* of P with respect to X_i.

12. FUNCTORIAL PROPERTIES OF K-DIFFERENTIALS

PROPOSITION 19. *Let*

$$\begin{array}{ccc} K & \xrightarrow{\rho} & K' \\ {\scriptstyle \eta}\downarrow & & \downarrow{\scriptstyle \eta'} \\ A & \xrightarrow[u]{} & A' \end{array}$$

be a commutative diagram of commutative ring homomorphisms, η (resp. η') *making*

A (resp. A′) *into a* K-*algebra* (resp. K′-*algebra*). *There exists one and only one* A-*linear mapping* $v : \Omega_K(A) \to \Omega_{K'}(A')$ *rendering commutative the diagram*

$$\begin{array}{ccc} A & \xrightarrow{u} & A' \\ {\scriptstyle d_{A/K}}\downarrow & & \downarrow{\scriptstyle d_{A'/K'}} \\ \Omega_K(A) & \xrightarrow[v]{} & \Omega_{K'}(A') \end{array}$$

$d_{A'/K'} \circ u$ is a K-derivation of A with values in the A-module $\Omega_{K'}(A')$; the existence and uniqueness of v then follow from Proposition 18 of no. 11.

The mapping v of Proposition 19 will be denoted by $\Omega(u)$; if there is a commutative diagram of commutative ring homomorphisms

$$\begin{array}{ccccc} K & \xrightarrow{\rho} & K' & \xrightarrow{\rho'} & K'' \\ {\scriptstyle \eta}\downarrow & & \downarrow{\scriptstyle \eta'} & & \downarrow{\scriptstyle \eta''} \\ A & \xrightarrow[u]{} & A' & \xrightarrow[u']{} & A'' \end{array}$$

it follows immediately from the uniqueness property of Proposition 19 that

$$\Omega(u' \circ u) = \Omega(u') \circ \Omega(u).$$

Since $\Omega_{K'}(A')$ is an A′-module, from $\Omega(u)$ we derive canonically an A′-*linear* mapping

$$\Omega_0(u) : \Omega_K(A) \otimes_A A' \to \Omega_{K'}(A') \tag{41}$$

such that $\Omega(u)$ is the composition of $\Omega_0(u)$ and the canonical homomorphism $i_A : \Omega_K(A) \to \Omega_K(A) \otimes_A A'$. For every A′-module E′, there is a commutative diagram

$$\begin{array}{ccc} \mathrm{Hom}_{A'}(\Omega_{K'}(A'), E') & \xrightarrow{\mathrm{Hom}(\Omega_0(u), 1_{E'})} & \mathrm{Hom}_{A'}(\Omega_K(A) \otimes_A A', E') \\ {\scriptstyle \phi_{A'}}\downarrow & & \downarrow{\scriptstyle \phi_A \circ r_A} \\ D_{K'}(A', E') & \xrightarrow[C(u)]{} & D_K(A, E) \end{array}$$

where $C(u)$ is the mapping $D \mapsto D \circ u$ (no. 7, Proposition 9) and r_A is the canonical isomorphism

$$\mathrm{Hom}(i_A, 1_{E'}) : \mathrm{Hom}_{A'}(\Omega_K(A) \otimes_A A', E') \to \mathrm{Hom}_A(\Omega_K(A), E');$$

this follows immediately from Proposition 19 and the definition of the isomorphisms ϕ_A and $\phi_{A'}$.

PROPOSITION 20. *Suppose that* $A' = A \otimes_K K'$, *with* $\eta' : K' \to A'$ *and* $u : A \to A'$ *the canonical homomorphisms. Then the* A'*-linear mapping*

$$\Omega_0(u) : \Omega_K(A) \otimes_A A' \to \Omega_{K'}(A')$$

is an isomorphism.

By virtue of the fact that in diagram (42) the vertical arrows are bijective, it reduces to proving that, for every A'-module E', the homomorphism $C(u) : D \mapsto D \circ u$ in diagram (42) is bijective (II, § 2, no. 1, Theorem 1). Now $\mathrm{Hom}(u, 1_{E'}) : \mathrm{Hom}_{K'}(A \otimes_K K', E') \to \mathrm{Hom}_K(A, E')$ is an isomorphism (II, § 5, no. 1, Proposition 1) and $C(u)$ is its restriction to $D_{K'}(A', E')$ and hence is injective; moreover, if $f : A' \to E'$ is a K'-linear mapping such that

$$f \circ u : A \to E'$$

is a K-derivation, it follows immediately from the fact that f is K'-linear and the fact that $f((x \otimes 1)(y \otimes 1)) = (y \otimes 1)f(x \otimes 1) + (x \otimes 1)f(y \otimes 1)$ for x, y in A, that f is a K'-*derivation*, the elements $x \otimes 1$ for $x \in A$ forming a generating system of the K'-module A'; this completes the proof that $C(u)$ is bijective.

From now on we confine our attention to the case where $\rho : K \to K'$ is the *identity mapping* of K; every K-algebra homomorphism $u : A \to B$ is therefore mapped to a B-linear mapping

$$\Omega_0(u) : \Omega_K(A) \otimes_K B \to \Omega_K(B). \tag{43}$$

On the other hand, we can consider the B-module of A-*differentials* $\Omega_A(B)$ since B is an A-algebra by means of u; the canonical derivation $d_{B/A} : B \to \Omega_A(B)$ being *a fortiori* a K-derivation, it factorizes uniquely into

$$B \xrightarrow{d_{B/K}} \Omega_K(B) \xrightarrow{\Omega_u} \Omega_A(B)$$

where Ω_u is a B-linear mapping (no. 11, Proposition 18). For every B-module E, there is a commutative diagram

$$\begin{array}{ccc} \mathrm{Hom}_B(\Omega_A(B), E) & \xrightarrow{\mathrm{Hom}(\Omega_u, 1_E)} & \mathrm{Hom}_B(\Omega_K(B), E) \\ \downarrow{\scriptstyle \phi_{A,B}} & & \downarrow{\scriptstyle \phi_{K,B}} \\ D_A(B, E) & \xrightarrow[j_u]{} & D_K(B, E) \end{array} \tag{44}$$

where j_u is the canonical injection (no. 7); this follows immediately from Proposition 18 of no. 11.

PROPOSITION 21. *The sequence of* B*-linear mappings*

$$\Omega_K(A) \otimes_A B \xrightarrow{\Omega_0(u)} \Omega_K(B) \xrightarrow{\Omega_u} \Omega_A(B) \to 0 \tag{45}$$

is exact.

It reduces to verifying that, for every B-module E, the sequence

$$0 \to \mathrm{Hom}_B(\Omega_A(B), E) \xrightarrow{\mathrm{Hom}(\Omega_u, 1_E)} \mathrm{Hom}_B(\Omega_K(B), E) \xrightarrow{\mathrm{Hom}(\Omega_0(u), 1_E)} \mathrm{Hom}_B(\Omega_K(A) \otimes_A B, E)$$

is exact (II, § 2, no. 1, Theorem 1); but by virtue of the fact that in the commutative diagrams (42) and (44) the vertical arrows are isomorphisms, it suffices to show that the sequence

$$0 \longrightarrow D_A(B, E) \xrightarrow{j_u} D_K(B, E) \xrightarrow{C(u)} D_K(A, E)$$

is exact, which is just Proposition 11 of no. 7.

We consider now the case where the K-algebra homomorphism $u: A \to B$ is *surjective*; if $\mathfrak{J}$ is its kernel, B is then isomorphic to $A/\mathfrak{J}$. We consider the restriction $d|\mathfrak{J}: \mathfrak{J} \to \Omega_K(A)$ of the canonical derivation $d = d_{A/K}$ and the composite A-linear mapping

$$d': \mathfrak{J} \xrightarrow{d|\mathfrak{J}} \Omega_K(A) \xrightarrow{i_A} \Omega_K(A) \otimes_A B.$$

Then $d'(\mathfrak{J}^2) = 0$, since, for x, y in $\mathfrak{J}$,

$$d'(xy) = d(xy) \otimes 1 = (x.dy + y.dx) \otimes 1 = dy \otimes u(x) + dx \otimes u(y) = 0$$

since $u(x) = u(y) = 0$. Hence we derive from d', by passing to the quotient, an A-linear mapping

$$\bar{d}: \mathfrak{J}/\mathfrak{J}^2 \to \Omega_K(A) \otimes_A B$$

and as $\mathfrak{J}$ annihilates the A-module $\mathfrak{J}/\mathfrak{J}^2$, $\bar{d}$ is a B-*linear* mapping.

PROPOSITION 22. *Let $\mathfrak{J}$ be an ideal of the commutative* K-*algebra* A, $B = A/\mathfrak{J}$ *and* $u: A \to B$ *the canonical homomorphism. The sequence of* B-*linear mappings*

$$\mathfrak{J}/\mathfrak{J}^2 \xrightarrow{\bar{d}} \Omega_K(A) \otimes_A B \xrightarrow{\Omega_0(u)} \Omega_K(B) \to 0 \tag{46}$$

is then exact.

Note that $\Omega_K(A) \otimes_A B$ is identified with $\Omega_K(A)/\mathfrak{J}\Omega_K(A)$ and that the image of $\bar{d}$ is the image of $d(\mathfrak{J})$ in this quotient module; the quotient of $\Omega_K(A) \otimes_A B$ by $\mathrm{Im}(\bar{d})$ is therefore identified with the quotient $\Omega_K(A)/N$, where N is the sub-A-module generated by $\mathfrak{J}\Omega_K(A)$ and $d(\mathfrak{J})$. Moreover, the composite mapping

$$A \xrightarrow{d_{A/K}} \Omega_K(A) \longrightarrow \Omega_K(A)/N$$

is a K-derivation (no. 7, Proposition 9) and, since it is zero on $\mathfrak{J}$ by definition of N, it defines, when passing to the quotient, a K-derivation $D_0: B \to \Omega_K(A)/N$.

Taking account of the uniqueness of the solution of a universal mapping problem, it reduces to proving that, for every B-module E and every K-derivation $D: B \to E$, there exists a unique B-linear mapping $g: \Omega_K(A)/N \to E$ such that $D = g \circ D_0$. But, the composite mapping $D \circ u: A \to E$ is a K-derivation (no. 7, Proposition 9) and hence there exists one and only one A-linear mapping $f: \Omega_K(A) \to E$ such that $f \circ d_{A/K} = D \circ u$. This relation shows already that f is zero on $d(\mathfrak{J})$; as also $\mathfrak{J}E = \{0\}$ since E is a B-module, f is zero on $\mathfrak{J}\Omega_K(A)$; hence f is zero on N and defines, when passing to the quotient, a B-linear mapping $g: \Omega_K(A)/N \to E$ such that $g \circ D_0 = D$; the uniqueness of g follows from the uniqueness of f.

It must not be thought that, even if $u: A \to B$ is an injective homomorphism, $\Omega_0(u): \Omega_K(A) \otimes_A B \to \Omega_K(B)$ is injective (Exercise 5). However we have the following proposition:

PROPOSITION 23. *Let* A *be an integral* K-*algebra*, B *its field of fractions and* $u: A \to B$ *the canonical injection. Then* $\Omega_0(u): \Omega_K(A) \otimes_A B \to \Omega_K(B)$ *is an isomorphism.*

Using the fact that in diagram (42) the vertical arrows are bijective, it reduces to proving that, for every vector space E over B, the mapping $C(u): D_K(B, E) \to D_K(A, E)$ is bijective. But this follows from the fact that every K-derivation of A into E can be extended uniquely to a K-derivation of B into E (no. 5, Proposition 5).

§ 11. COGEBRAS, PRODUCTS OF MULTILINEAR FORMS, INNER PRODUCTS AND DUALITY

In this paragraph, A *is a commutative ring with the trivial graduation. For a graded* A-*module* M *of type* **N**, M^{*gr} *will denote the graded* A-*module of type* **N**, *whose homogeneous elements of degrees* n *are the* A-*linear forms which are zero on* M_k *for all* $k \neq n$.

1. COGEBRAS

DEFINITION 1. *A cogebra over* A *(or* A-*cogebra, or simply cogebra if no confusion can arise) is a set* E *with a structure defined by giving the following:*

(1) *an* A-*module structure on* E;

(2) *an* A-*linear mapping* $c: E \to E \otimes_A E$ *called the coproduct of* E.

DEFINITION 2. *Given two cogebras* E, E′, *whose coproducts are denoted respectively by* c *and* c', *a morphism of* E *into* E′ *is an* A-*linear mapping* $u: E \to E'$ *such that*

$$(u \otimes u) \circ c = c' \circ u, \tag{1}$$

in other words, it renders commutative the diagram of A*-linear mappings*

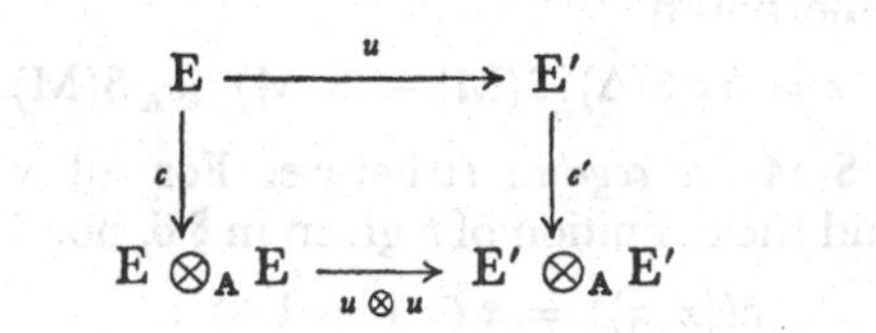

It is immediately verified that the identity mapping is a morphism, that the composition of two morphisms is a morphism and that every bijective morphism is an isomorphism.

Examples. (1) The canonical isomorphism $A \to A \otimes_A A$ (II, § 3, no. 5) defines an A-cogebra structure on A.

(2) Let E be a cogebra, c its coproduct and σ the canonical automorphism of the A-module $E \otimes_A E$ such that $\sigma(x \otimes y) = y \otimes x$ for $x \in E$, $y \in E$; the A-linear mapping $\sigma \circ c$ defines a new cogebra structure on E. With this structure E is called the *opposite* cogebra to the given cogebra E.

(3) Let B be an A-*algebra* and let $m: B \otimes_A B \to B$ be the A-linear mapping defining multiplication on B (§ 1, no. 3). The transpose ${}^t m$ is then an A-linear mapping of the dual B^* of the A-module B into the dual $(B \otimes_A B)^*$ of the A-module $B \otimes_A B$. If also B is a *finitely generated projective* A-module, the canonical mapping $\mu: B^* \otimes_A B^* \to (B \otimes_A B)^*$ is an A-module isomorphism (II, § 4, no. 4); the mapping $c = \mu^{-1} \circ {}^t m$ is then a coproduct defining a *cogebra* structure on the *dual* B^* of the A-module B.

(4) Let X be a set, $A^{(X)}$ the A-module of formal linear combinations of elements of X with coefficients in A (II, § 1, no. 11) and $(e_x)_{x \in X}$ the canonical basis of $A^{(X)}$. An A-linear mapping $c: A^{(X)} \to A^{(X)} \otimes_A A^{(X)}$ is defined by the condition $c(e_x) = e_x \otimes e_x$ and a canonical cogebra structure is thus obtained on $A^{(X)}$.

(5) Let M be an A-module and T(M) the tensor algebra of M (§ 5, no. 1); by II, § 3, no. 9 there exists one and only one A-linear mapping c of the A-module T(M) into the A-module $T(M) \otimes_A T(M)$ such that, for all $n \geqslant 0$,

$$(3) \qquad c(x_1 x_2 \ldots x_n) = \sum_{0 \leqslant p \leqslant n} (x_1 x_2 \ldots x_p) \otimes (x_{p+1} \ldots x_n)$$

for all $x_i \in M$ ($x_1 x_2 \ldots x_n$ denotes the product in the algebra T(M)). Thus T(M) is given a *cogebra* structure.

(6) Let M be an A-module and S(M) the symmetric algebra of M (§ 6, no. 1); the diagonal mapping $\Delta: x \mapsto (x, x)$ of M into $M \times M$ is an A-linear mapping to which there therefore corresponds a homomorphism $S(\Delta)$ of the A-algebra S(M) into the A-algebra $S(M \times M)$ (§ 6, no. 2, Proposition 3). On the other hand, in § 6, no. 6 we defined a canonical graded algebra isomor-

phism $h: S(M \times M) \to S(M) \otimes_A S(M)$; by composition we therefore obtain an A-*algebra* homomorphism

$$c = h \circ S(\Delta): S(M) \to S(M) \otimes_A S(M),$$

thus defining on $S(M)$ a *cogebra* structure. For all $x \in M$, by definition $S(\Delta)(x) = (x, x)$ and the definition of h given in § 6, no. 6 shows that

$$h((x, x)) = x \otimes 1 + 1 \otimes x.$$

It follows that c is the unique algebra homomorphism such that, for all $x \in M$,

$$c(x) = x \otimes 1 + 1 \otimes x. \tag{4}$$

As c is an algebra homomorphism, it follows that, for every sequence $(x_i)_{1 \leqslant i \leqslant n}$ of n elements of M,

$$c(x_1 x_2 \ldots x_n) = \prod_{i=1}^{n} (x_i \otimes 1 + 1 \otimes x_i) = \sum (x_{i_1} \ldots x_{i_p}) \otimes (x_{j_1} \ldots x_{j_{n-p}}) \tag{5}$$

the summation on the right hand side of (5) being taken over all ordered pairs of strictly increasing sequences (in some cases empty)

$$i_1 < i_2 < \cdots < i_p, \quad j_1 < j_2 < \cdots < j_{n-p}$$

of elements of $[1, n]$, whose sets of elements are complementary. The element $c(x_1 x_2 \ldots x_n)$ is an element of *total degree* n in $S(M) \otimes_A S(M)$ and its component of bidegree $(p, n - p)$ is

$$\sum_{\sigma} (x_{\sigma(1)} \ldots x_{\sigma(p)}) \otimes (x_{\sigma(p+1)} \ldots x_{\sigma(n)}) \tag{6}$$

where the summation is taken over all permutations $\sigma \in S_n$ which are *increasing* in each of the intervals $[1, p]$ and $[p + 1, n]$.

(7) Let M be an A-module and proceed with the exterior algebra $\bigwedge(M)$ as with $S(M)$ in *Example* 6; the diagonal mapping $\Delta: M \to M \times M$ this time defines a homomorphism $\bigwedge(\Delta)$ of the A-algebra $\bigwedge(M)$ into the A-algebra $\bigwedge(M \times M)$ (§ 7, no. 2, Proposition 2); on the other hand there is a canonical graded algebra isomorphism

$$h: \bigwedge(M \times M) \to \bigwedge(M) \,{}^g\!\otimes_A \bigwedge(M)$$

(§ 7, no. 7, Proposition 10), whence by composition there is an *algebra* homomorphism $c = h \circ \bigwedge(\Delta): \bigwedge(M) \to \bigwedge(M) \,{}^g\!\otimes_A \bigwedge(M)$, which can be considered as an A-module homomorphism $\bigwedge(M) \to \bigwedge(M) \otimes_A \bigwedge(M)$ and which therefore defines on $\bigwedge(M)$ a *cogebra* structure. It can be proved as in *Example* 6 that c is the unique algebra homomorphism such that, for all $x \in M$,

$$c(x) = x \otimes 1 + 1 \otimes x, \tag{7}$$

whence, for every sequence $(x_i)_{1 \leqslant i \leqslant n}$ of elements of M,

$$c(x_1 \wedge x_2 \wedge \cdots \wedge x_n) = (x_1 \otimes 1 + 1 \otimes x_1) \wedge \ldots \wedge (x_n \otimes 1 + 1 \otimes x_n)$$

where the product on the right hand side is taken in the algebra

$$\bigwedge(\mathrm{M}) \,{}^g\!\otimes_{\mathrm{A}} \bigwedge(\mathrm{M});$$

to calculate this product, consider, for every ordered pair of strictly increasing sequences $i_1 < i_2 < \cdots < i_p$, $j_1 < j_2 < \cdots < j_{n-p}$ of elements of $[1, n]$, whose sets of elements are complementary, the product $y_1 y_2 \ldots y_n$, where $y_{i_h} = x_{i_h} \otimes 1$ $(1 \leqslant h \leqslant p)$ and $y_{j_k} = 1 \otimes x_{j_k}$ $(1 \leqslant k \leqslant n - p)$ and the sum is taken over all these products. As the graded algebra $\bigwedge(\mathrm{M}) \,{}^g\!\otimes_{\mathrm{A}} \bigwedge(\mathrm{M})$ is anticommutative and the elements $x_i \otimes 1$ and $1 \otimes x_i$ are of total degree 1, by § 4, no. 6, Lemma 3 and Lemma 1,

$$(8) \quad c(x_1 \wedge x_2 \wedge \cdots \wedge x_n) = \sum (-1)^{\nu}(x_{i_1} \wedge \cdots \wedge x_{i_p}) \otimes (x_{j_1} \wedge \cdots \wedge x_{j_{n-p}})$$

ν being the number of ordered pairs (h, k) such that $j_k < i_h$ and the summation being taken over the same set as in (5). The element $c(x_1 \wedge \cdots \wedge x_n)$ is of *total degree n* in $\bigwedge(\mathrm{M}) \,{}^g\!\otimes_{\mathrm{A}} \bigwedge(\mathrm{M})$ and its homogeneous component of bidegree $(p, n - p)$ is equal to

$$(9) \qquad \sum_{\sigma} \varepsilon_{\sigma}(x_{\sigma(1)} \wedge \cdots \wedge x_{\sigma(p)}) \otimes (x_{\sigma(p+1)} \wedge \cdots \wedge x_{\sigma(n)})$$

the summation being taken over permutations $\sigma \in \mathfrak{S}_n$ which are *increasing in each of the intervals* $[1, p]$ and $[p + 1, n]$.

When in future we speak of $\mathrm{A}^{(\mathrm{X})}$, T(M), S(M) or $\bigwedge$(M) as *cogebras*, we shall mean, unless otherwise mentioned, with the cogebra structures defined in *Examples* 4, 5, 6 and 7 respectively.

(8) Let E, F be two A-cogebras and c, c' their respective coproducts. Let $\tau: (\mathrm{E} \otimes_{\mathrm{A}} \mathrm{E}) \otimes_{\mathrm{A}} (\mathrm{F} \otimes_{\mathrm{A}} \mathrm{F}) \to (\mathrm{E} \otimes_{\mathrm{A}} \mathrm{F}) \otimes_{\mathrm{A}} (\mathrm{E} \otimes_{\mathrm{A}} \mathrm{F})$ denote the associativity isomorphism such that $\tau((x \otimes x') \otimes (y \otimes y')) = (x \otimes y) \otimes (x' \otimes y')$ for x, x' in E and y, y' in F. Then the composite linear mapping

$$\mathrm{E} \otimes_{\mathrm{A}} \mathrm{F} \xrightarrow{c \otimes c'} (\mathrm{E} \otimes_{\mathrm{A}} \mathrm{E}) \otimes_{\mathrm{A}} (\mathrm{F} \otimes_{\mathrm{A}} \mathrm{F}) \xrightarrow{\tau} (\mathrm{E} \otimes_{\mathrm{A}} \mathrm{F}) \otimes_{\mathrm{A}} (\mathrm{E} \otimes_{\mathrm{A}} \mathrm{F})$$

defines a cogebra structure on the A-module $\mathrm{E} \otimes_{\mathrm{A}} \mathrm{F}$, called the *tensor product* of the cogebras E and F.

Let E be a cogebra and Δ a commutative monoid. A graduation $(\mathrm{E}_\lambda)_{\lambda \in \Delta}$ on the A-module E is said to be *compatible with the coproduct c* of E if c is a graded homomorphism of degree O of the graded A-module E into the graded A-module (of type Δ) $\mathrm{E} \otimes_{\mathrm{A}} \mathrm{E}$, in other words (II, § 11, no. 5) if

$$(10) \qquad c(\mathrm{E}_\lambda) \subset \sum_{\mu + \nu = \lambda} \mathrm{E}_\mu \otimes_{\mathrm{A}} \mathrm{E}_\nu.$$

In what follows, we shall most often limit our attention to graduations of type **N** compatible with the coproduct; a cogebra with such a graduation will also be called a *graded cogebra.* If F is another graded cogebra, a *graded cogebra morphism* $\phi: \mathrm{E} \to \mathrm{F}$ is by definition a cogebra morphism (Definition 2) which is also a *graded homomorphism of degree* 0 of graded A-modules.

Examples. (9) It is immediate that the cogebras $\mathsf{T}(\mathrm{M})$, $\mathsf{S}(\mathrm{M})$ and $\bigwedge(\mathrm{M})$ defined above are graded cogebras.

2. COASSOCIATIVITY, COCOMMUTATIVITY, COUNIT

Let E be a cogebra, c its coproduct, N, N′, N″ three A-modules and m a bilinear mapping of $\mathrm{N} \times \mathrm{N}'$ into N″. Let $\tilde{m}: \mathrm{N} \otimes_{\mathrm{A}} \mathrm{N}' \to \mathrm{N}''$ denote the A-linear mapping corresponding to m. If $u: \mathrm{E} \to \mathrm{N}$, $v: \mathrm{E} \to \mathrm{N}'$ are two A-linear mappings, we derive an A-linear mapping $u \otimes v: \mathrm{E} \otimes_{\mathrm{A}} \mathrm{E} \to \mathrm{N} \otimes_{\mathrm{A}} \mathrm{N}'$ and a composite A-linear mapping of E into N″:

$$(11) \qquad m(u, v): \mathrm{E} \xrightarrow{c} \mathrm{E} \otimes_{\mathrm{A}} \mathrm{E} \xrightarrow{u \otimes v} \mathrm{N} \otimes_{\mathrm{A}} \mathrm{N}' \xrightarrow{\tilde{m}} \mathrm{N}''.$$

Clearly we have thus defined an A-bilinear mapping $(u, v) \mapsto m(u, v)$ of $\mathrm{Hom}_{\mathrm{A}}(\mathrm{E}, \mathrm{N}) \times \mathrm{Hom}_{\mathrm{A}}(\mathrm{E}, \mathrm{N}')$ into $\mathrm{Hom}_{\mathrm{A}}(\mathrm{E}, \mathrm{N}'')$.

When E is a graded cogebra, N, N′, N″ graded A-modules of the same type and $\tilde{m}$ a graded homomorphism of degree k of $\mathrm{N} \otimes_{\mathrm{A}} \mathrm{N}'$ into N″, then, if u (resp. v) is a graded homomorphism of degree p (resp. q), $m(u, v)$ is a graded homomorphism of degree $p + q + k$.

Examples. (1) Take E to be the graded cogebra $\mathsf{T}(\mathrm{M})$ (no. 1) and suppose that N, N′, N″ have the trivial graduation. A graded homomorphism of degree $-p$ of $\mathsf{T}(\mathrm{M})$ into N (resp. N′, N″) then corresponds to a multilinear mapping of M^p into N (resp. N′, N″). Given a multilinear mapping $u: \mathrm{M}^p \to \mathrm{N}$ and a multilinear mapping $v: \mathrm{M}^q \to \mathrm{N}'$, the above method allows us to deduce a multilinear mapping $m(u, v): \mathrm{M}^{p+q} \to \mathrm{N}''$ called the *product* (relative to m) of u and v. Formulae (3) (no. 1) and (11) show that, for $x_1, \dots, x_{p+q}$ in M,

$$(m(u, v))(x_1, \dots, x_{p+q}) = m(u(x_1, \dots, x_p), v(x_{p+1}, \dots, x_{p+q})).$$

(2) Take E to be the graded cogebra $\mathsf{S}(\mathrm{M})$ (no. 1), preserving the same hypotheses on N, N′, N″. A graded homomorphism of degree $-p$ of $\mathsf{S}(\mathrm{M})$ into N then corresponds to a *symmetric multilinear mapping* of M^p into N (§ 6, no. 3). Then we derive from a symmetric multilinear mapping $u: \mathrm{M}^p \to \mathrm{N}$ and a symmetric multilinear mapping $v: \mathrm{M}^q \to \mathrm{N}'$ a symmetric multilinear mapping $m(u, v): \mathrm{M}^{p+q} \to \mathrm{N}''$, also denoted (to avoid confusion) by $u \cdot_m v$ (or even $u \cdot v$) and called the *symmetric product* (relative to m) of u and v. Formulae (6) (no. 1) and (11) show that, for $x_1, \dots, x_{p+q}$ in M,

$$(u \cdot_m v)(x_1, \ldots, x_{p+q}) = \sum_{\sigma} m(u(x_{\sigma(1)}, \ldots, x_{\sigma(p)}), v(x_{\sigma(p+1)}, \ldots, x_{\sigma(p+q)}))$$

the summation being taken over permutations $\sigma \in \mathfrak{S}_{p+q}$ which are increasing on each of the intervals $(1, p)$ and $(p + 1, p + q)$.

(3) Take E to be the graded cogebra $\bigwedge(\mathrm{M})$ (no. 1). Then we deduce similarly from an alternating multilinear mapping $u: \mathrm{M}^p \to \mathrm{N}$ and an alternating multilinear mapping $v: \mathrm{M}^q \to \mathrm{N}'$ an alternating multilinear mapping $m(u, v): \mathrm{M}^{p+q} \to \mathrm{N}''$, also denoted by $u \wedge_m v$ or $u \wedge v$ and called the *alternating product* (relative to m) of u and v. Formulae (9) (no. 1) and (11) show in this case that, for $x_1, \ldots, x_{p+q}$ in M,

$$(u \wedge_m v)(x_1, \ldots, x_{p+q}) = \sum_{\sigma} \varepsilon_{\sigma} m(u(x_{\sigma(1)}, \ldots, x_{\sigma(p)}), v(x_{\sigma(p+1)}, \ldots, x_{\sigma(p+q)}))$$

the summation again being taken over permutations $\sigma \in \mathfrak{S}_{p+q}$ which are increasing on each of the intervals $(1, p)$ and $(p + 1, p + q)$.

We return to the case where E is an arbitrary graded cogebra (of type **N**) and assume that the three modules N, N′, N″ are all equal to the underlying A-module of a *graded* A-*algebra* B of type **Z**, the mapping m being multiplication in B, so that $\tilde{m}: \mathrm{B} \otimes_{\mathrm{A}} \mathrm{B} \to \mathrm{B}$ is a graded A-linear mapping of degree 0. Thus a *graded* A-*algebra* structure is obtained on the graded A-module $\mathrm{Homgr}_{\mathrm{A}}(\mathrm{E}, \mathrm{B}) = \mathrm{C}$.

In particular, suppose that B = A (with the trivial graduation), so that $\mathrm{Homgr}_{\mathrm{A}}(\mathrm{E}, \mathrm{A})$ is the *graded dual* $\mathrm{E}^{*\mathrm{gr}}$, which thus has a graded A-algebra structure.

Let F be another graded cogebra c' its coproduct and $\phi: \mathrm{E} \to \mathrm{F}$ a graded cogebra morphism (no. 1); then the canonical graded morphism

$$\tilde{\phi} = \mathrm{Hom}(\phi, 1_{\mathrm{B}}): \mathrm{Homgr}_{\mathrm{A}}(\mathrm{F}, \mathrm{B}) \to \mathrm{Homgr}_{\mathrm{A}}(\mathrm{E}, \mathrm{B})$$

is a *graded algebra homomorphism*. For u, v in $\mathrm{Homgr}_{\mathrm{A}}(\mathrm{F}, \mathrm{B})$ and $x \in \mathrm{E}$,

$$(\tilde{\phi}(uv))(x) = (uv)(\phi(x)) = m((u \otimes v)(c'(\phi(x)))).$$

But by hypothesis $c'(\phi(x)) = (\phi \otimes \phi)(c(x))$, hence

$$(u \otimes v)(c'((x))) = (\tilde{\phi}(u) \otimes \tilde{\phi}(v))(c(x))$$

and therefore $\tilde{\phi}(uv) = \tilde{\phi}(u)\tilde{\phi}(v)$, which proves our assertion.

In particular, the graded transpose ${}^t\phi: \mathrm{F}^{*\mathrm{gr}} \to \mathrm{E}^{*\mathrm{gr}}$ is a graded algebra homomorphism.

Remark. Suppose that the E_p are *finitely generated projective* A-modules, so that the graded A-modules $(E \otimes_A E)^{*gr}$ and $E^{*gr} \otimes_A E^{*gr}$ can be canonically identified (II, § 4, no. 4, Corollary 1 to Proposition 4). If also the A-modules $A \otimes_A A$ and A are then canonically identified (II, § 3, no. 4), the linear mapping $E^{*gr} \otimes_A E^{*gr} \to E^{*gr}$ which defines multiplication in E^{*gr} can be called the *graded transpose of the coproduct c.*

PROPOSITION 1. *Let* E *be a cogebra over* A. *In order that, for every associative* A-*algebra* B, *the* A-*algebra* $\mathrm{Hom}_A(E, B)$ *be associative, it is necessary and sufficient that the coproduct* $c: E \to E \otimes_A E$ *be such that the diagram*

$$\begin{array}{ccc} E & \xrightarrow{c} & E \otimes_A E \\ {\scriptstyle c}\downarrow & & \downarrow{\scriptstyle 1_E \otimes c} \\ E \otimes_A E & \xrightarrow[c \otimes 1_E]{} & E \otimes_A E \otimes_A E \end{array} \tag{12}$$

be commutative.

Let B be an associative A-algebra and u, v, w three elements of $C = \mathrm{Hom}_A(E, B)$. Let m_3 denote the A-linear mapping $B \otimes_A B \otimes_A B \to B$ which maps $b \otimes b' \otimes b''$ to $bb'b''$. By definition of the product on the algebra C, $(uv)w$ is the composite mapping

$$E \xrightarrow{c} E \otimes E \xrightarrow{c \otimes 1_E} E \otimes E \otimes E \xrightarrow{u \otimes v \otimes w} B \otimes B \otimes B \xrightarrow{m_3} B$$

whilst $u(vw)$ is the composite mapping

$$E \xrightarrow{c} E \otimes E \xrightarrow{1_E \otimes c} E \otimes E \otimes E \xrightarrow{u \otimes v \otimes w} B \otimes B \otimes B \xrightarrow{m_3} B.$$

It follows that if diagram (12) is commutative, the algebra $\mathrm{Hom}_A(E, B)$ is associative for every associative A-algebra B. To establish the converse, it suffices to show that there exists an associative A-algebra B and three A-linear mappings u, v, w of E into B such that the mapping $m_3 \circ (u \otimes v \otimes w)$ of $E \otimes E \otimes E$ into B is *injective.* Take B to be the A-algebra T(E) and u, v, w the canonical mapping of E into T(E). The mapping $m_3 \circ (u \otimes v \otimes w)$ is then the canonical mapping $E \otimes E \otimes E = T^3(E) \to T(E)$, which is injective.

When the cogebra E satisfies the condition of Proposition 1, it is said to be *coassociative.*

Examples. (4) It is immediately verified that the cogebra A (no. 1, *Example* (1) the cogebra $A^{(X)}$ (no. 1, *Example* 4) and the cogebra T(M) (no. 1, *Example* 5) are coassociative. If B is an *associative* A-algebra which is a finitely generated projective A-module, the cogebra B* (no. 1, *Example* 3) is coassociative: for the commutativity of diagram (12) then follows by transposition from that of the diagram which expresses the associativity of B (§ 1, no. 3). Conversely, the same argument and the canonical identification of the A-module B with its bidual (II, § 2, no. 7, Corollary 4 to Proposition 13) show that if the cogebra B* is coassociative, the algebra B is associative. Finally, the cogebras S(M) and $\bigwedge$(M) (no. 1, *Examples* 6 and 7) are coassociative; this follows from the commutativity of the diagram

$$\begin{array}{ccc} M & \xrightarrow{\Delta} & M \times M \\ {\scriptstyle\Delta}\downarrow & & \downarrow{\scriptstyle 1_M \times \Delta} \\ M \times M & \xrightarrow[\Delta \times 1_M]{} & M \times M \times M \end{array}$$

the functorial properties of S(M) (§ 6, no. 2) and $\bigwedge$(M) (§ 7, no. 2), which give the corresponding commutative diagrams

$$(14)\quad \left\{ \begin{array}{ccc} S(M) & \xrightarrow{S(\Delta)} & S(M \times M) \\ {\scriptstyle S(\Delta)}\downarrow & & \downarrow{\scriptstyle S(\Delta_M \times 1)} \\ S(M \times M) & \xrightarrow[S(\Delta \times 1_M)]{} & S(M \times M \times M) \\ \\ \bigwedge(M) & \xrightarrow{\bigwedge(\Delta)} & \bigwedge(M \times M) \\ {\scriptstyle \bigwedge(\Delta)}\downarrow & & \downarrow{\scriptstyle \bigwedge(1_M \times \Delta)} \\ \bigwedge(M \times M) & \xrightarrow[\bigwedge(\Delta \times 1_M)]{} & \bigwedge(M \times M \times M) \end{array} \right.$$

and the existence and functoriality of canonical isomorphisms for symmetric and exterior algebras of a direct sum (§ 6, no. 6 and § 7, no. 7).

PROPOSITION 2. *Let* E *be a cogebra over* A. *In order that, for every commutative* A-*algebra* B, *the* A-*algebra* $\mathrm{Hom}_A(E, B)$ *be commutative, it is necessary and sufficient that the coproduct* $c: E \to E \otimes_A E$ *be such that the diagram*

$$(15)\quad \begin{array}{ccc} & E & \\ {\scriptstyle c}\swarrow & & \searrow{\scriptstyle c} \\ E \otimes_A E & \xrightarrow[\sigma]{} & E \otimes_A E \end{array}$$

(where σ is the symmetry homomorphism such that $\sigma(x \otimes y) = y \otimes x$) *is commutative* (in other words, it suffices that the cogebra E be *identical with its opposite* (no. 1, *Example* 2).

Let B be a commutative A-algebra and u, v two elements of $C = \mathrm{Hom}_A(E, B)$.

By definition of the product in C, uv and vu are respectively equal to the composite mappings

$$E \xrightarrow{c} E \otimes E \xrightarrow{u \otimes v} B \otimes B \xrightarrow{m} B$$

and

$$E \xrightarrow{c} E \otimes E \xrightarrow{v \otimes u} B \otimes B \xrightarrow{m} B.$$

It follows that if diagram (15) is commutative the algebra $\mathrm{Hom}_A(E, B)$ is commutative for every commutative A-algebra B. To establish the converse, it suffices to show that there exist a commutative A-algebra B and two A-linear mappings u, v of E into B such that $m \circ (u \otimes v): E \otimes E \to B$ is injective. Take B to be the algebra $S(E \oplus E)$ and u (resp. v) the composition of the canonical mapping $E \oplus E \to S(E \oplus E)$ and the mapping $x \mapsto (x, 0)$ (resp. $x \mapsto (0, x)$) of E into $E \oplus E$. If $h: S(E) \otimes S(E) \to S(E \otimes E)$ is the canonical isomorphism (§ 6, no. 6, Proposition 9) and $\lambda: E \to S(E)$ is the canonical mapping, then $h^{-1} \circ m \circ (u \otimes v) = \lambda \otimes \lambda$. Now $\lambda \otimes \lambda$ is injective, for $\lambda(E)$ is a direct factor of S(E) (II, § 3, no. 7, Corollary 5 to Proposition 7).

When the cogebra E satisfies the condition of Proposition 2, it is said to be *cocommutative.*

Examples. (5) It is immediate that the cogebra A (no. 1, *Example* 1) and the cogebra $A^{(X)}$ (II, § 11, no. 1, *Example* 4) are cocommutative. It follows from formula (5) of no. 1 that the cogebra S(M) is cocommutative. Finally, for an A-algebra B such that the A-module B is projective and finitely generated to have the property that the cogebra B^* (no. 1, *Example* 3) is cocommutative, it is necessary and sufficient that B be commutative; for (using the canonical identification of the A-module B with its bidual (II, § 2, no. 7)), this follows from the fact that the commutativity of diagram (14) is equivalent by transposition to that of the diagram which expressed the commutativity of B (§ 1, no. 3).

PROPOSITION 3. *Let* E *be a cogebra over* A. *In order that, for every unital* A*-algebra* B, *the* A*-algebra* $\mathrm{Hom}_A(E, B)$ *be unital, it is necessary and sufficient that there exist a linear form* γ *on* E *rendering commutative the diagrams*

$$(16) \qquad \begin{array}{ccc} E & \xrightarrow{c} & E \otimes_A E \\ & \searrow^{h'} & \downarrow{\scriptstyle \gamma \otimes 1_E} \\ & & A \otimes_A E \end{array} \qquad \begin{array}{ccc} E & \xrightarrow{c} & E \otimes_A E \\ & \searrow^{h''} & \downarrow{\scriptstyle 1_E \otimes \gamma} \\ & & E \otimes_A A \end{array}$$

where $c: E \to E \otimes_A E$ *is the coproduct and* h' *and* h'' *the canonical isomorphisms*

(II, § 3, no. 4, Proposition 4). *The unit of* $\mathrm{Hom}_A(E, B)$ *is then the linear mapping* $x \mapsto \gamma(x)1$ (where 1 denotes the unit element of B).

Let γ be a linear form on E rendering diagram (16) commutative. Let B be a unital A-algebra with unit element 1, $\eta: A \to B$ the canonical mapping and $v = \eta \circ \gamma$ the element of the A-algebra $C = \mathrm{Hom}_A(E, B)$. For every element $u \in C$, uv is the composite mapping

$$(17) \qquad E \xrightarrow{c} E \otimes E \xrightarrow{1_E \otimes \gamma} E \otimes A \xrightarrow{u \otimes \eta} B \otimes B \xrightarrow{m} B.$$

Then $uv = m \circ (u \otimes \eta) \circ h'' = u$. It is similarly proved that $vu = u$ and hence v is unit element of C. Conversely, let the A-module $A \oplus E$ be given a unital algebra structure such that $(a, x)(a', x') = (aa', ax' + a'x)$ for a, a' in A and x, x' in E. Let B denote the A-algebra thus defined and let C be the A-algebra $\mathrm{Hom}_A(E, B)$. Suppose that C is unital and let $e: x \mapsto (\gamma(x), \lambda(x))$ be its unit element (where $\gamma(x) \in A$ and $\lambda(x) \in E$). On the other hand let f be the element $x \mapsto (0, x)$ of C. An immediate calculation shows that fe is the element

$$x \mapsto (0, (h'')^{-1}((1_E \otimes \gamma)(c(x))))$$

of C. The condition $fe = f$ implies the commutativity of the second diagram of (16) and it is similarly seen that the condition $ef = f$ implies the commutativity of the first diagram of (16).

A linear form γ on E rendering diagrams (16) commutative is called a *counit* of the cogebra E. A cogebra admits *at most one* counit: for it is the unit element of the algebra $\mathrm{Hom}_A(E, A)$. A cogebra with a counit is called *counital*.

Examples. (6) The identity mapping is the counit of the cogebra A; on the cogebra $A^{(X)}$ (no. 1, *Example* 4) the linear form γ such that $\gamma(e_x) = 1$ for all $x \in X$ is the counit. On the cogebra $\mathsf{T}(M)$ (resp. $\mathsf{S}(M)$, $\bigwedge(M)$) the linear form γ such that $\gamma(1) = 1$ and $\gamma(z) = 0$ for z in the $\mathsf{T}^n(M)$ (resp. $\mathsf{S}^n(M)$, $\bigwedge^n(M)$) for $n \geqslant 1$ is the counit. Finally, let B be an A-algebra which is a finitely generated projective A-module and has a unit element e; then on the cogebra B^* (no. 1, *Example* 3) the linear form $\gamma: x^* \mapsto \langle e, x^* \rangle$ is the counit, for this form is just the transpose of the A-linear mapping $\eta_e: \xi \mapsto \xi e$ of A into B and by transposition the commutativity of diagrams (16) follows from that of the diagrams which express (using η_e) the fact that e is unit element of B (§ 1, no. 3); the same argument moreover shows that conversely, if the cogebra B^* admits a counit γ, the transpose of γ defines a unit element $e = {}^t\gamma(1)$ of B.

PROPOSITION 4. *Let* E *be a cogebra admitting a counit* γ *and suppose that there exists in* E *an element* e *such that* $\gamma(e) = 1$; *then* E *is the direct sum of the sub-A-modules* Ae *and* $E_\gamma = \mathrm{Ker}(\gamma)$ *and*

$$(18) \qquad \begin{cases} c(e) \equiv e \otimes e \pmod{E_\gamma \otimes E_\gamma} \\ c(x) \equiv x \otimes e + e \otimes x \pmod{E_\gamma \otimes E_\gamma} \quad \textit{for all } x \in E_\gamma. \end{cases}$$

The first assertion is immediate, for $\gamma(x - \gamma(x)e) = 0$ and the relation $\gamma(\alpha e) = 0$ implies $\alpha = 0$. Let $c(e) = \sum_i s_i \otimes t_i$, so that

$$e = \sum_i \gamma(s_i)t_i = \sum_i \gamma(t_i)s_i$$

by (16) and $1 = \gamma(e) = \sum_i \gamma(s_i)\gamma(t_i)$. Therefore

$$\begin{aligned}\sum_i (s_i - \gamma(s_i)e) \otimes (t_i - \gamma(t_i)e) = \sum_i s_i \otimes t_i - \sum_i e \otimes \gamma(s_i)t_i \\ - \sum_i \gamma(t_i)s_i \otimes e \\ + \sum_i \gamma(s_i)e \otimes \gamma(t_i)e\end{aligned}$$

which, by the above relation, is just $c(e) - e \otimes e$; this therefore proves the first relation of (18). On the other hand the decomposition of $E \otimes E$ as a direct sum

$$A(e \otimes e) \oplus ((Ae) \otimes E_\gamma) \oplus (E_\gamma \otimes (Ae)) \oplus (E_\gamma \otimes E_\gamma)$$

allows us to write, for $x \in E_\gamma$, $c(x) = \lambda(e \otimes e) + (e \otimes y) + z \otimes e) + u$ where $u = \sum_j v_j \otimes w_j$, y, z and the v_j and w_j belong to E_γ. The definition of the counit γ then gives $x = \lambda e + y = \lambda e + z$ and, as $\gamma(x) = 0$, necessarily $\lambda = 0$, $x = y = z$, whence the second relation of (18).

Remark. Let C be a counital coassociative A-coalgebra, B a unital associative A-algebra and M a left B-module. The A-bilinear mapping $(b, m) \mapsto bm$ of $B \times M$ into M defines an A-bilinear mapping

$$\mathrm{Hom}_A(C, B) \times \mathrm{Hom}_A(C, M) \to \mathrm{Hom}_A(C, M)$$

by the general procedure described at the beginning of this no. It is immediately verified that this mapping defines on $\mathrm{Hom}_A(C, M)$ a left module structure over the ring $\mathrm{Hom}_A(C, B)$.

3. PROPERTIES OF GRADED COGEBRAS OF TYPE N

PROPOSITION 5. (i) *Let E be a graded cogebra admitting a counit γ; then γ is a homogeneous linear form of degree 0.*

(ii) *Suppose further that there exists an element $e \in E$ such that $E_0 = Ae$ and $\gamma(e) = 1$. Then the kernel E_γ of γ is equal to $E_+ = \sum_{n \geqslant 1} E_n$, $c(e) = e \otimes e$ and*

$$c(x) \equiv x \otimes e + e \otimes x \pmod{E_+ \otimes E_+} \tag{19}$$

for all $x \in E_+$.

(i) It suffices to verify that $\gamma(x) = 0$ for $x \in \mathrm{E}_n$, for all $n \geqslant 1$. Since c is a graded homomorphism of degree 0,

$$c(x) = \sum_{0 \leqslant j \leqslant n} \Big(\sum_i y_{ij} \otimes z_{i,n-j}\Big) \tag{20}$$

with, for all j such that $0 \leqslant j \leqslant n$, y_{ij} and z_{ij} in E_j; applying (16) (no. 2) we obtain

$$x = \sum_{0 \leqslant j \leqslant n} \Big(\sum_i \gamma(y_{ij}) z_{i,n-j}\Big) = \sum_{0 \leqslant j \leqslant n} \Big(\sum_i \gamma(z_{i,n-j}) y_{ij}\Big),$$

whence, equating the components of degree 0 and degree n on the two sides

$$x = \sum_i \gamma(y_{i0}) z_{in} = \sum_i \gamma(z_{i0}) y_{in}$$

$$0 = \sum_i \gamma(y_{in}) z_{i0} = \sum_i \gamma(z_{in}) y_{i0}$$

and therefore $\gamma(x) = \sum_i \gamma(y_{in})\gamma(z_{i0}) = \gamma(0) = 0$.

(ii) Since $\mathrm{Ker}(\gamma)$ and E_+ are both supplementary sub-A-modules of $\mathrm{A}e = \mathrm{E}_0$ and $\mathrm{E}_+ \subset \mathrm{Ker}(\gamma)$ by (i), $\mathrm{E}_+ = \mathrm{Ker}(\gamma)$ (II, § 1, no. 8, *Remark* 1); the other assertions follow from Proposition 4 of no. 2.

PROPOSITION 6. *Let* E *be a graded cogebra over* A. *In order that, for every commutative* A-*algebra* B, *with the trivial graduation, the graded* A-*algebra of type* **Z** $\mathrm{Homgr}_\mathrm{A}(\mathrm{E}, \mathrm{B})$ (no. 2) *be anticommutative* (§ 4, no. 9, Definition 7), *it is necessary and sufficient that, if* σ_g *denotes the automorphism of the* A-*module* $\mathrm{E} \otimes_\mathrm{A} \mathrm{E}$ *such that*

$$\sigma_g(x_p \otimes x_q) = (-1)^{pq} x_q \otimes x_p$$

for $x_p \in \mathrm{E}_p$, $x_q \in \mathrm{E}_q$, *where* p *and* q *are arbitrary elements of* **N**, *the diagram*

$$\begin{array}{ccc} & \mathrm{E} & \\ {\scriptstyle c}\swarrow & & \searrow{\scriptstyle c} \\ \mathrm{E} \otimes_\mathrm{A} \mathrm{E} & \xrightarrow{\sigma_g} & \mathrm{E} \otimes_\mathrm{A} \mathrm{E} \end{array} \tag{21}$$

be commutative.

The proof is analogous to that of Proposition 2 of no. 2.

When the graded cogebra E satisfies the condition of Proposition 6, it is said to be *anticocommutative.*

Example. It follows immediately from formula (8) of no. 1 that for every A-module M, the graded cogebra $\bigwedge(\mathrm{M})$ is *anticocommutative.*

4. BIGEBRAS AND SKEW-BIGEBRAS

DEFINITION 3. *A graded bigebra* (resp. *skew graded bigebra*) *over a ring* A *is a set* E *with a graded* A-*algebra structure of type* **N** *and a graded* A-*cogebra structure of type* **N**, *with the same underlying graded* A-*module structure and such that:*

(1) *The* A-*algebra* E *is associative and unital.*

(2) *The* A-*cogebra* E *is coassociative and counital.*

(3) *The coproduct* $c: E \rightarrow E \otimes_A E$ *is a homomorphism of the graded algebra* E *into the graded algebra* $E \otimes_A E$ (resp. *graded algebra* $E\, {}^g\!\otimes_A E$ (cf. § 4, no. 7)).

(4) *The counit* γ *of* E *is a homomorphism of the graded algebra* E *into the algebra* A (*with the trivial graduation*) *such that, if* e *denotes the unit element of the* A-*algebra* E, $\gamma(e) = 1$.

If E is a graded bigebra whose graduation is *trivial*, E is called simply a *bigebra.* A graded bigebra is called commutative (resp. cocommutative) if the underlying algebra is commutative (resp. if the underlying cogebra is cocommutative); a skew graded bigebra is called anticommutative (resp. anticocommutative) if the underlying graded algebra is anticommutative (resp. if the underlying graded cogebra is anticocommutative).

It follows from Definition 3 and no. 2, Proposition 5 that, for a graded bigebra or a skew graded bigebra E,

$$(22) \quad \begin{cases} c(e) = e \otimes e \\ c(x) \equiv x \otimes e + e \otimes x \ (\text{mod. } E_+ \otimes E_+) \quad \text{for } x \in E_+ = \bigoplus_{n \geqslant 1} E_n. \end{cases}$$

If E and F are two graded bigebras (resp. two skew graded bigebras), a mapping $\phi: E \rightarrow F$ is called a *graded bigebra morphism* (resp. *skew graded bigebra morphism*) if: (1) ϕ is a graded algebra morphism (and hence maps the unit element of E to the unit element of F); (2) ϕ is a graded cogebra morphism such that, if γ and γ' are the respective counits of E and F, then $\gamma = \gamma' \circ \phi$.

Examples. (1) Let S be a monoid with identity element u, so that the algebra $E = A^{(S)}$ of the monoid S over A admits the unit element e_u (§ 2, no. 6); it has been seen on the other hand that E has canonically a coassociative cocommutative A-cogebra structure with a counit γ such that $\gamma(e_s) = 1$ for all $s \in S$ (no. 1, *Example* 4 and no. 2, *Examples* 4, 5 and 6). The formula $c(e_s) = e_s \otimes e_s$ giving the coproduct shows also immediately that c is an algebra homomorphism. Thus a *cocommutative bigebra* structure has been defined on E and E, with this structure, is called the *bigebra of the monoid* S *over* A.

If T is another monoid with unit element v, $f: S \rightarrow T$ a homomorphism such that $f(u) = v$ and $f_{(A)}: A^{(S)} \rightarrow A^{(T)}$ the A-algebra homomorphism derived from f (§ 2, no. 6), it is immediately verified that $f_{(A)}$ is a *bigebra homomorphism.*

(2) Let M be an A-module. The graded A-algebra (§ 6, no. 1) and graded A-cogebra (no. 1, *Example* 6) structures defined on S(M) define on this set a *commutative cocommutative graded bigebra* structure; for we have seen (no. 1, *Example* 6) that the coproduct on S(M) is an *algebra* homomorphism and it

follows from the definition of the counit γ (no. 2, *Example* 6) that $\gamma(1) = 1$ and that γ is an algebra homomorphism of E into A.

(3) Let M be an A-module. We see as in *Example* 2 that the graded A-algebra (§ 7, no. 1) and graded A-cogebra (no. 1, *Example* 7) structures on $\bigwedge(\mathrm{M})$ define on this set an *anticommutative anticocommutative skew graded bigebra* structure.

Remark. If M is an A-module such that $\mathrm{M} \otimes_{\mathrm{A}} \mathrm{M} \neq \{0\}$, the graded A-algebra (§ 5, no. 1) and graded A-cogebra (no. 1, *Example* 5) structures on $\mathsf{T}(\mathrm{M})$ *do not define* a bigebra structure, for in general

$$c(x_1 x_2 y_1 y_2) \neq c(x_1 x_2) c(y_1 y_2)$$

for four elements x_1, x_2, y_1, y_2 of M, as formula (3) of no. 1, shows.

5. THE GRADED DUALS $\mathsf{T(M)^{*gr}}$, $\mathsf{S(M)^{*gr}}$ AND $\bigwedge(\mathsf{M})^{*gr}$

From now on we return to the general conventions of the chapter on algebras, which will therefore be assumed (unless otherwise mentioned) to be associative and unital.

Let M be an A-module; the graded A-cogebra structures defined on $\mathsf{T}(\mathrm{M})$ (no. 1, *Example* 5), $\mathsf{S}(\mathrm{M})$ (no. 1, *Example* 6) and $\bigwedge(\mathrm{M})$ (no. 1, *Example* 7) allow us to define canonically on the graded duals $\mathsf{T}(\mathrm{M})^{*gr}$, $\mathsf{S}(\mathrm{M})^{*gr}$ and $\bigwedge(\mathrm{M})^{*gr}$ *graded algebra* structures of type **N**, by virtue of no. 2, Propositions 1 and 3 and the convention made on the graduation of the graded dual of a graded module (no. 1). Moreover, the graded algebra $\mathsf{S}(\mathrm{M})^{*gr}$ is *commutative* (no. 2, Proposition 2 and *Example* 5) and the graded algebra $\bigwedge(\mathrm{M})^{*gr}$ is *anticommutative* (no. 3, Proposition 6 and *Example*). In $\bigwedge(\mathrm{M})^{*gr}$ *every element of degree* 1 *is of zero square*; such an element is identified with a linear form f on M and its square is the alternating bilinear form $f \wedge f$ on M^2 such that

$$(f \wedge f)(x, y) = f(x)f(y) - f(y)f(x)$$

(no. 2, *Example* 3).

Let N be another A-module and u an A-linear mapping of M into N. We know that u defines canonically graded algebra homomorphisms

$$\text{(23)} \qquad \begin{cases} \mathsf{T}(u) : \mathsf{T}(\mathrm{M}) \to \mathsf{T}(\mathrm{N}) \\ \mathsf{S}(u) : \mathsf{S}(\mathrm{M}) \to \mathsf{S}(\mathrm{N}) \\ \bigwedge(u) : \bigwedge(\mathrm{M}) \to \bigwedge(\mathrm{N}) \end{cases}$$

(§ 5, no. 2, § 6, no. 2 and § 7, no. 2). It is immediately verified in formula (3) of no. 1 that $\mathsf{T}(u)$ is also a *cogebra morphism*. On the other hand, if Δ_{M} (resp. Δ_{N})

denotes the diagonal mapping $M \to M \times M$ (resp. $N \to N \times N$), there is the relation $(u \times u) \circ \Delta_M = \Delta_N \circ u$; it follows that $S(u \times u) \circ S(\Delta_M) = S(\Delta_N) \circ S(u)$

$$(\text{resp. } \bigwedge(u \times u) \circ \bigwedge(\Delta_M) = \bigwedge(\Delta_N) \circ \bigwedge(u)).$$

Using the definition of coproduct in $S(M)$ and $\bigwedge(M)$ (no. 1, *Examples* 6 and 7) and the functorial character of the canonical isomorphisms

$$S(M \times M) \to S(M) \otimes_A S(M)$$

and $\bigwedge(M \times M) \to \bigwedge(M) \,{}^g\!\otimes_A \bigwedge(M)$, it is seen that $S(u)$ and $\bigwedge(u)$ are also *cogebra morphisms*† (and hence in this case *bigebra* morphisms). It follows immediately that the *graded transposes* (II, § 11, no. 6) of the homomorphisms (23)

$$^tT(u) : T(N)^{*gr} \to T(M)^{*gr}$$

$$^tS(u) : S(N)^{*gr} \to S(M)^{*gr}$$

$$^t\bigwedge(u) : \bigwedge(N)^{*gr} \to \bigwedge(M)^{*gr}$$

are *graded algebra homomorphisms.*

We now note that the dual M^* of M is identified with the submodule of elements of degree 1 in $T(M)^{*gr}$ (resp. $S(M)^{*gr}$, $\bigwedge(M)^{*gr}$). It therefore follows from the universal property of the tensor algebra (§ 5, no. 1) and the universal property of the symmetric algebra (§ 6, no. 1) that *there exists one and only one graded algebra homomorphism*

$$\theta_T : T(M^*) \to T(M)^{*gr}$$

which extends the canonical injection $M^* \to T(M)^{*gr}$, *and one and only one graded algebra homomorphism*

$$\theta_S : S(M^*) \to S(M)^{*gr}$$

which extends the canonical injection $M^* \to S(M)^{*gr}$. On the other hand, the canonical injection of M^* in the *opposite* algebra to $\bigwedge(M)^{*gr}$ is such that the square of every element of M^* is zero; hence (§ 7, no. 1, Proposition 1) *there exists one and only one graded algebra homomorphism*

$$\theta_\wedge : \bigwedge(M^*) \to (\bigwedge(M)^{*gr})^0$$

which extends the canonical injection $M^* \to \bigwedge(M)^{*gr}$.‡ These homomorphisms

† This also follows from formulae (5) and (9) of no. 1.

‡ This injection is extended to a homomorphism into the opposite algebra to $\bigwedge(M)^{*gr}$ instead of a homomorphism into $\bigwedge(M)^{*gr}$ for reasons of convenience in the calculations.

are functorial: for example, for every A-module homomorphism $u: \mathrm{M} \to \mathrm{N}$, the diagram

$$\begin{array}{ccc} \mathsf{T(N^*)} & \xrightarrow{\mathsf{T}({}^t u)} & \mathsf{T(M^*)} \\ \Big\downarrow \theta_{\mathrm{T}} & & \Big\downarrow \theta_{\mathrm{T}} \\ \mathsf{T(N)^{*gr}} & \xrightarrow[{}^t\mathsf{T}(u)]{} & \mathsf{T(M)^{*gr}} \end{array}$$

is commutative, as follows immediately from the universal property of the tensor algebra (§ 5, no. 1); there are analogous commutative diagrams for θ_{S} and $\theta_{\wedge}$.

We shall find the homomorphisms θ_{T}, θ_{S} and $\theta_{\wedge}$ explicitly. For this we consider more generally a coassociative A-cogebra E with coproduct c and define by induction on n, for $n \geqslant 2$, the linear mapping c_n of E into $\mathrm{E}^{\otimes n}$ by $c_2 = c$ and

$$c_n = (c_{n-1} \otimes 1_{\mathrm{E}}) \circ c. \tag{24}$$

On the other hand we denote by $m_n: \mathrm{A}^{\otimes n} \to \mathrm{A}$ the canonical linear mapping such that $m_n(\xi_1 \otimes \xi_2 \otimes \cdots \otimes \xi_n) = \xi_1 \xi_2 \ldots \xi_n$ and note that, for $n \geqslant 2$,

$$m_n = m \circ (m_{n-1} \otimes 1_{\mathrm{A}}) \tag{25}$$

writing $m = m_2$. With this notation:

Lemma 1. (i) *In the associative algebra* $\mathrm{E}^* = \mathrm{Hom}_{\mathrm{A}}(\mathrm{E}, \mathrm{A})$, *the product of n elements* $u_1, u_2, \ldots, u_n$ *of degree* 1 *is given by*

$$u_1 u_2 \ldots u_n = m_n \circ (u_1 \otimes u_2 \otimes \cdots \otimes u_n) \circ c_n. \tag{26}$$

(ii) *Suppose also that the cogebra* E *is graded. Then, in the graded associative algebra* $\mathrm{E}^{*\mathrm{gr}} = \mathrm{Homgr}_{\mathrm{A}}(\mathrm{E}, \mathrm{A})$, *the product of n elements* $u_1, u_2, \ldots, u_n$ *of degree* 1 *is given by*

$$u_1 u_2 \ldots u_n = m_n \circ (u_1 \otimes u_2 \otimes \cdots \otimes u_n) \circ \delta_n \tag{27}$$

where $\delta_n: \mathrm{E} \to \mathrm{E}^{\otimes n}$ *is the linear mapping which maps each* $x \in \mathrm{E}$ *to the component of* $c_n(x)$ *of multidegree* $(1, 1, \ldots, 1)$.

Formula (26) is just the definition of the product in E^* for $n = 2$; to prove it by induction on n, observe that

$$\begin{aligned} u_1 u_2 \ldots u_n & \\ &= m \circ ((u_1 u_2 \ldots u_{n-1}) \otimes u_n) \circ c \\ &= m \circ ((m_{n-1} \circ (u_1 \otimes u_2 \otimes \cdots \otimes u_{n-1}) \circ c_{n-1}) \otimes u_n) \circ c \\ &= m \circ (m_{n-1} \otimes 1_{\mathrm{A}}) \circ (u_1 \otimes u_2 \otimes \cdots \otimes u_{n-1} \otimes u_n) \circ (c_{n-1} \otimes 1_{\mathrm{E}}) \circ c \\ &= m_n \circ (u_1 \otimes u_2 \otimes \cdots \otimes u_n) \circ c_n \end{aligned}$$

by virtue of (24), (25), II, § 3, no. 3, formula (5) and the relation

$$u_n = 1_{\mathrm{A}} \circ u_n \circ 1_{\mathrm{E}}.$$

When E is graded and the elements $u_i \in E^{*gr}$ homogeneous of degree 1, then by definition for *homogeneous* elements $x_i \in E$,

$$(u_1 \otimes u_2 \otimes \cdots \otimes u_n)(x_1 \otimes x_2 \otimes \cdots \otimes x_n) = 0$$

unless all the x_i are of degree 1, whence formula (27).

It follows from formulae (3), (5) and (7) of no. 1 and formula (24) that when E is taken to be one of the three graded cogebras $\mathsf{T}(M)$, $\mathsf{S}(M)$ and $\bigwedge(M)$, we obtain respectively by induction on n (using the fact that the coproduct is a graded homomorphism of degree 0), for $x_1, x_2, \ldots, x_n$ in M:

$$\text{when } E = \mathsf{T}(M), \quad \delta_n(x_1x_2\ldots x_n) = x_1 \otimes x_2 \otimes \cdots \otimes x_n$$

$$\text{when } E = \mathsf{S}(M), \quad \delta_n(x_1x_2\ldots x_n) = \sum_{\sigma \in \mathfrak{S}_n} x_{\sigma(1)} \otimes x_{\sigma(2)} \otimes \cdots \otimes x_{\sigma(n)}$$

$$\text{when } E = \textstyle\bigwedge(M), \quad \delta_n(x_1x_2\ldots x_n) = \sum_{\sigma \in \mathfrak{S}_n} \varepsilon_\sigma . x_{\sigma(1)} \otimes x_{\sigma(2)} \otimes \cdots \otimes x_{\sigma(n)}.$$

It suffices to note, for example when $E = \bigwedge(M)$, that in the expression $c_n(x_1x_2\ldots x_n) = (c_{n-1} \otimes 1_E)\left(\sum (-1)^\nu (x_{i_1}\ldots x_{i_p}) \otimes (x_{j_1}\ldots x_{j_{n-p}})\right)$ arising from formula (8) of no. 1, the only terms which can give a term of multidegree $(1, 1, \ldots, 1)$ are those for which $n - p = 1$ and hence

$$\delta_n(x_1x_2\ldots x_n)$$

is the term of multidegree $(1, 1, \ldots, 1)$ in the sum

$$\sum_{i=1}^{n} (-1)^{n-i} c_{n-1}(x_1 \ldots x_{i-1}x_{i+1} \ldots x_n) \otimes x_i$$

and this term is necessarily equal to

$$\sum_{i=1}^{n} (-1)^{n-i} \delta_{n-1}(x_1 \ldots x_{i-1}x_{i+1} \ldots x_n) \otimes x_i,$$

whence the result by the induction hypothesis.

Using Lemma 1, the product in $\mathsf{T}(M)^{*gr}$ of n linear forms $x_1^*, x_2^*, \ldots, x_n^*$ of M^* is given by

$$(28) \qquad \langle x_1^* x_2^* \ldots x_n^*, x_1x_2\ldots x_n\rangle = \prod_{i=1}^{n} \langle x_i^*, x_i\rangle$$

for $x_i \in M$ $(1 \leqslant i \leqslant n)$; the product of these n forms in $\mathsf{S}(M)^{*gr}$ is given by

$$(29) \qquad \langle x_1^* x_2^* \ldots x_n^*, x_1x_2\ldots x_n\rangle = \sum_{\sigma \in \mathfrak{S}_n} \left(\prod_{i=1}^{n} \langle x_{\sigma(i)}^*, x_i\rangle\right);$$

finally, the product of these forms in $\bigwedge(\mathrm{M})^{*\mathrm{gr}}$ is given by

$$(30)\qquad \langle x_1^* x_1^* \dots x_n^*, x_1 x_2 \dots x_n\rangle = \det(\langle x_i^*, x_j\rangle).$$

In each of the three cases, we have respectively

$$\theta_{\mathsf{T}}(x_1^* \otimes x_2^* \otimes \dots \otimes x_n^*) = x_1^* x_2^* \dots x_n^*$$

$$\theta_{\mathsf{S}}(x_1^* x_2^* \dots x_n^*) = x_1^* x_2^* \dots x_n^*$$

$$\theta_\wedge(x_1^* \wedge x_2^* \wedge \dots \wedge x_n^*) = x_n^* x_{n-1}^* \dots x_1 = (-1)^{n(n-1)/2} x_1^* x_2^* \dots x_n^*$$

and hence we deduce from (28), (29) and (30) the relations

$$(28\text{ bis})\qquad \langle\theta_{\mathsf{T}}(x_1^* \otimes x_2^* \otimes \dots \otimes x_n^*), x_1 \otimes x_2 \otimes \dots \otimes x_n\rangle = \prod_{i=1}^{n} \langle x_i^*, x_i\rangle$$

(in other words θ_{T} restricted to $\mathsf{T}^2(\mathrm{M}^*)$ is just the canonical homomorphism of II, § 4, no. 4)

$$(29\text{ bis})\qquad \langle\theta_{\mathsf{S}}(x_1^* x_2^* \dots x_n^*), x_1 x_2 \dots x_n\rangle = \sum_{\sigma \in \mathfrak{S}_n} \left(\prod_{i=1}^{n} \langle x_{\sigma(i)}^*, x_i\rangle\right)$$

$$(30\text{ bis})\qquad \langle\theta_\wedge(x_1^* \wedge x_2^* \wedge \dots \wedge x_n^*), x_1 \wedge x_2 \wedge \dots \wedge x_n\rangle = (-1)^{n(n-1)/2}\det(\langle x_i^*, x_j\rangle).$$

PROPOSITION 7. *Let* M *be a finitely generated projective* A*-module. Then the canonical homomorphisms* $\theta_{\mathsf{T}}:\mathsf{T}(\mathrm{M}^*) \to \mathsf{T}(\mathrm{M})^{*\mathrm{gr}}$ *and* $\theta_\wedge:\bigwedge(\mathrm{M}^*) \to (\bigwedge(\mathrm{M})^{*\mathrm{gr}})^0$ *are bijective. Also the graded dual* $\bigwedge(\mathrm{M})^{*\mathrm{gr}}$ *is then equal to the dual* $\bigwedge(\mathrm{M})^*$ *of the* A*-module* $\bigwedge(\mathrm{M})$.

Suppose first that M has a *finite basis* $(e_i)_{1 \leqslant i \leqslant m}$ and let $(e_i^*)_{1 \leqslant i \leqslant m}$ be the dual basis of M* (II, § 10, no. 4). Formula (28 bis) shows that, for every finite sequence $s = (j_k)_{1 \leqslant k \leqslant n}$ of n elements of the interval $[1, m]$ of **N**,

$$\theta_{\mathsf{T}}(e_{j_n}^* \otimes \dots \otimes e_{j_n}^*)$$

is the element of index s in the basis of $(\mathsf{T}^n(\mathrm{M}))^*$, *dual* to the basis of $\mathsf{T}^n(\mathrm{M})$ consisting of the $e_s = e_{j_1} \otimes \dots \otimes e_{j_n}$ (§ 5, no. 5, Theorem 1). Hence θ_{T} is bijective.

Similarly, formula (30 bis) shows that, for every finite subset H of $[1, m]$ with n elements, $(-1)^{n(n-1)/2}\theta_\wedge(e_{\mathrm{H}}^*)$ (notation of § 7, no. 8, Theorem 1) is the element of index H in the basis of $(\bigwedge^n(\mathrm{M}))^*$, *dual* to the basis of $\bigwedge^n(\mathrm{M})$ consisting of the e_{H}. Hence $\theta_\wedge$ is bijective.

Suppose now only that M is finitely generated and projective; then M is a direct factor of a finitely generated free A-module L, so that there exist two

A-linear mappings $M \xrightarrow{j} L \xrightarrow{p} M$ whose composition is the identity 1_M. We deduce a commutative diagram

$$\begin{array}{ccccc} T(M^*) & \xrightarrow{T({}^t j)} & T(L^*) & \xrightarrow{T({}^t p)} & T(M^*) \\ \downarrow{\scriptstyle \theta_T} & & \downarrow{\scriptstyle \theta_T} & & \downarrow{\scriptstyle \theta_T} \\ T(M)^{*gr} & \xrightarrow[{}^t T(j)]{} & T(L)^{*gr} & \xrightarrow[{}^t T(p)]{} & T(M)^{*gr} \end{array}$$

and an analogous commutative diagram where T is replaced by $\bigwedge$. The proposition then follows from the following lemma:

Lemma 2. Let

$$\begin{array}{ccccc} X & \xrightarrow{u} & Y & \xrightarrow{v} & X \\ \downarrow{\scriptstyle f} & & \downarrow{\scriptstyle g} & & \downarrow{\scriptstyle f} \\ X' & \xrightarrow[u']{} & Y' & \xrightarrow[v']{} & X' \end{array}$$

be a commutative diagram of sets and mappings such that $v \circ u$ and $v' \circ u'$ are the identity mappings of X *and* X' *respectively. Then, if g is injective* (resp. *surjective*, resp. *bijective*), *so is f.*

u is injective since $v \circ u$ is, hence, if g is injective, $u' \circ f = g \circ u$ is injective and therefore f is injective. Similarly v' is surjective since $v' \circ u'$ is; hence, if g is surjective, $f \circ v = v' \circ g$ is surjective and therefore f is surjective.

The last assertion of Proposition 7 follows from the fact that $\bigwedge(M)$ is then a finitely generated A-module (§ 7, no. 3, Proposition 6 and II, § 11, no. 6, *Remark*).

We now examine what can be said concerning the homomorphism θ_S when M is *projective and finitely generated.* Suppose first that M admits a finite basis $(e_i)_{1 \leqslant i \leqslant m}$. In the notation at the beginning of the chapter the A-module $S^n(M)$ admits as basis the family of elements e^α such that $|\alpha| = n$. Let u_α (for $|\alpha| = n$) denote the element of index α in the basis of $(S^n(M))^*$ *dual* to (e^α). The elements u_α, for $\alpha \in \mathbf{N}^m$, therefore form a basis of the algebra $S(M)^{*gr}$ and we shall obtain the multiplication table of this basis explicitly. We write

$$u_\alpha u_\beta = \sum_{\gamma \in \mathbf{N}^m} a_{\alpha\beta\gamma} u_\gamma \quad \text{with } a_{\alpha\beta\gamma} \in A.$$

Then by definition

$$a_{\alpha\beta\gamma} = \langle u_\alpha u_\beta, e^\gamma \rangle = m((u_\alpha \otimes u_\beta)(c(e^\gamma))),$$

where $m: A \otimes A \to A$ defines the multiplication on A and c is the coproduct of S(M). In other words, $a_{\alpha\beta\gamma}$ is just the coefficient of $e^\alpha \otimes e^\beta$ when $c(e^\gamma)$ is written in terms of the basis of $S(M) \otimes S(M)$ consisting of the $e^{\xi} \otimes e^{\eta}$, where ξ and η run through $\mathbf{N}^m$. But since c is an algebra homomorphism,

$$c(e^{\gamma}) = \prod_{i=1}^{m} (c(e_i))^{\gamma_i} = \prod_{i=1}^{m} (e_i \otimes 1 + 1 \otimes e_i)^{\gamma_i}$$

by formula (4) of no. 1; this gives

$$c(e^{\gamma}) = \sum_{\xi+\eta=\gamma} ((\xi, \eta))e^{\xi} \otimes e^{\eta} \tag{31}$$

where we write

$$((\xi, \eta)) = \prod_{i=1}^{n} \frac{(\xi_i + \eta_i)!}{\xi_i!\eta_i!} \quad \text{(cf. § 10, no. 4, formula (18)).} \tag{32}$$

Hence we obtain the multiplication table

$$u_{\alpha}u_{\beta} = ((\alpha, \beta))u_{\alpha+\beta}. \tag{33}$$

On the other hand, if $(e_i^*)_{1 \leqslant i \leqslant m}$ is the basis of M^*, dual to (e_i), it follows from formula (29 bis) that, for all $\alpha \in \mathbf{N}^m$,

$$\theta_S(e^{*\alpha}) = \alpha! u_{\alpha} \tag{34}$$

in the notation of § 6, no. 6. Hence the homomorphism θ_S is bijective if and only if the $\alpha! u_{\alpha}$ form a *basis* of $S(M)^{*gr}$, or also if the elements $\alpha!1$ are *invertible.*

PROPOSITION 8. *Suppose that the ring* A *is an algebra over the field* $\mathbf{Q}$ *of rational numbers; then, for every finitely generated projective* A*-module* M, *the homomorphism*

$$\theta_S : S(M^*) \to S(M)^{*gr}$$

is bijective.

It amounts to proving this when M is finitely generated and free; we pass from this to the general case using Lemma 2 as in the proof of Proposition 7.

Remark. Let M be an A-module and $\rho : A \to B$ a commutative ring homomorphism. Then there is a commutative diagram of graded B-algebra homomorphisms

$$\begin{array}{ccc} T((M^*)_{(B)}) & \longrightarrow & (T(M)^{*gr})_{(B)} \\ {\scriptstyle T(\upsilon_M)}\downarrow & & \downarrow{\scriptstyle \upsilon_{T(M)}} \\ T((M_{(B)})^*) & \xrightarrow[\theta_T]{} & T(M_{(B)})^{*gr} \end{array}$$

where the first row is a homomorphism composed of the homomorphism $\theta_T \otimes 1_B : T(M^*) \otimes_A B \to T(M)^{*gr}$ and the canonical isomorphism

$$T((M^*)_{(B)}) \to T(M^*) \otimes_A B$$

(§ 5, no. 3, Proposition 5). It is immediately verified, using formula (28) and the definition of the homomorphism υ_E (II, § 5, no. 4), that this diagram is *commutative*. When M is a *finitely generated projective* A-module, $M_{(B)}$ is a finitely generated projective B-module (II, § 5, no. 1, Corollary to Proposition 4) and all the homomorphisms of the above diagram are *bijective* (Proposition 7 and II, § 5, no. 4, Proposition 8). There are analogous commutative diagrams with T replaced by S or $\bigwedge$; the diagram for $\bigwedge$ also consists of bijective homomorphisms when M is projective and finitely generated (Proposition 7); if further A is an algebra over $\mathbf{Q}$, the diagram for S also consists of bijective homomorphisms (Proposition 8).

6. INNER PRODUCTS: CASE OF ALGEBRAS

Let $E = \bigoplus_{p \geqslant 0} E_p$ be a *graded* A-*algebra* of type $\mathbf{N}$ and P a graded A-module of type $\mathbf{Z}$; for every *homogeneous* element $x \in E_p$, *left* multiplication by x is an A-linear mapping $e(x)$ of E into itself which is *graded of degree p*. For every element $u \in \mathrm{Homgr}_A(E, P)$, the *right inner product of u by x*, denoted by $u \llcorner x$, is the element $u \circ e(x)$ of $\mathrm{Homgr}_A(E, P)$. We also write $(i(x))(u) = u \llcorner x$ and we see that $i(x)$ is a graded endomorphism of degree p of the graded A-module $\mathrm{Homgr}_A(E, P)$. If now $x = \sum_{p \geqslant 0} x_p$ is an arbitrary element of E (with $x_p \in E_p$ for all $p \geqslant 0$, $x_p = 0$ except for a finite number of values of p), we write $i(x) = \sum_{p=0}^{\infty} i(x_p)$, which is therefore an endomorphism of the A-module $\mathrm{Homgr}_A(E, P)$.

To remember which element, in the expression $u \llcorner x$, "operates" on the other, observe that the element x which "operates" on u is placed at the free end of the horizontal line in $\llcorner$.

The *associativity* of the algebra E goes over to the relation $e(xy) = e(x) \circ e(y)$ for x, y homogeneous; whence, by definition of $i(x)$,

$$i(xy) = i(y) \circ i(x) \tag{35}$$

first for x, y homogeneous and then, by linearity, for *arbitrary* x, y in E; this may also be written

$$(u \llcorner x) \llcorner y = u \llcorner (xy) \tag{36}$$

for x, y in E and $u \in \mathrm{Homgr}_A(E, P)$; as on the other hand clearly $i(1)$ is the identity mapping (since this follows from $e(1) = 1_E$) and $x \mapsto i(x)$ is A-*linear*, it is seen that the external law of composition $(x, u) \mapsto u \llcorner x$ ($x \in E$, $u \in \mathrm{Homgr}_A(E, P)$) defines, with addition, a *right* E-*module* structure on $\mathrm{Homgr}_A(E, P)$.

In particular we consider the case P = A, $\mathrm{Homgr_A}(E, P)$ being in this case the *graded dual* E^{*gr} of E; $i(x)$ is then the *graded transpose* of the A-linear mapping $e(x)$ (II, § 11, no. 6), in other words, for all x, y in E, $u \in E^{*gr}$,

$$\langle u \llcorner x, y\rangle = \langle u, xy\rangle. \tag{37}$$

With the convention at the beginning of the paragraph, note that, if $x \in E_p$, $i(x)$ is an endomorphism of E^{*gr} *of degree* $-p$.

For every homogeneous element $x \in E_p$, *right* multiplication by x is similarly denoted by $e'(x)$ and the element $u \circ e'(x)$ of $\mathrm{Homgr_A}(E, P)$, called the *left inner product of u by x*, by $x \lrcorner u$; we write $i'(x) = x \lrcorner u$ and $i'(x)$ is therefore a graded endomorphism of $\mathrm{Homgr_A}(E, P)$ of degree p; as above this definition can be extended to the case where x is an arbitrary element of E. As in this case $e'(xy) = e'(y)e'(x)$,

$$i'(xy) = i'(x) \circ i'(y) \tag{38}$$

which may also be written

$$x \lrcorner (y \lrcorner u) = (xy) \lrcorner u \tag{39}$$

and shows that the external law of composition $(x, u) \mapsto x \lrcorner u$ defines, with addition, a *left* E*-module* structure on $\mathrm{Homgr_A}(E, P)$. The associativity of E implies on the other hand that $e(x) \circ e'(y) = e'(y) \circ e(x)$ for x, y homogeneous in E, whence the relation

$$(y \lrcorner u) \llcorner x = y \lrcorner (u \llcorner x) \tag{40}$$

so that the two external laws of composition on $\mathrm{Homgr_A}(E, P)$ define on this set an (E, E)*-bimodule* structure (II, § 1, no. 14).

When we take P = A, $i'(x)$ is the graded transpose of $e'(x)$; in other words, for all x, y in E, $u \in E^{*gr}$,

$$\langle y, x \lrcorner u\rangle = \langle yx, u\rangle. \tag{41}$$

When the graded algebra E is *commutative*, obviously $u \llcorner x = x \lrcorner u$. When E is *anticommutative* and P = A, then for $x \in E_p$, $y \in E_r$ and $u \in E^*_q$, $yx = (-1)^{pr}xy$, whence, by (37) and (41), $\langle u \llcorner x, y\rangle = (-1)^{pr}\langle y, x \lrcorner u\rangle$. But as the two sides of this relation are zero except for $r = q - p$,

$$x \lrcorner u = (-1)^{p(q-p)} u \llcorner x.$$

Let F be another graded A-algebra and $\phi: E \to F$ an A-homomorphism of graded algebras; then $\tilde{\phi} = \mathrm{Hom}(\phi, 1_P): \mathrm{Homgr_A}(F, P) \to \mathrm{Homgr_A}(E, P)$ is a graded A-homomorphism of degree 0; by definition, for x, y in E and $u \in \mathrm{Homgr_A}(F, P)$

$$\begin{aligned}(\tilde{\phi}(u \llcorner \phi(x)))(y) &= (u \llcorner \phi(x))(\phi(y)) \\ &= u(\phi(x)\phi(y)) = u(\phi(xy)) = (\tilde{\phi}(u))(xy) = (\tilde{\phi}(u) \llcorner x)(y)\end{aligned}$$

or also

$$\tilde{\phi}(u \llcorner \phi(x)) = \tilde{\phi}(u) \llcorner x \tag{42}$$

and similarly

$$\tilde{\phi}(\phi(x) \lrcorner u) = x \lrcorner \tilde{\phi}(u). \tag{43}$$

In other words, when $\mathrm{Homgr}_A(F, P)$ is considered as an (E, E)-*bimodule* by means of the ring homomorphism $\phi: E \rightarrow F$, it is seen that $\tilde{\phi}$ is an (E, E)-*bimodule homomorphism* (or also an E-*homomorphism* of the (F, F)-bimodule $\mathrm{Homgr}_A(F, P)$ into the (E, E)-bimodule $\mathrm{Homgr}_A(E, P)$).

Examples. In particular the above may be applied when E is one of the graded algebras $\mathsf{T}(M)$, $\mathsf{S}(M)$ or $\bigwedge(M)$ for an A-module M and P an A-module (with the trivial graduation). To find explicitly the bimodule structures thus obtained, note that the elements of degree $-n$ of $\mathrm{Homgr}_A(\mathsf{T}(M), P)$ (resp. $\mathrm{Homgr}_A(\mathsf{S}(M), P)$, resp. $\mathrm{Homgr}_A(\bigwedge(M), P)$) are identified with the *n-linear mappings* (resp. *symmetric n-linear mappings*, resp. *alternating n-linear mappings*) of M^n into P. It suffices to express the products

$$f \llcorner (x_1 \otimes x_2 \otimes \cdots \otimes x_p)$$

(resp. $f \llcorner (x_1 x_2 \ldots x_p)$, resp. $f \llcorner (x_1 \wedge x_2 \wedge \cdots \wedge x_p)$) for every finite sequence $(x_i)_{1 \leqslant i \leqslant p}$ of elements of M and the analogues for the left inner product. It follows immediately from the definitions that

$$f \llcorner (x_1 \otimes x_2 \otimes \cdots \otimes x_p) = (x_1 \otimes x_2 \otimes \cdots \otimes x_p) \lrcorner f = 0 \tag{44}$$

if $p > n$ and that, for $p \leqslant n$, $f \llcorner (x_1 \otimes x_2 \otimes \cdots \otimes x_p)$ (resp.

$$(x_1 \otimes x_2 \otimes \cdots \otimes x_p) \lrcorner f)$$

is the $(n - p)$-*linear mapping* defined by

$$\begin{cases} (f \llcorner (x_1 \otimes x_2 \otimes \cdots \otimes x_p))(y_1, \ldots, y_{n-p}) \\ \qquad\qquad = f(x_1, \ldots, x_p, y_1, \ldots, y_{n-p}) \\ ((x_1 \otimes x_2 \otimes \cdots \otimes x_p) \lrcorner f)(y_1, \ldots, y_{n-p}) \\ \qquad\qquad = f(y_1, \ldots, y_{n-p}, x_1, \ldots, x_p). \end{cases} \tag{45}$$

For $p > n$, there are also in $\mathrm{Homgr}_A(\mathsf{S}(M), P)$ (resp. $\mathrm{Homgr}_A(\bigwedge(M), P)$) formulae (44) with $x_1 \otimes x_2 \otimes \cdots \otimes x_p$ replaced by $x_1 x_2 \ldots x_p$ (resp. $x_1 \wedge x_2 \wedge \cdots \wedge x_p$). For $p \leqslant n$, the same substitutions in (45) define the *symmetric* $(n - p)$-*linear mappings* $f \llcorner (x_1 x_2 \ldots x_p)$ and $(x_1 x_2 \ldots x_p) \lrcorner f$ (resp. the *alternating* $(n - p)$-*linear mappings*

$$f \llcorner (x_1 \wedge x_2 \wedge \cdots \wedge x_p) \quad \text{and} \quad (x_1 \wedge x_2 \wedge \cdots \wedge x_p) \lrcorner f).$$

When $n = p$, the above products are equal to the *constant* function on M equal to $f(x_1, \ldots, x_p)$.

If $u: \mathrm{M} \to \mathrm{N}$ is an A-module homomorphism, $\mathsf{T}(u): \mathsf{T}(\mathrm{M}) \to \mathsf{T}(\mathrm{N})$ is a graded A-algebra homomorphism, then it follows from what we have seen above that $(\mathsf{T}(u))^{\sim}$ is a $\mathsf{T}(\mathrm{M})$-*homomorphism* of the $(\mathsf{T}(\mathrm{N}), \mathsf{T}(\mathrm{N}))$-bimodule $\mathrm{Homgr}_{\mathrm{A}}(\mathsf{T}(\mathrm{N}), \mathrm{P})$ into the $(\mathsf{T}(\mathrm{M}), \mathsf{T}(\mathrm{M}))$-bimodule $\mathrm{Homgr}_{\mathrm{A}}(\mathsf{T}(\mathrm{M}), \mathrm{P})$, relative to the ring homomorphism $\mathsf{T}(u)$. There are analogous results for $(\mathsf{S}(u))^{\sim}$ and $(\bigwedge(u))^{\sim}$.

7. INNER PRODUCTS: CASE OF COGEBRAS

Let $\mathrm{E} = \bigoplus_{p \geqslant 0} \mathrm{E}_p$ be a *coassociative counital graded cogebra*. Then we know (no. 2, Propositions 1 and 3) that the graded dual $\mathrm{E}^{*\mathrm{gr}}$ has (with the convention on graduations made at the beginning of the paragraph) a *graded algebra* structure of type $\mathbf{N}$ over A, the product of two elements u, v of this algebra being defined by $uv = m \circ (u \otimes v) \circ c$, where $c: \mathrm{E} \to \mathrm{E} \otimes_{\mathrm{A}} \mathrm{E}$ is the coproduct and $m: \mathrm{A} \otimes_{\mathrm{A}} \mathrm{A} \to \mathrm{A}$ defines the multiplication. In other words, if, for $x \in \mathrm{E}$, $c(x) = \sum_i y_i \otimes z_i$, we can write (canonically identifying $\mathrm{A} \otimes_{\mathrm{A}} \mathrm{E}$ and E)

$$\begin{aligned}\langle x, uv\rangle = (uv)(x) &= \sum_i u(y_i)v(z_i) = v\Big(\sum_i u(y_i)z_i\Big) \\ &= v(((u \otimes 1_{\mathrm{E}}) \circ c)(x)) = \langle((u \otimes 1_{\mathrm{E}}) \circ c)(x), v\rangle.\end{aligned}$$

This can be interpreted by saying that, for all u homogeneous of degree p in $\mathrm{E}^{*\mathrm{gr}}$, the left multiplication $e(u): v \mapsto uv$ in $\mathrm{E}^{*\mathrm{gr}}$ is the *graded transpose* of the graded endomorphism of degree $-p$

$$i(u) = (u \otimes 1_{\mathrm{E}}) \circ c \tag{46}$$

of E; hence, in the above notation,

$$(i(u))(x) = \sum_i u(y_i)z_i.$$

Formula (46) also defines an element $i(u) \in \mathrm{Endgr}_{\mathrm{A}}(\mathrm{E})$ for every element $u \in \mathrm{E}^{*\mathrm{gr}}$; for all $x \in \mathrm{E}$ and all $u \in \mathrm{E}^{*\mathrm{gr}}$, we write

$$x \llcorner u = (i(u))(x) \tag{47}$$

so that, for u and v in $\mathrm{E}^{*\mathrm{gr}}$,

$$\langle x, uv\rangle = \langle x \llcorner u, v\rangle.$$

The element $x \llcorner u$ of E is called the *right inner product of x by u*.

Here again the element u which "operates" on x is placed at the free end of the horizontal line in $\llcorner$.

For any two elements u, v of E^{*gr},

(48) $$x \llcorner (uv) = (x \llcorner u) \llcorner v,$$

in other words

$$i(uv) = i(v) \circ i(u).$$

As above let $c(x) = \sum_i y_i \otimes z_i$, so that $x \llcorner (uv) = \sum_i (uv)(y_i) z_i$. If

$$c(y_i) = \sum_j y'_{ij} \otimes y''_{ij},$$

then

(49) $$x \llcorner (uv) = \sum_{i,j} u(y'_{ij}) v(y''_{ij}) z_i.$$

On the other hand, if $c(z_i) = \sum_k z'_{ik} \otimes z''_{ik}$, then

(50) $$(x \llcorner u) \llcorner v = \sum_{i,k} u(y_i) v(z'_{ik}) z''_{ik}.$$

Now, the coassociativity of E shows that (no. 2, Proposition 1)

(51) $$\sum_{i,j} y'_{ij} \otimes y''_{ij} \otimes z_i = \sum_{i,k} y_i \otimes z'_{ik} \otimes z''_{ik}$$

and the equality of expressions (49) and (50) follows from the fact that they are respectively the image of the left and right hand sides of (51) under the linear mapping f from $E \otimes E \otimes E$ to E such that $f(x \otimes y \otimes z) = u(x)v(y)z$.

We recall on the other hand (no. 2, Proposition 3) that the unit element of the algebra E^{*gr} is the linear form $e: x \mapsto \gamma(x).1$; hence

$$x \llcorner e = \sum_i \gamma(y_i) z_i = x$$

by virtue of the definition of counit. As the mapping $u \mapsto i(u)$ is linear, it is seen that the external law of composition $(u, x) \mapsto x \llcorner u$ defines a *right* E^{*gr}-*module structure* on E.

Similarly we define, for all $u \in E^{*gr}$, the endomorphism of E

(52) $$i'(u) = (1_E \otimes u) \circ c$$

and, for all $x \in E$, we write

(53) $$(i'(u))(x) = u \lrcorner x$$

and this element of E is called the *left inner product of x by u*. As above it is seen that the external law $(u, v) \mapsto u \lrcorner x$ defines a *left* E^{*gr}*-module structure* on E. Moreover, these two structures are *compatible*, in other words,

$$(54) \qquad (u \lrcorner x) \llcorner v = u \lrcorner (x \llcorner v)$$

for u, v in E^{*gr} (II, § 1, no. 14). With the same notation as above, the left hand side of (54) is $\sum_{i,j} u(z_i)v(y'_{ij})y''_{ij}$ and the right hand side is $\sum_{i,k} v(y_i)u(z''_{ik})z'_{ik}$; their equality follows from the fact that they are the respective images of the left and right hand sides of (51) under the linear mapping g of $E \otimes E \otimes E$ into E such that $g(x \otimes y \otimes z) = v(x)u(z)y$.

It is therefore seen that the two external laws of composition on E defines on this set an (E^{*gr}, E^{*gr})*-bimodule* structure.

When the cogebra E is *cocommutative*, then $u \lrcorner x = x \llcorner u$ for all $x \in E$ and $u \in E^{*gr}$; when it is *anticocommutative* (§ 4, no. 9) and $u \in E_p^*$ and $x \in E_q$, we can write $c(x) = \sum_{0 \leqslant j \leqslant q} \left(\sum_i y_{ij} \otimes z_{i,q-j}\right)$ with y_{ij} and z_{ij} in E_j for all j and then by hypothesis

$$\sum_i z_{ij} \otimes y_{i,q-j} = (-1)^{j(q-j)} \sum_i y_{ij} \otimes z_{i,q-j}.$$

By definition, $x \llcorner u = \sum_{0 \leqslant j \leqslant q} \left(\sum_i u(y_{ij}) z_{i,q-j}\right)$ and

$$u \lrcorner x = \sum_{0 \leqslant j \leqslant q} \left(\sum_i u(z_{i,q-j}) y_{ij}\right).$$

As $u(y_{ij}) = 0$ (resp. $u(z_{i,q-j}) = 0$ unless $j = p$ (resp. $q - j = p$), it is seen by the above that $u \lrcorner x = (-1)^{p(q-p)} x \llcorner u$.

Finally, let $\phi \colon E \to F$ be a *graded cogebra morphism*; then it has been seen (no. 2) that the graded transpose ${}^t\phi \colon F^{*gr} \to E^{*gr}$ is a *graded algebra homomorphism*; therefore, for $x \in E$, u, v in F^{*gr},

$$\begin{aligned}\langle \phi(x \llcorner {}^t\phi(u)), v\rangle &= \langle x \llcorner {}^t\phi(u), {}^t\phi(v)\rangle = \langle x, {}^t\phi(u){}^t\phi(v)\rangle = \langle x, {}^t\phi(uv)\rangle \\ &= \langle \phi(x), uv\rangle = \langle \phi(x) \llcorner u, v\rangle\end{aligned}$$

whence

$$(55) \qquad \phi(x) \llcorner u = \phi(x \llcorner {}^t\phi(u));$$

and similarly

$$(56) \qquad u \lrcorner \phi(x) = \phi({}^t\phi(u) \lrcorner x).$$

In other words, ϕ is an F^{*gr}*-homomorphism* of the (E^{*gr}, E^{*gr})-bimodule E into the (F^{*gr}, F^{*gr})-bimodule F, relative to the ring homomorphism

$${}^t\phi \colon F^{*gr} \to E^{*gr}.$$

Examples. In particular the above can be applied when E is one of the graded cogebras $\mathsf{T}(M)$, $\mathsf{S}(M)$ or $\bigwedge(M)$ for an A-module M (no. 1, *Examples* 5, 6 and 7). To find explicitly the bimodule structures thus obtained, we again identify a homogeneous element f of degree n in $\mathsf{T}(M)^{*gr}$ (resp. $\mathsf{S}(M)^{*gr}$, resp. $\bigwedge(M)^{*gr}$) with an *n-linear form* (resp. *symmetric n-linear form*, resp. *alternating n-linear form*, also called an *n-form*) on M^n. It suffices to express the products

$$(x_1 \otimes x_2 \otimes \cdots \otimes x_p) \llcorner f \text{ (resp. } (x_1 x_2 \ldots x_p) \llcorner f,$$

resp. $(x_1 \wedge x_2 \wedge \cdots \wedge x_p) \llcorner f$) for every finite sequence $(x_i)_{1 \leqslant i \leqslant p}$ of elements of M and the analogues for the left inner product. Now, definitions (46) and (52) and formulae (3), (6) and (9) of no. 1 give respectively:

$$(57) \quad \begin{cases} (x_1 \otimes x_2 \otimes \cdots \otimes x_p) \llcorner f = f \lrcorner (x_1 \otimes x_2 \otimes \cdots \otimes x_p) = 0 \\ (x_1 x_2 \ldots x_p) \llcorner f = f \lrcorner (x_1 x_2 \ldots x_p) = 0 \quad \text{for} \quad p < n. \\ (x_1 \wedge x_2 \wedge \cdots \wedge x_p) \llcorner f = f \lrcorner (x_1 \wedge x_2 \wedge \cdots \wedge x_p) = 0 \end{cases}$$

For $p \geqslant n$, we have respectively

$$(58) \quad (x_1 \otimes x_2 \otimes \cdots \otimes x_p) \llcorner f = f(x_1, \ldots, x_n) x_{n+1} \otimes \cdots \otimes x_p$$

$$(59) \quad (x_1 x_2 \ldots x_p) \llcorner f = \sum_\sigma f(x_{\sigma(1)}, \ldots, x_{\sigma(n)}) x_{\sigma(n+1)} \ldots x_{\sigma(p)}$$

$$(60) \quad (x_1 \wedge x_2 \wedge \cdots \wedge x_p) \llcorner f = \sum_\sigma \varepsilon_\sigma f(x_{\sigma(1)}, \ldots, x_{\sigma(n)}) x_{\sigma(n+1)} \wedge \cdots \wedge x_{\sigma(p)}$$

(where, in (59) and (60), the summations are taken over the permutations $\sigma \in \mathfrak{S}_p$ which are *increasing on each of the intervals* $[1, n]$ *and* $[n+1, p]$ of $\mathbf{N}$); and similarly

$$(61) \quad f \lrcorner (x_1 \otimes x_2 \otimes \cdots \otimes x_p) = f(x_{p-n+1}, \ldots, x_p) x_1 \otimes x_2 \otimes \cdots \otimes x_{p-n}$$

$$(62) \quad f \lrcorner (x_1 x_2 \ldots x_p) = \sum_\sigma f(x_{\sigma(p-n+1)}, \ldots, x_{\sigma(p)}) x_{\sigma(1)} \ldots x_{\sigma(p-n)}$$

$$(63) \quad f \lrcorner (x_1 \wedge x_2 \wedge \cdots \wedge x_p) = \sum_\sigma \varepsilon_\sigma f(x_{\sigma(p-n+1)}, \ldots, x_{\sigma(p)}) x_{\sigma(1)} \wedge \cdots \wedge x_{\sigma(p-n)}$$

(where, in (62) and (63), the summations are taken over the permutations $\sigma \in \mathfrak{S}_p$ which are *increasing on each of the intervals* $[1, p-n]$ *and* $[p-n+1, p]$ of $\mathbf{N}$).

8. INNER PRODUCTS: CASE OF BIGEBRAS

Let E be a graded bigebra (resp. skew graded bigebra) (no. 4, Definition 3); then the results of nos. 6 and 7 can be applied to define the right (resp. left) inner products $x \llcorner u \in E$ and $u \llcorner x \in E^{*gr}$ (resp. $u \lrcorner x \in E$ and $x \lrcorner u \in E^{*gr}$)

for all $x \in E$ and all $u \in E^{*gr}$. Thus an (E, E)-bimodule structure and an (E^{*gr}, E^{*gr})-bimodule structure are obtained on E. Further:

PROPOSITION 9. *Let* E *be a graded bigebra* (resp. *skew graded bigebra*). *For every element* x *of degree* 1 *in* E, *the left and right inner products by* x *are derivations* (resp. *antiderivations*) (§ 10, no. 2) *of the algebra* E^{*gr}.

In the notation of no. 6, for every homogeneous element x *of degree* 1 in a graded bigebra (resp. a skew graded bigebra) E,

$$c(x) = x \otimes 1 + 1 \otimes x,$$

by Proposition 5 of no. 3 and the fact that c is a homomorphism of degree 0. Suppose first that E is a graded bigebra. For all $y \in E$, by definition

$$\langle (uv) \llcorner x, y\rangle = \langle uv, xy\rangle = m((u \otimes v)(c(xy)))$$

and since c is an *algebra* homomorphism, $c(xy) = c(x)c(y)$. Let $c(y) = \sum_i s_i \otimes t_i$ with s_i and t_i in E; therefore

$$c(xy) = \sum_i (xs_i) \otimes t_i + \sum_i s_i \otimes (xt_i).$$

Hence $\langle (uv) \llcorner x, y\rangle = \sum_i u(xs_i)v(t_i) + \sum_i u(s_i)v(xt_i)$. But we may write

$$\sum_i u(xs_i)v(t_i) = m(((u \llcorner x) \otimes v)(c(y))) = \langle (u \llcorner x)v, y\rangle$$

and similarly

$$\sum_i u(s_i)v(xt_i) = m((u \otimes (v \llcorner x))(c(y))) = \langle u(v \llcorner x), y\rangle,$$

whence, returning to the notation $i(x)$ for the inner product.

$$(i(x))(uv) = ((i(x))(u))v + u((i(x))(v)) \tag{64}$$

which proves that $i(x)$ is a *derivation* in E^{*gr}.

Suppose now that E is a *skew* graded bigebra, that $u \in E_p^*$, $v \in E_q^*$ and $y \in E_r$; then we can write

$$c(y) = \sum_{0 \leqslant j \leqslant r} \Big(\sum_i s_{ij} \otimes t_{i,r-j}\Big)$$

where the s_{ij} and t_{ij} belong to E_j; by definition of the product in $E \,{}^g\!\otimes_A E$, then

$$c(xy) = c(x)c(y) = \sum_{0 \leqslant j \leqslant r} \Big(\sum_i (xs_{ij}) \otimes t_{i,r-j} + (-1)^j \sum_i s_{ij} \otimes (xt_{i,r-j})\Big)$$

whence this time

$$\langle (uv) \llcorner x, y\rangle = \sum_{0 \leqslant j \leqslant r} \Big(\sum_i u(xs_{ij})v(t_{i,r-j}) + (-1)^j \sum_i u(s_{ij})v(xt_{i,r-j})\Big).$$

Then also $\sum_{0\leqslant j\leqslant r}\left(\sum_i u(xs_{ij})v(t_{i,r-j})\right) = \langle(u \llcorner x)v, y\rangle$. On the other hand, $u(s_{ij}) = 0$ unless $j = -p$ and hence we may also write

$$\sum_{0\leqslant j\leqslant r}(-1)^j\left(\sum_i u(s_{ij})v(xt_{i,r-j})\right) = (-1)^p\langle u(v \llcorner x), y\rangle.$$

We therefore conclude that

$$(i(x))(uv) = ((i(x))(u))v + (-1)^p u((i(x))(v)), \tag{65}$$

in other words $i(x)$ is an *antiderivation* in E^{*gr}. The assertions relating to the left inner product by an element x of degree 1 in E are proved similarly.

Remarks. (1) Let E be a graded bigebra over A and N, N′, N″ three graded A-modules. Let m be an A-bilinear mapping of $N \times N'$ into N''; for $u \in \mathrm{Homgr}_A(E, N)$ and $v \in \mathrm{Homgr}_A(E, N')$, let $u.v$ denote the graded homomorphism $m \circ (u \otimes v) \circ c$ of E into N″. On the other hand, let $i(x)$ denote the (right or left) inner product by $x \in E$ in the A-modules $\mathrm{Homgr}_A(E, N)$, $\mathrm{Homgr}_A(E, N')$ and $\mathrm{Homgr}_A(E, N'')$. Then, if x is *of degree* 1,

$$(i(x))(u.v) = ((i(x))(u)).v + u.((i(x))(v))$$

for all $u \in \mathrm{Homgr}_A(E, N)$ and $v \in \mathrm{Homgr}_A(E, N')$.

Under the same conditions, if E is a skew graded bigebra and u is homogeneous of degree p, then

$$(i(x))(u.v) = ((i(x))(u)).v + (-1)^p u.((i(x))(v)).$$

The proofs are the same as in Proposition 9.

(2) The same argument as in the above proof proves, more generally, that for all $x \in E$, if $c(x) = \sum_j x'_j \otimes x''_j$, then, for all u, v in E^{*gr}, the "Leibniz formula"

$$(i(x))(uv) = \sum_j (i(x'_j))(u).(i(x''_j))(v)$$

holds. In particular, for every *primitive* element of a graded bigebra E, that is such that $c(x) = x \otimes 1 + 1 \otimes x$, $i(x)$ is a *derivation* of E^{*gr}.

PROPOSITION 10. *Let* E *be a graded bigebra* (resp. *skew graded bigebra*). *For every element f of degree* 1 *in* E^{*gr}, *the left and right inner products are derivations* (resp. *antiderivations*) *of the algebra* E.

Let $x \in E_p$, $y \in E_p$ ($p \geqslant 1$, $q \geqslant 1$). By Proposition 5 of no. 3, we can write

$$c(x) = x \otimes 1 + \sum_{1\leqslant j\leqslant p-1}\left(\sum_i x'_{ij} \otimes x''_{ij,p-j}\right) + 1 \otimes x$$

$$c(y) = y \otimes 1 + \sum_{1\leqslant k\leqslant q-1}\left(\sum_i y'_{ik} \otimes y''_{i,q-k}\right) + 1 \otimes y$$

where x'_{ij} and x''_{ij} belong to E_j, y'_{ik} and y''_{ik} to E_k. If E is a graded bigebra, the component of $c(xy) = c(x)c(y)$ belonging to $E_1 \otimes E$, is equal to

$$\sum_i x'_{i,1} \otimes x''_{i,p-1}y + \sum_i y'_{i,1} \otimes xy''_{i,q-1}$$

and hence by definition

$$(xy) \llcorner f = \sum_i f(x'_{i,1})x''_{i,p-1}y + \sum_i f(y'_{i,1})xy''_{i,q-1}$$
$$= (x \llcorner f)y + x(y \llcorner f)$$

and the right inner product by f is a *derivation*. If on the other hand E is a skew graded bigebra, the component of $c(xy)$ belonging to $E_1 \otimes E$ is equal to

$$\sum_i x'_{i,1} \otimes x''_{i,p-1}y + (-1)^p \sum_i y'_{i,1} \otimes xy''_{i,q-1}$$

and this time we obtain

$$(xy) \llcorner f = (x \llcorner f)y + (-1)^p x(y \llcorner f)$$

which shows that $i(f)$ is then an *antiderivation*. The argument is similar for the left inner product by f.

Examples. Propositions 9 and 10 apply in particular to the graded bigebra $\mathsf{S}(M)$ and the skew graded bigebra $\bigwedge(M)$. The inner products by elements of degree 1 in $\mathsf{S}(M)$ (resp. $\mathsf{S}(M)^{*gr}$) are *derivations* which *commute with one another*, since $\mathsf{S}(M)$ (resp. $\mathsf{S}(M)^{*gr}$) is commutative.

Similarly, the inner products by elements of degree 1 in $\bigwedge(M)$ (resp. $\bigwedge(M)^{*gr}$) are *antiderivations*, which are of *zero square*, for the square of an element of degree 1 in the algebra $\bigwedge(M)$ (resp. $\bigwedge(M)^{*gr}$) is zero.

9. INNER PRODUCTS BETWEEN T(M) AND T(M*), S(M) AND S(M*), $\bigwedge$(M) AND $\bigwedge$(M*)

The right inner product defines on $\mathsf{T}(M)$ (resp. $\mathsf{S}(M)$, resp. $\bigwedge(M)$) a right module structure over the algebra $\mathsf{T}(M)^{*gr}$ (resp. $\mathsf{S}(M)^{*gr}$, resp. $\bigwedge(M)^{*gr}$) (no. 7, *Examples*). Using the canonical homomorphisms θ_{T} (resp. θ_{S}, resp. $\theta_{\wedge}$) of no. 5, we derive

a right $\mathsf{T}(M^*)$-module structure on $\mathsf{T}(M)$

a right $\mathsf{S}(M^*)$-module structure on $\mathsf{S}(M)$

a left $\bigwedge(M^*)$-module structure on $\bigwedge(M)$.

The external law of any of these structures is also denoted by

$$(z^*, t) \mapsto i(z^*).t$$

(by an abuse of language); we also write $t \llcorner z^*$ instead of $i(z^*).t$ in the case of $\mathsf{T}(\mathrm{M})$ or $\mathsf{S}(\mathrm{M})$; on the other hand, we write $z^* \lrcorner t$ in the case of $\bigwedge(\mathrm{M})$ and say that this is a *left* inner product of t by z^*, since then we have a *left* $\bigwedge(\mathrm{M}^*)$-module law. For z^* homogeneous of degree n and t homogeneous of degree p, $i(z^*).t = 0$ if $p < n$ and, for $x_i \in \mathrm{M}$ $(1 \leqslant i \leqslant p)$, $x_j^* \in \mathrm{M}^*$ $(1 \leqslant j \leqslant n)$ and $p \geqslant n$, by virtue of formulae (58), (59) and (60) of no. 7,

$$(66)\quad i(x_1^* \otimes x_2^* \otimes \cdots \otimes x_n^*).(x_1 \otimes x_2 \otimes \cdots \otimes x_p) = \Big(\prod_{j=1}^{n} \langle x_j^*, x_j\rangle\Big) x_{n+1} \otimes \cdots \otimes x_p$$

$$(67)\quad i(x_1^* x_2^* \ldots x_n^*).(x_1 x_2 \ldots x_p) = \sum_{\sigma} \Big(\prod_{j=1}^{n} \langle x_j^*, x_{\sigma(j)}\rangle\Big) x_{\sigma(n+1)} \ldots x_{\sigma(p)}$$

$$(68)\quad i(x_1^* \wedge x_2^* \wedge \cdots \wedge x_n^*).(x_1 \wedge x_2 \wedge \cdots \wedge x_p) = (-1)^{n(n-1)/2} \sum_{\sigma} \varepsilon_\sigma \Big(\prod_{j=1}^{n} \langle x_j^*, x_{\sigma(j)}\rangle\Big) x_{\sigma(n+1)} \wedge \cdots \wedge x_{\sigma(p)}$$

where, the formulae (67) and (68), σ runs through the set of permutations $\sigma \in \mathfrak{S}_p$ which are *increasing* on the intervals $[1, n]$ and $[n+1, p]$.

We can also write, in the inner product notation,

$$\langle t \llcorner u^*, v^*\rangle = \langle t, \theta_{\mathsf{T}}(u^* v^*)\rangle \quad \text{for} \quad t \in \mathsf{T}(\mathrm{M}), u^*, v^* \quad \text{in} \quad \mathsf{T}(\mathrm{M}^*)$$

$$\langle t \llcorner u^*, v^*\rangle = \langle t, \theta_{\mathsf{S}}(u^* v^*)\rangle \quad \text{for} \quad t \in \mathsf{S}(\mathrm{M}), u^*, v^* \quad \text{in} \quad \mathsf{S}(\mathrm{M}^*)$$

$$\langle v^*, u^* \lrcorner t\rangle = \langle \theta_\wedge(u^* \wedge v^*), t\rangle \quad \text{for} \quad t \in \textstyle\bigwedge(\mathrm{M}), u^*, v^* \quad \text{in} \quad \textstyle\bigwedge(\mathrm{M}^*).$$

We leave to the reader the task of finding explicitly the analogous formulae for left inner products, this time using formulae (61), (62) and (63).

The above can be applied with M replaced by its dual M^*; M^* must then be replaced by the bidual M^{**} and $\mathsf{T}(\mathrm{M}^*)$, for example, thus has a right module structure over the algebra $\mathsf{T}(\mathrm{M}^{**})$. But the canonical mapping $c_{\mathrm{M}}: \mathrm{M} \to \mathrm{M}^{**}$ defines an algebra homomorphism $\mathsf{T}(c_{\mathrm{M}}): \mathsf{T}(\mathrm{M}) \to \mathsf{T}(\mathrm{M}^{**})$, by means of which $\mathsf{T}(\mathrm{M}^*)$ has a *right* $\mathsf{T}(\mathrm{M})$*-module* structure. Similarly $\mathsf{S}(\mathrm{M}^*)$ (resp. $\bigwedge(\mathrm{M}^*)$) has a *right* $\mathsf{S}(\mathrm{M})$*-module* (resp. *left* $\bigwedge(\mathrm{M})$*-module*) structure. The explicit formulae giving the external laws of these modules are derived immediately from the above by exchanging the roles of M and M^*. Note that, for all $x \in \mathrm{M}$, $i(x)$ is always a *derivation* (resp. *antiderivation of zero square*) of the graded algebra $\mathsf{S}(\mathrm{M}^*)$ (resp. $\bigwedge(\mathrm{M}^*)$).

PROPOSITION 11. *The canonical homomorphism* $\theta_{\mathsf{T}}: \mathsf{T}(\mathrm{M}^*) \to \mathsf{T}(\mathrm{M})^{*\mathrm{gr}}$ (resp. $\theta_{\mathsf{S}}: \mathsf{S}(\mathrm{M}) \to \mathsf{S}(\mathrm{M})^{*\mathrm{gr}}$, resp. $\theta_\wedge: \bigwedge(\mathrm{M}^*) \to \bigwedge(\mathrm{M})^{*\mathrm{gr}}$) *is a right* $\mathsf{T}(\mathrm{M})$*-module* (resp. *right* $\mathsf{S}(\mathrm{M})$*-module*, resp. *left* $\bigwedge(\mathrm{M})$*-module*) *homomorphism.*

We show first that, for $z^* \in \mathsf{T}(\mathrm{M}^*)$ and $t \in \mathsf{T}(\mathrm{M})$,

$$\theta_{\mathsf{T}}(z^* \llcorner t) = \theta_{\mathsf{T}}(z^*) \llcorner t. \tag{69}$$

Since M is a generating system of the algebra $\mathsf{T}(\mathrm{M})$, we need only prove (69) when $t = x \in \mathrm{M}$; moreover we can restrict our attention to the case where $z^* = x_1^* \otimes x_2^* \otimes \cdots \otimes x_p^*$, where the $x_j^* \in \mathrm{M}^*$, and then, by (66) with the roles of M and M* interchanged, $z^* \llcorner x = \langle x, x_1^* \rangle x_2^* \otimes \cdots \otimes x_p^*$. Therefore, for all $y_2, \ldots, y_p$ in M,

$$\begin{aligned}\langle \theta_{\mathsf{T}}(z^* \llcorner x), y_2 \otimes y_3 \otimes \cdots \otimes y_p \rangle &= \langle x, x_1^* \rangle \prod_{j=2}^{p} \langle y_j, x_j^* \rangle \\ &= \langle \theta_{\mathsf{T}}(z^*), x \otimes y_2 \otimes \cdots \otimes y_p \rangle \\ &= \langle \theta_{\mathsf{T}}(z^*) \llcorner x, y_2 \otimes \cdots \otimes y_p \rangle\end{aligned}$$

whence (69).

We prove secondly that for $z^* \in \mathsf{S}(\mathrm{M}^*)$ and $t \in \mathsf{S}(\mathrm{M})$,

$$\theta_{\mathsf{S}}(z^* \llcorner t) = \theta_{\mathsf{S}}(z^*) \llcorner t. \tag{70}$$

As above, we can limit ourselves to the case $t = x \in \mathrm{M}$. But further, here $i(x)$ is a *derivation* of $\mathsf{S}(\mathrm{M}^*)$ and a *derivation* of $\mathsf{S}(\mathrm{M})^{*\mathrm{gr}}$. Therefore (§ 10, no. 7, Corollary to Proposition 9) it suffices to verify (70) for $z^* = x^* \in \mathrm{M}^*$, since M* is a generating system of $\mathsf{S}(\mathrm{M}^*)$; but this is trivial, the two sides then being equal to $\langle x^*, x \rangle$. A similar argument proves the relation

$$\theta_{\wedge}(t \lrcorner z^*) = t \lrcorner \theta_{\wedge}(z^*) \tag{71}$$

for $z^* \in \bigwedge(\mathrm{M}^*)$ and $t \in \bigwedge(\mathrm{M})$: observe then that, for $x \in \mathrm{M}$, $i(x)$ is an *antiderivation* in $\bigwedge(\mathrm{M}^*)$ as well as in $\bigwedge(\mathrm{M})^{*\mathrm{gr}}$ and use § 10, no. 7, Corollary to Proposition 9. There is an analogous result for left inner products.

10. EXPLICIT FORM OF INNER PRODUCTS IN THE CASE OF A FINITELY GENERATED FREE MODULE

Let M be a finitely generated free A-module, $(e_i)_{1 \leqslant i \leqslant n}$ a basis of M and $(e_i^*)_{1 \leqslant i \leqslant n}$ the dual basis of M*. For every finite sequence $s = (i_1, \ldots, i_p)$ of elements of $[1, n]$, let $e_s = e_{i_1} \otimes e_{i_2} \otimes \cdots \otimes e_{i_p}$ (resp. $e_s^* = e_{i_1}^* \otimes \cdots \otimes e_{i_p}^*$). We know (§ 5, no. 5, Theorem 1) that the e_s form a *basis* of the A-module $\mathsf{T}(\mathrm{M})$ and the e_s^* a *basis* of the A-module $\mathsf{T}(\mathrm{M}^*)$. If s, t are two finite sequences of elements of $[1, n]$, let $s.t$ denote the sequence obtained as follows: if $s = (i_1, \ldots, i_p)$ and $t = (j_1, \ldots, j_q)$, $s.t$ is the sequence $(i_1, \ldots, i_p, j_1, \ldots, j_q)$ with $p + q$ terms. Then $e_{s.t} = e_s \otimes e_t$. It then follows from (66) that

$$\begin{cases} e_s \llcorner e_t^* = 0 & \text{if } s \text{ is not of the form } t.u \\ e_{t.u} \llcorner e_t^* = e_u. \end{cases} \tag{72}$$

Similarly, the symmetric algebra $S(M)$ has as basis the set of monomials e^{α} with $\alpha \in \mathbf{N}^n$ (§ 6, no. 6, Theorem 1) and $S(M^*)$ the set of monomials $e^{*\alpha}$ with $\alpha \in \mathbf{N}^n$; recall (no. 5) that u_α, for $|\alpha| = k$, denotes the element of the basis of $(S^k(M))^*$, dual to the basis $(e^{\alpha})_{|\alpha|=k}$ of $S^k(M)$; the u_α, for $\alpha \in \mathbf{N}^n$, therefore form a basis of $S(M)^{*gr}$. The definition of right inner product by e^{β} in $S(M)^{*gr}$ as the transpose of multiplication by e^{β} in $S(M)$ then shows that

$$\text{(73)} \qquad \begin{cases} u_\alpha \llcorner e^{\beta} = 0 & \text{if } \alpha \not\geqslant \beta \\ u_\alpha \llcorner e^{\beta} = u_{\alpha-\beta} & \text{if } \alpha \geqslant \beta. \end{cases}$$

Similarly, since $S(M)$ is here canonically identified with the graded dual of $S(M)^{*gr}$, $i(u_\beta)$ is the graded transpose of multiplication by u^{β} in $S(M)^{*gr}$ and hence from the multiplication table (33) (no. 5) of the basis (u_α) we deduce that

$$\text{(74)} \qquad \begin{cases} e^{\alpha} \llcorner u_\beta = 0 & \text{if } \alpha \not\geqslant \beta \\ e^{\alpha} \llcorner u_\beta = (\beta, \alpha - \beta) e^{\alpha-\beta} & \text{if } \alpha \geqslant \beta. \end{cases}$$

As for the right inner product of an element of $S(M)$ by an element of $S(M^*)$, the definition of this product (no. 9) and formula (34) of no. 5 allow us to deduce from (74) the formulae

$$\text{(75)} \qquad \begin{cases} e^{\alpha} \llcorner e^{*\beta} = 0 & \text{if } \alpha \not\geqslant \beta \\ e^{\alpha} \llcorner e^{*\beta} = \dfrac{\alpha!}{(\alpha-\beta)!} e^{\alpha-\beta} & \text{if } \alpha \geqslant \beta. \end{cases}$$

There are analogous formulae for the inner product of an element of $S(M^*)$ by an element of $S(M)$ interchanging the roles of M and M^* (since M^{**} is here identified with M).

Remark. Being given the basis $(e_i)_{1 \leqslant i \leqslant n}$ allows us to identify the algebra $S(M)$ with the polynomial algebra $A[X_1, \ldots, X_n]$ (§ 6, no. 6); formula (75) shows that the inner product by $e^{*\alpha}$ is just the differential operator $D^{\alpha} = D_1^{\alpha_1} D_2^{\alpha_2} \ldots D_n^{\alpha_n}$, where $D_i = \partial/\partial X_i$ for $1 \leqslant i \leqslant n$ (§ 10, no. 11, *Example*).

Consider finally the exterior algebra $\bigwedge(M)$, which has as basis the set of elements e_J, where J runs through the set of subsets of the interval $(1, n)$ of $\mathbf{N}$ (§ 7, no. 8, Theorem 1); similarly $\bigwedge(M^*)$ has as basis the elements e_J^*. It follows from formula (68) of no. 9 that

$$\text{(76)} \qquad \begin{cases} e_K^* \lrcorner e_J = 0 & \text{if } K \not\subset J \\ e_K^* \lrcorner e_J = (-1)^{p(p-1)/2} \rho_{K, J-K} e_{J-K} & \text{if } K \subset J \text{ and } p = \operatorname{Card}(K), \end{cases}$$

where $\rho_{K, J-K}$ is the number defined by formula (19) of § 7, no. 8. There are analogous formulae with the roles of M and M^* interchanged.

11. ISOMORPHISMS BETWEEN $\bigwedge^p(M)$ AND $\bigwedge^{n-p}(M^*)$ FOR AN n-DIMENSIONAL FREE MODULE M

PROPOSITION 12. *Let* M *be a free* A*-module of dimension* n; *let* $e \in \bigwedge^n(M)$ *be an element forming a basis of* $\bigwedge^n(M)$ *and let* e^* *be the element of* $\bigwedge^n(M^*)$ *such that* $\{(-1)^{n(n-1)/2}\theta_\wedge(e^*)\}$ *is the dual basis of* $\{e\}$ *in* $(\bigwedge^n(M))^*$. *Let* $\phi : \bigwedge(M^*) \to \bigwedge(M)$ *be the mapping* $z \mapsto z \lrcorner e^*$ *and* $\phi' : \bigwedge(M^*) \to \bigwedge(M)$ *the mapping* $z^* \mapsto z^* \lrcorner e$. *Let* ϕ_p (resp. ϕ'_p) *be the restriction of* ϕ (resp. ϕ') *to* $\bigwedge^p(M)$ (resp. $\bigwedge^p(M^*)$). *Then:*

(i) *The mapping* ϕ *is a left* $\bigwedge(M)$*-module isomorphism and the mapping* ϕ' *is a left* $\bigwedge(M^*)$*-module isomorphism; moreover the mappings* ϕ *and* ϕ' *are inverses of one another.*

(ii) *The mapping* ϕ_p *is an isomorphism of the* A*-module* $\bigwedge^p(M)$ *onto the* A*-module* $\bigwedge^{n-p}(M^*)$ *and the mapping* ϕ'_p *is an isomorphism of the* A*-module* $\bigwedge^p(M^*)$ *onto the* A*-module* $\bigwedge^{n-p}(M)$.

(iii) *If we write* $B(u, v^*) = \langle u, \theta_\wedge(v^*)\rangle$ *for* $u \in \bigwedge(M)$ *and* $v^* \in \bigwedge(M^*)$ *then, for* $u^* \in \bigwedge^p(M^*)$ *and* $v^* \in \bigwedge^{n-p}(M^*)$,

$$B(\phi'_p(u^*), v^*) = (-1)^{p(n-p)}B(u^*, \phi'_{n-p}(v^*)). \tag{77}$$

The fact that ϕ is $\bigwedge(M)$-linear and ϕ' is $\bigwedge(M^*)$-linear follows from the formulae $(u \wedge v) \lrcorner e^* = u \lrcorner (v \lrcorner e^*)$ and $(u^* \wedge v^*) \lrcorner e = u^* \lrcorner (v^* \lrcorner e)$ (no. 6, formula (37), using the fact that $\theta_\wedge$ is an isomorphism of $\bigwedge(M^*)$ onto the opposite algebra to $\bigwedge(M)^*$). On the other hand there exists a basis $(e_i)_{1 \leqslant i \leqslant n}$ of M such that

$$e = e_1 \wedge e_2 \wedge \cdots \wedge e_n \quad \text{and} \quad e^* = (-1)^{n(n-1)/2}e_1^* \wedge e_2^* \wedge \cdots \wedge e_n^*,$$

where (e_i^*) is the basis dual to (e_i). We write $I = (1, n)$; it follows from (76) that, for every subset J of I with p elements,

$$\begin{cases} \phi(e_J) = (-1)^{n(n-1)/2+p(p-1)/2}\rho_{J,I-J}e^*_{I-J} \\ \phi'(e^*_J) = (-1)^{p(p-1)/2}\rho_{J,I-J}e_{I-J}. \end{cases} \tag{78}$$

This proves that ϕ and ϕ' are bijective; moreover $\rho_{J,I-J}\rho_{I-J,J} = (-1)^{p(n-p)}$ (§ 7, no. 8, formula (21)); as the number

$$\frac{n(n-1)}{2} + \frac{p(p-1)}{2} + \frac{(n-p)(n-p-1)}{2} + p(n-p) = n(n-1)$$

is *even*, it follows that ϕ and ϕ' are inverses of one another. Finally, to prove (77), it suffices to take $u^* = e^*_J$ and $v^* = e^*_{I-J}$; the verification also follows from the definition of $\theta_\wedge$, formulae (78) and the relation $\rho_{J,I-J}\rho_{I-J,J} = (-1)^{p(n-p)}$ (§ 7, no. 8, formula (21)). Note that, for $u^* \in \bigwedge^p(M^*)$ and $v^* \in \bigwedge^{n-p}(M^*)$,

$B(\phi'_p(u^*), v^*)$ is, to within a sign, the coefficient of $u^* \wedge v^*$ with respect to the basis $\{e^*\}$ of $\bigwedge^n(M^*)$.

PROPOSITION 13. *With the hypotheses and notation of Proposition* 11, *for every endomorphism g of the* A-*module* M.

$$(\det g)\phi = \bigwedge({}^t g) \circ \phi \circ \bigwedge(g). \tag{79}$$

Clearly $\bigwedge({}^t g) = \theta_\wedge^{-1} \circ ({}^t\bigwedge(g)) \circ \theta_\wedge$; since $\bigwedge(g)$ is an endomorphism of the *algebra* $\bigwedge(M)$ and by definition, for all

$$z \in \bigwedge(M), \qquad \theta_\wedge(\bigwedge(g)(z) \lrcorner e^*) = \theta_\wedge(e^*) \llcorner \theta_\wedge(\bigwedge(g)(z)),$$

we deduce from formula (42) of no. 6 that

$$\begin{aligned}((\theta_\wedge^{-1} \circ ({}^t\bigwedge(g)) \circ \theta_\wedge) \circ \phi \circ \bigwedge(g))(z) &= \theta_\wedge^{-1}({}^t\bigwedge(g)(\theta_\wedge(e^*)) \llcorner z) \\ &= z \lrcorner (\bigwedge({}^t g)(e^*)) = (\det g)(z \lrcorner e^*) = (\det g)\phi(z)\end{aligned}$$

taking account of § 8, no. 4, Proposition 8.

COROLLARY. *For every automorphism g of* E,

$$\bigwedge({}^t g^{-1}) = (\det g)^{-1}\phi \circ (\bigwedge(g)) \circ \phi^{-1}. \tag{80}$$

12. APPLICATION TO THE SUBSPACE ASSOCIATED WITH A p-VECTOR

Let K be a field and E a vector space over K. Recall that with every p-vector $z \in \bigwedge^p(E)$ is associated a finite-dimensional subspace M_z of E, namely the smallest vector subspace M of E such that $z \in \bigwedge^p(M)$ (§ 7, no. 2, Corollary to Proposition 4).

PROPOSITION 14. (i) *The orthogonal of* M_z *in* E^* *is the set of* $x^* \in E^*$ *such that* $x^* \lrcorner z = 0$.

(ii) *The subspace* M_z *associated with z is the image of* $\bigwedge^{p-1}(E^*)$ *under the mapping* $\lambda_z : u^* \mapsto u^* \lrcorner z$ *of* $\bigwedge^{p-1}(E^*)$ *into* E.

Let N denote the image of λ_z. For $x^* \in E^*$ and $u^* \in \bigwedge^{p-1}(E^*)$,

$$\begin{aligned}\langle \theta_\wedge(x^*), u^* \lrcorner z\rangle = \langle \theta_\wedge(u^* \wedge x^*), z\rangle &= (-1)^{p-1}\langle \theta_\wedge(x^* \wedge u^*), z\rangle \\ &= (-1)^{p-1}\langle \theta_\wedge(u^*), x^* \lrcorner z\rangle.\end{aligned}$$

Therefore, for x^* to be orthogonal to N, it is necessary and sufficient that $x^* \lrcorner z$ be orthogonal to $\theta_\wedge(\bigwedge(E^*))$. Now, the latter condition is equivalent to saying that $x^* \lrcorner z = 0$; for let $(e_\lambda)_{\lambda \in L}$ be a basis of E; giving L a total ordering, it has been seen (§ 7, no. 8, Theorem 1) that the e_J, for J running through the set $\mathfrak{F}(L)$ of finite subsets of L, form a basis of $\bigwedge(E)$; it then follows from formula (30) of no. 5 that the elements $\theta_{\wedge}(e_J^*)$ are, to within a sign, the coordinate forms on $\bigwedge(E)$ relative to the basis (e_J); whence our assertion.

The orthogonal of N therefore consists of the $x^* \in E^*$ such that $x^* \lrcorner z = 0$ and the conclusion of (i) will therefore follow from (ii).

We show first that $N \subset M_z$. Let M be a vector subspace of E such that $z \in \bigwedge(M)$ and let $j: M \to E$ be the canonical injection; let μ_z denote the mapping $v^* \mapsto v^* \lrcorner z$ of $\bigwedge^{p-1}(M^*)$ into M; it follows from formula (60) of no. 7 that there is a canonical factorization

$$\lambda_z : \bigwedge^{p-1}(E^*) \xrightarrow{\bigwedge^{p-1}({}^t j)} \bigwedge^{p-1}(M^*) \xrightarrow{\mu_z} M \xrightarrow{j} E$$

which proves that $N \subset M$ and hence $N \subset M_z$ by definition of M_z. It remains to verify that $N = M_z$. Suppose the converse: there would then exist a basis $(e_i)_{1 \leqslant i \leqslant n}$ of M_z and an element $x^* \in E^*$ such that $\langle x^*, e_1\rangle = 1$, $\langle x^*, e_j\rangle = 0$ for $2 \leqslant j \leqslant n$ and such that x^* is orthogonal to N and hence $x^* \lrcorner z = 0$. We write $z = \sum_H a_H e_H$, where the sum is taken over the subsets of $[1, n]$ with p elements. By (68) (no. 9),

$$x^* \lrcorner e_H = 0 \quad \text{if} \quad 1 \notin H$$
$$x^* \lrcorner e_{(1) \cup H} = e_H \quad \text{if} \quad H \subset [2, n]$$

which shows that the relation $x^* \lrcorner z = 0$ implies $a_H = 0$ for $1 \in H$. But this is impossible, for z would then belong to $\bigwedge^p(M')$, where M' is the subspace of M generated by $e_2, \ldots, e_n$.

13. PURE p-VECTORS. GRASSMANNIANS

Let K be a field and E a vector space over K. A p-vector $z \in \bigwedge^p(E)$ is called *pure* (or sometimes *decomposable*) if it is non-zero and there exist vectors $x_1, \ldots, x_p$ in E such that $z = x_1 \wedge x_2 \wedge \cdots \wedge x_p$. For this, it is necessary and sufficient that the subspace M_z associated with z (which is always of dimension $\geqslant p$ for $z \neq 0$) be *exactly* of dimension p (since $\bigwedge^p(M_z)$ is then of dimension 1). In particular, every *non-zero* scalar, every non-zero element of $E = \bigwedge^1(E)$, every non-zero element of $\bigwedge^n(E)$, when E is of dimension n, is *pure.*

PROPOSITION 15. *Let* E *be a vector space of dimension n and let e be an element* $\neq 0$ *of* $\bigwedge^n(E)$ (*hence forming a basis of this vector space*). *Let* $\phi : \bigwedge(E) \to \bigwedge(E^*)$ *be the vector space isomorphism associated with e* (no. 11, Proposition 12). *If z is a pure element of* $\bigwedge^p(E)$, *then* $\phi(z)$ *is a pure element of* $\bigwedge^{n-p}(E^*)$ *and the subspaces associated with z and* $\phi(z)$ *are orthogonal.*

The cases $p = 0$ and $p = n$ are trivial. Suppose therefore that $1 \leqslant p \leqslant n - 1$ and let $z = x_1 \wedge \cdots \wedge x_p \neq 0$. Then there exists a basis $(e_i)_{1 \leqslant i \leqslant n}$ of E such that $e_i = x_i$ for $1 \leqslant i \leqslant p$ and $e = e_1 \wedge e_2 \wedge \cdots \wedge e_n$. It then follows from formula (78) of no. 11 that $\phi(z) = \pm e^*_{p+1} \wedge \cdots \wedge e^*_n$, whence the proposition.

COROLLARY. *If* E *is of dimension n, every non-zero* $(n - 1)$*-vector* E *is pure.*

PROPOSITION 16. *For an element $z \neq 0$ of $\bigwedge^p(\mathrm{E})$ to be pure, it is necessary and sufficient that, for all $u^* \in \bigwedge^{p-1}(\mathrm{E}^*)$,*

$$(81) \qquad (u^* \lrcorner\, z) \wedge z = 0.$$

The case $p = 0$ is trivial and we assume $p \geqslant 1$. If $z = x_1 \wedge \cdots \wedge x_p$, formula (68) (no. 9) with $n = p - 1$ shows that $u^* \lrcorner\, z$ is a linear combination of the x_i $(1 \leqslant i \leqslant p)$, whence (81). If on the other hand the subspace M_z associated with z is of dimension $> p$, consider a basis $(e_j)_{1 \leqslant j \leqslant n}$ of this subspace with $n > p$. It follows from no. 11, Proposition 13 that each of the e_j is of the form $u^* \lrcorner\, z$ for some $u^* \in \bigwedge^{p-1}(\mathrm{E}^*)$ and relation (81) therefore implies $e_j \wedge z = 0$ for $1 \leqslant j \leqslant n$. It follows that in the expression $z = \sum_{\mathrm{H}} a_{\mathrm{H}} e_{\mathrm{H}}$ (where H runs through the set of subsets of $[1, n]$ with p elements) all the coefficients a_{H} are zero, whence $z = 0$, contrary to the hypothesis.

The criterion of Proposition 16 is equivalent to writing conditions (81) when u^* runs through a *basis* of $\bigwedge^{p-1}(\mathrm{E}^*)$. In particular, suppose that E is of finite dimension n and let $(e_i)_{1 \leqslant i \leqslant n}$ be a basis of E. Conditions (81) are then equivalent to the conditions

$$(82\text{–}(\mathrm{J}, \mathrm{H})) \qquad \langle e^*_{\mathrm{J}}, (e^*_{\mathrm{H}} \lrcorner\, z) \wedge z \rangle = 0$$

for all subsets J, H of $[1, n]$ such that $\mathrm{Card}(\mathrm{J}) = p + 1$ and $\mathrm{Card}(\mathrm{H}) = p - 1$. Now, if I and I′ are two subsets of $[1, n]$ with p elements, formulae (76) of no. 10 and multiplication table (20) of § 7, no. 8 show that

$$\langle e^*_{\mathrm{J}}, \langle e^*_{\mathrm{H}} \lrcorner\, e_{\mathrm{I}}) \wedge e_{\mathrm{I}'} \rangle = 0$$

unless there exists an $i \in [1, n]$ such that $\mathrm{I} - \mathrm{H} = \{i\}$ and $\mathrm{J} - \mathrm{I}' = \{i\}$, in which case

$$(83) \qquad \langle e^*_{\mathrm{J}}, (e^*_{\mathrm{H}} \lrcorner\, e_{\mathrm{I}}) \wedge e_{\mathrm{I}'} \rangle = (-1)^{(p-1)(p-2)/2} \varepsilon_{i, \mathrm{J}, \mathrm{H}}$$

where $\varepsilon_{i,\mathrm{J},\mathrm{H}} = \rho_{\{i\},\mathrm{H}} \rho_{\{i\},\mathrm{I}'}$; it can then be said that for $i \in \mathrm{J} \cap \complement \mathrm{H}$, $\varepsilon_{i,\mathrm{J},\mathrm{H}}$ is equal to $+1$ if the number of element of J which are $< i$ and the number of elements of H which are $< i$ have the *same parity*, and -1 otherwise.

It follows immediately that if we write $z = \sum_{\mathrm{I}} a_{\mathrm{I}} e_{\mathrm{I}}$, where I runs through the set of subsets of $(1, n)$ with p elements, relation (82–(J, H)) is equivalent to the relation

$$(84\text{–}(J, H)) \qquad \sum_{i \in J \cup \complement H} \varepsilon_{i, J, H} a_{J-\{i\}} a_{H \cup \{i\}} = 0.$$

Relations (84) are called *Grassman's relations*; these are therefore necessary and sufficient conditions (when J describes the set of subsets with $p + 1$ elements and H the set of subsets with $p - 1$ elements of $(1, n)$) for an element $z \neq 0$ of $\bigwedge^p(E)$ to be *pure.*

Note that relations (84) are not independent. For example, for $n = 4$ and $p = 2$, Grassmann's relations reduce to the single relation

$$(85) \qquad a_{12}a_{34} - a_{13}a_{24} + a_{14}a_{23} = 0.$$

Let $D_p(E)$ be the subset of $\bigwedge^p(E)$ consisting of the *pure* p-vectors; clearly $D_p(E)$ is saturated with respect to the equivalence relation between u and v: "there exists $\lambda \in K^*$ such that $v = \lambda u$" and two elements u, v of $D_p(E)$ are equivalent under this relation if and only if the subspaces M_u and M_v of E which are associated with them are the same. Therefore we thus obtain a *canonical bijection* of the *set of p-dimensional vector subspaces* of E onto the *image* $\mathbf{G}_p(E)$ *of* $D_p(E)$ *in the projective space* $\mathbf{P}(\bigwedge^p(E))$ associated with $\bigwedge^p(E)$. The subset $\mathbf{G}_p(E)$ of $\mathbf{P}(\bigwedge^p(E))$ is called the *Grassmannian* of index p of the vector space E. When E is finite-dimensional and $(e_i)_{1 \leqslant i \leqslant n}$ is a basis of E, the Grassmanian of index p is the set of points of $\mathbf{P}(\bigwedge^p(E))$ for which a system of homogeneous coordinates (a_I) (relative to the basis (e_I) of $\bigwedge^p(E)$) satisfies Grassmann's relations (84).

When $E = K^n$, we sometimes write $\mathbf{G}_{n,p}(K)$ instead of $\mathbf{G}_p(K^n)$, so that $\mathbf{G}_{n,1}(K) = \mathbf{P}_{n-1}(K)$. The mapping $M \mapsto M^0$, which associates with every p-dimensional subspace of K^n the *orthogonal* subspace in E^* (identified with K^n under the choice of basis dual to the canonical basis of K^n) therefore defines a canonical *bijection* of $\mathbf{G}_{n,p}(K)$ onto $\mathbf{G}_{n,n-p}(K)$; Proposition 15 shows that this bijection is the *restriction* to $\mathbf{G}_{n,p}(K)$ of a canonical *isomorphism* of the projective space $\mathbf{P}(\bigwedge^p(K^n))$ onto the projective space $\mathbf{P}(\bigwedge^{n-p}(K^n))$.

APPENDIX

ALTERNATIVE ALGEBRAS. OCTONIONS

1. ALTERNATIVE ALGEBRAS

Let A be a commutative ring and F a (not necessarily associative) A-algebra. For any three elements x, y, z of F, we write

$$(1) \qquad a(x, y, z) = x(yz) - (xy)z$$

(*associator* of x, y, z); a is obviously an A-trilinear mapping of $F \times F \times F$ into F.

Lemma 1. *For all p, q, r, s in the algebra* F,

$$(2) \qquad a(pq, r, s) - a(p, qr, s) + a(p, q, rs) = a(p, q, r)s + pa(q, r, s).$$

The verification follows immediately from definition (1).

PROPOSITION 1. *For an* A-*algebra* F, *the following conditions are equivalent:*

(a) *For every ordered pair of elements x, y of* F, *the subalgebra generated by x and y is associative.*

(b) *The trilinear mapping $(x, y, z) \mapsto a(x, y, z)$ is alternating* (§ 7, no. 3).

(c) *For every ordered pair of elements x, y of* F, $x^2y = x(xy)$ *and* $yx^2 = (yx)x$.

Clearly (a) implies (c). We show that (c) implies (b): by definition (§ 7, no. 3) to prove (b), it suffices to verify that $a(x, x, y) = 0$ and $a(x, y, y) = 0$, which is precisely (c).

To prove that (b) implies (a), we use the following 4 lemmas:

Lemma 2. *Let* E *be an* A-*algebra such that the trilinear mapping $(x, y, z) \mapsto a(x, y, z)$ is alternating,* S *a generating system of* E *and* U *a sub-A-module of* E *containing* S *and such that $s\mathrm{U} \subset \mathrm{U}$ and $\mathrm{U}s \subset \mathrm{U}$ for all $s \in \mathrm{S}$. Then $\mathrm{U} = \mathrm{E}$.*

The set U′ of $x \in \mathrm{E}$ such that $x\mathrm{U} \subset \mathrm{U}$ and $\mathrm{U}x \subset \mathrm{U}$ is obviously a sub-A-module of E, which contains S by hypothesis. On the other hand, for x, y in U′ and $u \in \mathrm{U}$, by hypothesis

$$(xy)u = x(yu) + a(x, y, u) = x(yu) - a(x, u, y) = x(yu) - (xu)y + x(uy) \in \mathrm{U};$$

on passing to the opposite algebra, we have similarly $u(xy) \in \mathrm{U}$. Hence U′ is a *subalgebra* of E and, since it contains S, $\mathrm{U}' = \mathrm{E}$. Hence $\mathrm{EU} \subset \mathrm{U}$ and *a fortiori* $\mathrm{UU} \subset \mathrm{U}$, which proves that U is a subalgebra of E; as it contains S, $\mathrm{U} = \mathrm{E}$, which proves the lemma.

A subset H of F is called *strongly associative* if $a(u, v, w) = 0$ when *at least two* of the elements u, v, w belong to H.

Lemma 3. *Suppose that the mapping a is alternating. If* H *is a strongly associative subset of* F, *the subalgebra of* F *generated by* H *is strongly associative.*

As the set of strongly associative subsets of F is inductive, it suffices to prove that if H is a *maximal* strongly associative subset of F, H is then a *subalgebra* of F. As H is obviously a sub-A-module of F, it suffices to verify that for any two elements u, v of H, $\mathrm{H} \cup \{uv\}$ is also strongly associative, for by virtue of the definition of H, this will imply $uv \in \mathrm{H}$. Now, for all $z \in \mathrm{H}$ and all $t \in \mathrm{F}$, by (2)

$$a(uv, t, z) - a(u, vt, z) + a(u, v, tz) = 0$$

since H is strongly associative; as u, v, z are in H, also

$$a(u, vt, z) = a(u, v, tz) = 0,$$

whence $a(uv, t, z) = 0$. Using the fact that a is alternating, this shows that $a(p, q, r) = 0$ whenever at least two of the elements p, q, r belong to $\mathrm{H} \cup \{uv\}$ whence the lemma.

Lemma 4. *Suppose that the mapping a is alternating. Then, for all $x \in \mathrm{F}$, the subalgebra of* F *generated by x is strongly associative.*

$a(u, v, w) = 0$ whenever two of the three elements u, v, w are equal to x and it suffices to apply Lemma 3.

Lemma 5. *Suppose that the mapping a is alternating and let* X, Y *be two strongly associative subalgebras of* F. *Then the subalgebra of* E *generated by* $\mathrm{X} \cup \mathrm{Y}$ *is associative.*

Let Z be the set of $z \in \mathrm{E}$ such that $a(u, v, z) = 0$ for all $u \in \mathrm{X}$ and $v \in \mathrm{Y}$, this is obviously a sub-A-module containing X and Y since X and Y are strongly associative; by Lemma 2, it will suffice to verify that, for $u \in \mathrm{X}$ and $v \in \mathrm{Y}$, $u\mathrm{Z} \subset \mathrm{Z}$, $v\mathrm{Z} \subset \mathrm{Z}$, $\mathrm{Z}u \subset \mathrm{Z}$ and $\mathrm{Z}v \subset \mathrm{Z}$. Now, for u, u' in X, $v \in \mathrm{Y}$ and $z \in \mathrm{Z}$, by (2)

$$a(u'u, z, v) - a(u', uz, v) + a(u', u, zv) = a(u', u, z)v + u'a(u, z, v) = 0$$

by virtue of the fact that X is strongly associative and the definition of Z. But as X is strongly associative, $a(u', u, zv) = 0$ and since $u'u \in \mathrm{X}$, $a(u'u, z, v) = 0$ by definition of Z. Hence $a(u', uz, v) = 0$, which shows that $u\mathrm{Z} \subset \mathrm{Z}$. Applying (2) now with $(p, q, r, s) = (v, z, u, u')$, we obtain similarly $\mathrm{Z}u \subset \mathrm{Z}$. Interchanging the roles of X and Y and using the fact that a is alternating, we obtain $v\mathrm{Z} \subset \mathrm{Z}$ and $\mathrm{Z}v \subset \mathrm{Z}$; whence the lemma.

It now suffices, to prove that (b) implies (a) in Proposition 1, to take $\mathrm{X} = \{x\}$ and $\mathrm{Y} = \{y\}$, using Lemma 4.

Definition 1. *An algebra* F *is called alternative if it satisfies the equivalent conditions of Proposition* 1.

An associative algebra is obviously alternative. In no. 3 we shall give an example of an alternative algebra which is not associative.

If F is an alternative A-algebra, every A-algebra $\mathrm{F} \otimes_{\mathrm{A}} \mathrm{A}'$ obtained from F by extending the scalars (§ 1, no. 5) is an alternative A′-algebra, as follows from condition (b) of Proposition 1.

2. ALTERNATIVE CAYLEY ALGEBRAS

Proposition 2. *Let* A *be a ring,* F *a Cayley* A*-algebra, e its unit element, $s : x \mapsto \bar{x}$ its conjugation and* $\mathrm{N} : \mathrm{F} \to \mathrm{A}$ *its Cayley norm* (§ 3, no. 4).

(i) *For* F *to be alternative, it is necessary and sufficient that, for every ordered pair of elements x, y of* F, $x^2y = x(xy)$.

(ii) *If* F *is alternative, then* $N(xy) = N(x)N(y)$ *for all* x, y *in* F.

(iii) *Suppose that* F *is alternative. For an element* $x \in F$ *to be invertbile, it is necessary and sufficient that* $N(x)$ *be invertible in* Ae*; the inverse of* x *is then unique and equal to* $N(x)^{-1}\bar{x}$*; denoting this by* x^{-1},

$$x^{-1}(xy) = x(x^{-1}y) = y$$

or all $y \in F$.

The condition $x^2y = x(xy)$ is obviously necessary for F to be alternative (no. 1, Proposition 1). Conversely, if it holds for every ordered pair of elements of F, applying it to $\bar{x}$ and $\bar{y}$, it gives $\bar{x}^2\bar{y} = \bar{x}(\bar{x}\bar{y})$; applying the conjugation s to this relation, we obtain $yx^2 = (yx)x$, so that conditions (c) of Proposition 1 of no. 1 are satisfied.

Obviously $a(e, x, y) = 0$ for all x, y in F. If F is alternative, we therefore deduce from Proposition 1 (no. 1) that the subalgebra G of F generated by e, x and y is associative. As $\bar{x} = -x + T(x) \in -x + Ae$, $\bar{x} \in G$ and similarly $\bar{y} \in G$. Then $N(xy) = (xy)(\overline{xy}) = xy.\bar{y}.\bar{x} = N(y)x\bar{x} = N(y)N(x)$, using the fact that $N(y) \in Ae$. This proves (ii).

Finally we prove (iii). If $N(x)$ is invertible in Ae and we write $x' = N(x)^{-1}\bar{x}$, then $xx' = x'x = e$, for $N(x) = x\bar{x} = \bar{x}x$. Conversely if x admits a left inverse x'', then $N(x'')N(x) = N(e) = e$ by (ii) and $N(x)$ is invertible in Ae; further, as $x' = N(x)^{-1}\bar{x}$ is in the subalgebra generated by x and e, the elements x, x', x'' belong to the associative subalgebra generated by x, x'' and e and hence $x'' = x''(xx') = (x''x)x' = x'$, whence the uniqueness assertion. The formulae $x^{-1}(xy) = x(x^{-1}y) = y$ follow from the fact that x^{-1}, x and y are elements of the subalgebra generated by x, y and e, which is associative.

PROPOSITION 3. *Let* E *be a Cayley* A*-algebra,* γ *an element of* A *and* F *the Cayley extension of* E *defined by* γ *and the conjugation of* E (§ 2, no. 5, Proposition 5). *For* F *to be alternative, it is necessary and sufficient that* E *be associative.*

Let $u = (x, y)$, $v = (x', y')$ be two elements of F (where x, y, x', y' are in E). Then (§ 2, no. 5, formula (27))

$$(3)\quad \begin{cases} u^2v = ((x^2 + \gamma\bar{y}y)x' + \gamma\bar{y}'(y\bar{x} + yx), (y\bar{x} + yx)\bar{x}' + y'(x^2 + \gamma\bar{y}y)) \\ u(uv) = (x(xx' + \gamma\bar{y}'y) + \gamma(x'\bar{y} + \bar{x}.\bar{y}')y, y(\bar{x}'\bar{x} + \gamma\bar{y}y') + (y\bar{x}' + y'x)x). \end{cases}$$

Using the fact that $\bar{y}y$ and $\bar{x} + x$ are in Ae, examining these formulae shows that the associativity of E implies $u^2v = u(uv)$ and hence the fact that F is alternative (Proposition 2). Conversely, if F is alternative, the equation $u^2v = u(uv)$ applied when $y' = 0$ gives

$$(y\bar{x} + yx)\bar{x}' = y(\bar{x}'\bar{x}) + (y\bar{x}')x.$$

Now the left hand side is equal to $(yT(x))\bar{x}' = y(\bar{x}'T(x)) = y(\bar{x}'x + \bar{x}'\bar{x})$;

comparing with the right hand side, we obtain $(y\bar{x}')x = y(\bar{x}'x)$, which proves the associativity of E, since x, y and $\bar{x}'$ are arbitrary elements of E.

3. OCTONIONS

Let E be a quaternion algebra of type (α, β, γ) over A (§ 2, no. 5, *Example* 2) and let $\delta \in A$. The Cayley extension F of E by δ and the conjugation of E is called an *octonion algebra* over A and is said to be of type $(\alpha, \beta, \gamma, \delta)$. By Proposition 3 of no. 2, F is an *alternative* algebra. It has a basis $(e_i)_{0 \leqslant i \leqslant 7}$ of 8 elements, defined by

$$\begin{aligned} e_0 &= (e, 0), & e_1 &= (i, 0), & e_2 &= (j, 0), & e_3 &= (k, 0) \\ e_4 &= (0, e), & e_5 &= (0, i), & e_6 &= (0, j), & e_7 &= (0, k) \end{aligned}$$

where (e, i, j, k) is the basis of E defined *loc. cit.*; clearly e_0 (also denoted by e) is the unit element of F. If $u = \sum_{i=0}^{7} \xi_i e_i$ is an element of F (with the $\xi_i \in A$), formulae (23), (24) and (31) of § 2, no. 5, give for the conjugate, trace and norm of the octonion u

$$(4) \quad \begin{cases} \bar{u} = (\xi_0 + \beta\xi_1)e_0 - \sum_{i=1}^{7} \xi_i e_i \\ \mathrm{T_F}(u) = 2\xi_0 + \beta\xi_1 \\ \mathrm{N_F}(u) = \xi_0^2 + \beta\xi_0\xi_1 - \alpha\xi_1^2 - \gamma(\xi_2^2 + \beta\xi_2\xi_3 - \alpha\xi_3^2) \\ \qquad - \delta(\xi_4^2 + \beta\xi_4\xi_5 - \alpha\xi_5^2) + \gamma\delta(\xi_6^2 + \beta\xi_6\xi_7 - \alpha\xi_7^2). \end{cases}$$

Now let $u = (x, y)$, $u' = (x', y')$ and $u'' = (x'', y'')$ be three octonions (where the elements x, x', x'', y, y', y'' belong to E). Formulae (24) and (27) of § 2, no. 5 give

$$\begin{aligned} \mathrm{T_F}((uu')u'') &= \mathrm{T}(xx'x'') + \delta\mathrm{T}(\bar{y}'yx'') + \delta\mathrm{T}(\bar{y}''y\bar{x}') + \delta\mathrm{T}(\bar{y}''y'x) \\ \mathrm{T_F}(u(u'u'')) &= \mathrm{T}(xx'x'') + \delta\mathrm{T}(x''\bar{y}'y) + \delta\mathrm{T}(\bar{x}'\bar{y}''y) + \delta\mathrm{T}(x\bar{y}''y') \end{aligned}$$

(where T denotes trace in E and use is made of the fact that E is associative). As $\mathrm{T}(xy) = \mathrm{T}(yx)$ for all quaternions x, y (§ 2, no. 4, formula (17)), it follows that

$$(5) \qquad \mathrm{T_F}((uu')u'') = \mathrm{T_F}(u(u'u'')).$$

We study in particular *octonions of type* $(-1, 0, -1, -1)$; formulae (4) then simplify to

$$(6) \quad \begin{cases} \bar{u} = \xi_0 e_0 - \sum_{i=1}^{7} \xi_i e_i \\ \mathrm{T_F}(u) = 2\xi_0 \\ \mathrm{N_F}(u) = \sum_{i=0}^{7} \xi_i^2. \end{cases}$$

*If we take A to be the field $\mathbf{R}$ of real numbers, the octonions of type $(-1, 0, -1, -1)$ over $\mathbf{R}$ are called *Cayley octonions* (or *octaves*). It follows from Proposition 2 (ii) of no. 2 that every Cayley octonion $\neq 0$ is *invertible*.*

PROPOSITION 4. *Let* F *be an octonion algebra of type* $(-1, 0, -1, -1)$ *over* A. *There exists a vector space* V *of dimension* 3 *over the field with two elements* $\mathbf{Z}/2\mathbf{Z}$ *and a bijection* $\lambda \mapsto e'_\lambda$ *of* V *onto the basis* $(e_i)_{0 \leqslant i \leqslant 7}$ *such that*

$$(7) \qquad e'_0 = e_0, \qquad e'_\lambda e'_\mu = \pm e'_{\lambda+\mu}$$

for all λ, μ *in* V. *In order that* $e'_\lambda(e'_\mu e'_\nu) = (e'_\lambda e_\mu)e'_\nu$, *it suffices that, in* V, λ, μ, ν *be linearly dependent over* $\mathbf{Z}/2\mathbf{Z}$; *this condition is necessary if* $2 \neq 0$ *in* A.

We preserve the notation at the beginning of this no. It follows from formulae (33) of § 2, no. 5 that the set S consisting of the elements $\pm e_0$, $\pm e_1$, $\pm e_2$, $\pm e_3$ is stable under multiplication. Moreover, for x, y, y' in E, by formula (22) of § 2, no. 5,

$$(8) \quad (x, 0)(0, y') = (0, y'x), \quad (0, y')(x, 0) = (0, y'\bar{x}), \quad (0, y)(0, y') = (-\bar{y}'y, 0)$$

so that the set T consisting of the elements $\pm e_i$ $(0 \leqslant i \leqslant 7)$ is stable under multiplication; moreover, its multiplication table is *independent* of the ring A.

In particular, let A″ be the field $\mathbf{Z}/2\mathbf{Z}$ with two elements and let E″ be the quaternion algebra of type $(1, 0, 1)$ over A″ and F″ the algebra of octonions of type $(1, 0, 1, 1)$ over A″; let $(e''_i)_{0 \leqslant i \leqslant 7}$ be the basis of F″ formed as described above. Since $-e''_i = e''_i$, the set T″ of e''_i has 8 elements and is stable under multiplication; moreover, it follows immediately from the above that the mapping $\theta : \mathrm{T} \to \mathrm{T}''$ such that $\theta(e_i) = \theta(-e_i) = e''_i$ for $0 \leqslant i \leqslant 7$ is a homomorphism for multiplication. Moreover the quaternion algebra E″ is in this case commutative and hence F″ is associative (§ 2, no. 5, Proposition 5); further, conjugation in F″ is in this case the identity. Hence T″ is a *group* and formulae (8) show that it is *commutative*; these formulae and formulae (33) of § 2, no. 5 show that the square of every element of T″ is the unit. If V is used to denote the group T″ written additively, V can be given a unique vector space structure over $\mathbf{Z}/2\mathbf{Z}$, necessarily of dimension 3 since

$$\mathrm{Card}(\mathrm{V}) = 8 = 2^{\dim(\mathrm{V})}.$$

For all $\lambda \in \mathrm{V}$, let e'_λ then denote the element of $(e_i)_{0 \leqslant i \leqslant 7}$ such that $\theta(e'_\lambda) = \lambda$; then $e'_0 = e_0$; moreover, as θ is a homomorphism and the relation $\theta(x) = \theta(y)$ is equivalent to $x = \pm y$, $e'_\lambda e'_\mu = \pm e'_{\lambda+\mu}$. If λ, μ, ν are linearly independent over $\mathbf{Z}/2\mathbf{Z}$, they form a basis of V and hence all the elements e_i $(0 \leqslant i \leqslant 7)$ would belong to the subalgebra generated by e'_λ, e'_μ and e'_ν; when $2 \neq 0$ in A, $e'_\lambda(e'_\mu e'_\nu) = (e'_\lambda e'_\mu)e'_\nu$ is therefore impossible, for F would be associative and hence E would be commutative (§ 2, no. 5, Proposition 5), which contradicts relations (33) of § 2, no. 5. On the other hand, if λ, μ, ν are linearly dependent in V, the three elements e'_λ, e'_μ, e'_ν belong to a subalgebra with 2 generators of F, which is therefore associative (no. 1, Proposition 1); whence the conclusion.

Remark. As $\bar{e}'_\lambda = -e'_\lambda$ for $\lambda \neq 0$,

$$e'^2_\lambda = -e_0 \quad \text{for } \lambda \neq 0,$$
$$e'_\mu e'_\lambda = -e'_\lambda e'_\mu \quad \text{for } \lambda \neq 0,\ \mu \neq 0 \text{ and } \mu \neq \lambda.$$

EXERCISES

§ 1

1. Let E be an algebra over a (commutative) field K and L a commutative extension field of K. Show that, if the algebra $E_{(L)}$ admits a unit element, so does E (cf. II, § 3, no. 5, Proposition 6).

§ 2

1. Let E be an A-algebra with a basis of two elements e_1, e_2 and a unit element $e = \lambda e_1 + \mu e_2$.

(*a*) Show that the ideal $\mathfrak{a}$ of A generated by λ and μ is the whole ring A (in the contrary case, note that $E/\mathfrak{a}E = E \otimes_A (A/\mathfrak{a}) = \{0\}$ would hold contrary to the fact that $A/\mathfrak{a} \neq \{0\}$ and E is a free A-module).

(*b*) If α, β are two elements of A such that $\alpha\lambda + \beta\mu = 1$, show that the elements e and $u = \beta e_1 - \alpha e_2$ also form a basis of E and deduce that E is a quadratic A-algebra.

2. Show that every quadratic algebra over $\mathbf{Z}$ is isomorphic to an algebra of type $(0, n)$ with $n \in \mathbf{N}$, or an algebra of type $(m, 0)$ with m a non-square, or an algebra of type $(m, 1)$, where m is an integer which is not of the form $k(k - 1)$; moreover, no two of these algebras are isomorphic.

3. (*a*) Let K be a commutative field. Show that, for K to be the centre of a quaternion algebra of type (α, β, γ) over K, it is necessary and sufficient that one of the following cases hold: (1) $\beta\gamma \neq 0$; (2) $\gamma = 0, \beta \neq 0$ and $\beta^2 + 4\alpha \neq 0$; (3) $\beta = 0, \alpha \neq 0$ or $\gamma \neq 0$ and $2 \neq 0$ in K.

(*b*) Let K be a commutative field such that $2 \neq 0$ in K; show that the quaternion algebra over K of type $(1, \gamma)$ is isomorphic to the matrix algebra $\mathbf{M}_2(K)$ (consider the basis of this algebra consisting of the elements $\frac{1}{2}(1 + i)$, $\frac{1}{2}(1 - i)$, $(j + k/2\gamma)$, $\frac{1}{2}(j - k)$).

(*c*) Let K be a commutative field such that $2 \neq 0$ in K and let E be the quaternion algebra over K of type (α, γ). A quaternion $z \in E$ is called *pure* if $\bar{z} = -z$ (or also $T(z) = 0$). Show that if z is pure, so is every quaternion tzt^{-1}, where $t \in E$ is invertible.

(*d*) If u, v are any two quaternions, $uv - vu$ is pure. Deduce that if u, v, w are any three quaternions, then

$$(uv - vu)^2 w = w(uv - vu)^2.$$

(*e*) Show that if there exists in E a quaternion $z \neq 0$ such that $N(z) = 0$, there also exists a pure quaternion $z' \neq 0$ such that $N(z') = 0$. (Note, in the notation of formula (33) (no. 5), that, if α is not a square in K, the existence of $z \neq 0$ in E such that $N(z) = 0$ is equivalent to the existence of an element $y \in K + Ki$ such that $\gamma = N(y)$.)

5. Over a commutative field K in which $2 = 0$, every quaternion algebra E of type $(\alpha, 0, \gamma)$ is commutative and the square of every $x \in E$ belongs to K; the subalgebra $K(x)$ of E generated by an element $x \in K$ is therefore a quadratic algebra of type $(\lambda, 0)$ over K and E is a quadratic algebra over $K(x)$.

Show that the set C of $x \in E$ whose square is equal to the square of an element of K is a vector subspace of E whose dimension is equal to 1, 2 or 4. If C is of dimension 1 (in which case $C = K$), E is a field. If C is of dimension 2, there exists a quadratic algebra $K(x)$ contained in E which is a field and E has a basis over $K(x)$ consisting of two elements 1 and u with $u^2 = 0$; the set of $y \in E$ such that $y^2 = 0$ is an ideal $\mathfrak{a}$ of dimension 2 over K and $E/\mathfrak{a}$ is isomorphic to $K(x)$. Finally, if C is of dimension 4 (and hence equal to E), the set $\mathfrak{a}$ of $y \in E$ such that $y^2 = 0$ is an ideal of dimension 3 and $E/\mathfrak{a}$ is isomorphic to K; there exists in E a basis $(1, e_1, e_2, e_3)$ such that $e_1^2 = e_2^2 = e_3^2 = 0$, $e_1 e_2 = e_3$, $e_1 e_3 = e_2 e_3 = 0$; $Ke_3 = \mathfrak{b}$ is the only ideal of dimension 1 in E, $\mathfrak{b}$ is the annihilator of $\mathfrak{a}$ and $\mathfrak{a}/\mathfrak{b}$ is the direct sum of two ideals of $E/\mathfrak{b}$, of dimension 1, which annihilate one another.

6. Let K be a commutative field in which $2 \neq 0$ and E a quaternion algebra over K of type $(0, 0, \gamma)$.

(*a*) If γ is not a square in K, there exists no left (resp. right) ideal in E of dimension 1 over K; the set $\mathfrak{a}$ of $x \in E$ such that $x^2 = 0$ is a two-sided ideal of dimension 2 over K and $E/\mathfrak{a}$ is a field, a quadratic algebra over K.

(*b*) If γ is a square $\neq 0$ in K, there exists in E a basis $(e_i)_{1 \leq i \leq 4}$ with the following multiplication table:

	e_1	e_2	e_3	e_4
e_1	e_1	0	e_3	0
e_2	0	e_2	0	e_4
e_3	0	e_3	0	0
e_4	e_4	0	0	0

The set $\mathfrak{a}$ of $x \in E$ such that $x^2 = 0$ is a two-sided ideal of dimension 2, the direct sum of the two two-sided ideals Ke_3 and Ke_4 which annihilate one another; the latter are the only (left or right) ideals of dimension 1 in E. The quotient algebra $E/\mathfrak{a}$ is isomorphic to the product of two fields isomorphic to K.

(c) If $\gamma = 0$, the set $\mathfrak{a}$ of $x \in E$ such that $x^2 = 0$ is a two-sided ideal of dimension 3; $Kk = \mathfrak{b}$ is the only (left or right) ideal of dimension 1 in E; it is a two-sided ideal, the left and right annihilator of $\mathfrak{a}$. In the quotient algebra $E/\mathfrak{b}$, $\mathfrak{a}/\mathfrak{b}$ is the direct sum of two two-sided ideals of dimension 1 which annihilate one another; finally $E/\mathfrak{a}$ is a field isomorphic to K.

7. Let K be a commutative field in which $2 \neq 0$ and E the algebra over K with a basis of four elements $1, i, j, k$ (1 the unit element) with multiplication table

$$i^2 = j^2 = k^2 = 1, \qquad ij = ji = k, \qquad jk = kj = i, \qquad ki = ik = j.$$

Show that E is isomorphic to the product of four fields isomorphic to K (consider the basis of E consisting of the elements $(1 + \varepsilon i)(1 + \varepsilon' j)$, where ε and ε' are equal to 1 or -1).

The algebra E is the algebra (over K) of the product of two cyclic groups of order 2. Generalize this to the algebra of the product group of n cyclic groups of order 2 (cf. § 4, no. 1, *Example* 2).

8. The *quaternionic group* $\mathbf{Q}$ (I, § 6, Exercise 4) is isomorphic to the multiplicative group of the eight quaternions ± 1, $\pm i$, $\pm j$, $\pm k$ in the quaternion algebra of type $(-1, -1)$ over a field in which $2 \neq 0$. Show that the algebra of the group $\mathbf{Q}$ over a field K in which $2 \neq 0$ is isomorphic to the product of four fields isomorphic to K and the quaternion algebra of type $(+1, -1)$ over K (if c is the element of $\mathbf{Q}$ corresponding to the quaternion -1, the elements of $\mathbf{Q}$ can be written as $e, i, j, k, c, ci, cj, ck$; consider the basis of E consisting of the elements $\frac{1}{2}(e + c)$, $\frac{1}{2}(e - c)$, $\frac{1}{2}(e + c)i$, $\frac{1}{2}(e - c)i$, $\frac{1}{2}(e + c)j$, $\frac{1}{2}(e - c)j$, $\frac{1}{2}(e + c)k$, $\frac{1}{2}(e - c)k$).

9. (a) Show that the algebra E of the *dihedral group* $\mathbf{D}_4$ of order 8 (I, § 6, Exercise 4) over a field K in which $2 \neq 0$ is isomorphic to the product of four fields isomorphic to K and the algebra of matrices of order 2 over K. (If a, b are the two generators of $\mathbf{D}_4$ introduced in I, § 6, Exercise 4, the elements of $\mathbf{D}_4$ can be written as $a^i b^j$ with $0 \leqslant i \leqslant 3$, $0 \leqslant j \leqslant 1$; consider the basis of E consisting of the four elements $\frac{1}{2}(e + a^2)$, $\frac{1}{2}(e - a^2)$, $\frac{1}{2}(a + a^3)$, $\frac{1}{2}(a - a^3)$ and these four elements multiplied on the right by b; use Exercise 4.) Deduce that, if -1 is a square in K, the algebras over K of the *non-isomorphic* groups $\mathbf{Q}$ and $\mathbf{D}_4$ are *isomorphic*.

(b) Show similarly that the algebra F of the dihedral group $\mathbf{D}_3$ of order 6 over a field K in which $2 \neq 0$ and $3 \neq 0$ is isomorphic to the product of two fields isomorphic to K and the matrix algebra of order 2 over K (consider here the basis of F consisting of the elements $e + a + a^2$, $a + a^2 - 2e$, $a - a^2$ and these three elements multiplied on the right by b).

10. Let G be a group and H a normal subgroup of G. Show that, if $\mathfrak{a}$ is the two-sided ideal of the group algebra $A^{(G)}$, generated by the elements $ts - s$, where t runs through H and s runs through G, the group algebra $A^{(G/H)}$ is isomorphic to the quotient algebra $A^{(G)}/\mathfrak{a}$.

11. Let A be a commutative ring and S a commutative monoid with identity element e. Suppose that an external law $(s, x) \mapsto x^s$ is given on A with S as domain of operators, such that, for all $s \in S$, the mapping $x \mapsto x^s$ is an *automorphism* of the ring A and $(x^s)^t = x^{st}$ for all $x \in A$ and s and t in S (which implies that e is the identity operator for the external law considered). Then a multiplicative internal law of composition is defined on the A-module $A^{(S)}$, whose canonical basis is denoted by (b_s), by the relation

$$\Big(\sum_s x_s b_s\Big)\Big(\sum_s y_s b_s\Big) = \sum_s\Big(\sum_{tu=s} a_{t,u} x_t y_u^s\Big) b_s$$

where the $a_{s,t}$ belong to A.

Show that this law and addition on $A^{(S)}$ define a *ring* structure on this set, provided the $a_{s,t}$ satisfy the conditions

$$a_{s,t} a_{st,u} = a_{tu}^s a_{s,tu}$$

for all s, t, u in S. The element b_e is unit element of this ring if also

$$a_{e,s} = a_{s,e} = 1$$

for all $s \in S$; in that case A can be identified with a subring of $A^{(S)}$ and if C denotes the subring of A consisting of the elements invariant under all the automorphisms $x \mapsto x^s$, $A^{(S)}$ is a C-algebra, called the *crossed product* of the ring A and the monoid S, relative to the *factor system* $(a_{s,t})$. If the factor system $(a_{s,t})$ is replaced by $\left(\frac{c_s c_t^s}{c_{st}} a_{s,t}\right)$, where, for all $s \in S$, c_s is an invertible element of A and $c_e = 1$, the new crossed product obtained is *isomorphic* to the crossed product defined by the factor system $(a_{s,t})$.

If in particular A is taken to be a quadratic algebra over a ring C and S a cyclic group of order 2, say $\{e, s\}$, such that $x^s = \bar{x}$ for $x \in A$, show that every crossed product of A and S is a *quaternion algebra* over C.

12. Let I be a non-empty set. Show that the elements of the free magma M(I) can be identified with the sequences $(a_i)_{1 \leqslant i \leqslant n}$ (n an arbitrary integer $\geqslant 1$) where the a_i are either elements of I or a special symbol P (representing the "open brackets"), where these sequences satisfy the following conditions:

(i) The length n of the sequence is equal to twice the number of symbols P which appear in it, plus 1.

(ii) For all $k \leqslant n - 1$, the number of symbols P which appear in the subsequence $(a_i)_{1 \leqslant i \leqslant k}$ is $\geqslant k/2$.

(Use *Set Theory*, I, Appendix to associate with each sequence (a_i) an element of M(I); for example, to the sequence PPPxyzPxy corresponds the word $(((xy)z)(xy))$.)

¶ 13. Let A be a commutative ring and E an A-module. An A-module E_n is defined inductively by writing

$$E_1 = E, \qquad E_n = \bigoplus_{p+q=n} (E_p \otimes_A E_q) \quad \text{for } n \geqslant 2.$$

We write $ME = \bigoplus_{n=1}^{\infty} E_n$ and an A-algebra structure is defined on ME by setting, for $x_p \in E_p$ and $x_q \in E_q$, $x_p x_q = x_p \otimes x_q \in E_{p+q}$.

(*a*) Show that for every A-algebra B and every A-linear mapping $f: E \to B$, there exists one and only one A-homomorphism of algebras $ME \to B$ extending f. ME is called the *free algebra of the* A-*module* E.

(*b*) An integer $a(n)$ is defined by induction on n by the formulae

$$a(1), \quad a(n) = \sum_{p+q=n} a(p)a(q) \quad \text{if } n \geqslant 2.$$

Show that E_n is isomorphic to the direct sum of $a(n)$ modules isomorphic to $E^{\otimes n}$.

(*c*) Let $f(T)$ be the formal power series $\sum_{n=1}^{\infty} a(n)T^n$. Show that

$$f(T) = T + (f(T))^2.$$

*Deduce the formulae

$$f(T) = \frac{1 - \sqrt{1 - 4T}}{2} \quad \text{and} \quad a(n) = 2^{n-1}\frac{1.3.5\ldots(2n-1)}{n!}.*$$

(*d*) If I is any set and $E = A^{(I)}$, show that $\mathrm{Lib}_A(I)$ is identified with ME. There are canonical homomorphisms $M(I) \xrightarrow{w} \mathbf{N}^{(I)} \xrightarrow{l} \mathbf{N}$ whose composition is length in M(I); $\omega(m)$ is called the *multidegree* of $m \in M(I)$; the mapping ω defines on $\mathrm{Lib}_A(I)$ a graduation of type $\mathbf{N}^{(I)}$. Show that the set $(\mathrm{Lib}_A(I))_\alpha$ of homogeneous elements of multidegree $\alpha \in \mathbf{N}^{(I)}$ under this graduation is a free A-module of rank equal to

$$a(\alpha) = a(n)\frac{n!}{\alpha!} = 2^{n-1}\frac{1.3.5\ldots(2n-1)}{\alpha!}$$

where $\alpha! = \prod_{i \in I} (\alpha(i))!$ and $n = \sum_{i \in I} \alpha(i) = l(\alpha)$ is the length of α.

14. (*a*) Let M be a monoid whose law of composition is denoted by $\top$; suppose further that M is *totally ordered* by an order relation $x \leqslant y$ such that the relations $x < y$ and $x' \leqslant y'$, or $x \leqslant y$ and $x' < y'$, imply $x \top x' < y \top y'$. Show that if A is an integral domain, the algebra over A of the monoid M has no divisors of 0.

(*b*) Deduce that if A is an integral domain, every polynomial algebra $A[(X_i)_{i \in I}]$ is an integral domain.

(*c*) If A is an integral domain and f and g two polynomials in $A[(X_i)_{i \in I}]$ such that fg is homogeneous and $\neq 0$, show that f and g are homogeneous. In particular, the invertible elements of the ring $A[(X_i)_{i \in I}]$ are the invertible elements of the ring A.

15. Let A be a commutative ring and n an integer $\geqslant 2$ such that for all $a \in A$ the equation $nx = a$ has a solution $x \in A$. Let m be an integer $\geqslant 1$ and u a polynomial in A[X] of the form $X^{mn} + f(X)$ where f is of degree $\leqslant mn - 1$; show that there exists a polynomial $v \in A[X]$ of the form $X^m + w$, where w is of degree $\leqslant m - 1$, such that $u - v^n$ is zero or of degree $< m(n-1)$.

16. Let A be a commutative ring and $u = \sum_{k=0}^{m} a_k X^k$ a divisor of 0 in the ring A[X]. Show that if $v = \sum_{k=0}^{n} b_k X^k$ is an element $\neq 0$ of A[X] of degree n such that $uv = 0$, there exists a polynomial $w \neq 0$ of degree $\leqslant n - 1$ such that $uw = 0$. (Reduce it to the case where $b_0 \neq 0$; if $a_k v = 0$ for $0 \leqslant k \leqslant m - 1$, show that we can take $w = b_0$; if $a_k v = 0$ for

$$0 \leqslant k < p \leqslant m - 1$$

and $a_p v \neq 0$, show that $a_p b_0 = 0$ and therefore we can take $w = \sum_{k=0}^{n-1} a_p b_{k+1} X^k$.) Deduce that there exists an element $c \neq 0$ of A such that $cu = 0$.

17. Let A be a commutative ring. Show that, in the algebra A[X] of polynomials in one indeterminate over A, the mapping $(u, v) \mapsto u(v)$ is a law of composition which is associative and left distributive with respect to addition and multiplication in A[X]. If A is an integral domain, show that the relation $u(v) = 0$ implies that $u = 0$ or that v is constant and that, if $u \neq 0$ and the degree of v is > 0, the degree of $u(v)$ is equal to the product of the degrees of u and v.

18. Let K be a commutative field and K[X] the ring of polynomials in one indeterminate over K. Show that every automorphism s of the ring K[X] leaves K invariant (Exercise 14(*c*)) and induces on K an automorphism of this field; show further that necessarily $s(X) = aX + b$, with $a \neq 0$ in K

and $b \in K$ (cf. Exercise 17). Conversely, show that being given an automorphism σ of K and two elements a, b of K such that $a \neq 0$ defines one and only one automorphism s of K[X] such that $s|K = \sigma$ and $s(X) = aX + b$. If G is the group of all automorphisms of the ring K[X] and N the subgroup of G consisting of the K-*algebra* structure of K[X], show that N is a normal subgroup of G and that G/N is isomorphic to the automorphism group of the field K; the group N is isomorphic to the group defined on the set $K^* \times K$ by the law of composition $(\lambda, \mu)(\lambda', \mu') = (\lambda\lambda', \lambda'\mu + \mu')$.

19. Prove the polynomial identity in 8 indeterminates

$$(X_1^2 + X_2^2 + X_3^2 + X_4^2)(Y_1^2 + Y_2^2 + Y_3^2 + Y_4^2)$$
$$= (X_1Y_1 - X_2Y_2 - X_3Y_3 - X_4Y_4)^2 + (X_1Y_2 + X_2Y_1 + X_3Y_4 - X_4Y_3)^2 + (X_1Y_3 + X_3Y_1 + X_4Y_2 - X_2Y_4)^2 + (X_1Y_4 + X_4Y_1 + X_2Y_3 - X_3Y_2)^2.$$

(Apply formula (32) of no. 5 to a suitable quaternion algebra.)

20. Show that there exists no polynomial identity in 6 indeterminates

$$(X_1^2 + X_2^2 + X_3^2)(Y_1^2 + Y_2^2 + Y_3^2) = P^2 + Q^2 + R^2$$

where P, Q, R are three polynomials with coefficients in $\mathbf{Z}$ with respect to the indeterminates X_i and Y_i. (Observe that $15 = 3.5$ cannot be written in the form $p^2 + q^2 + r^2$, where p, q, r are integers.)

21. Let S be a multiplicative monoid with identity element e, satisfying condition (D) of no. 10 and further satisfying the two following conditions: (1) the relation $st = e$ implies $s = t = e$: (2) for all $s \in S$, the number n of terms in a finite sequence $(t_i)_{1 \leqslant i \leqslant n}$ of elements *distinct from* e and such that $t_1t_2 \ldots t_n = s$ is less than a finite number $\nu(s)$ depending only on s.

Show that, for an element $x = \sum_{s \in S} \alpha_s s$ of the total algebra of S over a ring A to be invertible, it is necessary and sufficient that α_e be invertible in A (reduce it to the case where $\alpha_e = 1$ and use the identity

$$e - z^{n+1} = (e - z)(e + z + \cdots + z^n)).$$

22. If K is a commutative field, show that in the ring of formal power series $K[[X_1, X_2, \ldots, X_p]]$ there is only a single maximal ideal, equal to the set of non-invertible elements. On the other hand, show that in the polynomial ring $K[X_1, \ldots, X_p]$ for $p \geqslant 1$ there always exist several distinct maximal ideals.

23. Let K be a commutative field; show that there exists no formal power series $u(X, Y) \in K[[X, Y]]$ such that, for an integer $m > 0$,

$$(X + Y)u(X, Y) = X^m Y^m.$$

24. Let K be a commutative field and let k be an integer such that $k \neq 0$ in K. Show that for every formal power series u of $K[[X]]$ with constant term equal to 1, there exists a formal power series $v \in K[[X]]$ such that $v^k = u$.

25. Let A be a ring, *commutative or otherwise*, and let σ be an endomorphism of A. On the additive product group $E = A^{\mathbf{N}}$ an internal law of composition is defined by writing $(\alpha_n)(\beta_n) = (\gamma_n)$, where $\gamma_n = \sum_{p+q=n} \alpha_p \sigma^p(\beta_q)$ (with the convention $\sigma^0(\xi) = \xi$).

(*a*) Show that this law defines, with addition on E, a *ring* structure on E, admitting as unit element e the sequence (α_n) where $\alpha_0 = 1$ and $\alpha_n = 0$ for $n \geqslant 1$. The mapping which associates with every $\xi \in A$ the element $(\alpha_n) \in E$, such that $\alpha_0 = \xi$, $\alpha_n = 0$ for $n \geqslant 1$, is an isomorphism of A onto a subring of E, with which it is identified. Then we write $\sum_{n=0}^{\infty} \alpha_n X^n$ instead of (α_n) and $X^p \beta = \sigma^p(\beta) X^p$ for all $\beta \in A$; if $u = \sum_{n=0}^{\infty} \alpha_n X^n \neq 0$, the smallest integer n such that $\alpha_n \neq 0$ is called the *order* of u and denoted by $\omega(u)$.

(*b*) If A is a ring with no divisor of 0 and σ is an isomorphism of A onto a subring of A, show that E is a ring with no divisor of 0 and that

$$\omega(uv) = \omega(u) + \omega(v)$$

for $u \neq 0$ and $v \neq 0$.

(*c*) For $u = \sum_{n=0}^{\infty} \alpha_n X^n$ to be invertible, it is necessary and sufficient that α_0 be invertible in A.

(*d*) Suppose that A is a field and σ an *automorphism* of A. Show that E admits a field of left fractions (I, § 9, Exercise 15) and that in this field F every element $\neq 0$ can be written uniquely in the form uX^{-h}, where u is an element of E of order 0.

¶ *26. (*a*) Let A and B be two *well-ordered* subsets of $\mathbf{R}$ (which are necessarily countable; cf. *General Topology*, IV, § 2, Exercise 1). Show that $A + B$ is well-ordered and that, for all $c \in A + B$, there exists only a finite number of ordered pairs (a, b) such that $a \in A$, $b \in B$ and $a + b = c$ (to show that a non-empty subset of $A + B$ has a least element, consider its greatest lower bound in $\mathbf{R}$).

(*b*) Let K be a commutative field. In the vector space $K^{\mathbf{R}}$, consider the vector subspace E consisting of the elements (α_x) such that the set (depending on (α_x)) of $x \in \mathbf{R}$ such that $\alpha_x \neq 0$ is *well-ordered*. For any two elements (α_x), (β_x) of E, we write $(\alpha_x)(\beta_x) = (\gamma_x)$, where $\gamma_x = \sum_{y+z=x} \alpha_y\beta_z$ (a sum which is meaningful by virtue of (*a*)). Show that this law of composition defines, with addition, a *field* structure on E; the elements of E are also denoted by $\sum_{t\in\mathbf{R}} \alpha_t X^t$ and called *formal power series with well-ordered real exponents*, with coefficients in K.$_*$

§ 3

1. Let Δ be a commutative monoid written additively with the identity element denoted by 0, all of whose elements are *cancellable*. Let E be a graded A-algebra of type Δ and suppose that E admits a unit element $e = \sum_{\alpha\in\Delta} e_\alpha$ where $e_\alpha \in E_\alpha$. Show that necessarily $e = e_0 \in E_0$.

§ 4

¶ 1. Let A be a commutative ring, E, F two A-algebras, M a left E-module, N a left F-module and G the algebra $G = E \otimes_A F$. For every ordered pair of endomorphisms $u \in \mathrm{End}_E(M)$, $v \in \mathrm{End}_F(N)$, there exists one and only one G-endomorphism w of $M \otimes_A N$ such that $w(x \otimes y) = u(x) \otimes v(y)$ for $x \in M, y \in N$ and we can write $w = \phi(u \otimes v)$, where ϕ is an A-linear mapping (called *canonical*) of $\mathrm{End}_E(M) \otimes_A \mathrm{End}_F(N)$ into $\mathrm{End}_G(M \otimes_A N)$. Suppose in what follows that A is a *field*.

(*a*) Show that the canonical mapping ϕ is injective (consider an element $z = \sum_i u_i \otimes v_i$, where the $v_i \in \mathrm{End}_F(N)$ are linearly independent over A and write $(\phi(z))(x_\alpha \otimes y) = 0$ for every element x_α of a basis of M over A and all $y \in N$).

(*b*) Suppose that M is a *finitely generated free* E-module. Show that the mapping ϕ is bijective in each of the following cases: (1) E is of finite rank over the field A; (2) N is a finitely generated F-module.

(*c*) Suppose that M is a *free* E-module and that there exist a $y \in N$ and an infinite sequence (v_n) of endomorphisms of the F-module N such that the vector subspace (over A) of N generated by the $v_n(y)$ is of infinite rank over A. Show that, if the mapping ϕ is bijective, every basis of M over E is necessarily *finite*.

(*d*) Suppose still that M is a *free* E-module. Show that, for the G-module $M \otimes_A N$ to be faithful, it is necessary and sufficient that the F-module N be faithful (consider a basis of E over A).

¶ 2. Let K be a commutative field, E, F two fields whose centre contains

K, M a left vector space over E and N a left vector space over F; let $G = E \otimes_K F$.

(*a*) Let $x \neq 0$ be an element of M and $y \neq 0$ an element of N. Show that $x \otimes y$ is free in the G-module $M \otimes_K N$ (use Exercise 1(*d*) and the transitivity of $\mathrm{End}_E(M)$ (resp. $\mathrm{End}_F(N)$) in the set of elements of M (resp. N) distinct from 0).

(*b*) Show that $M \otimes_K N$ is a free G-module. (If (m_α) is a basis of M over E and (n_β) a basis of N over F, show that the sum of the G-modules $G(m_\alpha \otimes n_\beta)$ is direct, using a method analogous to that of (*a*).)

§ 5

1. Let N be the **Z**-module **Z**/4**Z**, M the submodule 2**Z**/4**Z** of N and $j : M \to N$ the canonical injection. Show that $T(j) : T(M) \to T(N)$ is not injective.

2. Let I and J be two finite sets, $I_0 = I \cup \{\alpha\}$, $J_0 = J \cup \{\beta\}$, where α and β are two elements not belonging to I and J respectively, and suppose that I_0 and J_0 are disjoint. Let E be a vector space of finite rank over a commutative field K and let z be an element of $T^I_J(E)$. For all $i \in I$, let W'_i be the subspace of E^* consisting of the $x^* \in E^*$ such that $c^i_\beta(z \otimes x^*) = 0$ (c^i_β being the contraction of index $i \in I$ and index $\beta \in J_0$ in $T^I_{J_0}(E)$) and let V_i be the subspace of E orthogonal to W'_i. For all $j \in J$, let W_j be the subspace of E consisting of the $x \in E$ such that $c^\alpha_j(z \otimes x) = 0$ (c^α_j being the contraction of index $\alpha \in I_0$ and index $j \in J$ in $T^{I_0}_J(E)$) and let V'_j be the subspace of E^* orthogonal to W_j. Show that z belongs to the tensor product $\left(\bigotimes_{i \in I} V_i\right) \otimes \left(\bigotimes_{j \in J} V'_j\right)$ canonically identified with a subspace of $T^I_J(E)$. Moreover, if $(U_i)_{i \in I}$ is a family of subspaces of E, $(U'_j)_{j \in J}$ a family of subspaces of E^* such that z belongs to the tensor product $\left(\bigotimes_{i \in I} U_i\right) \otimes \left(\bigotimes_{j \in J} U'_j\right)$, identified with a subspace of $T^I_J(E)$, then $V_i \subset U_i$ and $V'_j \subset U'_j$ for all $i \in I$ and $j \in J$ (cf. II, § 7, no. 8, Proposition 17).

3. Let M be a finitely generated projective A-module. For every endomorphism u of M, let $\tilde{u}$ denote the tensor $\theta_M^{-1}(u) \in M^* \otimes_A M$ corresponding to it.

(*a*) For every element $x \in M$, show that $u(x) = c^1_2(x \otimes \tilde{u})$, where $x \otimes \tilde{u}$ belongs to $M \otimes M^* \otimes M = T^{\{1,3\}}_{\{2\}}(M)$.

(*b*) Let u, v be two endomorphisms of M and $w = u \circ v$; show that $\tilde{w} = c^2_3(\tilde{v} \otimes \tilde{u})$, where $\tilde{v} \otimes \tilde{u}$ belongs to $T^{\{2,4\}}_{\{1,3\}}(M)$.

(*c*) For every automorphism s of M and every tensor $z \in T^I_J(M)$, a tensor $s.z \in T^I_J(M)$ is defined by the condition that if $z = \bigotimes_{k \in I \cup J} z_k$ where $z_k \in M$ for $k \in I$ and $z_k \in M^*$ for $k \in J$, then $s.z = \bigotimes_{k \in I \cup J} z'_k$ with $z'_k = s(z_k)$ if $k \in I$,

$z'_k = {}^t s^{-1}(z_k)$ if $k \in J$. Show that the group $\mathbf{GL}(M)$ *operates* by the law $(s, z) \mapsto s.z$ on the A-module $T^I_J(M)$, the mappings $z \mapsto s.z$ being automorphisms of this A-module. Moreover, for every ordered pair of indices $i \in I$, $j \in J$,

$$c^i_j(s.z) = s.c^i_j(z).$$

If M is projective and finitely generated, show, in the notation of Exercise 2, that $s.\tilde{u} = (s \circ u \circ s^{-1})^\sim$.

4. Let M be a finitely generated projective A-module; for every bilinear mapping u of $M \times M$ into M, let $\tilde{u}$ denote the tensor $\theta_M^{-1}(u)$ in $T^{\{3\}}_{\{1,2\}}(M)$. Show that for the algebra structure on M defined by u to be associative, it is necessary and sufficient that the tensors $c^3_4(\tilde{u} \otimes \tilde{u})$ and $c^6_2(\tilde{u} \otimes \tilde{u})$ correspond under the canonical associativity isomorphism of $T^{\{4\}}_{\{1,2,3\}}(M)$ onto $T^{\{2\}}_{\{1,3,4\}}(M)$.

¶ 5. Let E be a vector space over a commutative field K and let T(E) be the tensor algebra of E.

(*a*) Show that T(E) contains no divisors of 0 and that the only invertible elements in T(E) are the scalars $\neq 0$. If $\dim(E) > 1$, T(E) is a non-commutative K-algebra.

(*b*) Let u, v be two elements of T(E). Show that, if there exist two elements a, b of T(E) such that $ua = vb \neq 0$, one of the elements u, v is a right multiple of the other. (Consider first the case where u and v are homogeneous; reduce it to the case where E is finite-dimensional and write $u = e_1u_1 + \cdots + e_nu_n$, $v = e_1v_1 + \cdots + e_nv_n$, where (e_k) is a basis of E. Conclude by arguing by induction on the smallest degree of the homogeneous components of ua.) Deduce that, if $\dim(E) > 1$, the ring T(E) admits no field of left fractions (I, § 9, Exercise 15).

(*c*) Show that for two non-zero elements u, v of T(E) to be permutable, it is necessary and sufficient that there exist a vector $x \neq 0$ in E such that u and v belong to the subalgebra of T(E) generated by x (use (*b*)). Deduce that, if $\dim(E) > 1$, the centre of T(E) is equal to K.

¶ 6. Let K be a commutative field and E, F two K-algebras each with a unit element which is identified with the unit element 1 of K. Consider the K-algebra $T(E \oplus F)$, where $E \oplus F$ is considered as a vector space over K; E (resp. F) is identified with a *vector subspace* of $T(E \oplus F)$ under the canonical injection. Let $\mathfrak{J}$ be the two-sided ideal of $T(E \oplus F)$ generated by the elements $x_1 \otimes x_2 - (x_1x_2)$ and $y_1 \otimes y_2 - (y_1y_2)$, where $x_i \in E$, $y_i \in F$, $i = 1, 2$; let $E * F$ denote the quotient algebra $T(E \oplus F)/\mathfrak{J}$; it is called the *free product* of the K-algebras E and F; let ϕ denote the canonical homomorphism of $T(E \oplus F)$ onto $E * F$ and α and β the restrictions of ϕ to E and F; α and β are K-*algebra* homomorphisms.

(*a*) Show that for every ordered pair of K-homomorphisms $u: E \to G$,

$v: \mathrm{F} \to \mathrm{G}$ into a K-algebra G, there exists one and only one K-homomorphism $w: \mathrm{E} * \mathrm{F} \to \mathrm{G}$ such that $u = w \circ \alpha$ and $v = w \circ \beta$ (which justifies the terminology).

(*b*) Let R_n be the vector subspace of $\mathsf{T}(\mathrm{E} \oplus \mathrm{F})$ the sum of the $\mathsf{T}^k(\mathrm{E} \oplus \mathrm{F})$ for $0 \leqslant k \leqslant n$ and write $\mathfrak{J}_n = \mathfrak{J} \cap \mathrm{R}_n$. Show that, if we consider a basis of E consisting of 1 and a family $(e_\lambda)_{\lambda \in \mathrm{L}}$ and a basis of F consisting of 1 and a family $(f_\mu)_{\mu \in \mathrm{M}}$, then a supplementary subspace of $\mathfrak{J}_n + \mathrm{R}_{n-1}$ is obtained in R_n by considering the subspace of R_n with basis the elements of the form $z_1 \otimes z_2 \otimes \cdots \otimes z_n$ where, either z_{2j+1} is equal to one of the e_λ for $2j + 1 \leqslant n$ and z_{2j} is equal to one of the f_μ for $2j \leqslant n$, or z_{2j+1} is equal to one of the f_μ for $2j + 1 \leqslant n$ and z_{2j} is equal to one of the e_λ for $2j \leqslant n$ (argue by induction on n).

(*c*) Let $\mathrm{P}_n = \phi(\mathrm{R}_n) \subset \mathrm{E} * \mathrm{F}$; then $\mathrm{P}_n \subset \mathrm{P}_{n+1}$, $\mathrm{P}_0 = \mathrm{K}$, $\mathrm{P}_h\mathrm{P}_k \subset \mathrm{P}_{h+k}$ and $\mathrm{E} * \mathrm{F}$ is the union of the P_n. Show that there is a canonical vector space isomorphism, for $n \geqslant 1$,

$$\mathrm{P}_n/\mathrm{P}_{n-1} \to (\mathrm{G}_1 \otimes \mathrm{G}_2 \otimes \cdots \otimes \mathrm{G}_n) \oplus (\mathrm{G}_2 \otimes \mathrm{G}_3 \otimes \cdots \otimes \mathrm{G}_{n+1})$$

where we write $\mathrm{G}_k = \mathrm{E}/\mathrm{K}$ for k odd, $\mathrm{G}_k = \mathrm{F}/\mathrm{K}$ for k even. In particular the homomorphisms α and β are injective, which allows us to identify E and F with sub-K-algebras of $\mathrm{E} * \mathrm{F}$. P'_n (resp. P''_n) is used to denote the inverse images in P of $\mathrm{G}_1 \otimes \mathrm{G}_2 \otimes \cdots \otimes \mathrm{G}_n$ (resp. $\mathrm{G}_2 \otimes \mathrm{G}_3 \otimes \cdots \otimes \mathrm{G}_{n+1}$); the *height* of an element $z \in \mathrm{E} * \mathrm{F}$, denoted by $h(z)$, is the smallest integer such that $z \in \mathrm{P}_n$; if $h(z) = n$ and $z \in \mathrm{P}'_n$ (resp. $z \in \mathrm{P}''_n$), z is called *pure odd* (resp. *pure even*). Then $\mathrm{P}_n = \mathrm{P}'_n + \mathrm{P}''_n$ and $\mathrm{P}'_n \cap \mathrm{P}''_n = \mathrm{P}_{n-1}$ for $n \geqslant 1$.

(*d*) Show that, if u, v are two elements of $\mathrm{E} * \mathrm{F}$ such that $h(u) = r$, $h(v) = s$, then $h(uv) \leqslant h(u) + h(v)$ and that $h(uv) = h(u) + h(v)$ unless u and v are pure and, either r is *even* and u, v *are not of the same parity*, or r is *odd* and u, v *are of the same parity*. (If for example r is even, $u \in \mathrm{P}'_r$, $v \in \mathrm{P}'_s$, show that $uv \in \mathrm{P}_{r+s-1}$ is possible only if $u \in \mathrm{P}_{r-1}$ or $v \in \mathrm{P}_{s-1}$.)

(*e*) Suppose that E has no divisors of zero and, in the notation of (*d*), suppose that r is even, u pure even and v pure odd. Show that $h(uv) = r + s - 1$, unless there exist two invertible elements x, y of E such that $uy \in \mathrm{P}_{r-1}$ and $xb \in \mathrm{P}_{s-1}$. (Let (a_i) (resp. (b_j)) be the basis of the supplementary subspace of $\mathfrak{J}_r + \mathrm{R}_{r-1}$ in R_r (resp. of $\mathfrak{J}_s + \mathrm{R}_{s-1}$ in R_s) considered in (*b*); write $u = \sum_i \phi(a_i)x_i$ and $v = \sum_j y_j\phi(b_j)$, where x_i and y_j are in E, and show that, if $uv \in \mathrm{P}_{r+s-2}$, then necessarily $x_iy_j \in \mathrm{K}$.) Treat similarly the other cases where $h(uv) < h(u) + h(v)$ when E and F are assumed to have no divisors of zero.

(*f*) Deduce from (*d*) and (*e*) that, when E and F have no divisors of zero, nor does $\mathrm{E} * \mathrm{F}$.

(*g*) Generalize the above definitions and results to any finite number of unital K-algebras.

¶ 7. Let E be a vector space of finite dimension n over a field K. For every integer $r \geqslant 1$, the symmetric group $\mathfrak{S}_r$ operates linearly on the tensor product $E^{\otimes r}$ by the action $(\sigma, z) \mapsto \sigma . z$ such that

$$\sigma . (x_1 \otimes x_2 \otimes \cdots \otimes x_r) = x_{\sigma^{-1}(1)} \otimes x_{\sigma^{-1}(2)} \otimes \cdots \otimes x_{\sigma^{-1}(r)}$$

for every sequence of r vectors $(x_i)_{1 \leqslant i \leqslant r}$ of E. Let $u_\sigma(z) = \sigma . z$. Show that for every sequence $(v_i)_{1 \leqslant i \leqslant r}$ of r endomorphisms of E,

$$(*) \qquad \operatorname{Tr}(u_\sigma \circ (v_1 \otimes v_2 \otimes \cdots \otimes v_r)) = \operatorname{Tr}(w_1)\operatorname{Tr}(w_2) \ldots \operatorname{Tr}(w_k)$$

where the endomorphisms w_j of E are defined as follows: σ^{-1} is decomposed into cycles, $\sigma^{-1} = \lambda_1 \lambda_2 \ldots \lambda_k$ and, if $\lambda_j = (i_1\ i_2 \ldots i_h)$, we write

$$w_j = v_{i_1} \circ v_{i_2} \circ \cdots \circ v_{i_h}.$$

(Reduce it to calculating the left hand side of (*) when each v_i is of the form $x \mapsto \langle x, a_h^* \rangle a_k$, where $(a_j)_{1 \leqslant j \leqslant n}$ is a basis of E and $(a_j^*)_{1 \leqslant j \leqslant n}$ the dual basis.)

¶ 8. Let A be an integral domain and K its field of fractions; for every A-module M, let $\tau(M)$ denote its torsion module (II, § 7, no. 10).

(*a*) Show that for the algebra T(M) to have no divisors of zero, it is necessary and sufficient that the A-module T(M) be torsion-free (observe that $T(M)_{(K)}$ is isomorphic to $T(M_{(K)})$ and use Exercise 5).

(*b*) Let $u : M \to M'$ be an injective A-module homomorphism. Show that $\operatorname{Ker}(T(u)) \subset \tau(T(M))$; if M' is torsion-free, then $\operatorname{Ker}(T(u)) = \tau(T(M))$ (cf. II, § 7, no. 10, Proposition 27).

(*c*) Give an example of a torsion-free A-module M such that $\tau(T(M))$ is non-zero (II, § 7, Exercise 31).

¶ 9. Let A be a commutative ring, F the free associative algebra over A of the set of integers $I = [1, n]$, where $n \geqslant 2$, and X_i $(1 \leqslant i \leqslant n)$ the corresponding indeterminates, so that a canonical basis of F over A consists of the products $X_{i_1} X_{i_2} \ldots X_{i_r}$, where $(i_1, \ldots, i_r)$ is an arbitrary finite sequence of elements of I.

(*a*) Let C_2 denote the set of commutators $[X_i, X_j] = X_i X_j - X_j X_i$ for $1 \leqslant i < j \leqslant n$. Assuming that C_{m-1} is defined, C_m denotes the set of elements $[X_i, P] = X_i P - P X_i$, where $1 \leqslant i \leqslant n$ and P runs through C_{m-1}. Let U be the graded subalgebra of F generated by 1 and the union of the C_m for $m \geqslant 2$. Show that, for $1 \leqslant i \leqslant n$, $[X_i, P] \in U$ for all $P \in U$ (confine it to the case where P is a product of elements of $\bigcup_m C_m$ and argue by induction on the number of factors).

(*b*) Suppose from now on that A is a *field*. For every integer $m \geqslant 0$, let U_m be the vector A-space of homogeneous elements of degree m in U (so that $U_0 = A$, $U_1 = \{0\}$); choose a basis $R_{m,1}, \ldots, R_{m,d_m}$ in each U_m, consisting of products of elements of $\bigcup_{k \leqslant m} C_k$. For every ordered pair of integers (m, k) such that $m \geqslant 2$, $1 \leqslant k \leqslant d_m$, let $E_{m,k}$ be the vector sub-A-space of U generated by the $R_{p,j}$ such that $p > m$ or $p = m$ and $j \geqslant k$; show that $E_{m,k}$ is a two-sided ideal of U and that, if $L_{m,k}$ is the left ideal of F generated by $E_{m,k}$, $L_{m,k}$ is a two-sided ideal of F.

(*c*) For every multiindex $\alpha = (\alpha_1, \ldots, \alpha_n) \in \mathbf{N}^n$, we write

$$X^\alpha = X_1^{\alpha_1} X_2^{\alpha_2} \ldots X_n^{\alpha_n}.$$

Show that the elements $X^\alpha R_{m,k}$ $(m > 0, 1 \leqslant k \leqslant d_m)$ form a basis of F over K. (To prove that these elements form a generating system of the vector space F, prove that every element $X_{i_1} \ldots X_{i_r}$ is a linear combination of them by induction on the degree r. To see that the $X^\alpha R_{m,k}$ are linearly independent, argue by *reductio ad absurdum* by considering the maximum degree with respect to one of the X_i of the monomials X^α which would appear in a linear relation between the $X^\alpha R_{m,k}$ and proceed by induction on this maximum degree; for this, reduce it to the case where K is infinite (considering if necessary $F \otimes_K K'$ for an extension field K' of K) and observe that when $X_i + \lambda$ is substituted in an $R_{m,k}$ for X_i, for some $\lambda \in K'$, the element $R_{m,k}$ is unaltered).

(*d*) Show that the two-sided ideal $\mathfrak{Z}(F)$ of F generated by the commutators $[u, v] = uv - vu$ of elements of F is the sum of the U_m for $m \geqslant 2$; $F/\mathfrak{Z}(F)$ is isomorphic to $A[X_1, \ldots, X_n]$ and hence is an integral domain.

(*e*) Show that the algebra U is not a free associative algebra by proving that the quotient $U/\mathfrak{Z}(U)$, where $\mathfrak{Z}(U)$ is generated by the commutators of elements of U, admits nilpotent elements $\neq 0$. (Consider the image z' in $U/\mathfrak{Z}(U)$ of the element $z = [X_i, X_j]$ of U; $z'^2 \neq 0$ but $z'^4 = 0$.)

10. Let A be a commutative ring,

$$M' \xrightarrow{u} M \xrightarrow{v} M'' \longrightarrow 0$$

an exact sequence of A-modules and n an integer >0. Let $L = \operatorname{Coker}(T^n(u))$ so that there is an exact sequence

$$T^n(M') \xrightarrow{T^n(u)} T^n(M) \to L \to 0.$$

For every subset J of the set $\{1, 2, \ldots, n\}$, we write

$$P_J = N_1 \otimes N_2 \otimes \cdots \otimes N_n, \quad \text{where } N_i = M' \text{ if } i \in J,\ N_i = M \text{ if } i \notin J$$

($P_\varnothing$ is therefore equal to $T^n(M)$) and

$$Q_J = N_1 \otimes N_2 \otimes \cdots \otimes N_n, \quad \text{where } N_i = M' \text{ if } i \in J,\ N_i = M'' \text{ if } i \notin J$$

($Q_\varnothing$ is therefore equal to $T^n(M'')$.) For every integer i such that $0 \leqslant i \leqslant n$, let $P^{(i)}$ denote the submodule of $T^n(M)$ generated by the union of the images of the P_J for all the subsets J such that $\text{Card}(J) = i$ and let $L^{(i)}$ be the canonical image of $P^{(i)}$ in L.

(*a*) Show that there exists a canonical surjective homomorphism

$$\bigoplus_{\text{Card}(J)=i} Q_J \to L^{(i)}/L^{(i+1)}$$

(cf. II, § 5, no. 6, Proposition 6).

(*b*) If the sequence $0 \to M' \xrightarrow{u} M \xrightarrow{v} M'' \to 0$ is exact and splits, show that the homomorphism defined in (*a*) is bijective.

§ 6

1. If M is a monogenous A-module, then $\mathsf{S}(M) = \mathsf{T}(M)$. Deduce an example of an injective homomorphism $j: N \to M$ such that $\mathsf{S}(j): \mathsf{S}(N) \to \mathsf{S}(M)$ is not injective (§ 5, Exercise 1).

2. Establish for symmetric algebras the properties analogous to those of Exercise 8(*a*) and (*b*) of § 5.

3. Let K be a field, B the polynomial ring $K[X_1, X_2]$ and $\mathfrak{p}$ the ideal of B generated by the polynomial $X_1^3 - X_2^2$.

(*a*) Show that $\mathfrak{p}$ is prime and therefore $A = B/\mathfrak{p}$ is an integral domain. Let a and b denote the canonical images of X_1 and X_2 in A; a and b are not invertible in A and $a^3 = b^2$.

(*b*) Let $\mathfrak{m}$ be the (maximal) ideal of A generated by a and b. Show that the symmetric algebra $\mathsf{S}(\mathfrak{m})$ is not an integral domain. (Observe that $\mathfrak{m}$ is identified with the quotient module A^2/N, where N is the submodule of A^2 generated by $(b, -a)$; conclude that $\mathsf{S}(\mathfrak{m})$ is isomorphic to the quotient ring $A[U, V]/(bU - aV)$ and show that in the polynomial ring $A[U, V]$ the principal ideal $(bV - aU)$ is not prime, by considering the product $b^2(bV^3 - U^3)$.)

¶ 4. Let A be a commutative ring and $\mathfrak{a}$ an ideal of A. The *Rees algebra of the ideal* $\mathfrak{a}$ is the subalgebra $\mathsf{R}(\mathfrak{a})$ of the polynomial algebra A[T] in one indeterminate, consisting of the polynomials $c_0 + c_1T + \cdots + c_nT^n$, where $c_0 \in A$ and $c_k \in \mathfrak{a}^k$ for every integer $k \geqslant 1$.

(*a*) Show that there exists one and only one surjective A-algebra homomorphism $r: \mathsf{S}(\mathfrak{a}) \to \mathsf{R}(\mathfrak{a})$ such that the restriction of r to $\mathfrak{a}$ is the A-linear mapping $x \to xT$.

(*b*) Suppose that the ring A is an integral domain. Let $a_1, \ldots, a_n$ be elements of A with $a_n \neq 0$; consider the A-algebra homomorphism

$$u: A[X_1, \ldots, X_n] \to A[T]$$

such that $u(X_i) = a_iT$. The kernel $\mathfrak{r} = \text{Ker}(u)$ is a graded ideal whose homogeneous component $\mathfrak{r}_m$ of degree m consists of the homogeneous polynomials f of degree m such that $f(a_1, \ldots, a_n) = 0$. If $\mathfrak{r}' \subset \mathfrak{r}$ is the ideal of $A[X_1, \ldots, X_n]$ generated by the homogeneous component $\mathfrak{r}_1$ of $\mathfrak{r}$, show that the A-module $\mathfrak{r}/\mathfrak{r}'$ is a torsion A-module. (To verify that, for all $f \in \mathfrak{r}_m$, there exists $c \neq 0$ in A such that $cf \in \mathfrak{r}'$, argue by induction on m, by writing

$$f(X_1, \ldots, X_n) = X_1 f_1(X_1, \ldots, X_n) + X_2 f_2(X_2, \ldots, X_n) + \cdots + X_n f_n(X_n)$$

where the f_i are homogeneous of degree $m - 1$; consider the polynomial $g(X_1, \ldots, X_n) = X_1 f_1(a_1, \ldots, a_n) + X_2 f_2(a_2, \ldots, a_n) + \cdots + X_n f_n(a_n)$, observe that $g \in \mathfrak{r}_1 \subset \mathfrak{r}'$ and form $a_n^{m-1} f - X_n^{m-1} g$.)

(*c*) Deduce from (*b*) that when A is an integral domain, the kernel of the homomorphism $\mathfrak{r}$ of (*a*) is the torsion module of $\mathsf{S}(\mathfrak{a})$. Deduce that, for $\mathsf{S}(\mathfrak{a})$ to be an integral domain, it is necessary and sufficient that $\mathsf{S}(\mathfrak{a})$ be a torsion-free A-module. In particular, if the ideal $\mathfrak{a}$ is a projective A-module, $\mathsf{S}(\mathfrak{a})$ is an integral domain and isomorphic to $\mathsf{R}(\mathfrak{a})$.

¶ 5. Let E be a vector space of finite dimension n over a field K.

(*a*) Show that for every integer m the symmetric power $S^m(E)$ and the vector space $S'_m(E)$ of symmetric contravariant tensors of order m have the same dimension $\binom{m+n-1}{n-1} = \binom{n+m-1}{m}$.

*(*b*) If K is of characteristic $p > 0$, show that the vector space $S''_p(E)$ of symmetrized tensors of order p has dimension $\binom{n}{p}$ (observe that if, in a tensor product $x_1 \otimes x_2 \otimes \cdots \otimes x_p$ of p vectors of E, $x_i = x_j$ for an ordered pair of distinct indices i, j, then the symmetrization of this product is zero; use the fact that if $(H_i)_{1 \leqslant i \leqslant r}$ is a partition of $\{1, 2, \ldots, p\}$ into non-empty sets, comprising at least two sets, the subgroup of $\mathfrak{S}_p$ leaving each of the H_i stable has order not divisible by p). On the other hand, the canonical image in $S^p(E)$ of the space $S'_p(E)$ of symmetric tensors of order p is of dimension n and is generated by the images of the tensors $x \otimes x \otimes \cdots \otimes x$ (p times), where $x \in E$ (use the same remark); the canonical mapping of $S'_p(E)$ into $S^p(E)$ is therefore not bijective although the two vector spaces have the same dimension.*

§ 7

1. Let A be an integral domain and K its field of fractions.

(*a*) Show that the exterior power $\bigwedge^2 K$ of the A-module K reduces to 0.

(*b*) For every A-module $E \subset K$, show that every exterior power $\bigwedge^p E$ is a torsion A-module for $p \geqslant 2$.

(*c*) If E is the A-module defined in II, § 7, Exercise 31, show that $\bigwedge^2 E$ is non-zero. Give an example of an integral domain A and an A-module E contained in the field of fractions K of A such that no exterior power $\bigwedge^p E$ reduces to 0.

2. For every ideal $\mathfrak{a}$ of a ring A and every A-module E, the exterior algebra $\bigwedge(E/\mathfrak{a}E)$ is isomorphic to $(\bigwedge E)/\mathfrak{a}(\bigwedge E)$.

3. Give an example of a ring A with the following property: in the A-module $E = A^2$, there exists a submodule F such that, if $j : F \to E$ is the canonical injection, $\bigwedge^2 j$ is identically zero even when neither $\bigwedge^2 E$ nor $\bigwedge^2 F$ reduces to 0 (II, § 4, Exercise 2).

4. Let E be a free A-module of dimension $n \geqslant 3$. Show that if $1 \leqslant p \leqslant n - 1$ the A-modules $\bigwedge^p E$ and $\bigwedge^{n-p} E$ are isomorphic; but if $2p \neq n$, there exists no isomorphism ϕ of $\bigwedge^p E$ onto $\bigwedge^{n-p} E$ such that for every automorphism u of E, $\phi \circ \left(\bigwedge^p u\right) = \left(\bigwedge^{n-p} u\right) \circ \phi$ (if (e_i) is a basis of E, consider the automorphism u such that $u(e_i) = e_i + e_j$, $y(e_k) = e_k$ for $k \neq i$, and give i and j all possible values).

5. Let K be a commutative field, E, F two vector spaces over K and u a linear mapping of E into F of rank r. Show that if $p \leqslant r$ the rank of $\bigwedge^p u$ is equal to $\binom{r}{p}$ and that if $p > r$, $\bigwedge^p u$ is identically zero (take in E a basis r vectors of which form a basis of a subspace supplementary to $u^{-1}(0)$ and the rest a basis of $\overset{-1}{u}(0)$).

¶ 6. Let K be a commutative field and E a vector space over K.

(*a*) For an element $z = \sum_{p=0}^{\infty} z_p$ $\left(\text{with } z_p \in \bigwedge^p E\right)$ of the exterior algebra $\bigwedge E$ of E to be invertible, it is necessary and sufficient that $z_0 \neq 0$ (prove it first when E is finite-dimensional and then pass to the general case, noting that for all $z \in \bigwedge E$ there exists a finite-dimensional subspace F of E such that $z \in \bigwedge F$).

(*b*) Suppose that K is such that $2 \neq 0$ in K. Show that, if E is infinite-dimensional or of even finite dimension, the centre of the algebra $\bigwedge E$ consists of the elements such that $z_p = 0$ for all *odd* indices p. If E is of odd dimension n, the centre of $\bigwedge E$ is the sum of the above subspace and $\bigwedge^n E$. In all cases, the centre of $\bigwedge E$ is identical with the set of elements of $\bigwedge E$ permutable with all the invertible elements of $\bigwedge E$.

7. For every A-module E, show that, if we write $[u, v] = u \wedge v - v \wedge u$ for any two elements of $\bigwedge E$, then for any three elements u, v, w of $\bigwedge E$, $[[u, v], w] = 0$.

¶ 8. Let E be a free A-module.

(*a*) Show that in $T^n(E)$, $\mathfrak{J}''_n$ (no. 1) is a free sub-A-module equal to the kernel of the endomorphism $z \mapsto \boldsymbol{a}.z$ and admits a supplement which is a free A-module; $\bigwedge^n E$ is therefore canonically isomorphic to $A''_n(E)$. If also in A the equation $2\xi = 0$ implies $\xi = 0$, then $A''_n(E) = A'_n(E)$.

*(*b*) Suppose that A is a field of characteristic $p > 0$ and that $p \neq 2$. Then the canonical image of $A'_p(E) = A''_p(E)$ in $\bigwedge^p E$ is zero.

(*c*) Suppose that A is a field of characteristic 2. Then $A'_n(E) = S'_n(E)$, $A''_n(E) = S''_n(E)$, but in general $A'_n(E) \neq A''_n(E)$.*

9. Let E be an A-module. The skewsymmetrization of a linear mapping g of $T^n(E)$ into an A-module F is by definition the linear mapping $\boldsymbol{a}g$ of $T^n(E)$ into F such that $\boldsymbol{a}g(z) = g(\boldsymbol{a}z)$ for all $z \in T^n(E)$; if $\phi : E^n \to T^n(E)$ is the canonical mapping, the n-linear mapping $\boldsymbol{a}g \circ \phi$ is also called the skewsymmetrization of the n-linear mapping $g \circ \phi$. Deduce from Exercise 8(*a*) that if E is free every alternating n-linear mapping of E^n into F is the skewsymmetrization of an n-linear mapping of E^n into F.

¶ 10. Let E be an A-module with a generating system of n elements. Show that every alternating n-linear mapping of E^n into an A-module F is the skewsymmetrization of an n-linear mapping of E^n into F. (Let a_j $(1 \leqslant j \leqslant n)$ be the generators of E and $\phi : A^n \to E$ the homomorphism such that $\phi(e_j) = a_j$ for all j, where (e_j) is the canonical basis of A^n. If $f : E^n \to F$ is an alternating n-linear mapping, so is $f_0 : (x_1, \dots, x_n) \mapsto f(\phi(x_1), \dots, \phi(x_n))$; show that if g_0 is an n-linear mapping of $(A^n)^n$ into F such that $f_0 = \boldsymbol{a}g_0$, g_0 can also be written as $(x_1, \dots, x_n) \mapsto g(\phi(x_1), \dots, \phi(x_n))$, where g is an n-linear mapping of E^n into F.)

¶ 11. Let K be a commutative field in which $2 = 0$, A the polynomial ring K[X, Y, Z] and E the A-module A^3/M, where M is the submodule of A^3 generated by the element (X, Y, Z) of A^3; finally, let F be the quotient A-module of A by the ideal $AX^2 + AY^2 + AZ^2$. Show that there exists an alternating bilinear mapping of E^2 into F, which is the skewsymmetrization of no bilinear mapping of E^2 into F, but that in $T^2(E)$ the module $\mathfrak{J}''_2$ is equal to the kernel of the endomorphism $z \mapsto \boldsymbol{a}z$ (consider $T^2(E)$ as a quotient module of $T^2(A^3)$ and show that the kernel of $z \mapsto \boldsymbol{a}z$ is the canonical image of the module of symmetric tensors $S'_2(A^3)$).

¶ 12. In the notation of Exercise 11, let B be the quotient ring

$$A/(AX^2 + AY^2 + AZ^2)$$

and let G be the B-module $E \otimes_A B$. Show that in $T^2(G)$, the module $\mathfrak{J}''_2$ is distinct from the kernel of the endomorphism $z \mapsto \boldsymbol{a}z$ (method analogous to that of Exercise 11).

13. (*a*) Let M be a graded A-module of type **N**. Show that there exists on the algebra $S(M)$ (resp. $\bigwedge(M)$) one and only one graduation of type **N** under which $S(M)$ (resp. $\bigwedge(M)$) is a graded A-algebra and which induces on M the given graduation. Show that if the homogeneous elements of even degree in M are all zero (in which case the graduation on M is called *odd*), the graded algebra $\bigwedge(M)$ thus obtained is *alternating* (§ 4, no. 9, Definition 7).

(*b*) Given a graded A-module of type **N**, consider the following universal mapping problem: Σ is the species of anticommutative graded algebra structure (§ 4, no. 9, Definition 7), the morphisms are A-homomorphisms of graded algebras (§ 3, no. 1) and the α-mappings are the graded homomorphisms of degree 0 of M into an anticommutative graded A-algebra. Let M^- (resp. M^+) denote the graded submodule of M generated by the homogeneous elements of odd (resp. even) degree. Show that the graded algebra $G(M) = \bigwedge(M^-) \otimes_A S(M^+)$ is anticommutative; a canonical injection j of M into this algebra is defined by writing $j(x) = x \otimes 1$ if $x \in M^-$, $j(x) = 1 \otimes x$ if $x \in M^+$. Show that the graded algebra $G(M)$ and the injection j constitute a solution of the above universal mapping problem.

$$G(M) = \bigwedge(M^-) \otimes_A S(M^+)$$

is called the *universal anticommutative algebra* over the graded A-module M.

(*c*) Prove the analogues of Propositions 2, 3 and 4 for the universal anticommutative algebra. Consider the case where M admits a basis consisting of homogeneous elements: find explicitly in this case a basis of the universal anticommutative algebra.

14. Let M be a projective A-module and let $x_1, \dots, x_n$ be linearly independent elements of M; for every subset $H = \{i_1, \dots, i_m\}$ of $m \leqslant n$ elements of the interval $[1, n]$, where $i_1 < i_2 < \dots < i_m$, we write

$$x_H = x_{i_1} \wedge x_{i_2} \wedge \dots \wedge x_{i_m}.$$

Show that when H runs through the set of subsets of $[1, n]$ with m elements, the x_H are linearly independent in $\bigwedge(M)$.

15. Let M be an n-dimensional free A-module. Show that every generating system of M with n elements is a basis of M.

§ 8

1. In a matrix U of order n, if, for each index i, the column of index i is replaced by the sum of the columns of index $\neq i$, the determinant of the matrix obtained is equal to $(-1)^{n-1}(n-1)\det U$. If in U, for each i, the sum of the columns of index $\neq i$ is subtracted from the column of index i, the determinant obtained is equal to $-(n-2)2^{n-1}\det U$.

2. Let $\Delta = \det(\alpha_{ij})$ be the determinant of a matrix of order n; for

$$1 \leqslant i \leqslant n-1$$

and $1 \leqslant j \leqslant n-1$, we write

$$\beta_{ij} = \begin{vmatrix} \alpha_{1j} & \alpha_{1,j+1} \\ \alpha_{i+1,j} & \alpha_{i+1,j+1} \end{vmatrix}.$$

Show that the determinant $\det(\beta_{ij})$ of order $n-1$ is equal to

$$\alpha_{12}\alpha_{13}\ldots\alpha_{1,n-1}\Delta.$$

3. Show the identity

$$\begin{vmatrix} x_1 & y_1 & \alpha_{13} & \alpha_{14} & \ldots & \alpha_{1n} \\ \lambda_1 x_2 & x_2 & y_2 & \alpha_{24} & \ldots & \alpha_{2n} \\ \lambda_1\lambda_2 x_3 & \lambda_2 x_3 & x_3 & y_3 & \ldots & \alpha_{3n} \\ \cdot & \cdot & \cdot & \cdot & \cdot & \cdot \\ \lambda_1\lambda_2\ldots\lambda_{n-1}x_n & \lambda_2\ldots\lambda_{n-1}x_n & \lambda_3\ldots\lambda_{n-1}x_n & \lambda_4\ldots\lambda_{n-1}x_n & \ldots & x_n \end{vmatrix}$$
$$= (x_1-\lambda_1 y_1)(x_2-\lambda_2 y_2)\ldots(x_{n-1}-\lambda_{n-1}y_{n-1})x_n.$$

Deduce the following identities:

$$\begin{vmatrix} 1 & 1 & 1 & \ldots & 1 \\ b_1 & a_1 & a_1 & \ldots & a_1 \\ b_1 & b_2 & a_2 & \ldots & a_2 \\ \cdot & \cdot & \cdot & \cdot & \cdot \\ b_1 & b_2 & b_3 & \ldots & a_n \end{vmatrix} = (a_1-b_1)(a_2-b_2)\ldots(a_n-b_n)$$

$$\begin{vmatrix} a_1b_1 & a_1b_2 & a_1b_3 & \dots & a_1b_n \\ a_1b_2 & a_2b_2 & a_2b_3 & \dots & a_2b_n \\ a_1b_3 & a_2b_3 & a_3b_3 & \dots & a_3b_n \\ \cdot & \cdot & \cdot & \cdot & \cdot \\ a_1b_n & a_2b_n & a_3b_n & \dots & a_nb_n \end{vmatrix} = a_1b_n(a_2b_1 - a_1b_2)\dots(a_nb_{n-1} - a_{n-1}b_n)$$

$$\begin{vmatrix} a_2a_3\dots a_n & a_3a_4\dots a_nb_1 & a_4\dots a_nb_1b_2 & \dots & b_1b_2b_3\dots b_{n-1} \\ b_2b_3\dots b_n & a_3a_4\dots a_na_1 & a_4\dots a_na_1b_2 & \dots & a_1b_2b_3\dots b_{n-1} \\ a_2b_3\dots b_n & b_3b_4\dots b_nb_1 & a_4\dots a_na_1a_2 & \dots & a_1a_2b_3\dots b_{n-1} \\ \cdot & \cdot & \cdot & \cdot & \cdot \\ a_2a_3\dots b_n & a_3a_4\dots b_nb_1 & a_4\dots b_nb_1b_2 & \dots & a_1a_2a_3\dots a_{n-1} \end{vmatrix} = (a_1a_2\dots a_n - b_1b_2\dots b_n)^{n-1}.$$

$$\begin{vmatrix} x & a_1 & a_2 & \dots & a_{n-1} & 1 \\ a_1 & x & a_2 & \dots & a_{n-1} & 1 \\ a_1 & a_2 & x & \dots & a_{n-1} & 1 \\ \cdot & \cdot & \cdot & \cdot & \cdot & \cdot \\ a_1 & a_2 & a_3 & \dots & x & 1 \\ a_1 & a_2 & a_3 & \dots & a_n & 1 \end{vmatrix} = (x - a_1)(x - a_2)\dots(x - a_n)$$

$$\begin{vmatrix} x & a_1 & a_2 & \dots & a_n \\ a_1 & x & a_2 & \dots & a_n \\ a_1 & a_2 & x & \dots & a_n \\ \cdot & \cdot & \cdot & \cdot & \cdot \\ a_1 & a_2 & a_3 & \dots & x \end{vmatrix} = (x + a_1 + a_2 + \dots + a_n)(x - a_1)(x - a_2)\dots(x - a_n)$$

(reduce the last determinant to the preceding one).

4. Calculate the determinant

$$\Delta_n = \begin{vmatrix} a_1 + b_1 & b_1 & b_1 & \dots & b_1 \\ b_2 & a_2 + b_2 & b_2 & \dots & b_2 \\ \cdot & \cdot & \cdot & \cdot & \cdot \\ b_n & b_n & b_n & \dots & a_n + b_n \end{vmatrix}$$

(express Δ_n in terms of Δ_{n-1}).

5. If the a_i and b_j are elements of a commutative field such that $a_i + b_j \neq 0$ for every ordered pair of indices (i, j), prove that

$$\det\left(\frac{1}{a_i + b_j}\right) = \frac{\sum_{i<j} (a_j - a_i)(b_j - b_i)}{\sum_{i,j} (a_i + b_j)}$$

("Cauchy's determinant").

6. Show that, if X is a matrix with n rows and m columns, Y a matrix with p rows and n columns and $Z = YX$ the product matrix with p rows and m columns, the minors of Z of order q are zero if $n < q$; if $q \leqslant n$, they are given by

$$\det(Z_{\mathrm{L,H}}) = \sum_{\mathrm{K}} \det(Y_{\mathrm{L,K}})\det(X_{\mathrm{K,H}})$$

where K runs through the set of subsets of $[1, n]$ with q elements (use formula (3) of § 7, no. 2).

7. Let $\Delta = \det(\alpha_{ij})$ be the determinant of a matrix U of order n; for each index i $(1 \leqslant i \leqslant n)$ let Δ_i denote the determinant obtained by multiplying the element α_{ij} in U by β_j for $1 \leqslant j \leqslant n$. Show that the sum of the n determinants Δ_i $(1 \leqslant i \leqslant n)$ is equal to $(\beta_1 + \beta_2 + \cdots + \beta_n)\Delta$ (expand Δ_i by the row of index i).

8. Let $\Delta = \det(\alpha_{ij})$ be the determinant of a matrix U of order n and σ a permutation belonging to $\mathfrak{S}_n$; for each index i $(1 \leqslant i \leqslant n)$ let Δ_i be the determinant obtained by replacing the element α_{ij} in U by $\alpha_{i,\sigma(j)}$ for $1 \leqslant j \leqslant n$; if p is the number of indices invariant under the permutation σ, show that the sum of the n determinants Δ_i $(1 \leqslant i \leqslant n)$ is equal to $p\Delta$ (same method as in Exercise 7).

9. Let A be a square matrix of order n, B a submatrix of A with p rows and q columns and C the matrix obtained by multiplying in A, each of the elements belonging to B by the same scalar α. Show that each term in the total expansion of $\det C$ is equal to the corresponding term in the total expansion of $\det A$ multiplied by a scalar of the form α^r, where $r \geqslant p + q - n$ and r depends on the term considered (perform a suitable Laplace expansion of $\det C$).

10. Let Γ, Δ be the determinants of two matrices U, V of order n; let H and K be any two subsets of $[1, n]$ with p elements and (i_k) (resp. (j_k)) the sequence obtained by arranging the elements of H (resp. K) in increasing order; let $\Gamma_{\mathrm{H,K}}$ be the determinant obtained by replacing in U the column of index i_k by the column of V of index j_k, for $1 \leqslant k \leqslant p$; similarly let $\Delta_{\mathrm{K,H}}$ be the determinant obtained by replacing in V the column of index j_k by the

column of U of index i_k, for $1 \leqslant k \leqslant p$. Show that, for every subset H of $(1, n)$ with p elements,

$$\Gamma\Delta = \sum_{\mathrm{K}} \Gamma_{\mathrm{H,K}} \Delta_{\mathrm{K,H}}$$

where K runs through the set of subsets of $(1, n)$ with p elements (use formulae (21) and (22) of no. 6).

11. Let Δ be the determinant of a square matrix X of order n and Δ_p the determinant of the square matrix $\overset{p}{\bigwedge} X$, of order $\binom{n}{p}$. Show that

$$\Delta_p \Delta_{n-p} = \Delta^{\binom{n}{p}}$$

(use formulae (21) and (22) of no. 6).

12. A matrix (α_{ij}) of order n is called *centrosymmetric* (resp. *skew-centrosymmetric*) if $\alpha_{n-i+1,n-j+1} = \alpha_{ij}$ (resp. $\alpha_{n-i+1,n-j+1} = -\alpha_{ij}$) for $1 \leqslant i \leqslant n$, $1 \leqslant j \leqslant n$.

(*a*) Show that the determinant of a centrosymmetric matrix of even order $2p$ can be expressed in the form of a product of two determinants of order p and the determinant of a centrosymmetric matrix of order $2p + 1$ in the form of a product of a determinant of order p and a determinant of order $p + 1$.

(*b*) Show that the determinant of a skew-centrosymmetric matrix of even order $2p$ can be expressed in the form of a product of two determinants of order p. The determinant of a skew-centrosymmetric matrix of odd order $2p + 1$ is zero if in A the relation $2\xi = 0$ implies $\xi = 0$; otherwise it can be expressed in the form of the product of $\alpha_{p+1,p+1}$ by two determinants of order p.

13. Let $\Delta = \det(\alpha_{ij})$ be a determinant of order n and Δ_{ij} the minor of order $n - 1$, the complement of α_{ij}. Show that

$$\begin{vmatrix} \alpha_{11} & \alpha_{12} & \dots & \alpha_{1n} & x_1 \\ \alpha_{21} & \alpha_{22} & \dots & \alpha_{2n} & x_2 \\ \cdot & \cdot & \cdot & \cdot & \cdot \\ \alpha_{n1} & \alpha_{n2} & \dots & \alpha_{nn} & x_n \\ y_1 & y_2 & \dots & y_n & z \end{vmatrix} = \Delta z - \sum_{i,j} (-1)^{i+j} \Delta_{ij} x_i y_j.$$

If $\Delta = 0$ and the α_{ij} belong to a *field*, show that the above determinant is the product of a linear form in $x_1, x_2, \dots, x_n$ and a linear form in $y_1, y_2, \dots, y_n$ (use Exercise 5 of § 5 and Exercise 8 of II, § 6).

Give an example where this result fails to hold when the ring of scalars A is not a field (take A to be the ring $\mathbf{Z}/(6)$ and $n = 2$).

14. Show the identity

$$\begin{vmatrix} 0 & 1 & 1 & 1 & \dots & 1 \\ 1 & 0 & a_1 + a_2 & a_1 + a_3 & \dots & a_1 + a_n \\ 1 & a_2 + a_1 & 0 & a_2 + a_3 & \dots & a_2 + a_n \\ 1 & a_3 + a_1 & a_3 + a_2 & 0 & \dots & a_3 + a_n \\ \cdot & \cdot & \cdot & \cdot & \cdot & \cdot \\ 1 & a_n + a_1 & a_n + a_2 & a_n + a_3 & \dots & 0 \end{vmatrix} = (-1)^n 2^{n-1} \sum_{i=1}^{n} a_1 a_2 \dots a_{i-1} a_{i+1} \dots a_n$$

(use Exercise 13).

15. Show that if the columns of a square matrix of order n over A are linearly independent, the rows of the matrix are also linearly independent (cf. II, § 10, Exercise 3).

16. Let E, F be two free A-modules of respective dimensions m and n. For a linear mapping $u : \mathrm{E} \to \mathrm{F}$ to be injective, it is necessary and sufficient that $m \leqslant n$ and that, if X denotes the matrix of u with respect to any two bases of E and F, there exist no scalar $\mu \neq 0$ such that the products by μ of all the minors of X of order m are zero.

17. Let

$$(*) \qquad \sum_{j=1}^{n} \alpha_{ij} \xi_j = \beta_i \qquad (1 \leqslant i \leqslant m)$$

be a system of m linear equations in n unknowns over a commutative ring A. Let x_j $(1 \leqslant j \leqslant n)$ be the columns of the matrix $X = (\alpha_{ij})$ of this system and $y = (\beta_i)$; suppose that in X all the minors of order $> p$ are zero but that $x_1 \wedge x_2 \wedge \cdots \wedge x_p \neq 0$. For the system $(*)$ to have a solution, it is necessary that $x_1 \wedge x_2 \wedge \cdots \wedge x_p \wedge y = 0$. Conversely, if this condition holds, there exist $n + 1$ elements ξ_j $(1 \leqslant j \leqslant n + 1)$ of A such that $\xi_{n+1} \neq 0$ and

$$\sum_{j=1}^{n} \alpha_{ij} \xi_j = \beta_i \xi_{n+1} \qquad (1 \leqslant i \leqslant m).$$

18. Let E and F be two free A-modules of respective dimensions m and n, $u : \mathrm{E} \to \mathrm{F}$ a linear mapping and X the matrix of u with respect to any two bases of E and F.

(*a*) If $m \geqslant n$ and there exists a minor of X of order n which is invertible, u is surjective.

(*b*) Conversely, if u is surjective, show that $m \geqslant n$ and that there exists a minor of X of order n which is non-zero. If further, in the ring A, the ideal generated by the non-invertible elements is distinct from A, there exists a minor of X of order n which is invertible (consider the exterior power $\bigwedge^n u$).

(*c*) Let B be a commutative ring and A the product ring B $\times$ B. Give an example of a surjective linear mapping of the A-module A^2 onto A, all the elements of whose matrix (with respect to the canonical bases of A^2 and A) are divisors of zero.

19. Let X be a matrix over a commutative field. For X to be of rank p, it suffices that there exist a minor of X of order p which is $\neq 0$ and such that all the minors of order $p + 1$ containing this minor of order p are zero (show that every column of X is then a linear combination of the p columns to which the elements of the minor of order p in question belong).

20. Let A be a commutative ring, M an arbitrary A-module and $U = (\alpha_{ij})$ a matrix of type (m, n) with elements in A.

(*a*) Suppose that $m = n$ and let $\Delta = \det(U)$; then, if $x_1, \dots, x_n$ are elements of M such that $\sum_{j=1}^{n} \alpha_{ij}x_j = 0$ for $1 \leqslant i \leqslant n$, then $\Delta x_i = 0$ for every index i.

(*b*) Suppose that m and n are arbitrary, that all the minors of U of order $> r$ are zero and that a minor of order r is invertible, for example that of the matrix (α_{ij}) where $i \leqslant r$ and $j \leqslant r$; then if a_i denotes the row of index i in U, the a_i of index $\leqslant r$ form a basis of the submodule of A^n generated by all the rows of U; for $i > r$, therefore $a_i = \sum_{k=1}^{r} \lambda_{ik}a_k$. Then let $y_1, \dots, y_m$ be elements of M; for there to exist elements $x_1, \dots, x_n$ of M satisfying the system of equations $\sum_{j=1}^{n} \alpha_{ij}x_j = y_i$ for $1 \leqslant i \leqslant m$, it is necessary and sufficient that $y_i = \sum_{k=1}^{r} \lambda_{ik}y_k$ for $i > r$.

¶ 21. Let K be an infinite commutative field, m, n, r three integers $\geqslant 0$ such that $r \leqslant n \leqslant m$, $\mathbf{M}_{m,n}(K)$ the vector K-space (of dimension mn) consisting of the matrices of type (m, n) over K and V a *vector subspace* of $\mathbf{M}_{m,n}(K)$ such that r is the greatest value of the rank of matrices $X \in V$. We propose to prove the inequality

$$(*) \qquad \dim(V) \leqslant mr.$$

(*a*) Show that we can assume that $m = n$ and that the matrix

$$X_0 = \begin{pmatrix} I_r & 0 \\ 0 & 0 \end{pmatrix}$$

belongs to V. Deduce that every matrix $X \in V$ is then of the form

$$\begin{pmatrix} X_{11} & X_{12} \\ X_{21} & 0 \end{pmatrix}$$

(where X_{11} is of order r) with the condition $X_{21}X_{12} = 0$ (use the fact that for all $\xi \in K$, the matrix $\xi X_0 + X$ is of rank $\leqslant r$).

(*b*) Deduce from (*a*) that, for two matrices X, Y in V,

$$X_{21}Y_{12} + Y_{21}X_{12} = 0.$$

(*c*) For every matrix $X \in V$, we write $F(X) = (X_{11}X_{12}) \in \mathbf{M}_{r,n}(K)$; f is a linear mapping and $\dim(V) = \dim(\operatorname{Im}(f)) + \dim(\operatorname{Ker}(f))$. On the other hand, let u_x denote the linear form $(Y_{11}Y_{12}) \mapsto \operatorname{Tr}(X_{21}Y_{12})$ on $\mathbf{M}_{r,n}(K)$. Show that the linear mapping $X \mapsto u_X$ of $\operatorname{Ker}(f)$ into the dual $(\mathbf{M}_{r,n}(K))^*$ is injective; complete the proof of (*) by noting that the image under $X \mapsto u_X$ of $\operatorname{Ker}(f)$ is contained in the orthogonal of $\operatorname{Im}(f)$.

¶ 22. Let F be a mapping of $\mathbf{M}_n(K)$ into a set E, where K is a commutative field, such that, for every triple of matrices X, Y, Z in $\mathbf{M}_n(K)$,

$$F(XYZ) = F(XZY).$$

We except the case $\mathbf{M}_2(\mathbf{F}_2)$; show then that there exists a mapping Φ of K into E such that $F(X) = \Phi(\det(X))$. (Using the Corollary to Proposition 17 (no. 9), show first that, for $U \in \mathbf{SL}_n(K)$, $F(X) = F(XU)$; deduce that, in $\mathbf{GL}_n(K)$, $F(X)$ depends only on $\det(X)$. On the other hand, if, for $r \leqslant n-1$, we write

$$X_r = \begin{pmatrix} I_r & 0 \\ 0 & 0 \end{pmatrix}$$

show that there exists a matrix Y of rank r such that $X_{r-1} = YX_r$ and $Y = X_rY$ and deduce that $F(X)$ has the same value for all matrices of rank $< n$.)

23. In every A-algebra E, if $x_1, \ldots, x_n$ are elements of E, we write

$$[x_1x_2\ldots x_n] = \sum_{\sigma \in \mathfrak{S}_n} \varepsilon_\sigma x_{\sigma(1)}x_{\sigma(2)}\ldots x_{\sigma(n)}.$$

Show that if E admits a finite generating system over A, there exists an integer N such that $[x_1x_2\ldots x_N] = 0$ for all elements x_j $(1 \leqslant j \leqslant N)$ of E.

24. Let X_i $(1 \leqslant i \leqslant m)$ be square matrices of the same order n over the commutative ring A. Show that if m is even then (Exercise 23)

$$\operatorname{Tr}[X_1X_2\ldots X_m] = 0$$

and if m is odd

$$\mathrm{Tr}[X_1X_2\ldots X_m] = m.\mathrm{Tr}(X_m[X_1X_2\ldots X_{m-1}]).$$

(If λ is the cyclic permutation $(1, 2, \ldots, m)$, C the cyclic subgroup of $\mathfrak{S}_m$ generated by λ and H the subgroup of $\mathfrak{S}_m$ consisting of the permutations leaving m invariant, note that every permutation in $\mathfrak{S}_m$ can be written uniquely as $\sigma\tau$ with $\sigma \in \mathrm{H}$ and $\tau \in \mathrm{C}$; then use the fact that the trace of a product of matrices is invariant under cyclic permutation of the factors.)

¶ 25. Let A be a commutative ring containing the field $\mathbf{Q}$ of rational numbers and let X be a square matrix of even order $2n \geqslant 2$ over A. Let L be the subset $\{1, 2, \ldots, n+1\}$ of $\{1, 2, \ldots, 2n\}$ and L′ its complement. Show the following formula ("Redei's identity"):

$$(*) \quad \det(X_{\mathrm{L},\mathrm{L}}) \det(X_{\mathrm{L}',\mathrm{L}'}) = \sum_{\mathrm{H},\mathrm{K}} (-1)^r \frac{1}{n\binom{n-1}{r}} \det(X_{\mathrm{H},\mathrm{K}}) \det(X_{\mathrm{H}',\mathrm{K}'})$$

where H runs through the set of subsets of n elements of $\{1, 2, \ldots, 2n\}$ containing 1 and K the set of subsets of n elements of L and $r = \mathrm{Card}(\mathrm{H} \cap \mathrm{L}')$. (Evaluate the coefficient of a term $x_{\sigma(1),1}x_{\sigma(2),2}\ldots x_{\sigma(2n),2n}$ in the right-hand side of (*) for an arbitrary permutation $\sigma \in \mathfrak{S}_{2n}$, noting that this coefficient can only be $\neq 0$ if $\mathrm{H} = \sigma(\mathrm{K})$; show that it can only be $\neq 0$ if $\sigma(\mathrm{L}) = \mathrm{L}$.)

26. (*a*) In the notation of Proposition 19 (no. 10), suppose that there exist two A[X]-linear mappings g_1, g_2 of M[X] into M′[X] such that $\psi' \circ g_2 = g_1 \circ \psi$; show that there then exists an A[X]-linear mapping g of M_u into $\mathrm{M}'_{u'}$ such that $\phi' \circ g_1 = \phi' \circ \bar{g}$. Further, if g_1 and g_2 are A[X]-isomorphisms of M[X] onto M′[X], g is an A[X]-isomorphism of M_u onto $\mathrm{M}'_{u'}$.

(*b*) In the notation of no. 10, the endomorphisms u, u' are called *equivalent* if there exist two isomorphisms f_1, f_2 of M onto M such that $u' \circ f_2 = f_1 \circ u$; they are called *similar* if there exists an isomorphism f of M onto M′ such that $u' \circ f = f \circ u$. Deduce from (*a*) that, for u and u' to be similar, it is necessary and sufficient that the endomorphisms $\mathrm{X} - \bar{u}$ and $\mathrm{X} - \bar{u}'$ (of the A[X]-modules M[X] and M′[X] respectively) be equivalent.

§ 9

1. Let K be a commutative field with at least 3 elements and A the algebra over K with a basis consisting of the unit element 1 and two elements e_1, e_2 such that $e_1^2 = e_1$, $e_1e_2 = e_2$, $e_2e_1 = e_2^2 = 0$; let A^0 be the opposite algebra to A. Show that there exists in A elements x such that $\mathrm{Tr}_{\mathrm{A}/\mathrm{K}}(x) \neq \mathrm{Tr}_{\mathrm{A}^0/\mathrm{K}}(x)$ and $\mathrm{N}_{\mathrm{A}/\mathrm{K}}(x) \neq \mathrm{N}_{\mathrm{A}^0/\mathrm{K}}(x)$.

§ 10

1. If A and B are K-algebras and α and β two K-homomorphisms of A into B, an (α, β)-derivation of A into B is a K-linear mapping d of A into B satisfying the relation

$$d(xy) = (dx)\alpha(y) + \beta(x)(dy) \quad \text{for} \quad x, y \quad \text{in} \quad A,$$

in other words, a derivation in the sense of Definition 1 (no. 2) where $A = A' = A''$, $B = B' = B''$, $d = d' = d''$, $\mu : A \times A \to A$ is multiplication, $\lambda_1 : B \times A \to B$ is the K-bilinear mapping $(z, x) \mapsto z\alpha(x)$ and $\lambda_2 : A \times B \to B$ the K-bilinear mapping $(x, z) \mapsto \beta(x)z$.

Suppose henceforth that A and B are *commutative fields* and that $\alpha \neq \beta$. Show that there then exists an element $b \in B$ such that d is the (α, β)-derivation $x \mapsto b\alpha(x) - \beta(x)b$. (It can be assumed that $d \neq 0$. Show first that the kernel N of d is the set of $x \in A$ such that $\alpha(x) = \beta(x)$ and then show that for $x \notin N$ the element $b = d(x)/(\alpha(x) - \beta(x))$ is independent of the choice of x.)

2. Let K be a commutative field of characteristic $\neq 2$, A and B two K-algebras and α a K-linear mapping of A into B; an α-derivation of A into B is a K-linear mapping d of A into B satisfying the relation

$$d(xy) = (dx)\alpha(y) + \alpha(x)(dy) \quad \text{for} \quad x, y \quad \text{in} \quad A,$$

in other words, a derivation on the sense of Definition 1 with $A = A' = A''$, $B = B' = B''$, $d = d' = d''$, μ multiplication in A, λ_1 the mapping $(z, x) \mapsto z\alpha(x)$ of $B \times A$ into B and λ_2 the mapping $(x, z) \mapsto \alpha(x)z$ of $A \times B$ into B.

Show that if B is a *commutative field*, an extension of K, and $d \neq 0$, there exists $b \neq 0$ in B such that

$$\alpha(xy) = \alpha(x)\alpha(y) + b(dx)(dy) \quad \text{for} \quad x, y \quad \text{in} \quad A.$$

If $b = c^2$ with $c \in B$, $c \neq 0$, there exist two homomorphisms f, g of A into B such that $\alpha(x) = (f(x) + g(x))/2$, $dx = (f(x) - g(x))/2c$. Otherwise, there exists a quadratic algebra (§ 2, no. 3) B′ over B, a $\mu \in B'$ such that $\mu^2 = c$, and a homomorphism f of A into B′ such that $\alpha(x) = (f(x) + \overline{f(x)})/2$, $dx = (f(x) - \overline{f(x)})/2$.

¶ 3. Let K be a field, commutative or otherwise, E the left vector space $K_s^{(N)}$ and let $(e_n)_{n \in N}$ be the canonical basis of this space. On the other hand let σ be an endomorphism of K and d a $(\sigma, 1_K)$-derivation (Exercise 1) of K into itself, in other words an additive mapping of K into itself such that

$$d(\xi\eta) = (d\xi)\sigma(\eta) + \xi(d\eta).$$

The elements of E are linear combinations of products $\alpha_n e_n$ and a multiplication is defined on E (**Z**-bilinear mapping of E $\times$ E into E) by the conditions

$$(\alpha e_n)(\beta e_m) = (\alpha e_{n-1})(\sigma(\beta)e_{m+1} + (d\beta)e_m) \text{ for } n \geqslant 1$$
$$(\alpha e_0)(\beta e_m) = (\alpha\beta)e_m.$$

Thus an (in general non-commutative) ring structure is defined on E with e_0 as unit element (identified with the element 1 of K). If we write $X = e_1$, then $e_n = X^n$ for $n \geqslant 1$, so that the elements of E can be written as

$$\alpha_0 + \alpha_1 X + \cdots + \alpha_n X^n,$$

with the multiplication rule

$$X\alpha = \sigma(\alpha)X + (d\alpha) \quad \text{for} \quad \alpha \in K.$$

This ring is denoted by $K[X; \sigma, d]$ and is called the *ring of non-commutative polynomials in* X *with coefficients in* K, *relative to* σ *and* d. For a non-zero element $z = \sum_j \alpha_j X^j$ of E, the *degree* $\deg(z)$ is the greatest integer j such that $\alpha_j \neq 0$.

(*a*) Show that if u, v are two elements $\neq 0$ of E, then:

$$uv \neq 0 \quad \text{and} \quad \deg(uv) = \deg(u) + \deg(v);$$
$$\deg(u + v) \leqslant \sup(\deg(u), \deg(v)) \quad \text{if } u + v \neq 0.$$

Moreover, if $\deg(u) \geqslant \deg(v)$, there exist two elements w, r of E such that $u = wv + r$ and, either $r = 0$, or $\deg(r) < \deg(v)$ ("Euclidean division" in E).

(*b*) Conversely, let R be a ring such that there exists a mapping $u \mapsto \deg(u)$ of the set R′ of elements $\neq 0$ of R into **N**, satisfying the conditions of (*a*). Show that the set K consisting of 0 and the non-zero elements $u \in R$ such that $\deg(u) = 0$ is a (not necessarily commutative) field. If $K \neq R$ and x is an element of R for which the integer $\deg(x) > 0$ is the smallest possible, show that every element of R can be written uniquely in the form $\sum_i \alpha_j x^j$ with $\alpha_j \in K$. (To see that every element of R is of this form, argue by *reductio ad absurdum* by considering an element $u \in R$ which is not of this form and for which $\deg(u)$ is the smallest possible.) Prove that there exist an endomorphism σ of K and a $(\sigma, 1_K)$-derivation d of K such that $x\alpha = \sigma(\alpha)x + d\alpha$ for all $\alpha \in K$. Deduce that R is isomorphic to a field or a ring $K[X; \sigma, d]$.

4. Let M be a graded K-module of type **N**; show that every graded endomorphism of degree r can be extended uniquely to a derivation of degree r of the universal anticommutative algebra G(M) (§ 7, Exercise 13).

* 5. Let K be a field of characteristic $p > 0$, B the field K(X) of rational functions in one indeterminate and A the subfield $K(X^p)$ of B. Show that the canonical homomorphism $\Omega_K(A) \otimes_A B \to \Omega_K(B)$ is not injective.*

§ 11

1. If V is a finitely generated projective A-module, so is End(V), canonically identified with $V \otimes_A V^*$ (II, § 4, no. 2, Corollary to Proposition 2). If every element $u \in \mathrm{End}(V)$ is associated with the A-linear form $v \mapsto \mathrm{Tr}(uv)$ on End(V), an A-linear bijection is defined of End(V) onto its dual $(\mathrm{End}(V))^*$. Identifying End(V) and its dual A-module under this mapping, the A-algebra structure on End(V) defines by duality (no. 1, *Example* 3) an A-cogebra structure on End(V), which is coassociative and has as counit the form $\mathrm{Tr}: V \to A$. In particular, for $V = A^n$, a cogebra structure is defined on $\mathbf{M}_n(A)$ for which the coproduct is given by $c(E_{ij}) = \sum_k E_{kj} \otimes E_{ik}$. This cogebra structure and the algebra structure on $\mathbf{M}_n(A)$ do not define a bigebra structure.

2. Show that in Definition 3 (no. 4), condition (3) (relative to graded bigebras) is equivalent to the following: the product $m: E \otimes E \to E$ is a morphism of the graded cogebra $E \otimes E$ into the graded cogebra E.

3. Show that if M is an A-module there exists on T(M) one and only one cocommutative bigebra structure whose algebra structure is the usual structure and whose coproduct c is such that, for all $x \in M$, $c(x) = 1 \otimes x + x \otimes 1$.

4. Let E be a commutative bigebra over A, m the product $E \otimes E \to E$, c the coproduct $E \to E \otimes E$, e the unit element and γ the counit of E.

(*a*) Show that for every *commutative* A-algebra B, the law of composition which associates with every ordered pair

$$(u, v) \in \mathrm{Hom}_{A\text{-alg.}}(E, B) \times \mathrm{Hom}_{A\text{-alg.}}(E, B)$$

the A-algebra homomorphism

$$E \xrightarrow{c} E \otimes E \xrightarrow{u \otimes v} B \otimes B \xrightarrow{m_B} B$$

(where m_B is the product in B) defines a monoid structure on $\mathrm{Hom}_{A\text{-alg.}}(E, B)$.

(*b*) In order that, for every commutative A-algebra B, $\mathrm{Hom}_{A\text{-alg.}}(E, B)$ be a group, it is necessary and sufficient that there exist an A-algebra homomorphism $i: E \to E$ such that $i(e) = e$ and the composite mappings $m \circ (1_E \otimes i) \circ c$ and $m \circ (i \otimes 1_E) \circ c$ are both equal to the mapping $x \mapsto \gamma(x)e$. i is then called an *inversion* in E; it is an isomorphism of E onto the opposite bigebra E^0.

5. Let E_1, E_2 be two bigebras over A. Show that on the tensor product $E_1 \otimes_A E_2$ the algebra and cogebra structures the tensor product of those on E_1 and E_2 define on $E_1 \otimes_A E_2$ a bigebra structure.

6. Let M be a graded A-module of type $\mathbf{N}$. Define a skew graded bigebra structure on the universal anticommutative algebra G(M) (§ 7, Exercise

13); deduce a canonical graded algebra homomorphism $G(M^*) \to G(M)^{*gr}$ and notions of inner product between elements of $G(M)$ and elements of $G(M^*)$.

7. With the hypotheses and notation of Proposition 11 (no. 9), prove the formula

$$(\det g)\phi^{-1} = \bigwedge(g) \circ \phi^{-1} \circ \bigwedge({}^tg).$$

Form this formula and (79) (no. 11) derive a new proof of the expression for the inverse of a square matrix using the matrix of its cofactors (§ 8, no. 6, formula (26)).

8. Let E be a free A-module of dimension n. For every A-linear mapping v of $\bigwedge(M)$ into an A-module G, we can write $v = \sum_{p=0}^{n} v_p$, where v_p is the restriction of v to $\bigwedge^p(M)$; we then set $\eta v = \sum_{p=0}^{n} (-1)^p v_p$, so that $\eta^2 v = v$. Let u be an isomorphism of M onto its dual M* and let Δ be the determinant of the matrix of u with respect to a basis (e_i) of M and the dual basis of M* (a determinant which does not depend on the basis of M chosen). If ϕ is the isomorphism of $\bigwedge(M)$ onto $\bigwedge(M^*)$ defined starting with $e = e_1 \wedge e_2 \wedge \cdots \wedge e_n$, show the formula

$$\eta^{n+1}\Delta\phi = \bigwedge({}^tu) \circ \phi^{-1} \circ \bigwedge(u).$$

9. The *adjoint* of a square matrix X of order n over A is the matrix $\tilde{X} = (\det(X^{ji}))$ of minors of X of order $n - 1$. Show that if X is invertible then $\det(\tilde{X}) = (\det X)^{n-1}$ and that every minor $\det(\tilde{X}_{H,K})$ of $\tilde{X}$ of order p is given by the formula

$$(1) \qquad \det(\tilde{X}_{H,K}) = (\det X)^{p-1}(X_{H',K'})$$

where H' and K' are the complements of H and K respectively in $[1, n]$ ("Jacobi's identities"; use relation (80) of no. 11).

10. From every identity $\Phi = 0$ between minors of an arbitrary invertible square matrix X of order n over A another identity $\tilde{\Phi} = 0$ can be derived called the *complement* of $\Phi = 0$, by applying the identity $\Phi = 0$ to the minors of the adjoint $\tilde{X}$ of X and then replacing the minors of $\tilde{X}$ by functions of the minors of X using identity (1) of Exercise 9. In this way show the following identity:

$$\det(X^{ih})\det(X^{jk}) - \det(X^{ik})\det(X^{jh}) = (\det X)\det(X^{ij,hk}) \quad \text{for} \quad i < j$$

where $X^{ij,hk}$ denotes the minor of X of order $n - 2$ obtained by suppressing in X the rows of index i, j and the columns of index h, k.

¶ 11. Let $\Phi = 0$ be an identity between minors of an invertible square matrix X of order n over a commutative ring A, $\tilde{\Phi} = 0$ the complementary identity (Exercise 10) and k an integer >0. Let Y be an invertible square matrix of order $n + k$ and let $\tilde{Y}_0$ be the submatrix of the adjoint matrix $\tilde{Y}$ formed by suppressing in $\tilde{Y}$ the rows of index $\leqslant k$ and the columns of index $\leqslant k$. If $\tilde{Y}_0$ is assumed to be invertible and the identity $\tilde{\Phi} = 0$ is applied to the minors of $\tilde{Y}_0$ and then each minor which appears in this identity (considered as a *minor of* $\tilde{Y}$) is replaced by its expression as a function of the minors of Y, using identity (1) of Exercise 9, an identity $\Phi_k = 0$ is obtained between minors of the matrix Y (valid if Y and $\tilde{Y}_0$ are invertible) which is called the *extension of order* k of the identity $\Phi = 0$.

In particular, Let $A = (\alpha_{ij})$ be an invertible square matrix of order $n + k$, B the submatrix of A of order k obtained by suppressing in A the rows and columns of index $>k$ and Δ_{ij} the determinant of the matrix of order $k + 1$ obtained by suppressing in A the rows of index $>k$ except that of index $k + i$ and the columns of index $>k$ except that of index $k + j$; if C denotes the matrix (Δ_{ij}) of order n, show the identity

$$\det C = (\det A)(\det B)^{n-1}$$

valid when A and B are invertible (show that it is the extension of order k of the total expansion of a determinant of order n).

State the identities obtained by extending the Laplace expansion (§ 8, no. 6) and the identity of Exercise 2 of § 8.

¶ 12. Let $X = (\xi_{ij})$ be a square matrix of order n over a commutative field K; let H be a subset of $[1, n]$ with p elements and H′ the complement of H in $[1, n]$; suppose that, for every ordered pair of indices $h \in \mathrm{H}$, $k \in \mathrm{H}'$,

$$\sum_i \xi_{ih}\xi_{ik} = 0.$$

Show that, for all subsets L, M of $[1, n]$ with p elements,

$$\rho_{\mathrm{M},\mathrm{M}'} \det(X_{\mathrm{L},\mathrm{H}}) \det(X_{\mathrm{M}',\mathrm{H}'}) - \rho_{\mathrm{L},\mathrm{L}'} \det(X_{\mathrm{M},\mathrm{H}}) \det(X_{\mathrm{L},\mathrm{H}'}) = 0.$$

(Consider the columns x_h of index $h \in \mathrm{H}$ in X as vectors in $\mathrm{E} = \mathrm{K}^n$ and the columns x_k^* of index $k \in \mathrm{H}'$ as vectors in E* and show that the $(n - p)$-form $x_{\mathrm{H}'}^*$ is proportional to $\phi_p(x_{\mathrm{H}})$.)

13. Let Γ and Δ be two determinants of invertible matrices of order n over a commutative ring and Γ_{ij} the determinant obtained by replacing in Γ the column of index i by the column of Δ of index j. Show that

$$\det(\Gamma_{ij}) = \Gamma^{n-1}\Delta.$$

(Expand Γ_{ij} by the i-th column and use Exercise 9.)

14. Let M be a free A-module of dimension $n > 1$. Show that every isomorphism ψ of $\overset{p}{\bigwedge}(\mathrm{M})$ onto $\overset{n-p}{\bigwedge}(\mathrm{M}^*)$ such that $\overset{n-p}{\bigwedge}(\check{u}) \circ \psi = \psi \circ \overset{p}{\bigwedge}(u)$ for every automorphism u of M, of determinant equal to 1, is one of the isomorphisms ϕ_p defined in Proposition 12 of no. 11 (proceed as in § 7, Exercise 4).

15. Let E be a free A-module of dimension n. Let z be a p-vector over E and z^* a $(p + q)$-form on E; z can be canonically identified (§ 7, Exercise 8) with the skewsymmetrization $\boldsymbol{a}z_1$ of a contravariant tensor z_1 of order p and z^* with the skewsymmetrization of a covariant tensor of order $p + q$. If, in the mixed tensor $z_1 . z^*$, the k-th contravariant index and the $(p + k)$-th covariant index are contracted for $1 \leqslant k \leqslant p$, show that the covariant tensor of order q thus obtained is skewsymmetrized and is canonically identified with the q-form $z^* \llcorner z$, up to sign.

16. Let E be a free A-module of dimension n. The *regressive product* of $x \in \bigwedge(\mathrm{E})$ and $y \in \bigwedge(\mathrm{E})$ with respect to a basis $\{e\}$ of $\overset{n}{\bigwedge}(\mathrm{E})$, denoted by $x \vee y$, is the element $\overset{-1}{\phi}(\phi(x) \wedge \phi(y))$, the isomorphism ϕ being relative to e. This product is only defined to within an invertible factor, depending on the basis $\{e\}$ chosen. Show that if x is a p-vector and y a q-vector, $x \vee y = 0$ if $p + q < n$ and that, if $p + q \geqslant n$, $x \vee y$ is a $(p + q - n)$-vector such that

$$y \vee x = (-1)^{(n-p)(n-q)} x \vee y.$$

The regressive product is associative and distibutive with respect to addition and defines on $\bigwedge(\mathrm{E})$ a unital algebra structure isomorphic to the exterior algebra structure. Express the components of $x \vee y$ as a function of those of x and y for a given basis on E.

17. Let K be a commutative field and E a vector space over K.

(*a*) Let z be a p-vector $\neq 0$ over E and let V_z be the vector subspace of E consisting of the vectors x such that $z \wedge x = 0$. Then $\dim(\mathrm{V}_z) \leqslant p$; if $(x_i)_{1 \leqslant i \leqslant q}$ is a free system of vectors of V_z, there exists a $(p - q)$-vector v such that $z = v \wedge x_1 \wedge \cdots \wedge x_q$. If E is finite-dimensional, then $\mathrm{V}_z = \overset{-1}{\phi}(\mathrm{M}_z^0)$, where M_z^0 is the orthogonal in E^* of the subspace M_z associated with z. For z to be a pure p-vector, it is necessary and sufficient that $\dim(\mathrm{V}_z) = p$ and then $\mathrm{V}_z = \mathrm{M}_z$.

(*b*) Let v be a pure p-vector, w a pure q-vector and V and W the subspaces of E associated respectively with v and w. In order that $\mathrm{V} \subset \mathrm{W}$, it is necessary and sufficient that $q \geqslant p$ and that there exist a $(q - p)$-vector u such that $w = u \wedge v$. In order that $\mathrm{V} \cap \mathrm{W} = \{0\}$, it is necessary and sufficient that

$v \wedge w \neq 0$; the subspace $V + W$ is then associated with the pure $(p+q)$-vector $v \wedge w$.

(*c*) Let u and v be two pure p-vectors and U and V the subspaces associated respectively with u and v. For $u + v$ to be pure, it is necessary and sufficient that $\dim(U \cap V) \geqslant p - 1$.

18. Let $z = \sum_H \alpha_H e_H$ be a non-zero pure p-vector of an n-dimensional vector space E, expressed using its components with respect to an arbitrary basis $(e_i)_{1 \leqslant i \leqslant n}$ of E. Let G be a subset of $(1, n)$ with p elements such that $\alpha_G \neq 0$; let $(i_h)_{1 \leqslant h \leqslant p}$ be the sequence of indices in G arranged in increasing order and $(j_k)_{1 \leqslant k \leqslant n-p}$ the sequence of indices in the complement G′ of G, arranged in increasing order. For every ordered pair (h, k) of indices such that $1 \leqslant h \leqslant p$, $1 \leqslant k \leqslant n - p$, let β_{hk} be the component α_H of z corresponding to the subset H of $(1, n)$ consisting of the $p - 1$ indices in G distinct from i_h and of index $j_k \in G'$; let X be the matrix (β_{hk}) with p rows and $n - p$ columns. Given an arbitrary subset L of $(1, n)$ with p elements such that $L \cap \complement G$ has $q \geqslant 1$ elements, show that $(\alpha_G)^{q-1}\alpha_L$ is equal, up to sign, to the minor of X of order q, consisting of the rows of index h such that $i_h \in G \cap \complement L$ and the columns of index k such that $j_k \in L \cap \complement G$. (Write z as being of the form $\alpha_G y_1 \wedge y_2 \wedge \cdots \wedge y_p$, where the vectors y_i are such that, in the matrix Y with n rows and p columns whose columns are the y_i, the submatrix consisting of the rows whose indices belong to G is the unit matrix.)

19. Over a finite-dimensional vector space E, let $z = x_1 \wedge x_2 \wedge \cdots \wedge x_p$ be a pure p-vector $\neq 0$; let v^* be a q-form $(q \leqslant p)$ and u^* an arbitrary element of $\bigwedge(E^*)$. Show that

$$(u^* \wedge v^*) \lrcorner z = (-1)^{q(q-1)/2} \sum_H \rho_{H,K} \langle v^*, x_H \rangle (u^* \lrcorner x_K)$$

where H runs through the set of subsets of $(1, p)$ with q elements, K is the complement of H in $(1, p)$ and x_H (resp. x_K) denotes the exterior product of the x_i such that $i \in H$ (resp. $i \in K$).

20. Let E be an n-dimensional vector space, z a pure p-vector over E, V the vector subspace of E associated with z, u^* a pure q-vector over E*, W′ the subspace of E* associated with u^* and W the subspace of E, of dimension $n - q$, orthogonal to W′. If $q < p$, $u^* \lrcorner z$ is a pure $(p - q)$-vector over E, which is zero if $\dim(V \cap W) > p - q$; if $\dim(V \cap W) = p - q$, $V \cap W$ is associated with $u^* \lrcorner z$.

21. (*a*) Show directly (without using the isomorphisms ϕ_p) that every $(n-1)$-vector over an n-dimensional vector space is pure.

(*b*) Let A be a commutative algebra over a field K with a basis consisting of the unit element 1 and three elements a_1, a_2, a_3 whose products with one another are zero. Let E be the A-module A^3 and $(e_i)_{1 \leqslant i \leqslant 3}$ its canonical basis; show that the bivector

$$a_1(e_2 \wedge e_3) + a_2(e_3 \wedge e_1) + a_3(e_1 \wedge e_2)$$

is not pure.

22. Let E be an ordered set in which every interval $[a, b]$ is finite. Let S be the subset of $E \times E$ consisting of the ordered pairs (x, y) such that $x \leqslant y$. Show that a coassociative cogebra structure is defined on the A-module $A^{(S)}$ of formal linear combinations of the elements of S by taking as coproduct

$$c((x, y)) = \sum_{x \leqslant z \leqslant y} (x, z) \otimes (z, y).$$

For this cogebra the linear form γ such that

$$\gamma((x, y)) = 0 \quad \text{if} \quad x \neq y \text{ in E}, \qquad \gamma((x, x)) = 1$$

is a counit.

23. Let C be a cogebra over a ring A. A sub-A-module S of C is called a *subcogebra* of C if $c(S) \subset \operatorname{Im}(S \otimes S)$.

(*a*) If C is coassociative (resp. cocommutative), so is every subcogebra. If γ is a counit of C, the restriction of γ to any subcogebra of C is a counit.

(*b*) Which are the subcogebras of the cogebra $A^{(X)}$ (no. 1, *Example* 4).?

(*c*) If (S_α) is a family of subcogebras of C, $\sum_\alpha S_\alpha$ is a subcogebra of C (the smallest containing all the S_α).

24. Let C be a cogebra over a ring A. A sub-A-module S of C is called a *right* (resp. *left*) *coideal* if $c(S) \subset \operatorname{Im}(S \otimes C)$ (resp. $c(S) \subset \operatorname{Im}(C \otimes S)$).

(*a*) A subcogebra (Exercise 23) is a right and left coideal. The converse is true if A is a field.

(*b*) Every sum of right (resp. left) coideals is a right (resp. left) coideal.

(*c*) If S is a right (resp. left) coideal of C, the orthogonal S^0 is a right (resp. left) ideal of the dual algebra C^*.

(*d*) If A is a field and $\mathfrak{I}'$ a right (resp. left) ideal of the dual algebra C^*, the orthogonal $\mathfrak{I}'^0$ in C is a right (resp. left) coideal of C. (Argue by *reductio ad absurdum* by assuming that there exists an $x \in \mathfrak{I}'^0$ such that $c(x) \notin \mathfrak{I}'^0 \otimes C$; write $c(x) = \sum_i y_i \otimes z_i$, where the z_i are linearly independent, and show that there would be $y' \in \mathfrak{I}'$ and $z' \in C^*$ such that $\langle y'z', x\rangle \neq 0$.)

Deduce that for a vector subspace S of C to be a right (resp. left) coideal, it is necessary and sufficient that S^0 be a right (resp. left) ideal of C^*.

(*e*) Deduce from (*d*) that if A is a field every intersection of right coideals (resp. left coideals, resp. subcogebras) of C is a right coideal (resp. left coideal, resp. subcogebra). For S to be a subcogebra of C, it is necessary and sufficient that its orthogonal S^0 be a two-sided ideal of the algebra C^*.

25. Let C be a cogebra over a ring A. A sub-A-module S of C is called a *coideal* if $c(S) \subset \mathrm{Im}(S \otimes C) + \mathrm{Im}(C \otimes S)$. If C is counital, with counit γ, a coideal S is called *conull* if $\gamma(x) = 0$ for $x \in S$.

(*a*) A right or left coideal is a coideal. Every sum of coideals is a coideal.

(*b*) If S is a coideal of C, the orthogonal S^0 is a *subalgebra* of the dual algebra C^*. If C is counital and S is conull, S^0 is unital.

(*c*) Suppose that A is a field and C a counital cogebra. Show that if E′ is a unital subalgebra of C^* with the same unit element as C^*, the orthogonal E'^0 in C is a conull coideal.

(*d*) If S is a coideal of C, show that there exists on C/S one and only one cogebra structure such that the canonical mapping $p: C \to C/S$ is a cogebra morphism. If C is counital and S is conull, C/S is counital.

(*e*) For every cogebra morphism $f: C \to C'$, $S = \mathrm{Ker}(f)$ is a coideal of C, $S' = \mathrm{Im}(f)$ a subcogebra of C′ and, in the canonical factorization

$$C \xrightarrow{p} C/S \xrightarrow{g} S' \xrightarrow{j} C'$$

of f, g is a cogebra isomorphism.

(26) Let C be a coassociative counital cogebra over a commutative field K.

(*a*) For every vector subspace E of C which is a left C^*-module under the law $(u, x) \mapsto u \lrcorner x$, the left annihilator of this C^*-module is a two-sided ideal $\mathfrak{J}'$ of the algebra C^*. Show that if E is finite-dimensional over K, $C^*/\mathfrak{J}'$ is finite-dimensional.

(*b*) Deduce from (*a*) that for every element $x \in C$, the subcogebra of C generated by x is finite-dimensional. (Note that the C^*-module E generated by x is finite-dimensional and that, if $\mathfrak{J}'$ is its left annihilator, $x \in \mathfrak{J}'^0$, the orthogonal of $\mathfrak{J}'$ in C.)

(27) Let B be an associative unital algebra over a commutative field K and let $m: B \otimes_K B \to B$ be the K-linear mapping defining the multiplication on B. The tensor product $B^* \otimes_K B^*$ (where B^* is the dual vector space of the vector space B) is canonically identified with a vector subspace of $(B \otimes_K B)^*$ (II, § 7, no. 7, Proposition 16).

(*a*) Show that for an element $w \in B^*$ the following conditions are equivalent: (1) ${}^t m(w) \in B^* \otimes B^*$; (2) the set of $x \lrcorner w$, where x runs through B, is a finite-dimensional subspace of B^*; (3) the set of $w \llcorner x$, where x runs through B, is a finite-dimensional subspace of B^*; (4) the set of $x \lrcorner w \llcorner y$, where x and y run through B, is contained in a finite-dimensional subspace of B^*.

(*b*) Let $B' \subset B^*$ be the set of $w \in B^*$ satisfying the equivalent conditions in (*a*). Show that ${}^t m(B') \subset B' \otimes_K B'$. (For $w \in B'$, write ${}^t m(w) = \sum_i u_i \otimes v_i$, where for example the u_i are linearly independent in B^*; show that then $v_i \in B'$ for all i using the associativity of B and arguing by *reductio ad absurdum*.) B' is therefore the *largest cogebra* contained in B^*, for which ${}^t m$ is the coproduct. B' is called the *dual cogebra* of the algebra B. When B is finite-dimensional, $B' = B^*$.

(*c*) Show that if B is a commutative field of infinite dimension over K, $B' = \{0\}$. (Note that the orthogonal of the set of $x \lrcorner w$, for same $w \in B^*$, x running through B, is an ideal of B.)

(*d*) Let B_1, B_2 be two K-algebras and $f: B_1 \to B_2$ a K-homomorphism of algebras. Show that ${}^t f(B_2') \subset B_1'$ and that ${}^t f$, restricted to B_2', is a cogebra homomorphism of B_2' into B_1'.

(*e*) If C is a coassociative counital cogebra over K, the canonical injection $C \to C^{**}$ of the vector space C into its bidual maps C onto a subspace of $(C^*)'$, the dual cogebra of the algebra C^* dual to C, and is a cogebra homomorphism of C into $(C^*)'$.

APPENDIX

1. Show that in alternative Cayley A-algebra (notation of § 2, no. 4) $T((xy)z) = T(x(yz))$. (Reduce it to proving that $a(x, y, z) = a(\bar{z}, \bar{y}, \bar{x})$.)

2. Let F be an alternative A-algebra, in which the relation $2x = 0$ implies $x = 0$. Show that, for all x, y, z in F, $x(yz)x = (xy)(zx)$ (Moufang's identity). (In the identity

$$(xy)z + y(zx) = x(yz) + (yz)x$$

successively replace y by xy and z by zx.)

3. Let F be an alternative Cayley A-algebra, in which the relation $2x = 0$ implies $x = 0$. Show that if $x \in F$ is invertible then, for all y, z, t in F,

$$(N(x) + N(y))(N(z) + N(t)) = N(x\bar{z} + y\bar{t}) + N(xt - (xz)(x^{-1}y))$$

(use Exercises 1 and 2).

HISTORICAL NOTE

(Chapters II and III)

(N.B. Numbers in brackets refer to the bibliography at the end of this Note.)

Linear algebra is both one of the oldest and one of the newest branches of mathematics. On the other hand, at the origins of mathematics are the problems which are solved by a single multiplication or division, that is by calculating a value of a function $f(x) = ax$, or by solving an equation $ax = b$: these are typical problems of linear algebra and it is impossible to deal with them, indeed even to pose them correctly, without "thinking linearly".

On the other hand, not only these questions but almost everything concerning equations of the first degree had long been relegated to elementary teaching, when the modern development of the notions of field, ring, topological vector space, etc. came to isolate and emphasize the essential notions of linear algebra (for example duality); then the essentially linear character of almost the whole of modern mathematics was perceived, of which "linearization" is itself one of the distinguishing traits, and linear algebra was given the place it merits. To give its history, from our present point of view, would therefore be a task as difficult as it is important; and we must therefore be content to give a brief summary.

From the above it is seen that linear algebra was no doubt born in response to the needs of practical calculators; thus we see the rule of three† and the rule of false position, more or less clearly stated, playing an important role in all the manuals of practical arithmetic, from the Rhind papyrus of the Egyptians to those used in our primary schools, by way of Āryabhaṭa, the Arabs, Leonard of

† Cf. J. Tropfke, *Geschichte der Elementar-Mathematik*, 1. Band, 2te Ausgabe, Berlin-Leipzig (W. de Gruyter), 1921, pp. 150–155.

Pisa and the countless "calculation books" of the Middle Ages and the Renaissance; but they never constituted more than a small part, for the use of practical men, of the most advanced scientific theories.

As for mathematicians proper, the nature of their research on linear algebra depends on the general structure of their science. Ancient Greek mathematics, as expounded in the *Elements* of Euclid, developed two abstract theories of a linear character, on the one hand that of magnitudes ([2], Book V; cf. Historical Note to *General Topology*, IV) and on the other hand that of integers ([2], Book VII). With the Babylonians we find methods much more akin to our elementary algebra; they know how to solve, and most elegantly ([1], pp. 181–183), systems of equations of the first degree. For a very long time, nevertheless, the progress of linear algebra is mainly confined to that of algebraic calculations and they should be considered from this point of view, foreign to this Note; to reduce a linear system to an equation of the type $ax = b$, it suffices, in the case of a single unknown, to know the rules (already, in substance, stated by Diophantus) for taking terms from one side to the other and combining similar terms; and, in the case of several unknowns, it suffices to know also how to eliminate them successively until only one is left. Also the Treatises on algebra, until the XVIIIth century, think that all is accomplished as far as the first degree is concerned, when they have expounded these rules; as for a system of as many equations as unknowns (they do not consider others) where the left hand sides are not linearly independent forms, they are content to observe in passing that this indicates a badly posed problem. In the treatises of the XIXth century and even certain more recent works, this point of view is only modified by the progress of notation, which allows writing systems of n equations in n unknowns, and by the introduction of determinants which allow formulae of an explicit solution to be given in the "general case"; this progress, the credit for which would have belonged to Leibniz ([7], p. 239) had he developed and published his ideas on this subject, is mainly due to the mathematicians of the XVIIIth and early XIXth centuries.

But we must first study various currents of ideas which, much more than the study of linear equations, contributed to the development of linear algebra in the sense in which we understand it. Inspired by the study of Appollonius, Fermat [4(*a*)], having conceived, even before Descartes [5], the principle of analytic geometry, has the idea of classifying plane curves according to their degree (which, having become little by little familiar to all mathematicians, can be considered to have been definitely grasped towards the end of the XVIIth century) and formulates the fundamental principle that an equation of the first degree, in the plane, represents a line and an equation of the second degree a conic: a principle from which he deduces immediately some "very beautiful" consequences relating to geometric loci. At the same time, he enunciates [4(*b*)] the classification of problems into problems with a single solution, problems which reduce to an equation in two unknowns, an equation in three unknowns, etc.; and he adds: the first consist of determining a point, the second a line or plane locus, the others a surface, etc. ("... *such a problem does not seek only a point or a line, but the whole of a surface appropriate to the question; here surfaces as loci have their genesis and similarly for the rest*", *loc. cit.*, p. 186; here already is the germ of n-dimensional geometry). This paper, formulating the principle of dimension in algebra and algebraic geometry, indicates a fusion of algebra and geometry in absolute conformity with modern ideas, but which, as has already been seen, took more than two centuries to penetrate into men's minds.

At least these ideas soon result in the expansion of analytic geometry which reaches its fulness in the XVIIIth century with Clairaut, Euler, Cramer, Lagrange and many others. The linear character of the formulae for transformation of coordinates in the plane and in space, which Fermat cannot have failed already to have perceived, is put in relief for example by Euler ([8(a)], Chapters II–III and Appendix to Chapter IV), who here lays the foundation of the classification of plane curves and that of surfaces according to their degree (invariant precisely because of the linearity of these formulae); he it is also (*loc. cit.*, Chapter XVIII) who introduces the word "affinity" to describe the relation between curves which can be derived one from the other by a transformation $x' = ax$, $y' = by$ (but without perceiving anything geometrically invariant in this definition which remains bound to a particular choice of axes). A little later we see Lagrange [9(a)] devoting a whole memoir, which long remained justly famous, to typically linear and multilinear problems of analytic geometry in three dimensions. Around about this time, in relation to the linear problem constituted by the search for a plane curve passing through given points, the notion of determinant takes shape, first in a somewhat empirical way, with Cramer [10] and Bezout [11]; this notion is then developed by several authors and its essential properties are definitively established by Cauchy [13] and Jacobi [16(a)].

On the other hand, whilst mathematicians had a slight tendency to despise equations of the first degree, the solution of differential equations was considered a capital problem; it was natural that, among these equations, linear equations, with constant coefficients or otherwise, should early be distinguished and their study contributed to emphasize linearity and related properties. This is certainly seen in the work of Lagrange [9(b)] and Euler [8(b)], at least as far as homogeneous equations are concerned; for these authors see no point in saying that the general solution of the non-homogeneous equation is the sum of a particular solution and the general solution of the corresponding homogeneous equation and they make no use of this principle (known however to d'Alembert); we note here also that, when they state that the general solution of the homogeneous linear equation of order n is a linear combination of n particular solutions, they do not add that these must be linearly independent and make no effort to make the latter notion explicit; it seems that only the teaching of Cauchy at the École Polytechnique throws some light ([14], pp. 573–574) on these points as on many others. But already Lagrange (*loc. cit.*) introduces also (purely by calculation, it is true, and without giving it a name) the adjoint equation $L^*(y) = 0$ of a linear differential equation $L(y) = 0$, an example typical of duality by virtue of the relation

$$\int z L(y)\, dx = \int L^*(z)\, y\, dx,$$

valid for y and z zero at the extremities of the interval of integration; more precisely, and 30 years before Gauss defined explicitly the transpose of a linear substitution in 3 variables, we see here the first example without doubt of a "functional operator" L^* the transpose or "adjoint" of an operator L given by means of a bilinear function (here the integral $\int yz\, dx$).

At the same time and again with Lagrange [9(*c*)], linear substitutions, in 2 and 3 variables at first, were in the process of conquering arithmetic. Clearly the set of values of a function $F(x, y)$, when x and y are given all integral values, does not change when a linear substitution with integral coefficients, of determinant 1, is performed on x and y; on this fundamental observation Lagrange founds the theory of representations of numbers by forms and that of the reduction of forms; and Gauss, by a step whose boldness it has become difficult for us to appreciate, isolates the notion of equivalence and that of class of forms (cf. Historical Note to I); on this subject, he recognizes the necessity of certain elementary principles relating to linear substitutions and introduces in particular the notion of transpose or adjoint ([12(*a*)], p. 304). From this moment onwards, the arithmetic study and the algebraic study of quadratic forms, in 2, 3 and later n variables, that of bilinear forms which are closely related to them and more recently the generalization of these notions to an infinity of variables were, right up to the present, to constitute one of the most fertile sources of progress for linear algebra (cf. Historical Note to IX).

But a perhaps still more decisive progress was the creation by Gauss, in the same *Disquisitiones* (cf. Historical Note to I), of the theory of finite commutative groups, which occur there in four different ways, in the additive group of integers modulo m (for m an integer), in the multiplicative group of integers prime to m modulo m, in the group of classes of binary quadratic forms and finally in the multiplicative group of m-th roots of unity; and, as we have already noted, it is clearly as commutative groups, or rather as modules over **Z**, that Gauss treats all these groups and studies their structure, their relations of isomorphism, etc. In the module of "complex integers" $a + bi$, he is later seen studying an infinite module over **Z**, whose isomorphism he no doubt perceived with the module of periods (discovered by him in the complex domain) of elliptic functions; in any case this idea already appears neatly in Jacobi's work, for example in his famous proof of the impossibility of a function with 3 periods and his views on the problem of inversion of Abelian integrals

[16(*b*)], to result soon in the theorems of Kronecker (cf. Historical Note to *General Topology*, VII).

Here, another current joins those whose course and occasional meanders we have sought to trace, which had long remained underground. As will later be expounded in more detail (Historical Note to IX), "pure" geometry in the sense understood in the last century, that is essentially projective geometry of the plane and space without using coordinates, had been created in the XVIIth century by Desargues [6], whose ideas, appreciated in their true value by a Fermat and put into practice by a Pascal, had then been buried in oblivion, eclipsed by the brilliant progress of analytic geometry; it was revived towards the end of the XVIIIth century, with Monge, then Poncelet and his rivals Brianchon and Chasles, sometimes completely and voluntarily separated from analytic methods, sometimes (especially in Germany) closely intermixed with them. Now projective transformations, from whatever point of view they are considered (synthetic or analytic), are of course just linear substitutions on the projective or "barycentric" coordinates; the theory of conics (in the XVIIth century) and later that of quadrics, with whose projective properties this school is principally concerned for a long time, are just that of quadratic forms, whose close connection with linear algebra we have already pointed out earlier. To these notions is added that of polarity: also created by Desargues, the theory of poles and polars becomes, in the hands of Monge and his successors and soon under the name of the principle of duality, a powerful tool for transforming geometric theorems; if it cannot be affirmed that its relation with adjoint differential equations was perceived during that period (they are indicated by Pincherle at the end of the century), then at least Chasles did not fail [17] to perceive its relation with the notion of reciprocal spherical triangles, introduced into spherical trigonometry by Viète ([3], p. 428) and Snellius as early as the XVIth century. But duality in projective geometry is only an aspect of duality of vector spaces, taking account of the modifications imposed when passing from the affine space to the projective space (which is a quotient space of it, under the relation "scalar multiplication").

The XIXth century, more than any period in our history, was rich in mathematicians of the first order; and it is difficult in a few pages, even restricting ourselves to the most salient features, to describe all that is produced in their hands by the coming together of these movements of ideas. Between the purely synthetic methods on the one hand, a species of Procrustean bed where their orthodox protagonists put themselves to torture, and the analytic methods related to a system of coordinates arbitrarily imposed on the space, the need is soon felt for a geometric calculus, dreamed of but not created by Leibniz and imperfectly sketched by Carnot: first appears addition of vectors, implicit in Gauss's work in his geometric representation of imaginary numbers and the applications he makes of this to elementary geometry (cf. Historical Note to *General Topology*, VIII), developed by Bellavitis under the name of "method of equipollences" and taking its definitive form with Grassmann, Möbius and Hamilton; at the same time, under the name of "barycentric calculus", Möbius gives a version of it suitable for the needs of projective geometry [18].

At the same period, and by the same men, the step, so natural (once engaged on this path), already announced by Fermat, is taken from the plane and "ordinary" space to n-dimensional space; indeed an inevitable step, since the algebraic phenomena which can in two or three variables be interpreted geometrically are still valid for an arbitrary number of variables; thus to impose, in using geometric language, the limitation to 2 or 3 dimensions, would be for the modern mathematician just as tiresome a yoke as that which always prevented the Greeks from extending the notion of number to ratios of incommensurable magnitudes. Hence the language and ideas relating to n-dimensional space appear almost simultaneously on all sides, obscurely in the work of Gauss, clearly in the work of the mathematicians of the following generation; and their greater or less assurance in using them was perhaps less due to their mathematical inclinations than to their philosophical or even purely practical outlook. In any case, Cayley and Grassmann, around 1846, handle these concepts with the greatest of ease (and this, says Cayley quite contrary to Grassmann ([22(*a*)], p. 321), "*without recourse to any metaphysical notion*"); Cayley is never far away from the analytic interpretation and coordinates, whereas in Grassmann's work from the start, addition of vectors in n-dimensional space and the geometric aspect take the upper hand, to result in the developments of which we shall speak in a moment.

Meantime the impulse given by Gauss was pushing mathematicians, in two different ways, towards this study of algebras or "hypercomplex systems". On the one hand, it was inevitable to try to extend the domain of real numbers otherwise than by introducing the "imaginary unit" $i = \sqrt{-1}$ and perhaps thus open up vaster domains just as fertile as that of the complex numbers. Gauss himself was convinced ([12(*b*)], p. 178) of the impossibility of such an extension, as long as one wants to preserve the principal properties of complex numbers, that is, in modern language, those which make it into a commutative field; and, either under his influence or independently, his contemporaries seem to have shared this conviction, which was only justified much later by Weierstrass [23] in a precise theorem. But, once multiplication of complex numbers is interpreted by rotations in the plane, then, if it is proposed to extend this idea to three dimensional space, (since the rotations in space form a non-Abelian group) non-commutative multiplications have to be envisaged; this is one of Hamilton's† guiding ideas in his discovery of quaternions [20], the first example of a non-commutative field. The singular nature of this example (the only one, as Frobenius was later to show, which can be constructed over the

† Cf. the interesting preface of his *Lectures on quaternions* [20] where he retraces the whole history of his discovery.

field of real numbers) somewhat restricts its import, in spite of or perhaps even because of the formation of a school of fanatical "quaternionists": a strange phenomenon, which was later reproduced around the work of Grassmann, and then by the vulgarizers who draw from Hamilton and Grassmann what is called "vector calculus". The abandoning a little later of associativity, by Graves and Cayley who construct the "Cayley numbers", opens up no very interesting path. But after Sylvester had introduced matrices and (without giving it a name) had clearly defined their rank [21], again it was Cayley [22(*b*)] who created the calculus of matrices, not without observing (an essential fact often lost sight of later) that a matrix is only an abridged notation for a linear substitution, just as Gauss denoted the form $aX^2 + 2bXY + cY^2$ by (a, b, c). This is just one aspect, the most interesting for us of course, of the abundant production by Sylvester and Cayley on determinants and everything connected with them, a production full of ingenious identities and impressive calculations.

Also (amongst other things) Grassmann discovers an algebra over the reals, the exterior algebra which still bears his name. His work, earlier even than that of Hamilton [19(*a*)], created in an almost complete moral solitude, remained for a long time little known, no doubt because of its originality, because also of the philosophical mists, in which it begins by enveloping itself and which for example at first deterred Möbius. Moved by preoccupations analogous to those of Hamilton but of greater import (and which, as he soon sees, are the same as those of Leibniz), Grassmann constructs a vast algebraico-geometric edifice, resting on a geometric or "intrinsic" conception (already more or less axiomatized) of n-dimensional vector space; among the more elementary results at which he arrives, we quote for example the definition of linear independence of vectors, that of dimension and the fundamental relation

$$\dim V + \dim W = \dim(V + W) + \dim(V \cap W)$$

(*loc. cit.*, p. 209; cf. [19(*b*)], p. 21). But it is especially exterior multiplication, then inner multiplication, of multivectors which provide him with the tools with which he easily treats first the problems of linear algebra proper and then those relating to the Euclidean structure, that is orthogonality of vectors (where he finds the equivalent of duality, which he does not possess).

The other path opened up by Gauss in the study of hypercomplex systems is that starting from the complex integers $a + bi$; after these follow quite naturally algebras or more general hypercomplex systems, over the ring $\mathbf{Z}$ of integers and over the field $\mathbf{Q}$ of rationals, and first of all those already envisaged by Gauss which are generated by roots of unity, then algebraic number fields and modules of algebraic integers: the former are the principal topic in the work of Kummer, the study of the latter was undertaken by Dirichlet, Hermite, Kronecker and Dedekind. Here, in contrast to what happens with algebras over the reals, it is not necessary to abandon any of the characteristic properties of commutative fields and attention was confined to the latter throughout the XIXth century. But linear properties and for example the search for the basis for the integers of the field (indispensible for a general definition of the discriminant) play an essential role at many points; and with Dedekind at any rate the methods are destined to become typically "hypercomplex"; Dedekind himself moreover, without setting himself the problem of algebras in general, is conscious of this character of his works and of what relates them for example to the results of Weierstrass on hypercomplex systems over the reals ([24], in particular vol. 2, p. 1). At the same time the determination of the structure of the multiplicative group of units in an algebraic number field, effected by Dirichlet in some famous notes [15] and almost at the same time by Hermite, was vitally important in clarifying ideas on modules over **Z**, their generating systems and, their bases (when such exist). Then the notion of ideal, defined by Dedekind in algebraic number fields (as a module over the ring of integers of the field), whilst Kronecker introduces in polynomial rings (under the name of "systems of modules") an equivalent notion, gives the first examples of modules over more general rings than **Z**; and in the work of the same authors, and then Hilbert, in particular cases the notion of group with operators is slowly isolated, and the possibility of constructing always from such a group a module over a suitably defined ring.

At the same time, the arithmetico-algebraic study of quadratic bilinear forms and their "reduction" (or, what amounts to the same, of matrices and their "invariants") leads to the discovery of the general principles on the solution of systems of linear equations, principles which due to the lack of the notion of rank, had escaped Jacobi.† The problem of the solution in integers of systems of linear equations with integral coefficients is attacked and solved, first in a special case by Hermite and then in all its generality by H. J. Smith [25]; the results of the latter are found again, only in 1878, by Frobenius, in the framework of a vast programme of research instituted by Kronecker and in which Weierstrass also participates; incidentally during the course of these works, Kronecker gives definitive form to the theorems on linear systems with real (or complex) coefficients, which are also elucidated, in an obscure manual, with the minute care characteristic of him, by the famous author of *Alice in Wonderland*; as for Kronecker, he disdains to publish these results and leaves them to his colleagues and disciples; the word "rank" itself is only introduced by Frobenius. Also in the course of their teaching at the University of Berlin Kronecker [26] and Weierstrass introduce the "axiomatic" definition of determinants (as an alternating multilinear function of n vectors in n-dimensional space, normed so that it takes the value 1 at the unit matrix), a definition equivalent to that

† Concerning the classification of systems of n equations in n unknowns when the determinant is zero, he says ([16(a)], p. 370): "*paullo prolixum videtur negotium*" (it could not be elucidated briefly).

derived from Grassmann's calculus and to that adopted in this Treatise; again during his courses Kronecker, without feeling the need to give it a name and in a still non-intrinsic form, introduces the tensor product of spaces and the "Kronecker" product of matrices (the linear substitution induced on a tensor product by given linear substitutions applied to the factors).

This research cannot be separated from the theory of invariants created by Cayley, Hermite and Sylvester (the "invariant trinity" of which Hermite later speaks in his letters) and which, from a modern point of view, is above all a theory of representations of the linear group. Here there comes to light, as the algebraic equivalent of duality in projective geometry, the distinction between series of cogredient and contragredient variables, that is vectors in a space and vectors in the dual space; and, after attention has been turned first to forms of low degree and then of arbitrary degree, in 2 and 3 variables, almost without delay bilinear, then multilinear forms are examined in several series of "cogredient" or "contragredient" variables, which is equivalent to the introduction of tensors; the latter becomes explicit and is popularized when, under the inspiration of the theory of invariants, Ricci and Levi-Cività, in 1900, introduce into differential geometry "tensor calculus" [28], which later came into great vogue following its use by the "relativist" physicists. Again the progressive intermingling of the theory of invariants, differential geometry and the theory of partial differential equations (especially the so-called problem of Pfaff and its generalizations) slowly leads geometers to consider alternating bilinear forms of differentials, in particular the "bilinear covariant" of a form of degree 1 (introduced in 1870 by Lipschitz and then studied by Frobenius), to result in the creation by E. Cartan [29] and Poincaré [30] of the calculus of exterior differential forms. Poincaré introduces the latter, in order to form his integral invariants, as the expressions which appear in multiple integrals, whilst Cartan, guided no doubt by his research on algebras, introduces them in a more formal way, but without failing to observe that the algebraic part of their calculus is identical with Grassmann's exterior multiplication (whence the name which he adopts), thus definitively restoring the work of the latter to its rightful place. The translation, into the notation of tensor calculus, of exterior differential forms, moreover shows immediately their connection with skew-symmetric tensors, which, once a purely algebraic point of view is adopted, shows that they are for alternating multilinear forms what covariant tensors are for arbitrary multilinear forms; this aspect is further clarified with the modern theory of representations of the linear group; and thus, for example, the substantial identity between the definition of determinants given by Weierstrass and Kronecker and that resulting from Grassmann's calculus is recognized.

We thus arrive at the modern period, where the axiomatic method and the notion of structure (at first vaguely perceived, and defined only recently) allow us to separate concepts which until then had been inextricably mixed, to formulate what was vague or left to intuition and to prove with proper generality theorems which were known only in special cases. Peano, one of the creators of the axiomatic method and also one of the first mathematicians fully to appreciate the work of Grassmann, gives as early as 1888 ([27], Chapter IX) the axiomatic definition of vector spaces (finite-dimensional or otherwise) over the field of real numbers and, in a completely modern notation, of linear mappings of one such space into another; a little later, Pincherle seeks to develop applications of linear algebra, thus conceived, to the theory of functions, in a direction it is true which has not been very fruitful; at least his point of view allows him to recognize "Lagrange's adjoint" as a special case of the transposition of linear mappings: which appears soon, still more clearly, and for partial differential equations as well as for differential equations, in the course of the memorable works of Hilbert and his school on Hilbert space, and its applications to analysis. It is on that occasion that Toeplitz [31], also introducing (but by means of coordinates) the most general vector space over the reals, makes the fundamental observation that the theory of determinants is not needed to prove the principal theorems of linear algebra, which allows these to be extended without difficulty to infinite-dimensional spaces; and he also indicates that linear algebra thus understood can of course be applied to an arbitrary commutative base field.

On the other hand, with the introduction by Banach, in 1922, of the spaces bearing his name,† spaces not isomorphic to their dual are encountered, albeit in a problem which is topological as much as it is algebraic. Already between a finite-dimensional vector space and its dual there is no "canonical" isomorphism, that is determined by its structure, which had long been reflected in the distinction between cogredient and contragredient. Nevertheless it seems beyond doubt that the distinction between a space and its dual was definitively established only after the work of Banach and its school; also in these works the importance of the notion of codimension comes to light. As for duality or "orthogonality" between the vector subspaces of a space and those of its dual, the way in which it is formulated today presents not just a superficial analogy with the modern formulation of the principal theorem of Galois theory (cf. *Algebra,* V) and with so-called Pontrjagin duality in locally compact Abelian groups; the latter goes back to Weber, who, in the course of arithmetical researches, lays in 1886 its foundations for finite groups; in Galois theory the "duality" between subgroups and subfields takes form in the work of Dedekind and Hilbert; and orthogonality between vector subspaces derives visibly, first from duality between linear varieties in projective geometry and also from the notion and properties of completely orthogonal varieties in a Euclidean space or a Hilbert space (whence its name). All these strands are reassembled in the contemporary period, in the hands of algebraists such as

† These are complete normed vector spaces over the field of real numbers or that of complex numbers.

E. Noether, Artin and Hasse and topologists such as Pontrjagin and Whitney (not without the ones influencing the others) to arrive, in each of these fields, at the stage of knowledge whose results are expounded in this Treatise.

At the same time a critical examination is made, which is destined to eliminate, on each point, the hypotheses which are not completely indispensable, and especially those which would close the way to certain applications. Thus the possibility is perceived of substituting rings for fields in the notion of vector spaces and, creating the general notion of module, of treating at the same time these spaces, Abelian groups, the particular modules already studied by Kronecker, Weierstrass, Dedekind and Steinitz and even groups with operators and for example of applying the Jordan-Hölder theorem to them; at the same time, with the distinction between right and left modules, we pass to the non-commutative, which arises from the modern development of the theory of algebras by the American school (Wedderburn, Dickson) and especially the German school (E. Noether, E. Artin).

Bibliography

1. O. Neugebauer, *Vorlesungen über Geschichte der antiken Mathematik*, Vol. I: Vorgriechische Mathematik, Berlin (Springer), 1934.
2. *Euclidis Elementa*, 5 vols., ed. J. L. Heiberg, Lipsiae (Teubner), 1883–88.
2 *bis*. T. L. Heath, *The thirteen books of Euclid's Elements*..., 3 vols., Cambridge, 1908.
3. Francisci Vietae, *Opera mathematica*..., Lugduni Batavorum (Elzevir), 1646.
4. P. Fermat, *Oeuvres*, vol. I, Paris (Gauthier-Villars), 1891: (*a*) Ad locos planos et solidos Isagoge (pp. 91–110; French transl. *ibid.*, vol. III, p. 85); (*b*) Appendix ad methodum... (pp. 184–188; French transl., *ibid.*, vol. III, p. 159).
5. R. Descartes, *La géométrie*, Leyde (Jan Maire), 1637 (= *Oeuvres*, ed. Ch. Adam and P. Tannery, vol. VI, Paris (L. Cerf), 1902).
6. G. Desargues, *Oeuvres*..., vol. I, Paris (Leiber), 1864: Brouillon proiect d'une atteinte aux éuénemens des rencontres d'un cône auec un plan (pp. 103–230).
7. G. W. Leibniz, *Mathematische Schriften*, ed. C. I. Gerhardt, vol. I, Berlin (Asher), 1849.
8. L. Euler: (*a*) *Introductio in Analysin Infinitorum*, vol. 2, Lausannae, 1748 (= *Opera Omnia* (1), vol. IX, Zürich-Leipzig-Berlin (O. Füssli and B. G. Teubner), 1945); (*b*) *Institutionum Calculi Integralis*, vol. 2, Petropoli, 1769 (= *Opera Omnia* (1), vol. XII, Leipzig-Berlin (B. G. Teubner), 1914).
9. J.-L. Lagrange, *Oeuvres*, Paris (Gauthier-Villars), 1867–1892: (*a*) Solutions analytiques de quelques problèmes sur les pyramides triangulaires,

vol. III, pp. 661–692; (*b*) Solution de différents problèmes de calcul intégral, vol. I, p. 471; (*c*) Recherches d'arithmétique, vol. III, pp. 695–795.

10. G. Cramer, *Introduction à l'analyse des lignes courbes*, Geneva (Cramer and Philibert), 1750.
11. E. Bezout, *Théorie générale des équations algébriques*, Paris, 1779.
12. C. F. Gauss, *Werke*, Göttingen, 1870–1927: (*a*) Disquisitiones arithmeticae, vol. I; (*b*) Selbstanzeige zur Theoria residuorum biquadraticorum, Commentatio secunda, vol. II, pp. 169–178.
13. A.-L. Cauchy, Mémoire sur les fonctions qui ne peuvent obtenir que deux valeurs égales et de signes contraires par suite des transpositions opérées entre les variables qu'elles renferment, *J. Ec. Polytech.*, cahier 17 (volume X), 1815, pp. 29–112 (=*Oeuvres complètes* (2), vol. I, Paris (Gauthier-Villars), 1905, pp. 91–169).
14. A.-L. Cauchy, in *Leçons de calcul différentiel et de calcul intégral, rédigées principalement d'après les méthodes de M. A.-L. Cauchy* by Abbé Moigno, vol. II, Paris, 1844.
15. P. G. Lejeune-Dirichlet, *Werke*, vol. I, Berlin (G. Reimer), 1889, pp. 619–644.
16. C. G. J. Jacobi, *Gesammelte Werke*, Berlin (G. Reimer), 1881–1891: (*a*) De formatione et proprietatibus determinantium, vol. III, pp. 355–392; (*b*) De functionibus duarum variabilium . . ., vol. II, pp. 25–50.
17. M. Chasles, *Aperçu historique sur l'origine et le développement des méthodes en géométrie* . . ., Bruxelles, 1837.
18. A. F. Möbius, *Der baryzentrische Calcul* . . ., Leipzig, 1827 (=*Gesammelte Werke*, vol. I, Leipzig (Hirzel), 1885).
19. H. Grassmann: (*a*) *Die lineale Ausdehnungslehre, ein neuer Zweig der Mathematik, dargestellt und durch Anwendungen auf die übrigen Zweige der Mathematik, wie auch auf die Statik, Mechanik, die Lehre vom Magnetismus und die Kristallonomie erläutert*, Leipzig (Wigand), 1844 (=*Gesammelte Werke*, vol. I, 1st Part, Leipzig (Teubner), 1894); (*b*) *Die Ausdehnungslehre, vollständig und in strenger Form bearbeitet*, Berlin, 1862 (=*Gesammelte Werke*, vol. I, 2nd Part, Leipzig (Teubner), 1896).
20. W. R. Hamilton, *Lectures on quaternions*, Dublin, 1853.
21. J. J. Sylvester, *Collected Mathematical Papers*, vol. I, Cambridge, 1904: No. 25, Addition to the articles . . ., pp. 145–151 (=*Phil. Mag.*, 1850).
22. A. Cayley, *Collected Mathematical Papers*, Cambridge, 1889–1898: (*a*) Sur quelques théorèmes de la géométrie de position, vol. I, pp. 317–328 (=*Crelle's J.*, vol. XXXI (1846), pp. 213–227); (*b*) A memoir on the theory of matrices, vol. II, pp. 475–496 (=*Phil. Trans.*, 1858).
23. K. Weierstrass, *Mathematische Werke*, vol. II, Berlin, (Mayer und Müller),

1895: Zur Theorie des aus *n* Haupteinheiten gebildeten complexen Grössen, pp. 311–332.
24. R. Dedekind, *Gesammelte mathematische Werke*, 3 vols., Braunschweig (Vieweg), 1930–32.
25. H. J. Smith, *Collected Mathematical Papers*, vol. I, Oxford, 1894: On systems of linear indeterminate equations and congruences, p. 367 (= *Phil. Trans.*, 1861).
26. L. Kronecker, *Vorlesungen über die Theorie der Determinanten...*, Leipzig (Teubner), 1903.
27. G. Peano, *Calcolo geometrico secondo l'Ausdehnungstheorie di Grassmann preceduto dalle operazioni della logica deduttiva*, Torino, 1888.
28. G. Ricci and T. Levi-Cività, Méthodes de calcul différentiel absolu et leurs applications, *Math. Ann.*, vol. LIV (1901), p. 125.
29. E. Cartan, Sur certaines expressions différentielles et le problème de Pfaff, *Ann. E.N.S.* (3), vol. XVI (1899), pp. 239–332 (= *Oeuvres complètes*, vol. II_1, Paris (Gauthier-Villars), 1953, pp. 303–396).
30. H. Poincaré, *Les méthodes nouvelles de la mécanique céleste*, vol. III, Paris (Gauthier-Villars), 1899, Chapter XXII.
31. O. Toeplitz, Ueber die Auflösung unendlichvieler linearer Gleichungen mit unendlichvielen Unbekannten, *Rend. Circ. Mat. Pal.*, vol. XXVIII (1909), pp. 88–96.

INDEX OF NOTATION

INDEX OF TERMINOLOGY

Printing and Binding: AZ Druck und Datentechnik GmbH, Kempten